HANDBOOK OF POLYMER TESTING

PLASTICS ENGINEERING

Founding Editor

Donald E. Hudgin

Professor
Clemson University
Clemson, South Carolina

1. Plastics Waste: Recovery of Economic Value, *Jacob Leidner*
2. Polyester Molding Compounds, *Robert Bums*
3. Carbon Black-Polymer Composites: The Physics of Electrically Conducting Composites, *edited by Enid Keil Sichel*
4. The Strength and Stiffness of Polymers, *edited by Anagnostis E. Zachariades and Roger S. Porter*
5. Selecting Thermoplastics for Engineering Applications, *Charles P. MacDermott*
6. Engineering with Rigid PVC: Processability and Applications, *edited by I. Luis Gomez*
7. Computer-Aided Design of Polymers and Composites, *D. H. Kaelble*
8. Engineering Thermoplastics: Properties and Applications, *edited by James M. Margolis*
9. Structural Foam: A Purchasing and Design Guide, *Bruce C. Wendle*
10. Plastics in Architecture: A Guide to Acrylic and Polycarbonate, *Ralph Montella*
11. Metal-Filled Polymers: Properties and Applications, *edited by Swapan K. Bhattacharya*
12. Plastics Technology Handbook, *Manas Chanda and Salil K. Roy*
13. Reaction Injection Molding Machinery and Processes, *F. Melvin Sweeney*
14. Practical Thermoforming: Principles and Applications, *John Florian*
15. Injection and Compression Molding Fundamentals, *edited by Avraam I. Isayev*
16. Polymer Mixing and Extrusion Technology, *Nicholas P. Cheremisinoff*
17. High Modulus Polymers: Approaches to Design and Development, *edited by Anagnostis E. Zachariades and Roger S. Porter*
18. Corrosion-Resistant Plastic Composites in Chemical Plant Design, *John H. Mallinson*
19. Handbook of Elastomers: New Developments and Technology, *edited by Anil K. Bhowmick and Howard L. Stephens*
20. Rubber Compounding: Principles, Materials, and Techniques, *Fred W. Barlow*
21. Thermoplastic Polymer Additives: Theory and Practice, *edited by John T. Lutz, Jr.*
22. Emulsion Polymer Technology, *Robert D. Athey, Jr.*
23. Mixing in Polymer Processing, *edited by Chris Rauwendaal*
24. Handbook of Polymer Synthesis, Parts A and B, *edited by Hans R. Kricheldorf*
25. Computational Modeling of Polymers, *edited by Jozef Bicerano*
26. Plastics Technology Handbook: Second Edition, Revised and Expanded, *Manas Chanda and Salil K. Roy*
27. Prediction of Polymer Properties, *Jozef Bicerano*
28. Ferroelectric Polymers: Chemistry, Physics, and Applications, *edited by Hari Singh Nalwa*
29. Degradable Polymers, Recycling, and Plastics Waste Management, *edited by Ann-Christine Albertsson and Samuel J. Huang*
30. Polymer Toughening, *edited by Charles B. Arends*

31. Handbook of Applied Polymer Processing Technology, *edited by Nicholas P. Cheremisinoff and Paul N. Cheremisinoff*
32. Diffusion in Polymers, *edited by P. Neogi*
33. Polymer Devolatilization, *edited by Ramon J. Albalak*
34. Anionic Polymerization: Principles and Practical Applications, *Henry L. Hsieh and Roderic P. Quirk*
35. Cationic Polymerizations: Mechanisms, Synthesis, and Applications, *edited by Krzysztof Matyjaszewski*
36. Polyimides: Fundamentals and Applications, *edited by Malay K. Ghosh and K. L. Mittal*
37. Thermoplastic Melt Rheology and Processing, *A. V. Shenoy and D. R. Saini*
38. Prediction of Polymer Properties: Second Edition, Revised and Expanded, *Jozef Bicerano*
39. Practical Thermoforming: Principles and Applications, Second Edition, Revised and Expanded, *John Florian*
40. Macromolecular Design of Polymeric Materials, *edited by Koichi Hatada, Tatsuki Kitayama, and Otto Vogl*
41. Handbook of Thermoplastics, *edited by Olagoke Olabisi*
42. Selecting Thermoplastics for Engineering Applications: Second Edition, Revised and Expanded, *Charles P. MacDermott and Aroon V. Shenoy*
43. Metallized Plastics, *edited by K. L. Mittal*
44. Oligomer Technology and Applications, *Constantin V. Uglea*
45. Electrical and Optical Polymer Systems: Fundamentals, Methods, and Applications, *edited by Donald L. Wise, Gary E. Wnek, Debra J. Trantolo, Thomas M. Cooper, and Joseph D. Gresser*
46. Structure and Properties of Multiphase Polymeric Materials, *edited by Takeo Araki, Qui Tran-Cong, and Mitsuhiro Shibayama*
47. Plastics Technology Handbook: Third Edition, Revised and Expanded, *Manas Chanda and Salil K. Roy*
48. Handbook of Radical Vinyl Polymerization, *edited by Munmaya K. Mishra and Yusef Yagci*
49. Photonic Polymer Systems: Fundamentals, Methods, and Applications, *edited by Donald L. Wise, Gary E. Wnek, Debra J. Trantolo, Thomas M. Cooper, and Joseph D. Gresser*
50. Handbook of Polymer Testing: Physical Methods, *edited by Roger Brown*

Additional Volumes in Preparation

Handbook of Polypropylene and Polypropylene Composites, *edited by Harutun G. Karian*

Polymer Blends and Alloys, *edited by Gabriel O. Shonaike and George P. Simon*

Star and Hyperbranched Polymers, *edited by Munmaya K. Mishra and Shiro Kobayashi*

Practical Extrusion Blow Molding, *edited by Samuel L. Belcher*

HANDBOOK OF POLYMER TESTING

PHYSICAL METHODS

edited by

ROGER BROWN
Rapra Technology Ltd.
Shawbury, Shrewsbury, England

MARCEL DEKKER, INC. NEW YORK • BASEL

Library of Congress Cataloging-in-Publication Data

Handbook of polymer testing: physical methods / [edited by] Roger Brown.
p. cm.— (Plastics engineering; 50)
Includes bibliographical references and index.
ISBN 0-8247-0171-2 (alk. paper)
1. Plastics—Testing—Handbooks, manuals, etc. 2. Polymers—Testing—Handbooks, manuals, etc.
I. Brown, Roger (Roger P.) II. Series: Plastics engineering (Marcel Dekker, Inc.); 50.
TA455.P5H352 1999
620.1'923'0287—dc21 98-45735
CIP

This book is printed on acid-free paper.

Headquarters
Marcel Dekker, Inc.
270 Madison Avenue, New York, NY 10016
tel: 212-696-9000; fax: 212-685-4540

Eastern Hemisphere Distribution
Marcel Dekker AG
Hutgasse 4, Postfach 812, CH-4001 Basel, Switzerland
tel: 44-61-261-8482; fax: 44-61-261-8896

World Wide Web
http://www.dekker.com

The publisher offers discounts on this book when ordered in bulk quantities. For more information, write to Special Sales/Professional Marketing at the headquarters address above.

Current printing (last digit)
10 9 8 7 6 5 4 3 2

PRINTED IN THE UNITED STATES OF AMERICA

Preface

It is essential for design, specification, and quality control to have data covering the physical properties of materials. It is also essential that meaningful data is obtained by using test methods relevant to the materials. The different characteristics and behavior of materials dictate that particular test procedures be developed, and often standardized, for each material type. Polymers, especially, have unique properties that require their own measurement techniques.

There is a wide range of polymers from soft foams to rigid composites for which separate industries have developed. Each has its own individual test methods and, for the major types of polymers, texts exist that detail these procedures. There are, however, many similarities between different polymer types and frequently it is necessary for laboratories to consider a spectrum of materials. Consequently, there are advantages in a book that comprehensively covers the whole polymer family, describing the individual methods as well as discussing the approaches taken in different branches of the industry.

Handbook of Polymer Testing provides in one volume that comprehensive coverage of physical test methods for polymers. The properties considered cover the whole range of physical parameters, including mechanical, optical, electrical, and thermal as well as resistance to degradation, nondestructive testing, and tests for processability. All the main polymer classes are included: rubbers, plastics, foams, textiles, coated fabrics, and composites. For each property, the fundamental principles and approaches are discussed and particular requirements and the relevant international and national standards for the different polymer classes considered, together with the most-up-to-date techniques.

This book will be of particular value to materials scientists and technologists, and to all those who need to evaluate a spectrum of polymeric materials, including students, design engineers, and researchers. Its structure allows reference for the main properties at both the general and the detailed level, thus making it suitable for different levels of knowledge.

Chapter 29 is based on material produced for the "Testing Knowledge Base" at Rapra Technology, Ltd. Extracts from British Standards were reproduced with the permission of BSI. Users of standards should always ensure that they have complete and current information. Standards can be obtained from BSI Customer Services, 389 Chiswick High Road, London W4 4AL, England.

The other contributors and I gratefully acknowledge the support, information, and helpful advice given by our colleagues during the preparation of this book

Roger Brown

Contents

Contributors

Mustafa Akay University of Ulster at Jordanstown, Jordanstown, Northern Ireland

Paul P. Ashworth Consultant, Withington, Manchester, England

Cyril Barry Consultant, Hay-on-Wye, Hereford, England

Freddy Boey Nanyang Technological University, Singapore

Frank Broadbent British Standards Ltd., London, England

Roger Brown Rapra Technology Ltd., Shawbury, Shrewsbury, England

John Dick Alpha Technologies, Akron, Ohio

Barry Evans Schlumberger Industries–Metflex, Blackburn, England

Martin Gale Rapra Technology Ltd., Shawbury, Shrewsbury, England

John Gearing Gearing Scientific, Ashwell, Hertfordshire, England

Xavier E. Gros Independent NDT Centre, Bruges, France

David Hands Consultant, Sutton Farm, Shrewsbury, England

Steve Hawley Rapra Technology Ltd., Shawbury, Shrewsbury, England

Ken Hillier British Vita, Middleton, Manchester, England

Ivan James Forncet, Wem, Shropshire, England

Dieter Kockott Consultant, Hanau, Germany

Peter Lewis Tun Abdul Razak Research Centre, Brickendonbury, Hertfordshire, England

Keith Paul Rapra Technology Ltd., Shawbury, Shrewsbury, England

Graham D. Sims National Physical Laboratory, Teddington, Middlesex, England

Alan Veith Technical Development Associates, Akron, Ohio

1
Introduction

Roger Brown
Rapra Technology Ltd., Shawbury, Shrewsbury, England

The physical properties of materials need to be measured for quality control, for predicting service performance, to generate design data, and, on occasions, to investigate failures. Without test results there is no proof of quality and no hope of successfully designing new products. The group of materials classed as polymers generally have complicated behavior, and as much or more than with any material it is critical that their properties be evaluated—and evaluated in a meaningful way. Their characteristics are such that methods used for other materials such as metals or ceramics will not usually be suitable. There are also distinct differences between the classes of materials that make up the polymeric group, from flexible fabrics and soft foams through solid rubbers and thermoplastics to very rigid thermosets and composites. Consequently, it is no surprise that particular procedures have been developed and standardized to suit the needs of each material class.

There are excellent texts that deal in great detail with test methods for rubber and for plastics, etc. In dealing with one field they recognize the unique requirements for each class of materials and emphasize the particular procedures that have been standardized in each industry. It is no criticism of such texts to say that by concentrating on a restricted scope they do not bring out the similarities and the common themes that run through the testing of all polymers. It is relatively recently that, rather than metallurgists and plastics technologists, etc., the material scientist or technologist has emerged with an important role. There is great need, for technical and commercial reasons, for many companies to consider and use a spectrum of materials. For these interests, a book that covers the fundamentals and the latest techniques for testing the whole polymer family will have many advantages to students, design engineers, researchers, and those who need to evaluate a wide range of products and materials.

Broadly stated, the scope of this book is the physical testing of polymers. Polymers have been taken to include rubbers, plastics, cellular materials, composites, textiles, and coated fabrics—all the materials generally considered to make up the polymer industry with the exception of adhesives. A great many adhesives are polymeric, but it is considered that treatment of adhesive testing does not fit well with physical testing of the main polymer classes and requires its own volume. The standardized adhesion tests for solid polymers adhered to themselves or other substrates are, however, included.

Physical testing is used in its literal sense and hence does not include chemical analysis. The distinction between physical and chemical is perhaps not completely clear-cut, in that aging and chemical resistance are generally considered as physical tests but clearly involve monitoring the effects of chemical changes. Thermal analysis, for example, straddles both camps, and particular techniques have either been included or excluded depending on their purpose.

The aim of this book is to present an up-to-date account of procedures for testing polymers, indicating the similarities and the differences between the approaches taken for the different materials. Within the restrictions mentioned above, it is intended to be comprehensive. Hence it sets out to cover all the physical properties from dimensional through mechanical, thermal, electrical, etc., to chemical resistance, weathering, and non-destructive. In addition to all these tests on the formed material or product, processability tests are also included. The focus is on testing materials rather than on finished products. Indeed, the vast number of tests, many ad hoc, devised for evaluating performance of the multitude of products made from polymers, would fill a volume, even supposing the subject could be coherently treated. Comment on product testing however is made where appropriate. It should also be noted that many tests used for products are adaptations of the normal material tests, for example stress–strain properties on plastic film products and geomembranes. Even more widely, it is the usual practice to cut test pieces for standard material tests from such products as hose, conveyor belting, and containers.

A rather novel structure has been used, which is designed to give a progressive path for the bulk of commonly measured properties, from background principles, through basic established practice, to the particular requirements of the different materials and then to less common and more advanced techniques. It hence allows ready reference at different levels and largely avoids the complications of dealing with the details of different procedures for several materials in one place.

The basic structure consists of five sets of chapters. Chapters 1 through 7 cover general topics, sample preparation, conditioning, accuracy, reproducibility, etc., all materials being considered together.

Processability tests differ from all the other physical properties included in the book by virtue of being concerned with properties of relevance to the forming of materials and not the performance of the finished material or product. Chapter 8 deals with the processability tests in two parts, for rubbers and plastics respectively.

Chapters 9 through 14 are resumes of the principles and the basic approaches taken for the more commonly tested parameters.

In Chapters 15 through 20 the particular requirements of each of the classes of polymeric materials covered are considered in more detail, including reference to the standardized procedures. The scope of properties covered is essentially the same as in Chapters 9 through 14.

The remaining chapters address selected topics. The topics have been chosen for one or more of three reasons: it is convenient to cover all polymer classes together; the parameters are not those most commonly measured; or the subject is of particular topical interest.

The practicality of this for the reader is that if the subject of interest is in Chapters 1 through 8 or 21 through 32 then selection of the relevant chapter will find the main coverage of the subject. For other properties, the procedures for a particular polymer class can be found by selection of the appropriate chapter in the group 15 through 20. If the principles of the more common tests and comparison of the approaches for different polymer classes are required, then consult Chapters 9 through 14. It is suggested that these chapters be read before the subsequent chapters, especially if the reader is relatively new to polymer testing. It is also essential that the requirements for test piece preparation, conditioning, and dimensional measurement covered in Chapters 5, 6, and 7 be considered in conjunction with all the procedures discussed later.

All reasonable effort has been taken to make the book integrated rather than a series of independent chapters by different contributors. Inevitably there will be some overlap and repetition, but it is believed that this, and the relative complexity of the structure, is outweighed by the confusion that could result from trying to weld discussion of common tests for contrasting polymer types. It is inevitable also that there should be differences in style adopted by the different authors, which perhaps illustrates that testing can be approached in more than one way.

The emphasis is on standard test procedures, which by definition are those that have become widely accepted. Where standardized methods exist, they should be used for quality control purposes and for obtaining general material property data, to ensure compatibility between results from different sources. It is counterproductive to invent alternative procedures when satisfactory and well-tried methods exist, and it prevents meaningful comparisons of data from being made. It has to be accepted that many standard methods have severe limitations for the generation of design data, but nevertheless they can often form a good basis for producing more extensive information.

Unfortunately, standard tests are not completely standard, in that different countries and organizations each have their own standards. The situation has been steadily improving in recent years as more national standards bodies adopt international methods, and this is a trend that we should all encourage. In this book the ISO (and for electrical tests the IEC) standards, together with those of two of the leading national English-language standards-making organizations, the ASTM and the BSI, are considered, plus the European regional (CEN) standards. In a great many cases British standards are identical with ISO standards, but ASTM standards are at very least editorially different. British standards will always be identical with CEN standards where these exist and, in turn, CEN standards are often identical with ISO.

It is not possible to claim that every type of test known for every property has been included, but, within the defined scope, any omission is by accident rather than design. It is also likely that not every standard from the standard bodies covered will have been referenced. Standards are continually being developed and revised, so that it can be guaranteed that between writing and publication there will have been some changes, thus it is essential that the latest catalogs from the ISO, etc., be consulted for the most up-to-date position.

The apparatus needed for tests is considered in conjunction with the test procedures, but in many cases it is not an easy matter to select from the range of apparatus available in differing levels of sophistication or, indeed, to be able to find any supplier at all. The Test Equipment and Services Directory (published by Rapra Technology, England, in hard copy and on CD) contains both advice on selection and a comprehensive guide to instrument suppliers.

2
Putting Testing in Perspective

Ivan James
Forncet, Wem, Shropshire, England

1 Philosophy

As a generality, technical people want to test to obtain knowledge, whereas commercial people will test only when there is some pressure to do so. In an age of cost-cutting and streamlining of production, it may seem that testing is an unnecessary expense, but the reverse is true, since alongside an awareness of cost has grown an increasing customer awareness of quality. The consequences that arise when testing is omitted are illustrated by the following examples.

Some years ago a colleague who served on several SI committees once attempted to buy a radio while he was abroad. The young assistant took one off the shelf, switched it on, and it didn't work. She then unpacked one from its box just as it had arrived from the factory, and that one didn't work either. Eventually she found one that did and was surprised when he declined to buy it. What astounded him was that these products could be made and packed without any testing whatever to prove fitness for purpose until they reached the point of sale.

However, testing the product may not be sufficient, since in a complex product such as a radio, the reliability of the assembly depends on the reliability of each of the individual components. This was brought home to a supplier who asked a buyer what failure rate he would accept. The curt answer "zero" failed to convince him and 0.1% was suggested. "No, zero," was the reply. "But that implies testing every component," said the supplier. "Exactly," answered the buyer. In practice, of course, virtually zero reject levels can be achieved without 100% testing by extremely tight control of the process, but the story illustrates the point. The increasing demand for product quality brings in its train a requirement for component reliability, and that implies component testing also.

Brown [1] has previously suggested that as well as these two reasons for testing, two others can be listed, namely tests to establish material properties for design data and tests to establish reasons for failure if a product proves to be unsatisfactory in service.

Polymers are complex materials, and aspects of their behavior are sometimes unexpected. For this reason, tests on polymers need to be well chosen and wide ranging in order to avoid embarrassing failures. It is important to establish early on that the grade of material chosen fully matches the design criteria for the product. For example, a plastic component, although initially of adequate strength, may on constant exposure to detergents suffer from environmental stress cracking.

Examination of failed products or components is related to this, and testing may reveal that the material did not meet the designer's specification or show that some important property, such as creep, has been overlooked. The coating applied to surgeon's gloves, for example, may be more important than the composition of the rubber.

Summarizing, then, and following the approach suggested by Brown, there are four main areas of testing, namely:

1. Quality control
2. Predicting service performance
3. Design data
4. Investigating failures

Before undertaking any tests, and before considering which properties to measure, it is essential to identify the purpose of testing, because the requirements for each of the purposes are different. Failure to appreciate this can lead to time-wasting tests that do not yield the required results. Similarly, a lack of understanding as to why another person is carrying out particular tests can lead to misunderstanding and argument, say, between the research department and the quality control department in a factory.

To the various attributes related to testing procedures, precision, reproducibility, rapidity, and complexity, may be added the ability for tests to be automated and the desirability for tests to be nondestructive. The balance of these various attributes, and the related cost, differs according to the purpose of the test that is undertaken. These will be considered in turn, but in all cases the precision and reproducibility must be appropriate to the tests undertaken.

2 Quality Control Tests

Nondestructive methods are advantageous and indeed essential when 100% of the output is being tested. The tests should be simple and inexpensive, and automation will probably aid the rapidity of testing. Tests related to product performance are preferred.

2.1 Tests Predicting Product Performance

The most important factor is that the tests relate to service conditions and to aspects of product performance. The tests should not be too complex, although rapidity and cheapness are less important than was the case with quality control. Nondestructive tests are not always appropriate when predicting product performance, as it may be necessary to establish the point at which failure occurs.

2.2 Tests for Producing Design Data

Usually test pieces are of a simple shape and a specified size, whereas the product envisaged may be of a different geometry and size. Data must be presented in a form that enables the designer to allow for changes in geometry, time scale, etc., which implies detailed and comprehensive understanding of material behavior and often multipoint data. It follows that data of this type are expensive to produce and that results are unlikely to be obtained with great rapidity. However, automation may be advantageous, particularly in the case of tests running for a long time (creep tests, say).

2.3 Tests for Investigating Failures

Some understanding of the various mechanisms of failure is necessary before suitable tests can be chosen. Tests need not be complex but must be relevant. For example, a simple measurement of product thickness may establish that there has been a departure from the specified design thickness. The radii of corners, moisture content of plastic, bubbles produced during molding, and a host of other factors may have contributed to the failure, and it is important to keep an open mind when carrying out tests of this type. The absolute accuracy of the test may not be important, but it is essential that it be capable of discriminating between the good and the bad product.

It will be clear from the above that the range of possible tests is very wide, and in any discussion of the philosophy of testing it is useful to classify groups of tests in some manner.

Several approaches are possible, for example, mechanical tests or electrical tests, or again, tests on flexible or on rigid materials, but in discussing the philosophy of testing a broader classification is needed. A useful approach is that given by Brown [1] in an earlier volume, namely,

Tests relating to fundamental properties
Tests relating to apparent properties
Tests relating to functional properties

He defines these using the example of strength. Fundamental strength of a material is that measured in such a way that the result can be reduced to a form independent of test conditions. Apparent strength is that obtained by a method that has completely arbitrary conditions so that the data cannot be simply related to other conditions. Functional strength is that measured under the mechanical conditions of service, probably on the complete product.

Fundamental properties are those relating to the underlying properties of the polymer—the refractive index of a transparent material is a simple example.

Apparent properties are closely related to fundamental properties but are not so tightly defined or controlled. An example is tear strength, where the standard methods yield results strongly dependent on test piece geometry.

Functional properties are usually related to a product. A measure of the resistance to ground of antistatic tires on castor wheels is a good example. The measured value depends on both the geometry of the product and the underlying resistivity of the material, but in the end it is the resistance at a specified voltage that is required.

This classification applies to all kinds of properties (mechanical, electrical, chemical, etc.) and is useful when considering what type of test is needed.

For example, when considering what force is needed to draw a stopper from a bottle, the answer depends both on the coefficient of friction and on the normal force, neither of which is known. In turn the normal force depends on the dimensions of the two components and the modulus of the stopper. However, for a quality control test (and a performance test) all that is needed is that limits be set on the upper and lower levels of force required to extract the stopper. Knowledge of the individual parameters is not necessary. This is an example of a functional test.

Staying with this frictional analogy, the inclined plane method of measuring friction would give an apparent coefficient of friction, since conditions are not tightly controlled (velocity, for example, cannot be specified) and there is little chance of relating the result to other conditions.

If, on the other hand, the requirement is to measure the coefficient of friction between two materials for design purposes, then shape, surface finish, normal load, velocity, temperature, cleanliness, and humidity all become important parameters needing to be controlled. Furthermore, this illustrates the shortage of truly fundamental tests in which the rules for extrapolating to other conditions are well known, as in this case it would probably be necessary to produce multipoint data.

These three types of test can be loosely related to the purpose of testing.

In establishing design data, it is mostly fundamental properties that are needed, but these are in short supply. Many thermal and chemical tests are fundamental in nature, but most mechanical tests give apparent properties. In the absence of established and verified procedures for extrapolating results to other conditions, multipoint data have to be produced at defined levels of all the parameters likely to influence the test result. Consequently, reliable tables of properties for designers are difficult and expensive to establish.

Standard test methods giving apparent properties are best suited to quality control, and only in relatively few cases are they ideal for design data. Quality control tests are the most easily established, and many existing methods fulfil this need. In seeking an improvement in test procedure it is not always a more accurate test that is required. Depending on the purpose for which testing is undertaken, it may be quicker or cheaper tests that are required, or most important of all, tests that are relevant to service performance.

For predicting service performance, the most suitable tests would be functional ones. For investigating failures, the most useful tests depend on the particular circumstances, but fundamental mechanical methods are unlikely to be needed.

3 Trends

Because of the different reasons for testing and the consequent difference in test requirements, developments in test methods do not follow one path. The basic themes are constant enough: people want more efficient tests in terms of time and money, better reproducibility, and tests more suited to design data and more relevant to service performance. However, the emphasis depends on the particular individual needs.

In recent years, the drive towards international standards has led to a close examination of long-established test methods, and it has been found that the reproducibility of many of the tests was poor. This in turn has not led to new tests but rather to the establishment of better standardization of test procedures. There has also been a growing realization of the need to calibrate test equipment with proper documentation of calibration procedures and results.

Where different test methods associated with different countries have been in use for a long time, it has sometimes been difficult to reach an acceptance of one method as a standard test procedure. In these cases it has been necessary to present 'Method A' and

'Method B' as equally acceptable. Similarly, different test conditions have been allowed, perhaps taking account of the difference between temperate and tropical conditions. Although in the local environment this may be quite satisfactory, it leads to difficulties if the results are presented in a database and used in a wider context. Figures presented in a database all need to be produced in exactly the same way, and consequently there has been a lobby for extremely tight standards, with no choice of method or test conditions, specifically to yield completely comparable data for presentation in a database. Admirable though this approach may seem, it has to be recognized that the freedom of action previously allowed enabled the tests to be used over a wider range of industrial conditions than that envisaged by those setting up databases.

Automation and, in particular, the application of computers to control tests and handle the data produced have brought about vast changes in recent years. It is not only a matter of automation saving time and labor; it also influences the test techniques that are used. For example, these developments have allowed difficult procedures to become routine and hence increased their field of application. There are many examples of tests that would not exist without certain instrumentation, thermal analysis techniques being one of the more obvious. Advances in instrumentation for an established test may change the way in which it is carried out but do not generally change the basic concept or change it to produce more fundamental data.

Whether automatic or advanced instrumentation really saves money is difficult to say. Initially the equipment costs more, but this is offset by a saving in labor. However, the old adage that if a thing can go wrong it will remains true, and maintenance costs of complex equipment are high. Finally, the calibration of such equipment can be difficult, and the software that so readily transforms the data can give rise to concern as to what has happened between the transducer and the final output.

While improvements in tests in respect of their usefulness for generating design data and predicting service performance are continually sought, the advances have perhaps been less dramatic. The fundamental tests needed for design are often very difficult to devise and are likely to be more expensive to carry out and required only by a minority. As with so many things, the advances can be related to commercial pressures and the amount of effort that is funded. Where better and more fundamental tests do exist they are not always used as often as they should be because of the cost and complexity involved.

There has been an increase in tests on products, which has resulted from a greater demand to prove product performance and from specifications more often including such tests as part of the requirements.

For the future, it is highly probable that the same themes will continue. The quality movement is still strong, and the generation of databases will probably ensure that greater compatibility is achieved. Certainly there will be further developments in instrumentation and the handling of data. It would be a brave person who predicted a surge in tests for better design data, but there are signs that the sophistication of markets will lead to wider needs in this direction.

4 Test Conditions

Under the broad heading of test conditions should be included the manner of preparation of the material being tested and its storage history, as well as the more obvious parameters such as test temperature, velocity of test, etc. While it is recognized that the result obtained depends on the conditions of the test, it is not always obvious that some of these conditions may have been established before the samples were received for testing. Sometimes the history of the samples is part of the test procedure, as in aged and unaged samples for example, but at other times it may not be at all clear that certain "new" samples are already several months old, with their intervening history unknown. Degradative influences such as the action of ozone on rubber samples cannot be compensated for, but standard conditioning procedures are designed, as far as possible, to bring the test pieces to an equilibrium state. The imposition of a standard thermal history before measuring the density of a crystalline polymer is a good example.

In some cases, conditioning may involve temperature only, but where the material is moisture sensitive it is likely that a standard atmosphere involving control of both temperature and humidity will be called for. Occasionally, other methods of conditioning, such as mechanical conditioning, are used, as will be discussed in a later chapter.

Even with careful conditioning, however, the results produced from specimens manufactured by different methods may vary, and if there is to be a controlled comparison it is important that the test pieces be prepared in exactly the same way. This is particularly important for figures being presented in databases. For example, laboratory samples of a rubber prepared on a mill may differ considerably from factory materials prepared in an internal mixer, and often these differences are not sufficiently emphasized in tables of data.

Equally, test piece geometry is important, and again, if comparison is to be made, a standard and specified geometry should be adhered to. Rarely is it possible to convert from one geometry to another, since polymers are complex materials and the influence of the various test parameters is often nonlinear. For example, it is difficult to scale up gas transmission results obtained on thin sheets to thick sheets of the same material.

For these various reasons the simulation of service behavior is at best difficult and often impossible. There are numerous examples of long-term tests over 20 or more years that have shown that artificial ageing using heat or other means yields results that are significantly different from those obtained with the passage of time. There has to be an awareness of the limitations of any test procedure and an acknowledgement that the results obtained apply only to the narrow range of conditions under which the test was performed. For these reasons, with important and complex products such as tires, it is often necessary to test them under the exact conditions under which they will be used.

Test procedures require careful attention to detail, as small and apparently innocent deviations can produce significant changes in results. This implies that the test conditions need to be accurately set initially and then monitored throughout the test. Sometimes it arises that when testing according to a published standard some deviation from the set procedure cannot be avoided (perhaps because of a limitation on the amount of material available). In these cases such deviations should always be recorded. In any test report it is important to state quite clearly which procedure has been followed.

5 Limitations of Test Results

However carefully tests are organized and carried out, there will be a limit to the accuracy attainable, an obvious fact sometimes hidden by the multidecimal results obtainable from pocket calculators. Furthermore, once interlaboratory variability is considered, the precision of a particular method may be lower than was anticipated.

The widespread availability of personal computers has greatly simplified statistical analysis of test results, and there is now a greater appreciation of the value of such analysis. To this end the expanded British Standard Application of Statistics to Rubber Testing [2], which is also applicable to testing other polymers, contains useful information and is recommended. It is not sufficient to present the results to a statistician once the tests have been done. Thought must be given to the design of the experiment in the very beginning, and if help from a statistician is needed it should be brought in at that stage. Because of the importance of applying statistical principles to test results, the subject is comprehensively covered in Chapter 3.

Brief mention was made of the precision of tests as judged by interlaboratory trials, and sometimes the quoted level of precision seems relatively poor. Usually the laboratories taking part in these trials are experienced, and the precision levels quoted should be representative of good practice. Poor figures may indicate snags with that method but, whatever the quoted levels, there is no reason to suppose that it is "only the others" who get divergent results. Interlaboratory comparisons sometimes lead to the elimination of poor test procedures, and so bring about improved accuracy, as will be discussed later.

However, no measurement is exact, and there is always some uncertainty. Calibration laboratories are required to make uncertainty estimates for all their measurements, and in the future it may be that all accredited testing laboratories will also have to do so. This involves estimating the uncertainty introduced by each factor in the measurement and is not at all easy to do. At the very least, it is essential to be conscious of the order of magnitude of the range within which the "true" result lies.

6 Sampling

Efficient sampling means selecting small quantities that are truly representative of a much larger whole, and the significance of test results is closely related to the efficiency of sampling.

Often, in the laboratory, one is limited by the amount of material available, and at least there is then the excuse that the tests relate only to the material available at the time. In a factory, where the whole output is available, the problem is a different one. Here the quality control manager has to decide not only what is adequate, but also what is reasonable, bearing in mind the production schedule and the profitability of the operation.

The frequency of sampling and the number of test pieces (or repeat tests) per item sampled depend on circumstances, and obviously financial considerations play an important part. Certain long-winded (and expensive) tests call for one test piece only, although if multiple tests are done the method may be quite variable. The use of a single test piece is hardly satisfactory, but it may be that multiple tests in numbers sufficient to increase precision are totally uneconomic. This is the dilemma that quality control managers (and the writers of specifications) have to face. In a continuous quality control scheme it may be that the number of test pieces at each point is less important than the frequency of testing.

Where multiple tests pieces are available, an odd number is advantageous if a median is to be taken, and five seems to be the preferred number. This is just about large enough to make a reasonable statistical assessment of variability. However, the current range of standard methods is not consistent, and numbers between one and ten or more may be called for.

The essence of efficient sampling is that the small quantity selected and tested (the sample) be truly representative of the much larger whole. The test pieces should be representative of the sample taken, the sample representative of the batch, and the batch representative of the wider population of material. In many cases, this information is not known to the tester, but there should be awareness of the limitations of the results in this respect, and the best possible practice should be followed in selection of samples and test pieces. This may include blending of several batches, randomizing the positions from which test pieces are cut, and testing on test pieces cut in more than one direction.

Care should be taken when sampling from production that items be taken at random, and that the time at which samples are taken does not always coincide with some factor such as a shift change.

7 Quality Control

Quality control embraces the monitoring of incoming materials, the control of the manufacturing processes, and checks of materials and products produced, so as to ensure and maintain the quality of the output from the factory. Physical testing methods are important in this regime, and most of the standardized test methods are intended for quality control use—it is probable that the majority of tests carried out are undertaken in the first place for quality assurance purposes. However, this book is about testing and is not a quality control manual, so discussion here is restricted to the quality control of the testing process.

Quality control is often thought of as applying only to products, since this affects the lives of the entire population. However, those of us that work in laboratories must recognize that correct and reproducible results are in a sense products, and that the application of quality control to test laboratories is designed to improve the general reproducibility of all test results.

Reliable results can only come from a laboratory where the apparatus, the procedures, and the staff are all subject to a quality assurance system. ISO 9000 standards are applied in a wide context to various companies, and their laboratories will be included under the general umbrella of such a system, but a more focused scheme for test and calibration laboratories may be found in ISO guide 25 [3] and national equivalents. These standards cover not only the calibration of equipment and the control of test pieces but also the training of staff, an item tending to be overlooked in the general context of quality control. The requirements listed set a high standard, and it has to be recognized that maintenance of this standard is time-consuming and difficult. In the UK the accreditation of laboratories is entrusted to the United Kingdom Accreditation Service (UKAS). Similar organizations may be found in other countries, and some of these bodies have mutual recognition agreements.

Undoubtedly the most expensive item in any system of laboratory control is the calibration of equipment. All test equipment should be calibrated, and every parameter relating to that machine requires formal calibration. For example, it is easy to see that the force scale and speed of traverse of a tensile machine need calibrating, but it is less obvious that the cutting dies for test pieces also need calibrating in order to ensure that the test pieces conform to specification.

Calibration is based on the principle of traceability from a primary standard through intermediate or transfer standards. A good example of a transfer standard would be boxes of certified weights that are not in general use but the sole purpose of which is to check the accuracy of those that are in use.

Obviously, at each stage of measurement there is some degree of uncertainty, and estimates of this uncertainty form part of the calibration procedure. It is perfectly acceptable for a laboratory to carry out its own calibrations, provided they maintain appropriate calibration standards and operate a suitable quality system. However, it is often more convenient to buy-in calibration services. Wherever possible the calibration laboratory used should be accredited (UKAS or equivalent).

Calibration of apparatus in the polymer industry has to some degree been hampered by the lack of definitive guidance, but a British standard has been developed covering the

Calibration of Rubber and Plastics Test Equipment [4]. This explains the principles of calibration and gives details of the parameters to be calibrated and the frequency required, together with an outline of the procedure to be used for all rubber test methods listed in the ISO system.

The ASTM gave the lead in conducting systematic interlaboratory trials, and this has been followed by the ISO and others. The variability obtained was far greater than was expected, and in some cases it was so bad that it was doubtful whether certain tests were worth doing at all. These interlaboratory comparisons and the drive towards improved quality led to an abandonment of the complacent attitude that had formerly existed and stimulated various initiatives to improve the situation.

On the whole, variability arises from malpractice rather than from a poorly expressed standard, but if an interlaboratory trial reveals an excessive variability it is first necessary to pinpoint the problem before a standards committee can correct it. Unfortunately this is a slow and expensive procedure.

The demand for higher quality has produced pressures to make laboratory accreditation commonplace, and as more laboratories reach this status it must be expected that reproducibility will improve. The calibration of test machines, training, documentation of test procedures, sample control, and formal audits all have an enormous influence, and the discipline involved in maintaining an accredited status helps to minimize mistakes and maintain reproducibility. International agreements undoubtedly widen the scope of accreditation schemes and ensure uniform levels of accreditation. This is found to have an influence on the standard of laboratories with a consequent improvement in interlaboratory comparisons.

The essential requirements of any piece of test equipment are that it should satisfy the requirements laid down in the standard relating to the test method under consideration and that it should be properly calibrated. Convenience of use or the cost of running the tests are not items that can be specified, but nevertheless they play a dominant role in the selection of equipment. Increasingly computer control and data handling are becoming standard.

The manipulation of data by computer is a particularly difficult operation to monitor, since in a busy laboratory it is only too easy to accept the software as correct in all circumstances. Obviously the accuracy of the quoted results depends not only on the accuracy of the original measurements but also on the validity of the data handling. Some standard bodies are now developing specifications giving rules and guidance on software verification.

These changes in the basic concepts of laboratory testing bring with them both advantages and disadvantages. While it is obvious that automation brings with it a saving in staff time, perhaps enabling measurements to continue with the apparatus attended only periodically rather than continuously, it is not clear what the effect on accuracy or reproducibility will be. Noncontact extensometers, for example, ensure that there are no unwanted stresses on the test piece, but the accuracy is related to the parameters built into the extensometer (e.g., the response time in following the recorded signal). It is important not to assume that more complex equipment necessarily means improved accuracy, although it is frequently true. A simple example may illustrate the difficulty. Increasingly doctors are using electronic sphygmomanometers to measure blood pressure. Here the end points still rely on the doctors' skill in detecting a pulse, but rather than reading the height of a mercury column, a pressure transducer gives a direct digital readout. This gives a degree of confidence that is absent from a mercury manometer, but the defects in the system are hidden. There is an obvious need for calibration, which may go unrecognized in a busy surgery, but also the question of linearity of response is crucial. It is easy to look critically at the equipment used in a different discipline, but the same principles should apply in our own laboratories.

References

1. Brown, R. P., *Physical Testing of Rubber*, Chapman and Hall, London, 1996.
2. BS 903, Part 2, Guide to the application of statistics to rubber testing, 1997.
3. ISO Guide 25, General requirements for the competence of calibration and testing laboratories, 1990.
4. BS 7825, Parts 1–3, Calibration of rubber and plastics test equipment, 1995.

3
Quality Assurance of Physical Testing Measurements

Alan Veith
Technical Development Associates, Akron, Ohio

1 Introduction

Measurement and testing play a key role in the current technologically oriented world. Decisions for scientific, technical, commercial, environmental, medical, and health purposes based on testing are everyday occurrences. The intrinsic value and fidelity of any decision depends on the quality of the measurements used for the decision process. Quality may be defined in terms of the *uncertainty* in the measured values for a specified test system; high quality corresponds to a small or low uncertainty. Quality is contingent upon whether the operational system is simple or complex. The equipment, the procedure, the operators, the environment, the decision process itself, and the importance of these decisions—are all part of the system. A lower quality can be tolerated for less important routine technical decisions than for decisions that have large commercial or financial implications. Measurement and testing for fundamental and applied research and development and also for producer–user transactions are important elements that are part of a larger organized effort that is frequently called a *technical project*. Measurement and testing play a key role in all technical projects and the assurance that the output data from any technical project are of the highest quality, consistent with the stipulated goals and objectives, is of paramount importance in technical project organization.

Quality for a test system has two major components: (1) how well the measured parameters relate to the properties that are involved in the decision process and (2) the magnitude of the uncertainty of the measured parameter value or values; the higher the uncertainty the lower the quality. The first component is usually more complex, since it involves scientific expertise and some subjective judgments. If the measured parameters are not highly related to the decision process properties, a *fundamental scientific uncertainty* exists. The second component, the *measurement uncertainty*, is somewhat easier to address;

it is assessed and controlled by the application of a number of statistical and technical disciplines that have been developed over the past several decades. Once a certain level of quality is established and maintained by the use of specified control techniques, it follows that a certain degree of assurance that this quality is achieved must be part of any ongoing project. The purpose of this chapter is to give some elementary background on how to assess, control, and assure the quality of physical property measurements.

The chapter begins with a brief description of the essential components of a technical project. Next is a short section on elementary statistical principles that reviews some of the necessary concepts used in the immediate sections that follow. This is succeeded by sections on the principles of measurement and calibration and then some basic sampling theory. A more extensive section on statistical analysis is next, which then leads to the principles of quality assessment and control . The chapter concludes with a discussion of the topics of precision, bias, and uncertainty in laboratory testing. The word *uncertainty* is used in two different senses in the chapter, first as a generic term as defined above and second as a particular or specific term that defines a range or interval for any point estimate of a measured value. This distinction is readily apparent in the sense of its use, and this topic is fully discussed in the last section and in some of the annexes. Annex A gives some general statistical tables useful for the various statistical analysis algorithms. Annex B describes a statistical model for the testing process, which is also discussed in the last section of the chapter. Annex C gives a procedure for evaluating accuracy and bias. Precision is expressed in terms of within-laboratory variation, called *repeatability*, and between-laboratory variation, called *reproducibility*, and since these are important concepts, the calculation procedure for these two precision parameters is given in Annex D. These annexes should be consulted as indicated and as needed in the various sections of the chapter.

There is no completely standardized or uniform terminology for the disciplines of quality assurance and statistical analysis. Attempts at harmonization are being addressed by standardization organizations world-wide with varying degrees of success. In this chapter an attempt is made to use a terminology that is consistent within the chapter and also with the current trends in statistical nomenclature, but the terms and symbols may be at variance with those from other sources. Since all concepts and symbols are fully defined, this should present no substantial problems for the reader.

In addition to references to specific literature sources, a bibliography is given at the end of the chapter for appropriate statistical textbooks and for ASTM and ISO standards that apply to quality assurance, statistical analysis, precision, bias and uncertainty, laboratory accreditation, and proficiency testing . The listing of standards is not exhaustive; only those that are anticipated to be worthwhile for the topic of this chapter are included. These standards, with some exceptions as noted, were developed by committees on statistics and quality as generic standards that apply in principle to all testing and measurement operations.

2 Defining Technical Projects

Testing operations are frequently organized on the basis of a technical project. Each project with its defined operational system should have a specific goal—the solution of a particular problem that requires measurements, the generation of test data, analysis, and technical decisions. A successful execution requires that the project be well organized with a number of steps that must be undertaken in a specified order. Figure 1 illustrates these steps in a flowchart diagram. The steps involve (1) planning and modeling of the project, both of these being closely linked and used in an interactive way, (2) selecting a sampling

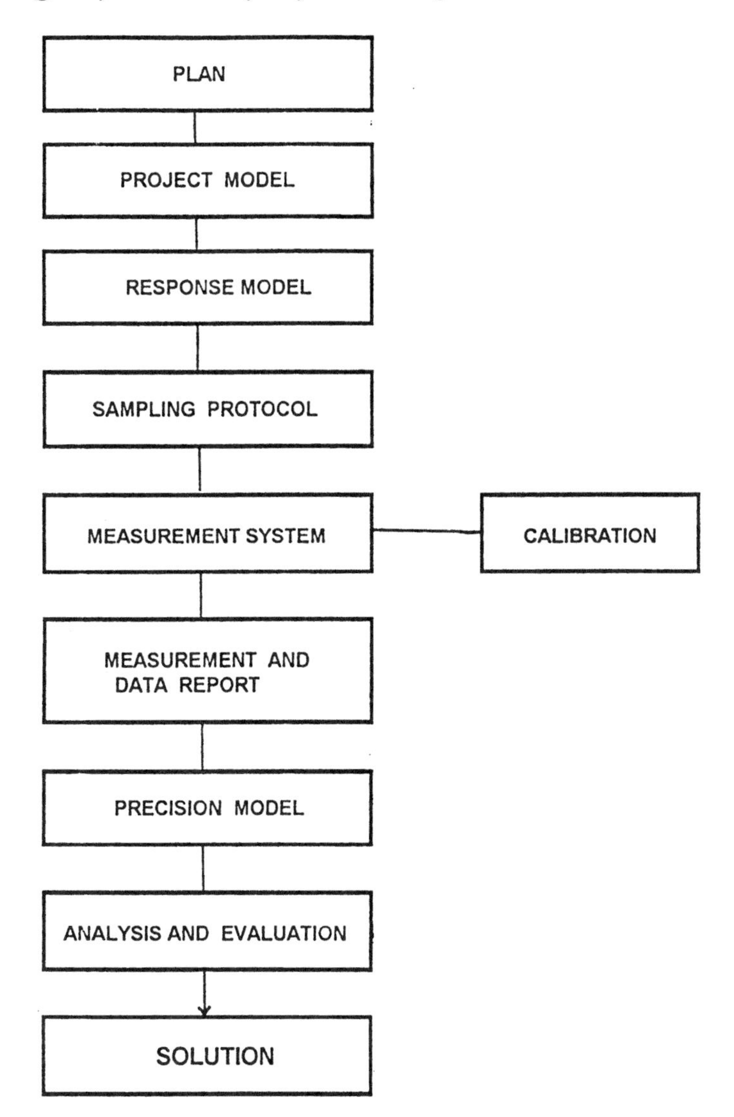

Figure 1 Flow diagram for technical project.

procedure and protocol, (3) setting up a defined capability measurement system with an appropriate calibration procedure, (4) performing the measurements and reporting the data, followed by (5) analysis and evaluation, which may make use of a test result uncertainty model to arrive at a solution. All of these must be of the highest quality for the successful execution of a complex project, especially if the project involves interlaboratory testing.

2.1 The Plan and the Models

A project plan and the required models, which are jointly evolved, are usually developed by a principal investigator in consultation with various participants who frequently are specialists in certain technical disciplines that are ancillary to the project. Topics that need attention are overall project organization, goals, and objectives, resources and constraints, performance criteria, the selection of the measurement methodology, and the selection of decision procedures. A set of well designed and coordinated standard operating procedures (SOP) must be selected and put into place. All the elements of the project required for a successful solution need to be specified.

A model may be defined as the *simplified representation* of a defined physical system. The representation is developed in symbolic form and is frequently expressed in mathematical equations and uses physical and/or chemical principles based on scientific knowledge, experimental judgment, and intuition, all set on a logical foundation. A model may be theoretical or empirical, but the formulation of an accurate model is a requirement for the successful solution of any problem.

Planning Model

A planning model is a generalized statement of the strategy to be used in the solution of a problem; it involves the selection of a coordinated set of SOPs and an assembly of the required system elements to arrive at a solution. Planning models are more descriptive in nature and are not as rigorous as response and analysis models.

Response Model

A response or analytical model is a mathematical formulation that describes one or more complex measurement operations that are part of a particular project. Once the test methodology has been selected, the response model should be constructed based on the performance parameters of the system and the independent variables that influence these performance parameters. This usually involves three steps: (1) formulation of the model, (2) calibration of the model, and (3) validation of the model. There are three important actions in the formulation: (i) identify the most important variables or factors that influence a selected response, (ii) develop a proper symbolic or mathematical representation for these variables, and (iii) identify the degree of interaction among the variables. Variables may be selected on the basis of theoretical principles or on an empirical approach using correlation and regression or principal components analysis techniques. Systematic experimental designs may be employed to formulate an empirical model; see Section 6.6.

The specified number and character of performance parameters and variables, i.e., the operational conditions, is defined as a *testing domain*. Simple domains for any project may require only a univariate model, while complex project measurement systems may require multivariate models. Some projects may have multiple response parameters, each of which may require a multivariate (independent) variable model. The calibration and validation operations are discussed below in Section 2.3.

Test Uncertainty Model

This is a model that may be used to relate the variation in the measured parameter(s) to sources of variation that are inherent in any testing operation. This is discussed in more detail in Section 8 on precision, bias, and uncertainty in laboratory testing.

2.2 Sampling Protocol

Sampling is a very important operation in any decision-making process based on measurement. The samples must be truly relevant and represent the population under consideration. Knowledge of the response and the uncertainty model are needed to be able to select a sampling protocol that will give the desired information about the process. The sampling plan should carefully describe how samples (number, type, etc.) are to be selected, and the operation of subsampling, if used, must also be documented. Procedures to protect or preserve samples during use and storage as well as the chain of custody for both the physical samples and the resulting data are required. See Section 5 for a discussion of basic sampling operations.

2.3 Measurement System

The appropriate methodology or test methods for conducting the selected measurements should be chosen and completely described as to written SOPs. Standard test methods should be used if at all possible, and if modifications are made on a standard method they should be clearly documented. See Section 4 for more details on the principles of measurement. The methodology should be validated as to its accuracy and sensitivity in measuring what is intended.

Model calibration, in distinction to test method calibration as discussed in Section 4, is essentially a fitting process, i.e., evaluating experimental model coefficients by regression and goodness-of-fit procedures; it should be completely described as to (i) how reference or other materials are used, (ii) the schedule for the initial and ongoing calibrations, and (iii) how the confidence in the calibration operation is to be quantified and expressed. It is important to perform the model calibration or coefficient evaluation process over a substantially large range of the operating or independent variables. This results in a more robust model with a broader applicable range. Validation is a process whereby the influence of the relevant input variables is observed for rational model output. The validation program should be based on input–output response generated from a separate testing and evaluation operation after all model development is concluded. The validation should also be conducted over the broadest range possible.

2.4 Test Data Report

Once the above issues have been addressed, the next step is to conduct the measurements and report the data. For simple projects the *as measured* data may be of primary concern, and if this is the case the averages or means for the data are given usually at some confidence level for a specified uncertainty interval. Analysis and other statistical operations, using any developed precision model for guidance, may be conducted on the raw data. However, the measured parameter(s) may be used as input to a mathematical expression that gives some derived parameter that is used to assess system performance and make decisions. A typical example is the use of one test or reference material as a control to express the performance of the tested candidate materials, usually as an index obtained by dividing the candidate measured parameter by the control measured parameter. When this is the case, the statistical parameters obtained by analysis of the raw data will not apply to these derived performance parameters or properties. Under these conditions, one must use propagation of error algorithms that give the variation or error of the derived property or parameter in terms of the measured parameter(s). A full discussion of this topic is given in Section 3 on statistical principles.

2.5 Analysis, Evaluation, and Solution

The techniques on statistical analysis as described in Section 6 may be used to analyze the data generated in project testing. For more sophisticated or complex analysis procedures, dedicated statistical software computer programs may be used. If derived expressions are necessary for performance, the formulas of Table 3.2 in Section 3.6 may be applied. Most projects will involve some form of iteration involving successive measurement operations to arrive at a satisfactory solution.

3 Elementary Statistical Principles

3.1 Test Data Variation Defined

Throughout this chapter data are assumed to be expressed on a continuous numeric variable scale; discrete or attribute data are not discussed. The general term *variation* is introduced in the usual classic textbook sense and may be described in terms of *deviations* of the output value from a *true* value that would be obtained for any measurement if, as in an ideal world, no variation existed. A true value may be (i) a theoretical value that can be calculated based on scientific principles; (ii) a theoretical value that exists but cannot be calculated from first principles: in its place a reference value obtained by a collaborative program by a standardization or other technical organization is adopted; this frequently uses standard reference materials; and (iii) an assigned value based on a reference test established by a recognized organization. Both of the later designations are called *conventional* true values.

All measurements are influenced to some degree by *perturbations* that influence the measurement process and generate these deviations. Each measurement may be perturbed by one or both of two types of deviation.

Random deviations : these are + and − differences about some central value that may be a true or a reference value; each execution of the test gives a specific difference, and the mean value of these differences is zero for a long run series of repetitive measurements.

Bias (or systematic) deviations: these are offsets or constant differences from the true value; these offsets may be + or − and are frequently unique for any particular set of test conditions.

Random and bias deviations are usually additive. There are four concepts that are applied to measurement variation: precision, bias, accuracy, and uncertainty. Frequently these are incorrectly used in an interchangeable manner. One of the main purposes of this chapter is to distinguish between these and show how each may be used correctly. Precision is defined in terms of the degree of agreement for repeated measurements under specified conditions and is customarily assumed to be caused by one or more random processes. High precision implies close or good agreement. Bias has not been addressed and investigated to the extent that has been devoted to precision. But as Section 8 of this chapter will show, bias is very important in testing and needs more attention to improve test quality. Accuracy is a concept that involves both precision and bias deviations. High accuracy implies that the sum of both types of deviation is low, or in an ideal situation, zero. The term uncertainty is of more recent origin and may be used in two different contexts as previously described. The relationship among these variation concepts is discussed in Section 8; in Annex B, which gives a statistical model for testing; and in Annex C on the evaluation of bias.

This section gives some of the more elementary statistical parameters that may be used to characterize and analyze data and discover the underlying relationships among variables that may be hidden by the overlaid variation or noise. Although everything discussed in this section is available in standard texts, a review of the more elementary statistical principles is presented to address the basic problems of measurement quality. This is given in what is hoped to be an orderly manner to make the chapter a reasonably self-contained document.

Today, with the widespread availability of computers and appropriate software, statistical analysis is within the reach of essentially anyone conducting a laboratory measurement operation. They may be used with spreadsheet programs for elementary analysis of small-to-intermediate databases, using the typical statistical algorithms built into these programs. Computers may also be used with dedicated statistical analysis programs for more complex and comprehensive analysis, especially for large databases. The use of computers both to acquire and to analyze data has evolved to the point where practically all test instruments are either equipped with a dedicated computer or connected to a personal computer and/or centralized computing system. These systems frequently are programmed to do elementary statistical calculations and print out means (averages) and standard deviations according to preset algorithms for each tested material or object.

However, rote application of packaged statistical routines without an adequate understanding of the principles involved can present a major problem in decision making. Statistical calculations and decisions should not be made blindly; they should always be accompanied by intuitive judgments based on well-established technical experience and common sense as well as an acceptable level of statistical competence. This approach is especially important when a small database is being considered.

3.2 Basic Probability Concepts

A simple but effective definition of probability is based on a frequency of occurrence concept. If an event A can occur in n_a cases out of N possible and equally probable number of cases, the probability $P(\mathrm{A})$ that the event will occur in any new trial is

$$P(\mathrm{A}) = \frac{n_a}{N} = \frac{\text{successful cases}}{\text{total number of cases}} \tag{1}$$

Probabilities are thus expressed as a value in the interval 0 to 1, where 0 indicates no probability and 1 indicates certainty. The following theorems are useful:

If $P(\mathrm{A})$ is the probability of an event, $1 - P(\mathrm{A})$ is the probability that the event will not occur.

If A and B are two events, the probability that either A or B will occur is

$$P(\mathrm{A\ or\ B}) = P(\mathrm{A}) + P(\mathrm{B}) - P(\mathrm{A\ and\ B}) \tag{2}$$

As a special case of the above, if A and B cannot occur simultaneously, i.e., they are mutually exclusive, then the probability that either A or B will occur is

$$P(\mathrm{A\ or\ B}) = P(\mathrm{A}) + P(\mathrm{B}) \tag{3}$$

If A and B are two events, the probability that events A and B will occur together is

$$P(\mathrm{A\ and\ B}) = P(\mathrm{A}) \times P(\mathrm{B}|\mathrm{A}) \tag{4}$$

where $P(\mathrm{B}|\mathrm{A})$ is the probability that B will occur assuming that A has already occurred.

A special case of the above is the situation where the two events A and B are independent (no influence of one event on the probability of the other event); then the probability of both A and B occurring is

$$P(\mathrm{A\ and\ B}) = P(\mathrm{A}) \times P(\mathrm{B}) \tag{5}$$

3.3 Data Distributions

Any individual data value belongs to or originates from some *population* of values that has certain properties that are collectively designated as a *distribution*. A population may be (1) a single (or a few) object(s) or a very limited mass of material, (2) a finite (but large) number of objects or a large mass of material or (3) a hypothetical infinite number of objects or mass of material; all three interpretations imply that the objects or material are generated by some identifiable process and usually have a recognized property range. Distributions describe the frequency of occurrence of individual data values about some specified central value: a mean or median as defined below. Distributions are characterized by a probability distribution function, which is expressed in terms of distribution parameters that relate to the central tendency and the dispersion of values about this central value.

The normal or gaussian distribution has a typical bell-shaped symmetrical frequency of occurrence below and above the mean. Nonnormal distributions have a nonsymmetric occurrence frequency, and such distributions can be made to approach normality by an averaging process. Thus a secondary population of individual means, each based on n values from a primary nonnormal population, approaches a normal distribution as n increases. This is called the *central limit theorem*. Nonnormal distributions can also be made to approach normality by the use of transformations. In any ongoing testing operation that has a known normal distribution, a number of actions can occur that may cause deviations from normality.

Presence of a number of outliers
Shifts in operating conditions of the system
Undetected cycles or transient instability in the operation

All three of these perturbations will be discussed in succeeding sections of this chapter.

3.4 Characterizing Distributions—Mean, Variance, Standard Deviation, and Range

Some elementary concepts of the distribution and its characterization will be discussed here. A more comprehensive computer analysis is listed in the Statistical Analysis section. A database that is presumed to have a normal distribution may be characterized by two types of statistical parameters: one that establishes its *central value* and one that characterizes the spread or *dispersion of values* around the central value.

Central value—the *mean* (or for special cases the median)
Dispersion—the *variance* and *standard deviation* (square root of the variance) and the *range*

The mean (frequently known as the arithmetic average), variance, and standard deviation can be realized in two ways: (1) as a *true parameter value* based on extensive measurement or other knowledge of the entire population, in which case these parameters are designated by the symbols μ, σ^2, and σ respectively or (2) as *estimates* of the true values based on samples from the population, in which case they are designated by the symbols $\bar{x}$, S^2 and S respectively. The equations to calculate the mean, variance, and standard deviation as estimates from a sample are

$$\bar{x} = \frac{x_1 + x_2 + \ldots + x_n}{n} \tag{6}$$

$$S^2 = \frac{\Sigma(x_i - \bar{x})^2}{n - 1} \tag{7}$$

$$S = \left[\frac{\Sigma(x_i - \bar{x})^2}{n - 1}\right]^{1/2} \tag{8}$$

where x_i = any data value, n = the total number of sample data values, and Σ indicates summation over all values. The degrees of freedom *df* for Eq. 7 and 8 is $(n - 1)$. The degrees of freedom is the number of independent differences available for estimating the variance or standard deviation from a set of data; it is one less than the total number of data values n, since one degree of freedom (of the total degrees equal to n) is used to estimate the mean.

The majority of statistical calculations and resulting decisions are based on sample estimates for $\bar{x}$, S^2, and S. In certain circumstances the true values are known and slightly different procedures are used for statistical decisions. The standard deviation gives an estimate of the dispersion about the central value or mean in measurement units. For a normal distribution the interval of $\pm\sigma$ about the mean μ will contain 68.3 percent of all values in the population; the $\pm 2\sigma$ interval contains 95.5 percent, and the $\pm 3\sigma$ interval contains 99.7 percent. A relative or unit-free indicator of the dispersion is the coefficient of variation:

$$\text{Coefficient of variation} = CV = \frac{S}{\bar{x}} \tag{9}$$

The coefficient of variation may be expressed as a ratio or as a percentage by multiplying the ratio by 100.

In most testing operations the mean is used in preference to the median, which is defined as the middle value of an ordered set of (ascending, descending) values. A median may however be used in special circumstances; in some respects it is a more robust statistical parameter, being less influenced by extreme values. An important use of the median is in tension testing, where the tensile strength very frequently has a nonnormal distribution. The use of the median for small test samples (only a few strength measurements per material) usually is a better indicator of the true strength, since extremely low values due to abnormal flaws are of little influence.

Pooled Estimates of Variance, Standard Deviation

The variance and its square root the standard deviation along with the range discussed below are measures of the spread of the data and as such indicate the precision in an inverse manner. It is frequently assumed that a substantial number of test replications (12 or more) have to be made to estimate a variance or standard deviation. There are certain situations where this number of replications is required for important testing operations, but there are alternative ways to get good estimates of these statistical parameters for any test without this large number of measurements for each test material by a procedure called pooling. There are two general approaches.

If a measurement of some property is being conducted on a number of materials that are similar in their characteristics and response to the measurement process, then a reasonable assumption can be made that, although the means may be different, the variance of the testing measurement is the same for each material. On this basis a simple testing plan to evaluate a number of materials can be used, with only a few measurements on each material. The pooling of these individual estimates will give a good overall testing variance estimate. A procedure is given in the Statistical Analysis section to determine if the variances obtained on a number of test materials can be considered as equal.

A second approach is the accumulation of individual variance estimates each with only a few degrees of freedom, on one or more reference materials or objects over a period of time. The reference material should have the same measured property magnitude and the same general test response as the experimental materials if the estimated variance is to be used for decisions on the experimental materials. Both approaches may be used for a more comprehensive effort.

The process of pooling or averaging the individual variance estimates (each with only a few degrees of freedom) is equivalent to a weighted average calculation. Thus the pooled variance is obtained from the sum of each individual variance estimate $S^2(i)$ multiplied by the number of values in its replicate set n_i divided by the sum of the individual number of values in each replicate set. The number of degrees of freedom *df* attached to the pooled variance is equal to the sum of the number of individual *df* of each replicate data set.

$$S^2(p) = \frac{n_1[S^2(1)] + n_2[S^2(2)] + \dots}{n_1 + n_2 + \dots} \tag{10}$$

Although it is not as important today with the ready availability of computers, there is a quick calculation algorithm to obtain a pooled variance, $S^2(p)$, from a sequence of duplicate measurements obtained under either of the approaches as listed above.

$$S^2(p) = \frac{\Sigma d^2}{2k} \tag{11}$$

where

d = difference between duplicate measurement values
k = number of duplicate sets measured ($df = k$)

Pooled standard deviations may be obtained as described in Eqs. 10 and 11 by taking the square root of the variance.

For small data sets (10 or less) another convenient way to obtain a measure of precision is the use of a multiplying factor for the range to give an estimate of the standard deviation. Equation 12 gives the expression for the standard deviation S and Table 1 gives the factors f_R for sample sizes of 2 to 10.

$$\mathrm{S} = w(f_R) \tag{12}$$

where

w = range = (max value − min value)
f_R = factor for standard deviation calculation

The Sampling Distribution of the Mean

If a random sample of size n is drawn from a normal population having a mean μ and variance σ^2, then the mean of the n values in the sample $\bar{x}_{(n)}$ is a random variable whose distribution has the mean μ and a variance of σ^2/n or a standard deviation of $\sigma/\sqrt{n}$, which is frequently called the standard error of the mean. Note that this parameter, which establishes the reliability of the mean $\bar{x}_{(n)}$, decreases as the square root of n; it is necessary to quadruple the values n in order to halve the standard error of the mean.

Table 1 Estimate of Standard Deviation from Range, w

Sample size n	Factor f_R
2	0.886
3	0.591
4	0.486
5	0.430
6	0.395
7	0.370
8	0.351
9	0.337
10	0.325

3.5 The *Z*-Distribution and the *t*-Distribution

The unique bell shape of the normal distribution curve may be characterized by an equation called the normal probability density function, which gives the probability of finding a given distribution value as a function of that value. To avoid having a separate equation for each measured parameter with its unique units, the function is adjusted to make the area under the distribution curve equal to one or unity, and this adjusted equation is called the *standard normal distribution.* The results of such calculations are given as tabular values. Using these tables, problems concerning the probabilities of occurrence of any value may be solved by the use of a random variable usually denoted by Z, associated with this distribution. Areas for a normal distribution for any variable x with known mean and variance are obtained by a simple transformation of origin and scale. This transformation converts the x variable with mean μ and standard deviation σ to the variable Z as given by

$$Z = \frac{x - \mu}{\sigma} \tag{13}$$

A distinction is made between the random variable Z and the standard normal distribution *tabulated values*, also equal to the right hand side of Eq. 13, which are designated by z. The expected value of Z is zero, and the standard deviation is 1. When sampling is conducted from a normal population that has been transformed into Z values, certain fractions or proportions of all of the values are contained within specified $-$ and $+$ multiples of Z. Just as in the case discussed previously for the standard deviation, 68.3% of all values drawn from the population are within the interval of $\pm Z$; 95.5% of all values within $\pm 2Z$ and 99.7% within $\pm 3Z$. When considering either actual x values or Z values, these intervals are called the one, two, and three sigma limits respectively, and they represent the long run outcomes of sampling.

An alternative way to express the occurrence of 68.3% of the Z (or x) values of the above population within the limits of $\pm Z$ (or $\pm\sigma$) is the use of the concept of probability. The statement—The probability that any selected population value x_i has a value in the range $\pm Z$ (or $\pm\sigma$) is 0.683—is equivalent to the above statement (for $\pm Z$) expressed in percentage format. With a standardized equation format, a table of probabilities may be generated that applies to any type of random normal variable. Probabilities are discussed in more detail below. Thus the probability that x will take on a value less than or equal to a is given by finding the probability P in a standardized normal distribution table for a value of $Z = (a - \mu)/\sigma$. Annex A Table A1 lists the values of Z and associated "areas" for each value of x expressed in terms of a difference from μ in σ units, where the area is equal to the probability that Z will take on a value less than or equal to the specified Z. These areas or probabilities are left-hand oriented, i.e., they begin at $-\infty$. From the table, the probability that a Z value is as low as -2.00 is 0.0228; only two chances out of a hundred.

A frequent use of Table A1 is for the selection of a value of z for a specified probability $P = \alpha$ to make certain statistical inferences. The z_α notation is used for this purpose, and two separate examples of its use are as follows

$$P(Z < -z_\alpha) = \alpha \tag{14}$$

$$P(Z > z_\alpha) = 1 - \alpha \tag{15}$$

The first of these expressions, Eq. 14, states that the probability that Z will fall in the region or area from $-\infty$ to $-z_\alpha$ is α. In the second application in the use of z_α, the probability that Z is equal to or greater than z_α is obtained by difference. First the probability that Z will fall in the range –infinity to +infinity is equal to the entire area under the curve or 1. The probability that Z will be equal to or less than z_α is equal to the area from $-\infty$ to z_α. The area to the right of z_α is the difference between these two areas or $1 - \alpha$.

Another application of the use of specific tabulated z values is in forming an interval. Intervals are discussed in more detail in Section 3.6 below. If α is divided into two equal regions at either extreme end of the distribution, then

$$P(-z_{\alpha/2} < Z < z_{\alpha/2}) = 1 - \alpha \tag{16}$$

which states that the probability that Z will be found in the region from $-z_{\alpha/2}$ to $z_{\alpha/2}$ is also $1 - \alpha$, since the original area has been cut in half and $-z_{\alpha/2}$ and $z_{\alpha/2}$ are defined as the two points on the z axis that cut off areas of $\alpha/2$ at each end. Equations 15 and 16 both have probabilities equal to $1 - \alpha$, but the z values are different in the two situations. Each of the two areas is equal to $\alpha/2$ at either end of the distribution. Equation 13 can be used for the mean of n sample values $\bar{x}_{(n)}$ from a population where x in the Z calculation expression is replaced by $\bar{x}_{(n)}$ and σ is replaced by $\sigma/\sqrt{n}$.

The Z-distribution and Eq. 13 are applicable when both the population mean and the variance are known. When the variance must be estimated from a sample, the z -distribution and the proportion of sampling values that fall within the – and + limits as given above no longer apply. In such circumstances a distribution called Student's t-distribution is used, and t is a random variable defined by Eq. 17.

$$t = \frac{\bar{x}_{(n)} - \mu}{S/\sqrt{n}} \tag{17}$$

The t-distribution is similar in shape to the normal or z-distribution, it has an expected mean of zero, but its variance depends on the degrees of freedom df associated with S, which is equal to $(n - 1)$. As n approaches infinity, the t-distribution approaches the normal or z-distribution and its variance approaches 1. Tables of t values, designated as t_α, at selected df are given for various probabilities $P = \alpha$, that the calculated value t (calc), as given by Eq. 17, will equal or exceed the tabulated t_α, which is usually called a critical t or t_α (crit). See Table A2 for tabulated t values, i.e., t_α (crit), at various df. Equation 17 applies to problems where the population mean is known or where a selected x value is to be compared to $\bar{x}_{(n)}$ for a decision on a significant difference.

3.6 Hypothesis Testing, Confidence Intervals, Tolerance Intervals

Hypothesis Testing

The act of testing an hypothesis is based on a procedure called *statistical inference*—decisions about sample data that are based on defined statistical principles. Statistical inference is necessary because samples contain incomplete information about any particular population, and any statement about population parameters has a certain level of confidence. This is usually expressed as the probability of making an error, i.e., of being wrong, when any assertion is made about the population. The assertion about a mean or variance is called a *statistical hypothesis*, and a procedure that permits agreement or disagreement about the hypothesis is called *a test of hypothesis*. Hypothesis tests are required because of *sampling error*, the variation of the calculated parameters in a series of samples, obtained from repeated testing, under identical operating conditions.

A particular hypothesis is tentatively adopted and evidence is sought to support it. If sufficiently strong evidence cannot be found, it is rejected. This may be illustrated by the following example using Eq. 13 above. If a population has $\mu = 10.0$ and $\sigma = 1.5$, and a value $x = 13.5$ is obtained, what can be concluded about the probability that such a value is really from the specified population? Evaluating Z gives

$$Z = \frac{x - \mu}{\sigma} = \frac{13.5 - 10.0}{1.5} = 2.3 \tag{18}$$

In making a decision about the likelihood that a Z of 2.3 could be obtained on the basis of chance alone, a *null hypothesis* is adopted that x is indeed from the specified population. Anticipating the possible outcome that the null hypothesis may not be true, a second hypothesis is adopted called the *alternative hypothesis*. These two are symbolically represented as H_0 and H_A respectively.

H_0 : x is a member of the specified population

H_A : x is not a member of the specified population

A criterion is set up to make a decision either (1) to accept the hypothesis as true with a selected probability of being incorrect, which is designated as a *level of significance*, or (2) to reject the null hypothesis and therefore accept the alternative hypothesis. A *Type I error* is made when H_0 is incorrectly rejected, i.e., H_0 is actually true but the sample-based inference procedure rejects it. A *Type II error* is made when H_0 is not rejected when in fact H_0 is not true but the inference procedure fails to detect this. The usual approach in inference testing is to select a certain probability or level of significance for Type I error. Statistical decisions about hypotheses are made on the basis of a *test statistic*; a parameter calculated from the sample whose sampling distribution can be specified for both the null and the alternative hypothesis at some selected level of significance α. At the selected α, the sampling distribution of this test statistic is used to define the *rejection region*. The rejection region is that set of values for which the probability is less than or equal to the specified α when the null hypothesis is true.

A maximum acceptable probability is selected, $P = \alpha$, for rejecting a true null hypothesis. This maximum or *critical probability* is customarily $P = 0.05$ or a one in twenty chance. If the calculated probability is equal to or less than the adopted critical probability or significance level, the null hypothesis is rejected. The general expression for the probability that a random normal variable x will take on a value between a and b is

$$P(a < x < b) = CA(b) - CA(a) \tag{19}$$

where

$P(a < x < b)$ = the probability that x will fall between a and b
$CA(a)$ = the cumulative area under the standard normal z-distribution curve up to the value a
$CA(b)$ = the cumulative area under the standard normal z-distribution curve up to the value b (both areas from − infinity)

In this example, the z value of 2.3 is equivalent to a, and b represents the probability that a random value will fall in the range −infinity to +infinity. Table A1 reveals that $CA(2.3) = 0.9893$. This is the probability of finding a z value in the range −infinity to 2.3. The cumulative area or probability b of finding a z value less than +infinity is of course exactly 1.0. Thus

$$P(a < x < b) = 1.000 - 0.9893 = 0.0107 = 0.011 \tag{20}$$

The calculated probability of 0.011 is substantially less than 0.05, and the null hypothesis is rejected and the alternative hypothesis accepted. Hypothesis testing may be applied to any statistical parameter (t-distribution, F-distribution, etc.) for which a sampling distribution may be calculated or otherwise evaluated.

An alternative method for reporting the results of significance calculations that has gained acceptance is to use the calculated probability as an indicator of the weight of evidence or the *strength* of an assertion about the parameter of interest. Instead of adopting a critical $P = \alpha$, with $\alpha = 0.05$ or other, and making a yes or no decision to reject the tentative null hypothesis, the calculations are made for P(calc), and this is used to indicate the decisiveness of the act of potential rejection of the null hypothesis. For this procedure, P is defined as the probability of committing a Type I error if the actual sample (measured) value of the statistic is used as the rejection value. It is the smallest level of significance to reject the null hypothesis based on the sample at hand and is usually called the *attained* or *empirical* significance level. Many statistical software programs give these calculated P values as output.

Confidence Intervals

The calculated values from any sample are considered as point estimates. Any such estimate may be close to the true value of the population (μ, σ or other) or it may vary substantially from the true value. An indication of the interval around this point estimate, within which the true value is expected to fall with some stated probability, is called a *confidence interval*, and the lower and upper boundary values are called the *confidence limits*. The probability used to set the interval is called the *level of confidence*. This level is given by $(1 - \alpha)$, where α is the probability as discussed above for rejecting a null hypothesis when it is true. In most circumstances, means are the most important point estimates, and confidence intervals for means are evaluated at some probability $P = (1 - \alpha)$ that the true population mean is within the stated confidence limits. This can be expressed for a population with a known standard deviation σ as given in Eq. 21.

$$P\left[\bar{x}_{(n)} - z_{\alpha/2}\left(\frac{\sigma}{\sqrt{n}}\right) < \mu < \bar{x}_{(n)} + z_{\alpha/2}\left(\frac{\sigma}{\sqrt{n}}\right)\right] = (1 - \alpha) \tag{21}$$

where

$\bar{x}_{(n)}$ = a mean evaluated from a sample of n values
$z_{\alpha/2}$ = the z value at $P = \alpha/2$
$\sigma/\sqrt{n}$ = the standard error (standard deviation of means of n)
μ = true mean of the population

This equation states that the confidence interval is defined as the difference between the two limits about the point estimate of the mean $\bar{x}_{(n)}$, i.e., from the lower limit $(\bar{x}_{(n)} - z_{\alpha/2}\sigma/\sqrt{n})$ to the upper limit $(\bar{x}_{(n)} + z_{\alpha/2}\sigma/\sqrt{n})$. This difference is designated as the $(1 - \alpha)$ 100% confidence interval. The true mean μ is a fixed number; it has no distribution and it is either in the interval or it is not. The interpretation of the value $(1 - \alpha)100\%$ is as follows—If a trial of 100 repetitions of the experiment of drawing a sample of n values from this population is conducted, then in the long run $(1 - \alpha)100\%$ of the intervals for each trial of 100, will contain the true value μ.

Confidence intervals may be alternatively formulated in terms of a factor k_{con} selected so that the calculated interval covers the mean μ a certain percent (proportion) of the time.

$$\text{Con Interval} = \pm k_{\text{con}} \frac{\sigma}{\sqrt{n}} \tag{22}$$

where Con Interval is the confidence interval at a selected P for means of samples of size n. As an example suppose that a sample of $n = 4$ is drawn from a population with $\sigma = 1.5$ and the estimated mean is 8.9. The 95% confidence interval ($\alpha = 0.05$), where 1.96 is the z value at $\alpha/2 = 0.025$ and $1.5/2 = 0.75$, is the standard error of means of 4, is given by

$$\begin{aligned} 95\%\ \text{Con Interval} &= \pm 1.96(1.5/2) = \pm 1.96(0.75) \\ &= \pm 1.47 \end{aligned}$$

or

$$\begin{aligned} 95\%\ \text{Con Interval} &= [8.9 - 1.47] \text{ to } [8.9 + 1.47] \\ &= 7.43 \text{ to } 10.37 = 2.94 \end{aligned} \tag{23}$$

For a situation where the standard deviation is not known but must be estimated from a sample in the same manner as the mean, the t-distribution applies, and the probability expression is

$$P\left[\left(\bar{x}_{(n)} - t_{\alpha/2}\frac{S}{\sqrt{n}} < \mu < \bar{x}_{(n)} + t_{\alpha/2}\frac{S}{\sqrt{n}}\right)\right] = 1 - \alpha \tag{24}$$

The lower limit is $\bar{x}_{(n)} - t_{\alpha/2}S/\sqrt{n}$ and the upper limit is $\bar{x}_{(n)} + t_{\alpha/2}S/\sqrt{n}$. As an example, if $\sigma = 1.8$ as calculated from the sample of $4(df = 3)$ and $\bar{x}_{(n)} = 8.9$ and $(1 - \alpha)100 = 95\%$ or $\alpha = 0.05$, the confidence interval would be found using the tabulated t value of 3.18 which is found at $P = 0.025$ for $df = 3$.

$$\begin{aligned} 95\%\ \text{Con Interval} &= [8.9 - 3.18(1.8/2)] \text{ to } [8.9 + 3.18(1.8/2)] \\ &= 6.04 \text{ to } 11.76 = 5.72 \\ &= \pm 2.86 \end{aligned}$$

This confidence interval is almost twice the previous example because the standard deviation as well as the estimated mean is only known to $df = 3$.

Tolerance Intervals

The word *tolerance* is used in a number of ways in testing and measurement technology. Engineering and design tolerances are usually designated as upper and lower limits on certain dimensions or other numerical factors for an object or product. Tolerances can also apply to the number of significant figures or digits to retain in a measurement. A third type of tolerance is concerned with the percentage of population values falling within some specified limits, and this type is considered here. In explaining this kind of tolerance it is important to distinguish it from confidence intervals or limits.

Confidence intervals provide a value for the region of uncertainty about an estimated *population parameter* (a point value), usually a mean, with a certain degree of confidence. Frequently it is desirable to obtain an *interval* which will cover a *fixed proportion or percentage of the population* with a specified confidence. Such intervals are called *tolerance intervals* and the two endpoints are called *tolerance limits*. Tolerance intervals can also be formulated in terms of a factor, k_{tol}, and the estimated standard deviation S

$$\text{Tol Interval} = \pm k_{\text{tol}}(S) \tag{25}$$

where k_{tol} is selected so that the interval will encompass a proportion p of the population with a confidence of $(1 - P)$. As an example if $\bar{x}_{(n)} = 14.0$, $S = 1.5$ and $n = 15$, the tolerance interval that will contain 99 percent of the population 95 percent of the time is given by referring to Table A3 where the confidence level is given by τ. Thus for $p = 0.99$, $\tau = 0.95$, and $n = 15$; the tabulated value for k_{tol} is 3.88, and

$$\text{Tol Interval} = \pm 3.88(1.5) = \pm 5.82$$

or

$$\bar{x}_{(n)} = 14.0 \pm 5.82 = 8.18 \text{ to } 19.82 \tag{26}$$

The distinction between confidence intervals and tolerance intervals is that confidence intervals refer to estimates of the population statistics (usually the mean) while tolerance intervals are concerned with proportions or fractiles of the population. Thus the term tolerance as used here should be distinguished from the frequently used tolerance in engineering design for dimensions and other factors in the construction or manufacture of some object or structure.

3.7 Propagation of Random Error in Testing

When measured parameters that have a certain *random variation* are used in mathematical calculations that express some derived property, the form of the mathematical relationship is important in determining the variation associated with the calculated property. The statistical technique that addresses this topic is called *propagation of error*. See Ku [1] as well as ASTM standard D4356 in the bibliography for background on the calculation algorithms as given here.

For any general relationship, where Y is some function of x_1 and x_2,

$$Y = \phi(x_1, x_2 \ldots) \tag{27}$$

the variance of Y, S_{Y}^2, is given by Eq. 28 in terms of the partial differential of the function with respect to x_1 times the variance of x_1, which is S_{x1}^2, plus the partial differential of the function with respect to x_2 times the variance of x_2, S_{x2}^2, etc.

$$S_{\text{Y}}^2 = \left[\frac{\partial\{\phi(x_1, x_2, \ldots)\}}{\partial x_1}\right]^2 S_{\text{x1}}^2 + \left[\frac{\partial\{\phi(x_1, x_2, \ldots)\}}{\partial x_2}\right]^2 S_{\text{x2}}^2 + \ldots \tag{28}$$

The variance for x_1, x_2, etc. refers to individual measurement values in any population. With simple linear relationships for the function, the differentials become constants, and the equation for S_Y^2 takes the form

$$S_Y^2 = k_1 S_{x1}^2 + k_2 S_{x2}^2 + \ldots \tag{29}$$

For the simplest linear form for two variables, a sum or difference relationship is given by

$$Y = x_1 + x_2 \qquad \text{or} \qquad Y = x_1 - x_2 \tag{30}$$

the value for S_Y^2 is

$$S_Y^2 = S_{x1}^2 + S_{x2}^2 \tag{31}$$

since the differentials are unity. Thus the act of adding or subtracting two measured values, each having a variance associated with its measurement, substantially increases the variance of the sum or difference. If both x_1 and x_2 have the same variance, the variance of the sum or difference is two times the individual variances.

With any functional form beyond a sum or difference, the variance of Y is influenced by the value for the differentials. For a ratio or quotient,

$$Y = \frac{x_1}{x_2} \tag{32}$$

the variance of Y is given by Eq. 33, and the evaluation of S_Y^2 has to be made at some selected values for x_1 and x_2.

$$S_Y^2 = \left(\frac{x_1}{x_2}\right)^2 \left(\frac{S_{x1}^2}{x_1^2} + \frac{S_{x2}^2}{x_2^2}\right) \tag{33}$$

Expressions for a number of functional relationships frequently encountered in calculating a derived property or parameter are given in Table 2. These expressions are intended to be used to evaluate the variance for Y in certain local regions as defined by the mean values used for x_1 and x_2 . Usually mean or average values for x_1 and x_2 will be used to evaluate the variance expressions, and the variance of the means (the standard error squared) of the

Table 2 Propagation of Error Expressions for Selected Functional Forms

Functional form	Expression for S_Y^2
1. $Y = k_1x_1 + k_2x_2$	$k_1^2\mathbf{S}_{x1}^2 + k_2^2\mathbf{S}_{x2}^2$
2. $Y = x_1/x_2$	$(x_1/x_2)^2[\{\mathbf{S}_{x1}^2/x_1^2\} + \{\mathbf{S}_{x2}^2/x_2^2\}]$
3. $Y = 1/x_1$	$\mathbf{S}_{x1}^2/x_1^4$
4. $Y = x_1/x_1 + x_2$	$(Y/x_1)^4(x_2^2\mathbf{S}_{x1}^2 + x_1^2\mathbf{S}_{x2}^2)$
5. $Y = x_1/(1 + x_1)$	$\mathbf{S}_{x1}^2/(1 + x_1)^4$
6. $Y = x_1x_2$	$(x_1x_2)^2[(\mathbf{S}_{x1}^2/x_1^2) + (\mathbf{S}_{x2}^2/x_2^2)]$ (*a*)
7. $Y = x_1^2$	$4x_1^2\mathbf{S}_{x1}^2$ (*a*)
8. $Y = (x_1)^{1/2}$	$1/4(\mathbf{S}_{x1}^2/x_1)$
9. $Y = \ln(x_1)$	$\mathbf{S}_{x1}^2/x_1^2$ (*a*)
10. $Y = kx_1^ax_2^b$	$Y^2[a^2(\mathbf{S}_{x1}^2/x_1^2) + b^2(\mathbf{S}_{x2}^2/x_2^2)]$ (*a*)
11. $Y = e^{x1}$	$e^{2x1}\mathbf{S}_{x1}^2$ (*a*)

(*a*) Expressions are approximate especially for small *n*.

x variables, designated as $S^2_{\bar{x}i}$, should be used, according to Eq. 34, where n is the number of values used for the mean for x_i .

$$S^2_{\bar{x}i} = \frac{S^2_{xi}}{n} \tag{34}$$

Under these conditions, wherever Y appears in the variance expression, a mean value for Y is to be used.

4 Principles of Measurement, Calibration, and Traceability

4.1 Measurement

Measurement is basically a comparison of an unknown to a known or standard object or value by way of a specified technical operation that generates data unique to the class of objects or material. Measurement theory describes how a measured parameter relates to a specific property. There are two general types of measurement (1) direct, where the measured parameter is the same as the property of interest and (2) indirect, where the measured parameter is related to the property of interest by some underlying theoretical relationship, usually a direct or linear one. Although most physical property measurements are Type 1, some physical properties are evaluated by a Type 2 process. Terminology needed for describing measurement technology includes the following terms.

Standard. (1) A recognized quantity, object, or criterion used for comparison, or (2) a protocol for some specified technical operation or other measurement goal, established by a standardization organization.

Measurement system. The entire collection of devices or instruments needed to make a specified type of measurement; it usually includes a data recording device (recorder, computer).

Technique. A technical operation using a measurement system that has been developed based on a specific scientific principle.

*Test method (*or *method).* The adaptation of a technique to a specific operation or process; this includes a written procedure for conducting the measurement(s), and the resulting data are defined as test results.

Protocol. A method and a complete set of definitive instructions to achieve a given testing or measurement goal; the instructions often include the sampling operation and other important ancillary information such as data treatment and reporting.

A number of secondary terms related to test methods are

Absolute method. A method where the test results are based on recognized physical or other standards or standard values directly derived from theory.

Relative or *comparative method.* A method where the test results are based on a comparison with the measurement of a reference material or object.

Standard method. A method that has been adopted by a standardization organization; it usually has broad technological acceptance and a stated precision.

Reference Method. A standard method with demonstrated high (or good) precision and frequently high (or good) accuracy.

4.2 Required Characteristics of a Measurement System

There are two major sets of conditions that need to be met. The first concerns the basic requirements of the measuring process. For any material or object class the requirements are that

The measurement system be in a stable condition
The measurements be independent and random
The measurements represent the population of direct interest

The importance of the first requirement is obvious. Unstable systems are really not acceptable for testing. The second requirement is the basis for conducting statistical tests where independence and randomness are assumed for probabilistic conclusions. When the first two of these specifications are met, the *measurement system* is in a state of statistical control. The third specification relates to how well the sampling is done.

The second set of conditions is related to this sampling operation and the test specimens derived from the samples. The sampling procedure should

Be conducted on a stable (nonchanging) population
Produce individual samples appropriately selected and independent of each other

The phrase "appropriately selected" refers to the different types of sampling that can be conducted; this topic is discussed in more detail in Section 5 on sampling principles. The importance of these sampling requirements is self evident; both are necessary for proper statistical analysis. When these sampling conditions are met, the *sampling system* is said to be in a stable condition or in statistical control.

The attainment of these required characteristics is often not straightforward. Conformance with the requirements is usually obtained by a twofold process. First, substantial experience with the system and attention to important technical details is required. Second, certain statistical diagnostic tests may be used, the most important being control charts, which are defined in Section 7, using standard or reference materials subjected to the same testing protocol as the experimental materials. Independence of individual measurements can be compromised if there is any carryover effect or correlation between one test measurement (sample or specimen) and the next measurement. Calibration operations, to be discussed later, can also be a source of problems in test measurement independence if they are not conducted in an organized or standardized manner.

4.3 Figures of Merit

The selection of a measurement system for a specified testing objective is usually made on the basis of certain figures of merit that apply to the following essential characteristics:

Precision and bias (uncertainty in test results)
Sensitivity
Useful range

and also to a number of desirable characteristics:

Low cost
Rapid testing and/or automated procedure
Ruggedness and ease of operation

Precision, Bias, Uncertainty

Good precision, which has been defined as the close agreement of measured values, is indicated by a low standard deviation. It is an essential requirement. Good precision and low or zero bias equate to good or high accuracy. See Annex C (to be discussed later) for more on precision, bias, and accuracy. The term *uncertainty* is also frequently used as a surrogate for accuracy in an inverse sense, i.e., low or acceptable uncertainty is equivalent to high accuracy. Section 8 contains an expanded discussion on these concepts.

Sensitivity

This is related to the ability to detect small differences in the measured property and/or the fundamental inherent property. Sensitivity has been defined in quantitative terms for physical property measurements by Mandel and Stiehler [2] as

$$\text{Sensitivity} = \frac{K}{s(\text{m})} \tag{35}$$

where

K = the slope of the relationship between the measured parameter m and the inherent property of interest Q, where $Q = f(\text{m})$
$s(\text{m})$ = the standard deviation of the measurement m

Sensitivity is high when the precision is good, i.e., $S(\text{m})$ is small, and when K is large. An example will clarify the factor K. The percentage of bound styrene in a butadiene-styrene polymer may be evaluated by a fairly rapid test, the refractive index. A curve of refractive index vs. bound styrene, with the styrene measured by an independent but more complex reference test, establishes an empirical relationship between the styrene and the refractive index. Over some selected bound styrene range, the curve has a slope of K, and this value divided by the precision standard deviation $s(\text{m})$, gives the sensitivity in this range. For polymers of this type the refractive index sensitivity may be compared to the sensitivity of alternative quick methods, such as density, by evaluating K and $s(\text{m})$ for each technique.

Useful Range

This is the range over which there is an appropriate instrument response to the property being measured. Appropriate response is expressed in terms of two categories, (1) the presence of a linear relationship between instrument output vs. level of the measured property, and (2) precision, bias (uncertainty), and sensitivity, at an acceptable level.

Ruggedness Testing

Frequently there is the need to determine if a test is reasonably immune to the perturbing effects of variation in the operating conditions such as ambient temperature, humidity, electrical line voltage, specimen loading procedures, and other ordinary operator influences. A procedure called *ruggedness testing* is conducted by making systematic changes in the factors that might influence the test and observing the outcome. Such testing is frequently conducted as a new test is being developed and fine tuned for special purposes or routine use. It can also be used to evaluate suspected operational factors for standardized methods if environmental or other factors for conducting the test have changed.

A series of systematic changes are made on the basis of fractional factorial statistical designs, which are discussed in more detail in Section 6. The early work was done by Plackett and Burman [3], Youden [4], and Youden and Steiner [5]. These designs are quite efficient. The most popular design evaluates the first-order or main effect of seven factors in eight tests or test runs. One important caveat in using these designs is that the second-order and all higher-order interactions of the seven factors are confounded with the main effects. See Section 6 for additional discussion on interaction and confounding. If there are any large interactions of this type, they will perturb the main effect estimates. However experience has shown that in the measurement of typical physical properties under laboratory conditions, first-order or main effects are usually much larger than interactions, so the use of these fractional designs has been found to be appropriate for ruggedness testing.

The Plackett–Burman statistical design for seven factors, A, B, C, D, E, F, and G that might influence the test outcome is given by Table 3, where -1 indicates the low level (value) of any factor, 1 indicates the high level of any factor, and y_i is the measured value or test result for any of the eight runs or combinations of factor levels. This design assumes that the potential influence of any factor on test response is a linear one. As indicated by the table, the design calls for the sequential variation of all seven factors across the eight test runs in a way that provides for an orthogonal evaluation of the effect of each factor.

The design is evaluated by a procedure that sums the eight y_i values in a specified way and expresses the results of the summing operation as the *effect* of each factor. Thus the effect of factor A, designated as $E(\mathrm{A})$, is given by Eq. 36 as the difference of two sums divided by $N/2$. The first sum is the total of the *products* obtained by multiplying each value of y_i by 1 for those rows (runs) that contain a 1 for column A, i.e., rows 1, 4, 6, and 7. The second sum of products is obtained in the same sense for all rows of column A that contain a -1, i.e., rows 2, 3, 5, and 8. The use of an expression analogous to factor A may be used for all other factors.

$$E(\mathrm{A}) = \frac{[\Sigma y_i \mathrm{A}(1) - \Sigma y_i \mathrm{A}(-1)]}{N/2} \tag{36}$$

where

$\Sigma y_i \mathrm{A}(1) =$ sum of y_i values for all runs (rows) that have 1 for factor A
$\Sigma y_i \mathrm{A}(-1) =$ sum of y_i values for all runs (rows) that have -1 for factor A
$N =$ total number of runs in the design ($= 8$); all sums are algebraic

The significance of the effects is evaluated on the basis of either (1) a separate estimate of the standard deviation of measurements of the same type (materials, conditions) as conducted for the Plackett–Burman design or (2) repetition of the design a second time to provide for two estimates of each factor effect.

If S is the separate estimate of the standard deviation of individual y_i measurements, then the standard deviation of the mean of four such measurements is $S/2$. If no real factor

Table 3 Plackett–Burman Ruggedness Test Design for Seven Factors (A to G)

Test run	A	B	C	D	E	F	G	Test result
1	1	1	1	−1	1	−1	−1	y_1
2	−1	1	1	1	−1	1	−1	y_2
3	−1	−1	1	1	1	−1	1	y_3
4	1	−1	−1	1	1	1	−1	y_4
5	−1	1	−1	−1	1	1	1	y_5
6	1	−1	1	−1	−1	1	1	y_6
7	1	1	−1	1	−1	−1	1	y_7
8	−1	−1	−1	−1	−1	−1	−1	y_8

effect exists, then the calculated effect $E(i)$, which is a difference between two means of four each, has an expected value of zero and a standard deviation of $(\sqrt{2}S/2) = S/\sqrt{2}$. If $E(i)$ is significant (a real effect), it should exceed zero by an amount greater than two standard deviations based on means of four, i.e., greater than absolute $2(S/\sqrt{2})$, provided that S is known at a certainty level of at least 18 to 20 *df*. The use of $2(S/\sqrt{2})$ as the interval to indicate significance is based on a $P = 0.05$ or 95% confidence level. If S is not based on at least 18 *df*, then the value of Student's t (double-sided test) at $P = 0.05$ should be substituted for 2 in this interval expression using the appropriate *df*.

If S is to be estimated from the ruggedness testing itself, the eight runs are repeated to produce two sets of estimates for each factor, where $E(i)$ = replication 1 value and $E'(i)$ = the second replication value. Since the standard deviation of $E(i)$ or $E'(i)$ is $S/\sqrt{2}$, each value of the difference $[E(i) - E'(i)]$ or d, for each of the seven factors, is an estimate of $(\sqrt{2}S/\sqrt{2}) = S$. Hence an estimate of S based on 7 *df* is

$$S = \left[\frac{\Sigma(d(\mathrm{A})^2 + d(\mathrm{B})^2 + \ldots + d(\mathrm{G})^2}{7} \right]^{1/2} \tag{37}$$

where

$$d(\mathrm{A})^2 = [E(i) - E'(i)]^2 \quad \text{for Factor A}$$
$$d(\mathrm{B})^2 = [E(i) - E'(i)]^2 \quad \text{for Factor B}$$

and so on for all factors. With two replications of the eight runs now available for estimating the influence of the seven factors, the mean effect is $E(i)_{\mathrm{m}}$ or $[E(i) + E'(i)]/2$ for each factor. For the value of any $E(i)_{\mathrm{m}}$ to be significant, it must exceed $2.37[\sqrt{2}S/\sqrt{8}] = 1.18S$, where S is of course given by Eq. 37 and 2.37 is the t value in the Student's distribution at 7 *df* and at the $P = 0.05$ or 0.95 significance level. Factors that are found to be significant in any test need to be investigated and the test procedure or protocol revised to reduce the sensitivity to those factors.

For certain test methods, especially those that are more fully developed, only a few factors may require an evaluation, perhaps 3 or 4. The Plackett–Burman seven factor design may still be used with the remaining factors, say E, F, and G, being dummy factors, i.e., factors that have no influence on the outcome. All eight test runs must be completed for any design independent of the actual number of factors evaluated. Alternatively the fractional factorial designs for 3 or 4 factors as given in Section 6 may be used.

4.4 Calibration and Traceability

Calibration is basic to all measurement systems. Two important definitions of calibration are (1) a set of markings on a scale graduated in the output parameter of a test device, or (2) the act of comparing a measured test value or response to an accepted *true*, *reference*, or *standard* value or object. The primary purpose of calibration is the elimination of bias or systematic deviation of measured values so that they correspond exactly with the true values. Once the bias is known from a calibration experiment, an adjustment of the testing device is made so that the true or accepted reference value is obtained from the test device to some suitable tolerance. Frequently standardization is used as a synonym for calibration.

The requirements for calibration include (1) an estimate of the accuracy (precision, bias) needed in the output measurement or response function, (2) the availability of documented calibration standards, (3) the presence of a state of statistical control for the test system, and (4) a fully documented protocol along with experienced personnel for the calibration. A realistic calibration schedule should be maintained. Decisions on the frequency of calibration are made by balancing the cost of the calibration vs. the risk of biased test output. When in a state of statistical control, repetitive instrument responses for each true or standard value should be randomly distributed about a mean, and when a series of true values are used, the response means vs. true values should give a linear least-squares regression relationship. This gives confidence intervals on the slope and the intercept and for selected points on the line in the calibration range. Annex C outlines procedures for evaluating bias that are equivalent to this type of calibration operation; see this for more details. See also Section 6 for background on regression analysis

Empirical relationships that appear to be linear can be tested for linearity by a number of approaches. Visual review of a plot is most appropriate to reveal any departures from linearity. A plot of residuals (residual = observed − computed response value) with respect to the level of the response should not show any correlation or systematic behavior. Such a review requires at least 7 pairs of data points (response levels) to be useful. A simple F test may also be employed. If S_{ir}^2 is the pooled variance for a set of repetitive instrument responses, each set of responses at one of a series of levels (true values) of the calibration standard, and S_{ff}^2 is the variance of points about the fitted function, when individual set response values (not means or averages) are used for the least-squares calculation, then the variance ratio

$$F(\text{calc}) = S^2_{ff} / S^2_{ir} \tag{38}$$

should not be significant, i.e., greater than F(crit) at $P = 0.05$ for the respective df values for both variances. See Section 6 for variance analysis procedures. If F(calc) equals or exceeds F(crit) under these conditions, there is some significant departure from linearity. This approach should be used with a sufficient number of points so that the df for each estimated variance is 8 to 10. When demonstrated nonlinearity exists, transformations may be used to linearize the response. See ISO11095 in the bibliography for more details on calibration.

Traceability

As the name implies, this is the ability to trace or accurately record the presence of an unbroken, identifiable and documented pathway from fundamental standards or standard values to the measurement system of interest. Traceability is a prerequisite for the assignment of limits of uncertainty on the output or response measurement, but it does not imply any level of quality, i.e., the magnitude of the uncertainty limits. Physical standards with certified values as well as calibration services are available from national standardization laboratories such as National Institute of Standards and Technology or NIST (formerly NBS) in the USA or from the corresponding national metrology laboratories for all developed countries. All of these standards are usually expressed in the SI units system.

5 Basic Sampling Principles

Since measurements must be performed on some selected object or material, the practice of selecting these items, which is usually called *sampling*, is an important operation in any technical project. For some testing operations the sample is what is tested. For other operations test portions or test specimens are prepared from the sample using a given protocol, and these are tested. Sampling is a very important operation in any decision-making process, and good sampling technique insures that all samples unquestionably represent the population under consideration. Only the most elementary sampling issues are addressed in this section. For more detailed information, various texts and standards on sampling and sampling theory should be consulted; see the bibliography for statistical texts and for standards on sampling.

5.1 Terminology

Sampling terminology systems vary to some degree among industrial and commercial operations, which frequently involve complex mechanical systems to draw samples or other increments from some large lot or mass of material. One of the important objectives in such operations is the elimination of bias in the samples. Increased sampling frequency can reduce the uncertainties of random sampling variation, but it cannot reduce bias. The terminology given here is that which applies more directly to laboratory testing, where the process of obtaining samples is reasonably straightforward. This type of sampling may be thought of as static as opposed to dynamic sampling of the output of a production line. Some important terms are:

Sample. A small fractional part of a material or some number of objects taken from a lot or population; it is (they are) selected for testing, inspection, or specific observations of particular characteristics.

Subsample. One of a sequence of intermediate fractional parts or intermediate sets of objects, taken from a lot or population, that usually will be combined by a prescribed protocol to form a sample.

Random sample. One of a sequence of samples (or subsamples), taken on a random basis to give unbiased statistical estimates.

Systematic sample. One of a sequence of samples (or subsamples), each taken among all possible samples by way of a selected alternating sequence with a random start; it is intended to give unbiased statistical estimates.

Stratification. A condition of heterogeneity that exists for a lot or population that contains a number of regions or *strata*, each of which may have substantially different properties.

Stratified sample. One of a sequence of samples (or subsamples), taken on a random basis from each stratum in a series of strata in a (stratified) population.

5.2 The Purpose of Sampling Plans

With every sampling plan or procedure, two characteristics need to be considered, the quality of the estimates of the properties of interest and the cost of conducting the sampling. Increasing the confidence in the estimates, which means a more extensive sampling plan, with special emphasis on bias reduction, increases the cost of the sampling. A sampling plan should generate an objective estimate of any measured parameter, and this is accomplished by using strict probabilistic or statistical sampling or some variant of such operations that gives objective decisions. However there are situations where this type of plan would be excessively costly for the importance level of the decisions to be made. In such cases an alternative approach using subjective elements or technical judgment is usually employed.

A comprehensive sampling plan will provide estimates of

Population (lot) mean and the confidence limits on the mean
Tolerance interval for a given percentage of the units of the population
The sample size or minimum number of tested units (samples, specimens) to establish the above intervals with a selected confidence or tolerance level

On a practical basis sampling is usually conducted according to the goals of a technical project. Since there is a wide range of projects from simple to complex, three generic types of sampling plans have been used.

Intuitive sampling. This plan is organized employing the developed skill and judgment of the sampler using prior information on the lot or population as well as past sampling and testing experience. The decisions made on the data generated by such a plan are based on a combination of the skill and experience of the tester buttressed by limited statistical conclusions. Strict probabilistic conclusions are not warranted.

Statistical sampling. This provides for authentic probabilistic conclusions. All three of the above population estimates may be obtained, hypothesis testing may be conducted, inferences may be drawn, and predictions may be made about future system or process behavior. Usually a large number of samples is needed if the significance of small differences is of importance. Conclusions from this type of sampling are usually not controversial. The statistical model chosen is important, and when the number of samples required is large, which may impose a testing burden, hybrid plans using some simplifying intuitive assumptions are frequently employed.

Protocol sampling. These are specified plans used for decision purposes in a given situation. Regulations (of the protocol) usually specify the type, size, frequency, and period of sampling in addition to the test methods to be used and other important test issues. A combination of intuitive and statistical considerations may be used, and the above population estimates may be obtained depending on the protocol. Testing for conformance with producer–user specifications for commercial transactions is typical for this approach.

5.3 Assuring the Quality of Samples and Testing

Quality is assured when a sampling procedure is developed based on the requirements of the testing operation and its decision process and when the samples are drawn according to the prescribed procedure. Such topics as sample homogeneity (or unintended stratification) and sample stability (conditioning or storage changes in the sample prior to testing) must be addressed. The sampling procedures, holding time, and other handling operations should be well documented. The test methods used for any program should be stable or in a state of statistical control and have demonstrated sensitivity as well as good precision in regard to the measured parameter. Sampling procedures for three types of *statistical* sampling are presented, followed by a section on the size of the sample that must be taken for a specified precision or confidence interval on the estimated population mean.

5.4 Basics of Statistical Sampling

Random Sampling

A lot or population may be thought of as some total number of N *units*. If the population is hypothetically of infinite size, then N equals infinity. The word unit refers to each object if the population is comprised of individual objects, or to some small defined incremental volume (or mass) for a bulk material. The important feature in simple *random sampling* is that each unit of a population (or lot) have an equal chance $1/N$ of being selected for testing.

Random sampling can be conducted in one of two ways: (1) *with replacement* of the selected units, under the conditions that the test operation does not change or consume the unit, or (2) *without replacement*, when the unit is changed in some way by the testing. For large populations there is no essential distinction between these two types of random sampling. For small populations (small N) a difference does exist. Since most physical testing might in some way change the sample (or test specimen prepared from the sample) the expressions given below are for the "without replacement" category.

A simple *random sample* is defined as n units drawn from the population where each unit has the same probability of being drawn. The ideal procedure for doing this is to identify all the units in the population, $i = 1, 2, \ldots, N$, and select units from a table of random numbers. For some sampling operations this may have to be modified according to the manner in which individual units can be identified. Each of the N potential or actual units has a value y_i, and the unbiased estimate of the true mean $\acute{Y}$ is given by the sample mean $\acute{y}$ as

$$\acute{y} = \frac{\Sigma y_i}{n} \tag{39}$$

where n is the number of units or size of the sample drawn from the population. The quantity n/N is referred to as the *sampling fraction*, and the reciprocal N/n is known as the *expansion factor*. The unbiased estimate of the variance of $\acute{y}$, designated as $S^2_{\acute{y}}$, is given by

$$S^2_{\acute{y}} = \left[\frac{N-n}{N}\right]\frac{S^2_{yi}}{n} \tag{40}$$

where

S^2_{yi} = the variance of the individual n units

The factor $[(N-n)/N]$, which can be expressed as $[1-(n/N)]$, reduces the magnitude of the variance of the mean by the sampling fraction when compared to an infinite population value. This reduction factor is called the *finite population correction factor* or fpc, and it indicates the improved quality of the information about the population when n is large relative to N. As n grows larger, the variance of the population mean decreases and becomes zero if $n = N$, since at this point the mean is known exactly. In many situations fpc has a minimal effect and is usually ignored if $n/N < 0.05$, and then $[(N-n)/N]$ is set equal to 1. The confidence interval at $P = \alpha$, for the estimated mean $\acute{y}$, is given by

$$\acute{y} = \pm t_{\alpha/2}\left\{\left[\frac{N-n}{N}\right]\frac{S^2_{\text{yi}}}{n}\right\}^{1/2} \tag{41}$$

If the sample size used to estimate S^2_{yi} is fairly large, $n \geqslant 30$, the value of 2 may be used for $t_{\alpha/2}$ to give a $P = \alpha = 0.05$ or 95% confidence interval.

Systematic Sampling

The actual process of drawing random samples can frequently be time consuming as well as costly, especially for large populations and large samples. An alternative procedure that is easier to conduct and that gives good estimates of the population properties is *systematic sampling*. This type of sampling is conducted as follows:

1. Identify all units of the population; $i = 1, 2, \ldots, N$.
2. Calculate the expansion factor N/n and round to nearest integer, call this ef.
3. Choose a random integer I_r such that I_r is in the interval from 1 to $(ef - 1)$.
4. Sequentially identify the i units to be included in the sample, i.e., $i = I_r$, $I_r + ef$, $I_r + 2ef$, etc., and select the sample based on this identification.

This sampling operation will give a sample of n units (± 1 due to rounding of ef) with each unit having an equal probability of being selected. Since the starting point is randomly selected, this type of sampling gives unbiased estimates as expressed in Eqs. 40 and 41 if the population has truly random variation. If there is any cyclic or transient trend nature in the population or in some aspect of the sampling, a biased estimate or biased sample may result due to any potential correspondence between the sample sequence and the cyclic trends.

Stratified Sampling

This type of sampling is a process generally applied to bulk lots or populations that are known or suspected of having a particular type of nonhomogeneity. These populations have *strata* or localized zones, and each stratum is usually expected to contain relatively homogeneous objects or material properties. However these properties may vary substantially among the strata, and samples are taken independently in each stratum. To apply stratified sampling techniques a mechanism must exist to identify all the strata in the lot or population. Once this is done the strata may be sampled by using *proportional allocation* where the sample fraction is the same for each stratum. Another approach is *optimum allocation* where the sample size or fraction may be increased in those strata with increased variance if this information is known beforehand.

The calculation as applied above for $\acute{y}$, the estimated population mean for random sampling, may be applied to each stratum and these individual values used to get a population mean based on all *stratum* values. Similarly the calculation for $S^2_{\acute{y}}$, the estimated population variance for random sampling, may be applied to each stratum and an analogous procedure used to get an overall variance based on all *stratum* values. If unequal samples are taken from the various strata, the overall population mean and variance values that represent the entire stratified population must be obtained on a weighted average basis; see Section 3.

5.5 Sample Size Estimation

The number of samples required to estimate the sample mean to some degree of precision or uncertainty is an important aspect of sampling. Increasing the sample size or the number of units tested to obtain a mean increases the precision or reduces the uncertainty. However the cost of this extra work increases linearly with sample size, while the precision increases at a much slower rate with the square root of the number of samples. Sample size problems are approached on the basis of the total uncertainty in the mean. There are two criteria that must be specified to calculate a sample size n.

The value of E, the maximum $\pm$ uncertainty (error) of estimation of the mean
The required level of confidence of the maximum uncertainty

To be able to calculate n, given some E, requires a value for the standard deviation of the individual unit measurements under a specified sampling and testing condition. The relationship between n and E is derived from the sampling distribution of the mean and the t-distribution as discussed in Section 3.4. Thus rearranging the usual form of the equation,

$$E = t_{\alpha/2} S_a / \sqrt{n} \tag{42}$$

where

$E = |\acute{Y} - \acute{y}|$ = the maximum (absolute) difference between the estimated mean $\acute{y}$ from the n samples and the true mean $\acute{Y}$

$t_{\alpha/2}$ = t value at a specified $P = \alpha$; i.e., at a $(1 - \alpha)100\%$ confidence level, where the *df* used for $t_{\alpha/2}$ is based on the *df* for S_a

S_a = the *applicable* standard deviation (among individual units tested), a function of the specific sampling and testing operation

This may be rearranged to solve for n to give

$$n = \left[\frac{t_{\alpha/2} S_a}{E}\right]^2 \tag{43}$$

An absolute difference is indicated, and the uncertainty E may be alternatively expressed as a plus and minus range E about the estimated mean $\acute{y}$. Thus the statement that the true mean will be contained in the range $\acute{y} \pm E$, will have a $(1 - \alpha)100\%$ confidence level.

The evaluation of a value for n requires information about the testing conditions in order to select a correct applicable standard deviation for any specified $(1 - \alpha)$. This may be clarified by consideration of the magnitude of two types of error, sampling error and measurement error, for a relatively simple sampling and testing operation. Table 4 illustrates four testing scenarios for sampling variance, S^2(sp), and measurement variance, S^2(m). The importance or significance of either component is determined in large part by the magnitude of the expected difference d for a simple comparison of the estimated means for two different or potentially different lots or populations.

A Type 1 scenario is encountered when the expected difference d is large compared to the known variances for both types of error; i.e., that $d \gg 4$ times the square root of the sum of the two variances. For this situation the variance of neither component is critical. Type 2 is typical of a less precise test where sampling variation is low. If the test is in control, and the variance known, the number of measurements needed can be calculated for any desired value for d.

Type 3 is characteristic of a relatively precise test measurement where several samples are required to give a good estimate of lot or population means needed to calculate d. A defined sampling program is required. Usually only a few measurements (one, two) need be made on any sample. Type 4 is the most complex, since both components are important or significant. This is unfortunately frequently encountered in much testing. A specified sampling plan with multiple samples is required as well as multiple measurements on each

Table 4 Four Testing Variation Component Scenarios

	Component[a]	
Type of scenario	S^2(sp)	S^2(m)
1	Not sig	Not sig
2	Not sig	Sig
3	Sig	Not sig
4	Sig	Sig

[a]Not sig = no significant or large variation component. Sig = a significant or large variation component.

sample. Substantial background knowledge in addition to a formal analysis of variance for such a test situation may be required for an efficient evaluation of d.

The uncertainty E needs to be selected in relation to d on the basis that $E \leqslant d$, where the preferred situation is that $E < d$. If d is small, then E needs to be made at least as small, and for constant test variance this requires that n be increased. The equations needed to calculate n for different testing scenarios with a selected E are given below. It is assumed that the standard deviation S_a is known with at least 30 *df* and thus $t_{\alpha/2} = 2$.

Type 1—Neither S(sp) nor S(m) Significant: For this case it is assumed that one or two measurements will yield values that can detect a normal expected difference d.

Type 2—Only S(m) Significant: The number of samples n_{T2} is

$$n_{T2} = \left[\frac{2S(m)}{E_{T2}}\right]^2 \tag{44}$$

and S_a is equal to S(m) the measurement error.

Type 3—Only S(sp) Significant: For this situation the number of samples n_{T3} is

$$n_{T3} = \left[\frac{2S(sp)}{E_{T3}}\right]^2 \tag{45}$$

Type 4—Both S(sp) and S(m) Significant: For this more complex situation the uncertainty E is a function of the combined variance terms and there is no unique solution for n. The relationship is expressed as

$$E = \left[\frac{S^2(sp)}{n_{sp}} + \frac{S^2(m)}{n_{sp}n_m}\right]^{1/2} \tag{46}$$

and several combinations of n_{sp} and n_m may give equal E. Values for n_{sp} and n_m have to be selected based on their respective variance magnitudes and the costs associated with sampling and measurement.

6 Statistical Analysis Techniques

Part 1 of this section on analysis gives a survey of some of the more frequently used elementary analysis procedures along with a brief review of experimental design. The basic purpose of this section is to give a simple overview of the statistical concepts used in the application of these analysis techniques. Since computer programs are readily available for a more comprehensive analysis, a brief review of typical software analysis algorithms is also presented in Part 2.

Part 1. *Elementary Analysis Procedures*

6.1 Preliminary Screening of Databases

Prior to a formal analysis, a database should be examined for any unusual characteristics of the data distribution. A database may have some number of outliers, an inherent nonnormal or skewed distribution, or a bimodal character due to the presence of two separate underlying distributions. Most tests for normality are intended for fairly large sample sizes of the order of 15 or more. Smaller databases may be reviewed for unusual characteristics by way of the usual statistical algorithms available with spreadsheets. Tests for normality are listed in Part 2.

Detecting Outliers

Outliers may be present in any size database. For a database of from several to 30–40 data values, an analysis may be conducted using spreadsheet calculations. Data values may be sorted from low to high and a plot of the sorted or ordered values will reveal any suspicious high or low values. Tietjen and Moore [6] described a test that can be used for a small database with a reasonable number of suspected extreme values or outliers (1 to 5). The suspected outliers may be either low or high, and the statistical test may be used when both types exist in the sample at the same time. The test is applicable to samples of 3 or more, and for sample sizes of 11 or more, as many as five suspected extreme values may be tested as potential outliers. The following procedure is used.

(1) The data values are denoted as $X_1, X_2, \ldots, X_n$. The mean of all values designated as $\bar{x}_{(n)}$ is calculated.
(2) The absolute residuals of all values are next calculated: $R_1 = |X_1 - \bar{x}_{(n)}|$, $R_2 = |X_2 - \bar{x}_{(n)}|$, etc.
(3) Sort the absolute residuals in ascending order and rename them as Z values, so that the Z_1 is the smallest residual, Z_2 is next in magnitude, etc.
(4) The sample is inspected for extreme values, low and high. The most likely extreme values or potential outliers (the largest absolute value residuals) are deleted from the sample (or database) and a new sample mean is calculated for the remaining $(n - k)$ values, with $k =$ the number of suspected extreme values. This new mean is designated as $\bar{x}_k$. The critical test statistic $E(k)$ is calculated according to

$$E(k) = \frac{\Sigma\{i \text{ to } (n-k)\}[Z_i - \bar{x}_k]^2}{\Sigma\{i \text{ to } n\}[Z_i - \bar{x}(n)]^2} \tag{47}$$

where

$n =$ sample size (original)
$k =$ number of suspected extreme values
$\bar{x}_{(n)} =$ mean for all (n) values in sample
$\bar{x}_k =$ mean for $(n - k)$ values

Critical values are given in Table A4 for the test statistic $E(k)$ at the $P = 0.05$ and the $P = 0.01$ levels, for sample size $n = 3$ to 30 and for selected numbers of suspected outliers k. If the calculated value of $E(k)$ is *less* than the tabulated critical value in the table, the suspected values are declared as outliers. This general approach may be used in an iterative manner until all potential outliers have been evaluated by the procedure.

The action to take when significant outliers have been identified is part of an ongoing debate in the data analysis community. One recommendation is that only data values with verified errors or mistakes should be removed. This ultraconservative approach overlooks the situation where outright errors are made but no knowledge is available that they are errors. The opposing view recommends removal if a data value is a significant outlier ($P = 0.05$ or lower). A middle ground position is to make a judgment based on a reasonable analysis considering technical and other issues that relate to the testing in addition to the importance of the decisions to be made.

6.2 Evaluating Differences among Precision Estimates

Precision Estimates

Although the evaluation of any significant differences in mean values is one of the most important statistical operations, the issue of precision, expressed inversely as a variance, needs to be addressed first because certain knowledge about precision variance estimates and their uniformity is required for a proper execution of such mean value statistical tests. Precision variance can be evaluated once a database has been surveyed for unusual distribution characteristics. Situations that might require a decision on potentially significant differences in precision variance and in mean values as well are

Possible undetected changes in the operation of a measurement system
Comparison of different technicians, instruments, laboratories for a common test
Comparison of two measurement systems
Comparison of different materials or test objects

These and other similar situations require an objective basis for decisions.

Two Variance Estimates

The statistical significance of apparent differences between two variances can be judged on the basis of a ratio of the two variances. The most frequently encountered situation is the comparison of two variance estimates, both estimates obtained from samples from two ostensibly different populations. The variance ratio, S_1^2/S_2^2, is called an *F*-ratio, and this ratio follows a sampling distribution called an *F*-distribution. The shape of these distributions depends on the number of degrees of freedom *df* associated with each of the variances (numerator, denominator). The statistical test to determine if the two variance estimates are equal is called an *F*-test; this is conducted by comparing the calculated value of the ratio *F*(calc) to a critical *F* value called *F*(crit) that would be obtained on the basis of chance at some probability level when there is no real difference, i.e., both estimates are drawn from the same population. There are two potential situations or cases: (1) from technical or operational conditions alone, one variance should have a larger value and (2) on a technical basis neither variance can be considered greater.

Case 1

In conducting an *F*-test for this situation, S_1^2 is assigned as the greater expected variance. The null hypothesis, H_0, that there is no difference in variance and the alternative hypothesis, H_A, that S_1^2 is larger than S_2^2, are designated symbolically as

$$H_0: S_1^2 = S_2^2 \qquad (48)$$
$$H_A: S_1^2 > S_2^2$$

These hypotheses are tentatively adopted and a sample is drawn from each population (1 and 2) and the variance calculated. The ratio, $S_1^2/S_2^2 = F(\text{calc})$ is evaluated. If this ratio is equal to or larger than *F*(crit), the ratio that would be expected by chance at a probability $P = \alpha$ (0.05 or other) for finding a value as large as *F*(calc) when the null hypothesis is true, the hypothesis of equality is rejected and the alternative hypothesis is accepted, i.e., that $S_1^2 > S_2^2$.

F-distribution tables are given for *F*(crit) values, that are equaled or exceeded a certain percentage of the time by chance alone, see Table A5. The *F*(crit) value cuts off an upper area under the *F* distribution curve that is equal to α, and if *F*(calc) falls in this cutoff region, then the null hypothesis is rejected. Tables of *F* values are arranged for different degrees of freedom *df* in the numerator and denominator, and *F*(crit) is usually listed for each *P* level as $F\ (df_n, df_d)$; where $df_n = df$ in the numerator, and $df_d = df$ in the denominator.

Case 2

In this situation there is no technical reason for expecting either variance to be greater than the other. A null hypothesis and an alternative hypothesis is adopted:

$$\begin{aligned} &H_0: S_1^2 = S_2^2 \\ &H_A: S_1^2 > \text{ or } < S_2^2 \end{aligned} \tag{49}$$

and after both variances are calculated the greater variance is placed in the numerator to evaluate *F*(calc). For making decisions at a $P = 0.05$ level, an allowance must be made for *F*(calc), to be greater than *F*(crit) if S_1^2 is greater than S_2^2 and conversely to be less than a different *F*(crit) at the other end of the *F* distribution, if S_1^2 is less than S_2^2. One half of the $P = 0.05$ rejection region is assigned to potential large values for *F*(calc) and one half to potential small values for *F*(calc). On this basis the $P = 0.05$ value for *F*(crit) would be found at a $P = 0.025$ upper tail *F* table, or conversely if the *F*(crit) value in a 95% confidence level upper tail *F* table were selected, the confidence level for making a decision would be $P = 0.10$, since there is a $P = 0.05$ probability for the *F*(calc) to be in either critical region.

One Variance Known

A less frequent situation is the case where one of the variances is a known or defined variance, represented by σ^2, while the other variance is from a sample and is equal to S^2. The ratio S^2/σ^2 has a sampling distribution known as a "chi squared over *df*", designated as (χ^2/df). For this application, chi-square χ^2 is a random variable that is given by the ratio of the product $S^2(df)$ to the known variance σ^2.

$$\chi^2 = \frac{S^2(df)}{\sigma^2} \tag{50}$$

To conduct a test of significance for the null and alternative hypotheses,

$$\begin{aligned} &H_0: S^2 = \sigma^2 \\ &H_A: S^2 \neq \sigma^2 \end{aligned} \tag{51}$$

an estimate S^2 of the variance is obtained and multiplied by the *df* (i.e., $n - 1$), with $n =$ number of data values associated with the sample variance estimate. The calculated ratio χ^2(calc) is found and compared to a critical ratio χ^2(crit) at a selected probability $P = \alpha$. Values for χ^2(crit) are obtained from chi-square tables at the applicable *df* for any particular application. Table A6 is a χ^2 table, and Table A7 is a (χ^2/df) table.

More Than Two Variances

For situations with more than two variances, one statistical test that may be employed is the Hartley *F*-max procedure. This is a test that may be applied to a balanced database, i.e., the same number of values in each of the data sets, each set constituting a certain level of a factor in a test program. The variance of each of *n* data sets to be tested for equivalent variance is evaluated, and the ratio of $S^2(\max)/S^2(\min)$ is calculated and designated as *F*-max(calc). The value of this is compared to *F*-max(crit) at a selected *P* level (0.05 or 0.01) in tables of the *F*-max distribution for *n* factor levels (data sets) and for the common *df* for each variance estimate. Table A8 is a Hartley *F*-max table. The null hypothesis is

$$\text{H}_0: \text{all } \sigma_i^2 \text{ equal}$$
$$\text{H}_A: \text{at least two } \sigma_i^2 \text{ not equal} \tag{52}$$

If F-max(calc) equals or exceeds F-max(crit), then a significant difference exists for the maximum vs. the minimum variance of the database. If F-max(calc) does not exceed F-max(crit), then all variance estimates are equivalent.

6.3 Evaluating Differences in Means

Two Estimates of a Mean

There are a number of situations where a decision is required about the statistical significance of the numerical difference between two means where the two means may be considered to come from different populations. The approach to this problem requires information about the means and the variance of each potential population. A number of distinctions with regard to the means and the variances are possible. For means: (1) the two means may be estimates, (2) only one mean is an estimate, the other is known. For variances: (1) both variances may be estimates, (2) one variance is estimated, one is known or (3) both variances may be known. Any of the options for means may be combined with any of the three options for variances. The most frequently encountered situation is both means and variances are estimates. This combination has two additional options: the two variance estimates are equal or the two are unequal.

Option 1. If both variances are assumed to be equal or they have been shown to be equivalent by an F-test, the separate estimates are pooled (see Section 3.3) to give a value for $S^2(\text{p})$. The null hypothesis and alternative hypothesis are

$$\text{H}_0: \mu_1 = \mu_2$$
$$\text{H}_a: \mu_1 \neq \mu_2 \tag{53}$$

The decision on statistical significance is based on the sampling distribution of the ratio of the difference between two means to the standard deviation of such differences as given by Student's t-distribution for the general case of unequal number of values for each mean.

$$t_{\alpha/2} = \frac{\bar{x}_{(n)1} - \bar{x}_{(n)2}}{[(S^2(\text{p})/n_1) + (S^2(\text{p})/n_2)]^{1/2}} \tag{54}$$

where

$\bar{x}_{(n)1}$ = mean of n_1 values, population 1
$\bar{x}_{(n)2}$ = mean of n_2 values, population 2
$S^2(\text{p})/n_1$ = variance of means (of n_1 values each), population 1
$S^2(\text{p})/n_2$ = variance of means (of n_2 values each), population 2
and $df = (n_1 + n_2 - 2)$

Since there is no information on whether either mean is potentially the greater, the calculated difference is indicated by $\bar{x}_{(n)1} - \bar{x}_{(n)2}$, which can be either positive or negative. The value for t designated as $t_{\alpha/2}$ is evaluated using Eq. 54. The statistical significance for a numerical difference is evaluated by comparing t(calc) to a critical value $t_{\alpha/2}$(crit) obtained at some level of probability $P = \alpha$. At this α, the value for $t_{\alpha/2}$(crit) at the indicated df of $(n_1 + n_2 - 2)$ cuts off an area in each tail of the t-distribution that equals $\alpha/2$, since t(calc) may be + or −, i.e., it may appear in either tail. The value for $t_{\alpha/2}$(crit) is selected from a table for the t-distribution at $P = \alpha/2$ and compared with t(calc). Table A2 gives critical values for t at various P levels. If $|t(\text{calc})| \geq |t_{\alpha/2}(\text{crit})|$, using absolute values, then the null hypothesis is rejected and the difference is declared significant at the $(1 - P)$ 100 confidence level. If $|t(\text{calc})|$ does not equal or exceed $|t_{\alpha/2}(\text{crit})|$, the difference is not significant at this level of testing for the selected $P = \alpha$ value.

Option 2. If the variances are not equal there are two choices: use a transformation and conduct the analysis on the transformed data, or conduct the normal t-test and use the df for the smaller sample to select $t_{\alpha/2}(\text{crit})$. The distribution of the t value so calculated is approximately the same as a true t distribution. This smaller sample df recommendation is made because with unequal variances the exact number of degrees of freedom cannot be determined by the usual procedure as given above.

Differences in Paired Means

It is not always possible to evaluate the variances of two populations being compared. Under such circumstances a test program may be conducted by using the "paired sample" technique. As an example, if the effect of a certain treatment on a particular polymer is being evaluated, a number of uniform polymer samples are prepared and divided into pairs. One sample of each pair is given the treatment and one is not, i.e., it is the control. Measurements are then conducted on a sufficient number of pairs p (six or more) for treated vs. untreated with some particular test. Inferences on the effect of the treatment or the difference in means of the two populations (treated vs. nontreated) are based on the mean value of the difference between n paired values. The distribution of the ratio of this mean difference to the standard deviation of these differences follows a Student's t-distribution.

$$t_{\alpha/2} = \frac{d_{(\text{av})}}{(S_{\text{d}}^2/n)^{1/2}} \tag{55}$$

where

$d_{(\text{av})}$ = the average difference in means, (treated, nontreated) for n paired values
S_{d}^2 = the variance (estimate) of all the individual differences
$df = p - 1$

The null and alternative hypothesis statement is

$$\begin{aligned} &\text{H}_0: d(\text{av}) = 0 \\ &\text{H}_\text{A}: d(\text{av}) \neq 0 \end{aligned} \tag{56}$$

The significance of $d_{(\text{av})}$ is found by the same procedure as for a normal or standard t-test. The variance S_{d}^2 is not the variance of either population but of a constructed population of *differences* between the two conditions, treated vs. nontreated.

6.4 Elementary Analysis of Variance

Variance analysis is a task that can be easily performed mechanically with the ready availability of computers. However to apply this analysis properly requires some understanding of the underlying concepts. Analysis of variance or ANOVA is essentially a linear separation of the individual components of total variation in any database. The analysis permits an objective decision on what part of the total variation is due to any variable or external factor of the system. An external factor may be an intended modification and thus constitute a treatment or a certain factor (uncontrollable environmental type) and thus impose unintended modification. The simplest analysis of variance is a one-way analysis, and this begins with a database generated by a testing program such as a series of treatments on a common material with some number of repeat tests for each treatment. This elementary one-way analysis procedure will illustrate the basic ANOVA concepts.

Basis of Variance Analysis

The database or data matrix illustrated below as Table 5 consists of j columns and i rows of data. The j columns may be various treatments of a uniform material or levels of an adjustable or independent variable, and the values in each column are replicates or repeated measurements for each of the treatment levels.

The measurements or observations x_{ij} are the dependent variable. Two basic assumptions are made; all potentially different populations for each treatment have a normal distribution, and the variance of all populations is equal even if one or more of the treatments are significantly different. This latter assumption may be checked by the use of the Hartley F-max test as previously described. Each data value x_{ij} may be expressed in terms of certain population *means* as given in Eq. 57, where each group as defined below represents a treatment.

$$x_{ij} = \mu + (\mu_j - \mu) + (x_{ij} - \mu_j) \tag{57}$$

where

$\mu =$ the grand mean, the mean of all values in the database
$(\mu_j - \mu) =$ the variation among the x_{ij} values attributed to the differences of its group mean μ_j from the grand mean μ; there are k differences
$(x_{ij} - \mu_j) =$ the variation attributed to the differences of the group mean μ_j from the individual i replicate values for each group (or treatment)

The equation may be rewritten by assigning symbols to the two differences.

$$\beta_j = \mu_j - \mu \text{ and } \varepsilon_{ij} = x_{ij} - \mu_j \tag{58}$$

The term β_j represents the series of differences, jth treatment mean minus the grand mean; if there are no significant differences for the treatments then $\beta_j = 0$ for all treatments. If one or more of the treatments are significant then $\beta_j \neq 0$ for the one or more treatments. The term ε_{ij} is a random difference, i.e., measured value x_{ij} minus the jth treatment mean. By subtracting μ from both sides we obtain

$$x_{ij} - \mu = \beta_j + \varepsilon_{ij} \tag{59}$$

This model equation states that the magnitude of any deviation of the dependent variable x_{ij} from the grand mean is the sum of two components of variation, β_j, the component due to any presumed real response effect of a treatment, and a second component ε_{ij}, a within-treatment difference or error. The long run mean value of all ε_{ij} equals zero, and the

Table 5 Layout of Data for Variance Analysis

Replicate number	Level of Independent Variable				
	1	2	3	...	j
1	x_{11}	—	—	...	—
2	x_{12}	—	—	...	—
⋮	⋮	⋮	⋮	...	⋮
i	—	—	—	...	x_{ij}

variance of ε_{ij} is equal to the basic test error variance in making a measurement. The two components are called the between-treatments source of variation and the within-treatment source of variation. The analysis answers the question, is the between-treatments variation significant when compared to the within-treatment variation?

The analysis begins by setting up the null and alternative hypotheses

$$\begin{aligned} &H_0: \beta_j = 0 \qquad \text{for } j = 1, 2, \ldots, k \\ &H_a: \text{At least one } \beta_j \text{ is not } 0 \end{aligned} \tag{60}$$

The between-treatment variation is calculated by noting that the number of β_j terms is j, and the variance among the j group means, $S^2_{\bar{x}}$, is given by

$$S^2_{\bar{x}} = \frac{\Sigma(\bar{x}_j - \bar{x}_{ij})^2}{k - 1} \tag{61}$$

where

$\bar{x}_j$ = the treatment mean (across all i values) for the jth treatment
$\bar{x}_{ij}$ = the grand mean across all i and j values
k = the number of treatments

In general, the standard error for any normal population mean is equal to $\sigma/\sqrt{n}$, and the variance is equal to σ^2/n, where n is the number of values that are used to calculate the mean and σ^2 is the population variance. For this analysis the term $S^2_{\bar{x}}$ is an estimate of the true variance σ^2_{β} among the k different means. If the null hypothesis is true, $S^2_{\bar{x}}$ can be used to evaluate the population variance. To evaluate the variance for individual population values, the calculated $S^2_{\bar{x}}$ must be multiplied by the number of test values (replicates) used for each of the means, designated as n_j, to give $n_j S^2_{\bar{x}}$, which is an estimate of $n\sigma^2_{\beta}$. The variance of the individual measurements for each of the k treatments is calculated by

$$S^2_{\bar{j}} = \frac{\Sigma(x_{ij} - \bar{x}_j)^2}{n_j - 1} \tag{62}$$

This variance is equal to the variance of ε_{ij}. The pooled or average variance for the k estimates of $S^2_{\bar{j}}$ is designated as S^2_p, and given that each treatment has the same number of replicates, it is calculated by

$$S^2_p = \frac{\Sigma(S^2_{\bar{j}})}{k} \tag{63}$$

We now have two estimates of the individual population variance; the first $n_j S^2_{\bar{x}}$ obtained from the variation in treatment means, and the second S^2_p the pooled variance obtained from the replicates. If there are no real effects of the treatments (all $\beta_j = 0$), both of these are estimating the underlying variance of the population. An F-test can be used to decide if the two estimates are equal. Using the hypotheses as given above and using the technically justified assertion that if the variances are not really equal the between-treatments variance should be larger than the within-treatment variance, the F-ratio is defined as

$$F(\text{calc}) = \frac{n_j S^2_{\bar{x}}}{S^2_p} \tag{64}$$

The degree of freedom df for the numerator is $k - 1$ and the df for the denominator is $(n_j - 1)k$. If F(calc) is equal to or greater than F(crit) at these respective df at a probability level $P = 0.05$ or less, then the larger variance is significantly greater than the lesser variance, with the implication that at least one value of β_j is not equal to 0.

ANOVA Calculations

The classical ANOVA calculations are not usually performed as given above but by shortcut methods that were developed to reduce the burden of calculations prior to the use of computers. An understanding of this approach can be gained by considering that the total variation in a database, as given above, is evaluated by calculating the total variance S^2_{tot} by

$$S^2_{\text{tot}} = \frac{\Sigma(x_{ij} - \bar{x}_{ij})^2}{kn - 1} \tag{65}$$

with x_{ij} and $\bar{x}_{ij}$ as defined above and the summation is over all values. The numerator of Eq. 65 is a sum-of-squares called the *total sum-of-squares* and represents all the variation in the database. It may be shown that this total sum-of-squares, the numerator, may be partitioned into the two components on the right-hand side of Eq. 66.

$$\Sigma(kn)(x_{ij} - \bar{x}_{ij})^2 = \Sigma(kn)(x_{ij} - \bar{x}_j)^2 + n\Sigma(k)(\bar{x}_j - \bar{x}_{ij})^2 \tag{66}$$

where

$\Sigma(kn)$ = summation over all kn values
$\Sigma(k)$ = summation over k values
$\bar{x}_j$ = mean of each of j treatments and other symbols as defined above

The first term on the right-hand side is called the *error sum-of-squares* and the second right-hand term is the *treatment sum-of-squares*. A one-way ANOVA is performed by calculating the total sum-of-squares $SS(\text{tot})$ by way of a short-cut expression that can be shown to give the values as defined by the left side of Eq. 66.

$$SS(\text{tot}) = \Sigma x^2_{ij} - C \tag{67}$$

where

Σx^2_{ij} = the individual measured values x_{ij} squared and summed over all values in the database
$C = T^2_{\text{all}}/kn$ = a constant called the correction term, where T_{all} is the grand total of all measured values in the database, k is as defined above, and the j has been dropped on n

The second ANOVA sum-of-squares is the treatment sum-of-squares $SS(\text{trt})$, given by Eq. 68, where ΣT^2_j is the sum of the squared totals for each of the k treatments.

$$SS(\text{trt}) = \frac{\Sigma T^2_j}{n} - C \tag{68}$$

The random variation or error sum-of-squares is obtained by difference:

$$SS(\text{error}) = SS(\text{tot}) - SS(\text{trt}) \tag{69}$$

The three sums-of-squares are used in a table with the following layout of Table 6. The sums-of-squares are divided by the appropriate degrees of freedom df to give variances that are called mean squares. As indicated in the previous exposition of variance analysis, the treatment mean square is divided by the error mean square to make a decision on the significance of the treatments.

Table 6 Analysis of Variance Table

Source of variation	df	Sum of squares	Mean square	F(calc)
Treatments	$k - 1$	SS(trt)	MS(trt) = SS(trt)/($k - 1$)	MS(trt)/MS(error)
Error	$k(n - 1)$	SS(error)	MS(error) = SS(error)/$k(n - 1)$	
Total	$kn - 1$	SS(tot)		

Although the analysis of variance is a powerful tool for data analysis, especially for more complex situations that are beyond the scope of this chapter, it needs to be supplemented by supplementary analysis tools. If F(calc) is found to be significant, this implies that at least one of the treatments is significantly different from some other treatment. If several treatments are used, it is appropriate to determine what the exact relationship is among the means for all the treatments. This type of analysis is called multiple comparisons. A typical statistical test of this type is the Duncan Multiple Range Test; see Part 2 of this section. This test makes all the pair-wise comparisons among the treatment means based on a statistical parameter called the least significant range, LSR. For each pair of means an LSR is calculated and compared to a critical LSR for the number of groups compared, the df for the error term, and a selected P.

6.5 Elements of Correlation and Regression Analysis

Correlation Analysis

Correlation analysis permits an objective evaluation of the degree of association between two variables. If a value of variable x is produced by a process that also produces a corresponding value of variable y, a correlation analysis will indicate if the association between the two variables is significant at some selected level of confidence. A significant correlation does not imply that y may be predicted from x or vice versa; it simply indicates that there is some significant association. Association does not imply a cause and effect relationship, i.e., that x physically generates y. The associated values of x and y may be the result of some unknown underlying cause and effect process that involves both variables.

Graphical techniques or xy scatter plots are important adjuncts to correlation analysis; they give a quick indication of any potential association between two variables. A series of such plots may be required for a number of variables. If simple review of such plots indicates the possibility of significant correlation, then an estimate of the correlation coefficient may be obtained that provides a quantitative index of the degree of association. It is assumed that the x and y values have a normal distribution. The correlation coefficient is a random variable with a defined distribution function that depends on the sample size and the population value (the true underlying value) of the correlation coefficient, designated as ρ. For any finite sample size (number of xy pairs) an estimate of the population value ρ is obtained; this estimate is designated as **r**. The population correlation coefficient has the following properties: (1) its value lies between $+1$, which indicates perfect positive linear association, and -1, which indicates perfect negative or inverse linear association; (2) a value of zero indicates no linear association (a curvilinear association may however exist); and (3) the population coefficient is symmetric with respect to x and y, i.e., x on y gives the same value as y on x.

Analysis begins by calculating the mean $\bar{x}$ of the x variable, $\acute{y}$, the mean of the y variable, and the deviations $(x_i - \bar{x})$ and $(y_i - \acute{y})$. The estimated correlation coefficient **r** is the ratio of the sum of the cross products of the deviations of x and y divided by the square root of the product of the sums of the squares of the same respective deviations.

$$\mathbf{r} = \frac{\Sigma(x_i - \bar{x})(y_i - \acute{y})}{[\Sigma(x_i - \bar{x})^2 \Sigma(y_i - \acute{y})^2]^{1/2}} \tag{70}$$

For any xy scatter plot of two variables that have a high degree of positive correlation, a large number of points will fall in the upper right quadrant 1 and the lower left quadrant 3 when the origin (center) of the quadrants is at the mean values $\bar{x}$ and $\acute{y}$. This clustering of the points insures that quadrant 1 will have positive and relatively large x deviations associated with similar positive and large y deviations. Quadrant 3 will have a similar situation with negative deviation values for both variables. When these deviation cross products are summed over all xy values, they will very nearly equal the square root of the product of the x and y deviations squared, and the ratio will be high or near 1. For a negative or inverse association, the points will cluster in the upper left quadrant 4 and the lower right quadrant 2. A strong association of this type will give negative cross products and a high negative ratio or **r** value approaching -1.

The significance of any calculated correlation coefficient is evaluated by adopting the usual two-sided hypothesis test,

$$\begin{aligned} &H_0: \rho = 0 \\ &H_a: \rho \neq 0 \end{aligned} \tag{71}$$

This hypothesis can be tested for any sample size n of 3 or greater by using a t-statistic given by

$$t_{(\alpha/2)} = \frac{\mathbf{r}}{[(1 - \mathbf{r}^2)/(n - 2)]^{1/2}} \tag{72}$$

where $t_{(\alpha/2)}$ is the value of a random variable that has a distribution that is approximately the usual t-distribution with $n - 2$ degrees of freedom. The correlation is considered significant if t(calc) is greater than a critical value $t_{(\alpha/2)}$(crit) at a level of significance designated by $P = \alpha$. Table A9 gives precalculated critical **r** values at $df = n - 2$.

Regression Analysis

Simple regression analysis is also concerned with the association between two variables x and y, but the goal of this analysis is to develop a mathematical relationship between the two variables to permit a prediction of y by knowing the value of x. The potential use of regression analysis implies that there is a correlation between the two variables and that the distinction between correlation and causation as discussed above applies to this analysis as well. The linear mathematical relationship is called a regression model; it explains or predicts the response or y variable in terms of the other (independent) variable designated as the x variable. Regression analysis is used to make inferences about the parameters of the regression model which is given by

$$y = \beta_0 + \beta_1 x + \varepsilon \tag{73}$$

This model, referred to as the regression of y on x, involves β_0, β_1, and ε defined as

β_0 = the intercept, the value predicted by the model when $x = 0$; it has no practical meaning if x cannot equal zero, but is necessary to specify the model
β_1 = the slope of the regression line, i.e., the change in y for unit change in x
ε = a random error term with population mean of 0 and variance of σ^2

A term called the *conditional mean*, defined by $\mu_{y|x}$, is a predicted value for the dependent variable y for some given x and is expressed as

$$\mu_{y|x} = \beta_0 + \beta_1 x \tag{74}$$

The regression model describes a line that is the locus of all values of the conditional mean, each conditional mean corresponding to *one* of a set of x values. Most regressions problems are concerned with selecting a set of x values that span a reasonable operational range and measuring y at each of these x values. Each of the observed or measured values of the response variable (at a given x) comes from a normal population with mean of $\mu_{y|x}$ and variance σ^2.

The purpose of a regression analysis is to use a set of measured or observed x and y values to estimate the parameters β_0, β_1, and the variance of the ε terms or σ^2 and to perform hypothesis tests and evaluate confidence intervals concerning β_1. Basic assumptions in this analysis are that the linear model is appropriate, that the ε error terms or deviations in the y variable are independent and normally distributed and that they have a common variance σ^2 at all x levels, and that the uncertainty variance in selecting each x level is small in comparison to σ^2. The analysis seeks to find estimates of β_0 and β_1 that produce a set of $\mu_{y|x}$ values that are a best fit to the data. The regression line may be written in an alternative format as

$$\hat{u}_{y|x} = b_0 + b_1 x \tag{75}$$

In this alternative format, $\hat{u}_{y|x}$ is an estimate of the mean of y for any given x, and b_0 and b_1 are estimates of β_0 and β_1 respectively. How well the estimates agree with the observed y is evaluated by the magnitude of the differences $y - \hat{u}_{y|x}$, which are called *residuals*. Small residuals indicate good fit, and the best fit is the line that gives the smallest combined magnitude for the squares or variance of the residuals.

The minimization of the squares of the residuals is called the least-squares criterion, which requires that the estimates of β_0 and β_1 minimize the sums as expressed by

$$\Sigma(y - \hat{u}_{y|x})^2 = \Sigma(y - b_0 - b_1 x)^2 \tag{76}$$

These values are obtained by means of "normal equations," a set of simultaneous equations of the form

$$\begin{aligned} b_0 n + b_1 \Sigma x &= \Sigma y \\ b_0 \Sigma x + b_1 \Sigma x^2 &= \Sigma xy \end{aligned} \tag{77}$$

Based on these equations the estimate for b_1 can be expressed as

$$b_1 = \frac{\Sigma(x - \bar{x})(y - \acute{y})}{\Sigma(x - \bar{x})^2} \tag{78}$$

where the numerator is called the corrected or mean centered sum of cross products and the denominator is called the corrected sum of squares. The expression for b_0 is given by

$$b_0 = \acute{y} - b_1 \bar{x} \tag{79}$$

The error sum of squares, designated by $SS\varepsilon$ is

$$SS\varepsilon = \Sigma(y - \hat{u}_{y|x})^2 \tag{80}$$

which describes the remaining variation in y after estimating the linear relationship y upon x. The degrees of freedom for $SS\varepsilon$ is $n - 2$, where n is the number of xy values, since two df are used for b_0 and b_1 and the mean square or variance estimate of σ^2 is

$$MS\varepsilon = \frac{SS\varepsilon}{df} = \frac{\Sigma(y - \hat{u}_{y|x})^2}{n - 2} \tag{81}$$

Statistical inferences on the value of β_1 are based on the assumption that the estimate b_1 is a random variable whose distribution is approximately normal with mean $= \beta_1$ and variance of $\sigma^2/\Sigma(x - \bar{x})^2$, which can be given in equivalent form as $\sigma^2/(n - 1)S_x^2$, where S_x^2 is the variance among the n values of x. Either of these expressions shows that the estimate b_1 has greater precision when—the population variance σ^2 is small, the sample size or number of xy pairs n is large, and the independent variable x has a wide range of values, i.e., S_x^2 is large. The standard error of the estimate b_1 is $[\sigma^2/(n - 1)S_x^2]^{1/2}$, hence the ratio

$$z = \frac{b_1 - \beta_1}{[\sigma^2/(n - 1)S_x^2]^{1/2}} \tag{82}$$

is a standard normal (z) variable. When the estimated variance $MS\varepsilon$ is substituted for σ^2, the ratio becomes a random variable with a t distribution with $n - 2$ degrees of freedom, and this may be used for hypothesis testing. Thus for testing whether β_1 equals a specific value β_1^*.

$$\begin{aligned} H_0&: \beta_1 = \beta_1^* \\ H_a&: \beta_1 \neq \beta_1^* \end{aligned} \tag{83}$$

the test statistic is

$$t_{\alpha/2} = \frac{b_1 - \beta_1^*}{[MS\varepsilon/(n - 1)S_x^2]^{1/2}} \tag{84}$$

Letting $\beta_1 = 0$ provides a test of the null hypothesis,

$$\begin{aligned} H_0&: \beta_1 = 0 \\ H_a&: \beta_1 \neq 0 \end{aligned} \tag{85}$$

and the confidence interval for b_1 is calculated as

$$b_1 \pm t_{\alpha/2}\left[\frac{MS\varepsilon}{(n - 1)S_x^2}\right]^{1/2} \tag{86}$$

Inferences on the model estimates of the response variable are also important. There are two different but related inferences: (1) inferences on the mean response, how well the model estimates the conditional mean at some x, and (2) inferences on prediction, how well the model predicts the value of the response variable y for a randomly chosen future x value. The point estimate for the first of these is $\hat{u}_{y|x}$, the estimated mean response for any x, and the estimate for the second case is $\acute{y}_{y|x}$, the predicted individual response value for any x. For a specified value x^*, the variance of the estimated *mean* is

$$S^2(\hat{u}_{y|x}) = \sigma^2\left[\frac{1/n + (x^* - \bar{x})^2}{(n - 1)S_x^2}\right] \tag{87}$$

and the variance for a single predicted value is

$$S^2(\acute{y}_{y|x}) = \sigma^2\left[\frac{1 + 1/n + (x^* - \bar{x})^2}{(n-1)S_x^2}\right] \tag{88}$$

Both of these variances have their minima when $x^* = \bar{x}$, i.e., the response is estimated with greatest precision when the independent variable is at its mean. At this location the conditional mean is $\acute{y}$ and the variance of the estimated mean is the familiar σ^2/n. Also note that $S^2(\acute{y}_{y|x}) > S^2(\hat{u}_{y|x})$, since a mean has greater precision than an individual value. When $MS\varepsilon$ is substituted for σ^2, the estimated variance is given by these two equations. Letting $\bar{x} = 0$ in the variance expression for $\hat{u}_{y|x}$ gives the variance for β_0, which can be used for hypothesis and confidence interval testing for β_0. This has applications for some regression problems where β_0 has some intrinsic meaning other than an arbitrary fitting constant.

One simple diagnostic test for simple linear regression that may give an indication that some of the essential assumptions have been violated is the residual plot. If the residuals $y - \hat{u}_{y|x}$ are plotted on the vertical axis and predicted values $\hat{u}_{y|x}$ or x values on the horizontal axis, the plot should give a horizontal band of points, centered on zero on the vertical axis, with relatively uniform height (vertical spread) across the entire horizontal span if there are no serious problems with the regression. If the height increases from end to end, this may indicate nonuniform variance across the x factor range. If the band curves up or down, left to right, this may indicate a nonlinear underlying true relationship and the model has not been properly specified, i.e., it may need a square term.

Multiple regression, where a dependent y variable is a function of a number of independent x variables or factors, is a widely used analysis technique for multivariable problems The use of a specialized simplified multiple regression analysis, where all the independent factors are orthogonal, is discussed in the next section on experimental design. A detailed discussion of customary multiple regression analysis, where orthogonality may or may not be present for all variables, is a topic that is beyond the scope of this chapter.

6.6 Design Of Experiments

Experiments have a wide variety of technical objectives, but all should have the same statistical objective to provide maximum information of the highest quality possible for the minimum cost. The design of experiments is the process of efficiently planning and executing a series of experiments with this objective in mind. Two types of planning are required, (1) effective organization as to test methodology and basic scientific principles, and (2) selection of an experimental protocol to generate test data with certain statistical characteristics. This second type of planning is emphasized in this section. The protocol is the set of instructions for assigning treatments or other experimental factors to measurement or observational units. The reliability of the measured parameters is assessed by the standard error of the estimates, and these errors may be reduced by increased sample size. However for relatively complex technical projects with numerous factors, this leads to excessive testing costs. One way to avoid this is the use of proven experimental designs that provide for efficient comparisons among treatments or factors and reduced residual or test variance.

Basic Design Concepts

In experimental design, *factors* or *treatments* are independent variables. The factors may be quantitative, with their values set at some *level* on a continuous scale, e.g., temperature, or qualitative, with the levels representing certain types or categories, e.g., reactor 1 vs. reactor 2. The *response* is the dependent or measured variable. The generic models for experiments are (1) *a fixed-effects model,* where only the selected levels of the factors in the experiment are of concern, (2) a *random-effects model*, where the factor levels chosen represent a sample from a larger population and conclusions from the program are applied to the population, and (3) a *mixed-effects model*, where both random and fixed effects factors are included in the design. Factors may be primary, i.e., the major factor(s) for evaluation that can have a direct influence, or secondary, i.e., factors that can have a less direct influence, e.g., environmental conditions, which may or may not be evaluated as secondary objectives.

In a typical experiment set up to evaluate three levels; 1, 2, and 3 for Factor A, a secondary factor may be designated as qualitative factor O, which of several operators conducts the test. The usual or classical experimental approach is to fix the level of O, i.e., select one operator and evaluate the response when Factor A is varied over the three levels (1, 2, and 3) with perhaps three replications for each level. This approach has 6 *df* for error estimation and nine total test measurements. Testing will evaluate the influence of A with the selected operator but give no indication of operator effects. A more comprehensive approach is to use a *block* experimental design. Select two diverse operators and for each operator evaluate the response for factor A at levels 1, 2, and 3 with two replications per level. The operators are the blocks, and the influence of factor A is evaluated independently in each block. This design also has 6 *df* for error with now a total of 12 measurements. The investment of 3 more measurements for the second design provides much more information. The influence of factor A is now evaluated for both operators, and any unusual influence or *interaction* of operators on factor A response can also be determined.

In complex experimental programs, *blocking* may be done for potentially perturbing factors that cannot be controlled, such as ambient temperature or humidity variations, by conducting several evaluations of the response in short time periods, where temperature and humidity are relatively constant. Dividing the total experiment into these *blocks*, or relatively uniform groupings, improves the quality of the estimated effects by reducing the standard error of the measurements.

Factorial Designs

A class of experimental designs called factorial designs is well suited to technical projects involving physical measurements. Factorial designs are defined as a group of unique combinations of levels across all the selected factors. When the factor levels are set at the values called for in a particular combination and a response measurement is made for this combination, this is called a (test) *run*. Each design has some number of specified runs, and the total layout or list of these runs is called the *design matrix*. A *complete factorial* design matrix is one where for each factor, all factor levels of the other factors appear in some combinations of the design matrix. Thus for three factors investigated at two levels each, a complete factorial design would require eight (2^3) response measurements or runs, each having a different combination of the two levels of the three factors. When the number of factors is large, a complete or full factorial requires too many test runs, and designs called *fractional factorials* are used where a certain fraction of the full factorial number of runs is selected on the basis that the design be balanced with respect to the number of selected levels of each factor.

Factorial designs may be divided into two types, (1) *screening designs* used to search for important factors or to rank factors in order of importance, and (2) *exploratory designs* used to explore and map out a region of technical interest in greater detail, thereby gaining empirical understanding. The simpler screening designs, used for a preliminary search for important factors in a system, have k factors each at two levels, designated as upper and lower levels. Exploratory designs also have k factors, but now each factor usually has at least three levels, upper, middle and lower; this permits the evaluation of nonlinear response relationships.

The designs as given in this section are analyzed in terms of model equations that simulate the system under study. The designs and the model equations are set up using special coded units for the independent variables (factors) of the design. Thus for the response variable y and two independent variables x_1 and x_2, the model equation that allows for the evaluation of any interaction between x_1 and x_2 is

$$y = b_0 + b_1x_1 + b_2x_2 + b_{12}x_1x_2 \tag{89}$$

where

b_0 = a constant; in system of units chosen it is the value of y when $x_1 = x_2 = 0$
b_1 = change in y per unit change in x_1
b_2 = change in y per unit change in x_2
b_{12} = an interaction term for specific effects of combinations of x_1 and x_2, it indicates how b_1 changes as x_2 changes 1 unit

The coded units are obtained by selecting for *each* factor a value that constitutes a *center of interest* or a *reference value*, and then selecting certain values that are below and above that center of interest by an equivalent amount. This is a straightforward process for quantitative factors but may not be possible for some qualitative factors that can exist at only two levels. In this case the center of interest is considered as theoretical or conceptual. The coded units for any x_i are defined by

$$x_i = \frac{V_E - CV_E}{SU} \tag{90}$$

with

V_E = selected factor value for x_i, in physical units
CV_E = center of interest factor value for x_i, in physical units
SU = scaling unit, i.e., change in physical units equal to 1 coded unit

When V_E is higher than CV_E by an amount equal to SU, then $x_i = 1$; when V_E is less than CV_E by an equal amount, $x_i = -1$, and when $V_E = CV_E$, $x_i = 0$. The center of interest values for all factors constitute the central point in the multidimensional *factor space* for the experiment. As indicated above, the constant b_0 is the value for y at this center in the factor space; it is the (grand) average of all responses and is an important analysis output parameter for these designs.

The design matrix for a 2^3 full factorial design is given in Table A10 along with an additional matrix called the *independent variables matrix* or *analysis matrix* that is used to calculate the effects of the independent variables; this matrix has an equal number of rows and columns. The analysis of factorial designs is a relatively straightforward hand calculator procedure that may be performed by using the analysis matrix generated from the design matrix as follows: (1) the first analysis matrix column consists of 1 values and is used to evaluate the grand mean; (2) the next three columns are the same as in the design matrix; (3) the remaining (interaction) columns are generated by multiplying respective values in the columns headed by x_i and x_j to give column entries for b_{ij}. The final column is generated in the same way for three factors. In general this operation is conducted for all factors up to k, for any 2^k design.

With this design the *effects* of the three factors x_1, x_2, and x_3 may be evaluated in terms of *effect coefficients*. There are three *main effect* coefficients, b_1, b_2, b_3; three *two-factor interaction* coefficients, b_{12}, b_{13}, b_{23}; and one *three-factor interaction* coefficient, b_{123}, as given in Eq. 91.

$$y = b_0 + b_1x_1 + b_2x_2 + b_3x_3 + b_{12}x_1x_2 + b_{13}x_1x_3 + b_{23}x_2x_3 + b_{123}x_1x_2x_3 \quad (91)$$

The coefficients are evaluated from the sum of the products obtained by multiplying certain column values or elements on each row, where (col b_i) indicates a specific row value in column b_i of the analysis matrix and y_i is the same specific row value for the response. The sum obtained over all N responses is divided by $N/2$, where N is the total runs in the design.

$$b_i = \frac{\Sigma[y_i(\text{col}\, b_i)]}{N/2}$$
$$b_{ij} = \frac{\Sigma[y_i(\text{col}\, b_{ij})]}{N/2}$$
$$b_{ijk} = \frac{\Sigma[y_i(\text{col}\, b_{ijk})]}{N/2} \quad (92)$$

The value for b_0 that is equal to the grand mean of y is given by

$$b_0 = \text{grand mean } y = \frac{\Sigma[y_i(\text{col}\, b_0)_i]}{N} \quad (93)$$

On the assumption of uncorrelated response measurements and equal variance at all response levels, the variance of the effect coefficients S^2_{bi} is given in terms of the variance of the individual response variance S^2_{yi} by

$$S^2_{bi} = \frac{4S^2_{yi}}{N} \quad (94)$$

Screening Designs

Seven screening designs with two to five factors are listed in Table A11. These are a collection of complete factorials or fractional factorials. The fractional factorials are designated as one-half replicate or one-fourth replicate, i.e., a one-half or one-fourth fraction of the full design. With the exceptions as noted below, all of these designs allow for the evaluation of two factor interactions that often are important in many technical investigations. Typically three factor interactions have no real significance in such programs. Thus any design that allows for direct calculation of two factor interactions is usually sufficient to give a good evaluation of any system.

All of the designs are *orthogonal* in the independent variables or factors, i.e., there is no correlation among these factors. This feature of orthogonality permits the use of the matrix as discussed above for easy analysis. The designs are *balanced*, i.e., for any factor level for factor i , the levels of all other factors appear at their upper and lower values the same number of times. The analysis for each design in Table A11 can also be conducted by multiple regression analysis with typical computer algorithms to give the values for all b coefficients and the other typical output parameters for multiple regression. One virtue of using a multiple regression analysis is the ability to evaluate the significance of the individual coefficients on the basis of t tests and thus obtain a model with only significant coefficients.

Table A11 gives *alias* and *confounding* information for the fractional or blocked designs. This information is needed for proper design set up, analysis, and interpretation. An alias exists in fractionated designs when the same sum of products $\Sigma y_i(\text{col}\, b_{ij})$ numerically evaluates the sum for two different or separate coefficients; thus no unique evaluation of each of the coefficients is possible. The aliases in the table are indicated by an equals sign. Design 3, the three-factor design conducted in two blocks, can be used to lay out a 2^3 design into two blocks as well as to use either of the blocks to evaluate the three main effects of a three-factor design with the indicated aliases. Thus each block is equivalent to a three-factor one-half replicate design. If it can be shown on technical grounds alone that no two-factor interactions are possible or are of negligible magnitude, then the two-block three-factor design may be used for blocking or for main effect evaluation. If this cannot be assured, then for all potential situations of this sort, designs with no "main effect—two-factor interaction aliases" must be used. Confounding implies a similar unresolved situation where certain higher-order coefficients are equivalent to block effects in the same sense as an alias. This is of lesser importance than aliases with two-factor interactions.

When assigning coefficient numbers to factors, certain features of the fractional designs can be used to give a layout as free from conflicting interpretations as possible. Design 5, a four-factor one-half replicate, has the two-factor interactions aliased with each other. Note that x_1 is part of each aliased combination or equivalency. If it can be ascertained in advance that one factor of the four can be guaranteed to have minimal interaction with the other three, then this one factor should be assigned as x_1. On this basis the left side of all the alias equations is zero or very close to zero, and the right-hand side becomes the real interaction if this interaction is determined to be significant. A similar but not exactly equivalent situation exists with regard to x_5 for the five-factor one-half replicate Design 7; x_5 appears in the alias equations with main effects for factors 1 to 4. Thus for this design the factor with the least likely probability of interacting with the other factors should be assigned as Factor 5.

Exploratory Designs

With the exception of the two-factor design, those designs illustrated in Table A12 consist of a core group of runs that is equivalent to a screening design with certain added runs that contain additional levels of the factors. The additional runs are located at the center of the design (center of interest) and at upper and lower levels beyond those that appear in a screening design. These added runs enable any nonlinear behavior to be evaluated and allow for main effect and two-factor interactions to be evaluated with enhanced reliability due to the increased number of runs. The center runs are replicated to give a small *df* estimate of error.

The two-factor design is a layout in the shape of a hexagon with two center points. The three-factor design is a full factorial augmented by lower and upper points selected to give an efficient estimation of the potential effects of the three factors. The center point is replicated four times. The four- and five-factor designs are built on a one-half replicate of the respective full factorial again augmented by lower and upper points for all factors plus replicated center points.

The selection of the physical units that represent the coded units of both screening and exploratory designs should be carefully thought out. In a screening design, the lower and upper physical unit levels should be as wide as possible to increase the sensitivity of the evaluation. The values should also be in the range of direct interest to the technical problem at hand. Exploratory designs, with perhaps five levels, also require careful selection of the range of values and the equivalence between physical and coded units.

Part 2. *Computer Software Procedures for Statistical Analysis*

6.7 A Typical Program

Although there are numerous computer statistical analysis products available, they all have many capabilities in common. The characteristics described here are a *basic* list for one of these products for a usual range of applications. They are listed without any explanation or definition of parameters and terms. Refer to [7] for a full description and explanation.

Initial Data Exploration

Descriptive statistics—mean, median, trimmed means, standard deviation and standard error, variance, minimum, maximum, range, interquartile range, skewness, kurtosis
Frequency statistics—outlier identification: boxplots, stem-and-leaf plots, and histograms
Frequency statistics—description: percentiles, probability plots, robust estimates or M-estimators, Kolmogorov–Smirnov and Shapiro–Wilk normality tests
Variance homogeneity—Levene's test for equality of variance

Cross-tabulations

For two-way and multi-way tables—Pearson's r, Pearson's χ-square, likelihood-ratio χ-square, Yates corrected χ-square, Spearman's rho, contingency coefficient, Goodman's and Kruskal's tau, eta coefficient, Cohen's kappa, relative risk estimate

Means and One-Way ANOVA

Mean values for subgroups and related univariate descriptive statistics are given for a number of dependent variables; a one-way ANOVA is performed

Means and *t* Tests

Independent (two) sample *t* tests, paired-sample *t* tests, one sample *t* tests

One-Way ANOVA

Typical one-way ANOVA with post hoc tests: LSD, Bonferroni, Duncan's, Sidak's, Scheffe, Tukey, Tukey's-b, R-E-G-W-F, R-E-G-W-Q, S-N-K, Waller–Duncan
Levene's homogeneity of variance

Simple Factorial Analysis

ANOVA for factorial experiments: cell means, ANOVA table, covariate coefficients, R and R^2, multiple classification table, eta and beta values

Bivariate Correlations

Pearson's correlation coefficient, Spearman's rho, Kendall's tau-b, univariate statistics, covariances and cross-products, outlier screening prior to analysis

Linear Regression

Estimates of linear regression coefficients, standard errors of coefficients, significance of coefficients, blocking of variables, residual calculation and residual analysis, standard ANOVA, weighted least-squares analysis

Curve Estimation

Models available: linear, logarithmic, inverse, quadratic, cubic, power, compound, S-curve, logistic, growth, exponential

For each model: regression coefficients, multiple R, R^2, adjusted R^2, standard error of estimate, ANOVA table, predicted values, residuals, prediction curves

Nonparametric Tests

Chi-square, binomial test, runs test, one-sample Kolmogorov–Smirnov test, Mann–Whitney U test, Moses test, Wald–Wolfowitz test, Kruskal–Wallis test, Wilcoxon signed rank test, Friedman's test, Kendall's W test, Cochran's Q test

Multiple Response Analysis

Frequencies and frequency tables calculated, multiple response cross-tabs given

7 Quality Assessment and Control

Although quality may be thought of as a recent or modern development, there is nothing new about the basic idea of a quality product. For centuries highly skilled artisans have achieved high levels of quality by a variety of techniques across a broad range of products and services. What is new is the current action to move up on the quality curve; to attempt to improve quality to higher levels by a series of quality improvement iterations until some desired level is attained.

There are three terms that are frequently used in discussing the generic concept of quality: *statistical process control* or SPC, *statistical quality control* or SQC, and *quality assurance* or QA. SPC and SQC address essentially the same goals and are often used interchangeably. Both operations are conducted in large part using *average* and *range* charts, in tandem with innovative technology for process improvement. Quality assurance is the act of insuring that a certain level of quality is attained in any process. In a production setting there are basically two ways to attain any quality level: (1) use rigorous inspection to cull out off-quality product, or (2) analyze and improve the process, using SPC/SQC techniques, to generate product to whatever target level is achievable or appropriate. Quality is assured when these techniques and other ancillary technical disciplines are used to attain the level of quality desired or needed.

As outlined in the introduction, the quality of a test system has two major components: (i) how well the measured parameters relate to the fundamental properties that are involved in the systems-related decision process and (ii) the magnitude of the uncertainty of the selected measured or derived parameter(s). In a production setting, system performance is related to achieving the properties of the product that satisfy the customer, and quality is related to this performance. In a laboratory setting where one or more test systems are of direct concern, performance is defined as the capability to obtain *acceptable output* (data) on selected reference or other materials or objects as well as on routine testing. Quality is defined by this performance, and it may have more than one level or grade depending on the system and the importance of any technical decision.

For any level of quality in a test system, acceptable output is determined jointly by (1) the mean, of a set of individual measurements, for a particular parameter, and (2) the variation of the individual repetitive measurements used to obtain the mean within each set. When both of these are evaluated over a period of time, they must be within certain limits, and the magnitude of these limits determines the quality level. Another quality term, *quality assessment* , is a process of discovering what these limits are and how they may be improved by reducing the inherent variation. Thus both production and test systems can be in control at various quality levels.

7.1 Basic Quality Assessment

Data used for technical decisions fall into one of two *categories*: (1) the test data themselves are the ultimate end point of the measurement process, and the data either meet some set of criteria or they do not; or (2) the test data are used as input to a more complex system, and this complex system may have other quantitative inputs; the operation of the system must also meet some set of criteria. In the first case, decision procedures are frequently developed on long-term experience with accepted test methods often developed by standards organizations, and the quality assessment is reasonably straightforward. In the second case, decisions are more difficult to make, since they depend on fundamental technical and scientific knowledge that may be in the process of development; often there is no extensive background testing experience. The guidance given here applies mainly to Category 1 decision quality assessment, although it may be applied to Category 2 situations after all issues concerning the fundamental scientific basis for the testing have been resolved. Quality assessment techniques may be classified as *internal* or *external* .

Internal Techniques

Some frequently used internal techniques are

Repetitive measurements
Internal test or reference samples (or objects)
Control charts
Interchange of operators and/or equipment
Independent measurements
Definitive (alternative) methods or measurements
Audits

The purpose of all of these techniques is to determine how a system performs based on the selection of one or more performance parameters. The first two techniques on repetitive measurement and the use of internal test or reference samples is the classical way to evaluate *precision*. However this can be a time consuming process if it is not carefully planned, and quality assessment frequently provides a way to minimize the number of measurements. The use of duplicate samples in routine testing, and the accumulation of this information over a time period, is another approach to evaluating precision. More details on this will be given in Section 8 on laboratory precision, which addresses both intralaboratory and interlaboratory operations.

The use of control charts is a documented way to interpret sequential test data and assess quality as well as to monitor or maintain quality. This is described in more detail below. The interchange of operators and equipment (if possible) can also provide valuable information about sources of variation and their influence on quality. Test data should be independent of such factors as operators and individual test machines of the same design. If output data are related to such factors, operator training and machine calibration operations need attention. Independent measurements and definitive methods are related; they both may be used to measure the same parameters by a different but equivalent method. Although independent measurements may not be available for some methods, the results of such testing may give information about any bias in a selected test method, although this topic is more rigorously approached by way of *external* techniques as described below. Audits by internal staff members of any standard operating procedure (SOP) may also assist in assessing and improving quality.

External Techniques

Some frequently used external techniques that may be used are

Collaborative or interlaboratory testing
Exchange of test samples (or objects)
External Reference Materials (RM)
Standard Reference Materials (SRM)
Audits

All of these techniques establish a relationship between a laboratory and the external world. Collaborative or interlaboratory testing consists of programs organized to test samples from one or more selected materials (or objects) that have some documented level of homogeneity and are supplied to a number of laboratories. Participation in collaborative programs allows a laboratory to determine how it stands in comparison to other participants in the program and the accepted reference value for the measured parameter(s). The exchange of test samples is a technique often used for producer–consumer situations to resolve any testing disagreements. Bias may be evaluated by the use of *reference materials*; these may be ad hoc reference materials or RMs, developed by a standardization committee or other recognized group, that have an accepted reference value; or they may be more formally developed *standard reference materials* or SRMs from various national standardization laboratories or bodies such as the National Institute for Standardization and Technology (NIST) in the USA.

By comparing the values obtained in any laboratory to the standard reference value the magnitude of any bias is clearly indicated. Biases may be dependent on a number of factors concerning the testing operation; calibration procedures, operator technique, and ambient laboratory conditions are some typical sources. See Annex C for more details on bias evaluation. Interlaboratory testing is used to evaluate *repeatability* or within-laboratory precision and *reproducibility* or between-laboratory precision. External audits by any number of accredited organizations are important and have been given increasing attention in the past decade as the interest in such standards as the ISO 9000 series listed in the bibliography has risen.

7.2 Basic Quality Control

Quality control consists of sustaining the level of performance or quality as assessed over a selected time period. The explanation of the concepts of quality control is often based on an industrial production process in which the quality can be or has been assessed by independent evidence perhaps of a specialized and more costly nature than the control testing that is to be established. Thus the state of statistical control as defined below is guaranteed by this independent evidence. Once this is in place, the operations to set up a control operation are straightforward. However in a laboratory setting independent evidence for a state of control may or may not be available. When it is not available, the process of establishing a bona fide quality control program must proceed by an iteration process.

After all steps have been taken to accurately install a specific testing system, a period of apparently stable operation is selected and the initial steps to establish the control system are taken. The initial control process is followed for an additional period of time during the testing operation, and the results are examined. If problems appear when specified analysis protocols are employed, the problem is resolved and a second quality control process is established. This procedure is repeated until no further improvement can be made with the technical evaluation procedures available. When this state is reached full statistical control is achieved within the scope of the testing technology.

All operating systems have one common characteristic—the output is inherently variable. When the variability present in any operating system is confined to *indeterminate random variation*, the system is in a state of *statistical control*. This type of variation is also called *common cause* variation. The magnitude of this random variation is a function of the complexity of the testing system and the technology available to detect and eliminate this unwanted variation. *Statistical control* is that system state after all sources of *determinate or assignable variation* have been eliminated by the tools available to the experimenter. Assignable variation is variation that can be traced to a specific cause such as poor calibration procedures, poorly trained operators, and faulty machine settings and similar problems. This type of variation is also called *specific cause* variation. The discovery and elimination of assignable variation is dependent on the skill and expertise of an experimenter and on the level of the technology available for searching out potential causes of specific variation.

The basic approach to both assessing and controlling variation of either sort is the *control chart* technique, as originally developed by Shewhart [8], which can be used (1) to assess the level of achievable quality in a series of repetitive application steps to discover and eliminate assignable causes of variation, and to (2) control the level of quality once a certain level has been established. Control charts can provide a clear indication of the repetitive nature of a measured parameter in the sense of evaluating the long-term variation and the short-term variation. The use of control charts is based on the assumption that once all sources of the most easily recognized assignable variation have been eliminated, the residual level of variation is represented by a normal distribution. It must be recognized that any level of residual variation may contain some components that may at some future time be identified as assignable variation. Quality assessment and control can be based on one of two types of data: (1) attribute data, which are frequently defined as *count data*, the number of defective items or units in a sample of specified size, or (2) variable or measurement data, which are expressed on a continuous scale. Although the basic ideas of quality assessment and control are the same for both types of data, the specific details of calculation are somewhat different. Since attribute data are typical of industrial production processes, the procedures for their use are not described here.

Control Charts

There are two basic types of control charts. One is a chart that illustrates the long-term variation of the process or system; it consists of a mean value (of a set of n measurements) designated as $\bar{x}_n$, for some measured parameter, plotted sequentially (hourly, daily, weekly). This is called an $\bar{x}_n$ chart (x-bar chart). A second type of chart is one that indicates the short-term variation in the set of n individual measurements for the average or mean $\bar{x}_n$. This usually takes the form of a range R, of two or more measured parameter values obtained in a specified time span (either side by side or within a brief specified time) also plotted sequentially as above; this is called an *R chart*. Both of these charts have certain characteristics or limits that are developed to aid in the interpretation of the $\bar{x}_n$ and R values as they are recorded in time.

Both types of charts have a *central line*, which can be defined as the *mean quality level*. In the $\bar{x}_n$ chart this equates to the mean value of the measured parameter set over some stable time period. In the R chart this central line is the average range also over some stable established time period. When sequential values of either $\bar{x}_n$ or R lie close to the central line, there is some degree of confidence that the system is in statistical control. More objective decisions on whether statistical control is achieved are made on the basis of limits on $\bar{x}_n$ and R.

The $\bar{x}_n$ chart. This has values and limits defined according the following:

$$\begin{aligned} \text{Central line} &= \text{mean of } \bar{x}_n \text{ values} = \bar{\mathbf{x}}_n \\ \text{Lower Control limit} &= LCL = \bar{\mathbf{x}}_n - 3\sigma_n = \bar{\mathbf{x}}_n - A_2\mathbf{R} \\ \text{Upper control limit} &= UCL = \bar{\mathbf{x}}_n + 3\sigma_n = \bar{\mathbf{x}}_n + A_2\mathbf{R} \end{aligned} \tag{95}$$

where

A_2 = a factor as given below for the range charts; when A_2 is multiplied by $\mathbf{R}$, the product approximates the 3 sigma limits
$\mathbf{R}$ = the mean of a number of R values

The central line is established as the mean of some large number of $\bar{x}_n$ values, designated as $\bar{\mathbf{x}}_n$, obtained during a period where the system was in a state of statistical control. The number of $\bar{x}_n$ values used to calculate $\bar{\mathbf{x}}_n$ should be at least 20 and preferably 30 or slightly more. The lower control limit is $\bar{\mathbf{x}}_n$ minus the value of three times the standard deviation among the 20 to 30 values (each with n measurements) used to obtain $\bar{\mathbf{x}}_n$. The upper control limit is $\bar{\mathbf{x}}_n$ plus the value of three times the same standard deviation.

The plus and minus three sigma limits may also be established on the basis of the variation in the single or individual measurements used for the mean $\bar{x}_n$. For an individual measurement standard deviation σ, based on 30 or more measurements under statistical control conditions (it is assumed that this gives a value sufficiently close to the true value to use the symbol σ), the standard deviation for $\bar{x}_n$ values, designated as σ_n, is given by the usual expression, $\sigma_n = \sigma/\sqrt{n}$, where n is the number of values in each set that are used to calculate $\bar{x}_n$.

For circumstances where less than 30 values are used for σ, the appropriate t value as well as the symbol S, should be substituted for 3 and σ in the LCL and UCL expressions. See Section 3. As indicated in the LCL and UCL expressions, another option is available for the three sigma limits. When the range of n single measurements is evaluated for each *set* of the 20 to 30 *sets* of values, an unbiased estimate of $3\sigma_n$ is obtained by using the mean range $\mathbf{R}$ among the 20 to 30 values, and an appropriate multiplying factor A_2, that depends on the number n .

The Range Chart. This type of chart is also defined in terms of a central line and upper and lower limits.

$$\begin{aligned} \text{central line} &= \text{mean of R values} = \mathbf{R} \\ \text{lower control limit} &= D_3\mathbf{R} \\ \text{upper control limit} &= D_4\mathbf{R} \end{aligned} \tag{96}$$

The values for the constants D_3 and D_4 depend on the number of values n in each range R. Table 7 gives the values for the typical quality control constants A_2, D_3, and D_4 for sets with various n values.

In addition to the control limits as given for both types of charts, very frequently lower and upper warning values are also given as the $\pm 2\sigma$ limits. These are calculated in the same fashion as the control limits in regard to sample size. The lower warning limit

Table 7 Control Chart Constants

Number in set	A_2	D_3	D_4
2	1.88	0	3.27
3	1.02	0	2.58
4	0.73	0	2.28
5	0.58	0	2.12
6	0.48	0	2.00
7	0.42	0.08	1.92
8	0.37	0.14	1.86

LWL and the upper warning limit UWL are defined as two-thirds of the control limits and are calculated using the LCL and the UCL by

$$LWL = 0.667(LCL) \text{ and } UWL = 0.667(UCL) \tag{97}$$

Control Chart Procedure

The following is a explanation of the step-by-step procedure to set up and maintain both an $\bar{x}_n$ chart and a range chart.

1. Decide on the objective of the control charts—what is being controlled?
2. Choose the measurement variable, if this is not the same as step 1.
3. Decide on the set or subgrouping size, i.e., on the value of n and on the frequency of sampling and/or measurement.
4. Establish worksheets or a computer format for data accumulation and calculation.
5. Calculate the *set* means $\bar{\boldsymbol{x}}_n$ and *set* mean ranges $\boldsymbol{R}$ for a period of known or presumed statistical control; calculate the central line and the LCL, UCL, LWL, and UWL for both charts and establish these on the charts.
6. Continue testing—do the charts indicate an "in control" situation, i.e., all points within the control limits? If not, search for the cause and take remedial action.
7. If control is not achieved in Step 6, repeat steps 1 to 6 as a second iteration after remedial action is taken.
8. Repeat steps 1 to 7 if needed, until statistical control is fully attained within the capability of the testing technology available to search for assignable causes.

7.3 Quality Improvement

The current trend in quality improvement for production processes is to design quality into a product or material, i.e., make it an inherent characteristic and not rely on intensive inspection to cull out the poor-quality items or lots of material. Improved quality can be achieved in a number of ways. A thorough scientific and theoretical understanding of the system under consideration is the best route to high quality. This is not possible with most real-world systems, so systematic empirical approaches must be used. W. E. Deming [9,10] was a pioneer in the fundamental quality improvement process. His efforts to assist Japan in achieving high quality levels for its products beginning in the 1950s are well recognized. J. M. Juran also has a distinguished record in quality assurance and control; see his text in the bibliography.

The quality level of production processes can be improved by using factorial design experimentation. Although such experimental designs have been used for various research and development programs for several decades, their application to industrial production processes was pioneered by G. Taguchi [11,12]. His work, which is in essence based on fractional factorial experimentation, has shown that these techniques can frequently be used (1) to select one production variable to minimize variation (by careful control) and select another variable to hold response on target, and (2) to create products that are less sensitive to the environmental conditions of the process. Cause-and-effect diagrams obtained as the output of brainstorming sessions with experienced personnel can often be used to good advantage to detect possible causes for poor quality. This type of information can lead to improvements in the SOPs for such testing.

Quality improvement can also be attained by a number of concerted efforts such as "ruggedness testing" procedures that evaluate the influence of operational factors, as outlined in Section 4. The use of comprehensive quality manuals and periodic audits and reviews is also helpful. When computers are used for data acquisition and processing, the verification and integrity of both the hardware and the software are essential to quality. All of these options for improving industrial production quality can be used with some minor modifications for improving the quality of laboratory testing and measurement programs.

8 Precision, Bias, and Uncertainty in Laboratory Testing

Although precision, bias, and uncertainty have been discussed in previous sections, an additional and more detailed and somewhat historically oriented discussion is required, especially when typical producer–user and other specification testing is considered which involves a comparison of interlaboratory or different location test results. It should be emphasized that all three of these parameters must always be defined or evaluated in terms of the testing domain to which they apply. Annex B should have been reviewed prior to addressing the topics in this section.

Precision

This concept, which is expressed in terms of the degree of agreement among repetitive test values, is probably as old as testing itself. The simplest precision testing domain is a typical analytical "bench test" in a specific location, usually a single laboratory with an uncomplicated instrument and one testing technician where replicated measurements can be obtained within an hour. Such well-established testing procedures usually have a reasonably good level of precision, and testing a number of objects or materials usually gives reliable data that permits distinguishing any real differences among the tested items. When more complex testing domains are encountered, precision may be less satisfactory.

There are two major domain categories for precision, (1) the ability to repeat within-laboratory test results, which is by nearly universal agreement designated as *repeatability,* and (2) the ability of different laboratories to get agreement on their test results, and this is designated as *reproducibility*. In the usual inverse relationship, high repeatability and high reproducibility imply that there is close or good repetitive test agreement. The most important circumstance where good precision is desired is in specification and other producer vs. user testing as conducted in various laboratories, frequently on a global basis. Although some degree of interlaboratory testing to evaluate reproducibility has been conducted since laboratories and technical trade initially came into widespread use, the first comprehensive interlaboratory testing was part of the action to introduce standardization into technology. In 1884 the American Society of Mechanical Engineers (as well as Civil and Mining Engineers) began conducting interlaboratory testing, and the results were a mass of conflicting data with very poor agreement. In Europe at about the same time similar activity was taking place. In Germany a voluntary standardization organization was created as the result of an international conference held in Munich in 1884.

As international trade has increased over the past several decades, those standardization organizations that develop test method and specification standards have made policy decisions that all test methods shall have, as part of the standard, a section on the typical precision that can be expected. The American Society for Testing and Materials, ASTM, took such action in 1976. Other national standardization organizations, such as the British Standards Institute, BSI, and the Deutsche Institut für Normung, DIN, in Germany, have also embraced the concept of test method precision. The International Standardization Organization, ISO, has adopted similar policies, and more than 30 ISO Committees are engaged in this work. To facilitate the work on evaluating precision for standard test methods, these organizations have developed guideline standards on how such evaluations shall be conducted, how the data are to be analyzed, and how the results are to be expressed. See the bibliography for ASTM and ISO standards on precision. Annex D gives the necessary background for evaluating precision; topics include organizing an interlaboratory test program (ITP), a review of the terminology used, the assumptions underlying the analysis, and the calculation algorithms for repeatability $\boldsymbol{r}$ and reproducibility $\boldsymbol{R}$. Although there may be some small differences in nomenclature when comparing the ASTM and ISO precision standards, the calculation algorithms for both are identical. Numerous precision evaluation programs have been conducted using these and similar guideline standards for the past four or more decades. In almost all fields of technical activity these precision evaluation programs have shown that many of the current and well-developed test methods show very poor interlaboratory or reproducibility precision.

Bias

The major reason for the customary poor reproducibility for many test methods is the existence of a nonnormal or *biased* between-laboratory data distribution. Bias exists because each laboratory has its own *testing culture*, a unique environment and way of conducting any test that is dependent on the operational conditions in the laboratory. This occurs despite the use of standardized testing methods. This biased output causes a laboratory to be almost always low or high compared to some reference value and to other laboratories. Annex C gives the needed steps to evaluate bias, provided that one or more reference materials or standards are available that have recognized (true) values.

One of the early pioneers in the analysis of ITP data, W. J. Youden, demonstrated more than thirty years ago the dominant influence of interlaboratory bias (or systematic error as he called it) in a series of publications [4,5,15]. He showed with simple graphical plotting techniques the unmistakable existence of bias. The existence of an essentially constant bias for any laboratory invalidates the customary assumption in ITP analysis that a random normal distribution adequately represents the between-laboratory variation. See Annex D.

Veith [13] reviewed the current state of precision testing using some ASTM test methods in the rubber manufacturing industry in 1987. Mooney viscosity (ISO 289, ASTM D1646), a widely used test for quality assessment of raw rubbers, gave reasonably good relative precision, Type 1 ($\boldsymbol{r}$) pooled values of 3.0 percent for several clear rubbers, and good pooled ($\boldsymbol{R}$) values of 3.8 percent on the same basis. See Annex D for the definition of Type 1 and 2 precision and ($\boldsymbol{r}$) and ($\boldsymbol{R}$). For a widely used rate-of-cure test, the oscillating disc curemeter (ISO 3417, ASTM D2084), the precision was substantially worse with Type 1 (***R***) values, which depend on the material being tested, in the range 20 to 81 percent. He also demonstrated that interlaboratory bias was responsible for the poor agreement.

Brown [14] reviewed the results of interlaboratory precision evaluation programs in ISO TC45 (Rubber Product Testing) in 1989. He found reasonable precision for hardness tests (ISO 48, ASTM D2240) with Type 1 (***r***) values in the range 3.0 to 6.0 percent, and similar (***R***) values in the range 6.0 to 11.0 percent. Tensile or stress–strain testing (ISO 37, ASTM D412), a test with widespread usage, gave Type 1 (***R***) values in the range of 8.0 to 32 percent. Many other common tests, such as compression set and temperature rise in flexometer testing, showed very poor precision. For compression set, Type 1 (***R***) values were in the range 26 to 32 percent. For temperature rise, Type 1 (***R***) values were in the range 80 to 97 percent. Brown called into question the wisdom of conducting some of the tests at all, considering the wide variation in interlaboratory results.

Uncertainty

The generic concept uncertainty has been used throughout the chapter because it is a word that effectively conveys the sense of ambiguity about a measured result. The alternative concept of *specific uncertainty* may be defined as "the estimate attached to a test result that defines or characterizes the range of values within which the true value of the measured property is asserted to lie." This definition is similar in principle to that for a confidence interval as discussed in Section 3.6, but it lacks the instructions on how to calculate the "range of values." The establishment of procedures to calculate this type of specific uncertainty is currently under development by standardization organizations, and the provisional standard ISO/CD 12102, which contains the above definition, is currently under review. The major problem in this effort is the development of a comprehensive method to express this *range* so as to encompass any selected testing domain with certain specific factors that influence the range. The remainder of this section is devoted to specific uncertainty, and for brevity the word "specific" is dropped.

There are a number of components to this uncertainty, each associated with particular testing or other operational factors that influence the uncertainty range. These components were previously addressed in Annex B on the statistical model for testing operations. As discussed in the annex, there are two categories that influence the deviations that perturb any measured value for an object or material: production variation and measurement variation. Within each of these categories there are two additional types of components that cause deviations about any measured value: random and bias. Current approaches to uncertainty concentrate on the measurement variation and ignore the production variation by implicitly assuming that this variation is or can be made to be negligible.

Uncertainty is frequently based on the concept of *error* populations, where for any of these populations "error" implies a deviation δ_i from some true or reference value μ_r that would be obtained for the measurement y_i in the absence of any type of perturbation by a physical cause. However the word error can be defined on a much broader basis, and *deviation* will be used rather than error.

$$\delta_i = y_i - \mu_r \tag{98}$$

There may be any number of causes $(1, 2, \ldots, i)$ that contribute a deviation component. This situation may be represented by Eq. B1 in Annex B, given here as Eq. 99.

$$y_i = \mu_0 + \mu_j + \Sigma(b) + \Sigma(e) + \Sigma(B) + \Sigma(E) \tag{99}$$

The reference or true value for any material or object class for a particular measurement process is the sum of the first two terms of Eq.. 99, thus

$$\mu_r = \mu_0 + \mu_j \tag{100}$$

A rearrangement of Eq. 99 using $\mu_r = \mu_0 + \mu_j$ shows that δ_i for any given measurement is

$$\delta_i = \Sigma(b) + \Sigma(e) + \Sigma(B) + \Sigma(E) \tag{101}$$

The terms $\Sigma(b) + \Sigma(e)$ contribute bias and random deviation components attributable to inherent or production process variation in the material (or object class), and the terms $\Sigma(B) + \Sigma(E)$ contribute bias and random components attributable to the operational conditions of the measurement. Most discussions of uncertainty assume that the magnitude of $\Sigma(b) + \Sigma(e)$ is negligible compared to $\Sigma(B) + \Sigma(E)$. However for any realistic appraisal of uncertainty this assumption may not be tenable. For the discussion to follow a negligible magnitude for $\Sigma(b) + \Sigma(e)$ will be assumed for the sake of simplicity.

Uncertainty evaluation is concerned with calculating a $\pm$*range* about any y_i value that has a high probability of including the reference or true value μ_r within the range. Just as in the case of a confidence interval, this range is obtained on the basis of the standard deviation of δ_i values for the defined testing domain. This standard deviation for δ_i is a composite standard deviation obtained as a special sum of the variances of all individual components contained in the four types of terms on the right-hand side of Eq. 101. Thus, omitting the terms $\Sigma(b) + \Sigma(e)$, a composite variance for δ_i may be defined as

$$\text{var}\,(\delta_i) = \Sigma\ \text{var}\,(B) + \Sigma\ \text{var}\,(E) \tag{102}$$

and expressed as a standard deviation, sd (δ_i)

$$\text{sd}\,(\delta_i) = [\text{var}\,(\delta_i)]^{1/2} \tag{103}$$

The bias components may be divided into two categories: (1) global or inherent bias, unique to the test and common to all locations and machines, and (2) local bias, unique to a particular location and/or machine. The set of particular terms as given in Annex B, Eq. B2, better illustrate bias components such as $B_{L,}$, a unique laboratory component, B_M, a machine component, B_{OP}, an operator component, as well as random components E_M, potential random differences among machines, and E_{OP}, potential random differences among operators. To these components which are local an additional potential global bias B_{Gbl} must be added. Uncertainty $\hat{\mathbf{u}}$ may be given in terms of sd (δ_i) as

$$\hat{\mathbf{u}} = k[\text{sd}\,(\delta_i)] \tag{104}$$

where k is a multiplying factor, and the measured value y_i with its uncertainty may be expressed as

$$y_i \pm \hat{\mathbf{u}} = y_i \pm k[\text{sd}\,(\delta_i)] \tag{105}$$

The key issue in calculating the uncertainty $\hat{\mathbf{u}}$ is evaluating sd (δ_i) and adopting a value for k. If all the components of sd (δ_i) are known on the basis of 18 to 20 *df* for each component, then a value of 2 may be used for k. The operation of insuring that all important components of sd (δ_i) are fully evaluated is complex and beyond the scope of this chapter. Guidance standards for evaluating sd (δ_i) are currently under development, and reference should be made to this current activity; see ISO/CD 12102 in the bibliography.

Improving Reproducibility Precision

Poor reproducibility precision has been the one of the major reasons for the establishment of laboratory *accreditation systems* and the organizations that administer such systems as well as the development of the ISO 9000 series of standards that address quality across a broad technical spectrum. Accreditation is defined as the procedure by which an authoritative body or organization gives formal recognition that a laboratory or other testing operation is competent to carry out specific tests or operations. *Proficiency* testing, which is defined as a method of evaluating laboratory testing performance in a collaborative interlaboratory test program, is being used on an increasing basis and is frequently a key part of an accreditation process. As part of an ongoing development in the quality assessment arena, the concepts of accreditation and proficiency testing are now being brought together under the concept of *conformity*, defined as the determination of whether a product or process conforms to a particular set of specifications. Conformity in its broadest sense applies to testing, certification, and management system registration. The goal of all accreditation, proficiency, and conformity effort is to bring about a substantial improvement in reproducibility. See the bibliography for a listing of the ISO 9000 standards and other related standards on accreditation, proficiency, and conformity as developed by ASTM Committee E36.

Veith [16,17] described attempts to address the common occurrence of poor reproducibility or bias by proposing a program called INTERCAL (INTERlaboratory CALibration), a technique for using previously established calibration curves, one for each laboratory, to correct for the essentially fixed interlaboratory bias deviations for a test in any particular laboratory. The correction for a laboratory (a calibration correction) is made in the same sense as a calibration correction or adjustment for a test instrument. The correction makes a laboratory agree with a reference or standard value; thus all laboratories have improved agreement with each other.

In a continuation of the exploration of interlaboratory bias [18], he developed a new approach to evaluating interlaboratory precision that clearly shows the consistent deviation patterns of those laboratories that have a substantial bias. This subsequent work, which was conducted for Mooney viscosity (ISO 289, ASTM D1646), involved an ITP with 25 participating laboratories and again showed the overwhelming influence of bias in poor reproducibility. Using special analysis techniques, he was able to show that the laboratories of this ITP could be grouped into three grades based on both their reproducibility and their repeatability: Grade A laboratories, with no evidence of bias, with good (r) and ($\boldsymbol{R}$), 2.2 and 2.4 percent respectively; Grade B laboratories, those biased laboratories that have good repeatability, that can be made to conform to the approximate level of Grade A for reproducibility by use of an INTERCAL correction; and Grade C laboratories, those poor reproducibility or biased laboratories that have unstable testing systems with unusually high within-laboratory variation that do not respond to the INTERCAL correction process.

Proficiency testing and other allied programs often called *cross-check* testing have the goals of improved reproducibility, but in most such programs the initiative to improve resides with the individual participants. This often results in a postponement of any steps to attain improved performance. Another approach to improved reproducibility is the detailed technical investigation of test methods to determine which of the steps in their execution are most prone to generate variability and how they may be made more rugged and immune to these procedural steps. Brown and Soekarnein [19] investigated the influence of some fourteen of the usual within-laboratory operational factors in hardness testing using dead load and spring loaded instruments as well as microhardness devices (miniaturized instruments able to measure hardness on smaller and thinner specimens). Although some of the selected factors were shown to be important (e.g., lower than allowed thickness), giving variations in the range of ± 1 or 2 points, there were no factors of substantial importance that could account for the interlaboratory deviations customary in many ITP.

Spetz [20,21,22] has reported on the development of improved measurement techniques concerned with aging as well as tensile testing. He has also described how testing ovens and other devices may be improved in their design and manufacture [23]. The major factors related to the elevated temperature ovens used for heating the test specimens are the distribution of the temperature in space (within the oven) and in time (the degree of control of a set temperature). Additional factors are the air velocity and air exchange rate in the oven. Substantial differences were found for these factors, and recommendations were made to reduce the current tolerances on these factors in the aging test method standard ISO 188. In the evaluation of five common test factors in tensile testing potentially responsible for poor agreement, four factors emerged as roughly of equal importance to attaining improvement agreement: calibration, specimen thickness measurement, specimen cutting technique, and using five rather than three test specimens. It was estimated that with proper attention to all four, a reduction in tensile strength reproducibility ***R*** of about 50% might be achieved.

References

[1] Ku, H. H., *Precision Measurement and Calibration*, NBS Special Publication 300—Vol. 1 (1969), U.S. Government Printing Office, Washington, D.C. (NBS is now the National Institute for Standards and Technology, NIST.)

[2] Mandel, J., and Stiehler, R. D., *Jour. Res. NBS, 53* (3), 155 (1954). See also [1] above.

[3] Plackett, R. L., and Burman, J. F., *Biometrica, 33*, 305 (1946).

[4] Youden, W. J., *Industrial and Engineering Chemistry, 51*, 794 (1959).

[5] Youden, W. J., and Steiner, E. H., *Statistical Manual of the Association of Official Analytical Chemists*, AOAC, P.O. Box 540, B. Franklin Station, Washington, D.C. 20044, 1975.

[6] Tietjen, G. L., and Moore, R. H., Some Grubbs type statistics for outlier detection of several outliers, *Technometrics, 14*, 583 (1977).

[7] SPSS (Base 7.0 for Windows), SPSS, Inc., 444 N. Michigan, Ave., Chicago, IL 60611.

[8] Shewhart, W. A., *Statistical Methods from the Viewpoint of Quality Control*, edited by W. Edwards Deming, Graduate School, Dept. of Agriculture, Washington D.C., 1939.

[9] Deming, W. E., *Quality, Productivity and Competitive Position*, MIT Press, Cambridge, MA, 1982.

[10] Deming, W. E., *Out of Crisis*, MIT Center for Advanced Engineering Study, Cambridge, MA, 1986.

[11] Blendell, A., Disney, J., and Pridmore, W. A., eds, *Taguchi Methods: Applications in World Industries*, IFS Publications, Bedford, U.K., 1989.

[12] Taguchi, G., *Taguchi Methods: Design of Experiments*, ASI Press, Dearborn, MI, 1993.

[13] Veith, A. G., Precision in Polymer Testing: An Important World-Wide Issue, *Polymer Testing, 7*, 239 (1987).

[14] Brown, R. P., Faith, Hope and Testing, *European Rubber Journal*, Jan./Feb. 1989.

[15] Youden, W. J., *ASTM Journal of Materials Research and Standards*, Jan., 9, 1963.

[16] Veith, A. G., *"INTERCAL"—A Proposed Program to Eliminate Poor Test Reproducibility in the Rubber Industry*, presented to ISO Technical Committee 45, October 1989, Kuala Lumpur, Malaysia, and to ASTM Committee D11, December 1989, Orlando, FL.

[17] Veith, A. G., *The INTERCAL Concept—A Test Case Using ASTM Test D-623 (Flexometer) Testing*, presented to ASTM Committee D11, June 1990, Philadelphia, PA, and to ISO Technical Committee 45, October 1990, The Hague, Netherlands.

[18] Veith, A. G., A New Approach to Evaluating Inter-Laboratory Testing Precision, *Polymer Testing, 12*, 113 (1993).

[19] Brown, R. P., and Soekarnein, A., An Investigation of the Reproducibility of Rubber Hardness Testing, *Polymer Testing, 10*, 117 (1991).
[20] Spetz, G., *Swedish Institute for Standardization Precision Report* 1988:30, ISBN91-7848-111-2, ISSN 0284-5172.
[21] Spetz, G., Improving The Precision Rubber Test Methods—Part 2—Aging, *Polymer Testing, 13*, 239 (1994).
[22] Spetz, G., *Improving the Precision Rubber Test Methods—Part 3—Tensile Test, Polymer Testing, 14*, 13 (1995).
[23] Spetz, G., Recent Developments in Heat Aging Tests and Equipment, *Polymer Testing, 15*, 381 (1996).

Bibliography

I. Statistical Texts

Freund, R. J., and Wilson, W. J., *Statistical Methods*, Academic Press, 1993.
McCuen, R. H., *Statistical Methods for Engineers*, Prentice-Hall, 1985.
Miller, I., and Freund, J. E., *Probability and Statistics for Engineers*, Prentice-Hall, 1990.
Juran, J. M., and Gryna, F. M., *Quality Planning and Analysis*, 3rd ed., McGraw-Hill, 1993.

II. Standards on Statistics and Quality

Most of the standards given below were prepared by committees devoted to the development of generic standards that apply in principle to all testing and measurement operations.

ASTM Standards for Statistical Applications

These standards are available from
American Society for Testing and Materials
100 Barr Harbor Dr, W. Conshohocken, PA 19428-2959
Fax 610-832-9555

General Standards developed by Committee E11 on Statistics and Quality
E0105—Probability Sampling of Materials
E0122—Sample Size to Estimate Quality of a Lot
E0177—Use of Terms Precision and Bias in ASTM Test Methods
E0178—Dealing with Outlying Observations
E0456—Terminology Relating to Quality and Statistics
E0691—Conducting an Interlaboratory Study to Determine Precision of a Test Method
E1169—Conducting Ruggedness Tests
E1325—Terminology Relating to Design of Experiments
E1402—Terminology Relating to Sampling
E4356—Establishing Consistent Test Method Tolerances

Standards Applicable to Rubber Testing developed by Committee D11 on Rubber
D4483—Determining Precision for Test Method Standards in Rubber and Carbon Black Manufacturing Industries
D4678—Preparation, Testing, Acceptance, Documentation and Use of Industry Reference Materials
D5406—Calculation of Producer's Process Performance Index

Standards Applicable to Plastics Testing developed by Committee D20 on Plastics
D1898—Sampling of Plastics
D4968—Flow Charts for Annual Review of Standards for Precision and Bias

Standards on Laboratory Accreditation developed by Committee E36 on Conformity Assessment (Laboratory and Inspection Agency Evaluation and Accreditation)
E548—General Criteria Used for Evaluating Laboratory Competence
E994—Calibration and Testing Laboratory Accreditation Systems: General Requirements for Operation and Recognition
E1224—Categorizing Fields of Capability for Laboratory Accreditation Purposes
E1301—Proficiency Testing by Interlaboratory Comparisons
E1322—Selection, Training and Evaluation of Assessors for Laboratory Accreditation Systems
E1323—Evaluating Laboratory Measurement Practices and the Statistical Analysis of the Resulting Data
E1579—Ensuring Data Integrity in Highly Computerized Laboratory Operations
E1580—Surveillance of Accredited Laboratories
E1738—Development of a Directory of Accredited Laboratories by an Accredited Body

ISO Standards for Statistical Applications

These standards are available from:

International Standards Organization
Case Postale 56, 1 rue Varemba
CH-1211 Geneve 20 Switzerland
Telefax: + 41 22 733 34 30

General Standards
ISO Guide 2—General Terms and Definitions on Standardization, Certification and Testing Laboratory Accreditation
ISO/IEC Guide 25—Requirements for Competence of Calibration and Testing Laboratories
ISO Guide 33—Use of Reference Materials (to validate a test system)
ISO Guide 35—Certification of Reference Materials: General and Statistical Principles

General Standards developed by Technical Committee TC69 on Statistics
ISO 2854—Statistical Interpretation: Estimating Means and Testing Variances
ISO 3534-1 to 3534-3 Statistical Nomenclature (Vocabulary) and Symbols
ISO 5725-1 to 5725-6 Accuracy (Trueness and Precision): Six Part Guideline Standard for Evaluating Precision and Accuracy under Various Conditions
ISO 7870—Control Charts—General Guide and Introduction
ISO 7873—Control Charts: Arithmetic Average with Warning Limits
ISO 8258—Shewhart Control Charts
ISO 11095—Linear Calibration using Reference Materials
ISO/CD 12102 (Draft)—Measurement Uncertainty

Standards Applicable to Rubber Testing: Developed by Technical Committee TC45 on Rubber
ISO TR 9272—Rubber, Rubber Products, Determination of Precision for Test Methods
ISO TR 9474—Rubber, Rubber Products, Determination of Accuracy and Bias of Chemical Test Methods

Standards on Quality Assessment and Control developed by Technical Committee TC176 on Quality Management and Assurance

ISO 9001—Quality Systems: Model for Quality Assurance in Design/Development, Production, Installation and Service

ISO 9002—Quality Systems: Model for Quality Assurance in Production and Installation

ISO 9003—Quality Systems: Model for Quality Assurance in Final Inspection and Test

ISO 9004—Quality Management and Quality System Elements—Guidelines

Annex A. Statistical Parameter Tables

Table A1 Cumulative Normal Distribution

z	X	Area	z	X	Area	z	X	Area
−3.25	$\mu - 3.25\sigma$	.0006	−1.00	$\mu - 1.00\sigma$	.1587	1.05	$\mu + 1.05\sigma$	.8531
−3.20	$\mu - 3.20\sigma$	.0007	− .95	$\mu - .95\sigma$	.1711	1.10	$\mu + 1.10\sigma$	.8643
−3.15	$\mu - 3.15\sigma$	.0008	− .90	$\mu - .90\sigma$	.1841	1.15	$\mu + 1.15\sigma$	.8749
−3.10	$\mu - 3.10\sigma$	.0010	− .85	$\mu - .85\sigma$	.1977	1.20	$\mu + 1.20\sigma$	.8849
−3.05	$\mu - 3.05\sigma$	.0011	− .80	$\mu - .80\sigma$	.2119	1.25	$\mu + 1.25\sigma$	.8944
−3.00	$\mu - 3.00\sigma$	.0013	− .75	$\mu - .75\sigma$	.2266	1.30	$\mu + 1.30\sigma$	.9032
−2.95	$\mu - 2.95\sigma$	.0016	− .70	$\mu - .70\sigma$	.2420	1.35	$\mu + 1.35\sigma$	.9115
−2.90	$\mu - 2.90\sigma$	.0019	− .65	$\mu - .65\sigma$	.2578	1.40	$\mu + 1.40\sigma$	.9192
−2.85	$\mu - 2.85\sigma$	.0022	− .60	$\mu - .60\sigma$	.2743	1.45	$\mu + 1.45\sigma$	.9265
−2.80	$\mu - 2.80\sigma$	.0026	− .55	$\mu - .55\sigma$	.2912	1.50	$\mu + 1.50\sigma$	.9332
−2.75	$\mu - 2.75\sigma$	.0030	− .50	$\mu - .50\sigma$	.3085	1.55	$\mu + 1.55\sigma$	.9394
−2.70	$\mu - 2.70\sigma$	.0035	− .45	$\mu - .45\sigma$	.3264	1.60	$\mu + 1.60\sigma$	.9452
−2.65	$\mu - 2.65\sigma$	.0040	− .40	$\mu - .40\sigma$	.3446	1.65	$\mu + 1.65\sigma$	.9505
−2.60	$\mu - 2.60\sigma$	.0047	− .35	$\mu - .35\sigma$	.3632	1.70	$\mu + 1.70\sigma$	.9554
−2.55	$\mu - 2.55\sigma$	.0054	− .30	$\mu - .30\sigma$	.3821	1.75	$\mu + 1.75\sigma$	.9599
−2.50	$\mu - 2.50\sigma$	.0062	− .25	$\mu - .25\sigma$	.4013	1.80	$\mu + 1.80\sigma$	.9641
−2.45	$\mu - 2.45\sigma$	.0071	− .20	$\mu - .20\sigma$	.4207	1.85	$\mu + 1.85\sigma$	.9678
−2.40	$\mu - 2.40\sigma$	.0082	− .15	$\mu - .15\sigma$	.4404	1.90	$\mu + 1.90\sigma$	.9713
−2.35	$\mu - 2.35\sigma$	.0094	− .10	$\mu - .10\sigma$	.4602	1.95	$\mu + 1.95\sigma$	.9744
−2.30	$\mu - 2.30\sigma$	.0107	− .05	$\mu - .05\sigma$	.4801	2.00	$\mu + 2.00\sigma$	.9772
−2.25	$\mu - 2.25\sigma$	.0122				2.05	$\mu + 2.05\sigma$	.9798
−2.20	$\mu - 2.20\sigma$	.0139				2.10	$\mu + 2.10\sigma$	.9821
−2.15	$\mu - 2.15\sigma$	.0158	.00	μ	.5000	2.15	$\mu + 2.15\sigma$	.9842
−2.10	$\mu - 2.10\sigma$	.0179				2.20	$\mu + 2.20\sigma$	.9861
−2.05	$\mu - 2.05\sigma$	.0202				2.25	$\mu + 2.25\sigma$	.9878
−2.00	$\mu - 2.00\sigma$	.0228	.05	$\mu + .05\sigma$	.5199	2.30	$\mu + 2.30\sigma$	.9893
−1.95	$\mu - 1.95\sigma$	.0256	.10	$\mu + .10\sigma$	.5398	2.35	$\mu + 2.35\sigma$	.9906
−1.90	$\mu - 1.90\sigma$	.0287	.15	$\mu + .15\sigma$	.5596	2.40	$\mu + 2.40\sigma$	.9918
−1.85	$\mu - 1.85\sigma$	.0322	.20	$\mu + .20\sigma$	.5793	2.45	$\mu + 2.45\sigma$	.9929
−1.80	$\mu - 1.80\sigma$	.0359	.25	$\mu + .25\sigma$	.5987	2.50	$\mu + 2.50\sigma$	.9938
−1.75	$\mu - 1.75\sigma$	.0401	.30	$\mu + .30\sigma$	.6179	2.55	$\mu + 2.55\sigma$	.9946
−1.70	$\mu - 1.70\sigma$	.0446	.35	$\mu + .35\sigma$	.6368	2.60	$\mu + 2.60\sigma$	.9953
−1.65	$\mu - 1.65\sigma$	.0495	.40	$\mu + .40\sigma$	.6554	2.65	$\mu + 2.65\sigma$	.9960
−1.60	$\mu - 1.60\sigma$	.0548	.45	$\mu + .45\sigma$	.6736	2.70	$\mu + 2.70\sigma$	.9965
−1.55	$\mu - 1.55\sigma$	.0606	.50	$\mu + .50\sigma$	.6915	2.75	$\mu + 2.75\sigma$	.9970
−1.50	$\mu - 1.50\sigma$	.0668	.55	$\mu + .55\sigma$	.7088	2.80	$\mu + 2.80\sigma$	.9974
−1.45	$\mu - 1.45\sigma$	.0735	.60	$\mu + .60\sigma$	.7257	2.85	$\mu + 2.85\sigma$	.9978
−1.40	$\mu - 1.40\sigma$	.0808	.65	$\mu + .65\sigma$	.7422	2.90	$\mu + 2.90\sigma$	.9981
−1.35	$\mu - 1.35\sigma$	.0885	.70	$\mu + .70\sigma$	.7580	2.95	$\mu + 2.95\sigma$	.9984
−1.30	$\mu - 1.30\sigma$	.0968	.75	$\mu + .75\sigma$	.7734	3.00	$\mu + 3.00\sigma$	.9987
−1.25	$\mu - 1.25\sigma$	.1056	.80	$\mu + .80\sigma$	.7881	3.05	$\mu + 3.05\sigma$	.9989
−1.20	$\mu - 1.20\sigma$	.1151	.85	$\mu + .85\sigma$	.8023	3.10	$\mu + 3.10\sigma$	.9990
−1.15	$\mu - 1.15\sigma$	.1251	.90	$\mu + .90\sigma$	.8159	3.15	$\mu + 3.15\sigma$	.9992
−1.10	$\mu - 1.10\sigma$	.1357	.95	$\mu + .95\sigma$	.8289	3.20	$\mu + 3.20\sigma$	.9993
−1.05	$\mu - 1.05\sigma$	.1469	1.00	$\mu + 1.00\sigma$	.8413	3.25	$\mu + 3.25\sigma$	.9994

Table A2 Critical Values for t

DF	Two-Sided Risk 10%	5%	2.5%	1%	0.5%
1	6.31	12.7	25.5	63.7	127
2	2.92	4.30	6.21	9.92	14.1
3	2.35	3.18	4.18	5.84	7.45
4	2.13	2.78	3.50	4.60	5.60
5	2.01	2.57	3.16	4.03	4.77
6	1.94	2.45	2.97	3.71	4.32
7	1.89	2.36	2.84	3.50	4.03
8	1.86	2.31	2.75	3.36	3.83
9	1.83	2.26	2.69	3.25	3.69
10	1.81	2.23	2.63	3.17	3.58
11	1.80	2.20	2.59	3.11	3.50
12	1.78	2.18	2.56	3.05	3.43
13	1.77	2.16	2.53	3.01	3.37
14	1.76	2.14	2.51	2.98	3.33
15	1.75	2.13	2.49	2.95	3.29
16	1.75	2.12	2.47	2.92	3.25
17	1.74	2.11	2.46	2.90	3.22
18	1.73	2.10	2.45	2.88	3.20
19	1.73	2.09	2.43	2.86	3.17
20	1.72	2.09	2.42	2.85	3.15
21	1.72	2.08	2.41	2.83	3.14
22	1.72	2.07	2.41	2.82	3.12
23	1.71	2.07	2.40	2.81	3.10
24	1.71	2.06	2.39	2.80	3.09
25	1.71	2.06	2.38	2.79	3.08
26	1.71	2.06	2.38	2.78	3.07
27	1.70	2.05	2.37	2.77	3.06
28	1.70	2.05	2.37	2.76	3.05
29	1.70	2.05	2.36	2.76	3.04
30	1.70	2.04	2.36	2.75	3.03
40	1.68	2.02	2.33	2.70	2.97
60	1.67	2.00	2.30	2.66	2.91
120	1.66	1.98	2.27	2.62	2.86
∞	1.64	1.96	2.24	2.58	2.81
One-Sided Risk	5%	2.5%	1.25%	0.5%	0.25%

Source: By permission from *Introduction to Statistical Analysis*, by W. J. Dixon and F. J. Massey. Copyright, 1951. McGraw-Hill Book Company, Inc.

Table A3 Factors for Computing Two-Sided Tolerance Intervals for a Normal Distribution

n/p	$\tau = 0.95$				$\tau = 0.99$			
	.90	.95	.99	.999	.90	.95	.99	.999
2	32.02	37.67	48.43	60.57	160.19	188.49	242.30	303.05
3	8.38	9.92	12.86	16.21	18.93	22.40	29.06	36.62
4	5.37	6.37	8.30	10.50	9.40	11.15	14.53	18.38
5	4.28	5.08	6.63	8.42	6.61	7.86	10.26	13.02
6	3.71	4.41	5.78	7.34	5.34	6.34	8.30	10.55
7	3.37	4.01	5.25	6.68	4.61	5.49	7.19	9.14
8	3.14	3.73	4.89	6.23	4.15	4.94	6.47	8.23
9	2.97	3.53	4.63	5.90	3.82	4.55	5.97	7.60
10	2.84	3.38	4.43	5.65	3.58	4.26	5.59	7.13
15	2.48	2.95	3.88	4.95	2.94	3.51	4.60	5.88
20	2.31	2.75	3.62	4.61	2.66	3.17	4.16	5.31
25	2.21	2.63	3.46	4.41	2.49	2.97	3.90	4.98

Source: Excerpted from "Experimental Statistics," NBS Handbook 91.

Table A4 Critical E_k Values (Tietjen–Moore)

	$a = 1\%$					$a = 5\%$				
$\downarrow n/k \rightarrow$	1	2	3	4	5	1	2	3	4	5
3	0.000					0.001				
4	0.004	0.000				0.025	0.001			
5	0.029	0.002				0.081	0.010			
6	0.068	0.012	0.001			0.146	0.034	0.004		
7	0.110	0.028	0.006			0.208	0.065	0.016		
8	0.156	0.050	0.014	0.004		0.265	0.099	0.034	0.010	
9	0.197	0.078	0.026	0.009		0.314	0.137	0.057	0.021	
10	0.235	0.101	0.048	0.016		0.356	0.172	0.083	0.037	0.014
11	0.274	0.134	0.064	0.030	0.012	0.386	0.204	0.107	0.055	0.026
12	0.311	0.159	0.083	0.042	0.020	0.424	0.234	0.133	0.073	0.039
13	0.337	0.181	0.103	0.056	0.031	0.455	0.262	0.156	0.092	0.053
14	0.374	0.207	0.123	0.072	0.042	0.484	0.293	0.179	0.112	0.068
15	0.404	0.238	0.146	0.090	0.054	0.509	0.317	0.206	0.134	0.084
16	0.422	0.263	0.166	0.107	0.068	0.526	0.340	0.227	0.153	0.102
17	0.440	0.290	0.188	0.122	0.079	0.544	0.362	0.248	0.170	0.116
18	0.458	0.306	0.206	0.141	0.094	0.562	0.382	0.267	0.187	0.132
19	0.484	0.323	0.219	0.156	0.108	0.581	0.398	0.287	0.203	0.146
20	0.499	0.339	0.236	0.170	0.121	0.597	0.416	0.302	0.221	0.163
25	0.571	0.418	0.320	0.245	0.188	0.652	0.493	0.381	0.298	0.236
30	0.624	0.482	0.386	0.308	0.250	0.698	0.549	0.443	0.364	0.298

Source: Reproduced from Tietjen, G. L., and Moore, R. H., "Some Grubbs Type Statistics for Detection of Several Outliers," *Technometrics*, Vol. 14, 1977, pp. 583–597. Symbols are defined as follows: n = sample size. k = number of suspected values.

Table A5 Part A Upper 2.5 Percent Points of F Distribution

Degrees of freedom for denominator	Degrees of freedom for numerator																		
	1	2	3	4	5	6	7	8	9	10	12	15	20	24	30	40	60	120	∞
1	647.8	799.5	864.2	899.6	922	937	948	957	963	969	977	985	993	997	1001	1006	1010	1014	1018
2	38.5	39.0	39.2	39.2	39.3	39.3	39.4	39.4	39.4	39.4	39.4	39.4	39.4	39.5	39.5	39.5	39.5	39.5	39.5
3	17.4	16.0	15.4	15.1	14.9	14.7	14.6	14.5	14.5	14.4	14.3	14.3	14.2	14.1	14.1	14.0	14.0	13.9	13.9
4	12.2	10.7	9.98	9.60	9.36	9.20	9.07	8.98	8.90	8.84	8.75	8.66	8.56	8.51	8.46	8.41	8.36	8.31	8.26
5	10.0	8.43	7.76	7.39	7.15	6.98	6.85	6.76	6.68	6.62	6.52	6.43	6.33	6.28	6.23	6.18	6.12	6.07	6.02
6	8.81	7.26	6.60	6.23	5.99	5.82	5.70	5.60	5.52	5.46	5.37	5.27	5.17	5.12	5.07	5.01	4.96	4.90	4.85
7	8.07	6.54	5.89	5.52	5.29	5.12	4.99	4.90	4.82	4.76	4.67	4.57	4.47	4.42	4.36	4.31	4.25	4.20	4.14
8	7.57	6.06	5.42	5.05	4.82	4.65	4.53	4.43	4.36	4.30	4.20	4.10	4.00	3.95	3.89	3.84	3.78	3.73	3.67
9	7.21	5.71	5.08	4.72	4.48	4.32	4.20	4.10	4.03	3.96	3.87	3.77	3.67	3.61	3.56	3.51	3.45	3.39	3.33
10	6.94	5.46	4.83	4.47	4.24	4.07	3.95	3.85	3.78	3.72	3.62	3.52	3.42	3.37	3.31	3.26	3.20	3.14	3.08
11	6.72	5.26	4.63	4.28	4.04	3.88	3.76	3.66	3.59	3.53	3.43	3.33	3.23	3.17	3.12	3.06	3.00	2.94	2.88
12	6.55	5.10	4.47	4.12	3.89	3.73	3.61	3.51	3.44	3.37	3.28	3.18	3.07	3.02	2.96	2.91	2.85	2.79	2.72
13	6.41	4.97	4.35	4.00	3.77	3.60	3.48	3.39	3.31	3.25	3.15	3.05	2.95	2.89	2.84	2.78	2.72	2.66	2.60
14	6.30	4.86	4.24	3.89	3.66	3.50	3.38	3.29	3.21	3.15	3.05	2.95	2.84	2.79	2.73	2.67	2.61	2.55	2.49
15	6.20	4.77	4.15	3.80	3.58	3.41	3.29	3.20	3.12	3.06	2.96	2.86	2.76	2.70	2.64	2.59	2.52	2.46	2.40
16	6.12	4.69	4.08	3.73	3.50	3.34	3.22	3.12	3.05	2.99	2.89	2.79	2.68	2.63	2.57	2.51	2.45	2.38	2.32
17	6.04	4.62	4.01	3.66	3.44	3.28	3.16	3.06	2.98	2.92	2.82	2.72	2.62	2.56	2.50	2.44	2.38	2.32	2.25
18	5.98	4.56	3.95	3.61	3.38	3.22	3.10	3.01	2.93	2.87	2.77	2.67	2.56	2.50	2.44	2.38	2.32	2.26	2.19
19	5.92	4.51	3.90	3.56	3.33	3.17	3.05	2.96	2.88	2.82	2.72	2.62	2.51	2.45	2.39	2.33	2.27	2.20	2.13
20	5.87	4.46	3.86	3.51	3.29	3.13	3.01	2.91	2.84	2.77	2.68	2.57	2.46	2.41	2.35	2.29	2.22	2.16	2.09
21	5.83	4.42	3.82	3.48	3.25	3.09	2.97	2.87	2.80	2.73	2.64	2.53	2.42	2.37	2.31	2.25	2.18	2.11	2.04
22	5.79	4.38	3.78	3.44	3.22	3.05	2.93	2.84	2.76	2.70	2.60	2.50	2.39	2.33	2.27	2.21	2.14	2.08	2.00
23	5.75	4.35	3.75	3.41	3.18	3.02	2.90	2.81	2.73	2.67	2.57	2.47	2.36	2.30	2.24	2.18	2.11	2.04	1.97
24	5.72	4.32	3.72	3.38	3.15	2.99	2.87	2.78	2.70	2.64	2.54	2.44	2.33	2.27	2.21	2.15	2.08	2.01	1.94
25	5.69	4.29	3.69	3.35	3.13	2.97	2.85	2.75	2.68	2.61	2.51	2.41	2.30	2.24	2.18	2.12	2.05	1.98	1.91
30	5.57	4.18	3.59	3.25	3.03	2.87	2.75	2.65	2.57	2.51	2.41	2.31	2.20	2.14	2.07	2.01	1.94	1.87	1.79
40	5.42	4.05	3.46	3.13	2.90	2.74	2.62	2.53	2.45	2.39	2.29	2.18	2.07	2.01	1.94	1.88	1.80	1.72	1.64
60	5.29	3.93	3.34	3.01	2.79	2.63	2.51	2.41	2.33	2.27	2.17	2.06	1.94	1.88	1.82	1.74	1.67	1.58	1.48
120	5.15	3.80	3.23	2.89	2.67	2.52	2.39	2.30	2.22	2.16	2.05	1.94	1.82	1.76	1.69	1.61	1.53	1.43	1.31
∞	5.02	3.69	3.12	2.79	2.57	2.41	2.29	2.19	2.11	2.05	1.94	1.83	1.71	1.64	1.57	1.48	1.39	1.27	1.00

Interpolation should be performed using reciprocals of the degrees of freedom.

Table A5 Part B Upper 5.0 Percent Points of F Distribution

Degrees of freedom for denominator	Degrees of freedom for numerator																		
	1	2	3	4	5	6	7	8	9	10	12	15	20	24	30	40	60	120	∞
1	161.4	199.5	215.7	224.6	230.2	234.0	236.8	238.9	240.5	241.9	243.9	275.9	248.0	249.1	250.1	251.1	252.2	253.3	254.3
2	18.5	19.0	19.2	19.3	19.3	19.3	19.4	19.4	19.4	19.4	19.4	19.4	19.4	19.5	19.5	19.5	19.5	19.5	19.5
3	10.1	9.55	9.28	9.12	9.01	8.94	8.89	8.85	8.81	8.79	8.74	8.70	8.66	8.64	8.62	8.59	8.57	8.55	8.53
4	7.71	6.94	6.59	6.39	6.26	6.16	6.09	6.04	6.00	5.96	5.91	5.86	5.80	5.77	5.75	5.72	5.69	5.66	5.63
5	6.61	5.79	5.41	5.19	5.05	4.95	4.88	4.82	4.77	4.74	4.68	4.62	4.56	4.53	4.50	4.46	4.43	4.40	4.36
6	5.99	5.14	4.76	4.53	4.39	4.28	4.21	4.15	4.10	4.06	4.00	3.94	3.87	3.84	3.81	3.77	3.74	3.70	3.67
7	5.59	4.74	4.35	4.12	3.97	3.87	3.79	3.73	3.68	3.64	3.57	3.51	3.44	3.41	3.38	3.34	3.30	3.27	3.23
8	5.32	4.46	4.07	3.84	3.69	3.58	3.50	3.44	3.39	3.35	3.28	3.22	3.15	3.12	3.08	3.04	3.01	2.97	2.93
9	5.12	4.26	3.86	3.63	3.48	3.37	3.29	3.23	3.18	3.14	3.07	3.01	2.94	2.90	2.86	2.83	2.79	2.75	2.71
10	4.96	4.10	3.71	3.48	3.33	3.22	3.14	3.07	3.02	2.98	2.91	2.85	2.77	2.74	2.70	2.66	2.62	2.58	2.54
11	4.84	3.98	3.59	3.36	3.20	3.09	3.01	2.95	2.90	2.85	2.79	2.72	2.65	2.61	2.57	2.53	2.49	2.45	2.40
12	4.75	3.89	3.49	3.26	3.11	3.00	2.91	2.85	2.80	2.75	2.69	2.62	2.54	2.51	2.47	2.43	2.38	2.34	2.30
13	4.67	3.81	3.41	3.18	3.03	2.92	2.83	2.77	2.71	2.67	2.60	2.53	2.46	2.42	2.38	2.34	2.30	2.25	2.21
14	4.60	3.74	3.34	3.11	2.96	2.85	2.76	2.70	2.65	2.60	2.53	2.46	2.39	2.35	2.31	2.27	2.22	2.18	2.13
15	4.54	3.68	3.29	3.06	2.90	2.79	2.71	2.64	2.59	2.54	2.48	2.40	2.33	2.29	2.25	2.20	2.16	2.11	2.07
16	4.49	3.63	3.24	3.01	2.85	2.74	2.66	2.59	2.54	2.49	2.42	2.35	2.28	2.24	2.19	2.15	2.11	2.06	2.01
17	4.45	3.59	3.20	2.96	2.81	2.70	2.61	2.55	2.49	2.45	2.38	2.31	2.23	2.19	2.15	2.10	2.06	2.01	1.96
18	4.41	3.55	3.16	2.93	2.77	2.66	2.58	2.51	2.46	2.41	2.34	2.27	2.19	2.15	2.11	2.06	2.02	1.97	1.92
19	4.38	3.52	3.13	2.90	2.74	2.63	2.54	2.48	2.42	2.38	2.31	2.23	2.16	2.11	2.07	2.03	1.98	1.93	1.88
20	4.35	3.49	3.10	2.87	2.71	2.60	2.51	2.45	2.39	2.35	2.28	2.20	2.12	2.08	2.04	1.99	1.95	1.90	1.84
21	4.32	3.47	3.07	2.84	2.68	2.57	2.49	2.42	2.37	2.32	2.25	2.18	2.10	2.05	2.01	1.96	1.92	1.87	1.81
22	4.30	3.44	3.05	2.82	2.66	2.55	2.46	2.40	2.34	2.30	2.23	2.15	2.07	2.03	1.98	1.94	1.89	1.84	1.78
23	4.28	3.42	3.03	2.80	2.64	2.53	2.44	2.37	2.32	2.27	2.20	2.13	2.05	2.01	1.96	1.91	1.86	1.81	1.76
24	4.26	3.40	3.01	2.78	2.62	2.51	2.42	2.36	2.30	2.25	2.18	2.11	2.03	1.98	1.94	1.89	1.84	1.79	1.73
25	4.24	3.39	2.99	2.76	2.60	2.49	2.40	2.34	2.28	2.24	2.16	2.09	2.01	1.96	1.92	1.87	1.82	1.77	1.71
30	4.17	3.32	2.92	2.69	2.53	2.42	2.33	2.27	2.21	2.16	2.09	2.01	1.93	1.89	1.84	1.79	1.74	1.68	1.62
40	4.08	3.23	2.84	2.61	2.45	2.34	2.25	2.18	2.12	2.08	2.00	1.92	1.84	1.79	1.74	1.69	1.64	1.58	1.51
60	4.00	3.15	2.76	2.53	2.37	2.25	2.17	2.10	2.04	1.99	1.92	1.84	1.75	1.70	1.65	1.59	1.53	1.47	1.39
120	3.92	3.07	2.68	2.45	2.29	2.17	2.09	2.02	1.96	1.91	1.83	1.75	1.66	1.61	1.55	1.50	1.43	1.35	1.25
∞	3.84	3.00	2.60	2.37	2.21	2.10	2.01	1.94	1.88	1.83	1.75	1.67	1.57	1.52	1.46	1.39	1.32	1.22	1.00

Interpolation should be performed using reciprocals of the degrees of freedom.

Source: This table is abridged with permission of Professor E. S. Pearson from E. S. Pearson and H. O. Hartley, *Biometrika Tables for Statisticians*, Vol. 1. The original computations appear in C. M. Thompson and M. Merrington, "Tables of the Percentage Points of the Inverted Beta (F) Distribution," *Biometrika*, **33** (1943), 73.

Table A6 Critical χ^2 Values

	Percentiles									
DF	0.5	1	2.5	5	10	90	95	97.5	99	99.5
1	0.000039	0.00016	0.00098	0.0039	0.0158	2.71	3.84	5.02	6.63	7.88
2	0.0100	0.0201	0.0506	0.1026	0.2107	4.61	5.99	7.38	9.21	10.60
3	0.0717	0.115	0.216	0.352	0.584	6.25	7.81	9.35	11.34	12.84
4	0.207	0.297	0.484	0.711	1.064	7.78	9.49	11.14	13.28	14.86
5	0.412	0.554	0.831	1.15	1.61	9.24	11.07	12.83	15.09	16.75
6	0.676	0.872	1.24	1.64	2.20	10.64	12.59	14.45	16.81	18.55
7	0.989	1.24	1.69	2.17	2.83	12.02	14.07	16.01	18.48	20.28
8	1.34	1.65	2.18	2.73	3.49	13.36	15.51	17.53	20.09	21.96
9	1.73	2.09	2.70	3.33	4.17	14.68	16.92	19.02	21.67	23.59
10	2.16	2.56	3.25	3.94	4.87	15.99	18.31	20.48	23.21	25.19
11	2.60	3.05	3.82	4.57	5.58	17.28	19.68	21.92	24.73	26.76
12	3.07	3.57	4.40	5.23	6.30	18.55	21.03	23.34	26.22	28.30
13	3.57	4.11	5.01	5.89	7.04	19.81	22.36	24.74	27.69	29.82
14	4.07	4.66	5.63	6.57	7.79	21.06	23.68	26.12	29.14	31.32
15	4.60	5.23	6.26	7.26	8.55	22.31	25.00	27.49	30.58	32.80
16	5.14	5.81	6.91	7.96	9.31	23.54	26.30	28.85	32.00	34.27
18	6.26	7.01	8.23	9.39	10.86	25.99	28.87	31.53	34.81	37.16
20	7.43	8.26	9.59	10.85	12.44	28.41	31.41	34.17	37.57	40.00
24	9.89	10.86	12.40	13.85	15.66	33.20	36.42	39.36	42.98	45.56
30	13.79	14.95	16.79	18.49	20.60	40.26	43.77	46.98	50.89	53.67
40	20.71	22.16	24.43	26.51	29.05	51.81	55.76	59.34	63.69	66.77
60	35.53	37.48	40.48	43.19	46.46	74.40	79.08	83.30	88.38	91.95
120	83.85	86.92	91.58	95.70	100.62	140.23	146.57	152.21	158.95	163.64

For large values of degrees of freedom the approximate formula $\chi^2_\alpha = \frac{1}{2}(z_\alpha + \sqrt{2n-1})^2$, where z_α is the normal deviate and n is the number of degrees of freedom, may be used.

Source: By permission from *Introduction to Statistical Analysis*, by W. J. Dixon and F. J. Massey. Copyright, 1951. McGraw-Hill Book Company, Inc.

Table A7 Critical (χ^2/df) Values

	Percentiles									
DF	0.5	1	2.5	5	10	90	95	97.5	99	99.5
1	0.000039	0.00016	0.00098	0.0039	0.0158	2.71	3.84	5.02	6.63	7.88
2	0.00501	0.0101	0.0253	0.0513	0.1054	2.30	3.00	3.69	4.61	5.30
3	0.0239	0.0383	0.0719	0.117	0.195	2.08	2.60	3.12	3.78	4.28
4	0.0517	0.0743	0.121	0.178	0.266	1.94	2.37	2.79	3.32	3.72
5	0.0823	0.111	0.166	0.229	0.322	1.85	2.21	2.57	3.02	3.35
6	0.113	0.145	0.206	0.273	0.367	1.77	2.10	2.41	2.80	3.09
7	0.141	0.177	0.241	0.310	0.405	1.72	2.01	2.29	2.64	2.90
8	0.168	0.206	0.272	0.342	0.436	1.67	1.94	2.19	2.51	2.74
9	0.193	0.232	0.300	0.369	0.463	1.63	1.88	2.11	2.41	2.62
10	0.216	0.256	0.325	0.394	0.487	1.60	1.83	2.05	2.32	2.52
11	0.237	0.278	0.347	0.416	0.507	1.57	1.79	1.99	2.25	2.43
12	0.256	0.298	0.367	0.435	0.525	1.55	1.75	1.94	2.18	2.36
13	0.274	0.316	0.385	0.453	0.542	1.52	1.72	1.90	2.13	2.29
14	0.291	0.333	0.402	0.469	0.556	1.50	1.69	1.87	2.08	2.24
15	0.307	0.349	0.417	0.484	0.570	1.49	1.67	1.83	2.04	2.19
16	0.321	0.363	0.432	0.498	0.582	1.47	1.64	1.80	2.00	2.14
18	0.348	0.390	0.457	0.522	0.604	1.44	1.60	1.75	1.93	2.06
20	0.372	0.413	0.480	0.543	0.622	1.42	1.57	1.71	1.88	2.00
24	0.412	0.452	0.517	0.577	0.652	1.38	1.52	1.64	1.79	1.90
30	0.460	0.498	0.560	0.616	0.687	1.34	1.46	1.57	1.70	1.79
40	0.518	0.554	0.611	0.663	0.726	1.30	1.39	1.48	1.59	1.67
60	0.592	0.625	0.675	0.720	0.774	1.24	1.32	1.39	1.47	1.53
120	0.699	0.724	0.763	0.798	0.839	1.17	1.22	1.27	1.32	1.36
∞	1.00	1.00	1.00	1.00	1.000	1.00	1.00	1.00	1.00	1.00

The values in the above tables were computed from percentiles of the F-distribution. Interpolation should be performed using reciprocals of the degrees of freedom.

Source: By permission from *Introduction to Statistical Analysis*, by W. J. Dixon and F. J. Massey. Copyright, 1951. McGraw-Hill Book Company, Inc.

Table A8 Critical Percentage Points of the Ratio S^2_{max}/S^2_{min} (Hartley)

	Upper 5 Percent Points										
υ \ k	2	3	4	5	6	7	8	9	10	11	12
2	39.0	87.5	142	202	266	333	403	475	550	626	704
3	15.4	27.8	39.2	50.7	62.0	72.9	83.5	93.9	104	114	124
4	9.60	15.5	20.6	25.2	29.5	33.6	37.5	41.1	44.6	48.0	51.4
5	7.15	10.8	13.7	16.3	18.7	20.8	22.9	24.7	26.5	28.2	29.9
6	5.82	8.38	10.4	12.1	13.7	15.0	16.3	17.5	18.6	19.7	20.7
7	4.99	6.94	8.44	9.70	10.8	11.8	12.7	13.5	14.3	15.1	15.8
8	4.43	6.00	7.18	8.12	9.03	9.78	10.5	11.1	11.7	12.2	12.7
9	4.03	5.34	6.31	7.11	7.80	8.41	8.95	9.45	9.91	10.3	10.7
10	3.72	4.85	5.67	6.34	6.92	7.42	7.87	8.28	8.66	9.01	9.34
12	3.28	4.16	4.79	5.30	5.72	6.09	6.42	6.72	7.00	7.25	7.48
15	2.86	3.54	4.01	4.37	4.68	4.95	5.19	5.40	5.59	5.77	5.93
20	2.46	2.95	3.29	3.54	3.76	3.94	4.10	4.24	4.37	4.49	4.59
30	2.07	2.40	2.61	2.78	2.91	3.02	3.12	3.21	3.29	3.36	3.39
60	1.67	1.85	1.96	2.04	2.11	2.17	2.22	2.26	2.30	2.33	2.36
∞	1.00	1.00	1.00	1.00	1.00	1.00	1.00	1.00	1.00	1.00	1.00

υ = degrees of freedom for evaluating each variance (equal for each group of values in each set) and k = number of variance estimates in the set for calculating the max/min variance ratio.

Source: Reproduced with the permission of Professor E. S. Pearson from E. S. Pearson and H. O Hartley, *Biometrika Tables for Statisticians*, Vol. 1, Table 31.

Table A9 Critical Values for Correlation Coefficient r

n^*	$P = .1$	.05	.02	.01
1	.9876	.9969	.9995	.9998
2	.900	.950	.980	.990
3	.805	.878	.934	.958
4	.729	.811	.882	.917
5	.669	.754	.832	.874
6	.621	.706	.788	.834
7	.582	.666	.749	.797
8	.549	.631	.715	.764
9	.521	.602	.685	.734
10	.497	.576	.658	.707
11	.476	.552	.633	.683
12	.457	.532	.612	.661
13	.440	.513	.592	.641
14	.425	.497	.574	.622
15	.412	.482	.557	.605
16	.400	.468	.542	.589
17	.388	.455	.528	.575
18	.378	.443	.515	.561
19	.368	.432	.503	.548
20	.359	.422	.492	.536
25	.323	.380	.445	.486
30	.296	.349	.409	.448
35	.274	.324	.381	.418
40	.257	.304	.357	.393
45	.242	.287	.338	.372
50	.230	.273	.321	.354
60	.210	.250	.294	.324
70	.195	.231	.273	.301
80	.182	.217	.256	.283
90	.172	.205	.242	.267
100	.163	.194	.230	.254

*For a total correlation, n is two less than the number of pairs in the sample; for a partial correlation, the number of eliminated variates also should be subtracted.

Table A10 Design and Analysis Matrix for Three-Factor, Two-Level Experiment

	Design Matrix			Response	Analysis Matrix							
Run	X1	X2	X3	y	b0	b1	b2	b3	b12	b13	b23	b123
1	−1	−1	−1	y1	1	−1	−1	−1	1	1	1	−1
2	1	−1	−1	y2	1	1	−1	−1	−1	−1	1	1
3	−1	1	−1	y3	1	−1	1	−1	−1	1	−1	1
4	1	1	−1	y4	1	1	1	−1	1	−1	−1	−1
5	−1	−1	1	y5	1	−1	−1	1	1	−1	−1	1
6	1	−1	1	y6	1	1	−1	1	−1	1	−1	−1
7	−1	1	1	y7	1	−1	1	1	−1	−1	1	−1
8	1	1	1	y8	1	1	1	1	1	1	1	1

Table A11 Screening Designs

Design No.

2 Factor

1.

Run No.	X1	X2
1	−1	−1
2	1	−1
3	−1	1
4	1	1

No Aliases

3 Factor (One Block)

2.

Run No.	X1	X2	X3
1	−1	−1	−1
2	1	−1	−1
3	−1	1	−1
4	1	1	−1
5	−1	−1	1
6	1	−1	1
7	−1	1	1
8	1	1	1

No Aliases

3 Factor (Two Blocks)

3.

Block	Run No.	X1	X2	X3
I	1	1	−1	−1
	2	−1	1	−1
	3	−1	−1	1
	4	1	1	1
II	5	1	1	−1
	6	1	−1	1
	7	−1	1	1
	8	−1	−1	−1

Aliases:
(I) b1 = b23, b2 = b13, b3 = b12
(II) −b1 = b23, −b2 = b13, −b3 = b12

(b 123 confounded with blocks)

4 Factor

4.

Run No.	X1	X2	X3	X4
1	−1	−1	−1	−1
2	1	−1	−1	−1
3	−1	1	−1	−1
4	1	1	−1	−1
5	−1	−1	1	−1
6	1	−1	1	−1
7	−1	1	1	−1
8	1	1	1	−1
9	−1	−1	−1	1
10	1	−1	−1	1
11	−1	1	−1	1
12	1	1	−1	1
13	−1	−1	1	1
14	1	−1	1	1
15	−1	1	1	1
16	1	1	1	1

No Aliases

Table A11 Screening Designs (*cont'd.*)

Design No. 4 Factor (1/2 Replicate)

5.

Run No.	*X1*	*X2*	*X3*	*X4*	
1	−1	−1	−1	−1	
2	1	1	−1	−1	
3	1	−1	1	−1	
4	1	−1	−1	1	
5	−1	1	1	−1	
6	−1	1	−1	1	Aliases:
7	−1	−1	1	1	b12 = b34, b13 = b24, b14 = b23
8	1	1	1	1	b1234 confounded with replicate

5 Factor (1/2 Replicate)

6.

Run No.	*X1*	*X2*	*X3*	*X4*	*X5*	
1	−1	−1	−1	−1	−1	
2	1	1	−1	−1	−1	
3	1	−1	1	1	1	
4	−1	1	1	1	1	
5	1	−1	1	−1	−1	
6	−1	1	1	−1	−1	
7	−1	−1	−1	1	1	
8	1	1	−1	1	1	
9	1	−1	−1	−1	1	
10	−1	1	−1	−1	1	
11	−1	−1	1	1	−1	
12	1	1	1	1	−1	
13	1	−1	−1	1	−1	
14	−1	1	−1	1	−1	
15	−1	−1	1	−1	1	
16	1	1	1	−1	1	No Aliases

5 Factor (1/4 Replicate)

7.

Run No.	*X1*	*X2*	*X3*	*X4*	*X5*	
1	−1	−1	−1	−1	−1	
2	1	1	−1	−1	−1	
3	−1	−1	1	1	−1	
4	1	−1	1	−1	1	
5	−1	1	1	−1	1	
6	1	−1	−1	1	1	Aliases:
7	−1	1	−1	1	1	b1 = −b25, b2 = −b15, b3 = −b45
8	1	1	1	1	−1	b4 = −b35, b5 = −b12

Table A12 Exploratory Designs

Design No. 2 Factor

1.

Run No.	*X1*	*X2*
1	0.00	1.00
2	0.87	0.50
3	0.87	−0.50
4	0.00	−1.00
5	−0.87	−0.50
6	−0.87	0.50
7	0.00	0.00
8	0.00	0.00

3 Factor

2.

Run No.	*X1*	*X2*	*X3*
1	−1	−1	−1
2	1	−1	−1
3	−1	1	−1
4	1	1	−1
5	−1	−1	1
6	1	−1	1
7	−1	1	1
8	1	1	1
9	−1.68	0	0
10	1.68	0	0
11	0	−1.68	0
12	0	1.68	0
13	0	0	−1.68
14	0	0	1.68
15	0	0	0
16	0	0	0
17	0	0	0
18	0	0	0

4 Factor

3.

Run No.	*X1*	*X2*	*X3*	*X4*
1	−1	−1	−1	−1
2	1	−1	−1	−1
3	−1	1	−1	−1
4	1	1	−1	−1
5	−1	−1	1	−1
6	1	−1	1	−1
7	−1	1	1	−1
8	1	1	1	−1
9	−1	−1	−1	1
10	1	−1	−1	1
11	−1	1	−1	1
12	1	1	−1	1
13	−1	−1	1	1
14	1	−1	1	1
15	−1	1	1	1
16	1	1	1	1

Table A12 Exploratory Designs (*cont'd.*)

Design No.	4 Factor – *cont'd*					
	Run No.	*X1*	*X2*	*X3*	*X4*	
	17	−2	0	0	0	
	18	2	0	0	0	
	19	0	−2	0	0	
	20	0	2	0	0	
	21	0	0	−2	0	
	22	0	0	2	0	
	23	0	0	0	−2	
	24	0	0	0	2	
	25	0	0	0	0	
	26	0	0	0	0	
	27	0	0	0	0	
	28	0	0	0	0	
	5 Factor					
4.	*Run No.*	*X1*	*X2*	*X3*	*X4*	*X5*
	1	−1	−1	−1	−1	1
	2	1	−1	−1	−1	−1
	3	−1	1	−1	−1	−1
	4	1	1	−1	−1	1
	5	−1	−1	1	−1	−1
	6	1	−1	1	−1	1
	7	−1	1	1	−1	1
	8	1	1	1	−1	−1
	9	−1	−1	−1	1	−1
	10	1	−1	−1	1	1
	11	−1	1	−1	1	1
	12	1	1	−1	1	−1
	13	−1	−1	1	1	1
	14	1	−1	1	1	−1
	15	−1	1	1	1	−1
	16	1	1	1	1	1
	17	−2	0	0	0	0
	18	2	0	0	0	0
	19	0	−2	0	0	0
	20	0	2	0	0	0
	21	0	0	−2	0	0
	22	0	0	2	0	0
	23	0	0	0	−2	0
	24	0	0	0	2	0
	25	0	0	0	0	−2
	26	0	0	0	0	2
	27	0	0	0	0	0
	28	0	0	0	0	0
	29	0	0	0	0	0
	30	0	0	0	0	0
	31	0	0	0	0	0
	32	0	0	0	0	0

Annex B. Statistical Model for Testing Operations

The purpose of this annex is to build a detailed model of the measurement process based on the concepts introduced in previous sections of the chapter. The mathematical model improves understanding and more clearly demonstrates how the various concepts relate to each other. In an ideal world all measurements would be obtained free of variation. In the real world, all measurements are perturbed to some degree by a *system-of-causes* that produces error or variation in the output of the instruments or machines used for the testing. There are two general *variation categories* for any system. These categories are defined by the character and source of *deviations* that perturb the observed values compared to what would be obtained under ideal conditions.

Production variation—deviations in certain properties that are inherent in the *process* that produces or generates a particular class of objects or materials being tested, or acquired storage or conditioning changes after such processes are complete.

Measurement variation—deviations in the *operation of instruments or machines* that evaluate certain properties for any class of objects or material.

As previously discussed, within each variation category deviations may be of two different types, (1) random, + and − differences about some central (true) value, or (2) bias or systematic differences, each being an offset or constant difference from a central value. Both types may occur in either category.

The system-of-causes is a function of the *domain* of the testing program, where domain is defined by the scope and organization of any testing program, i.e., test machines, materials, operators, locations, replication and sampling operations. These "cause systems" can vary from simple to very complex. The production process is broadly defined; it can be (1) the ordinary operation of a manufacturing facility, (2) a naturally occurring and ongoing process, or (3) some smaller scale processing or other procedure that generates a material or class of objects for testing. The discussion applies to both objects and materials.

Objects may be discrete manufactured items or test specimens generated by a particular preparation process. Materials may be tested in a direct manner, such as the modulus of a polymer, or in an indirect manner, such as the quality of a carbon black or other additive via a physical property in a standard formulation. In the case of direct testing for a bulk material, an appropriate sample taken from the lot is tested. In the case of indirect testing, the material tested is usually appropriately sampled and combined with other materials in a specified way, and the composite is tested. This composite testing may involve objects or test specimens for the measurement process. The treatment in this annex is developed in terms of random and bias deviations. The relationship between random and bias deviations and accuracy, which is a concept that includes both types of deviations, is developed in Annex C, which also describes how to evaluate bias.

General Model

For any established system-of-causes or testing domain, each measurement y_i can be represented as a linear additive combination of fixed or variable (mathematical) terms as indicated by Eq. B1. Each of these terms is an individual deviation or *component of variation*, and the sum of all component deviations is equal to the total variation observed in the individual measurement at any selected brief time period of testing for a test procedure. For the following development it is assumed that all participants test a selected number of classes of objects or different materials, drawn from a common lot, employ the same type of apparatus, use skilled operators, and conduct testing in one or more typical laboratories or test locations.

$$y_i = \mu_0 + \mu_j + \Sigma(b) + \Sigma(e) + \Sigma(B) + \Sigma(E) \tag{B1}$$

where

y_i = a measurement value, at time (i), using specified equipment and operators, at laboratory or location (q)
μ_0 = a general or constant term (mean value), unique to the type of test being conducted
μ_j = a constant term (mean value), unique to material or object class (j)
$\Sigma(b)$ = the (algebraic) sum of some number of component *bias deviations* in the *process* that produced material or object class (j)
$\Sigma(e)$ = the (algebraic) sum of some number of component *random deviations* in the *process* that produced material or object class (j)
$\Sigma(B)$ = the (algebraic) sum of some number of component *bias deviations* for measurement (i), generated by the *measurement system*
$\Sigma(E)$ = the (algebraic) sum of some number of component *random deviations* for measurement (i), generated by the *measurement system*

An alternative approach is to use a single μ term, i.e., μ_r, in place of the two terms $\mu_0 + \mu_j$, where both of the characteristics defined by μ_0 and μ_j are contained in the single term. This approach is taken in Section 8 on specific uncertainty. Equation B1 indicates that there are three groups of generic variation or deviation components: (1) constant terms (population mean values), (2) bias deviations, and (3) random deviations.

Specific Model Format

A more useful format is obtained when Eq. B1 is expressed in the format of Eq. B2, where the generic summations are replaced by a series of typical individual terms or components appropriate to interlaboratory testing on a number of different object classes or materials, over a particular time period.

$$y_i = \mu_0 + \mu_j + \Sigma(b) + \Sigma(e) + B_L + B_M + B_{OP} + E_M + E_{OP} \tag{B2}$$

The new terms are defined as follows.

B_L = a bias deviation term unique to laboratory or location (q)
B_M = a bias deviation term unique to the specific instrument or machine
B_{OP} = a bias deviation term unique to the operator(s) conducting the test
E_M = a random deviation in the use of the specific instrument or machine
E_{OP} = a random deviation inherent in the operator's technique

Other types of testing perturbations not included in Eq. B2 may exist, e.g., bias and random components due to temperature, time of the year, etc.

The $\mu_0+\mu_j$ Terms

In the absence of bias or random deviations of any kind, a number of materials or object classes would have individual measured test values given by the sum of the two terms, $\mu_0 + \mu_j$. The term μ_0 would be unique (a constant value) for whatever test was being employed; this term characterizes the overall value or magnitude of the measured parameter. Each material or object class would be characterized by the value of μ_j, which would produce a varying value for the sum $\mu_0 + \mu_j$ across the number of materials or object classes in the test program. The sum would be the "true" or unperturbed test value.

The Production Terms $\Sigma(b)+\Sigma(e)$

There will always be some bias and random variation in the materials or object classes produced by the process that generates them. This usually unknown number of bias and random variations is designated by $\Sigma(b) + \Sigma(e)$. The goal of some sampling plans may be the evaluation of either of these sources of variation. In other testing operations the goal may be to reduce such variation to the lowest possible level. Appropriate sampling and replication plans will reduce the random components to some selected level. However increased sampling and replication does not reduce bias components; such action merely enhances the fidelity of the magnitude of these effects. Reducing or removing bias requires special test programs to discover and eliminate the causes.

The Measurement Bias Terms

Bias deviations may be divided into two classes, *local* and *global*. A local bias is a fixed offset that applies to certain specific conditions within a testing domain, e.g., a single test machine or laboratory within a domain of many machines or laboratories. Such biases are what make one laboratory, location, or test instrument different in comparison to other laboratories, locations, and instruments. When the domain consists of a large number of machines or laboratories, the local biases may be *variable* (+ or −) deviations unique to each of these machines or laboratories, and the distribution may be either random with a zero mean in the long run or a nonrandom finite distribution with a nonzero mean. A global bias is either (1) a fixed offset that applies across the whole testing domain and is unique to a generic condition that is common within the domain or (2) an inherent deviation in a particular design of a test apparatus. Although more than one global bias may exist, global biases usually are not considered to have a distributional character.

Bias terms that are fixed under one system-of-causes may be variable under another system-of-causes and vice versa. As an example, consider the bias terms B_L and B_M, which apply to most types of testing. For a *particular laboratory* (with one test machine), both of these bias terms would be constant or fixed. For a *number of test machines*, all of the same design in a given laboratory, B_L would be fixed but B_M would be variable, each machine potentially having a unique value. For a domain consisting of *a number of typical laboratories,* each with one machine, both B_L and B_M would be variable for the domain, but both B_L and B_M would be fixed or constant for the system-of-causes in each laboratory.

The Measurement Random Terms

These are deviations or components that are frequently called *error*. Random deviations are + or − values that have an expected mean of zero over the long run. The distribution of these terms is assumed to be approximately normal but in practice it is usually sufficient if the distribution is unimodal. The value of each random term influences the measured y_i value on an individual measurement basis. However in the long run, when y_i values are averaged over a substantial number of measurements, the influence of the random terms may be greatly diminished or eliminated depending on the sampling and replication plan, since each term averages out to zero (or approximately zero) and the *mean* y_i is essentially unperturbed.

In ordinary testing the magnitude of the individual bias and random components or deviations are usually not known. Their *collective effect* influences each measured y_i value, and this collective effect is what is normally evaluated in ordinary variance testing.

Sample and Test Replication

There are three general types of sample replication procedures that apply to testing, where the word "item" refers to an object or a test sample (portion) of a bulk material.

Type 1. *Sample replication (m)* using the same test item with 1 to m repeated tests
Type 2. *Sample replication (n,1)* using n test items, each item being tested *one* time
Type 3. *Sample replication (n,m)* using n test items, each item being tested m times

For Type 1 replication, the sample size is 1, with m replicates; for Types 2 and 3 the sample size is n, also with $1, 2, \ldots m$ replicates. The scope of the sampling and replication plan needs to be clearly defined for any testing program. Replication Types 1 (with m tests) and 3 may be used for nondestructive testing; while Type 2 is the only type available for multisample destructive testing. Type 3 testing reduces the influence of the random production variation as well as the random measurement variation.

An example will better indicate the influence of random and bias deviations. Replicated testing of any type with only a few replicates (where n and m jointly or each equal to less than 10) gives a test result average value $Y_{(<10)}$, as indicated by Eq. B3, where the appearance of $\Sigma(e)$ and $\Sigma(E)$ indicates that these sums are not equal to zero. Frequently $\Sigma(E)$ and $\Sigma(e)$ are much less than $\Sigma(b)$ and $\Sigma(B)$.

$$Y_{(<10)} = \mu_0 + \mu_j + \Sigma(b) + \Sigma(e) + \Sigma(B) + \Sigma(E) \quad \text{(B3)}$$

Highly replicated testing, which is qualitatively defined as 10 or more measurements for both n and m, substantially reduces the perturbation of the random deviations. The reduction factor is the reciprocal of the square root of either n or m, and the use of 10 is given as an example only. It is assumed that this number for n and m reduces the random deviations to a negligible value compared to the magnitude of the biases. Using this gives a mean value $Y_{(>10)}$, which is perturbed by bias components only.

$$Y_{(>10)} = \mu_0 + \mu_j + \Sigma(b) + \Sigma(B) \quad \text{(B4)}$$

Equation B4 shows that ordinary highly replicated testing (usually Type 3) does not approximate the true value for any candidate if any production or measurement system bias deviations exist. The tester ordinarily does not know of the potential sources of this inherent process and measurement bias variation and thus no individual assignment of variation components can be made. These terms remain in their generalized format.

New Term, *M*(*j*)

With highly replicated testing programs (both production and test measurement replication), the average values obtained in any program are estimates of the value of a new combined term as given by

$$M(j) = [\mu_0 + \Sigma(b) + \Sigma(B)] + \mu j \quad \text{(B5)}$$

and $M(j)$ is the mean value for the material or class of objects being tested, for laboratory or location (q), for the specific equipment and operators used during the existing time period, and it contains bias components or potential bias components for all of these conditions. If all biases are fixed for any given program, the three terms in the bracket can be considered as a constant, and the average test value varies across the number of materials or object classes because of the varying value of μ_j. If there are variable biases, then both μ_j and the biases influence the average value for any candidate.

Evaluating Process and Measurement Variance

Equation B1 may be used to illustrate how the variance of individual measurements y_i may be related to the terms or components of the equation. Recall that μ_0 and μ_j are constants, $\Sigma(b)$ and $\Sigma(e)$ refer to the sum of bias and random components respectively for the production process, and $\Sigma(B)$ and $\Sigma(E)$ refer to the sum of bias and random components respectively for the test measurement operation. The magnitudes of the individual components are ordinarily not known, and the equation can be simplified by combining the bias and random components for both sources where $\Sigma(b, e)$ = sum of bias and random components for the *production process* and $\Sigma(B, E)$ = sum of bias and random components for the *measurement procedure*.

$$y_i = \mu_0 + \mu_j + \Sigma(b, e) + \Sigma(B, E) \tag{B6}$$

The variance of any individual measurement y_i, designated by $S^2(y_i)$, is

$$S^2(y_i) = [\Sigma\,\mathrm{Var}\,(b, e)] + [\Sigma\,\mathrm{Var}\,(B, E)] \tag{B7}$$

where

$[\Sigma\,\mathrm{Var}\,(b, e)]$ = a variance that is the sum of individual bias and random variances, for the *production* process
$[\Sigma\,\mathrm{Var}\,(B, E)]$ = a variance that is the sum of individual bias and random variances, for the *measurement* procedure

Eq. B7 can be written in simplified format as

$$S^2(y_i) = S^2(\mathrm{tot}) = S^2(\mathrm{p}) + S^2(\mathrm{m}) \tag{B8}$$

where

$S^2(\mathrm{tot})$ = total variance among the materials or object classes in a test program
$S^2(\mathrm{p})$ = variance due to the production process
$S^2(\mathrm{m})$ = variance due to the measurement operation

For the measurement situation where testing is nondestructive, and any sample may be tested more than one time, all three variance components can be evaluated. Table B1 for a typical testing scenario illustrates this. There are k materials or object classes tested; each has four samples ($n = 4$) taken at divergent locations in the production process; and each sample has two replicates ($m = 2$). Each pair of y_{ij} values constitutes a cell in the table. There are $(k) \times 8$ individual test values, and the variance for all of these values is $S^2(\mathrm{tot})$. The variance $S^2(\mathrm{m})$ is evaluated by taking the variance for each cell in the table (each cell has 1 *df*) and pooling this across all cells for all materials or classes. The variance $S^2(\mathrm{p})$ is evaluated by difference as

Table B1 Typical Layout for Type 3 Sampling–Replication Testing Program

Candidate material (or object class)	Sample No.			
	1	2	3	4
A	y_{11}, y_{12}	y_{21}, y_{22}	y_{31}, y_{32}	y_{41}, y_{42}
B	—	—	—	—
⋮	—	—	—	—
k	—	—	—	y_{ij}

$$S^2(\mathrm{p}) = S^2(\mathrm{tot}) - S^2(\mathrm{m}) \tag{B9}$$

This approach to production process and test measurement variance evaluation assumes that the replicate (within-cell) testing variance is equal in the long run for all cells. The value of $S^2(\mathrm{p})$ as obtained from this analysis is a collective value that represents all materials or object classes.

Annex C. Evaluating Test Measurement Accuracy and Bias

As clearly outlined in Annex B, a test result may have a number of bias components in addition to random error components. The number of such bias components and their magnitude is a function of the system-of-causes or the domain of the testing program. To discuss bias, the concept of accuracy is introduced as the *combined* effect of *random* and *bias* deviations. In the measurement of some property x, accuracy **A**, is defined on a relative scale from 0 to 100 percent by

$$\mathbf{A} = \left[1 - \frac{|\mu_\mathrm{r} - \bar{x}_n|}{\mu_\mathrm{r}}\right]100 \tag{C1}$$

where

μ_r = the true or *reference* value for x
$\bar{x}_n$ = the mean of n measurements on x, for test condition j
$|\mu_\mathrm{r} - \bar{x}_n| = \delta_j$ = an absolute difference, true vs. measured values, for test condition j

When the differential is zero, accuracy is 100 percent. To simplify notation the differential is designated by the symbol δ_j. Using the concepts developed in Annex B, δ_j is given by

$$\delta_j = \Sigma(b) + \Sigma(e) + \Sigma(B) + \Sigma(E) \tag{C2}$$

and the μ_r of Eq. C1 is equal to $\mu_0 + \mu_j$ in Eq. B1 of Annex B. The Eq. C2 expression applies to both local and global biases and to any combination thereof as defined in Annex A. The biases as evaluated in this annex may be of either type according to the particular conditions that exist for the testing domain. The value of δ_j will be a function of the number of measurements for $\bar{x}_n$, and thus it is a statistical estimate, and for any number of finite measurements it contains a random variation component. If δ_j is exactly known (no random e or E components), the differential is the exact or *true bias* for the operational conditions of measurement. In actual practice it is usually sufficient if the sum of all random components contained within δ_j is less than any anticipated bias components by a factor of the order of 0.1 or 0.01.

Biases are generated by any number of conditions that interfere with the measuring process for a test method. Typical causes for bias are unique testing factors due to test machines or instruments, operators, level of training or experience, and ambient conditions in a given testing location. Three categories of bias may exist:

A fixed bias $(B)_\mathrm{F}$ exists when the magnitude is independent of the level or value of the measured property.

A relative bias $(B)_\mathrm{R}$ exists when the magnitude is a function of the level or value of the measured property.

A composite bias $(B)_\mathrm{C}$ is equal to the sum $(B)_\mathrm{F} + (B)_\mathrm{R}$, which is also a function of the level of the measured property.

Expressing Bias

In order to evaluate bias for a particular testing domain designated as condition j, it is necessary that the true value, μ_r, for the measurand or measured property be available when conducting measurements. In chemical analysis it is often possible to have samples, of a material similar or identical to those under analysis, with known (true) levels of the constituents being analyzed. In physical testing, this situation is more difficult to attain, and more dependence must be placed on having *reference* standards or materials that have documented and accepted levels of the properties of interest. In this development of bias expressions it is assumed that reference materials or objects are available that can be defined as a *true value*. It is also assumed that several levels of μ_r are available, and that a linear relationship exists between the measured and the true values. Four cases will be described.

Case 1. Zero Bias. This is given as a basis for comparison. In the absence of any bias, there is a one-to-one correspondence between (1) the *long run* measured average $\bar{x}_n$, which is designated as $\bar{x}_{lr}$, and (2) the true value for condition j. A long run is defined as a mean value of $\bar{x}_n$ where n is sufficient to reduce the sum of random component deviations in Eq. C2, which is now designated as ε, to a negligible value compared to μ_r. Thus for Case 1,

$$(\bar{x}_{lr})_1 = \mu_r \pm \varepsilon \tag{C3}$$

where $(\bar{x}_{lr})_1$ indicates a *zero bias mean* and a plot of $(\bar{x}_{lr})_1$ vs. μ_r will yield a slope of 1 with a zero intercept. In the next three cases the mean of the $\pm\varepsilon$ terms is assumed to be zero.

Case 2. Fixed Bias. For a fixed bias, the value for $(\bar{x}_{lr})_2$ is

$$(\bar{x}_{lr})_2 = \mu_r + (B)_\mathrm{F} \tag{C4}$$

In an alternative form, the difference between the measured and the true value is

$$(\bar{x}_{lr})_2 - \mu_r = (B)_\mathrm{F} \tag{C5}$$

and the value $(B)_\mathrm{F}$ adds or subtracts a constant value to any measured $\bar{x}_{lr}$. For a plot of $(\bar{x}_{lr})_2$ vs. μ_r the slope is also unity, but now a nonzero intercept exists.

Case 3. Relative Bias. When test conditions produce only a relative bias, the linear expression for $\bar{x}_{lr}$ is

$$(\bar{x}_{lr})_3 = a\mu_r \tag{C6}$$

where a is the slope that deviates from unity, and the absence of an intercept indicates that the line goes through the origin. If we contrast Case 1 with Case 3 by subtracting Eq. C3 from Eq. C6, we get

$$(\bar{x}_{lr})_3 - (\bar{x}_{lr})_1 = a\mu_r - \mu_r = \mu_r(a-1) = (B)_\mathrm{R} \tag{C7}$$

The difference between $(\bar{x}_{lr})_3 - (\bar{x}_{lr})_1$ at any μ_r (for condition j) is equal to μ_r times the term $(a-1)$. Since Case 1 is zero bias, the actual absolute bias for Case 3 is this difference. If both sides of Eq. C7 are divided by μ_r we get the expression for relative bias

$$\frac{(\bar{x}_{lr})_3 - (\bar{x}_{lr})_1}{\mu_r} = a - 1 \tag{C8}$$

Case 4. Composite Bias. The composite bias is the sum of the fixed and the relative bias, both in absolute terms.

$$(\bar{x}_{lr})_4 - (\bar{x}_{lr})_1 = \mu_r(a-1) + (B)_\mathrm{F} \tag{C9}$$

This represents a plot that has a nonzero intercept and a slope that differs from unity.

Evaluating Bias

One Reference Material Only

If only one reference material is available, the bias can be evaluated only at the level of the reference material. Since both relative and fixed bias may be present, this bias is designated as composite bias, and it is given in absolute terms by

$$(B)_C = \bar{x}_n - \mu_r \tag{C10}$$

If n is at some level such that the sum of random deviations is not negligible, then the difference $(\bar{x}_n - \mu_r)$ divided by the standard error of this difference is assumed to have the typical t distribution to give

$$t(\text{calc}) = \frac{\bar{x}_n - \mu_r}{S/\sqrt{n}} \tag{C11}$$

where S is the standard deviation of individual measurements of x, n is the number of measurements for the mean $\bar{x}_n$, and the df is equal to $n - 1$. If $|t(\text{calc})| > t(\text{crit})$ at the selected level of confidence $(1 - P)$, then $(B)_C$ is declared significant. The confidence interval for $(B)_C$ is

$$\text{Confidence interval for } (B)_C = \pm|t(\text{calc})|\,\frac{S}{\sqrt{n}} \tag{C12}$$

Several Reference Materials

If several reference materials are available with a range of levels, a linear regression technique can be used to evaluate the bias. Again it is assumed that both types of bias may be present. The customary xy pair terminology for regression is used, and now the measured values are designated by $\acute{y}_n$ rather than $\bar{x}_n$ as used above, and x is used for the reference values $(\mu_r)_i$. The linear equation used to evaluate the estimates of the slope and intercept obtained from several xy pairs is

$$\acute{y}_n = a'(\mu_r)_i + b' = a'x + b' \tag{C13}$$

Estimated values for the slope a' and the intercept b' are obtained by way of regression calculations as given in Section 6.5. The fixed bias $(B)_F$ is equal to b', and the relative bias $(B)_R$ is equal to $a' - 1$. The composite bias is given by

$$(B)_C = (a' - 1)(\mu_r)_i + b' = (a' - 1)x + b' \tag{C14}$$

Confidence intervals for b' and a' may be calculated by special formulas that are different from but equivalent to the formulas as given in Section 6.5. The standard error of an estimate (variation of points about regression line) is S_e and is equal to

$$S_e = \left\{\frac{N-1}{N-2}[S_y^2 - (b')^2 S_x^2]\right\}^{1/2} \tag{C15}$$

with N equal to the number of xy points in the regression, S_y^2 the variance of the N values of $\acute{y}_n$, and S_x^2 the variance of the N values of x or $(\mu_r)_i$. The standard error of the slope, $S(a')$, is

$$S(a') = \frac{S_e}{(SS_{xx})^{1/2}} \tag{C16}$$

where SS_{xx} is the sum-of-squares for the x variable used in the calculation of S_x^2. The standard error for the intercept, $S(b')$, is

$$S(b') = S_e \left[\frac{\Sigma x_i^2}{N(SS_{xx})} \right]^{1/2} \tag{C17}$$

and Σx_i^2 is the sum of all N values of x squared. The confidence intervals for the slope a' and intercept b' are

$$\text{Confidence interval for } a' = \pm t(\text{crit})\, S(a')$$
$$\text{Confidence interval for } b' = \pm t(\text{crit})\, S(b') \tag{C18}$$

Knowledge of the confidence interval for a' permits an evaluation of the interval for $(a' - 1)$. The critical values for t are selected at $N - 2$ degrees of freedom, which indicates that several μ_r levels (approx. 6 or more) should be used for bias evaluations. If this many levels are not available, then bias will be evaluated with larger uncertainties.

Annex D. Calculating Repeatability and Reproducibility

Customarily, testing precision is expressed in terms of repeatability or within-laboratory variation and reproducibility or between-laboratory variation. Both of these are jointly evaluated by a collaborative interlaboratory test program or ITP, on several materials or object classes with a range of material or class properties. In its simplest form, an ITP consists of selecting and/or preparing homogeneous lots of several materials for evaluation. For each material that is designated as a *level* in the ITP, samples are drawn from the uniform lot and sent out to a number of labs (10 or more is recommended), and replicate tests are conducted during a specified time interval. Since the testing system should be "in control" or stable over an extended period, the usual repeatability evaluation is based on a Day 1, Day 2 replication, with several days between the Day 1, Day 2 replicates. Each of the replicated tests produces a test result that is usually the mean or median of some number of repeated determinations as specified by the test method, obtained in a short (one or two hour) time span. The repeatability for each material or class is given as a pooled or *average* value across all laboratories on the assumption that all laboratories are typical and have equal testing competence.

The reproducibility is presumed to apply to a group of typical laboratories as selected from the population of all laboratories equipped to conduct such testing. The precision results are characteristic of the materials or object class tested and the time of the testing. Another ITP on other materials or on the same materials at a different time may give different results. The analysis assumes that the system-of-causes that generates repeatability variations as well as reproducibility variations are both described as random normal populations. Table D1 gives the layout of the data obtained in a typical ITP with n replicates or test results in each cell. For each material or level a one-way ANOVA is performed.

The model used for the analysis is taken from Eq. B1 of Annex B with slightly different notation for y.

$$y_{ijk} = \mu_0 + \mu_j + \Sigma(b) + \Sigma(e) + \Sigma(B) + \Sigma(E) \tag{D1}$$

Table D1 Typical Analysis Layout for Inter-laboratory Testing Program

	Level or Material			
Laboratory	1	2	3	q
1	—	—	—	—
2	—	y_{ij1}, y_{ijk}	—	—
3	—	—	—	—
—	—	—	—	—
p	—	—	—	—

Number of labs $= p, (i = 1, 2, \ldots p)$
Number of materials or levels $= q\ (j = 1, 2 \ldots, q)$
Replicates in each cell (or lab–material combination) $= n$ (this is usually 2)
y_{ijk} = one replicate, i.e., a single test result (as defined by test method)
Example: Cell (ij) contains n_{ij} test results, y_{ijk} $(k = 1, 2, \ldots, n_{ij})$

where

y_{ijk} = a measurement value (test result), using specified equipment and operators, at laboratory or location (q)
μ_0 = a general or constant term (mean value), unique to the type of test being conducted
μ_j = a constant term (mean value), unique to level (material or object class) j
$\Sigma(b)$ = the (algebraic) sum of some number of component *bias deviations* in the *process* that produced level j
$\Sigma(e)$ = the (algebraic) sum of some number of component *random deviations* in the *process* that produced level j
$\Sigma(B)$ = the (algebraic) sum of some number of component *bias deviations*, for each test result, generated by the *measurement system*
$\Sigma(E)$ = the (algebraic) sum of some number of component *random deviations* for each test result, generated by the *measurement system*

A key assumption in any ITP is that each material sent out to all participating laboratories is homogeneous or uniform in its properties such that the random variation among the samples sent to all laboratories, i.e., the effect of the term $\Sigma(e)$, is negligible compared to the term $\Sigma(E)$. Each particular lot for any material sent out in an ITP has some characteristic unknown value for $\Sigma(b)$, but this is a constant for any ITP and thus does not contribute to within-laboratory or between-laboratory variability for any specific ITP. The more complex Eq. D1 may be simplified by the use of a combined term m_j

$$m_j = \mu_0 + \mu_j + \Sigma(b) + \Sigma(e) \tag{D2}$$

where m_j is the mean or *reference value* for material j across all laboratories in the ITP, and it contains production process bias and random components. Since $\Sigma(b)$ and μ_0 are constants and $\Sigma(e)$ is defined as essentially equal to zero, the variations in m_j will reflect variations in μ_j the true or reference value for material j. Using the newly defined m_j, a simplified model may be written as

$$y_{ijk} = m_j + (B_i)_j + (E_{ik})_j \tag{D3}$$

This is the classic model for interlaboratory testing, where $(B_i)_j$ represents the *combined bias* component in y_{ijk} for laboratory i, for material j, and $(E_{ik})_j$ represents the *random* deviation component in y_{ijk} for laboratory i, for material j. The bias $(B_i)_j$ is the sum of all potential sources of laboratory bias variation, and its value may be a distinguishing feature for any laboratory.

The two fundamental assumptions in the analysis of ITP data are that the random deviations defined by $(E_{ik})_j$ as well as the total deviations about m_j, which are defined by the sum $(B_i)_j + (E_{ik})_j$, are normally distributed. This requires a normal distribution for the $(B_i)_j$ values. A normal distribution for $(B_i)_j$ implies that the biases are transitory in the sense that for a repeat of the ITP, a substantially different $(B_i)_j$ value may be obtained including a sign reversal. If both $(B_i)_j$ and $(E_{ik})_j$ are normally distributed, then in the long run (many repetitions of the ITP) the mean values for these components will be zero to give the true value for m_j. Thus all laboratories would obtain the same long run *mean values*. If however the biases $(B_i)_j$ are unique to particular laboratories and do not substantially change from one ITP to the next, the distribution will not be normal, and the long run will give a value for each laboratory equal $m_j + (B_i)_j$, where $(B_i)_j$ is a mean value characteristic of material j and laboratory i.

Types of Precision

The simplest ITP, as discussed above, can be called a "classical analytical" type of test, where the material to be analyzed is ready for analysis upon receipt in the laboratory, i.e., no processing or other operation is required prior to making the intended measurements. In ASTM D4483, the precision guideline standard for Committee D11 on rubber testing, this type of precision evaluation is designated as *Type 1 precision*. In many industrial testing situations, more complex testing standards are used that require pretest processing of some sort prior to the testing proper. An example from the rubber manufacturing industry is the evaluation of various grades of carbon black by mixing the candidate carbon blacks into a standard rubber formulation and then testing this mixed and frequently cured compound for certain physical properties. In a producer–user scenario, the evaluation of the quality of a lot of carbon black is conducted by the consumer or user laboratory by such a mixing and testing operation. The results in the user laboratory must agree with the results obtained by the same operation in the producer's laboratory. The replication to establish a precision for such testing must include the compounding (necessary raw materials) and mixing operation as part of the test process. The type of precision that involves this more complex operation is designated as *Type 2 precision*. Thus it is important when an ITP is organized to consider fully what type of precision is used or required in the appropriate application of the test. The discussion below applies equally to either type of precision.

Within-Laboratory Variation

For each level, the cell variance S_{ij}^2 is calculated for each laboratory, and the pooled or mean variance across all laboratories is obtained as given by Eq. D4, where p is the number of laboratories in the ITP. This pooled value is an indicator of the general or overall *within-laboratory variance*.

$$S_r^2 = \Sigma \frac{S_{ij}^2}{p} \tag{D4}$$

The square root of this pooled variance, S_r, is called the *repeatability standard deviation.*

$$S_r = \left(\Sigma \frac{S_{ij}^2}{p}\right)^{1/2} \tag{D5}$$

Between-Laboratory Variation

A derived intermediate parameter is the *between-laboratory variance*, designated by S_L^2. For any level, this is evaluated from (1) the variance among the laboratory cell averages, designated by $S_{\acute{Y}}^2$ and (2) the repeatability variance S_r^2. The value for $S_{\acute{Y}}^2$ is calculated by

$$S_{\acute{Y}}^2 = \Sigma \frac{(\acute{y}_i - \acute{Y})^2}{p} \tag{D6}$$

where

$\acute{y}_i$ = cell *mean* at level j for laboratory i,
$\acute{Y}$ = grand *mean* all laboratories, for level j

and the summation is over all laboratories. S_L^2 is evaluated by

$$S_L^2 = S_{\acute{Y}}^2 - \frac{S_r^2}{n} \tag{D7}$$

and S_L^2 is an indicator of the inherent variation between the laboratories free of the influence of the within-laboratory variation contained in the term S_r^2/n. It is also used in the calculation of the reproducibility variance S_R^2 defined as the *total* variance among all the laboratory cell mean values for any level which now contains the within-laboratory variation. The reproducibility variance S_R^2 is given by

$$S_R^2 = S_L^2 + S_r^2 \tag{D8}$$

Substituting for S_L^2 gives

$$S_R^2 = S_{\acute{Y}}^2 - \frac{S_r^2}{n} + S_r^2 \tag{D9}$$

An alternative format after taking the square root is

$$S_R = \left[S_{\acute{Y}}^2 + S_r^2\left(\frac{n-1}{n}\right)\right]^{1/2} \tag{D10}$$

and S_R is defined as the *reproducibility standard deviation.*

Repeatability and Reproducibility

In the current standard precision practices for ASTM and ISO, repeatability and reproducibility are defined as intervals or upper and lower limits within which a measured value is expected to fall on the basis of the t distribution at some confidence level. From the development in Section 3.5, the absolute difference between two means $|\acute{y}_1 - \acute{y}_2|$ is given by

$$|\acute{y}_1 - \acute{y}_2| = t_{\alpha/2}\left[S_y\left(\frac{1}{n_1} + \frac{1}{n_2}\right)^{1/2}\right] \tag{D11}$$

where n_1 and n_2 are the number of respective values in each mean and S_y is the pooled standard deviation of individual values in each mean. If $n_1 = n_2$,

$$|\acute{y}_1 - \acute{y}_2| = t_{\alpha/2}\left[S_y\left(\sqrt{2/n}\right)\right] \tag{D12}$$

On the basis of the central limit theorem, the standard deviation of means $S_{\acute{y}}$ is given by

$$S_{\acute{y}} = \frac{S_y}{\sqrt{n}} \tag{D13}$$

and with slight rearrangement,

$$|\acute{y}_1 - \acute{y}_2| = t_{\alpha/2}\sqrt{2}(S_{\acute{y}}) \tag{D14}$$

Repeatability $\boldsymbol{r}$ is defined by Eq. D15 when $S_{\acute{y}}$ is set equal to S_r:

$$\boldsymbol{r} = |\acute{y}_1 - \acute{y}_2| = t_{\alpha/2}\sqrt{2}(S_r) \tag{D15}$$

where $|\acute{y}_1 - \acute{y}_2|$ represents the difference between two within-laboratory test results in a typical laboratory. Reproducibility $\boldsymbol{R}$ is defined by Eq. D16 when $S_{\acute{y}}$ is set equal to S_R:

$$\boldsymbol{R} = |\acute{y}_1 - \acute{y}_2| = t_{\alpha/2}\sqrt{2}(S_R) \tag{D17}$$

where $|\acute{y}_1 - \acute{y}_2|$ represents the difference between two between-laboratory test results selected from two typical laboratories.

For any confidence level, the values selected for $t_{\alpha/2}$ depend on the degrees of freedom *df* for S_r and S_R. The *df* for S_r for a ten-laboratory ITP will be 10, one for each cell, and the *df* for S_R will be 10 – 1 or 9. The respective $t_{\alpha/2}$ values at the 95% confidence level are 2.23 and 2.26. It has been customary in the precision analysis in both ASTM and ISO to assume that a large number of degrees of freedom are available in any ITP and to use 2.00 as an approximation of the high *df* value (ca. 20–25) for $t_{\alpha/2}$ at the 95% confidence level. When $t_{\alpha/2} = 2.00$, the customary expressions for $\boldsymbol{r}$ and $\boldsymbol{R}$ are

$$\boldsymbol{r} = 2.83\, S_r \tag{D18}$$

and

$$\boldsymbol{R} = 2.83\, S_R \tag{D19}$$

The usual interpretation for $\boldsymbol{r}$ is as follows. The *difference* between two test results obtained under the specified ITP repeatability conditions, on a material identical to or very similar to the level or material in the ITP, will exceed $\boldsymbol{r}$ on an average of not more than once in 20 cases or with a probability $P = 0.05$.

The interpretation for $\boldsymbol{R}$ is as follows. The *difference* between two test results obtained in two different laboratories under the specified ITP reproducibility conditions, on a material identical to or very similar to the level or material in the ITP, will exceed $\boldsymbol{R}$ on an average of not more than once in 20 cases or with a probability $P = 0.05$.

Both repeatability and reproducibility can be given on a relative or percentage basis by using the mean value for each level, thus for relative repeatability ($\boldsymbol{r}$) in percent

$$(\boldsymbol{r}) = \left(\frac{\boldsymbol{r}}{\acute{y}_{12}}\right) 100 \tag{D20}$$

where $\acute{y}_{12}$ = the mean of the two within-laboratory values $\acute{y}_1$ and $\acute{y}_2$ or the mean of a certain level of measurement, and ($\boldsymbol{R}$) may be expressed similarly as

$$(\boldsymbol{R}) = \left(\frac{\boldsymbol{R}}{\acute{y}_{12}}\right) 100 \tag{D21}$$

where now $\acute{y}_{12}$ is the mean of the two between-laboratory values $\acute{y}_1$ and $\acute{y}_2$ or the mean for a certain level of measurement.

If the observed difference for any given situation is greater than $\boldsymbol{r}$ or $\boldsymbol{R}$ in absolute measurement units or in percent units for $(\boldsymbol{r})$ and $(\boldsymbol{R})$, it is presumed that the test operation is unstable or not "in control" and some remedial action is warranted.

The repeatability (interval) may also be used to determine if significant differences exist between two potentially different materials tested in any given laboratory. If under repeatability conditions similar to the ITP of the precision program (same testing conditions, similar test levels) the difference between two individual test results on materials A and B exceed $\boldsymbol{r}$, then material A may be declared significantly different from B at the $(1 - 0.05)\,100 = 95\%$ confidence level.

4
Standardization

Paul P. Ashworth
Consultant, Withington, Manchester, England

1 Introduction

The main purpose of this chapter is to provide information about the committees of the International Organization for Standardization (ISO) that produce and are responsible for all international standards on polymers, including the physical testing of polymers. In order to put this information into a wider perspective, the first part of the chapter gives a brief history of international standardization, describes the aims and structure of the ISO, the preparation and types of international standards, aspects of regional standardization, and the membership of the ISO.

The development of methods of test for polymers and national and international standards are closely linked. Whilst testing of natural polymers predates modern work in standards as we understand it, the period between the start of the formation of national standards bodies at the beginning of this century, and of similar international bodies from the midpoint of the century, was a time when many test methods for synthetic polymers were being developed.

The international standards committees for textiles, rubber, and plastics were part of the first group of technical committees to be established by the ISO during its first year of operation in 1947, and these three committees remain right up to the present day as major contributors to the annual complement of standards published by it. Test method standards have always formed the largest proportion of these collections.

2 A Brief History of International Standardization

The vast majority of the world's countries, whether they be highly industrialized or essentially agrarian, have some organization responsible for standardization. The development of international standards was one response to the need that suppliers and customers had to speak and understand a common technical language.

The first recognized national standards body was created in the United Kingdom in 1901, when the Institution of Civil Engineers formed the Engineering Standards Committee, later to be called the Engineering Standards Association. This organization developed into areas other than engineering and in 1929 was granted a Royal Charter under which it was to operate, as it still does, as the British Standards Institution (BSI).

In Germany the Standards Committee of German Industry was formed in 1917. This became the Deutscher Normenausschuss, and finally the Deutsches Institut für Normung (DIN). In the United States the American Standards Engineering Committee, later to become the American National Standards Institute (ANSI), was established in 1918. In France, the Permanent Standardization Committee was formed in 1918, eventually to become the Association française de normalisation (AFNOR).

Other countries throughout the industrialized world, and eventually in most of the developing nations also, followed these early examples and established their own national standards bodies, with the result that at the present time there are more than 110 such bodies in existence.

The first real effort in international standardization lay in the field of uniformity of testing materials. A German, Professor Bauschinger, took the lead and set up the Dresden Conference of 1886 at which the International Association for Testing Materials (IATM) was formed. Twenty-eight nations had joined it by 1912, but it seems to have foundered some time after that.

At the beginning of the twentieth century, national standardization in the form of specifications had generally advanced further in electrical engineering than in other technologies. Following informal discussions between Professor Elihu Thomson of the USA and Colonel Crompton and Mr Le Maistre of the United Kingdom in Saint Louis in 1904, there was a gathering in London in June 1906 of 14 countries where the rules were established of the International Electrotechnical Commission (IEC). Lord Kelvin was the first president, with Mr Le Maistre, who was to remain in this office until 1953, as its first general secretary.

In his address to the first meeting of the IEC in that year, the Right Honourable A. J. Balfour, the former Prime Minister of the United Kingdom, summed up in a few words the raison d'être of international standardization:

> As I comprehend it, the chief desire of the Commission is so to arrange, by international agreement, the tests which are to be applied to different kinds of electrical machinery, so to describe the qualities of different machines, that, whilst the man who buys and the man who sells will know exactly what each, respectively, is doing, there will yet be the freest initiative left to both.

This is an excellent summary, considering that it was written about ninety years ago, when the aims of international standardization could hardly have been clear. It says precisely what standards are about, and if it is widened to cover all areas other than just electrical, it is just as valid now as it was then.

In April 1926, the 18 countries that by that time had set up standards bodies met to consider the extension of international collaboration to other fields. As a result, the International Federation of the National Standardizing Associations, the ISA, was formed, but this organization ceased to exist by the beginning of the Second World War in 1939. It will be seen from the above that the main brakes on the two attempts to achieve full international standardization activities in the first half of the twentieth century were the two world wars.

In 1944, following Anglo-American discussions on the coordination of allied standards for war purposes and postwar needs, a standards coordinating committee was set up under the United Nations banner. The committee met in New York in November 1945 and arranged a full-scale conference in London in October 1946 at which representatives of 25 standards bodies agreed to set up the International Organization for Standardization. The IEC became its electrical division, though with full technical and financial autonomy.

In the 50 years since its foundation, the membership of the ISO has increased steadily as firstly the developed, and more recently the developing, nations came to organize their national standardization activities in order to take part fully in the international field.

3 The Aims of the International Organization for Standardization

The scope of the ISO covers standardization in all fields except electrical and electronic engineering, which are the responsibility of the IEC. The international standardization system composed of the ISO and the IEC is the world's largest nongovernmental system for voluntary industrial and technical collaboration. The ISO's work results in international agreements that are published as International Standards. At the present time there are about 9,200 of these standards published and available.

The mission of the ISO is, in its own official statement, "to promote the development of standardization and related activities in the world with a view to facilitating international exchange of goods and services and developing cooperation in the sphere of intellectual, scientific, technological and economic activity."

The principal aims of the ISO can be listed under five headings, as follows:

Communication

The provision of a means of expression and of communication among all interested parties.

Economy in Effort, Materials, Energy

Overall economy in terms of human effort, materials, power, etc., in the production and exchange of goods. Standards promote economy through variety reduction. The nature of the standard will determine where the reduction lies, for example in a range of product sizes, a number of different types of equipment, or a variety of test methods.

Consumer Protection

The protection of consumer interests through adequate and consistent quality of goods and services. Standards provide for the quality of goods and services by defining those features and characteristics that govern their ability to satisfy given needs. Guides to quality assurance are a popular and frequent feature at the moment in both international and national standardization.

Safety, Health, Environmental Protection

Promotion of the quality of life. The setting of standards of safety implies the definition of what is acceptable as a reasonable level of risk in the foreseeable use or even misuse of a product. Safety aspects play a very important role in current standards work.

Removal of Technical Barriers to Trade

The breaking down of barriers to trade. World trading policy is traditionally based, or should be in an ideal world, on the assumption that prosperity depends on the widest possible market through unimpeded exports and imports. It is easier and more economical to manufacture to one international standard than to design products for a large number of different individual national standards. It is easier to indicate requirements for purchase too. This works to the advantage of the efficient and to the disadvantage of the inefficient. Used properly, international standards can bring great financial advantage. The same product can be marketed anywhere and, because this reduces unnecessary manufacturing variety, costs will come down and the item should become more competitive.

4 The Structure of the ISO

The ISO now comprises about 110 members, one from each participating country, who can involve themselves, if they so wish, as members of about 175 technical committees covering an enormous range of subjects.

A member body of the ISO is the national body "most representative of standardization in its country." It follows that only one such body for each country is accepted for membership in the ISO. There are lesser forms of membership, e.g., correspondent and subscriber, for smaller or less active countries, where full participation is not expected.

The nature of the ISO member bodies varies extensively. In many countries the member bodies are often a section of a government department, usually related to trade, industry, and quality. In other countries the member body may be entirely independent, funded for example by the sale of national standards and income derived from industry and/or government. Brief descriptions of two member bodies are given later in this chapter.

The international standards produced by the ISO are not legally binding on any member nation. Their implementation is entirely voluntary, but in order for them to have any value at all it is important that this implementation be as wide and as full as possible. The only means that the ISO has to try to ensure this implementation are persuasion and encouragement. However, the national standards bodies have an important role to play here. In some countries it has been policy for many years to adopt, totally unaltered, as a national standard, any International Standard on which that country had submitted a positive vote. Unfortunately, this policy is not followed in many countries, with the result that differing national standards effectively contribute to barriers to international trade. It is obvious, therefore, that for International Standards to become more effective they need to be adopted nationally by more and more countries.

The development of regional standardization bodies, which is described briefly later in this chapter, and in particular the rapid increase in the activities of the European Committee for Standardization (CEN), has meant that there is now greater control over national implementation. The members of CEN are obliged to implement nationally *all* European Standards, and since probably the majority of European Standards are based on, if not identical to, their ISO equivalents, this means that the relevant International Standards are automatically adopted nationally by the European countries.

With regards to the relationship between national standards and national legislation, practice varies enormously between countries, according to the connection between the national standards body and the government. In many countries, although they may be widely used, national standards are of a voluntary nature. In some cases, however, particularly in areas relating to safety, health, and consumer protection, national standards may be cited in the relevant national legislation, with the result that the products involved cannot be offered for sale without it being demonstrated in some form or other that they conform to the appropriate standard.

The policy committees of the ISO comprise the General Assembly, the ISO Council, and the Technical Management Board.

The ISO Council is the ultimate authority, comprising about 18 elected member bodies, with some permanent membership, but for all effective purposes all administration is by, and all effective authority resides in, the Technical Management Board, to which nominees from member bodies serving on the Council are appointed. The Board meets approximately three times a year.

The technical work of the ISO is carried out by technical committees (TCs), one for each field of activity. While these technical committees represent an overall view of the range of activities in the ISO, most of the detailed technical work is carried out in subcommittees (SCs) and, to an even greater degree, in working groups (WGs).

All ISO member bodies are eligible for membership in any technical committee or subcommittee and can then nominate delegations, with a delegation leader, to meetings. With full participating (P) membership they then have full voting rights; indeed, they have an obligation to vote at all stages; with observer (O) membership, they can attend meetings and comment, but not vote. *All* ISO members have an obligation to vote at the main consultation stage, the draft international standard (DIS). Subcommittee membership is open to all P- and O-members of the relevant technical committee, whilst O-members of a technical committee can be P-members of a subcommittee under its jurisdiction. The administration of these committees is the responsibility of an elected member body, which acts as the committee secretariat and nominates the chairman.

Working groups comprise a restricted number of individuals nominated from the P-membership of the senior committee for their expertise in that particular field. A working group is administered by a convener selected from among its membership who is responsible for all its activities, although a secretary may also be appointed if needed. The working group is the key area of activity within the ISO, in terms of the preparation of standards and discussions between experts.

Finally, the smallest and generally shortest-lived group is the ad hoc group, usually established from among the members present at a particular meeting to discuss and resolve a particular problem or aspect of work. The ad hoc group is then usually disbanded at the end of the meeting, or, at the latest, at the end of the following meeting.

Although all the technical work of the ISO is administered by the individual member bodies, there is a central office, ISO Central Secretariat, in Geneva, Switzerland, which is responsible for overall coordination, editing, printing, and distribution of published standards, and other related activities.

5 Preparation of International Standards

As with any organization involved in a wide level of consultation, there are many stages in the development and production of an international standard, which to an outsider might appear confusing. However, these procedures have been developed over many years and represent what is required in order to match the need to achieve full consultation of a very wide membership with the production of standards as quickly as possible. Needless to say, the ISO is endlessly involved in considering ways of speeding up standards production.

The stages and associated documents are given below, together with brief descriptive comments.

Stage 0. Preliminary Stage. Preliminary New Work Item (PWI)

At this stage documents can be looked at and discussed without being officially listed in the committee work programme.

Stage 1. Proposal Stage. New Work Item Proposal (NP)

The project is officially registered, and the "clock" governing the strict timetable for processing through the various stages is started.

Stage 2. Preparatory Stage. Working Draft (WD)

This is the most important stage in terms of discussing the contents of the draft and ensuring that it is in a fit state for ballot.

Stage 3. Committee Stage. Committee Draft (CD)

This is the voting stage for the committee, essentially the last stage when comments can be accepted and discussed.

Stage 4. Enquiry Stage. Enquiry Draft, i.e. Draft International Standard (DIS)

This is the full voting stage for the whole membership of the ISO, where final, minor, comments can be made.

Stage 5. Approval Stage. Final Draft International Standard (FDIS)

This is the stage where committee members give a final approval vote to the standard.

Stage 6. Publication Stage. International Standard (ISO)

The standard is published after final editing.

There are various ways in which these procedures can be short-circuited, depending on the circumstances.

The stages where technical input from national members is most important are all those stages except the first and the last. The preliminary stage is little used, although it may become more often used in the future, and the publication stage involves mainly the ISO Central Secretariat editors and the technical committee or subcommittee secretariat.

Three of the standards preparation stages require a vote in some form or another, with, at the enquiry and approval stages, a precise requirement for acceptability of the standard, in that a given proportion of the votes cast need to be approval votes. However, much of the work, especially in the early stages, depends on the achievement of consensus, which has an official ISO definition as follows:

> General agreement, characterized by the absence of sustained opposition to substantial issues by any important part of the concerned interests and by a process that

involves seeking to take into account the views of all parties concerned and to reconcile any conflicting arguments.

NOTE Consensus need not imply unanimity.

Target dates are becoming widely used by the ISO management as a means of ensuring a timely production of standards. Those currently in force are first working draft, six months; first committee draft, two years; Final Draft International Standard, three years. However in certain circumstances it is permitted to "stop the clock," for example to allow important interlaboratory work to be carried out for the validation of a test method.

The official languages of the ISO are English, French, and Russian. Nearly all standards are published in English and French, and some in Russian, although glossaries are generally in all three languages. Meetings generally are conducted in English, occasionally with interpretation into French; however there is an increasing tendency to work in English alone.

Standards are reviewed after a maximum period of five years. A decision is then taken, by majority vote of the committee, to confirm, amend, revise, or withdraw the standard.

6 Types of International Standard

There are probably four basic types of international standard, although such a division is not rigid and the types are open to various interpretations.

The first type is the *glossary of terms* or *vocabulary*. This is the basic "language" for each particular field of activity; it is of prime importance to ensure that all the nationalities involved in subsequent standardization work are saying, and meaning, the same thing. Glossaries are generally produced in the three official ISO languages, English, French, and Russian, although often other countries develop their own glossaries, based on the ISO text, in their own languages.

The second type is the *test method*. It goes without saying that, in order to facilitate international trade, it is essential that when a particular property or dimension is specified or described it is measured in the same way by all those involved in production, testing, quality control, and purchase.

In most of the fields of activity in the ISO, glossaries and basic test method standards are well established, and current efforts are devoted to revising and updating them and introducing new methods where more modern and sophisticated procedures have become available.

The third basic type of standard is the *material* or *product specification*. It could be said that this is by far the most important type of standard for fulfilling the aims of international standardization described earlier. However, it is also the area that leads to most problems and disagreements. It is obvious that throughout the whole world, with differing climates, social patterns, legislation, and markets, needs for material and product specifications vary enormously. While the ISO and its members realize that they have a commitment to produce all the specifications that they can, and on which they can obtain international agreement, there will be instances where a decision has to be taken to accept that there can never be agreement if a particular subject is an especially difficult one, and that efforts should be diverted to those projects where real progress can be made.

The final type of standard is the *code of practice*. This describes *how* to perform specific activities. Such codes are not very common in the ISO at the present time.

The prime duty of the ISO is to produce International Standards, but under certain circumstances Technical Reports (TRs) may be produced. The three types of Technical Report are as follows:

Type 1: for when a DIS cannot be approved. By majority decision of the committee, a Technical Report is produced, with a full explanation for so doing being given in the foreword.

Type 2: for when work is still under technical development. The project is taken up to the DIS stage; then if there is a majority vote in the committee a Technical Report is prepared. This is regarded as a "prestandard," and further comments are invited in the foreword. The Technical Report is reviewed after three years, with an allowance for one three-year extension, after which it must be converted into an International Standard or withdrawn.

Type 3: an informative document, not normative. This can be developed at any stage by a majority vote to go direct to publication. This is probably the most common type of Technical Report.

Types 1 and 2 must be reviewed after three years, with the aim of issuing an International Standard, but there is no such official requirement for a Type 3.

7 Regional Standardization

A recent development in international standardization has been the growth of interest in regional standardization, and of particular interest and sometimes concern to some members of the ISO has been the extensive increase in activity during the past 10 years of the European Committee for Standardization (Comité européen de normalisation) (CEN), prompted by government policies within the European Union.

In order to avoid unnecessary and possibly conflicting duplication of work, the governing bodies of both the ISO and the CEN have developed a series of guidelines intended to ensure as much cooperation between the two bodies as possible. Whilst it is recognized that there may be occasions when the particular requirements of Europe, or the need for the very rapid preparation of a standard, mean that a separate European Standard is needed, in most cases it has been accepted that the two bodies should work together, and the manner in which this is to be achieved was formulated in the "Vienna Agreement," This agreement lists four means of cooperation: cooperation by correspondence, cooperation through mutual representation at meetings, adoption of existing International Standards as European Standards, and cooperation by transfer of work and parallel approval of standards in the ISO and the CEN.

This last process is the one that is exercising many of the ISO committees at the moment. The procedure basically means that the DIS and FDIS stages are combined simultaneously with the similar stages in the CEN, and comments thereon referred to whichever is the lead body. Consequently a single version is finally published, which is the ISO Standard, the European Standard, and, by extension, the national standard of all the member bodies of the CEN. One major difference however is that for CEN members, but not for ISO members, a German text, in addition to English and French texts, is also required.

The preferred method is that the lead body in the work be the ISO rather than the CEN. It is only when particular expertise rests with CEN members that this body takes the lead in the work.

8 ISO Central Secretariat

Information about all aspects of the ISO, its structure, membership, published standards, and current projects, can be obtained from

ISO Central Secretariat
1, rue de Varembé
Case postale 56
CH-1211 Geneva 20
Switzerland

Tel +41 22 749 01 11
Fax +41 22 733 34 30
Internet central@isocs.iso.ch

The following publications issued by the ISO Central Secretariat provide useful information:

(a) *The ISO/IEC Directives:*

Part 1: Procedures for the Technical Work
Part 2: Methodology for the Development of International Standards
Part 3: Drafting and Presentation of International Standards

These outline the organization of and procedures for the work of technical committees, the rules for the drafting of standards, and the philosophy of standardization.

(b) *ISO Memento*

This lists the names and addresses of all ISO member bodies and the titles, scopes, and organization structure of ISO technical and administrative committees.

(c) *ISO Technical Programme*

This lists all documents that have reached the stages of committee draft and Draft International Standard, by technical committee.

(d) *ISO Catalogue*

This lists all published standards, by subject.

New editions of the *ISO Memento* and the *ISO Catalogue* are published annually, and the *ISO Technical Programme* appears every six months.

9 The Members of the ISO

There are currently 110 active member bodies in the ISO. These are the standards bodies of the participating countries, which, in addition to providing national input to the ISO, and sometimes regional standards work, are also generally responsible for monitoring the preparation of their own national standards.

The structures and methods of working of these organizations vary enormously. Many are government departments or parts of such; some are entirely independent of government. Two such independent bodies are described briefly below.

The officially recognized standardization body in the USA is the American National Standards Institute, the ANSI. Although all formal U.S. dealings with the ISO have to take place through the ANSI, the body that produces all the U.S. standards for polymer testing is the American Society for Testing and Materials, the ASTM. Its standards are generally collected together in separate volumes for particular subjects, produced annually, so that they are always relatively up-to-date. Whilst the members of the ASTM are active in the preparation of ISO standards for polymer testing, the ASTM standards are *not* identical to their ISO equivalents, although in many cases they are very similar.

In the United Kingdom the British Standards Institution (the BSI) is an independent body operating under a Royal Charter. More than half of its standards-writing income is derived from the sale of its standards; the remainder is provided by industry on a membership basis, matched almost equally by a government contribution. The BSI also derives considerable income from separate divisions responsible for certification and testing services. Since the greatest proportion of all British standards produced are now identical to either ISO or CEN standards, the committee structure of the BSI has been modified to reflect that of the international and regional bodies. The same committees are responsible for input to the ISO and the CEN as well as the preparation of national standards, even when these are additional to or different from the international ones.

Both ASTM standards and, particularly in the British Commonwealth countries, British Standards are widely used throughout the world, as are the DIN standards produced in Germany. Although ISO standards are increasingly recognized as the world-wide reference for testing purposes, the wide acceptance of standards from the major national standards bodies can sometimes lead to confusion. This can only be resolved by the adoption by *all* national bodies of ISO publications as their national standards.

Full details of all ISO member bodies are available from the ISO Central Secretariat, and all organizations are listed in the ISO Memento, published and updated annually.

Details of *some* of these member bodies are given in the appendix to this chapter. The selection has been made so that all those bodies have been included that are responsible for secretariat activities in the technical committees that deal with polymers (see later in this chapter). The addresses and other information relate to the *official* member bodies, i.e., the recognized standards bodies. However, it is often the case that a committee secretariat is the responsibility of either a separate department with a different address from the main body, or indeed of a separate organization altogether, working on a contract basis (this is especially the case with the ISO/TC 45 secretariat—see the next section of this chapter).

10 Standardization of Polymer Test Methods

Most polymer test methods, other than the very recent and those under development, have been converted to International Standards, and the three ISO technical committees responsible for the majority of these are ISO/TC 38 for textiles, ISO/TC 45 for rubber and rubber products, and ISO/TC 61 for plastics. All three of these committees were formed in 1947 during the first year of operation of the ISO, and each is responsible for several hundred published standards and similar numbers of current active projects. Their full structures are given below, and attention is drawn particularly to the scopes of the technical committees, which give details of the subjects covered.

Although each of these committees has subcommittees and working groups devoted specifically to test methods only, including physical test methods, this in not always the case, so the structures given below cover the *whole* of each technical committee, and not just those subsections that self-evidently deal with test methods.

The bodies responsible for the subcommittees and working groups are given in the lists as member bodies. However, in the case of working groups, those responsible are often the individual conveners. Detailed information about these can be obtained from the relevant member body listed, or from the secretariat of the technical committee or subcommittee responsible.

ISO/TC 38 "Textiles"

Secretariat: BSI

Scope

Standardization of

Fibers, yarns, threads, cords, rope, cloth and other fabricated textile materials; and the methods of test, terminology, and definitions relating thereto
Textile industry raw materials, auxiliaries and chemical products required for processing and testing
Specifications for textile products

Number	Responsible	Title
WG 9	PKN	Nonwovens
WG 13	—	Standard atmospheres for conditioning and testing
WG 14	BSI	Generic names for man-made fibers
WG 15	BSI	Pilling
WG 16	DIN	Abrasion resistance and seam slippage
WG 17	SAA	Physiological properties of textiles
WG 18	—	Low stress, mechanical and physical properties of woven apparel fabrics
SC 1	**BSI/ANSI**	**Tests for colored textiles and colorants**
WG 1	—	Light and weathering
WG 2	—	Washing
WG 3	—	Atmospheric contaminants
WG 4	—	Heat treatments
WG 5	—	Adjacent fabrics
WG 7	—	Color measurement
WG 8	—	Dry cleaning
WG 9	—	Carpets
WG 11	—	Characterization of dyestuffs
WG 12	DIN	ISO/CEN coordination group
SC 2	**ANSI**	**Cleansing, finishing, and water resistance tests**
WG 1	BSI	Dry cleaning
WG 2	ANSI	Finishing

Number	Responsible	Title
WG 3	BSI	Washing
WG 4	ANSI	Appearance retention
WG 7	BSI	Tumble drying
WG 8	ANSI	Water resistance tests
WG 10	DIN	Water vapor transfer test methods
SC 5	**ANSI**	**Yarn testing**
WG 2	SCC	Yarn strength
WG 3	DIN	Evaluation of twist tests
SC 6	**ANSI**	**Fiber testing**
WG 1	ANSI	Cotton fiber tests
SC 11	SNV	**Care labelling of textiles and apparel**
WG 1	AFNOR	Technical matters
WG 2	ANSI	Glossary
WG 3	SCC	Test methods and criteria
SC 12	**BSI**	**Textile floor coverings**
WG 1	AFNOR	Terminology
WG 2	BSI	Physical test methods
WG 3	BIS	Hand-knotted carpets and jute and coir products
WG 4	ANSI	Classification and grading
WG 5	ANSI	Static electrical propensity
WG 6	DIN	Appearance retention testing
WG 7	SNZ	Castor chair testing
WG 8	BSI	Soilability/cleanability testing
SC 19	**ANSI**	**Burning behavior of textiles and textile products**
WG 1	SCC	Terminology
WG 2	SAA	Apparel
WG 3	DIN	Curtains, drapes, and bedding items
WG 4	BSI	Industrial uses and tentage
WG 5	—	Floor coverings
WG 7	BSI	Vertical burning behavior tests
SC 20	**SABS**	**Fabric descriptions**
SC 21	**BSI**	**Geotextiles**
WG 1	SCC	Liaison with CEN/TC 189
WG 2	ANSI	Terminology, identification, and sampling
WG 3	UNI	Mechanical properties
WG 4	BSI	Hydraulic properties
WG 5	ANSI	Durability
SC 22	**TBS**	**Product specifications**

ISO/TC 45 "Rubber and rubber products"

Secretariat: DSM (see NOTE below)

Scope

Standardization of terms and definitions, test methods, and specifications for rubber in any form, rubber products (including their dimensional tolerances) and major rubber compounding ingredients.

By agreement with ISO/TC 61 "Plastics," coated fabrics, flexible cellular materials and hose, whether made of rubber or plastics, are also dealt with in ISO/TC 45. In the case of speciality hoses of strong user interest to other TCs, e.g., TCs 22 "Road vehicles," 44 "Welding and allied processes," 67 "Materials, equipment and offshore structures for petroleum and natural gas industries" and 131 "Fluid power systems," TC 45 will mutually agree with the user TC concerned on the appropriate method of establishing or revising the International Standards concerned.

Excluded

Rubber belting (dealt with by ISO/TC 41 "Pulleys and belts (including veebelts)")
Tires (dealt with by ISO/TC 31 "Tyres, rims and valves")
Certain speciality products, specifically those dealt with by ISO/TC 20 "Aircraft and space vehicles," ISO/TC 22 "Road vehicles," ISO/TC 121 "Anaesthetic and respiratory equipment," and ISO/TC 157 "Mechanical contraceptives," as well as the rubber seals dealt with by ISO/TC 131 "Fluid power systems."

Number	Responsible	Title
WG 1	ANSI	Chemical tests
WG 7	BSI	Flexible and semirigid cellular materials
WG 10	ANSI	Terminology
WG 13	BSI	Coated fabrics
WG 15	ANSI	Application of statistical methods
SC 1	**BSI**	**Hoses (rubber and plastics)**
WG 1	ANSI	Industrial, chemical, and oil hoses
WG 2	BSI	Automotive hoses
WG 3	BSI	Hydraulic hoses
WG 4	NNI	Hose test methods
SC 2	**SIS**	**Physical and degradation tests**
WG 1	BSI	Physical and viscoelastic properties
WG 3	DIN	Degradation tests
SC 3	**SCC**	**Rubber materials (including latex) for use in the rubber industry**
WG 1	BSI	General methods for sampling, mixing, and vulcanization
WG 2	BSI	Latex
WG 3	NNI	Carbon black
WG 4	DSM	Natural rubber
WG 5	AFNOR	Synthetic rubber
WG 6	DIN	Nonblack ingredients
SC 4	**DSM**	**Miscellaneous products**
WG 1	—	Rubber threads
WG 2	AFNOR	Sealing rings for pipes
WG 3	DIN	Rubber covered rollers
WG 4	ANSI	Rubber roof covering
WG 5	DSM	Gloves for medical applications
WG 6	ANSI	Generic marking of rubber products

NOTE DSM is the official standards body of Malaysia. However, the secretariats of ISO/TC 45 and ISO/TC 45/SC 4 are administered by

SIRIM Berhad
1, Persiaran Dato' Menteri
P.O. Box 7035, Section 2
40911 Shah Alam
Selangor Darul Ehsan
Malaysia

Tel +60 3 559 26 01
Fax +60 3 550 80 95
Internet http://www.sirim.my

ISO/TC 61 "Plastics"

Secretariat: ANSI

Scope

Standardization of nomenclature, methods of test, and specifications applicable to materials and products in the field of plastics.
Excluded:

Rubber, lac.

Number	Responsible	Title
SC 1	**BSI**	**Terminology**
WG 1	ANSI	Terms and definitions
WG 3	DIN	Symbols
SC 2	**AENOR**	**Mechanical properties**
WG 1	DIN	Static properties
WG 2	ANSI	Surface properties
WG 3	BSI	Impact properties
WG 4	ANSI	Dynamic properties
WG 5	BSI	Temperature-dependent mechanical properties
WG 6	UNI	Preparation of test specimens by machining
WG 7	ANSI	Fracture and fatigue properties
WG 8	BSI	Forms of data presentation
SC 4	**BSI**	**Burning behavior**
WG 1	ANSI	Ignitability and flame spread
WG 2	NNI	Smoke opacity and corrosivity
WG 3	AFNOR	Heat release
WG 4	BSI	Guidance on fire testing
SC 5	**SNV**	**Physical-chemical properties**
WG 1	SCC	Optical properties
WG 4	—	Permeability and absorption
WG 5	DIN	Viscosity
WG 8	AFNOR	Thermal analysis
WG 9	BSI	Melt rheology

Number	Responsible	Title
WG 10	NNI	Gas chromatography
WG 12	NNI	Ash
WG 13	—	Reference materials
WG 16	—	Electrical properties
WG 17	ANSI	Density
WG 18	ANSI	Extractable matter
WG 19	—	Determination of melting point of semicrystalline polymers
WG 21	IBN	Application of statistical methods
WG 22	JISC	Biodegradability
SC 6	**DIN**	**Aging, chemical, and environmental resistance**
WG 1	ANSI	Resistance to biological attack
WG 2	DIN	Exposure to light
WG 3	UNI	Various exposures
WG 4	DIN	Cracking and crazing
WG 5	UNI	Thermal stability
WG 7	ANSI	Basic standards
SC 9	**ANSI**	**Thermoplastic materials**
WG 6	BSI	Polyolefins
WG 7	ANSI	Styrene polymers
WG 8	AFNOR	Polyamides
WG 14	DIN	Polymer dispersions
WG 15	ANSI	Polycarbonate
WG 16	BSI	Cellulose ester
WG 17	ANSI	Thermoplastic polyesters
WG 18	ANSI	Preparation of test specimens
WG 19	UNI	Polymethyl methacrylate materials
WG 20	DIN	Polyvinyl chloride
WG 21	ANSI	Polyoxymethylene
WG 22	AFNOR	PTFE raw materials and products
WG 23	BSI	Polymers and copolymers of vinyl alcohol
WG 24	JISC	Polyphenylene ethers
SC 10	**SCC**	**Cellular plastics**
WG 11	BSI	Physical and mechanical properties
WG 12	DIN	Endurance properties
WG 13	—	Material and product specifications
SC 11	**JISC**	**Products**
WG 2	AFNOR	Decorative laminates
WG 3	BSI	Plastics films and sheets
WG 5	SNV	Polymeric adhesives
WG 6	UNI	Polymethyl methacrylate sheets
WG 7	DIN	Polycarbonate sheets
WG 8	JISC	Unplasticized polyvinyl chloride sheets
SC 12	**ANSI**	**Thermosetting materials**
WG 1	AFNOR	Thermoset molding compounds
WG 2	DIN	Phenolic resins
WG 5	AFNOR	Polyesters, epoxies, polyurethanes, and other resins
WG 10	DIN	Standards for designing with thermosets

Number	Responsible	Title
SC 13	**AFNOR**	**Composites and reinforcement fibers**
WG 11	BSI	General standards
WG 12	IBN	General methods of test for reinforcements
WG 13	AFNOR	Glass fibers
WG 14	JISC	Carbon fibers
WG 15	AENOR	Prepregs and molding compounds
WG 16	AFNOR	Composite materials

In addition to the three committees described in detail above, the following ISO technical committees can be regarded as having some connections with polymers and polymeric materials, and are listed for completeness. Their full subcommittee and working group details can be obtained from the relevant secretariats.

ISO/TC 6 "Paper"

Secretariat: SCC

ISO/TC 33 "Refractories"

Secretariat: BSI

ISO/TC 35 "Paints and varnishes"

Secretariat: NNI

ISO/TC 138 "Plastics pipes, fittings and valves for the transport of fluids"

Secretariat: NNI

ISO/TC 166 "Ceramic ware, glassware and glass ceramic ware in contact with food"

Secretariat: ANSI

ISO/TC 189 "Ceramic tiles"

Secretariat: ANSI

ISO/TC 206 "Fine ceramics"

Secretariat: JISC

11 Other ISO Committees

There are a number of committees within the ISO that may be regarded as service committees, providing useful and indeed invaluable information for users of other standards, particularly test methods.

A selective list is given below. Full details of the committee structures can be obtained from the relevant secretariats.

ISO/TC 12 "Quantities, units, symbols, conversion factors"

Secretariat: SIS

ISO/TC 19 "Preferred numbers"

Secretariat: AFNOR

ISO/TC 37 "Terminology (principles and coordination)"

Secretariat: ON

ISO/TC 46 "Information and documentation"

Secretariat: DIN

ISO/TC 47 "Chemistry"

Secretariat: NNI

ISO/TC 48 "Laboratory glassware and related apparatus"

Secretariat: GOST R

ISO/TC 57 "Metrology and properties of surfaces"

Secretariat: GOST R

ISO/TC 69 "Applications of statistical methods"

Secretariat: AFNOR

ISO/TC 92 "Fire tests on building materials, components and structures"

Secretariat: BSI

ISO/TC 135 "Non-destructive testing"

Secretariat: JISC

ISO/TC 145 "Graphical symbols"

Secretariat: DIN

Appendix—Selected Member Bodies of the ISO

It must be stressed that these addresses are subject to change over the years. The most recent information will be available from ISO Central Secretariat.

AUSTRALIA (SAA)

Standards Australia
1 The Crescent
Homebush—N.S.W. 2140

Postal address

P.O. Box 1055
Strathfield—N.S.W. 2135

Tel +61 2 9746 47 00
Fax +61 2 9746 84 50
Internet intsect@saa.sa.telememo.au

AUSTRIA (ON)

Österreichisches Normungsinstitut
Heinestrasse 38
Postfach 130
A-1021 Wien

Tel +43 1 213 00
Fax +43 1 213 00 650
Internet iro@tbxa.telecom.at

BELGIUM (IBN)

Institut belge de normalisation
Av. de la Brabançonne 29
B-1000 Bruxelles

Tel +32 2 738 01 11
Fax +32 2 733 42 64
Internet instituut.normalisitie@pophost.eunct.bc

CANADA (SCC)

Standards Council of Canada
45 O'Connor Street, Suite 1200
Ottawa, Ontario K1P 6N7

Tel +1 613 238 32 22
Fax +1 613 995 45 64
Internet info@scc.ca

CHINA (CSBTS)

China State Bureau of Technical Supervision
4, Zhichun Road
Haidian District
P.O. Box 8018
Beijing 100088

Tel +86 10 6 203 24 24
Fax +86 10 6 203 10 10

DENMARK (DS)

Dansk Standard
Kollegievej 6
DK-2920 Charlottenlund

Tel +45 39 96 61 01
Fax +45 39 96 61 02
Internet dansk.standard@ds.dk

FINLAND (SFS)

Finnish Standards Association SFS
P.O. Box 116
FIN-00241 Helsinki

Tel +358 9 149 93 31
Fax +358 9 146 49 25
Internet sfs@sfs.fi

FRANCE (AFNOR)

Association français de normalisation
Tour Europe
F-92049 Paris La Défense Cedex

Tel +33 1 42 91 55 55
Fax +33 1 42 91 56 56

GERMANY (DIN)

DIN Deutsches Institut für Normung
Burggrafenstrasse 6
D-10787 Berlin

Postal address

D-10772 Berlin

Tel +49 30 26 01-0
Fax +49 30 26 01 12 31
Internet postmaster@din.de

GREECE (ELOT)

Hellenic Organization for Standardization
313, Acharnon Street
GR-111 45 Athens

Tel +30 1 228 00 01
Fax +30 1 228 30 34
Internet elotinfo@elot.gr

INDIA (BIS)

Bureau of Indian Standards
Manak Bhavan
9 Bahadur Shah Zafar Marg
New Delhi 110002

Tel +91 11 323 79 91
Fax +91 11 323 40 62

ITALY (UNI)

Ente Nazionale Italiano di Unificazione
Via Battistotti Sassi 11/b
I-20133 Milano

Tel +39 2 70 02 41
Fax +39 2 70 10 61 06
Internet webmaster@uni.unicei.it

JAPAN (JISC)

Japanese Industrial Standards Committee
c/o Standards Department
Agency of Industrial Science and Technology
Ministry of International Trade and Industry
1-3-1, Kasumigaseki, Chyodu-ku
Tokyo 100

Tel +81 3 35 01 20 96
Fax +81 3 35 80 86 37

MALAYSIA (DSM)

Department of Standards Malaysia
21st Floor, Wisma MPSA
Persiaran Perbandaran
40675 Shah Alam
Selangor Darul Ehsan

Tel +60 3 559 80 33
Fax +60 3 559 24 97
Internet central@dsm4.gov.my

NETHERLANDS (NNI)

Nederlands Normalisitie-instituut
Kalfjeslaan 2
P.O. Box 5059
NL-2600 GB Delft

Tel +31 15 2 69 03 90
Fax +31 15 2 69 01 90
Internet [givenname].[surname]@nni.nl

NEW ZEALAND (SNZ)

Standards New Zealand
Standards House
155 The Terrace
Wellington 6001

Postal address

Private Bag 2439
Wellington 6020

Tel +64 4 498 59 90
Fax +64 4 498 59 94
Internet snz@standards.synet.net.nz

NORWAY (NSF)

Norges Standardiseringsforbund
Drammensveien 145 A
Postboks 353 Skøyen
N-0212 Oslo

Tel +47 22 04 92 00
Fax +47 22 04 92 11
Internet firmapost@norsk-standard.msmail.telemax.no

POLAND (PKN)

Polish Committee for Standardization
ul. Elektoralna 2
P.O. Box 411
00-950 Warszawa

Tel +48 22 620 54 34
Fax +48 22 620 54 34

RUSSIAN FEDERATION (GOST R)

Committee of the Russian Federation for Standardization, Metrology and Certification
Leninsky Prospekt 9
Moskva 117049

Tel +7 095 236 40 44
Fax +7 095 237 60 32

SOUTH AFRICA (SABS)

South African Bureau of Standards
1 Dr Lategan Rd, Groenkloof
Private Bag X191
Pretoria 0001

Tel +27 12 428 79 11
Fax +27 12 344 15 68
Internet postmaster@sabs.co.za

SPAIN (AENOR)

Asociación Española de Normalización y Certificación
Fernández de la Hoz, 52
E-28010 Madrid

Tel +34 1 432 60 00
Fax +34 1 310 49 76

SWEDEN (SIS)

SIS—Standardiseringen i Sverige
St Eriksgatan 115
Box 6455
S-113 82 Stockholm

Tel +46 8 610 30 00
Fax +46 8 30 77 57
Internet info@sis.se

SWITZERLAND (SNV)

Swiss Association for Standardization
Mühlebachstrasse 54
CH-8008 Zurich

Tel +41 1 254 54 54
Fax +41 1 254 54 74
Internet post@snv.snv.inet.ch

TANZANIA, UNITED REPUBLIC OF (TBS)

Tanzania Bureau of Standards
Ubungo Area
Morogoro Road/Sam Nujoma Road
Dar es Salaam

Postal address

P.O. Box 9524
Dar es Salaam

Tel +255 51 4 32 98
Fax +255 51 4 32 98

UNITED KINGDOM (BSI)

British Standards Institution
389 Chiswick High Road
GB-London W4 4AL

Tel +44 181 996 90 00
Fax +44 181 996 74 00
Internet info@bsi.org.uk

USA (ANSI)

American National Standards Institute
11 West 42nd Street
13th floor
New York, NY 10036

Tel +1 212 642 49 00
Fax +1 212 398 00 23
Internet info@ansi.org

5
Sample Preparation

Freddy Boey
Nanyang Technological University, Singapore

1 Introduction

Sample preparation forms an integral and important part of any experimental research work and is probably not fully appreciated in terms of its significance. Any analysis and interpretation of experimental results must ensure that the samples tested did not themselves contribute any variation. Unless one can be fully sure that the samples tested were prepared with consistent pretest or preconditioning properties, no meaningful interpretation of their tests can be made. Careful and consistent sample preparation for polymers and their composites is important because both physical and mechanical properties of almost all polymers and their composites vary significantly depending on the ways they are prepared. Different heating/cooling profiles, applied pressure, and even different types of atmosphere during preparation will result in significant differences in the sample properties. Care should also be taken to ensure that the variations of physical properties within the same sample should be minimized if not eliminated. Specifications and techniques for proper sample preparation are generally well covered for most polymers.

2 Preparing Plastics Samples

Probably the three most important variations to be avoided in sample preparation of polymers and their composites are the degree of crystallization, the glass transition temperature, and the void content of the polymers. To this extent, all three are separately discussed below.

2.1 The Degree of Crystallization

The physical and mechanical properties of all crystallizing polymers vary significantly with the degree of crystallization (DOC) within the sample, and in some cases also depending on the morphology of the polymer crystals. Both the degree of crystallization and the crystal morphology of the sample are in turn dependent on the thermal and pressure history of the sample, the profile, and the surface smoothness of the mold cavity. For example, higher cooling rates after melting will result in a lower DOC, and in some cases, a high enough rate of cooling may even result in a completely amorphous polymer. It should be noted that the nominal cooling rate (the difference in the soaking temperature and the quenching medium temperature, divided by the cooling time) is not a good comparison to ensure that samples are comparably prepared, unless the mold used is identical. A mold made of copper will have a higher cooling rate than an aluminum mold, for the same nominal cooling rate, due to the higher heat conductivity of the former. A rougher surface will induce a higher rate of crystal nucleation and result in smaller crystal sizes. In the case of their composites, crystallizing polymers may also experience different DOC for different reinforcement contents [1], all other things being equal. Table 1 compares the results of the DOC for a PPS composite with 20% glass fibers, after water quenching and annealing for 10 minutes, using three different types of mold material. In this case, due to the high degree of crystal nucleation of the resin material, the different mold materials used resulted in quite different crystallinity. The difference in the DOC obtained was due to the different rate of heat conduction for the mold material, with the copper mold giving the highest thermal conductivity and thus the lowest DOC.

For the above reasons, it is probably preferable that the samples be totally quenched to obtain a fully amorphous sample, and then to anneal the samples to achieve the correct crystallization profile. This is of course possible only for a polymer that can be quenched to its fully amorphous state. In some cases, the rate of crystallization for the polymers may be so high that some amount of crystallization is unavoidable. Most thermal analysis methods used for measuring crystallinity, such as the DSC or DMA methods, preclude the availability of data relating the heat of fusion or the storage modulus against the crystallinity [2]. Such data may not be always available especially for polymers that either do not fully crystallize or cannot be fully amorphous. An alternative method to determine the absolute crystallinity value would be to use the X-ray diffraction technique [3].

2.2 Glass Transition Temperature (T_g)

For polymers that go through significant glass transition phases, obtaining consistent sample T_g values becomes important. This is because the glass transition results in significant volumetric as well as mechanical strength changes. Since the glass transition temperature is a measurement of the transition of the amorphous phase of the polymer, its variation is more of a concern for noncrystallizing polymers. However, semicrystallized polymers, since they also contain some amount of amorphous phase, are also affected. In

Table 1 Effect of Mold Material on the Degree of Crystallinity

Mold material	Thermal conductivity, w/m-K	Degree of crystallinity, %
Copper (100%)	398	22
Aluminum 2014	192	27
Low carbon steel	120	34

instances where the polymer is prepared from a liquid base mixed with an initiator, which is the case for most thermosetting and elastomeric polymers, optimal curing to obtain the maximization T_g value of the samples is crucial. The final T_g value for a given sample is dependent on both its thermal and pressure history [4,5]. The latter arises because applied pressure in effect reduces the free volume of the polymer, thus reducing the volume shrinkage relaxation time, and so increasing the T_g. The thermodynamic equation relating both physical parameters has been shown [6] to be:

$$T_g = \exp \frac{e_v(p - p_o)}{V_v} \quad (1)$$

where p_o is the correlating pressure constant and e_v is the energy required for formation of a fluctuating void of minimum size of volume V_v.

Most thermosetting resins cure from a liquid mix to a glassy solid sample. The void growth in this case has been shown to be dominantly moisture driven [7]. Typically the gelation/curing process results in exothermic heat, which would be sufficient to convert the moisture within the sample into voids. As such, samples that contain fillers or fibers that have not been fully dried would provide both nucleating sites and the driving force for more voids, and significantly lower T_g. Hence application of pressure during preparation of composite materials would increase the T_g values more significantly than for homogeneous polymers, as has been shown by Boey et al. [8], which obtained an increase of almost 6°C when an additional pressure of 1.5 MPa was applied.

Proper curing of thermoset samples is also crucial to obtain optimal and consistent samples. Ideally, the samples should achieve maximum conversion or cross-linking during cure. For proper curing, both the heating rate and the time spent at the eventual cure temperature are important. The maximum permissible heating rate can be determined by use of temperature and time transformation curves [9]. Too rapid a heating rate may cause premature degradation leading to lower T_g values as well as low percentage conversion and as a result, nonoptimal samples. If possible, it is often better to cure at a lower heating rate. Thus instead of thrusting the samples into a preheated oven, one might put the samples first into the cold oven and then heating the oven to its required cure temperature.

2.3 Void Content

Voids exist in samples dominantly because of residual moisture content that is retained either within the liquid resin used or in the solid pellets that have not been thoroughly dried. Even a small amount of moisture can induce a large and significant amount of voids due to the high partial pressure of water. Since almost all polymer samples are either prepared from a liquid base or melted from granules/pellets into a mold, they inevitably experience a transition from liquid to solid phase. Thermoplastic specimens prepared by hot pressing of several single solid laminated sheets can also suffer from voids at the interfaces between the laminate sheets. Evolution of gases, or entrapped air during the process, may thus be frozen into the final sample piece. Void content has been shown to affect physical [8] and mechanical [10] properties of the sample, so they need to be minimized as much as possible. As a rule of thumb, a volume fraction of greater than 5% voids would probably affect the measured physical/mechanical properties of the specimens if tested in bulk. In the case of specimens cut out from homogeneous thin sheets or membranes, void fractions smaller than 5% could still affect the final test results adversely.

The main method of void reduction is the use of a simultaneous applied vacuum and pressure. For elastomeric samples prepared from a liquid base, the liquid mix is typically viscous, and to that extent the voids cannot be easily transported up to the surface. This is particularly so for highly contoured samples. Vacuum evacuation of the mix prior to casting/molding will help to reduce this problem. Alternatively, the cast can be completely placed into a vacuum box/oven for evacuation and subsequent pressing. Vacuum evacuation alone may not be sufficient for void reduction in the case of fiber reinforced polymers [11]. The fiber reinforcements typically allow for many more void nucleation sites, resulting in huge numbers of small voids, which are harder to remove than small numbers of large voids. In addition to vacuum evacuation, applied pressure is then required. This can be by means of a press or a pressurized autoclave vessel. The typical pressure applied can range from 500 kPa for less viscous and low fiber volume fraction samples to 1.2 to 2.0 MPa for highly viscous and high fiber volume fraction samples. Details of this vacuum/applied pressure method of sample preparation can be found in [10]. Table 2 shows the effect of an applied vacuum and autoclave pressure on the void content of some epoxy samples. Application of only vacuum brought the void content down to about 11%. A further application of pressure from 500 to 1500 kPa lowered the voids down to 6.5% and 2.5% respectively. Such a change in the void content will inevitably result in changes in the mechanical properties.

Measurements of void content can be determined basically by comparison of the densities of the resin and reinforcements separately, and the measured composite density. The void content is thus given as

$$V = 100\,\frac{T_d - M_d}{T_d}$$

where

M_d is the measured composite density
T_d is the theoretical composite density given as:

$$T_d = \frac{100}{R/D + r/d}$$

where

R = weight % of resin in the composite
r = weight % of reinforcement in the composite
D = resin density
d = reinforcement density

2.4 Specifications for Processing Samples

Several standards and specifications exist for the processing of test samples. The American Standards for Testing Materials (ASTM) probably provide the most comprehensive series

Table 2 Effect of Applied Vacuum and Pressure on the Flexural Modulus and Strength of Some Epoxy Samples

Applied vacuum, kPa	Applied pressure, kPa	Void, %
0	0	24.0
30	0	11.0
30	500	6.5
30	700	5.5
30	1000	3.0
30	1500	2.5

of standards [12–20]. The International Organization for Standardization (ISO) has identical specifications under its ISO 293, 294, 295, and 3167 publications [21–24]. Similar standards also exist in the British Standards under the series of publications numbered BS 2782 part 9 [25–28]. Some of the standards have been tabulated in Table 3, with its closest equivalents provided as well.

3 Producing Composite Samples

3.1 Autoclave Processing of Thermoset Composites

Production of good quality fiber composite samples requires the use of vacuum and pressure to minimize voids as well as to maximize the fiber volume fraction. The vacuum evacuation is made possible usually by means of a vacuum bagging method. The basic lay up for such a method is illustrated in Fig. 1. Simple flat shaped samples as well as samples with low degree of contours can be produced using an appropriate caul or molding plate. Aluminum material is good for the caul plate. For very smooth and flat surface finishes, float glass is the best material for use. For more contoured specimens, it is possible to use epoxy to cast the mold for the sample by producing a negative or female casting and then using this to cast the positive or male piece as the final mold. If the actual dimensional tolerances are not very stringent, RTV silicone rubbers (such as Dow Corning RTV 3110

Table 3 List of Some Standards for Processing Test Samples

Standard	Title	Equivalent
ASTM D1896-95	Design of Moulds for Test Specimens of Plastic Moulding Materials	
ASTM D957-95	Determining Mould Surface Temperatures of Moulds for Test Specimens of Plastics	
ASTM D3419-93	Practice for In-Line Screw Injection Moulding of Test Specimens from Thermosetting Compounds	
ASTM 3297-93	Practice for Moulding and Machining Tolerances for PTFE Resin Parts	
ASTM D3641-93	Practice for Injection Moulding Test Specimens of Thermoplastic Moulding	ISO 294-1995 BS 2782 (Method 910A-1996)
ASTM D1896-93	Practice for Transfer Moulding Test Specimens of Thermosetting Compounds and Extrusion	
ASTM D3138-92	Practice for Rubber—Preparation of Pieces for Test Purposes from Products	
ASTM D4703-93	Practice for Compression Moulding Thermoplastic Materials into Test Specimens	ISO 293-1986 BS 2782 (Method 901A-1988)
ASTM D5224-93	Practice for Compression Molding Test Specimens of Thermosetting Molding Compounds	ISO 295-1991 BS 2782 (Method 902A)
ISO 3167-1993	Plastics—Preparation and Use of Multipurpose Test Specimens	BS 2782 (Method 931A-1993)

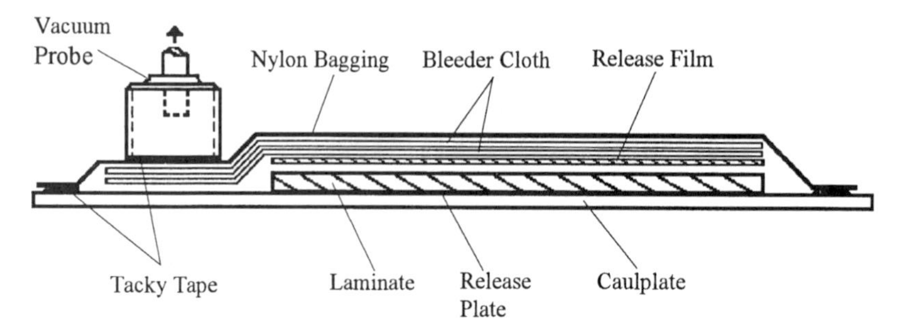

Figure 1 Schematic illustration of the lay-up of the specimens.

or 3120, both two-part silicone rubbers) have been effectively used for making such a contoured mold with the added advantage of not requiring additional release coatings.

The caul plate should be thoroughly cleaned with detergent or even better with acetone first. A release cloth is then applied onto the caul plate. To ensure thorough coverage, two coatings should be applied . Commercial release agents that can be used include those from Frekote (FRO, RRM 33 etc.), Dow Corning (Silicone Sprays) , General Electric (silicone sprays) and DuPont (VYDAX-PTFE type). Car polishing waxes and petroleum jelly are also some release agents that have been used. If the surface finish is not critical, a teflon release film can also be used to replace or in addition to the coating. Make sure that there are no wrinkles in the film and that the coverage of the film is sufficient for the whole lay up. The lay up is then done, whether by manual wet lay up or by using prepreg materials. In the case of the former, especially for room temperature curing types, it is important to ensure that the resin be carefully prepared. For example, the two parts of the resin should be separately measured and placed into two different containers. Mixing of the two parts should only be done just before the actual lay up. This is to provide sufficient time for the resin to permeate through the fiber laminates before it gels and becomes very viscous, when it would not be able to move and spread. For very large specimens that require a long time for lay up, it may be useful to put the two parts of the resins into a freezer to lower their temperature considerably. This way, even after mixing, the resin's ambient temperature would be low enough to slow the gelation process considerably.

For thermoset prepreg laminates that are to be stacked up, be careful about taking them out from the freezer and into a humid room. Very often, condensation of water vapor occurs, and if they are then inadvertently stacked together with the trapped water, unacceptable samples will result. To avoid excessive flow of the resin on the side, it is necessary to place a dam in the form of rubber or cork strips around the lay up. Allow for some space between the lay up and the dam, about 25–50 mm.

This is followed by a perforated release film to allow for transfer of the excess resin to the bleeder cloths and to separate the bleeder cloths from the actual laminate sample. Such films are readily available commercially, such as the FEP teflon 0.025 mm type film. However, a polypropylene or nylon sheet extensively perforated with a large pin will also suffice. Bleeder cloths are then added to absorb the excess resin, and also to provide a path for the vacuum to evacuate the volatiles and voids within the lay up. It is important to ensure that the bleeder cloth is extended continuously from the vacuum probe to the lay up. Industrial type absorbent paper tissue can also be used, but the better cloths would be the thick wool type cloths used for water filters, etc. The number of cloths is sometimes used to control the resin content of the cured sample. If maximum fiber content is required, the number of bleeder cloth layers should be such that the last layer is still dry after vacuum evacuation and applied pressure.

The whole configuration is then covered and sealed with a bagging bag. If bagging the whole set up is impossible or very difficult, a single bagging film can be used to cover the setup with its edges sealed onto the caul plate of a flat base using double-sided tape or a good sticky sealant. Make sure that some excess bagging material is provided for to allow for all the contours of the setup. If the curing is done at room temperature, simple polypropylene or high-strength polyethylene sheets should suffice. For hot curing, nylon bagging can be used up to about 180°C. For higher temperature curing, silicone rubber sheets can be used. While they are very costly, these silicone rubber sheets have the advantage that they are reusable. Special attention should be paid to what are termed pinhole defects in the vacuum bagging. Such pinholes can be detrimental since their occurrence often means that sufficient vacuum evacuation cannot be achieved, and thus the sample has to be scrapped. Testing of the bag can be done by effecting the vacuum in the bagging without the lay up to test the level of vacuum it achieves.

The vacuum evacuation is effected by means of a vacuum pump (or even a simple vacuum cleaner appropriately connected) via a vacuum probe, which is designed so that the excess resin will not enter into the vacuum pump to damage the machine. Vacuum evacuation can then commence. Typically evacuation down to 0.5 bars (about 50 kPa) would be sufficient. Pressurization of the sample is then effected by placing the whole configuration into an autoclave or any equivalent pressure vessel. If curing is to be done at elevated temperature, ensure that the temperature is raised to the appropriate level before pressurization. The vacuum should be maintained at all times. The pressure is then raised to about 10–12 bars (100–120 kPa) and maintained until the sample is fully cured. For hot curing samples, it is usually advisable to cool the sample down slowly to reduce residual stresses.

3.2 Vacuum Injection Molding

One method of sample preparation that avoids the tedious manual wet lay up method for fiber-reinforced composites, and is good also if very accurate placement of the fiber orientation is required, is the use of vacuum injection molding [29]. Figure 2 shows a simple but effective method for molding a flat panel from which samples can be cut out. The fibers are basically laid up without the resin and thus can be accurately placed in terms of their fiber orientation. If necessary, the fiber orientation can be accurately positioned by means of clamps at the ends of the fibers. A vacuum is then effected using a vacuum bagging method as mentioned in the previous paragraph. A separate resin transport channel is then opened and the resin is sucked in by the vacuum. A resin trap can be used to ensure that the resin does not inadvertently enter the vacuum pump. Once sufficient resin has been injected, the vacuum and resin hoses can be dismantled by quick disconnect valves and the whole sample panel with the vacuum bagging put into a heating oven or wrapped and heated with a heating blanket for full curing. In terms of process efficiency, vacuum injection molding works better with resin of lower viscosity. Samples of more contoured shapes, which cannot be produced using conventional methods, can be produced using this method. Figure 3 shows an example of a model smoke tunnel that has been produced using this method.

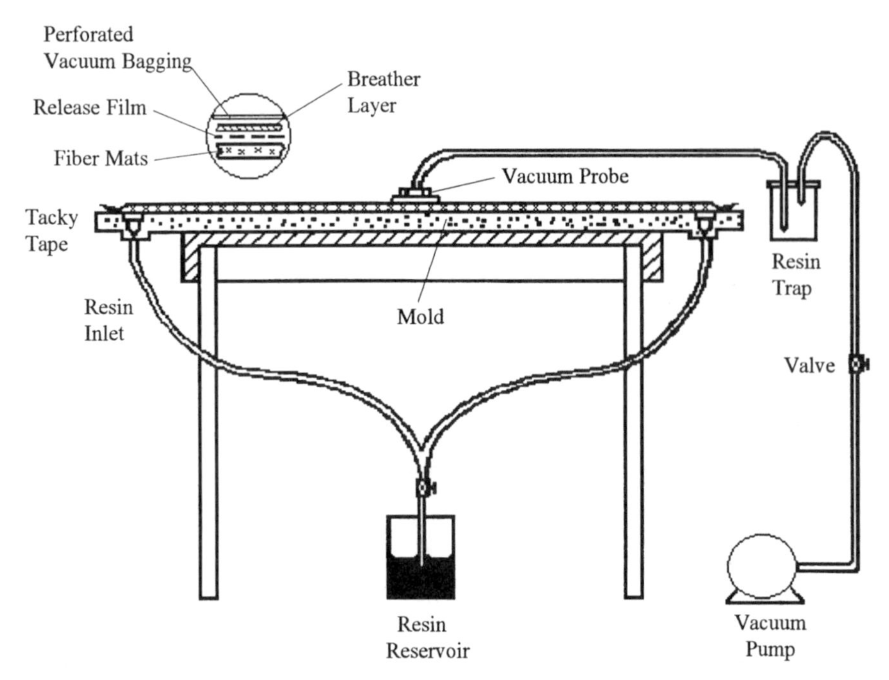

Figure 2 Schematic illustration of the whole VIM process, showing the mold, vacuum, and resin-inlet systems.

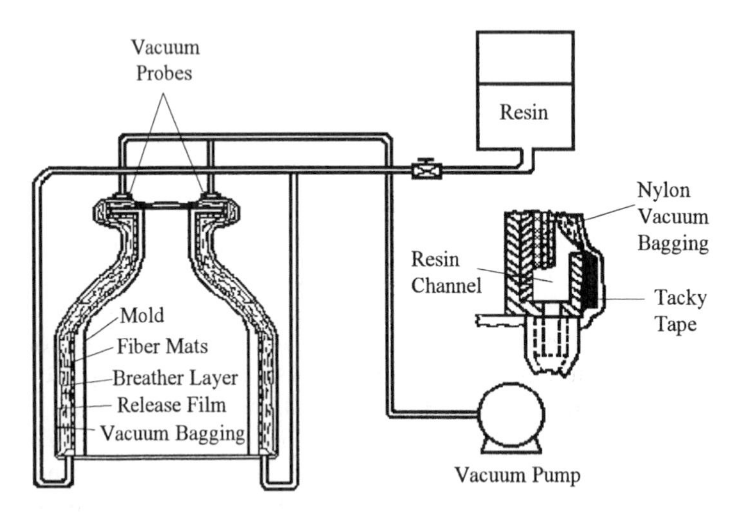

Figure 3 Schematic illustration of the complete mold, vacuum, and resin inlet set-up. The design of the resin inlet is enlarged.

3.3 Hot Forming of Thermoplastic Specimens

Thermoplastic composite samples are easier to produce in that it typically comes in the form of solid prepreg tapes that have no need for refrigeration. Heat is required during forming to melt and fuse the thermoplastic matrix together. Hot forming by matched die molding using a hot press with both platens heated is one of the typical methods for the production of thermoplastic composite specimens. Other methods include vacuum/diaphragm hot forming, where the laminate sheet is heated until it flows, and vacuum and if necessary additional pressure is used to form the sheet into shape [30].

Semicrystalline thermoplastic materials like PP basically contain two phases, the crystalline phase and the glassy or amorphous phase. The former has an ordered structure and significantly higher strength properties, but when heated significantly below its melting point it does not provide sufficient flow viscosity to be hot workable. The glassy phase, in contrast, has lower mechanical strength properties, but on heating beyond its glass transition temperature the flow viscosity drops significantly, and sufficiently to form without material damage. It is the reversible transition between the crystalline and the glassy phase that provides the mechanism for hot forming of the composite. In order for the materials to be hot workable, the composite is typically heated above its glass transition temperature. Both the forming temperature and induced forming strain may result in the crystallization of the work piece. In cases where there is a sufficient amount of the glassy phase present, the initial heating to above its glass transition temperature is sufficient. However, if the crystalline phase is dominant, it may be necessary to heat it to close to its melting point in order to melt the crystalline phase. The initial condition of the specimen piece is therefore preferably dominantly glassy. It is then annealed after forming to increase its crystalline phase.

The main process problem limiting the ability to form a thermoplastic composite specimen is the presence of its fiber constituent. For load bearing structures, the fiber volume content required is often more than 40–50%. While heating may vary the flow viscosity of the matrix, the fibers remain essentially unchanged in stiffness. Since the fibers typically bear most of the load in the composite, breakage or damage of the fibers is unacceptable during the hot forming process. Since the flow viscosity is dependent on the temperature that the specimen is heated up to, the heating temperature becomes an important parameter in sample preparation. It is important to note that inevitably due to the significant thickness of the laminate specimen piece, the specimen will experience differential heating across its thickness. The use of the Biot number method has been employed to estimate the thermal distribution across the thickness:

$$B_i = \frac{h_{\mathrm{eff}} t}{K} < 0.1 \tag{3}$$

$$h_{\mathrm{eff}} A(t_{\mathrm{h}} - T) = \frac{kA}{t}(t - T_{\mathrm{bot}}) \tag{4}$$

where

h_{eff} is the effective heat transfer coefficient (combining both convection and radiation effects)
t is the laminate thickness
k is the composite transverse heat conductivity
t_{h} is the heater temperature
T_{bot} is the bottom temperature
T is the surface temperature

Typically for thermoplastic composites, h_{eff} is estimated to be about 0.05–0.06 W/m^2 and k about 8×10^{-3} m/°C. For glass-reinforced PP thermoplastic composites, a 2.5 mm thick laminate has been found to give a temperature variation of about 7°C, sufficient to result in significant differences in both the flow characteristic of the matrix and the degree of crystallinity after cooling. Of course, heating at both sides would help to reduce this difference.

For thermoplastic prepreg laminates, if it should be required to stack the prepregs both sideways and vertically, a useful technique is to use a soldering iron to tack the prepregs that are side to side. Two or at most three tacks per pair of prepreg should suffice.

In the case of specimens having contoured or angular geometry such as a bend, hot forming over a die may result in higher stresses at the bends, giving rise to the thinning out of the sheet at the bends, which then increase the fiber volume fraction. More importantly, breakage of fibers at such bends becomes more likely. Typically, the glass fibers at the outer tensile surface break, but the glass fibers at the inner surface may actually be buckled due to the bending compressive rather than tensile stresses here. In some instances, it may be possible that failure may have occurred only at the outermost layer, so that removal of that layer would result in a good specimen. As in the case of thermoset curing, void reduction in also important during hot forming. Most often void occurrence is at the surfaces of the prepreg interlayer, the culprit more often than not being water. For plastics that can to be heated quickly above 100°C, this may be a simple way of eliminating the water at the surface. Otherwise, storage overnight in a dessicator or container with silica gel may be sufficient. If alcohol or acetone is used, make sure that sufficient time is given for the solvents to dry out of the surface, since such solvents often get absorbed into the surface, rather than just lying on the surface. If after such care has been taken there are still voids or defects at the interlayers, vacuum bagging using a high-temperature bagging material like nylon or silicone prior to forming, followed by forming into a flat sheet first, may be done. The single flat sheet can then be used to form whatever other shapes are required.

4 Preparing Rubber Test Samples

The production of rubber test pieces, unless they are taken from the finished products, involves the process of mixing, molding, and vulcanization to obtain a formed sheet before test pieces can be cut out. The processing variables for rubber processing can significantly affect the final physical and mechanical performance of the formed product. For this reason, it is always important to comply with an established specification. ASTM D3182-89 (reapproved in 1994) [31] provides a standard practice for producing standardized rubber sheets for testing. The standard is comparable to the ISO 2392 [32] and BS 1674 [33].

The ASTM D3182-89 provides a comprehensive listing of reference compounding materials required to prepare the rubber test compounds. The procedure for weighing of the compounds requires weighing accuracy down to ±0.1 g for the larger volume mill and internal mixer compounds, and to ±0.01 g for the miniature internal mixing compounds. In this regard, the ISO is even more stringent, requiring an accuracy of ±0.002 g for sulphur and other accelerators. Typically, the mixing is done by addition of all the compounding materials other than the rubber, carbon, and oil, into a mortar and mixed with a pestle, or mixed for about 10 minutes in a biconical blender with intensifier bar turning, or using a blender with intermittent 3-second mixing. The latter is done intermittently to avoid the stearic acid from melting, after which it will not be able to provide good dispersion. The carbon black should be dried by heating it at 125°C ±3 for one hour in an oven before being mixed into the compounding.

Standard mills can be used that have diameters of about 150 to 155 mm. The rolling speed should be about 0.4 rad/s, with the ratio between the fast and slow roll at 1.4 to 1. The standard internal mixers used should be about 1550–1600 cm^3 in volume. If smaller volumes are required, the miniature internal mixer to be used should have a capacity of about 120 cm^3 volume. The mixing procedure generally follows a procedure to ensure homogeneous mixing. The rubber is mixed with the compounds as it is banded on the slow roll. Carbon black is first progressively added evenly across the mill at a constant rate, with intermittent addition of oil. All other dry materials are then added slowly and evenly, again at a constant rate. The batch is thus passed through the mill several times as required. The ASTM D3182 actually requires the batch to be passed through an opening of 0.8 mm six times, and then through an opening of 6 mm four times, each time folding the batch piece on itself for the subsequent pass.

After mixing, the batch is then prepared for vulcanization by conditioning it at a relative humidity no greater than 55% for up to 24 hours. If a very stringent requirement is to be met, the conditioning can be by placing the sheeted compound in a closed container to prevent absorption of air moisture, or in an area controlled at 35% relative humidity.

The vulcanization equipment used is usually either a hot press with the two platens acting as the mold surface, or an actual mold piece heated up by a hot press. The required pressure should be no less than 3.5 MPa (500 psi) throughout the press platens. The molds used, or the sheet to be vulcanized, should not be less than 75 mm from the edges of the platens to ensure even heating. ASTM D3182 provides designs for a variety of multicavity molds, which can produce small tensile sheet or slab pieces. Control of the temperature is important and should be in the order of $\pm 0.5^\circ C$. The mold should first be heated in an enclosed press to the curing temperature for at least 20 minutes before the rubber batch is introduced into it. The press is quickly opened, with the rubber batch introduced into the mold, and then closed again in the minimum time possible. The time of cure is from when the pressure is then applied to the time the pressure is released. Once cure is completed, the vulcanized rubber piece should be immediately taken out and cooled to room temperature by water or a metal surface.

In some cases, test pieces are required from products already made. In this case, the vulcanization of the piece is not necessary, but the manner in which the test piece is produced from the product becomes important. ASTM D3183 provides the standard for this requirement. In the preparation of such pieces, it is necessary in many cases to separate the rubber of the product from the textile embedded within the product, such as rubber tires. If possible, the separation should not be done with a swelling liquid. If this is unavoidable, ASTM D3183 recommends the use of isooctane liquid. Cutting can be done using a knife lubricated with a mild soap solution. Uneven surfaces due to the textile previously embedded on the surface or surface patterns etc. can be removed by buffing through use of an abrasive wheel or abrasive flexible bands. Grit size during abrasion should be about No. 30 for rough buffing and 100–180 for final finishing.

5 Cutting Specimens

The cutting of specimens is often part of the process of their preparation, whether it is cutting from a rolled sheet or cutting from a molded piece. Most polymers and their composites can be machined in one form or the other, with the possible exception of silicone rubbers, which tend to tear as machining takes place. If dimension tolerances are important and cutting by a surgical knife or a pair of scissors is inadequate, the only option would be to injection or compression mold the test piece to its required dimensions. Particular care should be taken in the case of carbon-based fillers or fibers. Being conductive, the carbon particles that result from the cutting will readily short any nearby circuits if they are allowed to be airborne. These should be cut with a vacuum suction immediately collecting the resulting debris/dust. Cutting carbon, aramid and glass fibers should always be done with gloves to avoid painful sensitization of the machinist's skin.

The variations that can occur due to cutting are as follows:

Variations due to different edge cutting finishing. This is a particular problem for tensile testing of sheet type specimens, whether polymers or their composites. Results of the ultimate strength and the elongation have been known to vary widely for the same sample cut by different techniques, such as die cut, razor, rotary cutter, and shear cutter [34]. Such a problem can in part be resolved by ensuring that the cut edge is consistently sanded off with 80–220 grit grade sandpaper. In some cases for reinforced polymers, due to the usually wide disparity between the reinforcement and the matrix hardness, cutting techniques that can cut the reinforcements cleanly will vary with those techniques that cannot. Machining of composites with different fiber types may require different techniques. For example, Aramid fibers will require special cutting techniques such as water jet or laser cutting, while carbon/glass fibers can be saw machine cut. A thin transparent acrylic sheet can be used to be taped by double-sided adhesives onto the laminate to help reduce the damage to the saw blades during cutting. It is imperative especially for fiber-reinforced composite samples to be correctly aligned during cutting. To do this, a steel sheet backing can also be taped onto the laminate and secured magnetically onto the magnetic saw table once alignment is fully achieved.

Variations due to effects of lubricants/water used. Water, solvents, and lubricants can find their way into the polymers during cutting, and in doing so significantly affect their properties. Generally, it should not be necessary to use any liquid medium during cutting. For plastics, which have lower melting temperatures, lower-speed cutting may help to eliminate incipient melting. For plastics that are hard to cut because too flexible or having a tendency to chip off cutting edges, sandwiching the sheet between two acrylic sheets and cutting the sandwich laminates may help. Should it be very important to ensure that the edges cut do not experience any inadvertent heating that might change the properties at the edge, sandpaper (grade 80 is suitable) can be used to remove about 1 mm of the cut edge.

Standards specifying and recommending methods of machining and cutting can be found in [35] ASTM D882-1983 Tensile Properties of Thin Plastic Sheeting (with its equivalent given in [36], ISO 1184 (1983) Plastics—Determination of Tensile Properties of Films, and [37] BS 2782 Methods 326A-C, 1983 Determination of Tensile Strength and Elongation of Plastic Films), [38] ASTM 3297-1993 Preparation of Plastic Test Samples by Machining (with its equivalent given in [39], ISO 2818 (1994): Plastics—Preparation of Test Specimens by Machining, [40] ISO TC61 Determination of Tensile Properties, and [41] BS 2782 (Method 930A)-1996 Preparation of Test Specimens by Machining).

It is useful here to make mention of some problems that may be encountered in cutting of fibers and fabric. It is often not easy to try to obtain an accurate cutting of glass, carbon, kevlar, etc. fabric or textiles, the problem being that the fibers within the weave tend to move. To obtain an accurate cutting, and to ensure that the fibers within the fabric do not move after cutting, it is helpful to tape the fabric in the line where it is to be cut with masking tape, and then cut along the tape. Cutting should not be done with blades for cutting paper or knife cutters. A good tailor's scissors will be best for the job. It is also instructive to wear gloves when handling and cutting fibers, since they have a strong tendency to induce very irresistible itch on contact with human skin. Great care should also be taken to ensure that the fiber fabric pieces cut out are not contaminated during handling, since it is very important that they have good adhesion and spreading properties with respect to the resin used.

6 Cutting and Machining Fiber-Reinforced Composites

Machining and cutting for fiber-reinforced polymer composites present particular problems mainly due to the vastly different properties of the matrix and resin. Some of the problems encountered include contamination by cutting fluid or lubricant, induced fraying at the cutting edge or delamination during cutting, heat distortion, high tooling wear rate, etc. Industrially the best method of cutting fiber-reinforced composites has been the use of water jet cutting. This usually provides excellent cutting finishes with little or no distortion or damage during the cutting. Laser beam cutting and machining has also been successfully tried. However, this method usually leaves behind a damaged cutting surface of about 1–2 mm or even more.

Generally standard metal and woodworking cutting and machining tools can provide reasonably acceptable specimens, when touch-up finishing with sandpaper is included. However, for such facilities, the spindle speed should be modified to be significantly higher and the feed rate significantly reduced. For manual cutting using a bandsaw, a saw blade with stagger teeth can be used, but it must be sharp and run typically at speeds of about 20–30 m/s. A reverse type cutting where the heels rather than the hooks of the teeth of the saw blade are used to cut into the composite sample has been found to give better cutting finishes. If available, the use of circular diamond saws would be preferred. Routing for composites is usually done using high-torque, low-speed Buckeye routers. Generally, feed rate during cutting and trimming should be about 1 mm/s for thickness of 2.5 mm or less, and 0.5 mm/s for up to 10 mm thick.

Drilling of fiber-reinforced composites can also be properly done if special ceramic coated drills are used, such as titanium carbide drills, etc. Generally the types of drills that will give good results are the fluted twist drills with high helical angles and solid shank type drills. Good through hole finishing can be achieved by use of a backup piece; otherwise fraying at the bottom edges and even delamination may result. Aluminum would be a good and cheap backup piece. Typically, the backup piece should have the same thickness as the work piece. Drill speeds for carbide and similarly hard drills should be about 3,000 rpm, with a slower than conventional feed rate. If possible, avoid the use of fluid or lubricant.

The best method for close tolerance finishing is probably by sanding or grinding at about 20 m/s speed for 80–240 grit size paper. If perfectly flat surfaces are required, sandpaper can be taped or glued onto a float glass piece, and the specimen sandpapered on this surface.

References

1. Desio, G., and Rebenfield, J., *Appl. Polym. Sci.*, *39*, 825 (1990).
2. Jacob, J., Boey, F., and Chia, L., Thermal and non-thermal interaction of microwave radiation with materials, *Journal of Material Science*, *30*, 5321–5327 (1995).
3. Lee, T. H., Boey, F. Y. C., and Khor, K. A., X-ray diffraction analysis technique for determining polymer crystallinity in a semi crystalline thermoplastic PPS composite, *Polymer Composites*, *18*(6), 481–488 (1995).
4. Williams, M. L., Landel, R. F., and Ferry, J. D., *J. Amer. Chem. Soc.*, *77*, 3701 (1955).
5. DiBenedetto, A. T., *J. Polym. Sci. B: Polym. Phys.*, *25*, 1949 (1987).

6. Gibbs, J. H., and DiMarzio, E. A., *J. Chem. Phys.*, *28*, 373 (1958).
7. Kardos, J. L., Dudokovic, M. P., and Dave, R., *Advances in Polymer Science, 80: Epoxy Resins and Composites IV*, (K. Dusek, ed.), Springer, Berlin, 1986, p. 102.
8. Boey, F., Lee, T. H., and Sullivan-Lee, P., High pressure autoclave curing of composites: effect of high pressure on glass transition temperature, *J. Mater. Sci.*, *29*, 5985–5989 (1994).
9. Gilham, J. K., Time temperature transformation (TTT) state diagram and cure, in *The Role of the Polymeric Matrix in the Structural Properties of Composite Materials* (J. C. Seferis and L. Nicolais, eds.), Plenum Press, New York, 1983, pp. 1127–145.
10. Boey, F., Reducing the void content of a FRC using a vacuum injection molding method, *Polymer Testing*, *9*(6), 363–379 (1990).
11. Boey, F., and Lye, S. W., Void reduction in autoclave processing of thermoset composites, Part I: High pressure effects on void reduction, *Composites*, *23*(4), 261–265 (1992).
12. ASTM D1896-95, Design of moulds for test specimens of plastic moulding materials, American Society for Testing Materials Publication, Philadelphia, 1995.
13. ASTM D957-95, Determining mould surface temperatures of moulds for test specimens of plastics, American Society for Testing Materials Publication, Philadelphia, 1995.
14. ASTM D3419-93, Practice for in-line screw injection moulding of tet specimens from thermosetting compounds, American Society for Testing Materials Publication, Philadelphia, 1993.
15. ASTM 3297-93, Practice for moulding and machining tolerances for PTFE resin parts, American Society for Testing Materials Publication, Philadelphia, 1993.
16. ASTM D3641-93, Practice for injection moulding test specimens of thermoplastic moulding, American Society for Testing Materials Publication, Philadelphia, 1993.
17. ASTM D1896-93, Practice for transfer moulding test specimens of thermosetting compounds and extrusion materials, American Society for Testing Materials Publication, Philadelphia, 1993.
18. ASTM D3138-92, Practice for rubber—preparation of pieces for test purposes from products, American Society for Testing Materials Publication, Philadelphia, 1992.
19. ASTM D4703-93, Practice for compression moulding thermoplastic materials into test specimens, American Society for Testing Materials Publication, Philadelphia, 1993.
20. ASTM D5224-93, Practice for compression molding test specimens of thermosetting molding compounds, American Society for Testing Materials Publication, Philadelphia, 1992.
21. ISO 293, Plastics—Compression moulding test specimens of thermoplastic materials, International Organization for Standardization, 1986.
22. ISO 294, Plastics—Injection moulding of test specimens for thermoplastic materials, International Organization for Standardization, 1995.
23. ISO 295, Plastics—Compression moulding of test specimens for thermosetting materials, International Organization for Standardization, 1991.
24. ISO 3167, Plastics—Preparation and use of multipurpose test specimens, International Organization for Standardization, 1993.
25. BS2782, Part 9, Method 901A, Methods of testing plastics: Compression moulding test specimens of thermoplastic materials, British Standards Institute Publication, 1988.
26. BS2782, Part 9, Method 902A, Methods of testing plastics: Compression moulding test specimens of thermosetting materials, British Standards Institute Publication, 1992.
27. BS2782, Part 9, Method 910A, Methods of testing plastics: Injection moulding test specimens of thermoplastic materials, British Standards Institute Publication, 1996.
28. BS2782, Part 9, Method 931A, Methods of testing plastics: Preparation and use of multipurpose test specimens, British Standards Institute Publication, 1993.
29. Boey, F., and Liu, C. Y., Vacuum injection moulding for fibre reinforced composites, *Experimental Techniques*, Mar./Apr., 48–51 (1991).
30. O'Bradaigh, C. M., and Mallon, P. J., Effect of forming temperature on the properties of polymeric diaphram formed thermoplastic composites, *Composites Science and Technology*, *35*(3), 235–256 (1989).

31. ASTM D3182-89, Practice for rubber: Materials, equipment and procedure for mixing standard compounds and preparing standard vulcanized sheets, American Society for Testing Materials Publication, Philadelphia, 1994.
32. ISO 2392, Rubbers: Equipment and procedure for mixing and vulcanizing rubber test pices, International Organization for Standardization, 1993.
33. BS1674, Equipment and procedure for mixing and vulcanizing rubber test pieces, British Standards Institute Publication, 1976.
34. Patterson, C. D., Jr., *Materials, Research and Standards*, *4*(4), 169 (1964).
35. ASTM D882-1983, Tensile properties of thin plastic sheeting, American Society for Testing Materials Publication, Philadelphia, 1993.
36. ISO 1184-1983, Plastics—Determination of tensile properties of films, International Organization for Standardization, 1983.
37. BS2782, Methods 326A-C, 1983, Methods for testing plastics: Determination of tensile strength and elongation of plastic films, British Standards Institute Publication, 1993.
38. ASTM 3297-1993, Preparation of plastic test samples by machining, American Society for Testing Materials Publication, Philadelphia, 1993.
39. ISO 2818-1994, Plastics—Preparation of test specimens by machining, International Organization for Standardization, 1980.
40. ISO TC61, Determination of tensile properties, International Organization for Standardization, 1993.
41. BS2782, Method 930A-1996, Methods for testing plastics: Preparation of test specimens by machining, British Standards Institute Publication, 1983.

6
Conditioning

Steve Hawley
Rapra Technology Ltd., Shawbury, Shrewsbury, England

1 Introduction

Polymers are much more affected by variations in ambient conditions than are many other types of materials, such as metals, glasses, and ceramics. The ambient temperature and relative humidity conditions at the time of test can therefore be expected to influence the outcome of tests on polymers much more than for these other materials. For this reason it is necessary to specify the ambient conditions under which the tests are to be performed. In addition to the external conditions, the state of the material at the time of testing will also be important. The conditions prevailing prior to the commencement of testing may therefore be expected to have an influence on the outcome of the tests and so once again should be subject to controls.

Not all tests, nor all materials, are equally affected, and so the precise details of temperature and humidity will vary from test to test and from material to material. However, in every case, we can identify three phases during which the ambient atmosphere should be considered: storage, conditioning, and testing. Storage can be considered as that period of time between manufacture and the start of the process whereby the material is finally brought to the conditions required for the test. Conditioning is that phase during which the temperature and humidity are brought into line with the testing requirement. The testing phase is self-evidently that during which the actual test is performed.

2 Storage

Immediately after its formation a polymer is most likely to find its properties changing at their most rapid rate, in the absence of specifically aggressive environments. Thus, plasticized PVC and vulcanized rubbers in particular are well known to change their properties quickly at first and then to settle down at relatively stable values. Other examples include crystalline polymers that need time for their crystallinity to stabilize. Hydrophilic polymers like nylon and the cellulosics need to come to an equilibrium moisture content, and certain adhesives need to cure fully before their properties are optimized. Generally the process of change is approximately exponential, the first few hours being the most critical. It is wise to avoid testing material "hot off the press" if at all possible. After several days, the rate of change is generally imperceptible, and the effect of small differences in time between manufacture and testing becomes negligible.

In the case of vulcanized rubber, the recommended minimum period is 16 hours, as laid down in ISO 471 [1]. For plasticized PVC, when the hardness is to be measured, the time from manufacture to testing should be 7 ± 0.2 days as required by BS 2782 Method 365A [2].

At the other extreme, one would wish to avoid leaving test material for too long before it is tested if this is to represent unaged material. The longer the material is left before testing, the greater the chance of deterioration through accidental exposure to damaging environments such as heat, light, ozone, etc. There is, of course, usually a more pressing commercial reason for not leaving material too long before having it tested! It may be difficult in practice to know the length of time that has elapsed between manufacture and testing, nevertheless it must be considered good practice to try and standardize this. ISO 471 [1] again makes a very laudable attempt to provide reasonable criteria in suggesting a 4 week maximum for nonproduct tests and 3 months for product tests.

Irrespective of the time of storage, it is important that the polymer not be subjected during this period to high temperatures, ultraviolet light, or other potentially damaging environments. Many rubbers, for example, need to be protected from ozone and organic fluids; thermoplastics from stress-cracking fluids. During storage, provided harsh or aggressive environments are avoided, the polymer should suffer no significant deterioration. Only during the conditioning phase that precedes the testing will the atmospheric conditions need to be carefully controlled and monitored while the material is being brought into the desired state for the test.

3 Pretest Conditioning

Most test methods specify a conditioning period in a standard atmosphere prior to testing. Definitions for atmosphere, conditioning atmosphere, test atmosphere, and reference atmosphere may be found in ISO 558 [3], while ISO 554 [4] provides standard atmospheres for conditioning and testing. Both of these are general standards, not confined to polymer-based materials. The recommended atmosphere is 23°C/50% relative humidity, but 27°C/65% relative humidity is permitted for tropical countries and 20°C/65% relative humidity for certain special cases. Coated fabrics, for example, may require conditioning at 20/65, since this atmosphere has been traditionally used for fabrics.

Two levels of tolerance on temperature and relative humidity are given, these being ±2°C with 5% relative humidity and ±1°C with ±2% relative humidity where reduced tolerances are deemed to be necessary. (See Section 5.3 below for further consideration of these tolerances.)

ISO 291 [5] for plastics materials (or its equivalent, BS 2782: Part 0, Annex A [6]) and ISO 471 for rubbers [7] (BS 903 Part A35 [7]) follow the general standard in allowing the 23/50 and 27/65 atmospheres. The plastics standard does not permit the 20/65 option, but the rubber standard does because of the possible presence of a textile component. Where humidity is known to have little or no effect, the control of humidity may be waived.

ASTM D618 [8] designates 23/50 as the standard laboratory atmosphere but indicates six other conditions that may be used for particular purposes, neither 27/65 nor 20/65 being included in this list. For cellular materials BS EN ISO 845 [9] employs the 23/50 and 27/65 atmospheres as well as the uncontrolled humidity variants where the effect of humidity is not significant. The textiles and coated fabrics standard, BS EN ISO 2231 [10], lays down five standard atmospheres, which include all the ones previously mentioned. Interestingly, in this standard, Atmosphere A is the 20/65 condition, while 23/50 is "relegated" to the B atmosphere and 27/65 to the C atmosphere. The remaining two are 23° with no humidity control and 27° with no humidity control. 20° with no humidity control is not an option. Fiber-reinforced plastics generally use 23/50 (or just 23) as the preferred conditioning atmosphere, as for example in EN 62 [11] (equivalent to BS 2782 Method 1004 [12]), although other conditions may be encountered such as the use of 20°C, which is widely used in the metals industry.

The time required for conditioning is typically 3 hours (ISO 471) or 4 hours (ISO 291) for temperature alone, but up to a minimum of 96 hours where temperature and humidity are controlled. Where temperature and relative humidity are controlled, wide variations in conditioning time between types of material are observed. For example, 88 hours is a widely used minimum for thermoplastics, but cellular materials and fabrics coated on one side only with a polymer film may only need to be conditioned for a minimum of 16 hours. Reference should be made to the specific test method of interest to ensure compliance with its requirements.

Since conditioning for equilibrium moisture content can be very prolonged—and the conditioning times given in the standards are usually insufficient for true equilibrium to be attained—there are occasions when "accelerated" methods may be employed. Polyamides are a well-known class of plastics with an above-average sensitivity to moisture content, and it can take over a year for a 4 mm thick test piece to attain true equilibrium. ISO 1110 [13] provides an accelerated procedure for conditioning to reduce this time scale significantly by employing an elevated temperature and a slightly elevated relative humidity value.

4 Testing Conditions

The objective behind conditioning the test pieces at a given temperature (and perhaps humidity) is to bring them to an equilibrium condition under which the test itself is to be performed. However, standards are not always consistent in their application of the conditions. It is reasonably common to find that conditioning is to be performed in atmosphere 23/50, but that testing may be carried out in atmosphere 23. This is a sensible and pragmatic response, since it is more costly to have a large testing laboratory controlled to 23/50 rather than just 23 than it is to have a test chamber controlled to 23/50. Providing testing is performed reasonably quickly after removal from the controlled humidity environment, and the material being tested (or the test being applied) is not abnormally sensitive to moisture content, then no significant error is introduced.

However, instances of inconsistency over atmospheres used for conditioning and testing may be found in the standards literature. For example, in BS 4443 [14] for flexible cellular materials, in the former Method 1, now BS EN ISO 1923 [15], no specific conditioning requirement is stated, while Method 3 [16] gives no specific statement about the testing atmosphere to use, although there is the specific requirement to condition for 16 hours at 23/50. In Method 8 [17], the conditioning is to be at 23/50 and the test atmosphere is additionally defined as $23^\circ \pm 2^\circ$. (Similar situations arise in BS 4370 [18] for rigid cellular materials.) Clearly, a test temperature of 23°C would be understood by most people, if this were not explicitly stated, but it is unfortunate that these attentions to detail are sometimes missing in standards.

Where tests at other than ambient temperature are to be carried out, a list of preferred temperatures is provided in ISO 3205 [19]. This covers the temperature range from −269°C to +1000 °C and is not written specifically for polymeric materials. ISO 471 [7], reflecting the needs of elastomers in particular employs a smaller range of temperatures (−85° to +300°C) that do not always precisely match the more general ISO list, and the tolerances tend to be tighter. ASTM D618 [8] and D1349 [20] also provide lists of temperatures that, again, do not always accord with the ISO standards.

5 Apparatus for Conditioning

5.1 Air-Conditioned Rooms

Since test methods for polymers almost invariably require the test to be performed under reasonably tightly controlled temperatures, it is necessary for the testing laboratory to have appropriate air-conditioning if the precise requirements of the standard are to be met throughout the working day. Many tests involve testing to be performed overnight and over the weekend when laboratory staff are not present so that automatic temperature control then becomes essential. Specialists in heating systems should be consulted when considering the installation of air-conditioning in order to give as uniform a temperature throughout the working area as possible, especially taking into account the heat load from equipment. Even so, it is important for them to appreciate that the tolerances are to be maintained at all times and not just most of the time—requirements for the testing laboratory are much more stringent than those for the average office environment. Provided the laboratory does not have too many windows and outside doors (and even inside doors should be kept closed as much as possible), the provision of good temperature control need not be an excessively expensive option.

In some cases it may be necessary to include humidity control as well. This is invariably a more technically demanding requirement, which in turn carries a significant cost penalty. For this reason many laboratories in the polymer industry are not humidity controlled unless the laboratory is either regularly required to perform tests that are sensitive to moisture content, e.g., electrical tests, or is testing moisture-sensitive polymers, such as nylon. Where a temperature and humidity controlled room is to be provided, it is useful to have the controlled room situated within another room and with the minimum of windows and doors.

5.2 Enclosures

An alternative to having a temperature and humidity controlled room might be to use a cabinet that has these characteristics. Since polymers are generally slow to react to temperature and humidity changes, it may be sufficient to condition the test pieces at the appropriate temperature and humidity and then remove them to a less well controlled testing atmosphere and test them as quickly as possible. Careful reading of the standards of interest may be necessary to see if this is permitted and, where not, the agreement of the interested parties should be obtained to ensure that any deviation is acceptable. It is true to say that in many, though not all, situations, the temperature and humidity control requirements in standards are tighter than is strictly necessary for the outcome of the test, taking into account the inherent variability in the test method and the materials involved.

There are two types of humidity-controlled cabinet in general use: salt-tray cabinets and moisture-injection cabinets. The simpler are the salt-tray cabinets, in which saturated salts or standard solutions provide the appropriate humidity in the test space, which is at a controlled temperature. The design of these cabinets is critical if the humidity in particular is to be kept within the required tolerances throughout the working space. ISO 483 [21] contains useful information on the use of such cabinets, and although specifically written for plastics it is equally applicable to all polymeric materials. For the most normal test conditions for polymers, viz., 23°C and 50% relative humidity, a glycerol solution of refractive index 1.444 is proposed. The tolerance on the refractive index is very tight, being ±0.005 for a variation of ±5% relative humidity and ±0.002 for a relative humidity variation of ±2%.

The more sophisticated injection-humidity cabinet permits a wide variation in temperature and humidity to be created with a few simple settings of the controls. The humidity is measured by a suitable moisture sensor, such as a wet and dry bulb hygrometer or capacitive sensor, and this is used to control the injection of moisture into the chamber. Through the use of suitable control circuits it is also possible to cause such a chamber to cycle in temperature and/or humidity so that varying ambient conditions over wide extremes may be simulated for assessing such effects on the environmental resistance of polymer products.

5.3 Hygrometers

Dew point hygrometers are usually used as reference standards for measuring relative humidity, but for normal use in equipment and enclosures, capacitance/impedance instruments or wet and dry bulb thermometers are generally found. For the latter, platinum resistance thermometers are preferred, as they are very stable and robust. Even so, for these to operate accurately there should be an air flow over them of at least 3 m/s. BS 4833 [22] provides hygrometric tables for use with wet and dry bulb thermometers. Hair or paper hygrometers may prove of use in some instances because of the size and the relatively low cost, but it must be understood that they are generally very inaccurate and should never be used where precise humidity readings are needed or where long-term stability is required.

Typically, the tolerance on humidity control has been ±5% for the standard atmosphere and ±2% for tightly controlled atmospheres. In practice, however, there has probably been a good deal of lip service, or genuine ignorance, applied to these tolerances. It is now becoming recognized that ±2% is about at the limit of what the best calibration laboratories can achieve for the uncertainty of humidity sensor readings. On that basis, one would have to keep the humidity of the laboratory absolutely unchanging in order to ensure compliance with the ±2% humidity tolerance. ISO 291 [5], the general conditioning standard for plastics, is currently under revision, and a relaxation of the traditional tolerance values to ±5% and ±10% has been proposed, which probably reflects more reasonable practice for the typical testing laboratory.

5.4 Thermometers

There are many types of temperature-measuring instruments available, and, while the electronic versions are very widely used, the ordinary mercury-in-glass or alcohol-in-glass thermometers are still prevalent. There is more that can go wrong with a mercury-in-glass thermometer than is often appreciated, so there is a need to have them calibrated frequently, and the mercury thread should be examined for continuity before use. For accurate work the thermometer must be immersed to the correct depth. BS 593 [23] deals specifically with mercury-in-glass thermometers but does not cover the alcohol types, which may have to be used for certain low-temperature tests, since mercury freezes at about −39°C. Alcohol thermometers are not as accurate as mercury thermometers, however. Other standards covering the use of thermometers include ISO 653 [24] and 654 [25] and British standards BS 1704 [26] and 5074 [27].

The sensing elements for electronic thermometers vary, with the most widely used being thermocouples or platinum resistance thermometers. The latter are more stable and linear but are less robust than the former, and the temperature range they can cover is not as great since changing the metal combinations in the thermocouple enables very wide temperature ranges to be achieved. As noted for the liquid-in-glass thermometers, frequent calibration is still a requirement. The various parts of BS 1041 [28] give guidance into the selection and use of thermometers of various types, and there is also an ASTM manual on the use of thermocouples.

5.5 Apparatus for Elevated and Subambient Temperature

If a test is to be carried out at a nonambient temperature, then conditioning of the test pieces at that temperature takes place prior to the test being performed. Generally this occurs in a test chamber that is attached to the test machine itself and is an integral part of the machine. However, there are times when conditioning has to be carried out in a separate chamber and then the sample removed and tested as quickly as possible. An example of the latter is the low-temperature pendulum impact testing of plastic bars, where having the whole instrument in the cold chamber would risk the freezing up of the pendulum bearings, thereby influencing the outcome of the test.

The Rapra Guide to Test Equipment [29] offers comment on the types of enclosure available, and any particular requirement for a test will be indicated in the relevant section of this book. A number of standards have been published that deal with the subject of conditioning at elevated and subambient temperatures. ISO 3205 [19] lists the preferred test temperatures for use with all materials. As such it covers a range that is excessive for polymeric materials, as noted earlier, but it does provide a rationalization for choosing a consistent series of nonambient temperatures. ISO 3383 [30] on the other hand covers both elevated and subambient conditions for rubbers only. Within ASTM there are standards D832 [31] and D3847 [32], which cover requirements for low-temperature conditioning and testing.

When nonambient temperature conditioning is applied prior to testing, there is clearly the need to condition for long enough to allow temperature stabilization to be achieved. This time will depend on several factors, the major ones being temperature difference between the ambient and the conditioning temperature, the dimensions of the test specimen, the surface heat transfer coefficient, and the thermal diffusivity of the material. Fortunately, for normal conditioning purposes, these factors do not need to be known with great precision, and tables are available that give approximate conditioning times for various geometries, temperature gradients, and either gaseous or fluid heat exchange media. Tables 1 to 3 provide a reasonably comprehensive selection for the commonly encountered geometries.

However, one must be clear whether the test is looking to measure the effect of temperature alone or whether structural changes in the molecular morphology are also to be examined. This is generally a low temperature concern. Prolonged elevated-temperature conditioning is normally associated with heat aging effects. At low temperatures, polymers may crystallize or undergo other reversible structural changes, and these may take one or two orders of magnitude longer to reach equilibrium than the temperature. When undertaking such tests, therefore, the details of the test method or specification should be carefully followed, or quite different results may arise. Both these effects are considered further in Chapter 13

6 Mechanical Conditioning

This is a features generally associated with particular rubber tests, it being known from the work of Mullins [33,34] that elastomers containing fillers have their stress–strain behavior modified when they are deformed. Experience shows that repeated deformation under the same constraints leads to an equilibrium stress–strain curve being produced. Given sufficient time, the filler–rubber structure can rebuild, and there is a return towards the original stress–strain behavior. Standard test methods that require some form of mechanical conditioning prior to test indicate the precise conditioning required, and this should be followed if consistent and reproducible data are to be produced. Further information can be found in Ref. [35]

While mechanical conditioning is most commonly associated with elastomers, thermoplastics and composites are not excluded. Reference [36] for example looks at the influence of mechanical conditioning on the viscoelastic behavior of glass-fiber-reinforced epoxy resins.

Table 1 Thermal Equilibrium Times for Cylinders

			Time to 1°C off equilibrium (min)					
			Rubber		Crystalline plastic		Amorphous plastic	
Diameter (mm)	Height (mm)	Temperature (°C)	in air	in oil	in air	in oil	in air	in oil
64	38	−50	130	75	135	60	130	80
		0	95	60	100	45	95	65
		50	105	65	115	50	105	70
		100	130	80	140	60	130	85
		150	145	85	155	65	145	90
		200	155	90	165	70	155	95
		250	160	90	170	75	160	100
40	30	−50	75	35	85	30	75	40
		0	55	30	60	25	55	35
		50	60	30	70	25	60	35
		100	75	35	85	30	75	45
		150	85	40	95	35	85	45
		200	90	45	100	35	90	50
		250	95	45	105	40	90	50
37	10.2	−50	35	10	40	10	35	10
		0	25	10	30	10	25	10
		50	30	10	35	10	25	10
		100	35	10	40	10	35	10

Table 1 Thermal Equilibrium Times for Cylinders (*continued*)

Diameter (mm)	Height (mm)	Temperature (°C)	Time to 1°C off equilibrium (min)					
			Rubber		Crystalline plastic		Amorphous plastic	
			in air	in oil	in air	in oil	in air	in oi
		150	40	10	45	10	35	10
		200	40	10	50	10	40	15
		250	45	10	50	10	40	15
32	16.5	−50	45	15	50	15	45	20
		0	35	15	40	10	30	15
		50	35	15	45	15	35	15
		100	45	20	55	15	45	20
		150	50	20	60	15	50	20
		200	55	20	65	20	50	20
		250	55	20	65	20	55	25
29	25	−50	50	20	60	20	50	25
		0	40	15	45	15	40	20
		50	45	20	50	15	40	20
		100	55	25	60	20	50	25
		150	60	25	70	20	55	25
		200	65	25	70	25	60	30
		250	65	25	75	25	65	30
28.7	12.7	−50	35	10	40	10	35	15
		0	25	10	30	10	25	10
		50	30	10	35	10	30	10
		100	35	15	45	10	35	15
		150	40	15	50	10	40	15
		200	45	15	50	15	40	15
		250	45	15	55	15	40	15
25	20	−50	40	15	50	15	40	20
		0	30	15	35	10	30	15
		50	35	15	40	10	35	15
		100	45	15	50	15	40	20
		150	45	20	55	15	45	20
		200	50	20	60	15	50	20
		250	50	20	60	15	50	20
25	8	−50	25	5	30	5	25	10
		0	20	5	20	5	20	5
		50	25	5	30	5	25	10
		150	30	10	35	5	25	10
		200	30	10	35	10	30	10
		250	30	10	35	10	30	10
25	6.3	−50	20	5	25	5	20	5
		0	15	5	20	5	15	5
		50	20	5	20	5	15	5
		100	20	5	25	5	20	5
		150	25	5	30	5	20	5
		200	25	5	30	5	25	5
		250	25	5	30	5	25	5
13	12.6	−50	20	5	25	5	20	5
		0	15	5	20	5	15	5

Table 1 Thermal Equilibrium Times for Cylinders (*continued*)

Diameter (mm)	Height (mm)	Temperature (°C)	Time to 1°C off equilibrium (min)					
			Rubber		Crystalline plastic		Amorphous plastic	
			in air	in oil	in air	in oil	in air	in oil
		50	20	5	20	5	15	5
		100	20	5	30	5	20	10
		150	25	10	30	5	25	10
		200	25	10	30	5	25	10
		250	25	10	35	5	25	10
13	6.3	−50	15	5	20	5	15	5
		0	10	5	15	5	10	5
		50	15	5	15	5	15	5
		100	15	5	20	5	15	5
		150	20	5	20	5	15	5
		200	20	5	25	5	20	5
		250	20	5	25	5	20	5
9.5	9.5	−50	15	5	5	5	15	5
		0	10	5	5	5	10	5
		50	15	5	5	5	10	5
		100	15	5	5	5	15	5
		150	20	5	5	5	15	5
		200	20	5	5	5	15	5
		250	20	5	5	5	20	5

Table 2 Thermal Equilibrium Times for Flat Sheets

Thickness (mm)	Temperature (°C)	Time to 1°C off equilibrium (min)					
		Rubber		Crystalline plastic		Amorphous plastic	
		in air	in oil	in air	in oil	in air	in oil
25	−50	135	90	115	80	145	100
	0	95	75	80	65	105	85
	50	110	80	90	70	120	90
	100	140	90	115	80	150	100
	150	155	95	130	85	165	105
	200	160	100	135	85	180	110
	250	170	105	140	90	185	115
15	−50	70	35	60	30	80	40
	0	50	30	40	25	55	30
	50	60	30	45	30	65	35
	100	75	35	60	30	80	40
	150	80	40	65	35	90	40
	200	85	40	70	35	95	40
	250	90	40	75	35	100	45
10	−50	45	15	35	15	50	20
	0	30	15	25	15	35	15
	50	35	15	30	15	40	15

Table 2 Thermal Equilibrium Times for Flat Sheets (*continued*)

Thickness (mm)	Temperature (°C)	Time to 1°C off equilibrium (min)					
		Rubber		Crystalline plastic		Amorphous plastic	
		in air	in oil	in air	in oil	in air	in oil
	100	45	20	40	15	50	20
	150	50	20	40	15	55	20
	200	55	20	45	15	60	20
	250	55	20	45	20	60	20
8	−50	35	10	30	10	40	15
	0	25	10	20	10	30	10
	50	30	10	25	10	30	10
	100	35	10	30	10	40	15
	150	40	10	35	10	45	15
	200	40	15	35	10	45	15
	250	45	15	35	15	50	15
5	−50	20	5	20	5	25	5
	0	15	5	15	5	20	5
	50	20	5	15	5	20	5
	100	20	5	20	5	25	5
	150	25	5	20	5	25	5
	200	25	5	20	5	30	5
	250	25	10	20	10	30	5
3	−50	15	5	10	5	15	5
	0	10	5	10	5	10	5
	50	10	5	10	5	15	5
	100	15	5	10	5	15	5
	150	15	5	15	5	15	5
	200	15	5	15	5	20	5
	250	15	5	15	5	20	5
2	−50	10	5	10	5	10	5
	0	10	5	5	5	10	5
	50	10	5	5	5	10	5
	100	10	5	10	5	10	5
	150	10	5	10	5	10	5
	200	10	5	10	5	15	5
	250	10	5	10	5	15	5
1	−50	5	5	5	5	5	5
	0	5	5	5	5	5	5
	50	5	5	5	5	5	5
	100	5	5	5	5	5	5
	150	5	5	5	5	5	5
	200	5	5	5	5	5	5
	250	5	5	5	5	10	5
0.2	−50	5	5	5	5	5	5
	0	5	5	5	5	5	5
	50	5	5	5	5	5	5
	100	5	5	5	5	5	5
	150	5	5	5	5	5	5
	200	5	5	5	5	5	5
	250	5	5	5	5	5	5

Table 3 Thermal Equilibrium Times for Flat Strips

Width (mm)	Thickness (mm)	Temperature (°C)	Time to 1 °C off equilibrium (min)					
			Rubber		Crystalline plastic		Amorphous plastic	
			in air	in oil	in air	in oil	in air	in oil
25.4	12.7	−50	45	15	50	10	40	15
		0	30	10	35	10	30	10
		50	35	10	40	10	35	15
		100	45	15	55	10	40	15
		150	50	15	60	15	45	15
		200	50	15	60	15	50	15
		250	55	15	65	15	50	20
25.4	10.0	−50	35	10	45	10	35	10
		0	25	10	30	5	25	10
		50	30	10	35	10	30	10
		100	35	10	45	10	35	10
		150	40	10	50	10	40	10
		200	40	10	50	10	40	15
		250	45	10	55	10	40	15
25.4	9.5	−50	35	10	40	10	30	10
		0	25	10	30	5	25	10
		50	30	10	35	10	25	10
		100	35	10	40	10	35	10
		150	40	10	45	10	35	10
		200	40	10	50	10	40	10
		250	40	10	50	10	40	10
25.4	6.5	−50	25	5	30	5	25	5
		0	20	4	20	5	15	5
		50	20	5	25	5	20	5
		100	25	5	30	5	25	5
		150	30	5	35	5	25	5
		200	30	5	35	5	25	5
		250	30	5	40	5	30	10
25.4	5.0	−50	20	5	25	5	20	5
		0	15	5	20	5	15	5
		50	15	5	20	5	15	5
		100	20	5	25	5	20	5
		150	20	5	30	5	20	5
		200	25	5	30	5	20	5
		250	25	5	30	5	25	5
25.4	3.0	−50	15	5	15	5	10	5
		0	10	5	10	5	10	5
		50	10	5	15	5	10	5
		100	15	5	15	5	10	5
		150	15	5	20	5	15	5
		200	15	5	20	5	15	5
		250	15	5	20	5	15	5
25.4	2.0	−50	10	5	10	5	10	5
		0	10	5	10	5	5	5
		50	10	5	10	5	10	5
		100	10	5	10	5	10	5

Table 3 Thermal Equilibrium Times for Flat Strips (*continued*)

			Time to 1 °C off equilibrium (min)					
			Rubber		Crystalline plastic		Amorphous plastic	
Width (mm)	Thickness (mm)	Temperature (°C)	in air	in oil	in air	in oil	in air	in oil
		150	10	5	15	5	10	5
		200	10	5	15	5	10	5
		250	10	5	15	5	10	5
25.4	1.0	−50	5	5	5	5	5	5
		0	5	5	5	5	5	5
		50	5	5	5	5	5	5
		100	5	5	5	5	5	5
		150	5	5	10	5	5	5
		200	5	5	10	5	5	5
		250	5	5	10	5	5	5
15.0	15.0	−50	35	10	45	10	35	15
		0	30	10	35	10	25	10
		50	30	10	35	10	30	10
		100	40	10	45	10	35	15
		150	40	15	50	10	40	15
		200	45	15	55	15	40	15
		250	45	15	55	15	45	15
12.7	12.7	−50	30	10	35	10	30	10
		0	25	10	25	5	20	10
		50	25	10	30	10	25	10
		100	30	10	40	10	30	10
		150	35	10	40	10	35	10
		200	35	10	45	10	35	10
		250	40	10	45	10	35	10
12.7	10.0	−50	25	10	35	5	25	10
		0	20	5	25	5	20	5
		50	20	5	25	5	20	5
		100	30	10	35	5	25	10
		150	30	10	35	10	30	10
		200	30	10	40	10	30	10
		250	35	10	40	10	30	10
12.7	9.5	−50	25	10	30	5	25	10
		0	20	5	25	5	20	5
		50	20	5	25	5	20	5
		100	25	10	35	5	25	10
		150	30	10	35	10	30	10
		200	30	10	40	10	30	10
		250	35	10	40	10	30	10
12.7	6.5	−50	20	5	25	5	20	5
		0	15	5	20	5	15	5
		5	15	5	20	5	15	5
		100	20	5	25	5	20	5
		150	25	5	30	5	20	5
		200	25	5	30	5	25	5
		250	25	5	30	5	25	5
12.7	5.0	−50	15	5	20	5	15	5
		0	15	5	15	5	10	5

Table 3 Thermal Equilibrium Times for Flat Strips (*continued*)

Width (mm)	Thickness (mm)	Temperature (°C)	Time to 1 °C off equilibrium (min)					
			Rubber		Crystalline plastic		Amorphous plastic	
			in air	in oil	in air	in oil	in air	in oil
		50	15	5	156	5	15	5
		100	20	5	20	5	15	5
		150	20	5	25	5	20	5
		200	20	5	25	5	20	5
		250	20	5	25	5	20	5
12.7	3.2	−50	10	5	15	5	10	5
		0	10	5	10	5	10	5
		50	10	5	15	5	10	5
		100	10	5	15	5	10	5
		150	15	5	15	5	10	5
		200	15	5	20	5	15	5
		250	15	5	20	5	15	5
12.7	3.0	−50	10	5	15	5	10	5
		0	10	5	10	5	10	5
		50	10	5	10	5	10	5
		100	10	5	15	5	10	5
		150	15	5	15	5	10	5
		200	15	5	15	5	15	5
		250	15	5	20	5	15	5
12.7	2.0	−50	10	5	10	5	10	5
		0	5	5	10	5	5	5
		50	10	5	10	5	5	5
		100	10	5	10	5	10	5
		150	10	5	10	5	10	5
		200	10	5	15	5	10	5
		250	10	5	15	5	10	5
12.7	1.0	−50	5	5	5	5	5	5
		0	5	5	5	5	5	5
		50	5	5	5	5	5	5
		100	5	5	5	5	5	5
		150	5	5	10	5	5	5
		200	5	5	10	5	5	5
		250	5	5	10	5	5	5
6.35	12.7	−50	20	5	25	5	20	5
		0	15	5	20	5	15	5
		50	15	5	20	5	15	5
		100	20	5	25	5	20	5
		150	25	5	30	5	20	5
		200	25	5	30	5	20	5
		250	25	5	30	5	25	5
6.35	10.0	−50	20	5	25	5	15	5
		0	15	5	15	5	15	5
		50	15	5	20	5	15	5
		100	20	5	25	5	20	5
		150	20	5	25	5	20	5
		200	20	5	25	5	20	5

Table 3 Thermal Equilibrium Times for Flat Strips (*continued*)

Width (mm)	Thickness (mm)	Temperature (°C)	Time to 1 °C off equilibrium (min)					
			Rubber		Crystalline plastic		Amorphous plastic	
			in air	in oil	in air	in oil	in air	in oil
		250	25	5	30	5	20	5
6.35	6.5	−50	15	5	20	5	15	5
		0	10	5	15	5	10	5
		50	15	5	15	5	10	5
		100	15	5	20	5	15	5
		150	15	5	20	5	25	5
		200	20	5	25	5	15	5
		250	20	5	25	5	20	5
6.35	5.0	−50	15	5	15	5	15	5
		0	10	5	15	5	10	5
		50	10	5	15	5	10	5
		100	15	5	15	5	15	5
		150	15	5	20	5	15	5
		200	15	5	20	5	15	5
		250	15	5	20	5	15	5
6.35	3.0	−50	10	5	15	5	10	5
		0	10	5	10	5	10	5
		50	10	5	10	5	10	5
		100	10	5	15	5	10	5
		150	10	5	15	5	10	5
		200	10	5	15	5	10	5
		250	10	5	15	5	10	5
6.35	2.0	−50	10	5	10	5	10	5
		0	5	5	10	5	5	5
		50	5	5	10	5	5	5
		100	10	5	10	5	10	5
		150	10	5	10	5	10	5
		200	10	5	10	5	10	5
		250	10	5	10	5	10	5
6.35	1.52	−50	5	5	10	5	5	5
		0	5	5	5	5	5	5
		50	5	5	5	5	5	5
		100	5	5	10	5	5	5
		150	10	5	10	5	5	5
		200	10	5	10	5	10	5
		250	10	5	10	5	10	5
6.35	1.0	−50	5	5	5	5	5	5
		0	5	5	5	5	5	5
		50	5	5	5	5	5	5
		100	5	5	5	5	5	5
		150	5	5	5	5	5	5
		200	5	5	10	5	5	5
		250	5	5	10	5	5	5
4.0	12.7	−50	15	5	20	5	15	5
		0	10	5	15	5	10	5
		50	10	5	15	5	10	5

Table 3 Thermal Equilibrium Times for Flat Strips (*continued*)

Width (mm)	Thickness (mm)	Temperature (°C)	Time to 1 °C off equilibrium (min)					
			Rubber		Crystalline plastic		Amorphous plastic	
			in air	in oil	in air	in oil	in air	in oil
		100	15	5	20	5	15	5
		150	15	5	20	5	15	5
		200	15	5	20	5	15	5
		250	20	5	20	5	15	5
4.0	10.0	−50	15	5	15	5	15	5
		0	10	5	15	5	10	5
		50	10	5	15	5	10	5
		100	15	5	15	5	15	5
		150	15	5	20	5	15	5
		200	15	5	20	5	15	5
		250	15	5	20	5	15	5
4.0	6.5	−50	10	5	15	5	10	5
		0	10	5	10	5	10	5
		50	10	5	10	5	10	5
		100	10	5	15	5	10	5
		150	15	5	15	5	15	5
		200	15	5	20	5	15	5
		250	15	5	20	5	15	5
4.0	5.0	−50	10	5	15	5	10	5
		0	10	5	10	5	10	5
		50	10	5	10	5	10	5
		100	10	5	15	5	10	5
		150	10	5	15	5	10	5
		200	15	5	15	5	10	5
		250	15	5	15	5	10	5
4.0	3.0	−50	10	5	10	5	10	5
		0	5	5	10	5	5	5
		50	10	5	10	5	5	5
		100	10	5	10	5	10	5
		150	10	5	10	5	10	5
		200	10	5	15	5	10	5
		250	10	5	15	5	10	5
4.0	2.0	−50	5	5	10	5	5	5
		0	5	5	5	5	5	5
		50	5	5	10	5	5	5
		100	10	5	10	5	5	5
		150	10	5	10	5	10	5
		200	10	5	10	5	10	5
		250	10	5	10	5	10	5
4.0	1.0	−50	5	5	5	5	5	5
		0	5	5	5	5	5	5
		50	5	5	5	5	5	5
		100	5	5	5	5	5	5
		150	5	5	5	5	5	5
		200	5	5	5	5	5	5
		250	5	5	5	5	5	5

References

1. ISO 471, Rubber—Temperatures, humidities and times for conditioning and testing, 1995.
2. BS 2782, Method 365A, Determination of softness number of flexible plastics materials, 1976.
3. ISO 558, Conditioning and testing—Standard atmospheres—Definitions, 1980.
4. ISO 554, Standard atmospheres for conditioning and/or testing—Specifications, 1976.
5. ISO 291, Plastics—Standard atmospheres for conditioning and testing, 1977.
6. BS 2782, Part 0, Annex A, Plastics—Introduction—Standard atmospheres for conditioning and testing plastics, 1995.
7. BS 903, Part A35, Rubber—Temperatures, humidities and times for conditioning and testing, 1995.
8. ASTM D618, Standard practice for conditioning plastics and electrical insulating materials for testing, 1961.
9. BS EN ISO 845, Cellular plastics and rubbers—Determination of apparent (bulk) density, 1995.
10. BS EN ISO 2231, Rubber or plastics coated fabrics—Standard atmospheres for conditioning and testing, 1995.
11. EN 62, Glass reinforced plastics—Standard atmospheres for conditioning and testing, 1997.
12. BS 2782, Method 1004, Plastics—Glass reinforced plastics—Standard atmospheres for conditioning and testing, 1997.
13. ISO 1110, Plastics—Polyamides—Accelerated conditioning of test specimens, 1995.
14. BS 4443, Methods of test for flexible cellular materials, various dates.
15. BS EN ISO 1923, Cellular plastics and rubbers—Determination of linear dimensions, 1995.
16. BS 4443, Method 3, Flexible cellular materials—Determination of tensile strength and elongation at break, 1988.
17. BS 4443, Method 8, Flexible cellular materials—Determination of creep, 1988.
18. BS 4370, Methods of test for rigid cellular materials, various dates.
19. ISO 3205, Preferred test temperatures, 1976.
20. ASTM D1349, Rubber—Standard temperatures for testing, 1987.
21. ISO 483, Plastics—Small enclosures for conditioning and testing using aqueous solutions to maintain relative humidity at constant value, 1988.
22. BS 4833, Schedule for hygrometric tables for use in the testing and operation of environmental enclosures, 1986.
23. BS 593, Specification for laboratory thermometers, 1989.
24. ISO 653, Temperature measuring instruments—Long solid-stem thermometers for precision use, 1980.
25. ISO 654, Temperature measuring instruments—Short solid-stem thermometers for precision use, 1980.
26. BS 1704, Specification for solid-stem general purpose thermometers, 1985.
27. BS 5074, Specification for short and long solid-stem thermometers for precision use, 1974.
28. BS 1041, Temperature measurement, various dates.
29. Rapra, *Rapra Guide to Test Equipment*, Rapra Technology, n.d.
30. ISO 3383, Rubber—General directions for achieving elevated or sub-normal temperatures for test purposes, 1985.
31. ASTM D832, Practice for rubber—Conditioning for low temperature testing, 1992.
32. ASTM D3847, Practice for rubber—Direction for achieving sub-normal test temperatures, 1991.
33. Mullins, L., *Trans. IRI*, *23*, 280 (1948).
34. Mullins, L., RABRM Research Memo R.342, 1948.
35. Brown, R. P., *Physical Testing of Rubber* 3d ed., Chapman and Hall, 1996.
36. Donoghue R. D., et al., *Composites Sci. Technol.*, *44*(1), 43–55 (1992).

7
Mass, Density, and Dimensions

Roger Brown
Rapra Technology Ltd., Shawbury, Shrewsbury, England

1 Introduction

Mass, density, and dimensions have been grouped together largely for convenience, but there is an obvious connection between them in that density can be derived from a knowledge of dimensions and mass. They are also measurements used as essential parts of other physical tests. For example, density is used to calculate volume loss in an abrasion test, mass is an intrinsic factor in water absorption tests, and there are very few tests that do not in some way involve the measurement of dimensions.

Mass, dimensions, and density are also important in the factory, being factors in the costing of products and in quality control, from routine checking on dimensional accuracy of components to a simple control measure for the consistency of polymer compounds.

It follows that these measurements also have a connection in being amongst the most frequently used. Measurements that are made every day have a habit of being taken for granted, and this can certainly happen to the measurement of dimensions, resulting in unnecessary errors. When one considers that in, for example, the determination of tensile strength, any error in the measurement of the cross section results directly in an equivalent percentage error in the strength measurement, it is reasonable to devote considerable attention to the seemingly simple matter of measuring the width and thickness.

Mass and dimensions need no definition here, but it should be noted that in test methods a mass is often used to produce a force, and the term weight tends to be used indiscriminately. Using SI units there should not be any cause for confusion.

Density is mass per unit volume (at a defined temperature). Relative density is mass (of substance) compared with the mass of an equal volume of a reference substance (usually water) and being a ratio is dimensionless. It is relative density that is often the property measured, but in the usual units (Mg/m^3) it is normally adequate to take the density of water as 1. Furthermore, the determination is often made by observation of gravitational forces, but for convenience the forces are expressed in mass units. Relative density used to be commonly known as specific gravity, but this term is now deprecated and should not be used. Apparent density is the term used when the density of, for example, a powder is measured from mass and dimensions that include the voids between particles.

2 Measurement of Mass

The measurement of mass is quite simply a matter of weighing the test piece, sample, or whatever using an appropriate balance or scales. The accuracy needed varies according to the purpose, and it is essential that the instrument used be selected accordingly. In particular, it should be appreciated that reading to, for example, 1 mg is not the same as accurate to 1 mg, and standards are not always explicit.

3 Measurement of Density

The commonest method of determination is by weighing in air and water. The standard procedure for rubbers is given in ISO 2781 [1], method A, and specifies a test piece weighing a minimum of 2.5 g, which can be of any shape as long as the surfaces are smooth and there are no crevices to trap air. The test piece is weighed in air and then in water using a balance accurate to 1 mg. The best way of suspending the test piece is by means of a very fine filament, the weight of which can be included in the zero adjustment of the balance and its volume in water ignored. However, if smaller than standard test pieces are used, the effect of the filament could be significant. A top pan balance is not suitable, and it should be noted that the requirement is for accuracy to 1 mg not reading to 1 mg. It is permissible to wet the test piece with a liquid such as methylated spirit before weighing in water to eliminate air bubbles, and this is indeed common practice. The water then needs to be changed relatively frequently because of contamination by the alcohol. For most purposes the density is quoted to 0.01 Mg/m.

If the rubber is less dense than water, a less dense liquid of known density could be substituted, but it is more usual to attach a sinker to the test piece. The sinker can conveniently be a small piece of lead, but using an item like a paper clip to suspend the test piece leads to complications, as it will only be partly submerged. The weight of the sinker in water must be measured, and it is a common error among new technicians to make this weighing in air.

ISO 2781 also details a procedure (method B) for use when it is necessary to cut the sample into small pieces to avoid trapped air, as might happen with narrow bore tubing. The test piece comprises a number of small smooth pieces within the size 4 mm × 4 mm × 6 mm. These are weighed in a density bottle both with and without the remaining space filled with water. The bottle is also weighed without rubber both empty and filled with water. This is a more tedious procedure than method A and is generally only used as a last resort. Even with the test piece cut up, trapped air can still be a problem.

The British equivalent to ISO 2781 is BS 903, Part A1 [2], which is identical to the international method. Rather surprisingly, ASTM does not appear to have a specific method for density at the present time. There is however a section on density in the method on chemical analysis (D297) [3] and also a separate method for determining the density of rubber chemicals (D1817) [4].

Methods for plastics are given in ISO 1183 [5]. There is an immersion method and a pycnometer very similar to those for rubbers but with some differences in detail and emphasis, including very tight control of temperature, which reflects the closer tolerances needed on density for some plastics.

A third method in ISO 1183 utilizes two miscible liquids of different densities, one having a lower density and one a higher density than the test material. The test piece is introduced into a glass cylinder containing a quantity of the first liquid, and then the second liquid is buretted in until (after stirring) the test piece "floats" in equilibrium with the mixture. The density can be deduced from the relative quantities of the two liquids. A cruder version of this approach was previously specified, which simply used a series of liquids and a pair found for which the test piece just sinks in one and just floats in the other.

The fourth ISO 1183 method uses a density column, and this procedure is sometimes also used for rubbers if greater accuracy than that provided by the standard rubber methods discussed above is required. The principle of the method is that two miscible liquids of different densities can be run into a container so that a uniform density gradient from the bottom to the top of the container results. The container is normally a glass tube of not less than 40 mm diameter in a thermostatted jacket. This column can then be calibrated by floats of known density, which will come to rest at the depth in the column where their densities equal that of the immediate surrounding liquid. Small test pieces are then introduced into the column in the same manner and allowed to come to rest, their height in the column is then measured and their density deduced from a calibration graph. With care, a column will last several months, and the range of density in a single column would not normally be greater than 0.2 Mg/m but could be as little as 0.02 Mg/m. Ten minutes is suggested as the minimum time to allow test pieces to come to equilibrium, but a large number of samples can be tested at one time, and only a very small sample is required.

The British standards BS 2782, Methods 620 A-D [6], are identical to ISO 1183. ASTM D 792 [7] covers the displacement method and has two procedures, one for the displacement of water and one for that of other liquids. It is not clear why it has been split in this way, and it is notable that there is no mention of using a sinker, nor does it include a pycnometer method. The density gradient method is given in ASTM D 1505 [8] and is very similar to the ISO procedure. There is also an ASTM method for density of polyethylene by means of ultrasound, ASTM D 4883 [9]. This works on the principle of measuring sound velocity in the plastic, which correlates to density. The apparatus requires calibrating with reference materials but is claimed to give accuracies of 0.08% or better. The use of the method would mostly be in quality control, and it is questionable whether it should have been standardized. Essentially it describes the use of a commercial instrument with no apparatus details, not even the frequency used.

Because density is often used as a quality control check, particularly on batches of rubber compound, there has been a necessity to make measurements essentially in accordance with standards such as ISO 2781 but making the determinations as rapidly as possible. Hence various designs of "specific gravity balance" are in existence, primarily intended for rubbers, which to varying degrees automate the process. In one form of apparatus, the practical steps of weighing in air and water are taken, but the result may then be read directly from a scale calibrated in density. Completely automatic apparatus have also been produced.

If the density of cellular materials is attempted to be measured by immersion, then unless the test piece is completely covered in a solid skin the result will depend on the degree of water uptake into the cells. The standard method in ISO 845 [10] finds what it calls apparent or bulk density by weighing the test piece and measuring its dimensions. With cellular materials, distinction can be made between overall density, including any skins formed during manufacture, and core density with all skins removed, the former varying according to the percentage of area covered. With very low density foams the standard advises on making a correction to the weight determination for air buoyancy.

The same standard is given in BS EN ISO 845 [11], British, European, and ISO all being the same. There are five ASTM standards giving methods for density of cellular materials. They all use the weighing and measurement of dimensions principle. The method for rigid materials is sensibly contained in one standard, ASTM D 1622 [12], and this is essentially the same as ISO 845 (except that flexible materials are excluded). Methods for flexible materials are spread over ASTM D 1565 [13], ASTM D 1667 [14], ASTM D 3574 [15], and ASTM D 3575 [16] according to polymer type and cell structure. D1565 only makes measurements without any skin, D 1667 makes no mention of skin, and D3574 measures either core or a section (skin top and bottom).

The procedure of weighing and obtaining volume from measurement of dimensions could of course be used for solid rubbers and plastics, but accuracy is limited (as it is for foams) by the dimensions measurement, and it is inconvenient to go to great lengths to get a perfectly regular test piece.

The measurement of apparent or bulk density of powders is covered in ISO 60 [17] and ISO 61 [18]. The first is a procedure for powders that can be poured from a funnel. A funnel of the form shown in Fig. 1 is mounted vertically with its lower end 20–30 mm above the top of a measuring cylinder of 100 ml capacity and internal diameter 40–50 mm. With the lower orifice closed, 110–120 ml of well-mixed powder is poured into the funnel, and then the powder is allowed to flow into the measuring cylinder, assisted if necessary by being loosened with a rod. When the cylinder is full, a straight bladed knife is drawn across the top of of the cylinder to remove excess, and then the contents are weighed.

For materials that cannot be poured from a funnel, one uses a cylinder of 1000 ml capacity and internal diameter 90 ± 2 mm. A plunger of slightly smaller diameter and total mass 2300 g fits into the cylinder. Sixty grams of the powder is dropped, little by little, into the cylinder so that it is evenly distributed and has a level surface. The plunger is lowered onto the powder and rests there for 1 min before the height of the powder is measured. From the height of powder, the diameter of the cylinder, and the weight of the powder, the apparent density can be calculated.

The British methods, BS 2782, Methods 621 A and B [19,20], are identical to ISO 60 and 61. ASTM D 1895 [21] has three procedures. Methods A and B are similar to ISO 60 but have different funnel geometries, and slightly different results can be expected. The relation between method A and ISO 60 is given in an appendix. Method C is very similar to ISO 61.

ISO 1068 [22] measures the compacted bulk density of PVC resins with a cylinder method that uses a shaker to tamp down the material under a piston. BS 2782 Method 621D [23] is identical.

The bulk factor of a molding is defined as the ratio of the volume of a given mass of molding material to its volume in molded form—the ratio of the density of the molding to its apparent density before molding. ISO 171 [24], BS 2782, Method 621 C [25], and the procedure in ASTM D 1895 all require determination of apparent powder density and molded density by the appropriate methods discussed above.

For thin films and coated fabrics it is common to use a measure of mass per unit area rather than density. A test piece of given dimensions is cut and weighed, as for example in

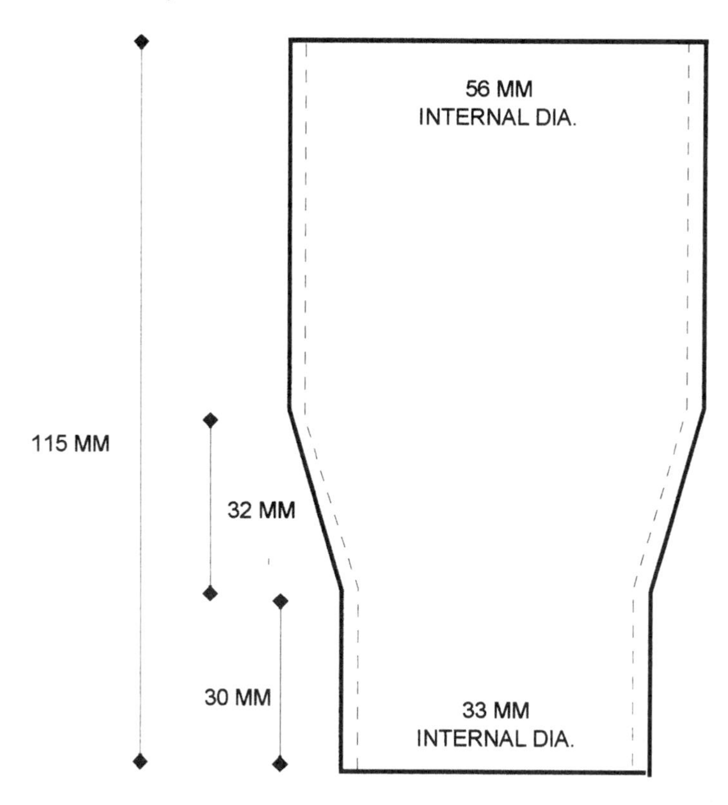

Figure 1 Form of funnel for apparent density.

ISO 2286 [26] for coated fabrics. For fibers it is convenient to measure mass per unit length, which is termed linear density (see Chapter 19).

4 Measurement of Dimensions

4.1 Standard Procedures

Virtually all physical test methods involve the measurement of test piece dimensions, and it has been common practice for each test method standard to specify the apparatus and means of making the dimensional measurements. Over the years the procedures became rationalized so that it was sensible that separate standards dealing specifically with dimensions should be produced, to which test method standards could refer.

For rubbers there is a general international standard, ISO 4648 [27], for test piece dimensions. It has four methods, dealing with dimensions less than 30 mm, dimensions over 30 mm, dimensions over 100 mm, and noncontact procedures respectively. BS 903, Part A38 [28], is identical.

The first method for dimensions under 30 mm specifies a gauge reading to 0.01 mm with a foot pressure of 22 ± 5 kPa (10 ± 2 kPa for hardness below 35 IRHD) acting on a plane flat foot. A table gives the nominal loading to achieve the pressure with various diameters of foot. The previous version of this standard had an annex that gave a diagram of a suitable apparatus involving a weight and a dial gauge with a lock such that the gauge spring pressure did not bear on the rubber. The more usual approach is to use an ordinary dial gauge minus the return spring but suitably loaded with a weight.

The previous edition also had a method intended for compression set test pieces, which was similar except that the force on the foot was 850 ± 30 mN, and the contact members were either domed surfaces or a spherical contact and a raised plane platform. This use of curved surfaces for compression set is based on the fact that after compression, particularly with nonlubricated test pieces, the rubber may well have concave surfaces. This does not happen if the test piece is lubricated, as is now the usual practice, and hence the curved surfaces were eliminated. However, if concave test pieces are encountered it may well be better to resort to the old method.

It is fairly obvious that the use of a different foot pressure would, with a soft and deformable material such as rubber, produce a different result. Hence, it is not permissible to use a dial gauge with a return spring, calipers, or a travelling microscope when this standard method is specified. The errors resultant on using different pressures have been reported by Clamroth and Dobroschke [29].

Dimensions such as the width of a dumbbell or the depth of a nick in a tear specimen will be less than 30 mm but could not be measured with a dial gauge. Because of the virtual impossibility of applying a known pressure, such measurements must be made in an essentially "contactless" manner. For low precision, calipers or a rule may suffice, but for readings to 0.01 mm a travelling microscope or projection microscope is most suitable, and this is specified in ISO 4648, Method D, and applies also to dimensions over 30mm. Projection microscopes also find use in examining profiles and for rapid swelling tests.

For the measurement of dimensions greater than 30 mm and up to 100 mm, ISO 4648 simply specifies vernier calipers capable of measuring with an error of not more than 1%, with the requirement that the test piece shall not be strained. In this context it may be noted that, although projection microscopes do not generally cover dimensions much greater than 30 mm, they can be used with a suitable jig to measure change in dimensions of a large test piece. The last method of ISO 4648 specifies a rule or tape for measuring dimensions greater than 100 mm with an accuracy of 1 mm.

There is not as yet a plastics equivalent to ISO 4648, but a draft is currently being circulated as a new work item. It is based on ASTM D5947 [30], which specifies micrometer calipers with or without a ratchet, a dead load dial gauge, and a motorized dial gauge. The standard specifies the instruments very fully, some of which specifications seem unnecessarily restrictive, and gives procedures for calibration. The procedures given for using the instruments go into astonishing detail.

By contrast, there is no advice on which method to use for a given purpose, and the standard oversteps its scope by discussing the effect of the force on the presser foot of a dial gauge on soft elastomers. One suspects that the authors have never read ISO 4648, and this perhaps illustrates that even in closely related industries test standards can diverge.

Rigid plastics do not have the problem of deforming under the measuring instrument, and in fact the reverse problem can occur using a dial gauge if the test piece being measured does not lie flat. For rigid materials a micrometer, which these days is generally digital, is most suitable for small dimensions. However, for softer plastic materials the same considerations apply as discussed for rubbers.

There are standards for dimensional measurement of plastic film, which are concerned with the product rather than for measuring test pieces. ISO 4591 [31] deals with gravimetric thickness, ISO 4592 [32] with length and width, and ISO 4593 [33] with thickness by mechanical scanning. The British standards [34–36] are identical.

In ISO 4591 a square or circle of material of area 100 cm^2 is cut from the film or sheet and weighed, and the density is determined so that the thickness can be calculated. In ISO 4582 procedures are given for measuring the length of a roll to 0.1 m and the width to 1 mm for widths over 100 mm or to 0.1 mm for widths less than 100 mm. The length measurement is made by laying the material on a flat surface 10 m long marked in meters and in 0.1 m for the last meter. The width is measured with a graduated scale, and to estimate tenths of a mm a magnifying glass with a graticule is employed.

In ISO 4593 three levels of accuracy of the measuring device are specified, 1 μm for thicknesses up to 100 μm, 2 μm from 100 up to 250 μm, and 3 μm for over 250 μm. The nature of the measuring device is not specified other than the foot diameter and force. For plane feet the diameter can be between 2.5 and 10 mm and the force between 0.5 and 1 N so that the tolerance on pressure is wide but the mean level is very low.

The international standard for dimensions of cellular materials [37] covers both rigid and flexible types. The European and British standards are identical. Measurements can be with a dial gauge, micrometer, sliding caliper, or tape depending on the size of the dimension, the type of cellular material, and the accuracy required. Advice is given on selection of instrument which distinguishes between the needs for rigid and flexible materials. The dial gauge procedure specifies a foot pressure of 100Pa, very much lower than for rubbers. Rather curiously, an annex gives an example of how this can be achieved that uses a screw micrometer together with an aluminum plate to apply the pressure and an electrical contact to detect when the micrometer tip touches the plate.

Methods for coated fabrics are akin to those for plastic film, there being methods for width and length of a roll [38], and the thickness [39]. Length is measured by progressively rolling out the fabric on a graduated surface and width with a scale wider than the fabric. If there is a selvedge of some kind then the useable width can be differentiated from the full width. Thickness is measured with a dial gauge and a choice of pressures, 2, 10, and 24 kPa, is given.

If a coated fabric is embossed or has an expanded layer, then sections can be cut and thickness determined with a microscope fitted with a graticule [40]. A similar approach can be taken for very thin coatings when it can be expedient to "pot" the sample in epoxy resin and microtome test pieces [41].

The dimensions of textile fibers are a critical factor in how the material processes and on the characteristics of the fabric produced. Hence dimensional measurement is more a material property concern than data required for other test methods. A variety of measurement techniques and terms are used and are detailed in Chapter 19.

For dimensions of rubber and plastics test pieces there has been considerable debate as to how many readings should be taken and what form of average should be used. ISO 4648 specifies at least three readings, taking the median. The draft ISO standard for plastics again specifies a minimum of three readings but takes the arithmetic mean. However, for a test such a tensile strength test there is merit in the notion that it is not the median thickness or width but the minimum thickness that is required. Practical difficulties probably make the adoption of the mean or median most sensible, but it can be expected that not all test methods will be totally in agreement.

No attempt will be made here to consider all the separate measurement clauses to be found in current test method standards. Until the ISO standards for the measurement of dimensions has become established long enough for all test methods to have been revised and reference it, each test method will have its own procedure, and there will not be universal agreement on detail. The essentials are to distinguish between a noncontact measurement and one applying a specified pressure, in the latter case to use the correct standard pressure, and to measure within the accuracy limits specified.

4.2 Non-Standard Methods

On-line inspection is a form of testing, and in this context dimensional measurements are those most often made. Apart from gauges, micrometers and so on, there are various optical, electrical, nuclear, and other methods that may have advantages in continuous production circumstances. Descriptions of the use of such techniques are extensively covered in journals and in manufacturers' literature. There is not space here to describe or review the individual techniques, but some fairly recent examples are given in Refs. 42–50. Methods having use in the laboratory have been reviewed in connection with swelling measurements [51,52].

Very often a great deal of dimensional information can be found by means of microscopy, which is such an important subject in its own right as to be in no way considered here as a branch of physical testing. For example, one would expect to employ a microscope to determine the thickness of a wax film on the surface of a rubber or to study the geometry of fibers or thin film, and much failure analysis involves detailed optical examination. There are inevitably a great number of special circumstances connected with polymers where an unusual type of dimensional measurement is required, such as the footprint area of tires or the crack length in fracture tests. A number of methods of interest will be mentioned in later chapters in conjunction with particular physical tests.

4.3 Surface Roughness

It is not often necessary to measure the surface roughness of polymer test pieces or products, and no standard methods exist. If measurements are attempted, either mechanical profiling as is standard with metals or possibly optical reflectance methods would be used. Generally, it is necessary to turn to methods established for metals, but only those that are suitable for, or can be adapted to take account of, the greater deformability of polymers, particularly rubbers and soft plastics, will be of potential value. One area where the surface finish is of great importance is in optical measurements, since light transmission and reflectance characteristics are very dependent on it. However, properties such as gloss and haze are measured rather than the surface roughness. Another area is friction, where the roughness of a surface may be measured to aid in the interpretation of the friction results. In rubber testing the surface finish of metals is of importance, for example on mold surfaces and compression set plates.

The ISO standards for surface finish are ISO468 [53] and ISO 4287, Parts 1 and 2 [54, 55]. The British equivalent is BS 1134. [56,57]. This British Standard is divided into two parts, the first concerning the method and instrumentation; the second forms a general explanation and is hence a good introduction to the subject. The parameter most often used to grade the roughness of a surface is Ra (previously known as CLA), the mean deviation of the surface profile above and below the center line. For example, for compression set of rubbers the arithmetic mean deviation (Ra) of the compression plates must be better than 0.2 m. There are, however, several other measures of texture covered by the standards.

4.4 Extensometry

The measurement of extension (or other mode of deformation) is an essential part of several tests, notably tensile or compression stress/strain properties and also thermal expansion. The range of measurement and the precision required depends not only on the type of deformation but on the material—there is clearly a big difference between a fiber reinforced composite and a soft rubber. Needless to say, the precision and range must be specified in the individual test method, and it is unlikely to be the same as that required for test piece dimensions. The method of measurement will also be to a considerable extent dependent on the test in question, and specific techniques may in some cases be given. Hence, the requirements for particular tests will be discussed in the relevant sections in later chapters.

4.5 Dimensional Stability

Dimensional stability can be taken to cover a fairly broad range of topics, including thermal expansion, shrinkage, softening point, and the effects of liquids, which overlaps with effects of temperature and the environment. This section will consider those methods that are specifically described as dimensional stability, leaving the others to the appropriate chapter later in the book.

Generally, vulcanized rubber is dimensionally very stable, which probably explains the lack of standard test methods for this property. Swelling in liquids (as well as thermal expansion) is not normally thought of as being a measure of dimensional stability, but nevertheless the effect of particular liquids can be very great. If a measure of dimensional change is required, other than by the standard swelling tests, the appropriate dimensions of a suitable sized test piece can be measured before and after an aging treatment by any of the methods mentioned in this chapter.

Although mold shrinkage of rubbers and plastics, i.e., the reduction in size of cooled molded articles compared to the mold dimensions, is principally a matter of thermal expansion, it is usual to make a direct measure of shrinkage by measuring a standard molded test bar. There must be any number of "standard" molds used in various rubber factories for this purpose, and the main essential is that the required accuracy be obtained. For example, to detect 0.1% shrinkage on a 10 cm bar requires a measurement to 0.1 mm. Shrinkage data has been given by Juve and Beatty [58] as well as a procedure for calculating shrinkage for different formulations. Standard procedures do exist for plastics and are referenced in Chapter 16.

An international standard method for dimensional stability of cellular materials is given in ISO 2796 [59]. The procedure is simply to measure a rectangular test piece before and after a conditioning period, which in this standard is a choice of various temperatures or temperature plus humidity.

The method for plastic film and sheet [60] is basically similar, with the refinements of putting gauge marks on the test piece and laying it on a bed of kaolin during exposure. A British method is specific to the stability of flexible PVC sheet [61] and involves conditioning in water at 100°C.

There do not appear to be ISO methods for coated fabrics, but there are national methods that indicate, as one might expect, that dimensional stability is of importance for these materials. There are three British standards: one covers stability to water immersion [62] and another to domestic washing [63]; the third is more unusual in that it measures the shrinkage of the material after it is unrolled and left unstrained for at least 24 hours [64].

When it comes to textiles, dimensional stability has attracted a lot of standards attention, and there are at least five international methods as well as national equivalents. The ISO methods cover washing and drying [65], free steam [66], cold water immersion [67], and dry heat [68]. The fifth standard covers the preparation, marking, and measuring of specimens and garments for dimensional change tests [69].

4.6 Dispersion

The dispersion of compounding ingredients in rubbers and plastics can have a large effect on physical properties, and a measure of dispersion can be used to judge the efficiency of mixing. The direct estimation of degree of dispersion is effectively a dimensional measurement using microscopy techniques and is just one example of the value of microscopy for fault diagnosis in polymer products.

Methods for plastics can be found in British standards BS 2782, Methods 823 A and B [70], which cover dispersion of carbon black in polyethylene, although they could be adapted for other materials. A thin section is examined under 100× magnification and comparison made with reference photographs. In method A the section is formed by melting and pressing and in method B by microtoming.

Dispersion measurements on rubber (most often carbon black dispersion) are normally made on cured material, although it is possible to prepare test pieces from some uncured materials. Probably the most widely used techniques are those described carefully by Medalia and Walker [71] and that form the basis of ASTM D2663 [72].

Examination of a torn surface with reflected light gives an overall picture of dispersion and is a useful rapid test in the control laboratory. To examine the dispersion of fine agglomerates of carbon black it is necessary to microtome sections using a freezing stage on a sledge microtome. The sections are examined by transmitted light, and either the percentage of agglomerates is estimated by counting or reference made to a standard chart.

Alternative approaches to the measurement of dispersion include the use of a stylus to measure roughness [73], electrical resistivity [74], and the analysis of the dark field image produced by a reflected light microscope [75]. Persson [76] has reported an improvement to the ASTM microscope method by using split field microscopy, and Richmond [77] describes a computer imaging technique. Belokur et al. [78] investigated the possibility of assessing dispersion from rheological measurements

Cembrola [79] has compared microscope, stylus, and resistivity methods and concludes that no one method is universally the best. A very comprehensive review of characterizing dispersions from every aspect has been given by Hess [80].

The optical method of observing a cut surface by reflected light is currently being standardized by ISO [81]. It includes the comparison against standard photographs using the split field technique first described by Persson (Fig. 2).

4.7 Cell Structure

Perhaps the most critical aspect of cellular materials is the structure of the cells—the size, shape, and proportion of open and closed cells. The size and shape of cells can be studied by microscopy or by projecting a very thin section, but to get numerical results is inevitably tedious. A British standard [82] determines cell count of flexible materials as the number of cells per linear 25 mm, which is considered a more convenient measure than cell size, particularly considering the variation in size for even uniform cell structures.

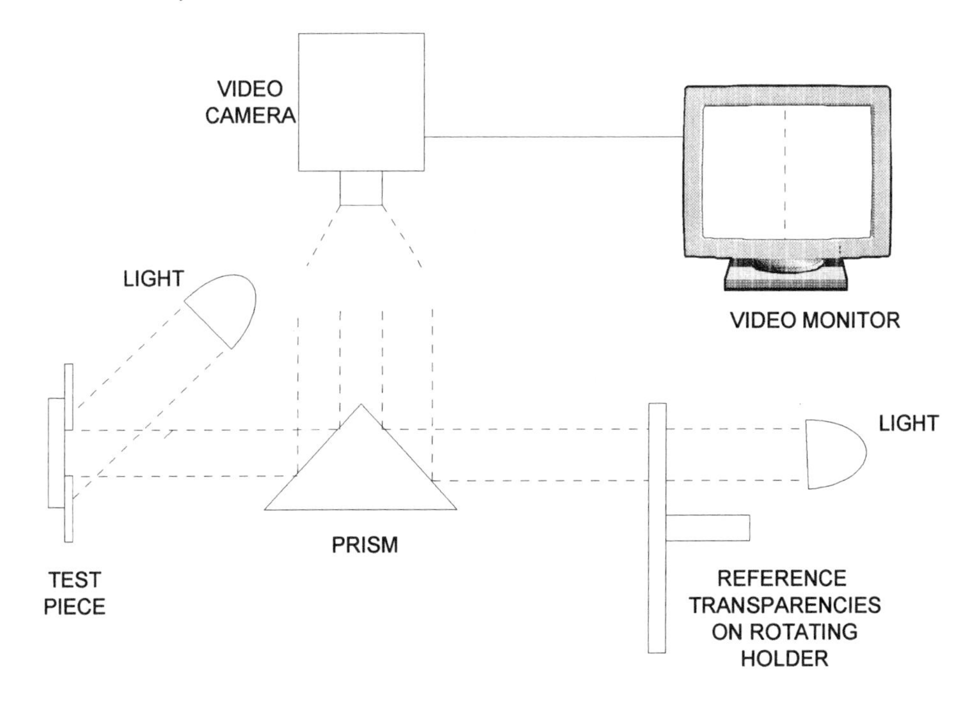

Figure 2 Split field dispersion method.

Image analysis techniques offer the possibility of detailed characterization much more rapidly. An account of the characterization of cellular structures has been given by Rhodes [83], who discusses various techniques including image analysis and optical microscopy. Image analysis has also been considered by Rhodes [84] and by Chaffanjon and Verhelst [85], while Sims and Khunniteekool [86] compare different methods of analysis of micrographs.

The volume percentage of open and closed cells can be estimated by determination of the surface area and geometrical volume followed by measurement of the impenetrable volume, either by a pressure variation method or by volume expansion. The detailed procedures for both methods, which are quite complicated, are given in ISO 4590 [87] for rigid materials.

The air flow permeability of cellular materials can be correlated with cell structure. A standard method of measuring air flow of flexible materials is given in ISO 4638 [88], based on detecting the pressure drop across a test piece through which air is passed under controlled conditions. A relatively simple quality control method is given in ISO 7231 [89], where the flow rate is measured at constant pressure drop.

References

1. ISO 2781, Determination of density, 1988.
2. BS 903, Part A1, Determination of density, 1980(88).
3. ASTM D297-93, Rubber products—Chemical analysis.
4. ASTM D1817-90, Chemicals—Density.
5. ISO 1183, Methods for determining the density and relative density of noncellular plastics, 1987.

6. BS 2782, Methods 620 A–D, Methods for determining the density and relative density of non-cellular plastics, 1991.
7. ASTM D792, Specific gravity (relative density) and density of plastics by displacement, 1991.
8. ASTM D1505, Density of (solid) plastics by density gradient technique, 1985.
9. ASTM D4883, Density of PE by ultrasound method, 1989.
10. ISO 845, Cellular plastics and rubbers—Determination of apparent (bulk) density, 1988.
11. BS EN ISO 845, Cellular plastics and rubbers—Determination of apparent (bulk) density, 1995.
12. ASTM D 1622, Rigid cellular plastics—Apparent density, 1993.
13. ASTM D 1565, Flexible cellular materials—Vinyl chloride polymers and copolymers (open cell foam), 1981.
14. ASTM D 1667, Flexible (closed cell) cellular materials—Vinyl chloride polymers and copolymers, 1976.
15. ASTM D 3574, Flexible cellular materials slab/bonded/molded urethane foams, 1995.
16. ASTM D 3575, Flexible cellular materials from olefin plastics, 1993.
17. ISO 60, Apparent density of material that can be poured from a specified funnel, 1977.
18. ISO 61, Determination of apparent density of material that cannot be poured from a specified funnel, 1976.
19. BS 2782, Method 621 A, Apparent density of material that can be poured from a specified funnel, 1978.
20. BS 2782, Method 621 B, Apparent density of material that cannot be poured from a specified funnel, 1978.
21. ASTM D1895, Apparent density, bulk factor and pourability of plastics materials, 1989.
22. ISO 1068, PVC resins—Determination of compacted apparent bulk density, 1975.
23. BS 2782, Method 621 D, Determination of apparent compacted bulk density of PVC resins, 1978.
24. ISO 171, Determination of bulk factor of moulding materials, 1980.
25. BS 2782, Method 621 C, Determination of bulk factor of moulding materials, 1983.
26. ISO 2286, Part 2, Determination of mass per unit area, 1988.
27. ISO 4648, Determination of dimensions of test pieces, 1991.
28. BS 903, Part A38, Determination of dimensions of test pieces, 1991.
29. Clamroth, R., and Dobroschke, P., Tech. Notes Rubb. Ind., No. 51, 1974.
30. ASTM D 5947, Physical dimensions of solid plastics specimens, 1996.
31. ISO 4591, Determination of average thickness of a sample, and average thickness and yield of a roll, by gravimetric techniques, 1992.
32. ISO 4592, Determination of length and width, 1992.
33. ISO 4593, Determination of thickness by mechanical scanning, 1993.
34. BS 2782, Method 631A, Determination of gravimetric thickness and yield of flexible sheet, 1993.
35. BS 2782, Method 632A, Determination of length and width of flexible sheet, 1993.
36. BS 2782, Method 630A, Determination of thickness by mechanical scanning of flexible sheet, 1994.
37. BS EN ISO 1923, Cellular plastics and rubbers—Determination of linear dimensions, 1995.
38. ISO 2286, Part 1, Determination of length, width and net mass, 1998.
39. ISO 2286, Part 3, Determination of thickness, 1998.
40. BS 3424, Method 28, Determination of coating thickness and thickness of an expanded layer, 1993.
41. BS 3424, Method 28B, Determination of coating thickness of thinly coated fabrics, 1993.
42. *Brit. Plast. Rubb.*, Jan. 1990, p. 15.
43. Braeuel, M., *Rubber World*, *202*, No. 1, 32 (1990).
44. Wullschleger, P., and Mannel, U., *Gummi. Fas. Kunst.*, *43*, No. 10, 538 (1990).
45. Spray, J. L., *Proceedings*, Antec 89, New York, 1–4 May, 1989.
46. Godden, P., *Mod. Plast. Int.*, *20*, No. 12, 59 (1990).

47. Tormala, S., and Rogers, J. H., Polymers, laminations and Coatings Conference, Boston, Sept. 4–7, Proceedings, Book 1, 1990, p. 457.
48. Spang, A., and Wuestenbergn, D., *Kunststoffe German Plastics*, *83*, No. 11, 14 (1993).
49. Larrison, D., and Whiteman, R., 148th ACS Rubber Div. Meeting, Cleveland, 17–20 Oct., Preprints, paper 80, 1995.
50. Reber, E. J., and Whiteman, R., Polymers, Laminations and Coatings Conference, Chicago, 27–31 August, Book 2, 1995, p. 457.
51. Brown, R. P., and Jones, W.L., *RAPRA Bulletin* (Feb. 1972).
52. Brown, R. P., *RAPRA Members Journal* (Aug. 1973).
53. ISO 468, Surface roughness, 1982.
54. ISO 4287, Part 1, Surface roughness terminology, surface and its parameters, 1984.
55. ISO 4287, Part 2, Surface roughness terminology, measurement of surface roughness parameters, 1984.
56. BS 1134, Part 1, Assessment of surface texture, methods and instrumentation, 1988.
57. BS 1134, Part 2, Assessment of surface texture, general information, 1988.
58. Juve, A. E., and Beatty, J. R., *Rubber World* (Oct. 1954).
59. ISO 2796, Cellular plastics, rigid—Test for dimensional stability, 1986.
60. ISO 11501, Determination of dimensional change on heating of film and sheeting, 1995.
61. BS 2782, Method 641A, Determination of dimensional stability at 100°C of flexible polyvinyl chloride sheet, 1983.
62. BS 3424, Method 20, Method for determination of dimensional stability to water immersion,1987.
63. BS 3424, Method 23, Method for determination of dimensional stability to domestic washing, 1987.
64. BS 3424, Method 36, Method for determination of dimensional changes on mechanical relation at zero tension, 1993.
65. ISO 5077, Determination of dimensional change in washing and drying, 1984.
66. ISO 3005, Determination of dimensional change of fabrics induced by free steam, 1978.
67. ISO 7771, Determination of dimensional change of fabrics induced by cold water immersion, 1985.
68. ISO 9866, Part 2, Determination of dimensional change in fabrics exposed to dry heat, 1995.
69. ISO 3759, Preparation, marking and measuring of fabric specimens and garments in tests for determination of dimensional change, 1994.
70. BS 2782, Methods 823 A & B, Methods for the assessment of carbon black dispersion in polyethylene using a microscope, 1978.
71. Medalia, A. I., and Walker, D. F., Technical Report RG 124 Revised. Published by Cabot Corporation.
72. ASTM D2663-89, Dispersion of carbon black.
73. Hess, W. M., Chirico, V. E., and Vegvari, P. C., *Elastomerics*, *112*, No. 1 (1980).
74. Boonstra, B. B., *Rubb. Chem. Technol.*, *50*, No. 1 (1977).
75. Ebell, P. C., and Hemsley, D. A., *Rubb. Chem. Technol.*, *54*, No. 4 (1981).
76. Persson, S., *Eur. Rubb. J.*, *160*, No. 9 (1978).
77. Richmond, B. R., ACS Rubber Division Spring Meeting, Denver, 18–21 May, Proceedings, 1993, p. 56.
78. Belokur, S. P., Skol, V. I., and Sapronov, V. A., *Int. Polym. Sci. Tech.*, *13*, No. 2 (1986).
79. Cembrola, R. J., *Rubb. Chem. Technol.*, *56*, No. 1 (1983).
80. Hess, W. M., *Rubb. Chem. Technol.*, *64*, No. 3, 386 (1991).
81. ISO CD 11345, Assessment of carbon black dispersion.
82. BS 4443, Method 4, Flexible cellular materials—Cell count, 1988.
83. Rhodes, M. B., *Low Density Cellular Plastics—Physical Basis of Behaviour*, Chapman and Hall, London, 1994, p. 56.
84. Rhodes, M. B., Polyurethanes '92, New Orleans, 21–24 October, Proceedings, 1992, p. 548.

85. Chaffanjon, P., and Verhelst, G., Polyurethanes World Congress, Nice, 24–26 October, Proceedings, 1991, p. 545.
86. Sims, G. L., and Khunniteekool, C., *Cell. Polym.*, *13*, No. 2, 137 (1994).
87. ISO 4590, Determination of volume percentage of open and closed cells of rigid materials, 1981.
88. ISO 4638, Polymeric materials, cellular flexible—Determination of air flow permeability, 1984.
89. ISO 7231, Polymeric materials, cellular flexible—Method of assessment of air flow value at constant pressure drop, 1984.

8
Processability Tests

John Dick
Alpha Technologies, Akron, Ohio
Martin Gale
Rapra Technology Ltd., Shawbury, Shrewsbury, England

1 Introduction

Most of this book is concerned with the properties of the finished material, i.e., as it is used in products. However, another important area of polymer testing is concerned with the tests needed to predict and control the properties of relevance to the various stages of processing. Collectively, these can be termed processability tests, and the majority are measures of the flow properties of the polymer melt and also the curing characteristics of thermosetting materials.

This chapter deals with processability tests for the two general polymer categories, rubbers and plastics. Not surprisingly, there are many similarities in the processing of the two types of material, but there are also many differences arising from their structural differences, differences in the processing methods and because of the separate development of the two industries. International and national standards for the two types of material are separate, and generally the apparatus used is specific to one material type. Consequently, rubbers and plastics are dealt with here sequentially, and it is left to the interested reader to contrast and compare the appoaches taken.

2 Introduction to Plastics Processability

The results of tests on molded test pieces and finished products can in many cases be influenced by the way the material has been processed. This in turn is influenced by the materials processability. Indeed there are very many cases of product failure resulting from processability changes, while misunderstandings and lack of information can also result in reduced output rates, high scrap rates, and general production inefficiencies.

A further need for processability testing is to ensure that polymer and formulation developments, substitutions and modifications do not result in processing problems and the subsequent difficulties already mentioned.

3 Thermoplastics Processing

Most thermoplastics processing involves screw machines, whether for extrusion (pipes, film, cables, etc.), blow molding (bottles, fuel tanks, etc.), or injection molding (from cocktail sticks to car bumpers). The processes in all cases involve a powder or pellet raw material being conveyed, melted, and pumped through a die or injected into a mold.

A processability difference between extrusion (including single-stage blow molding) and injection molding exists in that in extrusion the molten polymer is forced through a die into the atmosphere and then cooled to shape, while in injection molding the material is fully enclosed and cooled under pressure. Although extrusion is arguably the more demanding of a material's rheological properties, such data are important in injection molding for mold design while variations can cause mold filling variability, weld line weaknesses, shrinkage, warpage, and dimensional variations as well as final mechanical property deficiencies.

3.1 Thermoplastic Polymers

Many of the factors described above are a consequence of a number of polymer molecular structural features.

Polymers have long-chain molecules whose entanglement tends to be unravelled with orientation in the direction of flow during processing.

Polymers are produced with a range of average molecular weights, molecular weight distributions, and degrees of chain branching. Thus high-density polyethylene grades for large blow moldings will have high molecular weights giving high viscosities, while the opposite will be the case for melt spun fiber grade polypropylenes.

Some polymers are essentially amorphous (e.g., polystyrene, acrylonitrile butadiene styrene copolymer, polycarbonate, and polymethyl methacrylate) while others are semicrystalline (e.g., polyolefins and polyamides). The former tend to have a wide melting temperature range with a comparatively high melt strength, while semicrystalline polymers tend to have a narrow melting temperature range and frequently a low melt strength.

From a practical aspect, amorphous polymers tend to be less critical with regard to screw design and are favored by the processor for foams and profiles where a choice exists. Polyamide 66 and 6, as examples of semicrystalline polymers, have comparatively fluid melts that make profile extrusion very difficult.

Semicrystalline polymers need more attention to screw design to achieve uniform melting, more heat needs to be removed during cooling, and their high shrinkage is more likely to produce warpage and problems with dimensional tolerances. A consequence of these various effects is that an understanding of the rheology is important, particularly so with semicrystalline polymers.

3.2 Polymer Blends and Masterbatches

There has been a trend in recent years towards the use of polymer blends. In packaging films, several polyolefins may be blended together in an appropriate ratio to give particular film properties, e.g., toughness, gloss, fast heat seal, etc., at optimum cost. To prevent melt fracture in the extrusion of LLDPE film, it may be blended with 10–15% LDPE, or a fluoropolymer may be added as a processing aid. As a result the processor is confronted with numerous changes in rheological behavior that cannot be predicted from the supplier's individual data sheets.

Coextruded film or sheet edge trim and start-up scrap consisting of typically 3 to 5 different polymers (including adhesives) may need to be reprocessed as a homogenized blend. One cannot even assume that the processability will be unchanged when recycled monolayer scrap is used, particularly if the proportion of scrap to virgin is variable.

Any additive is a potential source of change to the overall processability from pellet feed to melt rheology. The not uncommon operator comment that one particular color always runs well and another always causes problems is more likely to be fact than superstition. Many products are colored by using masterbatches, whilst extruded products may contain antiblock and slip additives. Injection moldings may contain lubricants to aid mold release or provide product lubrication, e.g., bottle caps. In many cases such masterbatches contain low-viscosity polymer waxes to promote additive dispersion (particularly pigments) and masterbatch distribution throughout the natural polymer. There are in fact many additives, including fillers, flame retardants, and antioxidants, that can influence processability.

3.3 Typical Stages in Thermoplastics Processing

Figure 1 shows the typical stages involved in extrusion and injection molding. The properties that need to be measured to determine their influence on processing for each particular stage are indicated. For practically all stages, test methods exist, but only for a few have standards been established. The rheology of molten polymers influences many of the overall processes, including critical stages of flow in dies and molds. Fortunately, standard methods exist for capillary rheometry, the most important technique for such measurements.

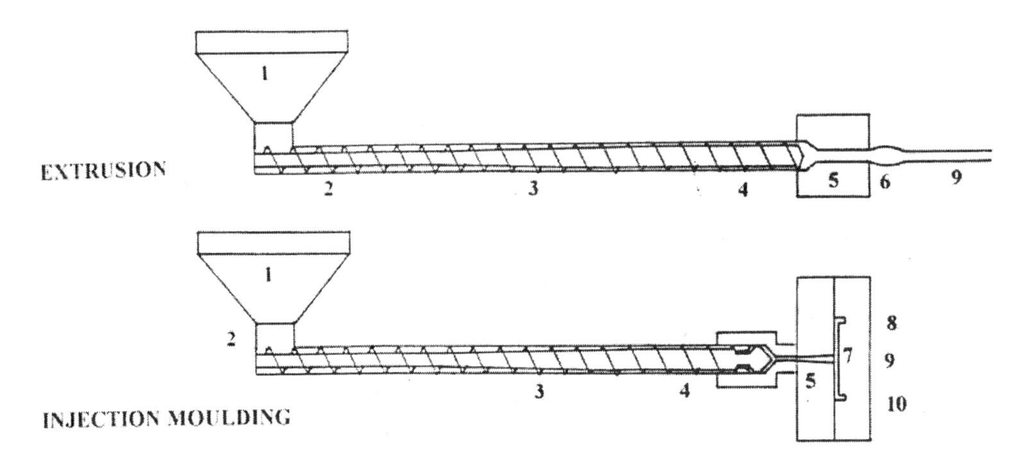

Figure 1 Typical stages in thermoplastics processing. (1) Solids flow; (2) screw filling and solids conveying; (3) melting; (4) polymer melt shearing/pumping; (5) shear flow in channels; (6) die swell; (7) elongational flow; (8) volume change under pressure; (9) thermal conduction; and (10) shrinkage.

The procedure used in this section on thermoplastics processability is to follow the material through the various stages through which it passes from pellet to product. It should be borne in mind that the pellets may not be a single material but a blend of different polymers, or the mix may contain pellets of masterbatched additives as described in Section 3.2.

It is probably appropriate to mention at this point that a number of feed materials may need drying before testing, because they either intrinsically take up moisture or contain additives that can promote this effect. Some polymers, notably polyamides and thermoplastic polyesters, will be degraded by processing if they are not dried beforehand under conditions specified by the supplier.

3.4 Particulate Properties

The particulate properties of polymers, blends, masterbatches, etc., are important from processability aspects for their influence on

Bulk handling including silo emptying
Auger metering
Particulate blending
Machine hopper flow
Screw channel filling
Extruder/molding machine screw conveying

Bulk handling, automatic blending, and hopper filling, etc., have become very widespread, and computer modelling of screws also needs particulate property data. Changes in the shape or bulk density of the feed pellets (or powder) can result in variable hopper feed rate and screw conveying. Hopper bridging and "rat holing" may occur. Overall the result may be variable dimensions and properties.

The measurable properties of concern are

Apparent (or bulk) density.
Friction of solid polymer particle or pellet against metal surfaces (termed external friction). This influences feed zone conveying rate as well as hopper flow.
Friction between solid polymer particles or pellets (termed internal friction). This also influences conveying as well as hopper flow, especially in grooved feed zones.

A particular problem is that a material that has excellent hopper flow may well exhibit poor machine screw conveying. For hopper flow, it might be expected that measurement of flow using a standard hopper would be a simple test that avoided the need for friction measurements, but unfortunately this is not the case.

Butters [1] has described a number of tests related to particle porosity, bulk density, and friction in relation to bulk storage and handling of polymers. However, in spite of the influence of these properties on uniformity of output rate and overall productivity, there is a general lack of standards and readily available testing equipment.

Apparent Density, Bulk Factor, and Pourability of Plastic Materials

These properties are covered partly or wholly by ASTM, ISO, and BS standards. ASTM D1895 [2] defines all three properties relevant to the standard as follows:

Apparent Density—the weight per unit volume of a material including voids inherent in the material as tested

Bulk Factor—the ratio of the volume of any given quantity of the loose plastic material to the volume of the same quantity of the material after molding or forming

Pourability—a measure of the time required for a standard quantity of material to flow through a funnel of specified dimensions

The ASTM standard uses a fairly simple apparatus which is described in sufficient detail that it can be made by most engineering workshops. It essentially consists of a cylindrical measuring cup mounted beneath a specified funnel such that a measured weight of material is allowed to flow from the funnel and overfill the cup, the excess being levelled off before weighing the known volume in the tared cup.

Method A has a funnel opening of 9.5 mm diameter.

Method B is more suited to pellets, having a larger funnel opening of 25.4 mm combined with a larger funnel and a larger cup.

Method C is more suited to materials in the form of coarse flakes, cut fibers, etc., and dispenses with the funnel. A hollow piston is placed on top of a weighed amount of material in a graduated cylinder. 2.3 kg of lead shot is then added to the piston and a reading taken from the graduated scale on the plunger after 1 minute.

ISO 60:1977 (E) [3], which is restricted to Apparent Density, uses essentially the same funnel and cylinder principle of the ASTM A and B test methods. However it uses one funnel for both pellets and powder, and the dimensions of both funnel and cylinder are different from the ASTM methods. On the other hand, ISO 61:1976 [4] is essentially the same as ASTM D1895, Method C, described above.

ASTM 1895 also has a pourability test in which the time taken for a weighed amount of material to flow out of the funnel of Method A is measured. This is obviously not suitable for plastics granules. A potential technique for overcoming the limitations of the standard test methods in making measurements applicable to plastics machinery hoppers has been evaluated by Boysen and Gronenbaum [5]. The method used seven hoppers 60 mm high with 45° sides and having discharge apertures of 3, 5, 10, 20, 25, and 30 mm. The system was aimed mainly at PVC powder compounds, but other materials were used in the investigations. The procedure replaced timed discharges of the standard methods with the establishment of the smallest aperture through which the material would freely flow. As plastics powders and granules flowing from hoppers into machine feed entries have minimal assistance from momentum and air entrainment, simple funnel tests should be treated with caution. Internal friction measurements made on annular shear cells described in Section 2.4.2 may be more appropriate. These measurements and other factors relevant to bulk storage and conveying have been reviewed by Butters [1]. This review also considers the effects of moisture and consolidation.

Polymer Particulate Friction

The laws of friction are as follows:

Amontons (1699): The frictional force F opposing motion is proportional to the normal force N (Fig. 2). The constant of proportionality is termed the coefficient of friction.

The coefficient of friction is independent of the apparent area of contact.

Coulomb (1785): The coefficient of friction is independent of the velocity between the two surfaces, provided that velocity is not zero.

It was not until 1934 that it was realized that the classical laws of friction were not obeyed by rubbers; and it has, hardly surprisingly, been shown that neither do plastics [6]. Further complications arise with polymers as a result of frictional heat, particularly at high velocities, and also from "rubbing-in" effects.

The position with regard to plastics processing was reviewed in the introduction of an article in 1981 by Huxstable, Cogswell, and Wriggles [7]. Most early measurements used the well-established sledge method but were limited in velocity and duration of test. Rotating cylinders and annuli overcame these objections, and of these the annular (or ring) shear cell was preferred. This approach has also been described by Gale [8], the instrument being based on earlier ones used for soil mechanics and silo design, etc. The general arrangement is shown in Fig. 3.

Polymer granules or powder are loaded into an annular trough that has a roughened bottom to prevent them sliding. The trough can be rotated at the required speed, whilst a stationary ring is pressed against the granules by a normal load, which can be varied. The ring can be smooth (for granule/metal friction) or have webs (for granule/granule friction).

The shear force is measured by a load cell restraining the ring from turning. The ring can be heated to simulate a hot barrel or screw surface.

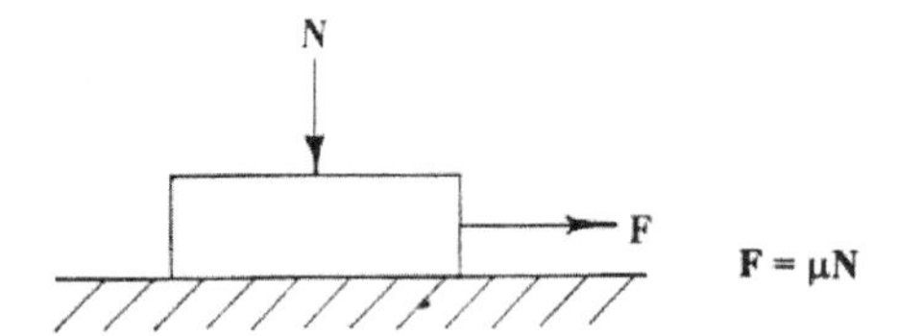

Figure 2 Coefficient of friction according to Amontons.

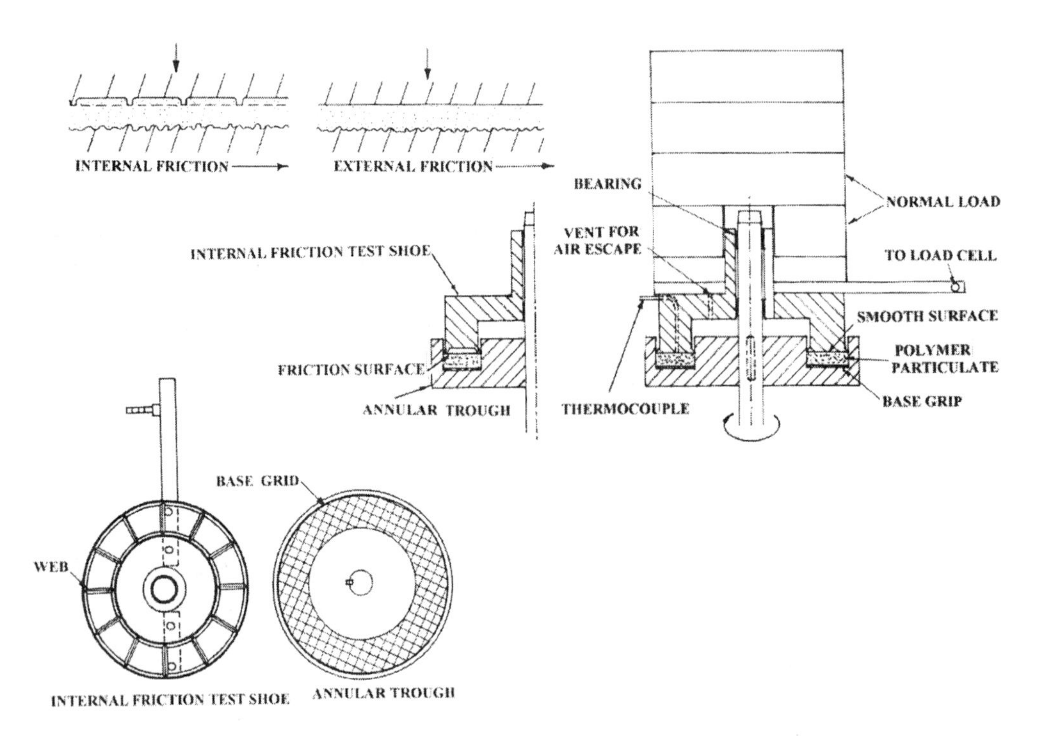

Figure 3 The annular shear cell.

The advantages of these instruments are that for feed zone properties they can operate at typical extruder velocities and normal loads. A further advantage is that the surface can be "rubbed in," i.e., be contaminated by polymer as has been found to be the real situation [9].

Huxstable et al. [7] produced graphs that showed the influence of temperature to be very significant. Measurements were recorded with a heating rate of 30°C/min; constant temperature results are difficult to obtain due to frictional heating. Influences of additives, particle size, and polymer type were also shown. The authors estimated that the coefficient of variation for friction was 4.5% and the temperatures at which friction events occurred were reproducible to within ±2°C. This raises the question, why are such instruments not more generally used? The answer is possibly that the test procedure is long and involved, with two runs necessary for each test, one for rubbing in and one for the test itself with cooling to ambient temperature between each run. Interpretation of the data may also be daunting, as coefficients of friction can have both peaks and troughs with increasing temperature.

Other rotating instruments include ones with three point loading [6], for which examples of correlation with extrusion behavior have been given [10] and an annular trough arrangement fitted to a Rheometrics mechanical spectrometer [11]. Fitting an annular shear cell to a torque rheometer is also possible [8].

3.5 Testing Melting Behavior

Melting in a screw is a very important stage of thermoplastics processing. The achievement of controlled and complete melting in an extruder or injection molding machine is dictated by the interrelationship between polymer and machine. Semicrystalline materials need more heat than amorphous polymers. For example, the total heat to process HDPE is three times that of polystyrene according to figures tabulated by Morton-Jones [12]. Thermal properties can be measured by DSC, but in a processing machine packing, friction, surface wetting/lubrication, etc., all play a part. In a screw processing machine such as an extruder, a general melting mechanism has been established [13–15] as shown in Fig. 4.

In practice melting may be completed earlier or later than the machine designer intended, which can often cause output rate instabilities and many types of product defects. As the melting rate also depends on operating temperatures, it cannot be easily predicted by measurement, although computer modelling is possible.

Measurements of bulk density and friction as described in Sections 3.4.1 and 3.4.2 have some relevance, as do the melt properties described in Section 3.6. However, apart from measurements of PVC gelation rates described in Section 3.7 using a torque rheometer, melting rate as a processability test appears to be quite unusual.

Measurements using a laboratory extruder are possible. The rapid cooling of an extruder and removal of the screw followed by examination of material removed from

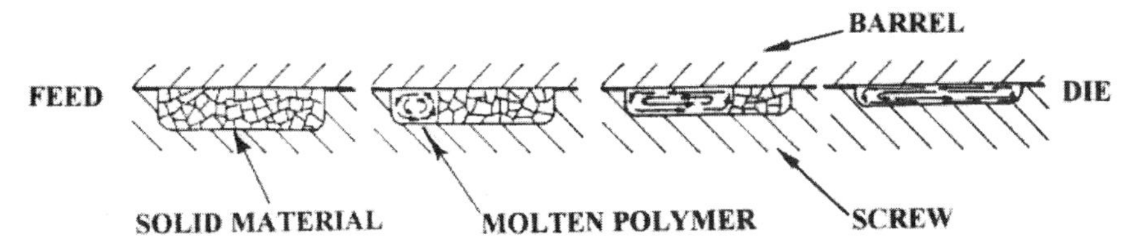

Figure 4 Polymer melting mechanism in a single screw machine.

the screw channel (as in Refs. 13, 14, and 15) is very time consuming and confined to research purposes. Instrumented laboratory extruders with pressure transducers in regularly spaced ports will give an indication of the material condition by both the overall pressure profile and the pressure waveform with screw rotation for each instrument [16].

An approximate measurement can be made by bleeding off small samples through bleed ports and examining their condition.

3.6 Properties of Molten Thermoplastics

Rheological Properties

The rheological properties of molten polymers influence many aspects of processing. In injection molding and extrusion they affect screw pumping efficiency, generation of mechanical heat, and die pressure (and hence output rate). They also influence die and mold weld lines, surface finish, product dimensions, etc.

Rheology of fluids is covered by a large number of books. For aspects of testing for processability of polymers, one by Cogswell [17] is of particular relevance. A very comprehensive book has been produced by Macosco [18]. There is also a useful journal article concerning polymer rheological testing by Delaney and Houlston [19].

Measurement of these properties in a meaningful way requires some understanding of the rather unusual rheological behavior of polymer melts. As polymers are viscous fluids, they continue to deform as long as a stress is applied. The energy applied is dissipated as heat, an important factor to be considered in processing behavior as rapid stressing can produce heat faster than it can be removed (as heat transfer is slow). This raises the melt temperature and consequently reduces its viscosity, which, although helpful to the processor, may not prevent overheating to unacceptable temperatures.

The ratio of shear stress to rate of strain defines the viscosity as

$$\eta = \frac{\tau}{\gamma}$$

If a liquid flows with a viscosity independent of stress level, its behavior is said to be Newtonian. However, polymers have long entangled chains, which in shear flow (which will be experienced in a screw channel or die) will become partly unravelled and aligned in the direction of shear. Cogswell uses the analogy of the more ordered state of spaghetti on a plate achieved by twirling a fork [17]. As a result of the molecular alignment the viscosity is reduced in varying degrees from the Newtonian ideal fluid and its behavior is termed pseudoplastic, i.e., it is shear thinning (Fig. 5).

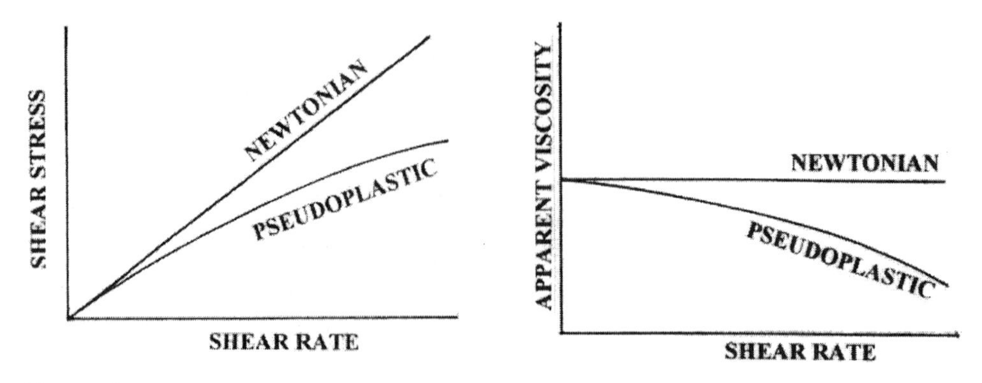

Figure 5 Rheological effects of increasing shear rate.

A useful representation is the graph of viscosity against shear rate, normally plotted on a log/log scale. As temperature changes this relationship to different degrees for different polymers, it is common practice to measure these properties at several temperatures. An example is shown in Fig. 6.

A further complication is an elastic effect, which is particularly relevant to die swell as described below. The combination of viscous and elastic effects can be represented by the Maxwell model, where a dashpot represents the viscous component and a spring represents the elastic component as shown in Fig. 7.

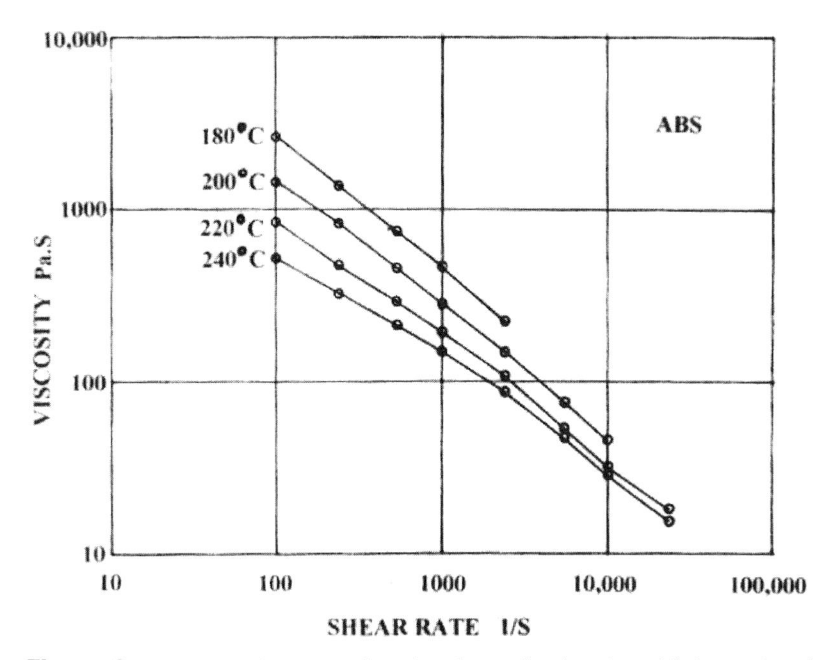

Figure 6 Capillary rheometer data for change in viscosity with increasing shear rate.

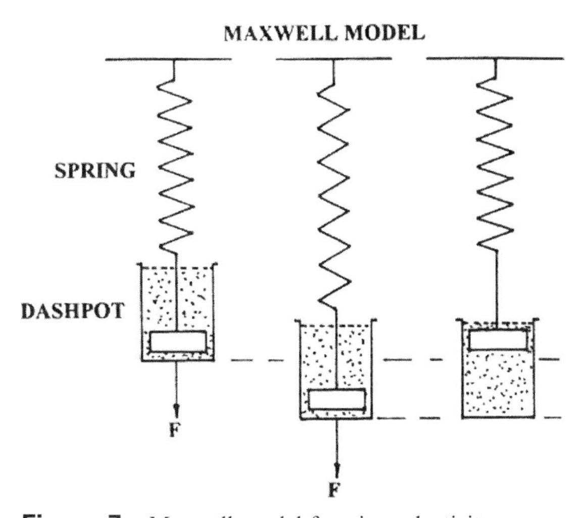

Figure 7 Maxwell model for visco-elasticity.

The spring obeys Hooke's law and extends instantly by an amount directly proportional to the force.

Although shear thinning complicates the measurement of polymer melt viscosity, it is fortuitous for polymer processing, as otherwise thermoplastics would be ten to a thousand times more difficult to extrude or injection mold.

Situations where the polymer is subjected to shear forces are the ones most commonly considered. Shear forces are imposed on polymer melts by rotating screws and by flow through dies. In theory the polymer touching the metal surfaces is stationary, and hence a velocity gradient occurs between fixed and moving parts, and between the center and edges of a die channel or mold runner as shown in Fig. 8.

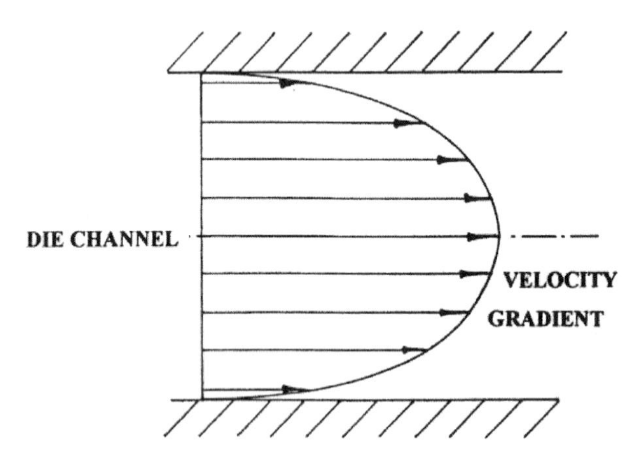

Figure 8 Velocity distribution across a die channel during flow of a viscous liquid.

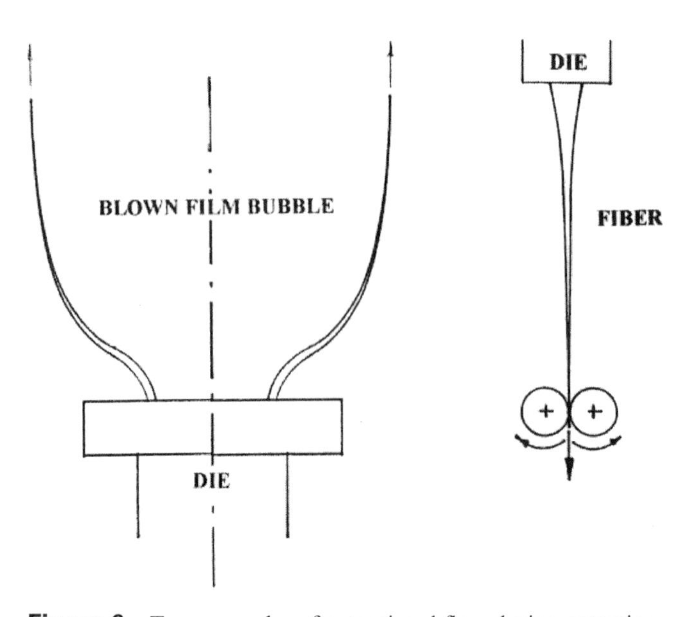

Figure 9 Two examples of extensional flow during extrusion.

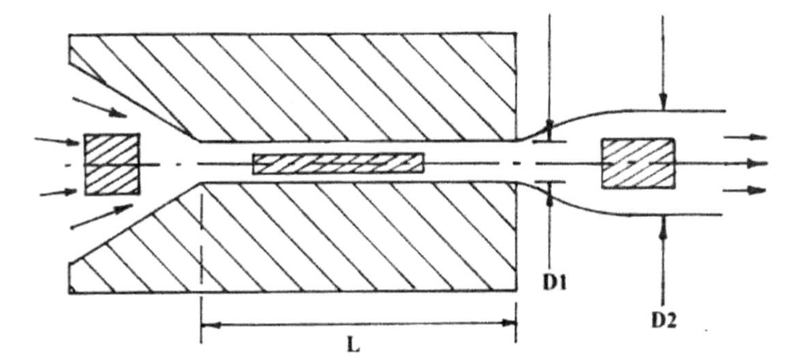

Figure 10 Mechanism of extensional flow effects in dies and die swell from elastic deformation.

$$\text{Die swell} = \frac{\text{extrudate thickness or diameter}}{\text{die gap or diameter}} = \frac{D2}{D1}$$

In addition to these shear flow situations, molten polymers are subjected to extensional flow. This occurs in film blowing, fiber drawing, and some injection mold filling situations where straightening and alignment of polymer chains may increase viscosity (Fig. 9).

An increase in extensional flow viscosity in film blowing, fiber spinning, and bottle blowing, etc., is usually advantageous, but the change in viscosity will depend very much on the polymer structure [17].

Extensional flow effects also occur at the entry and exits of dies (Fig. 10).

It is also in this region that the elastic component becomes important because the compressed polymer melt expands as it leaves the die. This increase in diameter or thickness is related to the elastic deformation of the polymer at the die inlet.

Die swell increases with shear rate and decreases with length/diameter (or thickness) ratio of the die aperture (L/D1). Hence higher output rates (and higher die pressures) give larger die swell.

Overall, for processability, tests are needed that will measure at least the following:

Viscosity/shear rate relationship over a range of shear rates and melt temperatures
Elongational viscosity over a range of shear rates and melt temperatures
Die swell

Capillary Flow Measurement of Shear Viscosity

In measuring viscosity it is assumed that the polymer behaves as a Maxwell model and measurements are made of an apparent viscosity that depends on such variables as temperature, pressure, stress, and time. Cogswell [17] suggested criteria for selecting a suitable test method. These included a strain rate up to $10{,}000\,\text{s}^{-1}$, to be appropriate for engineering purposes, conveniently measured to reproducibility of better than 5%, and to use a sample of less than 100 g. The capillary rheometer was considered to meet many of the requirements.

Shear stress and shear rate data can be calculated from measurements of melt pressure and volumetric flow rate from Poiseuille's law by

wall shear stress N/m^2 $$\sigma_s = \frac{PR}{2L}$$

wall shear rate S^{-1} $$\dot{\gamma} = \frac{4Q}{\pi R^3}$$

where

P = pressure drop N/m^2)
Q = volumetric flow rate (m^3/s)
R = capillary radius (m)
L = capillary length (m)

Capillary rheometers can usually measure over a range of shear stresses from 10^3 to 10^6 N/m^2 and shear rates from 0.001 to 500,000 s^{-1}.

These rheometers can be considered to be a form of heated syringe in which a plunger pushes polymer melted in a cylinder through a capillary die at its exit. The options for measuring pressure are

Apply a load with a set of known weights.
Apply an adjustable gas pressure to a smooth-fitting ball bearing piston
Apply a constant displacement and measure force with a load cell (a universal testing machine fitted with a compression jig can be used).
Apply a constant displacement and measure melt pressure with a melt pressure transducer.

The options for measuring volumetric flow rate are

Measure displacement of piston.
Weigh material issuing from capillary in unit of time.

The trend in commercial instruments has been towards calculating volumetric output rates from a preset range of piston speeds and measuring the resulting melt pressure with a melt pressure transducer for each piston speed.

The pressure drop through a capillary is shown in Fig. 11.

Measurement of shear viscosity by capillary rheometer is covered by two standards: ISO 11443 [20] and ASTM D3835 [21]. DIN 54811 and BS 2782, Part 7, Method 722B,1996 are the same as ISO 11443. The ISO standard gives options for a circular capillary die (Method A) and a rectangular slit die (Method B). See Fig. 12.

For both dies, the preset parameters can be the pressure (Method 1) or volume flow rate (Method 2). Thus a rectangular slit die with preset volume flow rate would be Method B2.

The advantage of the slit die is that with pressure transducers positioned along its length the entrance and exit pressure drop values can be determined, making it particularly

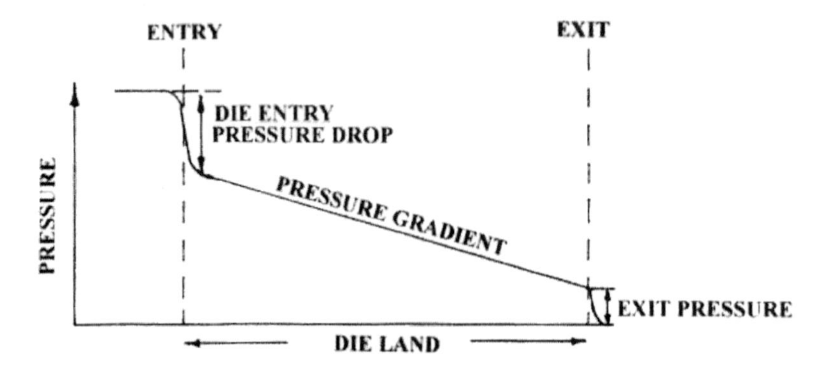

Figure 11 Pressure changes between a die entry and exit.

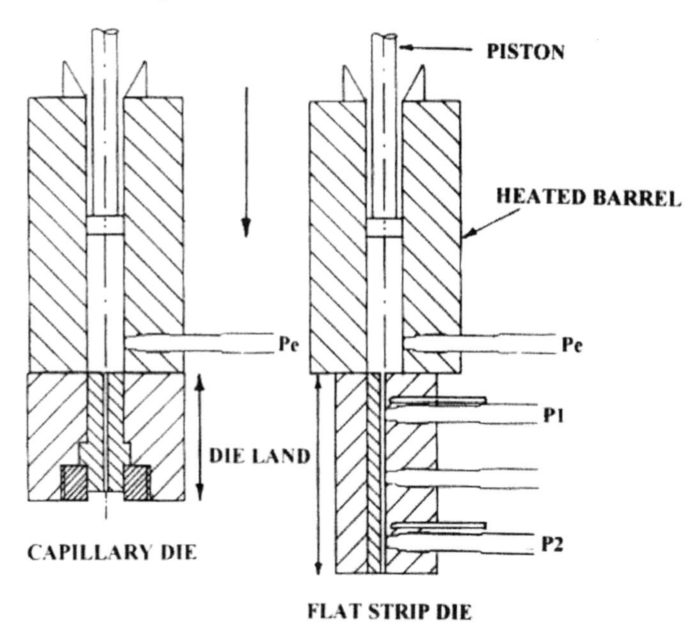

Figure 12 Capillary viscometer and flat strip die alternative.

suitable for computer-aided automatic measurements. With a circular capillary die, the end corrections can be made by making two experiments, one with a long capillary, e.g., L/D 20, and one with a short die, e.g., L/D 5. Using the data to apply the Bagley correction is described in the standard. The corrections convert the data from apparent shear stress to true shear stress. True shear rate can be calculated for both capillary and strip dies.

The standard specifies the apparatus in terms of a range of dimensions covering typical commercial instruments but gives values for surface hardness, dimensional tolerances, and roughness for barrel, piston, and dies (capillary and slit). Tolerances for piston diameter are also specified.

Temperature control and measurement and pressure transducer calibration is also covered, and test procedures with typical temperatures for specific polymers are given.

Two methods for measuring die swell are described (see Section 3.6.8 below).

ASTM D3835-95a [21] fulfils the same purpose and is very generally the same as ISO 11443. Unfortunately the standards differ a lot in detail, not so much in quantitative data but mainly in that one standard describes details that are left out of the other. Thus under definitions the ISO standard defines Newtonian and non-Newtonian fluids and the ASTM standard does not. The converse applies for "delay time."

Typical test temperatures only partly agree. The ISO method gives a range of 190 to 220°C for polyacetal while ASTM gives 190°C. ASTM die swell measurement procedures are different from the ISO procedures as described in Section 2.6.8. There is however good agreement on capillary tolerances, and the ASTM gives a procedure for measuring melt density.

Melt Flow Rate (MFR)

Melt flow rate (MFR) can also be referred to as melt flow index (MFI) or simply melt flow (MF). Thermoplastics materials data sheets normally contain this information (usually the only rheological information). Being a single value, it is particularly useful for specifications and making comparisons possible for consistency of quality. Under ideal conditions the MFR can be reproducible to within 3% and discriminate variations of 1% in molecular weight for molecular weights of about 100,000 [17].

The MFR is defined as the weight of polymer extruded through a specified die in 10 minutes under standard conditions of temperature and pressure. The numbers can often be clearly identified with the intended process, e.g., polypropylene with an MFR of 10 might be used for fibers or thin wall injection moldings, whereas for pipes the MFR might be 2. For low-density polyethylene thin films for freezer bags the MFR might be 7, but for heavy-duty agriculture sacks only 1.0 or less. The situation of having high values of MFR for lower viscosities and low values for high viscosities can sometimes cause confusion.

The use of MFR as a measure of consistency is acceptable, but as a measure of processability it can be very misleading. This is because the shear rates of the test are much lower than typical shear rates in extrusion and injection molding. Consequently, as a result of differences in molecular weight distribution, branching, etc., two polymers with similar MFRs can have widely differing viscosities at high shear rates, and consequently their processing behaviors may be quite different. The apparatus is shown in Fig. 13.

Fundamentally it is the same as the capillary rheometer, but the shear stress is kept constant from an applied load in the form of a weight, and instead of measuring the shear

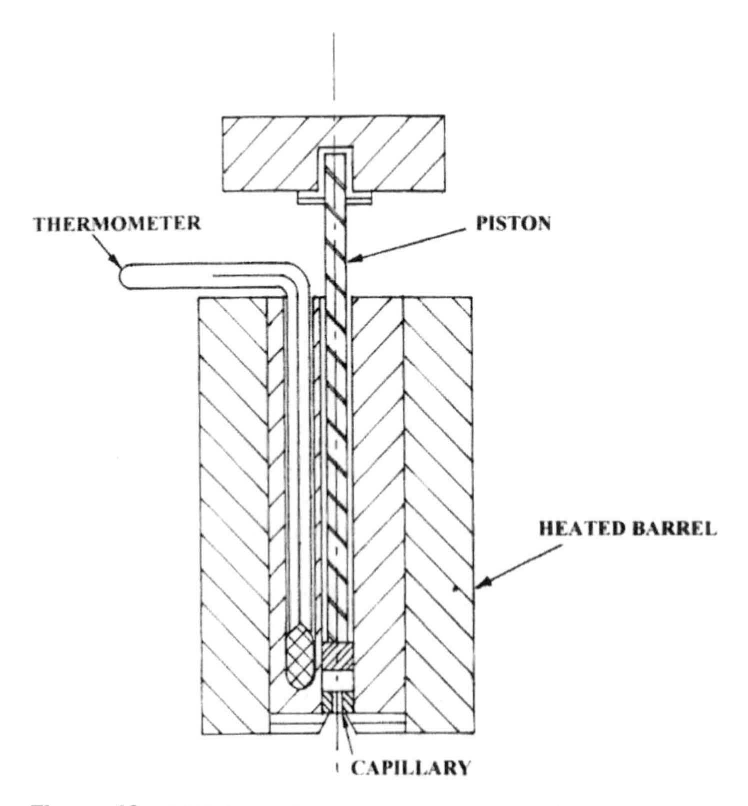

Figure 13 Melt flow rate apparatus.

rate, the weight extruded through a 2.095 mm die in a fixed time is measured. Compared with a capillary rheometer it is a low-cost instrument that is simpler to operate and has specified dimensions for piston and cylinder as well as the capillary.

The MFR test is covered by ISO 1133:1966 [22]. BS 2782, Method 720A, and DIN 53735 are the same as the ISO standard. ASTM D1238-95 [23] is "technically equivalent."

ISO 1133 lists standards for a number of polymers where MFR is specified as in the table.

ISO 1622-1:1985	Polystyrene	(PS)
ISO 1872-1:1986	Polyethylene and copolymers	(PE)
ISO 1873-1:1986	Polypropylene and copolymers	(PP)
ISO 2580-1:1990	Acrylonitrile butadiene styrene	(ABS)
ISO 2897-1:1990	Impact-resistant polystyrene	(HIPS)
ISO 4613-1:1985	Ethylene vinyl acetate	(EVA)
ISO 4894-1:1990	Styrene acrylonitrile	(SAN)
ISO 6402-1:1990	Various acrylonitrile copolymers	(ASA, AES, ACS)
ISO 7391-1:1987	Polycarbonate	(PC)
ISO 8257-1:1987	Polymethylmethacrylate	(PMMA)
ISO 99881-1:1991	Polyoxymethylene	(POM)

The test instrument is tightly specified with respect to dimensions, tolerances, and temperature control. The die is 2.095 mm dia ±0.005 mm by 8.00 mm long ±0.025 mm. In addition to procedure A in which the extrudate is cut and weighed, a procedure B covers automatic measurement of distance and time for the piston movement. Although the MFR is expressed as g/10 min the cutoff time is specified between 240 and 5 seconds depending on the MFR range, i.e., for high-viscosity polymers, which extrude slowly, cutoff time is 240 seconds, whereas for low-viscosity polymers it is between 5 and 15 seconds.

The standard contains a table of test temperatures and loads for each polymer. Polymers which have a very wide range of viscosities have up to three sets of test conditions to accommodate them. Consequently for materials such as ethylene vinyl acetate copolymer care must be taken to note the code letter (in this case B, D, or Z) in comparing different suppliers' literature.

Capillary Rheometer Attached to a Screw Extruder

Small plastics extruders are often used to produce test samples for color matching, gel counts, physical testing, fire tests, etc. A capillary die can be added and used as a rheometer using the method described in ASTM D5422 [24].

The screw extruder controls both the rate of extrusion and the melt temperature. The pressures are recorded by a pressure transducer and the output rates determined by cutting and weighing as for the MFR test. A minimum of two dies is necessary but three are recommended for calculating the Bagley correction factor. The ASTM standard tabulates die nozzle dimensions and lists typical screw-flight dimensions. Higher more regulated pressures can be achieved by adding a gear pump as shown in Fig. 14 and described below. A comparison of a slit die with a capillary die fitted to an extruder has been made by Padmanabhan and Bhattacharya [25] but aimed primarily at in-line measurements as described below.

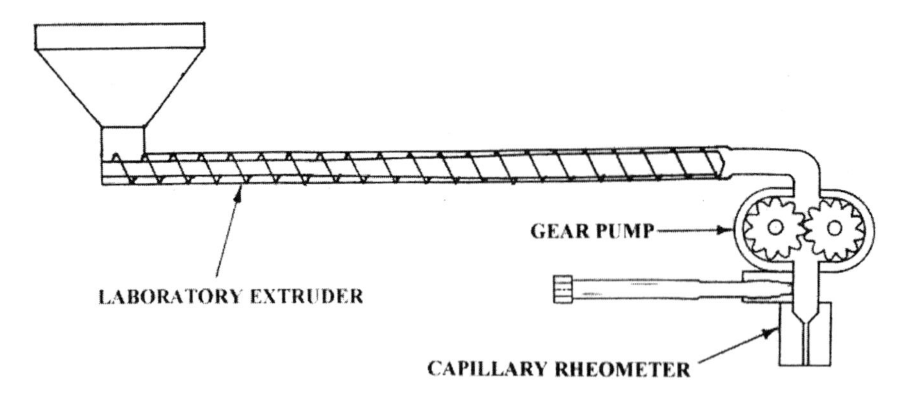

Figure 14 Capillary viscometer fed by an extruder and gear pump.

3.6.5 Measurement of Viscosity to Control Reaction Processes

The viscometers described above can be used for regular quality checks on samples taken from continuous polymerization units and twin-screw compounding extruders used for reactive processing and polymer blending. However, the time taken to carry out such off-line testing can be too long for corrective action to be taken. Furthermore, with computer control there are opportunities using on-line systems for integration of measurements into the complete process to give very tight control of processing. The basis is to divert a small stream of polymer through the instrument.

As there are a number of problems in the siting of the instruments, a variety of systems have been developed, each with its own advantages and disadvantages. By having the instrument attached directly to the machine, high temperatures and inaccessibility can be a problem, while the use of connecting pipes involves time delays and potential oxidation.

The continuous extrusion of a sample through a capillary can be used on the basis of that using an extruder described in Section 2.6. As the polymer is supplied to the instrument in a molten state at a controlled temperature, a gear pump can be used to provide both the required input pressure and controlled volumetric flow to the capillary. Kelly et al. [26] have described work using such an instrument in which several capillary dies covering a range of geometries were indexed through the melt stream, which provided a range of viscosity data and enabled Bagley corrections to be made.

According to Nelson [27], although this open system has advantages of easy installation, access, simplicity, and robustness, it has a number of disadvantages. The waste stream, although small, can cause housekeeping problems; volatile solvents and monomer may be present; and long time delays can occur between sampling and measurement due to the low capillary flow rate controlling flow time from the process to the instrument. The alternative systems described used a return pipe to avoid the waste and volatile problems, while a bypass around the capillary reduced sampling time. In one method, inlet and outlet gear pumps were used with a bypass valve closed during measurements, and in an alternative system three gear pumps were used, two to control flow through the capillary and one for the bypass. A further system used a double gear pump to feed dual capillaries measuring both shear and extensional viscosities.

Gear pumps have a disadvantage of generating heat and their output rates changing with melt temperature. An alternative to the capillary rheometer for on-line viscosity measurements has been described by Gogos et al. [28,29]. Measurement using a short

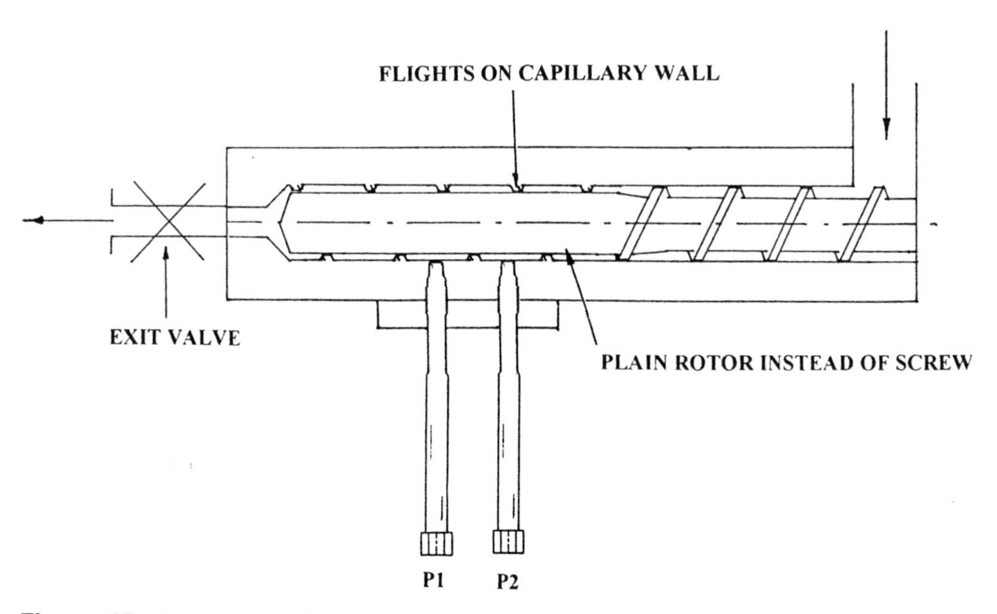

Figure 15 Viscometer using combined extrusion pressure development and measured pressure drop across a spiral stationary flight.

extruder screw with closed discharge avoids the need for both the generation of high pressures by a feed pump and the accurate measurement or control of throughput (Fig. 15).

The instrument uses a short extruder screw with a closed discharge. Calculation shows the pressure drop P to be a function of the geometry of the screw, the rotational speed, and the fluid viscosity. Hence the viscosity can be calculated by measurement of pressure change along the helical screw for a range of shear rates derived from several selected screw rotational speeds. The problem of having a waveform trace from a rotating screw is overcome by having the flight on the barrel surface combined with a closely fitting rotating cylinder. Unlike most of the gearpump fed capillary rheometers, the screw system measures the properties of discrete samples using a system of valves to empty and fill between tests. It avoids the need for end corrections while the screw will flush the system between samples and avoid the need for a bypass, but it is not fully continuous.

Rapid response off-line measurements can be made on the finished pellets by continuously diverting a sample stream into an extruder with a directly fed capillary as in Section 3.6.4 or with a gear pump between extruder and capillary die to provide pressures and volumetric control of output rate. According to Morrow and DeLaney [30], by using a low-level hopper and filling every 60 seconds an average response of 4–6 minutes was possible. Further advantages claimed were that the pellets/powder were more representative as not sampled from slow-moving material at the pipe wall, a second heating gave values that the customer would experience, and standard materials could be added periodically for calibration.

Extensional Flow Viscosity

Whereas the unravelling of tangled long-chain polymer molecules will usually cause shear flow viscosity to decrease with increasing shear rate, elongational viscosity is more likely to increase. This latter property is of particular importance in stretching circumstances, such as are found in blown film extrusion, blow-molding, and thermoforming. Extensional flow viscosity is also relevant to die swell. Extensional viscosity data by Sebastian and Dearborn [31] showed why LDPE and HDPE were better than PP from a processability aspect for blow molding and also why LDPE with a higher shear viscosity than LDPE was comparatively poor for a narrow molecular weight distribution but better when broader.

Macosko [18] showed, in simple diagrammatic form, seven of the most successful methods. Problems included the large strain required. These methods, which used solid specimens in extension, compression, and sheet stretching configurations, have been used for elastomers. However, with many thermoplastics there are the difficulties mentioned by Cogswell [17] of measurement, which requires the ends of a semiliquid sample to be held, and tendencies for samples to neck. The method is to take a strand or bar of polymer and, following heating to the test temperature, extend at a predetermined speed with simultaneous measurement of tension. There is a choice of pulling one end with one end fixed or pulling both ends. Presumably the advantage of pulling both ends by winding is that it eliminates a tendency to break at the fixed end.

Meissner and Hostettler [32] have described an improvement on an earlier double rotating clamp method in which metal conveyor belts were used at both ends. The test sample, a rectangular bar 60 × 7–8 × 2 mm, was supported on a cushion of inert gas.

Sebastian and Dearborn [31] used a strand of polymer, fixed at one end and the other end wound on a drum. This apparatus was based on one by Ide and White [33]. The specimen was immersed in a bath of silicone oil heated to the required temperature.

Rides et al. [34] referred to five papers using typically a force transducer and separate winding devices. In the same report a method was described based on a rotational rheometer (Fig. 16).

Test samples were extruded using a capillary rheometer. The problem of breaking at the clamp mentioned above was overcome by internally air-cooling it. The equipment was immersed in silicone oil, which acted as a heat transfer medium and minimized gravity and

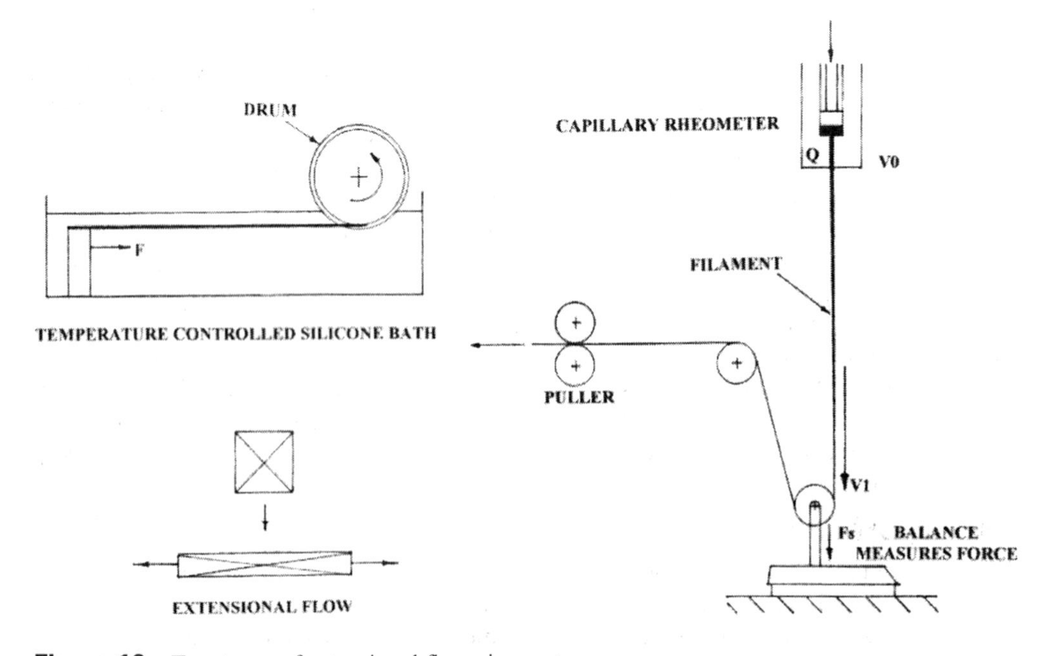

Figure 16 Two types of extensional flow viscometer.

buoyancy effects. A typical applied strain was 100% in 1 second, during which time the torque was measured. The fall in torque with time due to relaxation was then measured. There was good repeatability of results.

A comparatively simple system is to have a tension and reeling system as an addition to a capillary rheometer. An example of this type [35] is also shown in Fig. 16. The disadvantage of this method is that the measurement is not made under isothermal conditions.

The difficulties of direct measurement can be avoided, by using the "Cogswell calculations" of extensional viscosity from data obtained from a capillary viscometer using a plate die of "zero land length," which in reality will be about 1 mm long. In the measurements of shear viscosity described in Section 2.6, the pressure measurement represented the sum of the pressure drops at the inlet and outlet, which were mainly extensional flow effects, and the pressure drop along the capillary, which was a shear effect. It follows that comparisons between a zero-length capillary and a finite-length capillary enable the extensional viscosity to be calculated [36].

Rotational Viscometers

These normally have a sample between a fixed and a rotating surface. The surfaces may be two plates, but more usually one will be a cone. There is then a choice of oscillating and continuous rotation and a further choice of applied strain and measured torque or applied torque and measured strain. These instruments can be used to measure time-dependent effects, and although they have limited use for processability measurements, they are considered useful for molecular weight characterization [37]. Nevertheless, ASTM D4440-95a [38] lists a number of uses. As little as 3–5 g can be used for quality control, research and development, and establishment of optimum processing conditions. Graphical indication of molecular properties on melt processability and the effects of formulation additives that might affect processability or performance can be obtained.

Post-Extrusion Swelling or Die Swell

Obtaining the correct dimensions in extrusions and blow molding and in particular with extruded profiles is complicated by varying die swell between both different polymers and different grades of the same polymer. As a general rule, the longer the land length to thickness (or radius) ratio the less the die swell (up to a limit), and the higher the shear rate (increasing output rate or decreasing slot thickness or radius), the higher the die swell. This influences not only production equipment but also test instruments.

Die swell ratio is defined as the ratio of the diameter of the extrudate to the diameter of the capillary die (see Fig. 10 in Section 3.6). The increase in diameter can be expressed as a percentage of die diameter. Whether the measurements are at room temperature or test temperature is also part of the definition. Die swell can be measured using capillary rheometers from either measurement of diameter after the cooling of a piece of extrudate of specified length or by noncontact optical or photographic means of the hot extrudate as it issues from the die.

In ISO 11443 [20] the diameter of a cut sample no longer than 5 cm is manually measured or the diameter a distance of 10 mm below the die face is measured optically. ASTM D3835 [21] gives 6.25 mm distance from die face for measurement of cut extrudates and 25 mm from die face for optical measurements.

In an intercomparison study of tests in different laboratories, Allen and Rides [39] found that at a shear rate of 300 s^{-1} the contact measured results were 25–30% greater than those from optical measurements, but the latter method may have had less scatter.

However the different measurement distances of the optical sensors from the die face may have had a significant effect. Scatter could be reduced by using longer dies and smaller diameter dies (for a given extrusion rate). Use of entry cones also reduced scatter. The first conclusion was that with the ISO 11443 method a wide range of values was possible depending on the technique and testing conditions used.

Flow Defects

There are two generally recognized flow defects [17].

Sharkskin, which appears as a pattern of regular ridges positioned transversely to the direction of flow.

Melt fracture, which is a gross turbulence effect. In a strand extruded from a rheometer it will appear as a regular knotted or lumpy effect.

Sharkskin occurs at a lower shear rate than melt fracture, but the term melt fracture is often applied to all regular flow defects including sharkskin. Sharkskin occurs at a critical linear extrusion speed that can be raised by increasing melt temperature. Melt fracture starts at a critical shear stress and can be reduced significantly by reducing the die inlet angle, but, like sharkskin, it is also reduced by raising melt temperature. Linear low-density polyethylene is particularly prone to these defects, but they are minimized by the addition of special additives or blends with other polymers.

From a testing aspect, determination of critical shear rates and linear speeds can be made using a capillary rheometer by observing the conditions under which these defects occur at the relevant temperatures.

Data for Processability Computer Modelling

For prediction of mold filling and flow-through dies, computer modelling is increasingly used. In addition to viscosity, measurements may be needed for specific heat, shrinkage, thermal conductivity, and pressure–volume–temperature (PVT) relationship. A differential scanning calorimeter (DSC) can be used for specific heat measurements.

Measurement of shrinkage as a material property is covered by ASTM D955-89. In this test (which also covers transfer and compression molding), the shrinkage is derived from measurement of molded test pieces and the mold cavity dimensions. The moldings used are a bar 127 × 12.7 × 3.2 mm and a disc 102 mm dia × 3.2 mm thick.

Although this test is suitable for quality control and materials comparisons, processability computation frequently needs to take account of the shrinkage variables associated with the injection molding process, e.g., to minimize warpage. An empirical approach that provides guidelines for a number of polymers (including some with reinforcement) has been described by Fuzes [41]. This program covered a number of shapes, wall thicknesses, and tool temperatures, and it evaluated the influence of flow direction and type of gate.

There are two systems for measuring specific volume as a function of temperature and pressure [42]. The direct method uses a piston/capillary arrangement, and the indirect method has intermediate liquid (normally mercury) to pressurize, heat, and cool (Fig. 17)

Although in theory the indirect method should be better, Wendisch [42] concluded that there was little to choose between the two methods with regard to accuracy but that the direct method was easier from a practical aspect as no mercury was involved. Instruments are available for both methods. The indirect method was considered by Wendisch to be more suitable for research purposes [42].

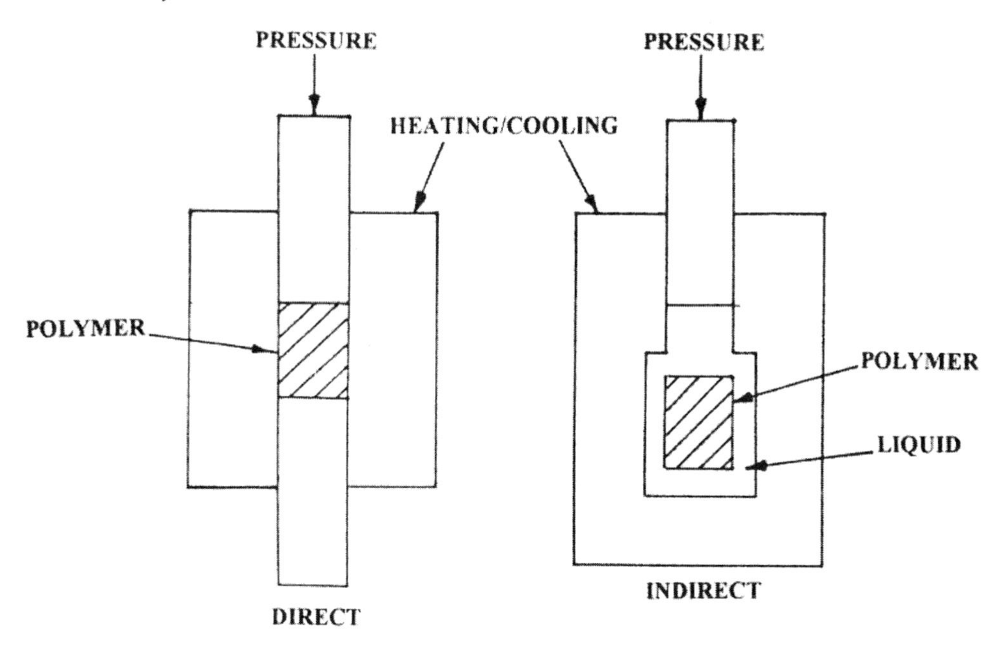

Figure 17 Arrangements of the "direct" and "indirect" PVT measuring systems.

Thermal conductivity measurement can be for steady state or unsteady state conditions. Steady state methods are more accurate for poor conductors. BS 874 [43] covers a guard ring method for solid and molten polymers. Coumar and Henry [44] have described an apparatus covering a temperature range of 30–300°C. For computer modelling, data may be needed for very high pressures as well as high temperatures. A transient linear source method by Lobo and Newman [45] takes 15–30 seconds, which avoids problems of degradation and chemical changes that might otherwise occur with the steady state methods. The various techniques were reviewed in the introduction of a paper by Sourour and Kamal [46]. Subsequently, apparatus has been described by Tan [47], Lobo and Newman [45], and Oehmke and Wiegmann [48]. This latter method used the direct PVT measurement apparatus in which a probe containing a heating wire and thermocouple was used to measure the increase in temperature as a known amount of heat was applied. Results were obtained from ambient to 400°C and 1600 bar pressure. Ejection temperature can be deduced from a DSC cooling trace, whilst the no-flow temperature can be measured using a cooling capillary rheometer.

3.7 Processability of PVC

PVC is often regarded as a special case. Set apart from most thermoplastics by its unusual particulate and subparticulate structures and the influence of the chlorine atom, it can be transformed by an exceptionally wide range of additives. As a result it can range from rigid to very flexible, from transparent to opaque, and from unfilled to heavily filled. From a processability aspect an extra dimension is the economy of in-house processing by powder mixing, often with direct conversion into the product, bypassing the compounding stage. This applies particularly to rigid PVC. As PVC cannot be processed into useful products without the addition of stabilizer and lubricant, but more often with at least five ingredients, a number of specific processability tests are of particular relevance. The processability testing of PVC has been reviewed by Bender and Booss [49].

Powder Properties

Powder and friction properties are covered in section 3.4. In ASTM D2395 [50] a test is described that measures the ability of PVC polymer to absorb plasticizer whilst remaining a free-flowing powder. The PVC polymer powder is mixed with liquid plasticizer in either a sigma blade mixer or a planetary mixer heated at 88°C and the torque is continuously recorded by a torque rheometer. The mix passes through a sticky state before becoming dry as the plasticizer is absorbed. The change in motor drive torque with time is used to measure the time taken for this absorbtion.

Fusion Time

This is equivalent to measuring the melting rate of other thermoplastics, but with PVC shear is as important as heat in the achievement of a homogenous melt. With unplasticized PVC powder blends in particular, direct extrusion into rigid products requires "fusion" to occur within certain limits to avoid overheating and subsequent degradation on the one hand, or underfused brittle products on the other. The torque rheometer graph of drive torque against time provides a fusion peak and a final steady melt torque. Information from the graph can be used as a measure of the compound's processability, although the

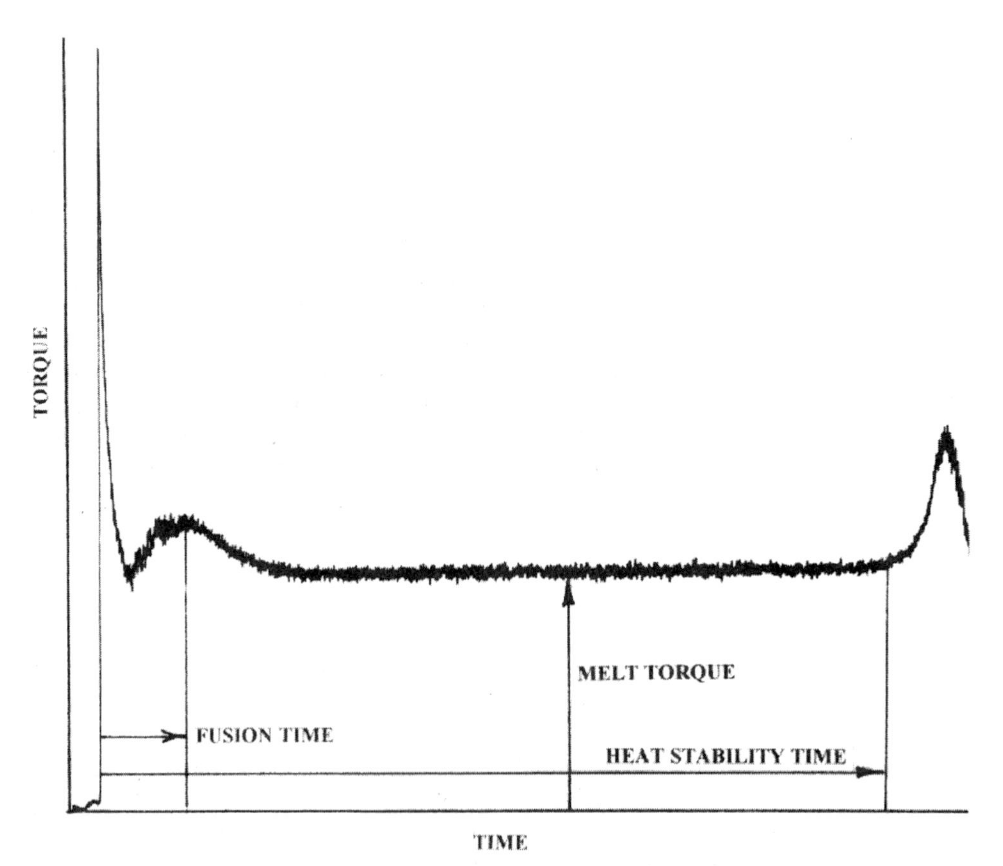

Figure 18 Typical torque rheometer trace for an unplasticized PVC powder compound.

values represent an inseparable combination of viscosity and lubrication. An example is shown in Fig. 18.

References to many papers concerned with this technique can be found in the Brabender bibliography [51]. ASTM D2538 [52] describes procedures for these tests, but although material consistency is not difficult to establish, processability predictions require a library of data based on comparisons with production experience.

Melt Flow Rate Test for PVC

An extension of the Melt Flow Rate test (described in Section 3.6) to accommodate PVC and rheologically unstable thermoplastics is covered by ASTM D3364 [53]. This standard covers semirigid and nonrigid (i.e., plasticized) PVC compounds. In so doing it addresses the potential problems with PVC that might otherwise occur from its limited thermal stability, elasticity, and time-dependent changes due to lubrication of the die. The die has the same 2.0 mm diameter of ASTM D1238 [23], but its length is about three times longer, and it has an included entry angle of 120°. Temperature is controlled at 175°C ± 0.1°C, noting that a 1°C temperature change can give a 19% change in flow rate. The applied load is 20 kg for some semirigid materials, which will cause problems with instruments not designed to accommodate this standard.

Orthmann [54] compared MFR with solution viscosities (K values) for PVC homopolymers and vinyl chloride/vinyl acetate copolymers. The polymer was milled with 0.5–1.0% tin stabilizer, granulated, and tested using a 10 kg load at 200°C for homopolymers and 155°C for copolymers. Samples were cut off every minute for the 9th to 13th minutes. MFR values for a number of materials ranged over three decades and did not correspond to K values in some instances, the MFR being the more sensitive measurement. No capillary details were given.

MFR values can be dependent on the PVC's level of gelation. This will depend on the shear and temperature history of the compound. An investigation by Parey and Zajchowski [55] showed that in the compounding process, increasing the extruder screw speed and increasing the extrusion temperatures both decreased the MFI, i.e., they increased the melt viscosity. A further complication with rigid PVC is wall slippage. Hegler et al. [56] concluded that the main influences on wall slippage were temperature, stress, and formulation.

Dynamic Thermal Stability

PVC compounds have a finite heat stability time during processing depending on their temperature, the final limit being "catastrophic" decomposition accompanied by a rapid release of hydrogen chloride. ASTM D2538 [57] gives two procedures. In the first the torque rheometer is run until the melt torque suddenly rises and the time from the maximum (fusion) torque to a sudden rise at decomposition is recorded. In the second method, small samples are removed from the mixing head at regular intervals, pressed into small plaques, and the time measured to reach a color end point. Color changes through pink, amber, brown, and finally black occur as PVC thermally degrades. Unlike the first procedure this test can be stopped before much chlorine is evolved, but quite obviously it is only suitable for clear, white, or light-colored compounds.

An alternative to the torque rheometer for dynamic heat stability testing of plasticized PVC is a two-roll mill. Measuring the time taken for the onset of sticking to the mill surface avoids serious chlorine evolution and can be used for any color, but it relies much on operator judgment.

3.8 Dispersion of Additives

The processor may use polymers in which the additives are already compounded or prepare a pellet blend of natural polymer and masterbatch. The additive concentration can vary from a few percent to about 80%, depending on the final concentration required. In many cases tests are carried out on the final product, but in some circumstances there is a need to check the quality of the compound or masterbatch before processing it into a final product.

In some cases samples are processed on a laboratory version of the production unit and the tests may be more concerned with materials quality than with processability. However, processability can become an issue when a masterbatch or compound introduces problems of achieving uniform coloring or specified electrical properties, etc. In extrusion, agglomerates can produce surface appearance defects, obstructions in small bore tubing (following "die drool"), voltage breakdown in cables, and failure to meet dispersion requirements of pipe, cable, and cistern standards.

With some notable exceptions, test methods in this general area are often without standards and passed around like good cookery recipes. BS 2782, Method 823A [58], includes an extrusion test in which the quality of carbon black dispersion is assessed by estimating the number of agglomerates in a fixed weight of compound. In this test $640\,g \pm 10\,g$ is extruded through a specified screen pack, and after purging the machine with natural polythene of the same MFR the screens are removed and the number of agglomerates counted under a microscope. The limit is 70. A variant of this test is to measure the melt pressure increase resulting from increasing clogging of the screens by an accumulation of agglomerates [59]. A method used for both carbon black and colors is to extrude masterbatch diluted with natural polymer or compound into blown film and and carry out an agglomerate count on a unit area laid on a light box [59]. For carbon black a final concentration of 1% masterbatch has been recommended [59]. Agglomerate counts by image analysis are possible both on microtomed sections [59] and on polymer streams passing a window [29]. The latter technique can be combined with on-line viscosity measurements described in Section 3.6. Incoming black technical compounds can be subjected to the normal materials quality tests using the microscopy techniques used in BS2782, Method 823B [60], and many similar product standards.

4 Thermosetting Plastics

For molding and casting processes the materials can be in the form of solid powders or pellets, doughs, or liquids. This group of materials in both of these processes exists in a liquid form during transformation into a solid product. More usually heat is applied such that the higher the temperature the faster the reactive change.

Bulk density and particulate flow tests can be performed as described in Section 3.4. Processability normally requires a measure of flow and rate of change from liquid to solid. However a problem normally exists in separating the two effects. Furthermore, the viscosity during the liquid stage may be continually changing. It typically decreases in the early stages as applied heat raises the temperature of the plastic and then increases as the cure process advances. If the reaction is significantly exothermic this will also influence both viscosity and onset of cure with time, and in turn will be influenced by test sample thickness.

4.1 Processability of Solid and Doughlike Thermosets

Thermal Flow and Cure Properties Using a Torque Rheometer

This test has some similarities with the measurement of fusion rate of PVC as described in Section 3.7. A test method is described in ASTM D3795 [61]. Unlike the cup flow tests described below, the method bears no similarity to the molding process, but this technique gives a better insight into the viscosity changes with time, particularly in the earlier part of the molding cycle.

A weighed amount is charged into the mixing chamber set at the appropriate test temperature and the instrument run at 40 rpm. The drive torque is recorded against time using either a chart recorder or a microprocessor. A typical chart recorder trace is shown in Fig. 19. Following melting of the thermoset, the torque falls as the viscosity falls with increasing material temperature. With the onset of cross-linking, the viscosity and hence the torque pass through a minimum and then rise, climbing to a peak where the material is no longer fluid. Beyond this point the torque falls as continued curing is accompanied by chopping and granulating by the rheometer so that the plastic is reduced to a crumb.

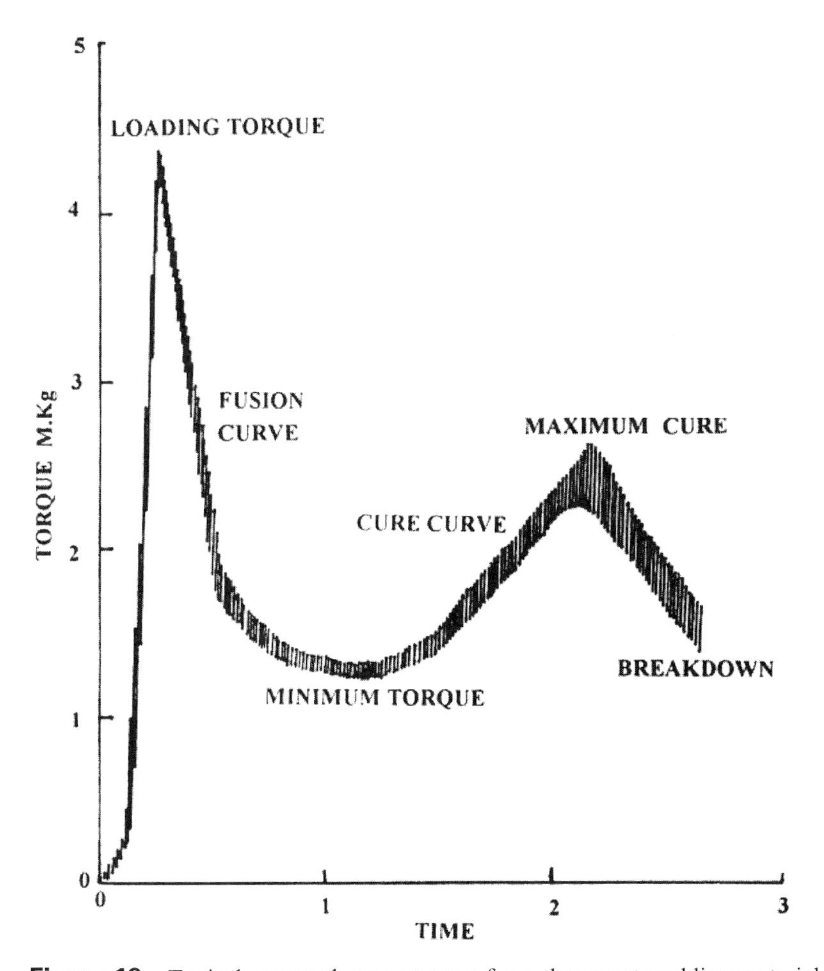

Figure 19 Typical torque rheometer trace for a thermoset molding material.

Consequently the graph beyond the peak is probably of little relevance, while stopping the instrument earlier and not emptying it extremely quickly could result in very difficult to remove blocks of hard plastic. ASTM D3795 lists head temperatures for eight different materials, including two temperatures for cross-linkable polyethylene, epoxy, and phenolic. Amino plastics are not in the list, although this type of instrument can be used for these materials as described by Paul [62].

Cup Flow Method for Thermosetting Molding Powders

This technique is essentially a model molding process. Its simulation of the real world of compression molding makes it a popular test for categorizing grades of molding powder, and it is probably as familiar in the field of phenolic molding powders as the MFR test is for thermoplastics. Regrettably, and in spite of its long history, the cup flow test, unlike the MFR test, has three quite different standards. BS 2782, Part 7, Method 720B [63], ASTM D731 [64], and DIN 53465 [65] are three versions of the same test. Although the three standards use a circular tumbler-shaped cup with the mouth larger than the base, the dimensions are quite different. ASTM D731 has the simplest design with a constant wall thickness throughout; DIN 53465 uses a cup that is nearly half as tall again with 60% of the wall thickness, while BS 2782, Method 720B, looks as though it were based on a real cup, being slightly smaller in diameter than the ASTM design but having a variable thickness base and tapering sides.

The BS, ASTM, and DIN designs are shown in Fig. 20. The BS standard according to its title is restricted to phenolic and alkyd molding materials, but the ASTM standard specifies temperatures for phenolic, melamine, urea, epoxy, diallyl phthalate, and alkyd molding materials. In all three standards the tool is mounted in a hydraulic press and the time taken for the ram to stop moving is measured, but the procedural details are different.

Measurement of Spiral Flow

Flow properties of thermosets can also be measured using transfer molding into a spiral flow tool as described in ASTM D3123 [66]. Material injected under specified transfer conditions flows spirally outwards along a groove with numbered distance marks until cured to an extent that flow ceases. Molding is carried out at 150°C. Length of spiral is measured to the nearest 25.4 mm (1 inch). The standard states that the test is suitable for

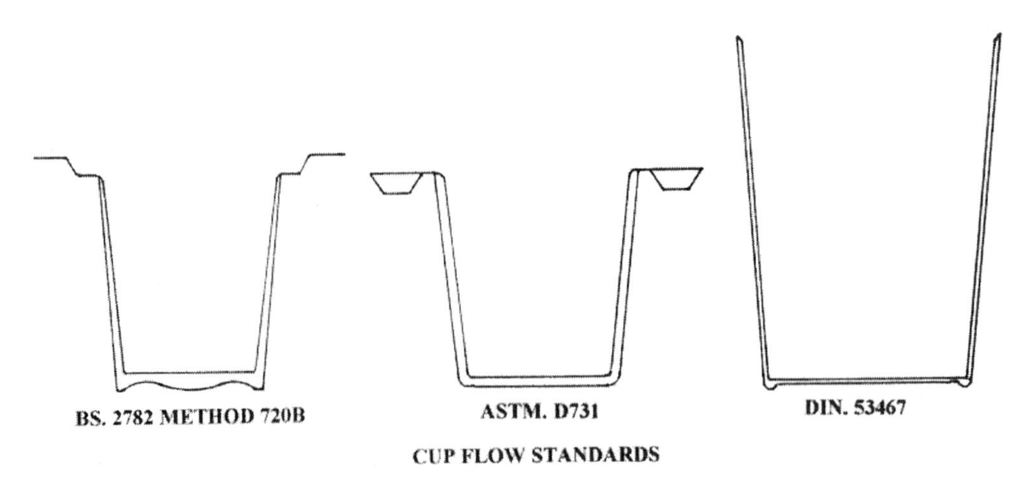

Figure 20 Flow cup designs for testing thermoset flow/cure behavior.

quality control testing as an acceptance criterion for low-pressure molding materials but that it is unsuitable for textile-reinforced compounds and comparing moldability of different molding compounds.

Resin Flow from Resin-Impregnated Glass Fabric

This test is covered by BS2782, Part 7, Method 721A [67]. Although similar to IEC249-3-1, BS4584, Part 11, and IEC249-3, it has a wider scope. The procedure assesses the flow of resin under heat and pressure from a fabric impregnated with polyester, epoxy, or silicone resins. It is primarily for glass fabrics but includes paper, cotton/synthetic fiber fabrics, glass mat and glass rovings, plus other resins such as phenolic, melamine, and polyamide. The method involves molding a multi-ply laminate from 100 mm square pieces cut from a roll of preimpregnated cloth fabric and following weighing, calculating the weight loss from the test piece.

4.2 Processability of Liquid Thermosets

Standards describe a surprisingly large number of different test methods that measure the time it takes for a thermosetting liquid composition to gel at a particular controlled temperature. The point at which the liquid gels is detected by a number of techniques ranging from operator judgment during manual stirring to automatic detection devices.

Measurement of Gel Time of Unsaturated Polyester Resins

Liquid polyester resins are often used at ambient temperatures, for applications such as glass fiber impregnation, for example, in the manufacture of boats, or for casting purposes. Curing is usually achieved by addition of peroxide initiator and accelerator immediately before use. Measurement of the time to gel, i.e., the time the operator has to impregnate the reinforcement or pour the castings, etc., is of particular relevance.

ISO 2535 [68] describes a method using a prescribed cure system with measurement of time from point of accelerator addition to onset of gelation at a temperature of 25°C. The resin is contained in a test tube with temperature control using a water bath. A glass rod is rotated at 1–2 rev/min by a torsion wire connected to a geared electric motor. The gel time is taken as the time taken for the wire to twist an amount corresponding to a liquid viscosity of 50 Poise for measuring this point and is shown in the standard.

BS 2782, Part 8, Method 835B [69] describes a manual method in which gelation is carried out at elevated temperatures achieved by heating with the vapor of a boiling liquid contained by a reflux condenser. Choice of temperatures of 81–82.5°C and 116–118°C is achieved using an appropriate alcohol with an intermediate vessel of liquid paraffin. The resin is stirred manually every 15 seconds until the operator detects gelation. Timing is from the immersion of the test tube holding the sample into the liquid paraffin to the observed gel point.

Method 835C [70], which uses a gel timer, applies to epoxide as well as to polyester resins. In this test the catalyzed resin is tested in an aluminum container of specified dimensions and placed in a water or oil bath at any one of seven recommended temperatures from 25° to 150°C or even higher if required. A plunger rod terminating in a disc is oscillated vertically by a motorized drive over a distance of 12.7 mm at a frequency of 6 s for gelation times between 5 and 20 min, and for long gel times the frequency is increased to between 6 and 60 s. Time is measured to when gelation causes resistance to movement to exceed the weight of the plunger.

Measurement of Gel Times for Phenolic Resins

Method 835D [71], which is suitable for phenolics and some other hardenable compositions including epoxies, measures the time for gelation of 1 g of material on a hot plate. During heating the molten resin or composition is kneaded by the operator with a palette knife following a defined procedure. Gel point is when the resin becomes stringy. This test is obviously rather messy and possibly operator dependent, but it has the advantage of ensuring a well-dispersed mix of hardener, reinforcement/filler, etc., without the need for very expensive test equipment such as a torque rheometer.

BS 2782, Part 8, Method 835A [72], for phenolic resins uses the general idea of ISO 2535 in which the sample is stirred with a glass rod in a test tube maintained at a constant temperature in a thermostatically controlled bath. In this method the suggested test temperatures are 130°C or 150°C (referring to methods 111A and 111B of BS 2782). Gel time is taken to the point where the manual stirring shows the resin to have become rubbery.

For gel time of carbon fiber-epoxy prepreg, ASTM D3532 [73] describes a method whereby a 6.3 square of prepreg is sandwiched between two microscope cover glasses placed on a hot plate at 121 or 177°C. The gel time is measured for a resin bead squeezed to the edge to become stringy. Manipulation is by wooden toothpick or fine glass rod.

Simultaneous Measurement of Gel Time and Rate of Cure

The above gelation tests do no more than measure the time to gelation. Time to reach complete cure may be independent of gelation time. Instruments such as the vibrating needle curemeter monitor changes after the gelation point. The VNC [74] replaces the slowly oscillating paddle with a high-frequency vibrating needle in which the amplitude of the constant force needle is measured, with measurement continued beyond the gel point. Thus in addition to gel time, viscous forces can be measured as cure progresses. By using a microprocessor the frequency of vibration can be changed to match the resonant frequency of the sample. As frequency will increase with increasing elasticity a plot of resonant frequency will monitor rate of cure. The instrument can be used for liquid thermosetting resins, sealants, cast elastomers, and polyurethane foams.

5 Viscosity Measurement for Liquid Polymers, Solutions, and Dispersions

A variety of standards exist, some of which apply to specific commercial instruments. All the following ISO standards have equivalent BS standards as listed in the references.

ISO 3219, 1993 [75], covers polyester resins as liquids, emulsions, or dispersions using a rotational viscometer, coaxial cylinder viscometer, and cone and plate system. ISO 2555 [76] refers to the Bookfield viscometer. ASTM D1824 [77] covers the use of this instrument for the measurement of apparent viscosity of plastisols and organisols at low shear rates. For high shear rates, ASTM D1823 [78] describes a method using a Burrell Severs A-120 viscometer. ISO 1628 [79] covers determinations of viscosity number and limiting viscosity for PVC, polyolefins, polycarbonate, thermoplastic polyester, and methyl methacrylate polymers, in parts 2 to 6 respectively.

Cellulose acetate is covered by ISO 1157 [80] and minimum film forming temperatures of polymer dispersions by ISO 2115 [81].

6 Introduction to Rubber Processability

An important goal in rubber compounding is to achieve specific product performance requirements while optimizing the cost of the compounding materials used. The rubber compounder is also concerned with measuring processing characteristics and predicting how a given compound will behave in the plant. The compounder often develops a formulation that will impart excellent performance characteristics to the rubber product at a low material cost; however, it must be processable to be useful.

Various processability tests are used to predict how well a rubber compound will process. However the "process" can vary greatly. Table 1 shows the diverse categories of rubber processes and some of the quality characteristics and concerns associated with each process. The natures of these processes vary because of differences in applied shear rates, temperatures, residence times, etc. Changing a given compound property can improve performance in one or more of these processes but could hurt the performance in another process. This is why it is important to look at all processes in a manufacturing operation when implementing compound changes.

Table 1 Rubber Processes with Example Quality Characteristics and Concerns

Mixing Quality	**Tire Building**
Filler incorporation time (BIT)	Building tack
% Filler dispersion	Green strength
Compound viscosity	Second step blow outs
Crumbly batches	
Lumpy stocks	**Bin Storage**
Slow mixing	Scorch
Milling	Green strength
Back rolling	Sticky slabs
Bagging	
Take off, mill release	**Compression/Transfer/Injection Molding**
Thickness	Backrinding
Extrusion	Appearance
Smoothness	Shrinkage
Dimensional stability	Cured hardness
Extrudate shrinkage	Porosity
Die swell	Mold release
Extrusion rate	Mold fouling
Pinholes	Mold flow/fill
Scorch	
Calendering	**Autoclave Cures**
Scorch	Appearance
Bare spots, holes	Shrinkage
Dimensional stability, width, thickness	Mandrel release
Heat blisters	
Trapped air	
Calender release (from rolls)	
Fabric penetration	
Liner release	

Experience has shown that the quality aspects for these greatly different processes are generally related to one or more of the ten fundamental rubber properties given below.

1. Viscosity
2. Shear thinning
3. Thixotropy
4. Green strength and extensional viscosity
5. Elasticity and V/E ratio
6. Tackiness (rubber to rubber) and/or bloom characteristics
7. Stickiness (rubber to metal or fabric) and/or surface lubricity
8. Filler dispersion
9. Cure properties (including scorch, cure rate, and ultimate state of cure)
10. Overcure stability (including reversion and marching modulus)

Viscosity is the resistance of a raw rubber or rubber compound to flow. Mathematically viscosity (η) is calculated from shear stress divided by shear rate as shown in Eq. 1.

$$\eta = \frac{\text{Shear stress}}{\text{Shear rate}} \tag{1}$$

Rubber viscosity can be measured by a rotational viscometer, a capillary rheometer, or an oscillating rheometer. These are discussed later in the chapter.

Shear thinning describes how much a rubber's measured viscosity decreases with an increase in the applied shear rate. By contrast, *Newtonian* fluids are materials that have a constant viscosity with increasing shear rate. (Water is an example of a Newtonian fluid.) Shear thinning is one of the forms of *non-Newtonian* behavior. Virtually all polymers display some degree of shear thinning. It is the degree of shear thinning (sometimes called *shear thinning index*) that is important in predicting the downstream process behavior for a rubber compound.

Thixotropy is the decline of viscosity with time for a rubber under conditions of steady state shearing. This decrease in viscosity can be related to filler loading and the destruction of filler agglomerates [82]. When the shearing is stopped, the viscosity can rise again or *recover* as the filler particles reagglomerate.

Green Strength is the tensile modulus and/or tensile strength of an *un*cured rubber compound. This property can relate to processes such as calendering, extrusions, and building. For example, this property is important for rubber compounds that are used in the construction of a tire on a second-stage tire building machine. Radial tires that are built with a rubber compound that has poor green strength may fail to hold air during expansion in the second stage of construction before cure. Elastomers with higher molecular weights and a tendency to crystallize on stretching will impart higher green strength to a rubber compound. Generally compounds based on natural rubber will have higher green strength than compounds based on most commonly used synthetic elastomers [83].

Elasticity is a quality that conforms with Hooke's law as follows [84]:

$$\sigma = E\gamma \tag{2}$$

where

σ = stress or force acting per unit area
γ = strain (displacement) as defined from change in length
E = the static modulus of elasticity

Perfectly elastic materials do not show any dependence on the rate of applied deformation but only on the magnitude of the deformation.

Uncured rubber compounds and raw elastomers in general will have differing degrees of both viscous and elastic quality. This combination of properties is called *viscoelasticity*. The elastic quality of uncured rubber is largely due to chain entanglements that occur with high molecular weight polymers. These entanglements restrict the movement of long molecular chains when a strain is applied. Two rubbers can have an identical viscosity but differ in elasticity. The rubber with higher elasticity will behave differently in a process, have greater resistance to a given deformation, be less thermoplastic in its behavior, be more "nervy" on the mill, and probably display greater die swell from an extrusion process. Many times the ratio of the viscous quality to the elastic quality is used to predict rubber processabillity. This is sometimes referred to as the *V/E* ratio. Often, a higher V/E means better processing characteristics for a rubber. When this ratio is calculated from sinusoidal deformation, it is referred to as the uncured *tangent* δ or *tan* δ.

Tackiness is defined as the ability of a rubber compound to stick to itself or to another rubber compound with only a moderate amount of applied pressure and a short dwell time [85]. Good rubber compound tack is necessary in building an uncured tire or constructing a conveyor belt by laying calendered plies on one another. Natural rubber commonly imparts a high level of building tack to a compound. Elastomers such as EPDM typically do not impart high tack to a rubber compound. In these compounds, ingredients called tackifiers are typically added to improve tack. Also, many compounding ingredients have limited solubility in rubber compounds. Some ingredients such as sulfur, accelerators, and/or antidegradants may separate from the compound under certain cooling conditions, exude to the rubber surface, and impart a surface *bloom*. This bloom can not only cause an appearance problem but also reduce or destroy building tack.

Stickiness is the ability of a rubber compound to stick to a nonrubber surface such as a metal or textile fabric surface. Excessive stickiness to metal can result in poor release from mills and problems in process equipment. On the other hand, the destruction of compound stickiness can cause slippage of the compound against metal surfaces in extruders or the rotors of an internal mixer. Various compounding additives, such as mill release agents or external lubricants, are used to control the level of stickiness. Note, compound stickiness and tackiness are not the same property.

Filler dispersion is a property that determines how well the filler partciles in a given rubber compound are dispersed as a result of the mixing process. This relates to carbon black dispersion as well as the dispersion of nonblack fillers such as silica, clay, calcium carbonate, titanium dioxide, etc. Also rubber curatives such as sulfur and accelerators can be poorly dispersed (commonly these ingredients are added late in the mixing cycle). Poor dispersion makes a mixed stock less uniform, and commonly the cured ultimate tensile strength will have more variability. Poor dispersion can affect other important cured physical properties such as abrasion, tear, and fatigue resistance, flexometer heat buildup, and other dynamic properties.

Cure properties of a compound are equally important in predicting rubber processability. *Time to scorch* is the time required at a set temperature for a rubber compound to develop incipient cross-links from the vulcanization process. When this scorch point is reached for a compound after a given heat history during factory processing, the compound can *no longer* be processed by milling, calendering, extruding, etc. Therefore measuring the scorch point is very important to determine whether a compound can be successfully processed in a particular factory operation. *Cure rate* and the *cure times* are also important characteristics. Cure rate measures how fast a compound increases modulus (builds cross-link density) during cure at a specified temperature. Cure times indicate how much time at a given temperature is required before a compound reaches a given state of cure; for example, how much time is required for the compound to reach 50 percent or 90 percent of the ultimate state of cure at a given cure temperature. These parameters help determine how long the uncured rubber article has to remain in a mold, press, autoclave, or salt bath. If the residence time is too short, poor cured physical properties will result. If the residence time is too long, some cured articles may display a deterioration in cured physical properties and/or reduction in production rates. Lastly, the *ultimate state of cure* is important as well. This can relate to the ultimate cross-link density and modulus that is achieved during the curing process.

Overcure stability is the steadiness of the compound's physical properties in the cured state under conditions of prolonged overcure. With thick article cures, some compounds are unavoidably exposed to overcure conditions. Certain compounds may not have good overcure stability. For example, many compounds based on natural rubber will decrease in modulus after peaking at the optimal cure when left at extended exposure times to the cure temperature. This decrease in modulus from overcure is called *reversion*. Also some compounds appear never to reach a cure plateau but continue to increase in modulus indefinitely at the cure temperature. This quality is commonly referred to as *marching modulus*. Obviously, the compound's overcure stability is very important in a rubber production facility.

To Lord Kelvin is attibuted the statement " . . . applying numbers to a process makes for the beginnings of a science." The ten fundamental rubber properties just discussed are all quantifiable. One or more of these fundamental properties relate to each of the eight different processes given in Table 1.

This next section reviews and compares the different processing tests that are used in the rubber industry today. These process tests either directly or indirectly measure one or more of the ten fundamental rubber properties just discussed.

For this review, process tests are categorized as follows:

1. Miniature internal mixers and extruders
2. Rotational viscometers
3. Capillary rheometers
4. Oscillating rheometers
5. Stress relaxation testers
6. Green strength tests
7. Tack testers
8. Carbon black dispersion tests

7 Miniature Internal Mixers and Extruders

Miniature internal mixers (MIMs) are used in the rubber industry to simulate the mixing process. MIMs can be used for testing raw rubbers or rubber compounds. They are described in ISO 2393 and ASTM D 3182 [86]. Because these MIMs consist of a jacketed stainless steel mixer head and bowl of relatively low volume (usually only 120 cm^3), the temperature control during mixing is quite good compared to a laboratory Banbury. Therefore these MIMs are used to measure such properties as raw rubber mastication profiles (breakdown energy) or carbon black incorporation times for mixed stocks. However, scale-up from a MIM to a factory Banbury shows a large step change because of the large difference in the ratios of mixer volume to mixer internal surface area. This scale-up affects a mixer's time–temperature profile and the dissipation of heat during mixing. MIMs are commonly used in compound development; but they are not as widely used in routine quality control of rubber compounds because their repeatability is often dependent on operator technique. Also, the MIM test time is long compared to some alternate test methods [87–88].

Laboratory extruders are also used to predict processing characteristics of a rubber compound. ASTM D2230 describes a commonly used laboratory extruder method using the Garvey die, which is designed to determine extrudability in the factory [89]. Figure 21 shows the cross section of the Garvey die. This ASTM standard uses a subjective rating system based on the smoothness and sharpness of the corners of the extrudates and the die swell. The extrusion rate can also be measured. One problem with the laboratory extruders as processability testers is that they require subjective evaluations of the extrudate quality.

8 Rotational Viscometers

Parallel plate or cone and plate rotational viscometers have been used to measure the viscosity of molten plastics and coatings for many years, but these types of viscometers are not commonly used to measure the viscosity of rubber. Because of the special nature of rubber, the Mooney viscometer has historically been used to measure and control the viscosity of raw rubbers as well as rubber compounds. The Mooney viscosity test is described by ISO 289, BS 903, Part A58, and ASTM D1646 [90].

The Mooney viscometer was developed by Melvin Mooney of the U.S. Rubber Co. in the 1930s and was adopted as the standard method for controlling the quality of GR-S (SBR) by the Technical Committee of the Rubber Reserve Co. in 1942 [91]. Since then, the Mooney viscometer has become the standard method for testing raw rubber as well as mixed stocks. A cross section of the Mooney viscometer is shown in Fig. 22. This method consists of rotating a special serrated rotor while embedded in a rubber sample within a

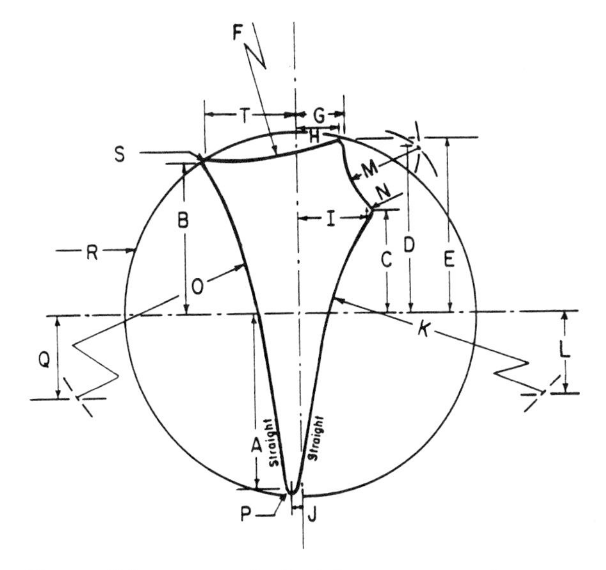

Figure 21 Garvey die for laboratory extrusion.

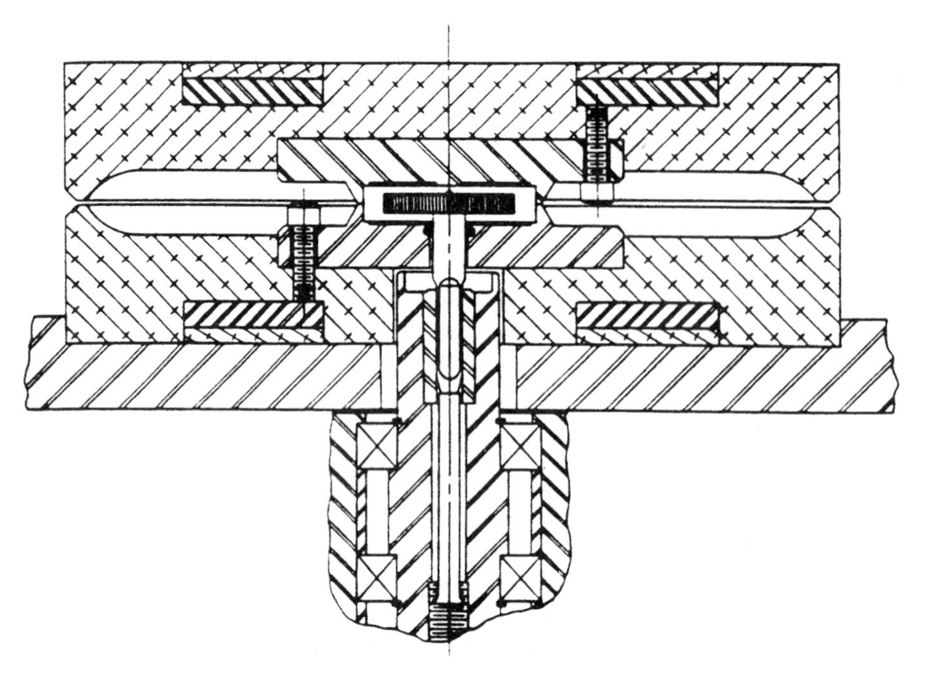

Figure 22 Cross section of Mooney viscometer dies, platens, and rotor.

sealed, pressurized, temperature-controlled cavity. The rotor turns at a standard rate of 2.0 revolutions/minute (0.21 rad/s). If the rate of rotation were higher, viscous heating might result. The special serrations on the rotor and cavity are necessary to prevent slippage [92]. Mooney viscosity results are reported in arbitrary Mooney units (MU), which are based on torque as defined by ISO 289 and ASTM D1646. (These standards define one Mooney unit equal to 0.083 N-m.) The Mooney viscometer is used to measure the viscosity of raw rubbers and compounds or to measure scorch characteristics of "final" mixed batches containing curatives.

Mooney viscosity is used by many synthetic rubber manufacturers as the single most important measure of their polymer's processing behavior. The Mooney test conditions that must be specified are as follows:

1. Mill massed specimen or unmassed specimen (U)
2. Test temperature
3. Size of the rotor (ML for large and MS for small)
4. Preheat time in minutes (or the delay after closing the dies but before moving the rotor)
5. Rotor running time in minutes

So UML 1 + 4(100°C) defines a test for a given rubber that was tested "as is" (unmassed) with a large rotor, at 100°C, a one-minute preheat, and four minutes of rotor running time.

The rubber industry, through ISO, ASTM, and other standards organizations, has reached a consensus on what the standard test conditions should be for 14 different classes of rubber, as shown in Table 2. A typical Mooney viscosity curve is shown in Fig. 23. As noted, the recorded Mooney viscosity typically reaches an initial peak when the rotor

Table 2 ASTM Standard Test Conditions for Mooney Viscosity Testing of Raw Rubber

Type rubber	Mooney test temperature, °C	Mooney running time, min
Natural rubber	100	4
BR	100	4
CR	100	4
IR	100	4
NBR	100	4
SBR	100	4
BIIR	100 or 125	8
CIIR	100 or 125	8
IIR	100 or 125	8
EPDM	125	4
EPM	125	4
Syn. rubber black masterbatch	100	4
IRM 241 butyl	100 or 125	8

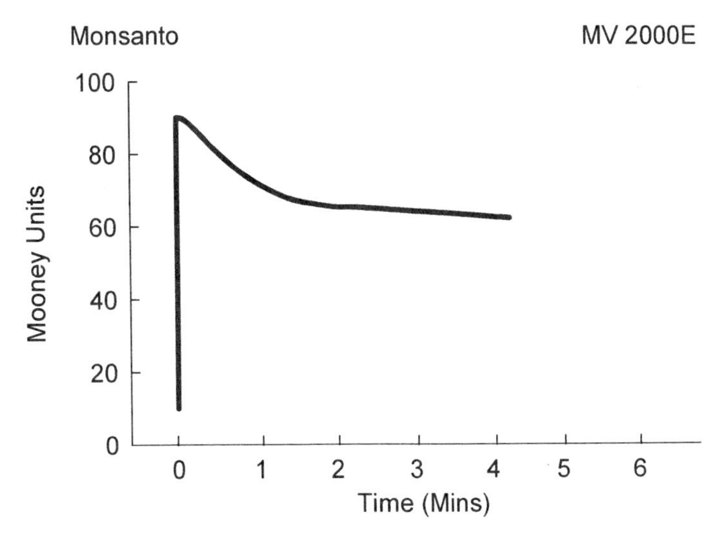

Figure 23 Typical Mooney viscosity curve for a raw rubber.

starts to rotate, but this viscosity normally drops with rotor running time denoting the thixotropic nature of the rubber compound or raw rubber being tested [93].

A Mooney scorch test is a method for measuring the scorch safety time (time to the onset of vulcanization at a given temperature) for a mixed stock containing curatives. This scorch is measured as the time required to reach a 5 MU rise above the minimum (t_5) when a large rotor is used, and 3 MU rise above minimum (t_3) when a small rotor is used. Figure 24 illustrates this measurement of scorch safety. Usually the Mooney viscometer is set at a temperature significantly higher than what is used to measure Mooney viscosity. Also, the cure rate is calculated as the "ASTM cure index" using Mooney scorch results and Equations 3 and 4 [94]:

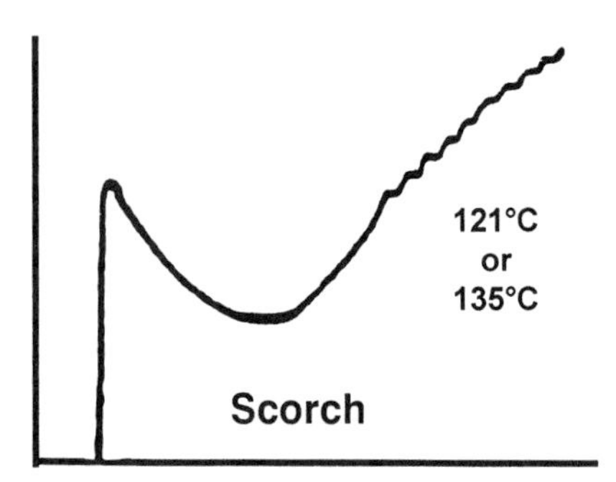

Figure 24 Typical Mooney scorch curve for a rubber compound with curatives.

$$\Delta t_S = t_{18} - t_3 \quad \text{for small rotor} \tag{3}$$

$$\Delta t_L = t_{35} - t_5 \quad \text{for large rotor} \tag{4}$$

A lower cure index represents a faster cure rate. There is a limit, however, to the cure information that a Mooney scorch test can provide as the cure advances beyond t_{18} (for a small rotor) and t_{35} (for a large rotor). This is because as the specimen cures and builds a cross-linked matrix, it loses flow properties and begins to slip and tear on the rotor [95].

9 Capillary Rheometers

Virtually all rubber compounds are non-Newtonian. This means that their measured viscosities will decrease with increasing shear rates. The slope of this viscosity drop is compound dependent. It is quite possible that a "compound 2" could have a lower viscosity than a "compound 1" at low shear rates, but compound 2 could cross over at higher shear rates where compound 1 has the lower viscosity [96]. This is illustrated in Fig. 25 [97]. The Mooney viscometer, when run at the standard rotational speed of 2 r/min, has a maximum shear rate of about $1\,s^{-1}$ (the shear rate changes across the rotor radius) [98]. The capillary rheometer is used because it can measure viscosity at much higher shear rates (at shear rates as high as $1000\,s^{-1}$).

ASTM D 5099 describes the capillary rheometer when testing rubber compounds. Figure 26 shows a schematic of a piston type capillary rheometer that is commonly used in rubber applications. A given quantity of rubber compound is packed into the heated barrel of the capillary rheometer. After a given preheat time, the drive system is started to push the rubber through a preselected die. The apparent shear rate is a function of the ram travel rate, which usually can be selected through the computer. The apparent shear stress is calculated from the resulting pressure measured by a pressure transducer near the die [99]. The apparent shear rate is an average shear rate, under the assumption that the rubber is Newtonian (which it is not). The "true" shear rate is determined by applying the Rabinowitsch correction (100–101]. Also, the apparent shear stress does not take into account the entrance/exit effects at the die. The "true" shear stress is calculated from the Bagley correction [102]. Viscosity is calculated by dividing "true" shear stress by "true" shear rate.

Some capillary rheometers (such as the Monsanto Processability Tester®) are equipped to directly measure the die swell of the rubber compound after it exits the die. This is done with special optical die swell detectors. *Running die swell* is measured at

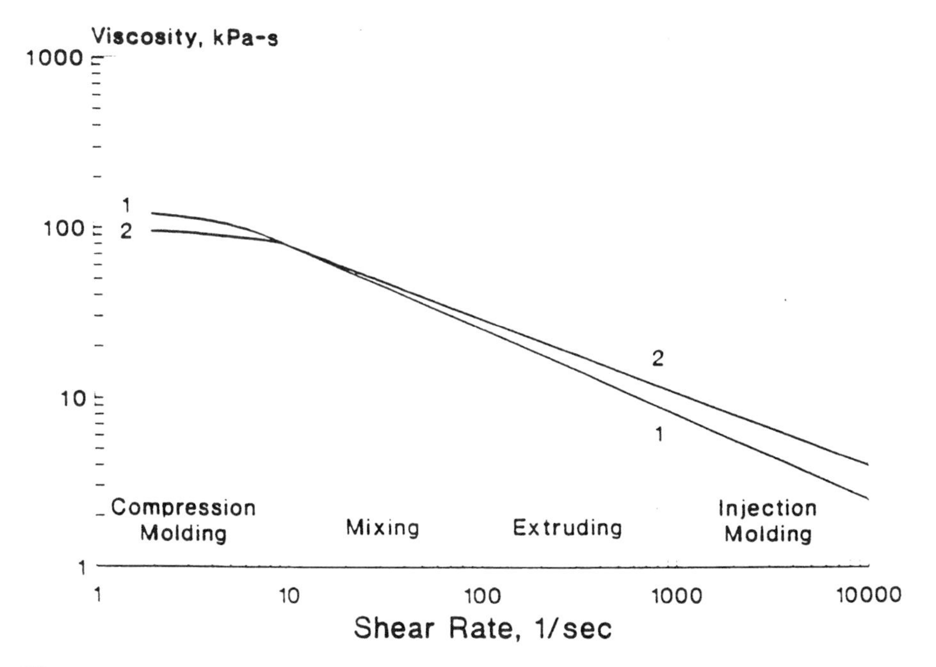

Figure 25 Typical response for viscosity of a rubber compound to changes in shear rate.

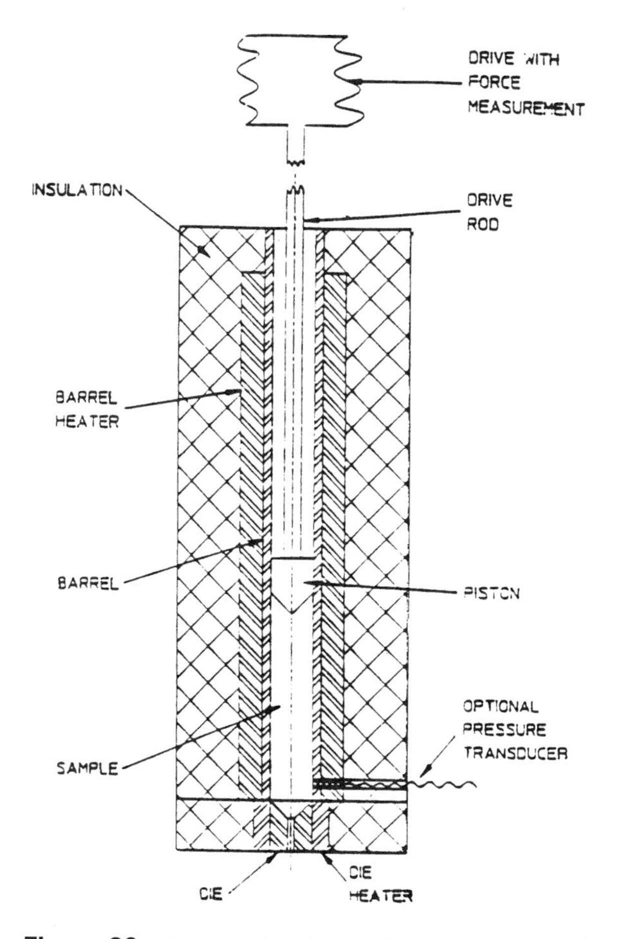

Figure 26 Cross sectional view of a piston type capillary rheometer.

different shear rates. *Relaxed die swell* can also be measured [103]. Compound die swell is a time and temperature-dependent property that correlates to the thickness of a calendered sheet, bagging on mills, and dimensions of extruded shapes [104].

10 Oscillating Rheometers

As discussed earlier, the Mooney scorch test can measure time to scorch and provide some information regarding the rate of cure. However, because the Mooney rotor rotates continuously, slipping and tearing of the specimen occur well before the ultimate state of cure is reached. It is not possible to observe the entire cure curve with the Mooney viscometer.

10.1 Oscillating Disk Rheometer

In 1963 Monsanto introduced the oscillating disk rheometer (ODR) [105]. In contrast to the Mooney viscometer, the ODR sinusoidally oscillates (not rotates) a biconical disk while embedded in a rubber sample in a sealed, pressurized cavity. Since the disk is oscillating instead of rotating, it can measure changes in the state of cure from scorch all the way to complete cure as shown in the ODR cure curve given in Fig. 27. This disk can oscillate with a strain amplitude of ± 1, 3, or 5 degrees arc. For compounds that cure to very high modulus, it is best to select a low strain amplitude to avoid slippage.

Unlike the other instruments we have discussed, the ODR gives useful information concerning not only a compound's processing behavior and scorch safety but also average and peak cure rate, final state of cure, and reversion resistance properties. Figure 28 shows a schematic of the ODR. The ODR is described in ISO 3417, BS 903, Part A60-2, and ASTM D2084 [106]. Since its introduction in 1963, over 5000 ODRs have been produced throughout the world.

10.2 Rotorless Curemeters

Even with the improved capabilities of the ODR, there were problems with the disk, which are listed here [107].

1. The disk functions as a heat sink, significantly increasing the time required for the specimen to reach the set temperature of the die cavity. This increases the total test time required and reduces the accuracy of cure studies.
2. Torque signals are measured at the bottom of the shaft that oscillates the disk. This makes it more difficult to measure accurately the dynamic properties due to the friction of the rotor and bearings.
3. At the end of a cure test, it is difficult to remove the sample because the rotor is embedded in the center of the cured sample.
4. The disk prevents effective use of film (either PET or Nylon 6,6) to protect the dies from being fouled. Thus cleaning is often required.
5. Automation is very difficult because of the requirement to remove the sample off the disk after the test is complete.

Because of these problems with the disk, rotorless curemeters were introduced in the 1980s. These new curemters apply a sinusoidal strain to the rubber specimen by oscillating the lower die (there is no rotor). An example of a rotorless curemeter die design is shown in Fig. 29 for the Alpha Technologies MDR 2000® Moving Die Rheometer. The reaction torque transducer is attached to the upper die and separated from the lower oscillating die. This separation produces a signal that is due only to the property of the rubber. This

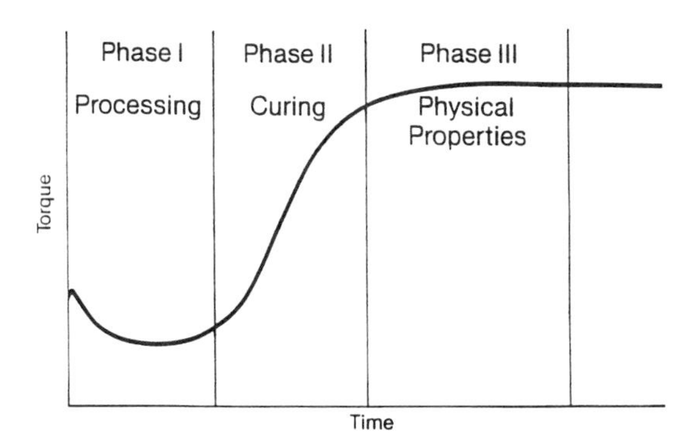

Figure 27 Typical ODR cure curve.

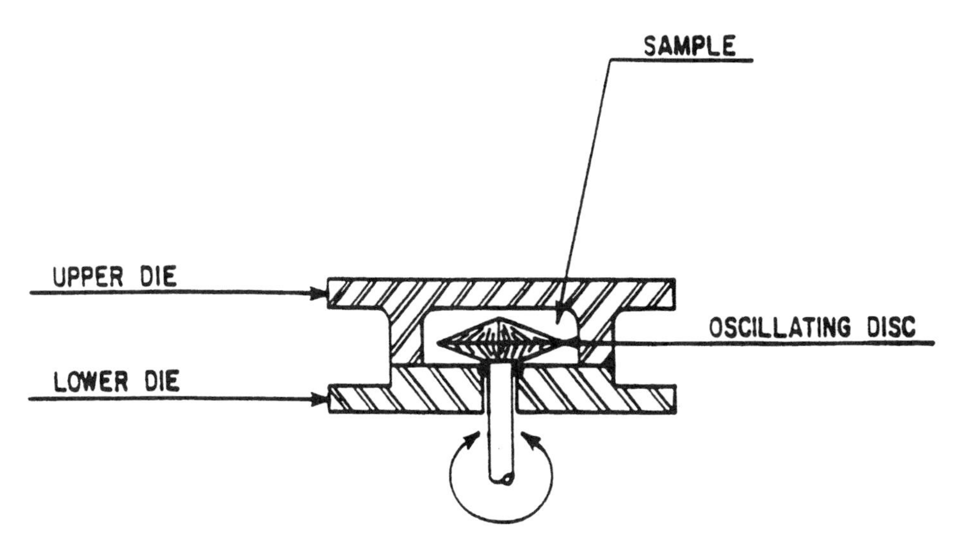

Figure 28 Cross section of ODR.

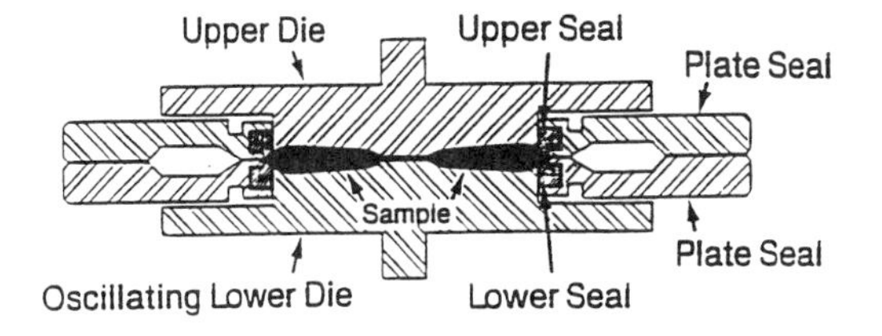

Figure 29 Schematic of MDR 2000 moving die rheometer test cavity.

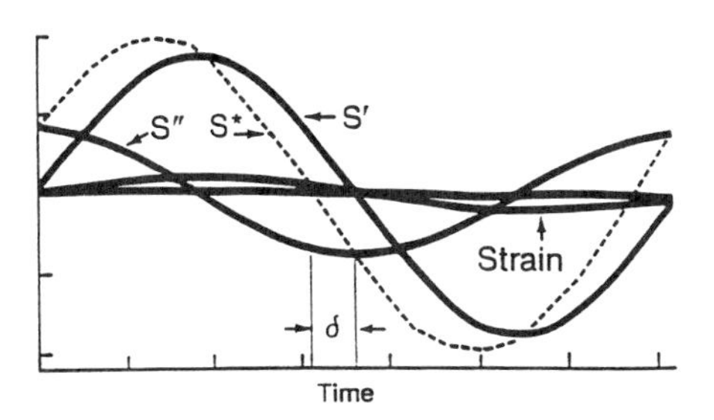

Figure 30 Torque response of MDR for a single cycle of die oscillation.

allows the measure of real dynamic properties. Because the frictional losses associated with the rotor were eliminated with the MDR, dynamic proeprties before, during, and after cure can be measured effectively. Figure 30 shows the elastic, viscous, and complex torque responses measured by the MDR with a rubber sample at a given sinusoidal strain. The dotted line in Fig. 30 indicates the complex torque (S*) response that the reaction torque transducer measures at the upper die. The peak strain and the peak complex torque occur at different times. This time or angle difference between the complex torque and the applied strain is noted by the phase angle shift δ. This difference is because of the visco-elastic nature of both uncured and cured rubber. The instrument software separates the complex torque into the elastic torque (S') and viscous torque (S''). Elastic torque (S') is in phase with the applied strain curve. The viscous torque (S'') is 90° out of phase with the applied strain. The parameter tan δ is calculated by dividing S'' by S'. A lower tan δ for a cured compound means greater resiliency [108,109].

Another feature of the rotorless curemeter die design (shown in Fig. 29) is significantly shorter temperature recovery time (the time required for the sample to reach the set temperature of the dies). A shorter recovery time produces a shorter cure test and better testing productivity compared to the ODR. Also, the MDR design gives significantly better test sensitivity in detection of compound variation than the ODR [110,111]. In addition, the MDR design gives significantly better test repeatability, better interlaboratory reproducibility, and better Gage R & R results than the ODR [112]. The MDR's near isothermal cure is more accurate and useful in cure kinetic studies [113,114]. The dynamic properties measured by the MDR provide the user with additional information that can be used to determine the assignable cause of variation in a rubber compound mixing process. Some of these dynamic property parameters are more sensitive to compound changes than the traditional ODR curemeter parameters [115].

ISO 6502 and ASTM D5289 describe rotorless curemeters used in the rubber industry [116]. The ASTM standard was published in 1993. This new technology is widely accepted by the rubber industry.

10.3 Rubber Process Analyzer

The rotorless curemeter technology has evolved into the Rubber Process Analyzer. This new rubber test instrument was commercially introduced by Monsanto in 1992 as the RPA 2000® [117]. Currently work is underway to establish this instrument as an ASTM standard. The RPA has the same die design as the MDR shown in Fig. 29. However, unlike other rotorless curemeters, the RPA moves the lower die with a special direct drive motor. Through a computer, the operator programs tests in which the oscillation frequency, strain amplitude, and/or test temperature are varied during a test. Special foil heaters enable rapid temperature increases while a forced air system reduces temperature rapidly. The range of applied strains, frequencies, and temperatures are given in Table 3. These features enable the RPA to function as a true dynamic mechanical rheological tester (DMRT) that can test raw polymers, masterbatches, and mixed stocks before, during, and after cure. Thus the RPA can be used as (1) a raw polymer tester, (2) a processability tester, (3) an advanced curemeter, and (4) a DMRT that measures after-cure dynamic properties at temperatures below the cure temperature [118]. The RPA can apply enough strain and frequency during oscillations to break up carbon black aggregate-aggregate networks of various rubber compounds and predict rubber factory processing behavior "down stream" [119]. The RPA can provide fundamental rheological properties such as storage modulus G′ and loss modulus G″ as well as dynamic complex viscosity η^* and real

Table 3 Commonly Used Range of RPA 2000 Test Conditions

Frequency	2 to 2000 cycles per minute
Strain	±0.1 to ±90° arc
Temperature	35° to 200°C

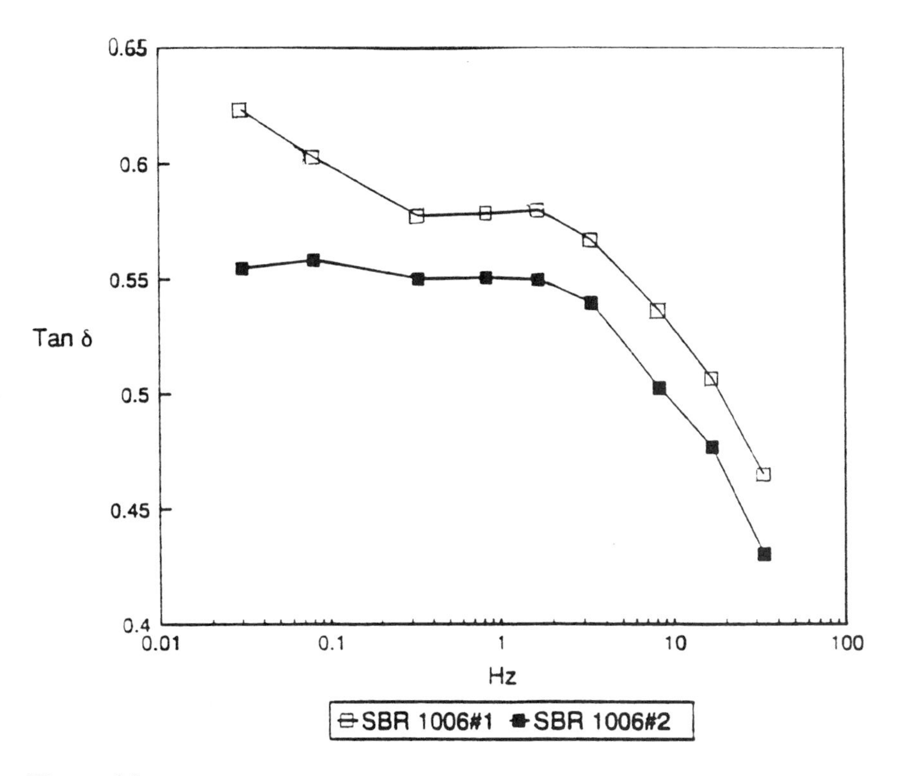

Figure 31 Comparison of two SBR 1006 polymers from different producers with the same tested Mooney viscosity values. Graph shows $\tan\delta$ response in an RPA frequency sweep at 100°C, 0.5 degrees arc.

dynamic viscosity η'. The RPA measures these properties with excellent repeatability because of its sealed, pressurized sample cavity [120].

Figure 31 shows the $\tan\delta$ response in an RPA frequency sweep on two SBR 1006 rubbers with the same Mooney viscosity. The RPA distinguishes polymers by their processing properties even though they may have the same Mooney viscosity. (Typically rubber lots can have the same Mooney viscosity but still process quite differently.) The RPA gives the viscoelastic profiles for raw rubbers, which relates directly to processing differences [121].

Figure 32 shows that the RPA can predict the quality of mix for a generic tire tread compound. As noted, the RPA uncured $\tan\delta$ correlates to the quality or state of mix. This RPA parameter also relates to percent carbon black dispersion [122].

Figure 33 shows that the RPA can be an advanced, rotorless curemeter. Both the S′ elastic torque and S″ viscous torque are shown during cure. The S″ cure curve is more sensitive to variations in oil and carbon black than the S′ cure curve [123].

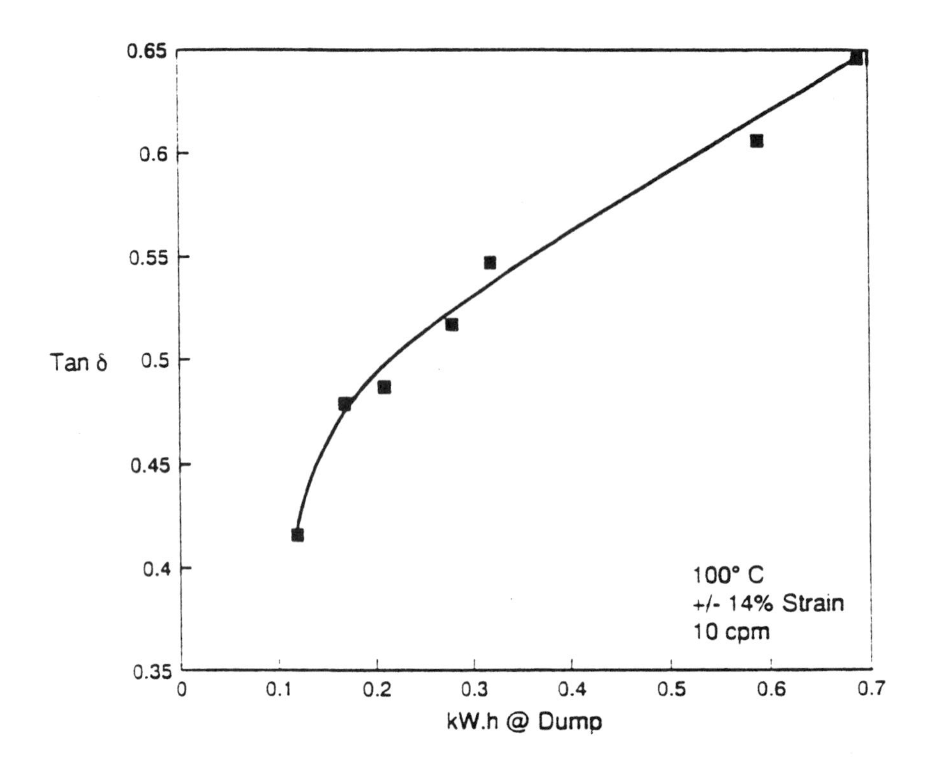

Figure 32 Uncured RPA tan δ response vs. Banbury kilowatt hour at dump for NR truck tread stocks. RPA test conditions: ±14% strain, 10 cpm, and 100°C test temperature.

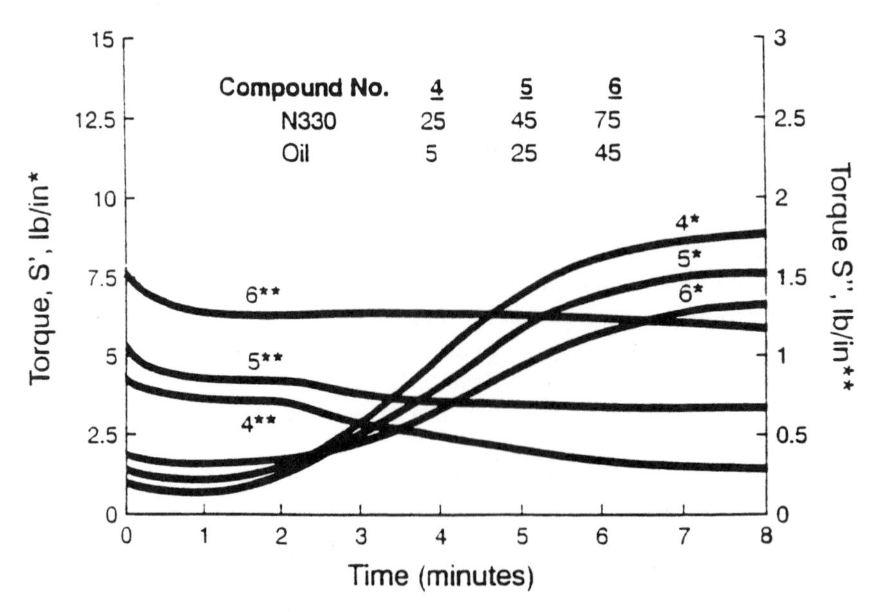

Figure 33 Changes in the S′ and S″ cure curves to indicated changes in oil and carbon black loadings in a vibration damping compound. Test conditions during cure: ±0.5° arc strain at 160°C, 100 cpm.

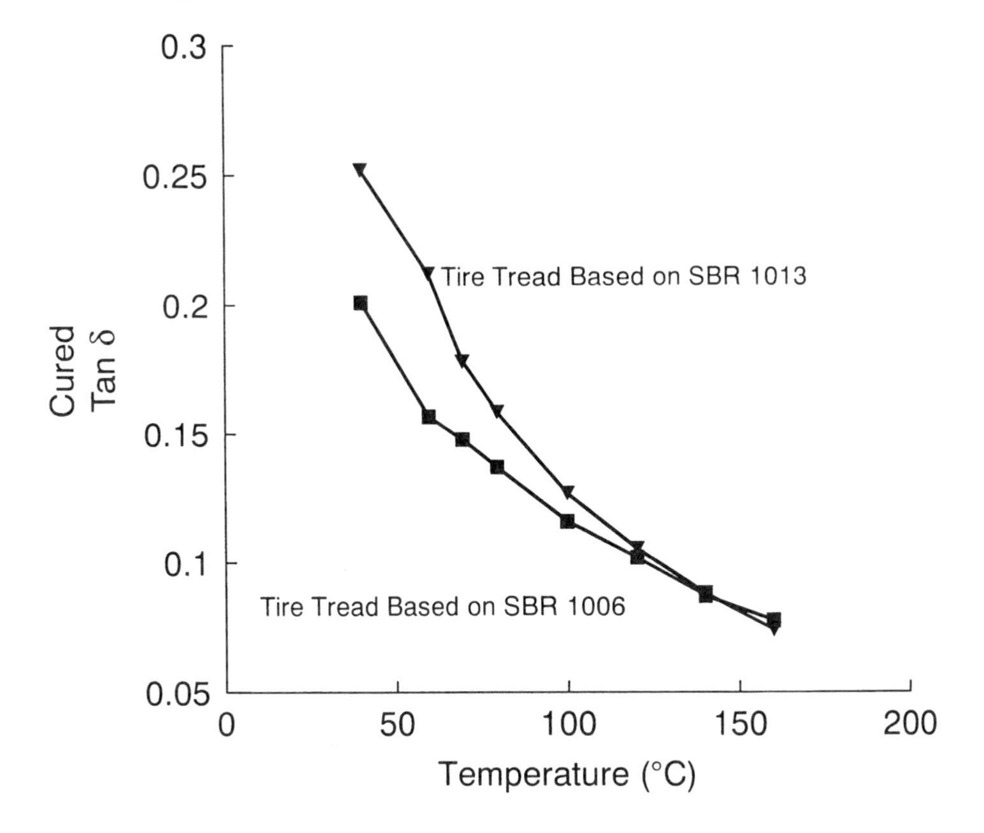

Figure 34 Comparison of RPA cured tan δ response in temprature sweep for a tread compound based on different hot emulsion SBRs as indicated. Test conditions: 100 cpm and $\pm 1.4\%$ strain. Cured at 170°C for 15 min.

Figure 34 shows the RPA tan δ response from a temperature sweep on two tire tread compounds cured *in situ* in the RPA. Because these two tread compounds are based on two different SBRs with different glass transition temperatures, the cured tan δ values, which are nearly identical at the cure temperature, are quite different at temperatures below the cure temperature. The RPA measures the curing characteristics of a tread compound, then is programmed to drop down automatically to a lower temperature and measure dynamic property measurements that relate to the tire rolling resistance. Thus the RPA is used to generate "real time" feedback on cured dynamic properties for a tire tread compound fresh from a Banbury mixing operation. This "real time" feedback provides enough time for corrective action before future mixes are made (if these dynamic properties fall outside of established specifications or warning limits) [124].

11 Stress Relaxation

Another very fast and effective method for measuring rubber processing properties is to perform a stress relaxation test. A stress relaxation decay curve can quickly quantify the viscoelastic properties of both raw rubbers and mixed stocks. The Maxwell model, shown in Fig. 35, illustrates this principle with a spring and dashpot in series [125]. A sudden

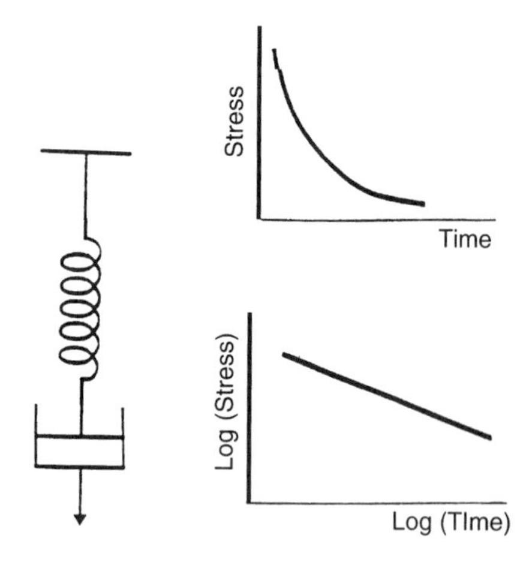

Figure 35 Maxwell model for viscoelastic behavior.

applied extensional deformation results in a characteristic stress relaxation curve as shown. A log–log plot of the resulting curve is often linear. Calculating the steepness of the slope in this log–log stress relaxation curve is a rapid method for quantifying viscoelasticity. When stress relaxation is fast, then there is a higher viscous quality relative to the elastic quality. When the stress relaxation is slow, then elasticity is the dominant quality. ASTM D6048 [126] (a general practice guide) has just been published that describes how stress relaxation measurements are done under the following test conditions.

1. Stress relaxation after sudden step shear strain [127,128].
2. Stress relaxation after sudden compression step strain [129] (this includes the German "Defo" method which is described in DIN 53514 and ASTM D6040)
3. Shear stress relaxation after cessation of steady shear flow (which includes Mooney stress relaxation [130] now described in ASTM D1646)
4. Shear stress relaxation after sudden stress application in a capillary rheometer [131].

The Alpha Technologies MV 2000® Mooney viscometer can perform a stress relaxation test after completing the viscosity test (see condition 3, above). For example, an ML 1+4 test can be done on a raw rubber sample followed by a two-minute stress relaxation test. The stress relaxation portion of the test is initiated when the Mooney rotor is suddenly stopped. The total test time would be seven minutes (or two minutes longer). Figure 36 shows two polymers that have the same Mooney viscosity yet have different stress relaxation profiles [132].

The RPA can also do a stress relaxation test. Figure 37 shows RPA stress relaxation decay curves from three truck tread stocks of the same formulation. Each curve indicates different qualities of mix caused by variations in mixing work history (variations in total energy at Banbury dump). As the energy at dump for these tread mixes increases, the peak torque decreases, the slope of the log–log plot of the stress relaxation curve becomes steeper, the regression line intercept decreases, the integrated area under the curve decreases, and the time to a given % drop decreases [133]. Therefore a simple stress relaxation test is a fast and effective way to measure variations in viscoelastic properties for both raw rubber and mixed stocks.

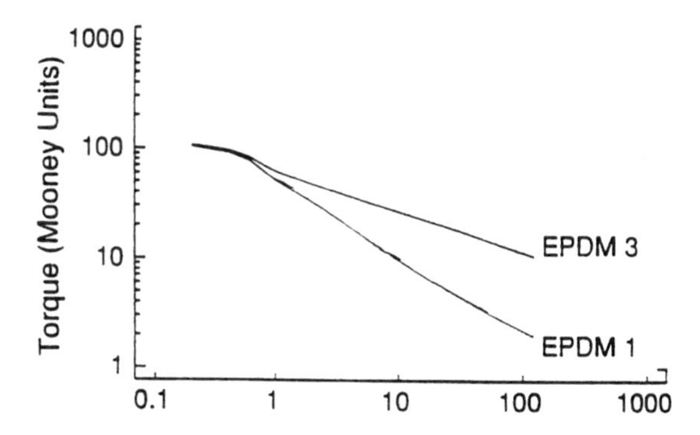

Figure 36 Mooney stress relaxation results on different production lots of EPDM polymers with the same Mooney viscosity.

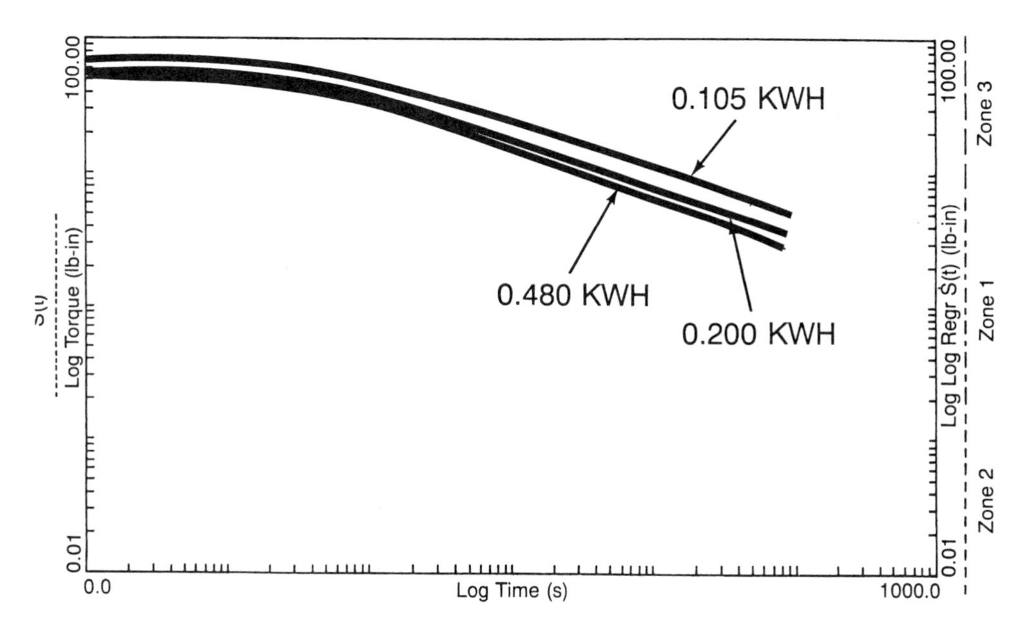

Figure 37 Comparison of RPA stress relaxation decay curves for three truck tread stocks with different qualities of mix caused by indicated variations in mixing work history (energy at Banbury dump).

Lastly, we should mention another category of tests called *compression plastimeters*. These tests are not stress relaxation methods, strictly speaking, but are related. They usually consist of measuring the resulting deformation of a cylindrically cut unvulcanized rubber specimen between two parallel plates from the application of a given force for a defined time period. This gives a measure of *plasticity*, which is defined as "ease of deformation." Also some of these test methods have recovery procedures. These procedures call for the specimen recovery height to be measured at a specified time period after the removal from the plastimeter. This specimen height recovery is related to the elasticity of the specimen. These methods were first introduced in the year 1924 [134]. They are usually simple to perform and are used in routine quality assurance testing for rubber comparisons of the same type. However, disadvantages of these methods are that they are performed at very low shear rates and sample preparation significantly affects test results. These shear rates are usually much lower than the shear rates seen in factory processing. ISO 2007 describes a rapid plastimeter method, while ISO 7323 and a similar method, ASTM D926, describe the Williams plastimeter procedure.

12 Green Strength Tests

Green strength tests are commonly performed on tensile testing instruments in accordance with International Standard ISO 9026. The unvulcanized rubber compound is sheeted off a laboratory mill at 2.2 mm thickness and placed in a mold. The sample is compressed for 5 minutes at 100°C with 2.5 MPa of platen pressure to assure that a smooth sheet without porosity results. Test pieces are dumbbells in the shape of ISO 37 type 1 or type 2. The grain direction (orientation direction) of the sheet is placed along the length of the dumbbell die, which cuts the test specimen. These specimens are then tested on a tensile tester with a separation rate of 100 mm/min. Typical stress–strain curves are shown in Fig. 38. Yield stress, maximum stress in MPa, and yield elongation in percent are usually reported. This ISO standard calls for the test to be carried out on a minimum of five test dumbbell speicmens and for reporting the median and range of these measurements.

13 Tack Tests

There are no standard ISO or ASTM test methods presently available for measuring tack or stick properties of an uncured rubber compound. Many rubber companies measure this property by their own proprietary methods. However, Monsanto Instruments Group introduced an instrument called the *Tel-Tak Tackmeter* in 1969 for measuring tack for tire and/or conveyor belt compounds.

Uncured Tel-Tak specimens were prepared by cutting rubber sheeted out from a mill at about 1.4 mm (0.06 in) thick. They were then placed over a piece of square woven fabric and pressed between a sheet of mylar on the top and cellophane on the bottom in a mold for 5 minutes at 100°C. After cooling, specimens were cut from the fabric side to 0.25 in. × 2 in. Two rectangular specimens are required for each test. Each test piece is mounted in the Tel-Tak instrument perpendicular to the other in order to assure the same contact area for each test.

After removing the protective mylar film, the Tel-Tak measured the maximum force in psi required to separate a $\frac{1}{4}$ inch wide rubber test specimen from another identical rubber specimen after these test pieces are touched together under a given load for a given dwell time that is defined as between 0.1 and 6.0 minutes and a selected contact pressure that is defined as between 16 and 32 psi. The rate of separation is 1 in./min. The tack (autoadhesion) is reported as the force required to separate the two identical rubber specimens. This test can be performed again where a stainless steel strip replaces the lower rubber specimen in order to measure the "stickiness" of the upper rubber specimen on contact with this polished stainless steel surface. Lastly the so called "true tack" is calculated by subtracting the stickiness value (rubber-to-metal) from the original rubber-to-rubber tack value.

Humidity, test temperature, and the age of the test samples all will affect tack measurements; therefore these variables should be controlled in order to assure good test repeatability. Testing in a temperature- and humidity-controlled room is advised. Also, optimal test dwell time and contact pressure should be selected and used as the standard test conditions. Variations in these properties will affect results.

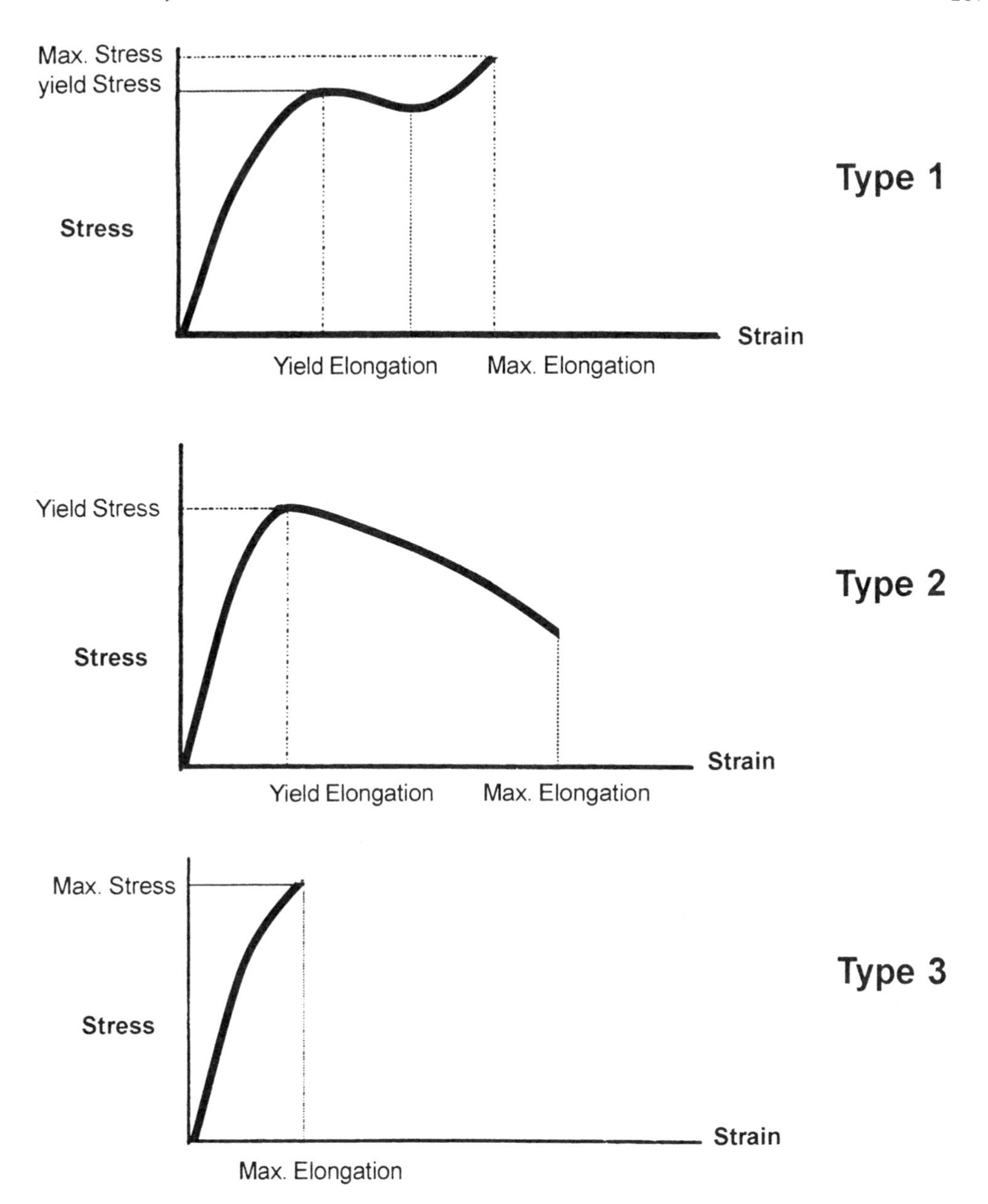

Figure 38

14 Dispersion Testers

ASTM D 2663 is the standardized test used in the rubber industry to measure the level of carbon black dispersion in a rubber compound. ASTM D 2663 describes three different methods for measuring dispersion.

ASTM D 2663, Method A, is a qualitative visual test method for rating mixed stocks against a set of standard photographs on a scale of 1 (for 70% dispersed) to 5 (for 99%

dispersed). This method calls for a visual examination of a freshly exposed surface from a torn vulcanized compound stock. The tear for this test may be initiated by a small cut. Also this standard describes how the surface of an uncured compound, without curatives, may be examined after being prepared using a sharp hot knife. However, unvulcanized compounds should be compressed in a mold for 5 minutes at 105°C to remove the air holes before testing.

ASTM D 2663, Method B, is a quantitative test method using a light microscope that determines the percentage area covered by black agglomerates in a microtomed section from the subject compound. The total cross-sectional area of all the agglomerates 5 μm or larger is counted. Since the content of carbon black is already known, the percent carbon black below 5 μm is expressed as the "percentage of carbon black dispersed." Percent dispersion above 99% is "very high," while at or below 95% it is "low." Preparing microtome specimens is somewhat labor intensive. This method requires a moderate amount of time to complete.

ASTM D 2663, Method C, is another quantitative method for measuring the level of carbon black dispersion using the trace of a stylus. A special dispersion analyzer system measures the roughness of a freshly exposed internal surface of the subject rubber compound. This "roughness" is measured from the trace of a fine stylus that has a tip radius of 2.5 μm. This roughness factor is based on the number average and average height of the "surface irregularities" from poorly dispersed carbon black agglomerates.

Lastly, ASTM D3313 for carbon black pellet hardness does *not* directly measure carbon black dispersion. This method does measure the individual carbon black pellet "hardness" or crush strength. Carbon black pellets that are too hard will not break up during mixing or disperse properly. Therefore by controlling the pellet hardness of carbon black, one cause of poor dispersion can be controlled. However, there are many other variables that influence dispersion as well. New sophisticated instruments are now available for measuring carbon black pellet crush strength.

References

1. Butters, G., *Plastics Pneumatic Conveying and Bulk Storage*, Applied Science Publishers, 1981.
2. ASTM D1895-89(reapproved 1990) Apparent density, bulk factor, and pourability of plastics materials, 1990.
3. ISO 60, 1977(E), and equivalent BS 2782, Part 6, Method 621A, Determination of apparent density of material that can be poured from a funnel, 1978.
4. ISO 61, 1976(E), and equivalent BS 2782, Part 6, Method 621B, Determination of apparent density of material that cannot be poured from a funnel, 1978.
5. Boysen, M., and Gronenbaum, J., *Plastverarbeiter*, *23*, No. 8, 549–550, (1972).
6. Klein, I., SPE, 34th ANTEC, April 1976, 444–449.
7. Huxstable, J., Cogswell F. N., and Wriggles, J. D., *Plast. Rubb. Process.* Appln. 1, No. 1, March 1981, 87–93.
8. Gale, G. M., SPE, ANTEC, 669–671, (1981).
9. Sneider, K., *Kunstoffe*, *59*, No. 2, 97, (Feb. 1969).
10. Riley, D. W., and Klein, I., SPE, 37th ANTEC, 153-155, (1979).
11. Emmanuel, J. A. O., and Schmidt, L. R., SPE ANTEC, 672–675, (1981).
12. Morton-Jones, D. H., *Polymer Processing*, Chapman and Hall, 1989.
13. Street, L. F., *Int. Plast. Engng*, *1*, No. 6, 289, (July 1961).
14. Maddock, B. H., *SPE J*, *15*, No. 5, 383, (May 1959).
15. Tadmor, Z., Duvdevani, I. J., and Klein, I., SPE ANTEC, 813, (1965).
16. Steward, E. L., SPE ANTEC, 73–76, (1987).

17. Cogswell, F. N., *Polymer Melt Rheology, A Guide for Industrial Practice*, George Godwin/John Wiley, 1981.
18. Macosko, C. W., *Rheology, Principles, Measurements and Applications*, VCH Publications, 1993.
19. Delaney, D., and Houlston, S., *Materials World*, 518–520, (Sept. 1996).
20. ISO 11443, Determination of the fluidity of plastics using capillary and slit die rheometers, 1995.
21. ASTM D3835-95a, Determination of properties of polymeric materials by means of a capillary rheometer.
22. ISO 1133, 1991(E), Plastics—Determination of the melt mass flow rate (MFR) and the melt volume flow rate (MVR) of thermoplastics. BS2782, Part 7, Method 722B, Plastics—Determination of the fluidity of plastics using capillary and slit die rheometers, 1996.
23. ASTM D1238-95, Flow rate of thermoplastics by extrusion plastometer.
24. ASTM D5422-93, Measurement of properties of thermoplastics materials by screw-extrusion capillary rheometer.
25. Padmanabhan, M., and Bhattacharya, M., *Rheol. Acta.*, *33*, 1, 71 (1994).
26. Kelly, A. L., Coates, P. D., Fleming, D. J., and Dobbie, T. W., *Plast. Rubb. Composites Processing and Applications*, *25*, 7, 313 (1996).
27. Nelson, B. J., Proc. 41st Annual ISA Analysis Div. Symposium, Vol. 29, 1996.
28. Gogos, C. G., Esseghir, M., Todd, D. B., Sebastian, D., and Garritano, R., SPE ANTEC, 1542–1545, 1993.
29. Yu, D. W., Esseghir, M., and Gogos, C. G., SPE ANTEC, 136–144, 1995.
30. Morrow, V., and DeLaney, D. E., SPE ANTEC, 1090–1094, 1996.
31. Sebastian D. H., and Dearborn, J. R., *Polym. Eng. Sci.*, *23*, No. 10, 572–575, (July 1983).
32. Meissner, J., and Hostettler, J., *Rheologica Acta*, *33*, 1–21, (1994).
33. Ide, Y., and White, J. J., *Appl. Polym. Sci.*, *22*, 1061, (1978).
34. Rides, M., Brown, C. S., Allen, C. R. G., Ferriss, D. H., and Gibbs, P. A. J., NPL Report CMMT(B)57. Extensional viscoelastic behaviour of high density polyethylene melts. National Physical Laboratory, Teddington, UK.
35. Rosand Precision Ltd. Technical data RH7 Capillary Rheometer.
36. Cogswell, F. N., *Polym. Eng. Sci.*, *12*, No. 1, 64–73, (Jan. 1972).
37. Rheometer Selection Guide Rh-01/96, Rosand Ltd., Stourbridge, UK.
38. ASTM D4440-95a, Rheological measurements of polymer melts using dynamic mechanical procedures.
39. Allen and Rides, NPL Report CMMT(A)9. Intercomparison of extrudate swell measurements using extrusion rheometers for a polyethylene melt, Jan. 1996.
40. ASTM D955, Standard test method for measuring shrinkage from mold dimensions of molded plastics, 1989.
41. Fuzes, L., *Int. Polym. Sci. Technol.*, *18*, 3, T32 (1991).
42. Wendisch, P., *Kunststoffe*, *86*, 11, 1730 (1996).
43. BS 874, Parts 1–3, Methods for determining thermal insulating properties, 1986–7.
44. Coumar, K. G., and Henry, K., SPE ANTEC, 1012 (1990).
45. Lobo, H., and Newman, R., SPE ANTEC, 802 (1990).
46. Sourour, S., and Kamal, M. R., *Polym. Eng. Sci.*, *16*, 7, 480 (1976).
47. Tan, V., *Adv. Polym. Technol.*, *11*, 1, 69 (1991).
48. Oehmke, F., and Wiegmann, T., SPE ANTEC, 2240 (1994).
49. Bender H., and Booss H J. Kunststoffe 73 No. 8 447-450 (1983) (eng. transl. 28–30).
50. ASTM D2396-94, Powder mix time of poly(vinyl chloride) (PVC) resins using a torque rheometer.
51. Brabender OHG, Bibliography—Application bulletins concerning tests on plastics materials etc. Duisburg.
52. ASTM D2538-95, Fusion of poly(vinyl chloride)(PVC) compounds using a torque rheometer.

53. ASTM D3364-94, Flow rates for poly(vinyl chloride) (PVC) and rheologically unstable thermoplastics.
54. Orthmann H J. Kunstoffe 58, No. 2 Feb 1968 159-162 (eng transl 21-23)
55. Parey J and Zajchowski S. Plastverarbeiter 32 No 6 1981 724-6
56. Hegler R P Mennig G and Weber G. Kunststoffe 73 No7 353–357 (1983) (eng. transl. 9–11)
57. ASTM D2538-95, Practice for fusion of poly(vinyl chloride) (PVC) compounds using a torque rheometer.
58. BS 2782, Part 8, Method 823A, Methods for the assessment of carbon black dispersion in polyethylene using a microscope, 1978.
59. Technical Report S-131. Cabot Corporation, Billerica, MA.
60. BS 2782, Part 8, Method 823B, Assessment of pigment dispersion in polyolefin pipes and fittings, microtome method, 1978.
61. ASTM D3795-93, Thermal flow and cure properties of thermosetting plastics by torque rheometer.
62. Paul, K T., Joint conference Brit Soc Rheol and PRI Loughborough, 49 (1975).
63. BS 2782, Part 7, Method 720B, Cup flow of phenolic and alkyd molding materials 1979.
64. ASTM D731-95. Molding Index of thermosetting molding powder.
65. DIN 53465, Determination of the closing time of thermosetting molding materials: cup flow test method.
66. ASTM D3123-94, Spiral flow of low-pressure thermosetting molding compounds.
67. BS 2782, Part 7, Method 721A, Determination of resin flow from resin impregnated fabric, 1988 (see also IEC C249-3-1 and BS 4584 Part 11).
68. ISO 2535-1974(E), Plastics—Unsaturated polyester resins—Measurement of gel time at 25°C.
69. BS 2782, Part 8, Method 835B, Determination of gelation time of polyester resins (manual method), 1980.
70. BS 2782, Part 8, Method 835C, Determination of gelation time of polyester and epoxide resins using a gel timer, 1980.
71, BS 2782, Part 8, Method 835D, Determination of gelation time of thermosetting resins using a hot plate, 1980.
72. BS 2782, Part 8, Method 835A, Determination of gelation time of phenolic resins, 1980.
73. ASTM D3532-76 (reapproved 1995), Gel time of carbon fiber-epoxy prepreg, 1995.
74. The Rapra Scanning VNC, Information Sheet 1, An introduction to cure monitoring, Rapra Technology, Shawbury.
75. ISO 3219, Plastics: Polymers in the liquid, emulsified or dispersed state: Determination of viscosity with a rotational viscometer working at defined shear rate, 1993.
76. ISO 2555, Resins in the liquid state or as emulsions or dispersions: Determination of apparent viscosity on the dispersions: determination of apparent viscosity by the Brookfield method, 1989.
77. ASTM D1824-95, Apparent viscosity of plastisols and organosols at low shear rates.
78. ASTM D1823-95, Apparent viscosity of plastisols and organosols at high shear rates by extrusion viscometer.
79. ISO 1628, Parts 1–6 and equivalent BS 2782, Methods 732A to 732F, 1991; ISO 1628-1, 1984; BS method 732A, Guidelines for the standardization of methods for the determination of viscosity number and limiting viscosity of polymers in dilute solutions, Part 1, General conditions. ISO 1628, Parts 2–6 are as follows: Determination of viscosity number and limiting viscosity number, Part 2, (BS method 732B)—Poly(vinyl chloride) resins. Part 3, (BS method 732C)—Polyethylenes and polypropylenes. Part 4, (BS method 732D)—Polycarbonate moulding and extrusion materials. Part 5, (BS method 732E)—Poly(alkylene terephalate). Part 6, (BS method 732F)—Methyl methacrylate polymers.
80. ISO 1157, and equivalent BS2782, method 733C, Determination of viscosity number and limiting viscosity number for cellulose acetate, 1990.
81. ISO 2115-1996 and equivalent BS2782, method 740C, Polymer dispersions—Determination of white point temperature and minimum film forming temperature.

82. Dick, John S., and Pawlowski, Henry, Applications of the rubber process analzyer in characterizing the effects of silica on uncured and cured compound properties, Presented at Rubber Div. Meeting, ACS, Montreal, Canada, May 4–8, 1996, p. 8.
83. ISO, Raw rubber or unvulcanized compounds—Determination of green strength, ISO 9026, International Organization for Standardization, 1988.
84. Blow, C. M., *Rubber Technology and Manufacture*, Newnes-Butterworths, 1971, p. 54.
85. Rhee, C. K., and Andries, J. C., Factors which influence autohesion of elastomers, *Rubber Chem. and Tech., 54*, 101 (1981).
86. *1995 Annual Book of ASTM Standards*, Section 9, "Rubber," Vol. 09.01, p. 456.
87. Morton, Maurice, *Rubber Technology*, 2d ed., Van Nostrand Reinhold, New York, 1973, p. 91.
88. Sezna, John, The use of processability tests for quality assurance, *Rubber World*, p. 23 (January 1989).
89. *ASTM Book of Standards*, p. 391.
90. *ASTM Book of Standards*, p. 325.
91. Symposium on Rubber Testing, ASTM Special Technical Publlication No. 74, Fiftieth Annual Meeting of ASTM, Atlantic City, New Jersey, June 16–20, 1947, p. 36.
92. Brown, R. P., *Guide to Rubber and Plastics Test Equipment*, 3d ed., RAPRA Technology Limited, Shawbury, Shrewsbury, Shropshire, England, 1989, p. 22.
93. Ibid., p. 22.
94. *ASTM Book of Standards*, p. 332.
95. Dick, J. S., *Compounding Materials for the Polymer Industries*, Noyes Publications, Park Ridge, New Jersey, 1987, pp. 115–116.
96. Morton, M., op. cit., p. 89.
97. Sezna, John, Instruments in rubber processes, Presented at Ft. Wayne Rubber Group Educational Symposium, November 8, 1994.
98. Brown, R. P., *Physical Testing of Rubber*, 2d ed., Elsevier Applied Science Publishers, London, 1986, p. 93.
99. *ASTM Book of Standards*, pp. 794–796.
100. Armstrong, Robert C., *Dynamics of Polymeric Liquids*, John Wiley, New York, 1987, pp. 527–528.
101. Sezna, John, *Monsanto MPT Applications Manual.*
102. Ibid., p. 4.
103. Monsanto Processability Tester MPT, Alpha Technologies LP, 1994.
104. MPT—Processability Tester, Alpha Technologies L. P., brochure.
105. Decker, G. E., Wise, R. W., and Guerry, D., An oscillating disk rheometer for measuring dynamic properties during vulcanization, *Rubber Chem. Tech., 36*, 451 (1963) and *Rubber World, 147*, No. 3, 68 (1963). Also presented at the Cleveland Rubber Division, ACS, October 17–19, 1962.
106. *ASTM Book of Standards.*
107. DiMauro, P. J., de Rudder, J., and Etienne, J. P., New rheometer and Mooney technology, *Rubber World* (January 1990).
108. Sezna, J. A., Pawlowski, H. A., and DeCoinck, D., New test results from rotorless curemeters, 136th Meeting of the ACS Rubber Division, Fall 1989, pp. 6–12.
109. Dick, John S., The optimal measurement and use of dynamic properties from the moving die rheometer for rubber compound analysis, *Rubber World* (January 1994).
110. Ibid.
111. Stanich, G. J., 3M evaluation of Monsanto Rheometers (ODR 100S, ODR 2000 and MDR 2000), July 31, 1991. Later published in *Rubber World.*
112. Sezna, J., and Dick, J., The use of rheometers for process control, *Rubber & Plastics News* (April 13 and 27, 1992).

113. Dick, John S., and Pawlowski, Henry, Applications for the curemeter cure rate in rubber compound development and process control, Presented at the Rubber Division, ACS, Philadelphia, May 2–5, 1995.
114. Dick, John S., and Pawlowski, Henry, Alternate instrumental methods of measuring scorch and cure characteristics, *Polymer Testing*, *14*, 45–84 (1995).
115. Dick, J. S., Optimal measurement, use of dynamic properties from the MDR for compound analysis, *Rubber World*, 19–25 (January 1994).
116. *ASTM Book of Standards*, p. 803.
117. Pawlowski, H., and Dick, J., Viscoelastic characterization of rubber with a new dynamic mechanical tester, presented at the Akron Rubber Group, April 23, 1992 (later published in *Rubber World*, June, 1992).
118. Dick, John S, and Pawlowski, Henry, Applications for the rubber process analyzer, *Rubber & Plastics News* (April 26, May 10, 1993).
119. Dick, John S., and Pawlowski, Henry, Rubber characterization by applied strain variations using the rubber process analyzer, *Rubber World* (January 1995).
120. Alpha Technologies RPA Training Course Manual, Part 2, Unit 5, p. 30.
121. Dick, J., and Pawlowski, H., *Rubber & Plastics News* (January 1995).
122. Ibid.
123. Ibid.
124. Dick, John S., and Pawlowski, Henry, Applications of the rubber process analyzer in predicting processability and cured dynamic properties of rubber compounds, presented at the Denver, Colorado meeting, May 18–21, 1993, p. 18.
125. Ferry, John D., *Viscoelastic Properties of Polymers*, 3d ed., John Wiley, New York, 1980, p. 15.
126. ASTM D6048, Standard practice for stress relaxation testing of raw rubber, unvulcanized rubber compounds and thermoplastic elastomers, New ASTM standard prepared in Stress Relaxation Task Group to D11.12, Subcommittee on Processability.
127. Pawlowski, H., and Dick, J., Measurement of the viscoelastic properties of elastomers with a new dynamic mechanical rheological tester, presented at the Rubber Div. ACS at Philadelphia, May 2–5, 1995, Paper 50.
128. Dick, John S., and Pawlowski, Henry, Applications for stress relaxation from the rubber process analyzer in the characterization and quality control of rubber, presented at the Rubber Division, ACS Meeting in Cleveland, October 17–20, 1995, Paper 26, p. 2.
129. Schramm, G., Rubber testing with the Defo-Elastometer, *Kautshuk + Gummi, Kunststoffe*, 1987, Vol. 40.
130. Burhin, H., Spreutels, W., and Sezna, J. A., MV 2000 Mooney viscometer—Mooney relaxation measurements of raw and compounded rubber stocks, presented at the Rubber Division ACS in Detroit, Oct. 17–20, 1989.
131. Sezna, John, MPT versus SRPT stress relaxation, Monsanto Tech Service Notes, June 3, 1981.
132. Burhin, H., Spreutels, W., and Sezna, J. A., MV 2000 Mooney viscometer—Mooney relaxation measurements of raw and compounded rubber stocks, presented at the Rubber Division ACS in Detroit, Oct. 17–20, 1989.
133. Dick, John S., and Pawlowski, Henry, Applications for stress relaxation from the rubber process analyzer in the characterization and quality control of rubber, presented at the Rubber Division, ACS Meeting at Cleveland, Oct. 17–20, 1995, Paper 26, p. 17.
134. Williams, I., *Ind. Eng. Chem.*, *16*, 362 (1924).

Appendix Description of Cited ASTM and ISO Standards

ASTM Standards

D 926—Standard test method for rubber property—Plasticity and recovery (parallel plate method), *Annual Book of ASTM Standards*, Vol. 09.01, 1996, p. 163.

D 1646—Standard test methods for rubber—Viscosity, stress relaxation, and prevulcanization characteristics (Mooney viscometer), *Annual Book of ASTM Standards*, Vol. 09.01, 1996, p. 328.

D 2084—Standard test method for rubber property—Vulcanization using oscillation disk cure meter, *Annual Book of ASTM Standards*, Vol. 09.01, 1996, p. 365.

D 2230—Standard test method for rubber property—Extrudability of unvulcanized compounds, *Annual Book of ASTM Standards*, Vol. 09.01, 1996, p. 394.

D 2663—Standard test methods for carbon black—Dispersion in rubber, *Annual Book of ASTM Standards*, Vol. 09.01, 1996, p. 417.

D 3182—Standard practice for rubber—Materials, equipment, and procedures for mixing standard compounds and preparing standard vulcanized sheets, *Annual Book of ASTM Standards*, Vol. 09.01, 1996, p. 460.

D 3313—Standard test method for carbon black—Individual pellet crush strength, *Annual Book of ASTM Standards*, Vol. 09.01, 1996, p. 516.

D 5099—Standard test methods for rubber—Measurement of processing properties using capillary rheometry, *Annual Book of ASTM Standards*, Vol. 09.01, 1996, p. 804.

D 5289—Standard test method for rubber property—Vulcanization using rotorless cure meters, *Annual Book of ASTM Standards*, Vol. 09.01, 1996, p. 815.

D 6048—Standard practice for stress relaxation testing of raw rubber, unvulcanized rubber compounds and thermoplastic elastomers, to be published in *Annual Book of ASTM Standards*, Vol. 0.901, 1998.

D 6049—Test for rubber property—Measurement of the viscous and elastic behavior of unvulcanized raw rubbers and rubber compounds by compression between parallel plates, to be published in *Annual Book of ASTM Standards*, Vol. 0.901, 1998.

ISO Standards

ISO 37—Rubber, vulcanized or thermoplastic—Determination of tensile stress–strain properties (1994).

ISO 289—Rubber, unvulcanized—Determinations using a shearing-disc viscometer—Part 1, Determination of Mooney viscosity; Part 2, Determination of prevulcanization characteristics (1994).

ISO 2007—Rubber, unvulcanized—Determination of plasticity—Rapid plastimeter method (1991).

ISO 2393—Rubber test mixes—Preparation, mixing and vulcanization—Equipment and procedures (1994).

ISO 3417—Rubber—Measurement of vulcanization characteristics with the oscillating disc curemeter (1991).

ISO 6502—Measurement of vulcanization characteristics with rotorless curemeters (1991).

ISO 7323—Rubber, raw and unvulcanized compounded—Determination of plasticity number and recovery number—Parallel plate method.

ISO 9026—Raw rubber or unvulcanized compounds—Determination of green strength.

9
Strength and Stiffness Properties

Roger Brown
Rapra Technology Ltd., Shawbury, Shrewsbury, England

1 Introduction

1.1 General

In a general definition, strength and stiffness properties cover everything from hardness through tensile stress–strain and including dynamic stress–strain tests, creep, and stress relaxation. For convenience, the definition is restricted in this chapter to the tests commonly referred to as short-term or static stress–strain; i.e., those where the effects of long times or cycling are ignored. In particular, time dependent properties are covered in Chapter 11 and dynamic tests in Chapter 21. Clearly, these distinctions are somewhat arbitrary.

Test methods for most of the short-term properties have been standardized and are very commonly used, particularly for quality control. Generally, they are carried out with test piece geometries and test conditions which owe more to experimental convenience than to fundamental considerations. Because the mechanical properties of polymers vary with time, temperature, and test conditions, these standard tests cannot provide a complete characterization of a material, but they are nevertheless of great practical value. Ideally, properties would be measured as a function of geometry, time, etc., but this is not usually viable because of cost considerations. It is necessary to appreciate the limitations of single-point measurements, to select the most relevant test and conditions for the purpose, and not assume that the result will be valid if applied to different conditions. If there were simple relationships for the variation of properties with test piece geometry, speed of test, etc., then single-point measurements could be extrapolated. Unfortunately, such universal conversions do not exist, and the relationships are often complex.

1.2 Modes of Deformation

Mechanical tests are carried out using a variety of modes of deformation of which tension, compression, flexure and simple shear are the most common. Others include torsion, biaxial tension, and hybrid configurations, as in indentation hardness or the drape of fabrics. The mode of deformation should be chosen as that most relevant to the intended application, but the choice may also be influenced by experimental convenience and the form of test piece available. This to a large extent explains why hardness is so commonly used as a measure of the modulus of rubber: it makes little demand on test piece preparation and is very quick and cheap to perform. Similarly, tensile properties are much more commonly measured than shear because they are relatively easier to perform. The most popular mode of deformation also depends on the class of material. Flexural tests are common for rigid plastics but virtually never used for solid rubbers because their stiffness is too low for the result to be meaningful. Perhaps incongruously, the flexibility of coated fabrics is of importance and bears no relation to the tensile stiffness. Although in principle stiffness in the different modes of deformation are related, the relationships are generally complex, and conversions are approximate even for homogeneous materials. This is well illustrated by an account of the relationships between the various moduli of rubbers [1].

1.3 Rate of Deformation

The stress–strain characteristics of polymers are dependent on the strain rate, so that the deformation speed selected is important, the apparent stiffness increasing with increasing strain rate. As a general rule the stiffer materials are tested at lower speeds than the flexible materials, although it will be appreciated that impact tests are a special case. Hence, tensile properties of rubbers are obtained at 500 mm per min, and the most rigid composites at perhaps 1 mm per minute. With many plastics the transition from ductile to brittle behavior occurs within the practical range of testing speeds, and a considerable change in properties is observed for a modest change in speed. It should be noted that the strain rate will change if the test piece dimensions are changed at constant testing speed. The standard test methods often allow a choice of test piece size, testing speed, and even test piece geometry, which conspires to make comparison of data more difficult.

This discussion of the speed of testing assumes that the tests are made at constant rate of traverse (of the testing machine moving member), and this is far and away the most common for polymer testing. The alternative of constant rate of loading is rarely met with.

Properties also change with temperature, and it is possible to relate time and temperature effects approximately by use of the Williams, Landel, and Ferry (WLF) equation [2].

2 Hardness

2.1 Significance

The term hardness has been applied to scratch resistance and to rebound resilience, but for polymers it is taken to refer to a measure of resistance to indentation. The mode of deformation under an indentor is a mixture of tension, shear, and compression, and hardness is by no means a fundamental property. The result depends on the indentor geometry and the degree of indentation as well as the time of indentation after which the measurement is made. Regardless of the arbitrary nature of the test, it is attractive because of its cheapness and apparent simplicity.

Although hardness tests are applied to most classes of polymer materials, they are far more common or important for some materials than for others. Hardness is almost inevitably included in properties of rubbers, fairly often for plastics, and only rarely for composites. With cellular materials a test more equivalent to compression stiffness is termed hardness, while the indentation of fabrics is not practical. Because of the range of stiffness involved across these materials, the methods applied vary considerably.

Despite the complexity of the deformation, approximate relations between hardness and modulus have been derived in some cases. By far the most effort in this direction has been for rubbers. The statistical theory of these gives the relationship for indentation by a rigid ball in the form

$$\frac{F}{E} = KP^{3/2}R^{1/2}$$

Scott [3] produced an empirical relationship (for indentations up to 0.8 of the ball diameter) as

$$F = KGR^{0.65}P^{1.35}$$

or

$$\frac{F}{E} = 1.9P^{1.35}R^{0.65}$$

Expressions have also been derived for a flat ended cylinder:

$$P = k_2\left(\frac{F}{E}\right)d^{-1}$$

and a cone:

$$P = k_3\left(\frac{F}{E}\right)^{0.5}$$

where

F = force
K = numerical constant
R = radius of ball
P = depth of identation
d = diameter of cylinder
k_1, k_2, and k_3 are constants, k_3 involving the angle of the cone

A truncated cone of fairly small angle behaves roughly like the cylinder.

Briscoe et al. [4] have quoted relationships for a number of indentor geometries and considered the effect of geometry in detail.

2.2 Forms of Test

The indenting force can be applied in three ways:

(a) Application of a constant force, the resultant indentation being measured
(b) Measurement of the force required to produce a constant indentation
(c) Use of a spring resulting in variation of the indenting force with depth of indentation

Considering the case of a ball, method (b) is attractive because, for rubber at least, the measured force should be proportional to the modulus. When standard methods were formulated, the measurement of force would have been much more of a complication that it is with modern force transducers, and the method has not been seriously adopted. Portable instruments, usually called durometers, always use a spring loading system, which does enable a much closer to linear relationship between indentation and log modulus. However, because springs are not considered precision measuring elements, standard reference methods use weights to apply a constant force.

A variety of indentor geometries are used, notably a ball, a truncated cone, and a pyramid. The pyramid shape is derived from methods developed for metals and is applied to rigid plastics, whereas a ball or truncated cone can be applied to both soft and relatively hard materials.

The normal tests use indentors with dimensions of the order of mm, but there are also micro tests that are scaled down by approximately an order of magnitude and allow thinner test pieces to be used and, on rigid materials, produce less damage. With rubbers hardness test are essentially nondestructive. A review of micro tests has been given by Lopez [5].

If test pieces are too thin, the base material has an effect, and different results may also be obtained on curved surfaces. Measurements made on nonstandard test pieces are sometimes referred to as apparent hardness measurements.

The total number of hardness methods that are used is considerable, and a frequently asked question is how the different scales are related. Because of the arbitrary nature of the methods, any relation is at best approximate, but a number of conversions have been established. Computer software has been produced that includes all those that have been published [6].

3 Tensile Stress–Strain Properties

3.1 Significance

Other perhaps than hardness, measurement of tensile stress–strain properties is the most common mechanical measurement on most polymer materials. The principle is simple enough. Stretch a test piece until it breaks and measure the force and elongation at various stages. Even when the application of the material is in shear or compression, tensile properties are commonly measured as a general guide to quality, which probably owes much to the relative convenience and simplicity of this geometry. Because the results are at least to some degree dependent on test piece geometry, the tensile properties measured are generally considered arbitrary rather than absolute.

The basic parameters measured are strength, elongation at break, and modulus, but even these can have different significance depending on the material. Plastics may yield before failing such that the strength at break is not the maximum stress attained and the elongation figure has very little practical meaning. Because polymers generally do not have linear stress–strain curves, a number of different measures of stiffness have been adopted, and hence there are several definitions of modulus and yield points. These vary to an extent depending on the material class and hence will be considered in Chapters 15 to 20. For those less familiar with polymeric materials it is important to appreciate that the quoted result will be dependent on the detailed procedure used and that comparisons can only sensibly be made when the procedures and the definitions of the parameters are the same.

3.2 Test Piece Geometry

Test pieces are most often in the form of two-dimensional dumbbell shapes or flat strips, but there are exceptions, such as ring test pieces, sometimes used for rubbers, and three dimensional dumbbells, for certain types of plastic. Dumbbells have the basic attribute of being a way of concentrating the stresses so that failure takes place in the narrow portion and not preferentially where the test piece is gripped, although this is not always successful in practice. For textiles a dumbbell would mean that some threads at the ends were not supported, and no advantage over a flat strip has been found for many plastics films. Particular variations on these shapes have been developed for composite materials to cope with their particular properties.

The detailed shape of dumbbells varies between standards, and there is also a range of sizes. The shapes are to a degree arbitrary, but in some cases they result from theoretical predictions or practical experiments with the aim of optimizing stress distribution. The different sizes largely result from the need to cater for circumstances where only small amounts of material are available. Because results are likely to depend on the geometry and the size, comparisons should strictly only be made where the same test piece has been used.

3.3 Test Apparatus

The tensile machine (or universal test machine used in the tensile mode) is essentially the same for all materials but clearly varies in capacity. The basic elements are grips to hold the test piece, a means of applying a strain (or stress), a force-measuring element, and an extensometer. A specification for machines suitable for rubbers and plastics is given in ISO 5893 [7], and the content largely applies to other materials. This standard was produced with the intention that test method standards would refer to it and hence obviate the need to include a description of a complex engineering instrument.

Commercial machines exist in a variety of capacities and degrees of sophistication, and a choice has to be made in the light of the materials to be tested and the nature of the work undertaken. For routine tensile measurements on flexible materials a simple and low capacity machine may be quite adequate. Where a variety of tests are undertaken it becomes sensible to choose a "universal" machine, i.e., one that is capable of being adapted for a variety of purposes and offering a wide range of facilities—operation in compression, wide speed range, etc. Essentially, the load capacity, the cross-head travel, and the speed range must be adequate for the materials tested, and these cover a wide range for the different types of polymers. Capacities range from 500 N for rubbers up to 25 kN for very stiff materials and compression tests; cross-head travel may be small for rigid plastics but up to 1 m for rubbers; speeds range from 0.1 to 500 mm/min.The physical size must also be such as to have sufficient space within the frame to accomodate jigs and environmental cabinets to be used and products that are to be tested.

Modern machines all use some form of electrical transducer to measure force that is essentially free from inertia and relatively stiff (i.e., deflecting very little). They allow through amplification a large range of force to be covered with one load cell and readily allow recording or automatic handling of the output data. Modern machines also all use a data capture system and a computer to store the force and displacement data in digital form. Software in various degrees of sophistication allows manipulation of the data and automatic control of the testing process. As a result, hard copy ouput is via a normal printer instead of chart recorders, which could introduce significant inertia errors.

The test piece must be gripped in some way whilst the strain is applied, and a considerable number of grip designs have been developed to cope with the range of materials from very soft foams to rigid reinforced plastics. The objective is to hold the test piece in correct alignment, sufficiently firmly to prevent slippage but to avoid crushing. For rubber dumbbells and other flexible materials a self-tightening mechanism is generally favored, such as the Gavin design using rollers moving in inclined slots. Probably the most widely used for plastics are wedge action grips, which are again self-tightening. Some strong and brittle materials are best held with vicelike chucks that in some cases include a bolt passing through a hole in the test piece end. An alternative for many circumstances is pneumatic or hydraulic grips where grip pressure can be adjusted before test and is maintained throughout the test. Friction of the test piece with any grip can be modified by a surface pattern or the insertion of emery paper. Rubber ring test pieces are held by a pair of pulleys on roller bearings with one or both pulleys rotated to ensure that the ring is uniformly stretched.

The elongation of rings and strip test pieces where slippage is totally prevented can be measured by grip separation, but for dumbbells some form of extensometer is necessary. A variety of contact types (i.e., where the extensometer is physically attached to the test piece) have been used based on strain gauge elements, LVDTs or potentiometers.The design varies according to the magnitudes of extension to be measured, and generally different extensometers would be required for flexible, semirigid, and very stiff materials. Noncontacting extensometers overcome a number of the problems of the contacting types and are now the preferred option if funds allow.

The first commercial noncontact extensometers were optical extensometers that used either visible or infrared light to illuminate targets on the test piece. The essential difference between the optical and contact types is in the method of following the extension as illustrated in Fig. 1; thereafter they both use some form of transducer to measure the movement, which largely dictates the range and sensitivity.

The principle and use of an optical noncontact extensometer available commercially has been described in some detail [8]. Two photoelectric sensing devices automatically follow, by means of a servomechanism, contrastingly colored gauge marks on the test piece. The separation of the auto followers is measured by some form of transducer, and the resulting electric signal is fed to a recorder. It is apparent that, in addition to the advantages given above, such a system can be used with very weak polymer films and could contribute to increased efficiency and time saving. An evaluation of optical extensometers was made by Hawley [9].

These devices eliminate the problems of contact but do introduce limitations of their own, such as the marks sometimes affecting the test piece and certain colors and surfaces being difficult to mark successfully. There is also the effort of applying the marks and, although in principle they will operate through an oven window, distortion of light may prevent this.

The next development was the laser extensometer, which uses reciprocating or rotating mirrors to sweep a laser beam through an angle between two marks on the test piece. The angle is calibrated against the distance and corrections made for the changes in beam path length with changing angle (Fig. 2). Laser extensometers are relatively cheap to produce compared to the optical type and can be used through the window of an oven, but the problems of diffraction in the glass are more severe. There may also be difficulties with cord specimens or if the marks distort. Accuracy at low strains is limited by the measurement of angle, but a useful advantage is that the gauge length need not be known.

The most recent development is the video extensometer (Fig. 2), which claims to overcome all of the disadvantages of the contact, optical, and laser types. A video camera

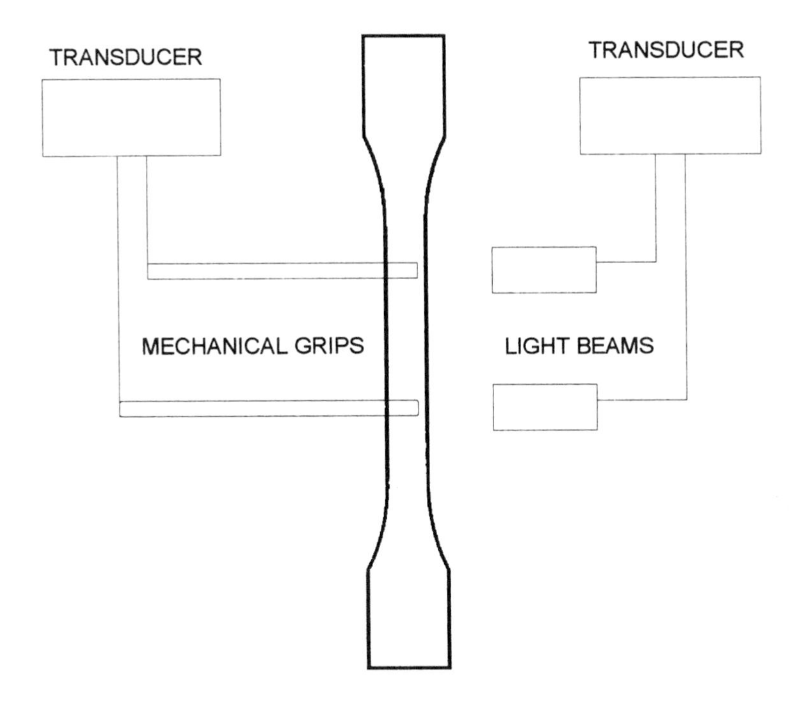

Figure 1 Optical and contact extensometers.

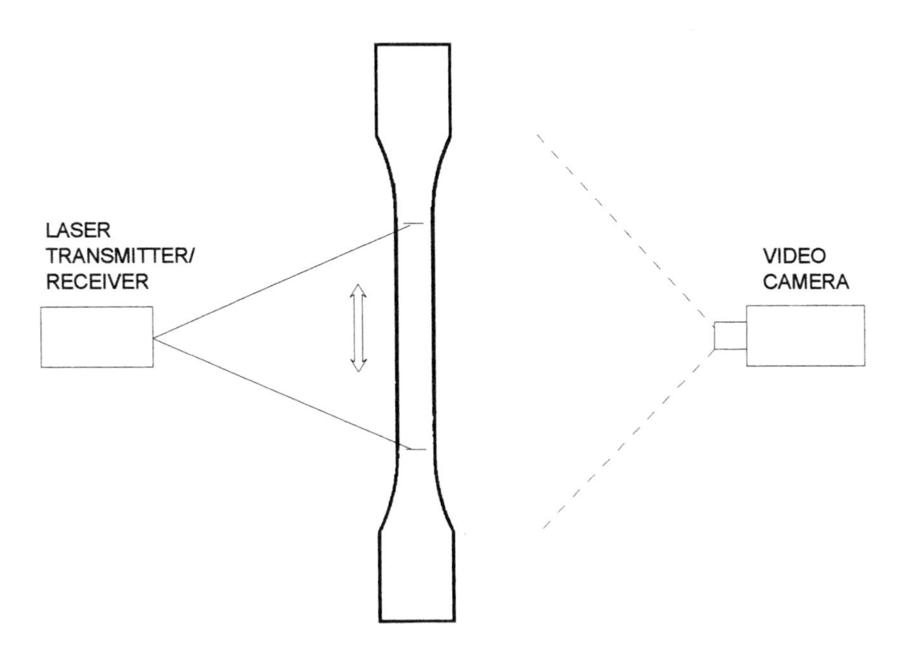

Figure 2 Laser and video extensometers.

produces an image of the test piece together with its gauge marks, which is fed into a computer. Range may be varied by change of lens, and resolutions down to a few micrometers can be achieved. Calibration can be carried out using an accurately marked calibration test piece so that distortion by glass can be taken into account. While a camera, an interface, and special software are needed, the system is in one sense simple in that there are no moving parts. It also has the unique advantage of producing an image of the test piece, so that there is a visual record of the mode of fracture.

The choice of extensometer depends on the strain range of interest, the accuracy needed, and of course the price. A noncontact type is generally preferred for convenience and lack of interference with the test piece. A laser system will probably be the most suitable for general testing of rubbers, but a higher-priced optical instrument would be preferred for lower strains. The video system shows promise for dealing with everything, but at a fairly high price.

4 Compression Stress–Strain

4.1 Significance

Measuring stress–strain characteristics in compression is clearly appropriate when the product is to be stressed in that manner in service. There are many such applications for rubbers, and yet tests in compression are less common than in tension, which is particularly surprising as in many ways they are easier to carry out. Probably the lack of compression facilities and accurate measurement of cross-head movement on older tensile machines was a factor.

The largest use of cellular materials is in compression, and this is reflected in the greater use of this type of test. Whilst the more usual type of compression test is performed by compressing a test piece between larger area platens, for cellular materials tests are also made with indentors smaller than the test piece area. Such tests are effectively a large-scale hardness test. For impact situations, an indentor fitted with an accelerometer is dropped under gravity onto the test piece or product and the dynamic response characterized by recording deceleration against time of impact.

Plastics are also frequently subjected to compressive stresses, although in many applications shear or tensile stresses may be more important. For rigid materials the geometry can be such that buckling under compression is a possible problem, and this is reflected in the different test piece shapes as discussed below.

For all materials (other than fabrics, for which the concept is not relevant) the basic parameter is a measure of stiffness or modulus derived from the stress–strain curve. As with tensile tests, because the stress–strain relation is generally not linear, care must be taken to compare only measures of stiffness defined in the same way. With rigid foams and plastics there are additionally measures of yield or strength.

4.2 Test Piece Geometry

Standard compression tests on rubbers and foams are carried out on test pieces with height significantly smaller than compressed area, so there is no question of buckling. A disc or short cylinder is most common. However, the ratio of height to area is important, together with whether the test piece is lubricated.

In theory there are two conditions under which a test piece can be compressed: either with perfect slippage between the test piece and compressing members of the apparatus, or with complete absence of slip. If there were perfect slippage, every element of the test piece would be subjected to the stress and strain, and a cylindrical test piece would remain a true cylinder without barrelling. For rubber under these conditions the stress and strain are approximately related by

$$\frac{F}{A} = G(\lambda^{-2} - \lambda) = \frac{E}{3}(\lambda^{-2} - \lambda)$$

where F = compression force, A = initial cross-sectional area, E = Young's modulus, G = shear modulus, and λ = ratio of compressed height to initial height.

If there is complete absence of slippage, stress and strain are not uniform throughout the test piece, and barrelling takes place on compression. The relation between stress and strain is then dependent on the shape factor of the test piece. The stress–strain relationship can then be expressed as:

$$E_c = E(A + BS^n)$$

where S is the shape factor defined as the ratio of the loaded cross-sectional area to the force free area (Fig. 3). For a disc this is

$$S = \frac{\text{diameter}}{4 \times \text{thickness}}$$

and for a rectangular block it is

$$S = \frac{ab}{2h(a + b)}$$

For a disc $A = 1$, $B = 2k$, and, for natural rubber at least, $n = 2$. k is a numerical factor that varies with modulus, and its values have been tabulated by Lindley [10].

The most common situation, both in practice and in experiment, is for the rubber to be bonded to metal plates or held between surfaces that effectively eliminate slip. In this situation the effect of shape factor means that the thinner the rubber the stiffer it appears, and this property is much exploited in the design of rubber mounts and bearings.

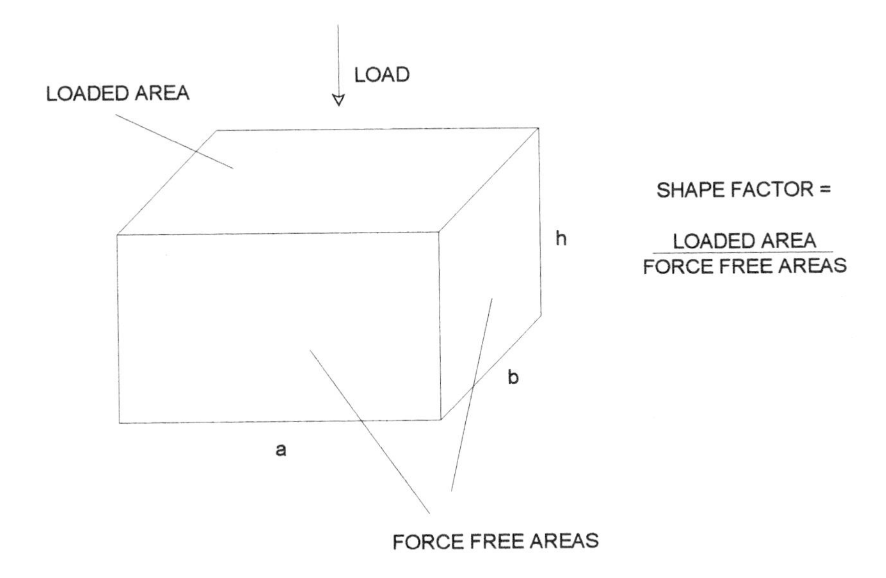

Figure 3 Shape factor.

When the shape factor becomes very high such that E_c approaches the bulk modulus K, the effective modulus is less than expected due to appreciable bulk compression and can be estimated from

$$\frac{E_c}{1 + \dfrac{E_c}{K}}$$

Standard tests for plastics often use test pieces, which may be right prisms, cylinders, or tubes, with the height greater than the diameter or width. The slenderness ratio is, however, chosen to avoid buckling at the strains likely to be reached during the test. It can be reasoned that the more slender test piece will improve the accuracy of modulus measurements where the material is very stiff and and the strains small. Also, with this geometry, failure is often achieved by the test piece shearing at an angle to its length. With a very slender test piece, failure would occur through buckling, and for a squat test piece failure by "bursting apart" would only be achieved at extremely high forces.

As mentioned above, large scale indentation tests are also carried out on foams with the indentor smaller than the the test piece area when the test piece is simply a convenient slab of the product. Dynamic impact tests measuring deceleration can use a test piece either smaller or larger than the indenting platen to suit the situation being emulated. Compression tests have also been made on plastics using a pair of bars acting on a strip test piece in the so-called plain strain compression test to investigate behavior at very large stresses.

4.3 Test Apparatus

The usual approach is to use a "tensile" or "universal" testing machine that is equipped with a compression load cell. The compression takes place between parallel steel plates, which may incorporate a self-alignment mechanism. Rather than a universal machine, an apparatus working in compression only is equally satisfactory if the volume of work justifies a dedicated apparatus. An alternative is to use a compression cage in a tensile machine effectively to reverse the motion of the machine. However, these are now seldom seen, as they can introduce considerable friction errors, and there is always difficulty in alignment. As will be appreciated, for stiff materials the forces and hence machine capacity required can be large.

Strain is normally measured by movement of the compression platens by means of a convenient transducer. Machine cross-head movement can be used if the machine is sufficiently stiff in relation to the test piece to avoid the introduction of errors.

5 Shear Stress–Strain

5.1 Significance

Tensile and compressive forces are normal to the plane on which they act, but shear forces are parallel to the plane. Simple shear can be represented by planes sliding parallel to a given plane by an amount proportional to their distance from that plane. In Fig. 4 the shear stress is

$$\tau = \frac{F}{lxw}$$

where w is the width (not shown). The shear strain is

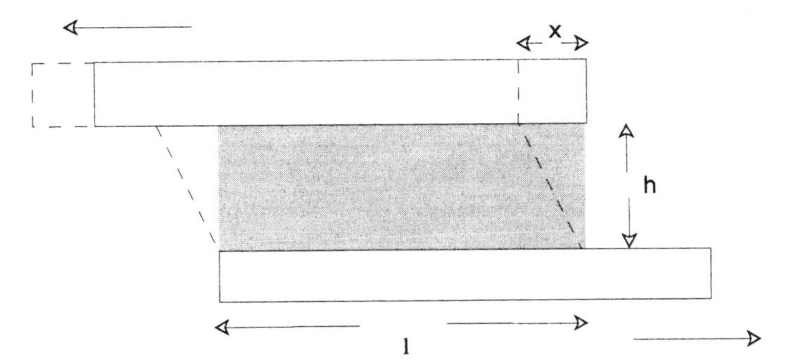

Figure 4 Simple shear.

$$\gamma = \frac{x}{h}$$

The shear stress–strain relationship is in many cases linear to greater strains than in tension or compression.

Pure shear is represented in Fig. 5 and is defined as a homogeneous strain in which one of the principal extensions is zero and the volume is unchanged. If the extension ratio $\lambda_1 = \alpha$ while $\lambda_2 = 1$, then λ_3 is $1/\alpha$.

Shear stresses are involved in a great many applications but, despite this, even less shear testing is carried out than compression testing. Whilst a shear test is no more difficult than a tension test, there is the added complexity for the most common geometry of having to bond a test piece to rigid members in order to apply a shearing force.

The parameters measured are the shear modulus and the shear strength, but not necessarily both in the same test. Shear modulus is usually measured at relatively modest strains where the stress–strain relationship is essentially linear.

5.2 Test Piece Geometry

There are a number of loading systems that give rise to shear stresses. In this section we shall consider variations on the lap shear, punch shear, torsion, and four-point loading.

For rubbers and foams the most usual approach is based on the lap, or sandwich, shear geometry, which is essentially the form of many mountings. There can be one, two, or four elements, as shown in Fig. 6, but clearly the four-element design is the most stable. In this geometry, there will be increasing bending strains as the thickness of the elements is increased, and standard methods limit the thickness/area ratio to ensure that bending is insignificant. The test pieces can be formed by bonding during molding or adhered afterwards. The test can be used to measure modulus or to measure strength. In the latter case this may be the strength of adhesion to the metal plates. For strains where the shear stress–strain relationship is linear,

$$\frac{F}{A} = G\gamma$$

where F = force, A = area, G = shear modulus, and γ = shear strain.

Relations between stress and strain for other shear and shear/compression configurations are given by Freakley and Payne [11].

Surprisingly, this geometry is not commonly found for quasistatic tests on plastics but has been standardized for dynamic tests.

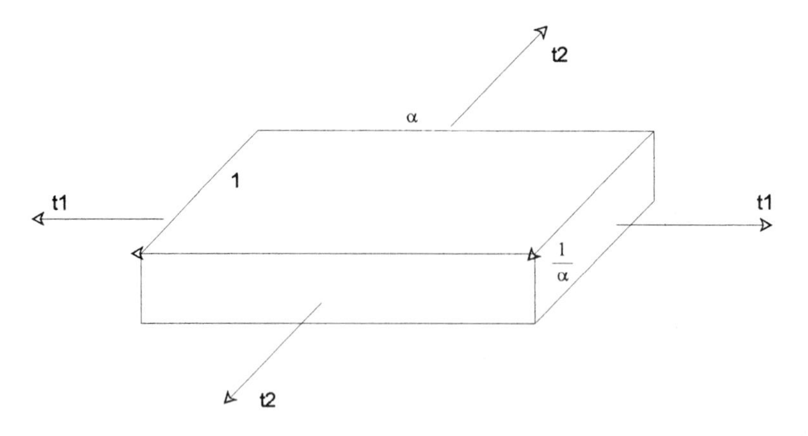

Figure 5 Pure shear.

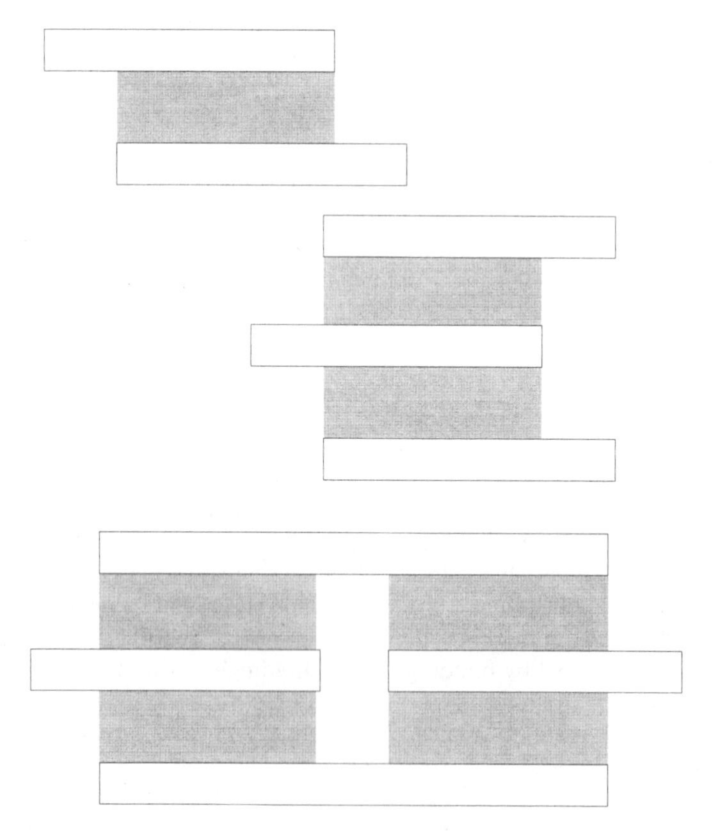

Figure 6 Shear test pieces.

Punch shear geometry is relatively popular for plastics to measure shear strength, although it can be used for any material. The approximation to pure shear conditions are achieved by a punch bearing on a sheet of material supported by a die. The smaller the difference between the internal diameter of the die and the external diameter of the punch, the nearer the approximation.

Some engineering components operate in torsion, and shear modulus can be measured in this mode. In practice, it is largely restricted to measuring stiffness of rubbers, flexible plastics, and coated fabrics as a means of characterizing low-temperature perfomance. A strip geometry is used when force and deflection are related by

$$\tau = \frac{kbt^3 G\theta}{l}$$

where τ = applied torque, k = shape factor, b = width of test piece, t = test piece thickness, G = shear modulus, θ = angle of twist, and l = effective length of test piece.

Stress–strain relationships for other torsional configurations can be found in Payne and Scott [12].

Bending tests are considered in the next section, where it will be mentioned that the geometry has to be defined to make shear stresses negligible. Conversely, the geometry can be chosen to make shear dominant, and this is the object of the so-called interlaminar shear strength test for fiber-reinforced plastics. The span is reduced to six times the test piece thickness to encourage shear failure. Shear can also be induced in directionally reinforced materials by suitable arrangement of the orientation of the reinforcement relative to the direction of straining in a tensile test.

A close approximation to pure shear in rubbers can be achieved by stretching a strip with length much greater than width normal to its length.

5.3 Test Apparatus

Shear tests with the sandwich type geometry, punch shear tests, and tests where shear is induced from straining in tension are carried out using a tensile test machine with appropriate jigs and grips to mount and strain the test piece. Strain can be measured with a transducer or by cross-head movement in a similar way to compression tests. Bending tests also use a tensile machine with a specially designed bending jig in accordance with the relevant standard.

Tests in torsion clearly require a device to apply a torsional strain. Examples of very simple devices used for low-temperature tests can be found in Chapters 15, 16, and 20, and apparatus for dynamic tests in Chapter 21.

6 Flexural Stress–Strain

6.1 Significance

Flexing, or bending, occurs by intent or accident with all the materials although the extent is clearly dependent on the stiffness. With rigid plastics, measuring modulus and strength in flexure is almost as commonly practiced as tensile tests, one reason for its popularity being that a strip is easier to produce than a dumbbell and there are no gripping problems. As well as flexural or bending tests, the term cross-breaking strength is also found, although its origin is not clear to this author.

Flexural tests are also common for rigid foams, but rubbers and flexible foams are not stiff enough to make the normal tests sensible. When such materials are apparently deformed by bending, as for example in shaped door seals, the material is also deformed in shear tension or compression, and the most appropriate test is a "compression" test on the actual product. With fabrics, coated fabrics, films, and thin rubber sheet, the flexibility may be of importance, for example the "handle" of fabrics, and tests can be devised where essentially a comparison is made using an ad hoc geometry and no attempt is made to calculate stresses or moduli.

When a beam is bent, a continuous gradation of stress occurs from a maximum tensile stress on one surface through a neutral axis to a maximum compressive stress on the other surface. It is the maximum tensile stress and strain that are calculated. Because of the geometry differences and the fact that in bending tests the surface stress rather than a homogeneous stress is considered, values for strength and modulus cannot be simply equated with those from normal tensile tests, although in theory they are equal.

The test piece can be flexed in the form of a simple cantilever or by three or four point loading (Fig. 7). Three-point loading is the most popular, but four-point loading has the advantage that the stress is constant over the whole of the span between the two inner supports, rather than a maximum occuring opposite the central support. Cantilever loading has found favor in simple tests where deadweight loading is used.

Approximate relationships between stress and strain applicable to small strains are given below, but a detailed discussion with more generally applicable formulae was given by Heap and Norman [13].

For a rectangular beam in three-point loading the flexural stress (maximum fiber stress) is given by

$$\sigma_F = \frac{3FL}{2bh^2}$$

where F is the force at midpoint. L is the span, b the width, and h the thickness.

The modulus in flexure is given by

$$E_b = \frac{L^3 F}{4bh^3 Y}$$

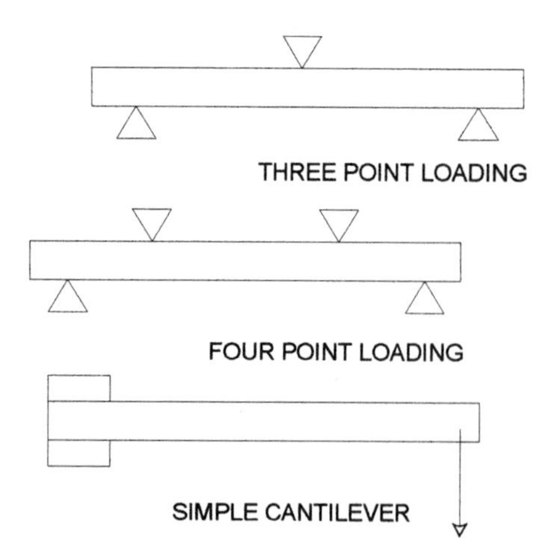

Figure 7 Flexural configurations.

where F/Y is the slope of the initial linear force deflection curve. In four-point bending of a rectangular beam,

$$\text{Flexural stress} = \frac{6F_x L_1}{bh^2}$$

where F_x = force on each bearing point (half total force) and L_1 is the distance between one outer bearing point and the adjacent inner support.

For a rectangular beam as a simple cantilever,

$$\text{Flexural stress} = \frac{6FL}{bh^2}$$

6.2 Test Piece Geometry

Standard methods use rectangular beam test pieces. The geometry of the beam is chosen to make shear stresses and flexure across the width unimportant. For three-point loading a span-to-depth ratio of 16 is generally satisfactory but does vary with the material characteristics. The quite different situation of deliberately introducing shear forces to measure interlaminar strength was discussed in Section 5.

6.3 Test Apparatus

Flexural tests in three- or four-point loading can be carried out using a tensile or a compression testing machine in conjunction with a suitable loading jig. Deflection is usually measured by cross-head movement. The loading jigs need to be stiff in comparison with the test piece and give accurate alignment. The test piece is supported and the load applied through circular section rods, the radii of which should be such as not to cause damage to the test piece; they are carefully specified in standard methods.

Apparatus for cantilever beam tests comprises a clamp to hold one end of the test piece and a means of attaching weights to the other end together with a scale to indicate deflection.

7 Tear Tests

7.1 Significance

Tear tests are applicable to flexible materials and thin plastic films. For rigid plastics and composites, a parallel can be drawn with the various crack growth tests, which are covered in Chapter 22. In a normal tensile test, taken to break, the force to produce failure in a nominally flawless test piece is measured. In a tear test, the force is not applied evenly but concentrated on a deliberate flaw or sharp discontinuity, and the force to produce continuously new surface is measured. This force to start or maintain tearing will depend in rather a complex manner on the geometry of the test piece and the nature of the discontinuity.

Tearing can occur in a wide range of products and is involved in fatigue and abrasion processes as well as the catastrophic growth of a cut on application of a force. Given this importance and the different results obtained from different geometries, it is not surprising that a considerable number of tear tests have been devised. The different tests in part reflect the different stress concentrations found in different products, but in many cases they are somewhat arbitrary. In consequence, the measured tear strength is not an intrinsic property of the material, and it can be difficult to correlate directly the results of laboratory tests with service performance.

For rubbers, a fracture mechanics approach can be taken that uses the concept of energy of tearing, which is the energy required to form unit area of new surface during tearing. The tearing energy is a basic material property and independent of test piece geometry. In theory, if the tearing energy and the elastic characteristics of the material are known, the force needed to tear a given geometry can be predicted. For certain geometries the relationship between force and energy is relatively simple, but to date it has been ignored in the standard methods and the arbitrary tearing force has been reported.

Cutting can be considered as the precursor to tearing, and if the cutting takes place while there is other stress on the material, it can be viewed as a sharp object assisting tearing. Cutting involves both the strength properties of the material and the friction, so that if a stress is applied whilst cutting, friction is much reduced and with it the force needed to cause cutting. Generally, cutting or puncture tests operate under ad hoc conditions intended to relate to the stress and geometry conditions of service.

7.2 Test Piece Geometry

Distinction can be made between the force to initiate a tear and the force to propagate a tear. Both are important, as even when a tear has started, for example because of an accidental cut, the resistance to propagation will determine whether the damage becomes catastrophic. The discontinuity at which a stress concentration is produced is formed either by a cut, a sharp reentry angle, or both. Most standard test pieces involve an artificially introduced cut, and only in a method with a sharp angle and no cut would any measure of tear initiation force be possible.

By far the most common form of geometry is where the tear is induced at right angles to the direction of applied force; three examples of variants are shown in Fig. 8 In this type of geometry the stresses at the tip of the tear are essentially tensile. The examples are of a crescent tear, which is used with an initial cut for rubbers and plastic film, an angle tear, which can be with or without an initial cut, and is again used for rubbers and plastics, and a strip with central cut, which is the basis of the Delft test for rubbers. The angle tear without a cut is the only geometry in general use where an initiation force is measured, but it requires the angle of the cutter to be very carefully maintained to get consistent results. The Delft geometry lends itself to small test pieces for where material is limited.

The other common approach is the trouser tear, also illustrated in Fig. 8, where the stresses must include shear forces. This form of test piece is used with variations for rubbers, plastics, and fabrics. For rubbers the trouser tear has the advantage of being particularly convenient for calculating tearing energy. It also allows the course of tear propagation to be followed and is a relatively easy shape to cut. There can be difficulties in highly extensible materials due to excessive leg extension and a variation is to reinforce the legs with a textile.

Cutting or puncture tests are, as mentioned previously, normally made with geometries intended to simulate service conditions.

7.3 Test Apparatus

Most of the normal tear tests are carried out using a tensile machine with suitable grips. The tearing force can rise rapidly, and in trouser type tests it can rise and fall during the

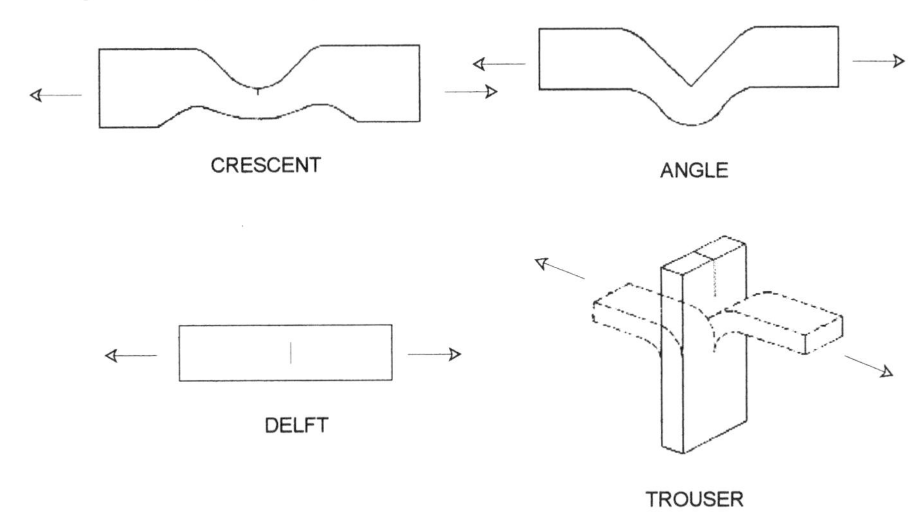

Figure 8 Tear test geometry.

test; therefore the response characteristics of the apparatus are particularly important. Puncture tests require a machine arranged to act in compression.

An alternative approach is to apply a force by means of a pendulum, which is the basis of the widely used Elmendorf apparatus. Essentially, part of the energy stored in the pendulm is used to produce the tearing (and any deformation of the test piece), and the magnitude of this is indicated by the energy lost compared to there being no test piece. The incentive for this approach originally was probably to have a relatively simple and cheap stand-alone apparatus.

8 Impact Tests

8.1 Significance

The basic characteristics of an impact test are that stresses are applied by subjecting the test piece to a sudden blow; hence the speed of the test (the strain rate) is relatively high compared to the methods covered in previous sections. Impact is clearly something that can occur in service, and the test methods that have been developed seek to simulate this and to provide a relatively simple means of achieving high strain rates. It is notable that considerable effort has been applied to impact tests, often with arbitrary geometries, but very little to high-speed testing in the standard tensile and shear geometries, etc.

Impact tests are principally applicable to rigid plastics but are also used on rigid foams and, with suitable geometry, on flexible sheet materials. A general distinction can be made between pendulum tests, in which the arm of a pendulum strikes the test piece in a bending geometry, tensile tests, which also use a pendulum but have the test piece arranged so that it is stressed in tension, and falling weight tests, where the impact is made with a loaded striker falling under gravity. Occasionally tests are made with a projectile fired from a gun arrangement.

The pendulum tests with the test piece stressed in bending is only sensible with rigid materials. Often the test piece is notched to provide a stress concentration. Tensile tests are also only really applicable to rigid materials and are far less commonly used. A stress concentration is provided by using a dumbbell shape. Falling weight tests can in principle be applied to any material, but with more flexible films the result is rather a puncture test. Falling weight tests are readily applied to many products as well as sheet test pieces.

A second distinction is between the basic tests, where the energy to cause failure is measured, and instrumented tests, where the apparatus includes transducers that allow the stress–strain curve during the impact to be derived. In the uninstrumented tests the magnitude of the result is very much geometry dependent, and there is a strong argument to say that the important factor is whether the failure was brittle or ductile under the conditions of the test. The speed of the test, the temperature, and the notch geometry all have an important bearing on whether the failure is brittle. In the absence of the ability to change speed or temperature, testing at different sharpness of notch is often used to distinguish differences in materials. With an instrumented test, considerably more information is gathered, and generally these tests are much preferable though they incur a large cost penalty. Instrumented apparatus usually allow testing over a range of temperatures.

A rather different use of an instrumented falling weight test is for cushioning materials, such as foams, where the impact is made not to break the test piece or product but to determine the force reduction, which is a measure of the cushioning performance.

8.2 Test Piece Geometry

Test pieces for the pendulum tests in bending mode are strips or bars that are either held as a simple cantilever (Izod tests) or subjected to three-point loading (Charpy tests). Various dimensions are used, with and without various notch geometries. These geometries are, other than the notch radius, arbitrary, and results are only comparable between pieces of the same geometry. Tensile impact tests use small dumbbells, usually with the waisted region formed with a single radius. Falling weight tests can in principle be performed on any test piece shape, but the normal standard is a flat sheet or the actual product.

8.3 Test Apparatus

Impact test apparatus are normally specifically made to comply with the relevant standard method. The pendulum impact tests use a heavy pendulum of specified velocity and energy to strike the test piece held in a clamp or positioned on supports. The energy remaining after the impact is registered by a pointer against a scale. The equipment for Izod and Charpy type tests is essentially similar, and it is possible with change of pendulum and test piece support to do both with the same base apparatus. Similar apparatus, but not complying with the standards, do exist, usually to work with smaller test pieces at lower energy.

Tensile impact can also be carried out with a pendulum apparatus equipped with a device to hold the test piece so that on impact by the pendulum the test piece is strained in tension. There are two types of test piece holder, one where the test piece is held on the frame of the apparatus and the other where it is held in the head of the pendulum. In both approaches corrections have to be made for the energy used in moving part of the test piece support.

A falling weight apparatus is basically a vertical frame with guides down which a striker falls and impacts the test piece held horizontally at the bottom. The heights of fall, striker geometry, and mass are all specified by the standards. Clearly, it is essential that the guides impart minimal friction and that there be a release mechanism for the weighted striker that operates smoothly.

Instrumented impact apparatus can in principle use any of the standard forms but, although some pendulum machines have been instrumented, the falling weight mode is by far the most common. The usual type of instrumentation consists of a load cell in the falling striker connected to a suitable signal conditioning and data capture system. The natural frequency of the striker assembly must be high in comparison to the effective frequency of the impact, and the energy of the striker must be such that the reduction in speed caused by the impact is relatively small. The load cell needs to be positioned as close to the tip of the striker as possible, and with very brittle materials resonant oscillations can occur in the load cell, necessitating the use of a filter in the data conditioning circuit.

For tests to provide a measure of cushioning performance, a common approach is to have an accelerometer in the falling striker and to use the peak deceleration or some integral of the deceleration over the time of impact as the performance measure.

References

1. Brown, R. P., *Physical Testing of Rubber*, Chapman and Hall, 1996.
2. Ferry, J. D., *Viscoelastic Properties of Polymers*, John Wiley, 1970.
3. Scott, J. R., *Trans. IRI*, 11 (1935).
4. Briscoe, B. J., Sebastian, K. S., and Adams, M. J., *J. Phys. D*, *27*, 6, 1156 (1994).
5. Lopez, J., *Polym. Test*, *12*, 4, 437 (1993).
6. Rapra. Convert software, Rapra Technology, n.d.
7. ISO 5893, Rubber and plastics test equipment, tensile, flexural or compression types (constant rate of traverse), 1997.
8. Bennett, F. N. B., *Polym. Test*, *1*, 2 (1980).
9. Hawley, S., *Polym. Test*, *2*, 1 (1981).
10. Lindley, P. B., *Nat. Rubb. Tech. Bul.*, No. 8, 3d ed., 1970.
11. Freakley, P. K., and Payne, A. R., *Theory and Practice of Engineering with Rubber*, Applied Science Publishers, 1978.
12. Payne, A. R., and Scott, J. R., *Engineering Design with Rubber*, Maclaren and Sons, 1960.
13. Heap, R. D., and Norman, R. H., *Flexural Testing of Plastics*, Plastics Institute, 1969.

10
Fatigue and Wear

Roger Brown
Rapra Technology Ltd., Shawbury, Shrewsbury, England

1 Introduction

Fatigue and wear are two aspects of a material's ability to survive repeated stressing. Fatigue could be defined as any change in the properties of a material caused by the prolonged action of stress or strain, but this general definition would then include creep and stress relaxation and even wear. Here fatigue is taken to cover only changes resulting from repeated cyclic deformation which means, in effect, long term dynamic testing.

The general definition of wear is the loss of material by any means, but it is usually taken to be loss caused by rubbing together of two surfaces, and the word is used in that sense here. Abrasion is then essentially another word for wear, and the two are commonly used indiscriminately. One aspect of the wear process is fatigue on a micro scale.

Fatigue and wear are clearly important properties with respect to durability. One may be important whilst the other is insignificant, as in a bearing suffering from wear but not being subjected to macro deformations. In other cases the two occur at the same time as in car tires. Wear is closely interrelated with friction, and the two can be studied in the same experiment, although standard methods for polymeric materials address them separately. Friction is covered in detail in Chapter 23. Fatigue and wear are measured in quite separate experiments, but the connection is worth bearing in mind.

2 Fatigue

2.1 Types of Fatigue Test

The essence of a fatigue test is that that a stress or deformation is applied repeatedly. The result of the fatiguing is a change in stiffness and/or a loss of mechanical strength and ultimately rupture. The manner of breakdown will vary according to the geometry of the component, the type of stressing, and the environmental conditions. The mechanisms that may contribute to the breakdown include thermal degradation, oxdation, and attack by chemical agents such as ozone as well as the basic propagation of cracks.

Fatigue tests for polymers can be classified according to the mode of deformation used or by the measure of degradations used. An important distinction can be made between those where the stressing is intended to produce cracking or rupture without subjecting the test piece to significant rise in temperature and those where the prime aim is to cause heating of the specimen by the stressing process. The vast majority of tests are in the former category, as most applications are not supposed to involve significant fatigue heating. Tests in the second category are normally for rubbers where heat build up is an expected hazard for products such as tires.

Although even a small change of stiffness may be important to an application, it is very rarely measured. It could be done by continuing a dynamic stress–strain test and monitoring change in modulus (see Chapter 21), but this would be expensive in that it would tie up an expensive apparatus for long periods. Alternatively, modulus could be measured at intervals after dynamic cycling on a separate apparatus. The lack of such tests is probably because for many products it is the loss of strength shown by cracking and/or complete rupture that is considered to be the important factor.

In principle any mode of deformation can be used in a fatigue test, including combinations of different modes, and many are. For convenience, tests are categorized here into those using flexing or bending and the rest. Flexing or bending tests are by far the most common, probably because it is an easy geometry on which to base an apparatus and is representative of many applications of rubbers, coated fabrics, and textiles.

Another important distinction is between tests using strain cycles and those cycling between set stresses. Almost all standard and widely used test methods use strain cycles, again mostly because of the relative simplicity of the apparatus. Where there is a change in modulus or set in the material, it is obvious that the effects will be very different, and ideally the test would simulate the mode in service. A third alternative is cycling to constant energy, but this is rarely done. It can be sensible to match service by superimposing the dynamic cycles onto a static preload or strain. This also avoids zero strain in the cycle, which can be difficult to control precisely and in rubbers can be particularly severe.

An unfortunate confusion of terminology in rubber testing should be noted. The term flexometer has traditionally been generically applied to apparatus for heat buildup tests, whereas such apparatus generally work in compression or shear or a combination of the two.

2.2 Significance

Because the manner of breakdown and the rate at which it occurs in a fatigue test depend on so many factors, notably the mode of stressing, whether stress or strain cycles are used, the frequency, the shape of the deformation cycle (e.g., smooth sinusoid or sharp pulse), and the environment, it is clear that a single test can only relate to the actual conditions used. One of the problems with the traditional standard tests is that the results are specific to the particular conditions, and these are arbitrary.

Although the standard test conditions are arbitrary, they were generally chosen to represent the deformations met in service. The bending and compression/shear rubber tests correspond to tire sidewall flexing and the bulk deformations of a tire respectively. Other bending tests were intended to relate to the flexing of shoe materials or belting, and the composite actions of some tests for fabrics are intended to simulate the complex stressing that such materials can be subject to.

Nevertheless, if materials are to be evaluated in more than a simple comparative manner, then tests are needed as a function of at least the amplitude of deformation. Perhaps because they are "newer" materials, it is more common to find this being allowed for in procedures for plastics and composites. Amplitude effects can also be studied in the more recently introduced tension fatigue test for rubbers. This allows the application of fracture mechanics concepts to the results, as will be discussed in a later section.

A further consideration is that the range of fatigue lifetimes that are observed for a given set of test conditions is very large in comparison to the more simple mechanical tests such as tensile strength—as much as decades of time. In consequence, a large number of replicate test pieces are required to yield reliable statistics.

2.3 Flexing and Bending Tests

Perhaps the first consideration in a bending test is the severity of the flexing action, which can vary between bending the material sharply through 180° to over a large radius and small amplitude. Not surprisingly, more severe flexing is used for thin flexible materials such as fabrics than for rigid plastics. For fabrics and rubbers at least, the deformation is always intended to be a constant strain amplitude rather than a stress amplitude. The stress realized for a given bending action will depend on the thickness of the test piece, and often it is ill defined so that it also varies with the modulus of the material tested. In many tests the exact strain reached at the nominally zero level is uncertain.

Distinction can be made between those measurements that count the cycles to the initiation of cracks and those that determine the rate of crack growth and in which the end point is taken as rupture. The distinction is important because a material may commence to crack easily but be resistant to growth or vice versa.

A great variety of flexing tests have been devised for rubbery materials, although many are no longer in common use. The majority were developed for particular products such as belting, footwear, and coated fabrics and as such had limited applicability. An interesting review is given by Buist and Williams [1].

Flexing in its basic form is taken to be simple bending in one direction. Composite modes of flexing deformation are used for fabrics in an attempt to simulate what happens in service. This is usually a mixture of twisting and compression, which results in a sort of crumpling action. With such tests there is no way of determining exactly what stresses are developed.

2.4 Other Modes of Deformation

After bending, tension is probably the most common mode of flexing. Its use for rubbers involving fracture mechanics principles is considered in Section 2.6. For plastics, care is taken to keep deformations within the elastic region.

Shear, compression, or torsion cycles with constant stress or constant strain amplitude and various cycle shapes are all perfectly feasible but not commonly used except for the heat buildup tests below. A procedure for plastics taken from metals testing involves rotating a cylindrical test piece with its ends constrained by bearings that are misaligned. The result is that each element of the test piece goes through a sinusoidal cycle from tension to compression.

With these geometries it is relatively easy to superimpose the cyclic deformation onto a static strain or stress. Also, it can be arranged to increase the stress or strain amplitude with time. With such a range of possibilities it becomes clear that the number of variations available for fatigue tests is legion.

2.5 Head Buildup

In a so called heat buildup test the prime object is to induce a temperature rise in the test piece, to measure its magnitude, and to study its effects. The heating is a result of the viscoelastic nature of polymers: some of the work done in stressing the material is dissipated by viscous forces between molecules and converted to heat energy. Compression, shear, or some combination of them is normally used with relatively bulky test pieces. The tests are almost exclusively applied to rubbers and were generally intended to be relevant to tires. The geometries used are inevitably arbitrary and involve superimposing dynamic strain or stress cycles onto a prestress or strain.

2.6 Treatment of Fatigue Data

In most of the traditional standard fatigue tests the number of cycles is recorded to reach a given measure of fatigue life under set conditions when the only possibility is to compare different materials. In principle, any of the tests could be carried out over a range of different peak strains or stresses. If peak stress or strain is then plotted against cycles to failure, a so-called S–N or Wohler curve (Fig. 1) is produced. A feature of this curve is that on reducing the stress or strain towards a particular value the fatigue life increases rapidly giving rise to the concept of a limiting stress or strain below which (in the absence of chemical degradation) the material has an infinite fatigue life. Because of the need to run many replicate test pieces to counter scatter of results and to run tests for long periods of time, to characterize a material in this way is an extremely expensive exercise.

The more traditional fatigue tests do not lend themselves to a rigorous theoretical treatment of the energy balance equations that govern crack formation and propogation. However, by suitable choice of test geometry it is possible to apply fracture mechanics principles [2,3] (see Chapter 22) to obtain more fundemental material data. For example, in the tensile fatigue test for rubbers the rate of crack growth is related to the tearing energy available as a result of the elastic work done on the test piece. The tearing energy parameter is defined as the decrease in elastic strain energy caused by the growth of a crack to produce a unit increase in area:

$$T = -\left(\frac{\partial W}{\partial A}\right)_l$$

Where W is the strain energy and A is the area of one of two crack faces formed. The subscript l refers to constant length so that no work is done by the external forces on extending the test piece. The minus sign is a convention to make T positive as W decreases as A increases.

The tensile test is of the single edge notched (SEN) geometry when

$$T = 2kW_0c$$

where λ is a slowly decreasing function of extension ratio, W_0 is the strain energy density (total elastic strain energy divided by test piece volume) at maximum deformation, and c is crack length. Theoretically, k is found from

$$k = \frac{\pi}{\sqrt{\lambda}}$$

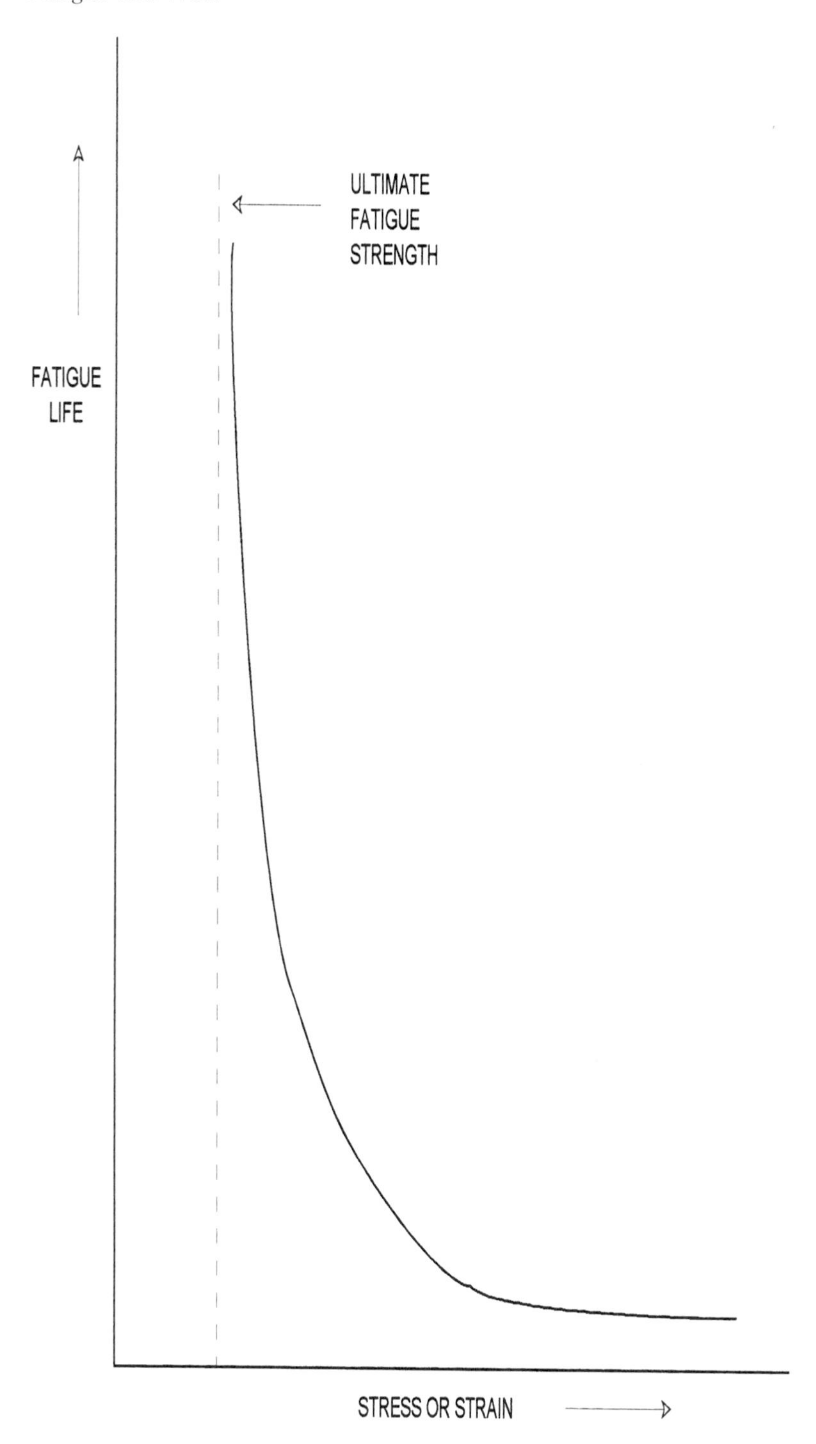

Figure 1 Fatigue life as a function of stress or strain.

where λ is the strain expressed as an extension ratio. Similar expressions can be found for other geometries.

By determining crack growth rate at different peak strains (and hence values of T), a curve of crack growth rate against T can be constructed. The fatigue response curve can be expressed as

$$\frac{dc}{dn} = f(T)$$

It is then possible to derive an expression for the number of cycles to failure [4], which for the case of intermediate tearing energies and no deliberate crack introduced (i.e., only naturally occuring flaws) is

$$N = \frac{1}{(\beta - 1)B(2kW_0)^{\beta} c_0^{\beta-1}}$$

The value of c_0, the size of a naturally occuring flaw, is typically between 20 and 60 μm.

3 Wear

3.1 Abrasion Mechanisms

The mechanisms by which wear occurs are complex, involving friction, local adhesion fatigue, and tearing. The relative importance of these factors depends on the material in question and the experimental conditions, particularly the nature of the abradant. More than one mechanism may be involved in any case, although one may predominate. Conventional materials such as metals can be abraded by cutting with a harder object but also by adhesion, where welds are formed by contact of minute asperities on the two surfaces under high local pressure. Rubber, in contrast, can slide over smooth surfaces with negligable abrasion despite high friction, and its abrasion is related to high local deformations caused by the interaction of asperities.The literature should be consulted for detailed discussions of wear mechanisms [5–11].

The consequence of the complex nature of abrasion for testing is that the test must essentially reproduce the conditions of service if good correlation is to be obtained. Even comparative results may be invalid if the predominant abrasion process in the test is different from that in service. The complexity of abrasion and the variety of service circumstances are largely the reasons for there being many abrasion tests in existence. Failure to appreciate the necessity of matching service conditions accounts for a common belief that all laboratory abrasion tests are useless except for quality control. In some instances it is virtually impossible to reproduce in the laboratory, with a reasonable degree of acceleration, the complex conditions of service, but for many products meaningful results can be obtained by careful choice of apparatus and abradant, the latter not necessarily that given in a standard method.

3.2 Types of Abrasion Test

A very large number of abrasion tests have been developed and a not inconsiderable number standardized at one time or another. The possible geometries by which the test piece and abradant can be rubbed together is legion, and some well-known configurations are shown in Fig. 2. It is not sensible to make any rigid classification, but in type (a) the test piece is reciprocated linearly against a sheet of abradant (or alternatively the abradant could be moved past a stationary test piece); in (b) the abradant is a rotating disc with the

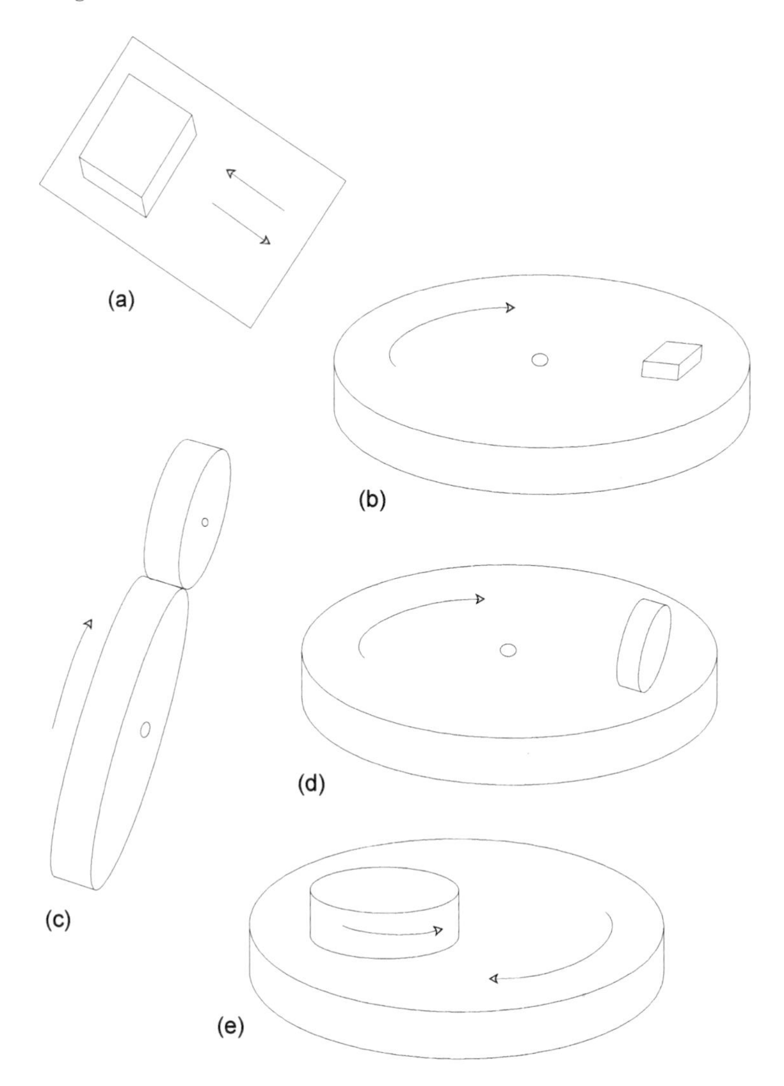

Figure 2 Abrasion configurations.

test piece held against it; in (c) both test piece and abradant are in the form of wheels (either of which could be the driven member); in (d) the abrasive wheel is driven by a rotating flat test piece; and in (e) both the test piece and the abradant are rotated in opposite directions.

In all these examples the abradant can be termed solid or fixed. A distinction can be made between tests with a solid abradant and those using a loose abradant. A loose abradant can be used in the manner of a shot blasting machine, or the test piece can be tumbled in it. This is a logical way to simulate the wear of conveyor belts, tank linings, etc. A loose abradant can also be used between two sliding surfaces, and this situation occurs in practice through contamination and as a result of accumulated wear debris. A car tire is an example of a solid abradant, the road, and a free flowing abradant in the form of grit particles.

Scratch tests can also be viewed as a form of abrasion test, although they are often seen as a measure of hardness. They are not favored for hardness, probably because of poor reproducibility, and in the abrasion sense they are probably best considered as a measure of mar resistance, but even here they find little application.

Abradants can consist of anything that will cause wear. Loose materials are likely to be the sharp particles from which solid abradants are manufactured, or sometimes the product to be used in service. The most common solid abradants are abrasive wheels (vitreous and resilient), abrasive papers or cloth, and metal "knives."

The prime factor in choosing an abradant is its relevance to service, but it also has to be available in a convenient form and, for anything but ad hoc tests, it is essential that it be reproducible. In consequence of these considerations, abrasive wheels and papers or cloths predominate where cutting by sharp asperities is to be simulated. The abrasive wheel is probably the most convenient, because of its low cost, its mechanical stability, and the ease with which it can be refaced to maintain a consistent surface. Abrasive papers and cloths are cheap and easy to use but are not so readily refaced and will deteriorate in cutting power more quickly. Although basically low in cost, both wheels and papers are a considerable expense when bought as standard reference materials. Materials such as textiles or smooth metal plates are more relevant for some applications, but they abrade relatively slowly, and if conditions are accelerated they give rise to excessive heat buildup.

3.3 Test Conditions

Whatever the geometry with a fixed abradant, there is relative movement, or slip, between the abradant and the test piece, and the degree of slip is a critical factor in determining the rate of wear. In Fig. 2a there is 100% slip, and the rate of slipping is the same as the rate of movement of the test piece or abradant. In 2c, the degree of slip can be varied by changing the angle between the wheels, and in 2d by varying the distance of the wheel from the center line of the test piece.

Another distinction between geometries is that in most the test piece is continuously and totally in contact with the abradant and the heat generated is not dissipated, resulting in a rise in temperature. Temperature rise is an important factor in obtaining correlation between laboratory and service, and an excessive rise results in degradation of the test piece. In Figs. 2c and 2d, each part of the test piece surface is only in contact with the abradant for part of each cycle. It is very difficult to control the temperature during such a test, and despite its importance it is rarely measured.

The contact pressure between the test piece and the abradant is a critical factor in determining the wear rate, and together with the actual rate of slip it influences the heat generated. Wear rate can be approximately proportional to pressure, but sometimes step changes are observed due to a change in the wear mechanism that may be due to temperature change. Rather than consider factors such as degree of slip and contact pressure individually, the power required to move the test piece over the abradant can be used as a measure of the severity of a test.

In many forms of test, different grades of abradant can be used, although only one may be standardized. Clearly, the wear rate will be affected by any change in the nature of the abradant. This includes wear of the abradant itself during the test, the effect of lubricants or loose abradants introduced, wear debris between the surfaces, and clogging of the abradant. Clogging or smearing of the abradant is a common problem and will invalidate the test. Wear debris can be removed by continuously brushing the test piece or by the use of air jets fed from an uncontaminated supply. With wheels it is normal to reface at intervals with a standardized procedure. Papers can be cleaned, but generally they have to be replaced before abrasive power reduces significantly. One purpose of using standard reference materials at intervals is to allow compensation for change in abradant. In some tests, new abradant is fed to the test piece continuously throughout the test, and this type is advantageous if high temperatures, which cause smearing, are likely. Smearing can also be reduced by introducing a powder between the test piece and the abradant and is a feature of some tests. Few tests allow the introduction of a liquid lubricant, but this may be relevant to service.

It cannot be overemphasised that test conditions must be chosen extremely carefully to match those found in the product application if there is to be any chance of obtaining a correlation between laboratory test and service.

3.4 Significance

It will be clear that measured abrasion resistance is not a fundamental property but relates only to the specific conditions of the test. The complex conditions of most product applications means that obtaining a high degree of correlation with a laboratory test is very difficult. For this reason a great deal of abrasion testing is carried out with the intention of only making comparisons between reasonably similar materials. Apart from being a way of compensating for changes in the abradant, standard reference materials are also a means of providing a norm by which the performance of any new material can be judged. Given that wear rate depends critically on test conditions, and that it is not easy to maintain a precisely reproducible abradant, the standard reference material route seems very attractive. However, there is something of a conflict in abrasion testing between putting trust in the consistency of a standard material or the consistency of a standard abradant. The belt and braces approach is to use standard abradants but also to run standard reference materials—and hope that it is possible to demonstrate which is at fault if a change is observed.

In most abrasion tests the measure of degree of wear is the loss in weight of the test piece. This is, however, not necessarily the the significant factor in practice. Generally volume loss is more relevant, and it is common practice to convert weight losses to volume. For coatings, laminates, and fabrics, it is likely that the loss of thickness or the complete loss of the surface layer is more important. In mar resistance tests the appearance of scratches and the change in surface gloss are relevant factors.

Weight or volume loss can be related to unit distance travelled over the abradant, per 1000 cycles or whatever is convenient. If comparison is then made with a reference material, an abrasion resistance index can be defined by

$$\text{Abrasion resistance index} = \frac{\text{Loss of standard material}}{\text{Loss of materiall under test}} \times 100\%$$

With inhomogenous materials it may be necessary to take into account the change of wear rate with time of abrasion.

Specific wear rate is the abrasion loss per unit surface area abraded.

References

1. Buist, J. M., and Williams, G. E., *India Rubb. World, 127*, 320, 447, 567 (1951).
2. Kinlock, A. J., and Young, R. J., *Fracture Behaviour of Polymers*, Elsevier Applied Science, 1983
3. Williams, J. G., *Fracture Mechanics of Polymers*, Ellis Horwood, 1984.

4. Lake, G. J., *Rubb. Chem. Technol.*, *68*, 3, 435 (1995).
5. Lancaster, J. K., *Plastics and Polymers* (Dec. 1973).
6. Shallamach, A., *Progress in Rubber Technology*, Elsevier, 1984.
7. Gent, A. N., and Pulford, C. T. R., *J. Appl. Polym. Sci.*, *28*, 3 (1983).
8. Thavamani, P., and Ghowmick, A. K., *Rubb. Chem. Tech.*, *67* (1994).
9. Ahagon, A., *Int. Polym. Sci Tech.*, *23*, 6 (1996).
10. Zhang, S. W., Wang, Deguo, and Yin, Weihua, *J. Mat. Sci.*, *30*, 4561 (1995).
11. Briscoe, B. J., and Tabor, D., in *Polymer Surfaces* (D. T. Clark and W. J. Feast, eds.), John Wiley, New York, 1978.

11
Time-Dependent Properties

Roger Brown
Rapra Technology Ltd., Shawbury, Shrewsbury, England

1 Introduction

All properties are time-dependent if you consider that the results of all the strength and stiffness tests depend on the rate of applying strain, dynamic test results depend on frequency, and properties change with time of aging. However, for the purpose of this chapter the term is used to cover the long-term effects of an applied stress or strain.

Creep is the measurement of the increase of strain with time under a constant applied stress; stress relaxation is the measurement of change of stress with time under constant strain; and set is a measure of the recovery after the removal of an applied stress or strain. For all of these properties there are two distinct causes, one physical and one chemical. The physical effect is due to the viscoelastic nature of polymers, which means that the response to an applied stress or strain is not instantaneous but develops with time. The chemical effect is due to the aging of the material under environmental influences. The two causes cannot be totally separated as both will always occur. However, at short times, normal ambient temperatures and in the absence of other environmental effects the physical effects will dominate, but at longer times, higher temperatures and the introduction of other degrading agents the chemical effects will increase.

The two causes of the phenomena give rise to essentially two ways in which creep, stress relaxation, and set tests are used. They can be intended essentially to measure the physical viscoelastic properties, or they can deliberately be used as a measure of aging resistance. The general practice as to which aim is being sought depends on the polymer type and the particular property. Hence creep tests are very often used, particularly with plastics, to obtain the magnitude of the physical effect; stress relaxation of rubbers is commonly used as an aging test, and set of rubbers is most often a quality control test including both physical and chemical effects.

If a creep test is continued for long enough, complete failure of the test piece can be induced. Such a test is termed static stress rupture or static fatigue and is essentially a creep test with the bother of measuring strain removed.

2 Creep

2.1 Significance

The long-term stress–strain behavior of polymers is generally more important than short-term properties where the product is expected to sustain a stress or strain in service. Creep is clearly the most relevant where the product or component is to be subjected to a more or less constant stress. This is the case for a great many uses of rigid plastics and for such products as rubber mountings. Hence, creep data is often an essential design factor for plastics but is only used for other polymer types when particular applications are in mind.

Creep data is usually obtained for a number of different stresses, as creep modulus will only be independent of stress over limited ranges. It may also be important to obtain data as a function of temperature. Commonly, isochronous stress–strain curves are derived from the creep curves at different stress levels as a useful way of displaying the information.

There can be confusion with definitions of creep used in different industries. Creep strain is defined in plastics standards [1], using tensile creep as an example, as the increase in length produced at any time during the test, expressed as a ratio or percentage of the original unstressed length. In the rubber standards [2] a creep increment is defined as the increase in strain that occurs in a specified time interval expressed as a ratio of the original unstressed length. Thus creep increment will be the same as creep strain if the beginning of the interval is zero. The rubber standards also define creep index, which is the increase in length over a time interval expressed as a ratio of the strain at the beginning of the time interval. This latter definition of creep has been traditionally used in the rubber industry. Creep modulus is the ratio of initial stress to creep strain, but again the rubber standards differ, giving a creep compliance as the ratio of the increase in strain that occurs in a specified time interval to the stress applied. There is also a "scientific" definition of creep which subtracts the instantaneous or elastic strain from the total strain to give the creep strain. By suitable choice of time intervals the rubber definition of creep increment can approximate to this. If the deformation shows a linear dependence when plotted against log time then a creep index rate can be defined that is often quoted in percent per decade.

2.2 Types of Test

Creep tests can be carried out using any mode of deformation, and logically the mode most relevant to the application would be chosen. For plastics, creep in tension is by far the most common, but creep in flexure is also internationally standardized. By contrast, creep of rubbers is most commonly carried out in compression with tests in shear also being standardized. These differences reflect the different applications of these materials where creep is likely to be important.

2.3 Apparatus

The basics of a creep apparatus are simple enough: a means of applying a load and a means of monitoring the deformation with time. In practice, considerable care has to be taken to design an apparatus that gives the necessary accuracy and stability over long times, and the sophistication increases as the deformations to be measured become smaller. The principle of a tension measurement is shown in Fig. 1. The force must be applied smoothly and without overshoot, and the mechanism must be such that the line of action of the applied force remains coincident with the axis of the test piece as it creeps. The extensometer used must not only be capable of the required sensitivity but also be stable for what may be considerable periods of time. The temperature similarly has to be controlled throughout the test.

It is instructive to consider how the features required for creep testing of plastics, for example, differ from those for testing metals. This illustrates some of the factors and raises the point that it is generally not satisfactory to use 'modified' metals testing apparatus for polymers. The cost of generating creep data is inevitably high. It is essential, therefore, that the data collected be accurate if full advantage is to be made of the cost reducing benefits of interpolation between stress levels and extrapolation (in time). Plastics are comparatively flexible with quite large operational strains, which introduces various considerations in the design of plastics creep machinery.

(a) The comparatively low loads applied means that frictional forces, to remain negligible, must also be low.

(b) The specimen, being flexible, cannot easily align itself with the load via passive (universal) couplings. Alignment should preferably be achieved by the use of slide/guide systems or other active parallel motion systems. The axiality of load should be such that the strain on opposing faces of the specimen should not differ by greater than 1%.

(c) When the load is, for instance, applied through a lever arm, the arm should include radiused segments or other means of ensuring that the arm ratio remains sensibly constant at large specimen deformation.

(d) The sensitivity of plastics to temperature is much greater than that for steel. The thermal expansion coefficient is very much higher and also the moduli of a typical plastic will change by about 4% for every 1°C change in temperature. Creep testing of plastics should be carried out at a temperature that is constant to within $\pm 1^\circ$C.

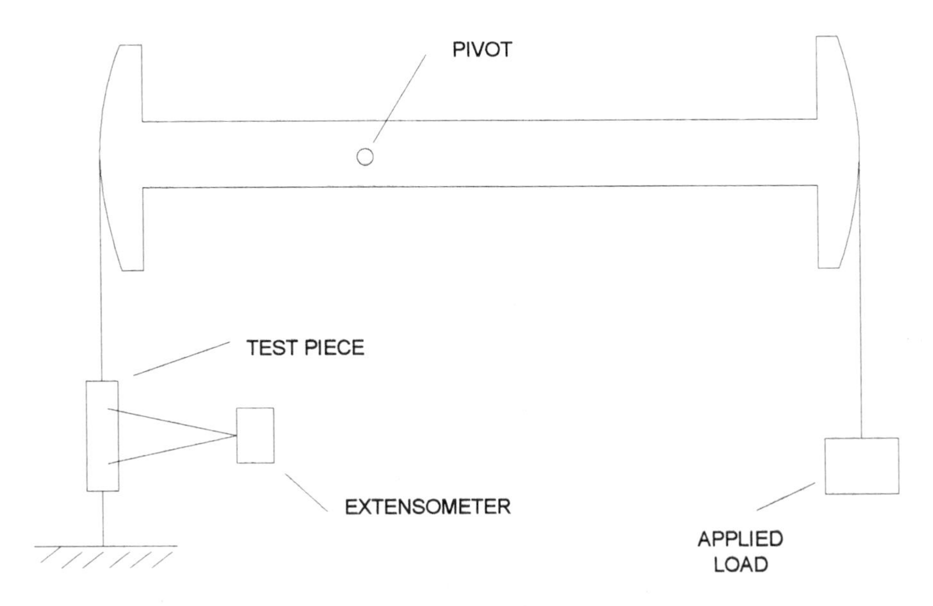

Figure 1 Principle of tensile creep measurement.

(e) The extensometry required for strain measurements need not be as sensitive as in the creep measurement of steel. Over-sensitive devices will merely record the ripple due to temperature fluctuations. However, it is important that the extensometer design be such as to provide great stability over long times.

In common with the creep testing of other materials, the plastic specimen must be loaded as quickly and smoothly as possible, that is, without "bounce." This also applies to the removal of load.

An apparatus specifically designed for plastics has been described by Wright [3]. This makes use of a moire fringe extensometer, which is essentially digital in operation and hence does not introduce any errors due to drift in sensitivity with time. Comparatively less sophisticated apparatus will be adequate for materials of lower stiffness and for shorter times of test.

3 Stress Rupture

With a high enough stress, complete failure, or rupture, will be virtually instantaneous, while with suitably low stresses, in the absence of other effects, the time to failure will be effectively infinite. In between there are clearly levels of stress that will cause failure in measurable time scales. By choice of loading, a creep test, without the need to measure strain, will yield stress rupture times. Such tests are not commonly carried out under normal atmospheric conditions, but where they are used, tension is the usual mode of stressing.

On certain products such as pipes or hose this type of test is more common and standardized with the stressing mode being the application of internal pressure. This is simulating the actual use conditions of the product.

Stress rupture tests on test pieces are very important under conditions where, in addition to the stress, the atmosphere is chosen to accelerate failure. The best known type of test is a test of the so-called environmental stress cracking of plastics, where the aggressive atmosphere is a chemical that causes cracking when the material is in a strained state. These tests are usually considered as a form of chemical resistance test and are covered in Chapter 14. Ozone cracking of rubber, also an environmental resistance test, is another example.

4 Stress Relaxation

4.1 Significance

The practical significance of stress relaxation is the opposite of creep, i.e., it is relevant when the material is subjected to a constant strain and the change in stress is of interest. This is particularly important in many sealing applications using rubbers and also in springs. The position is that stress relaxation tests are widely standardized and fairly often used for rubbers but only rarely found for plastics.

Stress relaxation is expressed as the change in force in the test piece as a percentage of the initial force, and there is not the confusion of definitions as with creep. Relaxation is not critically dependent on the applied strain for the strain levels usually of interest but is very dependent on temperature.

There are two distinct practical uses of stress relaxation data, the decay in stress that reduces sealing efficiency, and the use of stress relaxation as a general guide to aging performance of a material. The former type of test is sometimes called a sealing force test. Quite clearly the distinction can be somewhat woolly in that when producing sealing force data it might well be of interest to make measurements at elevated temperatures and/ or in the presence of liquids to be met in service. As mentioned at the beginning of the chapter, the cause of relaxation can be both physical and chemical, which considerably complicates the interpretation of data when significant times and elevated temperatures are involved. The introduction of chemicals that may swell or extract from the polymer further complicates the situation.

Whereas many sealing force measurements are made at normal ambient temperature and the decay measured is primarily physical, when stress relaxation is used to measure aging resistance the conditions are deliberately chosen to make the chemical effects dominant.

4.2 Types of Test

The types of stress relaxation test standardized (for rubbers) split into the two intended purposes. Seals are commonly stressed in compression or in a complex mode of stressing that compression can represent. Hence, sealing force tests are usually made in compression. The bulk of such test pieces slows oxidative aging and also the uptake of any fluid, but swelling and aging effects do have to be considered in long-term applications.

Stress relaxation tests as a guide to aging are inevitably carried out using tensile test pieces and test conditions involving elevated temperatures. These tests are considered in more detail in Chapters 14 and 15, but it is relevant to point out here the distinction between continuous and intermittent relaxation. The former is the normally defined stress relaxation situation, decay of force under a constant strain. Intermittent relaxation refers to a form of aging test where the test piece is strained at intervals for measurement of stress but aged in the unstrained condition. The term is something of a misnomer as the stress may in fact increase with time rather than decay due to stiffening of the material caused by the aging process.

4.3 Apparatus

Because of the expense involved in having a force measuring head for every test piece, the usual form of apparatus utilizes jigs that can in turn be placed under a measuring head to determine the stress. The general principle is shown in Fig. 2. The test piece is held between two platens, and at intervals the force is measured by applying a very small additional strain. This small strain is sufficient to result in the top platen being just separated from the body of the jig, and the force is transmitted via the central rod to a force measuring device.

Although the principle of a jig such as that illustrated schematically in Fig. 2 is fairly simple in practice, there are many difficulties. The two essential problems are to provide an efficient and reproducible detection system for the point at which the extra compression is applied and to prevent the platens tilting while not introducing appreciable friction

The force measuring head, together with provision for applying the extra compression, can be a beam balance, as used in the well-known Lucas apparatus, a universal tensile machine, or a specially designed electronic load cell unit [4]. The point at which the small amount of extra compression has been applied can be detected by breaking an electrical circuit. An early apparatus used a load cell attached to an arbor press. The operator manually lowered the press until the break in the electrical circuit was indicated visually by the extinction of a light. The Lucas apparatus has a similar detection system; the balance weights are adjusted manually until the force exerted by the beam just overcomes the force exerted by the test piece. Both of these approaches involve a somewhat

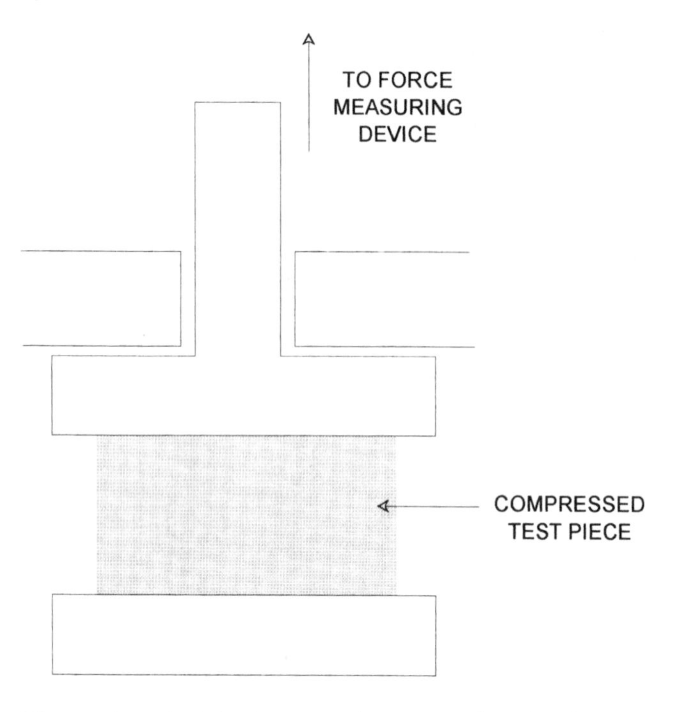

Figure 2 Principle of compression stress relaxation jig.

delicate operation. The use of a tensile machine affords some reduction in experimental difficulty but a very slow speed must be used to avoid overshoot as the increase in compression is very small.

One apparatus developed at RAPRA [5] reduces operator dependence as far as possible. A load cell is driven onto the jig by a pneumatic ram; at the moment when a very small additional compression on the test piece is detected electrically, the ram is automatically stopped, the force reading digitally recorded and the ram reversed. This measuring head can be used with a variety of jig designs including Lucas jigs.

The jigs previously specified in an ASTM standard utilized ball bushings for the load application to slide in, and the electrical contact was made through a circular plate. In principle, this arrangement gives good lateral stability, but it introduces friction that may be excessive, and the large electrical contact area is not conducive to a clean break. The Lucas jigs use a ball contact to make and break the electrical circuit but, as they have no lateral constraint, they are prone to tilting and can hence only be used with rings having a large diameter-to-height ratio. The RAPRA jigs use a ball electrical contact but provide lateral support through circular leaf springs with high lateral but low vertical stiffness, which eliminates friction. A small correction is made to the measured force for the vertical spring stiffness.

An alternative approach to detecting slight overcompression with an electrical circuit is to arrange for a preset amount of overcompression. Simple jigs of this type relying on a mechanical stop will probably produce excessive overcompression and be somewhat variable in use. The actual amount of overcompression of the test piece will depend on the stiffness of the load cell, and errors can be introduced by any lack of parallelism of the plates and plunger.

5 Set

5.1 Significance

Measurement of set is effectively restricted to rubbers and flexible cellular materials, where it has traditionally been paid rather more attention than stress relaxation and creep tests. Its popularity has a lot to do with the simple apparatus required. If set is measured on plastics it is usually made by following recovery after removal of load in a creep test.

The practical significance of the recovery after application of a stress or strain is obvious and is clearly important in such applications as seating and flooring. At first sight, set appears to be an important parameter in the effectiveness of a seal, but generally it is the force exerted by the seal that matters rather than the amount it would recover if released. Set has a general correlation with relaxation, but this is not always good enough to use it as a substitute.

Most measurements of set on rubbers are made not to obtain data to relate to a practical application but as a quality control test to indicate the state of cure. The test conditions are then arbitrary and usually involve straining at an elevated temperature. Short-term tests do not necessarily correlate well with long-term performance [6].

A schematic representation of a set measurement in compression is given in Fig. 3. The amount of set is partially transient and partially permanent. In the standard tests the

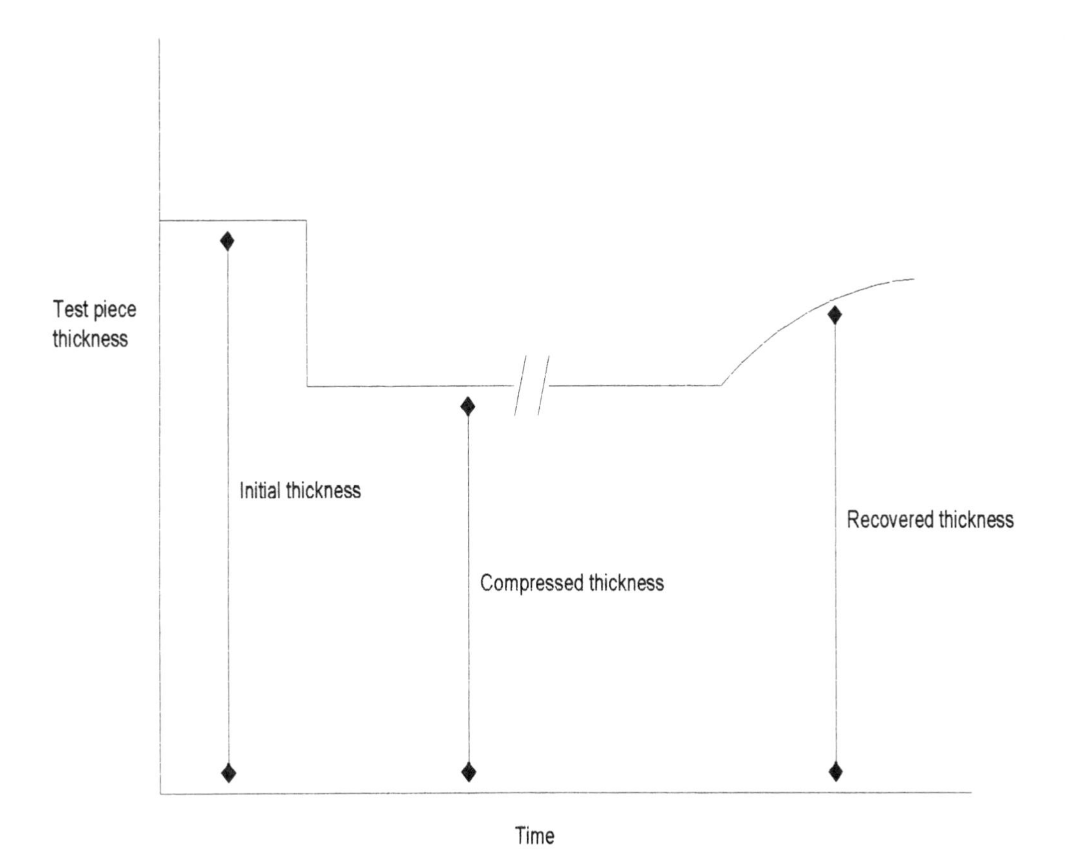

Figure 3 Diagramatic illustration of compression set.

final measurement is made at an arbitrary and relatively short time after release of the test piece, in many cases before all transient set has recovered. The term permanent set is sometimes applied to the result, but the proper meaning of this term is the set remaining after an infinite recovery time.

5.2 Types of Test

Set tests are made in either tension or compression, although any other mode of deformation could be used in principle. For quality control tests the choice of mode can be according to the convenience of the test piece available, although compression seems to be more often favored and does mean that a more bulky test piece, which is more likely to be undercured, is used. For tests intended to simulate service the most relevant mode of deformation would be used, and again this is more likely to be compression.

Set can be measured after the test piece has been subjected to constant stress or constant strain, but the latter is by far the most widely used, and constant stress measurements are virtually never seen for quality control purposes.

If set is to be used as a measure of performance, it is necessary to disregard the conditions specified in standard tests and to test under conditions relevant to service. As well as mode of deformation, this may involve recovery at the test temperature and testing as a function of test temperature and recovery time.

5.3 Apparatus

For tests under constant strain in compression the apparatus consists simply of smooth platens and spacers of appropriate thickness together with a dial gauge or other means of measuring thickness. The equivalent for tension tests is a rod or other guide fitted with a pair of grips and a suitable length measuring device. Despite such simplicity care has to be taken in construction to achieve the necessary tolerances.

Constant stress measurements are usually made using a jig incorporating a calibrated spring to apply the specified stress.

References

1. ISO 899-1, Plastics—Determination of creep—Tensile creep, 1993.
2. ISO 8013, Rubbers—Determination of creep in compression and shear, 1988.
3. Wright, D. C., *Rapra Bulletin*, *25*, 6, 133 (1971).
4. Aston, M. W., Fletcher, W., and Morrell, S. H., Proc. 4th Int. Conf. on Fluid Sealing, Philadelphia, 1969.
5. Brown, R. P., and Bennett, F. N. B., *Polym. Test*, *2*, 2 (1981).
6. Moakes, R. C., *Rapra Members Journal* (Sept./Oct. 1975).

12
Effect of Temperature

Roger Brown
Rapra Technology Ltd., Shawbury, Shrewsbury, England

1 Introduction

The effect of temperature can be separated into short-term and long-term effects. Short-term effects are generally physical and reversible when the temperature returns to ambient, while long-term effects are generally dominated by chemical change and are not reversible. The long-term effects at elevated temperature are usually referred to as the results of aging.

All physical properties of polymers change as temperature is varied, some to a greater extent than others. Hence, in principle any property could be used to monitor short-term sensitivity to temperature change, but clearly some will be more convenient and more useful than others. Because polymer properties are temperature dependent, it cannot be overemphasized that to fully characterize a material, properties should be measured over a range of temperatures. Those chosen would sensibly be the properties of interest for the application in question. In practice, the simpler mechanical tests are frequently selected to give a general indication of temperature effects because of experimental simplicity and their broad relevance, regardless of whether they are the most sensitive or the most relevant. These remarks on selection of properties apply equally to measuring the aging effects of long-term exposure.

Certain types of test to measure short-term effects receive special attention. Thermal expansion, glass transition point, softening point, and melting point are thought of as separate properties although they are particular cases of the effect of temperature. Low-temperature resistance can be measured in many ways, but many of the procedures in common use have been developed specially for reasons of experimental convenience.

2 Thermal Analysis

One of the most significant revolutions in testing has been the development of thermal analysis techniques. Thermal analysis is essentially the study of one or more properties of a material as functions of temperature, and a thermal analyzer is an apparatus that allows the automatic monitoring of the property with change in temperature. The breakthrough was the development of instrumentation to achieve this that made it possible to produce a large quantity of data quickly and economically. By thermal analysis a scan of property change over a wide temperature range can be made with less effort than it would take to achieve just a few points by conventional experiment. Apparatus is available to measure a considerable number of different properties.

Thermal analysis is not really one subject, because the information gained and the purposes for which it can be used are quite varied. The main "truly thermal" technique is differential scanning calorimetry (DSC). The heat input and temperature rise for the material under test are compared with those for a standard material, both subjected to a controlled temperature programme. In power compensation DSC the difference in heat input to maintain both test pieces at the same temperature is recorded. In heat flux DSC the difference in heat input is derived from the difference in temperature between the sample and the reference material. Heat losses to the surroundings are allowed but assumed to depend on temperature only.

The parameter measured is change in enthalpy, the quantity of heat absorbed or released. Changes of enthalpy accompany physical and chemical changes, and hence DSC is used for such applications as measuring transition points and studying cure or decomposition as well as determination of specific heat.

The general principles of DSC are outlined in a draft ISO standard, ISO 11357-1 [1] for plastics, and this contains a bibliography giving references to more detailed information. There are many other papers describing applications of thermal analysis to polymers, including those by Brazier [2] and Richardson [3]. Failure analysis and quality control using thermal analysis techniques has been discussed by Song and Ehrenstein [4] and DSC as an injection molding trouble-shooting tool by Thompson [5]. Some applications to plastics of modulated DSC, in which Fourier transform of the modulated heat flow is performed, are given by Sauerbrunn et al. [6], and the combination of scanning thermal microscopy and modulated DSC for characterizing polymer blends by Leckenby [7]. A classic reference is Thermal Characterisation of Polymeric Materials edited by Turi [8].

Dynamic mechanical thermal analysis measures damping and dynamic moduli and is covered in Chapter 21. Thermal mechanical analysis measures deformation of a test piece such as the dimensional change due to thermal expansion (also called thermodilatometry) and indentation at the softening point of the material.

Thermogravimetry measures the change in weight of a sample and is primarily a chemical analysis method.

3 Thermal Expansion

The cofficients of linear and volume expansion are defined by

$$\alpha = \frac{1}{l} \cdot \frac{\delta l}{\delta T} \qquad \beta = \frac{1}{V} \cdot \frac{\delta V}{\delta T}$$

where

l = length, V = volume, and T = temperature.

For an isotropic and homogeneous material the coefficients are related by

$$\beta = 3\alpha$$

The coefficient of linear expansion can be measured on a fairly large test piece as an average over tens of °C to reasonable precision using a suitable gauge. For flexible materials a travelling microscope can be used, and it is convenient to lay the test piece on smooth rollers in a glass sided bath. Precision is limited by the sensitivity of the measuring gauge and the effects of friction if the test piece is horizontal. Mold shrinkage is predominately a thermal expansion (or rather contraction) effect and can be measured by the difference in length of the mold cavity and the molded part, usually with a long bar test piece.

Thermal mechanical analysis affords a method of obtaining precision results on a small test piece as a function of temperature. A procedure for plastics using DMA is given in a draft ISO standard, ISO DIS 11359-2 [9].

The classical method for volume expansion is to use a liquid in a glass dilatometer. The test piece is placed in a chamber and covered with a known mass of liquid. As the temperature is raised the increase in volume is measured by the rise of the liquid up a graduated capillary. The expansion is calculated after making corrections for the expansion of the liquid and the container. Although simple in principle, dilatomers require very careful construction, calibration, and operation and the procedure is lengthy. Automatic dilatometers and thermodilatometers have been devised using various transduceres to measure the capillary height, but it can be more expedient if the facility is available to measure linear expansion by TMA in three directions. Methods for polymers have been given by Griffiths [10], Landa et al. [11], Tyagi and Deshpande [12] and Kim and Lim [13].

4 Transition Temperatures

A crystalline solid will melt when heated, and this change of state is known as a first-order transition. It is accompanied by step changes in volume and heat content as well as the obvious change from a solid to a liquid. Softening point is the temperature at which a somewhat arbitrary stress condition results in a predefined reduction in modulus.

A glass transition temperature is where an amorphous material changes from a glassy to a rubbery state and is generally regarded as a second-order transition. There is not a step change in volume or heat content, but modulus and many other properties show a more or less abrupt change. Because polymeric materials have a range of molecular weights, the transition does not take place at a single temperature but over a short temperature range, and the sharpness of the change depends on the material. A blend of two materials may show two glass transition points.

Because many physical properties, mechanical, thermal, and electrical, change by various degrees, there are potentially a number of different tests that can be used to measure the glass transition point. Frequency and temperature are interdependent: increasing frequency is equivalent to decreasing temperature, so that the measured glass transition will vary according to the effective frequency of the test chosen. A "slower" test yields a higher transition temperature than a "faster" one.

Probably the most convenient and most widely used methods are specific heat using DSC (see Section 2 in this chapter and Section 5 in Chapter 24) and dynamic mechanical analysis (see Chapter 21). However, a range of methods have been tried and reported on in the literature, from volume change to refractive index. For rubbers a measure can be obtained from the low temperature tests discussed in Section 5. Practically, the glass transition point is extremely important because of the dramatic change in properties that generally marks the point at which the suitability of the material for an application changes. In this context a test relevent to service and performed at an appropriate frequency will be the most sensible to use. An example is impact testing of rigid plastics, which detects the onset of brittleness at higher temperatures than a slower mechanical test.

Melting point is only applicable to materials sufficiently crystalline to show a fairly sharp change of state from solid to liquid, which in practice is limited to a few types of plastic. For these materials, tests have been standardized based on somewhat diverse techniques. The ring and ball method involves observing the temperature at which a steel ball sinks through a cast ring of the test material. The capillary tube method and the hot plate method are mostly used for polyamides. In the former, a sample is ground to a powder and sealed into a capillary tube which is heated. The melting is based on observing the point at which the powdered sample, having first coalesced, starts to transmit light. In the hot plate method, a granule of material together with a few drops of silicone oil is placed between a microscope slide and a cover glass on a heated plate. The cover, oil, and granule are arranged so that the melting point when the granule no longer supports the cover is readily observed. For materials that exhibit birefringence, the optical method can be used. When viewed between crossed Nicol prisms the crystalline material transmits light, but on melting and becoming noncrystalline it no longer does so.

Tests for softening point have been standardized for plastics with various terms having been used, including plastic yield, heat distortion point, deformation under load, and temperature of deflection under load. They are all based on applying an arbitrary load to the test piece and observing the temperature at which a defined degree of deformation has occurred. There are two commonly used approaches, to indent the test piece with a weighted rod or to apply the load by bending, but there are also tests that operate in tension. Generally, the conditions have been chosen so that the measured softening point marks where the material is about to suffer a drastic fall in stiffness and could no longer be considered a self-supporting solid. These tests are practically useful both for quality control and to indicate the temperature limit of a material's ability to support any load. However, it has to be appreciated that the results are completely dependent on the arbitrary test conditions used and that the material will be unsuitable for many applications well before the measured softening point is reached.

5 Low-Temperature Tests

The simple definition of a low-temperature test is the measurement of a property at subambient temperatures. Indeed, short-term low-temperature perfomance can sensibly be determined by measuring chosen properties as a function of temperature—and the same applies to above ambient temperatures. Primarily for experimental convenience, a number of specific low-temperature tests have been devised and standardized for rubbers, coated fabrics, and flexible plastics.

They are are based on one of three characteristics that are important for these flexible materials but that change dramatically as temperature is lowered:

Stiffness
Brittleness
Recovery

All these forms of test are used for rubbers but only stiffness and brittleness for the other types of material.

Additionally, some rubbers stiffen at low temperatures by partial crystallization. This is a gradual process continuing over many days or weeks but is most rapid at a particular temperature characteristic of each polymer, for example −25°C for natural rubber. Hence tests are also made to measure the effects of crystallization by changes of stiffness or recovery after periods of "aging" at a low temperature.

The obvious ways to measure the temperature at which the ability to recover from a deformation is lost are by the loss tangent from dynamic tests or by compression or tension set measurements. Dynamic analyzers are an excellent way of characterizing low-temperature characteristics, stiffness as well as viscous loss, but they are a relatively modern invention and expensive. It is, however, a little surprising that rebound resilience tests have not been commonly used. Set is quite often used, with compression set being favored over tension set. A particular form of recovery test developed and standardized for measurement of low-temperature behavior is the so-called temperature retraction test. This consists of stretching a dumbbell test piece, placing it in a bath at −70°C, and allowing it to retract as the temperature is raised—in a sense a variation on tension set.

Although torsion tests are little used to measure modulus at ambient temperature, they have proved convenient for measuring stiffness change at low temperatures, mostly because of relative experimental simplicity in devising means of straining in a temperature-controlled bath. Two types of apparatus are commonly used, the Clash and Berg, where strain is applied by a pulley, cords, and weights, and the Gehman, which uses a torsion wire. The Clash and Berg is still used for flexible plastics, but the Gehman is standardized for rubbers and coated fabrics. This is perhaps an example of the traditions of the particular industry being maintained.

These tests are not noted for their good reproducibility and there is a strong argument to use tensile modulus or dynamic modulus now that suitable apparatus is more widely available. Another approach is to measure hardness, but difficulties with icing up of moving parts of the apparatus restricted its appeal. Hardness is, however, now used for detecting crystallization (see below).

The simplest way to approach detection of brittleness point is quickly to bend the material while holding it at the low temperature, and a considerable number of tests of this type have been devised and some standardized. The degree of strain is often not precisely defined, and these tests tend to be operator-dependent, so that reproducibility can be poor. A more satisfactory approach is to impact a test piece at low temperatures. Adaptations of falling weight impact tests are tests are commonly used for rigid plastics, but for flexible materials a procedure in which a strip test piece held as a simple cantilever and impacted by a striker is very widely standardized.

Brittleness temperature measurement is essentially measuring the point at which a flexible material has lost its principal characteristic and hence is no longer suitable for its purpose. It has the disadvantage that it gives no information on behavior at other temperatures.

With at least three types of low-temperature test and several variants of each in use, it is not surprising that comparisons and correlations between them are sought. As mentioned at the beginning of this chapter, the glass transition point will depend on the effective frequency of test and hence before any equivalence of results can be expected compensation has to be made for different speeds. Generally, reasonably linear correlation can be expected between tests of the same type or between measures that are at roughly the same point in the stiffening process, but correlations between, for example, brittleness point and the temperature at which stiffness has doubled are likely to be less successful.

The various types of test do not have equal relevance to a given product, and wherever possible the procedure should be chosen to provide information relevant to the particular application.

In principle any of the tests could be used to study crystallization of rubbers by conditioning the test pieces for much longer times than normal, but in practice the favored method is change in hardness, one reason being that unvulcanized materials can be tested. Because crystallization is more rapid in the strained state, a particular type of compression set test has also been standardized for rubbers.

A practical point that should be considered in low-temperature tests is the medium used to maintain the temperature. Differences can occur between a liquid and a gaseous medium where crystallizing rubbers are concerned simply because of the difference in time needed to reach equilibrium. More generally, a liquid medium must be selected that does not affect the material being tested by swelling or extraction.

6 Heat Aging

The longer term effect of elevated temperatures is to hasten the process of degradation of the material. There are two distinct purposes for which heat aging tests are used: to measure changes at the elevated service temperature and to estimate the degree of change that would take place over much longer times at normal ambient temperature. Predicting service life from accelerated tests is extremely difficult and has to involve measurements after aging at a series of temperatures. A great many aging tests are simply made at one temperature either purely for for quality control or on a comparative basis.

The principle of heat aging tests is straightforward: the material is held at an elevated temperature for a period and the changes in selected properties are measured. Because the change with time is not usually linear, and with thermosetting materials may even change direction, rather more information is obtained if measurements are made after a number of exposure times. For complete characterization, and particularly for predictive purposes, tests should be made at several temperatures. Unfortunately, such complete characterization is both time-consuming and expensive.

Aging is carried out in thermostatically controlled ovens, which may be of the usual cabinet type or consist of a number of cells in a heated block. For plastics, relatively little consideration appears to have been given to details of the oven other than control of temperature, but for rubbers other factors that can influence results are carefully specified, and it would seem reasonable to suppose that these could be significant for plastics also.

Cabinet ovens are generally less expensive than cells and have the advantage of being able to accomodate large test pieces and test jigs, but control of air flow and velocity in them is more difficult. If different materials are exposed in the same oven there is the possibility of migration of constituents, and this is one incentive for using cell ovens. Whether one uses cabinet or cell type ovens, the air exchange rate should be controlled, as it affects the degree of degradation; three to ten changes per hour are specified for rubbers. The air velocity also affects the degradation rate, and a distinction is made between ovens with low velocities which are dependent on the exchange rate, and those with high velocities generated by a fan. It is easy to see that the cell type oven with air entering the bottom and being exhausted at the top of each cell is more amenable to control of exhange rate and velocity, although not for high velocities. Test pieces must be arranged so that there is air flow on all sides. The materials from which the oven is constructed must not have an effect on aging, and for rubbers the absence and copper and copper alloys is specified.

Temperature should be controlled as closely as possible, as even 0.5°C can be significant, but ±1 or ±2 is commonly specified. In a cabinet oven, only the central part is likely to remain in tolerance, and calibration is needed to define the useful volume.

If the test is to simulate service conditions, then the temperature of the test is chosen accordingly, but care still has to be taken with effects of geometry, air flow, etc. Where the test is intended to be accelerated, the main problem is that although a high temperature can give rapid change, it can also produce misleading results because reactions take place at the test temperature that do not occur at ambient. Factors in addition to oxidative degradation include evaporation of plasticizers and further cross-linking.

As with tests to measure the short-term effect of temperature, the properties chosen to monitor the effect of aging should preferably be those most relevant to service because the degree of degradation seen will be very property dependent. In practice the simpler mechanical tests such as of tensile properties are most commonly used because of experimental convenience. The geometry of the test piece is significant, because the limiting factor to the rate of degradation, particularly at higher temperatures, is the rate of oxygen diffusion onto the material. This means that degradation will be greater on the outside than in the bulk, and surface properties will show greater change than bulk properties. It is undesirable to compare results from test pieces of markedly different size, and caution is needed in extrapolating results from thin test pieces to large products.

One variation on heat aging tests is to measure plasticizer loss by using an absorbant material such as charcoal and determining weight loss. Other tests for particular materials have been devised, for example thermal stability of PVC can be measured by the rate of evolution of hydrogen chloride. For rubbers, aging in oxygen at elevated pressure is sometimes used to accelerate the test further.

Procedures to extrapolate accelerated test results to ambient temperatures and also to estimate maximum service temperatures are discussed in Chapter 29.

References

1. ISO DIS 11357, Plastics—Differential scanning calorimetry—Part 1, General principles.
2. Brazier, J., *Rubb. Chem. Tech.*, *153*, 3 (1980).
3. Richardson, M. J., *Polym. Test.*, *4*, 2–4 (1984).
4. Song, J., and Ehrenstein, G. W., Antec 93, Proceedings, Vol. 1, 1993, pp. 1023–1031.
5. Thompson, S. L., Antec 94, Proceedings, Vol. 1, 1994, pp. 766–769.
6. Sauerbrunn, S. R., Crowe, B. S., Foreman, J. A., and Reading, M. Antec 93, Proceedings, Vol. 1, 1993, pp. 1014–1016.
7. Leckenby, J., *Brit. Plast. Rubb.* (May 1995).
8. Turi, E. A. (ed.), *Thermal Characterisation of Polymeric Materials*, Academic Press, 1997.
9. ISO DIS 11359-2, Plastics—Thermomechanical analysis, Part 2, Determination of coefficient of linear expansion and glass transition temperature.
10. Griffiths, M. D., *Rapra Members J.*, July, 187 (1974).
11. Landa, Yu. I., Merkel, N. D., Nevelskii, I. V., and Sukharina, N. N., *Int. Polym. Sci. Tech.*, *12*, 5, T/90-1 (1985).
12. Tyagi, O., and Deshpande, D. D., *J. Appl. Polym. Sci.*, *37*, 7, 2041 (1989).
13. Kim, C. S., and Lim, K. L., *J. Polym Sci. Phys.*, *28*, 11, 2019 (1990).

13
Environmental Resistance

Roger Brown
Rapra Technology Ltd., Shawbury, Shrewsbury, England

1 Introduction

The common factor to the subjects considered in this chapter is that they are all concerned with evaluating the resistance of materials to exposure to some environmental agent other than temperature (and air). Durability involves many factors: mechanical, thermal, and electrical stresses including creep, fatigue, and abrasion. These can be difficult enough to measure and are greatly complicated by the effects of nonambient temperatures, but when the effects of environments are superimposed, the problems can increase in a quantum leap.

The main environmental agents in addition to temperature can be grouped as chemicals, weathering, biological attack, radiation, and fire. Chemical resistance includes gases and liquids with the latter including the special cases of water and environmental stress cracking. The permeation of gases and liquids as opposed to their degradation effects are considered in Chapter 30.

Many environmental tests are accelerated because if they were not they would offer no advantage over a field trial. This introduces the problem seen with heat aging tests of satisfactorily extrapolating from short-term tests to long-term service conditions. When several environmental factors are present simultaneously, this difficulty is greatly increased. Extrapolation of accelerated environmental resistance data is discussed in Chapter 29.

2 Moist Heat and Steam Tests

Polymers vary considerably in their resistance to water, but materials with hydrolyzable bonds such as some polyurethanes can be very susceptible to degradation in water. Significant increase in the rate of degradation has been reported for materials generally considered water resistant when high humidity is added to high temperatures. Standard tests for exposure to high humidity are not common, but this may be largely because it is thought easier to carry out tests in liquid water for temperatures below boiling point. The two are not equivalent, as in moist air there will be a greater supply of oxygen and the effects are likely to increase with increasing humidity level. The principle of a moist heat test is similar to heat aging tests with a humidity cabinet replacing the oven.

Tests in steam at 100°C or above would be considered very severe but would be relevant for products such as hoses intended to be used in such conditions. A particular test procedure would be given in the product specification, but one general approach would be to use an autoclave. In designing such tests and interpreting results, consideration should be given to the oxygen levels present, as this could have a large effect on the degree of degradation found.

3 Effect of Liquids

The effect of a liquid on a polymer may be swelling, chemical reactions, or extraction of constituents of the material—or indeed all three. Exposure of polymers to chemicals is generally termed a chemical resistance test, although for rubbers they are more frequently referred to as swelling tests, volume swell tests, or oil aging (because standard grades of oil are frequently specified as the liquid).

While the principle of chemical resistance tests is the same for all materials—exposure of test pieces to the liquid and measurement of changes in properties—there are significant differences in the way in which such tests are viewed for rubbers and plastics. For plastics, the approach is generally to take a liquid or liquids relevant to service and to measure change in properties of interest. Such tests may be made to characterize a material's resistance by using a wide range of liquids or to investigate specifically the effect of a particular liquid for a given application. There is also a most important branch of chemical resistance of plastics known as environmental stress cracking, where the test piece is exposed in the strained condition that is considered in a separate section below.

For rubbers, exposure to liquids is very commonly specified using standard oils or fuels with the prime property change measured being increase (or sometimes decrease) in volume. Volume change is taken as a general indication of resistance, although quite often changes in tensile properties are measured as well. The use of standard oils and fuels reflects the widespread use of rubbers in such fluids and the great difference in rubber polymers in their resistance to hydrocarbons. For particular purposes other chemicals and properties are of course also used.

For any polymer the change in mass or volume indicates the degree of absorption taking place, but it is necessary also to measure other properties to quantify the effect this has on strength, modulus, etc. For some products a decrease in volume (extraction) can be more serious than swelling. If there is low swelling it may be hiding a significant chemical effect and a large deterioration in physical properties.

Another use of swelling tests on rubbers quite unrelated to chemical resistance is to estimate the degree of cure, higher swelling indicating a lower cross-link density. The cross-link density is estimated from the Flory–Rehner equation:

$$\frac{1}{M_c} = \frac{\log_e(1 - V_r) + V_{r-} + \mu V_r^2}{\rho V_1 (V_r^{1/3} - \frac{1}{2} V_r)}$$

where M_c = number average molecular weight of network chains, V_r = the volume fraction of rubber in the swollen material, μ = a solvent–rubber interaction constant,

ρ = density of the network, and V_1 = molar volume of the swelling liquid. The concentration of effective cross-links is $1/2M_c$.

The absorption of a liquid into a polymer is a function of time as illustrated in Fig. 1. In curve (a) a maximum equilibrium absorption (volume or mass increase) is reached. In curve (b) there is a decrease before equilibrium is reached illustrating the effect of extraction of soluble matter. If extraction was greater than swelling, the curve would show a reduction in volume (or mass) and reach a negative equilibrium level. For some materials, a continued increase in uptake can occur as shown in curve (c) because of oxidation due to oxygen not being totally excluded.

Figure 2 illustrates how a particular absorption can result from either a high maximum absorption but slow rate of absorption or a lower maximum absorption with a more rapid approach. It makes clear the importance of either taking readings as a function of time or, if one reading only is taken, making sure that it does not lie on the early part of the curve where the degree of absorption is still changing. The time to reach equilibrium is a function of the polymer and the liquid, and also the thickness of the test piece. Equilibrium time is roughly proportional to the square of test piece thickness.

The most usual tests involve total immersion of the test pieces in the liquid; care needs to be taken to ensure that the liquid is in contact with all surfaces, and that there is a sufficient volume of liquid, at least fifteen times the volume of the test pieces being often specified. Where service involves exposure on one surface only, for example with coated fabrics, it is sensible to test in that manner and a suitable exposure jig is used.

When the test piece is removed from the liquid, the volume or mass will decrease with time as the liquid permeates back out. For volatile liquids this dictates that the time between removal and test is critical and it may be necessary to extrapolate back to "zero" time. It is also possible to dry the test piece and by further weighing estimate the amount of matter extracted, but this is liable to be inaccurate. The physical properties after drying can be measured and may be particularly relevant for some applications.

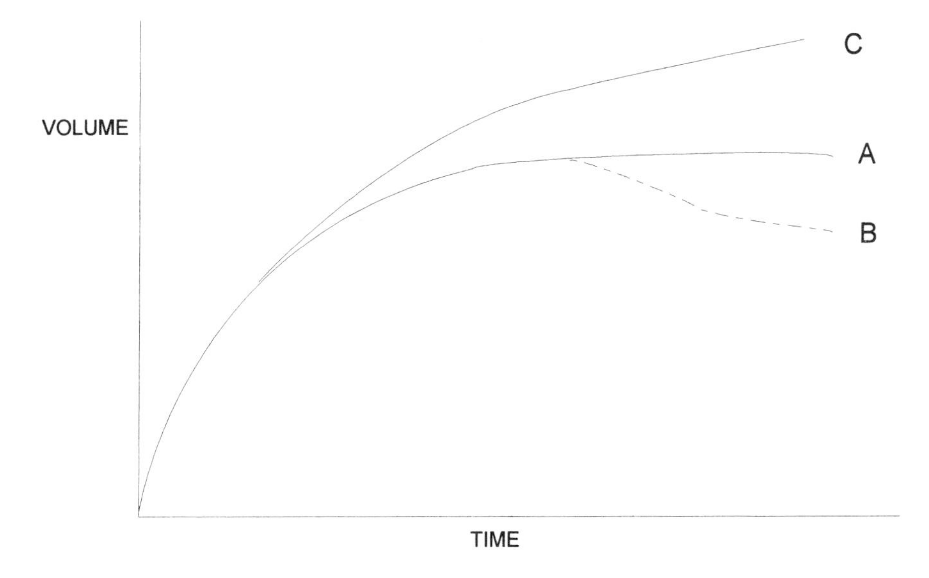

Figure 1 Time–swelling curves.

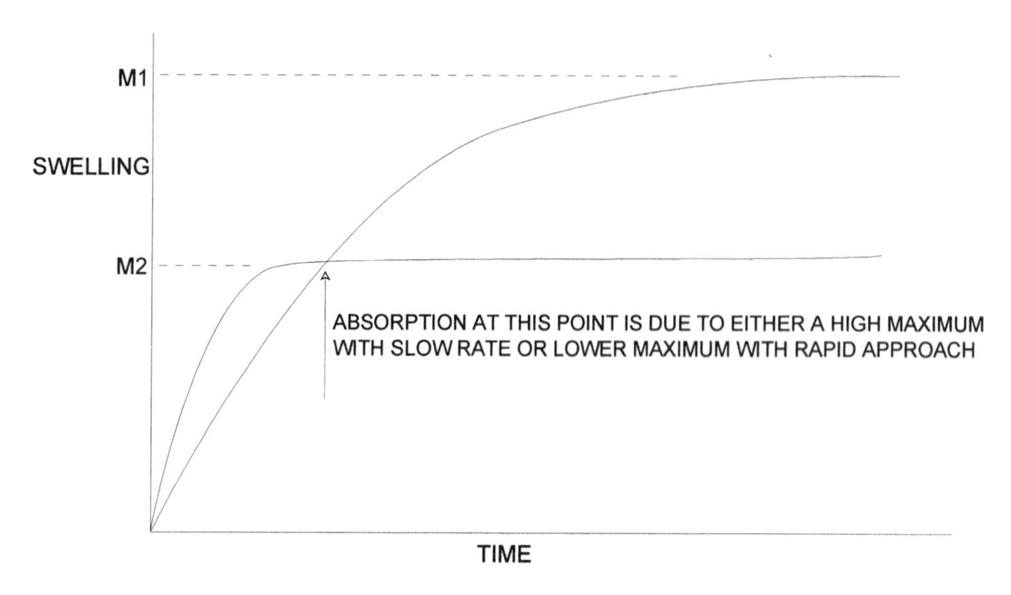

Figure 2 Swelling rate.

Water is essentially one particular chemical, but tests specifically for water exposure have been standardized for all the material types. Water is absorbed very slowly in most solid polymers, which means that long times are required to reach equilibrium unless a test piece with a large surface area to volume ratio is used. One approach is cut the material into very small pieces, and this is done for rubbers with exposure to water vapor.

Using water vapor means that absorption is related to the humidity used, and temperature has to be controlled very tightly. Also the water must be pure or there will be a reduction in absorption. Indeed, the equilibrium absorption measured is substantially the same as that which would be obtained in an aqueous solution which would maintain the test humidity.

There are several water absorption tests specified for plastics with various geometries and conditions. Generally they use times that would not result in equilibrium, and care needs to be taken in comparing results, particularly if surface area to volume ratios are different. Coated fabrics and textiles are normally relatively thin and hence there is less problem with very long times being needed to reach equilibrium.

4 Environmental Stress Cracking

Liquids which have no significant effect on an unstrained plastic test piece may cause severe cracking if the test piece is exposed in a strained condition. This is known as environmental stress cracking and is responsible for a large number of failures of plastics components in service. The problem is particularly severe because many people are unaware of the problem, and because only very small amounts of liquid, such as detergents and solvents, are required.

It follows that it is most important that chemical resistance studies of plastics include environmental stress cracking tests, and several have been standardized. These will be detailed in Chapter 16, but they can be classified into constant stress and constant strain methods. Simple bent strip methods are convenient for quality control, but for more comprehensive evaluation control of stress or strain at several levels is necessary. In constant strain methods the strain is relaxed on cracking, while with constant stress it is maintained and crack propagation as well as initiation can be studied.

5 Effect of Gases

The effect of oxygen is a very special case of the effect of gases and has been covered under heat aging in Chapter 12. Oxygen is also present to various degrees in liquid exposure tests and is an important consideration in weathering tests. With one notable exception, very little testing of the effect of other gases on polymers is carried out compared to tests with liquids. If the effect of a particular gas needs to be studied it would be necessary to devise a test rig comprising, for example, a chamber through which the gas is circulated. Alternatively, for vapors and gases readily obtainable in the liquid state, the liquid could be used.

The notable exception is the exposure of rubber, including relevant foams and coated fabrics, to ozone. Many rubbers are susceptable to cracking when exposed in the strained condition to ozone in very small quantities—this is in fact a case of environmental stress cracking discussed above for plastics. Ozone levels in the atmosphere are generally no more than the odd part per hundred million, but this is quite sufficient to cause severe damage. Accelerated tests are usually carried out in special ozone cabinets at levels between 25 and 200 parts per hundred million, and the procedures are described in Chapter 15.

6 Weathering

All polymers are affected by the environment in which they are used or stored, although the rate and degree of change is dependent on the polymer and indeed the particular grade or formulation. The most common environment is that termed weathering: exposure to light, ambient temperature, precipitation, and possibly atmospheric contaminants.

It follows that from the first introduction of polymeric materials a need was perceived to measure their resistance to weathering. The obvious first approach was to expose samples or products to the natural elements and to measure the change in key properties with time. For a material with reasonable resistance, such natural weathering will clearly involve long exposure periods before sufficient data can be collected. For this and other practical reasons, artificial weathering tests were developed that could simulate the natural conditions and at the same time offered the possibility of accelerating the degradation process.

Despite all the effort that has been expended on developing and evaluating weathering tests, the present situation is far from ideal or even satisfactory. As well as problems with reproducibility there is very often failure to obtain correlation with service. At the root of all difficulties and problems with weathering tests is the simple fact that there is no such thing as consistent, let alone standard, weather. The light intensity, wavelengths, temperature, and moisture levels vary throughout each day, from season to season and from place to place. Furthermore, these variations are quite large.

The problem is compounded by interaction and synergy between the degrading agents. Although light is often seen as the main agent in weathering tests, it is quite possible, for example, for temperature to be the predominant factor. The net result is that we are trying to characterize material behavior for what is effectively an infinite number of permutations of conditions.

There is no letup to the magnitude of the problem when we consider that different polymers will react in different ways to the various permutations. This explains the circumstances often reported where a correlation with service is found for one material but the same test procedure fails to do likewise for other materials.

A yet further complication is that the observed effect of weathering depends on the property chosen to monitor the degradation. As simple examples, there may be gross color change but negligible effect on mechanical properties, or relatively large change in microhardness (a surface effect) compared to change in a bulk property such as tensile modulus.

With so many problems it can be argued that weathering tests will never be successful, but needing to know how well a material will stand up to exposure is so important that more tests rather than fewer will be carried out. The detailed procedures, the options available, and the limitations are fully discussed in Chapter 28.

The vast majority of weathering tests are carried out on plastics. For rubbers the most catastrophic cause of deterioration when exposed to the atmosphere out of doors is ozone, but it has been found much more convenient to to use laboratory ozone tests as discussed above. Oxygen, temperature, moisture, and sunlight all affect rubbers, but at normal ambient temperatures the rate of degradation by oxygen is slow and the effects of temperature are usually evaluated by accelerated heat aging tests (Chapter 12). The main reason why weathering tests are less common on rubbers is that the effect of light is much less important than it is with plastics. Degradation is generally restricted to the surface layer and is of most consequence for coated fabrics and very thin-walled products. The exception is where change of color in nonblack rubbers is considered important.

7 Biological Degradation

There are two sides to attack by living organisms, when it is unwelcome and when it is wanted. Although polymers and/or additives can prove attractive to living organisms, unrequired attack is relatively rare. Nevertheless, there can be serious problems in tropical countries, and there has been considerable concern that products such as rubber pipe seals in temperate climates are susceptible. Increasingly, biodegradable polymers have been introduced that are specially formulated so that they are broken down by microorganisms relatively quickly and hence their disposal after use causes no environmental problems.

Exposure to living organisms is a biological rather than a physical test and is normally entrusted to specialist laboratories. A British standard for pipe joint rings [4] includes a requirement for resistance to microbiological degradation, and there is an ISO standard for resistance of plastics to fungi and bacteria [5].

Evaluation of biodegradable materials has attracted considerable attention very recently, and the ISO has embarked on producing a number of test methods that are at draft stage at the time of writing. This range of ISO draft standards [6–10] reflects the various procedures for measuring the degree of degradation and the different conditions under which degradation may take place. Tests can be separated into those carried out in an aqueous environment, with either aerobic or anaerobic conditions, and composting methods. For the aqueous environment, the aerobic tests use either measurement of released carbon dioxide, oxygen demand, biomass, or dissolved organic carbon. The anaerobic test measures conversion of carbon to carbon dioxide and methane, while the composting test analyses released carbon dioxide. In addition to these draft standards, there are new work proposals for anaerobic biodegradation under high solids digestion conditions and a standard giving a test scheme for acceptance of biodegradable materials. The ISO drafts were doubtless influenced by a series of ASTM standards that had been established earlier [11–15] although they are not the same and illustrate the great variety in the detailed techniques being used in what is a relatively new area of measurement.

Itavaara and Vikman [16] give an overview of biodegradability test methods and distinguish between screening tests and tests that simulate in situ conditions. Screening tests are those in aqueous conditions where the authors suggest that for packaging materials compost or soil microbes be added in addition to the more usual sewage sludge. Model studies on laboratory scale composting methods have been described by Gross et al. [17]. Clearly, the difficulties with composting tests simulating in situ conditions are that they can be laborious and expensive, and it is not easy to produce a standard compost, but there is a strong argument that they are the only way to prove the degradability of a material. This appears to be the view taken in the ISO new work item for a test scheme for final acceptance.

8 Radiation

Radiation is taken here to mean atomic and nuclear particles, i.e., gamma rays, electrons, neutrons, etc. The intensity of such radiation at the earth's surface is not high enough to affect polymers significantly, and hence radiation exposure tests are only required in connection with applications in nuclear plants and possibly where radiation is used to induce cross-linking. The specialist equipment needed to carry out irradiation tests is clearly not generally available, and the limited interest means that there has not been wide-scale standardization of test procedures. Nevertheless, there is an IEC guide to determining the effects of ionizing radiation on insulating materials written in four parts [18–21] and an ASTM recommended practice for exposure of polymeric materials to various types of radiation [22].

The exposure to radiation clearly needs a very specialized facility, but the methods used to monitor changes are the same as for any other aging test. There is also a parallel with other aging tests in that the effects are often dependent on dose rate as well as total dose, and this has to be carefully considered when designing tests.

9 Fire

The unique properties of the various polymers have made them essential or highly desirable for a vast range of applications, but unfortunately they will burn. This has given rise to considerable concern, particularly with certain plastics and foams. Great effort has been expended on improving fire resistance by means of retardants, and not surprisingly fire testing is extremely important for many products. In recent years there has been a high level of activity in the development of test apparatus and procedures that are more relevant to real fire situations. The subject is dealt with in detail in Chapter 27, but two preliminary points can be made here.

It is necesssary to distinguish between small-scale and large-scale tests and to define clearly which aspect of burning is being evaluated, for example ease of ignition, rate of burning, smoke production, etc. Large-scale tests are necessary to evaluate the performance of a material or product in most real fire situations.

It is fairly easy to invent small-scale tests that are confusing as to which aspect of fire they are meant to evaluate and may give dangerous impressions because of ill-conceived presentation of results. Consequently, the established standards should be adhered to and the limitations of small-scale tests in terms of predicting performance appreciated.

References

1. Flory, P. J., *J. Chem. Phys.*, *18*, 108 (1950).
2. Flory, P. J., and Reyner, J., *J. Chem. Phys.*, *11*, 521, (1943).
3. Kraus, G., *Rubb. World*, 67, (1956).
4. BS 2494, Specification for elastomeric joint rings for pipework and pipelines, 1990.
5. ISO 846, Determination of behaviour under the action of fungi and bacteria—Evaluation by visual examination or measurement of change in mass or physical properties, 1978.
6. ISO CD 14851, Evaluation of the ultimate aerobic biodegradability of plastics materials in an aqueous medium—Method by determining the oxygen demand in a closed respirometer.
7. ISO CD 14852, Evaluation of the ultimate aerobic biodegradability of plastics materials in an aqueous medium—Method by analysis of of released carbon dioxide.
8. ISO CD 14853, Evaluation of ultimate anaerobic biodegradation of plastics materials in an aqueous system—Method of analysis of carbon conversion to carbon dioxide and methane.
9. ISO CD 14854, Evaluation of ultimate aerobic biodegradation of plastics materials in an aqueous system—Method by determining the oxygen demand, biomass and DOC production in a closed respirometer.
10. ISO DIS 14855, Evaluation of the ultimate aerobic biodegradability and disintegration under controlled composting conditions—Method by analysis of released carbon dioxide.
11. ASTM D 5338, Determining the aerobic biodegradation of plastics materials under controlled composting conditions, 1992.
12. ASTM D 5209, Determining the aerobic biodegradation of plastics materials in the presence of municipal sewage sludge, 1992.
13. ASTM D 5271, Determining the aerobic biodegradation of plastics materials in an activated sludge wastewater treatment system, 1993.
14. ASTM D 5210, Determining the anaerobic biodegradation of plastic materials in the presence of municipal sludge, 1992.
15. ASTM D 5247, Determining the aerobic biodegradability of degradable plastics by specific microorganisms, 1992.
16. Itavaara, M., and Vikman, M., *J. Environmental Polym. Degrad.*, *4*, 1 (1996).
17. Gross, R. A., Gu, Ji-Dong, Eberiel, D., and McCarthy, S. P., *J. M. S.—Pure Appl. Chem.*, *A32*(4), 613 (1995).
18. IEC 544, Part 1, Guide for determining the effects of ionising radiation on insulating materials—Radiation interaction, 1994.
19. IEC 554, Part 2, Procedures for irradiation and test, 1991.
20. IEC 554, Part 3, Test procedures for permanent effects, 1979.
21. IEC 554, Part 4, Classification system for service in radiation environments, 1985.
22. ASTM E 1027, Exposure of polymeric materials to ionizing radiation, 1992.

14
Other Physical Properties

Roger Brown
Rapra Technology Ltd., Shawbury, Shrewsbury, England

1 Introduction

Having dealt with the mechanical properties, durability, and the effect of temperature and the environment, there is a number of somewhat diverse physical property measurements remaining. For convenience, these are collected together for introductory treatment in this chapter. The topics covered are electrical, optical, and thermal properties, friction, permeability, and staining.

In many senses these have absolutely nothing in common, but in two respects there are features that they share. Whereas the mechanical properties and durability tests are very widely applicable to most applications of polymers, the parameters considered here could be termed specialist in that they will only be of interest for particular products. This results in tests to measure these parameters being carried out relatively infrequently and in many cases only by specialist laboratories. The second aspect is that the methods used differ little regardless of which polymer type is being tested. This is in part due to the relatively infrequent use and hence the lack of particular procedures developing in the different branches of the polymer industry, but also because for these parameters the differences in form and stiffness of the polymer types is of relatively little significance.

Having said that these properties are only of interest in particular applications, it will be appreciated that for a given application one of these physical properties can be the factor with overriding importance. For example, a cable covering is nothing if it is not insulating and a balloon is useless if it has high permeability.

In the same way, although the test methods are essentially similar for all types of polymer, there are characteristics of polymers that differ from materials in general that have to be taken into account when the test procedure is formulated.

2 Thermal Properties

The thermal properties are conductivity, diffusivity, and specific heat. Other properties are sometimes included under this title but thermal expansion, transition points, low temperature properties, and heat aging are more properly the effects of temperature (Chapter 12). Thermal analysis in all its various forms is also a study of the effect of temperature rather than measurements concerning the transport of heat, although thermal analysis techniques can be used to measure thermal transport properties.

The first, and obvious, area of interest in thermal properties is for applications involving thermal insulation. The polymer type most frequently involved is foams. For plastics and rubbers there is also a need for transport properties, particularly diffusivity, in the prediction of processing behavior. Most processes for forming these materials involve heat, often for quite short times, and the rate at which heat is transferred can be critical.

Compared to materials in general, polymers, and particularly foams, have relatively low thermal conductivity, which has to be taken into account when considering the sensitivity of test procedures. When data is required in connection with polymer processing, it is preferably obtained at processing temperatures, which can involve the polymer going into the molten state. Clearly, this places a considerable restriction on the design of measuring apparatus.

Although test methods specifically for polymers have been developed, very few are standardized, and for insulation applications, methods for insulating materials in general are normally used. Test procedures are considered in detail in Chapter 24.

3 Electrical Properties

Electrical properties are of concern in in electrical and electronic components, such as cables, plugs, and sockets, and also in products to give protection against high voltages such as gloves and casings. The interest in electrical properties is, however, more widespread in that they can be of importance in any product where electricity could be a hazard. A particular area is the dissipation of electric charge, which can affect the specification of such things as flooring, coverings, and clothing.

When considering properties to be measured and the test methods, a distinction can be made between materials that are conducting or semiconducting (antistatic) and those that are insulating. Although there is a grey region at the higher end of semiconducting and the lower end of insulating, test methods can be divided into the two areas.

The main tests on insulating materials are resistance or resistivity (the reciprocal of conductivity), power factor, permittivity, electric strength, and tracking resistance. For conducting and antistatic materials, only resistivity and surface charge are useful, because such materials would not be used in situations requiring low dielectric loss or high electric strength. With insulating materials a distinction can be made between surface and volume resistivity, as the surface is often more conductive than the bulk, but with conducting materials such distinction is not sensible.

Electrical property measurement is relevant to all the polymer types, although most applications are for plastics or rubbers. Rubbers hold a particular interest because of the unique combinations of electrical properies and low stiffness that can be achieved. With respect to testing, the relative softness of rubbers and foams has to be considered when choosing electrode systems. In addition to the resistance of the polymer, there is a contact resistance at the electrode/test piece interface that can be very high, and this is a complication in testing and an important factor in the choice of electrodes.

The most frequently used properties are well standardized nationally and internationally. Previously, slightly differing standard methods were used for rubbers and plastics, but there has been a trend for them to be combined, with provision made to account for the softer materials. Mostly the methods are a subset from the methods for materials in general, but there are particular procedures for conducting rubbers. For less common properties, general methods are used, and for antistatic behavior there is an increasing number of product tests being introduced. Test procedures are considered in detail in Chapter 25.

4 Optical Properties

Optical parameters, which are properties of the surface, gloss, reflectance, and color, are in principle relevant to all polymer materials, but it is not frequently that rubbers and foams are used where these aesthetic qualities are important. Not that these properties are entirely a matter of appearance, because color and reflectance are important in terms of solar energy reflected and hence the temperature rise in an exposed material. Roofing sheet is a particular example, but generally the color would only need to be approximately matched. Gloss is also important in terms of glare from a surface and would be controlled in such products as artificial sports surfaces. Surface texture will affect gloss and reflectance and how a surface looks, but it is not in itself an optical property.

The optical parameters that are properties of the bulk of the material, transmittance, refractive index, birefringence, and haze are only applicable to transparent or translucent materials. In practice these measurements are essentially restricted to plastics, and their relevance is the same as for any other material that transmits light.

Visual inspection is obviously an optical matter that includes noninstrumental assessment of aspects of appearance such as gloss, color matching against standard colors, and detection of flaws. Optical microscopy is an extension of visual inspection that makes examination of small details of appearance and structure possible. Visual inspection is also a nondestructive technique and this aspect is considered further in Chapter 32. Microscopy is an art and science of its own, and in-depth treatment of it is beyond the scope of this book.

Optical properties of polymers, both of the surface and the bulk, are measured in the same way as for any other material, and in most cases they present little in the way of particular problems or special considerations. Some methods have been standardized specifically for polymers, although it is not always obvious why this was necessary. Test procedures used are covered in Chapter 26.

5 Friction

Friction is an important factor in a surprising number of applications from tires and bearings to bottle tops and footballs; in some cases the problem is too little friction and in others too much. Despite its importance, friction is relatively infrequently measured as a general material property; most measurements are made in the context of a particular product.

Relatively crude measures of friction can be easily obtained with inclined planes and simple towed sleds, but accurate and complete characterization of a material requires very well designed apparatus and is time-consuming. The correlation of measured friction values with product performance is notoriously difficult in many cases. One difficulty is that friction is sensitive to just about all experimental factors including temperature, velocity, and stiffness of the apparatus. Further, it can change significantly with quite small variations in the surface conditions of texture and contaminants. The limited value of measurements made on simple apparatus and at one set of conditions is not always appreciated.

More simple test methods for certain polymer types have been standardized for many years: the inclined plane for fabrics and the simple towed sled for plastic films. But it is only fairly recently that more comprehensive procedures have been recognized in standards. There is no reason why such a comprehensive method should not be written for use with all materials, but the tendency has been for standardization to be carried out within particular industries. One reason for this is that there are a considerable number of tests for particular products, mostly developed with the aim of simulating the particular conditions of use. The measurement of friction is comprehensively covered in Chapter 23.

6 Permeability

Polymers are used a great deal as barriers to liquids and gases in such areas as damp-proof membranes, food packaging, and balloons. For these applications, the permeability will be a prime consideration in the selection of materials.

Permeability measurement becomes increasingly difficult and time-consuming as the rate of permeation decreases. The basic methods are based on direct measurement of the amount of substance permeated, using weight, volume, or pressure change, but at the lower permeation levels these techniques reach their limit of sensitivity and become very time-consuming and experimentally difficult. More sensitive and/or rapid methods using more sophisticated detection systems were then developed for quality control in particular industries or to enable very low permeation rates to be measured, particularly with low-pressure differentials.

Polymers are newer than many materials, and test methods for most properties have origins in those developed for more traditional materials. Permeability tests have generally been first developed for polymers. Basic methods have long been standardized, those for liquids being based on weight change and those for gases on either pressure or volume change. The more advanced techniques have as yet been largely ignored by standards committees, probably because of their specialist interest. There has been a tendency for the methods to be developed in the separate industries, but the essential methodology is the same for plastics and rubbers. The procedures used are covered in detail in Chapter 30.

7 Staining

Additives such as antioxidants can cause staining of any surface that comes in contact with or near to the polymer material. The problem is mostly found with rubber or flexible cellular seals and is an important consideration in such consumer products as cars and kitchen appliances. The compounder has to achieve adequate environmental resistance without an unacceptable degree of staining. No staining would be ideal, but in practice some staining may have to be tolerated. To ensure that levels are acceptable, tests to produce and measure staining are often included in specifications.

Staining can occur by direct contact, on the area surrounding contact, and by water that has leached constituents from the polymer compound. The staining can also be intensified by heat and light, and one or both of these is usually included in the test procedure. There had been confusion over terms used to describe the cause of staining until an international standard was published. In ISO 3865,

Contact stain is the stain which occurs on the surface directly in contact with the rubber.

Migration stain is the stain which occurs on the surface surrounding the contact area.

Extraction stain is the stain caused by contact with water containing leached-out constituents of the rubber.

Penetration stain is the staining of a veneer layer of an organic material bonded to the rubber surface.

It should be noted that the stain on the surface directly in contact with the rubber is always contact stain even if the stain has to be intensified by exposure to light after removal of the rubber.

Details of the experimental procedures for rubbers are given in Chapter 15, and these could be used for other polymer types if required.

15
Testing of Rubber

Peter Lewis
Tun Abdul Razak Research Centre, Brickendonbury, Hertfordshire, England

1 Introduction

1.1 Requirements for Rubber

Small-scale laboratory tests for rubber are used in material and product specifications and for such activities as compound development, component design, factory quality control, and life prediction. Many of these tests have been shaped by the distinguishing features of rubber, rubber products, and their means of manufacture. Such features can conveniently be grouped as follows:

Properties

A very low Young's modulus, ranging typically from 1 to 20 MPa (depending on composition and test strain)
Very high elongation at break, in some instances in excess of 1000%
A viscoelastic response to deformation, with modulus and damping properties strongly influenced by temperature and frequency
The Gough Joule effect, shown as an increase in modulus with an increase in temperature and the retraction of stressed rubber on heating
An ability of some elastomers to undergo strain-induced crystallization
The susceptibility of unsaturated rubbers to ozone attack and subsequent cracking in the stretched state

Manufacture

The addition of a wide variety of compounding ingredients to facilitate processing, effect vulcanization, and modify or improve vulcanizate properties, calling for tests for measurement of their incorporation, dispersion, and performance in the rubber compound

A wide range of manufacturing operations calling for measurement of properties as diverse as hot tear strength (for moldings extracted from heated molds), compression modulus (as a measure of porosity), coefficient of friction (for surface-treated components such as latex gloves and windscreen wiper blades) and adhesion strength (for rubber-to-fabric or rubber-to-metal bonds)

Applications

A wide range of finished products from disposable singe-use surgeons' gloves to structural bearings designed for at least 50 years service

Static and dynamic deformation in tension, shear, compression, and torsion

A wide range of operating temperatures down from near the glass transition up to several hundred degrees

Among pressures influencing the development and usage of tests have been the needs of small manufacturers and custom compounders having limited testing facilities, the insistence of many end-users for engineering data and reliable prediction of long-term performance, and, more recently, the rapid growth of the thermoplastic rubber sector. Pressure has also been exerted by major customers, notably the motor, civil engineering, and health-care industries.

1.2 Test Development

In his 1950 review of the history of rubber testing, Buist [1] observed that prototypes of many of the tests then in use had been tried with varying degrees of success before 1900, and there is no doubt that technologists of that period would still recognize many of the features of the present-day rubber laboratory. The value of tensile stress–strain properties as control measures was appreciated in the late nineteenth century; the principles of the indentation hardness test were established before the First World War; and the accelerated air-oven aging test dates back at least 80 years.

A measure of the development—and importance—of the physical test has been the growth of national and international standards in the testing sector. The ASTM was an early starter in this area, and by 1948, committee D11 (which exists to this day) had issued some 80 standards, comparable to the current number. In the UK most of the formative work took place in the 1930s, leading to the first issue of BS903 in 1940. By 1950 this standard contained 27 physical tests; it now has over 50.

On the international scene, ISO/TC45 (Rubber and Rubber Products) began its work in 1948, yet by 1957 it had only published three test methods for rubber, a reflection of the difficulties of test harmonization. However, 20 years later, the number had risen to over 40, and now there are about 70 in the ISO series. These have helped to make ISO/TC45 one of the most productive of ISO technical committees.

1.3 Popularity of Rubber Tests

The uptake of standard tests can conveniently be gauged by their usage in national, international, and company specifications for materials and products, although the use of tests for internal quality control work and for R & D activities should never be underestimated. In specifications, small-scale laboratory tests fall into three broad categories.

Classification and control, headed by hardness, density, tensile stress–strain properties, accelerated aging resistance, and compression set

Polymer selection and environmental requirements, headed by oil resistance, low-temperature properties, and ozone resistance

Special requirements, with examples being stress relaxation, creep, wear, compression stress–strain behavior, dynamic properties, staining, and metal corrosion.

For the first category, there is close agreement between national and company or sector-interest specifications, but differences do often emerge in the second category, with companies favoring tests for fairly widespread use irrespective of the material or product concerned. Particular examples of this trend noted in some automotive rubber specifications are low temperature testing, ozone resistance, and tear strength; these can sometimes almost be as popular as tensile strength, hardness and accelerated aging.

A recent survey by Coveney and Jamil [2] of U.K. manufacturers of rubber engineering components has confirmed the popularity of the traditional test method and has revealed a continuing reluctance to adopt some of the standard methods introduced to aid the generation of design data; these include several of the tests available for dynamic testing.

1.4 Sources of Information

The rubber industry has appreciated the need for guidance on the selection and use of the rubber test methods. Attention must be drawn in particular to the books published by Brown [3] and Smith [4]; the first of these will appeal to the professional, as it discusses and compares a whole range of standard tests, while the second gives a general grounding suitable for readers having a limited knowledge of rubber. Mention should also be made of reviews prepared for special interests such as engineering design [5].

In the U.K., the BSI has introduced two standard guides to assist users of the growing BS903 series of test methods. These are BS903, Part 1, on the selection and use of tests, and BS903, Part 2, on the use of statistics in rubber testing.

2 Strength and Stiffness Properties

2.1 General

Short-term stress–strain testing is widely practised in the rubber industry, especially in the form of indentation hardness, tensile strength, and elongation at break. Applications range from quality control and measurement of the state of cure to material specification and a convenient means of monitoring aging resistance.

2.2 Hardness

The International Rubber Hardness Degree (IRHD) scale is described in ISO 48 (BS903, Part A26, and ASTM D1415). Three methods are available for the dead-load test over the hardness range of 10 to 100 IRHD. The normal and high hardness procedures require a test piece thickness of 8–10 mm, whilst the softer L procedure needs a thickness of 10–15 mm. Such thicknesses can be obtained by plying thin sheets or test pieces, but for many finished products the dead-load instruments are unsuitable. The standard therefore includes a micro test using a scaled-down indentor, and this requires a test piece of 2 mm thickness, although it can be undertaken on sections down to 1 mm thick.

The Shore A scale, or as it is sometimes called, durometer hardness, enjoys considerable success in North America and in particular is widely accepted by the automotive industry; the scale is an integral part of the classification system for elastomeric materials used in automotive applications (ASTM D2000). Shore hardness has also become the industry standard for the rapidly growing thermoplastic rubber sector, where advantage can be taken of the complementary Shore D scale for harder grades and for rubber-modified plastics. Seven Shore scales are described in ASTM D2240, while Shore A and D hardness are also detailed in the international standard for pocket hardness meters, ISO 7619 (BS903, Part A57). The latter also specifies a pocket meter based on the IRHD scale.

Correlations are frequently sought between the IRHD and Shore A scales [3]. There is approximate numerical equivalence for vulcanized natural rubber and competitive synthetic rubbers, but there can be significant differences for some elastomers. In some product and material specifications both scales are included as acceptable alternatives, but this practice is not widely pursued.

For engineering design—and in some accompanying specifications—hardness has been used as a means of assessing Young's modulus and related moduli. Correlations between IRHD and shear or compression modulus have been established, with closer agreement being obtained for a series of related rubber compounds [6]. Muhr and Thomas [7] make use of the extensive test data in a series of engineering data sheets to show that the low-frequency dynamic shear modulus calculated from hardness agrees with direct modulus measurements to within 30%, provided the shear strain matches the deformation in the hardness test. However, these authors also show that the estimated shear modulus is inappropriate for calculating compression modulus unless the compression strain is very low.

Design engineers are generally sceptical of the value of hardness as a design parameter and are among the most critical of the reproducibility of measurements. Key factors governing test errors are operator variations, test conditions, the level of appropriate calibration, and test piece thickness [8,9]. There remains a surprising lack of appreciation of the importance of the latter, with many users assuming that because of their size the Shore A and IRHD pocket hardness meters can be used for comparatively thin test pieces. ISO 7619 actually specifies a minimum thickness of 6 mm, although a plied-up test piece can be adopted. Bassi et al. [10] have considered the influence of thickness on Shore A hardness and the need for a correction to achieve closer agreement between measurements.

ISO 48 includes methods for measuring the hardness of curved surfaces, and in the related ISO 7267 (BS7442, Part 3) this principle has been extended specifically for rubber-covered rollers, with the IRHD method being joined by the Shore A and Pusey and Jones hardness scales.

2.3 Tensile Stress–Strain Properties

Tensile strength and elongation at break are measured not only on specially prepared test sheets but also on products as diverse as surgeons' gloves and laminated structural bearings. Accordingly, the test method ISO 37 (BS903, Part A2) is intended for wide appeal, with a choice of four dumbbells ranging from type 1 having an overall length of 115 mm down to type 4 with a length of only 35 mm. Two ring-shaped test pieces are also specified, one with an internal diameter (ID) of 44.6 mm and the other with an ID of only 8 mm. The small ring and the two smallest dumbbells are intended primarily for use with products, and it is hoped that this provision will discourage a proliferation of other test pieces in various industrial sectors. The equivalent ASTM D412 permits use of strip test pieces for products such as tubing, where dumbbells and rings are impractical. All the standard test pieces can be used for strength, elongation at break, "modulus" (stress) at given elongations and, in the case of thermoplastic rubbers, tensile strength and elongation at yield.

The specified minimum of three test pieces is commonly employed, but in a study of test precision Spetz [11] advocates the use of five test pieces to improve test reproducibility. He also draws attention to the importance of thickness measurement and calibration. Care too should be taken to ensure test piece cutters are fully maintained. Some tolerance in cutter sharpness can often be accepted in tests at standard laboratory temperature, but a slightly blunt cutter can seriously impair tensile strength measured at elevated temperatures [12]. Cutter flaws not only reduce the temperature at which strength begins to fall rapidly but also increase the temperature band over which the fall occurs. With some vulcanized rubbers, the critical temperature range can embrace the laboratory temperature and so lead to significant scatter in results.

Special care should also be exercised when thin latex films are tested. Elongation at break can exceed 1000%, slippage from the grips can occur, especially with lubricated surfaces, and there is sometimes risk of grip tearing. The standard ISO 37 test pieces are sometimes found to be unsuitable for latex applications, and for this reason it has been necessary to specify a dumbbell of higher than normal tab width to test length-width ratio in the CEN specification for surgeons' rubber gloves (EN 455). Thickness measurement is especially critical with latex films. Poor reproducibility in an interlaboratory test program on condoms was attributed in part to variation in recorded thickness [13]. Any small misalignment between a dial gauge measuring foot and base plate will certainly be reflected in measurements on films that can be as thin as 150 mμ. Reproducibility can be improved by applying an adhesive to the thread of the foot to ensure parallelism to the plate or even by bonding a second foot to the original one. However, a much easier way is simply to compute the thickness from the weight of the test piece. Weighing instead of direct measurement reduced the coefficient of variation (CV) for thickness from 9.3% to 1.8% and in turn halved the CV for tensile strength from 10.7% to 4.7% [14].

2.4 Tear Strength

ISO 34 (BS903, Part A3) contains as many as five test methods for tear strength, a reflection not only of the general importance of the property but also of the difficulty of matching laboratory tear tests to product behavior. The methods comprise the trouser, crescent, Delft, and nicked and unnicked angle test pieces, none of which gives comparable results to another. ASTM D624 specifies crescent test pieces with and without tab ends (Dies A and B respectively), the unnicked angle test piece (Die C), and the trouser test piece (Die T).

The trouser test piece, which has its origins in fundamental work on the fracture mechanics of tearing, is the preferred ISO method principally on the grounds that the test result is not sensitive to nick length—one of the shortcomings of the other methods—and is less influenced by test piece shape. Trouser tear strength is also less sensitive to tensile modulus, provided leg extension is negligible. Even so, nearly twenty years after its standardization in ISO 34, the method is still not widely used and much less specified than the alternatives. A major deterrent is test variation, even though much of the variation is intrinsic to the material being tested. Attempts have been made to constrain the crack path and to reinforce the legs with fabric for the combined purpose of minimizing leg extension and reducing knotty tearing behavior (see the annex to ASTM D624). Neither technique is attractive, and Clamroth and Kempermann [15] have demonstrated that poor repeatability remains even when a groove has been inserted to encourage controlled tear propagation. In their studies, the smallest coefficient of variation was given by the crescent test piece, a feature later confirmed by Warhurst, Slade and Ochiltree [16], who also showed that reproducibility was influenced by thickness, straining rate, and test sheet grain.

The applicability of each method to service behavior remains questionable, and choice becomes often a matter of convenience. The Delft test piece is just 60 × 9 mm in overall dimensions, half the size of the other test pieces, and so it is particularly suitable for use with products. Tear strength is certainly relevant to mold stripping, where good hot tear resistance is usually demanded, to applications where there is risk of failure through a crack growth process, and to the sometimes rough handling of products during installation. Silicone and nitrile rubbers often need special attention, since attempts to optimize tear strength can lead to some sacrifice in other properties such as resistance to compression set.

In spite of the lack of universal agreement on the choice of methods, there is still interest in the levels of correlation among the different standard methods. Warhurst et al. [16] report slightly higher results for the unnicked angle test piece than for the crescent test piece, with both of them giving a resistance at least three times higher than the trouser test piece. Work on a series of black-reinforced natural rubber vulcanizates at the Tun Abdul Razak Research Centre [6] concluded that the crescent test piece generally gave higher results than the unnicked angle test piece. Adding a 1 mm nick to the angle test piece reduced tear strength by at least half, and by up to two-thirds with gum vulcanizates. Anyone interested in a comparison of the methods should also study the precision statement given at the end of ISO 34, Part 1.

In nicked tear test pieces, the result is strongly affected by the size and quality of the nick. When any comparisons are made, it should be remembered that some old published data for crescent tear strength will be for a previous standard nick size of 0.5 mm, and in most instances this will be reflected in a higher test value.

2.5 Other Strength Measures

Conventional tensile and tear tests are not particularly suitable tools for a detailed investigation of strength-related properties of small finished components or over the thickness of a bulky product, unless the item can be sectioned to size. An alternative approach, particularly suitable for diagnostic work, involves the use of a puncture test [17]. An indentor of specified geometry is pressed into the component, and the force to cause rupture is measured at a specified penetration or over a range of depths. One suitable application is the study of aging across the thickness of a bulky product [18].

The puncture force F at rupture is related to the tearing energy Tc by the equation

$$Tc = F\,\frac{1-\lambda_c}{2\pi r_0}$$

where λ_c is the compression ratio and r_0 is the surface crack radius after rupture.

Good agreement has been reported between tearing energies derived from the test and those from trouser tear measurements [18].

2.6 Measurement of Modulus

Tensile modulus, or more correctly the stress at a given elongation, is much less frequently specified than tensile strength and elongation at break, but it is conveniently measured at the same time and is usually recorded as M100 or M300 (the stresses at 100 or 300% elongation), often as a measure of cure or filler reinforcement. One advantage tensile modulus has over hardness as a measure of stiffness is that it can easily be used to determine grain and other anisotropic affects.

A more valuable measure, especially of the state of cure, is "relaxed modulus," the term used to describe the tensile stress at 100% elongation after one minute's relaxation. The test was once specified as a measurement of the vulcanizing characteristics of Standard Malaysian Rubber but has since been replaced for this purpose by curometry. However, the test, which is described in the now withdrawn BS1673, Part 4, remains useful as a means of recording differences in modulus when comparisons are made of some modulus-dependent properties, such as fatigue life at a specified elongation. The test can also be adapted to measure physical cross-link density in vulcanized rubber [19]. The inverse measurement of elongation at a given tensile stress remains a standard method in ASTM D1456.

Some engineers are surprised that a standard test for determining Young's modulus is not available, despite acceptance of its limitation to low strains. An ASTM method conducted in flexure was withdrawn in 1995. For load-bearing applications more appropriate properties are shear modulus and compression modulus, from both of which some estimate of Young's modulus can be made.

A test method for shear modulus is described in ISO 1827 (BS903, Part A14). A bonded quadruple shear test piece, which may or may not have previously been mechanically conditioned, is deformed to a maximum shear strain of 30%, and the result is reported as shear modulus at 25% strain. A second method in the same standard involves loading the test piece so that adhesion strength between rubber and substrate can be determined. No corresponding ASTM method is available.

Compression modulus is determined in accordance with ISO 7743 (BS903, Part A4). Two procedures are available, one using lubricated plates to facilitate test piece slippage and the other using a bonded test piece. In both instances the test piece is compressed by 25% and the modulus recorded at 10% and 20% compressive strains. Lubrication ensures that the resulting modulus is independent of test piece geometry, whereas bonding means that the modulus depends on the shape factor of the test piece, that is, the ratio of one loaded face to the total force-free area. The shape factor for the larger compression set disc used in the standard test is 0.58. The compression modulus at other shape factors can be calculated from the Young's or shear modulus of the rubber. ISO 7743 is intended to assist engineering design of load-bearing applications, but it has also been recommended as a means of detecting porosity in finished components such as seals.

Equipment for determining compression modulus has been designed for ease of use in routine testing [20]; a test piece is squashed between two plates to a predetermined strain (typically 10%) and the force recorded after one minute.

Dynamic modulus can be determined by a range of methods, and attention is drawn to ISO 1856 and its UK equivalent, BS903, Part A24. In the USA, the Yersley oscillograph is specified in ASTM D945. For detailed guidance on the determination of dynamic properties, see Chapter 21. For a test methodology on rubber properties of engineering design, reference should be made to Gregory's review [21].

3 Fatigue and Wear Processes

3.1 General

Fatigue crack growth, abrasive wear, and thermal degradation resulting from heat buildup can be grouped together as dynamic failure mechanisms, although with some nonstrain crystallizing rubbers fatigue cracking can also occur under static strain conditions. These properties are relevant to such applications as tires, conveyor and transmission belts, engine mountings, and other springs.

3.2 Fatigue Cracking

Among the better known methods for measuring fatigue resistance is the long established De Mattia flex cracking test, which remains a standard despite the availability of improved, less subjective procedures. The test for crack initiation is described in ISO 132, BS903, Part A10, and ASTM D430. It involves the flexing of a grooved test piece at 5 Hz and the measurement of the number of cycles to a given state of cracking in the groove with the aid of a series of graded photographs or crack descriptions. At face value, the test has several shortcomings: the strain range is indeterminate, opportunities for varying the test conditions are severely limited, the test is unsuitable for finished products given the need for a molded groove, and there is a highly subjective element in the visual assessment of damage. Nonetheless, the De Mattia machine has proved suitable for some comparative testing and for compound evaluation. In ASTM D430 there is provision for the testing of dumbbells in an extension mode, as well as alternative Scott and DuPont flexing machines; these are seldom used outside the USA.

More appropriate for design purposes is the tension fatigue test described in ISO 6943 (BS903, Part A51, and ASTM D4482). An ISO 37 type 1 or type 2 dumbbell or a 44.6 mm ID ring is cycled to a predetermined maximum extension at a frequency between 1 and 5 Hz, and the number of cycles to test piece rupture is recorded. The fatigue life can be reported or plotted as a function of the maximum test strain, the corresponding maximum stress, or the strain energy density, the choice depending on the particular service application. One merit of the test is that it is based on a fundamental fracture mechanics approach, which enables laboratory fatigue behavior to be related to product performance [22]. Dumbbell test pieces are generally favored in the USA and much of Europe, with ring test pieces finding some popularity in the UK. Broadly comparable results are obtained with the two test pieces, with dumbbells proving to be especially valuable for investigating anisotropic effects.

The current standard fatigue test is confined to test pieces that pass through zero strain during cycling, whereas many products are designed to operate under prestrained or nonrelaxed conditions. With strain-crystallizing rubbers, imposition of a minimum strain significantly enhances fatigue life. Nor does a standard test method exist for fatigue cracking in compression. However, Stevenson [23] has demonstrated that, for a compressed test piece as for one in tension, a relationship exists between tearing energy and crack growth rate, confirming the general applicability of the fracture mechanics approach.

Some of the problems with crack growth fatigue relate to the analysis and interpretation of test results [24]. It is still not always appreciated that fatigue life is intrinsically more variable than tensile strength, a reflection of the sensitivity of crack growth to tearing energy and flaw size. For natural rubber, which in fatigue terms is generally regarded as a well-behaved elastomer, the ratio of the highest to the lowest test result will usually be in the order of 2, while for some nonstrain crystallizing rubbers, for example SBR, the ratio can be as high as 10, with little improvement being forthcoming from any refinements in ingredient dispersion and uniformity of cure. ISO 6943 recommends that a simple median be recorded as a measure of central tendency, while ASTM D4482 opts for a geometric mean and states that the logarithm of fatigue life normalizes distribution, reducing standard deviation at multiple extension levels.

Insufficient attention is often paid to stress softening and the development of set during cycling. Both these changes can enhance fatigue-to-failure life at constant deformation and can result in extremely long fatigue lives, which in turn lead to an overestimate of the mechanical fatigue limit for crack initiation. Thermoplastic rubbers are especially susceptible to this behavior.

3.3 Crack Growth

The relationship between crack growth rate, strain input, and fatigue life has been appreciated for over 30 years [25], yet until comparatively recently crack growth measurements outside the R & D sector were essentially confined to the De Mattia test (ISO 133, BS903, Part A11) and tests using specialized flexing equipment such as the Ross machine (ASTM D1052). In ISO 133, the number of cycles for an inserted cut to grow by a specified amount or the crack growth after a specified number of cycles is recorded. Such measurements do not overcome the previously mentioned shortcomings of the De Mattia machine.

Some advance was made possible with the standardization of the Texus flex machine in ASTM D3629. A pierced T-shaped grooved test piece is bent or flexed at a given frequency and adjustable flexing angle to vary severity. Normally the number of cycles to a 5- or 10-fold increase in cut length is recorded. The claims for the method are modest, with results being shown to relate qualitatively to tire cut growth.

Disappointing correlations between conventional fatigue cracking tests and tire sidewall performance prompted Young and coworkers [26,27] to seek a method for measuring crack growth over a range of strain inputs. A shear test piece was favored to facilitate measurement of tearing energy and minimize the dependence on crack length.

A similar pursuit was made by Kim and Lee [28] and by Upadhyay and Warrach [29], who found they could make considerable savings in test times with a procedure still sufficiently sensitive to detect the antifatigue cracking activities of some antidegradants. Good agreement with De Mattia crack growth behavior was reported.

Most interest in crack growth behavior has focused on tires, but Nichols and Pett [30] have reported agreement between cut growth measurements and the service performance of automotive power steering hose.

Researchers at the Bayer company [31] in Germany have developed a "tear" analyzer for crack growth measurements. A 1 mm lateral incision is made into a strip test piece, which is then cycled through a predetermined strain regime. A video camera is used to measure crack growth, and the test conditions can be varied to match service environment in terms of frequency, temperature, and straining. The rate of crack growth can then be plotted against such parameters as stress, strain amplitude, tearing energy, and strain energy. Excellent correlation between laboratory crack growth characteristics and tire endurance has been claimed.

A fatigue and cut growth measurement has also been designed by the Tun Abdul Razak Research Centre [32] primarily for use with the ISO 37 type 2 dumbbell test piece.

3.4 Heat Generation Tests

Flexometer tests are used to determine thermal stability under dynamic straining conditions. Measurements include temperature rise after a specified period of cycling, set and creep, and in some instances the time or number of cycles to failure in the form of thermal runaway or test piece destruction. In contrast to fatigue cracking tests, heat buildup tests call for a comparatively thick test piece so that heat hysteresis is converted into a temperature rise.

The principles of flexometer testing are described in ISO 4661/1 (BS903, Part A49). The best known test machine is the Goodrich compression flexometer described in ISO 4661/2, (BS903 A40 and ASTM D623). An alternative is the rotary flexometer of ISO 4661/2; a specific example, the Firestone flexometer, is specified in ASTM D623.

3.5 Wear Testing

Wear is important in applications as diverse as tires, conveyor belting, footwear, and windscreen wiper blades, yet most of us would accept that the rubber industry still awaits a laboratory test or series of tests that can predict service performance with any confidence. Over the years, there certainly has been no shortage of small-scale tests for abrasion resistance and indeed no shortage of claims or reports of satisfactory correlation with the behavior of various products, but strong doubts remain, and some sectors, not least the tire industry, conclude that there is no substitute for a service trial or a simulated product test, for example, a road trial on tires held at a small slip angle to accelerate wear.

Thirty years ago, Bulgin and Walters [33] concluded that the prediction of service wear needed detailed information on service conditions and laboratory tests that involved both cutting and frictional abrasion processes. Twenty years later, Muhr and Roberts [34] reached a similar conclusion, drawing attention to the complexity of wear and the contribution made by such parameters as friction, temperature, lubrication, the abrasive surface and smearing, as well as compound hardness and resilience. Yet many argue that the principal stumbling block is really an incomplete understanding of the factors determining the service wear of products. For tires, these will include the road surface and geometry, climatic conditions, wheel geometry, load, traffic conditions, and driver response [35,36].

Despite the accepted complexity of abrasion, there is currently only one ISO test method specific for rubber. This is the rotating cylindrical drum device, or as it is more popularly known, the DIN abrader, described in ISO 4649. A loaded cylindrical test piece is traversed along an abrasive cloth attached to a rotating drum, and the mass loss is measured after a specified length of travel. Advantages of this device include the use of a test piece small enough to be cut from a product or a comparatively thin sheet and a much reduced risk of abrasive contamination caused by debris or smearing. The main problem is often the effort needed to ensure the abrasive cloth meets and maintains the mass loss limits for the standard comparison rubber. Confusion can also arise with the alternative means of expressing the result, either as a relative volume loss, compared with the standard rubber, or the more straightforward abrasion resistance index, which is the ratio of the volume loss of a black-reinforced standard rubber to the volume loss of the test rubber.

The DIN abrader is quite widely used in Europe but finds little support in the USA, where the standard method is the PICO test described in ASTM D2228. In this method, abrasion is caused by two tungsten carbide knives, with a dusting powder applied to engulf abraded rubber particles.

In BS903, Part A9, the ISO method is one of four standard tests, the others being the DuPont, the Akron, and the Taber or rotary platform double head type machine. Like the ISO method, the DuPont test is one of continuous abrasion against a replaceable abrasive paper, this time of silicon carbide rather than of aluminum oxide, but there is no built-in procedure to avoid abrasive smearing and progressive changes in abrasive power other than a rotation of test pieces to even out differences. The Akron method involves the wear of a molded rubber disc against a loaded free-running abrasive wheel, with severity altered by adjusting the slip angle between disc and wheel. The Taber test also uses abrasive wheels and is suitable for testing sheets down to 2 mm in thickness. Both methods are susceptible to abrasive wheel contamination.

The British Standard offers three standard rubbers in addition to the DIN standard formulation. All results for the three non-ISO methods are expressed as an abrasion resistance index.

4 Time-Dependent Properties

4.1 Creep

Resistance to creep is relevant to load-bearing applications such as structural bearings, but the measurement has not, until comparatively recently, had a high profile in many product specifications, even though there has been a British Standard, currently BS903, Part A15, for over 50 years. One reason for this has been the specification of particular rubber types known to have generally low creep properties, and therefore it has been sufficient to confine test requirements to controls on state of cure and level of fillers; these include strength and compression set. However, there is now a shift towards the inclusion of a creep requirement both in specifications and in development work, in response (a) to product performance specifications that are no longer oriented towards particular rubbers and (b) to the manufacture of compounds for earthquake isolation, where the need for high damping can call for some sacrifice in creep resistance.

ISO 8013, now dual numbered as BS903, Part A15, provides test procedures for creep in shear and in lubricated or bonded compression. The tests require a bonded double shear test piece or a standard compression set disc. There is no current ASTM equivalent. Creep can be expressed as a percentage of the original dimension or as a percentage of the test strain, the custom in the rubber industry, and if a linear plot against logarithmic time is obtained, the result—a measure of physical creep—is conveniently expressed as creep per decade. Using a test similar to that described in BS903, Part A15, Derham and Waller [37] have found remarkably good agreement between the long-term creep of a building mounting and the laboratory prediction from the sum of the physical creep and the chemical creep estimated from high-temperature tests.

Derham [38] has investigated many of the factors affecting test performance. Test piece history is important, especially with filled rubbers, since prestressing reduces the creep rate and there need not be a full recovery. Temperature control is also essential, since thermal cycling can significantly increase creep.

ISO 8013 is confined to static strain conditions and can seriously underestimate the creep that occurs under dynamic loading. The creep rate in cycled rubber is higher than that predicted by a simple Boltzmann superposition, but linearity is still observed between creep and logarithmic time or the logarithm of the number of cycles, as long as a physical mechanism applies [39,40]. The increase is most striking with strain-crystallizing elastomers such as natural rubber.

4.2 Stress Relaxation

Stress relaxation is an important design consideration for sealing applications from O-rings to pipe joints for water, gas, and petroleum pipelines. However, like creep, it has been a relative latecomer in material and product specifications, again largely because of a previous reliance on suitably resistant rubber types. The development of performance-based specifications and growing concern about the long-term resistance of both traditional and newer materials have strengthened the need for stress relaxation testing.

Compression stress relaxation is described in ISO 3384 (BS903, Part A42). The principal procedure uses Type A or Type B compression set test pieces compressed by 25% against lubricated surfaces, but ring test pieces are specified for tests conducted under fluids.

The linearity between physical stress relaxation and logarithmic time is the basis for short-term test requirements in product specifications such as BS 2494 and its ISO equivalents for pipe sealing rings. A typical requirement is a maximum level of relaxation between the starting point of 30 minutes after loading and 168 h, the time gap being equivalent to about 2.5 decades. In the absence of longer-term testing, there has been a tendency to specify especially tight limits on physical relaxation in order to provide some insurance against the initially slower chemical stress relaxation after prolonged exposure. Such exacting requirements, typically less than 5% per decade, cannot be achieved by the new thermoplastic rubbers, where 8% per decade is a more typical stress relaxation rate, and this disparity has added pressure for a test able to measure—and analyze—both physical and chemical relaxation.

Recent work in this area has been spearheaded by Birley and colleagues [41–43] at Loughborough University, principally for the specification of pipe sealing rings. Attention has been drawn in particular to the importance of test temperature, temperature control, and loading rate. To assist the interpretation of test data, they have proposed that at the end of the relaxation test, the test piece should be allowed to recover under the same conditions of time and temperature [43]. The recovery rate is always slower than the relaxation rate, and lack of recovery is indicative of the secondary relaxation caused by oxidation and network changes.

As with creep, stress relaxation testing has essentially been confined to static conditions and thus may be inappropriate, certainly as a design tool, for seals undergoing some mechanical cycling. Dynamic relaxation is higher than static relaxation under otherwise equivalent conditions, especially in the presence of high filler levels and at high test strains. The change is most pronounced in strain-crystallizing rubbers, but an increase has also been observed in reinforced SBR vulcanizates.

Davies et al. [44] have found that

$$\sigma_N = \sigma \frac{(N - k)^{-s}}{1 - k}$$

where σ_N is the stress after N loading cycles, σ is the stress on the first loading cycle, k is a constant, and s is the negative slope of the stress relaxation line.

4.3 Compression Set

Compression set is one of the most widely measured properties of rubber, featuring in a range of material and product specifications and finding use as a sensitive measure of the state of cure in both conventional thermoset rubbers and newer dynamically vulcanized thermoplastic rubbers. Comparative ease of determination has made compression set a popular test for sealing applications, although it is increasingly accepted that the property should not be used in place of the often more relevant stress relaxation. However, it can be relevant to seals where sufficient bodily movement occurs to open up a leakage channel before recovery of shape can take place.

The standard test is detailed in ISO 815 (BS903, Part A6) and is applicable to low, ambient and elevated temperatures. It is confined to constant strain conditions, usually 25%, and set is reported as a percentage of the compression strain. The same disc test pieces used for creep and stress relaxation are specified, with the smaller 12.5 mm diameter, 6 mm thick Type B disc being especially valuable for samples taken from finished products. Plied test pieces are permitted for sheet material and thin products, and these are highlighted in the generally similar ASTM D395 test method. Because of difficulties in cutting standard test pieces from small pipe sealing rings, the specification BS2494 permits the additional use of discs 3.5 mm in height and 5 or 7 mm in diameter. Such scaling down should not be encouraged for general use, given the reduction of the test strain to under one millimeter, but the changes are acceptable provided tests are comparative, and they provide an opportunity to test the finished component instead of a test sheet of nominally comparable state of cure.

Lubrication of the test pieces is now standard practice in order to eliminate one obvious source of variation. The more uniform flattening of the test piece also eases measurement of thickness after release from compression. However, there remain specifications in which set is determined in the absence of lubricants. It has also become common practice with general-purpose rubbers to measure compression set after just one day at 70°C, which for sulfur-vulcanized elastomers can be a sensitive measure of the state of cure. Higher test temperatures are specified for special-purpose and speciality synthetic rubbers, but the one-day test has remained popular, not least as a classification criterion and grade requirement in such specifications as ASTM D2000 and the British Standard series of material specifications for individual rubber types. Tests seldom last more than seven days, and recovery is usually confined to the standard 30 minutes after release, during which time the test piece cools to standard laboratory temperature if taken from an oven. The short-term nature of the test and the absence of isothermal conditions during recovery has been questioned by Birley and other workers [43].

ASTM retains a constant stress method for compression set, and one is also specified for electrical mats in BS 921. In these tests the set is expressed as a simple percentage of the original test piece thickness, and thus it is not unusual to see specification limits and test results that are numerically much smaller than those obtained in an equivalent constant deformation test. Such differences can sometimes lead to confusion and indeed alarm when confronted by users much more familiar with the tests of ISO 815.

4.4 Tension Set

The determination of set in tensile strain is much less commonly specified than that in compression, although in principle it is a particularly straightforward procedure: a strip, dumbbell or ring test piece of known reference length is stretched to a given extension, exposed in this condition to a combination of temperature and time, and then released for a specified period before measurement of the reference length. One clear advantage for the test, as say a measure of the state of cure, is that test pieces can be cut from even the thinnest of latex films, whereas the minimum ply thickness of a laminated compression set test piece is 2 mm.

The standard test is described in ISO 2285 (BS903, Part A5) and it offers a range of test pieces and test strains from 15% up to 300% if the elongation at break of the test rubber allows. Based on the experience with compression set and general-purpose rubbers, the recommended test conditions are one day at 70°C unless other times and temperatures are specified. The 30 minutes recovery time is also retained for general purposes. However,

ISO 2285 provides alternative recovery procedures in recognition of the influence these can have on the test results. The first involves release immediately on removal from any test oven and recovery at standard laboratory temperature, reminiscent of the usual procedure for compression set. The second involves release and recovery of the test piece at the test temperature before cooling for measurement of reference length. This operation often accelerates recovery from any reversible source of set and thus the results are generally smaller. The third procedure involves cooling the test piece in its stretched state for 30 minutes and then proceeding as normally. An increase in set on changing to this procedure can be indicative of a "frozen in" source of set, such as a thermally labile cross-link or crystallization.

Tension set under constant load is described in ISO 12244. A test piece is subjected to a tensile stress of 2.5 MPa (or 1 MPa for very soft rubbers), and the resulting elongation is measured after both 30 seconds and 60 minutes. The first measurement is comparable with the determination of strain at a given stress described in ASTM D1456, while the difference between the two readings will be a measure of tensile creep. Tension set is then determined when the test piece is unloaded at the end of the hour. A major difference between ISO 12244 and ISO 2285 is the speed of the test, with set in the former being measured just 10 minutes after release. The constant load test is also confined to standard laboratory temperature and thus will be of limited value, although suitable for measuring state of cure and for quality control of thin-walled products.

5 Determination of Temperature Effects

5.1 General

Resistance to temperature can be gauged in two ways. The first is the instantaneous effect of a temperature change on properties, which at low temperature is generally dependent on the rubber's glass transition temperature and at elevated temperatures on the thermoplasticity and ease of strain-induced crystallization. The other is the effect of storage at a given temperature on stability, manifested as thermal or thermal-oxidative aging at elevated temperatures and the crystallization of stereo-regular rubbers at low temperatures. It is interesting to observe that surprisingly little attention is made to the actual measurement of properties *at* high temperatures, compared with the measurement at standard laboratory temperature after test pieces have been aged at elevated temperature. In some contrast, at low temperatures there is far more attention on the immediate effect of temperature change than on the frequently underestimated effect of medium-to-long-term storage under cold conditions.

5.2 Elevated Temperatures

High-Temperature Testing

It is not always appreciated that many rubbers renowned for their outstanding heat resistance are mechanically quite weak at elevated temperatures and indeed often at temperatures well below those used to characterize thermal stability. The most noticeable changes are a loss in tensile and tear strengths and a decrease in modulus. Yet such properties are seldom determined at temperatures other than the standard laboratory temperature, often as much through sheer habit as through a reluctance to install high-temperature enclosures around tests. The consequence is that high-temperature tests, other than those for accelerated aging and compression set, are rarely called up in material and product specifications. The emphasis in technical literature and data sheets is also on heat endurance or simply accelerated aging, with little attention paid to high-temperature measurements, with the exception of flexometer testing and occasionally creep or stress relaxation.

It should be noted that the standard test methods for tensile stress–strain properties, tear strength, rebound resilience, and other dynamic properties provide for high-temperature measurements, preferably at the recommended temperatures of ISO 471.

Heat Resistance and Resistance to Thermal-Oxidative Aging

Heat resistance and accelerated aging tests are widely specified for both products and materials, finding use for type approval, quality control, and prediction of long-term performance. Nowadays the two terms are used synonymously, but some authorities still regard heat resistance tests as measures of stability at the anticipated maximum working temperatures. All are agreed accelerated aging involves the use of an elevated temperature to increase the rate of thermal or thermal-oxidative degradation in order to assess stability at ambient temperatures or temperatures below the test temperature.

The most widely employed test method is air oven aging in a "normal" or cell oven at temperatures of 70°C and above. The test is described in ISO 188 (BS 903, Part A19) and in ASTM D573. A test tube enclosure method is described in ASTM D865, and this, like the cell oven, is intended to prevent cross-contamination between different materials, particularly from the volatilization of antidegradants. Also available in ISO 188, but much less extensively used, is the oxygen pressure or oxygen bomb test in which test pieces are exposed to an atmosphere of oxygen under an increased pressure (typically 2100 kPa at a temperature of 70°C) to accelerate the rate of oxidative aging. As might be expected, this test increases the role of oxidation in the thermal aging and so is particularly valuable as a measure of antioxidant or prooxidant activity. An alternative is an air bomb test of the type described in ASTM D454, which specifies a pressure of 500 kPa and a temperature of 125°C.

Pressure aging tests have been specified for cable sheathing and insulation but are considered too far removed from the working environment of most rubber products to gain widespread acceptance.

Today's air aging oven is still closely reminiscent of the Geer aging oven developed at the beginning of the century [1]. However, the rubber industry is still taking stock of its value for the prediction of service performance and of the main sources of test error [45]. Royo [46] has drawn attention to the importance of air flow rate between 3 and 15 changes/hour, and in standard tests there is now a provision for using a high exchange rate oven, although the user must still decide which is the more appropriate, especially as the effect of air change varies from compound to compound. In more recent work on test precision, Spetz [47] has highlighted the importance of strict temperature control, concluding that present standard temperature tolerances should be halved, for example from 1°C to 0.5°C at 100°C. The need for such control has already been recognized in the standard test (ISO 2930) for the determination of the plasticity retention index of technically specified natural rubber, with a tolerance of ±0.5°C being specified at the aging temperature of 140°C.

Improving test reproducibility is normally identified with the widespread use of the air oven aging test in specifications and in quality control work, with tests being completed in a few days. The value of aging tests for predicting long-term service performance is considered in Chapter 29, but it is appropriate to review features of specific importance to rubber. Key considerations are choice of test conditions, the property being measured, and the influence of test piece thickness.

Evidence of reliable correlations between accelerated aging and product endurance is still needed to convince many manufacturers and end users of the value of undertaking medium-term aging tests at moderately elevated temperatures, even when it is accepted that a few weeks test exposure is still very short when related to a product designed for perhaps 50 years service life. In many respects, progress has been hampered by the continued use of material specifications such as ASTM D2000; the latter standard has undergone very little change over the years, despite the increasing demands of the automotive industry, with a 70 hour aging test remaining supreme. Some motor companies have pressed for longer test periods, and 1000 h (6 weeks) is now specified for a range of company specifications [48], usually, although not exclusively, at a single aging temperature. For an improved classification of heat resistance, the Society of Automotive Engineers (SAE) has gone one stage further by introducing a test procedure for the determination of a continuous upper temperature limit; this is defined as the aging temperature at which the test compound will retain 50% of its initial tensile strength after 1008 hours exposure [49]. This procedure obviously calls for a considerable commitment in terms of time and equipment, but the SAE considers it essential for materials intended for automotive products expected to last for 100,000–125,000 miles.

The SAE approach is simply an extension of the conventional aging test, inasmuch as tensile strength and elongation at break are retained as particularly sensitive measures of thermal and thermal oxidative aging and are determined at standard laboratory temperature irrespective of the intended maximum working temperature of the rubber. Tensile strength is especially suitable for natural rubber, which can soften as well as harden on aging as a result of chain scission alongside oxidative cross-linking and postvulcanization. Elongation at break is a more appropriate measure for rubbers undergoing stiffening on aging.

For many applications, such as structural bearings, modulus stability is a more important consideration than retention of strength or extensibility, provided serious degradation has not occurred. However, the prediction of long-term modulus—or hardness—from accelerated tests remains especially difficult. This results in part from the strong temperature dependence of oxidative chain scission and cross-linking. With natural rubber, high-temperature aging favors scission reactions and softening, whereas at temperatures closer to ambient, cross-linking and stiffening are encouraged. Tests at high temperatures can therefore seriously underestimate the stiffening that can occur, albeit very slowly, at atmospheric temperatures. Until comparatively recently this complication was largely ignored on the grounds that strength retention was a sufficiently good measure of compound stability in combination with other properties and any control on composition. The uptake of rubber in structural mountings and bearings has shifted attention directly to modulus stability because of the strong dependence of performance on bearing stiffness and in consequence the specification of particularly tight design limits on such properties as shear modulus.

Early work by Stenberg [50] revealed not only differences between static and dynamic moduli on aging but also the complication caused by the diffusion control of oxidation. Inspection of bulky products recovered after many years' service has confirmed time and time again that oxidative aging is essentially confined to the surface, with the changes occurring in the bulk being much more consistent with milder anaerobic aging. Thus tests conducted on standard thin sheets (usually no more than 2 mm thick) can easily overestimate the role of oxidative aging. One solution is to age a simple laminate of standard sheets and then test each to obtain a profile of aging from the surface to the center. This technique was pursued by Knight and Lim [51] at comparatively high aging temperatures and later by Fuller et al. [52] at moderately elevated temperatures for an assessment of the stability of compounds for seismic bearings. Japanese workers [53] also concerned with earthquake protection adopted a similar approach except that a thick block was spliced into 2 mm thick sections after aging. The limiting feature in each of these studies was that information was necessarily confined to tests undertaken on thin sheets and so favored tensile stress–strain behavior, irrespective of its relevance to the performance of the intended product. Lindley and Teo [54] used the more novel approach of determining the puncture strength at various depths in the aged block, and they used this to show an arrhenious relationship between the depth of oxidation and the aging temperature. One advantage of the puncture test is that parallel tests can be conducted on products recovered from service without the need for sectioning. One shortcoming is that a puncture can provide a channel for subsequent oxidation if the test rubber is returned to the aging oven for additional measurements. Care must be taken to ensure that any further punctures are made some distance from previous ones.

The puncture test is a semidestructive method in that not all the rubber is destroyed. Nondestructive methods have been developed by Dinzburg and Bond [55] for the determination of stiffness in a bending mode, and by Brown [56] who used tensile stress relaxation and modulus retention. The latter approach has since been standardized in ISO 6914 (BS903, Part A52) and comprises three procedures—one is a continuous relaxation test conducted at 50% elongation and the others are intermittent strain tests in which the test piece is relaxed during the air oven aging periods and only strained for stress measurements, in one case at the aging temperature and in the other more conveniently at standard laboratory temperature. The continuous relaxation test is a direct measure of degradative scission reactions and the results are amenable to structural analysis. The intermittent tests are measures of the combined effects of scission and cross-linking reactions, with the standard laboratory temperature stress measurement being reminiscent of the modulus measured in the conventional tensile stress–strain tests. The standard test uses a 1 mm thick test piece to minimize the effect of diffusion control on oxidation, especially at high temperatures.

A novel means of assessing anaerobic thermal stability has been found to be the standard curemeter test (e.g., ISO 3417, BS903, Part A60, and ASTM D2084). The time to reach a given degree of reversion is measured over a range of temperatures and the activation energy estimated in this way can be used as a guide for the extrapolation of anaerobic modulus changes to ambient temperatures [52]. The curemeter can also be used to assess the role of "marching modulus curing" at service temperatures.

This chapter is concerned with the physical testing of rubber, but thermal aging is primarily a chemical process, and therefore it should be recognized that aging can usefully be monitored by chemical methods, especially those sensitive to surface changes and able to give a more precise measure of diffusion-controlled oxidation. These techniques include chemiluminescence [57] and a range of thermal analytical methods such as DSC and mass change [58]. Thermal methods have been used to characterize polymeric roofing membranes and are expected to enter the specification area when more data are available.

5.3 Low-Temperature Testing

Short-term Effects

The most popular low-temperature test for rubber is impact brittleness (ISO 812, BS903, Part A25, and ASTM D2137). Test pieces held at low temperatures are impacted by a striker and, if they break, new sets tested at progressively higher temperatures until no breakages occur. The result is expressed as the brittleness temperature or, as more normally in specifications, as pass or fail at a given low temperature. Because of a relatively high 2 m/s striker velocity at impact, the brittleness temperature is much higher than the normally quoted glass transition temperature (*Tg*), and so with natural rubber with a *Tg* in the order of −70°C, the brittleness temperature is close to −55°C. For many products the test is a relatively severe one but it discriminates between rubbers varying in *Tg* and is a useful measure of the effects of added plasticizers. The measurement is specified as the low temperature classification criterion in ISO 4632, the international equivalent of ASTM D2000. It is also specified as a grade requirement throughout both these material specifications.

In Scandinavia the temperature retraction (or TR) test is reported to be especially popular and to give the best overall level of reproducibility [59]. The test described in ISO 2921, BS903, Part A29, and ASTM D1329 involves first cooling an extended test piece to a temperature at which it does not recover and then measuring the temperature at which a specified retraction occurs.

Correlations have been established between T10 (the temperature at which 10% retraction occurs) and brittleness temperature for broadly similar rubbers and between T70 and low-temperature compression set. The test can also be adapted to give a measure of crystallization.

Modulus changes at low temperatures can be monitored by the torsional modulus or Gehman test of ISO 1432 (BS903, Part A13) and by measurement of hardness.

Crystallization

A time-dependent stiffening can occur in stereo-regular rubbers through low-temperature crystallization [60]. The phenomenon is most closely identified with natural rubber and chloroprene rubber, the former at temperatures in the range −10°C to −40°C and the latter in the range of +10°C to −20°C. It is a reversible process, with the hardening disappearing when the rubber is heated, but the stiffening can sometimes persist at ambient temperature and so can easily be confused with the normally irreversible age hardening. A clear distinction needs to be made, with the support of properly specified tests, because compounding to minimize the risk of aging can actually detract from resistance to crystallization. Making a distinction between crystallization hardening and the more immediate glass hardening is just as important, since a plasticizer added to lower glass transition temperature may facilitate crystallization.

The best known standard measure of crystallization is hardness increase described in ISO 3383 and BS903, Part A63. Surprisingly there is no ASTM equivalent. The test involves measuring hardness at the test temperature after test piece conditioning and then again after one or more storage periods. Any increase is a direct measure of crystallization (or any other time-dependent cause of stiffening), whereas any difference between the initial hardness measurement and hardness determined at standard laboratory temperature is more a reflection of glass stiffening (once test errors have been eliminated). Crystallization resistance can be expressed in terms of the time to a given hardness increase or to half the overall increase where a hardness ceiling is found, but in specifications it is commonplace to find a maximum increase after seven days at −10°C (for chloroprene rubber) or −25°C (for natural rubber).

Strain accelerates crystallization and can be used in testing to shorten exposure times or allow completion of the process in a reasonable period [60]. Stress relaxation in tension can be used for this purpose, but complete decay of stress can occur before full crystallization and therefore the degree of stiffening can be underestimated. This limitation also applies to compression set, for which there is a low-temperature provision in ISO 815, and to a high compression strain recovery test of Russian origin described in ISO 6471.

Crystallization tests are conducted at a constant low temperature and so do not take account of the diurnal and seasonal thermal cycling that may occur in some products. It has been reported that thermal and mechanical cycling disrupts crystallization and results in a lower stiffness increase than a constant temperature test would indicate [61]. In consequence, products used for long periods at low temperatures, for example bridge bearings in cold climates, may not undergo the extent of hardening predicted in a laboratory test. The reproducibility of low-temperature hardness and compression set tests have also been questioned and for these reasons withdrawn from the ASTM and ISO specifications for bridge bearings. An alternative test is now being sought for this application, preferably one offering a direct measurement of shear modulus.

While there is now acceptance that crystallized rubber can yield on deformation, crystallization can cause a significant increase in shear modulus, and it reestablishes itself much more rapidly than it has developed initially [62].

6 Environmental Properties

6.1 General

Rubbers are used in a wide variety of service environments and share many of the requirements and features of other materials discussed in this book, although resistance to light aging is generally much less serious a problem than in plastics because of the extensive use of carbon black as a filler or pigment. Attention is therefore drawn to other chapters on environmental resistance and weathering. This section focuses on properties more closely identified with rubbers.

6.2 Ozone Cracking

Rubbers having main chain unsaturation can undergo ozone cracking when test pieces are exposed to atmospheric traces of ozone at an elongation above a characteristic threshold strain [63]. For most diene rubbers this strain is typically around 5% in the absence of any protective agent, although under dynamic strain conditions it can be reduced to less than 1%. Antiozonant materials are added to rise the threshold strain above the maximum strain encountered in service life or alternatively, as is usually the case in dynamically strained applications, to reduce the rate of crack growth.

Laboratory ozone resistance tests involve the exposure of stretched test pieces, usually strips or dumbbells, to a specified combination of ozone concentration and temperature. The standard concentrations permitted in ISO 1431/1 (BS903, Part A43) are 25, 50, 100, and 200 pphm ozone.* For most purposes, 50 pphm is considered suitable and is now found in a wide range of material and product specifications. The lowest concentration is intended for applications used under low-severity conditions, whereas 100 and 200 pphm are used to ensure that there is sufficient protection for badly polluted areas or to ensure the use of an inherently resistant rubber type.

---

*The rubber industry expresses ozone concentration as parts per hundred million parts of air, whereas environmental agencies have traditionally used parts per billion (thousand million). ASTM D1149 (the corresponding American test method to ISO 1431/1) expresses and specifies ozone concentration in terms of partial pressure to accommodate differences in barometric pressure between test laboratories. 50 mPa is equivalent to 50 pphm ozone.

In spite of these guidelines, opinions still vary on the appropriate choice of concentrations. Some authorities have argued for test concentrations even lower than 25 pphm in order to improve the degree of correlation with service exposure in the region of 1 pphm ozone, whereas others insist the concentration should be high enough to discriminate between compounds within a few days as well as to ensure sufficient antiozonant for long-term protection. It is certainly important that the test be severe enough to provide a clear distinction between a rubber having a high threshold strain and one that is resistant simply because of a very low crack growth rate. Thus chloroprene rubber is more inherently ozone resistant than natural rubber not because it has a higher threshold strain but because its rate of crack growth can be less than 1/20th that of natural rubber [63].

The test temperature should be chosen to match service conditions, with more than one test becoming desirable if the application is to encounter a range of temperatures. The choice of temperature (and its control) is especially important where waxes are used as part of the protective system, since temperature governs the solubility and diffusion coefficient of the wax in the rubber [64]. Some users are still unaware that an increase in temperature reduces the levels of wax available for surface blooming, and that a lower temperature reduces the rate at which the wax migrates to the surface to form a bloom or to repair a damaged one. Most test equipment is designed for use at temperatures above ambient, with 40°C now being a preferred temperature for routine purposes. Hill and Jowett [65] have argued for low-temperature testing since a wax designed for summer conditions and hot climates may fail to provide protection in winter and cold regions.

The usual test strain for static exposure is 20%, which is not only representative of a range of products but sufficiently high to ensure the use of a suitably resistant compound. A range of test strains, involving several test pieces or a multistrain test piece, is necessary for the measurement of the threshold strain. The standard test for dynamic strain conditions (ISO 1431/2 and BS903, Part A44) involves cycling test pieces from zero strain to a specified maximum at a frequency of 0.5 Hz. The usual maximum strain is 10%, but a range of elongations can be used to determine the time to first crack as a function of maximum strain. ISO 1431/2 also provides an intermittent dynamic procedure comprising periods of cyclic and static strain exposures. The combined cyclic and static sequence is intended to simulate behavior in products from tire sidewalls to engine mountings, but the standard has generated very little interest. Many manufacturers and end users are not prepared to invest in dynamic test equipment in the absence of detailed guidance on the selection of straining conditions and test periods. As a result the dynamic ozone test seldom features in product specifications, and there is a continued lack of information on the behavior of antiozonants under the standard test conditions. It can reasonably be argued that establishing correlations for static strain tests and service exposures is difficult enough without the complication of dynamic exposure. ASTM D3395 has two methods for dynamic ozone testing—one using a strip cycled to 25% elongation and the other using a fabric belt over which the test piece is applied.

In the standard test, ozone resistance is assessed by the visual inspection of cracks, both for the determination of the time to first crack at a given strain and for the estimation of threshold strain. The accuracy of these measurements is also dependent on the frequency of observation, for which the standard times are 2 h, 4 h, 8 h, 16 h, and then daily. For a more accurate measure of the threshold, ISO 1431/1 recognizes the value of graphical presentation to assist in the interpretation interpretation of the results. One procedure involves the plotting of the logarithm of the test strain against the logarithm of the time to first crack. The resulting curve sometimes approximates to a straight line, but this should never be assumed, and extrapolation to times beyond the longest test period is discouraged since it would seriously underestimate any limiting value in the threshold strain [66].

A limiting threshold strain—one that no longer falls with time—is often less evident in prolonged atmospheric exposures than it is in accelerated ozone tests, and this is one reason why its value has been questioned. However, a gradual decay in the threshold can sometimes be traced to losses of antiozonants by other means than simple ozone attack [64]. These include depletion by thermal-oxidative aging and by leaching. Neither of these is anticipated in short-term standard tests, and for long-term protection there is a strong case for assessing antiozonant activity after a period of leaching or accelerated air oven aging. Such a need has been anticipated in some product specifications, but no test procedures have so far been developed.

Ozone cracking is essentially a visual process, but there have been attempts to introduce quantitative methods of assessing crack growth as alternatives to simple inspection. One is stress relaxation [67] to reflect the gradual decrease in the cross-sectional area of the test piece, but this is insensitive in the region of the threshold strain where cracks are isolated. Less dependent on crack density and size is fatigue life or tensile strength [68], although this necessarily calls for the destruction of the test pieces.

Wet titration, electrochemical, and instrumental methods are used for the determination of ozone concentration, and variations within and between these have been shown to be responsible for some lack of agreement among laboratories. ISO 1431/1 and ISO 1431/2 have now been in existence for many years without a universally agreed reference method. A UV absorption reference method will form the basis of the forthcoming ISO 1431/3.

Provision for assessing ozone cracking and resistance outdoors is given in ISO 4665.

6.3 Resistance to Liquids

Liquids can swell rubber, leach out extractible materials, and react chemically to modify or degrade the compound. The results are changes in mass, dimensions, physical properties, and resistance to aging, and each of these is accommodated in the standard test ISO 1817 (BS903, Part A16) and its U.S. equivalent ASTM D471.

The temperature and the time of test piece immersion are key variables, but the most important factor is the choice of liquid, given the many standard fuels, oils, and simulated service liquids listed in the standards. The standard reference oils are used primarily for classification purposes, with resistance to the high volume swell oil No.3 (IRM 903) being a key classification criterion in ASTM D2000 and its ISO and BS equivalents. Resistance to the higher aniline point reference oil No.1 is also a popular measurement, featuring throughout ASTM D2000 as a grade requirement and often being specified alongside oil No.3 in product standards [48]. Oils No.2 (IRM 902) and No.5 are essentially for special purposes. Among the most commonly used standard simulated fuels is a 50/50 *iso*-octane/toluene mix (Fuel C).

The principal application, although by no means the principal aim, of ISO 1817 is to ensure the use of a suitably oil-resistant rubber for products coming into contact with swelling oils and fluids. The test method can, however, give misleading information on the performance of many hydrocarbon rubbers. Immersing a thin test piece of natural rubber in a standard oil at an elevated temperature leads to rapid and extensive swelling, yet many applications of natural rubber, notably engine mounts, are used in an oily environment quite satisfactorily. This better-than-expected performance is a reflection of the much greater resistance of relatively thick products and the strong viscosity dependence of oil absorption. A viscous oil penetrates rubber much more slowly than a low-viscosity one of similar chemical composition, even though both these would eventually lead to the same level of swelling [69].

It is important to note that the standard test is also designed for measuring resistance to water, aqueous solutions, and other inorganic liquids. Resistance to distilled water is specified for a range of products from pipe sealing rings to plant linings and marine applications. While distilled water reduces one source of test variation, especially on a global scale, it can be unrepresentative of the service environment, and therefore it is essential that the quality and composition of the water in contact with the application should be considered, say with respect to the presence of traces of prooxidant materials such as copper, manganese, and iron.

6.4 Contact with Other Materials

Staining of Light-Colored Surfaces

Rubber can contain discoloring antidegradants (mainly amines), hydrocarbon oils, and occasionally vulcanization residues capable of causing the staining of light-colored surfaces, for example other rubbers, paints, and lacquers, in contact with or nearby the rubber product. A test method for measuring staining resistance is described in ISO 3865 (BS 903, Part A33), which includes procedures for contact stain (on the precise area of contact), migration or "halo" stain (the area immediately surrounding the contact area), penetration stain (from the rubber compound to an applied surface lacquer), and extraction stain (the stain caused by materials leached out from the rubber).

Corrosion and Adhesion of Metals

Some rubber compositions, for example those containing reactive chlorine, can promote the surface corrosion of metals in contact with them, and in certain applications this can be sufficient to impair component performance. To guard against the use of such materials, a test method is described in ISO 6505 (BS903, Part A37). The procedure is similar to that for contact stain, in that a sandwich of rubber and test surface, in this case a specified metal, is stored under load in a temperature-controlled environment for a given period. The measurement is a visual one and includes an indication of the ease of separation of the test rubber and the metal at the conclusion of the test.

The British Standard includes a second method in which pure zinc is exposed to the volatiles from the rubber in a water-saturated atmosphere at 50°C. The zinc is weighed before and after exposure, the mass loss being recorded as a measure of corrosion.

References

1. Buist, J. M., in *History of the Rubber Industry* (P. Schidrowitz, and T. R. Dawson, eds.), W. Heffer and Sons, 1952.
2. Coveney, V. A., and Jamil, S., RubberCon '95, Göteborg, Sweden, 1995.
3. Brown, R. P., *Physical Testing of Rubber*, Chapman and Hall, 1996.
4. Smith, L. P., *The Language of Rubber*, Butterworth Heinemann, 1993.
5. Muhr, A. H., *J. Nat. Rubb. Res.*, *7*(1), 14 (1992).
6. Engineering Data Sheets, EDS 1-50, Malaysian Rubber Producers' Research Association, 1979–1987.
7. Muhr, A. H., and Thomas, A. G., *NR Technology*, *20*, 27 (1989).
8. Brown, R. P., *Polymer Testing*, *10*, 117 (1991).

9. Spetz, G., *Polymer Testing*, *12*, 351 (1993).
10. Bassi, A. C., Casa, F., and Mendichi, R., *Polymer Testing*, *7*, 165 (1987).
11. Spetz, G., *Polymer Testing*, *14*, 13 (1995).
12. Bell, C. L. M., Stinson, D., and Thomas, A. G., *Rubb. Chem. Technol.*, *55*, 66 (1982).
13. Goodchild, I. R., and Williams, M. J., *Rubber Developments*, *45*, 91 (1992).
14. Pendle, T. D., *Rubber Developments*, *49*, 45 (1996).
15. Clamroth, R., and Kempermann, Th., *Polymer Testing*, *6*, 3 (1986).
16. Warhurst, D. M., Slade, J. C., and Ochiltree, B. C., *Polymer Testing*, *6*, 463 (1986).
17. Stevenson, A., and ab Malek, Kamarudin, *Rubb. Chem. Technol.*, *67*, 743 (1994).
18. ab Malek, Kamarudin, and Stevenson, A., *J. Nat. Rubb. Res.*, *7*, 126 (1992).
19. Fletcher, W., Gee, G., and Morrell, S. H., *Trans. Instn. Rubber Ind.*, *28*, 85 (1952).
20. Yeoh, O. H., *Polymer Testing*, *7*, 121 (1987).
21. Gregory, M. J., *Polymer Testing*, *4*, 211 (1984).
22. Lake, G. J., *Rubb. Chem. Technol.*, *68*, 435 (1995).
23. Stevenson, A., *Polymer Testing*, *4*, 289 (1984).
24. Royo, J., *Polymer Testing*, *11*, 325 (1992).
25. Lake, G. J., and Lindley, P. B., *Rubber J.*, *146*(10), 24; *146*(11), 30 (1964).
26. Young, D. G., Kresge, E. N., and Wallace, A. J., *Rubb. Chem. Technol.*, *55*, 428 (1982).
27. Young, D. G., *Rubb. Chem. Technol.*, *58*, 785 (1985).
28. Kim, S. G., and Lee, S. H., *Rubb. Chem. Technol.*, *67*, 649 (1994).
29. Upadhyay, N. B., and Warrach, W., *Rubber World*, *203*, No. 1, 38 (1990).
30. Nichols, M. E., and Pett, R. A., *Rubber World*, *211*, No. 6, 27 (1995).
31. Sumner, A. J. M., Kelbch, S. A., and Eisele, U. G., *Rubber World*, *213*, No. 2, 38 (1995).
32 Tun Abdul Razak Research Centre 1996, Universal Fatigue and Cut Growth Test.
33. Bulgin, D., and Walters, M. H., Proc. Int. Rubb. Conf., Brighton, Maclaren, 1967, p. 445.
34. Muhr, A. H., and Roberts, A. D., in *Natural Rubber Science and Technology* (A. D. Roberts, ed.), Oxford Science Publications, 1988.
35. Ambelang, J. C., *Tire Sci. Technol.*, *1*, 39 (1973).
36. Veith, A. G., *Tire Sci. Technol.*, *14*, 201 (1986).
37. Derham, C. J., and Waller, R. A., *Consulting Engineer*, *39*(7), 49 (1975).
38. Derham, C. J., *J. Materials Sci.*, *8*, 1023 (1973).
39. Derham, C. J., and Thomas, A. G., *Rubb. Chem. Technol.*, *50*, 397 (1977).
40. Pond, T. J., *J. Nat. Rubb. Res.*, *4*, 93 (1989).
41. Birley, A. W., Fernando, K. P., and Tahir, M., *Polymer Testing*, *6*, 85 (1986).
42. Tahir, M., and Birley, A. W., *Polymer Testing*, *7*, 3 (1987).
43. Prabhu, A. M., Birley, A. W., and Sigley, R. H., *Polymer Testing*, *10*, 39 (1991).
44. Davies, C. K. L., De, D. K., and Thomas, A. G., *Prog. Rubb. Plast. Technol.*, *12*, 208 (1996).
45. Clamroth, R., Tobisch, K., Barczewski, H., and Wundrich, K., *Polymer Testing*, *13*, 129 (1994).
46. Royo, J., *Polymer Testing*, *3*, 113, 121 (1982).
47. Spetz, G., *Polymer Testing*, *13*, 239 (1994).
48. Ford Motor Company, Engineering Materials Specifications, Volumes 5A and 5B, 1995.
49. Klingensmith, W., *Rubber World*, *206*, No. 5, 16 (1992).
50. Stenberg, B., *Polymer Testing*, *2*, 287 (1981).
51. Knight, G. T., and Lim, H. S., Proc. Int. Rubb. Conf., Kuala Lumpur, 5, 1975, p. 57.
52. Fuller, K. N. G., Ahmadi, H. R., and Pond, T. J., ASME PVP Conference, Montreal, July 1996.
53. Fujita, T., Ishida, K., Mazda, T., Nishikawa, I., Muramatsu, Y., Hamanaka, T., Yoshizawa, T., and Sueyasu, T., *Rubber World*, *211*, No. 3, 37 (1994).
54. Lindley, P. B., and Teo, S. C., *Plast. Rubb. Mat. Applns.*, *2*, 82 (1977).
55. Dinzburg, B., and Bond R., *Rubber World*, *201*, No. 4, 20 (1990).
56. Brown, R.P., *Polymer Testing*, *1*, 59 (1980).
57 Mattson, B., and Stenberg, B., *Prog. Rubb. Plast. Technol.*, *9*, 1 (1993).
58. Paroli, R. M., and Delgado, A. H., *Rubber World*, *214*, No. 4, 27 (1996).

59. Spetz, G., *Polymer Testing*, 9, 27 (1990).
60. Stevenson, A., *Kautschuk Gummi Kunststoffe*, *37*, 105 (1984).
61. Coe, D., Lachmann, C., and Howgate, P., *Eur. Rubb. J.*, *170* No. 5, 34 (1988).
62. Pettifor, J. D., and Coveney, V. A., Arctic Rubber Conf. Tampere, Finland, 1989.
63. Braden, M., and Gent, A. N., *J. Appl. Polym. Sci.*, *3*, 90, 100 (1960).
64. Lewis, P.M., *NR Technology*, *3*(1), No. 1 (1972).
65. Hill, M. L., and Jowett, F. *Polymer Testing*, *1*, 259 (1980).
66. Lake, G. J., *Polymer Testing*, *11*, 117 (1992).
67. Ganslandt, E., and Svensson, S., *Polymer Testing*, *1*, 81 (1980).
68. Lake, G. J., and Thomas, A. G., Proc. Int. Rubb. Conf., Brighton, 1967, p. 525.
69. Muniandy, K., Southern, E., and Thomas A. G., in *Natural Rubber Science and Technology*, (A.D. Roberts, ed.), Oxford Science Publications, 1988.

16
Particular Requirements for Plastics

Steve Hawley
Rapra Technology Ltd., Shawbury, Shrewsbury, England

1 Strength and Stiffness Properties

1.1 Hardness

Although hardness is much less used in the characterization of plastics than it is for rubbers, there are, nevertheless, several methods of test frequently encountered. The principles behind hardness testing have been considered in Chapter 9.

The most popular scale is probably the Shore durometer hardness, which has two main variants. The durometer is a small, hand-held instrument with an indentor of given geometry that is pressed into the surface of the material to be measured under a spring of given stiffness. The amount of penetration of the indentor is measured by a suitable scale marked directly in hardness degrees. Both the common Shore variants are standardized in ISO 868 [1]. For soft plastics, the Shore A scale is used (see Fig. 1a). In this method the spring-loaded indentor consists of a truncated cone of included angle 35° and diameter at the flat of 0.79 mm. As mentioned in the previous chapter, the Shore A scale and the IRHD scale of ISO 48 [2] are essentially the same over the normal operating range. The Shore D scale (Fig. 1b), which is more suited to typical thermoplastics materials, has a sharper indentor of included angle 30° with only a slightly rounded (0.1 mm radius) tip.

The test is also standardized as BS 2782 Method 365B [3], which is identical to the ISO method, and ASTM D2240 [4], on which the ISO method was originally based. In both methods the test piece to be measured must be 6 mm or more thick (3 mm is allowed in ASTM for the D scale) to avoid the hardness value being affected by the harder surface on which the test material is resting. To avoid edge effects, measurements should not be made nearer to any edge than 12 mm. It is clear, therefore, that a standard hardness test requires a relatively large and flat sample of material. Where these criteria are not met, only apparent hardness can be quoted, and it is important that deviations from standard

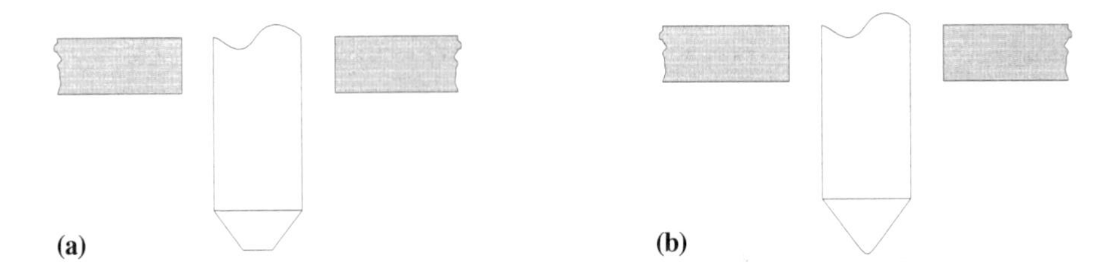

Figure 1 (a) Indenter for Type A durometer. (b) Indenter for Type D durometer.

conditions should be recorded along with the observed value if true comparisons are to be made between materials, measuring instruments, operators etc.

Another potential source of disagreement between measurements is the time of application of the load before the reading is taken. The ISO (and hence BS) standard sets this time at 15 seconds, although an "instantaneous" reading may be estimated by making the reading after nominally a 1 second application of load. In the ASTM standard the preferred time is 1 second but others may be used by agreement.

For guidance on the use and the calibration of durometers reference should be made to BS 903, Part A57, which is covered in Chapter 15.

A test more frequently found in Continental Europe than in the U.K. or the U.S.A. is the ball indentation hardness test, which is standardized in ISO 2039, Part 1 [5], also dual numbered as BS 2782, Method 365D [6]. In this a 5 mm diameter hardened steel ball is pressed into the test surface under a specified load so that the indentation is between 0.07 and 0.10 mm for method A, or between 0.15 and 0.35 mm for method B. The time of application of the load is 30 seconds, and a minimum test piece thickness of 4 mm is recommended. Unlike the Shore or IRHD scales, where hardness is directly related to the penetration of the indentor, the ball indentation hardness is given by:

$$\text{ball indentation hardness} = \frac{\text{applied load}}{\text{surface area of impression}} \tag{1}$$

In ISO 2039, Part 2 [7] the Rockwell hardness test is described. This is based on the same principle as the ball indentation hardness, but due to the severity of the test in terms of the load applied and the ball diameter through which the load is applied to the test piece, the test is really only suited to the harder thermoplastics and thermosets. Several hardness scales are defined according to the ball and load used. Table 1 illustrates the defined scales in the ISO standard.

The test piece is required to be at least 6 mm thick. In procedure A, the minor load is applied for 10 sec and then the major load for 15 sec. The hardness is read off the scale 15 sec after removal of the major load, while the minor load is being applied. Note that this is

Table 1 Rockwell Scales (ISO 2039-2)

Rockwell scale	Minor load (N)	Major load (N)	Indentor diameter (mm)
R	98.07	588.4	12.7 ± 0.015
L	98.07	588.4	6.35 ± 0.015
M	98.07	980.7	6.35 ± 0.015
E	98.07	980.7	3.175 ± 0.015

a different procedure to the previously mentioned tests, in which the hardness is measured while the major load is still being applied. A variant on this, which is applied to the R scale only, is to take the reading just before the major load is removed, thus making it a similar procedure to ball indentation. The Rockwell alpha value is then taken as 150 minus this indentation reading. Fett [8] has shown that ball indentation hardness H and Rockwell alpha R_α are correlated through the expression

$$H = \frac{(448.6)^{1.23}}{(150 - R_\alpha)} \tag{2}$$

The Rockwell test is standardized also in BS 2782, Method 365C [9], and ASTM D785 [10]. The British standard is identical (dual numbered) and the ASTM technically equivalent, although this does add the even more severe scale K to the list given in Table 1.

For flexible plastics there is a softness measurement standardized in BS 2782, Method 365A [11]. Technically this is in agreement with BS 903, Part A26, and ISO 48, which were dealt with more fully in Chapter 15, although the expression of the result is somewhat different. The softness number is the penetration expressed in hundredths of a millimeter; there is no secondary conversion of this value to a hardness number as there is for the IRHD scale for rubber. The test piece is required to be between 8 and 10 mm thick for the standard test to apply, and no measurement should be made nearer than 10 mm from any edge. Nonstandard dimensions can be used, but these must be stated along with the softness number obtained. The standard contains a special note concerning plasticized PVC, which is known to vary in hardness with time after molding. To minimize the effect of this the softness measurement must be made 7 ± 0.2 days after molding.

Other hardness tests that are sometimes applied to plastics include Barcol, Brinell, and Vickers. The former is normally applied to composites and is described more fully in Chapter 18. Brinell and Vickers are not standardized for plastics, being designed for metals, although values for plastics using these scales can be found in the literature. The Brinell method is standardized for use with metals in ISO 6506 [12], BS EN 10003 [13], and ASTM E10 [14].

Like the Rockwell test, the Brinell method uses a spherical indentor, but here the diameter of the impression is measured rather than the depth of penetration, the hardness relationship being given by the expression

$$HB = \frac{2F}{\pi D^2[1 - (1 - \{d/D\}^2)^{1/2}]} \tag{3}$$

in which

F = applied load (kg)
D = diameter of the indentor (mm)
d = diameter of the impression made (mm)

The Vickers method on the other hand uses a right diamond pyramid on a square base as the indentor, and the mean diagonal of the impression is measured, the hardness relationship here being given by the expression

$$HV = \frac{2F \sin(\theta/2)}{d^2} \tag{4}$$

in which

F = applied load (kg)
d = mean diagonal width of the impression (mm)
θ = apex angle of the pyramid (= 136°)

The Vickers test is standardized for use with metals in ISO 6507 [15], BS 427 [16], and ASTM E92 [17].

As noted for rubbers, micro tests using scaled-down indentor sizes and smaller loads are being used, and special mention should be made of the micro Vickers test [18], which has been applied successfully to plastics (see Ref. [19] for a review) in the determination of crystallization effects, anisotropy, the effect of weathering and heat aging, the study of polymer blends, etc.

Many other tests have found niche uses, but these have not been widely adopted or standardized at national or international level for use with plastics. Among these, mention may be made of the Knoop micro hardness test, which is similar to the Vickers test but with the diamond indentor having diagonal lengths in the ratio 7:1. This gives rise to smaller indentations, making the shattering of brittle materials less likely. The TNO test uses a polished sapphire pyramid as the indentor but is otherwise based on the Vickers principle. Tests like Moh's scale (using natural minerals) and pencil hardness (Kohinoor test) rate materials according to their scratch resistance. Clearly there is a qualitative relationship between the hardness of a surface and its scratch resistance, but many other factors are involved also, and correlations between these "hardness" tests and standard tests cannot be expected to be very high. Similarly the schleroscope is sometimes described as measuring hardness, when in fact it measures resilience. Special tests have been devised for coatings and films that include pendulum or rocking hardness tests.

Magdanz [20] has given an overview of hardness testers of various types, while Guevin [21] focuses on techniques for coating hardness. A more general review on equipment covering all branches of materials science can be found in Ref. [22]

1.2 Tensile Properties

The tensile testing of plastics materials is covered by the various parts of ISO 527. The disparate standards that once characterized the testing of plastics in their various forms have now been drawn together under one number. Different parts of the standard refer to different forms of plastic, such as general molding and extrusion compounds, films, general purpose composites, high-performance composites, etc. All of these are linked by Part 1 [23] which sets out the general principles to be applied, whatever the specific form of material to be tested.

There are many terms applied to tensile testing, and often the terminology is used casually or imprecisely, which leads to ambiguity and confusion. The following terms are used in ISO 527, and these should be taken as the standard definitions for all plastics tensile testing (note that the term "modulus" tends to be used rather differently in rubber testing, as pointed out in the previous chapter):

Gauge length: the initial distance between the gauge marks on the central part of the test specimen
Speed of testing: the rate of separation of the grips of the testing machine during test
Tensile stress (engineering): the tensile force per unit area of the original cross section within the gauge length, carried by the test specimen at any given moment

Tensile stress at yield: first stress at which an increase in strain occurs without an increase in stress. Also referred to as yield stress
Tensile stress at break: the tensile stress at which the test specimen ruptures
Tensile strength: the maximum tensile stress sustained by the test specimen during a tensile test
Tensile stress at x% strain: the stress at which the strain reaches the specified value x expressed in percentage. It may be measured, for example, if the stress–strain curve does not exhibit a yield point. In this case x must be defined either in the relevant product standard or agreed upon by the interested parties
Tensile strain: the increase in length per unit original length of the gauge. It is used for strains up to the yield point; for strains beyond this limit see nominal tensile strain below
Tensile strain at yield: tensile strain at the yield stress
Tensile strain at break: the tensile strain at the tensile stress at break, if it breaks without yielding
Tensile strain at tensile strength: the tensile strain at the point corresponding to the tensile strength, if this occurs without or at yielding
Nominal tensile strain: the increase in length per unit original length of the distance between the grips (grip separation). It is used for strains beyond the yield point. It represents the total relative elongation that takes place along the free length of the test specimen
Nominal tensile strain at break: the nominal tensile strain at the tensile stress at break, if the specimen breaks after yielding
Nominal tensile strain at the tensile strength: the nominal tensile strain at the tensile strength, if the specimen breaks after yielding
Modulus of elasticity in tension: the ratio of stress difference to the corresponding strain difference. These strains are defined in the standard as being 0.05% and 0.25%. Also known as Young's modulus. This definition is not applicable to films (or rubber as noted earlier)
Poisson's ratio: the negative ratio of tensile strain in one of the two axes normal to the direction of pull to the corresponding strain in the direction of pull within the initial linear portion of the longitudinal versus normal strain curve

Some of these terms are illustrated in Fig. 2 While a number of them are rather cumbersome and pedantic, they do make clear precisely what is being considered. The new standard has no definition for offset yield stress or proportional limit because the idea of trying to specify the extent of the initial linear portion of the stress–strain curve has been abandoned. It was in any case untenable for many plastics materials and of little practical value for the others.

Because plastics are temperature and time dependent, by virtue of their viscoelasticity, both parameters must be defined where comparisons between materials are to be made. The test temperature is typically the standard one as noted in Chapter 6, although non-standard conditions are used to measure the effect of temperature. The test is normally carried out at one of the standard test speeds, chosen from a set of values given in the standard. These recommended speeds are shown in Table 2. Note that for modulus measurement the test speed is normally 1 mm/min.

It is worth noting the definitions for tensile strain and nominal tensile strain. This distinction was not made in previous editions of the standard.

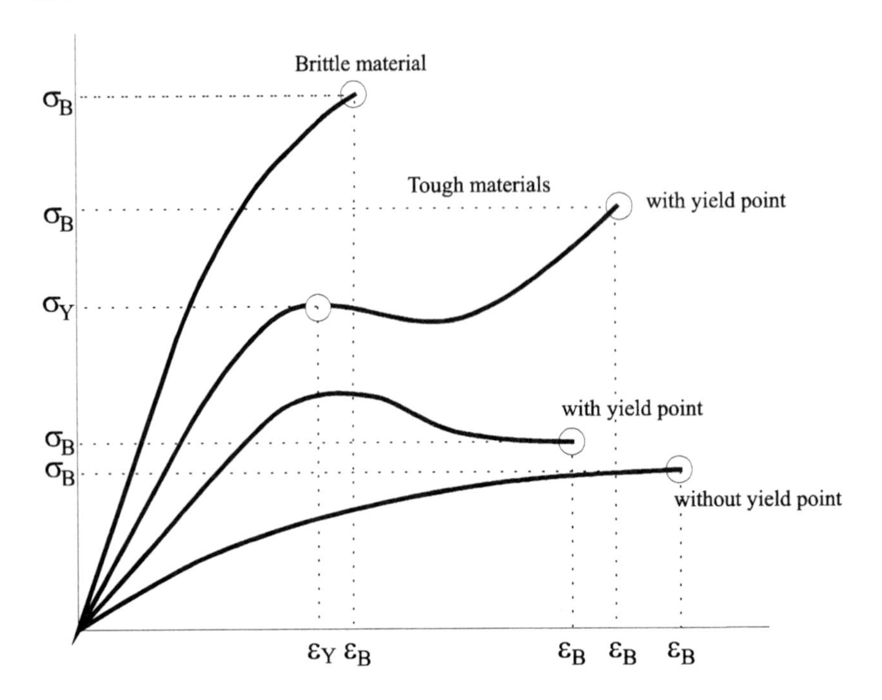

Figure 2 Typical stress–strain curves.

Table 2 Recommended Testing Speeds (ISO 527-1)

Speed (mm/min)	Tolerance (%)
1	±20
2	±20
5	±20
10	±20
20	±10
50	±10
100	±10
200	±10
500	±10

Tensile strain measures the increase in length of an initial gauge length that is contained within the parallel portion of the dumbbell. Up to a point close to the yield point, the change within this region tends to be reasonably uniform. At the yield point a neck forms in the parallel portion and the strain in the necked region is quite different and much higher than that in the unnecked region. What happens is that the highly disordered polymer chains in the unnecked region are transformed into highly ordered patterns in the necked region at the boundary, which moves along the test piece. As the test proceeds, the boundaries of the neck move towards the gripped tab ends. If the gauge length is being monitored by means of an extensometer, erratic behavior can occur at and beyond yielding. It is not unknown for the neck to start outside the gauge length. If this happens the gauge length can apparently decrease as the neck forms, causing measured strain to go backwards until one edge of the neck reaches the nearest gauge mark. Strain then increases again, but once the neck has advanced as far as the second gauge mark there is almost no additional strain within this highly strained necked region, and so the elongation appears to come to a stop even though the grips of the test machine are still moving apart and increasing the overall length of the test piece. Depending on where the test piece breaks with regard to the position of the necks relative to the gauge marks, inconsistent elongation at break data can result.

To counter these problems, and in recognizing that elongation has little practical meaning beyond the yield point, the nominal tensile strain is now defined that takes account only of the movement of the grips. Because of the nonlinear extension of the dumbbell, the tab ends extending by a smaller percentage than the parallel portion, due to their larger area, and because of any grip slippage and movement in the grip linkages, load cell, and testing frame, this is normally a less precise estimate of extension. Beyond the yield point, however, the magnitude of this error is small, and the benefit of not having interpretational problems with the behavior of the gauge length far outweigh the error.

The modulus of elasticity in tension, or Young's modulus, sometimes also referred to as E-modulus, has now been defined in strict relation to two (arbitrary) values of strain, these being 0.05% and 0.25% of the gauge length. There are two ways of determining the slope of the stress–strain curve between these two points: by simply fitting a straight line between the two points, or by finding the best fit straight line between the points (which may not pass through either of them). The former method is used when analogue data has been produced, e.g., from a graph drawn by a plotter or chart recorder, and the latter where the data has been captured digitally and can be processed by computer. In principle the two could give different results, although the difference is likely to be small and masked by the variation between test pieces.

The use of 0.05% strain as the lower limit for determining modulus is a very stringent requirement and needs an extensometer accurate to at least 1 micrometer. There are also implications for the test piece in terms of flatness or warping and for the precision of the gripping mechanism which, frankly, are unrealistic in some cases. It may be that some of these points will be addressed at the next revision of the standard.

Poisson's ratio has now been defined within the framework of the standard. This requires a very precise strain measurement for either the change in width or the change in thickness. The change in width is usually the easier to measure, and hence is the one most frequently measured. For a homogeneous material the Poisson's ratio in the width direction would be equal to the Poisson's ratio in the thickness direction.

The individual parts of ISO 527 define the standard test pieces to be used. In Part 2 [24], which deals with plastics for molding and extrusion, there are six types in all, arranged in three pairs. The normal dumbbell (Fig. 3a) has a minimum length of 150 mm, a width at the widest part of 20 mm, and a width along the parallel portion of 10 mm. The preferred thickness is 4 mm. The two variants of this type differ in the length of the parallel portion. For type 1A, the radius of the shoulder joining the parallel portion to the tab end is from 20 to 25 mm making the parallel portion 80 mm long. For type 1B the radius is 60 mm or more to make the parallel portion 60 mm long. Type 1A is preferred where the test piece is directly molded and is identical to the multipurpose test specimen of ISO 3167 [25]. Type 1B is for use where the test piece is to be machined. The advantage of having the 80 mm parallel portion along the dumbbell is that many plastics test methods have a test piece of this length and with a cross section of 10×4 mm so they can be prepared by simply cutting away the tab ends of the dumbbell.

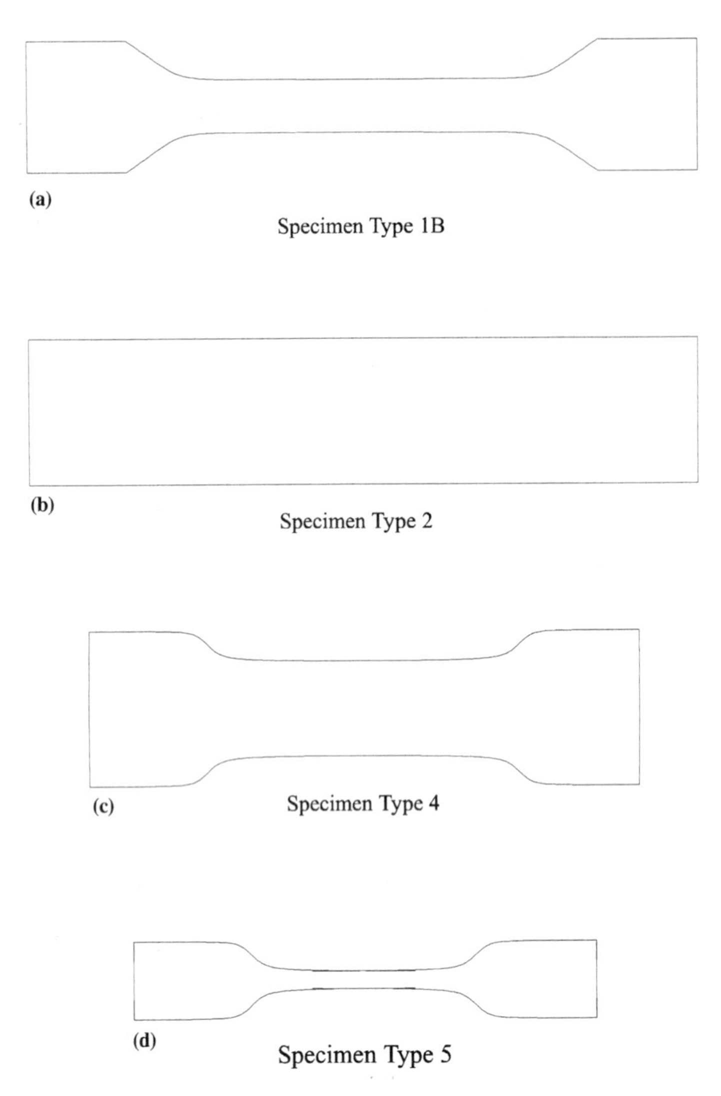

Figure 3 Specimen types. (a) Type 1B. (b) Type 2. (c) Type 4. (d) Type 5.

Where the standard type 1A or 1B specimen cannot be used for whatever reason, the other types may be substituted. Specimens 1BA and 1BB are proportionally scaled versions of the type 1B, the size ratios being 1:2 for the type 1BA and 1:5 for the type 1BB. The speed of testing is also to be scaled in the same proportion, so that the nominal rate of strain remains the same; for modulus, however, 1 mm/min remains the norm. Very soft, highly extensible plastics may be tested with the type 5A or 5B specimen, these being identical to the type 2 and 3 of ISO 37, which were covered in the previous chapter.

Part 3 [26] covers the tensile testing of films and sheeting and replaces the earlier ISO 1184. The preferred test piece is now the parallel strip, 10 to 25 mm wide and having an overall length of at least 150 mm. This test piece has been given the type 2 designation (Fig. 3b). Other test pieces that may be used are the type 5 (Fig. 3d), which is identical to the "rubber" type 1 dumbbell of ISO 37, the type 1B as described above, and the type 4 (Fig.

3c), which is an extra wide dumbbell of 25.4 mm width in the parallel portion. Other dimensions in this dumbbell have strong connections with imperial units. All of these test pieces are intended for special purposes: the type 5 for highly extensible films, the type 1B for rigid sheets, and the type 4 for "other types of flexible thermoplastic sheet." The type 4 looks particularly useful for films where breakage at the grip is a problem, since the extra width of the tab end reduces the chance of failure at the grip edge. The normal speeds of testing within the range 5 to 500 mm/min are allowed, but a speed of 300 mm/min is also permitted, while the 10 and 20 mm/min are not used.

A particular practical problem with films is the measurement of extension. They are normally far too flimsy to bear the weight of an extensometer, and so some form of noncontacting system is almost essential. The high extensibility of some types of film material can also lead to difficulty with the test piece out-stretching the movement of the tensile testing machine. In these circumstances it is permitted to reduce the initial 100 mm distance between the grips, as normally used, to 50 mm. Apart from the type 5 dumbbell, which has a 25 mm gauge length, the other test pieces all use 50 mm for the initial gauge length.

Tensile tests for cellular materials and composites have their own particular requirements, and these are covered in subsequent chapters.

The standards in BS 2782 covering tensile testing are dual numbered with the corresponding ISO. Thus Method 321 [27] corresponds to ISO 527-1, Method 322 [28] to ISO 527-2, and Method 326E [29] to ISO 527-3 (Methods 326 A to C having been recently superseded). A standard not covered by ISO, however, is Method 327A [30], which covers the tensile testing of PTFE products in various forms such as tube, rod, sheet, and tape.

Greater diversity of test pieces exists when ASTM standards are considered, although the test temperature and speeds are encompassed by the ISO version. The general test methods are given in D638 [31] for general purpose plastics and D882 [32] for films and sheeting. There is also a "metricated" version of D638 (D638M), which has test pieces corresponding to the type 1 and 5 dumbbells of ISO 527. There is provision for the use of microtensile test pieces (Fig. 4) given in ASTM D1708 [33], where only small quantities of material are available, but only for use where there is a history of data using this test piece; otherwise the very small type V specimen of D638 is recommended.

The effects of fillers on tensile properties have been investigated by Godard and Bomal [34], Canova [35], Skelhorn [36], and Meddad et. al. [37] amongst others. Where cross-head movement is used to estimate extension, the correction for the finite compliance of the test machine has been reported by Turek [38]. The effect of crystalline and amorphous phases in a semicrystalline polymer has been reported by Carraher [39]. Some of the

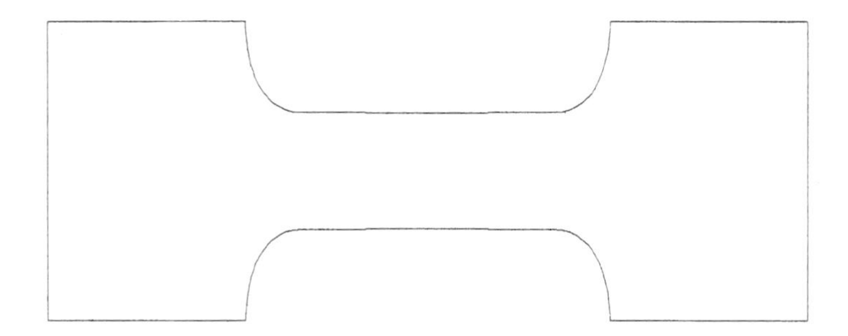

Figure 4 The microtensile test piece.

problems associated with applying data from a tensile test to finite element modelling have been discussed by Boyce et al. [40] and by Rackowitz [41], even when comparing data from data sheets. The effect of molecular weight on tensile properties has been reported by Kennedy et al. [42] and by Shick [43].

1.3 Compressive Properties

The measurement of compressive properties of plastics is covered by ISO 604 [44]. Unlike with the tensile test a single document is able to cover a wide range of materials, since test geometries and other conditions tend to be less variable between types of plastic than is the case for tensile testing. Nevertheless, there are types of materials that the standard does not purport to cover, and these include textile reinforced plastics, cellular materials, and sandwich constructions involving cellular materials.

As the compression test is—in principle at least—only a reverse tensile test, it is not surprising that many of the same definitions apply, with a few obvious textual changes (such as decrease in length rather than increase in length) for definitions of strain. Because the test is performed by loading the test piece between suitable flat metal anvils, the shape of the test piece can be quite varied, and the standard permits the use of right prisms, cylinders, and tubes. Unlike the tensile test piece, the test piece for compression testing must have very flat and parallel end faces, lathes or milling machines being recommended for their preparation. Parallelism to 0.025 mm normal to the long axis of the test piece is required. Of significance in the compression test, but not in the tensile test, is the problem of buckling. As the test piece is compressed there is a tendency for it to buckle rather than to compress uniformly. Short, squat test pieces are much less prone to buckling than tall, slender, test pieces. For this reason there are certain restrictions placed on the ratio of key dimensions. The two key dimensions are the test piece length (the direction along which the compressive force is applied) and the critical dimension at right angles to this. The relationship between the parameters that govern buckling is given as

$$\varepsilon_c^* \leq 0.4\left(\frac{x}{l}\right)^2 \tag{5}$$

where

ε_c^* is the maximum nominal compressive strain experienced during the test
l is the length of the test piece
x is the diameter if the test piece is a cylinder, the outer diameter if it is a tube, or the thickness (i.e., the smaller lateral dimensions) if it is a prism

For measuring compressive modulus, the x/l ratio should not be less than 0.08, while for other tests it should not be less than 0.4. The smaller ratio for the modulus test is permitted since the strain range required for the test is so small and hence the point of buckling is unlikely to be reached. As with other ISO test methods, preferred specimen sizes (Fig. 5) are given, which in this case are

for modulus: length = 50 mm, wdith = 10 mm, thickness = 4 mm
for strength: length = 10 mm, width = 10 mm, thickness = 4 mm

It is noted in the standard that these test pieces can be prepared from the multipurpose test piece given in ISO 3167 [25].

It is in the nature of the test for much smaller deformations to be observed before the end point is reached than is the case for tensile testing. Accordingly, there is no require-

Figure 5 Compression test pieces. (a) For modulus. (b) For strength.

ment to have high test speeds specified, and the preferred speed should be chosen from the list: 1, 2, 5, 10, or 20 mm/min. The choice depends on the nature of the material and the measurement being made. For modulus, the speed should be the closest in value from this list to 5% of the specimen length. For strength tests on brittle materials, the speed should be the closest to 10% of the specimen length, and for ductile materials, to 50% of this length. Thus for the standard test pieces, the most appropriate speed for modulus and for the strength of brittle materials is 1 mm/min; for the strength of ductile materials it is 5 mm/min.

The corresponding standard to the above ISO test method is BS 2782, Method 345A [45], which is dual numbered with it. In ASTM the standard test method is D695 [46], or the metricated D695M version, which gives rather more detail than does the ISO standard over tooling, to effect the compression in a controlled and systematic way. The standards also define the details of suitable lateral support when thin specimens are to be tested. The preferred test piece is a cylinder or prism of length twice the principal diameter or width.

Brief mention may be made here of a compression test that measures the cohesion between layers of laminated tube as given in BS 2782, Method 346A. This is covered more fully in Chapter 18.

1.4 Flexural Properties

Flexural or bending tests are frequently applied to plastics materials, and the test is standardized internationally as ISO 178 [47]. It is widely used for a variety of different plastics types, although cellular materials and sandwich structures are excluded from its scope. The ISO test covers only the three-point bending method (Fig. 6), although a four-point bending test is under consideration. Definitions similar to the corresponding tensile and compression standards apply in the main, although in bending tests, the stress and strain are not uniform throughout the thickness. They change from the tensile form on the outer surface opposite the loading point, through zero at the neutral plane, to the compression form on the surface adjacent to the loading point. For an isotropic material, the neutral plane is along the center of the test piece. Stress and strain are therefore defined in terms of the outer surfaces only and the midpoint of the span.

Figure 6 Schematic test arrangement for three-point bend test.

One term unique to the flexural test is the stress at the conventional deflection. Generally with ductile materials the test piece does not reach a point of fracture but simply keeps bending until eventually it slips from the outer supports. The conventional deflection is defined as 1.5 times the test piece thickness, which for the standard span of 16 times the thickness equates to a strain of 3.5%. The stress at this point forms a useful, if arbitrary, characteristic for ductile materials, which occurs before the peak in the force–deflection curve is reached.

The modulus of elasticity in flexure is defined in an analogous way to the tensile and compressive tests, with the strains of 0.05% and 0.25% again taken as the limiting values between which it is determined. This places very severe constraints on the accuracy of the test equipment and the test piece itself. For the standard test piece, 0.05% strain corresponds to a deflection of the outer surface of only 0.08 mm from its starting point. This does not make much allowance for any backlash in the test jig, lack of flatness in the test piece, or lack of alignment in any of the three loading bars.

In theory, the flexural modulus and the tensile modulus should be the same. In practice this is only approximately so, because plastics materials are seldom (if ever) isotropic throughout the thickness. Differential cooling rates, variations in extrusion or injection rates, changes in flow patterns, etc., all contribute to nonuniform properties throughout the thickness. When coupled with the nonuniform stresses already mentioned, it is hardly surprising that inconsistencies with the tensile test arise.

The same test speeds as for the tensile test are permitted, although in practice these tend to be limited to the slowest end of the range. The speed closest to 1% strain per min is specified, and this translates into 2 mm/min for the preferred test piece. The latter is a bar of length 80 mm and cross section 10×4 mm; this can be taken from the center of the multipurpose test specimen of ISO 3167 [25]. When the preferred test piece cannot be obtained, other specimen sizes are permitted. The length of the test piece should be (20 ± 1) times the thickness, and the width increases from 10 mm to 50 mm as the thickness varies from 3 to 50 mm. Below 3 mm thickness, a width of 25 mm is used. These figures apply to plastics other than textile or long-fiber reinforced plastics, for which wider samples are typically used at a given thickness. Chapter 18 deals with these more fully.

For most tests, the span should be set to (16 ± 1) times the thickness, although for soft thermoplastics (to avoid indentation from the supports) and certain fiber-reinforced materials (to avoid delamination), higher span-to-thickness ratios may be needed. For very thin test pieces a lower ratio may be needed in order to keep the forces generated within the measuring range of the equipment.

The ISO standard covers only rectangular test pieces in three-point bending, for which the stress and strain relationships respectively are

$$\sigma_F = \frac{3FL}{2bh^2} \tag{6}$$

$$\varepsilon_F = \frac{6hs}{L^2} \tag{7}$$

in which:

F = force at midspan
L = span
b = test piece width
h = test piece thickness
s = deflection of the surface of the test piece at midspan

From these expressions we obtain the expression for modulus as

$$E_f = \frac{L^3}{4bh^3}\, slope \tag{8}$$

where *slope* is the slope of the force–deflection curve between the reference strains of 0.05% and 0.25%.

A more accurate expression for the stress, which takes into account the horizontal component of the flexural moment, is given by

$$\sigma_F = \frac{3FL}{2bh^2}\left(1 + \frac{4s^2}{L^2}\right) \tag{9}$$

Since s is typically very much less than L, the second term in parentheses makes only a very small contribution to the stress and can be ignored. This correction is not covered in the standard.

For a circular rod, the expressions for stress and modulus are

$$\sigma_F = \frac{8FL}{\pi D^3} \tag{10}$$

$$E_f = \frac{4L^3}{3\pi D^4}\, slope \tag{11}$$

where D is the diameter of the rod.

In four-point bending, the stress for a rectangular beam and for a circular rod become respectively

$$\sigma_F = \frac{12FL_1}{bh^2} \tag{12}$$

$$\sigma_F = \frac{64FL_1}{\pi D^3} \tag{13}$$

where F is the total force (i.e., on both bearing faces) and L_1 is the span between the inner and outer bearing faces.

The equivalent British standard can be found in BS 2782, Method 335A [48], which is identical to the ISO standard. The corresponding ASTM standards are D790 [49] and its metricated version D790M. The test follows very much the same lines as the ISO test except for the provision of a four-point loading test in addition to the normal three-point test. The advantage of the four-point test is that the curvature between the inner loading noses tends to be much more circular than in the three-point test, making it conform more accurately to the underlying mathematical principles on which the flexural test is based. A further test is D747 [50], in which a strip of material is clamped in a vice and the load applied through a pivot point at the end of the vice, where the free length of the test piece starts. The greater the angle of bend, the lower the stiffness and hence modulus of the material. The test is better suited to estimating relative modulus than in making absolute determinations.

The effect of fillers on flexural strength has been reported by Shi and Nedea [51] as well as by others (see [34–37]).

1.5 Shear Properties

The measurement of shear strength and modulus is relatively little used in the context of general plastics testing, although the tests are more frequently applied to rubbers (Chapter 15) and composites (Chapter 18). While there are no general ISO test methods, there are BS and ASTM tests.

BS 2782, Methods 340A and B [52], cover the determination of shear strength for molding material and for sheet material respectively. For Method 340A (Fig. 7), the test pieces are molded discs 25.3 ± 0.1 mm in diameter and 1.6 ± 0.1 mm thick, while for

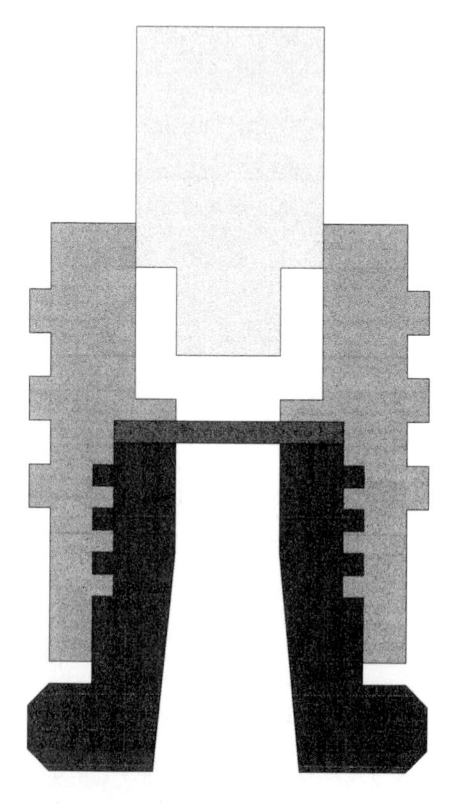

Figure 7 Shear strength jig.

method B a rectangular bar test piece is used of length 32 mm and 6.4 ± 0.2 mm wide by the thickness of the sheet under test. However, if the sheet is more than 6.35 mm it is machined on one surface only to reduce the thickness to 6.10 ± 0.25 mm.

The test is carried out in compression by placing the test piece in a special bolster into which sits a close-fitting punch that bears against the test piece surface. The jig is placed in a universal testing machine and the load on the punch is increased so as to cause failure in the test piece within 15 to 45 seconds from first application of the load.

For this test the calculation of strength for Methods A and B is respectively given by the expressions

$$S = \frac{F}{\pi DT} \tag{14}$$

$$S = \frac{F}{2.096BT} \tag{15}$$

in which S is the shear strength, F is the force at break, D is the diameter of the punch, B is the mean width of the test piece, and T is the mean thickness of the test piece.

ASTM D732 [53] follows the same pattern as Method A above, but permits any thickness between 0.125 mm and 12.5 mm to be tested, and the test piece is drilled centrally to locate a guide pin on the punch.

Jawad [54] has used the determination of shear moduli to examine molecular processes in liquid crystalline copolymers, while Hedner et al. [55] describe a test for general use with thermoplastics. Boyce et al. [40] compare the large strain compression, tension, and shear behavior of polycarbonate.

1.6 Tear Strength

The tear strength measurements on plastics invariably are applied to films or sheeting, there being two internationally standardized methods of test. In the first part of ISO 6383 [56], the trouser tear geometry (Fig. 8) is used, which is essentially the same as that described for rubbers. A test piece 150 mm by 50 mm is cut along the center of its long axis, from one end to halfway down its length. The two "legs" so formed are gripped in the stationary and moving jaws of a universal testing machine and pulled apart at 200 mm/min (although for the time being a speed of 250 mm/min is also permitted).

Typically, an irregular wavelike trace results (Fig. 9), and the standard defines the tearing force as the mean force after the first 20 mm and last 5 mm of the tearing trace is ignored. This tearing force is then normalized by dividing it by the film or sheet thickness to produce the tearing resistance value.

As with other mechanical tests, the properties of the film or sheet may vary significantly with direction, the machine or longitudinal direction often generating an easier tearing path at lower force than the transverse or cross direction. Indeed, getting the test piece to tear along the length of the specimen can be the greatest difficulty associated with the test.

Part 2 [57] of the same standard specifies the Elmendorf method. Here the test piece is held in the jaws of a pendulum, which is released and causes the initial cut in the specimen to propagate across it (Fig. 10). The energy absorbed by this tearing process is reflected in, and hence measured by, the height to which the pendulum swings once tearing is complete. Two test pieces are permitted, the rectangular and the constant radius specimen. As their names imply the rectangular test piece is rectangular, of sides 75 mm by 63 mm with a 20 mm long slit cut in the middle of the longer side and parallel to the shorter side, and the

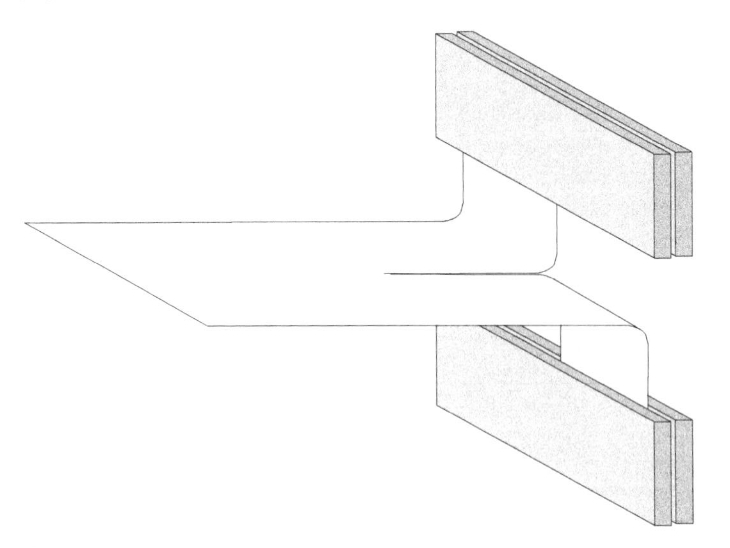

Figure 8 Trouser tear test geometry.

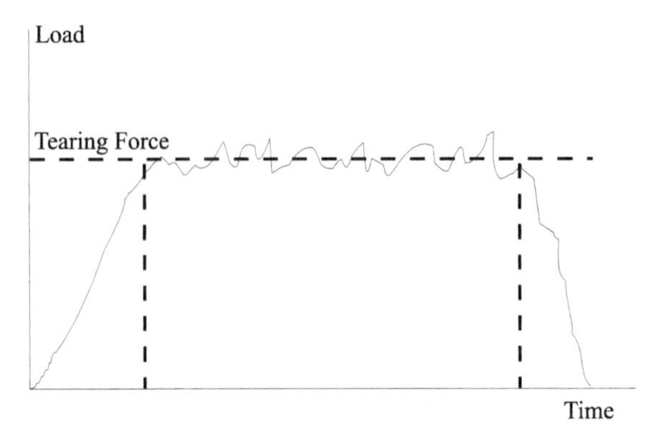

Figure 9 Typical tearing force graph.

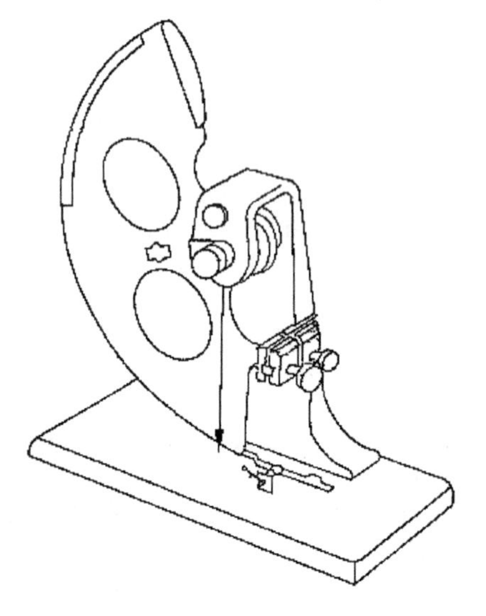

Figure 10 Elmendorf tear tester.

constant radius test piece is of the same 75 mm length with the same 20 mm slit, but the edge facing the cut is circular of radius 43 mm rather than straight. This means that if the tear propagates at an angle to the motion of the pendulum, the tear length remains constant, and so, in principle, should the tearing energy.

Although the test is essentially energy based, it is a tearing force that is calculated from the scale reading and the conversion factors provided by the manufacturer. Interestingly, in this part to the standard, the tearing resistance is this tearing force and not the tearing force divided by the test piece thickness as it is in Part 1.

In British standards, the same methods are standardized in the dual numbered BS 2782, Method 360B [58] and Method 360A [59]. BS 2782 Method 360C [60] and ASTM D1004 [61] use the angle tear test piece similar to that detailed in ISO 34 from the previous chapter, although the dimensions are not identical. The standard test speed for this is 250 mm/min for the BS and 51 mm/min for the ASTM, so differences in test results between them can be expected. D1922 [62] uses an Elmendorf pendulum with the constant radius test piece similar to the ISO standard, while D1938 [63] is based on the trouser tear geometry tested at a grip separation of 250 mm/min. D2582 [64] uses a novel falling dart arrangement (Fig. 11) which is intended to simulate snagging hazards. A weighted carriage is released from a standard height. Fastened to the side of the carriage and protruding horizontally from it is a cylindrical tearing probe with a truncated cone at the tearing end. This falls against the test film, which is clamped to a curved holder adjacent to the tower down which the carriage falls, so that the distance from the film to the tower decreases the further down the tower the carriage falls. After falling 508 mm, the probe just touches the surface of the film and thereafter penetrates and rips down the film, the length of tear so produced being a measure of the tear resistance of the test film.

1.7 Impact Strength

The impact strength of plastics falls into two basic categories: the pendulum tests and the falling weight tests. Each of these is then further subdivided into more specific classes.

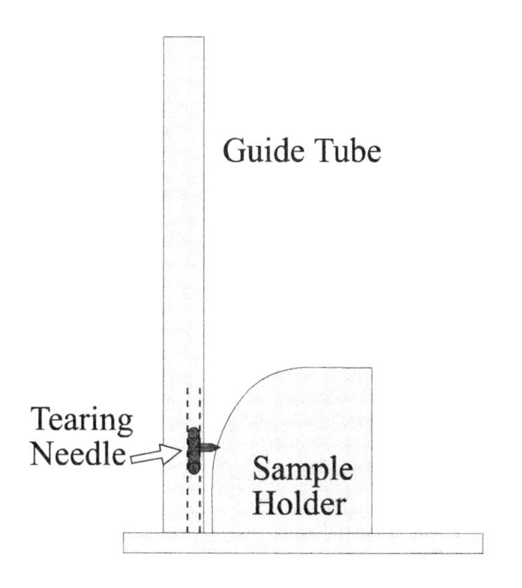

Figure 11 ASTM film tear test.

Charpy Test

The Charpy test is detailed in ISO 179 [65], the identical BS 2782, Method 359 [66], and the related ASTM D256, Method B [67]. The following discussion is based on the ISO test method, with comments on the ASTM variant at the end.

In the Charpy test the test piece is supported as a horizontal beam and is broken by a single swing of a pendulum, the line of impact being midway between the supports. Both notched and unnotched test pieces may be tested, and the test piece may be oriented in the edgewise or the flatwise direction. The edgewise test is illustrated in Fig. 12.

The directionality of the test is best understood in relation to the test piece dimensions themselves. The standard test bar is $80 \times 10 \times 4$ mm, which can be cut from the centre parallel portion of the multipurpose test specimen [25]. In the flatwise test, the direction of the pendulum at impact is in the 4 mm direction of the test piece, so that bending takes place over the 80×10 mm surface. In the edgewise test, it is the 80×4 mm plane that is bent, and the pendulum travels in the direction of the 10 mm dimension. The edgewise test is now the preferred form of geometry for most testing purposes. In former times it was the flatwise test that was typically used, and the edgewise test was reserved for investigating the effect of fibre reinforcements on impact strength. Now the flatwise test is reserved for investigating surface effects such as might occur when the material is weathered by UV light or exposed to chemicals.

For laminated test pieces, tests may be performed both flatwise and edgewise, and for each of these there exists the possibility of having the laminations parallel or normal to the direction of blow. All these are permitted, and a suitable coding scheme is defined to enable the options chosen in a given test to be defined very succinctly.

The test can be performed using either unnotched or notched test pieces. Three types of notches are standardized (Fig. 13), the preferred one having a radius at the notch base of 0.25 mm (the type A notch). A blunt 1.0 mm (type B notch) and a very sharp 0.1 mm (type C) notch are also covered. Notches of different base radius are useful for more extensive characterization of plastics than a simple quality test or data sheet entry, in that they enable an estimate of the notch sensitivity of the plastic to be investigated. The flatwise test can also be performed notched or unnotched, except that here the notched test has two notches machined across the 4 mm direction and directly opposite

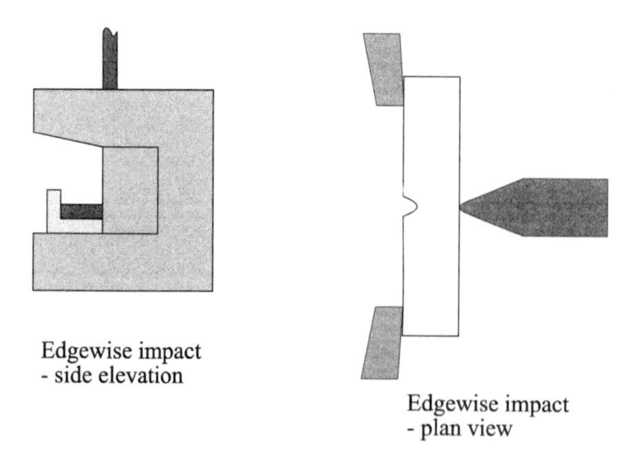

Figure 12 Charpy impact geometry. (a) Edgewise impact—side elevation. (b) Edgewise impact—plan view.

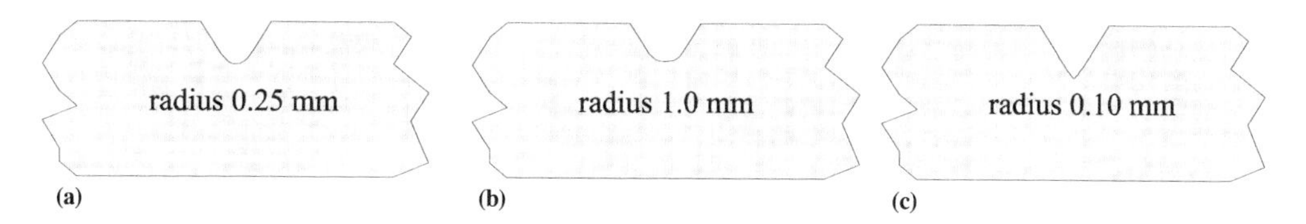

Figure 13 Notch types. (a) Type A notch. (b) Type B notch. (c) Type C notch.

each other to give a 6 mm width to the test piece between the notches. All three types of notch may be used in the flatwise test.

The standard test piece is suitable for general purpose testing, although where sheet material is to be tested it is permissible to test the full thickness of the sheet up to a maximum of 10.2 mm. Above this the sheet should be machined on one surface to reduce the thickness to 10 mm. Where the sheet is reinforced in some way the reinforcement must be regularly distributed and be of only one type. Thin samples are not suited to this test as buckling of the test piece can occur when tested edgewise or bending without failure when tested flatwise. For long-fiber reinforced plastics alternative geometries are permitted, but these are covered in Chapter 18.

The ISO standard covers two pendulum lengths, giving different velocities at the point of impact. The one most frequently used has an impact velocity of 2.9 m/s, and five pendulums having energies of 0.5, 1.0, 2.0, 4.0, and 5.0 J are specified. The larger machine, having an impact velocity of 3.8 m/s, has pendulums of energy 7.5, 15.0, 25.0, and 50.0 J.

The impact strength is the energy removed from the pendulum as a result of work done in breaking the test piece divided by the cross-sectional area of the test piece in the direction of swing. In fact, because the test piece is bending during the impact event, there is a deformation volume rather than simply an area and so test pieces of different size do not give results that are proportional to the cross-sectional area but rather to some indeterminate volume. For this reason, results obtained from test pieces of different size cannot be compared.

The ASTM test follows the same principles but differs in certain details. The standard test specimen is based on imperial units, so that the preferred test piece is 127 mm long by 12.7 mm wide by a thickness of 12.7, 6.4, or 3.2 mm. The details of the apparatus used are the same as for the Izod test, which is covered in the next section; but the form of expressing the result is different, being based on the energy normalized with respect to the length of the notch only, and not on the area behind the notch. This only serves to add to the difficulty in making comparisons between data obtained by the ASTM standard, with its different test piece sizes and impact conditions, to that of the ISO standard.

A new part to ISO 179 is under preparation that will cover the instrumentation of the Charpy pendulum so that force–time (and by integration, force–deflection) curves can be obtained. This allows for a fuller characterization of the impact behavior of the plastic than can be derived from only the energy to break of the typical test. There has been an instrumented version of the falling weight impact test (see later in this section) for several years and the same principles apply to both.

Nakamura et al. [68] have used the instrumented Charpy test to examine the effect of silica fillers on epoxy resins, while Wang et al. [69] have used the same technique for examining RIM parts. An examination of fillers in polypropylene has been reported by Jancar and DiBenedetto [70]. Trantina and Oehler [71] discuss the application of Charpy (and Izod) tests to the prediction of impact resistance for use in design calculations. Sharpe and Boehme [72] have used a small Charpy test to investigate dynamic fracture toughness measurements.

Izod Test

The Izod test is notionally very similar to the Charpy test, except that the test piece is clamped at one end below the notch, or at the center of the specimen if it is unnotched, and struck by a pendulum close to the other end. It is therefore a cantilever bending test (Fig. 14). Traditionally the Izod test has been more favoured in North America, while the Charpy test has been more popular in Europe. The test details are given in ISO 180 [73], BS 2782, Method 350 [74] (which is identical to it), and ASTM D256, Method A [67] (also methods C, D, and E).

Considering the ISO standard first, the standard test piece is the ubiquitous 80 × 10 × 4 mm test piece so widely used in ISO tests, and three variants are permitted: unnotched, notched with a 0.25 mm radius notch (type A), and notched with a 1.0 mm radius notch (type B). These match the same conditions as for Charpy tests, but the type C notch is undefined for Izod tests. The test is almost always carried out edgewise, although where laminated plastics are to be tested it is possible to test flatwise as well and using the same parallel or normal arrangements as for Charpy tests.

Unlike the Charpy test, the notched Izod is capable of being tested either with the notch on the same side as the point of impact, which is the normal way, or on the opposite side, when it is called the reverse notched test. Thus in the normal test the side containing the notch is placed under tension and the notch fulfils its purpose as a stress concentrator. In the case of the reverse notch, it is the unnotched face that is under tension, and no stress concentration occurs; in fact the notch is placed under a compressive deformation. This arrangement is possible in the Izod test because the pendulum strikes the test piece at a point remote from the notch, and the advantage of having the reverse notch is that the test piece is otherwise identical. For the Charpy test, the cross section of the unnotched test piece must be greater than the notched test piece.

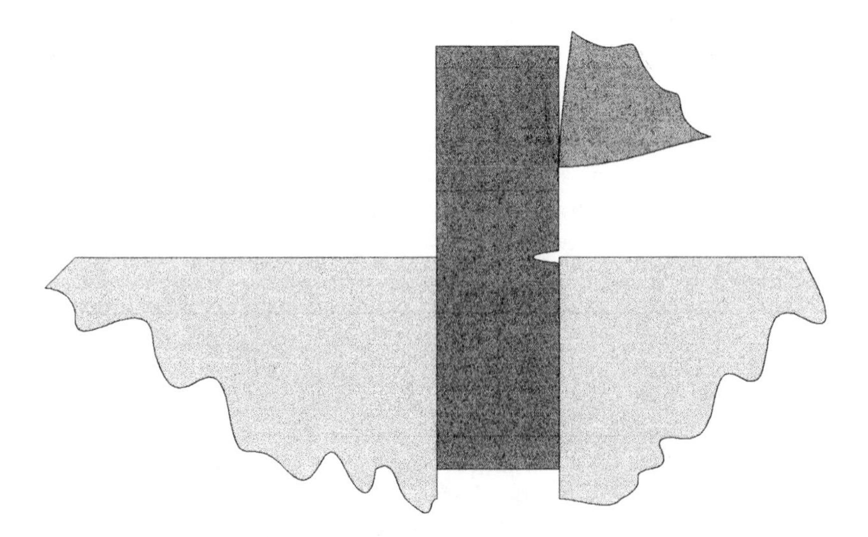

Figure 14 Izod impact geometry.

The impact velocity for the test is 3.5 m/s, and pendulums of energy 1.0, 2.7, 5.5, 11.0, and 22.0 J are used. As for the Charpy test, the energy absorbed by the impact should be between 10% and 80% of the capacity of the pendulum.

Certain plastics can give results that vary according to clamping pressure, a problem from which the Charpy test is free, and the standard recommends that when testing such materials some means of standardizing and recording the clamping pressure should be used. However, it gives no advice on which plastics are so affected nor on how to determine whether the effect is significant or not.

The test is often applied to plastics at subambient temperatures, but it is far from ideal for this. Again, the Charpy test should be preferred. There are serious practical problems in carrying out the test with the apparatus itself at the low test temperature due to icing of the bearings, etc. It is therefore common practice to soak the test piece at the test temperature and then quickly remove it and test it. However, the test piece must be clamped into a large metal heat source, the clamping vice, at a point adjacent to the critical notch region where the bending takes place. The actual test temperature is therefore quite indeterminate and likely to be variable from test piece to test piece.

The ASTM test follows the same principles but differs from the Charpy test in certain details. Again, the test is based on imperial units with the impact resistance characterized by the length of the notch, rather than the area behind the notch. The details of the apparatus itself mirror very closely the requirements of the ISO test method, which was largely derived from the ASTM standard. Method A covers the normal test procedures, which are applied to materials having an impact resistance in excess of 27 J/m. For lower values than this, Method C is applied, which attempts to make a correction for the energy required to "toss" the test piece. This involves carrying out a secondary test on the broken test piece, wherein the halves of the test piece are reassembled and the energy value obtained when this broken test piece is impacted is taken to be the energy absorbed in accelerating the initially stationary test piece. Since this energy is not due to the impact event as such, it is then subtracted from the apparent energy obtained during the first impact event, when the test piece was unbroken. Objections to the scientific principles behind this idea can be raised, and it has never found acceptance within the ISO community.

Method D deals with the estimation of notch sensitivity by having the test carried out at two notch radii, 0.25 mm and 1.0 mm. The ratio of the difference in the two energy values to the difference in notch radii is then taken as the index of notch sensitivity. Where the 1.0 mm radius leads to test pieces which do not break, a 0.5 mm radius notch may be substituted.

Method E covers the reversed notch test.

Fu et al. [75] have investigated the toughening of polyethylene by calcium carbonate using the Izod test, while Grocela and Nauman [76] have tried to derive quantitative models for Izod to predict strength for impact modified polystyrene. The Izod test has been used to investigate the toughening mechanism of low molecular weight polybutadiene in polystyrene [77,78]. Weier and Hemenway [79] describe the use of the instrumented Izod test on PVC/acrylic composites.

Tensile Impact

Both the previous two methods require the test piece to be sufficiently rigid for buckling of the specimen under test to be negligible. For thinner section materials and for those exhibiting a high elongation before fracture, the tensile impact test may be the only viable pendulum method. The test is standardized in ISO 8256 [80], BS 2782, Method 354A & B [81], and ASTM D1822 [82], which has the metricated version D1822M.

There are two basic types of tensile impact test: the specimen-in-bed type (illustrated schematically in Fig. 15) and the specimen-in-head type. Method A of ISO 8256 covers the first of these and Method B the second. Two pendulum lengths are given in the standard, one of which gives an impact velocity of 2.8 m/s and the other of 3.7 m/s. The former is applied to pendulums having an energy of 2 and 4 joule, while the latter is applied to pendulums having an energy of 7.5, 15, 25, or 50 joule.

For Method A the test piece is clamped into a suitable holder mounted onto the bed of the apparatus. One end of the holder is rigidly mounted on the bed and the other, the cross-head, is free to move along the bed. The test piece forms a bridge between them. A pendulum is released, and at the bottom of its swing it makes contact with the arms of the cross-head. Kinetic energy is transferred to the test piece, which extends to rupture, and the absorbed energy is determined from the height of swing of the pendulum. However, some energy is also expended in tossing the cross-head, and so a correction must be applied for this. The correction is a constant for a given pendulum and cross-head and can be determined from the equation

$$E_q = \frac{E_{\max}\mu(3+\mu)}{2(1+\mu)} \tag{16}$$

where

$$\mu = \frac{1}{4}\frac{m_{\mathrm{cr}}}{E_{\max}}\left(\frac{gT}{\pi}\right)^2(1-\cos\alpha) \tag{17}$$

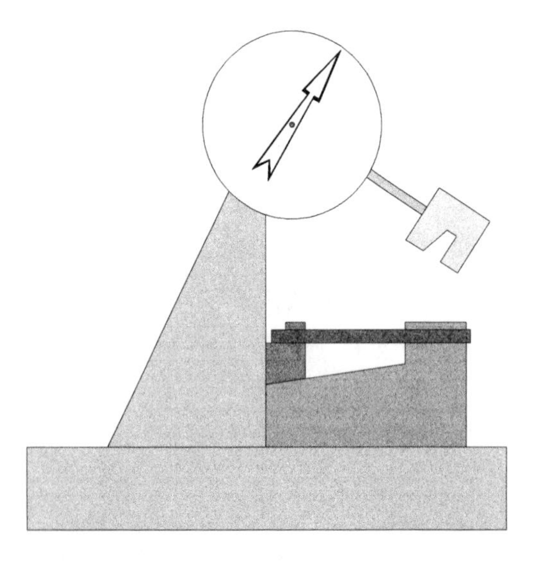

Figure 15 Tensile impact—specimen-in-bed type.

E_q is the energy correction due to the plastic deformation of the kinetic energy of the cross-head
$E_{\max}$ is the maximum impact energy of the pendulum
m_{cr} is the mass of the cross-head
g is the acceleration due to gravity
T is the period of the pendulum
α is the angle between the positions of the maximum and minimum height of the pendulum

The desired energy to rupture the test piece is then simply the difference between the uncorrected energy read from the maximum swing of the pendulum after impact and the above correction energy. As for the Charpy and Izod tests, the result is normalized with respect to the area of the test piece cross section, although unlike the other pendulum impact tests, there are several types of test piece that are used, Fig. 16 illustrating these.

For Method B the test piece is clamped into the compound head of the pendulum, which is released from its raised position. As it reaches the lowest point of its swing, the rear of the pendulum strikes rigid supports on the frame of the apparatus and is arrested. The front of the pendulum continues its swing, extending and rupturing the test piece. As for Method A, corrections must be applied to the energy read from the swing of the pendulum to compensate for the cross-head bounce energy. In this case, the correction is added to the reading from the pendulum, because immediately after impact the two halves of the pendulum are traveling in opposite directions. The correction for a given apparatus is determined by means of a special calibration procedure, which is detailed in the standard.

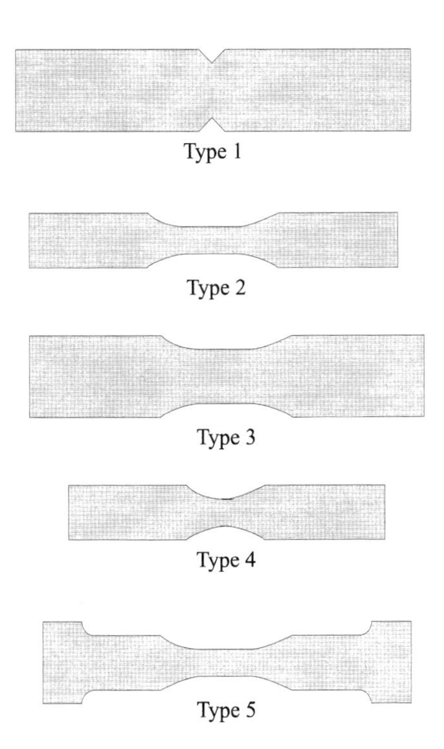

Figure 16 Test pieces. (a) Type 1. (b) Type 2. (c) Type 3. (d) Type 4. (e) Type 5.

BS 2782, Method 354A, is identical to Method A of the ISO standard, while 354B is identical to Method B. ASTM D1822 and its metricated equivalent are essentially Method B of the ISO standard, the specimen-in-head geometry being that favored in North America while the specimen-in-bed geometry has been typically favored in Europe.

Tensile impact has been used to characterize the effect of molecular weight on impact and stress–strain properties [83], while Dijkstra et al. [84] have used tensile impact to investigate the toughening effects of rubber in nylon 6.

Falling Dart Methods

The traditional falling dart methods require a large number of test specimens, because for each drop there can be only two outcomes: the test piece fails according to some agreed criteria, or it passes. The amount by which it passes or fails cannot be judged. Results must therefore be analyzed statistically in order to quantify the mean energy or mass or height that causes failure. With the newer methods, piezoelectric or resistive transducers are built into the dart, so that the force during impact can be monitored directly and a quantitative result obtained for each test piece tested.

The general test for plastics is covered in ISO 6603, Part 1 [85], which is dual numbered in BS 2782, Method 353A [86]. Test pieces of preferred size 60 mm square (or circular) by 2 mm thick are supported on an annular base of inside diameter 40 mm, and the dart with a 20 mm diameter striker is released from a preferred height of one meter. The test piece may be clamped or unclamped on the support, although the standard indicates that different results are likely to occur from these two techniques, and it is permitted also to use a 10 mm diameter striker.

Two methods of analysis are covered. The preferred method is the "staircase" method in which the mass of the dart is varied in given increments according to whether the test piece previously tested passed or failed. If it passed, the mass is increased to increase the probability of failure next time, and if it failed, the mass is decreased to decrease the probability of failure. At least 30 test pieces are required plus an additional 10 used as preliminary specimens to select a suitable starting mass and increment. In Method B, the "statistical" or "probit" method, a minimum of 40 test pieces is also required, although in practice 60 or more tend to be needed. Here 10 test pieces are tested under given conditions and the percentage of failures recorded. The mass is then altered and a further 10 are tested and so on until at least three results are obtained with percentage failures greater than 0% and less than 100% with at least one result greater than 50% and at least one result less than 50%.

For both tests it is permitted to vary the height rather than the mass, although this is not the preferred way to carry out the test as impact velocity is changing along with the impact energy.

The mean impact strength and standard deviation are determined by means of a rather complex calculation for the staircase method or for the statistical method by plotting the percentage passes (or failures) against impact parameter (energy, mass, or height according to the requirements) on probability paper and finding the best fit straight line. The parameter that corresponds to the 50% failure probability is the mean value, and the difference between the 50% and the 16% (or the 84%) probabilities is the standard deviation.

The method detailed in ASTM D5420 [87] is somewhat unusual compared to other impact standards in that the test piece, which is placed on a support plate having a circular hole of given size, has a striker resting upon it, and the striker is then impacted by the falling weight. This is the so-called Gardner impact test, and a number of variations in geometry are allowed. It uses the "staircase" approach to varying the energy of impact, with drop height rather than the drop mass being varied. ASTM D5628 [88] is rather more conventional and follows the same pattern as ISO 6603, Part 1, albeit with different dart shapes and drop height.

Films and sheeting are tested by similar methods, although the test piece diameter tends to be rather larger, typically 125 mm or so, as does the impacting striker. The standardized tests are given in ISO 7765-1 [89], and the identical BS 2782, Method 352E [90], and in ASTM D1709 [91]. There is very little difference between these standards, although the ASTM method does permit either the staircase or the probit method of analysis, while the ISO and BS allow only the staircase method to be used. The drop height is 0.66 m for method A and 1.5 m for method B.

Instrumented impact tests are cited in ISO 6603, Part 2 [92], for general purpose plastics testing, and ISO 7765, Part 2 [93], for films and sheeting. These tests are echoed in the corresponding BS 2782 methods, 353B [94] and 352F [95] respectively, which are identical to the corresponding ISOs. The essential difference to the noninstrumented variant is that some load-sensing transducer is built into the dart; it is preferable to have this transducer as close to the point of impact as possible to reduce interference from "ringing" effects as the force wave sweeps up the dart from the moment of contact. Figure 17 gives a schematic illustration of the dart arrangement. The transducer is generally a resistive or piezoelectric device, the latter being preferred as it has a higher natural frequency and is therefore capable of recording faster transitions without attenuation of the signal.

Much of the detail concerning test pieces and geometry are as given in ISO 6603, Part 1. It is noteworthy that for the film and sheeting tests the same 40 mm span is used as for the general test and not the much larger span of 125 mm as in the noninstrumented test. Much in these standards is devoted to the requirements for the instrumentation—the frequency response of the transducer and the bandwidth of the amplifier to ensure that attenuation or distortion of the signal generated by the impact event are not significant. Unlike the simple falling weight test, where the energy of the dart is arranged to be similar to the impact energy of the material being tested, for the instrumented option the dart should have a large excess of energy so that the reduction in velocity during impact is small (less than 20%).

In addition to the impact energy, this test is also capable of delivering the peak force, the energy to peak force, and the displacement at peak force. The shape of the force–displacement curve can be instructive in characterizing the material's behavior as well, and

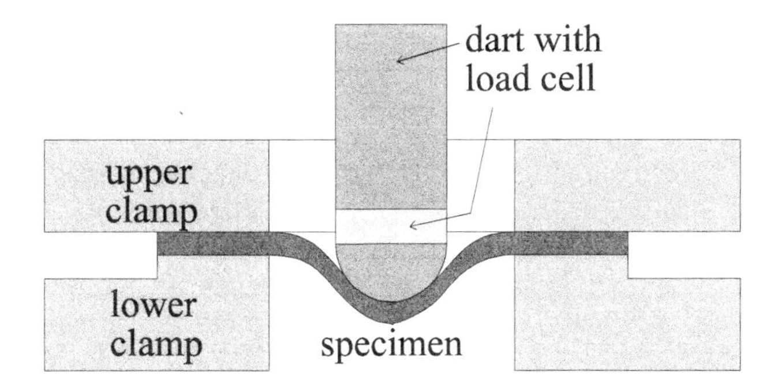

Figure 17 Instrumented impact geometry.

the standards give several examples of the type of curve that may be observed, as well as various failure criteria that can be applied. Clearly, therefore, much more data can be derived from this test than is possible from the simple test. It also requires far fewer test pieces: 10 is the norm, but for quality purposes 5 may be used. The negative side, of course, is that the apparatus is much more expensive and cannot be put together with a minimum of workshop facilities, as can the simple test.

2 Fatigue and Wear

2.1 Fatigue

Despite its importance in the context of choosing the right material and then designing to minimize the effects of fatigue, the standards literature is almost barren of fatigue test methods for plastics. Other materials (cf. Chapter 15) have a number of standards from which to choose. A lot of fatigue testing of plastics is carried out, of course, but these are frequently of an ad hoc nature and many take metals test methods as their basis. Reference may be made to standards such as BS 3518 [96] and ASTM E1150 [97] for metals testing for valuable information relating to terminology, data handling, and test methods. A very important factor to be borne in mind when adapting a metals test, however, is that plastics are typically much higher damping, lower thermal conductivity materials than metals so viscous heating can be a serious problem, making low frequency testing more or less mandatory. This has significant cost implications to data gathering on plastics materials.

The one standardized test is given in ASTM D671 [98], which applies repeated bending moments to the test piece under a defined peak load. Two test pieces are described, to be selected according to the material thickness and stress range over which measurements are to be made. Both specimens are of triangular form with a rectangular cross section, providing a uniform stress distribution over their respective test spans. Details of machining the rather complex shapes are provided, as these can have a significant influence on the outcome of the test. Having indicated that frequency can affect the results, the test is in fact only performed at 30 Hz. There is a useful appendix defining the various terms used in fatigue testing and details are given of constructing the *S–N* curve (Fig. 18), which is the applied stress versus number of cycles to failure very often used in the fatigue characterization of materials.

One of the problems associated with many fatigue tests is the time consuming process of measuring the length of the fatigue cracks that are formed. Riemslag [99] reports a method based on image analysis of a video scan of the crack during the fatigue test, while Trotignon [100] describes a universal fatigue testing machine that allows temperature, environment, and frequency to be varied in bending, torsion or tension in either stress or strain control.

2.2 Wear

There is currently only one standard test method in ISO devoted to the measurement of abrasion resistance for plastics materials, this being the "Taber" test standardized in ISO 9352 [101]. In this test a specimen is clamped onto a turntable that rotates at either 60 rev per min (where the electrical mains frequency is 50 Hz) or 72 rev per min (where the electrical mains frequency is 60 Hz) and weighted abrasive wheels press against the surface of the test piece. The center of rotation of the test piece is offset from the line of contact of the wheels, and so as the former is rotated this causes the wheels to rotate and to wear an annular track into the specimen surface. Figure 19 gives an illustration of the test piece

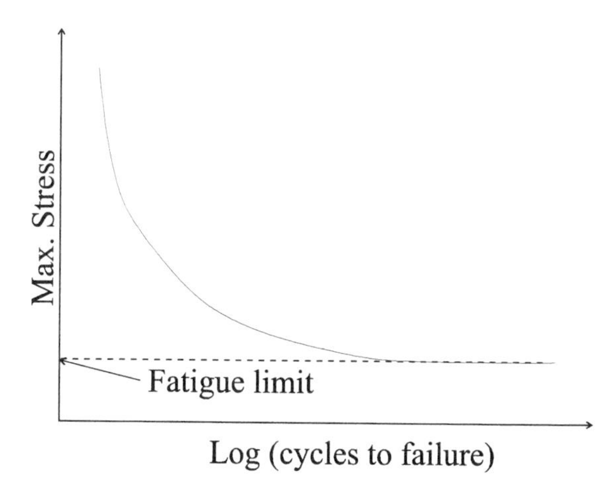

Figure 18 S–N curve for fatigue.

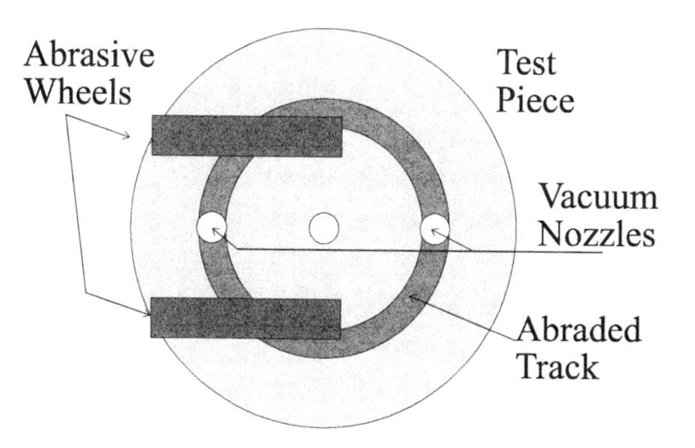

Figure 19 Taber abrasion arrangement.

arrangement. Abrasion resistance is typically then measured by weight loss after a given number of cycles, although for certain purposes loss of thickness, gloss, transparency, or other parameter can be far more sensitive indicators of deterioration. Standard weights applied to the arms of the apparatus to alter the load under which the wheels press against the plastics surface are 250 gm, 500 gm, and 1000 gm.

The ISO standard employs either standard abrasive papers, which are bonded to rubber wheels, or the Calibrase/Calibrade type wheels, in which the abrasive particles are compounded into the rubber (type CS) or vitrified (type H) base. These give a wide range of abrasive severity which can be checked against a standard zinc plate. Like many test method standards, this one gives a wide range of options that the user may select. It relies on the parameters to be used in a given situation being stated in the material or product specification. Where this is not available, agreement between the interested parties needs to be sought, and the details selected should be recorded so that the result can be checked in the future by adopting the same parameters.

A further ISO document, ISO 6601 [102], related to wear exists, although this is really a terminology document for both friction and wear, and ISO 6691 [103] is concerned with the classification and designation of plastics for bearings.

BS 2782 has an identical method to ISO 9352, which is Method 370 [104], and so nothing further need be said on this. ASTM on the other hand has three test methods that are applied to plastics. D673 [105] measures the mar resistance of plastics by having a controlled amount of carborundum grit of given size fall onto the inclined surface of the test piece from a controlled height (Fig. 20). The effect is measured by the loss of gloss as measured by a glossmeter or hazemeter (see Chapter 26). Clearly the method is only appropriate for use where a polished surface is important to the function to which the plastic is applied. ASTM D1044 [106] covers the use of the Taber Abraser. The test uses the same hazemeter as is quoted in D673 and so is only suited to transparent plastics. In ASTM D1242 [107] two methods are standardized; one has the abrasive grit falling onto a rotating surface, which is passed under a rotating weighted foot to which the test piece is mounted, pressing the grit against the plastic surface. In the second, a rotating bonded abrasive belt is pressed against the specimen surface as it is drawn uniformly across the belt's surface (Fig. 21). With this arrangement, several test pieces can be mounted to run at the same time. Both tests measure the volume loss of the test pieces from the recorded weight change and the previously measured density.

No abrasion test is capable of giving predictive data for the design engineer to use, and all are best suited to comparing materials for ranking purposes. For this reason the great advantage of the Taber test is its versatility compared to a number of other standard laboratory tests (cf. ISO 4649 described in Chapter 15). By varying the weight on the specimen, the number of cycles, and the type of abrasive wheel, the severity of the test may be altered over a wide range. In addition to the types of wheels previously described, tungsten carbide wheels are also available, which can provide a very severe abrasion regime. The major disadvantage to the Taber test is the relatively large and flat test

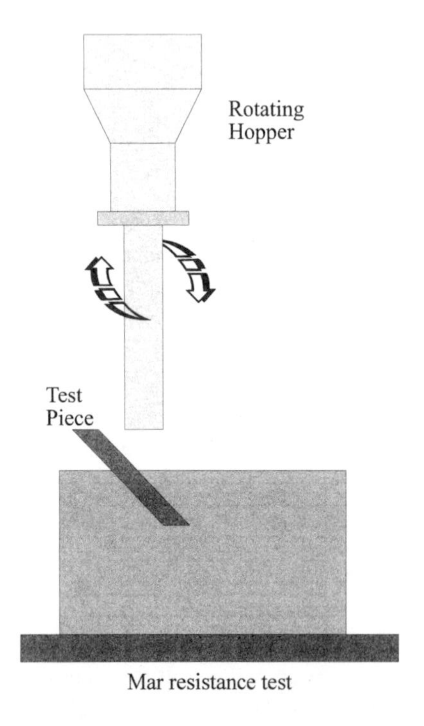

Figure 20 Mar resistance test.

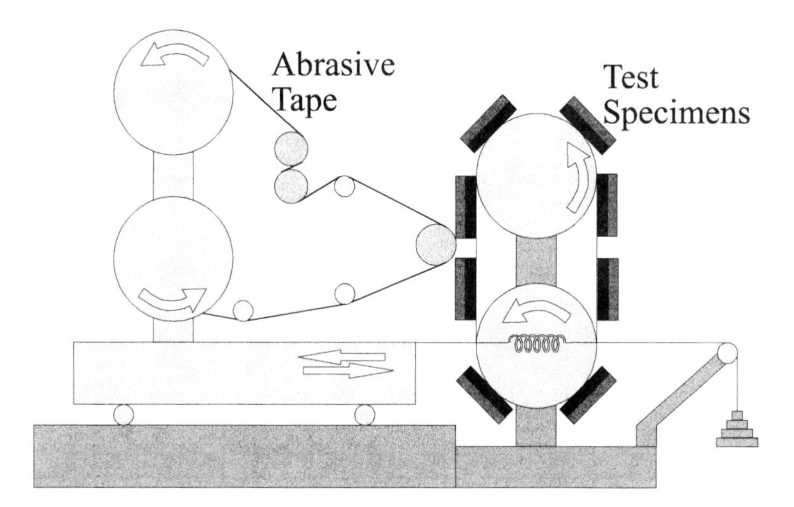

Figure 21 Bonded paper abrasion tester.

piece required, being about 100 mm square or round. When product testing, for instance, this requirement can sometimes make the Taber impossible to carry out.

Mention has already been made in Section 1 to the use of Moh's scale and pencil hardness, which are somewhat more concerned with mar resistance than they are with hardness as such, although harder materials tend to give better abrasion resistance, generally speaking.

3 Time-Dependent Properties

3.1 Creep

In a creep test, the load on the test piece is kept constant, and the dimensional changes that this brings about are monitored as a function of the time. Internationally, there are two standards available for the evaluation of creep properties of plastics: ISO 899-1 [108], which deals with tensile creep, and ISO 899-2 [109] (formerly ISO 6602), which deals with flexural creep. In both of these standards the initial stress, i.e., the stress based on the original (unloaded) dimensions of the test piece is assumed (as it is for the corresponding short-term tensile and flexural strength tests), and the strain is similarly defined as for the short-term strength tests, except, of course, that in this case, it is the strain as a function of time that is of paramount interest (Fig. 22a). The term "creep strain" is used to differentiate this type of strain from the short-term strain induced by classical tensile and flexural tests. The modulus is defined as the ratio of the initial stress to the creep strain and is referred to as the "creep modulus." Since the stress is constant and the strain increases with time, it follows that the creep modulus decreases as time increases. Other properties of interest when conducting creep tests are

The isochronous stress–strain curve, which is a cartesian plot of stress against creep strain at a specific time after the application of the load (Fig. 22b).

The time to rupture, which is the period of time that elapses from the point at which the test piece is fully loaded to the time it ruptures.

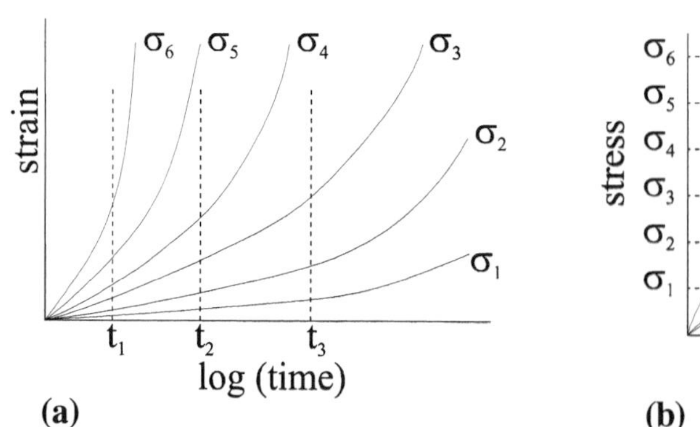

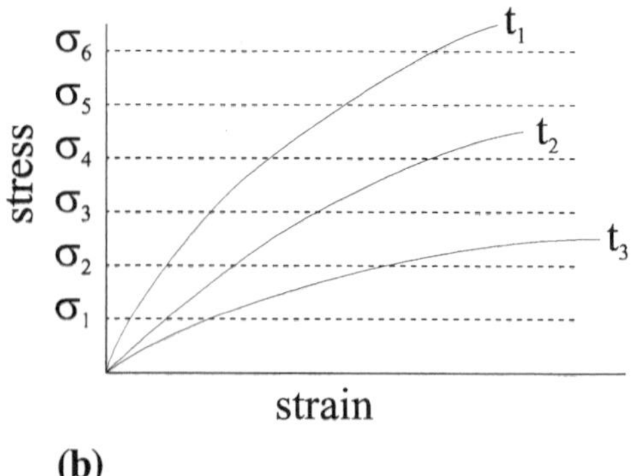

Figure 22 (a) Creep curves. (b) Isochronous stress–strain curves.

The creep-strength limit, which is the initial stress that will just cause rupture or a specified strain in a specified time (under the stated environmental conditions).

The recovery from creep, which is the decrease in strain at any given time after completely unloading the test piece. It is expressed as a percentage of the strain just prior to the removal of the load.

The tensile standard uses the tensile dumbbells standardized in ISO 527-2 and requires that the load applied be maintained within ±1% of the desired load. The loading mechanism must be designed both to apply the load smoothly and without overshoot and to do this consistently. In creep testing it is essential to avoid any shocks at any time during the test. The strain must be measured to an accuracy of 0.01 mm, and the elapsed time to 0.1%. Contactless optical extensometers are recommended, especially for creep rupture tests. The standard states that strain gauges are only suitable where they can be attached by adhesive, and only when the adhesion quality is constant throughout the test.

In the flexural test, the normal flexural properties (ISO 178) test piece is used, and the same requirements for the loading and timing device as for the tensile test apply. The accuracy of the deflection measuring device must be within 1% of the final deflection.

The standards note that the test results are dependent on the thermal history of the test pieces as well as the prevailing temperature and, for certain plastics, the relative humidity during conditioning and testing. Examples of the effect are shown in the appendices to the standards.

It is frequently convenient to take measurements at nominally equal logarithmic time intervals, and even more convenient if this can be done automatically. A suggested framework for this is given in the standards as in the table. In practice, however, where manual logging of data is undertaken, it is generally more convenient to work in whole numbers of days and weeks than in arbitrarily round numbers of hours.

1 min	3 min	6 min	12 min	30 min
1 hr	2 hr	5 hr		
10 hr	20 hr	50 hr		
100 hr	200 hr	500 hr	etc.	

Because creep testing is, by definition, a multipoint measurement, its results are frequently presented graphically. In its simplest form the creep strain is plotted as a function of time (or more generally $\log_{10}$ time), and since it is usual to perform the test at several levels of initial stress, the graph generally consists of several lines corresponding to each level of stress (Fig. 22a). From this basic plot the creep modulus versus time can be constructed, since for any given creep curve the ratio of the stress value to the creep strain at time t gives the creep modulus at time t. The isochronous stress–strain curves can be deduced by selecting any time(s) of interest and determining the creep strain for each level of stress applied, and then plotting the stress value against the corresponding creep strain at this time (Fig. 22b). At least three different stress levels and preferably more are required for the construction of the isochronous curves. The value of the isochronous curve is that it shows the equivalent stress–strain behavior of the material at a fixed instant in time, while the more normal tensile or flexural test shows the material's behavior with time as well as strain changing—some of the strain in the tensile/flexural test is due to creep induced by the previously applied stress. Of particular, although not exclusive, interest is the isochronous curve's ability to indicate the stress–strain properties at relatively long (days, weeks, or even years) time scales. These data are especially important to designers of plastics products and components.

If the test is carried out for long enough, the test piece eventually breaks and the time to creep rupture can be determined. Plotting stress against time to creep rupture allows the designer to estimate the probable longevity of his product, provided the data were derived at similar temperatures to service conditions. Here then is a further dimension to add to the experimental programme. The creep tests should be carried out not only at several stress levels but also at several temperatures (and for certain materials, several moisture contents) if a full understanding of the material response is to be gained. Clearly, this involves a considerable investment of time and resource, so it not surprising that extensive creep data are harder to come by than the "normal" tensile and flexural data.

These ISO standards have not been adopted into BS 2782, the UK having raised a number of technical objections down the years that have never been fully accepted. The British Standard that lays down the creep testing requirements for plastics is BS 4618, Section 1.1 [110], which covers not only tensile and flexural creep testing but also creep lateral contraction ratio (Poisson's ratio), which is not covered in the ISO standards. It does not, however, encompass creep rupture tests. BS 4618 is more a standard giving the philosophy and general methodology for undertaking creep testing than in laying down specific requirements. Unlike the ISO standards it provides a more detailed classification system for the strain measuring device. Category A extensometers have a minimum detectable increment of 0.002% strain, Category B have 0.008% and Category C, 0.02%. The size of a comprehensive evaluation for a plastic is made much clearer in this standard than in the ISO standards, which still tend to be modeled on the style of the routine "quality control" type of document like ISO 527 and 178.

Within ASTM the general test method is contained in D2990 [111], which allows for compressive creep as well as tensile and flexural creep. The details are in general agreement with the ISO and BS standards, but much more extensive data handling and extrapolation information, including time–temperature superposition methods, are presented in the appendix. Where creep rupture tests are being performed, it is recommended that stress levels giving failures after nominally 1, 10, 30, 100, 300, and 1000 hours be employed.

A creep rupture test specific to plastic pipes is given in ASTM D1598 [112], in which a length of pipe is pressurized by gas or liquid as appropriate and the time to failure is recorded. Pressure stability for up to a 100 hour test duration must be ±0.5%. Above this a ±1% stability is required. As for the other test methods, temperature control must be to ±2°C.

Over a limited range of stresses, the effect of stress level on creep rate is linear, but as the stress levels are raised this ceases to be true. The physical aging of the plastic during the creep test can also cause interference in the interpretation of data in the linear viscoelastic region. A paper by Read [113] discusses these issues and gives methods for analyzing creep curves in glassy polymers, while Yakushev et al. [114] describe the use of laser interferometry for measuring creep strains at low deformations. The determination of creep rupture data can be very time-consuming, and in a paper by Challa and Progelhof [115] the possibility of deriving creep rupture data from tensile creep data is discussed. While the standard test methods employ constant loading techniques, Brueller [116] notes that in practice loads are often applied through some kind of spring system so that the load decreases as the creep increases. His paper describes a numerical technique for quantifying this effect where the "normal" creep behavior for the material is known, and in a paper by Hornberger [117] the creep of actual plastics moldings are compared with predicted values based on a knowledge of the creep behavior of the plastic.

3.2 Relaxation

Stress relaxation is the other side of the viscoelastic coin in which the deformation is held constant and the stress (or force) is monitored over time. It is relatively little used in the characterization of plastics, cf. rubber, creep being considered a more useful parameter for plastics. Nevertheless, its occurrence can be of practical significance, as for example in the use of plastic fasteners, where generally a fixed deformation is applied when the fastening process is completed. The initial "tightness" of the fastening will decrease with time as the built-in stress decays. There are no longer any standard test methods for the determination of stress relaxation in plastics, ASTM D2991 having been discontinued.

Grzywinski and Woodford [118,119] report on the use of stress relaxation tests for the determination of design data for polycarbonates, in trying to determine long-term tensile and creep behavior from a 24 hour relaxation test, while Tsou et al. [120] examine the stress relaxation of several types of plastic film in bending and tension.

4 Physical Properties

Carbon black is an effective absorber of ultraviolet light and is thus often used at the level of a few percent in polyethylene compounds to provide UV protection. However, to be fully effective, the distribution of the carbon black must be uniform. BS 2782, Method 823A-B [121], provides two methods for assessing this: Method A is intended for use with compound only, while Method B is intended for use only with extrusions and mouldings. In Method A, a small sample of the material is squeezed between two microscope slides that have been heated to either 170° or 210°C, according to the specification requirements, so as to produce a film of about 20 to 30 micron thickness. This film is then examined under a ×100 magnification and its appearance compared with standard photomicrographs and a visual assessment made as to the degree of dispersion of the black. In Method B, a microtome section of the material, 10 to 20 micron in thickness, is taken and examined in the same way as for Method A.

When colored plastics materials are brought into intimate contact with other materials for any period of time, there can be a tendency for the colorant to "bleed" from the one material to the other. ISO 183 [122], and the dual numbered BS 2782, Method 542A [123] standard, provides a test method for assessing this tendency. The test material is placed in contact with an acceptor material under a constant pressure at a given temperature for a given time, 72 hours being the standard test period. The precise details are either to be agreed between the interested parties or found in the appropriate material/product specification. However for PVC, 50°C is the recommended temperature. Acceptor materials include white filter paper or a plasticized PVC of defined composition that is given in the standard. After exposure the acceptor material is examined against both a white and black background to see if any color bleeding has taken place.

5 Effect of Temperature

5.1 Thermal Expansion

There are currently no standards tests for measuring the thermal expansion of plastics in ISO or in BS 2782. In ASTM, test method D696 [124] uses a relatively thick test piece, which is placed in a chamber whose temperature can be controlled. This is shown schematically in Fig. 23. The expansion of the sample is transmitted to a remote dial gauge via a quartz rod that has a very low expansion coefficient. This same technique is applied in modern thermomechanical analyzers (TMA), but the dial gauge is replaced by linear displacement transducers or other electronic devices capable of detecting smaller dimensional changes. In turn this allows thinner specimens to be tested and permits wider temperature ranges to be examined. There are developments within the ISO to provide a standard for these types of instruments.

The problem of measuring the expansion of thin samples has been tackled by Tong et al. [125], who use both a capacitance change and a Fabry-Perot laser interferometric method to measure directly the change in thickness of polyimide films. Pottiger and Coburn [126] on the other hand make volume expansion measurements via pressure–

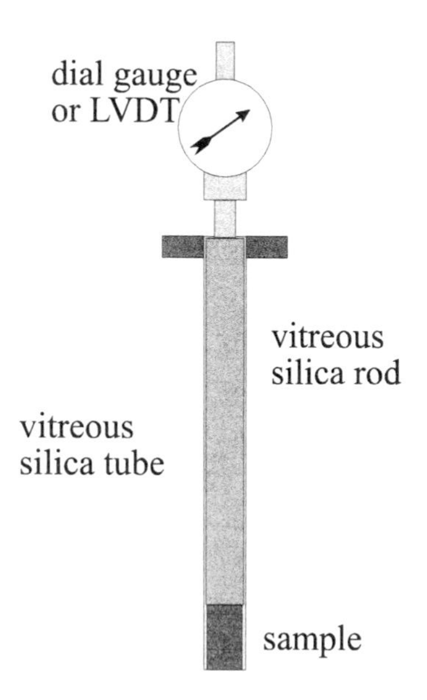

Figure 23 Thermal expansion apparatus.

volume–temperature (PVT) apparatus and then calculate the thickness change from the measured thermal expansions of the length and width directions of the film.

The measurement of volume expansion is generally performed using a liquid-in-glass dilatometer. A small sample of the plastic is placed in the dilatometer and covered with a known mass of the test liquid (often mercury). It is essential that the fluid have no effect on the plastic under test. The dilatometer is then placed in a temperature-controlled bath, and after reaching thermal equilibrium, the height of the fluid in the dilatometer stem is recorded. By repeating the test at several temperatures, and from a prior knowledge of the expansion of the fluid and the dilatometer, the expansion of the plastic may be deduced. It is a tedious technique, requiring great care in calibrating and operating the apparatus, but it is capable of very precise work if all the due precautions are taken. The method is no longer standardized in the ISO, the BS or the ASTM (method D864 having now been withdrawn).

5.2 Shrinkage

The dimensions of the test bar or plaque do not match, precisely, the mold from which it was made—shrinkage takes place, either through differential thermal expansion between the mold material and the polymer or through microstructural changes in the polymer after molding. ISO 2577 [127] and BS 2782, Method 640A [128], with which it is dual numbered, is a method for determining shrinkage in thermosetting materials. For compression molding, the mold cavity is 120 mm long by 15 mm wide by 10 mm deep, while for injection molding, a square of side 120 mm with a depth of 4 mm is used. The test material is molded as most suitable and the test pieces so formed stored in one of the standard atmospheres of ISO 291 (see Chapter 6) for between 16 and 72 hours. The dimensions of the test pieces are measured to the nearest 0.02 mm and the shrinkage then calculated as the difference between the mold dimension and the test piece dimension relative to the mold dimension. In some instances the post-shrinkage is required also, in which case the test pieces are placed in an oven set to 80°C if the material is a urea-formaldehyde molding compound or 110°C otherwise for a period of either 48 or 168 hours. The post-shrinkage is then the difference in the final dimension and the initial dimension of the test piece as molded, relative to the initial dimension.

ISO 11501 [129] (or the identical BS 2782, Method 643B [130]) is for the determination of shrinkage in film or sheeting when heated. The 120 mm square test piece, after measurement, is placed on a kaolin bed in a tray within an air-circulating oven set to the desired temperature for the material under test. An annex to the standard gives appropriate temperatures that have been found suitable for a range of materials and range from 70°C for plasticized PVC to 175°C for polypropylene. The test piece is left for preferred times of 5 minutes for nonshrink film and sheeting or 30 minutes for film and sheeting that is intended to be thermoshrunk or thermoformed. After removal from the oven the test pieces are conditioned for 30 minutes in the same atmosphere as used for the initial measurements and their measurements retaken and the shrinkage calculated for each of the two principal axes of the sample.

In ISO 3521 [131], dual numbered as BS 2782, Method 644A [132], the total volume shrinkage of polyester or epoxide casting resins is determined as the change in density that occurs from the moment of mixing to that after curing and conditioning at 23°C. Where the curing takes place at elevated temperature, the reactants are heated to the desired temperature, mixed, and then poured into a test tube. A sinker of known volume is suspended in the mixture and the change in buoyancy with time is monitored and extrapolated to zero time, which corresponds to the moment of mixing the components. The density of the mixture can thus be deduced. For resins that cure at ambient temperature, the densities of the components at this temperature are determined separately and that of the mixture calculated on an assumed additivity of volumes. After an appropriate post-curing operation and conditioning at ambient temperature, the density of the cast resin is determined by one of the methods outlined in Chapter 7.

The dimensional stability of plasticized PVC sheet is covered in BS 2782, Method 641A [133], and involves the measurement in the change in length, on heating for 15 minutes in water at 100°C, of a 250 mm long strip of material cut from the sheet. The test piece is required to be flat and straight during the measurement, and the standard recommends placing the strip in a 1.5 mm deep groove, which is slightly wider than the strip itself, alongside of which is a graduated rule marked in 0.5 mm divisions.

BS 2782, Method 643A [134], is a method for determining the shrinkage on heating of films to be used in shrink-wrap applications. A test piece of sufficient length to give a free length of 100 mm after looping around a wire frame and of width 10 mm is fully immersed in a fluid bath at the required temperature and left for 20 seconds. After this it is removed and immediately placed on a horizontal, absorbing, surface and its free length remeasured. Five test pieces are tested from each of the two principal directions of the film and the mean values reported separately. It is essential to place the films horizontally as soon as possible on removal from the fluid to avoid them stretching under their own weight. For PVC films the recommended temperature is 90°C, while for low density polyethylene it is 115°C.

ASTM D955 [135] follows a similar procedure to that of ISO 2577 and provides molding details for compression, injection, and transfer molding, the test piece in each case being in the form of a bar 127 mm by 12.7 mm by 12.7 mm. For injection molding, however, the preferred thickness is only 3.2 mm because of the difficulty in molding thick sections. The standard covers both thermoplastics and thermosets.

ASTM D2732 [136] is applied to films and sheeting not more than 0.76 mm thick. The test pieces consist of squares 100 mm on a side that are placed in holders that keep the test piece flat but allow intimate access of the immersion fluid to the full surface area of the sample. The test pieces are immersed in a suitable fluid at the agreed temperature for the material (determined by the interested parties or the material specification) for a period of 10 seconds and then removed and quickly plunged into a fluid bath at ambient temperature for a further 5 seconds for cooling. The final dimensional measurements must be made within 30 minutes of removal from the cooling bath.

In contrast to previous methods, where the changes in dimensions are recorded for test pieces that are unrestricted, ASTM D2838 [137] covers the measurement of the buildup of force in a specimen when it is restrained from shrinking on heating. It covers two procedures, both applicable to film or sheeting of less than 0.8 mm in thickness. Procedure A measures the maximum force exerted by the test piece, which is totally restrained as it is rapidly heated to a specific temperature, while Procedure B permits a controlled amount of shrinkage to take place before the specimen is restrained. The jig for the test consists of an L-shaped frame, to the top of which is attached a cantilever arm with strain gauges bonded close to the attachment. The other end of the cantilever is vertically above the base of the L, and a test piece can be fastened across these ends to form a bridge (Fig. 24). In Procedure A, the test piece is fastened so as to give the minimum achievable loading on the strain gauges, while in Procedure B marks are placed on the test piece at greater than the separation of the arms so that a known amount of "slack" in the test piece occurs. The jig is then placed smoothly and quickly into the fluid bath set to the desired temperature,

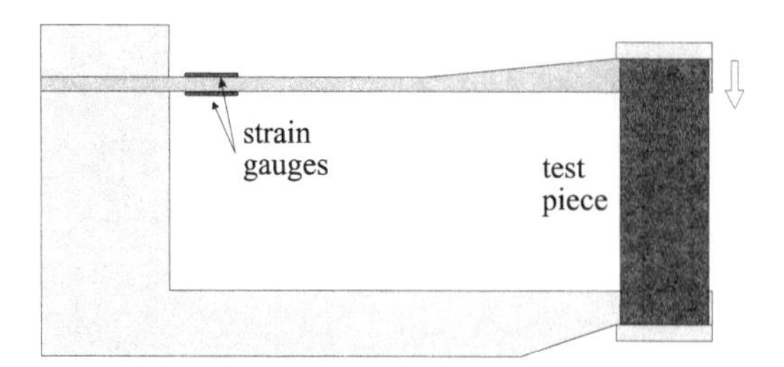

Figure 24 Film shrinkage test.

and the change in resistance of the strain gauges as the sample shrinks is recorded by means of a resistance bridge circuit. The plateau or peak value of force is recorded.

5.3 Melting and Softening Points

Bending Methods

A very popular test method for measuring the softening point of plastics is the "temperature of deflection under load" test, also known as the heat distortion or HDT test, standardized in ISO 75 [138]. The standard is now published in three parts, Part 1 covering general principles, Part 2 for plastics and ebonite, and Part 3 for reinforced plastics. Part 3 will not be considered further in this chapter.

The test consists of subjecting a bar-shaped test piece to a given bending stress while its temperature is increased at a uniform rate (Fig. 25). The temperature at which a certain deflection is reached is taken as the end point of the test. The test is essentially a flexural test, and so the equations for bending (Section 1) apply. The flexural stress to be applied is chosen from a list of three specified in Part 2: 1.80 MPa (Method A), 0.45 MPa (Method B), and 8.00 MPa (Method C), the last being a new addition to cater for thermoplastics of very high softening point. The high stress level enables a test result to be obtained using the traditional silicone fluids, since at 1.8 MPa the HDT temperature can be so high that the oil decomposes very rapidly.

Traditionally, the test has been performed edgewise—that is, with the load applied across the thickness, with bending taking place across the width—on a test piece of length between 110 and 130 mm, width between 9.8 and 15.0 mm, and thickness between 3.0 and 4.2 mm. The span for this geometry is 100 mm. With the move to try and have as many tests as possible performed using the standard multipurpose test specimen—of dimensions $80 \times 10 \times 4$ mm—a new variation has been introduced that allows the test to be performed flatwise on this test piece with a span of 64 mm, mirroring the normal flexural test. In this geometry the test piece is also more stable when the load is applied, although for the lowest stress level, achieving the low applied force can be something of a problem.

The test piece is placed in the air-circulating oven or temperature-controlled fluid bath and the load applied for five minutes, after which the dial gauge, or other deflection measuring device, is zeroed and the temperature raised at 120°C per hour. The temperature at which the standard deflection is reached is recorded. The standard deflection varies according to the size of the test piece in the bending direction, being between 0.32 and 0.36 mm for the flatwise test piece and between 0.21 and 0.33 mm for the edgewise test piece. Part 2 of the standard gives tables of standard deflections for the range of thickness

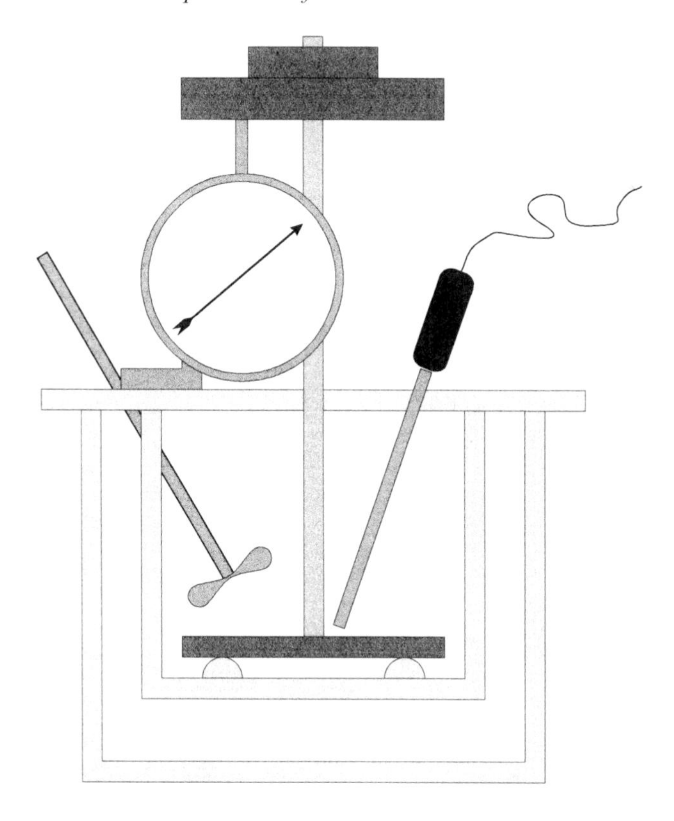

Figure 25 Heat distortion test.

(or width) allowed in the method, the standard deflections being equivalent to a strain of 0.2% in the surface of the test piece.

The heat distortion test is also covered in BS 2782, Method 121 [139], Methods 121A–C [140], and Method 121D [141]. Being identical to the corresponding parts of ISO 75, they are not considered further here. The corresponding ASTM standard is D648 [142], which is also technically equivalent to the ISO standard.

Penetration Methods

A qualitatively similar test, and one frequently encountered, is the Vicat Softening Temperature (VST), which is standardized in ISO 306 [143]. Instead of a bar, the test piece consists of a small plaque at least 10 mm square (or diameter) and of thickness between 3 and 6.5 mm, against which a weighted, circular, flat-ended pin of area 1 square millimeter is pressed. The test assembly is placed in a suitable fluid bath or oven, which is increased in temperature at a uniform rate (Fig. 26). The VST is the temperature at which the pin has penetrated by 1 mm. Loads of either 10 or 50 N may be used, and the heating rate may be either 50 or 120°C per hour. This gives four variations in the test, which are distinguished by the letter A for the 10 N load and B for the 50 N load, followed by the rate of heating. Thus B120 corresponds to the test conditions of 50 N and 120°C/h heating rate. There is no simple correlation between the various options and the VST value obtained, nor between VST and HDT on the same material, so it is important when trying to compare data that the same conditions be applied.

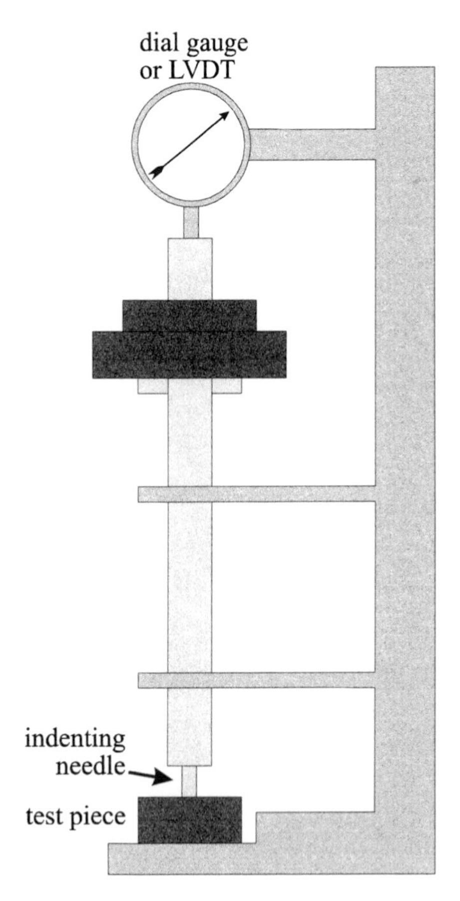

Figure 26 Vicat test.

Where thinner material than the required minimum thickness must be used, it is permissible to ply up the test piece, although the standard warns that different results can be observed in such cases.

The displacement measuring device, generally a dial gauge or a displacement transducer, is usually set up to read positive numbers as the pin penetrates the test piece, for ease of operation. It can be disconcerting, therefore, sometimes to find the reading becoming more negative rather than positive as the test progresses. The most probable cause of this is the state of molecular orientation in the test material. Highly oriented samples will relax and increase in thickness as the plastic is heated and may do so more rapidly than the pin is penetrating the sample, at least in the earlier stages of the test. Such annealing processes may significantly distort the VST value, and it may be necessary for the interested parties to agree on a preannealing procedure to be applied prior to the test being run. Monitoring the displacement of the pin as a function of temperature, preferably via logging equipment, can be a useful exercise in trying to resolve differences of opinion over the VST value of a given material. Similar annealing considerations apply also to the HDT test.

BS 2782, Methods 120A to E [144], excluding 120C, are identical to the corresponding methods in ISO 306 and need not be considered further. Method 120C [145] is peculiar to the UK and is applied to polymethyl methacrylate (PMMA) sheet. It follows the same methodology as the standard Vicat test under a load of 10 N and a heating rate of 50°C per hour, but the end point is the temperature at which the pin has penetrated by 0.10 mm. Stacking of thin sheet to make a minimum of 2.5 mm test piece thickness is permitted, and thinner sheet may be tested than is allowed by the more conventional test. In ASTM the equivalent test method is D1575 [146], which follows the standard ISO procedure.

A further type of softening point test is covered in BS 2782, Method 124A [147]. In this test a steel ball is allowed to penetrate a preformed disc of resin, located in a ring, whose temperature is steadily increased until penetration occurs (Fig. 27). A fixed quantity of freshly broken lumps of material is weighed into the ring, which is placed in an oven set to some 10 to 20°C higher than the softening point of the resin. Once the resin has melted, the ring is removed and placed on a metal surface, excess resin being removed by a hot knife. After cooling, a steel ball is placed centrally on the resin surface using a guide ring for location, and the whole is placed in a jig that is immersed in a heat exchange fluid, glycerol being often used. The fluid is heated at the rate of 5°C per minute, and the softening point is taken as the temperature at which the resin, or the ball, reaches the lower plate of the jig, which is some 25 mm below the upper plate on which the resin samples are placed. Despite the apparent crudeness of the test, it is capable of giving reproducibilities of the order of 1°C.

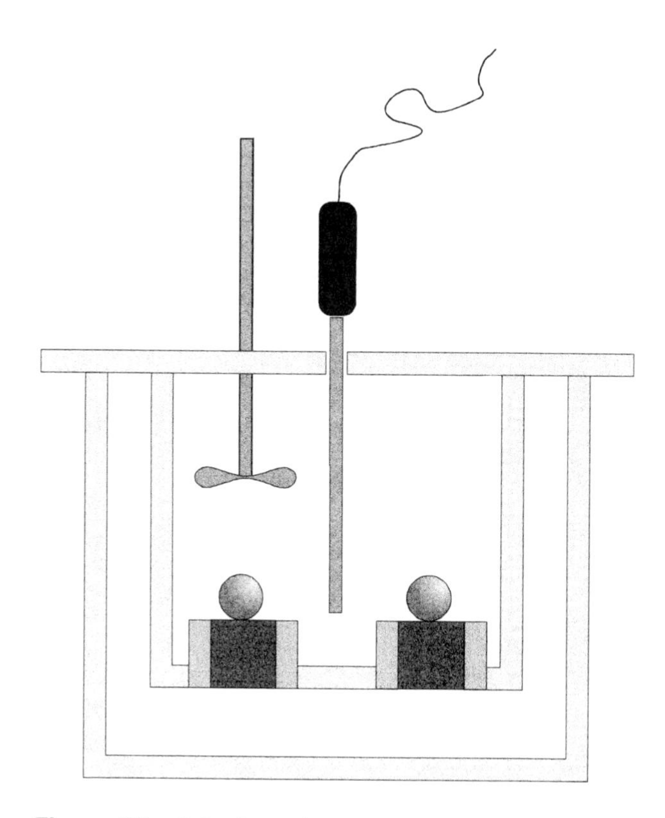

Figure 27 Softening point of resins.

Melting Point Methods

The HDT and VST tests may be applied to both amorphous and semicrystalline plastics. For the latter, however, a crystalline melting point can also be determined by using methods such as those detailed in ISO 1218 [148] or ISO 3146 [149]. ISO 1218 is specific to polyamides and incorporates two methods. In the first, a capillary tube containing a sliver of the test material, and a thermometer, are placed in a heated metal block with viewing and illumination ports at the side to enable the capillary tube to be inspected through a suitable lens. The block may be heated quite rapidly to start with as long as the rate of heating is reduced to 120°C per hour once the temperature has reached about 10° below the expected melting point. The temperature at which the sharp edges of the test piece disappear is taken as the melting point.

In the second method a small sample of the material is placed in the well of a hot stage containing a quantity of low-viscosity silicone oil (Fig. 28). The test piece is prepared by a special cutter of diameter 1.6 mm from a granule that will pass through a 0.80 mm sieve but is retained on a 0.63 mm sieve. The disk so formed is cut into quarters to form the test piece. A cover slip is placed over the test piece to form a wedge-shaped gap, and the oil is present in sufficient quantity to form a meniscus between the cover slip and the cell floor. The hot stage is heated in the same way as the previous method and, as the test piece begins to melt, the meniscus of the oil moves across the field of view of a low power lens

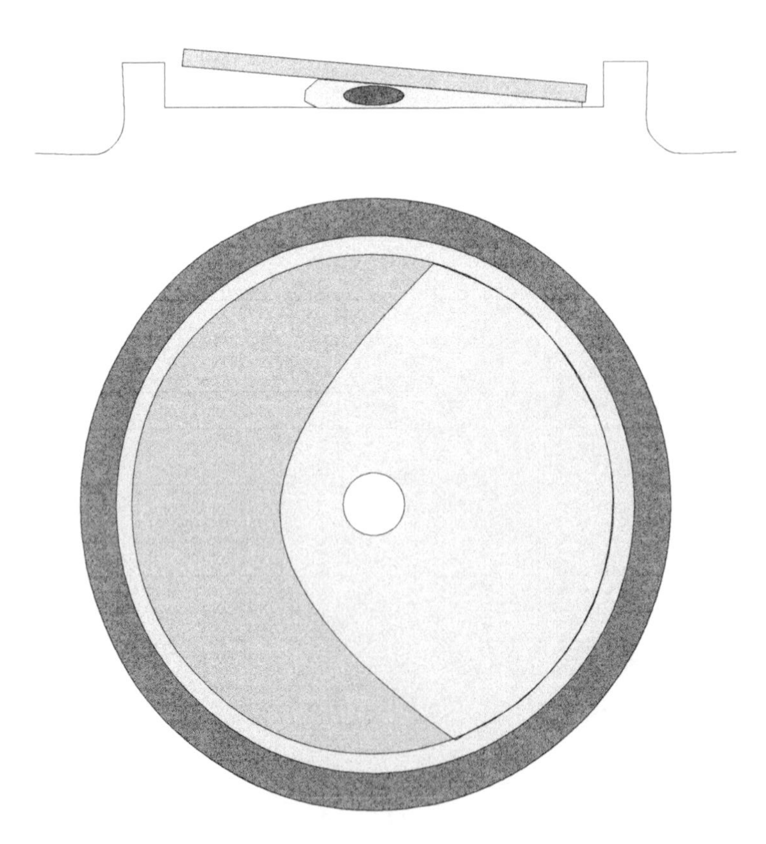

Figure 28 Melting point of nylon.

used to observe the progress of the test. This method only works for polyamides because of their relatively sharp melting points and low melt viscosities.

BS 2782, Methods 123A [150] and B [151] are technically similar to the capillary tube and optical methods of ISO 1218. In Method 123A, as well as the same type of capillary tube apparatus as that shown in the ISO standard, an electrically heated Thiele tube may be used for uncured thermosetting resins that do not have a sharp melting point. ASTM D789 [152] uses the same optical method as ISO 1218, Method B. ASTM D2116 [153] and D1457 [154] both use the differential scanning calorimetry (DSC) method for determining melting behavior. Thermal analysis is covered more fully in Chapter 24.

ISO 3146 is a more general standard consisting of four methods. Method A is the same as the capillary tube method of ISO 1218. Method B utilizes the fact that many crystalline polymers are birefringent. A hot stage microscope is used, the test sample being prepared from the material to hand by heating a small amount, typically a few milligrams, to slightly above its melting point and pressing it between the glass slide and a cover slip. This produces a thin film, which is allowed to cool slowly. The test piece so prepared is then heated at a suitable rate and viewed between crossed polars, the crystalline regions in the sample allowing some light to reach the eyepiece. As the melting temperature is reached, the rate of heating is reduced to between 1 and 2°C per minute, and the temperature at which a totally dark field is achieved is taken as the melting temperature. Method C, which is split into two parts according to the type of instrument used, employs thermal analysis techniques, either differential thermal analysis (DTA), given in Method C1, or DSC, given in Method C2. ISO 3146 is dual numbered in British Standards as BS 2782, Method 125 A–C2 [155].

5.4 Low Temperature Brittleness and Flexibility

Torsional Methods

The main torsional test method is the Clash and Berg test standardized in ISO 458, Parts 1 [156] and 2 [157], which are dual numbered in BS 2782 as Methods 153A [158] and B [159]. ASTM D1043 [160] is technically equivalent to Part 1.

The Clash and Berg test consists of taking a rectangular strip of material of uniform thickness and applying a known torque, through a weight and pulley arrangement, to one end, while the other end is kept fixed (Fig. 29). The angular deflection of the pulley is a measure of the stiffness of the material—the greater the deflection, the lower the stiffness. By immersing the test piece in a temperature-controlled bath, the effect of temperature on the stiffness can be determined.

The standard test piece is 70 ± 10 mm long by 6.0 to 6.4 mm wide. The thickness may vary from 1 to 5 mm. The heat transfer medium in which the apparatus is placed must be chosen to have no detrimental effect on the plastic under test, and several alcohols, acetone, silicone oils, etc. have been found to be suitable in most cases. Cooling of the medium may be by mechanical refrigeration or the addition of coolants such as liquid nitrogen or dry ice (solid carbon dioxide). Efficient stirring of the medium is essential to ensure uniform temperature distribution. The test piece is conditioned at the test temperature for three minutes before the torque is applied, and readings are made 5 seconds after applying the torque to ensure a consistent effect due to the creep of the test piece. It is normal practice to start at the lowest temperature of interest and increase this rather than the reverse.

Two methods are defined: Method A, which allows the angular deflection to vary between 10° and 100°, and Method B, which constrains the angle to between 50° and 60°.

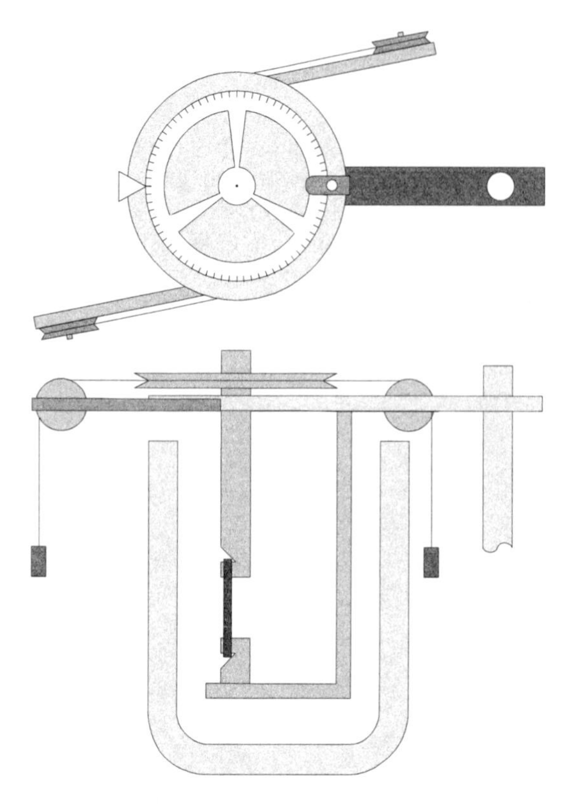

Figure 29 Clash and Berg apparatus.

The latter requires the torque to be constantly adjusted and so is more inconvenient to apply, but it keeps the strain in the test piece much more constant, and therefore the nonlinear stress–strain behavior of the material causes less interference in the estimation of the temperature effect.

The apparent torsional modulus of elasticity is given by the expression

$$T = \frac{917LM}{bd^3 \mu \theta} \tag{18}$$

where

L is the span (mm)
M is the applied torque (N · mm)
b is the width of the test piece (mm)
d is the thickness (mm)
μ is the dimensionless ratio given by the expression

$$\mu = 5.33 - 3.36 \frac{d}{b}\left(1 - \frac{d^4}{12b^4}\right) \tag{19}$$

θ is the angular deflection (°)
T is in units of MPa

In Part 2 of ISO 458, which is limited to vinyl chloride polymers, a modified version of Method A from Part 1 is applied. The angle is constrained to between 55° and 65°, and three test specimens are required: specimen 1 of between 1.8 and 2.0 mm thickness for the high stiffness region of 300 MPa; specimen 2 of between 2 and 5 mm thickness for the stiffness region of 23 MPa; and specimen 3 of between 4 and 5 mm thickness for the low stiffness region of 4 MPa. In all cases the specimen length is 60 ± 2 mm, the width is 6.3 ± 0.1 mm, the span is fixed at 40 mm, and the torque to be applied to each test piece is calculated for each of the torsional stiffness values of 300, 23, and 4 MPa. Tests are performed on each of the test pieces so as to give a valid result (i.e., deflection angles of 60 ± 5°) at each of at least three temperatures separated by 5°C. A graph is plotted of the logarithm of the stiffness against temperature, and the temperatures at which the torsional stiffness is 300, 23, and 4 MPa are recorded.

The ASTM method is almost identical to the general ISO test method, except that the allowed angular deflection can be in the range 5° to 100°, and the test piece is much more tightly controlled at a length of 63.5 ± 0.025 mm and width of 6.35 ± 0.025 mm. There appears to be no good reason for applying such tight tolerances to these dimensions. Thickness is allowed to vary between 1 and 3 mm.

A test very similar to the Clash and Berg is the Gehman test. This uses a torsion wire rather than weight and pulley arrangement to effect the torque. However, despite its greater simplicity, and the ease with which several test pieces can be tested in rapid succession, it has never caught on for plastics testing and is confined to the testing of rubbers. The interested reader is referred to Chapter 15 for further details.

Extension and Bending Methods

BS 2782 contains three methods that are specific to flexible vinyl compounds. In Method 150B [161], the Clash and Berg apparatus is used as noted in the previous section, but the result of the test is the temperature at which the standard test piece deflects by 200° under an applied torque of 0.057 N·m. The test piece is 65 mm long, with the span (the free length between the clamps) set to 38 ± 0.5 mm. The width can be between 6.2 and 6.4 mm and the thickness is 1.30 ± 0.08 mm. A graph of angular deflection against temperature is plotted, and the temperature at which the deflection equals 200° is read off. This value is then corrected by subtracting 0.5°C for each 0.025 mm of thickness of the test piece above 1.30 mm or adding 0.5°C for each 0.025 mm of thickness of the test piece below 1.30 mm.

In BS 2782, Method 150C [162], the amount of extension produced at a temperature of −5°C in a test piece placed under a tensile stress of 10.3 MPa is recorded. The test piece is 180 mm long (with a span of 127 mm) by 6.3 mm wide and of a thickness not in excess of 0.9 mm. The test piece is immersed in the heat exchange fluid at the set temperature for between 30 and 60 seconds according to thickness, and then the load on the upper clamp is increased uniformly at a rate equivalent to 20.7 ± 0.4 MPa per minute. The amount of extension at the point when the stress (based on the original cross-sectional area of the test piece) reaches 10.3 MPa is recorded.

Finally in this group there is BS 2782, Method 151A [163], which measures the lowest temperature at which none of a set of three test pieces cracks or fractures when wound round a specified mandrel (Fig. 30). The test is only carried out to the nearest 5°C. The test pieces are 100 mm long by 4.8 mm wide and are cut from a sheet of thickness 1.27 ± 0.08 mm. Three test pieces are placed in the guides of the instrument and the free end clamped to the mandrel. The guides are aligned at an angle of 68.5° to the axis of the mandrel. The test pieces and mandrel are immersed in the heat exchange fluid of industrial methylated spirit, which is cooled by either liquid nitrogen or dry ice. After 10

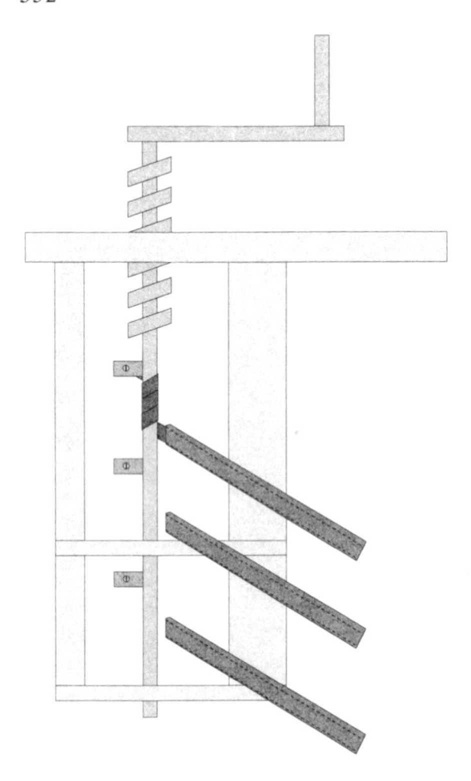

Figure 30 Cold bend test.

minutes' immersion at the test temperature, and while still immersed, the mandrel is rotated three full turns at the rate of 1 turn per second. The test pieces are then removed and inspected for signs of fracture or cracking. The test is repeated on fresh test pieces at 5°C increments until the lowest temperature at which no failures are observed in the three test pieces simultaneously tested. This is the cold bend temperature of the compound.

Impact Methods

ISO 974 [164] and ASTM D746 [165] both describe the determination of a brittleness temperature by means of an impact test, although there are some technical differences between them. In the ISO standard, the test piece is 20 mm long by 2.5 mm wide by 1.6 mm thick. These may be notched across the thickness in the middle of the length if desired. The test pieces are clamped so that an 8.5 mm length protrudes beyond the edge of the rear clamp face and are struck by a hammer traveling at 2 m/sec at a point 2 mm beyond this same point (Fig. 31). The rear clamp face has a cylindrical profile of radius 4 mm, so that, as the test piece is bent under the impact of the hammer, it bends around this radius. The clamped test piece is immersed in the heat exchange fluid, which is brought to the test temperature by the addition of a suitable coolant and left for 3 minutes to condition. After this time, the hammer is released, and following the impact the test piece is removed and examined; failure is defined as complete separation into two or more pieces.

At least four temperatures must be applied, and a minimum of 100 test pieces must be tested, with no fewer than 10 test pieces used at any one temperature. The temperatures are chosen so that percentage failures of between 10% and 90% inclusive are obtained. As

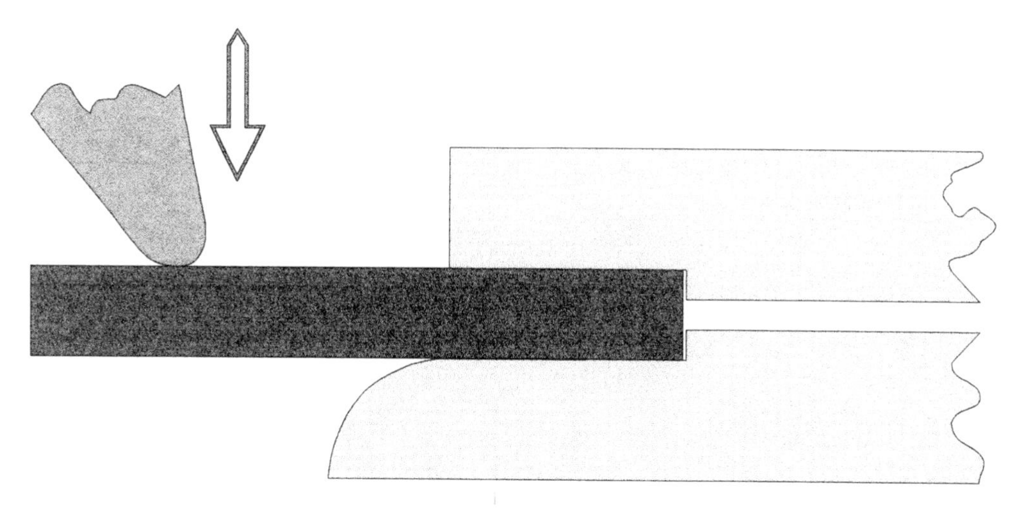

Figure 31 Impact brittleness test.

the fracture event is statistically determined, the temperature at which 50% of the test pieces can be expected to fail can be found by plotting percentage failure against test temperature on probability paper. The temperature at which the best fit straight line drawn through the experimental points meets the 50% probability line is the brittleness temperature. Alternatively, a calculated temperature can be found using the expression

$$T_{50} = T_{\mathrm{h}} + \Delta T\left(\frac{S}{100} - \frac{1}{2}\right) \tag{20}$$

where

T_{50} is the required brittleness temperature
T_{h} is the highest temperature at which all the test pieces failed
ΔT is the (uniform) temperature increment between successive tests
S is the sum of the percentage of failures at each temperature from the temperature corresponding to no failures down to and including T_{h}

It should be noted that these different methods of analysis may lead to slightly different results.

In ASTM D 746 two types of apparatus and three types of test piece may be used. The type B apparatus is as the ISO, while the type A apparatus uses a clamp without the radiused support on the rear clamp face. The hammer strikes the test piece 7.87 mm away from the clamped edge of the test piece, and the free length of the test piece is 25.4 mm. The type I test piece is used with the type A apparatus and consists of a rectangular strip 31.75 mm long by 6.35 mm wide by 1.91 mm thick, while the type II test piece is a modified T-50 dumbbell and is also used with the type A apparatus. The modified T-50 dumbbell has a rectangular tab end nominally 6.35 mm square with a strip 25.4 mm long and 2.54 mm wide protruding from the center of one of the 6.35 mm edges. The type III test piece is used only for the type B apparatus and matches the ISO test piece.

The test methodology and method of analysis is the same as for the ISO standard, although for routine testing against a specification for the material, it is permissible to test at the single specification temperature using 10 test pieces. Not more than 5 shall fail as a pass requirement. One additional point raised in the ASTM standard is the recommended use of a defined clamping torque to avoid excessive deformation of the sample.

For film and thin sheeting, a very simple impact test is given in BS 2782 150D [166], which uses a test machine called the Williams Cold Crack Apparatus. The apparatus consists of a spring-loaded plunger having a hammer at the free end which impacts against an anvil on which is placed the looped test piece (Fig. 32). The design of the apparatus is such that the impact velocity is 2 m/sec, the same as that used in the impact brittleness test. The test uses approximately 20 test pieces 32 mm long by 6.5 mm wide of thickness up to 1 mm. The upper thickness limit is determined by the inherent stiffness of the material being tested. The same 50% failure criterion is used as in ISO 974, a failure in this test being defined as a crack at the apex of the loop that extends over at least half of the width. Any cracks elsewhere are ignored. The looped test piece is placed in the heat exchange fluid at the test temperature and left for one minute to condition, after which the plunger is released and the test piece examined for failure. If the test piece did fail, the temperature is raised by 2°C, and if it passed, the temperature is lowered by 2°. At the end of the test, there must be at least four broken and four unbroken test pieces from the test sequence, and not less than 12 test pieces can be used.

ASTM D1790 [167] employs a similar approach to the Williams cold crack test except that the impact is effected by a pivoted metal arm 267 mm long falling from a vertical position onto the looped test piece placed on the anvil. The test piece is 146 mm long by 50.8 mm wide, the free ends being stapled together some 12.7 mm from the ends so that a loop is formed. The brittleness temperature defined in this test is that at which 80% of test pieces would be expected to fail with a probability of 95%. The analysis is performed in the same statistical way as the ISO 974/ASTM D746, except that the 80% failure line is

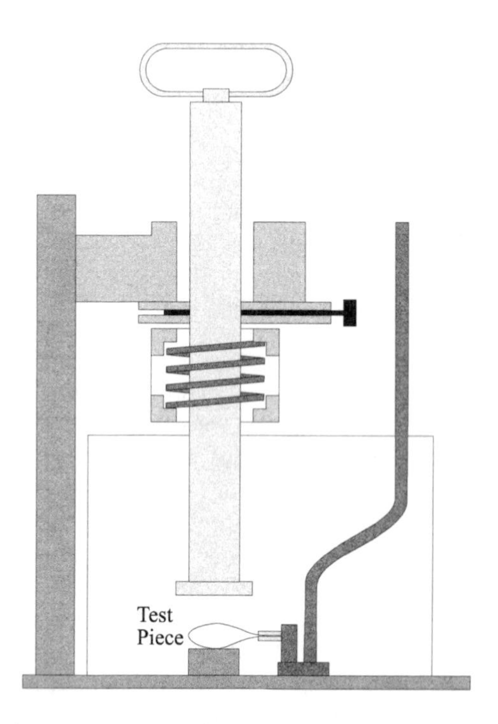

Figure 32 Williams cold crack apparatus.

used in place of the 50% failure line. Based on this type of analysis it is not clear what the significance of the 95% confidence limit is.

5.5 Heat Aging

General Methods

A general standard for assessing the effect of heat aging on properties of plastics is given in ISO 2578 [168], which is dual numbered in BS 2782 as Method 135 [169], and follows, albeit much more briefly, the standard given in the various parts of IEC 216, Part 1 of which [170] provides general guidance on procedures and evaluation. This standard is dual numbered as BS 5691. The IEC document is very comprehensive and a valuable source of advice on the subject. A number of characterizing parameters are used to quantify the aging behavior of the plastic, and the following terminology is defined.

Thermal endurance graph — (or Arrhenius graph) is the graph in which the logarithm of time to reach a specified end point in a thermal endurance test is plotted against the reciprocal thermodynamic (absolute) test.

Temperature index (TI) — the number corresponding to the temperature (°C) derived from the thermal endurance relationship at a given time. 20,000 hours is the normally chosen time.

Relative temperature index (RTI) — the temperature index of a test material obtained at the time that corresponds to the known temperature index of a reference material when both materials are subjected to the same aging and diagnostic procedures in a comparative test.

Halving interval (HIC) — number corresponding to the temperature interval (°C) that expresses the halving of the time to end point taken at the temperature of the TI or the RTI. It is a measure of the slope of the thermal endurance graph.

Threshold value — the value corresponding to a percentage of the initial value of the property under investigation at which the aging test is stopped and time to failure calculated.

Time to failure — the time required at the exposure temperature for a specimen either to fail the proof test or to reach the threshold value of the characteristic under investigation, whichever is the shorter.

In the test, a series of test pieces is aged at a variety of temperatures, and at suitable intervals a number are withdrawn, they are tested for whatever property has been selected to monitor the aging process, and the change in the property relative to the "unaged" value calculated. This change in property is plotted against the time for each of the test temperatures, and suitable curves are drawn through the points temperature (Fig. 33a). A horizontal line drawn at the threshold value through these curves determines the end point

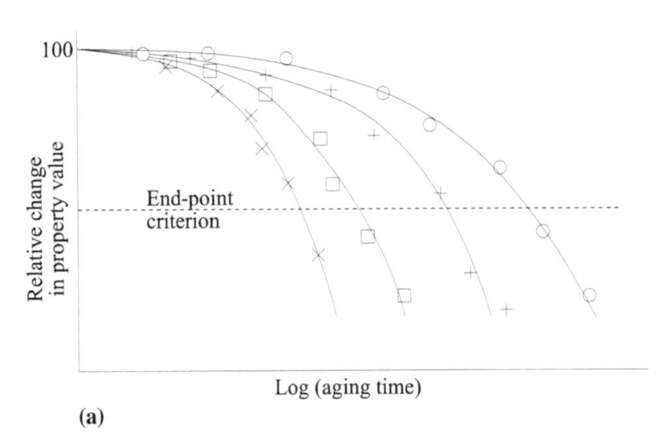

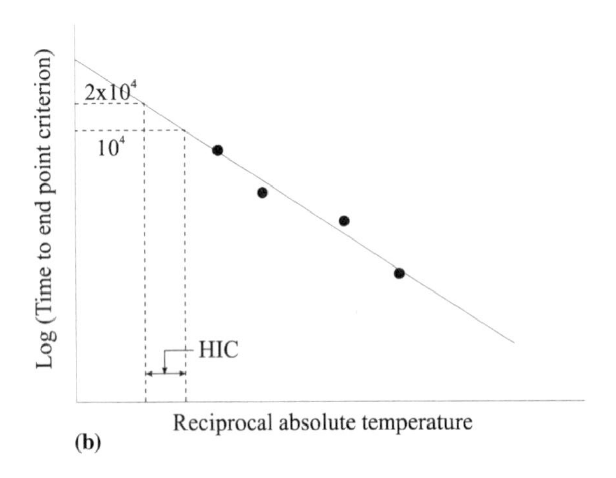

Figure 33 (a) Change of property on aging graphs. (b) Graphical determination of temperature index.

time for each temperature, and then the logarithm of this time is plotted against reciprocal absolute temperature (the Arrhenius plot) to enable the TI value to be derived temperature (Fig. 33b).

The standard recommends using at least three to four temperatures, but this really is the absolute minimum for a viable TI value to be calculated, and at each temperature there must be sufficient results to define clearly the change in property value down to and beyond the threshold limit so that this may be estimated by interpolation and not extrapolation. It is clear that the determination of TI is a time-consuming and expensive process. It must also be borne in mind that the calculated TI value may not equate with any precision to the true service temperature at which a product may be used. In practice it is likely that several properties may be influential in determining the viability of the product, and each of these will almost certainly have a different TI value. Nevertheless, the TI value remains a useful indicator for comparing the relative long-term effects of thermally induced aging in plastics.

ASTM D3045 [171] is similar in content, and D 794 [172] is also a general but extremely brief introduction to the subject.

ISO 11248 [173] follows a similar approach to evaluating the relative resistance of thermosets to exposure to high temperatures, although its title might suggest that the tests themselves were performed at elevated temperature. The characterizing tests are flexural properties (ISO 178), tensile properties (ISO 527), and compression properties (ISO 604); temperature increments of 25°C are employed. BS 2782, Method 136 [174] dual numbers this standard.

In BS 2782, Method 132B [175], the blistering propensity of thermosets is examined by heating a suitable molded test piece in a liquid bath at a rate of 120°C/hr. At a temperature some 10°C below the expected blister temperature, a test piece is removed and visually examined for blisters. The process is repeated on separate test pieces at every 5° increment in temperature until the examined test piece shows signs of blistering.

In the oxidation induction time method of BS 2782, Methods 134A and B [176], a DSC or DTA apparatus is used. A small sample, of the order of 15 mg, is held at the stated test temperature under a flow of oxygen. Once the antioxidant is consumed and the polymer begins to degrade, there will be either a temperature change relative to the reference cell (DTA apparatus) or a relative heat flux (DSC apparatus). The time taken for the onset of this reaction is taken as the induction time at the stated test temperature. In Method A, nitrogen is first used to purge the cell and polymer of all oxygen and to establish uniform test conditions before oxygen is introduced. In Method B, the oxygen is introduced at the start of the test.

ASTM D1870 [177] deals with general aging in which a selected property or properties are measured after aging for a time and at a temperature that are agreed upon by the interested parties or defined in the specification being used. Temperatures between 50° and 150°C are covered and air velocities within the range 100 to 250 m/sec; for any given test this must be maintained within the oven to ±10% of the nominal rate or better.

General Mechanical Tests

In BS 2782, Methods 131C and D [178], test pieces are heat aged and then crushed in a universal testing machine. For method C, test pieces are molded cylinders or cubes with the principal dimensions set to 10 mm; for method D, cubes of side 10 mm are cut from the sheet or molding. Duplicate test pieces are heated to 135°C for 17 hours, followed by a further 6 hours at 170°C, after which they are immersed in a fusible metal bath at 400°C for 30 minutes and then cooled to ambient temperature in a desiccator. The compression test is applied so as to give failure in 30±15 seconds.

General Tests Based on Weight Loss

ISO 176 [179] consists of two methods for quantitatively determining the loss in mass of a plastic in the presence of activated carbon when it is subjected to given temperature and time conditions. In Method A, the test material is in direct contact with the carbon. This method is particularly suited to materials that must be tested at low temperatures because they flow at higher temperatures. Method B, on the other hand, has the test material placed in a wire cage to prevent direct contact with the carbon. In both methods the standard test piece is a disk 50 mm diameter by 1 mm thick, although other dimensions are allowed where other factors make this necessary. For Method A, the test pieces are weighed and the thickness recorded before they are placed in a metal container containing a bed of 120 cm^3 of activated carbon. A further 120 cm^3 of carbon is then placed on top of the first test piece, and then the second test piece is placed on this. The process is repeated for the third test piece, and then activated carbon is added over this. Finally, the container is covered, placed in an oven at 70°C, and left for 24 hours. At the end of this time the test pieces are removed, and all traces of carbon are carefully brushed off before the test pieces are reweighed and the percentage weight loss calculated. In Method B, the process is very similar, except that the test pieces are placed in fine wire mesh cages, and a temperature of 100°C is used.

In ISO 177 [180], the mass loss that occurs when plasticizer migrates from the test substance to a reference absorbent is determined. The standard absorbents are the rubber defined in Annex B of ISO 4649 [181], polyethylene without additives, and polyvinyl acetate without plasticizer. The standard points out that where a particular application will bring the plastic under test into intimate contact with some other material, this material may be substituted as the backing medium. The test is performed in an analogous way to ISO 176, except that the stack now consists of alternating layers of backing medium and test sample. Onto the upper layer of backing material an aluminum foil is placed, and then a rubber sheet for evening out the applied pressure. Finally a glass plate is placed over the assembly and a 5 kg weight positioned on this before the assembly is placed in an oven set to 70°C and left for 24 hours. The test pieces and the backing disks are reweighed after conditioning, and the amount of migration from the test plastic into the backing material determined. If required the test can be continued for longer times to build up a knowledge of the progress of the migration with time.

ISO 176 is dual numbered as BS 2782, Method 465A and B [182], while ISO 177 is dual numbered as BS 2782, Method 465C [183]. ASTM D1203 [184] is very similar to ISO 176, although here the test piece is only 0.25 mm thick.

Tests for Specific Polymers

The thermal stability of polyvinyl chloride is the subject of a number of standards, reflecting the commercial importance of this material. When heated, PVC tends to degrade by loss of hydrogen chloride (dehydrochlorination) leading to discoloration as well as loss of mechanical properties. Various techniques for assessing this loss of HCl have been devised and are standardized in ISO 182-1 to -4 [185–188] which are dual numbered as BS 2782, Methods 130A to D [189–192]. In Part 1 a Congo red indicator paper is used, suspended in a closed test tube containing a sample of the compound. The test tube is placed in a fluid bath at the required temperature (either 180° or 200°C according to the type of PVC under test), and the time it takes the indicator paper to turn from red to blue is noted. The standard permits the use of universal indicator paper, for which the color change corresponding to a final pH of 3 is required. The standard points out that other substances may be evolved from the PVC that could interfere with the action of the HCl on the Congo red and that comparisons between compounds of very different composition should be made with caution. A similar situation applies to Part 2, which employs a pH meter to monitor the evolution of the HCl. The test material is placed in a dehydrochlorination cell, which is immersed in the heated fluid bath, and a steady stream of dried, carbon-dioxide-free nitrogen passed over it. The stream of nitrogen is bubbled through an absorbing solution of sodium chloride, which contains the electrodes of the pH meter, this being used to monitor continuously the increasing acidity of the solution. The time taken for the pH to change from an initial value of 6.0 (adjusted by the addition of sodium hydroxide) to 3.8 is taken as the stability time. Test temperature is as for Part 1. Parts 3 and 4 use the same apparatus as Part 2, but with a conductance meter or potentiometer replacing the pH meter respectively. In this case the absorbing liquid is either demineralized water alone (Part 3) or an aqueous solution of potassium nitrate (/sulfate), and the end point is taken when the conductivity of the water has changed by 50 μS/cm, or the concentration of chloride ions is 1.585×10^{-4} mol/L. International round-robin tests have shown that the three methods given in Parts 2 to 4 inclusive give comparable results.

ISO 305 [193] is applied to PVC and related compounds. It monitors the discoloration of PVC when heated either in an oil bath (Method A), which essentially excludes air, or in an air circulating oven (Method B). The standard points out that the two methods may not give stability times that correspond to each other because of this difference, and so the results from the methods may be compared to assess the effect that oxygen has. In Method A, the test piece consists of disks 14 mm diameter by 1 mm thick, and for Method B, squares of side 15 mm and 1 mm thick. The disks for Method A are placed into test tubes that contain an aluminum block, and a cylinder of aluminum is then placed over the test piece to sandwich it. The test tube assembly is placed in the oil bath set to the stated temperature (180°C if not otherwise agreed) and left for a suitable period of time. A series of test tubes is similarly immersed and each is removed at periods of time, cooled, and the test piece removed and clipped to a card along with its exposure time. The color of the sequence of test pieces is then measured against either an agreed color chart or a suitable photometer to determine the time taken for a measurable change in color to be observed. In the oven method a similar sequence is followed, except that the test pieces are typically placed in aluminum foil on a rack in the oven. It is important that the temperature in the oven be uniform so that an even color change will take place wherever in the oven the test pieces are placed, and the standard defines a checking procedure to ensure compliance with this requirement. As the standard points out, this technique, while being very simple, can be confounded by the initial color of the compound, and it may be necessary to apply one of the methods in ISO 182.

ASTM D2115 [194] is of similar content but is not technically equivalent.

ISO 1599 [195] monitors the viscosity loss in dilute solution of cellulose acetate. The solution viscosity is determined according to ISO 1157 [196] on the compound both before and after a prescribed compression molding cycle, and the loss in viscosity is then expressed as the ratio of the difference in viscosity before and after molding to the viscosity before molding, expressed as a percentage.

ISO 3671 [197] uses an oven set to 55°C to measure the weight loss of aminoplastic molding materials. A known weight of molding material, in a weighing bottle, is placed in the oven at the set temperature, and after 3 hours it is removed, placed in a desiccator to cool for at least an hour, and then reweighed. The percentage loss in mass relative to the initial mass is then calculated.

ISO 4577 [198] covers the thermal oxidative stability of polypropylene. Test pieces are mounted on a rotating drum located in a forced air circulating oven set to the preferred temperature value of 150°C, although where the effect of temperature on the stability is required, temperatures down to 100°C may be used (Fig. 34). The failure time in this standard is taken as the time in days for the test pieces to show localized crazing, crumbling, and/or discoloration.

BS 2782, Method 122A [199], measures the indentation of a weighted cylinder into a flexible PVC compound at a specified temperature and time. Test pieces 13 mm diameter by 1.27 mm thick are cut from sheet material and placed under a guided 3 mm diameter plunger having a load of 9.8 N applied to it. The initial thickness and final thickness after 24 hr at 70°C are measured with a device reading to 0.002 mm.

BS 2782, Method 131B [200], describes the determination of extensibility after heat aging of polyvinyl chloride sheet. This standard uses Method 150C, previously described,

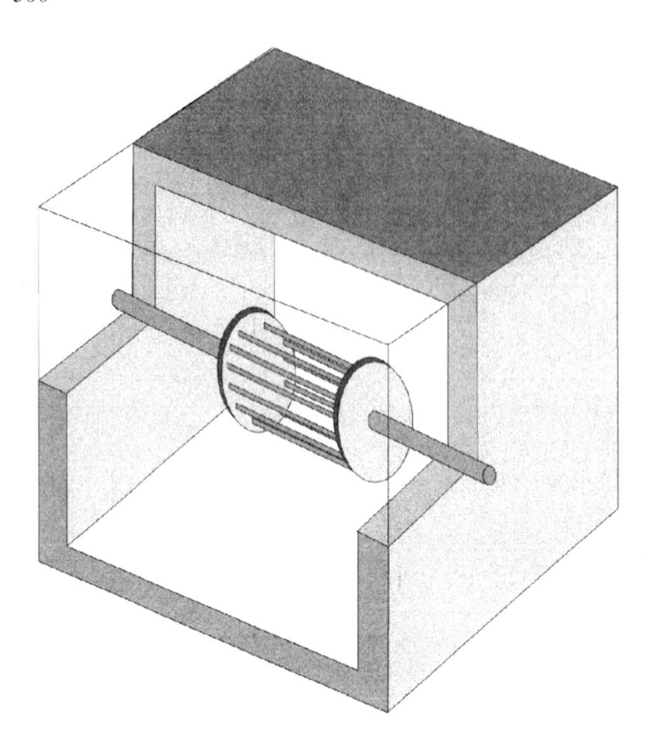

Figure 34 Oven aging of polypropylene.

but at room temperature and on material aged at 100°C for 24 hours to Method 465B (ISO 176).

ASTM D3012 [201] is used to assess the thermooxidative stability of polypropylene compounds using a biaxially rotating specimen holder. The test pieces are cut from compression molded plaques 1.25 mm thick and are 50 × 10 mm in size. Five test pieces are attached to the holder by metal clips lined with a fluoropolymer, and the test is conducted at 150°C unless otherwise agreed. The holder rotates through both a vertical and a horizontal axis in an oven having a high flow rate of air (1 m/s) to ensure a uniform aging of each test piece. Each day the test pieces are inspected for signs of failure, defined in this test as discoloration or crumbling.

ASTM D1457 [154] is a specification for PTFE-based compounds that contains a thermal stability index test based on the difference in specific gravity between test pieces before and after sintering under specified conditions. The principle of the test is said to be based on molecular weight change, but it is not clear how density measurements taken on two samples prepared by different techniques necessarily equate to loss of molecular mass on prolonged heating.

5.6 Physical Tests at Nonambient Temperature

When it is more appropriate to perform a mechanical or other test at a temperature other than ambient, then clearly this may be done even if the "standard" test specifies a test temperature of 23°C. Where nonambient temperatures are to be used, it is desirable to use one or more of the preferred temperatures given in ISO 3205 (see Chapter 6).

6 Environmental Resistance

6.1 Effect of Fluids

Environmental Stress Cracking

A plastic material that is resistant to a given chemical in an unstressed state may crack when exposed under stress to the same chemical. This stress may arise not only from externally applied forces but also from internal stresses due to molding operations. Most plastics suffer from this phenomenon of environmental stress cracking, and it is the cause of many failures in service. Several test methods have been developed to quantify this effect.

ISO 4599 [202] and its identical UK version, BS 2782, Method 832A [203], use a bent strip to generate the stressed state (Fig. 35). An indicative property, such as flexural strength or tensile elongation at break is chosen to measure the effect of the fluid, and an appropriate test piece is clamped to a former of known radius prior to being brought into contact with the test fluid at the required temperature. After the specified or agreed time has elapsed, the test piece is tested and the change in property compared to the "unaged" value determined. By using formers of differing radii, different strains can be generated and the minimum strain at which an agreed failure criterion is reached can be found.

The formers (and clamps) used must be completely resistant to the fluid being tested, and radii between 30 and 500 mm are found to give an adequate range of strains for most plastics over thicknesses from 2 to 4 mm. The preferred test temperatures are 23°, 40°, and 55°C, and the preferred durations are 22 to 24 hours for the short-term test and 1,000 hours for the long-term test, although other values may be used. A number of failure criteria are laid down in the standard to facilitate easy identification for specification purposes, and the table here illustrates these.

Designation	Indicative property	Criterion of failure
A1	State of surface (by visual examination)	Cracks or crazes around the extended edges
A2	State of surface (by visual examination)	Cracks or crazes in the extended surface
A3	State of surface (by visual examination)	Any other observations such as color change
B1	Tensile stress at rupture	80% of the value obtained on unprestrained, unexposed test pieces
B2	Flexural stress at maximum load	80% of the value obtained on unprestrained, unexposed test pieces
B3	Percentage elongation at break	50% of the value obtained on unprestrained, unexposed test pieces
B4	Charpy impact strength (unnotched)	50% of the value obtained on unprestrained, unexposed test pieces
B5	Tensile impact strength	50% of the value obtained on unprestrained, unexposed test pieces
B6	Any other agreed property	As agreed by interested parties

As well as measuring the strain to cause a given failure mode, the relative stress cracking factor may be determined, in which the ratio of the failure strain in the test fluid to that produced by a reference fluid is found.

In ISO 4600 [204] or its equivalent, BS 2782, Method 831A-B [205], a pin or a ball of known oversize is pressed into a reamed hole in the plastic, which is then brought into

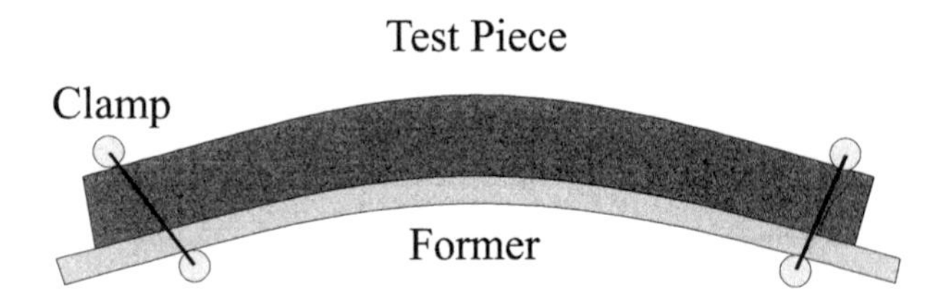

Figure 35 Stress cracking—bent strip test.

contact with the test fluid (Fig. 36). Several tests are performed using balls or pins of progressively increasing oversize. The failure limit is the smallest overstrain to produce failure in a given time and at a given temperature, the failure being defined by one of the following criteria:

1. Visible cracks observable to the naked eye (Method A)
2. A 5% reduction in maximum tensile force compared to zero overstrain (Method B1)
3. A 5% reduction in maximum flexural force compared to zero overstrain (Method B2)
4. A 20% reduction in tensile elongation at break compared to zero overstrain (Method B3)

The relative stress cracking factor may be determined as for ISO 4599.

Unlike the previous two standards, which apply a given strain to the test piece, ISO 6252 [206], which is dual numbered as BS 2782, Method 833A-C [207], uses a dumbbell-shaped test piece placed under constant tensile stress and immersed in the test medium until it breaks, time to failure being recorded (Fig. 37). In Method A the tensile stress leading to rupture in 100 hours is determined by interpolating points obtained at various stress levels that encompass the 100 hour point. In Method B the time to failure at a specified stress is found. In Method C the tensile stress versus time to failure is plotted graphically. The test piece normally used for this test is the general purpose plastics dumbbell of ISO 527, but with its dimensions scaled down by a factor of 2. The preferred test temperatures are 23° and 55°C, but other recommended temperatures include 40°, 70°, 85°, and 100°C.

Note may be made of BS 4618 Section 1.3.3 [208], which is a guidance document on the generation of data for design purposes.

ASTM D1693 [209] is a test specifically for polyethylene and is generally known as the Bell Telephone Test. It consists of bending a strip of plastic into which has been cut a

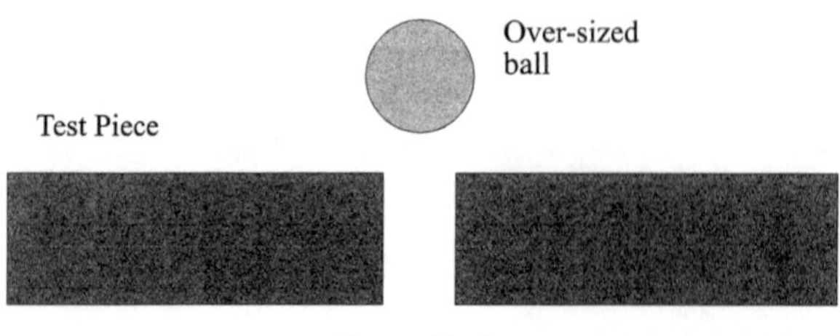

Figure 36 Stress cracking—pin or ball method.

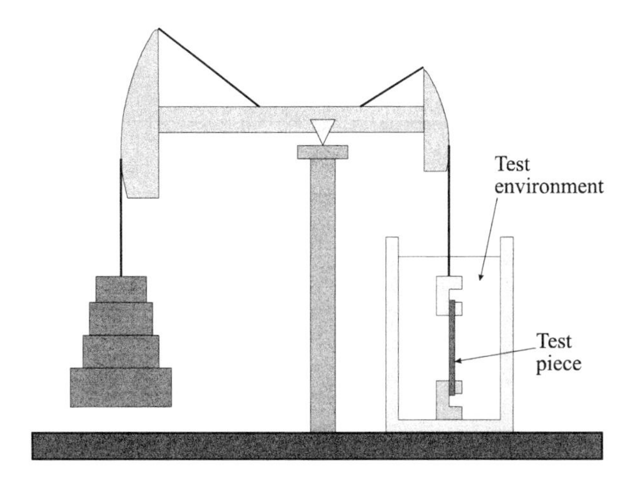

Figure 37 Stress cracking—constant tensile stress.

notch of given depth and immersing the strip in the test fluid at a given temperature and for a given time, or until the time to failure (Fig. 38). The test piece is examined to see if cracks have developed from the notch; extensions to the notch are not deemed to be failures, since true stress cracks normally run perpendicular to the notch.

Although the test uses very simple apparatus, there are a number of practical aspects that require careful control if reproducible results are to be obtained. The test piece is of a precisely controlled width (13.00 ± 0.08 mm) and must be prepared with sharp cutters that give square edges—bevelled edges are to be avoided. The notch must be made to a controlled depth with a sharp razor blade of specified length, which is effected through the use of a notching jig, and the test piece is then bent in a controlled manner and transferred to the specimen holder to maintain the controlled bend throughout the test. Again a jig is used to effect the bending and the transfer to the holder, which is simply a "U" channel, typically of brass. Different notch depths are used according to the thickness of the test piece.

Two test temperatures are specified, 50°C for Conditions A and B and 100°C for Condition C (Conditions A and B refer to test pieces of thickness 3.0 to 3.3 and 1.75 to 2.00 respectively), with the test fluid being nonylphenoxy poly(ethyleneoxy)ethanol, which has the trade name Igepal CO-630. This is used either as a 10% solution in water for Conditions A and B or undiluted for Condition C.

Water and Moist Heat

The standard test for assessing water absorption of plastics is given in ISO 62 [210], which defines four methods of test. It is applicable to a wide range of plastics in various forms but is not suitable for cellular plastics.

In Method 1, triplicate test pieces are dried in an oven at 50°C for 24 hours and then weighed after cooling in a desiccator before being placed in a container containing distilled water and left for a further 24 hours at 23°C. The tolerance on the temperature is either the normal ±2°C or the much tighter ±0.5°C, according to the material/product specification. After immersion, the test pieces are removed, dried with a cloth or filter paper, and then reweighed. It is normal to express the result simply as the weight change in milligrams,

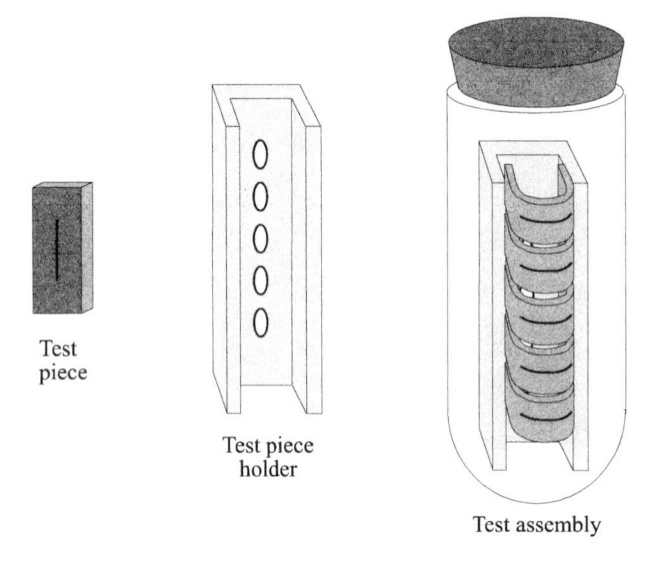

Figure 38 Stress cracking—Bell telephone test.

although sometimes percentage mass change or mass change per unit surface area are found.

Method 2 follows the same procedure as Method 1, except that a final step is added whereby the test pieces are returned to the oven at 50°C for a further 24 hours to redry them; they are weighed a third time after being cooled to ambient temperature. This method enables a quantitative estimate to be made of any water-soluble matter present in the plastic.

Methods 3 and 4 mirror Methods 1 and 2 except that the test temperature is 100°C (boiling distilled water) rather than 23°C.

BS 2782, Method 430A-D [211], is identical to ISO 62, while ASTM D570 [212] is technically similar, except that seven test conditions are specified:

1. 24 hr at 23°C
2. 2 hr at 23°C
3. 2 hr at 23°C followed by a further 22 hr at 23°C
4. Repeated weighing until an equilibrium weight is attained
5. 2 hr in boiling water
6. ½ hr in boiling water
7. 48 hr at 50°C

ISO 4611 [213] (which is identical to BS 2782, Method 551A [214]) deals with a broader range of water effects and covers three specific areas: damp heat, water spray, and salt mist.

In the damp heat test, the test pieces are subjected either to static temperature and humidity conditions of 40°C at 93% relative humidity or to 12 + 12 hour cycling between 23°C at >95% relative humidity and 40°C (or 55°C) at 93% relative humidity. For the water spray test a mist of distilled water at 40°C is sprayed over the surface of the plastic at the rate of 125 to 250 ml per hour per square meter of surface, while for the salt spray test the distilled water is replaced by salt solution at 35°C at a concentration of 50 gram per liter.

The effect of the exposure is measured by means of the mass change, changes in dimensions and/or appearance, or some change in a physical property such as a strength or modulus or an optical or electrical characteristic.

In ASTM D756 [215] there are seven environmental procedures that may be applied, the effects of which are limited to the measurement of mass or dimensional changes. In Procedure A a test cycle of 24 hours at 60°C at 88% relative humidity followed by 24 hours at 60°C dry heat is used. In Procedure B only dry heat at 60°C for 72 hours is applied. Procedure C specifies a test cycle of 24 hours at 70°C at 70% relative humidity followed by 24 hours at 70°C dry heat. Procedure D uses a test cycle of 24 hours at 80°C over water followed by 24 hours at 80°C dry heat. Procedure E is more complex, with 24 hours at 80°C at 70% relative humidity followed by 24 hours at either −40°C or −57°C (as required in the material/product specification), then 24 hours of dry heat at 80°C followed by a further 24 hours at either −40°C or −57°C. In Procedure F the test cycle is 24 hours at 38°C at 100% relative humidity followed by 24 hours of dry heat at 60°C. Finally in Procedure G a test cycle of 24 hours at 49°C at 100% relative humidity followed by 24 hours at 49°C dry heat is used.

Other Fluids

ISO 175 [216] provides a methodology for carrying out any general fluid immersion test on a plastic material. The properties that can be used to monitor the effect include mass, appearance, dimensions, or some physical property—mechanical, optical, thermal, electrical, etc. The property or properties should be chosen to reflect, where possible, the use to which the plastic is to be put, and, as with most tests on plastics, comparisons should only be made where test pieces have the same dimensions and geometry and where the same state of internal stress applies. The preferred test temperatures are 23°C and 70°C, but where other temperatures are more appropriate they may be used, ideally the selection being made from the list of preferred test temperatures given in ISO 3205. The standard test durations are 24 hours for the short duration test, 7 days for the standard duration test, and 16 weeks for the long duration test. However, where it is desired to plot the changing value of the selected property (properties) as a function of time, a standard scale of times is provided as follows:

1 - 2 - 4 - 8 - 16 - 24 - 48 - 96 - 168 hours
2 - 4 - 8 - 16 - 26 - 52 - 78 weeks
1.5 - 2 - 3 - 4 - 5 years

Where a graphical representation is made, it is frequently found to be convenient to use either a logarithmic or a square root time scale rather than a linear time scale.

An annex to the standard provides a list of chemicals that might be used in a general evaluation, but the use of chemicals known to come into contact with the product, such as brake fluids, special oils, and proprietary deicing liquids, should always take precedence over general fluids.

BS 2782, Method 830A [217], is identical to ISO 175, while ASTM D543 [218], Practice A, is very similar. (Note that Practice B in D543 is rather similar to ISO 4599.)

Mention can also be made of BS 4618, Section 4.1 [219], which is a general procedure similar to ISO 175, and to BSs 1763 [220] and 2739 [221], in which their respective Appendixes D give a procedure for assessing the tendency of the compound to stain in the presence of sulfur-containing chemicals. ASTM D1712 [222] gives a more general sulfur-staining test that is not confined to PVC, while ASTM D2151 [223] considers the effect on PVC of chemicals used in rubber formulations. This is done by pressing a standard rubber against the PVC under a weight at elevated temperature and then examining the plastic for signs of staining.

6.2 Effect of Gases

In comparison with the effects of liquids on plastics, very little testing is carried out on the effect of gases other than air aging tests, which were covered in the previous section of this chapter. Ozone, which is of particular concern to many elastomers (see Chapter 15), has no special effect on most plastics, beyond the normal oxidative degradation processes, as these are fully saturated in the main. The main concern of the effect of gases is through their permeability, a subject dealt with in Chapter 30.

6.3 Biological Effects

Most plastics materials are not seriously affected by microorganisms, so that they are often regarded as being rather "eco-unfriendly," although the use of certain types of additives, plasticizers, for example, may render a compound more susceptible to attack than are the base polymers alone. ISO 846 [224] gives a methodology for the use of five strains of fungi using one of two methods. In method A, the plastic is exposed to fungal spores in the presence of an incomplete nutritive medium, so that the fungi can only grow at the expense of ingredients in the plastic. This test is therefore sometimes referred to as a "test for growth." In method B, a complete nutritive medium is supplied so that the spores can grow whether the plastic supplies nutrients or not. The secreted metabolic products from the fungi are deposited on the plastic surface and may then attack it. On the other hand, the plastic may inhibit the growth of the fungi, in which case the test is referred to as a "test for fungistatic effect."

The effect of the fungi may be assessed by purely visual means, or by means of the quantitative loss of some property like mass, tensile strength, flexural modulus, etc.

Because of the pathogenic hazards associated with this type of test, the standard makes plain that the test must be performed only by staff competent in the handling of microorganisms.

Because of the natural resistance of most polymers to biological degradation, there is an increasing interest in preparing polymers with enhanced biodegradability. We need test methods that can evaluate the biodegradability of plastics in quantitative terms. This subject is beginning to be addressed at the international level, and there are three documents being prepared within ISO/TC61 that seek to achieve this. At the time of writing these were being circulated for their first public comments and as such may well undergo considerable revision before they are published. For information these are ISO/CD 14851 [225], ISO/CD 14852 [226], and ISO/CD 14855 [227].

BS 4618, Section 4.5 [228], provides guidance in the exposure of plastics to soil burial, but it is only a general guidance document and does not lay down precise procedures for evaluation programmes. ISO 846 is referenced in this standard, and a useful bibliography is provided. Within ASTM, G21 [229] and G22 [230] cover the effects of fungi and bacteria respectively.

References

1. ISO 868, Plastics and ebonite—Determination of indentation hardness by means of a durometer (Shore hardness), 1985.
2. ISO 48, Rubber, vulcanized or thermoplastic—Determination of hardness (hardness between 10 IRHD and 100 IRHD), 1994.
3. BS 2782, Method 365B, Determination of indentation hardness by means of a durometer (Shore hardness), 1992.
4. ASTM D2240, Rubber property—Durometer hardness, 1995.
5. ISO 2039-1, Plastics—Determination of hardness—Part 1: Ball indentation hardness, 1993.
6. BS 2782 Method 365D, Determination of hardness by the ball indentation method, 1991.
7. ISO 2039-2, Plastics—Determination of hardness—Part 2: Rockwell hardness, 1987.
8. Fett T., *Material Prüfung*, *14*(5), 151 (1972) .
9. BS 2782, Method 365C, Determination of Rockwell hardness, 1992.
10. ASTM D785, Rockwell hardness of plastics and electrical insulating materials, 1993.
11. BS 2782, Method 365A, Determination of softness number of flexible plastics materials, 1976.
12. ISO 6506, Metallic materials—Hardness test—Brinell test, 1981.
13. BS EN 10003-1, Metallic materials—Brinell hardness test, Part 1 Test method, 1995.
14. ASTM E10, Test method for Brinell hardness of metallic materials, 1993.
15. ISO 6507, Metallic materials—Hardness test—Vickers test, Part 1 (1982) HV 5 to HV 100, Part 2 (1983) HV 0.2 to less than HV 5, Part 3 (1989) less than HV 0.2.
16. BS 427, Method for Vickers hardness test and for verification of Vickers hardness testing machines, 1990.
17. ASTM E92, Test method for Vickers hardness of metallic materials, 1982.
18. ISO 4516, Metallic related coatings—Vickers and Knoop microhardness tests, 1980.
19. Lopez. J., *Polymer Testing*, *12*, 437–458 (1993).
20. Magdanz, H., *European Coatings Journal*, *3*, 142–146 (March 1996).
21. Guevin, P. R., *J. Coatings Technol.*, *67*, No. 840, 61–65 (1995).
22. *Materials World*, *2*(2), 105–111 (1994).
23. ISO 527-1, Plastics—Determination of tensile properties—Part 1: General principles, 1993.
24. ISO 527-2, Plastics—Determination of tensile properties—Part 2: Test conditions for moulding and extrusion plastics, 1993.
25. ISO 3167, Plastics—Multipurpose test specimens, 1993.
26. ISO 527-3, Plastics—Determination of tensile properties—Part 3: Test conditions for films and sheets, 1995.
27. BS 2782, Method 321, Determination tensile properties, general principles, 1994.
28. BS 2782, Method 322, Determination tensile properties, test conditions for moulding and extrusion plastics, 1994.
29. BS 2782, Method 326E, Determination tensile properties, test conditions for films and sheets, 1996.
30. BS 2782, Method 327A, Determination tensile strength and elongation at break of polytetrafluoroethylene (PTFE) and products, 1993.
31. ASTM D638, Tensile properties of plastics, 1995.
32. ASTM D882, Tensile properties of thin plastic sheeting, 1995.
33. ASTM D1708, Tensile properties of plastics by use of micro-tensile specimens, 1995.
34. Godard, P., and Bomal, Y., Filplas 92 (19th–20th May), Conference Proceedings, Manchester, Paper 7. 51, 1992.
35. Canova, L. A., Antec '93 (9th–13th May). Conference Proceedings, New Orleans, LA, Vol. II, 1993, pp. 1943–1949. 012.
36. Skelhorn, D. A., Antec '93 (9th–13th May). Conference Proceedings, New Orleans, LA, Vol. II, 1993, pp. 1965–1970. 012.
37. Meddad, A., Fellahi, S., Pinard, M., and Fisa, B., Antec '94 (1st–5th May). Conference Proceedings, San Francisco, CA, Vol. II, 1994, pp. 2284–2288. 012.

38. Turek, D. E., *Polym. Engng. Sci.*, *33*, No. 6, 328–333 (1993).
39. Carraher, C. E., *Polym. News*, *19*, No. 4, 112–113 (1994).
40. Boyce, M. C., Arruda, E. M., and Jayachandran, R., *Polym. Engng. Sci.*, *34*, No. 9, 716–725 (1994).
41. Rackowitz, D. R., *Plast. Des. Forum*, *19*, No. 4, 28–31 (1994).
42. Kennedy, M. A., Peacock, A. J., and Mandelkern, L., *Macromolecules*, *27*, No. 19, 5297–5310 (1994).
43. Shick, K. S., Antec '94 (1st–5th May), Conference Proceedings, San Francisco, CA, Vol. II, 1994, pp. 1895–1902. 012.
44. ISO 604, Plastics—Determination of compressive properties, 1993.
45. BS 2782, Method 345A, Determination of compressive properties, 1993.
46. ASTM D695, Compressive properties of rigid plastics, 1991.
47. ISO 178, Plastics—Determination of flexural properties, 1993.
48. BS 2782, Method 335A, Determination of flexural properties, 1993.
49. ASTM D790, Flexural properties of unreinforced and reinforced plastics and electrical insulating materials, 1995.
50. ASTM D747, Apparent bending modulus of plastics by means of a cantilever beam, 1993.
51. Shi, G. Z. H., and Nedea, C., Antec '94 (1st–5th May). Conference Proceedings, San Francisco, CA, Vol. II, 1994, pp. 2488–2493. 012.
52. BS 2782, Methods 340A and B, Determination of shear strength of moulding material (340A), Determination of shear strength of sheet material (340B), 1978.
53. ASTM D732, Shear strength of plastics by punch tool, 1993.
54. Jawad, S. A., *Polym. Int.*, *33*, No. 4, 373–376 (1994).
55. Hedner G., Selden, R., and Lagercrantz, P., *Polym. Engng. Sci.*, *34*, No. 6, 513–518, (1994).
56. ISO 6383-1, Plastics—Film and sheeting— Determination of tear resistance—Part 1: Trouser tear method, 1983.
57. ISO 6383-2, Plastics—Film and sheeting— Determination of tear resistance—Part 2: Elmendorf method, 1983.
58. BS 2782, Method 360B, Determination of tear resistance of plastics film and sheeting by the trouser tear method, 1991.
59. BS 2782, Method 360A, Determination of tear resistance of plastics film and sheeting by the Elmendorf method, 1991.
60. BS 2782, Method 360C, Determination of tear resistance of plastics film and sheeting by the initiation method, 1991.
61. ASTM D1004, Initial tear resistance of plastic film and sheeting, 1994.
62. ASTM D1922, Propagation tear resistance of plastic film and thin sheeting by pendulum method, 1994.
63. ASTM D1938, Tear-propagation resistance of plastic film and thin sheeting by a single tear method, 1994.
64. ASTM D2582, Puncture-propagation tear resistance of plastic film and thin sheeting, 1993.
65. ISO 179, Determination of Charpy impact strength, 1993.
66. BS 2782, Method 359, Determination of Charpy impact strength, 1993.
67. ASTM D256, Determining the pendulum impact resistance of notched specimens of plastics, 1993.
68. Nakamura Y., Yamaguchi, M., and Okubo, M., *Polym. Engng. Sci.*, *33*, No. 5, 279–284 (1993).
69. Wang, K. J., Sue, H. J., and Berg, J. W., Antec '94 (1st–5th May), Conference Proceedings, San Francisco, CA, Vol. II, 1994, pp. 1808–1812. 012.
70. Jancar, J., and DiBenedetto, A. T., Antec '94 (1st–5th May), Conference Proceedings, San Francisco, CA, Vol. II, 1994, pp.1710–1712. 012.
71. Trantina, G. G., and Oehler P. R. (1994). Antec '94 (1st–5th May). Conference Proceedings, San Francisco, CA, Vol. III, 1994, pp. 3106–3110. 012.
72. Sharpe, W. N., and Boehme, W., *J. Test. Eval.*, *22*, No. 1, 14–19 (1994).

73. ISO 180, Determination of Izod impact strength, 1993.
74. BS 2782, Method 350, Determination of Izod impact strength, 1993.
75. Fu Q., Wang, G., and Shen, J., *J. Appl. Polym. Sci.*, *49*, No. 4, 673–677 (1993).
76. Grocela, T. A., and Nauman, E. B., *Polymer*, *34*, No. 11, 2315–2319 (1993)
77. Piorkowska, E., Argon, A. S., and Cohen, R. E., *Polymer*, *34*, No. 21, 4435–4444 (1993).
78. Cheng C., Peduto, N., Hiltner, A., Baer, E., Soskey, P. R., and Mylonakis, S. G., *J. Appl. Polym. Sci.*, *53*, No. 5, 513–525 (1994).
79. Weier, J. E., and Hemenway, C. P., Antec '93. (9th–13th May), Conference Proceedings, New Orleans, LA, Vol. II, 1993, pp. 1690–1697. 012.
80. ISO 8256, Determination of tensile-impact strength, 1990.
81. BS 2782, Method 354A and B, Determination of tensile-impact strength, 1991.
82. ASTM D1822, Tensile-impact energy to break plastics and electrical insulating materials, 1993.
83. Hosoda S., and Uemura, A., *Polym. J.* (Jpn), *24*, No. 9, 939–949 (1992).
84. Dijkstra, K., ter Laak, J., and Gaymans, R. J., *Polymer*, *35*, No. 2, 315–322 (1994).
85. ISO 6603-1, Determination of multiaxial impact behaviour of rigid plastics—Part 1: Falling dart method, 1985.
86. BS 2782, Method 353A, Determination of multiaxial impact behaviour by the falling dart method, 1991.
87. ASTM D5420, Impact resistance of flat, rigid plastic specimens by means of a striker impacted by a falling weight (Gardner impact), 1993.
88. ASTM D5628, Impact resistance of flat, rigid plastic specimens by means of a striker impacted by a falling dart (tup or falling mass), 1995.
89. ISO 7765-1, Plastics film and sheeting—Determination of impact resistance by the free-falling dart method—Part 1: Staircase methods, 1988.
90. BS 2782, Method 352E, Determination of impact resistance by the free-falling dart method (staircase method), 1996.
91. ASTM D1709, Impact resistance of plastic film by the free falling dart method, 1991.
92. ISO 6603-2, Determination of multiaxial impact behaviour of rigid plastics—Part 2: Instrumented puncture test, 1989.
93. ISO 7765-2, Plastics film and sheeting—Determination of impact resistance by the free-falling dart method—Part 2: Instrumented puncture test, 1994.
94. BS 2782, Method 353B, Determination of multiaxial impact behaviour by the instrumented puncture test, 1991.
95. BS 2782, Method 352F, Determination of impact resistance by the free-falling dart method (instrumented puncture test), 1996.
96. BS 3518, Part 1, Methods of fatigue testing—Part 1: Guide to general principles, 1993.
97. ASTM E1150, Definitions of terms relating to fatigue, 1993.
98. ASTM D671, Flexural fatigue of plastics by constant amplitude of force, 1993.
99. Riemslag, A. C., *J. Test. Eval.*, *22*, No. 5, 410–419 (1994).
100. Trotignon, J. P., *Polym. Test.*, *14*, No. 2, 129–147 (1995).
101. ISO 9352, Plastics—Determination of resistance to wear by abrasive wheels, 1995.
102. ISO 6601, Plastics—Friction and wear by sliding—Identification of test parameters, 1987.
103. ISO 6691, Thermoplastics for plain bearings—Classification and designation, 1989.
104. BS 2782, Method 370, Determination of resistance to wear by abrasive wheels, 1996.
105. ASTM D673, Mar resistance of plastics, 1993.
106. ASTM D1044, Resistance of transparent plastics to surface abrasion, 1994.
107. ASTM D1242, Resistance of plastics materials to abrasion, 1995.
108. ISO 899-1, Plastics—Determination of creep behaviour—Part 1: Tensile creep, 1993.
109. ISO 899-2, Plastics—Determination of creep behaviour—Part 2: Flexural creep, 1993.
110. BS 4618, Section 1.1, Recommendations for the presentation of plastics design data—creep, 1970.
111. ASTM D2990, Tensile, compressive and flexural creep and creep-rupture of plastics, 1995.

112. ASTM D1598, Time-to-failure of plastic pipe under constant internal pressure, 1986.
113. Read, B. E., *J. Rheol.*, *36*, No. 8, 1719–1736 (1992).
114. Yakushev, P. N., Peschanskaya, N. N., Shpeizman, V. V., and Ioffe, A. F., *Int. J. Polym. Mat.*, *20*, No. 3/4, 245–250 (1993).
115. Challa, S., and Progelhof, R. C., Antec '93 Conference Proceedings, New Orleans, LA, 9th–13th May, Vol. II, 1993, pp. 1421–1424. 012.
116. Brueller, O. S., *Polym. Engng. Sci.*, *33*, No. 2, 97–99 (1993).
117. Hornberger, L. E., Antec '92, Plastics: Shaping the Future. Volume 2. Conference Proceedings, Detroit, MI, 3rd–7th May, 1992, pp. 2064–2066. 012.
118. Grzywinski, G. G., and Woodford, D. A., *Mat. Des.*, *14*, No. 5, 279–284 (1993).
119. Grzywinski, G. G., and Woodford, D. A., *Polym. Engng. Sci.*, *35*, No. 24, 1931–1937 (1995).
120. Tsou, A. H., Greener, J., and Smith, G. D., *Polymer*, *36*, No. 5, 949–954 (1995).
121. BS 2782, Methods 823A and B, Methods for the assessment of carbon black dispersion in polyethylene using a microscope, 1978.
122. ISO 183, Plastics—Qualitative evaluation of the bleeding of colorants, 1976.
123. BS 2782, Method 542A, Qualitative evaluation of the bleeding of colorants, 1979.
124. ASTM D696, Coefficient of linear thermal expansion of plastics, 1991.
125. Tong H. M., Saenger, K. L., and Su, G. W., *Polym. Engng. Sci.*, *33*, No. 22, 1502–1506 (1993).
126. Pottiger, M. T., and Coburn, J. C., Antec '93, Conference Proceedings, New Orleans, LA, 9th–13th May, Vol. II, 1993, pp. 1925–1927. 012.
127. ISO 2577, Plastics—Thermosetting moulding materials—Determination of shrinkage, 1984.
128. BS 2782, Method 640A, Determination of shrinkage of test specimens in the form of bars of compression moulded thermosetting molding materials, 1979.
129. ISO 11501, Plastics—Film and sheeting—Determination of dimensional change on heating, 1995.
130. BS 2782, Method 643B, Determination of dimensional change on heating of film and sheeting, 1996.
131. ISO 3521, Plastics—Polyester and epoxy casting resins—Determination of total volume shrinkage, 1976.
132. BS 2782, Method 644A, Determination of total volume shrinkage of polyester and epoxide casting resins, 1979.
133. BS 2782, Method 641A, Determination of dimensional stability at 100°C of flexible polyvinyl chloride sheet, 1983.
134. BS 2782, Method 643A, Shrinkage on heating of film intended for shrink wrapping applications, 1976.
135. ASTM D955, Measuring shrinkage from mold dimensions of molded plastics, 1989.
136. ASTM D2732, Unrestrained linear thermal shrinkage of plastic film and sheeting, 1983.
137. ASTM D2838, Shrink tension and orientation release stress of plastic film and sheeting, 1995.
138. ISO 75, Plastics—Determination of temperature of deflection under load—Part 1: General test method; Part 2: Plastics and ebonite; Part 3: High-strength thermosetting laminates and long-fibre-reinforced plastics, 1993.
139. BS 2782, Method 121, General test method for determination of temperature of deflection under load of plastics, 1994.
140. BS 2782, Method 121A-C, Determination of temperature of deflection under load of plastics and ebonite, 1994.
141. BS 2782, Method 121D, Determination of temperature of deflection under load of high strength thermosetting laminates and long fibre reinforced plastics, 1994.
142. ASTM D648, Deflection temperature of plastics under flexural load, 1995.
143. ISO 306, Plastics—Thermoplastic materials—Determination of Vicat softening temperature (VST), 1994.
144. BS 2782, Methods 120A, B, D, E, Determination of Vicat softening temperature of thermoplastics, 1990.

145. BS 2782, Method 120C, Determination of 1/10 Vicat softening temperature of thermoplastics, 1990.
146. ASTM D1525, Vicat softening temperature of plastics, 1995.
147. BS 2782, Method 124A, Determination of softening point of synthetic resin (ring and ball method), 1992.
148. ISO 1218, Plastics—Polyamides—Determination of "melting point", 1975.
149. ISO 3146, Plastics—Determination of melting behaviour (melting temperature or melting range) of semi-crystalline polymers, 1985.
150. BS 2782, Method 123A, Determination of the melting point of synthetic resins (capillary tube method), 1976.
151. BS 2782, Method 123B, Determination of the melting point of polyamides, 1976.
152. ASTM D789, Determination of relative viscosity, melting point and moisture content of polyamide (PA), 1994.
153. ASTM 2116, FEP-Fluorocarbon molding and extrusion materials, 1995.
154. ASTM D1457, Polytetrafluoroethylene (PTFE) molding and extrusion materials, 1992.
155. BS 2782, Method 125A to C2, Determination of melting behaviour (melting temperature or melting range) of semi-crystalline polymers, 1991.
156. ISO 458-1, Plastics—Determination of stiffness in torsion of flexible materials—Part 1: General method, 1985.
157. ISO 458-2, Plastics—Determination of stiffness in torsion of flexible materials—Part 2: Application to plasticised compounds of homopolymers and copolymers of vinyl chloride, 1985.
158. BS 2782, Method 153A, Determination of stiffness in torsion of flexible materials (general method), 1991.
159. BS 2782, Method 153B, Determination of stiffness in torsion of flexible materials (method for vinyl chloride compounds), 1991.
160. ASTM D1043, Stiffness properties of plastics as a function of temperature by means of a torsion test, 1992.
161. BS 2782, Method 150B, Determination of cold flex temperature of flexible polyvinyl compound, 1976.
162. BS 2782, Method 150C, Determination of low temperature extensibility of flexible polyvinyl chloride sheet, 1983.
163. BS 2782, Method 151A, Determination of cold bend temperature of flexible polyvinyl chloride extrusion compound, 1984.
164. ISO 974, Plastics—Determination of the brittleness temperature by impact, 1980.
165. ASTM D746, Brittleness temperature of plastics and elastomers by impact, 1995.
166. BS 2782, Method 150D, Cold crack temperature of film and thin sheeting, 1976.
167. ASTM D1790, Brittleness temperature of plastic sheeting by impact, 1994.
168. ISO 2578, Plastics—Determination of time–temperature limits after prolonged exposure to heat, 1993.
169. BS 2782, Method 135, Determination of the time–temperature limits after prolonged exposure to heat, 1993.
170. IEC 216-1, Guide for the determination of thermal endurance properties of electrical insulating materials—Part 1: General guidelines for ageing procedures and evaluation of test results, 1990.
171. ASTM D3045, Practice for heat aging of plastics without load, 1992.
172. ASTM D 794, Determining permanent effect of heat on plastics, 1993.
173. ISO 11248, Plastics—Thermosetting moulding materials—Evaluation of short-term performance at elevated temperatures, 1993.
174. BS 2782, Method 136, Determination of the short-term performance of thermosetting moulding materials at elevated temperatures, 1994.
175. BS 2782, Method 132B, Determination of blister temperature of thermosetting materials, 1976.

176. BS 2782, Method 134A and B, Determination of the oxidation induction time of thermoplastics, 1992.
177. ASTM D1870, Standard practice for elevated temperature aging using a tubular oven, 1991.
178. BS 2782, Method 131C and D, 131C: Crushing strength after heating, heat resistance of thermosetting moulding material. 131D: Crushing strength after heating, heat resistance of thermosetting laminated sheet or mouldings, 1978.
179. ISO 176, Plastics—Determination of loss of plasticizers—activated carbon method, 1976.
180. ISO 177, Plastics—Determination of migration of plasticizers, 1988.
181. ISO 4649, Rubber—Determination of abrasion resistance using a rotating cylindrical drum device, 1985.
182. BS 2782, Method 465A and B, Determination of loss of plasticizers (activated carbon method), 1979.
183. BS 2782, Method 465C, Determination of migration of plasticizers, 1990.
184. ASTM D1203, Volatile loss from plastics using activated carbon methods, 1994.
185. ISO 182-1, Plastics—Determination of the tendency of compounds and products based on vinyl chloride homopolymers and copolymers to evolve hydrogen chloride and any other acidic products at elevated temperatures—Part 1: Congo red method, 1990.
186. ISO 182-2, Plastics—Determination of the tendency of compounds and products based on vinyl chloride homopolymers and copolymers to evolve hydrogen chloride and any other acidic products at elevated temperatures—Part 2: pH method, 1990.
187. ISO 182-3, Plastics—Determination of the tendency of compounds and products based on vinyl chloride homopolymers and copolymers to evolve hydrogen chloride and any other acidic products at elevated temperatures—Part 3: Conductometric method, 1993.
188. ISO 182-4, Plastics—Determination of the tendency of compounds and products based on vinyl chloride homopolymers and copolymers to evolve hydrogen chloride and any other acidic products at elevated temperatures—Part 4: Potentiometric method, 1990.
189. BS 2782 Method 130A, Determination of the thermal stability of polyvinyl chloride by the Congo red method, 1991.
190. BS 2782, Method 130B, Determination of the thermal stability of polyvinyl chloride by the pH method, 1991.
191. BS 2782, Method 130C, Determination of the thermal stability of polyvinyl chloride by the conductometric method, 1993.
192. BS 2782, Method 130D, Determination of the thermal stability of polyvinyl chloride by the potentiometric method, 1993.
193. ISO 305, Plastics—Determination of the thermal stability of polyvinyl chloride, related chlorine-containing homopolymers and copolymers and their compounds—Discoloration method, 1990.
194. ASTM D2115, Standard practice for oven heat stability of polyvinyl chloride compositions, 1992.
195. ISO 1599, Plastics—Cellulose acetate—Determination of viscosity loss on moulding, 1990.
196. ISO 1157, Plastics—Cellulose acetate in dilute solution—Determination of viscosity number and viscosity ratio, 1990.
197. ISO 3671, Plastics—Aminoplastic moulding materials—Determination of volatile matter, 1976.
198. ISO 4577, Plastics—Polypropylene and propylene copolymers—Determination of thermal oxidative stability in air—Oven method, 1983.
199. BS 2782, Method 122A, Determination of deformation under heat of flexible polyvinyl chloride compound, 1976.
200. BS 2782, Method 131B, Determination of extensibility after heat ageing of flexible polyvinyl chloride sheet, 1983.
201. ASTM D3012, Thermal-oxidative stability of propylene plastics using a biaxial rotator, 1995.
202. ISO 4599, Plastics—Determination of resistance to environmental stress cracking (ESC)—Bent strip method, 1986.

203. BS 2782, Method 832A, Determination of resistance to environmental stress cracking (ESC) by the bent strip method, 1991.
204. ISO 4600, Plastics—Determination of resistance to environmental stress cracking (ESC)—Ball or pin impression method, 1992.
205. BS 2782, Method 831A and B, Determination of resistance to environmental stress cracking (ESC) by the ball or pin impression method, 1993.
206. ISO 6252, Plastics—Determination of resistance to environmental stress cracking (ESC)—Constant tensile stress method, 1992.
207. BS 2782, Method 833A to C, Determination of resistance to environmental stress cracking (ESC) by the constant tensile stress method, 1993.
208. BS 4618, Section 1.3.3, The presentation of plastics design data—Environmental stress cracking, 1976.
209. ASTM D1693, Environmental stress cracking of ethylene plastics, 1995.
210. ISO 62, Plastics—Determination of water absorption, 1980.
211. BS 2782, Method 430 A to D, Determination of water absorption, 1983.
212. ASTM D570, Water absorption of plastics, 1995.
213. ISO 4611, Plastics—Determination of the effects of exposure to damp heat, water spray and salt mist, 1987.
214. BS 2782, Method 551A, Determination of the effects of exposure to damp heat, water spray and salt mist, 1988.
215. ASTM D756, Determination of weight and shape change of plastics under accelerated service conditions, 1993.
216. ISO 175, Plastics—Determination of the effects of liquid chemicals, including water, 1981.
217. BS 2782, Method 830A, Determination of the effects of liquid chemicals, including water, 1986.
218. ASTM D543, Evaluating the resistance of plastics to chemical reagents, 1995.
219. BS 4618, Section 4.1, The presentation of plastics design data—Chemical resistance to liquids, 1972.
220. BS 1763, Thin PVC sheeting (calendered, flexible, unsupported), 1975.
221. BS 2739, Thick PVC sheeting (calendered, flexible, unsupported), 1975.
222. ASTM D1712, Resistance of plastics to sulfide staining, 1989.
223. ASTM D2151, Staining of poly(vinyl chloride) compositions by rubber compounding ingredients, 1995.
224. ISO 846, Plastics—Determination of behaviour under the action of fungi and bacteria—Evaluation by visual examination or measurement of change in mass or physical properties, 1978.
225. ISO/CD 14851, Biodegradability—Aerobic biodegradability—Oxygen demand in closed respirometer.
226. ISO/CD 14852, Biodegradability—Aerobic biodegradability—Carbon dioxide release (aqueous).
227. ISO/CD 14855, Biodegradability—Aerobic biodegradability—Carbon dioxide release (composting).
228. BS 4618; Section 4.5, The presentation of plastics design data—The effect on plastics of soil burial and biological attack, 1974.
229. ASTM G21, Determining the resistance of synthetic polymeric materials to fungi, 1990.
230. ASTM G22, Determining the resistance of plastics to bacteria, 1976.

17
Cellular Materials

Ken Hillier
British Vita, Middleton, Manchester, England

1 Introduction

Most polymers can be produced in a cellular or foamed form, and there are generally two main reasons for doing this. The first is usually associated with a particular property that gives to the material a useful characteristic unobtainable in the solid material. Second, the reason for making a polymer cellular in structure can be purely economic in nature. Of course both reasons could be important.

Cellular polymers can be produced in a wide range of densities (3–900 kg/m^3) [1] and it can be a useful approximation to assume that the mechanical properties of the material will increase as the density increases. This does not always apply for all properties, but for the mechanical values a power law relationship is generally found to apply [2]:

$$X_f = X_s \phi^n$$

where X_f is the foam property, X_s is the property of the solid polymer, and ϕ is the density. The exponent n depends on the property under consideration but usually lies in the region $1.0 < n < 2.0$.

A polymer type has its own set of physical and chemical properties that can set it apart from other polymers. For example, some are intrinsically rigid in nature, whereas others are more flexible. Some may exhibit good energy absorption characteristics, whereas others have a resistance to moisture penetration. These properties make them ideal for, say, insulation, cushioning, crash pads, or floatation applications respectively.

Take for example a rigid foam. It will be made up of a complex matrix of a solid polymer phase and a gas phase. Although the physical structure is usually complex, which is to say that the solid and gaseous phases will comprise cells, struts, cell walls, pinholes, and other irregularities, it is generally true to say that the chemical makeup of both the solid and gaseous phases is homogeneous in the fully cured foam. A rigid foam then produced as a thermal insulation medium relies on the thermal insulation properties of the enclosed gas in the cells. It is usually assumed that the nature of the gas phase is homogeneous. There is an optimum size associated with the cell to give the minimum thermal conductivity. If the cells are too large or interconnecting and thermal transfer is allowed by convection, then the thermal conductivity will increase. A solid form of the polymer will of course have minimal thermal insulation properties. The nature of the polymer then is important in several ways. It must have properties that are appropriate for the final application while maintaining an appropriate degree of thermal insulation. Many different types of polymeric cellular thermal insulation are in use today. The properties required are very diverse, and the choice of polymer will depend on the application.

As another example we may take flexible foam designed for domestic seating or cushioning. A solid rubber or plastic without any cellular form will have very little in the way of cushioning properties, at least as far as domestic upholstery is concerned. If cells are introduced without any consideration given to the interconnecting nature of those cells, then if those cells are completely closed and nonconnecting the cushioning characteristics will be poor and the feel will approximate to sitting on a balloon. A properly formulated foam with the correct degree of interconnecting cell networks will produce a foam with good cushioning and fatigue properties. Only flexible polyurethane seems to have the correct flexible and durability properties to fulfil the requirements of this application.

Another example would be those polymers designed for floatation. Here the cells are designed to be nonconnecting, and the polymer itself is preferably hydrophobic. A noncellular polymer may well float, but its buoyancy properties would not be important. It is only when the upthrust of a fully enclosed cell is considered that the excellent buoyancy characteristics of, say, cellular polystyrene or PVC can be understood.

With almost all cellular polymers there is a trade-off between acceptable foam physical properties and the unit mass of the cellular material. Usually the physical property under consideration shows a smooth curve relationship with the density of the product. This is not always the case, however, and care should always be exercised when adopting this rule. The relationship between the thermal conductivity of rigid foams and density is an example of where this rule can break down (Fig. 8).

Some polymers are easier to foam than others. Indeed, it was not until methods were found to circumvent the inclusion of cells in the early history of the phenol formaldehyde polymer that it gained any commercial significance. The development of foamed phenolic resins only became important much later when a specific need arose to produce rigid foam with reduced flammability. This consideration also led to the development of polyisocyanurate foams and carbodiimide foams. On the other hand, the polypropylene family of polymers, although having a tonnage far exceeding that of phenol formaldehyde resins, is

Table 1 Choice of Polymeric Cellular Insulant

Polymer	Important property	Thermal insulation
Rigid polyurethane	Cell walls impermeable	Excellent
Rigid polystyrene	Cell walls impermeable Water resistant	Good
Flexible polyolefin	Cell walls impermeable	Fair
Flexible polyurethane	Open celled	Poor
Phenol formaldehyde	Low flammability	Good

only now beginning to exploit the new markets being made available to foamed polypropylene. This is because the formation of foamed polypropylene has never been easy up until recently, due largely to its melt rheology properties. Recent developments have been aimed at overcoming this problem.

Some polymers can be used to produce foam possessing a wide range of properties. For example, polyurethane can be made hard or soft, flexible or rigid, at high or low densities. Polyurethane has an exceptional range of physical property variation and provides an unprecedented example to which many of the other cellular polymer test methods can be compared. Indeed, many of the methods used for the polyurethane family of polymers are common to the other polymeric foams. This subject is extensive, and a chapter such as this cannot hope to be exhaustive. However, it is hoped that most major physical property measurements are covered or at least guidance given to the reader as to where details can be obtained.

Most cellular materials are anisotropic in nature. Usually this means that the foam cells are orientated in the direction in which the foam was formed. The cells are usually elongated. This means that the physical properties are different when measured in the direction of the longest cell dimension or perpendicular to it. Most of the properties described in this chapter are measured on pieces cut perpendicular to the direction in which the foam rose, although for some foams, e.g., rigid insulation, the hardnesses are quoted in both directions. When using a cellular polymer, care should always be exercised in ensuring that the direction of cut is consistent throughout a product. For example, a settee or lounge chair should be constructed only with flexible polyurethane cut in the same direction with respect to foam rise.

Foam Types

Cellular polymers can generally be divided into three classes, rigid, semirigid, and flexible. Rigid foams find uses such as thermal insulation, packaging, and structural moldings. Semirigid foams are used in hard-wearing padding, moldings, and crash padding. Lastly, flexible foam can be used as a filling or cushioning material, in packaging, and in filters. Some authors [3] have tried to classify foams in a slightly different way. Just two groupings are used, Rigid, and Semirigid-to-Flexible. The first group is then subdivided into Tough-Rigid (e.g., polystyrene, PVC, and polyurethane) and Brittle Rigid (e.g., phenolic and urea types). The second group is also subdivided into Mainly Closed Cell (polyethylene and PVC) and open Cell (polyurethane). The distinction between the categories, whichever system is used, is not always easy to define, and when looking at foams in general it may be that the choice as to which type is chosen will depend on factors other than the physical properties at ambient conditions. Price and durability of course will also be prime factors.

Conditioning

As with all polymer testing, the conditioning of the specimen is very important to ensure good, meaningful, reproducible results (see also Chapter 6). In some cases, cellular materials, with their large surface areas, tend to be even more susceptible to, say, atmospheric moisture than the corresponding solid polymer counterpart. Usually the test method employed will stipulate the conditioning method and time to be used, and these tend now to be laid down as for example in Standard Atmospheres for Conditioning [4]. With some test methods, where the conditioning is extremely important in relation to the property, e.g., the thermal conductivity, then a great deal of care is necessary to ensure reproducible results.

Scope

Many tests commonly applied to plastics in the form of solids, elastomers, coatings, and sealants may of course also be applied to cellular plastics. For example, notched impact strength tests can be applied to high-density structural foams, and it is very common to quote Shore hardness values for skin hardnesses of rigid and semirigid foam moldings. It is the intention here, however, to restrict the discussion to tests derived exclusively for plastics containing a cellular structure.

Dimensions and Density

Measurement of the dimensions and density of rigid, semirigid, and flexible foams is now standardized and described in BS EN ISO 845, 1995 [5] and ISO 1923 [6]. ASTM D 1622, 93 [7] is technically equivalent (Chapter 7).

2 Rigid Foam

2.1 Compressive Strength

The compressive strength of a rigid foam is measured by applying a force to the specimen using a moving platen and noting the force at a specified deflection or the maximum force. The cell structure of a rigid foam has a significant influence on the foam properties. In particular, the anisotropy of the foam can result in large differences in the hardness values measured in the direction of foam rise compared to those measured in the perpendicular direction. Usually the results and specification values for the compressive strength of a rigid foam is expressed in both directions. This property usually follows a power relationship [1]:

$$L = K\rho^n$$

where

L = load
K = constant
n = constant
ρ = density

Figure 1 shows a typical force vs. deformation curve.

BS 4370, Part 1, Method 3 [8], which is related to ISO 844, 1985 [9] , gives a result for either the compressive strength or the compressive stress at 10% compression of the foam. The test specimen should where possible be a cube of 50 mm side, and this material is subjected to increasing compression at a fixed rate of 10% of the thickness per minute until the specimen is reduced to 90% of the original thickness.

The maximum force is noted or the force at 10% compression. Plying up of specimens to the required thickness is not permitted. The compressive strength is calculated from the equation

$$C_m = \frac{10^3 F_m}{S_0}$$

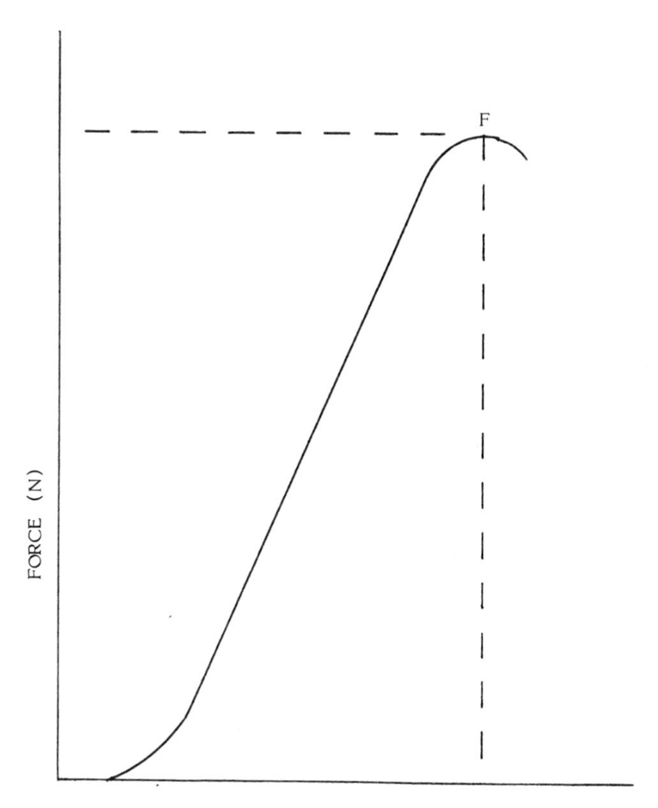

Figure 1 Force-deformation curve for rigid foam.

where

C_m = compressive strength
F_m = maximum force in N
S_0 = cross-sectional area

The compression strength is recorded if the maximum stress is observed before 10% compression is reached. If the maximum is not reached below 10% compression, then this value is defined as the compressive stress at 10% compression.

The American Standard ASTM D 1621 [10] and ISO 844 are technically equivalent. Test specimens can have cross-sectional areas of between 25.8 and 232 cm^2 with minimum thickness of 25.4 mm and maximum thickness no greater than the width or diameter of the specimen. The compressive strength is determined in a similar fashion to that described in BS 4370.

For some rigid foams, standards exist that define the types of foam available and where they can be used. For example in BS 3837, Part 1, 1986 [11], expanded polystyrene board can be designated by the appropriate structure, grade, and type. Typical examples are given as

Cut board, Standard duty, Type N
Moulded board, High duty, Type A

Five grades are defined: Standard (SD), High (HD), Extra High (EHD), Ultra High (UHD), and Impact Sound (ISD). Type A differs from Type N in that it is flame retarded to pass the horizontal burn test BS 4735 [12]. The properties expected for these various grades are specified in tables such as that partially depicted in Table 2.

The test methods mostly follow British Standards, but some are more closely related to the ISO tests. Care must be taken to ensure that the correct sample size is used. The determination of water absorption by diffusion is based on the Swiss Standard SIA 279 : Part 5.07 [13] (see Section 2.6 below). Similarly the properties of extruded board are specified in BS 3837, Part 2, 1990 [14]. BS 3927, 1986 [15], specifies rigid phenolic foam for thermal insulation in the form of slabs and profiled sections. The material is classified as types A, B, and C, which differ principally in thermal conductivity, water vapor permeability and apparent water absorption. Thermal conductivity is measured by methods described in BS 4370, Part 2, Method 7 [16] or Appendix B of BS 874 [17] .

BS 4840, Part, 1 [18] for polyurethane foam specifies that the two types of foam (LD and HD) should have minimum compressive strengths of 210 kPa (LD) or 2100 kPa (HD).

2.2 Cross-Break Strength

BS 4370, Part 1, Method 4, shows how the cross-break strength in flexure of rigid cellular specimens can be measured when subjected to a three-point loading as illustrated in Fig. 2.

The test pieces are 200 mm long by 35–70 mm wide and 50 mm thick. These are measured accurately and placed on two parallel supporting bars 150 mm apart and a load applied centrally at a rate of 100 mm/minute by means of a bar moving essentially parallel to the line of contact of the supports.

The load at the point of fracture is recorded and the mean result taken from five determinations derived from the equation

$$C_b = \frac{1500FL}{BD^2}$$

where

C_b = cross-break strength
F = maximum force in N
B = width of specimen in mm
D = thickness of the specimen in mm
L = distance between the lines of contact of the supports in mm

Table 2 Expanded Polystyrene Boards—Specification for Boards Manufactured from Expandable Beads

Property	Grade			
	SD	HD	EHD	UHD
Thermal conductivity (W/(m°K))	0.038	0.035	0.033	0.033
Compressive strength (kPa)	70	110	150	190
Cross-break strength (kPa)	140	170	205	275
Dimensional stability (% linear change at 80°C)	1.0	1.0	1.0	1.0

Source: BS 3837, Part 1, 1986.

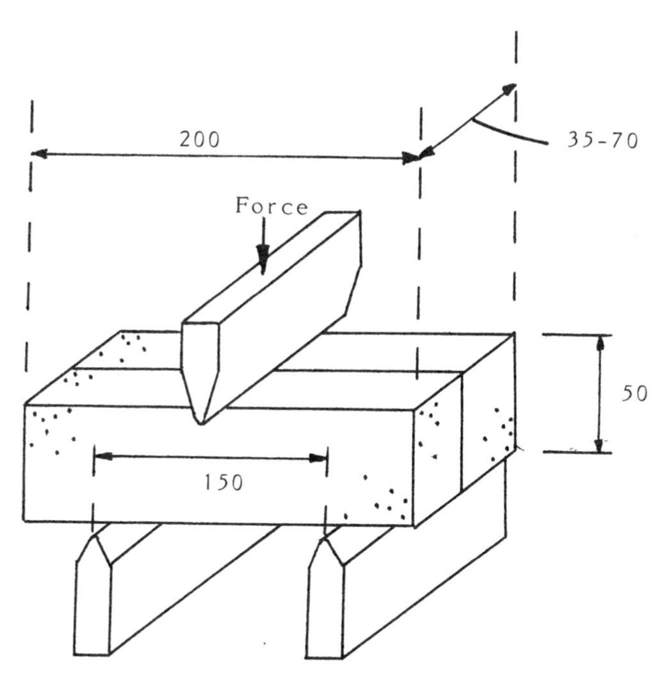

Figure 2 Determination of cross-break strength

With this test method it is permitted to use plied-up samples, but only if these are not less than 12 mm thick.

2.3 Modulus of Elasticity

BS 4370, Part 4, Method 14 [19], which is related to ISO 1209–2 [20], describes a specimen ratio of length to thickness much larger than that used in BS 4370, Part 1, Method 4 above and hence the strength in flexure will differ significantly from the cross-break strength. The specimen supports are two parallel cylindrical bars set 300 mm apart, each having an edge radius of 15 mm. The load bar is of the same dimensions as the supports, and when it is located centrally it is used to apply to the specimen an increasing load at a rate of 20 mm/minute. Specimen dimensions are 350 × 50 × 25 mm, and plied-up samples are not permitted. The force/deformation curve is recorded and a tangent drawn to the steepest part of this curve such that the deformation can be estimated. The test is stopped when the specimen fractures or when the tensile strain in the lower surface reaches 0.05. With the standard test piece this occurs when the deflection reaches 30 mm. The specimen is inspected for any signs of crushing, and if this is found the flexural strength is not calculated. Modulus of elasticity E (in kPa) is calculated from the expression

$$E = \frac{L^3 F_T \times 10^6}{4bd^3 x_T}$$

where

F_T = the force in kN
x_T = the corresponding deformation in mm
L = the span in mm
b = the test specimen width in mm
d = the test specimen thickness in mm

The flexural strength R (in kPa) is calculated from the equation:

$$R = \frac{1.5 F_R L \times 10^6}{bd^2}$$

where

F_r is the maximum force applied in kN

2.4 Tensile Strength

In contrast to the similar specimen shapes and dimensions used in Europe and the USA (see Section 3.2 below), when testing the tensile strength of flexible cellular polymers the two continents differ quite widely in the methods used to determine this property for rigid materials. The ISO 1926 [21] and BS 4370, Part 2, Method 9 standards use dumbbell-shaped specimens, the size of which can be increased if the thickness of the material is greater than 12.5 mm.

The cross-sectional shape in the region of fracture is rectangular and preferably a square of side 12.5 mm. It is important that no forces be applied to the specimen on clamping into the tensile testing machine that will result in the specimen breaking too far from the center of the narrow part of the test piece. To ensure that the probability of this happening is reduced, a specially shaped clamping device is used. If fracture does occur away from the central narrow region of the test piece, then these results are discarded and other samples tested. After measuring the thickness and width of the narrow part of the test piece, it is placed in the holder and the force applied in the longitudinal axis direction. The specimen is extended at a rate of 10 mm/minute until it breaks, and the tensile strength is calculated using the equation

$$S = \frac{F}{bd} \times 10^3$$

where F is the maximum force measured before break, b is the width (in mm), and d is the thickness (in mm).

With ASTM D 1623 [22], the principle of the test is the same as that of the ISO method, but the cross section of the specimen is circular, has a cross-sectional area of 1 square inch (645 mm^2), and is only 108 mm long.

The sample is preferably prepared using a lathe and the conically shaped ends fit into specially shaped holders that make the fracture within the gauge marks more probable. The standard speed of test machine jaw separation should be between 3 to 6 minutes to rupture with a suggested rate of cross-head movement set at 1.3 mm/minute for each 25.4 mm of test section gauge length. This test method also can be used to determine the tensile adhesion properties.

2.5 Shear Strength

The shear strength of a cellular product, as described in ISO 1922, 1981 [23], which is equivalent to BS 4370, Part 2, Method 6, is determined by applying a shear stress to a $250 \times 50 \times 25$ mm parallelepiped specimen by means of applying the force via two mild steel supports bonded to the largest surfaces (Fig. 3). This test method is limited to cellular products not above 100 kg/m^3 density.

The method of adhesion of the plates to the foam surfaces and the type of adhesive used are all-important in this test and, depending on the nature of the polymer, epoxy- or polyester-based adhesives are recommended.

After the requisite conditioning period, which can be used also to encompass the adhesive curing period , the shear stress is applied in a direction parallel to the longitudinal axis of the specimen by placing the specimen and supports vertically between the movable grips of a machine capable of separating the grips at a rate of 1 mm/minute. The resulting force-deflection diagram is recorded.

The shear strength q is calculated from the formula

$$q = \frac{1000F_{\mathrm{m}}}{lb}$$

where

l = the initial length of the specimen
b = the initial width of the specimen
F_{m} = the maximum force applied to the specimen

The shear modulus G is calculated from the formula

$$G = \frac{1000d\theta}{lb}$$

where

θ = the slope of the linear portion of the force-deflection diagram expressed in N/mm
d = the thickness of the specimen

Shear properties of sandwich core materials can be determined by a similar technique described in ASTM C273–94 [24]. In this method the support plates may be adhered directly to the facings of the sandwich or the core materials. The recommended speed of jaw separation is 0.50 mm/minute.

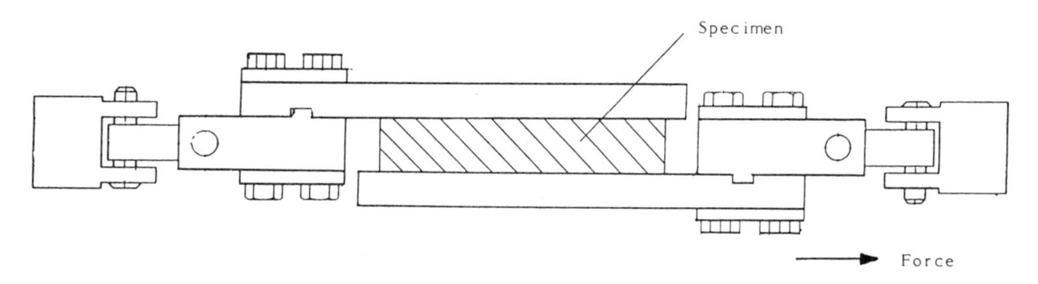

Figure 3 Determination of shear strength.

2.6 Water Vapor Transmission

BS 4370, Part 2, Method 8 is related to ISO 1663, 1981 [25] and describes the determination of water vapor transmission rate (μg/(m^2 · s)), water vapor permeance (ng/(m^2 · s · Pa)) and water vapor permeability (ng/(m · s · Pa) for rigid cellular materials that have thicknesses of between 10 mm and 70 mm. The first two properties are specific to the specimen thickness, whereas the permeability is a property of the material if this is homogeneous. A table is included in this British Standard giving conversion factors for all three properties. The method described is suitable for materials having a water vapor transmission rate of between 200 to 3000 μg/(m^2 · s).

The five specimens of measured thickness are sealed using wax into the open mouths of glass or metal beakers that contain the desiccant calcium chloride (Fig. 4). The minimum exposed surface area should be 32 cm^2 and if the test is designed to measure permeability the thickness should be 25 mm. The metal template is used to obtain a good seal with the wax after the specimens are in place in the beakers. The beakers and test specimens are weighed and transferred to a constant humidity and temperature environment. This environment may be either 38°C with a relative humidity gradient of 0% to 88.5% or 23°C with a humidity gradient of 0% to 50%. The beakers are removed from the chamber after intervals of about 24 hours and reweighed. The contents are shaken to mix the desiccant and then placed back in the chamber. The daily observed weight is plotted against time, and when three consecutive points are obtained that lie on a straight line, excluding the initial reading, the test is terminated.

The results are calculated using the equations

$$W_t = \frac{W}{3600A} \qquad P_e = \frac{1000W}{3600AP} \qquad P_y = \frac{1000Wd}{3600AP}$$

where W is the rate of mass change determined from the slope of the plot (in μg/h), A is the exposed area (in m^2), and P is the water vapor pressure difference under the chosen environmental conditions (5860 Pa at 38°C and 1400 Pa at 23°C).

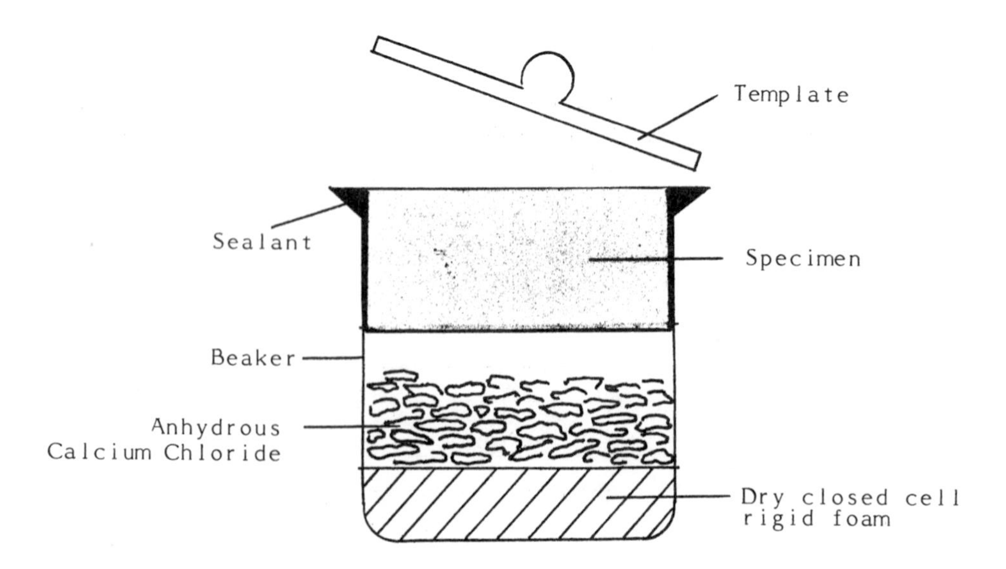

Figure 4 Determination of water vapor transmission.

BS 4840, Part 1, specifies that both types of polyurethane slab foam (Type 1, LD and HD) should have a maximum water vapor permeability normal to the major plane of the slab of 5.5 ng/(Pa · s · m). BS 5241, Part 2 [26], specifies that the five types of sprayed-up rigid polyurethane and polyisocyanate foam (for thermal insulation or buoyancy applications) should have maximum values of 3.5, 5.5, or 8.5 ng/(Pa · s · m).

BS 3837, Part 2, describes a method for the determination of the water absorption by diffusion for expanded polystyrene board. This method is based on the Swiss Standard SIA 279, Part 5.07, and involves subjecting the specimen (500 × 500 × 50 mm) to the environment above a small reservoir of water maintained at 50°C with the other side of the foam at 0°C. After 28 days the increase in mass is expressed as the water absorption (Fig. 5).

2.7 Water Absorption

BS3837, Part 2, Appendix E is specifically concerned with expanded polystyrene board, whereas the related ISO 2896, 1987 [27] covers all rigid cellular materials. The water absorption is determined by measurement of the buoyant force of a specimen when immersed in distilled and deaerated water for four days. The specimen, which should have a volume of at least 500 cm^3 (150 × 150 × 75 mm), is weighed and placed in the stainless steel mesh cage (Fig. 6), the buoyant force of which has been determined previously.

This cage is designed so that it can be suspended from a balance and appropriate weights added to allow the buoyant force to be measured. The cage is then immersed in water so that the water surface is approximately 50 mm above the top surface of the specimen. Air bubbles are removed from the surface of the specimen and the containing vessel is covered with a nonpermeable film to prevent evaporation. After four days the apparent mass is again determined and the water absorption, expressed as a percentage by volume, is calculated. For some rigid foam materials, for example certain grades of phenol-formaldehyde foam (BS 3927) , it is important to ensure that the surface of the water is maintained at approximately 50 mm above the top face of the specimen throughout the four-day test duration. Rigid foams with high water absorption rates will tend to absorb water quickly at the beginning of this test, and care must be exercised particularly at the beginning of the four-day period to prevent the top specimen surface becoming clear of the water surface.

A great deal of attention must be paid to any swelling that might occur during the course of this determination, and BS 3837 has two procedures that can be applied to

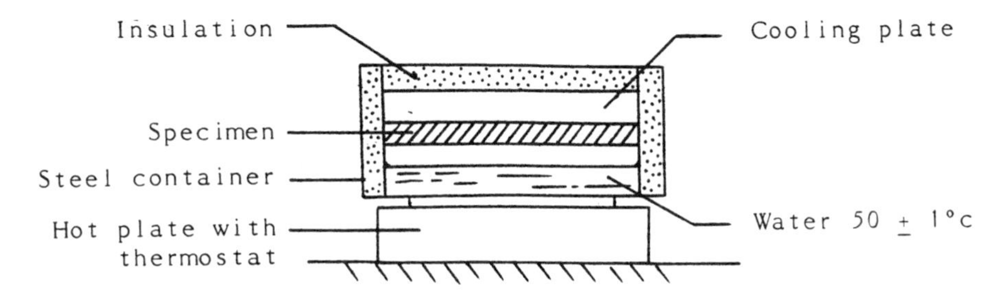

Figure 5 Determination of water absorption by diffusion.

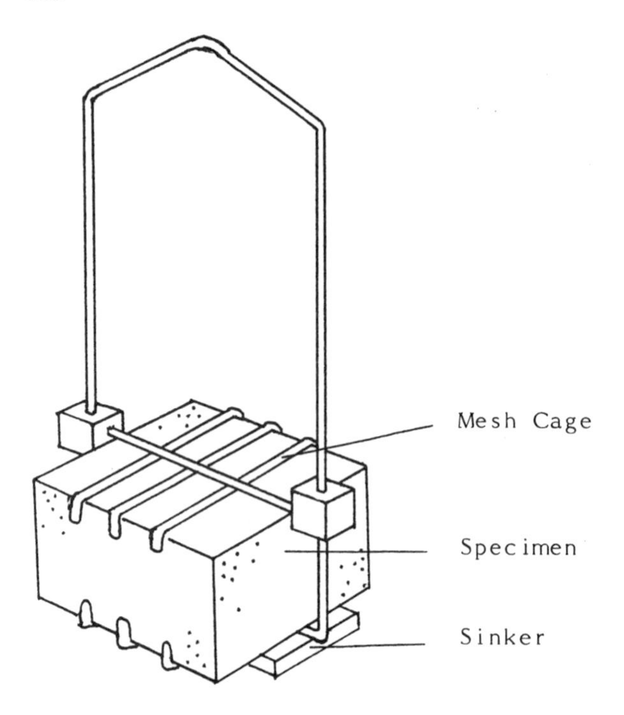

Figure 6 Apparatus for water absorption.

materials that exhibit uniform swelling and nonuniform swelling. A correction for the water present in the cut surfaces is also applied.

BS 3927 for phenol-formaldehyde foam uses the same basic technique, but the specimen size at 50 mm cube is different, and the period of immersion is set at seven days. Similarly BS 4840, Part 1, for rigid polyurethane foam in slab form uses the same sample size and immersion period.

BS 5617 [28] is a specification for urea-formaldehyde foam systems suitable for thermal insulation of cavity walls. Appendix C describes a method for water absorption using a 100 × 100 × 40 mm sample floated on the surface of the water for 24 hours.

The water absorption of flexible foams is much more difficult to measure than that of rigid foams, and when this property is requested it is usually in relation to a specific application need. There is no standard test method.

2.8 Closed Cell Content

The ASTM D 2856–94 [29] method uses an air pycnometer and follows the same basic principles as ISO 4590, 1981 [30] but differs significantly in experimental detail. The long established BS 4370, Part 2, Method 10 has been superseded by BS EN 4590, 1995 [31].

These test methods are based on the determination of the volume of closed cells in a cellular plastic by the application of Boyle's law to two enclosures in the pycnometer, one of which contains the test specimen (Fig. 7). By equalizing the pressure in the two chambers it is possible to determine the apparent volume of the specimen.

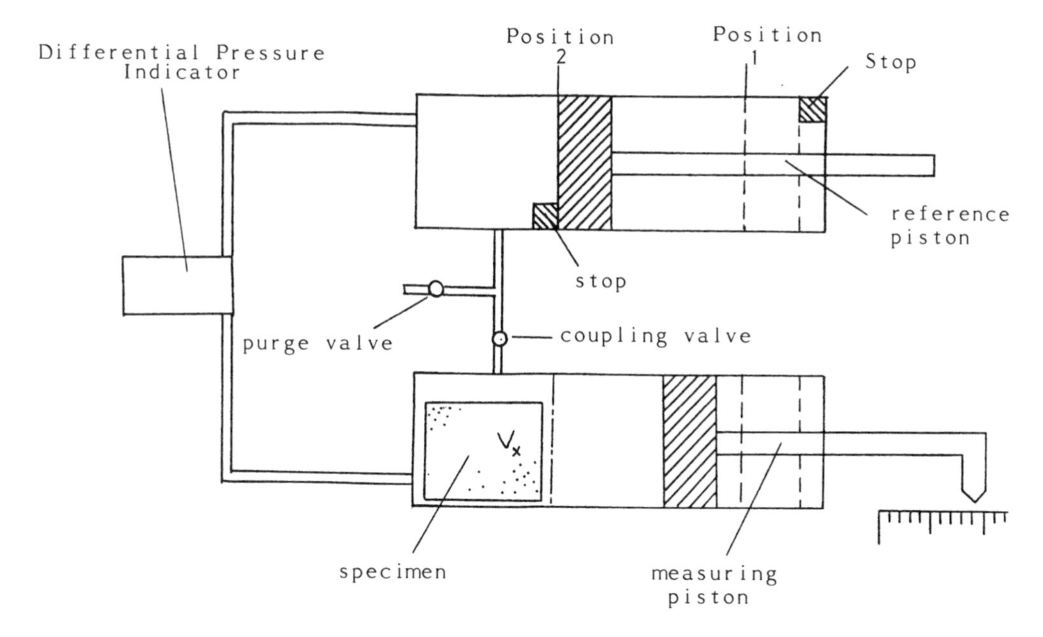

Figure 7 Air pycnometer.

Five test specimen cubes of 2.5 cm side are prepared and one placed in the sample cup. In order to maintain equal pressure on both sides of the differential pressure indicator, any movement of one piston from Position 1 must be duplicated by a similar movement of the second piston when the apparatus is empty. If both pistons are advanced to Position 2 with the specimen in position and the coupling valve closed, the pressure in the chambers will not be equal. The pressure is equalized by withdrawing the measuring piston by an amount proportional in volume to V_x and the apparatus calibrated to give this amount directly in cubic centimeters.

2.9 Friability

The friability of a rigid foam is not an easy property to determine, and it is seldom used as a quality control measurement. However, for certain materials such as phenol-formaldehyde foam it can be a useful tool in formulation work to ensure that the product is suitable for the application area. In certain instances the test method is best adapted to the foam being tested, for example the time duration of the test may be shortened if the material is being abraded too harshly.

The friability of the rigid cellular material is determined by measuring the degree of disintegration as a percentage mass loss when the foam is subjected to a grinding process. This grinding process as described in BS 4370, Part 3, Method 12 [32] is achieved by the use of a ball mill of specified dimensions (260 mm diameter, 90 mm long) containing 3.8 kg of 19 mm diameter porcelain balls. The mill is rotated during the test at a rate of 62 revolutions per minute. Four specimen cubes of side 20 mm are weighed and then loaded into the mill, which is then revolved for 60 revolutions. The specimens are then removed, vacuum cleaned, and reweighed. The friability (%) is then simply calculated from the original and final weights.

2.10 Thermal Conductivity

The thermal conductivity of a cellular polymer varies with temperature and, amongst other things, the density of the foam [2,33]. The nature of the specimen surfaces is also important. An example of how the thermal conductivity of rigid EPS foam [34] changes with respect to these variables is shown in Fig. 8.

The study of the thermal insulation behavior of cellular polymers is complex, and further reading of the cited references is recommended. There are essentially two ways in which the thermal conductivity can be determined, both ways being described in BS 4370, Part 2, Method 7A & 7B, 1993.

The thermal conductivity k is defined by the equation

$$k = \frac{Hd}{\Delta T}$$

where H is the rate of heat flow per square meter (in W/m^2), d is the mean specimen thickness (in meters), and ΔT is the difference in temperature (in K) between the hot and cold faces of the specimen. By definition then the thermal conductivity of a homogeneous material is the rate of heat flow, under steady conditions, through unit area, per unit temperature gradient in the direction perpendicular to that area.

Method 7A is based on the absolute guarded hot plate method described in BS 874, 1986 for determining the steady state thermal transmission properties of insulating materials in the temperature range −20°C to 100°C.

This procedure is similar to those given in ISO 2582 [35], DIN 52612 [36], and ASTM C–177 [37]. Two foam specimens of size 300 × 300 × 30–50 mm are placed in contact with the opposite faces of an electrically heated hot plate [33,38]. The edges of the hot plate and the samples are guarded by heated collars that minimize heat loss from these areas. Steady-state conditions are established when the electrical input to the hot plate equals the heat flow through the specimen. The thermal conductivity is then calculated.

Method 7B describes a heat flow meter method for the determination of the "apparent" thermal conductivity. This is a much faster method of determining this property and is the usual method employed. It uses a heat flow meter as for example described in BS 874, 1973, Appendix C. An international standard, ISO 8301, 1991 [39] exists that defines the use of heat flow meters and the calculation of the heat transfer properties of specimens. The

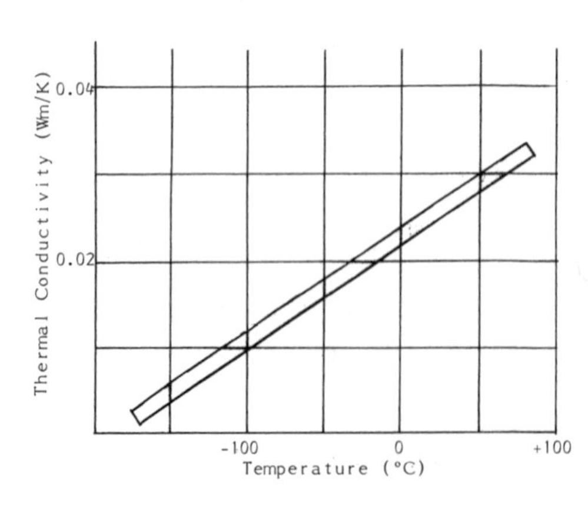

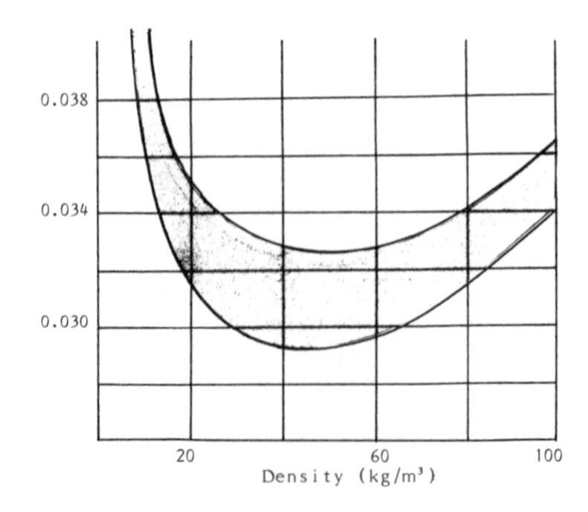

Figure 8 Variability of thermal conductivity of EPS foam with temperature and density.

bibliography included in ISO 8301 is useful as a starting point if thermal conductivity is to be studied in more depth. The specimen is placed between two surfaces whose temperature is accurately controlled and the heat flow measured by some electronic means. The apparatus is calibrated using materials of similar k value that have been determined using the guarded hot plate method. The material specification should give the mean temperature requirement. In the absence of this, the mean temperature of the test should be 10°C, and the minimum temperature difference between the hot and the cold plates should be 15°C. Commercially available machines can be purchased to measure the thermal conductivity of foams, and one such piece of apparatus is shown schematically in Fig. 9 [40]. The thickness is measured automatically when the sample is placed into the machine.

2.11 Thermal Stability

The influence of temperature on rigid cellular materials can be studied using several available standards covering dimensional stability (BS 4370, Part 1, Method 5), compressive creep (ISO 7616 [41]), and the determination of the coefficient of linear thermal expansion (BS 4370, Part 3, Method 13). The first two of these tests are relatively easy to perform, and the dimensional stability in particular tends to be quoted widely.

2.12 Dimensional Stability

Dimensional stability is a particularly important property of rigid foams, especially if used as insulating materials that can be exposed to extremes of temperature and humidity. It is

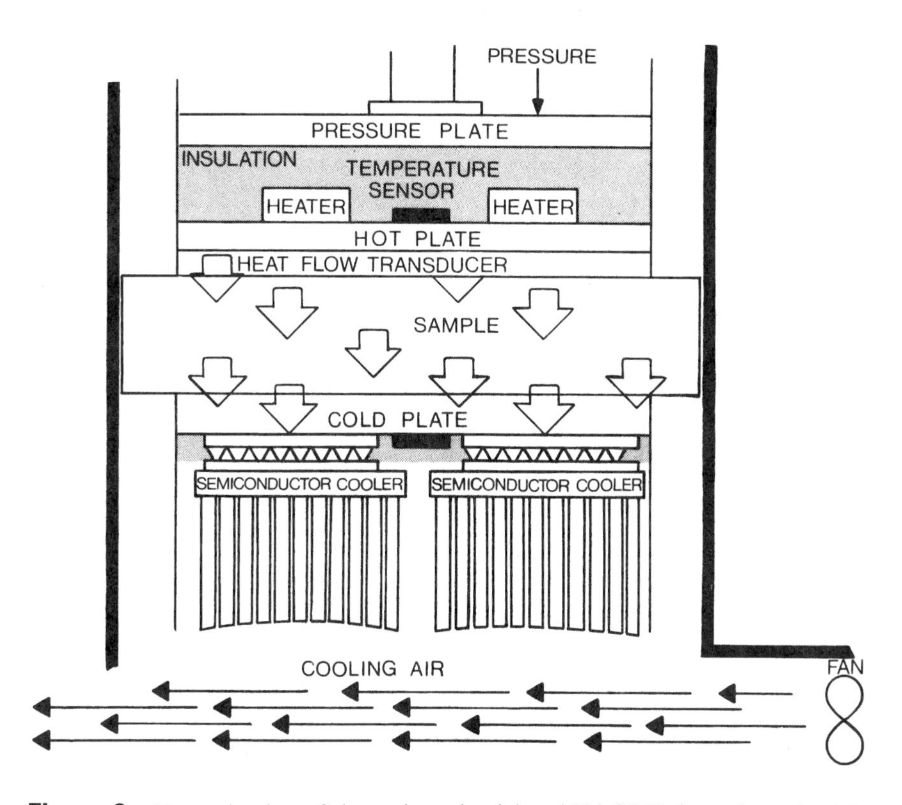

Figure 9 Determination of thermal conductivity. ANACON thermal conductivity analyzer.

best to test the foam at the specified conditions of use, but failing this the BS 4370, Part 1, Method 5 (Method 5A being related to ISO 2796 [42]) suggests fourteen temperatures for use in dry conditions ranging from −55°C to 150°C and at two temperatures of 40°C and 70°C between 90–100% relative humidity. For polyurethane foams it is not unusual to specify requirements at −25°C (freezer or cold store insulation) to 70–100°C (low oven temperatures). Expanded polystyrene usually has dimensional stability figures quoted at 70–95°C.

Two specimen sizes are specified (100 or 250 mm square and 25 mm thick (Method 5A) or 250 × 50 × 25 mm (Method 5B)) with the square samples being preferred. The foam, which should not have any skinned surface, is measured accurately for width (three places), length (three places) and thickness (five places) and then conditioned in the usual way. The specimens are then laid horizontally in a test chamber on a rigid wire mesh that allows free circulation of air. Exposure conditions are chosen from one to fourteen days and then, after 1–3 hours in the original conditioning atmosphere, the specimens are remeasured. The test results are expressed as the percentage change in linear dimensions. ASTM D2126 [43] describes a similar test method.

2.13 Compressive Creep under Specified Load and Temperature Conditions

The ISO 7616 test is designed for use on building thermal insulation materials. The specimens are rectangular parallelepipeds with a square base of 50 mm and a thickness of not less than 20 mm although 50 mm is recommended. The sample is placed between two flat plates that can be loaded to give stresses of 20 or 40 kPa for 48 hours and the thickness measured under these standard conditions. The sample is then subjected to 70°C or 80°C for 2 or 7 days and the thickness again measured. The compressive deformation under standard and elevated temperatures is calculated as a simple percentage of the thicknesses compared to the original thickness. Care should be taken to ensure that the stress is applied to the foam in the direction in which it would be exerted in use.

3 Flexible Foam

Most of the test methods outlined in this section for flexible foams are described in ISO tests or the related BS4443 [44]. These methods usually can be applied to flexible foams in general (The BS standard encompasses "flexible cellular materials of polymeric origin"), although in some instances the standards include only certain polymer types, for example ISO 2439, 1980 [45] includes only latex, urethane, and open-cell PVC. Care must be exercised when using the test methods for foams not covered in the scope of the test method. ASTM D 3574–95 [46] relates specifically to urethane foams.

Some of the methods described in BS 4443 have recently been superseded by BS EN ISO standards.

3.1 Hardness

There are several ways to measure the hardness of a flexible foam, and the chosen method appears to be mostly dictated by tradition or geographic location rather than by technical considerations. The indentation load deflection (ILD), as with all the "hardness" determinations, is generally accepted to be a measure of the load-bearing capability of the foam. Up until about 10 years ago in the UK the hardness of a foam could be described with reference to BS 3667 [47]. This technique indented a foam of dimensions 10 × 10 × 3 inches. BS 4443 has now replaced this earlier method and uses a test piece of dimensions 380 × 380 × 50 mm. At one time, some foam-making organizations published both hardness values, which entailed testing to both techniques and cutting two sets of test pieces. Because these two techniques employed samples of different dimensions, there was no true correlation between the two sets of values when determined for the same type of foam. Fortunately this situation has now been virtually eliminated, and BS 4443 is the accepted hardness measurement. All ILD methods indent a foam sample using a pressure foot of specified dimensions at a specified speed. The force required to give a certain indentation is recorded, Fig. 10.

Two procedures described in BS4443, Part 2, 1988 [48] are suitable in particular for latex, PVC, and polyurethane foams. One is faster to carry out and can be used as a quality control method (Procedure A). Procedure B can be used to determine the load to give indentations of 25, 40, and 65% deflections and hence the sag factor can be determined. In addition, by measuring the load for specified indentations of the foam on loading (as with Procedure B), followed by measuring the indentations on unloading, a measure of the foam hysteresis can be determined. Hysteresis is a measurement of the energy absorbed by a foam when subjected to a deformation.

For BS 4443 a test machine is required that is capable of indenting the test specimen between a supporting ventilated surface and the indentor. The indentor should be movable in the vertical plane with a uniform speed of 100 ± 20 mm/minute. This machine should be capable of precisely measuring the force required to give the specified deformation. The accuracy of all the measurement parameters are specified in the test method, and machines of the required type are available from several sources. An example of one such machine is shown in Fig. 11.

The indentor is of a flat circular shape with a diameter of 200 mm. This is usually mounted on a ball joint that allows it to conform to the shape of the sample but that does not allow any vertical movement. Molded or shaped samples can be measured on specially constructed base plates that conform closely to the base of the article. At the start of the test a small force (5N) is applied to the specimen, which is 380 × 380 × 50 mm in size, and the thickness measured. Plied up samples may be used if the sample is below the specified

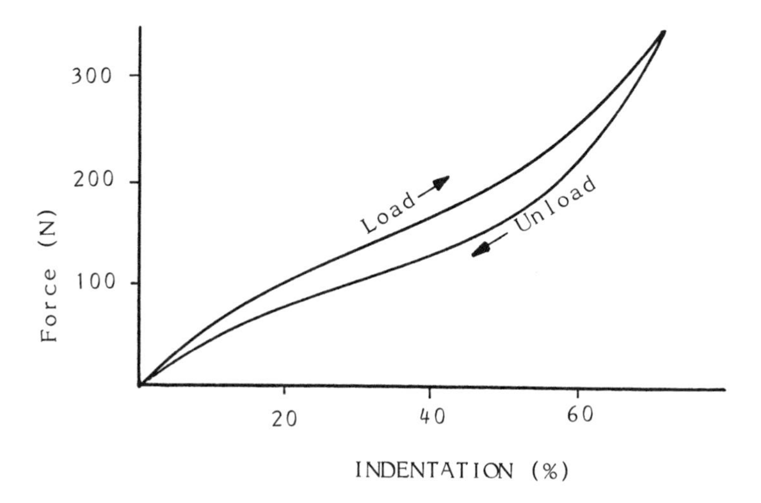

Figure 10 Force-deformation curve for flexible foam.

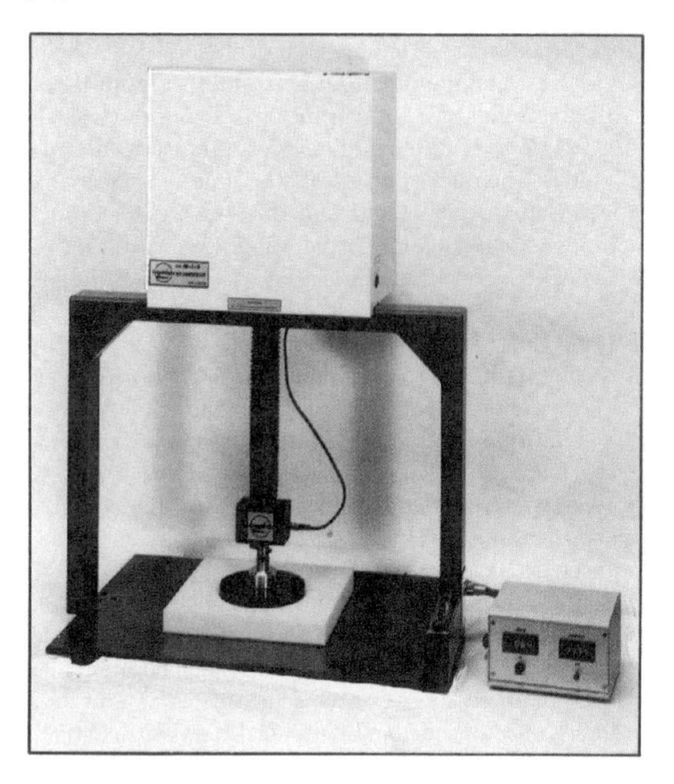

Figure 11 Hampton identometer for hardness testing.

thickness.The specimen is then indented to 70% of the original thickness three times. This procedure conditions the foam and helps eliminate undue errors caused by various factors including the elimination of any closed cells that may have been present in the foam. For Procedure A, immediately after the third indentation, the foam is indented to 40% of the original thickness and, after a period of 30 seconds, the force required to produce this indentation is measured. It is important to note that the same test specimen should not be remeasured immediately for hardness if at all. If this is unavoidable, the specimen should be left until the following day before being measured again. With polyurethane foam it is only after leaving the material for a considerable time that consistent results can be obtained. This depends on, among other things, the hysteresis characteristics of the foam, Fig. 10.

In Procedure B, immediately after the third unloading cycle, the force is measured after indenting to 25, 40 and 65% of the original specimen thickness. Once again in this procedure the foam is maintained at the required indentation for 30 seconds to allow consistency of results to be obtained.

ISO 2439 is related to BS4443, Part 2, with the procedures being the same. A faster quality control method is also included that does not require 30 seconds to elapse between compression to 40% of original thickness and measurement of the force.

The American test method for indentation force deflection (IFD), ASTM D 3574–95, describes a procedure similar to the British Standard, but it differs in several ways that must be recognized if direct comparisons are to be made. D 3574–95 is specifically for "Flexible Cellular Materials—Slab, Bonded, and Molded Urethane Foams." Specimen size is 380 × 380 × 20 mm, with specimens below 20 mm being plied up. The apparatus has an indentor foot of 323 cm^2 (similar to but not quite the same as the European standard), and the speed with which the indentation is carried out is about 50 mm/minute. The specimen is preflexed twice to 75–80% of the original thickness at a rate of 240 mm/minute and then allowed to rest for about six minutes after this preflex. The thickness of the sample is then measured while applying a force of 4.5 N. The force required to indent the foam to 25% and 65% is then measured after allowing the foam to remain indented at each value for 60 seconds.

A second hardness test is also described in ASTM D–3574; it is used to determine the indentation residual deflection force (IRDF). For certain upholstery applications it is useful to know how a given section of seating will compress under the weight of an average person. The test is preferably carried out on the complete manufactured article, and the position of the test area is selected by agreement between the furniture manufacturer and the purchaser. Using similar procedures described in the IFD test, the thickness of the article is determined by applying forces of 4.5 N, 110 N, and 220 N. IRDF values are usually given in centimeters, and the original thickness of the specimen must be also stated to be meaningful. An older name for this test was the indentation residual gauge load (IRGL).

An alternative method of measuring the load-bearing capabilities of the foam is the compression load deflection (CLD) technique, which is carried out on a much smaller sample than required with the ILD technique, Fig. 12. This test measures the compression stress–strain characteristics of the foam and arguably gives a better understanding of the intrinsic load-bearing characteristics of the foam than the ILD measurement. This is because the ILD measurement is influenced by several factors over and above those that influence the CLD. These factors include the thickness of the sample, the dimensions of the sample, the tensile properties of the material, and the shape of the indentor. It is becoming more common to use the CLD as a quality assurance test rather than the ILD, since these extraneous factors can then be ignored. The usefulness of ILD measurements however, in the gauging of how the material performs in a cushioning application, remains valid.

In the test method BS4443, Part 1, 1988 there are two CLD methods described that can be used for testing flexible materials with intercommunicating cells with densities up to

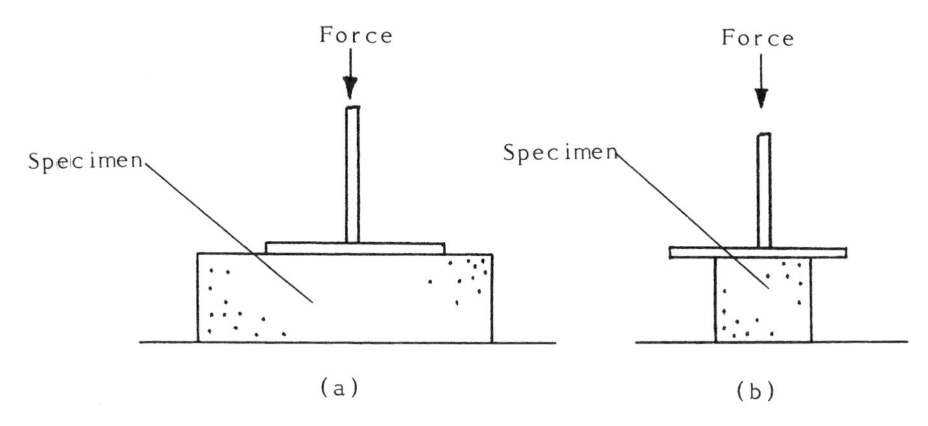

Figure 12 Measurement of (a) indentation load deflection (ILD) and (b) compression load deflection (CLD).

250 kg/m^3 (Method 5A) and for testing materials at higher densities, usually of the non-communicative type, e.g., the cellular rubbers.

In Method 5A, the test specimen shape is a right parallelepiped with a minimum width-to-thickness ratio of 2 : 1. Usually the size is cut to 100 × 100 × 50 mm, which incidentally is a very convenient library sample size to be retained in the laboratory. The thickness of the sample should not be less than 10 mm, and plied-up samples are allowed. The method stipulates that in any plying-up procedure the thickness of any sheet must be greater than 10 cell diameters. The test machine can be the same as that used to determine the ILD; the speed of indentation and the preflexing techniques are identical to BS4443, Part 2. After the third preflexing cycle, the force required to produce 25%, 40%, 50%, and 65% compression strain is measured, preferably by an autographic recording device. The compression stress–strain characteristics CC at each of the compression strains are calculated from the equation

$$CC = \frac{1000F}{A} \quad \text{(kPa)}$$

where

F = force in Newtons
A = area of the test specimen (mm^2)

In Method 5B, the method of test for the higher density materials is similar but the sample size , although still a parallelepiped, is reduced to 40 mm square sides for the load-bearing surfaces with a width-to-thickness ratio of 4 : 1. The area of the load-bearing surface should not be less than 1600 mm^2. The specimen is compressed at a rate of 5 mm/minute, and a preflexing procedure is carried out whereby the sample is compressed to the specified strain three times before the stress is determined on the fourth cycle.

The ISO 3386/1 [49] CLD standard for cellular flexible foam up to 250 kg/m^3 is related to BS4443 with the test procedures being identical. In addition to the CC characteristics, a compression stress value (CV_{40}) is given prominence and is calculated using the equation

$$CV_{40} = \frac{1000F_{40}}{A}$$

where F_{40} is the force, in Newtons, recorded in the fourth loading cycle for 40% compression. ISO 3386/2 [50] is used for materials above 250 kg/m^3 and is also related to BS 4443. It differs from ISO 3386/1 in that compression stress values have been deleted and cylindrical test pieces are not permitted.

ASTM D-3574 allows smaller sizes to be used with the minimum load-bearing surface area being 2500 mm^2 and the thickness 20 mm. The test consists of measuring the force necessary to produce a 50% compression. Preflexing conditions and procedures for compression are the same as in the IFD test.

3.2 Tensile Strength and Elongation at Break

The tensile strength of a cellular material is determined by measuring the uniformly applied force required to break the sample. The test piece is usually dumbbell-shaped (Fig. 13) and is cut or stamped (clicked) out of a sheet of foam with a test piece cutter. The rectangular cross section specimen should be prepared with care, since any snags or imperfections may cause premature fracture of the material under test. The cutting out of the material may also be useful in obtaining an indication of the "clickability" of the foam,

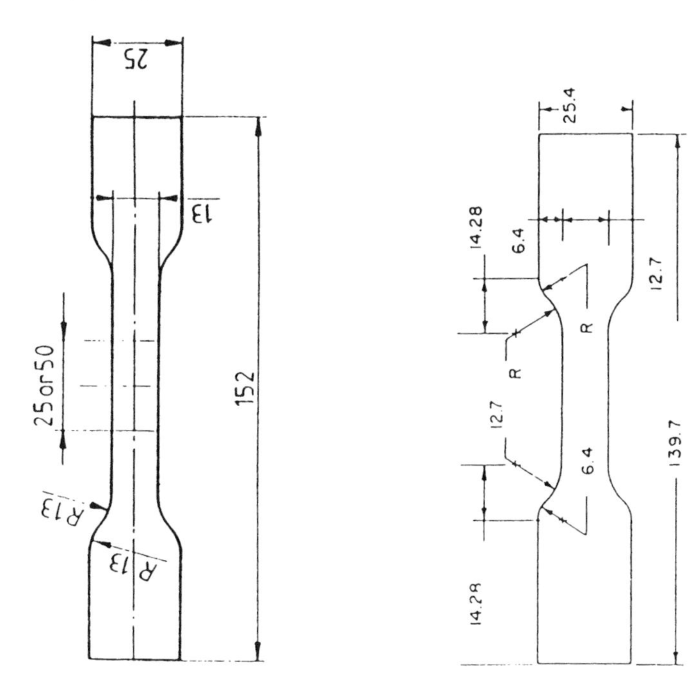

Figure 13 Tensile strength dumbbell specimens (dimensions in mm). (a) ISO 1798. (b) ASTM D3574.

or in other words the tendency of the foam to weld when subjected to a cutting force. There are no common test methods available to determine this characteristic, which is important in many foam fabrication techniques.

With ISO 1798 [51] the test piece (Fig. 13a) is placed symmetrically in the jaws of a machine capable of separating at a speed of about 500 mm/minute. The force is noted at the point at which the foam breaks together with the separation between the gauge marks. Tensile strength is calculated from the equation

$$\text{Tensile strength} = \frac{F}{A}$$

where F is the breaking force in newtons and A is the average initial cross-sectional area between the gauge marks.

Elongation at break is expressed as a percentage of the original gauge length and is calculated from the formula

$$\text{Elongation at break} = \frac{L - L_0}{L_0} \times 100$$

where L = gauge length at break and L_0 = the original gauge length.

With ASTM D 3574, the test specimen (Fig. 13b) is of a very similar shape to that used in the ISO standard although its overall dimensions are a little smaller.

The BS 4443, Part 1 standard is identical to the ISO method, although an additional comparative elongation at break technique is described.

3.3 Tear Strength

The tear resistance of flexible cellular polymers can be determined using EN ISO 8067, 1995 [52], which is for materials having a thickness of greater than 24 mm. The specimen (Fig. 14) is placed in the jaws of a machine capable of measuring the force at which rupture of the foam takes place when subjected to a jaw separation speed of 50–500 mm/minute.

When it is necessary to keep the cut in the center of the block while tearing, a sharp blade is permitted. When the tear has travelled 25 mm along the test piece, the test is terminated, and the maximum force is recorded as the tear strength. The equation used is

$$R = \frac{F}{d}$$

where F is the maximum force in newtons and d is the original thickness in meters.

BS 4443, Part 7, Method 17 [53] describes a similar procedure to that of ISO 8067 [54] but is designed to test those materials with an integral skin. The sample is precut to allow the material to tear by application of the force in a direction perpendicular to the skin, but in this case it is not permitted to help the tear develop with the aid of a sharp blade. The jaw separation rate is 100 mm/minute.

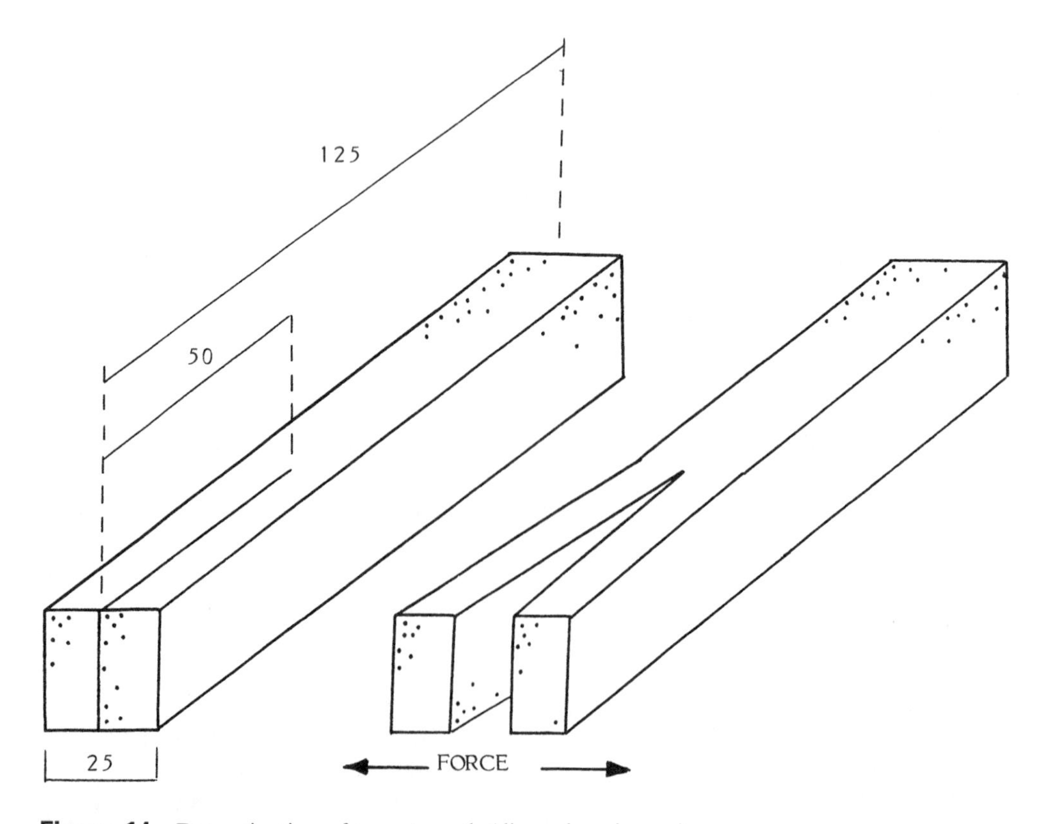

Figure 14 Determination of tear strength (dimensions in mm).

ASTM D 3574 is also very similar to the ISO standard although the specimen dimension is a little shorter as is the initial cut length. Jaw separation speeds are allowed in the range 45–540 mm/minute.

3.4 Porosity/Airflow

The airflow or porosity of a cellular plastic is very important, as it is intimately linked to many of the foam's properties. Flexible foams for cushioning tend to have intercommunicating cells (open cells), and if closed cells exist in their structure then comfort and fatigue characteristics can be adversely affected. Conversely some flexible foam applications, such as energy absorption, packaging, and buoyancy, rely on a degree of closed cell nature. Flexible foam gasketing applications can also call for a degree of closed cell nature although here it is sometimes also important that foam resilience be retained. Reticulated foams (fully open celled) find uses in the filtration of gases and liquids.

Rigid foams tend to be closed cell in nature, and this characteristic is necessary for some of the typical application areas such as thermal insulation and floatation aids. Some rigid foams do exist however with an open cell structure such as the phenol-formaldehyde-based horticultural and floral foams.

Considering the importance of the open or closed cell nature of a foam (and hence its porosity) it is perhaps rather surprising that the measurement of this property is generally neglected in terms of quality control procedures. It is only when a secondary application or process dictates the use of a certain foam porosity that the property is measured. In the flexible foam market it is also found that specific company test methods are in widespread use as opposed to international or national standards.

For a flexible foam, the openness of the structure is usually determined by blowing air through a sample of standard thickness at a set pressure differential. The rate of airflow measured is used to give an indication of porosity. When a constant pressure differential is used, the test methods can be useful quality assurance techniques. For a more in-depth understanding of the cellular material, techniques employing changing the airflow and measuring the resulting pressure differentials are in some instances considered more useful (ISO 4638 [55]).

Methods for determining airflow are described in ISO 7231 [56] using techniques involving either positive or negative pressure differentials with respect to atmospheric pressure. The test piece size is $51 \times 51 \times 25$ mm, which is designed to give a snug fit into the apparatus specimen holder cavity, which has dimensions of $50 \times 50 \times 25$ mm. After inserting the specimen into the cavity of the test apparatus (Fig. 15), a constant pressure of 125 Pa is applied across the foam and the airflow noted. The airflow value is expressed in cubic decimeters per second. Fig. 15 shows the apparatus for applying a pressure above atmospheric using an air pump.

The BS4443, Part 6, Method 16 [57] and ASTM D3574 are similar to the ISO 7231 method and need not be considered separately. The American Standard however does have slightly different apparatus dimensions and does not describe a higher than atmospheric pressure method.

ISO 4638 describes a method for determining the airflow permeability K, which is given by D'Arcy's law:

$$u = \frac{q_v}{A} = \frac{K \Delta p}{\eta \delta}$$

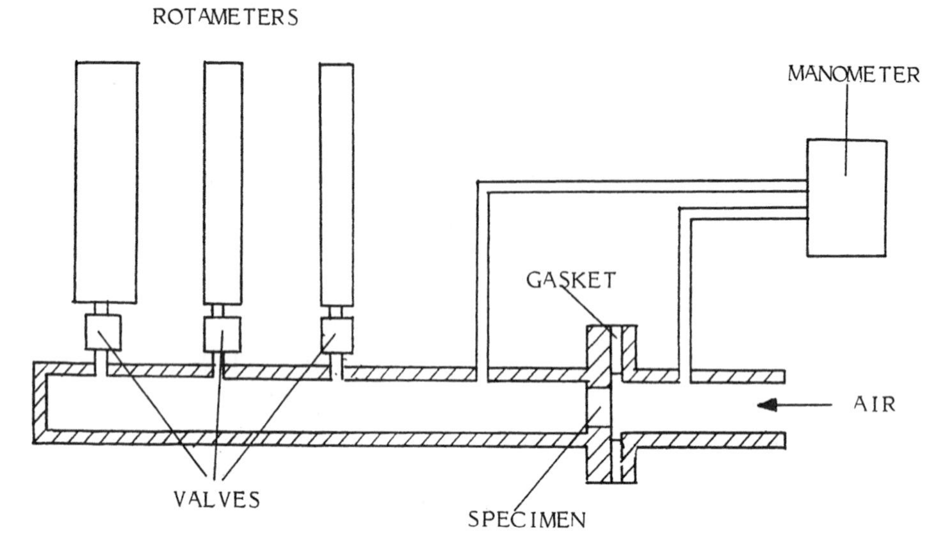

Figure 15 Measurement of airflow by applying pressure above atmospheric. ISO 7231.

where

u = linear airflow velocity
q_v = volumetric airflow rate
A = right cross-sectional area
K = flow permeability
Δp = pressure drop
η = dynamic viscosity
δ = test piece thickness

The airflow permeability can be used to gain an insight into the structural or physical properties of the cellular materials. The apparatus used is shown in Fig. 16. It comprises a cell into which the specimen can be placed and a method by which a steady flow of air through the specimen can be achieved. This steady airflow can be controlled to give different values, and hence K can be deduced.

It is found that porosity is important for many foam applications, and hence it is no surprise that many companies have their own test methods and specifications for this property. In some instances this derives from the fact that the end use employs different sizes and shapes of the foam or fabricated foam product. For example the Engineering Standard GME 60 286 [58] of Adam Opel AG/Vauxhall Motors Limited uses a similar apparatus to that described in ISO 7231, although the pressure differential is generally lower than the ISO standard at 10 or 20 Pa. For gasketing type applications, the porosity test can take the form of merely determining how the foam can contain a set pressure within a closed vessel. The test method described in Ford Engineering Specification WSD–M99D57 [59] and ESA–M4D200–B [60] consists of a 2 gallon paint pressure tank with a 6″ long pipe extending from the top, to which is attached a 1″ pipe flange with a 1″ opening to the atmosphere (Fig. 17). The tank is fitted with a pressure gauge and the means of introducing an air pressure into the pot of 5 psi. The foam gasketing sample is bolted between the pipe flange plates, and the pot raised to a pressure of 5 psi. After turning off the air pressure supply, the pressure within the pot is measured over 30 minutes, and usually a residual pressure of about 1.5 psi is required to pass the test.

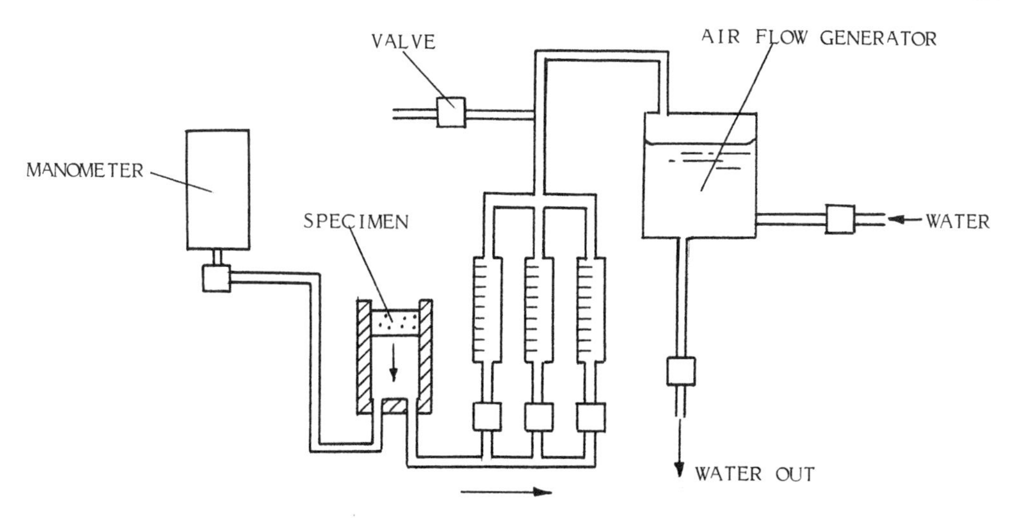

Figure 16 Measurement of airflow permeability. ISO 4638.

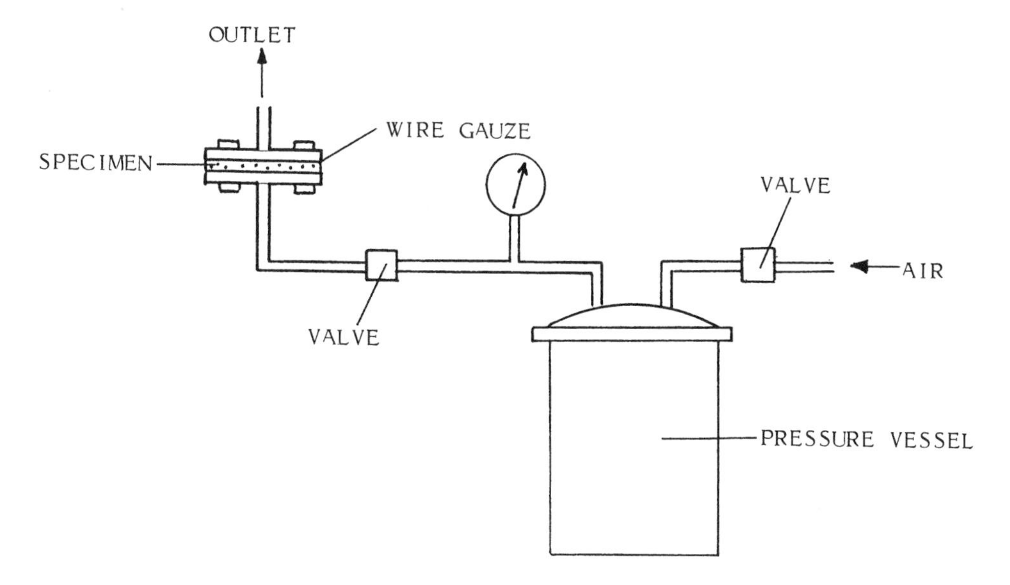

Figure 17 Measurement of foam porosity (Ford method).

3.5 Durability (Fatigue)

The durability of a flexible foam may be determined by using a shear or a pounding force on the foam. Tests described here are designed to assess the suitability of foam for use in upholstery and are therefore primarily aimed at cellular latex or polyether urethane types of foam. A dynamic fatigue test using a roller shear at constant force is described in ASTM D 3574, Test I_2. However a much more popular fatigue test for cellular upholstery materials is determined by constant load pounding.

BS EN ISO 3385 [61] uses three test pieces (right parallelepipeds 380 × 380 × 50 mm) that are each subjected to a constant load of 750 N during one load cycle at a rate of 70 per minute. The load is applied by way of an indentor of diameter 250 mm (Fig. 18) that is rigidly fixed to its guide but allowed movement in the vertical plane as the specimen and indentor approach their nearest position. The indentor surface is smooth but not polished. The machine is capable of oscillating either the platen or the indentor so that the constant load is attained at one time during the cycle.

The indentor support mounting is constructed so that the indentor force is carried by it for most of the cycle time, but for a short period of time, no more than 25% of the total cycle time, the full force of the indentor is carried by the test piece alone. For a manually adjusted piece of apparatus, as shown in Fig. 18 , the position of the indentor is changed so that the time requirement for full indentor support is kept within the specified test requirements. This is necessary as the test specimen softens throughout the test. A force-measuring device such as mounting the platen on a load cell is required to determine the applied load.

The thickness and hardness of the test piece are measured, and it is then placed centrally beneath the indentor. The stroke is adjusted to equal the thickness of the test piece and the position of the indentor and platen adjusted to ensure that the correct load is applied. This is usually ensured by the indentor just being lifted in its mounting.

Usually 80,000 load cycles are applied, and the specimen is removed and allowed to rest for 10 minutes before remeasuring the thickness and the hardness. In ISO 3385 the recommended hardness test is ISO 2439 using 40% indentation including the preliminary indentation procedure, although the original thickness of the specimen is used to calculate the 40% indentation level. The test report should include both the thickness loss and the hardness loss of the samples. When these losses have been determined, the suitability of the material can be gauged by reference to BS 3379 [62] "'Flexible polyurethane cellular materials for loadbearing applications" [12]. In-use fatigue testing is a subject that is found to change and evolve with end-use variety, and it is recommended that the fatigue test be chosen after thorough investigation of end-use market. Specific tests are available

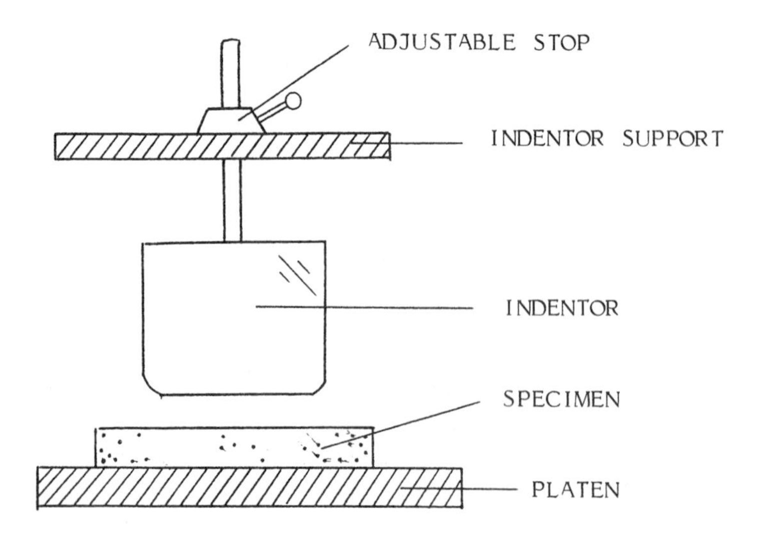

Figure 18 Determination of fatigue.

for upholstery [63], carpet underlay [46] and shoe soles. A specification for flexible cellular material is given in ISO 5999 [64].

Dynamic cushioning performance is not covered in this section, but the ISO 4651 [65] gives a good introduction to the methods, as does Ref. 3 in Chapter 9.

3.6 Compression Set

The techniques described in EN ISO 1856, 1996 [66], DIN 53572 [67], and ASTM 3574, Test D, are all very similar, and only a general description is given here. The U.S. and German specifications test three samples, whereas the others recommend five. Specimens are right parallelepipeds of square load-bearing surfaces of side 50 mm and width-to-thickness ratio of not less than 2 : 1. When testing thin samples of less than 10 mm they can be plied-up without cement so that their thickness will be greater than 25 mm. When very thin specimens need to be tested, in the order of about 1 mm, the use of glass slides between the slices of foam is essential, since there can be considerable interpenetration of the cell structures of adjacent layers.

A compression device is used consisting of at least two flat polished plates larger in dimensions than the test specimens. Spacers and clamps are used to give the required degree of compression and to ensure that the plates are held parallel to each other during the test (Fig. 19). If many samples are to be tested, then two tiers can be used. After the conditioning period, the sample thicknesses are measured and then the samples are compressed within the apparatus to the specified degree, which is usually 50%, 75%, or 90%. The deflected samples and the apparatus are placed within 15 minutes into a circulating oven maintained at 70°C and 5% relative humidity . The humidity requirement (only specified in ASTM D 3574) is achieved by placing the oven in a conditioned room at 23°C and 50% relative humidity. After 22 hours at this temperature the apparatus is removed and the specimens removed within one minute and then allowed to recover for between 30 and 40 minutes on a wooden (or other low thermal conductivity) surface. The thickness is then measured again and the compression set determined from the equation

$$\mathrm{CS} = \frac{T_o - T_r}{T_o} \times 100$$

where

T_o = original thickness of the material
T_r = thickness after recovery

The ISO specification also describes methods that allow the specimens to be compressed at the conditioning temperature for 72 hours and also compression under specified conditions. The ASTM test also gives the equation used to express the compression set in terms

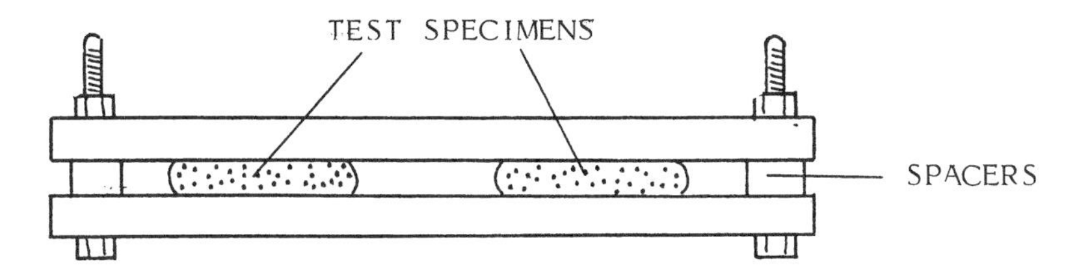

Figure 19 Determination of compression set.

of the original deflection. This last reporting technique is more commonplace with polyurethane molded material.

After the introduction of combustion-modified polyurethane foam into the upholstery and bedding market, a wet compression set method, BS 4443, Part 7, Method 18, has become much more widely used. This is carried out by using a humidity cabinet set at 40°C and 95–100% relative humidity. Apart from recommending the use of a stainless steel compression apparatus, the only major difference from EN ISO 1856, 1996 is that the compression is carried out at 70% of the original thickness as opposed to the more normal 75%. This slight discrepancy arises because of the original test methods devised by the automotive trade.

3.7 Windscreen Fogging

Problems associated with the deposition of "fog" on automotive windscreens have been with us for many years, but only relatively recently has this problem assumed real importance, as most of the major car companies in the USA, Europe, and Japan now ask for compliance to their own fogging tests for most interior trim components. The problem is not specifically related to cellular polymers, and indeed much of the interior trim can be seen not to be cellular. It could also be true that most of the windscreen fogging is derived from air external to the car interior, e.g., exhaust fumes from traffic.

The automotive company tests are all relatively similar, but great care must be exercised to ensure that the test conditions are as specified. Usually the test requires small samples of the specimen to be placed in a beaker, the mouth of which is covered with a glass plate . The beaker and contents are placed in a heated oil bath at a specified temperature (e.g., 75°C for Volvo/Saab, 6 hours; 90°C for General Motors, 3 hours; 100°C for Ford, 3 hours) and the glass plate kept in contact with a condenser at an ambient temperature usually set at 20°C. After the specified time, the glass plate is removed and the degree of fogging measured with a glossmeter. More recent developments, especially with German automotive companies, measure the degree of fogging by weighing the residue deposited on an aluminum foil placed at the mouth of the beaker. Calibration of the apparatus is usually carried out using a standard liquid such as DIDP. Reproducibility of these tests is not easy, and great care needs to be exercised in many aspects, especially with respect to the cleanliness of the glass and beakers. Details of some test methods are given in Ref. 68.

3.8 Rebound Resilience

This property of rebound resilience is very rarely specified, but it is a useful, fast, simple test that can be carried out quantitatively to assign the foam type to conventional or high-resilience types. The test is described in ASTM D 3574–95, Test H and consists of dropping a 16 mm diameter steel ball of weight 16.3 g onto the specimen, which preferably is in the form of the end product, or onto the foam at a minimum thickness of 50 mm. Plied-up samples are permitted without the use of cement.

The ball is released within a vertical clear plastic tube of internal diameter 38 mm so that it falls without rotation. The drop height is 500 mm, and the tube is marked at 25 mm intervals with a full circle. Rebound resilience is reported in percent. For a discussion on ball rebound resilience see Ref. 2.

3.9 Sound Absorption

Cellular materials are used in many acoustical insulation applications, and the subject is covered in depth by Lauriks [2]. Usually the foam is of the flexible or semirigid variety, and a useful selection of references is also given in "Flexible Polyurethane Foams" [63]. Cellular polymers are themselves rather poor materials for reducing sound transmission, but they usually are much better in absorbing sound waves of certain frequencies. Open celled varieties of foam (especially reticulated foam) are particularly effective in this respect. It is now common practice to use foam in the form of anechoic wedges when wishing to construct chambers whose walls are designed to be sound deadening.

The normal incidence sound absorption coefficient of a material can be determined using the method described in BS EN ISO 9614 [69] (BS 4196, Part 5 [70]). In this type of determination, a plane sound wave is made to be reflected by the sample mounted in a standing wave tube. The single-frequency wave hits the sample at normal incidence, and the amplitudes of the incident and reflected waves are measured over a series of chosen frequencies. Usually the frequencies of interest lie between 50 and 300 Hz, although some automotive specifications will cover much wider ranges (see Ford Engineering Specification WSK-M2D406-A).

3.10 Cell Count

BS 4443, Part 1, Method 4 defines the cell count as the number of cells per linear 25 mm of the flexible cellular material. Cell count has a large influence on the physical properties of the foam (see Ref. 1 in Chapter 3), although only in limited circumstances is it specified. It is a property that is not easy to measure accurately and in many instances will be somewhat dependent on the tester. A 25 mm cloth counting glass is used to count the actual number of cells against the counting edge of the glass. In practice it is useful to mark the surface of the foam with ink of a contrasting color to that of the foam. In this way the cut edges of the foam are enhanced, thus making counting easier.

3.11 Cushioning

Both flexible and rigid foams are used in packaging, and BS 1133, Section 12, 1986 [72] can be used as a good introduction to the subject. The design and application of cushioning materials is outside the scope of this work. BS 1133 gives a comparison of the cushioning efficiency of such cellular polymers as polyethylene, polyurethane chipfoam, polystyrene, and ethylene vinyl acetate.

3.12 Buoyancy

Although given here under the flexible foam section, this test applies equally to the more rigid types of foam also. The flexible foam types are usually of the thermoplastic, predominantly closed cell variety.

BS EN 396, 1994 [73] titled "Lifejackets and Personal Buoyancy Aids—Lifejacket 150," primarily describes the lifejacket itself in terms of materials of construction and the buoyancy of the complete article. However Annex C describes a method for determining the foam buoyancy by weighing the specimen in air and then when submerged in water at 20°C in a preweighed cage. The difference in weight is then expressed as a percentage volume change. Annex H goes on to describe a test for measuring the compressibility of inherently buoyant material. The specimens are 100 × 100 mm of the appropriate thickness and are placed under a flat metal plate and compressed at a speed of 200 mm/minute until a load of 500 kPa is reached. This is repeated after decompression four more times. The specimen is then kept under the metal plate so that it is only just weighted by the plate to remain under water. The load required to achieve this is recorded as the original buoyancy. The specimen is allowed to dry for seven days and the compression cycle repeated without water a further 500 times. The specimen is then returned to the atmosphere for three days and the buoyancy then measured as before. The loss of buoyancy is given as a percentage.

4 Semirigid Foam

Semirigid foam is usually tested using methods normally applied to rigid or flexible foams. In some instances they are tested for ultimate elongation (as a percentage to BS903, Part A2 [74]). The surface hardness is also important in some molded part applications, and that property can be measured using standard Shore A hardness procedures.

References

1. Klempner, D., and Frisch, K. C., *Polymeric Foams*, Hanser, 1991.
2. Hilyard, N. C., and Cunningham, A., *Low Density Cellular Plastics*, Chapman and Hall, 1994.
3. Saechtling, H., *International Plastics Handbook*, Hanser, 1987.
4. Brown, R. P., *Handbook of Plastics Test Methods*, Elsevier Science, 1988.
5. BS EN ISO 845, Cellular plastics and rubbers—Determination of apparent (bulk) density, 1995.
6. ISO 1923, Cellular plastics and rubbers—Determination of linear dimensions, 1981.
7. ASTM D 1622-94, Compressive properties of rigid cellular plastics.
8. BS 4370, Part 1, Rigid cellular materials, 1988.
9. ISO 844, Cellular plastics—Compression test of rigid materials, 1985.
10. ASTM D, 1621-94, Compressive properties of rigid cellular plastics.
11. BS 3837, Part 1, Expanded polystyrene boards. Specification for boards manufactured from expandable beads, 1986.
12. BS 4735, Laboratory method of test for assessment of the horizontal burning characteristics of specimens no larger than 150 mm × 50 mm × 13 mm (nominal) of cellular plastics and cellular rubber materials when subjected to a small flame.
13. SIA 279, Part 5.07, Water absorption by diffusion.
14. BS 3837, Part 2, Expanded polystyrene boards. Specification for extruded boards, 1990.
15. BS 3927, Rigid phenolic foam (PF) for thermal insulation in the form of slabs and profiled sections, 1986.
16. BS 4370, Part 2, Method 7, Rigid cellular materials, 1993.
17. BS 874, Determining thermal insulating properties with definitions of thermal insulating terms, 1973.
18. BS 4840, Part 1, Rigid polyurethane (PUR) foam in slab form, 1985.
19. BS 4370, Part 4, Method 14, Determination of flexural properties, 1991.
20. ISO 1209–2, Cellular plastics, rigid-flexural tests—Part 2, Determination of flexural properties, 1990.
21. ISO 1926, Cellular plastics–Determination of tensile properties of rigid materials, 1979.
22. ASTM D, 1623–78, Tensile and tensile adhesion properties of rigid cellular plastics.
23. ISO 1922, Cellular plastics—Determination of shear strength of rigid material, 1981.
24. ASTM C273–94, Shear properties of sandwich core materials.
25. ISO 1663, Rigid cellular materials, 1981.
26. BS 5241, Part 2, Rigid polyurethane (PUR) and polyisocyanurate (PIR) foam when dispensed or sprayed on a construction site, 1991.

27. ISO 2896, Cellular plastics, rigid—Determination of water absorption, 1987.
28. BS 5617, Urea-formaldehyde (UF) foam systems suitable for thermal insulation of cavity walls with masonry or concrete inner and outer leaves, 1985.
29. ASTM D, 2856-94, Open-cell content of rigid cellular plastics by the air pycnometer.
30. ISO 4590, Cellular plastics-Determination of volume percentage of open and closed cells of rigid materials, 1981.
31. BS EN 4590, Cellular plastics-Determination of volume percentage of open and closed cells of rigid materials, 1995.
32. BS 4370, Part 3, Rigid cellular materials, 1988.
33. Woods, G., *The ICI Polyurethanes Book*, ICI Polyurethanes and John Wiley, 1990.
34. Styrocell Technical Manual STY 5.1, 2nd ed.
35. ISO 2582, Cork and cork products—Determination of thermal conductivity—hot plate method, 1978.
36. DIN 52612.
37. ASTM C-177.
38. *Revista de Plasticos Modernos*, Núm. 436, Oct. 1992.
39. ISO 8301, Thermal Insulation—Determination of steady-state thermal resistance and related properties—Heat flow meter apparatus, 1991.
40. ANACON Thermal Conductivity Analyser, ANACON, Thame, Oxfordshire.
41. ISO 7616, Cellular plastics, rigid—Determination of compressive creep under specified load and temperature conditions, 1986.
42. ISO 2796, Cellular plastics, rigid—Test for dimensional stability, 1986.
43. ASTM D 2126, Dimensional stability, 1987.
44. BS 4443, Part 1, Flexible cellular materials, 1988.
45. ISO 2439, Polymeric materials, cellular flexible—Determination of hardness (indentation technique), 1980.
46. ASTM D 3574–95, Flexible cellular materials—Slab, bonded, and molded urethane foams.
47. BS3667 (withdrawn), Methods of testing flexible polyurethane foam.
48. BS 4443, Part 2, Flexible cellular materials, 1988.
49. ISO 3386/1, Polymeric materials, cellular flexible—Determination of stress–strain characteristic in compression—Part 1, Low-density materials.
50. ISO 3386/2, Polymeric materials, cellular flexible—Determination of stress–strain characteristic in compression—Part 2, High-density materials.
51. ISO 1798, Polymeric materials, cellular flexible—Determination of tensile strength and elongation at break.
52. EN ISO 8067, Flexible cellular polymeric materials—Determination of tear strength, 1995.
53. BS 4443, Part 7, Flexible cellular materials.
54. ISO 8067, Flexible cellular materials—Determination of tear strength.
55. ISO 4638, Polymeric materials, cellular flexible—Determination of air flow permeability.
56. ISO 7231, Polymeric materials, cellular flexible—Method of assessment of airflow value at constant pressure-drop.
57. BS 4443, Part 6, Flexible cellular materials.
58. GME 60 286, Test method for determining the air permeability of trim materials, GM Engineering Standards Europe.
59. WSD-M99D57, Polyurethane foam, pressure sensitive adhesive, low permeability, Ford Engineering Material Specification.
60. ESA–M4D200–B, Polyurethane foam, flexible—restricted flow, Ford Engineering Material Specification.
61. BS EN ISO 3385, Flexible cellular polymeric materials—Determination of fatigue by constant-load pounding.
62. BS 3379, Flexible polyurethane cellular materials for load bearing applications, 1991.
63. Flexible polyurethane foams, the Dow Chemical Company, 1991.

64. ISO 5999, Polymeric materials, cellular flexible—Polyurethane foam for load-bearing applications excluding carpet underlay—Specification.
65. ISO 4651, Cellular rubbers and plastics—Determination of dynamic cushioning performance.
66. EN ISO 1856, Polymeric materials, cellular flexible—Determination of compression set, 1996.
67. DIN 53572, Determination of the compression set after constant deformation.
68. Volvo 417–01, Saab 1082, General Motors GME 60 326, Method B, Ford FLTM BO 116–3.
69. BS EN ISO 9614, Acoustics—Determination of sound power levels of noise sources using sound intensity.
70. BS 4196, Part 5, Precision methods for determination of sound power levels for sources in anechoic and semi-anechoic rooms, 1981.
71. WSK–M2D406–A, Polyurethane foam, acoustic grade, Ford Engineering Specification.
72. BS 1133, Section 12, Packaging code. Methods of protection against shock, 1986.
73. BS EN 396, Lifejackets and personal buoyancy aids—Lifejacket 150, 1994.
74. BS 903, Methods of testing vulcanised rubber.

18
Particular Requirements for Composites

Graham D. Sims
National Physical Laboratory, Teddington, Middlesex, England

1 General

Polymer matrix composites (PMCs), or fiber-reinforced plastics (FRPs), provide a wide range of properties and behavior. Materials with discontinuous fibers are slightly stiffer than conventional unreinforced plastics, whereas the fully aligned continuous fiber systems can record exceptionally high specific properties (property divided by density), exceeding those of competing materials such as steel and aluminum. There are a virtually infinite number of materials, and material formats that can be combined to form a composite material, as shown in Table 1.

A range of generic materials exists that are commercially viable at each property or technology level (see Table 2). However, new methods of manufacture (e.g., thermoforming) and new products (e.g., stitched fabrics and co-mingled tows containing the reinforcing glass-fiber and a polymer fiber that will be melted to form the matrix) challenge existing design and test methods.

Characteristics of PMCs influencing the design and choice of test methods include the following.

1. The properties are anisotropic to varying degrees (i.e., mechanical, thermal, and electrical properties vary with direction in the material). The highest anisotropy is illustrated by the properties of a fully aligned 60% (by volume) carbon fiber/epoxy laminate, where the properties parallel with the fiber direction can be thirty times greater than in the perpendicular direction, whereas in a molded short-fiber system the ratio of properties in perpendicular directions may only be a factor of two. The fibers themselves may have even higher anisotropy (e.g., carbon and aramid fibers).
2. The ability to vary intentionally the fiber properties, format, volume fraction, and orientation and to choose different resins and additives result in an exceptionally

Table 1 Constitutent Materials

Constituent	Types
Fibers	glass, carbon, aramid, polyethylene, polypropylene
Matrices	polyester, epoxy, phenolics, vinylesters, polyimides
Additives	fillers, coupling agents, fire retardants
Sandwich cores	polymer foams (PVC, PU), balsa, honeycombs (aluminum, nomex)

wide range of materials (e.g., Young's modulus E varies from 1 to 250 GPa). A wide range of Poisson's ratio values are possible compared with the values obtained for homogeneous materials (e.g., 0.02 to 1.5 depending on fiber orientation, test direction, and materials).

3. For short fiber systems, the fiber orientations obtained are less controllable, but techniques are being developed to control fiber orientation during injection molding to some degree.
4. The variety of materials possible results in specimens having a range of behaviors. Thus, even in a single test, such as interlaminar shear, unacceptable failure modes may occur depending on the material characteristics.
5. The material is often produced by bringing together the fiber and matrix while manufacturing the final product. However, there is an increasing volume of composite materials supplied as premixes, such as aerospace preimpregnates (prepregs), sheet molding compounds (SMC), glass mat thermoplastic (GMT), and co-mingled glass and polypropylene fibers in tows and fabrics.

Table 2 Simplified Composite Material Classification System

Class	Fiber format	Typical products	Property
A	Continuous	Unidirectional prepreg, aligned, unidirectional fabrics, pultruded (aligned) rods, filament hoop wound cylinders	Highest anisotropy, layered structure, properties in transverse and through-thickness properties low and similar
B	Fabrics, combination of formats	Pultrusions (structural, general purpose), aligned SMC (XMC) and GMT, combinations of fabrics, mats and/or aligned	Reduced anisotropy, layered structure, through-thickness properties lowest
C	Mat (chopped, swirled)	Sheet molding compound (SMC), glass mat thermoplastics (GMT), chopped strand mat (CSM)	"Isotropic" in-plane, layered structure, through-thickness properties lowest
D	Discontinuous short (<7.5 mm) and long (>7.5 mm)	Injection molded thermoplastics. Bulk and dough molding compounds (BMC, DMC)	Nominally random in and through the plane, but flow induced anisotropy and layering

2 Material Systems

A wide range of constituent materials (see Table 1) is used in practical systems having two main constituent components (i.e., fiber and matrix) together with additives, and for sandwich structures, core materials. In addition, the fiber format can be short discontinuous (<7.5 mm length), long discontinuous (≥ 7.5 mm length), discontinuous mat (chopped at 25–50 mm), continuous mat (swirled), continuous aligned, and fabrics and mixtures of these formats (e.g., in pultruded profiles). A simplified classification system is given in Table 2.

The different combinations of components and formats lead to a wide variety of behavior and properties. Consequently, the test methods used must allow for these differences. The scope of multipart or multispecimen standards provides guidance as required on the appropriate test method or specimen design, so that, for example, a range of specimen designs are available in the tensile test discussed below.

3 Standards Bodies for Specialist Composite Test Methods

The principal sources of standard test methods for PMCs are the ISO (International Standards Organisation) series produced by the ISO TC61/SC13 committee and the CEN (Comité Européen de Normalisation) series produced by the CEN TC249/SC2 committee. The parent Technical Committees (ISO TC61 and CEN TC249) are responsible for all test methods for plastics, many of which are applicable to PMCs along with other classes of plastics. It is mandatory to publish CEN standards in Europe and to replace national European standards of the same scope, such as BSI (U.K.), DIN (Germany), and NF (France). CEN committees will adopt many ISO standards covering plastics and some PMCs. However, many new or updated standards covering more recently developed PMCs are being produced by collaborative action by ISO and CEN under the Vienna agreement.

A long established series of test methods is published by the D30 committee of the ASTM (American Society for Testing and Materials). In addition, other standards are produced by the Japan Industrial Standards (JIS) body, which are being replaced by ISO methods when published, and by various trade bodies such as the Society of Automobile Engineers (SAE). In Europe, the EN Aerospace series is used for aerospace applications. The unofficial CRAG recommendations [1] from the U.K. were a welcome first attempt to draw together a consistent set of test methods. Further details of these methods are given in several reviews published by Sims [2–5].

4 Test Panel and Specimen Manufacture

Composite materials are fabricated using a wide range of industrial processes ranging from hand layup used for the majority of trade moldings to autoclave molding typically used in the Grand-Prix and aerospace industries. The fabrication of panels for preparation of test specimens is covered by ISO 1268 [6], which is being revised as a multipart document covering most techniques to reflect the range of industrial processes used. The molding of thermoplastics and thermoset based composites reinforced with short fibers is done in a manner similar to conventional plastics using ISO 293 [7], 294 [8] and 295 [9]. These standards, and parts, are listed in Table 3.

The preparation of specimens is an important part of the test procedure. Some directions are included as annexes to the test method standards. A specialist document

Table 3 Test Panel Preparation Standards

Standard	Title
ISO 1268, Part 1	General principles
ISO 1268, Part 2	Contact and spray-up molding
ISO 1268, Part 3	Wet compression molding
ISO 1268, Part 4	Molding of preimpregnates
ISO 1268, Part 5	Filament molding
ISO 1268, Part 6	Pultrusion molding
ISO 1268, Part 7	Resin transfer molding
ISO 1268, Part 8	Molding of SMC/BMC
ISO 1268, Part 9	Molding of GMT/STC
ISO 1268, Part 10	Injection molding of BMC/DMC
ISO 293	Compression molding test specimens of thermoplastic materials
ISO 294	Injection molding test specimens of thermoplastic materials
ISO 295	Compression molding test specimens of thermoset materials

covering the machining of composites and specimen preparation is under study in CEN TC249/SC2/WG5 (c.f. the basic machining procedures for unreinforced plastics in ISO 2818 [10]).

5 Material Characterization

As explained earlier, due to the combination of components possible, and the final processing by the fabricator, it is important that the material tested be fully characterized. The properties to be measured include the following.

5.1 Component Fractions (e.g., Fiber, Matrix and Void)

For glass fiber or other inert fiber systems, the resin and fiber fractions can be obtained by burning off the resin at ~650°C for approximately 1 hour according to ISO 1172 [11]. The 1997 revision of this standard included additional clauses covering determination of soluble fillers. ISO 7822 [12] is used for determining void fractions, which for a CSM material widely used for trade moldings could be typically 5% compared to negligible for autoclaved prepregs used in the electrical and aerospace industry.

For carbon fiber systems, there is some loss of fiber with the resin burnoff method, so an alternative method based on chemical digestion, ISO/DIS 14,127 [13], has been developed. This standard also includes void fraction determination, but for high-quality aerospace materials, where the voidage is less than 1%, the method is not sufficiently accurate. However, a voidage level of less than 1% is generally acceptable and need not be measured.

5.2 Fiber Format

Fiber layout can be confirmed, if known, by careful examination of the residue from a resin burnoff test. Alternatively, polished through-thickness sections can be examined under a microscope.

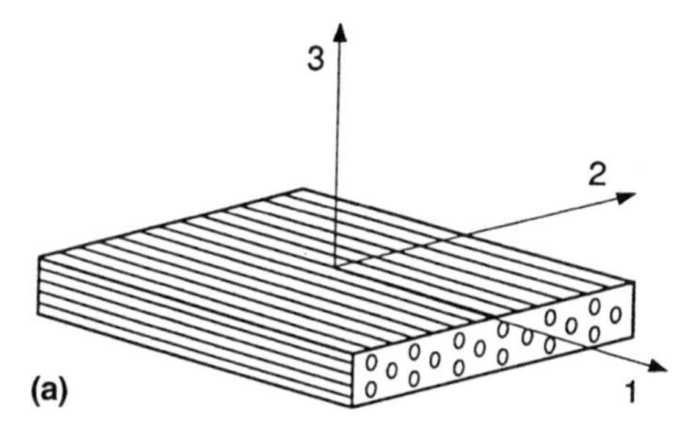

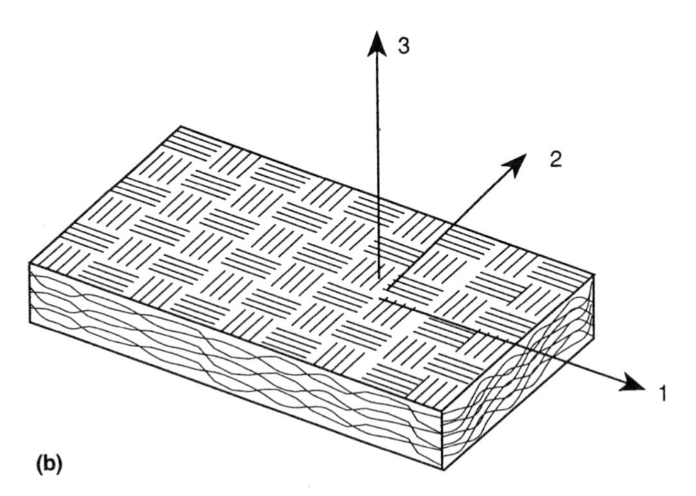

Figure 1 Axes of symmetry. (a) Unidirectionally reinforced systems. (b) Fabric reinforced systems.

These examinations are used to define the type of fiber formats, principal directions, and degree of anisotropy, as shown in Fig. 1. For some systems (e.g. pultruded profiles) two or three fiber formats will be used together.

5.3 State of Cure

It is important when conducting expensive test programmes that the constituents be checked for quality prior to use and that the fabricated materials be checked after manufacture, or on receipt if already fabricated. The techniques used vary from the Barcol hardness test (see below) to the more rigorous DMA [dynamic mechanical analysis: ISO 6721 (ten parts) [14]) or DSC [differential scanning calorimetry: ISO 11,357 (7 parts) [15]) test methods. There is some concern that these tests do not give the same values due to subtle variations between different manufacturers' equipment.

The simple and nondestructive Barcol hardness test has the added advantage that it can be conducted in situ on the factory floor (see Section 6). The other techniques are more costly, require a greater degree of operator training, and are not practical for in situ factory use. However, they are suitable for checking incoming material as preimpregnates or laminates.

6 Hardness

The only specialist use of hardness testing for composite materials is related to the assessment of cure using the Barcol hardness test available as EN 59 [16]. The test is frequently used as a low-cost quality check on the cure of laminates such as hand layup CSM materials used in marine and general purpose commercial moldings. Care should be taken to avoid cracking thin laminates.

7 Strength and Stiffness

The strength and stiffness of a composite material are the properties that are most generally recognized as being anisotropic. The test methods frequently require the properties to be measured in two mutually perpendicular directions depending on the material being evaluated.

7.1 Tension Properties

The redraft of ISO 527, Tensile test for plastics [17] has resulted in two complementary parts for composites. Part 4, based on and replacing ISO 3268, covers "isotropic and orthotropic" materials, and Part 5 covers "unidirectional" materials. These parts were drafted by the author working in cooperation with colleagues from other countries. The remaining three parts are Part 1, General principles, Part 2, Plastics and moulding materials, including short fiber-reinforced materials, and Part 3, Films.

The general principles given in Part 1 have been applied in the specialist parts. One of the two dumbell specimens in Part 2, for plastics, is available in ISO 527 Part 4 [18] for "isotropic" composites. This normally machined dumbell specimen is only successful for the molded materials in Class D of Table 2. For other classes of composites, unwanted failures normally occur at the grip but shear failures occur at the shoulders of the dumbell. In addition, to the dumbell specimen in Part 4, there is a straight specimen, 250 mm long × 25 mm wide × 2–10 mm thick, which may be tabbed or untabbed. There is an allowance in Part 4, if high scatter is obtained due to the 25 mm wide specimen being too narrow to be representative (e.g., for coarse reinforcements), to use a 50 mm wide specimen.

The main technical requirement was to obtain a harmonized version for the totally new Part 5. Inputs were included mainly from ASTM D3039, JIS 7073, CRAG 300, EN 61, and EN aerospace methods (2561, 2597, and 2747). EN ISO 527, Part 5 [19] was parallel voted as a CEN standard and should, with Part 4, replace EN61 and possibly the Aerospace EN standards.

Obtaining agreement between the different series of standards (e.g., ISO, EN, ASTM, and JIS) on specimen sizes for the unidirectionally reinforced materials covered in EN ISO 527, Part 5, was a notable first step towards harmonization of test methods, particularly as the 0.5 inch wide ASTM D 3039 specimen had been in extensive use for many years and contributed to many databases. The specimen sizes agreed for Part 5 are 250 mm × 15 mm × 1 mm in the 0° direction (parallel to fiber axis) and 250 mm × 25 mm × 2 mm in the 90° direction. These sizes are now the same for all series except EN Aerospace for glass fiber systems.

Several other aspects of the test method were decided during the drafting of the new standard following agreement on the specimen sizes. For example, a glass fiber fabric/epoxy, aligned at ±45° to the specimen axis, was chosen as the preferred tabbing material, but alternatives were allowed providing it was shown that the strength was at least equal to

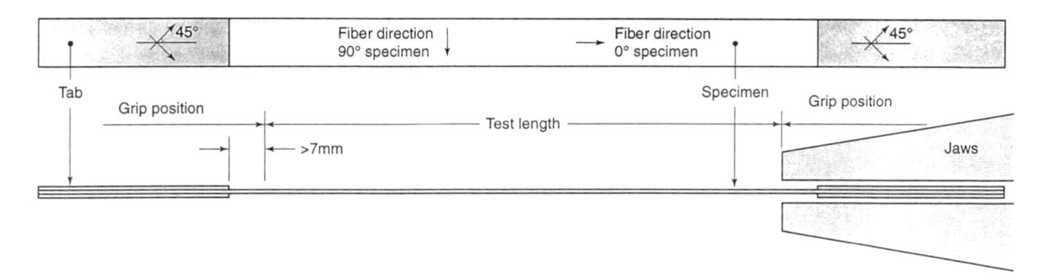

Figure 2 Tabbed tensile specimen.

that obtained with the preferred tab material and that there was no increase in scatter. A tab angle of 90° was adopted as in EN 2561 and CRAG [3d edition], with additional support from Japan, whose round-robin [RR] results were inconclusive regarding the preferred tab angle.

The modulus is calculated as the chord value between 0.0005 (=0.05%) and 0.0025 (=0.25%) strain to agree with Part 1 of ISO 527. However, the EN Aerospace series still uses load-based criteria (c.f. allowable strain design limits used) to determine the modulus (i.e., between 0.1 and 0.5 of the maximum load). The specimen preparation details are given now in a "normative" annex, prior to development of a separate standard on this aspect, as proposed in a CEN study item. The remaining differences between various versions of the tensile test to be tackled in the next revisions are given in Table 4.

Particularly when testing aligned materials in the fiber direction, it is necessary to use the minimum grip pressure sufficient to avoid slippage, as otherwise crushing of the specimen within the grips can occur due to the low through-thickness and transverse strengths, compared to the axial strength. This also applies to a lesser degree to other composite materials. A particular feature of Parts 4 and 5 for composites is that an informative annex has been added recommending an alignment check to be undertaken due to low failure strains involved, particularly for the 90° specimen of fully aligned materials.

Table 4 Remaining Differences Between Tensile Test Methods

Parameter	Options	Standards
Modulus measurement	Strain based (EN ISO 0.05%–0.25% or ASTM 0.1%–0.3%)	ISO, EN General, ASTM, JIS, CRAG
	Load based	EN Aerospace
Materials scope	All fibers covered by one standard	ISO, EN General, ASTM, JIS, CRAG
	Fiber specific, several versions needed	EN Aerospace
Application area	All	ISO, EN General, ASTM, JIS, CRAG
	Aerospace only	EN Aerospace
Tab angle	90 degrees	ISO, EN General, JIS, CRAG, EN Aeropsace
	Tapered (7–90 degrees)	ASTM

Test equipment requirements mainly relate to the need for a tensile capacity of 100 to 200 kN to test the higher-performance laminates. For strain measurements, clip gauge extensometers are commonly used as being more economical than strain gauges.

A similar approach to harmonization was used for other test methods in the CEN/ISO programme. These items were prepared by the author as convenor of the CEN TC249/SC2/WG5 and were CEN led "parallel processing" ballots. Technical harmonizers were nominated by ASTM and JIS for each method to cooperate with the CEN working group [WG5]. In all cases the approach and conditions represented in ISO 527–5 were introduced into the new documents as appropriate using the same clause structure and clause content whenever applicable, for example, in clauses on tab material and specimen preparation. This has made it easier to prepare the documents to a consistent standard and for users to understand the content of the clauses; and to reduce costs in specimen preparation and testing by using the same procedure.

7.2 Compression Properties

Compression testing is particularly difficult for PMCs due to the occurrence of macro- and microbuckling modes. The rectangular prism specimen in ISO 604 is not suitable for laminated composites, as these will split vertically between the laminate layers when loaded axially in the plane of the laminations. Variations on these designs are suitable for compression, and tensile testing, in the through-thickness direction (i.e., direction 3 in Fig. 1).

Most composites are available as relatively thin laminated sheets, so that tensile type specimens 2 mm to 4 mm thick are not suitable for compression tests, as they fail by Euler column buckling. Three approaches have been used to avoid specimen buckling;

1. Increased specimen thickness, limited by machine capacity.
2. Short test span (i.e., column height equal to the "gauge" length): this is the most popular.
3. Support jig on faces of specimen, for example as used in ASTM D695; however the level of support and load carried by the jig is unknown.

In practice, a short gauge length, 10 mm for a 2 mm thick specimen of unidirectionally reinforced material, or 25 mm for a 4 mm thick fabric or mat reinforced specimen, is most frequently used. A framework has been developed in order to rationalize the many test methods and the even greater number of loading fixtures/jigs. The final failure will frequently be by local microbuckling, or kinking, of bundles of fibers. A similar failure occurs in individual carbon and aramid fibers, which have a fibrillar structure.

The approach taken is to validate the quality of the test by requiring the specimen to be strain gauged on both faces and for the test to be completed, or terminated, within a buckling strain of less than 10% of the axial strain. By concentrating on the quality of the test, a range of jigs can be used, thus allowing the continuing development of jigs and ensuring that established jigs are used correctly.

The compression standard EN ISO/DIS-F 14,126 [20] was based on ISO 8515 with input from prEN 2850, ASTM D3410, CRAG 400, and ASTM D695 (modified). Both shear face loaded and end loaded jigs, as in ISO 8515, are allowed (see Fig. 3). The jig used must be reported, either an established jig from the annex or any other that is used. The standard allows a disposition, requested by the U.S.A., that providing the first five specimens fail correctly within the 10% criteria, further specimens of the same batch, tested at the same time by the same operator, equipment etc, to not require the strain to be

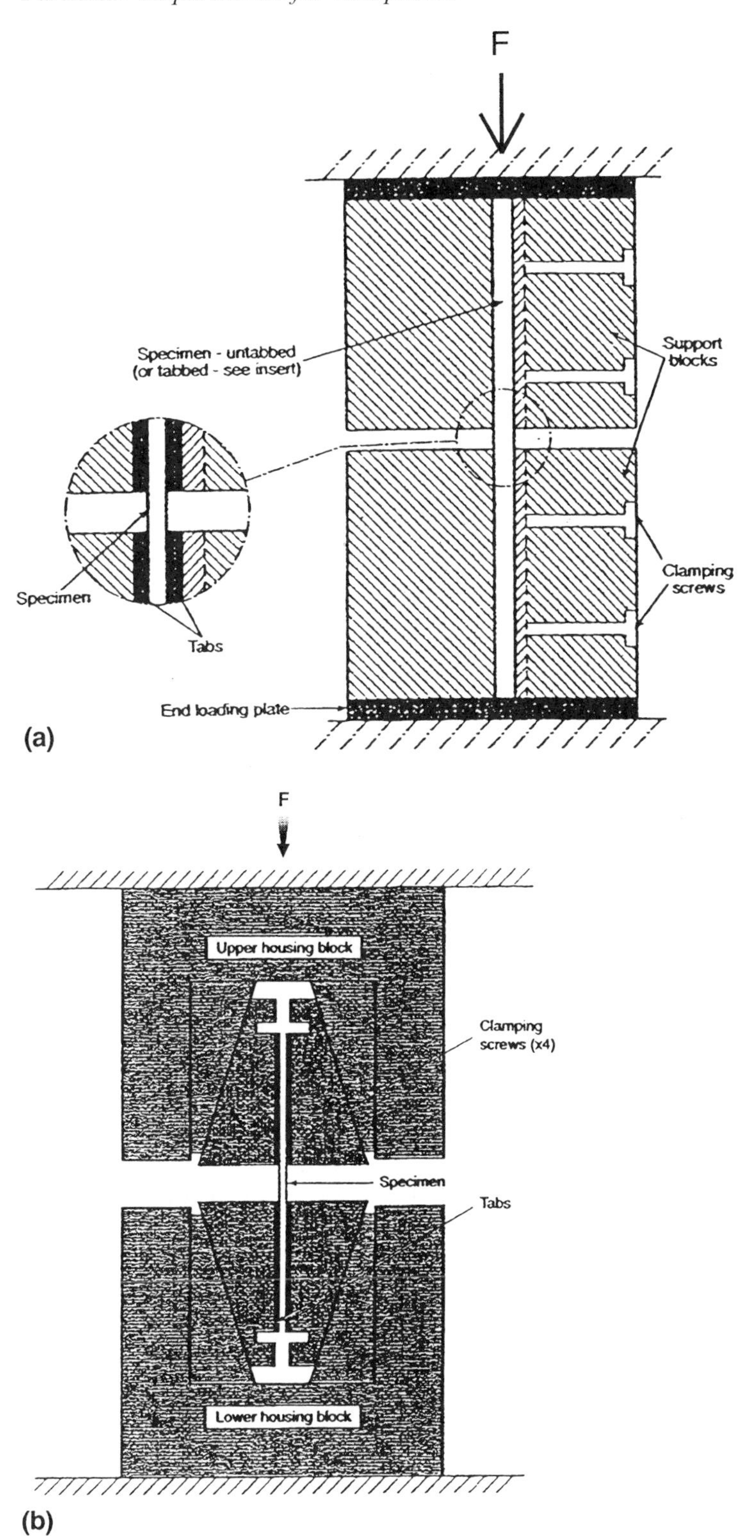

Figure 3 Compression specimens and loading jigs.

measured on both faces. It should be noted that the ASTM D-30 committee sent back their completed revision of ASTM D 3410, shortly before publication, in order to achieve greater harmonization with the CEN/ISO standard (i.e., to use 10 mm width and gauge length rather than 12.7 mm as proposed initially (~0.5 inch)).

A second specimen size, 25 mm wide by 25 mm gauge length, is applicable for coarser fabrics, mats, and multidirectional layups but requires a 4 mm or greater thickness, based on ASTM D3410 guidelines, to avoid premature failure by Euler buckling. There was no general support for including the machined specimen described in prEN 2850. This specimen requires a separate special layup and careful machining and polishing of the cut faces and radii. This is not a practical situation for general engineering use when a finished test panel is supplied to obtain a set of properties. Also, results from this method do not always show an improvement over other methods.

7.3 Three- and Four-Point Flexure

Flexure testing is a cheap and easy method of obtaining data that are almost equivalent to data from axially-loaded tests. However, there are several difficulties in applying this "simple" method. In a flexure test a proportion of the measured deflection occurs in shear, rather than in flexure, resulting in a modulus 1% lower than that achieved by a tensile test. The effect is proportional to the ratio of the tensile Young's modulus to the through-thickness shear modulus. For composite materials this ratio is much higher (e.g., 5–50) than for polymers (e.g., 2.5), resulting in a larger error in the modulus of the order of 10% at the same span (i.e., 16 × specimen thickness.) In order to obtain an equivalent or comparable value, the span must be increased in proportion. The span/thickness ratio is 40 for a unidirectionally reinforced carbon fiber specimen.

The other difficulty concerns the tendency for different failure modes to occur. For example, materials with mat reinforcement will fail in tension, while a fully unidirectional specimen will often fail on the compression face through local microbuckling assisted by the local compression and shear loads at the central loading roller(s).

The three and four point flexure standard EN ISO 14,125 [21] was based on ISO 178 with input from ASTM D790, JIS K7074, EN 63, CRAG 200, and EN Aerospace 2562 and 2746. Composite materials are divided into four classes related to their structure and ratio of Young's modulus to through-thickness shear modulus, as in Table 2. Care is required in determining the failure mode. A four-point loading arrangement has been included to delay the initiation of compression failure by reducing the load on the central loading roller(s). In some cases, failure of the compression face can be avoided by use of a 0.2 mm thick pad beneath the center loading roller, following Japanese practice. However, flexure tests would not be recommended for generating design data.

7.4 Shear Test Methods

A range of shear tests exists to satisfy all the requirements for in-plane and some through-thickness shear properties. Although torsion of tubes or rods is often recommended, it is rarely used in practice and has not been standardized. Equally, the rail shear test only appears as an ASTM guide [ASTM D 4255]. It is not favored, since a large specimen is required and the test suffers from premature failure initiated at the bolt holes. Several other methods that have been standardized are discussed below.

1. The *apparent interlaminar shear test* uses a short beam flexure loading mode to measure the shear strength along the plane of lamination. The analysis of the test is

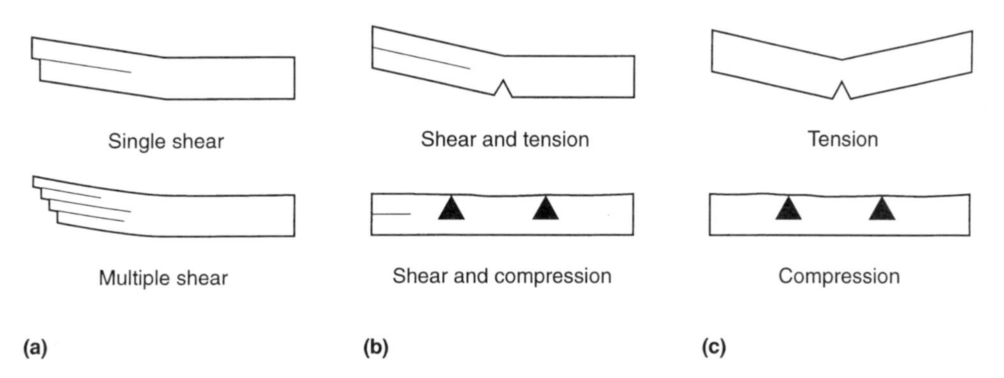

Figure 4 Interlaminar shear test by short beam flexure—failure modes.

simplified by assuming that the material is homogeneous, and the apparent strength determined is not recommended for use in design. EN ISO 14,129 [22] is based on ISO 4585 with input from ASTM D2344, CRAG 100, EN 2377, and EN 2563. The only significant change was to use a 2 mm thick standard specimen to take into account the existing preferences when testing aerospace materials, with other thicknesses being accommodated by the application of the existing scaling rules. An increased emphasis is placed on obtaining an interlaminar shear failure, as other failure modes frequently occur, such as tension for mat reinforced materials or compression for fabric reinforced materials, as shown in Fig. 4. Modulus is not measured by this technique.

2. For the measurement of *in-plane shear properties* of unidirectional and fabric reinforced specimens the ±45° tensile test can be used. The EN ISO 14,130 [23] standard was based on ASTM D3518 with input from prEN 6031 and CRAG 101. It applies to all fibers, and in common with the ASTM revision the test will be terminated at 5% shear strain. This strain limit results in a shortening of the normally excessive test duration for tougher matrix systems that have a very flat load-time response. It also minimizes fiber rotation and heating effects. The peak load obtained before or at 5% shear strain is taken as the shear strength. Both modulus and strength data are obtained using specimens and procedures similar to EN ISO 527–4. Strain gauges set at ±45° are normally used to measure the shear strain. The shear modulus is measured between 0.001 (=0.1%) and 0.005 (=0.5%) shear strain (i.e., twice the strain used in the tensile and compressive test values).

It has been reported that the results obtained are dependent on the number of layers, or shearing interfaces, in the specimen. This may have technical and harmonization implications for the Airbus Industries/prEN 6031 version of this test method, which is the only one using 1mm thick specimens as opposed to 2 mm in all others. It is recommended that the number of shearing faces (e.g., interfaces) should be kept constant when individual ply thickness varies from the normal 0.125 mm plies. The test is not suitable for a wide range of PMCs, which prompted the recent development of the plate-twist test described below.

3. The *double notch shear test* (e.g., Iosipescu) has been standardized as ASTM D 5379 [24] and can be used for both strength and modulus measurements (see Fig. 5). Modulus is relatively straightforward once the small strain gauges are applied in the center of the specimen. The measurement of shear strength needs additional care as, for some composite materials and some directions, the failure is initiated by localized tensile stresses rather than shear stresses. Redesign of the loading jig in the recommended ASTM drawing

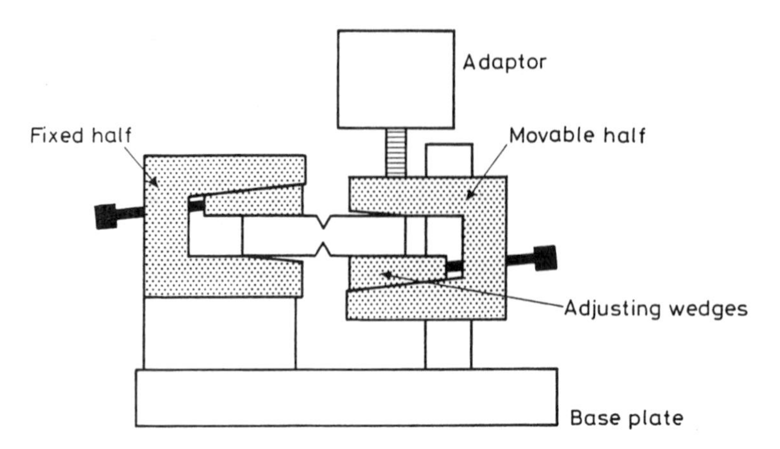

Figure 5 Double notch test.

for this standard has been proposed by NPL to reduce the tendency for unsymmetrical loading of the specimen.

4. The *plate-twist test for in-plane shear modulus* requires only that a square plate be supported on two diagonal points and loaded on the other diagonal by two further points. The test is similar to a flexure test and has the same benefits of low loads, large (linear) specimen deflections, cheap specimens, and simple test equipment (see Fig. 6). The test has been standardized for plywood for many years and has featured in many reviews over the last 25 years.

Recently, a revised analysis developed at NPL allowed the loading and support points to be positioned in-board of the actual specimen plate corner, with a major effect on the ease and speed of the test. The in-plane shear modulus by plate twist, ISO/DIS 15,310 [25], was at the DIS ballot stage in 1997. Precision data obtained in a round-robin exercise organized by the National Physical Laboratory supported the original New Work Item (NWI) submission by the U.K. No special layup is required for the specimen, but as the shear loading is developed in flexure, the test material should appear "homogeneous" through the thickness and have orthotropic symmetry. Therefore for a 0°/90° layup there should be several layers rather than three. Figure 7 shows data obtained for a single set of test plates tested at several sites.

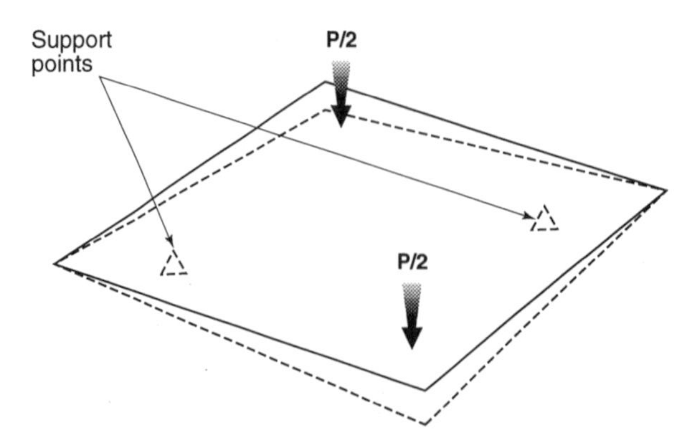

Figure 6 Plate-twist geometry.

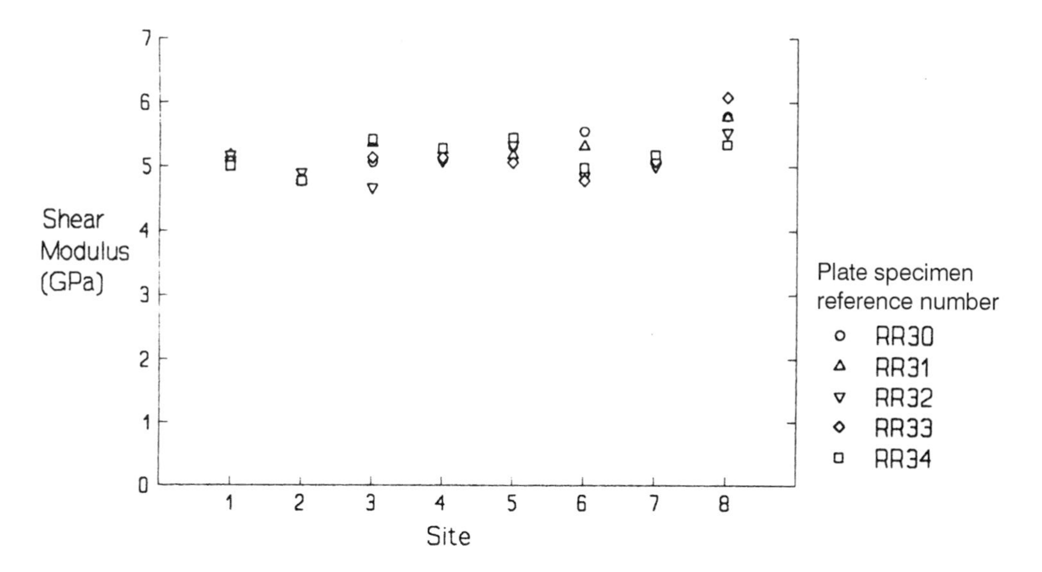

Figure 7 Plate-twist test precision data.

7.5 Impact and Fracture Toughness Test Methods

The measurement of "toughness" properties is one area where the particular nature of composites (i.e., poor interlaminar strength of laminated composites) affects both the use of existing tests and the creation of new test methods.

Beam Impact

In general, the notched Charpy (ISO 179) [26] and Izod (ISO 180) [27] tests are not meaningful for composites, and ISO 179 recommends that only unnotched specimens should be tested. The difficulty with these tests is that in the notched condition the majority of specimens tested perpendicular to the plane of the test panel delaminate at the root of the notch. This reduces the specimen to a thinner version of the unnotched specimen, which as described above for the interlaminar shear test (see Section 5.4) is susceptible to compression-initiated failures under complex local loads. Specimens cut in the plane of the laminate or sheet will be less susceptible to delamination at the notch tip, and crack growth will be possible from the notch tip. However, other compression/shear failure modes are still possible in some composites, and they will not be loaded in this direction in most applications.

Puncture Impact

More recently, the trend has been to test plastics, reinforced and unreinforced, by the falling weight or falling dart test using either Part 1 (uninstrumented) or Part 2 (instrumented) of ISO 6603 [28]. When tested on the 40 mm diameter support, the failure can be limited and influenced by the proximity of the edge of the support ring. The 100 mm diameter support added during the latest revision in 1997 is much less likely to influence the failure. This difference in failure behavior also applies to unreinforced plastics. The instrumented tests allow load, displacement, and energy values to be obtained at different points throughout the test, such as at the peak, at any particular event, and at test completion (i.e., test specimen puncture).

Compression-After-Impact (CAI)

In the aerospace industry an application representative test has been developed specifically for composites that is not required for a metal. The test assesses the damage tolerance of a panel (e.g., an aircraft wing skin) that has been impacted to a small degree compared to the puncture test above. The impact energy applied is considered to be equivalent to a tool dropped by a maintenance engineer or stones thrown up from a runway during take-off or landing. The impact causes delamination between the ply layers, plus some associated fiber and matrix damage, without necessarily showing any blemish on the impacted surface. The damage is described as barely visible impact damage (BVID). This damage occurs in composites due to their poor properties in the through-thickness direction. The effect of the damage is assessed by compression of the test panel, with all edges supported, until the initial damage propagates and specimen failure occurs by buckling.

The damage state after the impact phase was initially assessed using ultrasonic C-scan to measure the total projected delaminated area. This area does not accurately reflect the delaminations, which tend to form a helix of "petals" between different plies. Recently, input from aircraft maintenance requirements has promoted a "dent" criterion related to inspection of the impacted surface. This approach is based on the external damage that can be found during inspection and, although the amount of damage will vary depending on the material used and support conditions, the component must remain serviceable within the design strain. However, the dent size criterion was found in a U.S.A. programme to be inconsistent as a failure criterion and further study to define this test is required.

The test is expensive to undertake because it uses a relatively large and thick specimen. It is also rather arbitrary in nature in combining two aspects (i.e., impact and compression). It would be useful to perform the first phase using ISO 6003–2 (Instrumented plate impact) using the 100 mm diameter support included in the current revision, so that a standard procedure is used and extra data quantifying the impact (e.g., energy absorbed, peak load, and displacement) can be obtained. The EN Aerospace version used by Airbus, originally developed by Boeing, requires extra material to obtain confirmation of critical energy level currently based on a dent (c.f. delaminated area measured by ultrasonic C-scanning), although the actual dent size used for the criterion varies in different drafts.

Although the test was developed for use in the aerospace industry and is used in their materials specifications, the data are being offered to other industries. Some tests of this type are relevant to other types of composites (e.g., delamination of CSM materials in chemical tanks or boat hulls), but current standardization is orientated towards the aerospace industry. The tests have been shown to be less satisfactory in circumstances different from the initial problem (e.g., thicker material, sandwich materials). Researchers in Japan and the U.K. have proposed a smaller specimen, which lowers the material cost and allows a lower test machine capacity to be used in the compression phase of test.

Fracture Toughness Tests

Fracture toughness tests for isotropic materials normally use edge-notched flexure beams and compact tension specimens. These methods are, as for the impact tests, only suitable for the injected or compression molded materials, which do not have a strongly laminated structure. Laminated composite materials have their primary failure path between the layers, and a new set of test geometries has been developed.

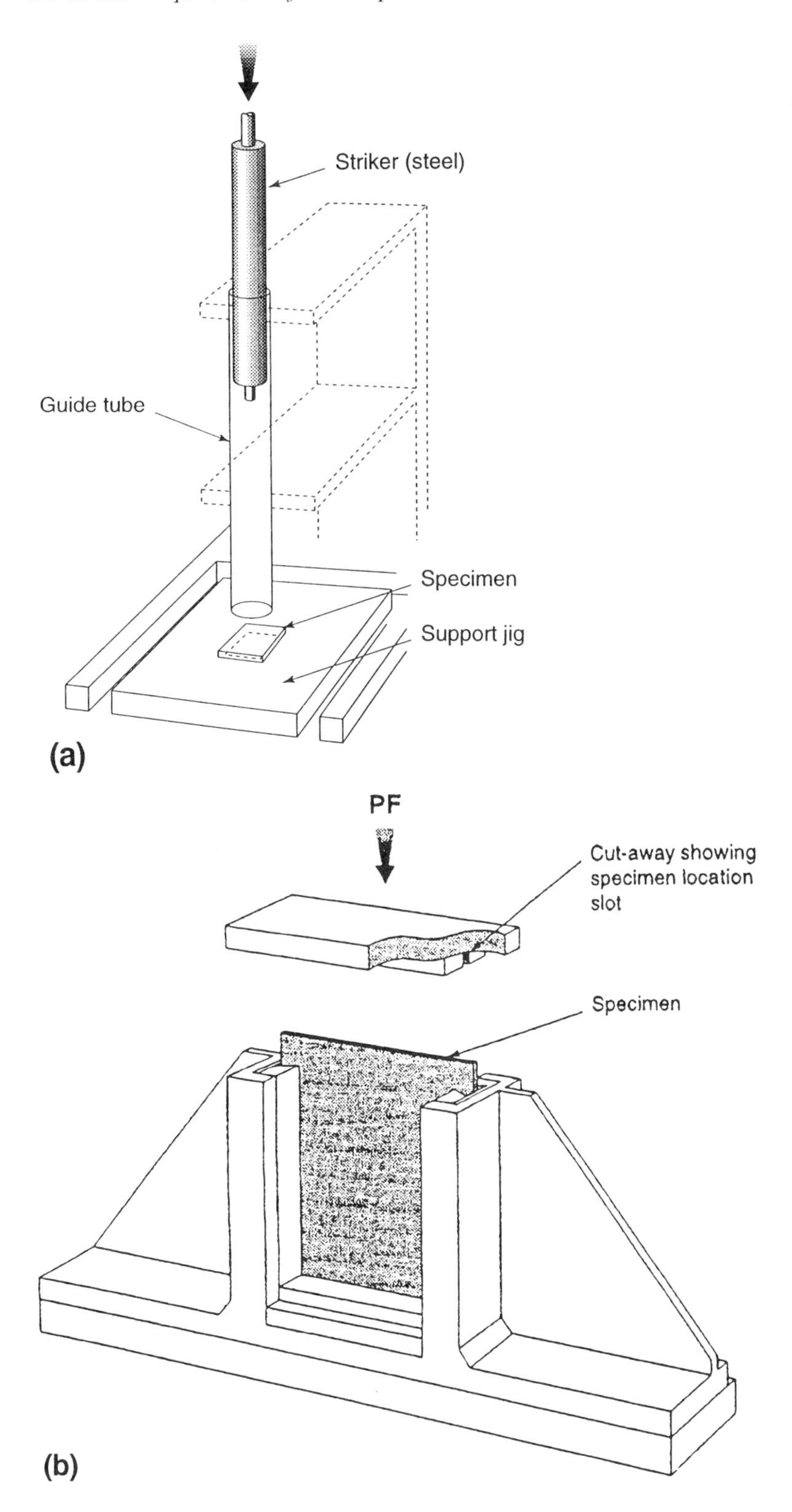

Figure 8 Compression after impact jigs.

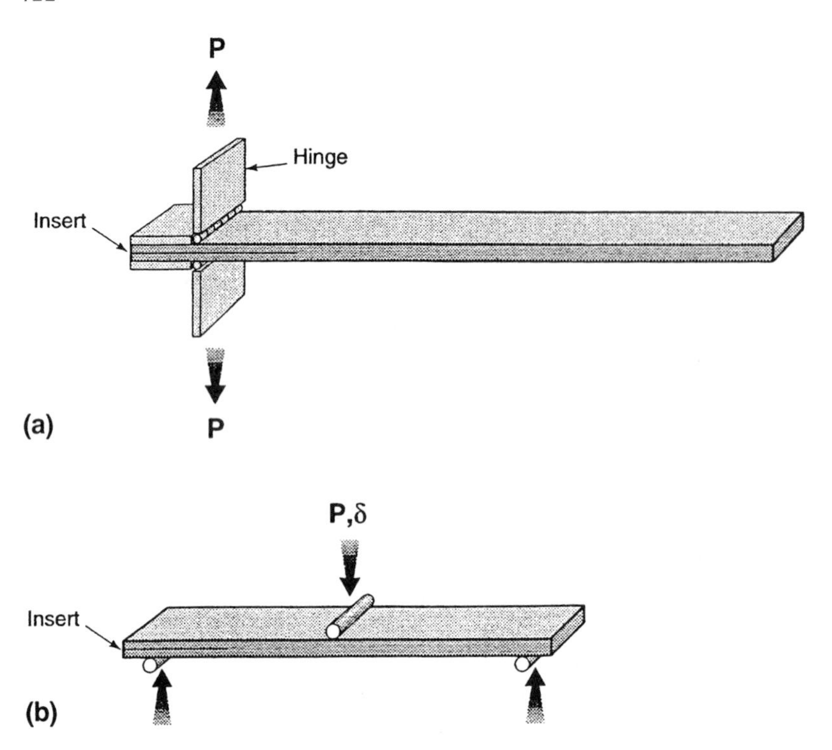

Figure 9 Composite fracture toughness test modes (a = Mode I, b = Mode II).

Mode I opening mode tests propagate the crack along the interlayer plane using the procedure being standardized as ISO/DIS 15,024 [29] (see Fig. 9). A thin film (e.g., PTFE) is laminated into the composite on the midplain to act as a starter notch.

Agreement had been obtained internationally for Mode I test conditions, whereas in 1997 international round-robins were being organised by the prenormalisation VAMAS programme to compare competing Mode II shear mode methods. Shear modes considered are end-notched flexure (ENF) under three-point bending, stabilized ENF, four-point versions of ENF, and end-longitudinal split (ELS).

8 Fatigue, Creep, and Wear

As composites are proposed in more infrastructure applications, such as bridges, there is an increased demand for long-term data suitable for reliable prediction of lives as long as 120 years.

8.1 Fatigue Testing

Fatigue testing of PMCs covered by ISO/CD 13,003, General procedures [30] uses the standard test geometries covered above, including for determination of a reference short term or static ultimate strength. Additional recommendations are made in annexes for tension and flexure test modes, but the approach given can be used for any test geometry and mode. For glass fiber reinforced systems, the standard recommends that the ultimate strengths be measured at a similar rate to that used in the fatigue test (i.e., failure in 0.25

seconds rather than the usual 30–90 seconds) to allow for the rate dependence of the properties of these systems.

The maximum fatigue frequency that can be used depends on the material under test. For glass fiber reinforced systems the self-generated (or autogenous) heating limits the test frequency to 2–5 Hz in tension. Higher frequencies can be used for flexure due to the smaller volume of material under the maximum stress. Higher frequencies can also be used for carbon fiber based systems because of the lower strains applied and the improved heat dissipation due to the higher thermal conductivity of the test material.

Tensile tests normally use load control with failure defined by specimen separation, whereas the flexural loading is undertaken under displacement control. In this latter case the fatigue failure in composites normally results in well-distributed damage causing "softening" of the specimen and a falloff in load. This behavior is prompting the development of a failure criterion related to the loss of specimen stiffness (e.g., 20%). Experimental techniques have been developed to measure specimen stiffness continuously during the fatigue test.

8.2 Creep Tests

Specific creep tests have not been developed for composites at the coupon level. Data have been obtained using several of the standard test geometries described above, but creep is normally considered less of a problem than fatigue. The creep tests for plastics, ISO 899 [31], can be used for guidance with the standard composite specimens.

8.3 Wear Testing

Wear testing, when conducted, would follow standard procedures, while being aware of the anisotropic nature of the material under test. For example, for a unidirectional material, wear along the fibers would be different from across the ends of the fibers.

9 Effect of Temperature

It is recommended for all plastics that the reference temperatures in ISO 3025 [32] are used. This will improve the quality of data collected by allowing easier comparison of different materials.

The standard mechanical tests, as described in Section 7, can normally be undertaken with care for composites as a function of temperature. The difference between fiber- and matrix-dominated properties can result in different temperature dependencies. Changes in the residual thermal stresses present can occur both between the fibers and the resin, and between layers, in particular between 0° and 90° orientated unidirectional layers. Care needs to be taken in assessing the failure mode, particularly in flexural and compressive tests where there can be changes, particularly at elevated temperatures, due to the matrix providing a lower degree of support to the fibers, thus encouraging compression failure.

Tabbing of high-performance materials needs additional consideration at elevated temperatures to ensure the tab material and adhesive are acceptable at the test temperatures. Some test fixtures, as used for compression testing, are more difficult to use at nonambient temperatures. Additional care needs to be taken on strain measurement, especially if using bonded strain gauges.

It should be noted that the determination of the temperature of deflection under load for PMCs uses ISO 75–3 [33]. Unlike Part 2, which uses fixed applied loads, Part 3 uses a load equivalent to a proportion of the maximum flexure load in order to allow for the wide range of initial stiffness found in PMCs. Some inconsistencies have been noted compared to Part 2 data for reinforced and unreinforced versions of the same polymer that have resulted in some users preferring to use for PMCs Part 2 rather than the correct Part 3. The new ISO 10,350–2 [34] database standard for comparable data for materials selection will recommend in the second CD draft that Part 2 of ISO 75 should be used for the measurement of the heat distortion temperature until this problem is resolved.

In addition, a specific tests for composites ASTM 2733 [35] uses the double notched specimens (c.f. single lap joint) to assess the shear strength at elevated temperatures. An extension of this test is in an internal aerospace test [36], which uses "thermal spiking" to assess performance. In this test a sample is conditioned at the aerospace standard 70°C/85%RH for two weeks prior to a rapid excursion for a maximum of one minute to a higher service temperature. The specimen is then conditioned under the standard conditions for a further two weeks prior to the evaluation of the weight gain.

10 Environmental Resistance

At the material level, the environmental tests available would be mainly the same as for unreinforced plastics, although this topic is being studied by CEN TC249/SC2/WG5. Specific tests are either associated with particular industry sectors (e.g., aerospace) or product area (e.g., process and chemical plant).

Specific tests in the aerospace industry include the assessment of humid atmospheres on physical and mechanical properties. In prEN 2823 [37] the normally used conditions are 70°C and 85% RH, although 95% RH is also available. Some users consider both these options are more severe than normal operating conditions and may unnecessarily penalize the material properties that can be used in design. ASTM D 5229 [38] specifically provides for measuring moisture uptake in polymer matrix composites.

In addition to the ISO 175 [39] standard for chemical degradation assessment for all plastics, the aerospace standard, EN 6063 [40] uses microscopic assessment of microcracks following immersion in the test environment. A particular test, EN 6030 [41], used in this industry is for the assessment of the effect of chemical paint stripper using the ±45° tensile test for shear properties.

Additional environment tests at the product level (e.g., see BS 4994 and BS 7676—being replaced by EN equivalents) use a single sided plate or a deformed section of a pipe. These tests are appropriate to the application when exposure will be one-sided on a nonhomogeneous layup (e.g., use of resin rich and tissue surfaces). The combined application of load and the aggressive environment is considered most representative of actual service conditions.

11 Composite Material Database

Following the success of ISO 10,350 [34] on comparable data for materials selection for plastics, it was agreed in 1996 to prepare two parts. Part 1 covers unreinforced plastics and plastics reinforced by fibers shorter than 7.5 mm, and Part 2 covers reinforced plastics with fibers greater than 7.5 mm. These database standards quote the appropriate international standard, normally ISO as discussed above. The original standard has been adopted widely leading to a greater use of the referenced ISO standards. The wide range of material properties found in composite materials requires that the correct method (e.g., Part 4 or 5 of ISO

527) for the material under evaluation be used. The data obtained still retains the quality of equivalent accuracy and relevance as embodied in the original standard, which was able to use a single specimen in each case because of the greater similarity of these materials.

References

1. P. T. Curtis (1988), CRAG test methods for the measurement of engineering properties of fibre reinforced plastics, 3d ed., RAE TR 88012 .
2. G. D. Sims (1990), Standards for polymer matrix composites, Part II, Assessment and comparison of CRAG test methods, NPL Report DMM(A)7.
3. G. D. Sims (1991), Development of standards for advanced polymer matrix composites, *Composites 22*, 4, 267–274.
4. G. D. Sims (1994), International standardisation of coupon and structural element test methods, ECCM—Composites: Testing and standardisation 2, Hamburg.
5. G. D. Sims (1994), The UK contribution to the international standardisation of composite test methods, BPF Congress, Birmingham.
6. ISO 1268, Composites and reinforced plastics—Preparation of test plates.
7. ISO 293, Compression moulding test specimens of thermoplastic materials.
8. ISO 294, Injection moulding test specimens of thermoplastic materials.
9. ISO 295, Compression moulding test specimens of thermoset materials.
10. ISO 2818, Plastics—Preparation of test specimens by machining.
11. ISO 1172, Textile glass reinforced plastics—Determination of loss on ignition.
12. ISO 7822, Textile glass reinforced plastics—Determination of void content; loss on ignition, mechanical degradation and statistical counting methods.
13. ISO/DIS 14,127, Composites—Determination of resin, fibre and void content for composites reinforced by carbon fibre.
14. ISO 6721, Plastics—Determination of dynamic mechanical properties.
15. ISO 11,357, Plastics—Differential scanning calorimetry.
16. EN 59, Glass reinforced plastics: Measurements of hardness by means of a Barcol impressor.
17. BS EN ISO 527, Plastics—Determination of tensile properties.
18. BS EN ISO 527, Part 4, Plastics—Determination of tensile properties—Test conditions for isotropic and orthotropic fibre-reinforced plastic composites.
19. BS EN ISO 527, Part 5, Plastics—Determination of tensile properties—Test conditions for unidirectional fibre-reinforced plastic composites.
20. EN ISO/FDIS 14,126, Fibre-reinforced plastic composites—Determination of the in-plane compression strength.
21. BS EN ISO 14,125, Fibre-reinforced plastic composites of the flexural properties.
22. BS EN ISO 14,129, Fibre-reinforced plastic composites—Determination of the in-plane shear stress/shear strain, including the in-plane shear modulus and strength, by the $\pm 45^\circ$ tension test method.
23. BS EN ISO 14,130, Fibre-reinforced plastic composites—Determination of apparent interlaminar shear strength by short-beam method.
24. ASTM D 5379, Shear properties of composite materials by the V-notched beam method.
25. ISO/DIS 15,310, Reinforced plastic—Determination of in-plane shear modulus by the plate twist method.
26. ISO 179, Plastics—Determination of Charpy impact strength.
27. ISO 180, Plastics—Determination of Izod impact strength.
28. ISO 6603-2, Determination of multi-axial impact behavior by the instrumented puncture test.
29. ISO 15,024, Determination of Mode I delamination resistance of unidirectional fibre-reinforced polymer laminates using the double cantilever beam.

30. ISO/DIS 13,003, Fibre reinforced plastics: Determination of fatigue properties under cyclic loading.
31. ISO 899–1, Plastics—Determination of creep behaviour—Part 1: Tensile creep.
32. ISO 3025, Preferred test temperatures.
33. ISO–75, Plastics—Determination of temperature of deflection under load—Part 2, Plastics and ebonite.
34. ISO 10,350–2, Plastics—Acquisition and presentation of comparable single-point data—Part 2, Long fibre-reinforced plastics.
35. ASTM 2733, Interlaminar shear strength of structural reinforced plastics at elevated temperatures.
36. Unpublished—Determination of the influence of thermal spikes on fibre reinforced composites.
37. prEN 2833, FRP—Determination of the effect of exposure to humid atmosphere on physical and mechanical characteristics.
38. ASTM D 5229, Moisture absorption properties and equilibrium conditioning of polymer matrix composite materials.
39. ISO 175, Plastics—Determination of the effect of liquid chemicals, including water.
40. prEN 6063, FRP—Determinatrion of material degradation due to chemical products.
41. prEN 6030, FRP—Determination of the mechanical degradation due to chemical paint strippers.

TC = Technical committee SC = Subcommittee WG = Working Draft
CD = Committee Draft DIS = Draft International Standard
F-DIS = Formal (final) vote on DIS

19
Textile Polymers

Frank Broadbent
British Standards Ltd., London, England

1 Scope

The study of textile materials covers an immense range, from fiber chemistry and fiber formation through fiber and filament processing; it embraces a multitude of different techniques for the preparation and production of fashion fabrics and highly specialized technical fabrics and their eventual finishing, recycling, and physical properties. Some of these considerations are fiber specific, whereas some are more related to the method of fabric formation. For present purposes the scope is limited to only those aspects of the polymer that are specifically related to synthetic or more properly manufactured fibers and takes no account of purely fabric construction-related properties. To do otherwise would clearly be beyond the bounds of this present publication, especially when one considers the range of textile polymers available, and more so if naturally occurring fibers and regenerated fibers were to be included. The range and classification of textile fibers is illustrated in Fig. 1

The term "polymer" encompasses a wide range of textile materials when one considers the generic definition of the term given in *Textile Terms and Definitions* [1]: "A large molecule built up by the repetition of small chemical units".

A similar definition appears in ASTM D 4466 [2]: "A macromolecular material formed by the chemical combination of monomers having either the same or different chemical composition".

The different types of textile polymer chains can be further classified into atactic, isotactic, and syndiotactic.

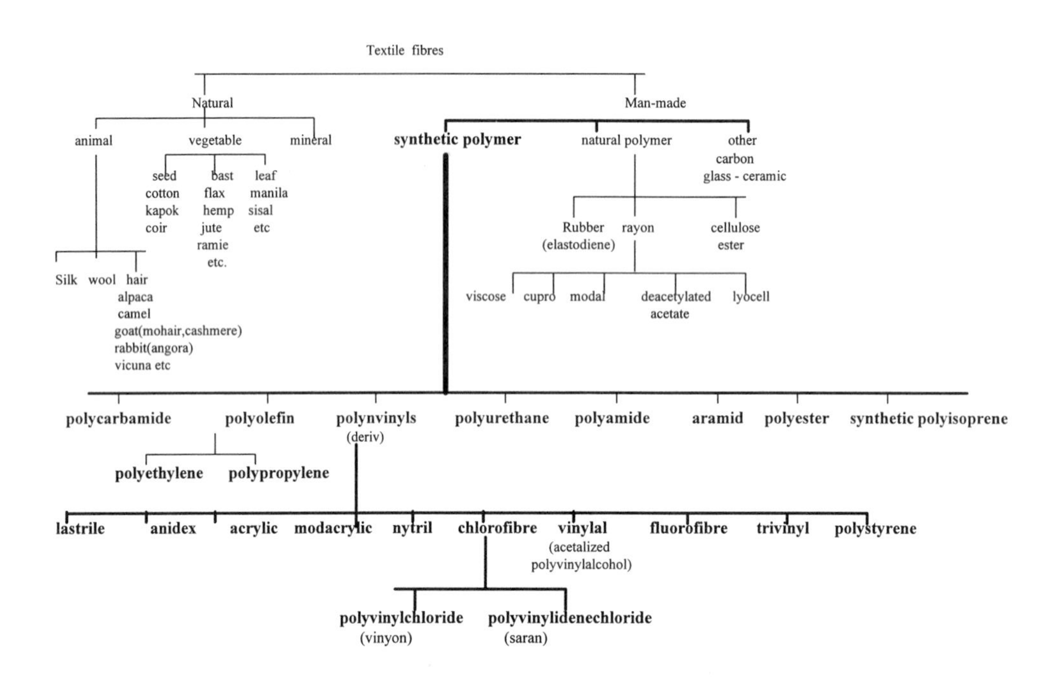

Figure 1 Classification of textile fibers.

2 Orientation and Crystallinity

For the present, probably the most significant feature of textile polymers is their anisotropism. Indeed, it might even be argued that if a polymer (or perhaps more accurately a polymer product) does not exhibit anisotropism, it is not in the family of products that can be described as textiles. So perhaps the logical place to start our investigation of textile polymers is with the investigation of isotropy.

Molecular orientation and crystallinity are of course not the same thing, but the effect they have on the way a polymer performs is often difficult to distinguish. So it is important to identify the degree of molecular orientation as well as the degree of crystallinity, as this provides useful information in relation to a number of performance and process parameters, such as tensile modulus, elasticity, moisture relations, and dye uptake.

2.1 Optical Birefringence of Fibers

The birefringence of textile fibers provides a measure of the molecular orientation of the polymer chain of the fiber. It might also be something of an indicator of the degree of crystallinity of the polymer chain. Birefringence is the difference between the refractive indices of the fiber in a direction parallel to the fiber axis and in a direction at right angles to the fiber axis. Details of how this refractive index is obtained are discussed elsewhere [3], but briefly the fiber is mounted in a liquid whose refractive index is known (or has been previously determined) and the fiber is then examined in plane-polarized monochromatic light until the fiber, in the direction of viewing, becomes invisible. Of course if the liquid used is operative in both the parallel (||) and in the perpendicular ($\perp$) directions, then the fiber can be said to be optically isotropic, as the refractive indices in both directions are the same, and the birefringence value will be 0.000. It follows that the greater the difference in the refractive indices ($\Delta\eta = \eta\| - \eta\perp$), the greater the degree of molecular orientation. In the great majority of fibers this value is positive, but an exception is to be found in the acrylics, where a negative value for birefringence is sometimes obtained [4].

Of course with most types of bicomponent fiber, the measurement of birefringence is not possible because of the interference of the different polymer groups along the fiber axis. Also, with deeply dyed fibers, the results if obtainable are unreliable because of the interference from the dye molecule.

The study of fiber birefringence is not however a purely academic exercise but has definite practical applications, some of which are

1. Examination of changes in molecular orientation [3]
2. Fiber identification [4]

It follows that as birefringence rises so too will the difference in the optical reflectance of colour when viewed from different angles and also the perception of color will change as the degree of drawing increases.

These factors can be of extreme importance in color matching or in trying to reproduce a specification against very tight performance factors, such as infrared reflectance in disruptively patterned military equipment.

2.2 Crystallinity

Crystallinity is not a subject that is routinely investigated, and accordingly there is no recognized standard to which one might refer. Indeed, a number of techniques are used, such as X-ray diffraction and synchrotron studies [3].

Table 1 Birefrigence Values of Common Textile Fibers

Fiber type	$\eta\|$	$\eta\perp$	$\eta\| - \eta\perp$	Nominal tenacity values in cN/tex	Standard regain %
Cotton	1.59[a]	1.525[a]	0.065	27 to 36	7.0 to 8.5[b]
Silk	1.59[a]	1.54[a]	0.05		11.0[b]
Triacetate	1.48[a]	1.47[a]	0.01	13	3.5[b]
Acrylic	1.52[a]	1.51[a]	0.01	24	1.5[b]
Aramid					3.5 to 7.0[b]
Glass	1.55[a]	1.55[a]	0.00	63	0.00[b]
Modacrylic	1.54[a]	1.53[a]	0.01	27	0.4 to 3.0[b]
Nylon	1.57[a]	1.51[a]	0.06	45	4.5[b]
Nylon 6.6 & HT	1.58[a]	1.52[a]	0.06	48 to 80	
Polyester	1.72[c]	1.54[c]	0.18[c]	61 to 80[c]	0.3[b]
Polypropylene	1.56[a]	1.51[a]	0.05	45 to 80	0.00[b]
Polyethylene	1.56[a]	1.51[a]	0.05		0.00[b]
Viscose	1.55[a]	1.53[a]	0.02	21 to 32	11.0[b]

[a] *Source*: AATCC Test Method 20, 1990 (as referenced by ASTM D 276–87) [4] [5].
[b] *Source*: ASTM D 1909–86 (90) [6].
[c] *Source*: Hoechst Trevira; values depend upon specific tenacity type.

Crystallinity has been defined [1] as three-dimensional order in the arrangement of atoms and molecules within a chemical phase, but this definition seems overly simplistic depending on the molecular weight of the chemical compound, and it now seems more likely that sites of crystallinity within a polymer have boundaries with amorphous or noncrystalline sites that are sometimes indiscernible, thus perhaps giving rise to such terms as paracrystalline.

However, once the polymer is drawn and subsequently redrawn, the degree of orientation of the crystallites becomes more fixed [7], unless or until the structure is subjected to a time–temperature equivalence that exceeds the original energy equilibrium, thus allowing shrinkage tension to develop [8].

3 Fiber Dimensions and Dispersion

3.1 Fiber Dimensions

The physical dimensions of textile fibers (the term fiber will be used to encompass filaments unless expressly stated otherwise) are critical to almost every characteristic of the resulting fabric and also to their performance in processing, with the possible exception of electrostatics. They include

1. Fineness
2. Length (in the case of staple) and variation in length
3. Cross-sectional shape
4. Crimp

3.1.1 Fineness

With some textile fibers, whose cross section is almost circular, such as glass, carbon, wool (excluding lambs' wool), high modulus polyethylene, polynosic, some regenerated cellulosics such as Lyocell, fineness can be described by reference to the fiber diameter, which on the whole ranges from about 9 or 10 μm up to around 140 μm (Boron), with the bulk of the synthetics falling into the 30 to 40 μm range (as a very rough guide).

The expression of fiber fineness by reference to fiber diameter tends to be in the minority in the textile industry as a whole, simply because the yarn measuring or counting systems tend to be direct (i.e., mass per unit length) systems such as Tex or metric or Denier, and it is simpler to translate fiber data to yarn data when the system on which each is measured is the same. In SI units, which is the favored metrology system internationally, the yarn counting or measuring method is Tex (g/km) or submultiples thereof. For the sake of completeness, however, the methods of measuring or estimating fiber diameter are

1. Normal methods of microscopy as described in ASTM D 578, ISO 1888, 1979, and ISO 11567, 1995 (Methods B and C)
2. Projection microscope techniques such as the method used for measuring wool fiber diameter (see ASTM D 2130 and ISO 137).
3. FDA techniques (i.e., fiber diameter analyser) which operate by a light scanning non-microscopical technique, or laser diffractometry.
4. Indirect techniques such as airflow resistance and calculations based on measured linear density and fiber density

When measuring the "diameter" of filament yarns continuously from a package or bobbin it is crucial that the yarn tension be kept as low as possible, but, above all, constant, as excessive variations in yarn tension may result in consequential variations in yarn diameter.

Each of these methods operates on the assumption that the fiber is circular in cross section, though in many cases it is not truly circular and may be flattened or trilobal or elliptical, which again is another reason for not using diameter as a measure of fiber fineness. However in some cases such as glass fiber processing the diameter is of importance for machine processing or laying-up reasons concerned with the space available rather than anything else.

Indirect measurements of fiber diameter based on the fiber (or filament) linear density and the fiber density are based on the physical proportions of

$$\text{mass} = \text{volume} \times \text{density}$$

and

$$\text{volume} = \frac{\text{mass}}{\text{density}}$$

then

$$\text{fibre volume} = \text{cross sectional area} \times \text{length}$$

and

$$\text{cross-sectional area} = \pi \text{r}^2$$

then

$$\pi r^2 l = \frac{\text{mass}}{\text{density}}$$

$$r^2 = \frac{\text{mass}}{\text{density} \times \pi l}$$

then

$$\text{fiber diameter } (d) = 2\sqrt{\frac{\text{mass}}{\rho\pi \text{l}}}$$

If we then substitute for mass and length the linear density (g/km = tex), we have the basis for the equations in ISO 11567 (methods A and C).

It can be seen from this that different fiber types of the same linear density but of different densities can have different fiber diameters, so a fiber with a higher density will have a smaller diameter than a fiber of the same linear density but a lower density. This highlights another area where fiber (filament) diameter is also important, and that is in calculating cover factor. Cover factor is simply a number that gives an indication of the extent to which the area of fabric is covered by one set of threads or as a measure of the relative density of packing (i.e., looseness or tightness) of the yarn in the fabric. It is calculated, in the case of woven fabrics, from the equation

$$\left.\begin{array}{l}\text{Cover factor (warp)}\\ \textit{or}\text{ (weft)}\end{array}\right\} = \frac{\text{threads/cm.} \times \sqrt{tex}}{10}$$

or, in the case of weft knitted fabrics, from

$$\frac{\sqrt{tex}}{\text{stitch length (mm)}}$$

These equations were developed later from original work by Pierce [9] and by Pollitt [10] based on ring spun cotton yarn, where it was shown that the yarn diameter d in inches was proportional to $(28\sqrt{N})^{-1}$, where N was the English cotton count (i.e., No. of lengths of 840yards/pound-weight = Number English cotton count [= NeCC]).

However, as shown above, filament diameter can be different if the polymer density is different, so the above equations for cover factor can in many cases be only a guide and may need to be modified for density if comparisons are to be made. Also, of course, in the case of staple fiber yarns particularly, twist factor can have a marked effect on the yarn density and hence on yarn diameter, but as a first approximation the yarn diameter of a staple fiber yarn spun on the cotton system can be found from

$$\frac{\sqrt{tex}}{678.6}$$

The microscopy and projection microscope techniques are notoriously subject to error, suffering as they do from operator error, calling for considerable dexterity by the operator in cutting and manipulation, and at present I am not aware of any statistical precision data on the methods (although see ASTM D 578). It is unlikely that reproducibility R will ever approach a level enjoyed by some laboratories in repeatability r tests.

Another technique for measuring fiber diameter indirectly is to measure the airflow resistance of a standard mass of fiber (sufficiently well opened to avoid hard lumps or plugs of fibers) that is compressed into a standard volume. It is sometimes assumed that this is a direct measure of the fiber diameter, but this is a misconception. It is in fact a measure of the specific surface of the fiber, which is commonly described as

$$\frac{\text{surface area}}{\text{volume}}$$

If the fiber is circular in cross section, the specific surface is simply

$$\frac{\pi dL}{\pi \left|\frac{d}{2}\right|^2 L} = \frac{4}{d}$$

So the specific surface is inversely proportional to the fiber diameter. By using this data to deduce the fiber diameter and by knowing the weight of the compressed plug, we can derive the linear density of the fiber by inserting the value for the density of the fiber (cited in standards such as ISO 11567 in g/cm^3 or in dtex as in ISO 1973: 1995, but commonly measured in mg/cm$^3 \times 10^{-5}$). This is the principle in ISO 10306 for the determination of cotton fiber maturity and its linear density, but the principle can be equally applied to synthetic fibers, given appropriate calibration. For fibers of circular cross section, the plot of airflow rate to fiber diameter, for each particular fiber type, would not be linear but approximately parabolic, as shown below, as the airflow rate will be inversely proportional to the square of the specific surface of the fiber, i.e., the surface area per unit volume of material. The air flow rate for fibers of different densities will of course differ, given that the compressed plug weight is the same, for reasons given above.

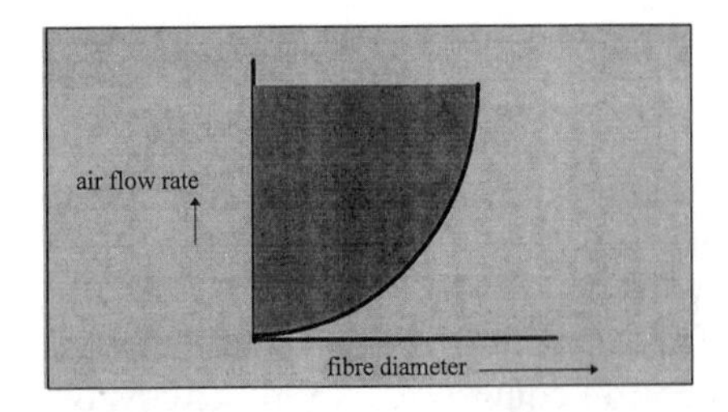

Figure 2 Theoretical increase in air-flow with increases in fiber diameter.

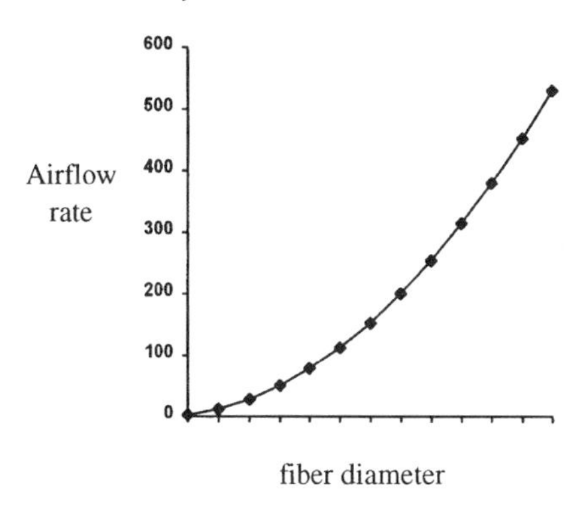

Figure 3 Practical increase in air-flow rate with increases in fiber diameter.

However, as stated above, the description of fiber fineness by reference to its diameter tends to be in the minority, and usually fiber fineness is expressed in terms of linear density. There have been many systems of yarn measurement throughout the world, not to mention the variations in the cotton spinning, worsted, and woollen industries, but for present purposes only the SI system will be used (tex = g/km), as this is universally recognized and is the system employed in international standards; after 50 years of gestation it is now being widely used across the globe.

It may at this point be opportune, before embarking on a study of how linear density is measured, to explain why fiber fineness is important.

Yarn manufacturers and fabric producers are concerned with yarn uniformity (apart that is from the specialist and fancy yarn producers who go to great lengths to produce yarns with random periodicities and what would otherwise be regarded as faults). Yarn uniformity is important to the yarn producer not only because he can sell it at a premium but also because the yarn contains fewer thin places, which cause the yarn to break in processing. Every time this happens, a spindle has to stop, and production is lost through downtime. Similarly, during weaving, when a warp yarn breaks, the weaving machine is stopped, or if it is not stopped and production continues, a fabric is produced with a fault, which attracts a lower sale price. *Quality therefore is not an altruistic virtue but a necessity that goes straight to the bottom line.*

Everything else being equal, (i.e., fiber length, fiber length variation, interfiber friction, flexural rigidity), the yarn uniformity is directly proportional to the number of fibers in the cross section of the yarn per unit length. How the staple yarn producer ensures that the number of fibers per cross section of yarn is uniform along the length of the yarn is an issue of processing technology that is beyond the scope of this book.

If for example a yarn of 25 tex is spun from a fiber of 1.25 dtex, the number of fibers per cross section would be on average

$$\frac{25 \text{ tex} \times 10}{1.25 \text{ dtex}} = 200$$

If however the linear density of the fiber was 2 dtex, then clearly the number of fibers in the cross section would only be 125, and as a consequence the yarn regularity would be worse (i.e., the CV% would be higher).

The coefficients of variation of linear density for short staple yarns specified in ASTM D 2645–95 are 5% for carded yarns and 4% for combed yarns. ISO 10290 does not differentiate between carded or combed yarns and has the CV% for linear density as 4% for both, but this is clearly the result of optimism. A method for the indirect measurement of yarn regularity is given in ASTM D 1425–96 and is concerned with the measurement of the unevenness of tow, top sliver, roving, and yarn produced from staple fibers and filament yarns, by means of continuous runs using capacitance testing equipment. Unevenness can be measured in U% (i.e., mean deviation %) or in CV%, depending on the type of integrator used by the instrument.

As mentioned above, the usual or common means of describing fiber fineness is by reference to its mass/unit length or linear density. The standard SI unit for this measure is tex, which is the number of grams per unit length of 1,000 m (i.e., g/km), or its multiples such as kilotex (for the extremely coarse measures such as ropes > 4 mm ϕ), which is kg/km, or for practical purposes g/m or its submultiples such as decitex [dtex] for the finer yarns and fibers, which is g/10 km or mg/10 m. For a review of these counting systems see ASTM D 861 or ISO 1144 and ISO 2947.

The standard techniques for measuring linear density are gravimetric method and resonance methods, and descriptions of these techniques are given in ASTM D 1577, ISO 1973, ISO 2060, ISO 10120, and ASTM D 1907.

Gravimetric Method. This method determines the *average* linear density of the sample, irrespective of the variability of the density of fibers within the sample. It is therefore unsuitable for fiber blends in which the linear densities of the fibers are markedly different. This does sometimes happen, and blends of fibers having different crimp rigidities are sometimes processed to permit changes in yarn properties or fabric properties to be developed later, but generally speaking the staple yarn spinner will try to blend fibers of the same linear density, as this creates fewer problems in production.

In ASTM D 1577–90 the average linear density is determined by taking specimen bundles of fiber from a sample (see later). A specimen bundle of fibers is composed of around 400 to 800 fibers (the requirement being to select sufficient fibers so that the bundle, before cutting, weighs 0.5 mg to 7.5 mg). In ISO 1973, 1995, ten sets of 50 fiber bundles are selected from precombed bundles, providing that the balance is sufficiently precise for such small amounts to be weighed accurately to within ±1%. But as one of the acknowledged variabilities in the test is the accuracy of weighing, a larger bundle size is probably preferable. The problem with this of course is that the actual number of fibers has to be counted later in the test, so when the linear density of the fiber is very low, as it would be when testing microfibers (1.3 dtex or less or 1.0 dtex or less depending on your particular point of view), the counting of the fibers becomes very tedious and time-consuming.

The fiber bundle is combed using a *combsorter* so that all the fibers are aligned and the short ends and broken fiber removed. Of course if filament is being tested they can be arranged in parallel rows without the aid of a combsorter.

Any crimp is removed by pretensioning the fiber, but care is needed to avoid introducing strain into the fiber as this could affect the end result of the test; see Fig. 4.

While in this position (i.e., free of crimp and in parallel alignment) the fibers are cut to a selected length using a suitable fixed die cutter. The specimen bundle is then weighed to the nearest 5 µg (= 0.005 mg) and the fibers in the specimen bundle are counted.

The average linear density of the fibers in the specimen bundle is determined according to ASTM D 1577–90 from the formula:

$$\text{dtex} = \frac{1000W}{Ln}$$

where

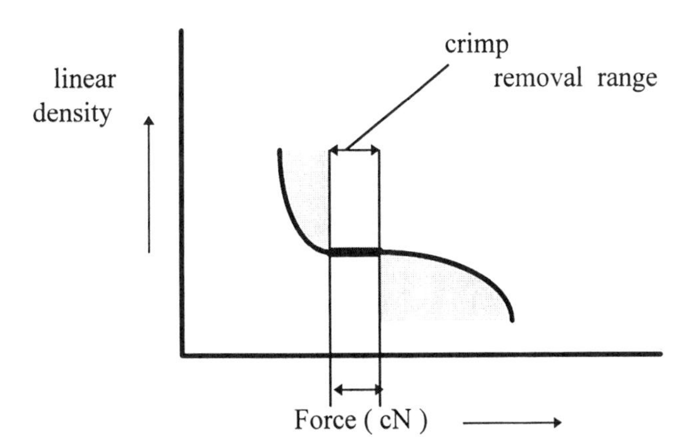

Figure 4 Pretensioning for crimp removal.

W is the specimen bundle weight in mg

L is the cut length of the fibers in mm

n is the number of fibers in the bundle

Of course this is slightly wrong as the formula actually gives the linear density in tex and not dtex.

[As mg/mm = g/m, and tex = g/m $\times 10^3$, then dtex = g/m $\times 10^4$.]

To calculate the fiber linear density in dtex from these measurements, the equation should be

$$\text{dtex} = \frac{W}{Ln} \times 10,000$$

and not 1,000 as in ASTM D 1577–90.

In ISO 1973, 1995, the average linear density of the fiber (filament) is calculated as

$$\rho_{l,b} = \frac{m_b}{n_f . l_f} \times 10^4$$

where

$\rho_{l,b}$ is the mean linear density of the fibers

m_b is the mass in milligrams of the fiber bundle

n_f is the number of fibers in the bundle

l_f is the length of millimetres of the individual fibers in the bundle

So if the length of fibers in the bundle is 50 mm and the number of fibers in each bundle is 50, and if the weight of the bundle is 0.385 mg, the mean linear density of the fibers in the bundle in dtex would be

$$\frac{0.385\,\text{mg}}{50 \times 50\,\text{mm}} = 0.000154 \times 10^4 = 1.54\,\text{dtex}$$

The reliability of the result is determined in ISO 1973 by calculating the 95% confidence interval of the mean and expressing this as a percentage of the mean value. If this is greater than 2%, the number of bundles tested is increased. The confidence interval is calculated using Student's ***t*** factor taken from statistical tables, but of course it is much easier if one simply uses the tables and equation in ISO 2602.

The linear density of filament yarns may be determined by

1. The skein or hank method (see ISO 2060)
2. Cutting lengths of filament to a prescribed length
3. On-line linear density measurement (as marketed for example by Microsensors Inc, USA)

In any event, and whichever method is employed, it is essential that the tension applied to the filament be sufficient to straighten the test piece without causing any measurable strain. For most purposes this will be between 4 mN/tex to 6 mN/tex (i.e., approximately 0.4 g/tex to 0.6 g/tex).

Another important consideration is to ensure that the test is conducted with the fibers in moisture equilibrium approached from the dry side of the hysteresis curve (see later). Alternatively, the oven-dry weight may be taken and an allowance made for the commercial moisture regain. If the yarn is sized or contains other finishes, either an allowance can be made for these or the yarn can be scoured and desized prior to testing.

The gravimetric methods use the values of length and mass of the test piece to calculate the linear density as

$$\text{Linear density (tex)} = \frac{m \times 10^3}{L}$$

where:

m is the weight of the test piece or skein in grams (as measured or agreed)

L is the length of the test piece or skein in meters

Instruments for measuring the filament linear density continuously use capacitance of the filaments to operate the sensor, similar in principle to the Uster and Keisoki eveness testers. The instruments are usually accurate to between 1% and 3%, but as capacitance is affected by relative humidity and temperature, these parameters need to be strictly controlled. Otherwise the accuracy of the results will be somewhat spurious, even when calibration is on line, as the strict control of relative humidity within an environment where the temperature varies by more than ±2°C around an ambient temperature of over 25°C is not realistic, as can be seen from any psychometric chart.

Resonance (Vibroscope) Method. The principle underlying the operation of the vibroscope is that of a vibrating string, the natural fundamental frequency f (in Hz) of which is given by

$$f = \frac{1}{2l}\sqrt{\frac{T}{M}}$$

where

l is the length of the string

M is the mass per unit length of a perfectly flexible string

T is the tension in the string

However we are not of course dealing with a perfectly flexible fiber but one that has an element of stiffness. The above formula should therefore be modified by a correction factor that is proportional to the root of the Young's modulus and the square of the fiber radius. As this correction factor is in the majority of cases less than 3%, the above equation can be said to be accurate to ±3% (on average).

From the above equation it can be seen that the mass per unit length M can be found from

$$M = T\left[\frac{1}{2lf}\right]^2$$

since

$$\sqrt{\frac{M}{T}} = \frac{1}{2lf} \quad \text{and} \quad \frac{M}{T} = \left[\frac{1}{2lf}\right]^2$$

It is on this basis that the mass per unit length of a fiber or filament is determined using the Vibroscope. Although the instrument is capable of being used by relatively unskilled operators, some care is needed to avoid confusing the natural frequency of vibration with the third harmonic frequency (see ASTM D 1577).

The method is also the subject of ISO 1973, 1995. Both ASTM D 1577 and ISO 1973 admit that it is necessary first to calibrate the instrument by determining the linear density of the fiber gravimetrically, but if the fiber type and source remain constant, the instrument can be used as a very quick and easy means of verifying or checking the linear density without going through the laborious technique of gravimetric determination.

Admittedly ISO 1973, 1995 does require the CV% of the individual readings to be determined, as well as the confidence interval of the mean, but by programming the data tables and basic equations from ISO 2602 into a relatively inexpensive electronic calculator, the computation becomes very simple and need only take a few minutes, even allowing for the inputting of the data.

3.1.2 Fiber Length and Fiber Length Dispersion

It has been said previously that fiber fineness is the single most important factor ruling yarn regularity. This is true, but it is not the whole truth. Fiber length or more precisely the variation in fiber length is equally important, at least in terms of staple fiber yarns, because it is the variability of fiber length that is the crucial element in the control of fibers during the yarn preparation and spinning stages of production. When the percentage dispersion of fibers in a fiber diagram is large, the result is the production of a random periodicity in the yarn caused by incomplete control of the fibers in the drafting zones. To go into this in any detail is beyond the scope of this book, but it does give some indication of why it is necessary to know something about the degree of fiber dispersion in the staple fiber raw material.

Effective fiber length, or more approximately, staple length, is a feature that goes to the root of staple fiber yarn strength (*all other things being equal*), because the longer the fiber the greater the number of points of interfiber contact in the yarn and the greater the yarn strength, until of course an optimum is reached (Fig. 5).

A more realistic and practical graphical representation associating fiber length with practical upper count limits (i.e., the upper linear density limit to which a fiber will spin commercially) must take account also, as has been said already, of the fiber fineness. See Table 2, which is based on a short staple ring spun system.

The values in Table 2 are shown graphically in Fig. 6, expressed in the form of English cotton counts, and in Fig. 7, expressed in terms of Tex values.

A number of terms are used to describe fiber length variability, such as uniformity index,uniformity ratio, and upper half mean length, but these are terms that probably go beyond the confines of the present volume, and for the moment it seems necessary only that the reader gains a brief understanding of the importance of measuring the various aspects

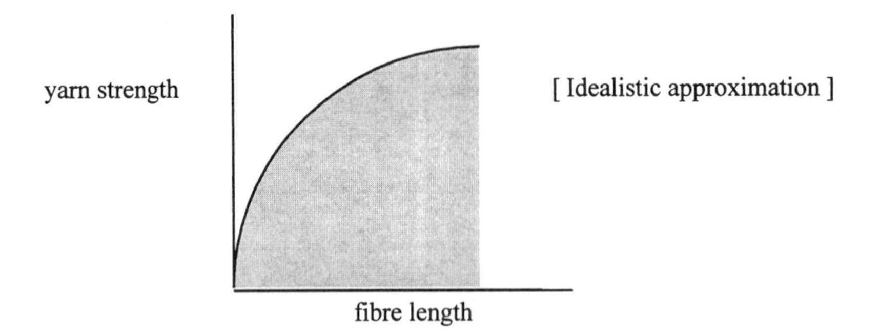

Figure 5 Potential increase in yarn strength with increasing fiber length.

of fiber length for processability and quality reasons. An outline of the various methods used to measure fiber length and its variabilty are given below, but it may nevertheless be useful to know why a knowledge of fiber length and its variation is important to the staple yarn producer.

First of all, the spinner needs to know what machine settings he must adopt in order to achieve the maximum amount of fiber control and the least degree of fiber damage. Secondly, he needs to know what the baseline is at the start of operations so that quality checks can be conducted to assess what amount of fiber damage is taking place. Third, he needs the information in order to assemble together a blend of materials that can be reproduced and from sources that are compatible and economic.
The proper discussion of these topics is not within the scope of this present volume, but this perhaps gives some idea of the importance of the topic to the yarn producer.

Table 2 Ring Spinning Limits Based on Fiber Length and Fiber Fineness

Linear density in mtex (A)	Effective length in mm (B)	Count range factor $(B)/(A) \times 10^3$	Spinning limit Tex	Spinning limit NeCC
310	15	48	60	10
290	21	72	60	10
230	22	96	30	20
248	24	96	25	25
190	27	142	15	40
240	28	116	13	45
203	30	148	15	40
176	32	182	12	50
160	32	200	7	80
146	34	232	6.5	90
146	36	246	4.9	120
140	40	286	4.2	140

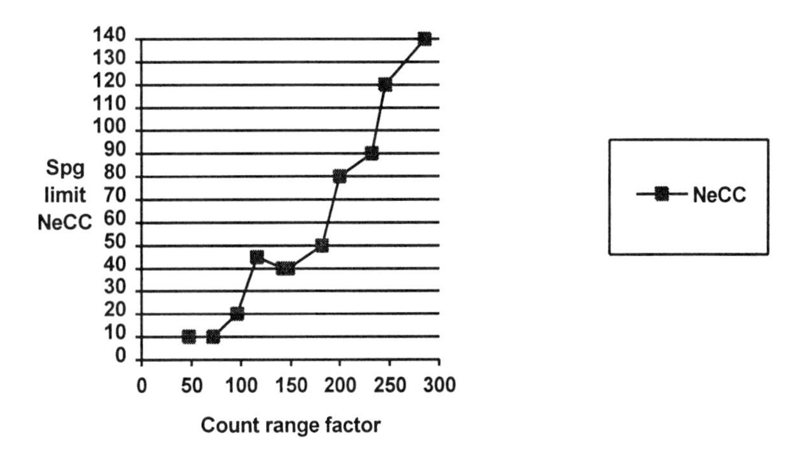

Figure 6 Illustration of upper count limit (in NeCC) using only fiber length and fineness factors.

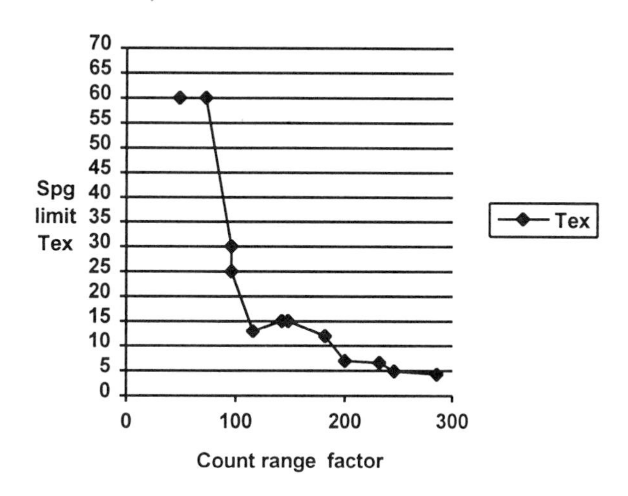

Figure 7 Illustration of upper count limit (in Tex) using only fiber length and fineness factors.

The standard methods used to provide data on the various length parameters and dispersion or variability of length are outlined below:

ASTM D 5103–95 describes a method to measure the length manually of 50 fibers. A fiber diagram is plotted of fiber length against cumulative frequency, as illustrated in Table 3.

Table 3 Cumulative Frequency Table

Length group, mm	Number of fibers	Cumulative number of fibers, %
42	1	2
41.5	1	4
41	1	6
40.5	1	8
40	2	12
39.5	1	14
39	2	18
38.5	6	30
38	12	54
37.5	7	68
37	3	74
36.5	2	78
36	4	86
35.5	2	90
35	1	92
34.5	1	94
34	1	96
33.5	1	98
33	1	100
Total	50	

The data in Table 3 are expressed in a cumulative frequency curve as shown in Fig. 8.

ISO 6989 is somewhat similar to ASTM D 5103 but more realistically measures 500 fibers and also permits the use of the WIRA fiber length machine. The modal length (i.e., the central length of the most numerous class length) and the mean length are calculated. Either a histogram or a cumulative frequency diagram can be prepared (see also ASTM D 3661, below).

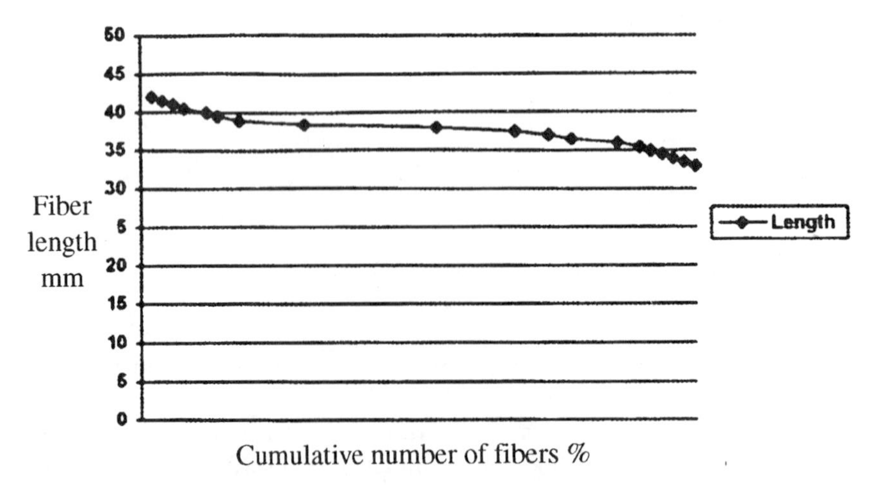

Figure 8 Cumulative frequency curve for fiber length.

ASTM D 3660–90 employs a comb sorter to separate the fibers, which are then arranged on a fiber array board. Fiber length is then measured and plotted against cumulative frequency. The method is somewhat laborious and requires considerable dexterity. ASTM admit that the method is really only suitable for R&D and arbitration purposes.

ASTM D 3661–90 uses the WIRA comb sorter (as in method C in ISO 6989) and is suitable for synthetic polymer fibers having lengths of 25 mm upto 250 mm. One hundred fibers are measured. Some dexterity is necessary in handling the fibers and laying them into the fiber guide. Again a staple length distribution plot is made of fiber length against cumulative frequency.

ASTM D 1447–89(94) specifies the use of the Zellweger Uster Digital Fibrograph. This measure the length and length uniformity of short staple fibers. In this method the fibers are held on combs at random positions along their length (as they would be in practice by the rollers of a drafting system), and the frequency and length of the fibers is scanned photoelectronically. The method is regarded as having acceptable levels of (*R*) and (*r*). Details of the critical differences between mean values are given in ASTM D 1447. As the majority of fiber terms relating to length uniformity, such as uniformity index, span length, and floating fiber percent, are universally related to the data provided using the Fibrograph, this now tends to be the preferred option. ISO 4913, 1981, somewhat cryptically, does not actually refer expressly to the Fibrograph, but instead requires a "Testing instrument, suitable for optically scanning beards of cotton fibers," but it is nevertheless the Fibrograph that is (presumably) intended.

ASTM D 3937 provides a method for determining the crimp frequency of synthetic staple fibers, but the between-laboratory variation is known to be great. It is, however, suitable for in-house QC operations.

The above is only a brief résumé of the methods available for investigating the length and length uniformity of staple fiber, but it may be useful to look a little more closely at comb sorter diagrams, cumulative frequency diagrams, and the operation of the digital Fibrograph.

A fuller understanding of the practical application of these data can really only be gained by a study of the principle of roller drafting and short staple spinning technology generally, which is beyond the scope of the present volume.

4 Density

The main practical use of fiber density is in fiber identification.

The determination of average fiber mass presents of course no experimental problems, but this is not true of determining the average fiber volume, as this does present one or two difficulties.

First, a bundle of fibers will entrap air both between the fibers and also on the fiber surfaces, since we are not dealing with cylindrical specimens but with fibers which may have very complex cross-sectional shapes and helices. If all this entrapped air is not displaced and remains trapped by the fiber, the liquid displacement will be too high (i.e., the assumed fiber volume will be too high) and the calculated density will as a consequence be too low.

The second difficulty is one of liquid absorption by the fiber. This of course results in a smaller measured displacement of liquid, giving a calculated density that would be too high. This latter consideration is most important, and it is essential to ascertain first that the liquids used in the test are inert chemically to the polymer being tested or are not absorbed by the fiber during the test.

Table 4 Approximations of General Fiber Densities

Fiber type	Density g/cm^3 (=mg/mm^3) DRY
glass	2.47–2.57
polyethylene	0.92–0.95
nylon 6	1.14
nylon 6.6	1.14
nylon 11	1.04
polyester	1.32–1.39
triacetate	1.3
modacrylic	1.28–1.42
aramids	1.37–1.42
acrylics	1.17–1.19

Three methods are common, viz, the liquid displacement method, the sink–float method and the density gradient column method. Each of these is a common, standard technique and is fully described in ISO 10119, 1992 (for the determination of the density of carbon fiber), and also in ASTM D 276–87 (reapproved in 1993), which in fact also refers to ASTM D 1505, ASTM D 792, and AATCC, Method 20 (1990) (Fiber identification), each of which deals with the above techniques. ISO 10119 is a very good and concise description of the techniques. However the measurement liquids specified in ISO 10119 of ethanol, methanol, acetone, tricloroethane, and carbon tetrachloride, although suitable for carbon fibers, are not at all suitable for the general range of textile polymers, with the exception perhaps of ethanol and methanol. ASTM D 276–87 recommends the use of *n*-Heptane for universal application, except, of course for the olefins, such as polyethylene. A range of typical fiber densities is given in Table 4.

In making determinations of fiber density it is essential that either the % regain of the fiber be known or that the determination be conducted with the fiber at oven-dry weight, as this could affect the measured density, particularly for the hygroscopic fibers.

5 Moisture Relations, Temperature Effects, and Thermal Properties

I remember my old tutor at college more than thirty years ago saying, "Don't forget to sell the water." Moisture relations in textiles, however, is concerned with more than just the commercial regain allowances specified in ISO/TR 6741-4 or ASTM D 1909, important though they are.

Traditionally the conditioning and testing atmosphere for textiles has been 20 ± 2°C and 65 ± 2% r.h. Quite why is lost in the mists of time, but it is thought that in Lancashire (England), where the industrial revolution and large-scale production of textiles began, the temperature in the "ring room" or in the weaving shed was commonly around 68°F and a frequent ambient humidity would be around 65% r.h., but this is itself a guess, as the means of measuring relative humidity, even into the 1950s, was notoriously inaccurate, if not even whimsical. This historical sourcing of the now oft-referred-to conditions of 68/65, or, to give the more up-to-date metric version. 20/65, is however not established with any degree of certainty.

Moisture *regain*, incidentally, being a concept unfamiliar in conditioning bulk polymers, is the *amount of water contained in a textile expressed a percentage of the oven-dry mass*. Its measurement is specified in ISO 6741–1, 1989, and in ASTM D 2654–89. The standard atmosphere for the conditioning and testing of textiles is given in ISO 139, 1973, and in ASTM D 1776–90. The commercial allowances or permitted regains are however not necessarily the same thing as the moisture equilibrium of the textile in the standard atmosphere, and these are specified in ISO 6741–4 and in ASTM D 1909–86.

An interesting study was undertaken in 1994 by the British Textile Technology Group (BTTG) into "the influence of environmental conditions on the testing of textile materials," and it was published in 1995 by the British Standards Institution as PD 6577. It will, we hope, be discussed widely. Its broad conclusions are reproduced here:

> It is true that cotton and linen are slightly stronger and wool more extensible at high humidity.These "improvements" in properties however are balanced by the detrimental weakening effect of high humidity for example on the strength of viscose, acetate and wool.
>
> A consideration of all the above factors leads inevitably to the conclusion that the 50% r.h./23°C atmospheric condition should be recommended for conditioning and testing textiles
>
> The current textile humidity permitted range of ±2%, is generally regarded as extremely difficult to achieve, first, because this degree of control is difficult to retain over the whole area of a laboratory and to avoid short time excursions. Small changes in temperature produce large drifts in humidity (% r.h.), e.g. an increase in temperature of 1°C produces a decrease in r.h.% of about 4%.
>
> Second, the uncertainty in measurement of relative humidity using commonly available electronic hygrometers is at least ±2% of the reading and is often greater due to non-linearity, hysteresis and drift [see Pragnell, R.F. SIRA Ltd "The relationship between humidity calibration standards and standards for environmental testing," Humidity seminar, June 1988].
>
> Recognized authorities in the field of humidity measurement hold the view that control of a laboratory to ±2% is impossible; control to ±5% could be achieved with care, but ±10% is typical for most 'controlled' laboratories . . .

This seems to take no account of variations in atmospheric pressure, but the suggested tolerance of ±10% seems not unreasonable given that if the dry bulb temperature is 20°C and the wet bulb depression is 4.2°C at 1013 hPa, the r.h.% will be 65%, but if the wet bulb depression changes to 3.2°C, at the same pressure, the r.h. % will be 72.9%, i.e., a change of 7.9%. This of course assumes that wet bulb temperatures can be measured to an accuracy of 0.1°C. It seems clear therefore that the traditional standard atmosphere of 20°C ± 2°C and 65% r.h. ±2% r.h. assumes that it is possible in the laboratory to maintain a constant barometric pressure, a dry bulb temperature that does not vary by more than ±2°C and that the wet bulb depression is constant at 4.2°C. Is this feasible?

> Temperature control on the other hand is quite different. Thermometers can be accurately calibrated to 0.1°C. The laboratory in which the work was conducted was controlled with ease to better than ±0.5°C; well within the specified range.

The overall recommendations of the report, in summary form are that

> (a) the standard atmosphere for conditioning and testing textiles should be changed from (65 ± 2)% r.h. and (20 ± 2)°C to (50 ± 5)% r.h. and (23 ± 2)°C. (b) preconditioning from the dry-side should always be performed before conditioning and testing textiles made from hygroscopic polymers [the precise definition of "hygroscopic" in

this context is not included]; and (c) the standard tropical atmosphere should be withdrawn.

The above report and recommendations will no doubt prove controversial in the industry and in international trade for some time to come, but irrespective of the logic that the report claims, there is certainly some truth in the charge that many industries and not only the textile industry, have for many years tacitly accepted the tight tolerances on relative humidity specified in industrial standards without looking too closely at whether they were certifiable .

Moisture relations in textile polymers are concerned mostly with the hydrophilic fibers, such as the cellulose groups of cotton, flax, hemp, jute, viscose, modal, and acetates. This is not to say however that the essentially hydrophobic fibers, which fall into the synthetic polymer group (see Fig. 1) have zero moisture imbibition or are totally unaffected by moisture, although this will be true for some of the fibers in this group.

So what is the range of water sorptions that can be expected at the standard conditioning atmosphere, and what is the difference, if any, between these moisture regains and those agreed as commercial regains for trade purposes? Table 5 gives some indications, the second column showing the expected moisture regain in equilibrium with the standard atmosphere for conditioning and testing (as cited in ASTM D 276–87(93)), and the third column showing some of the commercial regains cited in ISO 6741–4.

Reference was made above in the extract from the BTTG report to "pre-conditioning from the dry-side." By this is meant from the dry side of the hysteresis curve. This is because, particularly for the more hygroscopic fibers, there is a disparity in equilibrium

Table 5 Comparison of Equilibrium Regains at Standard Atmosphere with Commercial Allowances

Fiber type	% regain at 65% r.h. & 21°C	% regain cited in ISO 6741-4: 1987
Cotton	7.4	raw 8.5
Flax	12.0	12.0
Viscose	13.3	13.0
Acetate	5.0	BISFA 9.0
Triacetate	2.5	BISFA 7.0
Nylon 6	4.0	BISFA 5.75
Nylon 6.6	4.2	BISFA 5.75
Nylon 11	1.0	BISFA 3.5
Aramids		
Nomex ®	6.5	US 3.5, 4.5 or 7.0
Kevlar ®	3.5	depending on end use
Acrylic	1.8	BISFA 2.0
Modacrylic	1.3	
Spandex	1.3	
Polyester	0.1 to 0.4 depending on type	BISFA 1.5
Polyolefins		
Polyethylene	0.0	
Polypropylene	0.1	BISFA 2.0
Glass	0.1	
Fluorocarbon	0.0	

regain, at any given relative humidity, depending upon whether the fiber was initially dryer than its equilibrium regain, or whether it was wetter than its equilibrium regain. Traditionally this has been illustrated as shown in Fig. 9.

The reality however is not quite so simplistic, and the following graphical representation gives perhaps a clearer idea of what is happening during the desorption/absorption of moisture during the fibers' approach to equilibrium, although much will depend upon the individual fiber types' heats of sorption. The subject is far too complex to be dealt with here, and readers are directed to more specialized studies of the subject, such as Mackay and Downes [11], Watt and McMahan [12], A.B.D. Cassie [13], and Morton and Hearle [7].

These references are concerned with the mechanisms through which moisture is absorbed by the fiber. The effect it has on fiber properties however varies from fiber type to fiber type, and although the effects are wider than merely the effect on tensile properties, it is on these properties that the standard methods of test are mainly concerned. Many of the standard tests previously cited, such as ISO 5079, ISO 2062, ASTM D 2256, and ASTM D 3217, all make provision for conducting the tests with the fiber or yarn in the wet rather than the conditioned state.

Of course it goes without saying that all tests for determining the fiber linear density, diameter, and density should be conducted on specimens conditioned in the standard atmosphere for conditioning and testing from the dry side of the hysteresis curve.

Figure 11 gives an approximation of the effect of moisture on the tenacity of a range of textile fibers. Because of the scales involved, it is not reasonable to show breaking extensions in the same histogram, so these are shown separately.

The values in Fig. 11 are in some cases maximums and in some cases averages only, as they are for illustration purposes only. It can for example be seen that in the wet condition the tenacity of cotton rises whereas that of viscose drops dramatically, whilst polyester and modacrylic are largely unaffected.

Relative breaking extensions in the dry and wet state for the above range of fibers is given in Fig. 12.

Despite the extreme importance of the thermal properties on the processing and performance of textile polymers, particularly in filament processing and finishing, there are surprisingly few standard methods of test dealing with the subject. This is perhaps due to the complexity of the subject in terms of the effects of temperature on oriented chain molecules and influence of moisture on the polymer. However, a standard method that is available is ASTM D 5591–95, which is concerned with the thermal shrinkage force of yarn and cord. The instrument specified is the Testrite thermal shrinkage force tester

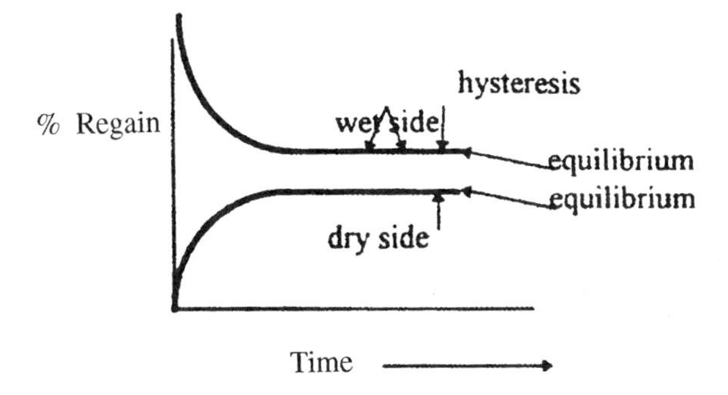

Figure 9 Traditional illustration of moisture hysteresis.

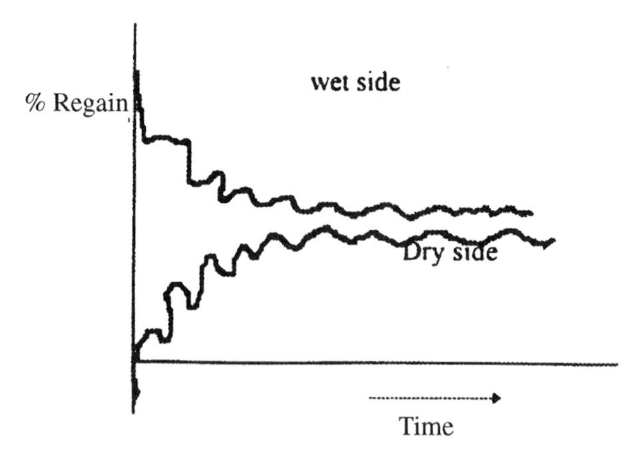

Figure 10 Probable approach to equilibrium, (exaggerated), allowing for fiber heats of sorption.

(manufactured by Testrite of Halifax, W. Yorkshire, England). The ASTM method is applicable to the measurement of the thermal shrinkage force of yarns and cords whose shrinkage force at 177°C ± 2°C in air does not exceed 20 N. It is applicable to nylons, polyesters and aramid yarns.

Thermal shrinkage force or thermally induced contraction is critical to many commercial operations, such as tire production, transmission belts, and certain types of conveyor belting and is dependent on the annealing and drawing processes during filament manufacture.

The test consists of allowing a specific length of yarn to relax under zero tension for 12 to 28 h in the standard atmosphere for conditioning and testing. It is then placed in the instrument under a pretension of 19.6g/dtex, clamped into position, and exposed to a dry

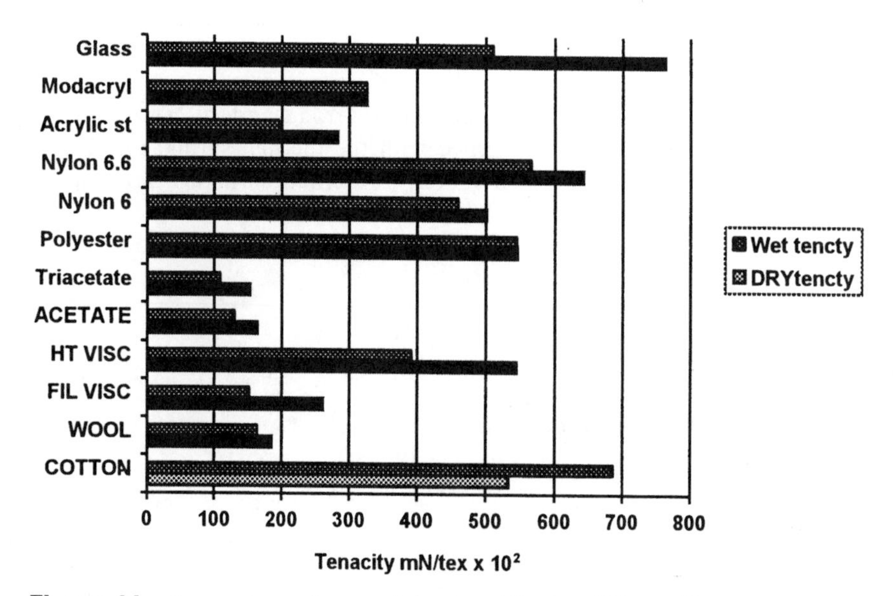

Figure 11 Comparison of wet and dry tenacities for different fibers.

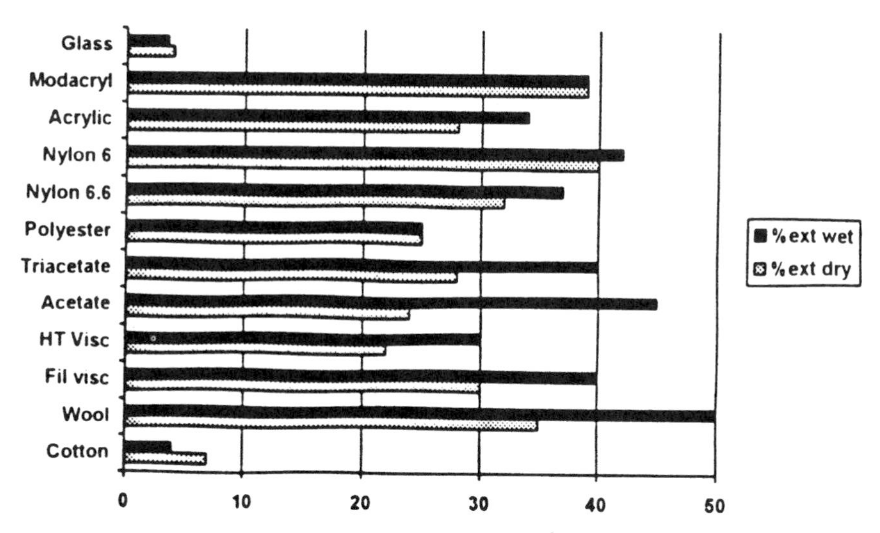

Figure 12 Effect of moisture on breaking extensions of different fibers.

heat temperature of 177°C ± 2°C for 120 s ± 5 s. The force or increase in tension caused by the thermal shrinkage is read directly from the instrument to the nearest 0.1 N.

This method of test of course only provides data on thermal shrinkages under the specific conditions of test, and if the temperature or time of exposure, or both, or the pretensioning value change, the results can be considerably different. On a polyester for example, as the temperature increases above 177°C at 120 s, the percentage thermal shrinkage may rise from around 11% to 14 or 15% at 200°C for the same exposure and pretensioning force. Generally speaking, the lower the pretensioning force the higher will be the thermal shrinkage for the same time/temperature exposure. However, these are broad generalizations, and the reader is directed to the original works by Ribnick [14], Dismore and Statton [15], and Dumbleton and Buchanan [16] et al.

The virtue of ASTM D 5591–95 lies in its use as a means of assessing the quality or continuity of supply, rather than in its use as a predictor of performance under different conditions.

Important reference works on the subject can be found in *The Setting of Fibres and Fabrics* (J.W.S. Hearle and L.W.C. Miles, eds.), Merrow, Herts, UK, 1971; *Fibres, Films Plastics and Rubbers* by W.J. Roff and J.R. Scott, Butterworths, London, 1971; *Fibres from Synethetic Polymers*, by I. Marshall and J.R. Whinfield, Elsevier, Amsterdam, 1953; *Advances in Fibre Science* (S.K. Mukhopadhyay, ed.), The Textile Institute, 1992.

6 Electrical Properties

The accumulation of an electrical charge on a textile polymer product (filament, yarn, fiber, or fabric) causes many problems in processing, from carding to spinning, beaming, weaving and knitting, to cutting, sewing and making-up. It also causes problems (now seemingly known by the new breed of "management" as "challenges") during wear and can be a cause of the attraction of particles of dust and dirt. It can also be a source of danger, since if the charge is high enough it can discharge into air, causing sparks. If this takes place in a flammable atmosphere, an explosion may ensue [see EN 1149–1].

The tendency of textile polymers to accumulate static electrical charges is related to their electrical resistance. With solid materials, the electrical resistance is defined as the resistance between opposite faces of a 1 m cube. With textile polymers, for reasons similar to those given earlier, it is more convenient to express the resistance of fiber filaments by

$$\text{Resistance (R)} = R_s \frac{l}{NT} \times 10^5$$

where

R_s is the mass specific resistance (= specific resistance in $\Omega \times$ density; see IEC 250)
l is the distance between the ends of the test piece
N is the number of fibers or filaments
T is the linear density in tex

(see *J. Text. Inst. 44*, T 117 [1953]).

Results are plotted in log R due to the very wide range of values encountered. Electrical resistance however is related to the relative permittivity of the material.

It is not possible to predict the propensity of a textile fiber or filament to accumulate a static electrical charge from a knowledge only of its chemistry. Its morphology and crystalline structure, as well as the other materials with which it comes into contact and any finish that may have been applied, all play a significant part.

A technique for the measurement of the electrostatic propensity of textile filaments can be found in ASTM D 4238–90.

The method does not directly measure the actual charge developed on the material as might be done using a Faraday cylinder; instead it records only a voltage that is proportional to the voltage developed on the material. In that sense it is arbitrary and relative only.

The electrostatic propensity of a textile polymer is defined in ASTM D 4238 as the capacity of a nonconducting material to acquire and hold an electrical charge by induction (via corona discharge) or by triboelectric means (i.e., rubbing with another material).

A test specimen that has been brought into equilibrium with the test atmosphere (presumably from the dry side of the hysteresis curve) is clamped to a metal disc (which is grounded). It is essential that the test specimen be as flat as possible against the disc and that there be no wrinkles, creases, snarls, or knots present, as they would seriously affect the result. The metal disc and test specimen are then covered by a metal frame exposing an area of test specimen 32 × 32 mm. A high-voltage corona discharge electrode is positioned 15 mm above the surface of the test specimen. A detecting electrode pin is also mounted above the test specimen to measure the charge developed on the test specimen. Before the test commences, the test specimen is discharged of any latent static charge by an ionizing bar. With both the corona discharge electrode and the detecting electrode in place, the metal disc is rotated at 1720 rpm and a voltage of 10 kV is applied to the discharge electrode. The maximum voltage is attained after about 15 s to 30 s and is recorded on an oscilloscope. After the maximum voltage is recorded the discharge electrode is moved away from the test specimen and time started. After 10 min, or when the measured voltage has been reduced to one-half, the test recording is terminated and the electrostatic decay half-life is reported. The electrostatic potential of the test specimen is calculated by multiplying the measured voltage by the calibration factor, which is determined separately using polyethylene film.

What might be regarded as a serious drawback to the method is that the linear density of yarns tested must be at least 340 dtex or folded (plied) to a resultant dtex of at least 340.

On a more general note, it must be pointed out that from a practical point of view, if the relative humidity is greater than 50%, the propensity of fibers or filaments to develop an electrical charge is much reduced, and the voltage developed will be lower, or no charge will be accumulated, or the half-life of the charge that is developed will be reduced.

It has been shown [17] that with most textile materials the propensity of the material to hold a charge falls rapidly with an increase in moisture regain, but perhaps this is to be expected.

At the beginning of this section it was stated that the propensity of a textile polymer to accumulate static electrical charges is related to their electrical resistance. It has also been said that electrical resistance is related to the relative permittivity of the material. Relative permitivity or permittivity is sometimes (now incorrectly) called dielectric constant, as this used to be the common term before SI units became the fashion. Relative permittivity is therefore often used as a shorthand method of indicating the static electrical propensity of fibers and filaments. Its measurement is discussed in IEC 250.

With dielectrically anisotropic materials such as textile filaments or fibers, it is important to state the direction of the electric field, and consequently it might in some cases be advisable to make measurements of permittivity in more than one direction. Permittivity also varies with the electrical frequency, and so it is advisable to make measurements of permittivity over as wide a range of frequencies as may be relevant to the material. The results should be plotted as permittivity against $\log_{10}$ frequency. A typical plot is illustrated in Fig. 13.

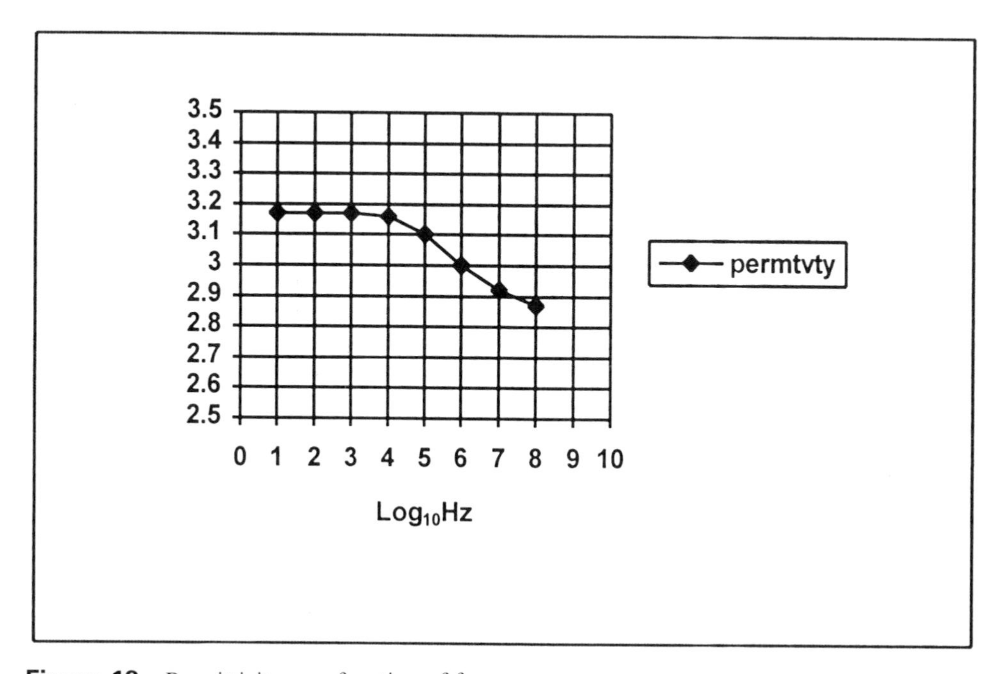

Figure 13 Permittivity as a function of frequency.

Relative permitivity ε' is the ratio of the capacitance between pairs of electrodes with the material as the dielectric to the capacitance between the electrodes without the material (air being the dielectric). So

$$\varepsilon' = \frac{Cx}{Cv}$$

where

Cx is the capacitance with the material

Cv is the capacitance without the material

Relative permittivity also varies with the temperature, as well as frequency, so it is also advisable to determine the permittivity over a range of temperatures relevant to the material operating range, but obviously below the material softening temperature. The results can then be plotted as temperature over $\log_{10}$ frequency.

7 Time-Dependent Properties

As will be seen in Section 8, the duration and speed of the application of a load on a fiber or filament has a profound effect on the way and the extent to which they deform.

This might seem at odds with Professor Hearle in his *Physical Properties of Textile Fibres* (The Textile Institute, 3rd ed., 1993, p. 359 at 16.4.3) where he states, "As a first approximation we can say that the breaking extension is independent of the rate of loading." However, as Hearle says, it is a first approximation, and as approximations go, it is not very approximate, as he himself later shows. In fact the extension at break is much dependent on the rate of loading and also on the manner in which the load is applied.

In this respect, textile polymers can almost be described in the same way as isotropic polymers in terms of their viscoelastic behavior, but not quite. The same relative criteria apply, such as elastic deformation, primary creep, yield, secondary creep, creep recovery, isothermal, or adiabatic deformation.

The subject of textile polymer rheology, however, although growing apace, cannot so far predict precisely how a drawn polymer will behave in a given set of stress-inducing conditions, apart from first approximations. So we are left, as is invariably the case, with a "suck it and see" scenario. In fact the Foreword to BS 4029, 1978, acknowledges the situation in the following words:

> Although the data on tensile elastic recovery are not normally specified for commercial purposes, they may be required in the assessment of the suitability of a given type of fiber for particular purposes and are often quoted in manufacturers publications and the technical press . . . It is recognised that values for the tensile elastic recovery of a fiber may vary considerably with the conditions of test, and no single set of conditions will define adequately the behaviour of the fiber. Consequently the tensile elastic recovery should be measured for at least two widely differing conditions of extension or relaxation time, one of which is a preferred set of conditions, the other being appropriate to the purposes of testing.

This does not mean, however, that we are approaching the problem without any supporting knowledge or data, but the test history, degree of crystallinity, orientation, and polymer weight all influence the results and need to be given their proper consideration. To expect to be able to predict their viscoelastic behavior, even in the same polymer group, if two of these criteria are different is stretching credulity. (*Note*: Credulity is not included in Figure 1 in Section 1).

Theoretical and academic considerations apart, the standard methods that are available for investigating the elastic properties of textile fibers are in ASTM D 1774–93 and BS 4029, 1978, although one hopes that at some time in the future the ISO may publish ISO 15602, which might address the problem of standardizing the test criteria for international comparison purposes. See later.

ASTM D 1774–93 is concerned with the measurement of the elastic behavior of fibers by assessing their ability to recover from strain-induced stress and to recover their original dimensions after extension. Procedures are also described for crimped and uncrimped fibers. The fiber is tested at three levels of extension, viz, 2%, 5%, and 10%. A CRE type instrument is used in the test, equipped with a constant or continuous system for plotting the load/extension and a planimeter or automatic integrator for measuring the area under the load–extension curve. Normally 10 specimens are tested at each extension. The specimens are conditioned in the standard atmosphere for conditioning and testing prior to the test, and the more hygroscopic materials such as rayon, acetates, and nylon are conditioned from the dry side of the hysteresis curve by first preconditioning. The gauge length selected must of course be the same for the fibers tested, but where possible gauge lengths of 127 mm or 265 mm are recommended. The test specimens are extended at a standard rate of 10% per minute, so clearly the results between 127 mm gauge length and 265 mm gauge length tests will not be comparable. The test specimen is first extended 2%, extension and then the cross-head is stopped for 60 s to allow the stress induced in the specimen to decay (B–G in Fig. 14). The cross-head is then returned to the original gauge length at the same speed as the rate of extension. The test specimen is then allowed to relax for 3 min, allowing creep recovery to take place. The test specimen is then again extended (at the same rate) to its original level of extension. At this point it will commonly be found that when the second cycle is initiated there will be a certain amount of fiber extension without any appreciable load being recorded (A–D in Fig. 14). This extension or elongation is the amount of permanent deformation suffered by the test specimen and is commonly expressed as a percentage of the total extension (A–E).

So permanent deformation % = (AD × 100)/AE.

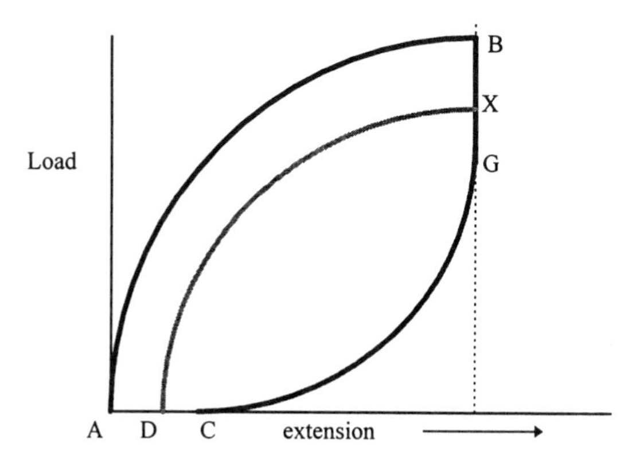

Figure 14 Stylized illustration of the effect of mechanical conditioning on elastic recovery.

However, many fibers will not suffer any permanent deformation at this degree of elongation, and indeed ASTM D 1774 acknowledges that "Determination of the permanent deformation after 2% extension will be inaccurate if the gage length used was less than 5 inches (127 mm) due to the difficulty of measuring the small extent of irrecoverable deformation."

The test is repeated on fresh specimens at 5% extension and 10% extension (providing of course that this does not exceed the breaking extent of the fiber, in which case the test results would be lost).

The tensile strain recovery and the work recovery % are also calculated. The procedure for crimped specimens is much the same, except that the test specimen is first pretensioned to take out the crimp and the extension shown on the recording chart eliminated by rezeroing the recorder. The location of the cross-head is locked in the memory at this point, so that on recovery the cross-head does not reenter the crimp range.

BS 4029, 1978, specifies a method of test for the measurement of tensile elastic recovery of all types of single filaments or fibers when subjected to tensile strain applied by a CRE type machine. No special pretreatments such as mechanical cycling or relaxation of internal stresses by immersion in boiling water are specified, although it is acknowledged that they may be required for some special purposes. The determination of elastic recovery from a given stress (specific stress) is not included.

Before describing this test in more detail it might be of interest to compare the various definitions relevant to elastic modulus given in ASTM D 4848-94a, BS 4029, 1978, and Textile Terms and Definitions [1].

Immediate elastic recovery (BS 4029). The immediate change in elongation experienced by a fiber during a loading cycle when, after holding at a defined elongation for a defined time, the applied tension is reduced to the pretension value.

Residual elongation (BS 4029). The nonrecovered change in elongation experienced by a fiber during defined loading cycles.

Delayed deformation (ASTM D 4848). Deformation that is time-dependent and exhibited by material subject to a continuing force (=Creep).

Elastic limit (ASTM D 4848). The greatest stress that can be applied to a material without permanent deformation (cf. yield point).

Discussion. Elastic limit is a property of a material, whereas yield point is a specific point on a stress–strain curve.

Elasticity (ASTM D 4848). That property of a material by virtue of which it tends to recover its original size and shape immediately after removal of the force causing deformation.

Elongation (ASTM D 4848). The ratio of the extension of a material to the length of the material prior to stretching (cf. extension).

Extension (ASTM D 4848). The change in length of a material due to stretching.

Extension recovery cycle (ASTM D 4848). In tension testing, the continuous extension of a specimen, with a momentary hold at a specified extension, followed by a controlled rate of return to zero extension.

Immediate elastic recovery (ASTM D 4848). Recoverable deformation that is essentially independent of time, that is, occurring in (a time approaching) zero time after removal of the applied force. (See Delayed deformation and compare with (delayed) elastic recovery.)

Tensile hysteresis curve (ASTM D 4848). A complex load–elongation or stress–strain curve obtained under either of two conditions:

(1) When a specimen is successively subjected to the application of a load or force less than that causing rupture, and the removal of the load or force according to a predetermined procedure.

(2) When a specimen is stretched less than the breaking elongation and allowed to relax by removal of the strain according to a predetermined procedure. (= tensile recovery curve, work recovery curve, and strain recovery curve.)

Yield point (ASTM D 4848). In a stress–strain curve, the point beyond which work is not completely recoverable and permanent deformation takes place. (cf. elastic limit).

Yield point (TT & D). The point on the stress–strain curve corresponding to the elastic limit.

Elastic limit (TT & D). The greatest strain that a material is capable of sustaining without any permanent strain remaining after complete release of the stress.

Elastic recovery (TT & D). The immediate reduction in extension observed in a material when, after being held at a defined elongation for a given time, the applied force is removed.

Elasticity (TT & D). That property of a material by virtue of which it tends to recover its original size and shape immediately after removal of the force causing deformation.

Creep (TT & D). The time-dependent increase in strain resulting from the continuous application of a force. (Note: *Creep tests are usually carried out at constant load and constant temperature*.)

Creep recovery (TT & D). The time-dependent decrease in strain following removal of stress.

Stress decay (BS 5421, Part 2, 1978). The decrease in tensile force that occurs when a specimen is held at a constant extension, expressed as a percentage of the original force at that extension.

In BS 4029 the test specimen is extended at a constant rate of 25% ± 5% per minute (of the original gauge length), compared to 10%/minute in ASTM D 1774. It is admitted that elastomeric yarns (of the type covered by BS 5421) may be required to be extended at the rate of 100% min (in fact in BS 5421, Part 2, 1978, which is a series of tests for PU thread, the gauge length is 50 mm and the cross-head speed is 500 mm/min, thus giving a rate of extension of 1,000%/min, and not 100%/min as stated in BS 4029).The gauge length adopted in BS 4029 is 200 mm for filament and 20 mm or 50 mm for staple. The test specimens are pretensioned at 0.5 cN/tex (untextured) and 2.0 and 1.0 cN/tex for textured polyester and nylon or acrylics and acetates respectively in filament form, whereas staple nylon and polyester are pre-tensioned at 1.0 cN/tex and acrylics at 5.0 cN/tex. Viscose, cupro, modal, acetate, and triacetate are pre-tensioned at 0.5 cN/tex. Due to their low modulus, elastanes are pre-tensioned at only 0.01 cN/tex.

Table 6 Comparison of Test Parameters Between ASTM and BS in Measuring Elastic Recovery of Filaments

Parameter	BS 4029	ASTM D 1774
Gauge length	200 mm (fil) 20 mm or 50 mm (staple)	127 mm or 265 mm (filament only)
Extension rate	25% or 100% per min	10%/min
Initial extension	2%	2%
2nd optional extension	5%	5%
3rd & other optional extension	20% or 100%	10%
Dwell time	30 s or 1 min or 5 min	1 min
Zero tension relaxation time	1 min	3 min
Reports	Immediate elastic recovery and residual elongation after 1 min	% permanent deformation % tensile strain recovery % work recovery

The procedure is similar to ASTM D 1774 but contains more options, which are best illustrated in Table 6.

Clearly the results from the two methods will not be comparable, but if the test criteria are known this at least gives the technologist a chance to interpret the results in some sort of meaningful way. Unfortunately it is common practice for some fiber manufacturers and some parts of the trade press to fail to specify the methods of test used or the test parameters employed. One can understand and sympathize with a commercial reluctance to not give too much away, but if the motives are for some other reason, it is well worth remembering the statement that "the days are gone when traders and commercial people can hide behind an ambiguity." So when one sees a tabular comparison of fiber data of various fibers from a range of manufacturers, without the commonality of test methods, go gentle into that good night.

8 Tensile Properties and Toughness

If everyone liked the same thing we would have run out of it, whatever it was, long ago. So it is with textile polymers. There are an almost infinite number of end uses, each of which demands a particular combination of fiber/yarn/fabric/composite-construction properties, but one of the most common features of textile polymer products is that they are all *orthotropic*, i.e., they have elastic properties with considerable variation of strength in two directions perpendicular to one another. This element of elasticity has a profound impact on fiber toughness and also on the time dependent properties of the polymer product.

Tensile strength is not the most important element of a textile fiber. Important it is, but there are other factors of equal importance, such as flexibility, handle, dye-uptake, and moisture absorption/desorption. In fact, in many cases, even though the material specification is designed with a high emphasis on tensile properties, quite often the rigours of use come nowhere near the product's tensile capabilities.

Which particular property or combination of properties are the most important will depend upon the end use to which the product will be put. Nevertheless, tensile properties are important, to some degree or other, in almost every case, if only because of other properties, which are themselves related to the tensile modulus, such as tear and burst strengths.

Fibre tensile properties have their own importance, however, long before they are embodied in any end product. For example, the spinning limits illustrated in Section 3.1.2 take account only of fiber linear density and effective length.

Fibre quality index however, which is a numerical value that indicates the processabilty of cotton, is given by

$$\frac{2.5 \times \text{uniformity ratio} \times \text{tenacity (gf/tex)}}{\text{micronaire value}}$$

where uniformity ratio is the ratio between two span lengths (commonly 50% and 2.5%) expressed as a percentage of the longer span length. Span length is determined using the Fibrograph instrument described earlier. The value 2.5% span length is the extent exceeded by only 2.5% of the fibers. What this fiber quality index illustrates is that in addition to the spinning limit being dependant on fiber length and fineness and fiber uniformity, it is also dependant upon fiber tenacity. This is quite obvious when one realizes that with the finer yarns in particular, the greater the proportion of the spinning tensions (forces) fall on individual fibers. So if end-break rates are to be kept to acceptable limits (say 5/100 spindle hours or less), the fiber tenacity must be sufficient to withstand these forces. It is also equally obvious that the yarn strength cannot be greater than the sum of the strengths of the individual fibers.

Before embarking however on a study of the methods for the physical testing of textile polymer fibers it may be worthwhile first to establish a few related definitions for terms that differ from testing bulk polymers.

Tenacity [1]. The tensile force per unit linear density corresponding with the maximum force on a force/extension curve.

Note 1. In testing textile fibers and yarns, tensile force is normally measured in Newtons (or multiples or submultiples thereof) and linear density in tex (or decitex). Thus

$$\text{tenacity} = \frac{\text{maximum tensile force (N)}}{\text{linear density (tex)}}$$

Note 2. For the purposes of this definition, the linear density is that measured before a test piece is extended, not that applying at the point of maximum force. Because of the relatively high extensibility of most textile materials, the extension at maximum force should always be stated.
Note 3. The expression "ultimate tensile stress" is sometimes used in textiles as a synonym for tenacity, but in some other disciplines "ultimate tensile stress" and tenacity relate to the maximum tensile force per unit cross-sectional area of the test piece (Fig. 15).

Breaking strength : tensile strength [1]. The maximum tensile force recorded in extending a test piece to breaking point.

Tensile strength at break [1]. The tensile force recorded at the moment of rupture.
Note. The tensile strength and the tensile strength at break may be different if, after yield, the elongation continues and is accompanied by a drop in force resulting in tensile strength at break being lower than tensile strength.

Extension at break; breaking extension. The extension percentage of a test specimen at breaking point.

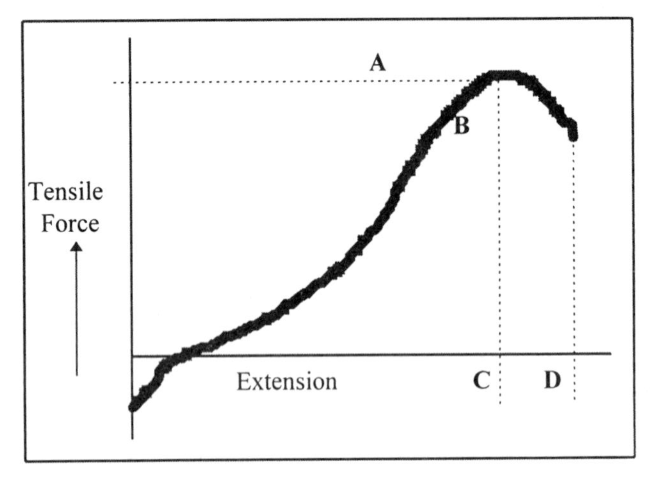

Figure 15 Illustration of key tensile parameters. A, maximum tensile force; B, point of severance or rupture; C, extension at maximum force; D, extension at break.

In engineering and testing bulk polymers, the term stress is taken to be the ratio of applied force to the cross sectional area. The SI unit of stress is the newton per square meter (N/m^2). This is also numerically equivalent to the pascal (Pa), albeit that the Pa is usually employed as a unit of compression stress (i.e., pressure) rather than tensile stress.

However, as explained earlier in Section 3, textile fibers invariably have irregular cross sections, making the area of cross section difficult to measure, except indirectly from its mass and density (see Section 3.1). Therefore instead of employing the traditional unit of stress, the term specific stress or mass stress is employed. This is given by

$$\text{specific stress} = \frac{\text{force (N)}}{\text{linear density (tex)}}$$

Textile fibers are intrinsically variable materials. When subjected to an increasing tensile force they will of course break at their weakest point. Statistically there is a greater chance of a weak point occurring in a long fiber than in a short one, so the mean breaking force of a series of tensile tests on longer fiber specimens will be lower than the same series of tests conducted on shorter fibers.

The elongation at break of the vast majority of fibers will be greater if the force is applied over a longer period of time. Exceptions might be found in carbon, boron and similar nonviscoelastic materials. Therefore the force required to break the fiber will vary depending upon the rate at which the force is applied. If the fiber is shock loaded, i.e., the fiber is extended to its breaking extension instantaneously, the force recorded will be higher than if the breaking extension is achieved over a long period of time. Fiber manufacturers therefore who wish to emphasise the high breaking force of a particular fiber may tend to conduct tests at a high extension rate, thus taking advantage of this fact. Naturally, the reverse side of the coin is that whilst extending the specimen to rupture at a high speed results in a high breaking force being recorded, the extension at break is correspondingly lower. Conversely, if the specimen is extended at an exceedingly slow rate, the extension at break will be proportionally higher.

Another feature of creating stress–strain curves of textile fibers that can be misleading is by plotting force against actual elongation and not specifying or taking account of the fiber linear density. This is illustrated in Fig. 16.

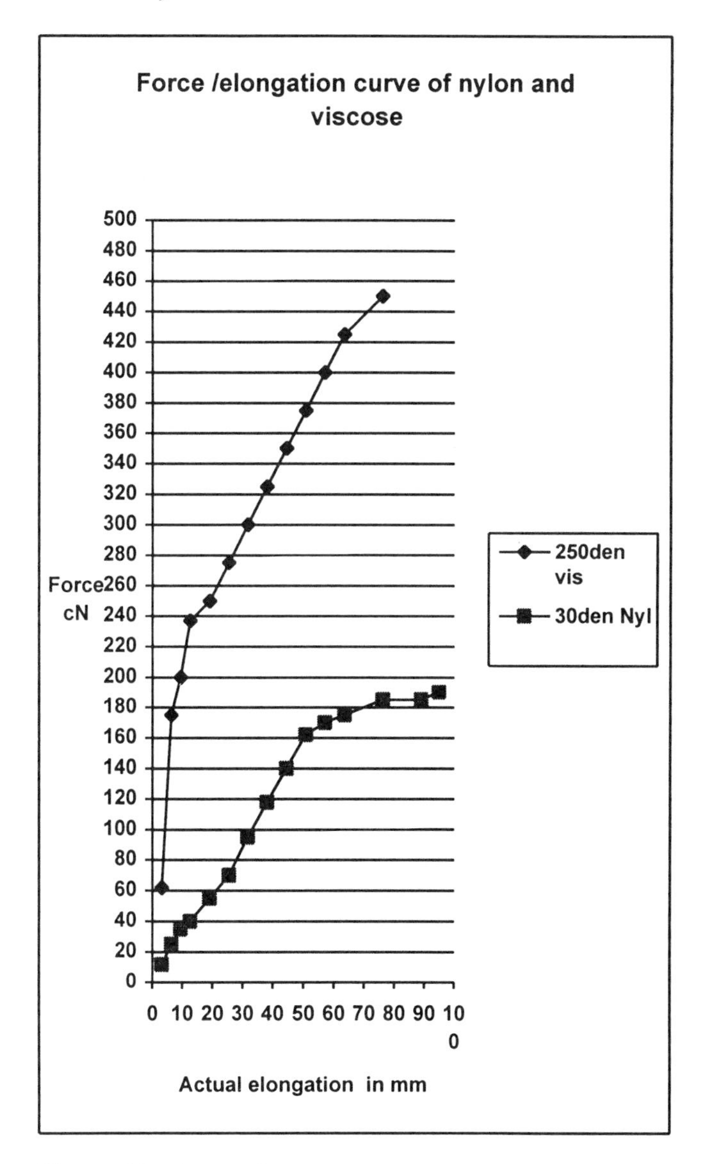

Figure 16 Force /elongation curves for nylon and viscose.

However, when the applied force is expressed in terms of the linear density of the fiber at the commencement of the test and the elongation is expressed in terms of the percentage elongation of the fiber, the true story emerges, as shown in Fig. 17 where the force/elongation curve is transferred from Fig. 16.

Tenacity or tensile strength at break is also not necessarily the best indicator of fitness for purpose, even when tensile strength is crucially important. The work done in extending the specimen to breaking point may be of greater importance, as for example in motor vehicle web restraints, car seat belts, climbing and mooring ropes, and safety harnesses. In these cases the work of rupture or toughness is the more important issue. The work of rupture is the energy absorbed by the fiber, filament or product up to the point of rupture and is measured in joules. It is the total area bounded within the stress–strain curve and is

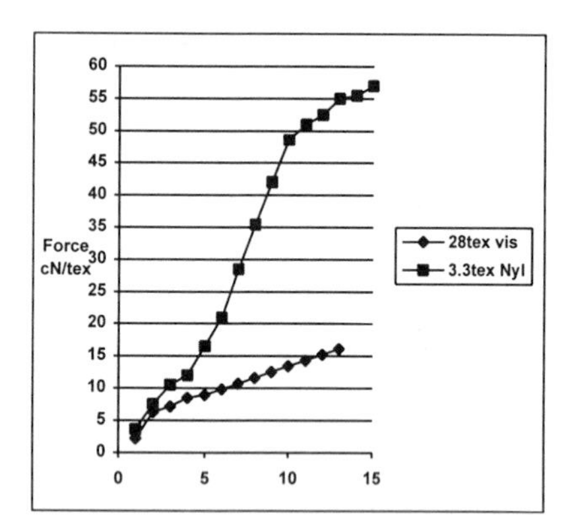

Figure 17 Force/linear density vs. extension curves for the yarns shown in Figure 16.

the work done within the stress–strain curve (as work done is force × displacement) as illustrated in Fig. 18.

However, all other things being equal, the work of rupture will depend on the fiber linear density and, because of the reasons mentioned above, on the gauge length of the specimen. Therefore, standardizing gauge length, in order to compare different types of material, the term specific work of rupture is employed and is given by

$$\text{specific work of rupture} = \frac{\text{work of rupture}}{\text{linear density} \times l}$$

where

work of rupture is in N · m
linear density is in g/Km (tex)
and l is in m

So specific work of rupture can actually be expressed as

$$\frac{\text{N} \cdot \text{m}}{\text{tex} \times \text{m}} = \frac{\text{N}}{\text{tex}}$$

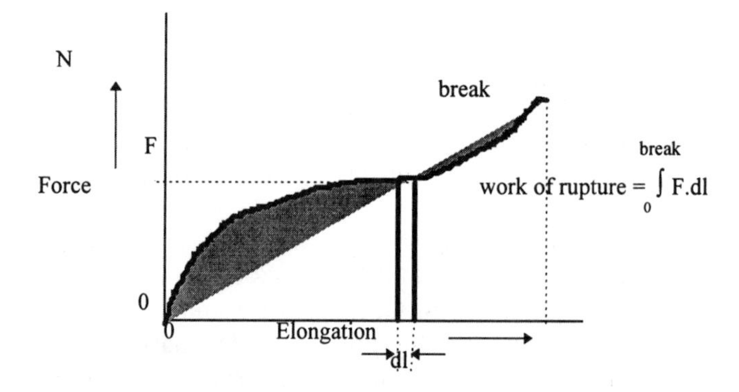

Figure 18 Force/elongation curve for illustration of work of rupture.

Numerically N/tex is equivalent to kJ/g, although in ASTM D 3822–95 (see later) work of rupture is in J/tex.

Of course if Hooke's law were to be followed throughout the test, the area under the curve would be

$$\frac{\text{F} \times \text{elongation}}{2} \qquad \text{or} \qquad \tfrac{1}{2}(F \times \text{elongation})$$

In order to define simply whether the work done in extending the specimen to rupture exceeds or is lower than this ideal, the term work factor has been adopted. This is given by

$$\text{work factor} = \frac{\text{work of rupture}}{\text{force at break} \times \text{elongation at break}}$$

So for materials having the same tenacity, the work of rupture will be higher if the work factor exceeds 0.5 or lower if the work factor is less than 0.5.

This is illustrated graphically in Fig. 19.

Effect of Crimp

As was illustrated in Section 3.1.1, the removal of fiber crimp requires a force, and care is needed in doing so in order to avoid introducing strain into the fiber. This is of course equally true when conducting tensile tests. If a fiber containing crimp is tested without the crimp being first removed by a suitable pretension, the force elongation curve will register a force without any apparent elongation. This is because the force is employed in removing the crimp. This is illustrated in Fig. 20.

The preceding brief summary of the important principles that need to be considered in tensile testing fiber polymers and filaments will we hope be useful to the reader in looking at the detail of the following standard methods of test.

In *ISO 5079, 1995, Textiles—Fibres—Determination of breaking force and elongation at break of individual fibers*, the terms "rupture" and "break" are differentiated. Consequently separate terms are used for "elongation at break" and "elongation at rupture" as illustrated in Fig. 21.

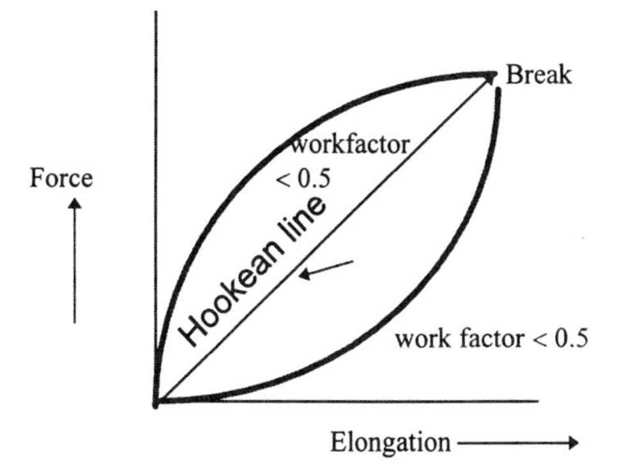

Figure 19 Force/elongation curve illustrating work factor.

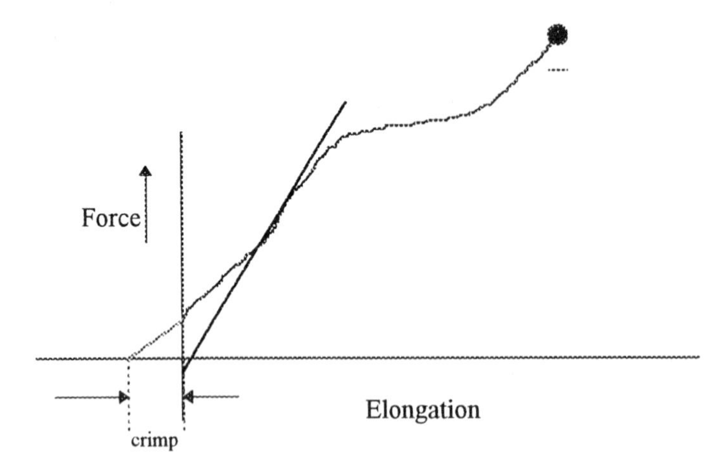

Figure 20 Illustration of the effect of crimp on stress/strain curve.

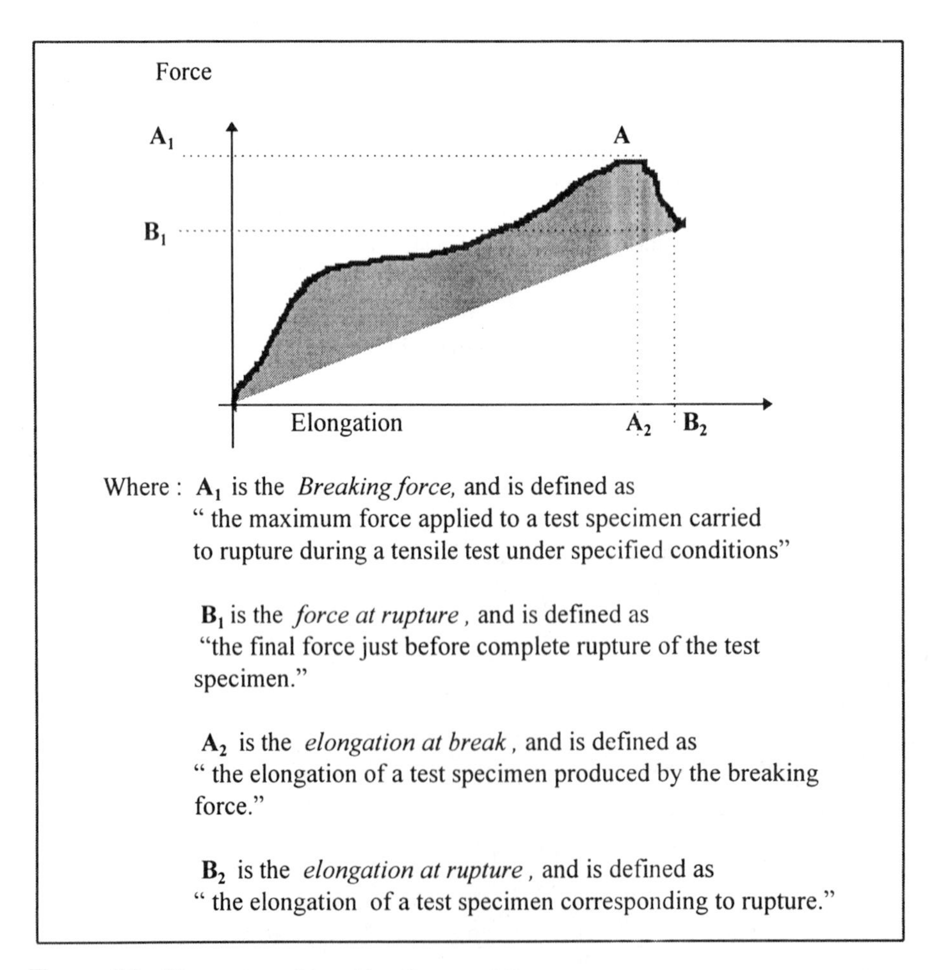

Figure 21 Illustration of breaking force and force at rupture.

It might be considered somewhat pedantic, but one might argue that the above definition of *elongation at break* gives a misguided view and that perhaps a more accurate definition might be that the elongation at break is the elongation that under the particular conditions of the test (i.e., initial gauge length and cross-head speed) develops in the test specimen sufficient tension to cause the test specimen to break.

A pretension is applied that varies with the type of fiber and whether the test is conducted wet or dry and is given in a tabulated format without explanation or reference to the crimp rigidity of the different synthetic fiber types. The responsibility for this aspect appears to have been left to interested parties whoever they may be. The tabulated data is reproduced in Table 7.

The linear density of the fibers tested is determined in accordance with ISO 1973 (see earlier in Section 3).

Individual fibers are extended to rupture at constant rates of extension that may vary between 5 and 20 mm/min. For reasons explained above, results of tests conducted at different rates of extension will not of course be directly comparable, but these options are available to enable fibers of different elastic modulus to be tested.

The standard requires specimens whose elongation at break is less than 8% to be extended at a rate of 50% per minute. The normal nominal gauge length is 20 mm. Consequently if the fiber extension at break is, say, 3.5% to 4%, the cross-head traverse speed will be 10mm/min. This would result in a time to break of 4.5 s. The standard alternatively specifies that if the extension at break is equal to or greater than 8% the elongation rate shall be 100%/minute. Therefore in this case if the extension at break were, say, 10%, the cross-head traverse speed would be 20 mm/min and the time to break would be 6 s.

At least 50 fibers are tested, and the mean breaking force is reported along with the mean elongation at break and their 95% confidence intervals.

ASTM D 3822-95—Method of test for tensile properties of single textile fibers (although the words "fiber" and "filament" are used interchangeably throughout this method), provides directions for measuring the breaking force and elongation at break of single fibers and for calculating the (breaking) tenacity, initial modulus, chord modulus, tangent modulus, tensile stress at specified elongation, and breaking toughness. It should be noted however that the tensile stress referred to is specific stress and not stress in the classical engineering sense (see earlier). Breaking force is expressed in mN and tenacity in mN/tex as is chord modulus and initial modulus. Again a CRE type instrument is used, but the rate of extension is 10% per minute for fibers whose breaking extension is less than 8%, 60% per minute for fibers whose breaking extension is between 8% and 100% and 240% per minute for fibers whose breaking extension is greater than 100%. The standard

Table 7 Pretensioning Forces for Different Fiber Types

Fiber type	Pretension[a] in cN/tex
Cellulose man-made fibers	
test in conditioned state	0.6 ± 0.06
test in wet state	0.25 ± 0.03
Polyester fibers	
linear density < 2 dtex	2.0 ± 0.2
linear density ≥ 2 dtex	1.0 ± 0.1

[a]A higher pretension, e.g. to remove crimp, may be applied subject to agreement between the interested parties.

does acknowledge however that "for the optimum degrees of comparability, tensile properties of filaments should be measured at the same rate of extension" (see earlier and contrast with ISO 5079).

The gauge length selected is 10 mm, or 250 mm where applicable, but for the reasons given earlier it is acknowledged that "when comparisons are to be made between different fibers . . . it is advisable to use the same gage length for all tests, selecting it to accommodate the shortest fibers of interest."

A pretension of 25 mN/tex to 100 mN/tex is used to remove crimp in crimped fibers. At least 20 acceptable fiber breaks are conducted (i.e., fiber breaks not at the jaw–fiber interface or at the tab–fiber interface or where there is no slippage of the fiber in the jaws of the instrument or in the end tabs). End tabs are a means of connecting the fiber ends in an epoxy resin bound to a cardboard "tab." This technique is common where it is difficult to grip the fiber directly without crushing it at the jaw interface, which would result in a jaw interface break not caused by a natural tensile weakness.

The breaking force, tenacity, elongation at break, initial modulus, chord modulus, and toughness (i.e., work of rupture in J/g) is calculated for each individual test result and the mean value reported.

It may at this point be useful to look at the determination of initial modulus in a general sense, before illustrating the approach in ASTM D 3822, which at first sight may appear unusually complex.

Initial modulus. This term is intended as "initial Young's modulus" and is the ratio of specific stress (N/tex) to fractional strain along the axis of the stress–strain curve in the Hookean region, with starting point being zero stress and zero strain. This is illustrated in simple geometric terms as in Fig. 22.

$$\text{Initial Young's Modulus} = \tan\theta = \frac{\text{A.A}_1}{0.\text{A}_1} = \frac{\text{B.B}_1}{0.\text{B}_1} = \frac{\text{C.C}_1}{0.\text{C}_1}$$

It is not always an easy matter however to determine the point of zero strain or indeed where the Hookean region in the stress–strain curve might be.

The first prerequisite is obviously to establish if there is an obvious yield point, as this would clearly establish the upper extent of the Hookean region. The part of the stress–strain curve approaching the yield point and having the maximum gradient is taken to be the Hookean region, and the zero starting point of this gradient line, extended back to the

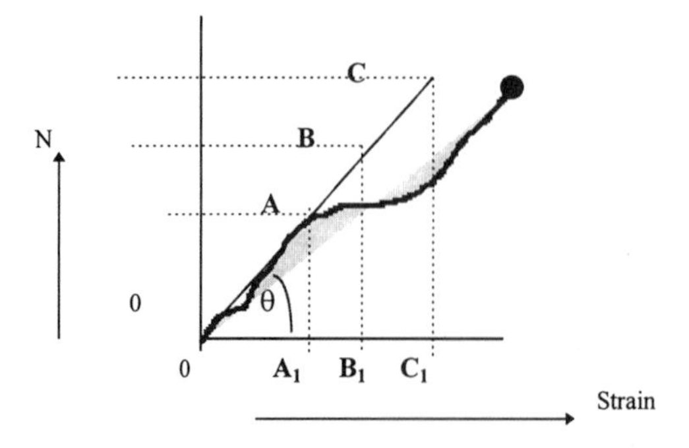

Figure 22 Geometry for Young's modulus.

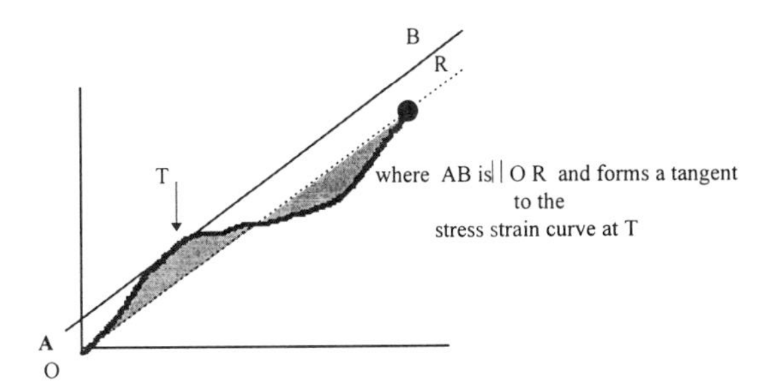

Figure 23 Determination of yield point according to Meredith.

base extension line, will intersect at the point that represents the zero extension point from which strain is measured.

Determination of Yield Point

There are basically two approaches to the determination of yield point. The first was suggested by Meredith [18] as in Fig. 23. The second was suggested by M.J. Coplan in U.S. Air Force Technical Report 53–21, as in Fig. 24.

This is the underlying theory supporting the determination of initial modulus in ASTM D 3822, Appendix XI, where there are two illustrations, one of a material with an obvious Hookean region and one of a material with no obvious Hookean region, as in Fig. 25.

ASTM D 2256 says at Appendix X1, "In the case of a yarn exhibiting a region that obeys Hooke's law, a continuation of the linear region of the curve is constructed through the zero-force axis.

This intersection point B is the zero elongation point from which strain is measured. The initial modulus can be determined by dividing the force at any point along the line BD by the corresponding strain measured from point B."

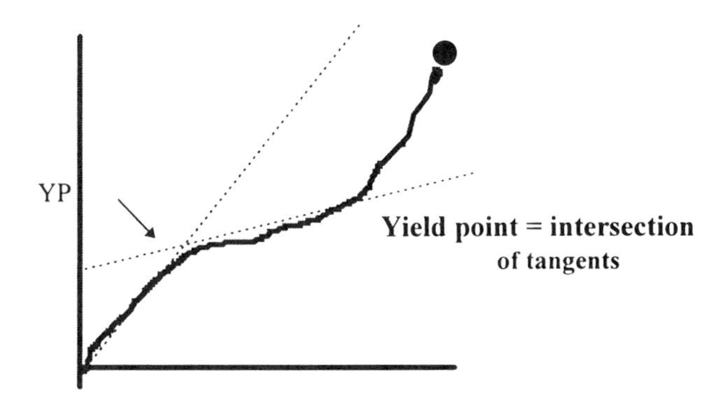

Figure 24 Determination of yield point according to Coplan.

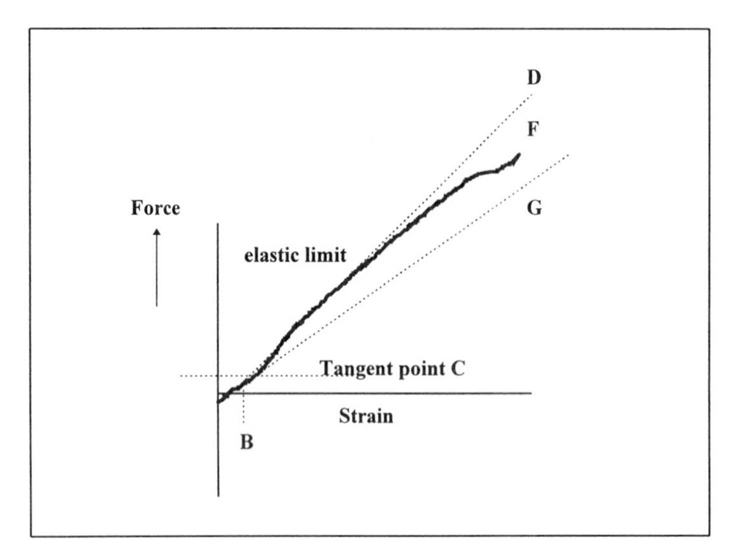

Figure 25 Determination of initial modulus according to ASTM D 2256.

As a practical illustration, Fig. 26 may be interesting, taken from a viscose yarn of 33 tex.

Work of Rupture

As explained earlier, the work of rupture is *represented* by the area under the stress–strain curve. However, the proportions and limits of the stress–strain curve, and hence the area it encompasses, will be dependent upon the speed of jaw separation and cross-head load range in the vertical axis, and the chart width and chart recording speed. Also of course, as

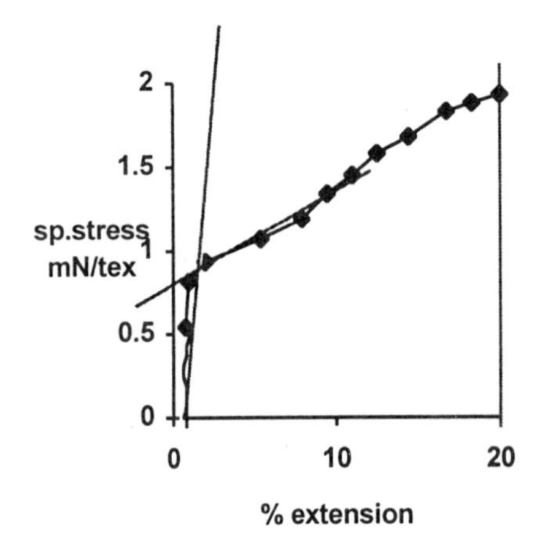

Figure 26 Stress/strain curve for 33 tex viscose yarn.

illustrated earlier in Fig. 17, the linear density must be taken into account, or otherwise the stress–strain curve will not be material representative. Consequently in ASTM D 3822, the breaking toughness (*Tu*) in J/g is calculated using the following equation:

$$Tu = \frac{A_c \times F_{fs} \times R_s}{W_c \times C_s \times D_L \times L_e}$$

where

A_c is area under stress-strain curve in mm^2
F_{fs} is the full scale force range in mN
R_s is testing speed rate in mm/min
W_c is recording chart width in mm
C_s is the recording chart speed in mm/min
D_L is the fiber linear density in mtex
L_e is the effective specimen length in mm

ASTM D 2102–94 is concerned with determining the tensile properties of single man-made textile filaments taken from yarns and from crimped or uncrimped tows. If however the yarns have been subjected to a bulking or texturing process the method is not applicable. Again the method covers the measurement of breaking force and elongation at break as well as tenacity, initial modulus, and toughness.

Single fiber (filament) specimens are extended to rupture at a constant rate of extension of 10% per minute for filaments whose elongation at break is less than 8%, at 60% per minute for filaments whose breaking elongation is between 8% and 100% and at 240% per minute for filaments whose breaking elongation is greater than 100%. This gives the variable times to break of around 50 s down to 25 s. See ASTM 2256 later. If the filament is crimped, a pretension of 0.3 gf /tex to 1 gf/tex is used to remove crimp.

Linear density of the filament is determined using a vibroscope (described earlier in ASTM D 1577 or ISO 1973 in Section 3).

The breaking toughness can be reported directly from the test instrument if equipped with a program for integration. Otherwise it is calculated in gf·cm/tex from the equation

$$Tu = \frac{V}{T}$$

where

T = linear density in tex
V = work done in g · cm/cm length, and

$$V = \frac{A \times S \times R}{G \times W \times L}$$

where

A is the area under the force--elongation curve in cm^2
S is the full-scale load in gf
R is the cross-head speed in cm/min
G is the (corrected) effective specimen length in cm
W is the chart width in cm
L is the chart speed in cm/min

The initial modulus is determined in a manner similar to ASTM D 3822 and described above.

ASTM D 2256–95 provides a method for the determination of the tensile properties of monofilament, multifilament, and spun yarns, wither single, plied, or cabled, providing that the yarns do not extend more than 5% when the tensile force is increased from 5 mN/tex to 10 mN/tex. The method covers the measurement of breaking force and elongation at break, tenacity (at break), initial modulus, and breaking toughness. Options are included for knotted and looped forms (for which see also ISO 2062).

Generally speaking this single strand method is more accurate than a skein method (for which see ISO 6939 later) and is more economic in terms of material. It provides a result that is invariably higher, in terms of the sum of the breaking forces of the same number of ends as a skein test. Most types of yarn can be tested by this method, although some modification of clamping techniques may be necessary for a given yarn depending on its structure and composition. Special clamping adaptions may be necessary with high modulus yarns such as glass, aramid, or extended chain polyolefin.

A CRE type instrument is recommended with a variable speed drive so that a time to break can be standardized at an average of 20 s $\pm$ 3 s. This has been consistently found to minimize the differences in test results between CRE and CRT type instruments and has been recommended from time to time by ISO/TC 38.

If for example the breaking extension of the filament is known to be around 16% to 20% and is extended at 50% per min, the time to break will be around 20 s. So if the initial gauge length is 250 mm, the jaw separation rate will need to be around 125 mm/min. If however the jaw separation rate were 100 mm/min and the initial gauge length were 250 mm, the specimen would be extended at 40% per min and the time to break would be around 27 s.

For ready reference these relations can be reduced to

$$T = \frac{E_b \times 60}{P_{er}}$$

where

T is time to break in seconds
E_b is the % extension at break
P_{er} is the % extension rate per minute

Similarly,

$$P_{er} = \frac{CHS \times 100}{G_L}$$

where

CHS is the cross-head speed in mm per minute
G_L is the initial gauge length in mm

and

$$\frac{E_b \times 60 \times G_L}{CHS \times 100} = T$$

or

$$CHS = \frac{P_{er} \times GL}{100} = \frac{E_b \times 60 \times G_L}{T \times 100}$$

So for a filament with an extension at break expected of around 30%, an initial gauge length of 250 mm, and a required time to break of $20\,\text{s} \pm 3\,\text{s}$, the rate of jaw separation (*CHS*) per minute would need to be

$$CHS = \frac{30 \times 60 \times 250}{20 \times 100} = 225\,\text{mm/min}$$

It is for these reasons that a variable speed drive is necessary in order to comply with ASTM D 2256, unless the option of mutual agreement between concerned parties is invoked of using an operating speed fixed at $120 \pm 5\%$ extension per minute. As the gauge length is still 250 mm, this would give a rate of jaw separation of

$$CHS = \frac{P_{\text{er}} \times G_{\text{L}}}{100} = \frac{(120 \pm 5) \times 250}{100} = 287.5 \text{ to } 312.5\,\text{mm/min}$$

In fact ASTM D 2256 refers to a cross-head speed of 290 to 310 (i.e. 300 ± 10) mm/min, but this is clearly nominal.

The initial modulus and tenacity (breaking) are calculated in the same way as in ASTM D 2101–94.

The instruction for determining breaking toughness is similar to that given in ASTM D 2101 in terms of the equations used, but the following notes are added, viz, "When using the force–elongation curves, draw a line from the point of maximum force of each specimen perpendicular to the elongation axis. Measure the area bounded by the curve, the perpendicular and the elongation axis by means of an integrator or a planimeter, or cut out the area of the chart under the force–elongation curve, weigh it, and calculate the area under the curve using the weight of the unit area."

Of course one of the disadvantages to a standard time to break, as shown, is that the speed at which the specimen is extended varies, and as previously explained this gives rise in itself to consequential variations in breaking extension and tenacity. The advantage of the one perhaps is more than outweighed by the other.

Knot breaking strength

When a textile fiber or filament is bent into a loop or formed into a knot and a tensile stress imposed, its breaking strength will be lower than if the fiber were tested in the straightened condition. This is a measure of the bending modulus or stiffness of the fiber. In its looped or knotted state, there is a part of the circumference of the fiber that is under strain. If the knot strength or loop strength is expressed as a percentage of its tensile strength, it is found that the value is lowest for those fibers (filaments) which have the lowest elongation at break. Nylon for example will have a percentage loop breaking force of around 82% of its tensile strength, whereas fiberglass will have a percentage loop strength of around only 8% to 9%.

ASTM D 2256 acknowledges that "the reduction in breaking force due to the presence of a knot or loop is considered a measure of the brittleness of the yarn."

Although ASTM D 2256 includes options for conducting the test in loop or knotted configurations, the recommended methods for determining the tenacity of yarns in knotted and loop configurations are in ASTM D 3217–95 and BS 1932, Part 2, 1989.

ASTM D 3217–95 includes "the measurement of the breaking tenacity of man-made fibers taken from filament yarns, staple or tow, either crimped or uncrimped and tested in either a double loop or as a strand formed into a single overhand knot."

A CRE type instrument is used and operated at a rate of extension of 100% of the nominal gauge length (25 mm) per minute. Linear density is determined previously using a Vibroscope (see earlier).

The loop and knot configurations are shown in Figs. 27 and 28.

BS 1932, Part 2, 1989, knot strength ratio is defined as the ratio of the knot breaking strength (as measured in BS 1932, Part 2) to the yarn breaking strength (as measured in accordance with ISO 2062).

The loop strength ratio on the other hand is defined as the ratio of the loop breaking strength as measured in accordance with BS 1932, Part 2 to *twice* the yarn breaking strength measured in accordance with ISO 2062.

The knot configuration in BS 1932, Part 2, as shown in Fig. 29, is for a Z twist yarn, and a different configuration is shown for an S twist yarn, but no configuration is shown for the loop strength test.

Glass Yarns

The tensile testing of glass yarns is described in ISO 3341, 1984. It is intended for yarns having a diameter of less than 2 mm or a linear density lower than 2 ktex but can be used for yarns exceeding these values. Dynamometers of CRL, CRE, or CRT type are permitted. A gauge length of 500 mm is used, although a nominal gauge length of 250 mm may be used by agreement. It is however acknowledged that in this case the test results may be slightly higher (by which is presumably meant the breaking strength rather than the extension at break—see above). The test is conducted at a speed to produce a break in 20 ± 3 s or, if CRE is used, at 50 mm/min.

Consequently, if the time to break T is to be 20 s and the gauge length G_L is 500 mm, and assuming the extension at break is 3.5%, the jaw separation rate (CHS) can be calculated as described earlier, i.e.

$$T = \frac{E_b \times 60 \times G_L}{CHS \times 100}$$

and

$$CHS = \frac{E_b \times 60 \times G_L}{T \times 100} = \frac{3.5 \times 60 \times 500}{20 \times 100} = 52.5\,\text{mm/min}$$

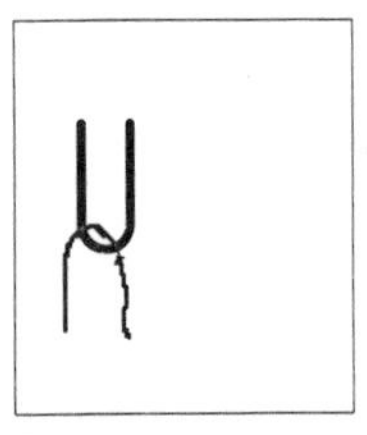

Figure 27 Loop configuration.

Figure 28 Knot configuration.

Figure 29 Knot configuration for Z twist yarn according to BS 1932, part 2.

Obviously if the jaw separation rate were fixed at 50 mm/min, the time to break would be slightly longer, 21 s, but still within the permitted tolerance of 20 ± 3 s.

Crimp Rigidity and Crimp Frequency

Some mention has already been made of the necessity of pretensioning fibers prior to tensile testing in order to remove crimp and thereby obtain a straight fiber before the extenuating load is applied. It should not however be assumed from this that fiber crimp is a nuisance. Far from it. It is in fact essential to the overall cohesive power of staple fibers in forming a yarn and is vital if a card web is not to collapse (albeit that the introduction of hooks during carding is greatly helpful in this respect). In synthetic filament processing it is the amount and strength of the in-built crimp in the filaments that governs the bulking potential of the yarn and thereby the handle and elasticity of the resulting fabric.

There are basically two standard methods for measuring crimp rigidity, BS 6663, 1986, and ASTM D 4031–95. In the first of these there is Method A and Method B, but Method B is essentially the same as Method A except that the test hank is first relaxed in hot water at 75°C for 10 minutes. Method A is essentially the H.A.T.R.A. crimp rigidity test described by Booth [19] and criticized by G. S. Wray [20]. The instrument is now marketed by SDL of Didsbury, Stockport, Cheshire, U.K.

BS 6663 does acknowledge that the test as written is only suitable for nylon yarns processed by twist/detwist, false twist, stuffer box, and edge crimped methods and "does not characterize the yarn completely but is a useful quality control check for throwsters and knitters." This last acknowledgement by the BS means that although the test may not be suitable for predicting how the yarn will behave when in fabric form, (which was Wray's criticism) it is useful as a comparative in order to ensure that yarns with widely disparate crimp rigidities are not used together in the same fabric, unless of course (no pun intended) that was the intention.

The Method A in BS 6663, 1986 consists very simply of wrapping a skein of yarn on a wrap reel. The skein is suspended from a hook, and two weight pieces are applied to the lower end to exert a total pulling force equivalent to 0.9 mN/dtex. The skein is suspended in this way in a cylinder of water. This total weight exerting a force equivalent to 0.9 mN/dtex is made up of two weight pieces a "lightweight" equivalent to 0.0176 mN/dtex and a "heavyweight" equivalent to 0.8824 mN/dtex, (e.g., 150 g/1667 dtex). After 2 minutes the heavyweight is removed and the length of the hank is measured (it is in practice using the commercial instrument zeroed at 0%). After a further 2 minutes (under the influence of only the lightweight), the contracted length is measured (it is in fact read directly from a scale as %contraction) and the % contraction calculated. The basis of the % contraction is

$$\% \text{ contraction} = \frac{L_1 - L_2}{L_1} \times 100$$

where

L_1 is the length at 2 minutes under the total force of 0.9 mN/dtex
L_2 is the contracted length after 2 minutes relaxation under the lightweight force of 0.0176 mN/dtex

This % contraction is referred to as the crimp rigidity.

The method in ASTM D 4031–95 is related to BS 6663 but cannot be said to be the same test, as the conditions of loading the yarns are different, dry heat and hot water options are used, and the actual loading of the yarns with weight pieces is different. Consequently the results from the two methods, i.e., the BS and the ASTM, should not be compared. In fact even the terminology is different. The following definitions appear in ASTM D 4031-95a:

Bulk shrinkage. A measure of potential stretch and power of stretch yarns or a measure of bulk of textured-set yarns.

Crimp contraction. An indicator of crimp capacity or a characteristic of a yarn's ability to contract under tension.

Crimp development medium. For testing of textured yarn, an environment that allows the temporary set of fiber crimp to be overcome and that allows the filaments to assume their permanently set configuration.

Crimp recovery. A measure of the ability of a yarn to return to its original crimped state after being subjected to tension.

Skein shrinkage. A measure of true or intrinsic yarn shrinkage not including crimp contraction.

In *Textile Terms and Definitions* [1] the following terms relating to crimp are included:

Crimp, latent. A crimp that is potentially present in specially prepared fibers or filaments and that can be developed by a specific treatment such as by thermal relaxation or by tensioning and subsequent relaxation.

Crimp frequency. The number of full waves or crimps in a length of fiber divided by the straightened length.

Crimp recovery. A measure of the ability of a yarn to return to its original crimped state after being subjected to tension. (*Note*: This definition is the same as that in ASTM D 4031.)

Crimp retraction; crimp contraction. The contraction in length of a previously textured yarn from the fully extended state (i.e., where the filaments are subsequently straightened), owing to the formation of crimp in individual filaments under specified conditions of crimp development. It is expressed as a percentage of the extended length.

Crimp rigidity test. A form of crimp rigidity test used in the UK for the testing of false-twist textured nylon yarns.

Crimp stability. The ability of a textured yarn to resist the reduction of its crimp by mechanical and /or thermal stress.

Note: Crimp stability is normally expressed as the ratio of values of crimp retraction measured before and after a specified mechanical and/or thermal treatment of the yarn.

In ASTM D 4031–95a, a skein of yarn is subjected to a crimp development medium using three options of loading. As the crimp is developed and shrinkage occurs in the yarn, the skein length becomes shorter. The lengths of the skein under specified tension forces are used to calculate bulk shrinkage, crimp contraction, skein shrinkage, and crimp recovery.

The options for loading the skeins consist of using light loads equivalent to 0.04 mN/tex to 0.98 mN/tex that extend but do not remove the crimp, to heavy loads of 8.8 mN/tex that remove the crimp but do not elongate the yarn. As with BS 6663, tables provide a list of weights to be used and the options available.

The crimp development conditions are as in Table 8.

For each specimen the following measurements are made:

1. Length before development under light load C_b
2. Length before development under heavy load L_B
3. Length after development under light load C_a
4. Length after development under heavy load L_a
5. Length with light load on a developed specimen after removal of heavy load C_c

The results are calculated as

$$\text{Crimp contraction before development to } 0.1\% = \frac{(L_b - C_b) \times 100}{L_b}$$

$$\text{Crimp contraction after development to } 0.1\% = \frac{(L_a - C_a) \times 100}{L_a}$$

$$\text{Skein shrinkage to } 0.1\% \text{ after development} = \frac{(L_b - L_a) \times 100}{L_b}$$

$$\text{Crimp recovery to } 0.1\% = \frac{(L_a - C_c) \times 100}{(L_a - C_a)}$$

$$\text{Bulk shrinkage to } 0.1\% = \frac{(C_b - C_a) \times 100}{C_b}$$

As observed in relation to BS 6663, the values so obtained are true only under the crimp development conditions and loads used and are useful for comparative purposes only. However, ASTM D 4031, unlike BS 6663, does include precision data.

Table 8 Crimp Development Conditions

All crimp parameters calculated	Dry heat at 120°C and 0.04 Mn/tex to 0.44 mN/tex for textured polyester (option A)
Bulk shrinkage	In water at 82°C and 0.13 mN/tex for textured Nylon yarns (option C)
Bulk shrinkage	In water at 97° at 0.13 mN/tex (option C)

Crimp frequency is defined in ASTM D 3937 as "in man-made staple fibers, the number of crimps or waves per unit length of extended or straightened fiber." Compare this with the Textile Institute definition given above.

ASTM D 3937–90 provides a method for determining the crimp frequency of man-made staple fiber irrespective of how the crimp was inserted or developed. Equivalent ISO and BS methods are not available.

The test consists of placing specimens of staple fiber on a plush or short pile (e.g., raised and cropped) surface, which should be of a contrasting color in a well illuminated, draft-free environment. Each fiber is arranged separately from its neighbor so that the crimps along the entire length of the fiber can be counted. After the crimps are counted, the fiber is straightened, without causing extension to the fiber, and its uncrimped length is measured. Crimp frequency is reported as the number of crimps per unit of extended (i.e., straightened) fiber length. The ASTM defines crimp for the purposes of the test as being "characterized by a change in the directional rotation of a line tangent to the fiber as the point of tangent progresses along the fiber. Two changes in rotation constitute one unit of crimp."

To illustrate this, a figure shows stylized crimped fibers with various degrees of crimp frequency, ranging from 4 to 23.

Naturally, before the test commences, the fiber sample is first conditioned in the standard atmosphere for conditioning and testing. There is no provision in the standard for preconditioning from the dry side of the hysteresis curve, but with the more hydrophobic man-made fibers this should not present any difficulties. The ratio of precision between laboratories and within a laboratory is in the region of 2.05 to 3.48, depending upon whether the results are made in single or multimaterial tests. These sort of figures are in keeping with fine detail working requiring a degree of manual dexterity. The ratio of within-laboratory precision to single operator precision is of course, as one normally expects, considerably better at 1.08 to 1.95 (or, very approximately, half the above ratios between laboratories), again depending upon whether results relate to single or multimaterial tests.

9 Frictional Properties

Next to fiber toughness and flexibility, fiber (yarn/filament) friction is arguably the most important element of a fiber or filament, as it affects every process between raw staple or filament supplier right through to the finished garment and beyond. Somewhat anachronistically however there has only been a relatively small amount of research data published on the subject. Why is this? Partly perhaps because it is known to be such a difficult subject to study and partly because there is a related subject that gathers more of the funds available and that is supported by centuries of experience, viz, "handle." What research data that is available dates from mostly around the 1950s and sound though it is, it of course predates the advantages that modern technology can bring to its study.

There are two aspects to the study of fiber friction: interfiber friction or cohesion, and the friction between the fiber or filament and solids, such as ceramics, chromes, PTFE, stainless steels, brass and other alloys, etc.

Interfiber friction is important in the development of staple fiber yarn strength, in the amount of energy needed during the opening process, in ensuring that there is sufficient fiber cohesion to hold together the web at the card, in all the drafting processes, whether they be roller or apron or gill type, in coiling, in allowing withdrawal of sliver at the creel of subsequent operations, in needle-punched felts and stitch bondeds, and in the formation of nonwovens generally.

The frictional properties of yarns and filaments in contact with solids, affect such things as balloon tension in ring spinning and two-for one twisting, tension control and winding-on tension during winding, reed abrasion in weaving, needle transfer, and heating during knitting, needle breaks and heating during sewing operations and other instances of abrasion and frictional heating during yarn transfer processes.

Surprisingly, there are no standard methods of measuring fiber or yarn friction published by the ISO, and the methods that are available are ASTM D 4120–93; ASTM D 2612–93a; ASTM D 3412–95, and ASTM D 3108–95.

ASTM D 2612–93a provides a method for measuring the fiber cohesion force required to cause initial separation of fibers in a bundle of fibers. This observed cohesive force is converted to cohesive tenacity based on the linear density of the fiber specimen. The method is not advised for trade purposes but as an in-house quality check, since the coefficient of variation of the test using man-made fibers is in the region of 21%, or 15% with a single operator. The ASTM method advises the conditioning of the test sample to equilibrium with the standard atmosphere from the dry side of the hysteresis curve by first preconditioning the test sample. Clearly, if the test sample were to be conditioned from the wet side, the results would be totally uncomparable with tests conducted on preconditioned specimens.

The specimen consists of sliver or top mounted (initially on a paper carrier to prevent stretching or fiber separation during mounting) between the jaws of a CRE tensile testing machine, and a force /extension curve is developed. From this is determined the cohesive force to the nearest 0.1 gf from the maximum point of the curve. The drafting tenacity is calculated by

$$DT = \frac{F \cdot L}{1{,}000M}$$

where

DT = drafting tenacity in mgf/tex
F = cohesive force in gf
L = length of specimen in mm
M = mass of specimen in g

ASTM D 4120–93 describes the measurement of fiber cohesion as the dynamic cohesive force required to maintain drafting in rovings, slivers, or top when they are subjected to stress induced by passing between pairs of drafting rollers of different surface speeds. It provides an indication of the ability of fibers to hold together by measuring the force required to slide fibers in a direction parallel to their length. It is acknowledged that higher drafting forces are encountered when the fibers have a predominance of trailing hooks. Consequently the fiber cohesion will appear higher in first passage drawframe sliver than in card sliver, where the predominance of hooked fiber are leading on being withdrawn from the sliver can. Also, the drawframe sliver is more compact than card sliver, thus giving rise to a greater number of interfiber points of contact and a higher fiber cohesion. Again the sample is conditioned from the dry side of the hysteresis curve. The apparatus consists of a cohesion meter (manufactured and supplied by Rothschild of Zurich, Switzerland), a tensiometer, and a tachometer for measuring the drafting roller speed. The drafting rollers are set at 1.4 × the staple or fiber length as determined according to ASTM D 3660 or ASTM D 3661 (see earlier in Section 3). After balancing the tensiometer at slow speed, the instrument is operated at 5 m/min until 50 readings are obtained of the drafting force in gf.

Of course one problem associated with this technique is that the slow operation speed of 5 m/min does not reflect the reality of processing speeds of 300 m/min, at which other factors, such as roller heating, fiber heating, turbulence, and development of electrostatic charges, will radically affect the actual drafting forces necessary to overcome interfiber friction, but as an in-house quality check for processability the method has many advantages.

ASTM D 3412–95 measures the frictional properties for both filament and staple yarns using a technique similar to that described by Lindberg and Gralen in 1948 [21].The method has been used with yarns in the range from 1.5 tex to 200 tex, but it is accepted that this range may be exceeded. A length of yarn is moved at a known speed in contact with itself or another yarn of similar construction at a specified twist/wrap angle. The input and output tensions are measured, and from this data the coefficient of friction is calculated. The method is not suitable for acceptance testing but might find some uses as an in-house quality test. The apparatus is somewhat complex and can give rise to fluctuations due to changes in r.h./% and pulley bearing friction. The calculation of the coefficient is based on a variation of Amonton's law, which is dealt with later. A schematic illustration of the apparatus and the yarn path is shown Fig. 30.

The instrument is first calibrated by running yarn through the system without the 'snarling' twist at W, and the input and output tensions are measured through the yarn tension gauges illustrated. In fact this twist range W is not actually snarling twist in the technical sense, but the expression gives an idea of the yarn configuration. The difference between these input and output tensions is then designated as ΔT. The apex angle $\propto$ is calculated as:

$$\propto = 2\arctan\left[\frac{H}{V - W}\right]$$

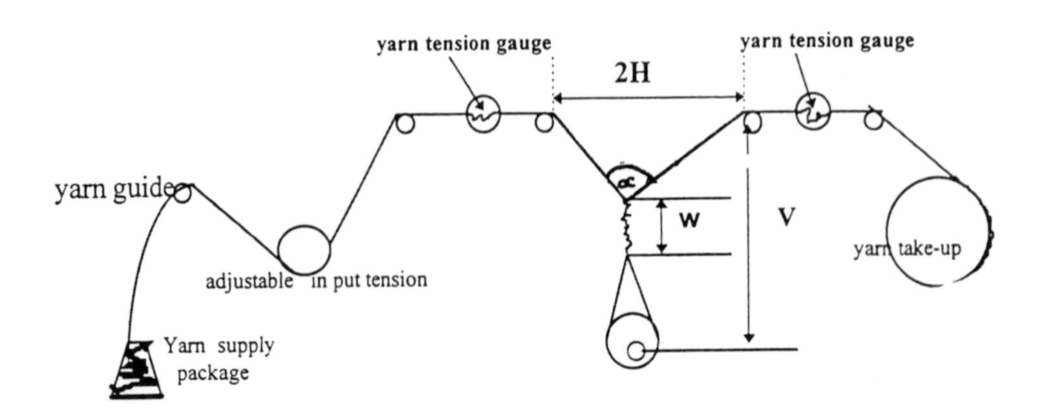

Figure 30 Schematic illustration of apparatus and yarn path for yarn/yarn friction test.

The coefficient of friction is calculated, after having run the instrument with snarling twist at W, as

$$\mu = \frac{\ln \dfrac{T_2 - \Delta T/2}{T_1 + \Delta T/2}}{2\pi n_{\alpha}}$$

where

μ = coefficient of friction
ln = log to base e (2.718)
T_1 = mean input tension
T_2 = mean output tension
ΔT = zero twist tension (see above)
n = number of wraps
α = apex angle (see above)

The capstan method, which is included as an alternative option, is much simpler, and in my view probably more reliable, and this is to some extent borne out by the interlaboratory precision data included in the standard, where the critical differences for the capstan method are approximately half those using the twisted strand method. The capstan method consists of winding a tube of fixed diameter with the yarn under test and then a single strand of the same yarn is pulled over it, one end being connected to a tension gauge and the other to a weight. The arrangement is illustrated in Fig. 31.

The coefficient of friction is then calculated according to Amonton's law, i.e.

$$T_2 = T_1 e^{\mu\theta}$$

$$\text{Therefore } \mu = \frac{\log_e(T_2/T_1)}{\theta}$$

The further application of Amonton's law will be dealt with later when looking at friction with solids.

Polymer friction of the bulk polymers and that of the textile polymers follow roughly the same laws, namely that Coulomb's third law of friction (i.e., that kinetic friction is independent of the speed of sliding) and Amonton's first law (that the frictional force is independent of the area of contact) just do not apply, especially with the thermoplastics. Instead we are faced with the discovery that with a steady increase in the speed of sliding the frictional force can increase until it reaches a point where it can drop dramatically if the friction coefficient and the linear speed are high enough. But this will of course depend very much on the particular fiber type. One illustration of this was found by D. G. Lyne

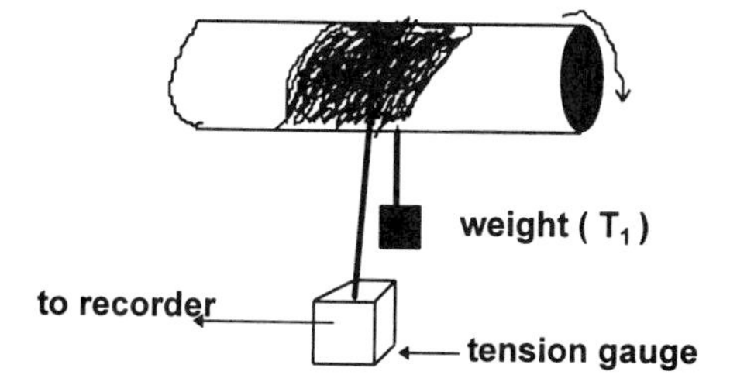

Figure 31 Schematic illustration of yarn/yarn capstan friction test.

[22], who ran an acetate yarn over a guide at various speeds and found that the value of T_2 increased with yarn speed (and consequently that the value of μ_k rises). In fact this particular reference is quoted by ASTM in D 3108–95. The resulting graph is shown in Fig. 32.

ASTM D 3108–95 is concerned with the determination of the coefficient of friction of yarn to solid materials. A yarn speed of 100 m/min is used, which, considering the speed of yarn withdrawal, winding on speeds, and weft insertion rates of over 1,000 m/min for projectile looms and over 2,000 m/min for air-jet and water-jet looms, is somewhat unrealistically low from the point of view of reflecting reality. Also as will be seen below, a single yarn speed is not going to provide any data on what happens to the μ_k when the yarn speed is increased. Standard wrap angles of 180° (3.14 radians) and 360° (6.28 radians) are used, but it is acknowledged that other angles of wrap may be used provided that with yarns of high coefficients these wrap angles are not exceeded, as otherwise the tension necessary for maintaining the yarn speed may exceed the yield strength of the yarn. It is recommended however that the angle of wrap should not be lower than 1.57 radians. The method can be said to be suitable for yarns in the range 10 tex to 80 tex and for yarns having values of μ_k between 0.1 and 0.5, but yarns outside this range may also be tested.

The method consists of running a length of yarn at the designated speed in contact with single or multiple friction surfaces whose angle of contact with the yarn are known. The yarn input and output tensions are measured and the coefficient of friction calculated according to the common capstan equation. It should at this point be noted that this coefficient of friction μ_k is only true for the particular set of materials used by the instrument and in the condition in which they are maintained, as a frictional coefficient for a fiber or yarn on its own cannot be expressed, because it is axiomatic that for friction to exist there must be another material. So when the coefficient μ_k is quoted for any friction testing system, the test parameters should also be quoted. For example, in the case of ASTM D 3108 it would be advisable in my view to state : "μ_k – 100 m/min–chromed steel 4 μm to 6 μm roughness" and preferably the regain percentage of the yarn at the beginning and end of the test, as it has been found (see J. A. Morrow [23] and King [24]) that as % regain rises, so does μ_k, not linearly but more in line with the graph shown in Fig. 33 (as an approximate mean representation). The condition of that other material and whether fiber lubricants are used will obviously affect the results. Nevertheless, yarn

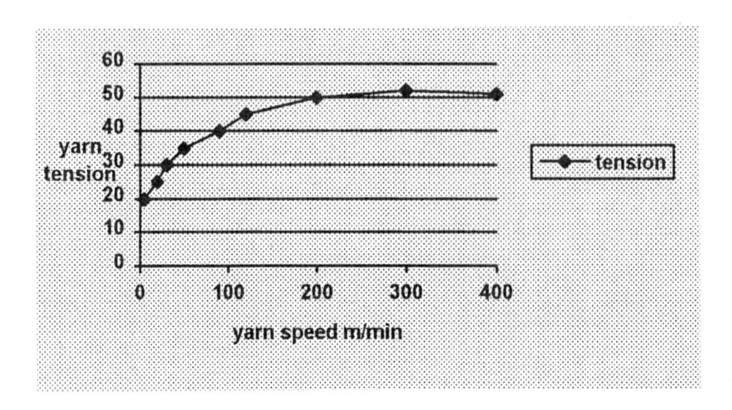

Figure 32 Illustration of yarn tension with yarn speed.

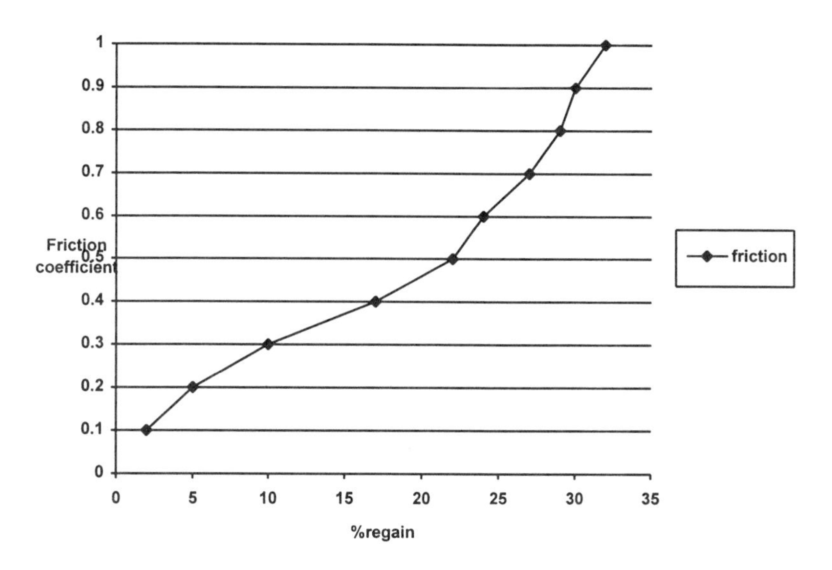

Figure 33 Illustration of rise in μ_k with increase in regain.

friction is crucial to almost every operation in which yarns are employed, either because of their effect on yarn guide or needles in terms of wear or heating effects.

A schematic of a typical yarn friction tester is shown in ASTM D 3108 and is in my view far superior to the arrangement shown in PD 6527, 1990, published by the BSI. An approximation of the ASTM schematic is shown in Fig. 34.

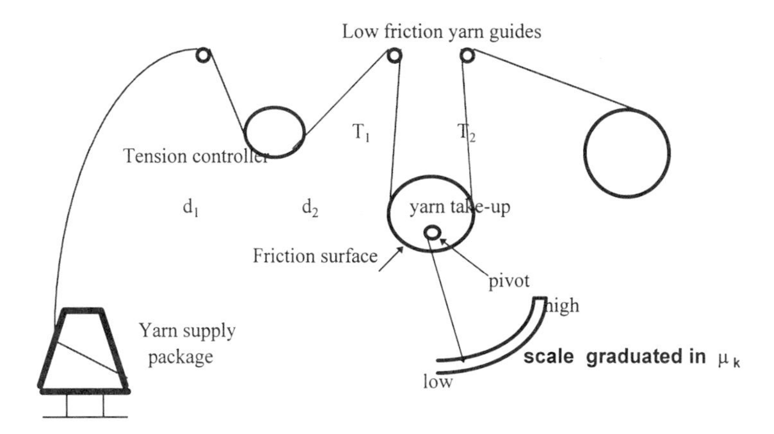

Figure 34 Schematic apparatus and yarn path for yarn to solid friction test.

10 Color

Probably more than any other single item of industrial and consumer interest (motor vehicles included), the color and maintenance of color in textile products is, if not paramount, of extremely high importance. In fact, in the case of consumer textiles, from clothing to household textiles, the aesthetics of color and the ultimate selection of the product is based quite often entirely on the color or colors of the product. Color is not always imparted to the textile in the yarn or fabric stages but is often introduced during the manufacture of the fiber. When color is introduced in this way, it is usually almost permanent. Nevertheless, the fiber manufacture and the fiber buyer need to assure themselves that the color will be retained to an acceptable level of density or depth when exposed to various agencies, not the least of which is daylight, and UV in particular.

Of course the retention of color is not the only measure of performance after exposure to the selected agency, but the assessment of other physical parameters is dealt with elsewhere. It is however often assumed that changes in color of a textile after exposure to the relevant agency result only in a loss of aesthetic appeal, and changes in other properties are often not investigated. In fact when one looks at the specimen size employed in ISO 105 BO2 for example, it is clear that post-exposure physical testing was not even envisaged by the writers of that standard.

The importance placed on the retention of color is emphasised by the number of methods or agencies against which textile colored products are tested. In the ISO 105 series there are a total of 72 published standards, which, allowing for the few that are concerned with general principles, blue cloths, grey scales, adjacent fabrics, and instrumental assessment of standard depth, etc., leaves about 57 or so methods concerned with assessing color changes against different agencies, or, to put it into a well known phrase or saying, there are approximately 57 different varieties of color fastness test for textiles.

Some of these agencies will be of particular concern to elastomer and polymer composites in particular, such as ISO 105, Parts S01, S02, and S03, which are concerned with color fastness to hot air,sulfur monochloride and open steam vulcanization, and ISO 105, Part X10, which is concerned with the assessment of the migration of textile colors into PVC coatings.

The equivalent U.S. methods are published by the American Association of Textile Chemists and Colorists (AATCC) in bound volumes, rather than as individual publications, which is the general norm with ISO publications.

Clearly the color measurement and color fastness testing of textile polymers is a subject of its own and deserves far greater attention than can be given to it here, and the reader is directed to publications such as Peters and Freeman on *Colour Chemistry* or *Colour Physics for Industry* by Roderick McDonald, both published by the Society of Dyers and Colourists. There are however a few things that should be said here.

The first is that ISO 105 needs to make greater provision for the testing of textiles in the UV absorption bands, and the second is that provision also needs to be made for larger test specimens in order to allow for post-exposure assessment of changes in properties other than color. This could be of particular importance where the textile product is itself a safety critical item.

11 Sampling

So far nothing has been said of sampling. This is not because sampling textile materials is simple and straightforward, but because it is complex and of extreme importance and consequently deserves the honor of being the last word. Sampling is akin to calibration. If the sample is biased, so too will be the test results.

Staple Fiber Sampling

The Introduction to ISO 1130, "Textile fibers—Some methods of sampling for testing," is quite succinct and deserving of repetition. It says, "No single technique of sampling can be devised that will serve in all circumstances. Sampling from a bale of cotton, for example, presents problems quite different from those encountered in sampling from a consignment of yarn packages, while sampling from a card web is again different from either."

ISO 1130, 1975 provides a selection of methods for dealing with, among other things, sampling from bulk consisting of several bales of fiber, sampling from a bulk consisting of drawn fiber, such as sliver, rovings, and yarns, and sampling methods for loose fibers. Not all of these techniques are said to be applicable to man-made staple, but in general, the techniques are suitable and valid. The standard deals with the subject of zoning and cut-square methods.

The object in every case is to obtain a sample that is as representative of the total population as is economically or commercially viable, without introducing fiber length or fiber extent bias.

The mathematical and technological theory underlying these methods is not dealt with in ISO 1130, nor will it be here, since the subject would take more space than is justified in a publication of the present type.

Sampling for quantitative analysis of textiles is dealt with briefly in ASTM D 629–95 and in even briefer form in AATCC Test Method 20A–1989, "Fiber analysis—Quantitative." However, more detailed advice on sampling methods is given in ASTM D 4271–88, "Standard practice for writing statements on sampling in test methods for textiles," and also in ASTM D 3333–95, "Standard practice for sampling man-made staple fibers, slivers or tow for testing."

BS 2545, 1965(1990), "Methods of fiber sampling for testing," is essentially the basis for ISO 1130 (which is rather obvious when one reads the foreword to the BS).

When the problem is one of sampling yarn packages, one can then treat each package as an identifiable individual element of the population. The number of packages selected to form the sample can then be determined by reference to ISO 2859–1, 1989, "Specification for sampling plans indexed by acceptable quality level (AQL) for lot-by-lot inspection," or by ISO 2859–2,1985, "Specification for sampling plans indexed by limiting quality (LQ) for isolated lot inspection," or by ISO 8422, 1991, "Specification for sequential sampling plans, or, for inspection by variables," ISO 3951, 1989, "Specification for single sampling plans indexed by acceptable quality level (AQL) for lot-by-lot inspection." The required number of test specimens to conduct the necessary tests can then be selected from representative lengths from each package.

Sampling of cloth presents different problems altogether, and this highlights the ever-present problem, namely that of making the sample large enough to be sure of representing the population without destroying the commercial value of the product. On the other hand, the sample should not be so small that it would allow defects to escape the chance of being included in the sample, especially if that defect could find itself included in a product that was intended to provide personnel protection, e.g., safety lines or safety harnesses, lifting slings etc.

The frequency of sampling also requires to be given careful consideration when sampling a textile element with a safety connection, and in this respect the process technology associated with the textile product needs to be reviewed.

However, all this requires a combination of knowledge of the process technologies, sampling techniques, and risk analysis, which is clearly beyond the scope of this chapter.

References

1. *Textile Terms and Definitions*, 10th ed., The Textile Institute, Manchester, UK., 1995.
2. ASTM D 4466–94—Terminology for multicomponent textile fibers
3. Greaves, P., *Microscopy of Textile Fibres*, Bios Scientific Publishers in association with the Royal Microscopal Society.
4. ASTM D 276–87 (93)—Identification of fibers in textiles.
5. AATCC Test Method 20–1990 (as referenced in ASTM D 276).
6. ASTM D 1909–94—Table of commercial moisture regains for textile fibers.
7. Morton, W. E., and Hearle, J. W. S., *Physical Properties of Textile Fibres*, The Textile Institute, Manchester, U.K., 1993.
8. Mukhapadhyay, S. K., ed., *Advances in Fibre Science*, The Textile Institute, Manchester,U.K., 1992.
9. *J. Text. Inst.*, March (1937).
10. *J. Text. Inst.*, *40* (1949).
11. *J. Text. Inst.*, *60* (1969).
12. *Text. Res. J.* (1966).
13. *Trans. Faraday Soc.*, *36* (1940).
14. *Text. Res. J.*, *38* (1969).
15. *J. Pol. Soc.*, Part C.13 (1966).
16. *J. Pol. Soc.*, A-2 (1968).
17. *J. Text. Inst.*, T 702 (1949).
18. *J. Text. Inst.*, *36*, T 107 (1945).
19. Booth, J. E., *Principles of Textile Testing*, Newnes-Butterworth in association with The Textile Institute, 1968.
20. *Report on the 1966 Conference on Bulk, Stretch and Texture*, Text. Inst. and Ind., July/August 1966.
21. *Text. Res. J.*, *18* (1948).
22. *J. Text. Inst.*, *46* (1955).
23. *J. Text. Inst.*, *22* (1931).
24. *J. Text. Inst.*, *41* (1950).

The full titles and references of the standards referred to in the text are listed below in numerical order, in the sequence ISO, ASTM, AATCC, BS, EN.

ISO 105, BO2, 1994—Colour fastness to artificial light, Xenon arc fading lamp test.

ISO 105, Part S01, 1993—Colour fastness to vulcanization , Hot air.

ISO 105, S02, 1993—Colour fastness to vulcanization , Sulfur monochloride.

ISO 105, S03, 1993—Colour fastness to vulcanization, Open steam.

ISO 105, Part X10, 1993—Assessment of migration of textile colors into polyvinylchloride coatings.

ISO 137, 1975—Wool—Determination of fiber diameter—Projection microscope method.

ISO 139, 1973—Textile—Standard atmospheres for conditioning and testing.

ISO 1130, Textile fibres—Some methods of sampling for testing.

ISO 1144, 1973—Textiles—Universal system for designating linear density (tex system).

ISO 1888, 1979—Textile glass—Determination of the average diameter of staple fibres, or continuous filaments constituting a textile glass yarn—Cross section method.

ISO 1973, 1995—Textile fibres—Determination of linear density—Gravimetric method and vibroscope method.

ISO 2060, 1994—Textiles—Yarn from packages—Determination of linear density (mass per unit length) by the skein method.
ISO 2062, 1993—Textiles—Yarn from packages—Determination of single end breaking force and elongation at break.
ISO 2602, 1980—Statistical interpretation of test results—Estimation of the mean—Confidence interval.
ISO 2859-1, 1989, Specification for sampling plans indexed by acceptable quality level (AQL) for lot-by-lot inspection.
ISO 2859-2, 1985, Specification for sampling plans indexed by limiting quality (LQ) for isolated lot inspection.
ISO 2947, 1973—Textiles—Integrated conversion table for replacing traditional yarn numbers by rounded values in the Tex system.
ISO 3951, 1989, Specification for single sampling plans indexed by acceptable quality level (AQL) for lot-by-lot inspection.
ISO 4913, 1981—Textiles—Cotton fibres—Determination of length (span length) and uniformity index.
ISO 5079, 1995—Textile fibres—Determination of breaking force and elongation at break of individual fibres.
ISO 6741-1, 1989—Textiles—Fibres and yarns—Determination of commercial mass of consignments, Part 1, Mass determination and calculations.
ISO/TR 6741-4, 1987—Textiles—Fibres and yarns—Determination of commercial mass of consignments, Part 4, Values used for the commercial allowances and the commercial moisture regains.
ISO 6989, 1981—Textile fibres—Determination of length and length distribution of staple fibres (by measurement of single fibres).
ISO 8422, 1991, Specification for sequential sampling plans.
ISO 10119, 1992—Carbon fibre—Determination of density.
ISO 10120, 1991—Carbon fibre—Determination of linear density.
ISO 10290, 1993—Textiles—Cotton yarns—Specifications.
ISO 10306, 1993—Textiles—Cotton fibres—Evaluation of maturity by the air-flow method.
ISO 11567, 1995—Carbon fibre—Determination of filament diameter and cross-sectional area.
ISO 15602 (not yet published)—Monofilament yarns—Methods of test.
ASTM D 276–87 (93)—Identification of fibres in textiles.
ASTM D 578–90—Standard specification for glass fiber strands.
ASTM D 629–95—Quantitative analysis of textiles.
ASTM D 792–91—Test methods for density and specific gravity (relative density) of plastics by displacement.
ASTM D 861–95—Use of the Tex system to designate linear density of fibres, yarn intermediates, and yarns.
ASTM D 1425–96—Test method for unevenness of textile strands using capacitance testing equipment.
ASTM D 1447—Length and length uniformity of cotton fibres by fibrograph measurement.
ASTM D 1505–85—Test method for density of plastics by the density-gradient technique.
ASTM D 1577–90—Linear density of textile fibres.
ASTM D 1774–93—Elastic properties of textile fibres.
ASTM D 1776–90—Conditioning textiles for testing.
ASTM D 1907–89—Yarn number by the skein method.
ASTM D 1909–94—Table of commercial moisture regains for textile fibers.
ASTM D 2101–94 (discontinued in 1995)—Tensile properties of single man-made textile fibers taken from yarns and tows.
ASTM D 2102–90—Shrinkage of textile fibers.
ASTM D 2130–90—Diameter of wool and other animal fibers by microprojection.
ASTM D 2256–95—Tensile properties of yarns by the single-strand method.
ASTM D 2612–93—Fiber cohesion in sliver and top in static tests.

ASTM D 2645–95—Yarns spun on the cotton or worsted systems.
ASTM D 2654–89—Moisture in textiles.
ASTM D 3108–95—Coefficient of friction, yarn to solid material.
ASTM D 3217–95—Breaking Tenacity of Man-Made Textile Fibres in Loop or Knot Configurations
ASTM D 3333—95, Standard practice for sampling man-made staple fibres, slivers or tow for testing.
ASTM D 3412–95—Coefficient of friction, yarn to yarn.
ASTM D 3660–90 (discontinued 1995)—Staple length of man-made fibers, average and distribution (fiber array method).
ASTM D 3661–90 (discontinued 1995)—Staple length of man-made fibers, average and distribution single fiber length machine method.
ASTM D 3822–95—Test method for tensile properties of single textile fibers.
ASTM D 3937–94—Test method for crimp frequency of man-made staple fibers.
ASTM D 4031–95—Test method for bulk properties of textured yarns.
ASTM D 4120–93—Test method for fiber cohesion in roving, sliver, and top in dynamic tests
ASTM D 4238–90—Test Method for Electrostatic Propensity of Textiles
ASTM D 4271–88(93), Standard practice for writing statements on sampling in test methods for textiles.
ASTM D 4848–95—Force, deformation and related properties of textiles.
ASTM D 5103–95—Length and length distribution of man-made staple fibers (single-fiber test).
ASTM 5591–95—Thermal shrinkage force of yarn and cord using the testrite thermal shrinkage force tester.
AATCC Test Method 20A, 1989, Fiber Analysis—Quantitative.
BS 2545,1965 (1990)—Methods of fibre sampling for testing.
(BS) PD 6527, 1990—Methods of test for textiles (other than those methods already published as British Standards).
(BS) PD 6577, 1995—The influence of environmental conditions on the testing of textile materials.

20
Coated Fabrics

Barry Evans
Schlumberger Industries–Metflex, Blackburn, England

1 Introduction

What is a coated fabric? To state the obvious, it is a fabric or textile with a coating or deposit of polymeric material. To state the accepted [1], it is a material composed of two or more layers, at least one of which is of a textile material (woven, knitted, or nonwoven) and at least one of which is a substantially continuous polymeric film, bonded closely together by means of an added adhesive or by the adhesive properties of one or more of the component layers. A precise definition is difficult because the combinations and varieties are vast and the end uses so diverse. This prompts a number of questions. Is the coating present in a continuous film, or is it printed or sprayed on,dipped or laminated? Is it on one side or both sides of the material, or is it sandwiched in the center, off center, or is it multi-ply? Is the fabric facing inside or outside? Is it a fashion item, decorative or functional, like waterproof clothing? Is it lightweight or heavyweight, woven, nonwoven, or knitted? Is it there as a carrier or support for the coating, or is it structural and reinforcing as with inflatable structures? The end application often dictates the construction, and some of the uses are listed as follows: airbags, footwear, protective clothing, upholstery, inflatables, tarpaulins, gloves, bags, and luggage to name but some. Why coat a fabric at all? Essentially, it is to impart some property that it does not have on its own. Some of the earliest coated fabrics were rubber coatings made by Amazonian Indians who directly spread tree latex on cloth, to impart waterproofing. In Manchester in the 1880s an industry grew up [2] based on spreadcoating, involving the early rubber pioneers Hancock and Macintosh , the "mac" still being synonymous with rainwear to this day. A range of products is illustrated in Hancock's personal narrative published in 1857 [3]. Early history has been reviewed by Wooton [4] in the introductory chapter of his book.

2 Scope

Industrial uses of textile composites include conveyor belting, tires, and hoses. These are not considered here specifically, other than that several of the test methods may well be applicable where the textile element is of relevance. These are products in their own right and have been extensively reviewed elsewhere [4, 5]. It is the intention in this chapter to review the test methods available for coated fabrics, their practicability and their application in the areas of quality control, design, and specifications relevant to product end use. Some knowledge of processing is assumed, but clarification is given as necessary.

3 Background

A coated fabric behaves differently from both its textile part and the elastomer and can be classed as a composite, and therefore properties are difficult to predict. Since the base textile has an orientation, warp or weft in a woven cloth, so then does the coated fabric.Even with nonwoven or knitted substrates, the properties of the combination are anisotropic, that is, directionally different. Properties are generally measured in two directions designated longitudinal and transverse, and sometimes diagonally, that is, on the bias. Test specimens are conditioned before testing [6] in a standard atmosphere of controlled temperature and humidity. Since moisture can affect the base textile, depending on its nature (cotton, rayon, polyamide, polyester, etc.) and whether it is fully encapsulated by the coating, moisture equilibrium is approached from the dry side. It is considered to have been reached when there is no increase in mass over successive weight measurements. Most tests are carried out a minimum of 16 hours after final processing has elapsed unless done as part of an in-house production check during manufacture. In the case of a made-up product, or a customer test requirement, normally three months is the maximum time lapse expected. Some products however may have a life expectancy of ten years, or be stored for a lengthy period before being called into service. In such cases specific arrangements will need to be reached with the customer.

4 Roll Characteristics

Coated fabrics are often supplied for product fabrication in roll form and therefore commercially must conform to an agreed standard between the supplier and customer. It is normal practice to measure (and state) the length, width, and net mass of a roll [7] and also the thickness, total weight, and coating weight [8] of the composite, if required.Test methods are specified for these parameters. In essence, the length is determined by progressively rolling out the coated fabric end to end, without tension, along a precalibrated table (minimum five meters in length), marked off in meters with the final meter in centimeters, or using a drum and a calibrated measuring wheel. Tension control during processing or fabrication is important, particularly where knitted or extensible fabrics are used [9], since this can affect the dimensional stability. Width is normally measured using a calibrated millimeter scale wider than the fabric. The "useable width" is often quoted, and for samples taken for further testing,these are selected so that the full width of the coated fabric is represented. Textiles are often produced with a reinforced edge or selvedge, and may on occasions be dyed or finished, necessitating the use of a tenter process, which leaves pinholes down each side. Some manufacturers may remove this edge after processing by trimming, in which case sampling is straightforward since the width as presented is trimmed. Others may leave the product as processed. The term "useable width" has therefore been defined as "that width of coated fabric excluding the selvedge which is consistent in its properties, uniformally finished and free of unacceptable flaws" [7]. It is worth noting at this point that most product specifications include a test specimen selection diagram as an appendix for assistance.

The mass of the roll is determined by weighing using any suitable means and deducting the mass of the tube or holder from the gross mass recorded. From the mass, width, and length so determined, the mass per unit area can be calculated if required. Normally this weight would be obtained from a smaller cut piece, representative of the roll taken on line during processing as part of a quality control check. The method used [7] involves a balance and a standard-sized cutter giving a test piece of 100 square centimeters. Calculation of grams per square meter (gsm) is therefore easily carried out. Samples can be taken across the width, the results averaged, and corrections made to the processing conditions if remedial action needs to be made due to variation edge to center during manufacture. Coating mass can be determined if the weight of the fabric is known. If not,then the cloth has to be carefully removed from the coating using a suitable solvent system or stripping medium. Often the fabric can be removed by moistening with liquid, but if the coating needs to be dissolved away, then Soxhlet extraction or immersion and agitation may be used. It is clearly important to use a solvent which does not affect the fabric. Drying, reconditioning, and weighing the fabric enable the calculation to be done. The polymer coating will also dictate the type of solvent to be used. Safe methods of working and the use of personal protective clothing equipment may be required,and certain solvents may need other precautionary measures. Safety statements are now being written into standards to ensure good working practices.

5 Thickness

Thickness measurements [10] are carried out using a dial gauge of the deadweight type, where a presser foot moves vertically, exerting a known pressure on the material placed between the foot and a baseplate anvil. Screw micrometers or ratchet micrometers are not suitable because coated fabrics compress easily, unlike metals where these devices are used satisfactorily. Dial gauges must be capable of measurement to 0.02 mm, and the foot must not be less than 9 mm in diameter and be able to exert a pressure to the material of 2, 10, or 24 kPa. If coated fabrics have an embossed surface , these surfaces are not to be avoided during measurement. Again, product specifications usually dictate where the sampling shall be conducted. A strip of material from across the width of a roll can be measured by this method at ten equally spaced points.The presser foot is normally in position for ten seconds before the reading is taken. If an expanded or foam layer is part of the coating it is practice to use a large diameter foot. When the precise thickness of the coating or skin coating/expanded layer is required, samples can be viewed through a $\times$ 50 microscope, fitted with a calibrated graticule whilst supported in an appropriate holder [11]. For very thinly coated fabrics,perhaps applied by transfer coating, a similar technique can be used [12], but it may be practical to "pot" the sample in an epoxy resin and microtome a section for viewing and measuring. In many instances it is important to know the thinnest layer of coating above the thickest part of the textile(usually where one yarn crosses another), as this dictates the physical performance of the coated fabric in use. Often the mean, maximum, and minimum thicknesses are reported.

6 Tensile Strength and Elongation at Break

The strength of a coated fabric is an important property, and the end use dictates the construction of the product, since the textile element mainly imparts this strength feature. For example, the properties needed for sailcloth or inflated structures are different from those for upholstery or garments. Nevertheless, the method of measurement is the same for all, utilizing a universal test machine; but dumbell geometry test pieces are not used, as with polymer testing and the cross-sectional area is not measured or needed for the computation of stress. For coated fabrics and textiles, stress is the force applied to the test specimen, and the maximum force is that recorded in extending the test piece to breaking point. This is the accepted breaking strength value. Two methods are normally used to determine this tensile/ breaking strength [13]. The first is the strip test method, where the strength and elongation at break can be measured, the full width of the test piece being held in the jaws (see Fig. 1). The second is the grab test, where only the central part is utilized (see Fig. 2). The method is simple ,with the test piece clamped between two jaws, one fixed, one movable, then being extended at a constant rate until it breaks.

The normal extention rate for coated fabrics is 100 ± 10 mm/minute. The two methods require different sample dimensions and different jaws, and they vary in jaw separation distance. The strip method samples where possible are "frayed" down to 25 mm width [14] i.e., threads are removed by hand from the edges; or 50 mm width [13] and are of sufficient length to allow a jaw edge separation of 200 mm. This is often reduced to 100 mm for materials with greater than 75% extensibility. Where fraying is not possible, the sample is cut parallel to the yarns. Test samples having a woven support are cut wider using scissors and then frayed, to 50 mm. It is worth noting that incorrect preparation of samples can influence results. Knitteds and nonwovens are cut with care to the final dimension to obtain tidy edges. However, this may lead to inaccuracies, and it may be more beneficial to conduct a bursting strength test, which can be carried out regardless of fabric construction. It must be mentioned that the results obtained from tensile and burst testing are not directly comparable. If difficulties are encountered because of weave distorsions, or if the textile is not visible, then the grab test method may be preferable. Test pieces can be mounted with a known pre-tension [14] or in a slack condition if deformation occurs. In the strip test,the jaw width is not less than the width of the test piece and preferably of width at least 60 mm. The jaw edges are radiused to prevent cutting of the

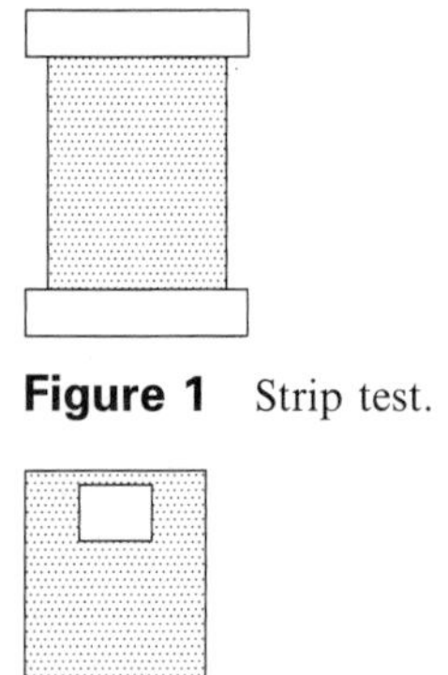

Figure 1 Strip test.

Figure 2 Grab test.

samples. Should slippage or breakage occur in the jaws during the test, then the results should be discarded and the test repeated. Emery paper, felt, leather, rubber or other such materials may be used to assist grip between the test piece and the jaw face. In the case of the grab test, the test piece size is 100 mm by 150 mm and the jaw separation is set at 100 mm, or 75 mm if specified [14]. The clamps have a face of 25 × 25 mm and the back clamp is 25 × 50 mm with the test piece being held centrally on the narrowest side. The test pieces are then extended at a constant rate of 100 mm/minute to the breaking point. Tests are normally carried out in both the longitudinal and transverse directions and may also be performed on wet specimens if required. Sometimes, for example with knitted products, the method may not be satisfactory because the test piece necks down badly or does not give a reproducible result. In these cases an alternative breaking strength is determined by burst testing.

7 Seam Strength

Often with garments or shoes where bonding or sewing takes place, the strength of the seam or its ability to resist seam opening needs to be known. A conventional grab test is therefore carried out first and the values recorded. A seamed test piece is then made up by lapping, bonding, or sewing [15, 16] and the seam arranged to be placed centrally in between the upper and lower jaws of the tensile machine. The value at break can then be expressed as a percentage of that of the original parent fabric.

8 Elongation and Tension Set

Whilst knowledge of the breaking strength is valuable, it may also be useful to know how far the material stretches under a particular load and whether, if this load is removed, the product recovers. Samples are normally tested in the two directions, but with most coated fabrics extensibility is more noticeable in the bias direction. Sampling in this case requires the use of a left and right hand bias sample to eliminate bow and skew [17]. The degree of nonrecovery of elongation is known as tension set or permanent set and may be measured after stretching under a given load or to a fixed elongation. The method of extensibility under load can either be done as a dead load test, where the piece is clamped and fixed at one end, and a constant load applied to the other [18], or a load can be progressively applied until the required percentage elongation is reached, held for a specified duration and then released at a predetermined rate. With extensible or nonwoven fabrics, where products are required for furnishings or footwear, elongation and set are often specified. The method of dead load application to a known extensibility is used as an in-house quality control test to monitor the laying on of stretch fabrics to a polymer base coat on transfer coating lines. The dead load test could also give an indication of creep if the extension were monitored over time, creep being the gradual increase in deformation in a material subjected to a constant force.

9 Bursting Strength

Tensile or breaking strength is sometimes difficult to measure if the substrates are extensible. For some applications a burst test may be more representative of end use properties and is often quoted. Two methods are the most frequently used, one using a conventional tensile machine with a ring-clamped test piece pressed against a moving steel ball, and the other using a diaphragm burst tester operated using hydraulic pressure. The bursting attachment for the tensile machine comprises an upper and lower clamp, having concentric grooves and crowns that intermesh with the test piece clamped between these plates but being exposed at an aperture of 45 mm diameter. The center portion pushes against a steel ball of diameter 25 mm at a rate of 5 mm/second until it ruptures. The bursting strength is then calculated from the force of rupture *F* and the internal cross-sectional area *A* of the test piece, i.e., F/A. In the diaphragm method the aperture size is different (31 mm or 35.7 mm), and the pressure to rupture is applied using a rubber diaphragm mounted below the exposed clamped test piece. Hydraulic fluid is introduced behind the rubber diaphragm at a known rate, and the burst pressure at rupture is measured using a pressure gauge. The upper clamp is then removed and the pressure now indicates the force to distend the diaphragm. This "correction factor" is subtracted from the burst pressure at rupture to give the actual burst pressure of the test piece. The two methods may not however give the same results . Current thinking is to increase the size of the steel ball to 38 mm diameter to improve the reproducibility. Revision of ISO3303 will follow [19].

10 Tear Resistance

This property has relevance to the intended end use, or performance criteria. It can be influenced by many things, for example, the type and construction of the yarn or substrate, the polymer coating,and even how the coating has been applied. Factors affecting tear resistance have been studied and reported on by Eichert [20] for PVC coated fabrics but are applicable to most other polymers. The test methods have developed over several years dependant on the parent industry, from paper, leather, and leather cloth, and these tests have subsequently been adopted or adapted for coated fabrics. As novel substrates develop however, testing is still evolving in an attempt to be more representative of the conditions or type of tear initiation, again perhaps influenced by end use. Testing must always be relevant, and whilst the value of the tear force is of interest, it is of little consolation if the product has actually torn in use. The following methods are in use or being developed: trouser or single tear; double or tongue tear; wing tear; trapezoidal tear; ballistic pendulum; puncture or snag tear; tack tear; and wounded burst tear. Most of the test methods have a basic commonality with only the geometry of the test piece changing, being historically industry specific. Recourse should be made to the referenced standard for the detail. In essence the test piece has a slit cut into it, along some part of its length, and the two resulting parts are mounted in the jaws of a tensile testing machine and moved apart at a fixed rate. The tear force is determined in the two directions, along and across the material, and recorded for interpretation. Results obtained in the two directions will probably not be equal in value. The single and double tear are described in ISO4674 [21]. Even the most simple type, the trouser tear (see Fig. 3), has evolved, since the early specimens had a 5 mm diameter hole punched at the end of the slit, so that the tear "broke out," allowing an initial reading to be made [22]. This reading represents the force required for tear initiation, subsequent readings being the force to propagate the tear. The rate of tear is normally 100 ± 10 mm per minute. The main problems encountered in carrying out tear testing are that sometimes the tear does not propagate in the direction of the jaw traverse, tearing towards the sample edge. In the tongue or double slit test (see Fig. 4), the tongue may stretch and a tensile effect occur, or threads may pull out rather than break. Under these conditions an alternative specimen shape may be chosen or a larger test piece taken and the procedure repeated. Nonwoven and knitted substrates are often tested using larger samples than those initially specified in the method. High-strength products are tested using large trouser and trapezoidal test pieces [23] (see Fig. 5), lighter

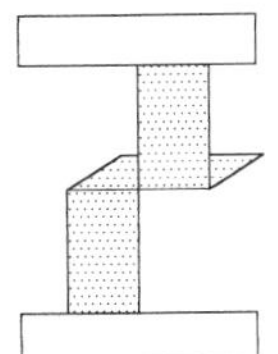

Figure 3 Trouser tear.

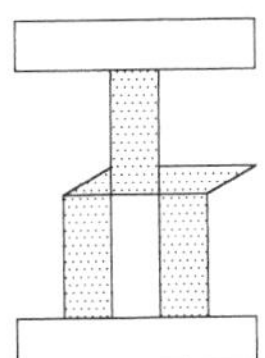

Figure 4 Tongue tear.

weight ones single or wing rip [24] (see Fig. 6). The tear test is generally recorded on an autographic trace so that the chart can be analyzed. As each thread is torn through, the trace rises and falls, producing a series of peaks and troughs. The mean, maximum or median can be determined as required. Interpretation of traces has generated much discussion and argument over the years, including whether a minimum force should also be quoted; the controversy continues.

Wet testing is sometimes done following the soaking of the test pieces in water containing a wetting agent. Another popular method is the falling pendulum (Elmendorf) type tester [25], but other pendulum devices are equally suitable [26]. In this test the coated fabric is held between two clamps, one fixed and one movable, and the tear is propagated by a falling pendulum carrying a graduated scale to indicate the force needed to tear the specimen. The method is used for lightweight coated fabrics and textiles. Tack-tear resistance is described in ASTM D751 as a simulated furnishings test where the coated fabric will be nailed or tacked in place during use [27]. Two other methods have been developed

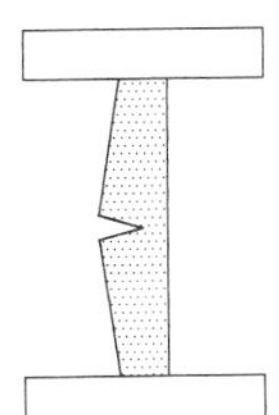

Figure 5 Trapezoidal tear.

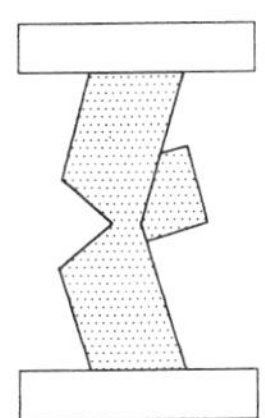

Figure 6 Wing tear.

for tear, based on puncture tests; one uses a ring clamp held in a tensile machine , and a piercing instrument penetrates it [28]. The other [29] uses a burst test machine, where the sample has a 3 mm diameter hole punched in its center, is held between two clamped rings, and a steel ball is pushed through it. This test is known as a wounded burst tear and yields only the lowest values of the material. A method also exists for the determination of the crush or cutting resistance of fabrics coated with rubber or plastics, and an assessment is made of the coating for possible cutting or penetration. Whilst it is not strictly tear testing, this test is mentioned for completeness [30].

11 Adhesion

Adhesion of the polymer coating to the substrate is important because loss of adhesion leads to delamination and often failure of the product in use. Products can be coated on both sides, as for example with tarpaulins, single coated, as with garments or upholstery, or the polymer can be sandwiched between fabrics, creating a waterproofing /bonding layer as in double texture materials. Success or failure of the composite is dependent upon the adhesion of the polymer to the substrate, and knowledge of the level of adhesion is vital. Materials are often tested in strip form (having first begun the separation process by hand) by clamping the two separated ends of the strip in the jaws of a pulling device and peeling them apart at a fixed rate.The rate of separation is important and is considered in some detail by Wake in his discussion of peel performance and other aspects of adhesion [31]. The rate of jaw separation has been set at 100 mm/minute for a dynamic test in ISO2411 [32]. In general terms, the slower the rate of separation, the lower the adhesion value obtained, and the higher the adhesion to the substrate, the lower the tear strength.This is one reason why it is not advisable to compare results from the deadweight or static test with the dynamic methods from tensometers. Sample preparation is important, and the early test standards on adhesion described in some detail the exact method to be followed. [33]. These were often polymer specific, e.g., PVC, nitrocellulose, or polyurethane, and treatment varied dependent on whether the coating was thick or thin. For thick films, often the strength of the coating exceeded the adhesion, and the coating could therefore be cut through down to the fabric, then manually stripped, prior to testing. Sample sizes were 75 $\times$ 200 mm, then trimmed to 50 $\times$ 200 mm. Thin coatings of PVC were often recoated with a plastisol in several coats pregelled at 100°C, to obtain a satisfactory coating, and then fully gelled at higher temperature prior to stripping. However, this may have favorably affected the true adhesion values if the original coating was not sufficiently gelled initially. Another technique is to combine the two coated faces of two samples together using a suitable adhesive cement. Again, the adhesive carrier, usually a solvent, can affect the result if it is not completely removed prior to testing. After drying, this is cut down through to one fabric layer and then peeled apart initially prior to fixing the fabric in one jaw and the double coating and fabric in the other. Alternatively, instead of gluing the two coats together, a cotton fabric could be used instead of the second sample, thus forming a double texture, which could then be similarly prepared. With all these methods the use of additional bonding or bonded materials could influence the results. Some lacquer coats or surface finishes may also prevent the application of secondary coatings. Despite the reservations, however, these techniques are essential, particularly with thin films, where the adhesion needs to be known, but the coating may not be strong enough to effect a peel, without itself breaking. Once prepared and initially separated the test pieces are marked 50 mm apart [32], mounted in the jaws and pulled at a rate of 100 $\pm$ 10 mm/minute until 100 mm of coated fabric has been peeled. The separation forces can be recorded on an autographic trace to monitor the variation. The maximum and minimum values will be shown as peaks and troughs. Where these are clearly defined, a midvalue point can be determined and a mean adhesion value quoted. For less well-defined traces, interpretation becomes a problem, and recourse can be made to ISO6133 for guidance [34]. In these cases a maximum and minimum value are often stated. Adhesion values for coated fabrics are quoted in newtons per 50 mm width. Within ASTM/ISO test methods, only the dynamic method is described. For quality control purposes, where a maximum adhesion value is required to be established or perhaps the minimum load to effect separation is requested , then a static or dead load method is used. In the dead load method, one end of the sample is in a fixed jaw and the other end is attached to a jaw to which additional loads can be added. The whole unit is hung in a vertical plane and loads added smoothly to determine the maximum load to effect separation at a rate not exceeding 5 mm in 5 minutes. As a variant, a dead load proof test can be used, where a given load is applied for a specified time and the specimen must not be stripped more than a given amount. The static method employs a spring balance fitted with a grip. The 50 mm wide sample is inserted, hung vertically, and the other end of the sample attached to a second grip that can be pulled until separation occurs. The lower jaw can then be fixed in position. Separation will cease once the adhesion force is balanced by the tension in the spring and the value can be read. Both these methods [35] have the advantage of being simple to do and inexpensive; they are however being superseded by the dynamic test.

12 Resistance to Damage by Flexing

Flex performance of coated fabrics tends to be product specific and leads to it being specified as resistance to a set number of flexings without damage, or flexing to a known point and the damage must be of a certain severity. Sizes of test pieces vary depending on the method and whether further performance tests need to be carried out. Durability can also be assessed on a time basis. The difficulty is that the methods may not be representative of practice, since the test machines have been adapted from other industries; for example the De Mattia for rubber test pieces and the Schildknecht for plastic. Some others have been devised to be more product oriented, such as the SATRA Vamp flex, and Ballyflex for footwear or a Crumple flex test for garments. All have some drawbacks, mainly that the test pieces are small, and cannot be easily folded or shaped into position, but flex performance has been viewed as a mark of quality despite these limitations. Blocking, adherence, or surface finish can affect results. Ventilation and temperature control may also be needed to prevent localized heating; wet testing can also be carried out but this may lead to delamination. The methods are individual in their actions and comparisons between them are not meaningful.

12.1 De Mattia Method

Historically, the De Mattia method has been used for rubber testing and measures cracking (ISO 132) or crack growth (ISO 133) and utilizes a grooved molded rubber sample [36]. The apparatus comprises a pair of flat grips capable of reciprocation in a vertical plane towards each other. In the open position, the grips are 70 mm apart, and when closed are approximately 13 mm apart. The sample is therefore capable of being repeatedly flexed through 180°, normally at a rate of 5 times per second. For coated fabrics [37] the test piece is folded over on itself twice longitudinally, with the coating outermost, from a size of 125 × 37.5 mm to 125 × 12.5 mm. The sample is conditioned and then mounted in the grips so that an outward fold is created at the midpoint, once the grips are moved towards each other. The apparatus can then be set in motion and stopped frequently to assess damage or when the required number of cycles has been completed. Damage is assessed by comparing flexed and unflexed material and graded as a deterioration in appearance but without the use of magnification. This damage is reported as none, slight, moderate, and severe. Cracking is viewed using a ×10 eyepiece and is graded initially on appearance with a description of penetration into the coating/middle layer/base fabric exposure/ or total cracking through the product. The number of cracks of the lowest grade (or nil cracking), the length of the cracks in mm, and delamination if examined more closely is also reported. Other tests can be done to assess further deterioration such as adhesion or hydrostatic head if specified.

12.2 Schildknecht Method

Instead of using grips, a cylindrical sample made from material 105 mm long by 50 mm wide is mounted around two 25.4 mm cylinders set vertically apart, and held using hose clips at the upper and lower points so that the clips are 36 mm apart when the cylinders are in the fully extended position. One cylinder is capable of a reciprocating motion to give a stroke length when in motion of 11.7 ± 0.35 mm. Larger test pieces can be utilized of dimensions 105 × 65 mm, if further testing is needed, clamped again at 36 mm between clips but such that the free length of the unclamped test piece is 44 mm when the cylinders are fully extended. The rate of flexing is 500 times per minute (8.3 ± 0.4 Hz). This method [37] is useful for lightweight constructions and produces a concertina type of flex representative of clothing folding at elbows, knees, etc., in garments. Assessment is as described above in the De Mattia method. These two methods produce a unidirectional flex, but more often than not products are stressed in more than one direction in use. The crumple flex test was devised to try to address this situation.

12.3 Crumpleflex Method

This test requires a larger test sample but allows for further post testing to assess deterioration; in this way it removes the subjective evaluation required by other forms of flexing. A large test piece 220 × 190 mm is sewn into a cylindrical shape of length 190 mm and diameter 64 mm, with the coating on the outside. The product is then mounted between two shouldered discs and held in position with clamp rings. One of the discs is capable of oscillating through 90° at a rate of 200 twists per minute. The other moves inward with a stroke of 70 mm at a rate of 152 strokes per minute thereby inducing compression. Maximum extention between the discs is 152.4 mm, and they are mounted on the same axis, normally horizontal. After the machine has been set in motion and has completed the specified number of cycles, the sample can be stripped from the machine and examined for breakdown of the coating. Performance is again assessed as in the flex methods above, but more frequently it is supplemented by postflex testing for water penetration resistance. The method can also be carried out at low temperatures. This test is now specified for water resistant clothing [38] followed by hydrostatic head testing to ISO1420 [39].

12.4 SATRA Vamp Flex Method

Some methods have been developed specifically for end product use but have been adopted for coated fabrics. This test was originally devised for footwear and is called up in ISO4643 [40] for PVC industrial boots and is part of BS 3424 Pt 9 [41]. A 64 mm square sample is mounted symmetrically between two V clamps in the fully open position, the V clamps being at an angle of $40^\circ \pm 1^\circ$. The V clamps are 28.5 ± 2.5 mm apart open and 9.5 ± 1mm closed. One clamp is moved towards the other at a frequency of 5 ± 0.5 Hz, with a stroke of 19 ± 1.5 mm. The clamping forms a downward double crease centrally parallel to the face of the clamps similar to that of a shoe upper. The test can be used for coated fabrics and for subsequent testing. Flex assessment can be carried out on a pass/fail basis or a grading as before. For low temperature testing (-5°C) the frequency is changed to 1.5 ± 0.15 Hz due to frictional heat buildup, raising the temperature of the material. This method has an advantage for thicker coatings.

12.5 Ballyflexometer Method

A test piece is fixed in a folded position along its length between a stationary clamp and an upper reciprocating clamp that repeatedly dips vertically through an angle of 22.5° at a rate of 100 ± 5 cycles per minute. A shaped guide in the upper clamp creates a running crease in the material, which moves vertically upwards and downwards. At the lowest point (22.5° below horizontal), the three edge fold center of the crease is 2 mm above the top of the lower clamp. Some abrasion can occur between meeting surfaces of the folded specimen, and sometimes crease formation is not consistant. V shaped guides are fitted in the bottom clamp to overcome this. The method is described in BS4161 Pt3 [42] as a test for gas meter diaphragms, but it was originally used for testing shoe components. As with earlier methods, the samples can be tested at low temperatures.

13 Abrasion Resistance

Many coated fabrics are subjected to abrasion during their lifetimes which results in wear or deterioration, damage and lack of performance. Whilst it may be difficult to predict lifetimes, these simulated tests can be used for comparative purposes and quality control, and often further testing can be done when relevant. Two methods have been widely used over the years, the WIRA (Martindale) tester, and the rotary platform double head abraider (Taber), using abrasive wheels. Both methods are described in ISO5470 [43]. Wear assessment can be visual, based on the number of cycles or on the weight loss of the coating against a specific abradant. The Martindale employs a circular sample (38 mm diameter) mounted in a specimen holder, which is subjected to a specific fixed load. The holder fits above a base plate covered with a square of abradant. The sample rests on the abradant and is then cycled backwards and forwards in a Lissajous motion thus producing even wear. Normally the abradant is silicon carbide paper or wool worsted mounted over felt. If further tests need to be carried out and larger test pieces are required, then the roles can be reversed and the abradant put in the holder with the specimen as the base platform. This has the advantage in that the resultant specimen can be employed for post abrasion water penetration resistance testing.

The rotary abrader applies two abrasive wheels under controlled pressure to a circular sample mounted on a rotating table or platform. The wheels are free to rotate under the friction exerted by the moving specimen, and being mounted at an acute angle they induce wear. The weight loss of the coating can be determined after a fixed number of cycles (rotation is 60 revolutions per minute), and is reported in mg/revolution. Visual inspection against a known predetermined standard is useful for internal quality control. A further variation of these methods is contained in ISO5981 [44], determining the flex abrasion. This method is a visual assessment of wear after subjecting a specimen to mild abrasive rubbing of a foot pressing/resting against a changing series of folds produced in a coated fabric using a rubbing machine. Scuffing and snagging can also be determined by exposing the coated product to gravel chippings in a rotating drum [45]. Again visual assessment is made. This type of method is being reconsidered for rubber abrasion. One further property in this area of testing is color fastness to wet and dry rubbing, an assessment being made against a grey scale or discoloration of the rubbing medium. The equipment is a basic crockmeter used in textile testing where the specimen is mounted and gripped on a fixed flat bed over a glass plate, and a pivotal arm whose head is a brass peg rests on the material, the two surfaces, peg and sample, being parallel to each other. The peg is equipped with a screwed ring, and a piece of bleached cotton (or emery if required) is held by the ring over the peg. The arm/peg arrangement applies a total mass of 1000 g against the coated fabric and is reciprocated in a straight line over a distance of 100 mm at a rate of 0.25 cycles per second. After a given number of rubs (normally 20 cycles), the sample is checked for wear, damage, print loss and color fastness [46].

The next sections contain some methods that are common to rubber, textiles, and plastics and will therefore not be discussed at length but are highlighted for the purpose of completeness and possible comparison.

14 Accelerated Aging/Service Testing

Properties of coated fabrics can alter or change with exposure to humidity, heat, cold, fatigue, and chemicals, and tests have been devised to simulate these conditions and monitor behavior.

14.1 Oven Methods

Oven aging tests are used to observe and assist the measurement of deterioration due to changes in environmental conditions from a known baseline. Samples can be exposed in a standard oven to temperature for a fixed period [47]; weight loss can be determined, if it is relevant, and plasticizer absorbed on activated carbon in the case of PVC, for instance [48].Changes in tensile strength, elongation, adhesion, discoloration, softness, hardness, etc. can be measured and comparisons made with the original results. For possible use in tropical climates, high humidity can be introduced as a variant. This is useful to assess polymers sensitive to hydrolytic attack, as with polyester polyurethanes, for example. Some products, if left in contact with themselves either in roll form or in boxes, may adhere or block, and this can be simulated in an oven [49] by stacking the possible variations in surface contact, face to face, face to back, or back to back, sandwiched between glass plates and with weight applied. The ability to separate them after exposure and cooling determines the degree of blocking. Volatility and the ability to fog a glass plate to simulate windscreen fogging in automobiles is another specialized heat test referenced here for interest [50].

14.2 Ozone Resistance

For completeness, resistance to ozone cracking under static conditions, which can be classed as ageing, will be considered at this point. The principle of ozone generation, measurement of concentration, and verification has been discussed elsewhere, and rubber-coated fabrics are not exempt from cracking unless suitably formulated. Test pieces are exposed to ozone under specified conditions [51], the sample being clamped over a mandrel. The time for the first cracks to appear can be noted, or the sample can be exposed for a fixed period and pass/fail criteria used. Temperature and ozone concentration can be varied as specifications dictate.

15 Resistance to Liquids

Resistance to penetration by liquids, for protective clothing, has moved on from the simple test of pouring the medium into a cone-shaped sample of the product over a beaker and observing penetration over a period of time [52]. Health and safety legislation regarding chemicals handling, and greater awareness of the potential dangers involved in some of the tests, has now led to the ISO adding a caveat regarding safety to all its standards. ISO6450 describes the procedure for determining the resistance to liquids [53]. In service coated fabrics are rarely totally immersed, so this test is carried out exposing one surface only to the liquid medium. The test is performed using the apparatus and test liquids prescribed in ISO1817 [54]. Mass loss can be determined if necessary, but normally changes in physical properties are measured before and after exposure and differences expressed as a percentage of the original. The method can be used for other test liquids if chemical resistance needs to assessed, due care being taken regarding safety.

16 Resistance to Penetration by Water

Whilst protective clothing may be subjected to droplets or splashes in practice, and need to be resistant to the chemical medium, weatherwear has much longer time exposure to the elements, and testing therefore needs to reflect durability and fitness for purpose, as with fisherman's clothing for example. The resistance to water penetration is basically a pressure test on a specimen of coated fabric, but these are not designed as bursting tests. Due to the diversity of coatings and substrates, the pressure tests can be divided into low- and high-pressure methods, and those with small and large test pieces, depending on the previous testing done to whichever specification. Testing is time consuming, and multi-headed machines have been devised, but the purpose of the test is to look for water penetration, and sometimes it is difficult to monitor large numbers of test pieces. Pressure rise has to be controlled, and therefore test times can be lengthy if statistical information is needed. A more realistic and simpler test for quality control is to subject samples to a rapid rise in pressure which can be maintained.

Normally the sample is clamped in a ring and subjected to an increasing pressure of water from above or below depending on the method and which face is to be tested. Once a predetermined pressure is rea ched it can be held until water comes through or the pressure can be continually increased and noted at the moment of leakage. This is used as the basis of a pass/fail test. Suitable equipment is described in ASTM 751, BS 3424, and ISO1420. These pressure tests are called up in a range of specifications, often preceded by abrasion, flex, or ageing tests. Sometimes it may only be wettability or shower rating that needs to be determined, and a funnel and showerheads placed above an inclined sample are one simple method of assessment [55]. If dimensional stability is required, a piece of material can be measured, immersed fully in water for a given period of time, removed and allowed to dry, and then remeasured. The change in dimensions can then be expressed as a percentage of the original as a shrinkage or an expansion [56]. Taken one stage further, stability to washing in a domestic machine can also prove useful [57].

17 Permeability

As coated fabrics have become more lightweight for outerwear, materials have been developed that "breathe" and yet remain waterproof. Due to the difference between the size of water vapor molecules and water droplets, coatings have been produced that allow perspiration to pass through but remain resistant to rain. Tests for water vapor permeability have been developed as an aid to assessing breathability, and a standard has been written for these products [58]. The principle is straightforward; the coated fabric is placed over the top of a dish containing water, supported if necessary, and sealed in place with adhesive or self-adhesive tape. The whole is accurately weighed and then placed in a controlled environment, on a rotating table that ensures a fixed amount of air movement above the material. The dish is weighed periodically to determine the rate of weight loss or transmission. From the dimensions of the dish (area of the face) the water vapor permeability (WVP) is then calculated in grams per square meter per 24 hours. If the test is conducted using a standard fabric whose WVP is measured at the same time, an index value can be calculated so that meaningful comparisons can be made. It also serves to determine if the test is meaningful, or if something has altered during exposure. If the transmission rates of volatile liquids need to be measured, a similar technique can be used to that described in ISO6179 [59]. In this method the dish/vessel is inverted so that the liquid is in contact with the sample at all times. Elevated temperatures can be used if needed. The transmission rate is expressed in grams per square meter per hour. For transmission of gases, the test piece is clamped in a closed cell with a vector or carrier gas on one side and the tracer gas on the other. An analyzer system then detects and measures the concentration of the tracer gas in the vector gas as it permeates through the material, and the rate can then be determined for the gas in question [60]. Where porosity or airflow resistance needs to be determined, perhaps with perforated products, air permeability is normally a sensible method. A rate of airflow is measured through a material that produces a fixed pressure drop across a specimen of known test area and a rate of volume per area per second is then calculated [61]. Permeability may be a desirable feature, and it may have been purposely introduced by perforation, as for example in automotive seating and lining materials.

One property, however, that may not be desirable, particularly with inflatable coated fabrics, is lateral leakage or wicking. This can occur in gaseous or liquid media where the textile part of the coated fabric allows passage along or through the material by capilliary action. Penetration of the polymer into or between the yarns can minimize this effect, and some forms of textile treatment are available to reduce the incidence.Tests have been devised for this and are described in BS3424 [62].

18 Flexibility

Stiffness, rigidity, drape, handle, are all terms that describe the product, but it is a difficult property to quantify. Several test methods have been developed to determine flexibility; ISO5979 is one such method [63]. A rectangular test piece 600 $\times$ 100 mm is folded back on itself lengthwise on a flat surface, and the ends are superimposed and then held under a steel bar. The height of the loop produced is measured; the higher the loop the stiffer the product. This test has been incorporated into a CEN standard together with a method of measuring bending length [64]. In this latter test, a rectangular piece of coated fabric 200 × 25 mm wide is moved under a graduated scale until the edge of a platform is reached. On continuing, the strip overhangs the edge, bending under its own weight until the end reaches an inclined plane. At this point a length can be read on the graduated scale, which is the bending length. The test can be repeated for the opposite face and the longitudinal and transverse values measured, and a mean determined. Knowing the weight and the bending length, it is possible to calculate a flexural modulus, which can then be used comparatively.

19 Low-Temperature Performance

Flexibility can also be measured at low temperatures as a bending test [65]. Samples are kept at low temperature for a specified period and then bent through 180^0 and the surface examined for damage (cracks) and graded accordingly.The maximum recommended thickness of sample is 2.2 mm. The test can be used to assess the temperature of cracking or to pass or fail a specification.

If more specific information is required, then a cold crack temperature can be established [66]. Here, specimens are held in a loop form in a liquid coolant and impacted between a plunger and an anvil at various temperatures and assessed for cracking. A temperature is established where, from five samples, there are more passes than failures, and this is reported as the cold crack temperature. However, the test only uses a short cooling period, which may give misleading results with some materials. Other low-temperature impact tests have been developed [67] where the specimen is clamped in a cantilever beam in a bath of liquid and struck at right angles with a striker at a velocity of 1.8–2.1 m/second. The product is then visually examined at 5× magnification by folding it 180° around a 6 mm mandrel. Again, this can be a pass/fail test. This method is also used for brittleness temperature by impact for plastics [68]. When further tests need to be done such as water vapor or hydrostatic head, after the low temperature exposure another method may be adopted [69]. Samples are first folded into four (across and lengthways) and exposed to −30°C under 4 kPa pressure for 48 hours. The piece is then unfolded whilst still at this low temperature, examined for cracks and delamination, then subjected to hydrostatic head pass/fail criteria at 25 kPa water pressure.

20 Product Specifications

The major coated fabrics test methods have been outlined and reviewed, and these form the basis of standardization for product specifications. There also exist a variety of chemical and other tests [70], several specific to government or military establishments. Some may now have been discontinued, and others may be hidden away in annexes. They do however offer an insight into specific industry applications. Specifications exist for upholstery [71], tarpaulins [72], and waterproof clothing [73] at the ISO level and have been expanded at the national level [74].

21 Summary

The use of coated fabrics is growing all the time as newer polymers are developed and further applications are found. Tents and awnings have developed into architectural features and air-supported structures. Footwear has become fashioned into sports and leisurewear, and marine applications range from oil booms and inflatables/liferafts to hovercraft skirts. Airbags, lifejackets, safety clothing, and weatherwear are all demanding newer test methods for their particular properties. As applications develop, standards will need to match them. These must be meaningful, applicable, and reproducible, but more importantly, if they can be accepted universally, then specifications will have much more value.

References

1. BS 3546, Coated fabrics for use in the manufacture of water penetration resistant clothing, Part 4, 1991.
2. Buist, J. M., Plenary lecture, Rubbercon 96, 1996.
3. Hancock, T., Personal narrative of the origin and progress of the caoutchouc or India rubber manufacture in England, 1857.
4. Wake, W. C., and Wooton, D. B., *Textile Reinforcement of Elastomers*, Applied Science Publishers, 1982.
5. Evans, C. W., *Developments in Rubber and Rubber Composites*, Applied Science Publishers, 1983.
6. ISO 2231, Fabrics coated with rubber or plastics—Standard atmospheres for conditioning and testing, 1991.
7. ISO 2286–1, Determination of roll characteristics, Part 1, Determination of length, width, and net mass, 1998.
8. ISO 2286–2, Determination of roll characteristics, Part 2, Determination of mass per unit area, 1998.
9. BS 3424, Part 20, Method 23, Determination of dimensional changes on mechanical relaxation at zero tension, 1987.
10. ISO 2286–3, Determination of roll characteristics, Part 3, Determination of thickness, 1998.
11. BS 3424, Part 25, Method 28, Determination of coating thickness and thickness of an expanded layer, 1993.
12. BS 3424, Method 28b, Determination of coating thickness of thinly coated fabrics, 1993.
13. ISO 1421, Determination of tensile strength and elongation at break, 1998.
14. ASTM D 751, Standard test methods for coated fabrics, 1995.
15. ASTM D 1683, Standard test method for failure in sewn seams of woven fabrics, 1990.
16. ASTM D 751, Seam strength, 1995.
17. BS 3424, Part 21, Method 24, Determination of elongation and tension set, 1993.
18. ISO 7617–1, Annex B, Plastics coated fabrics for upholstery, Part 1, Specification for PVC coated knitted fabrics, 1994.
19. ISO 3303, Determination of bursting strength, 1995.
20. Eichert, U., Eurofabric 92, 19th European Akzo Symposium on Broad Woven Industrial Fabrics.
21. ISO 4674, Part 1, Determination of tear resistance, 1998.
22. BS 3424, Method 7C, Single tear, 1973.
23. EN 1875–3, Determination of tear resistance, Part 3, Trapezoid tear, 1997.
24. ISO 13937–3, Textiles—Tear properties of fabrics, Part 3, Determination of tear force of wing shaped test specimens.
25. ASTM D 1423–83, Tear resistance of woven fabrics by falling pendulum (Elmendorf).
26. ISO 4674–2, Determination of tear resistance, Part 2, Ballistic pendulum method, 1998.
27. ASTM D 751, Tack tear, 1995.
28. ASTM D 751, Puncture resistance, 1995.
29. BS 3424, Methods for test for coated fabrics—Wounded burst test (in preparation).
30. ISO 5473, Determination of crush resistance, 1997.

31. Wake, W. C., and Wooton, D. B., *Textile Reinforcement of Elastomers*, Chapter 5, Applied Science Publishers, 1982.
32. ISO 2411, Determination of coating adhesion, 1998.
33. BS 3424, Method 9, Determination of coating adhesion, 1973.
34. ISO 6133, Rubber and Plastics—Analysis of multi-peak traces obtained in determination of tear strength and adhesion strength, 1998.
35. BS 3424, Determination of coating adhesion, Methods 9A and 9C, Deadweight and static methods, 1973.
36. ISO 132 and ISO 133, Determination of flex cracking and determination of crack growth (De Mattia), 1983.
37. ISO 7854, Determination of resistance to damage by flexing, 1995.
38. BS 3546, Part 5, Immersion suits, annex D, 1995.
39. ISO 1420, Determination of resistance to penetration by water, 1992.
40. ISO 4643, Plastics moulded footwear—Polyvinyl chloride industrial boots—specification, Annex C.
41. BS 3424, Part 9, Determination of resistance to damage by flexing, 1996.
42. BS 4161, Gas meters, Part 3, 1977.
43. ISO 5470, Determination of abrasion resistance, Parts 1 and 2, 1997.
44. ISO 5981, Determination of flex abrasion, 1997.
45. BS 3424, Part 31, Method 34, Determination of resistance to scuffing and snagging, 1990.
46. BS 3424, Method 16, Determination of fastness of colour to wet and dry rubbing, 1973.
47. ISO 1419, Accelerated ageing and simulated service tests, 1995.
48. BS 3424, Part 12, Method 14A, Loss of volatile matter on heating of plasticised PVC coated fabrics, 1990.
49. ISO 5978, Determination of blocking resistance, 1995.
50. ISO DIS 6452, Determination of fogging resistance of trim materials in the interior of automobiles, 1996.
51. ISO 3011, Determination of resistance to ozone cracking under static conditions, 1997.
52. BS 3424, Method 19, Determination of resistance to penetration by liquids, 1973.
53. ISO 6450, Determination of resistance to liquids.
54. ISO 1817, Vulcanised rubber—Resistance to liquids—Method of test.
55. BS 3424, Part 26, Method 29D, Resistance to water penetration and surface wetting, 1990.
56. BS 3424, Part 17, Method 20, Determination of dimensional stability to water immersion, 1987.
57. BS 3424, Part 36, Method 39, Determination of the dimensional stability of coated fabrics to domestic washing, 1993.
58. BS 3546, Part 4, Specification for water vapour permeable coated fabrics, 1991.
59. ISO 6179, Rubber, vulcanised—Rubber sheets and rubber coated fabrics—Determination of transmission rate of volatile liquids (gravimetric technique), 1989.
60. ISO 7229, Measurement of gas permeability, 1997.
61. BS 3424, Part 16, Method 18, Determination of air permeability or air flow resistance, 1996.
62. BS 3424, Method 21, Determination of resistance to wicking and lateral leakage to air, 1986.
63. ISO 5979, Determination of flexibility—Flat loop method, 1994.
64. EN 1735, Determination of flexibility, Method 2, Bending length, 1994.
65. ISO 4675, Low temperature bend test, 1995.
66. BS 3424, Part 8, Determination of cold crack temperature, 1983.
67. ISO 4646, Low temperature impact test, 1989.
68. ISO 974, Plastics—Determination of the brittleness test by impact, 1997.
69. BS 3546, Part 5, Specification for coated fabrics for immersion suits, Annex F, 1995.
70. BS 3424, Methods of test for coated fabrics, Appendix A, 1973.
71. ISO 7617, Plastics coated fabrics for upholstery, Parts 1, 2, and 3.
72. ISO 8095, PVC coated fabrics for tarpaulins, 1995.
73. ISO 8096, Coated fabrics for water resistant clothing, Parts 1, 2, and 3.

74. BS 3546, Coated fabrics for use in the manufacture of water penetration resistant clothing; Part 2, Specification for non water vapour permeable coated fabrics, 1993. Part 4, Specification for water vapour permeable coated fabrics, 1991. Part 5, Specification for coated fabrics for immersion suits, 1995.

21
Dynamic Mechanical (Thermal) Analysis

John Gearing
Gearing Scientific, Ashwell, Hertfordshire, England

1 Introduction

DMTA (or DMA)—the exciting of a material with a periodic stress and monitoring of the resultant strain—has become a commonly used technique for both scientists and engineers who need to know the viscoelastic properties of a material with respect to temperature, humidity, vibration frequency, dynamic or static strain amplitude, or other parameter against time. This chapter will attempt to introduce its principles, cover a short history of the technique relating it to other mechanical tests, and discuss its application to a wide range of polymers and other materials.

There are two rather different uses for DMTA. The chemist will often want to know the morphology including glass transition(s) and melt(s) of a newly synthesized few grams of polymer or copolymer with a quick thermal scan, whereas the engineer or physicist will wish to maintain temperature equilibrium and check the performance in conditions similar to the end use of a component made from this material. Environmental effects such as static strain, humidity, or solvent ingress can be fundamental to the expected life of the material, either in a detrimental way (e.g., solvents into polycarbonate) or beneficially (as in the case of hair conditioners and hand creams). Both the chemists' and the engineers' needs will be addressed here with the compromises necessary to attempt to achieve both in one apparatus, which should also be affordable, look good, and give traceable results for a long time with minimum servicing.

ISO 2856 covers the dynamic testing and separates it into two broad classes, free vibration and forced vibration. A complementary technique worth mentioning here is rebound resilience (ISO 4662/BS903/ASTM D1054/D2632), which is in effect a drop weight impact test (or pendulum strike) and measures the rebound height of a small spherical steel ball on a well-clamped rubber specimen. Its main appeal is its simplicity in a half cycle of deformation, and the ratio of the rebound height to the drop height gives an indication of the resilience, which is approximately equal to tangent delta (the angle between the applied stress and strain in a forced vibration system). Different types of these include Luepke, Schob, Zerbini, Dunlop Tripsometer, Goodyear-Healey, Shore Scleroscope, Bashore, and the ADL tester. (See Brown [142] or Payne and Scott [143] for more details.)

Free vibration methods such as the torsion pendulum are covered by ISO 4663 and are limited to very low strains and frequencies, and are in much less frequent use these days than the forced vibration nonresonant systems on which this chapter will focus. The early Du Pont DMA and German Myrenne used input energy to maintain the resonant oscillation amplitude, but the main limitations were variable frequency according to the sample size (which had to be glassy or plastic) or one frequency only (1 Hz) respectively.

Forced nonresonant DM testing is covered by ISO 6721–5, 2856, 4664; ASTM D2231; and DIN 53513, and this chapter will particularly look at the variation of properties as the temperature, stress amplitude, or frequency of vibration is varied in a wide range of test geometries so that almost any nonfluid material can be loaded and measured.

2 History

During the 1930s Lessig [106] at Goodrich in Akron developed the Goodrich flexometer, which became an ASTM standard (D623) for tire rubbers, and was modernized in the 1990s [107] to operate from 5 to 40 Hz and from room temperature up to 120°C only. This later paper by Askea shows an interesting comparison with the large servohydraulic systems, which almost always give very low values of modulus and where the accurate control of very small phase angles is not as easy as with the electrodynamic systems studied in this chapter. Dynamic strains and frequencies can be swept at a particular static strain. During the 1980s this lab also purchased a smaller DMTA to enable them to scan from −150°C to 500°C for their research.

Roelig [1] in Germany did some DM work on rubbers during 1940, and one of his early apparatus arrived in the Admiralty Research Establishment, Holton Heath (U.K.) shortly after the war. This was a large apparatus used for looking at properties at a constant temperature. During the 1940s and 1950s, plastic was considered a brittle curiosity, so apart from some further DM work on rubbers by groups including Alexandrov and Lazurkin [2], and at MRPRA [79] and RAPRA [78] in the UK, one of the first references to measuring the DM properties of plastics is Schmieder and Wolf in 1953 [3]. Nolle published a fine 3D picture of SBR rubber properties with respect to frequency and temperature as early as 1950 [4], and other workers who produced classic volumes shortly afterwards were Nielsen [5], Tobolsky [6], Ferry [7], and McCrum/Read/Williams [8].

Takayanagi [9] published DM work on crystalline polymers in 1965 and from this work developed the Rheovibron. Early users of this apparatus included Murayama, Dumbleton, Williams, and Bell, followed by MacKnight (pu), Karasz (block copolymers), and Wilkes (collagen) during the early 1970s [10–16]. By 1981, Wedgewood/Seferis [108] were producing accurate data and published the classic guide to overcoming the many inherent problems. At this time Heijboer was active [20] and Davies [17] at Leeds and Wetton [18] at Manchester were building their own. RAPRA were using a large Keelavite, and various groups were buying the Rheovibron from Toyo Baldwin of Japan, which was designed for thin fibers (e.g., Imperial College, London, Liverpool, and Strathclyde Universities). This gave information at 3.5 and 11 Hz, and if resonances could be avoided also at 35 and 110 Hz, with a maximum tan delta of only 1.7 and temperature limit of

250°C. The early ones had tube electronics and you had manually to try to maintain a tension as the sample relaxed during a thermal scan.

A Torsional Braid Analyser or Torsion Pendulum (ISO 4663 or BS903/A31) had been developed for metals to study grain boundaries and crystal dislocations, e.g., Zener [21]. Curing polymer systems on glass or metal braids was being researched by many groups in the U.S.A. such as Lewis and Myers at Lehigh University in 1958 and other groups in Japan and Germany who found that good data could be obtained with a simple stop watch! A recent review by Gillham and Enns elegantly covers all this early TBA history [19], and it is still one of the simplest and most sensitive methods for monitoring from a liquid via gelation through to a glassy solid. Very low strains and frequencies in the 0.1 to 10 Hz region limit its use to materials characterization rather than to obtain engineering data. Brabender-Lonza, Gillham, and Myrenne have manufactured many of these.

The Yerzley Oscillograph (ASTM D945) has been used for rubber testing by monitoring the decay of the sample subjected to a dynamic oscillation by a cantilever. Terms such as Yerzley hysteresis and resilience seem peculiar to this apparatus, and since large deformations were necessary, the nonlinearity suggests that only an average modulus could be obtained.

In 1971, Rheometrics started to publish results on polymer melts and then solids and built up quite a following with their step isotherm approach, where quite extensive manual adjustment was needed at each temperature. Their RSA solids analyzer was introduced in the late 1980s, and in 1994 their parent company bought PL Thermal Science Division and hence the PL DMTA and other thermal techniques too.

1975 saw Du Pont Instruments launch the 980 DMA as a thermal analyzer complement to their DSC, TGA, and TMA. This had two long arms that gave a natural resonant frequency that would change with the sample dimensions and as the sample softened. However once the glass transition was reached (for an unfilled polymer) the two arms could not be prevented from trying to oscillate independently, so little data could be obtained in the leathery or rubbery regions. Hundreds of these were sold, and Boeing insisted that all their composite suppliers had to measure T_g on them. After many improvements, such as removing the 28 Hz resonance of the first polymeric feet(!), strengthening the rather flimsy steel arms in the 982, and offering a forced or resonant frequency option, this unit finally died in the early 1990s. TA Instruments have launched a much improved version in 1996.

In 1976 the Rheovibron agent in Germany designed a larger unit for one of his tire customers, and the first Gabo Eplexor was born. These were mainly sold in Germany, and it was soon realized that 150 N was needed to compress a glassy 10 × 10 mm solid cylinder by 2 microns, to give data through T_g, so a range of larger units were developed. An Eplexor first came to the U.K. in 1995. These currently use a 32 bit computer to control every parameter and can be used for millinewton forces up to 10k newton for the larger highly filled specimens, with every conceivable clamping option. As the tire companies relied more and more on this data for QC, the first DM autosampler was delivered in 1995, allowing 50 prepared samples to be measured and then run unattended over the weekend or overnight, in effect trebling the capability of one operator and apparatus, and allowing any inhomogeneities of the samples to be seen, as well as giving very accurate engineering data on a statistically meaningful number of samples. As we went to press this was still the only autosampled DMA on the market, but I am sure that others will follow suit soon! Options include 1100°C for glass/ceramics in 4 point bend.

On a similar time scale, Metravib in France, IMASS in the USA (also the Rheovibron importer) and Iwamoto (Japan) developed larger engineers' apparatus, each with its own following, mainly in the country of origin. MTS (USA), Dartec, Keelavite and ESH (GB) and Schenk (Germany) have produced large servohydraulic apparatus for DM testing but usually at room temperature only, or with such a large oven that hours are needed for equilibration at other temperatures.

The late 1970s saw Polymer Laboratories develop their DMTA using dual cantilever bending, which works well for most small samples from −150°C to the onset of melt. Shear, tensile, torsion, and simple compression options followed, as did the complementary Di-electric Thermal Analyser (DETA), and computers were used from 1982 to both control and analyze the data. Seiko Instruments copied this and tried to patent it, and others such as Netzsch, Perkin-Elmer, and TA Instruments looked very closely at this before launching their own. For comparative data and fast thermal scans they all can give good data, but for absolute modulus numbers most systems need to consider the frame compliance, sample end corrections, and relative dimensions, and hence only a limited range of sample dimensions can be used for accurate measurement of modulus in a particular mode of deformation.

Perkin-Elmer and Mettler produced various vibrating TMAs during the 1970s and 1980s, and these would show a massive increase in wave amplitude as the glass transition approached, giving some limited information. In the late 1980s Perkin-Elmer produced the DMA7 and updated many of their older TMAs at the same time. This unit could be supplied with glass fittings and reach 1000°C; it is suitable for small samples and as a thermal scanning DMA.

3 Principles

The response of a sinusoidal stress signal applied to a material will depend on the viscoelastic nature of that material, hence a sample of spring steel will be almost totally elastic, whereas a tub of grease or honey will be predominantly viscous. Hooke's "True theory of elasticity" said, "The power of any spring is in the same proportion with the tension thereof. i.e. if one power stretch will bend it one space, two will bend it two, three will bend it three and so forth." Hence force $F = k(\Delta x)$, where delta x is the displacement. Euler refined this to include the cross-sectional area A and the original length L, but it was not until about 1800 that Thomas Young published it more widely as:

$$\frac{F}{A} = E\frac{\Delta x}{L} \qquad \text{which has now become} \qquad \sigma = E\gamma$$

where sigma is the stress, or force per unit area, E is the tensile or Young's modulus, and gamma is the strain, or extension divided by the original length.

Also in the late 17th century Newton theorized that "The resistance from the want of lubricity in the parts of a fluid is, other things being equal, proportional to the velocity with which the parts of the fluid are separated from one another". The want of lubricity is now known as viscosity, and it was left for two centuries until Poisson, Stokes, Poiseulle and others expressed this as, "The stress used in deforming a liquid is in direct proportion to the rate of strain imposed thereon." Thus

$$\sigma = \eta\dot{\gamma}$$

where the proportionality constant for a Newtonian fluid is η, the viscosity, sigma is the stress, and gamma dot (or $d\gamma/dt$) the strain rate. Polymers will be in between these two extremes, with C fiber filled fully cured composites being close to entirely elastic (but

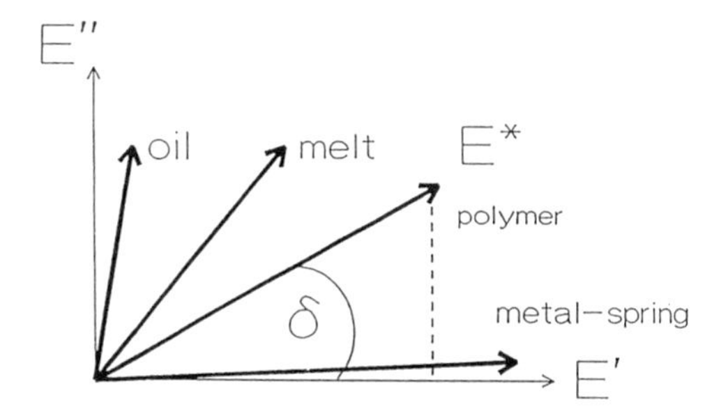

Figure 1 Formulas: $E^* = E' + iE''$; $\tan\delta = E''/E'$.

highly anisotropic!), followed by glassy plastics, leathery materials, rubbers, foams, gels and then molten polymers, as can be seen in Fig. 1.

In the solids that we are dealing with in this chapter, the elastic moduli will be more important than the viscosity. Various models have been suggested over the years to describe polymers using springs and dashpots, and combining Maxwell's series model with Voigt-Kelvin's parallel one gives a good approximation (Fig. 2).

From Hooke's law, the six independent components of the stress tensor can be expressed as a function of the six components of the strain tensor in a symmetrical matrix of order 6 with 21 modulus components for a general anisotropic sample of material. For an isotropic body, there are only two independent components. The mode of deformation will determine which modulus will be measured.

The bulk modulus K ($= 1/H$, the reciprocal of the bulk compliance) can be measured in compression with a very low height-to-thickness (h/t) ratio and unlubricated flat clamp surfaces. In pure compression with a high h/t ratio, and lubricated clamps, the compressive modulus E ($= 1/D$, the reciprocal of the compressive compliance) will be measured. Any intermediate h/t ratios will measure part bulk and part compressive moduli. Hence it is vital for comparing samples to use the same dimensions in thermal scans and the same h/t ratio when accurately isotherming and controlling static and dynamic strains and frequencies.

Young's modulus E ($= 1/D$, the compliance) is commonly measured in tension, and with the free length-to-width (l/w) and length-to-thickness (l/t) ratios greater than 10 will give accurate data. Note that the static stress will always have to exceed the dynamic one if

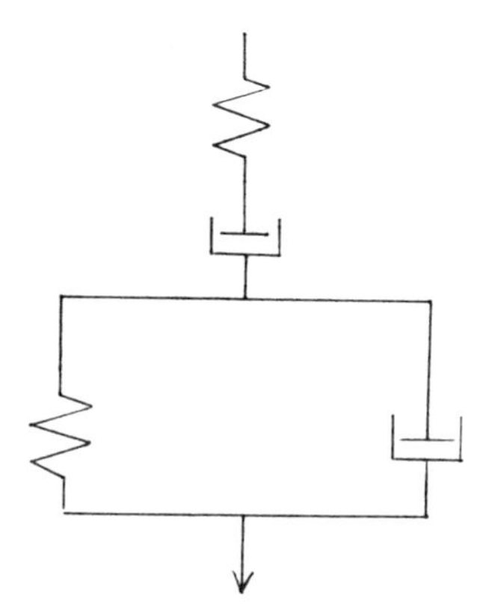

Figure 2 Four-component model.

the sample must remain in tension (as is the case of thin films or fibers). Hence a constant static stress can become a problem when the material softens dramatically, as will happen at T_g (the glass transition, where the modulus can drop from 1000 to 1 MPa.) One way around this is to reduce the static load in proportion to the modulus, or to apply a constant static strain with superimposed dynamic strain.

For an elastomer it is quite important to record how it was cooled, as very different results may be seen for samples cooled under stress, and cooled and allowed to contract (or expand). Some experimenters use a so-called holding stress—just enough to prevent the sample buckling—and others swear by putting a 10% strain on the elastomer at room temperature and maintaining it during the cooling to begin the thermal scan. Certainly very different results will often be achieved if the static force is applied during the cooling cycle, compared to just a very small holding stress. During the thermal scan, in some apparatus it is possible to relax the sample completely in between the step temperature measurements to allow for the sample to reach temperature equilibrium before reapplying the static and dynamic loads.

Many experienced rubber testing laboratories will start at the highest temperature of interest and then sweep frequency and/or static and dynamic strain before moving to the next lower temperature, equilibrating, and then repeating this until the lowest temperature of interest is reached. This way, no undue damage is applied to the sample by working it in the leathery region with the very large forces needed to achieve the desired strains. Of course if the l/w and l/t ratios are low, then a tension/compression experiment can be performed about any static stress including zero.

If measured in three-point bending mode (for rigid materials only), Young's modulus will again be measured, but as the l/t ratio goes below 10, some shear will be measured too, giving a lower value. For polymers to scan from glassy plastic to the onset of melt, the dual- or single-cantilever bending mode is preferred by those looking for transitions (and with small force apparatus), since the sample will be measurable over this four to five decades of modulus, even though it is a mixed modulus that is recorded with the typical experimental l/t of between 3 and 5. One variety of four-point bend will give Young's modulus (where it is in effect a wide central clamp of a three-point bend, allowing pure tension only between the two center clamps) as shown in Fig. 3.

Other variants have been tried, and one particularly clever bend mode was developed by MacInnes [29], who measured from the liquid state to the glassy state in an enclosed polymer bag in the form of a thin disc. The outside ring is fixed, and the center is oscillated up and down. One advantage over the TBA is that all the moisture is contained in the bag, which can be vitally important for environmental studies or food material research. Of course this would give good relative information and reasonably good tan delta values, but engineering modulus is usually less important for foodstuffs, where the subjective relative taste/texture properties such as crispness or crunchiness are difficult to define! Because of the inhomogeneity of many foodstuffs and their dependance on the water content, this

Figure 3 (a) Three-point bend. (b) Four-point bend.

does allow the water content to be maintained constant, which was always a problem with the bending or shear experiments we made at Reading or Loughborough in the 1980s.

For simple shear measurements of rubbers or elastomers, the shear modulus G ($= 1/J$, the shear compliance) will be determined with one or two samples clamped about an oscillating center part, and for torsional shear a cylindrical (or better, ring-shaped) sample will be mounted between two parallel plates. An immediate problem can be seen with the torsional shear test, if the samples are likely to have a varying G with strain amplitude (i.e., the sample is nonlinear, as most filled polymers will be), since the central part of the sample has zero strain imposed on it, increasing to a maximum at the outside edge. Hence a narrow ring-shaped or hollow cylindrical-shaped sample will give more accurate data if torsion has to be used, since the strain is then a small variation about the mean radius of the ring (Fig. 4a).

The shear modulus of a glassy plastic is impossible to measure in the clamps shown in Fig. 4a without adhering the sample into the clamp (but then you are also measuring the adhesive too), and the leathery region (about 100 MPa) is the limit for shear on the smaller 10 N force apparatus. One group at 3M research used very long, thin samples in the double shear setup of Fig. 4a and were able to measure from the glassy state through T_g, but in effect this was dual cantilever bending with the sample ends glued into the clamps. The four-point bend approach can be used for shear measurements in the glassy region (without adhesives) where the clamps alternate, as can be seen in Fig. 4b.

These moduli are related for an isotropic rubber with temperature as seen in Fig. 5.

These are most easily represented by the equation $E^* = E' + iE''$, where E' is the ratio between (the amplitude of the in-phase stress component/strain, σ/ε) and E'' is the loss modulus (the amplitude of the out-of-phase component/strain amplitude). Similarly for G^* and K^* and the ratio between the Young's modulus E^* and the shear modulus G^* includes Poisson's ratio υ, for an isotropic linear elastic solid with a uniaxial stress. (Poisson's ratio is more correctly defined as minus the ratio of the perpendicular strain to the plane strain, or $-\varepsilon_{22}/\varepsilon_{11}$ for one orthogonal direction 22 which equals the 33 strain if the sample is isotropic.)

$$E = \frac{9KG}{(3K + G)} \qquad G = \frac{E}{2(1 + \upsilon)}$$

$$K = \frac{E}{3(1 - 2\upsilon)} \qquad \upsilon = \frac{3K - 2G}{2(3K + G)}$$

Poisson's ratio itself is a complex quantity, as there is a phase lag between the lateral motions and the in-plane stress and strain of a dynamically stressed system. Most isotropic plastics and rubbers have Poisson's ratios of 0.3 and 0.5 approximately respectively, but it

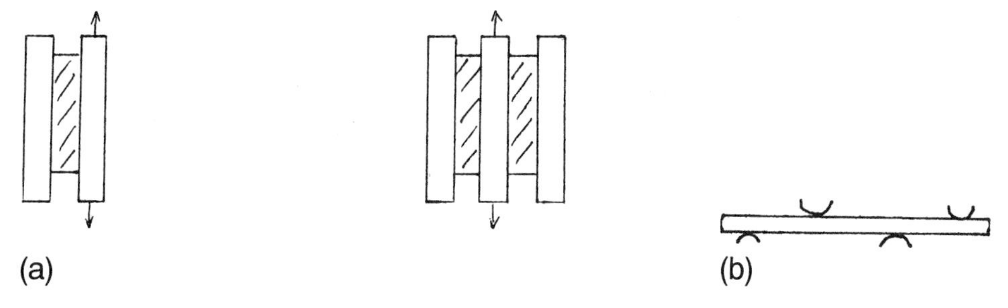

Figure 4 (a) Simple shear clamps. (b) Four-point bend for shear.

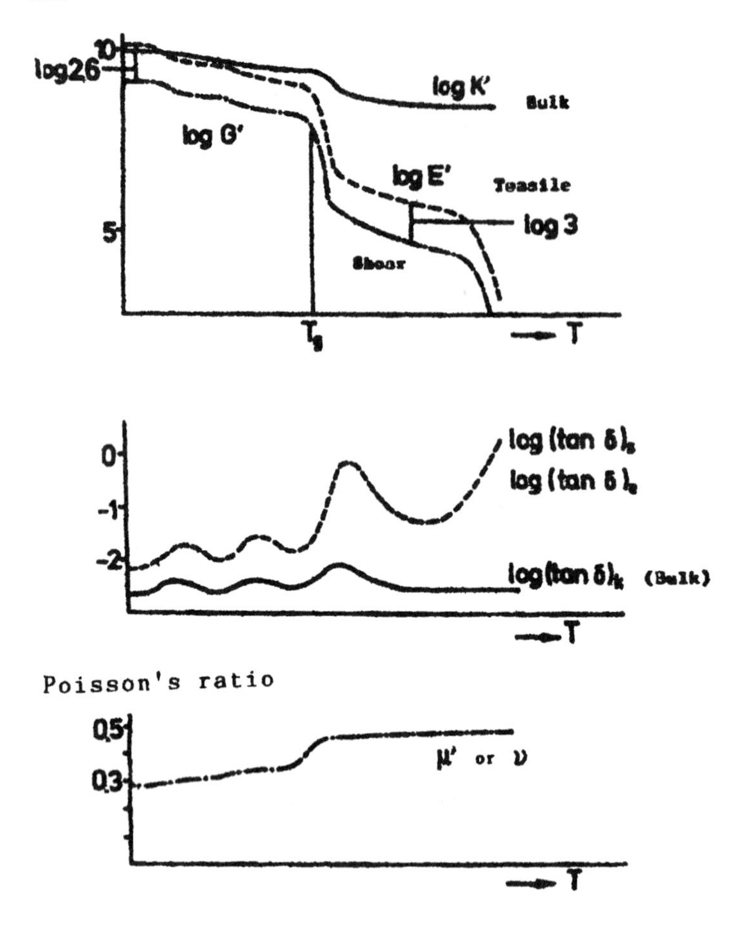

Figure 5 Shear modulus for an isotropic rubber.

is dangerous to assume these values for new materials, and particularly for structural components, since these are often designed to have different properties in the various directions of strain. Lempriere [30] studied Poisson's ratios for nonisotropic materials such as composites, showing that there are thermodynamic constraints limiting the moduli and that the product of all three ratios should be less than 0.5. They are constrained to be within the range $-1 < \nu < 0.5$. For example, Summerscales [31] discusses these ratios for composites, and some new structural foams have been developed with negative Poisson's ratios (i.e., higher shear moduli than Young's moduli in either the plastic or rubbery region or in both); see for example Evans [22].

Compliance* is the reciprocal of the modulus* and favored by some scientists, and the ceramicists have used Q, the mechanical quality factor or cot delta ($= 1/\tan\delta$).

For completeness, the equation of motion of the torsion pendulum in free oscillation is also given here:

$$M\frac{d^2x}{dt^2} + \frac{S''}{\omega}\cdot\frac{dx}{dt} + S'x = 0$$

can be solved to give the stiffnesses

$$S' = m\omega^2\left(1 + \frac{\Lambda}{4\pi^2}\right), \qquad S'' = \frac{m\omega^2}{\pi}, \qquad \text{and} \qquad \tan\delta = \frac{\Lambda}{\pi(1 + \Lambda/4\pi^2)}$$

(ω is the angular frequency and Λ is the log decrement).

4 Effects of Temperature and Frequency

A polymer may be scanned at constant temperature over a wide frequency range, but there are severe limitations with regards to one's patience at the low end (as well as the time-dependent properties of the polymer) and mechanical capabilities at the higher end, so that in practice few researchers scan below 0.001 Hz or above 1000 Hz, which is limited by machine resonances. This more precise interpretation [7] is hence often impractical, so that a wide temperature scan at a few multiplexed frequencies is easier to achieve in a few hours. By scanning at say 2°C/min and measuring at 0.3, 3, and 30 Hz, or at 0.5°C/min and 0.1, 1, 10, and 100 Hz, a clear picture of the various molecular relaxations and any crystalline melting can be seen. From the temperature shift of the loss tangent peaks with frequency, the activation energy can be calculated for that particular transition. Any crystalline parts melting will show peaks that are at the same temperature, just as a few percent free sulphur in a rubber compound will show up near 122°C, the melting point of sulphur, and indeed this or a metal with a low melting point such as indium (at 156°C) could be used as a temperature calibrant in a composite sample (Fig. 6).

The thermal scan, being faster and more convenient than step isotherming, is often the preferred way of evaluating new compounds. Since many industrial vibrations are in this frequency region, good experimental data may be obtained for the stress engineer as long as clamping errors and static and dynamic loads are considered. Many people prefer to use modulus rather than compliance, but notice that the peak tan delta is the same in Fig. 7 whichever measurement is used. As tan delta is also the ratio of kE''/kE', where k is the geometry constant, it is also noted that a misshapen or uneven sample will still give an accurate tan delta curve, whereas there can be large errors in measured modulus, particularly in bending, where k is proportional to the cube of the t/l ratio.

Tan delta responds in a systematic way to an increase in the amorphous phase volume fraction, as can be seen in the example of zinc chloride addition to poly(vinyl alcohol), which reduces the crystallinity [23] Fig. 8.

As frequency is increased, the temperature at the tan δ and E'' peaks shifts to higher temperatures, because more energy is being converted into heat at the higher frequencies, where the material is acting as a glass, because there is no time to respond to the applied stress; whereas at very low frequencies the molecular motion can follow the applied stress in a near-equilibrium, rubbery manner. The diffusional motion of the molecules thus coincides with the applied frequency and produces a peak, because being away from equilibrium the entropy production

$$\Delta S_{\mathrm{ir}} = \frac{\Delta Q_{\mathrm{ir}}}{T}$$

The temperature rises because of the heat

$$\Delta Q_{\mathrm{ir}} = C_{\mathrm{p}}\,\Delta T$$

It is clear from the above that (unlike in DSC) we must define which glass transition we are measuring—and commonly this is peak tan delta (1 Hz), or peak $\log E''$ (1 Hz). In the case

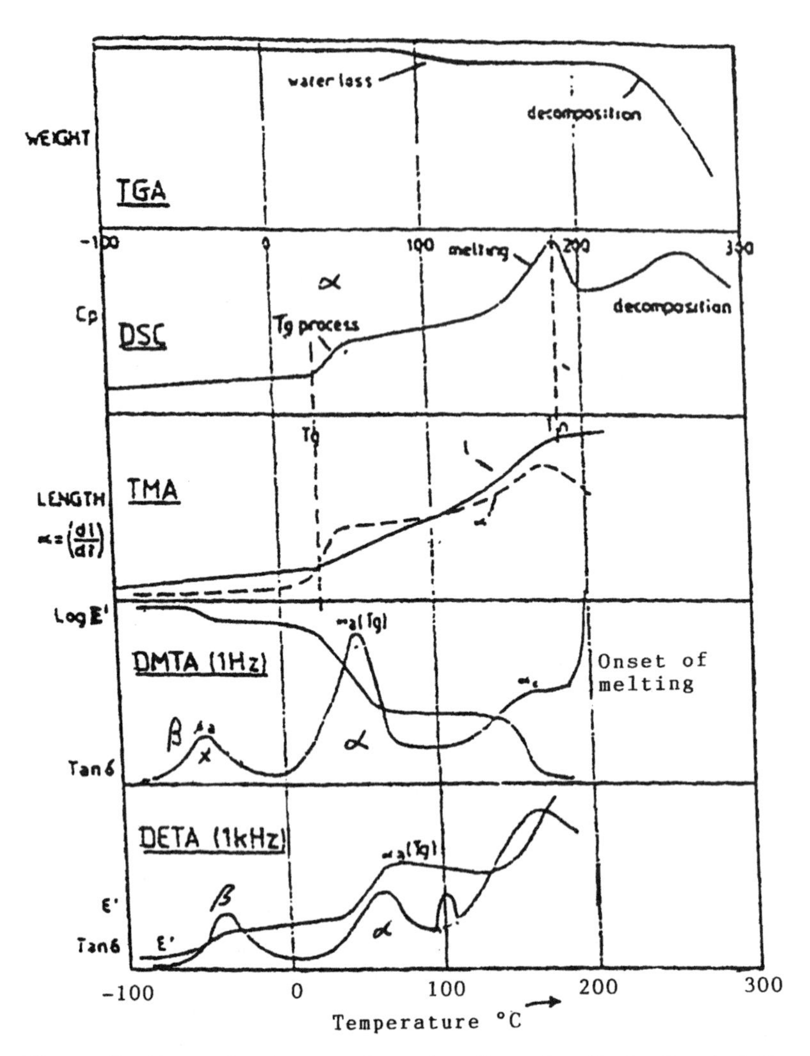

Figure 6 Comparison between the various thermal methods of analysis for a typical semicrystalline polymer. Notice that the beta and alpha peaks in the DMTA trace of tangent delta give much more information than the DSC. For most geometries the expansion given by TMA is also available.

of polypropylene the E'' peak is much clearer than the tan delta one (which is on a steeply rising baseline), and other groups have chosen the drop-off point of the E' curve, particularly where the software can calculate this more or less automatically. In this case, whether it is the log or linear E' curve (or E^* curve) must be defined as well as the frequency. Old DMAs used to plot lin E', whereas most DMTA users plot the log E', since it will show the three or four decade drop more clearly into the rubbery region. For brittle materials, or where there were problems going through T_g, the lowest T_g temperature is often recorded, which is normally the E' drop-off point. At 1 Hz this will often be within a few degrees of the DSC measured T_g (corrected to zero rate).

At large strains and frequencies above a few Hz, the heat buildup can be quite large, whereas at low frequencies and small strains it is negligible. Tire rubbers are highly filled with carbon black to help dissipate this heat energy as well as to increase the modulus.

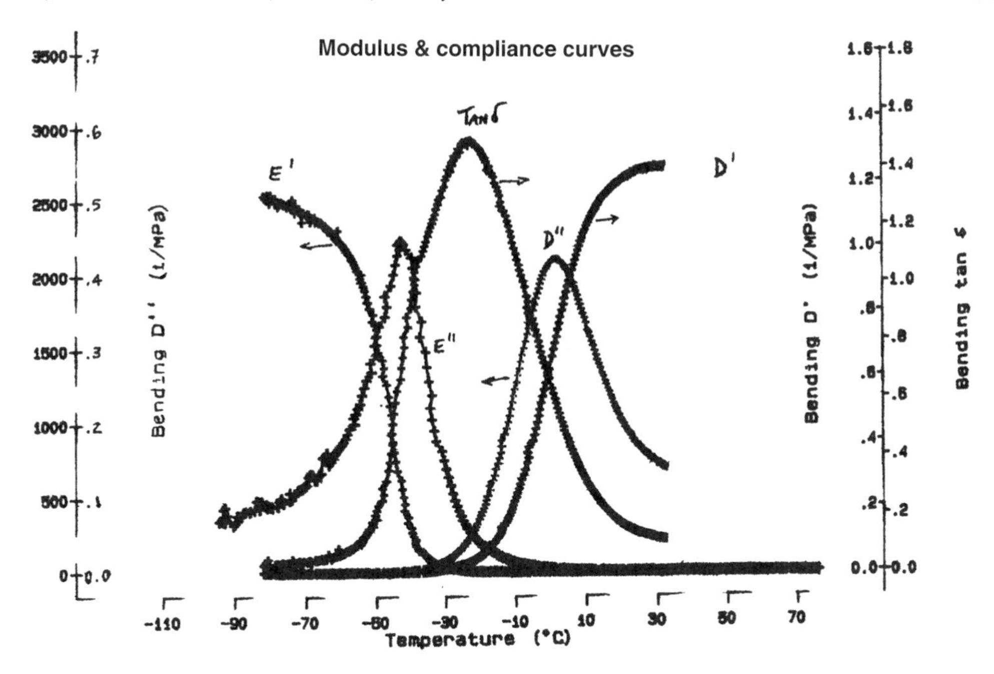

Figure 7 DMTA Head: standard 300°C AL Loughborough. Polyisobutylene; 1 Hz, bending mode strain (×4 operator: Gary Foster; DIM: 5.00, 11.160, 1.760; filename: polyisobutylene; Dec/07/1990).

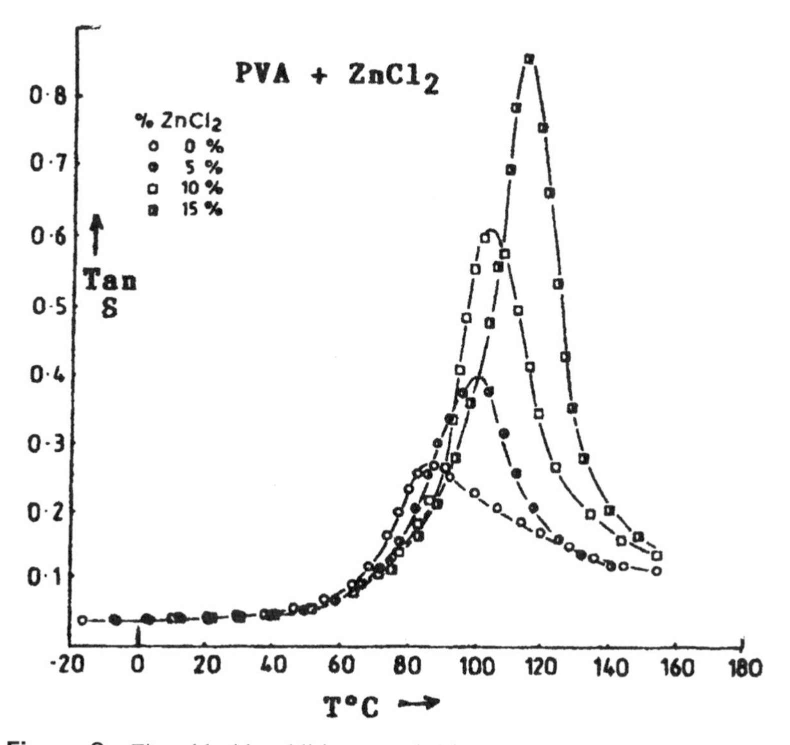

Figure 8 Zinc chloride addition to poly(vinyl alcohol) reduces the crystallinity.

Here is an example of a foamed polymer measured in shear where the activation energy (enthalpy) can be calculated from the Arrhenius equation

$$\Delta H^* = \frac{R \ln(f_2/f_1)}{(1/T_1 - 1/T_2)}$$

(Fig. 9). Notice that the peaks become broader at the higher frequencies, so if one is looking for a couple of very close transitions then running at a lower fequency will help separate them. In this case the heating rate has to be slow enough to allow a reading to be made every one or two degrees, and many DMAs will not allow you to run slower than 0.1 degrees per minute, which is still too fast for a 0.001 Hz run! Hence the step isotherm approach must be used for these very low frequencies, and reliable overnight running is the norm.

For a more complete analysis of this see for example Wetton [24]. The relaxation process in polymers can also be approached from the measurement of a wide range of frequencies at a given temperature, but in practice this is impossible, so approximations have been made where measurements over 3 or 4 decades only are used. Such second-order approximations by Ferry [7] and Schwarzl/Staverman [25] show that the relaxation spectrum can be represented from the storage modulus by

$$H(\tau) = AE'\left(1 - \frac{d \log E'}{d \log \omega - 1}\right)$$

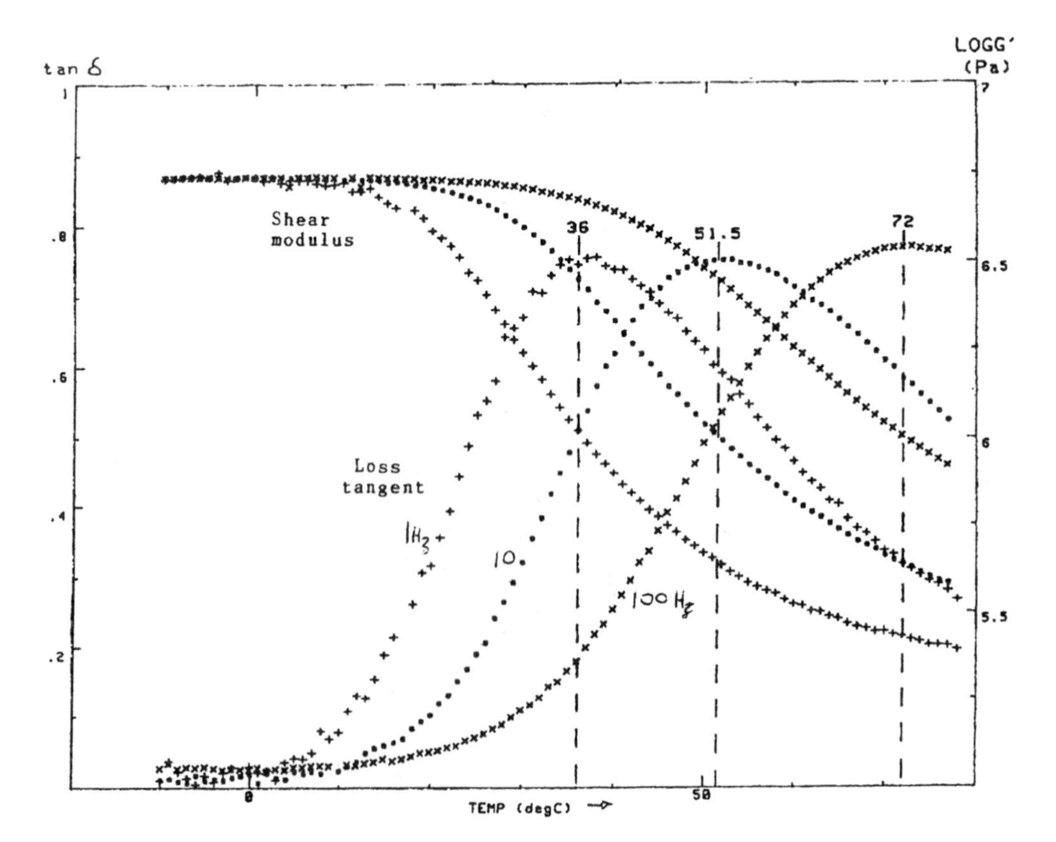

Figure 9 A typical foam that is good at absorbing low-frequency vibrations.

and

$$H = BE''\left(1 - \frac{d\log E''}{d\log \omega}\right) \text{ from the loss modulus,}$$

where A and B are functions of the slope of the $\log H(\tau)$ against $\log(\tau)$ curve.

Williams, Landel, and Ferry described an empirical equation to represent the shift of time and temperature of the viscoelastic parameters, so that a very wide frequency range can be estimated from a narrow range of measurements, limited by the resonance of the apparatus at the high end and by one's patience and sample degradation/change at the low frequency end [26].

Some of the most elegant recent work on this has been performed by Townend [27], (who shows that a careful temperature scan measurement at one frequency for the rubbers he was working with can predict properties at any other temperature or frequency); and Starkweather [32], who derived master curves from 10^{-4} to 10^{8} for PMMA, polystyrene, polyvinylacetate, and some ethylene/propylene copolymers; and Matsuoka [131] with his theory of intermolecular cooperative relaxation to predict the frequency dependence of DMTA data; and Oyadiji [136], who measures from 1 Hz to 1 kHz and then predicts up to 10^{11} Hz, showing some discrepancies above 10^{8}, which are still being investigated. Personally, I have often used WLF but would normally only believe two to three decades up and down from the four decades that I have measured! Whether using square wave excitation and fast fourier analysis one can measure at even higher frequencies remains to be seen. Some apparatus are already offering this option.

Hartman [84] suggests that a rule of thumb for polymers generally is 7°C per decade of frequency, but rather than accept another generalization (which is never always true, as can be seen in Fig. 9), I would recommend setting up a 0.1, 1, 10, 100 Hz run at 0.5°C/min for the typical samples that your lab will be running and measuring your own rules of thumb!

5 Applications

5.1 Nonpolymers

Before looking at polymers in some detail, it is perhaps worth mentioning a few other materials that have been successfully analysed by DMTA. For food researchers, to whom taste, touch, and texture are so important, DMTA has made great strides in understanding the fundamental food physics related to staling, cooking, crispness, and moisture content. Blanshard and Lillford's book on the glassy state in foods [123] contain 31 papers from the April 1992 Nottingham conference, where many world famous researchers were present from both the food and polymer industries. The three main schools of thought were polymers + water, water + ingredients, and water + ingredients + temperature. (Bon appetit!)

Since the early DM work on metals with unautomated torsion pendulum systems [21], few papers have been published, perhaps because of the limited temperature range available. In 1989 Duncan presented some DM traces on common metals to 800°C [124]. Glass has had a couple of papers recently. Of the major glass companies in Europe, only Schottglas use this technique to my knowledge, and of the few published papers in the open literature Hill [35] and Roeder et al. [36] are examples. Hill showed that liquid–liquid phase separation prior to crystal nucleation and real time studies of the kinetics of crystallization could be studied in glass ceramics, and activation energies could be matched to crystal growth. Transitions seen were due to ion hopping, rotation of nonbridging oxygen molecules, and segmental backbone motion. Glass ionomer dental cements were measured at the LGC, London, in the 1980s by Nicholson [144]. Other work above 600°C has been done in the Canadian defence research labs at Medicine Hat using the glass parts and clamps of the DMA7. For piezoelectric materials, mechanical and electrical properties are particularly important, but most of this work that I have come across was in the kHz region (e.g., Ref. 130)

Bone has been looked at by Deckmann and Gabriel [125], and this will be an increased area of interest as biocompatible polymers for coating implants are researched [127]. The effects of moisture and creams on the epidermis and hair have been studied by many cosmetic companies, and it is quite dramatic to watch the modulus drop a decade as a drop of liquid is applied to the sample in the DMTA at room temperature and constant humidity! Paper has had several groups publishing, including Munoz et al. on the quality of handsheet papers from eucalyptus pulp [126], and for some years the research departments at Amcor, Kimberley Clark, English China Clays, and Sanyo Kokusaku Pulp have used DMTA for mechanical testing of papers. Tobacco leaf parenchyma shows a T_g at about 8°C (peak log E'' @ 1 Hz), and the mechanical properties of the shredded leaf are important to prevent the tobacco from falling out of the cigarette in the high-speed production process [132].

Explosives mixtures were analyzed by Baker [128] and Hoffman et al. [129], and DMTA has been a QC tool at several propellants and explosives factories for some time, usually with at least one secondary temperature measurement device (since some of the early temperature programmers had a habit of running away) to switch off the heating power, if the temperature of these quite large explosive specimens exceeded the safe point!

5.2 Polymers

Polymers are normally classified into four main architectural types: linear (which includes rigid rod, flexible coil, cyclic, and polyrotaxane structures); branched (including random, regular comb-like, and star shaped); cross-linked (which includes the interpenetrating networks (IPNs)); and fairly recently the dendritic or hyperbranched polymers. I shall cover in some detail the first three types, but as we went to press very little DM work has been performed yet on the hyperbranched ones, which show some interesting properties. (Compared to linear polymers, solutions show a much lower viscosity and appear to be Newtonian rather than shear thinning [134].) Johansson [135] compares DM properties of some hyperbranched acrylates, alkyds, and unsaturated polyesters and notes that the properties of his cured resins so far are rather similar to conventional polyester systems.

Amorphous Polymers

When looking at amorphous homopolymers, DMA has about 1000 times the sensitivity of differential scanning calorimetry, which can show very little below the glass transition. Even the new modulated DSCs are still about 100 times less sensitive, but they are able to distinguish certain overlapping phenomena. Whether modulated DMA will ever be of interest we shall have to wait and see. (The temperature ramp is modulated at perhaps 0.1 Hz and amplitude of a few degrees on top of the 5°C per minute ramp.) (Fig. 10.)

From absolute zero four transitions are normally described. The δ transition is possibly due to local mode motion of chain molecules because of cooperative torsional motion

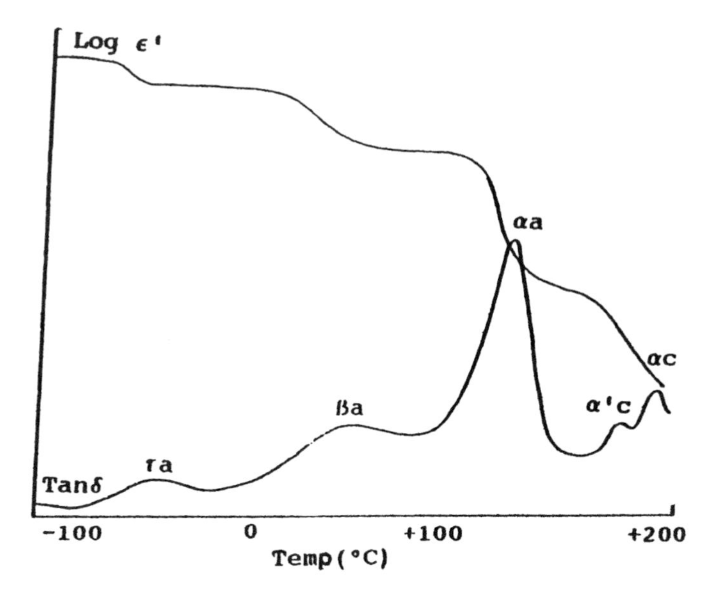

Figure 10 Typical relaxation processes in a semicrystalline polymer. α_a is the T_g relaxation process. The a and c subscripts refer to amorphous and crystalline phases.

about bonds but staying in the same potential energy well [28]. The γ transition could be cooperative angular motion to give local configurational changes such as single bond flips. The β transition is related to side chain motions in the molecules and so can be seen to be much larger in a poly(cyclohexyl methacrylate) than in poly(methyl methacrylate), where the methyl group is so much smaller. This can be of particular importance when looking at curing systems such as epoxy paints, where the size of this transition (between −80 and −50°C) can give a good indication of the degree of cure *without* heating the sample above room temperature, and hence curing further the sample that one is trying to measure! In the case of polyethylene only, Sha et al. [68] and Khanna et al. [69] have argued that the beta transition is the glass transition.

The α transition in most amorphous polymers is the glass transition T_g, where the modulus can change by a factor of one thousand or more as the glassy plastic softens into the leathery phase before becoming a rubber. This large change means that it is usually more practical to plot the log (rather than linear) modulus. The sensitivity of the DMTA technique means that trace materials can be identified, such as the 3% poly(butyl acrylate) (of MW approx 100,000) used to coat $CaCO_3$ particles in the sample of HIPS (high impact poly(styrene)) shown in Fig. 11, whereas at lower molecular weights pba was much more soluble in the ps phase and did not show as a discrete peak at all.

For an extensive review of crystalline and part-crystalline polymers, I recommend Boyd's [33] 1985 paper, which looks at low, medium, and high-crystallinity polymers and has a further 112 references. The problem here is that the heat history of the sample can dramatically affect the data, and it is important to note what sample preparation has been made before making such a test, and after the first thermal scan, how the sample is cooled to perform a second. In a part-amorphous, part-crystalline polymer, as well as the transitions for the amorphous phase described above, the various crystalline parts will show melting points, and these are often referred to as $(\text{alpha})_c$, and the T_g would be $(\text{alpha})_a$. Note that as $T_{melt}s$ are independant of frequency of vibration, any $(\text{alpha})_c$ peaks

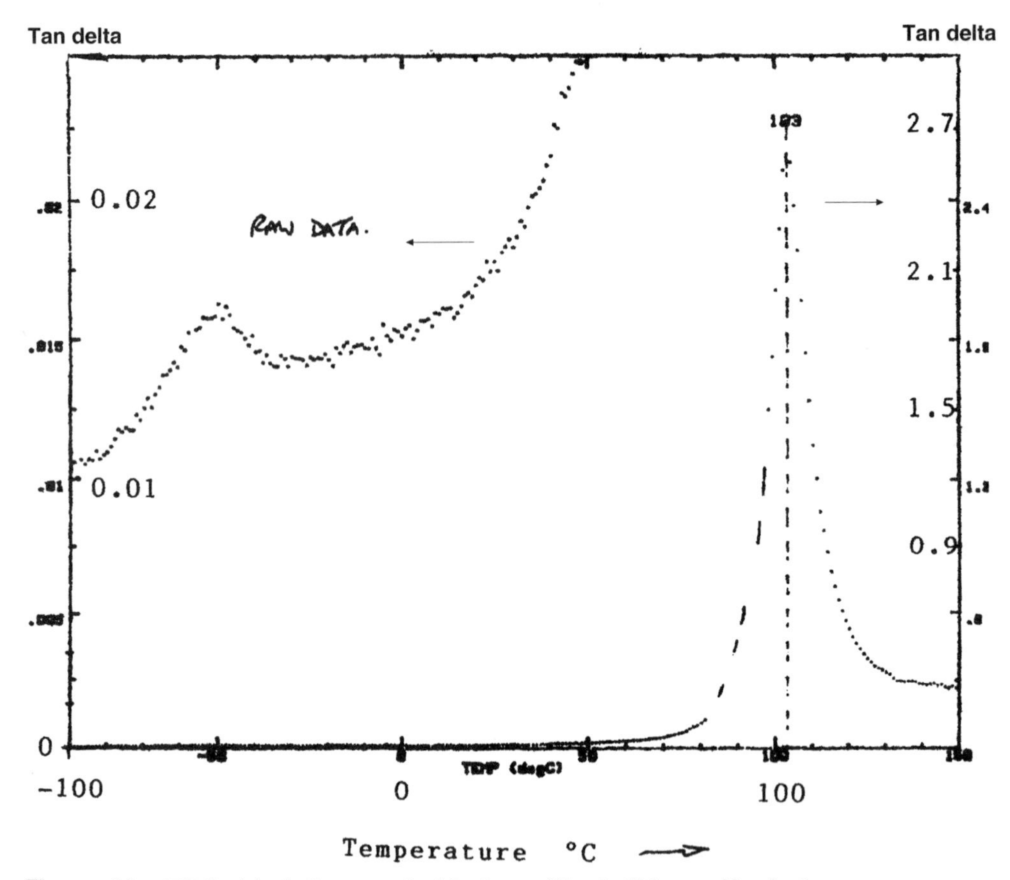

Figure 11 HIPS with $CaCo_3$ coated with pba as filler (×120 magnification).

will occur at the same temperature, but the peaks due to the amorphous region will be shifted with frequency. Khanna [67] estimated the crystallinity by DMTA of a whole range of semicrystalline polymers and compared them to values obtained by x-ray diffraction and the less accurate DSC.

As well as heat- and time-induced crystallization, one can also come across materials that show a large stress-induced crystallization, such as the copolymers reported by Okoroafor and Rault [34] based on poly(THF/amides). As here, the structure of many polymers will be affected by the static stress applied, a relatively trivial case being the humble rubber band, which can have a 100-fold variation in modulus according to its static stress! (Fig. 12.)

Copolymers

Many copolymers have been studied using the technique, and the degree of compatibility and number of phases present can be clearly seen in four rather simplified generic cases (Fig. 13).

For many years polymer blends have been seen as means to combine the best properties of the two homopolymers, or as means of improving the fabrication method and at a lower cost by incorporating fillers. Miscible blends can be seen to be one continuous phase with resultant properties between those of its two components. Immiscible blends can

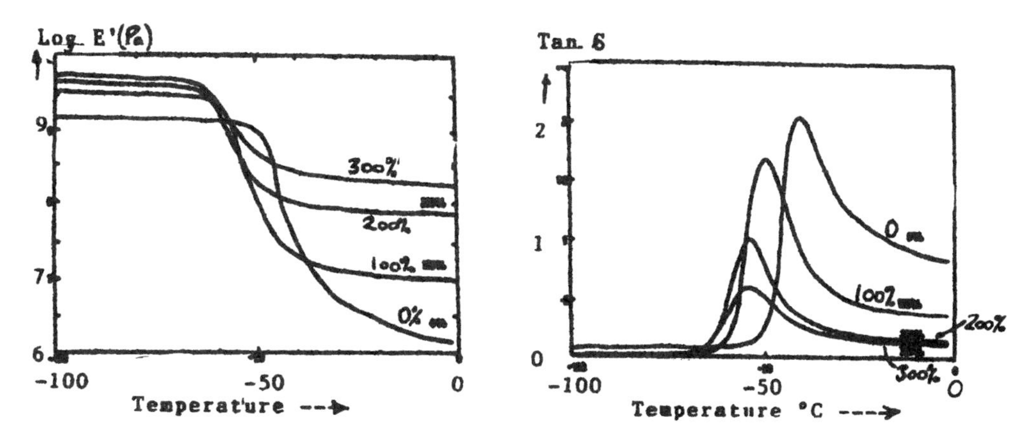

Figure 12 Natural rubber at 300%, 200%, 100%, and 0% extension, showing that it is three times stiffer in the glassy state and 100 times stiffer at 0°C in the rubbery state.

sometimes give good mechanical performance because of good interfacial adhesion between the phases. Examples are PVC/PMMA blends, which combine the good fire resistance of PVC with the higher T_g of PMMA, and have a higher notched impact strength than either polymer alone. Hence these have been used for aircraft interiors [48].

For further reading on polymer blends, I recommend a selection of papers, including ppo/sbs/ps (Chiu and Hwung) [49], pmma/pc (Eastmond et al.) [50–53], peek/p(etheri-

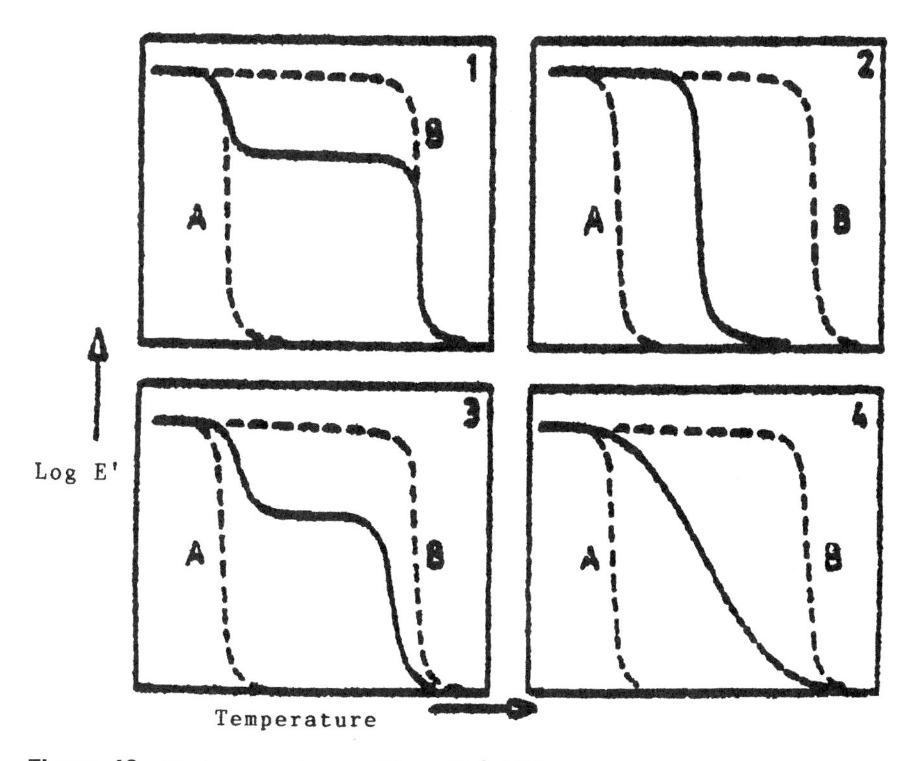

Figure 13 Various copolymers. (From Ref. 146.)

mide) (Crevecoeur and Groeninckx) [54], lignin/pva or /pmma (Ciemniecki and Glasser) [58, 59], basic features of the glassy state (Benavente et al.) [60], p(amide)/pp (Utracki and Sammut) [61], p(caprolactone)/bisphenol A (Defieuw et al.) [62], lignin/pu (Kelley et al.) [63], pvdf/pv(pyrrolidone) (Pizzoli et al.) [64], p(ester)/pva (Bucknall et al.) [65], aromatic p(esters) (Diaz-Calleja et al.) [66], pvdf/ps/pva (del Rio and Acosta) [133], ps/pvp with strengthened interfaces (Edgecombe et al.) [120] and p(lactic acid)/pcl (Lostocco and Huang) [115], (Iannace et al.) [116]; and for elastomers, e.g., poly(HFPO) (Ding et al.) [70], poly(hydroxy ether siloxanes) (Haidar et al.) [71], SBR and PBDs (Wetton et al.) [72], EPDM/rubber blends (Radusch et al.) [73], filled rubbers/wet skid behavior (Gabler et al.) [76].

Highly filled rubbers (as used for tires) have been studied for many years in the U.S.A. and in Germany, with Continental Gummi using the technique from 1976 and Goodyear (Luxemburg) shortly after. Many papers have been submitted at the ACS Rubber Division meetings (in the U.S.A.), and at the Tyre Tech meetings in Europe and Singapore. In Europe, RAPRA and MRPRA (both in the U.K. [78, 79]) have studied DM properties on very large samples (using up to 20 kN and 5 meganewton respectively) and also on very small chemists' samples using millinewtons of force. Much of their work has been in confidence for clients, of course, but papers are regularly submitted at international conferences by their engineers.

Problems of heat buildup in large specimens and scragging are beyond the scope of this chapter, but one of the reasons for the 50% carbon black in tires is to dissipate the large buildup of heat generated when any rubber sample is vibrated at frequencies above a few hertz. Some DMTS apparatus allow up to three thermometers to be placed in or on the sample to measure this, whilst other workers prefer to run down in temperature to prevent overworking the rubber samples, particularly at constant strain.

Rubber samples have been molded onto metal clamp parts directly (to avoid use of adhesives) where they are to be measured in single or double shear. For compression measurements a 10 mm diameter cylinder × 10 mm high is commonly used, and for tension the typical sample dimensions will be 30 × 2 × sample thickness, which after clamping gives a free length of about 20 mm. Samples can be prepared with metal ends to aid clamping and particularly where many samples are to be run in the autosampler overnight. Test routines have been covered by DIN 53 513, ISO 6721–5, 2856, and 4664, and ASTM D-2331, and further rubber DM reading could include works by R. Wetton [137], Stopp and Heine [138], Lake and Thomas [139] and Wisch and Meinecke [140].

Table 1 shows an example of engineering data for a filled rubber.

Coatings

Many curing studies have been made both on TBAs and on DMTA for thin coatings on rigid substrates, and the beauty of this technique compared to a viscometer is that one can monitor the full experimental cure from a low-viscosity liquid, through the gelling and right up to the glassy solid. Since many skin effects can affect the bulk properties of the coating, it is interesting to look at different thicknesses academically, or for a real application make sure that the studied thickness is similar in size to the end-use coating. During the 1980s there was an unofficial competition as to who could see the T_g of the thinnest coating, and this was probably won by Wetton et al., who were able to see a 0.3 micron rubber coating [37]. This was done in dual cantilever on a thin metal strip, and the glass transition temperature varied with the thickness,various suggestions being put forward for this, including reduced entropy because of surface constraints (Fig. 14a).

Table 1 Typical Data from the Gabo Explexor DMTS

Moduli and tan δ

Nr:	E' (MPa)	E'' (MPa)	tan δ	$\|E^*\|$ (MPa)	Frequ. (Hz)	Temp. (°C)	Time (sec)
1	36.362	5.008	0.1377	36.705	10.000	99.6	27
2	34.174	5.392	0.1578	34.597	10.000	99.8	55
3	30.870	5.498	0.1781	31.356	10.000	100.1	85
4	27.607	5.370	0.1945	28.125	10.000	100.0	118
5	24.927	5.170	0.2074	25.458	10.000	100.0	146
6	22.246	4.741	0.2131	22.745	10.000	99.9	174
7	19.857	4.238	0.2134[a]	20.304	10.000	99.9	203
8	17.542	3.650	0.2080	17.918	10.000	100.0	244
9	15.853	3.145	0.1984	16.162	10.000	100.1	277
10	15.721	3.121	0.1986	16.028	10.000	99.9	311

Load values

Nr:	Dyn. force (N)	Dyn. stress (MPa)	Dyn. strain (%)	Preforce (N)	Prestress (MPa)	Prestr. (%)	L0 (mm)	Lm (mm)
1	1.49	0.045	0.099	55.52	1.673	15.00	6.41	5.45
2	2.24	0.068	0.158	58.03	1.749	15.05	6.39	5.43
3	3.26	0.098	0.252	57.61	1.736	15.04	6.39	5.43
4	4.63	0.139	0.397	56.48	1.702	15.01	6.38	5.43
5	6.72	0.203	0.633	57.69	1.739	15.05	6.38	5.42
6	9.59	0.289	1.003	57.66	1.738	15.04	6.38	5.42
7	13.65	0.411	1.586	57.46	1.732	15.04	6.38	5.42
8	19.22	0.579	2.512	55.80	1.682	15.02	6.38	5.42
9	27.73	0.836	3.983	55.43	1.670	15.00	6.38	5.42
10	27.63*	0.833*	4.002	55.01	1.658	15.02	6.38	5.42

[a]Maximum tan delta at between 1 and $1\frac{1}{2}$.

*Example of a highly filled rubber—strain sweep at a constant static compression of 15%. Notice that the strain sweep stopped at 4%, as the maximum force of 25 N (+10%) was met.

As many cure experiments take minutes or even hours, it can be instructive to monitor at several multiplexed frequencies. Note here that the apparent cure at 100 Hz happens long before the 0.01 Hz (Fig. 14b). It should also be run once over a several day isotherm, in case there are any "ambient drift" problems such as have been reported with a sinusoidal baseline of a 24 hour periodicity, suggesting that the warmer/colder day/night ambient lab conditions were causing some recording errors in the instrumental electronics. For very fast curing systems using ultraviolet light, for example, a high frequency must be used, since the cure may have happened in less than one second! 100 Hz on a modern DMTA could give a good indication (but older ones were limited to a 0.25 Hz sampling acquisition rate so were no good for these faster cures), although dielectric, infrared, and fluorescence polarization have been successfully used too [46].

To study much thinner coatings will probably require the vibrating AFM (atomic force microscope) or nano indenting technique, which use similar principles, but where the theory states that as long as the surface coating is softer than the substrate, and one only indents 10% into the coating, then the loading and unloading curves give one the modulus of the coating in situ. From the 300 nm on a DMA this new technique has certainly taken the thinnest coating down to the tens of nanometers region or at least a factor of 10

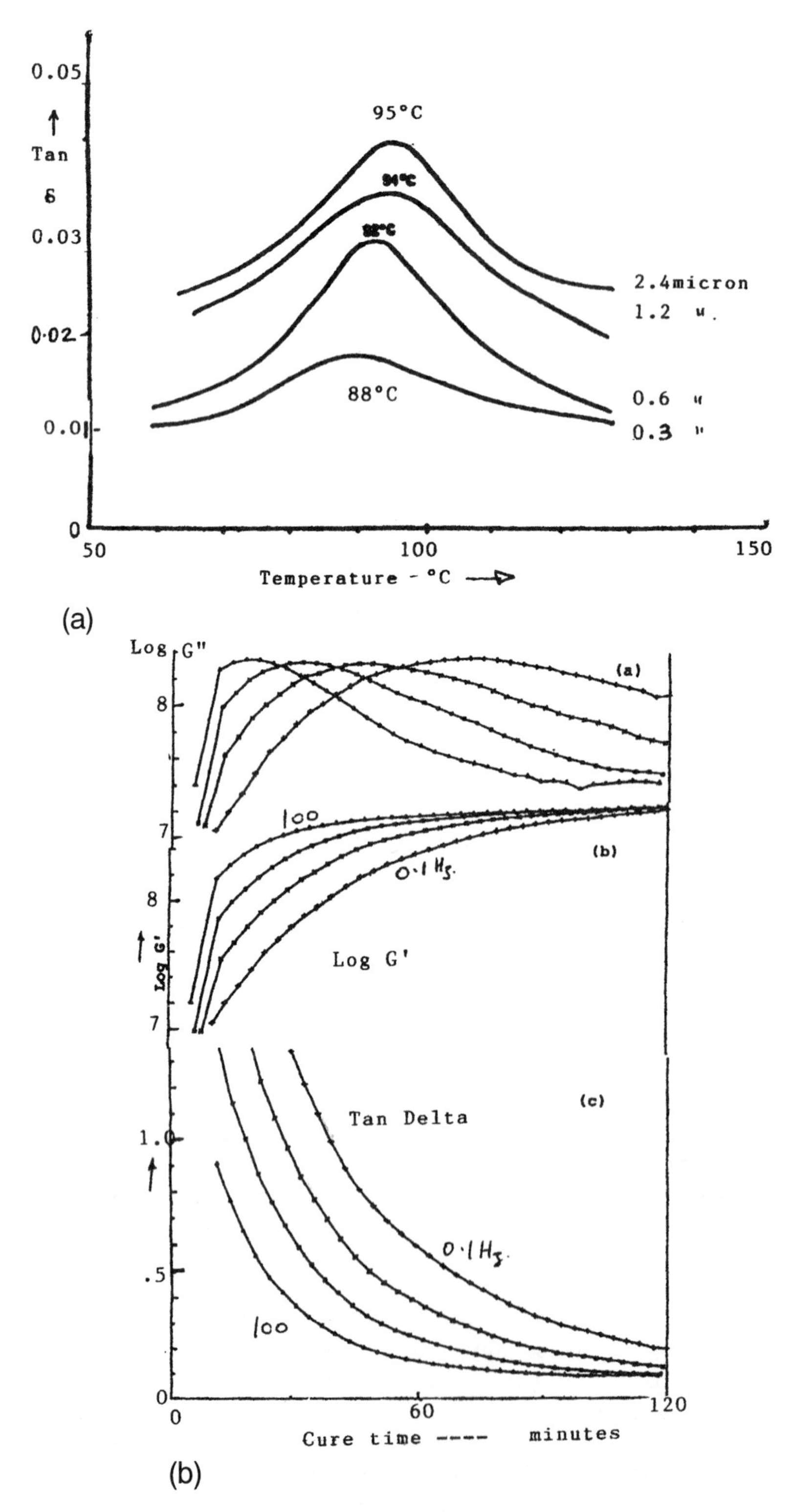

Figure 14 (a) Very thin PVC coatings on 0.5 mm steel. (b) Epoxy cure monitored at 0.1, 1, 10, and 100 Hz.

improvement. The modulus of the coating(s) in situ as well as their hardness can be accurately measured, whereas all that the DMA could give is the overall modulus of the coating on the substrate. Groups active in this area include Pollock [38], Smith [39], Reading [145] and Briscoe [88] in the U.K., Brotzen [89], Colton [90], Oliver and Pharr [91], and Doerner and Nix [92] in the U.S.A., and Baker [93] and Burnham [90] in Germany. Kajiama et al. [113] recently looked at ps and ps/pvme polymers of 200 nm and 25 nm thickness and compared the results with bulk material ones, showing good agreement at high molecular weights, but anomalies at lower ones. Already some groups of researchers are making DSC and DMA measurements on *molecules* on the surface of various materials using these very fine-pointed probes. From the meganewtons of civil engineering antiearthquake rubbers to the nanonewtons of an AFM is quite a range for this versatile technique!

Hill has recently reviewed the DM method for coatings [85, 86] and suggests that running the DM thermal scan into the rubbery region for an unpigmented paint film until the E' curve increases again gives one E'_{min}, which is directly related to cross-link density, from 4×10^6 Pa for a lightly cross-linked one to 2×10^8 Pa for a very highly cross-linked one. He defines the moles of elastically effective network chains per cubic centimeter of film as

$$\nu = \frac{E'_{min}}{3RT}$$

where R is the gas constant and T the temperature in degrees Kelvin, assuming that the G' is one third of the E'.

Perera (at CORI in Belgium) has done a lot of work on aging phenomena in coatings, although most of it, being for clients, is unpublished. In a recent paper [109] he suggests that physical aging is associated with conformational arrangements, increased molecular packing, and densification and cites a further 24 references. Towards the goal of totally aqueous paints, Heuts et al. are trying to reduce the use of cosolvents to lower the MFT (minimum film formation temperature) by a two-stage emulsion polymerization, using DMTA to monitor the mechanical properties and presence of inhomogeneous polymer distributions [110]. Richard has investigated the role of core chains and particle/particle interfaces [111].

Optical fiber coatings (which are usually multilayer) can be studied using a simple clamp modification so that the circular sample is gripped at three points (rather than two, which would cause slippage), by inserting a small V notch into one side of the clamp frame. Groups in Australia and the U.K. used this very successfully in the 1980s. Of secondary interest could be the observation of the ion peaks in the glass itself, and these vary in temperature according to the ion present (see start of this application section and Fig. 15b).

If asked to look at the degree of cure of a coating, I would recommend starting at -100°C, since the size of the beta transition will be directly related to the freedom of the side chains, and hence smaller for a more complete cure. For epoxy systems this proves much more effective than heating up through T_g (to try to use this temperature as the indicator), since the heating scan causes more cure to be given to the coating, and hence it is then difficult to estimate what the original cure was. Practically, if you are heating from -100°C to room temperature a few times, it is usually worth taking every fifth run to over (+)100C, so that the liquid nitrogen tubing has a chance to dry out, to avoid problems with blocked pipes from the slightly wet liquid nitrogen that is often used.

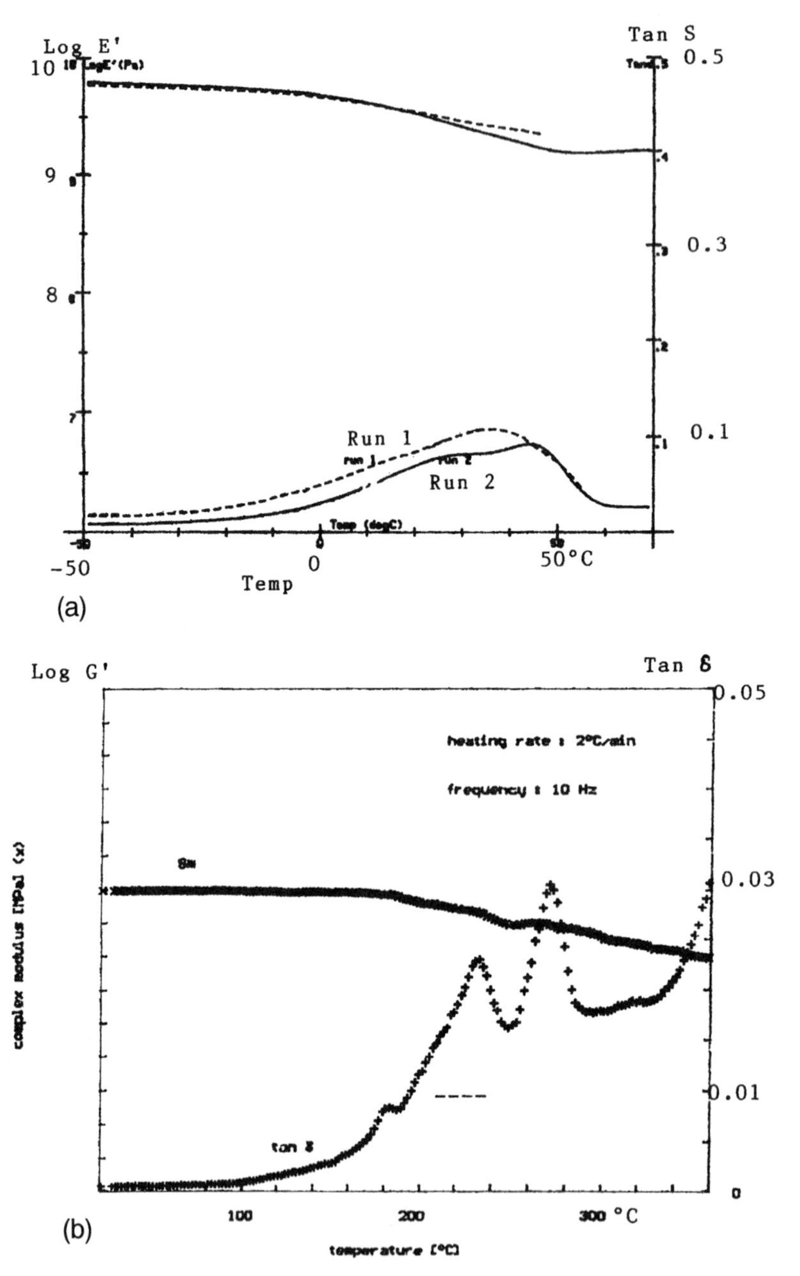

Figure 15 (a) Optical fiber coating run in dual cantilever bending with notched clamp. Total thickness 150 microns. Cure of the coating can be seen between run 1 and run 2. (b) Peaks due to ion-hopping in glass.

Adhesives

Adhesive cure is another example that can be monitored successfully by DMTA, and many papers have been published here by Eldridge and Ferry in 1954 [40] and other groups in the USA such as Winter, Foley, Schroeder, Dale, and Prime [41–45]. These looked respectively at the viscoelasticity at the gel points of polymers including PDMS, elastomer latexes, Gelva^{TM} pressure-sensitive adhesives (PSA), and effect of pressure on cure of acrylate/methacrylate copolymers. McArdle et al. looked at the surface initiated redox polymerization in adhesives at room temperature after modification with Cu salts [81]. IAZs (interfacial accomodation zones) [82] that could decouple the adhesive from the substrate have been studied by Dickie and Morman [83] for epoxies and silicones on zinc, aluminum, and stainless steel substrates. Curves such as Fig. 16, showing the performance according to frequency or temperature, have become industry standards for PSAs.

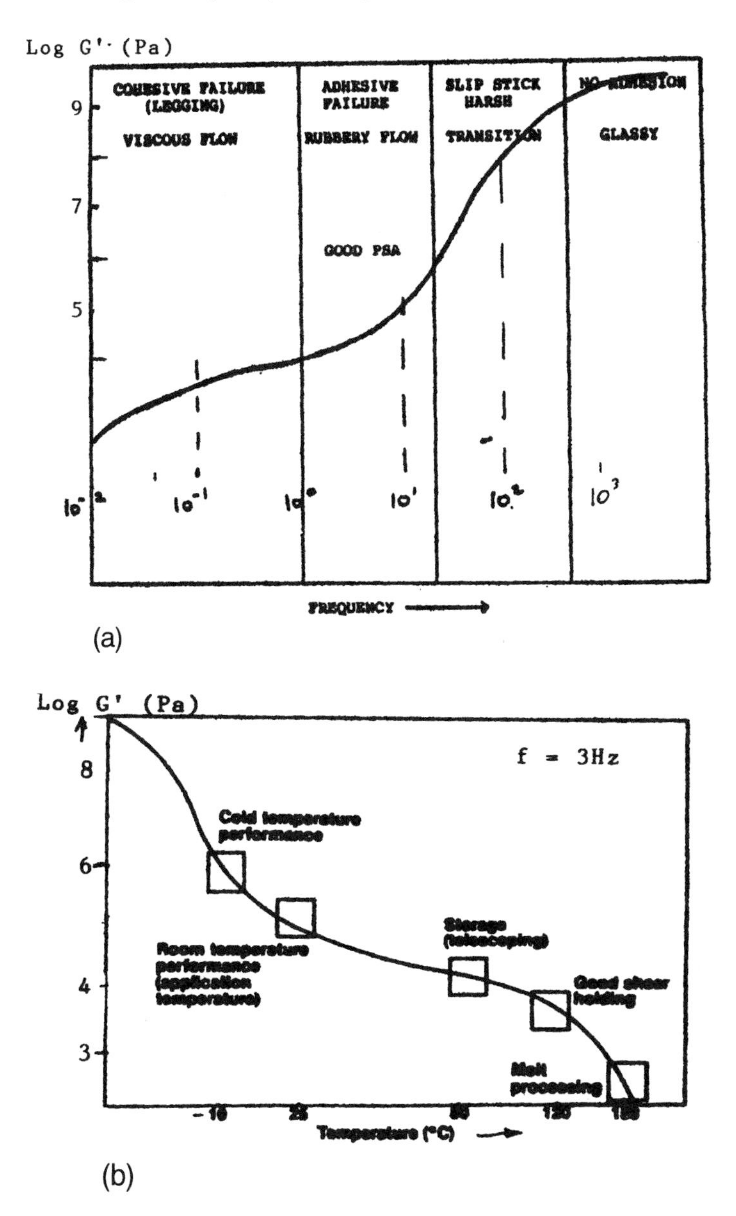

Figure 16 (a) A good pressure-sensitive adhesive with G' of 10^4 to 10^6 at room temperature. A mirror image of this curve at 1 Hz vs. T would show the operating temperature for best use. (b) Temperature performance at low frequency.

Bitumens and sealants have been extensively studied too, although many of the bitumen testing apparatus have to be much larger, since the roadstone filler can be 3/4 inch (19 mm) aggregate! Huet (France) [74], Francken (Belgium) [77], and the British TRRL published early work here, and some more recent work is by Leahy et al. [75] and by Kopsch [87]. Review papers on sealants by Lowe [47] and Malik [114] are useful, and one of the recent catastrophic spacecraft failures could possibly have been averted if the science of heating and cooling the structural sealants had been more closely studied before the event.

Thermosets and Composites

The Du Pont DMA and PL DMTA were both used extensively here for looking at prepregs, and the finished 4 to 32 ply composite material, but care is needed to overcome certain idiosyncracies of each apparatus. In particular the measured modulus in a dual or single cantilever bend mode is always well below the static tensile test one because of end errors in the clamps. It is estimated that a free (l/t) ratio of greater than 200 is needed to measure the E' accurately in this mode. Since the longest length that the apparatus can handle is often only 20 mm, real composite specimens are too thick to measure thus. Larger and later systems can overcome this by using three-point bend to give a truer value of modulus. For single ply systems expect to see a different T_g according to the orientation [112], although almost all practical composites will be multilayer and usually with plies at different angles, hence a single glass transition can be seen and quoted.

Beware of heating rate effects, which were due to larger metal clamps giving an apparent second T_g peak a few degrees later, as well as the expected thermal lag showing a higher temperature. One way to search for such anomalies is to do a series of experiments on similar samples at a range of heating rates from 0.1 to 20°C/min (which is a very good discipline for the new trainee anyway!). In my experience, for a polymer that is normally not heat-conducting, the error in T_g thus measured is about half the increase in temperature per minute. The side benefit of this experience on your normal materials is that when an urgent run needs to be made at the last minute, you have a lot more confidence in the temperature correction needed to allow for the faster than normal heating rate. For heat-conducting rubbers such as those used in tires, the correction is smaller.

As well as the study of curing mechanisms, the relative performance of the fibers, and any coatings on the fibers to improve the bonding of the fibers in the matrix, have been well studied. From the 1970s WLF studies on cure of epoxies [80], there have been many new developments both for fibers and resins and coatings to improve the interface region. In 1986 Robeson summarized the fundamental aspects of polymer/polymer miscibility and reported on the large amount of work in the previous ten years [102], and Don and Bell [99] have studied epoxy/pc blends following the conflicting data of Bucknall et al. [100] and Di Liello et al. [101] to try to improve the ductility and toughenability.

Carfagna et al. [103] and Giamberini et al. [104] have reviewed the liquid crystalline epoxies and reacted rigid rod epoxy terminated monomers with curing agents such as aromatic amines for higher T_gs and aliphatic acids for lower ones. In the U.S.A., Langlois et al. [105] describe their results using a bis-propargyl monomer to produce a T_g of 260°C (tan delta peak at 1 Hz). Recent papers include producing an improved modulus into the rubbery region using biphenol epoxies cured by catechol novolac rather than phenol novolac [96]; improved toughness using PEI-modified epoxies [97] or acrylate-terminated polyurethanes [98]; zero shrinkage on cure of a phenolic resin [121]; interfacial

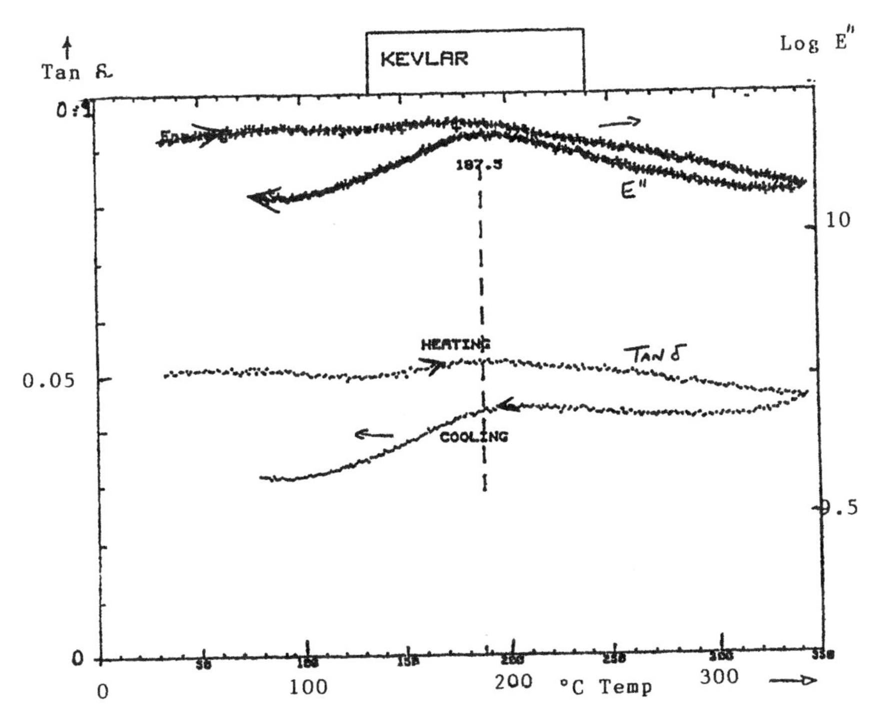

Figure 17 This shows the cure of a thin coating on a Kevlar fiber whilst being heated to 350°C. On cooling, one can see the extra stiffness from the cured coating (epoxy) and the T_g of the coating in the 187–190°C region. A small bundle of fibers used as the monofilament of ~ 8 μm diameter was *too* thin. Diameter of the bundle was approx. 50 μm.

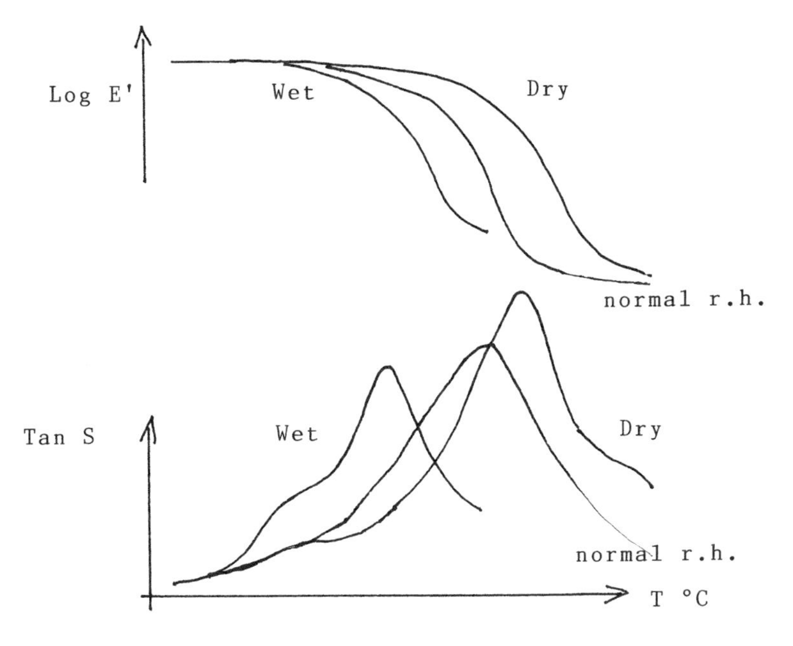

Figure 18 Three runs on three differently conditioned samples of an epoxy at 6°C/minute and 1 Hz.

aspects by using core-shell rubber particles with different surface chemistry [119]; and nanocomposites using silicate layers in poly(ε caprolactone) [122].

Aging and moisture effects are important, witness the apocryphal stories about early parts of aircraft made from epoxy composites that had crossed the equator and on returning to the Northern hemisphere aerodrome could have the moisture poured out of them! Aging has already been mentioned under coatings (Perera [109]) and others. Recent reviews of aging of polymers include Simon and Plazek [117] for a polyetherimide, and Hay [141] on polymer blends. The moisture absorbance is critical for several polymers including nylons and epoxies as it is for foods, and some DM apparatus offer relative humidity control as well as the thermal oven. If this is not available, then crude dry, wet, and normal humidities can be compared very simply in a standard DMTA using dried Si gel, and a little boat of kitchen foil filled with water respectively, as can be seen in Fig. 18.

References

1. Roelig, H., *Rubber Chem. Tech.*, *18*, 62 (1945).
2. Alexandrov, A., and Lazurkin, Y., *Acta Phys.-chim. URSS*, *12*, 647 (1940).
3. Schmieder, K., and Wolf, K., *Kolloid Zeit*, *134*, 149 (1953).
4. Nolle, A., *J. Polymer Science*, *5*, 1 (1950).
5. Nielsen, L., *Mechanical Properties of Polymers*, Reinhold, New York, 1962.
6. Tabolsky, A., *Properties and Structures of Polymers*, John Wiley, New York, 1960.
7. Ferry, J., *Viscoelastic Properties of Polymers*, John Wiley, New York, 1961.
8. McCrum, N. G., Read, B., and Williams, G., *Anelastic and Dielectric Effects in Polymeric Solids*, John Wiley, London, 1967, republished (unabridge) by Dover in 1991.
9. Takayanagi, M., Viscoelastic behaviour of crystalline polymers, Prc. 4th Intl Congress Rheology 1, 161–187 (1965).
10. Murayama, T., Dumbleton, J., and Williams, M., Viscoelastic properties of oriented nylon 6,6 fibers, part 111, Stress relaxation and DM properties, *J. Macromol. Sci.*, *B1*, 1–14 (1967).
11. Dumbleton, J., Murayama, T., Bell, J., On the DM behaviour of drawn PET fibers, *Kolloid Zu. Z. fur Polymere*, *228*, 54–58 (1968).
12. Kajiyama, T., and MacKnight, W., Relaxations in polyurethanes in the Tg region, ACS Pol. Preprints, *10*, 1, 65–71 (1968).
13. Illinger, J., Schneider, N., and Karasz, F., Low temperature DM properties of pu-polyether block co-polymers, *Pol. Eng. Sci.*, *12*, 1, 25–29 (1972).
14. Nguyen, A., Vu, B., and Wilkes, G., Melting and DM props of reconstituted collagen, ACS Pol. Preprints, Vol. 13, 2, 705–709 (1972).
15. Murayama, T., Dumbleton, J., and Williams, M., Viscosity of oriented PET, *J. Pol. Sci.*, *A2*, 6, 787–793 (1968).
16. Murayama, T., and Bell, J., Relation between network structure and DM properties of a typical amine cured epoxy polymer; *J. Pol. Sci.*, *A2*, Vol. 8, 437–445 (1970).
17. Gibson, A. G., Davies, G., and Ward, I., DM behaviour and longitudinal thickness measurements on ultra-high modulus linear polyethylene: A quantitative model for the elastic modulus, *Polymer*, *19*, 683–693 (1978).
18. Wetton, R., and Allen, G., *Polymer*, *7*, 331 (1966).
19. Gillham, J., and Enns, J., On the cure and properties of thermosetting polymers using TBA, *Trends in Pol. Sci.*, *2*, 12, 406 (1994).
20. Heijboer, J., *Brit. Polymer J.*, *1*, 3 (1969).
21. Zener, C., Elasticity and Anelasticity of Metals, Chicago Univ. Press (1948).
22. Evans, K. E., *Endeavour*, n.s., *15*, 4, 1991.
23. Rowe, A. M., Ph.D. thesis, Loughborough University of Technology, 1982.
24. Wetton, R., DMTA of polymers and related systems, Chapter 5 of *Developments in Polymer Characterisation* (Dawkins, J., ed.), Elsevier Applied Science, 1986.

25. Schwarzl, F., and Staverman, A., *Die Physik der Hochpolymeren* (Stuart, ed.), Springer Verlag, Berlin, 1956.
26. Williams, M., Landel, R., and Ferry, J., *ACS*, *77*, 3701 (1955).
27. Townend, D., A computerised technique for producing WLF shifted data from a single frequency. Temperature scan without graphical manipulation, NATAS Meeting, Kingston, Ont., August, 1982.
28. Saito, N., Okano, K., Iwayanagi, S., and Hideshima, T., *Solid State Physics*, *14* (Seitz and Turnbull, eds.), Academic Press, New York, 1963.
29. MacInnes, M., Private communication. See also Ref. 123 (The glassy state in foods), pp. 223–248.
30. Lempriere, B., Poisson's ratio in orthotropic materials, *AIAA J.*, *6*(11), 2226–2227 (1968).
31. Summerscales, J., and Fry, S., Poisson's ratio in fibre reinforced polymer composites with high void content, *J. Material Science Letters*, *13*, 912–914 (1994).
32. Starkweather, H. W., Jr., DM studies of the glass transition in the frequency domain. NATAS, Ohio, Paper 11 (1986).
33. Boyd, R. H., Relaxation processes in crystalline polymers: Experimental behaviour—A review. *Polymer*, 26, March, 323–347 (1985).
34. Okoroafor, E., and Rault, J., Cryodilation of thermoplastic PEBA elastomers. *J. Pol. Sci. B*, *29*, 1427–1437 (1991).
35. Hill, R., and Gilbert, P., High temperature DMTA of a Li Zn silicate glass-ceramic. *J. Am. Cer. Soc.*, *76*(2), 417–425 (1993).
36. Roeder, E., Mayer, H.-J., and Liedhegener, R., Elastizität u. mech. Dämpfung eiens unidirectional SiC kurzfaser-verstätkten Alkali-Kalk-Silicatglases. *Mat.-Miss. u. Werkstofftech.*, *25*, 244–251 (1994).
37. Wetton, R., Stone, M., and Gearing, J., Mech. properties of very thin pol. coatings. Proceedings of NATAS 9/85, San Francisco (1985).
38. Pollock, H. M., Nanoindentation from ASM handbook, Volume 18, *Friction Lubrication and Wear Technology* (OH, USA).
39. Smith, J. F., Measurement, Testing and optimisation of wear-resistant coatings, The NanoTest 550, *Tribotest J.*, *6* (1995).
40. Eldridge, J., and Ferry, J., Studies of the cross-linking process in gelatin gels, *J. Phys. Chem.*, *58*, 922 (1954).
41. Winter, H., and Chambon, F., Analysis of lin. viscoelasticity of a cross-linking pol. at the gel point. *J. Rheol.*, *30*(2), 367 (1986) and *J. Rheol.*, *31*(8), 683–697 (1987).
42. Foley, K., and Chu, S. G., Rheological characterisation of elastomer latexes for PSAs, *Adhesives Age*, *9*, 24 (1986).
43. Schroeder, J., Madsen, P., and Foister, R., Structure/property relationships for a series of crosslinked aromatic/aliphatic epoxy mixtures, *Polymer*, *28*, May, 929–940 (1987).
44. Dale, W., Haynes, J., Paster, M., and Alstede, E. (Monsanto Inc., Spec. Resins Division, Indian Orchard, MA 01151), Measurement of fundamental mech. properties of Gelva(tm) PSA and their relationships to industry standard testing (1985/86).
45. Gipstein, E., Prime, R. B., and Allen, R., The effect of pressure on the polymerisation/cure of acrylate/methacrylate copolymers, NATAS, Paper 113 (1991).
46. Scarlata, S., Ors, J., and Enns, J., Determination of the phase transition of a pol. film by fluorescence polarization, NATAS, 12 (1987).
47. Lowe, G. B., DMTA of sealants for glazing, Thiokol Chem. Seminar, 9/88.
48. Paul, D., and Barlow, J., *J. Macro Sci—Rev. Macro Chem.*, c18, 109 (1980).
49. Chiu, H.-T., and Hwung, D.-S., Miscibility on polyphenylene oxide and SBS triblock copolymer blends, *Eur. Pol. J.*, *30*(10) (1994).
50. Eastmond, G., Jiang, M., and Malinconico, M., Morphologies and properties of polyblends: 1 Blends of pmma and a chlorine containing polycarbonate, *Polymer*, *24*, Sept., 1162–1170 (1983).

51. Eastmond, G., Jiang, M., and Malinconico, M., Two blends of pmma and a chlorine containing PC—Effect of added copolymer, *Brit. Pol. J.*, *19*, 275–285 (1987).
52. Sakellariou, P., Eastmond, G., and Miles, I., Interfacial activity of pc/pmma graft copolymers in pc/pmma blends: Effect of copolymer concentration, *Polymer*, *32*(13), 2351–2362 (1991).
53. Sakellariou, P., Eastmond, G., and Miles, I., Effect of homopolymer molecular weights, *Polymer*, *33*, 21, 4493–4504 (1992).
54. Crevecoeur, G., and Groeninckx, G., Binary blends of peek and poly(ether-imide) miscibility, crystallization behaviour and semicrystalline morphology, *Macromolecules* (in press 1/91 when received by me).
55. Sperling, L., *Interpenetrating Polymer Networks and Related Materials*, Plenum Press, New York, 1981.
56. Hourston, D., and Huson, M., Semi- and fully Interprenetrating pol. networks based on pu-polyacrylate systems. XI. The influence of polymerisation temperature on morphology and properties, *J. Appl. Pol. Sci.*, 1753–1762 (1992).
57. Decurtins, S., Hourston, D., Polyamide-pu IPNs prepared by synthesis in the melt, *J. Appl. Pol. Sci.*, 365–375 (1988).
58. Ciemniecki, S., and Glasser, W., Multiphase materials with lignin: 1. Blends of hydroxypropyl lignin with pmma, *Polymer*, *29*, June, 1021–1029 (1988).
59. Ciemniecki, S., and Glasser, W., 2. Blends of hydroxypropyl lignin with poly(vinyl alcohol), *Polymer*, *29*, 1030–1036 (1988).
60. Benavente, R., Perena, J., Perez, E., Bello, A., and Lorenzo, V., *Basic Features of the Glassy State* (J. Colmenero and A. Alegria, eds.), World Scientific, Singapore, 1990, pp. 490–500.
61. Utracki, L., and Sammut, P., Rheological response of polyamide/pp blends: Part 3: Solid state behaviour, *Plastics, Rubber & Composites Proc. & Applications*, *16*, 4, 221–229 (1991).
62. Defieuw, G., Groeninckx, G., and Reynaers, H., Miscibility, crystallization and melting behaviour, and semi-crystalline morphology of binary blends of polycaprolactone with poly(hydroxy ether of bisphenol A), *Polymer*, 462–489 (1989).
63. Kelley, S., Ward, T., Rials, T., and Glasser, W., Engineering plastics from lignin. XVII. Effect of molecular weight on pu film properties. *J. Appl. Polymer Sci.*, *37*, 2961–2971 (1989).
64. Pizzoli, M., Scandola, M., and Ceccorulli, G., Viscoelastic relaxations of polyvinylidenefluoride/polyvinylpyrrolidone blends, European Symposium on Polymer blends, Strassbourg 9/1987.
65. Bucknall, C., Davies, P., and Partridge, I., Phase separation in styrenated polyester resin containing a poly(vinyl acetate) low profile additive, *Polymer*, *26*, January (1985).
66. Diaz-Calleja, R., Ricard, E., and Guzman, J., Influence of static strain on DM behaviour of amorphous networks prepared from aromatic polyesters, *J. Polymer Science*, Polymer Physics Edn. 23 (1986).
67. Khanna, Y., Estimation of polymer crystallinity by DM techniques, *J. Appl. Sci.*, *37*, 2719–2726 (1989).
68. Sha, H., Harrison, I., and Zhang, X., A DMTA study of polyethylenes, 1990 Natas conference ref BA2.
69. Khanna, Y., Turi, E., Taylor, T., Vickroy, V., and Abbott, R., DM relaxations in polyethylene, *Macromolecules*, *18*, 1302 (1985).
70. Ding, J.-F., Khan, A., Proudmore, M., Mobbs, R., Heatley, F., Price, C., and Booth, C., Elastomers based on polyhexafluoropropylene oxide crosslinked via triazine groups, *Macromol. Chem. Phys.*, *195*, 3137–3148 (1994).
71. Haidar, B., Hedrick, J., Russel, T., and Hofer, D., Synthesis and properties of segmented poly(hydroxyether siloxane) CA07 (1990).
72. Wetton, R., Foster, G., and Corish, P., Elastomer microstructure/Dynamic property correlations, *Polymer Testing*, *10*, 175–188 (1991).
73. Radusch, H.-J., Laemmer, E., and Richter, C., Weichmachung dynamischer Vulkanisate auf der Basis von Polyolefin-Kautschuk Blends, *Polymerwerkstoffe*, 137–146 (1994).

74. Huet, C., Etude par une méthode d'impedance due comportement viscoelastique des materiaux hydrocarbones, Thesis for Docteur Ingenieur at the Fac. of Sciences of the University of Paris, with English summary (1965).
75. Leahy, J., Drake, J., and Birkenshaw, C., Structural characteristics of peat bitumen and peat petroleum bitumen blends, and their consideration as potential road binder materials, *J. Mat. Sci.*, 3688–3692 (1990).
76. Gabler, A., Straube, E., and Heinrich, G., Korrelationen des Nassrutschverhaltens fussgefüllter Vulkanisate mit ihren viskoelastischen Eigenschaften, *KGK (Kautschuk Gummi Kunstoffe)*, *46*, 12/93, 941–948, with English summary (1993).
77. Frankcen, L., and Moraux, C., Influence des fillers sur les caracteristiques de la consistance des mastics bitumineux, *Bituminfo*, December, 7–18 (1985).
78. RAPRA, Shawbury, Shropshire GB SY4 4NR, (44) 1939 250 383, fax 251 118. http:/www.rapra.net.
79. MRPRA, Brickendonbury, Herts, GB SG13 8NL, (44) 1992 584 966, Fax 554 837.
80. White, R. P., Jr., Time temperature superpositioning of viscosity time profiles of 3 high temperature epoxy resins, *Polymer Eng. Sci.*, *14*, 50, January (1974).
81. McArdle, C., Burke, J., and McGettrick, B., Characterisation of cure distribution in surface initiated redox polymerisation by DM and di-electric thermal analyses, *Plastics, Rubber Composites Processing Applications*, *16*, 245–253 (1991).
82. Knollman, G., and Hartog, J., *J. Adhesion*, *17*, 251 (1985).
83. Dickie, C., and Morman, K., DM response of adhesively bonded beams: Effect of environmental exposure and interfacial zone properties, *Pol. Eng. Sci.*, *30*, 4, 249–255 (1990).
84. Hartman, B., Chapt. 2 of *Sound and Vibration Damping in Polymers*, ACS Symp Series 424 (1990).
85. Hill, L. W., Structure/property relationships of thermoset coatings, *J. Coatings Techn.*, *64*, 808, 29 (1992).
86. Hill, L. W., DM and tensile properties, *ASTM Manual 17*, 46, 534–546.
87. Kopsch, H., Zum Verhalten von polymermodifizierten Bitumen bei tiefen Temperaturen (On the condition of pol modified bitumen at low temperatures), *Bitumen*, *1*, 10–13 (1990).
88. Brisco, B., Sebastian, K., and Adams, M., The effect of indenter geometry on the elastic response to indentation, *J. Phys. D. Applied Physics*, *27*, 1156–1162 (1994).
89. Brotzen, F., Mechanical testing of thin films, *Int. Materials Review*, 39, 1, 24–45 (1994).
90. Burnham, N., Colton, R., Pollock, H., Interpretation of force curves in force microscopy, *Nanotechnology*, *4* (1993).
91. Oliver, W., and Pharr, G., An improved technique for determining hardness and elastic modulus using load and displacement sensing indentation experiments, *J. Material Research (MRS)*, *7*, 6, 1564–1583 (1992).
92. Doerner, M., and Dix, W., A method for interpreting the data from depth sensing indentation insts., *J. Mat. Res.*, *1*(4), 601 (1986).
93. Baker, S. P., The analysis of depth sensing indentation data, *MRS Symp. Procdgs.*, 308, 209–216 (1993).
94. Watanabe, J., and Tominga, T., *Macromolecules*, *26*, 4032–4036 (1993).
95. Carswell, R., Cormack, P., Sherrington, D., and Moore, B., Solid phase synthesis of liquid crystalline oligopeptides, RSC Polymer Symposium Wilton (1996).
96. Ochi, M., Shimizu, Y., Tsuyuno, N., Nakanishi, Y., and Murata, Y., Effect of network structure on thermal and mechanical properties of cured epoxy resin containing mesogenic groups, *American Chem. Society, PMSE*, *74*, Spring, 20–21 (1996).
97. Girard-Reydet, E., Pascault, J., Sautereau, H., and Vicard, V., TP modified epoxy networks/ Influence of cure schedule and introduction of block co-polymers on morphologies and mechanical properties, *ACS PMSE*, *74*, Spring, 24–25 (1996).
98. Zielinski, J., Vratsanos, M., Laurer, J., and Spontak, R., Phase separation studies of heat-cured ATU-flexibilized epoxies, *Polymer*, 37, 1, 75–84 (1996).

99. Don, T.-M., and Bell, J., Studies of reactions in epoxy resin/polycarbonate blends, *ACS PMSE*, *74*, Spring, 127–128 (1996).
100. Bucknall, C., Partridge, I., Jayle, L., Nozue, I., Fernyhough, A., and Hay, I., *Pol Preprints (ACS)*, *33*(1), 378 (1992).
101. Di Liello, V., Martuscelli, E., Musto, G., Ragosta, G., and Scarinzi, G., *J. Pol. Sci. B, Pol. Physics*, *32*, 395 (1994).
102. Robeson, L., Fundamental aspects of polymer/polymer miscibility: Influence of the molecular structure and specific interactions, At Leuven 10/86 VCV Workshop on New Polymeric Materials—Advances in Pol. Alloys and Composites.
103. Carfagna, C., Amendola, E., and Giamberini, M., Anisotropic epoxy based networks, *ACS PMSE*, *74*, Spring, 129–130 (1996).
104. Giamberini, M., Amendola, E., and Carfagna, C., *Mol. Cryst. Liq. Crystals*, *266*, 9–22 (1995).
105. Langlois, D., Smith, M., Benicewiz, B., and Douglas, E., Phase behaviour of liquid crystal thermosets, *ACS PMSE*, *74*, Spring (1996).
106. Lessig, E. T., *Industrial Eng. Chemistry* (Anal Edn) *9*, 582–588, December (1937).
107. Askea, D., DM properties from the BFGoodrich Flexometer 11. ACS rubber division meetings 5/93, Paper 63 and 10/95, Paper 100.
108. Wedgewood, A., and Seferis, J., Error analysis and modelling of non-linear stress-strain behaviour in measuring DM properties of polymers with the Rheovibron, *Polymer*, *22*, July, 966–991 (1981).
109. Perera, D., Schutyser, P., de Lame, C., and Van den Eynde, D., On film formation and phys. ageing in organic coatings, *ACS PMSE*, *73*, Fall, 187–188 (1995).
110. Heuts, M., le Febre, R., van Hilst, J., and Overbeek, G., Influence on film formation of acrylic dispersions, *ACS PMSE*, 140–441 (1995).
111. Richard, J., Dynamic viscoelastic properties of polymer latex films: Role of core chains and particle/particle interfaces, *ACS PMSE*, Fall, Chicago, *73*, 41–42 (1995).
112. Gearing, J., and Stone, M., The DM Method for the characterisation of solid composite polymers, *Polymer Composites*, *5*, 4, 312–319 (1984).
113. Kajiyama, T., Tanaka, K., Ge, S.-R., and Takahara, A., Interfacial characteristics of pol. blend ultrathin films based on scanning viscoelasticity microscopy, *ACS PMSE*, Fall, 269–270 (1996).
114. Malik, T., Sealant rheology and its practical measurements, *ACS PMSE*, Fall meeting, Orlando, 291–292 (1996).
115. Lostocco, M., and Huang, S., The synthesis and characterisation of polyesters derived from L-lactide and variably-sized poly (caprolactone), *ACS PMSE*, Fall, 379–380 (1996).
116. Iannace, S., Ambrosio, L., Huang, S., and Nicolais, L., *J. Pure Appl. Chem.*, A32, 881 (1995).
117. Simon, S., and Plazek, D., Physical ageing of a polyetherimide: Volume, enthalpy and creep, *ACS PMSE*, Fall, 423–424 (1996).
118. McKenna, G., Schultheisz, C., and Leterrier, Y., Volume recovery and phys. ageing: Dilatometric evidence for different kinetics. 9th Intl. Conf: Deformation, Yield & fracture of Polymers 4/1994 Cambridge. 31/1–31/4 (1994).
119. Bagheri, R., and Pearson, R., Interfacial aspects in rubber-roughened epoxies, *ACS PMSE*, Fall, 473–474 (1996).
120. Edgecombe, B., Frechet, J., Xu, Z., Jandt, K., and Kramer, E., Novel styene-based copolymers for the strengthening of interfaces between immiscible polymers using hydrogen bonding interactions, *ACS PMSE*, Fall, Orlando, 478–479 (1996).
121. Ishida, H., and Low, H., Polybenzoxazines: Expanding phenolic resin with structural applications, *ACS PMSE*, Fall, 115–116 (1996).
122. Krishnamoorti, R., and Giannelis, E., Polymer layered silicate nanocomposites, *ACS PMSE*, Fall, Orlando, 46–47 (1996).
123. Blanshard, J., and Lillford, P. (eds.), *The Glassy State in Foods*, Nottingham Univ. Press, 1993.

124. Duncan, J., Gearing, J., Wetton, R., Fisher, J., and Blow, A., DMTA of metals. NATAS 10/1988 paper 78. Lake Buena Vista (Fl USA) (1988).
125. Deckmann, H., and Gabriel, E., Microstructure and DM properties of bones and implants, Gabo Eplexor Appn 22. (Ahlden).
126. Munoz, M., Santamaria, A., Guzman, G., Dynamic viscoelastic measurements of handsheet papers, *Tappi J.*, 217–221 (1990).
127. West, J., Hem, D., and Hubbell, J., Polymers in medicine: manipulating wound healing, *ACS PMSE*, Fall, 411 (1995).
128. Baker, F., and Privett, G., DM studies of nitrocellulose/nitroglycerine mixtures, *Polymer*, 28, June, 1121–1126 (1987).
129. Hoffman, D. M., Matthews, F., and Pruneda, C. D., DM&TA of crystallinity development in Kel-F 800 and TATB/Del-F 800 plastic bonded explosives. Paper 33.
130. Haerdtl, K., Electrical and mechanical losses in ferroelectric ceramics, *Ceramics Intl.*, *8*, 4, 121–127 (1982).
131. Matsuoka, S., Predicting frequency dependence of DMTA data, NATAS 1992, p. 76, and *Relaxation Phenomenon in Polymers*, Hanser/Oxford Univ. Press, New York, 1992.
132. Jadraque, D., and Perena, J., Rheology in tobacco. Coresta Congress, Jerez de la Frontera, Spain 10/1992.
133. del Rio, C., and Acosta, J. L., Thermal transitions, microstructure and miscibility in Ternary Polyblends based on polyvinylidene-fluoride, *Polymer International*, SCI 30, 47–53 (1993).
134. Dvornic, P., Uppuluri, S., and Tomalia, D., Rheological properties of starburst polyamido-amine dendrimers in concentrated ethylene diamine solutions, *ACS PMSE*, Fall, Chicago, 131–132 (1996).
135. Johannson, M., and Hult, A., Synthesis, characterisation and curing of resins based on hyperbranched structures, *ACS PMSE*, Fall, 178–179 (1996).
136. Oyadiji, S., Characterisation of the aggregate complex modulus properties of a fibre/wire reinforced composite viscoelastic pipe, ECCM7 Composite Materials Conference, London 5/96. Inst. of Materials (Woodhead Publishing, Ltd.) pp. 167–172.
137. Wetton, R., Design of elastomers for damping applications—Chapter 2 in *Elastomers: Criteria for Engineering Design*, Applied Science, Barking, GB.
138. Stopp, R., and Heine, S., Parameter determination for the design of rubber parts and gaskets, SAE 930123, Detroit 3/1993 Conf.
139. Lake, G., and Thomas, A., Strength properties of rubber—Chapter 15 in *Natural Rubber Science and Technology*, Oxford Science Publishers, 1988.
140. Wisch, C., and Meinecke, E., The static and DM properties of filled elastomers with respect to C black loading and the stress-strain behavior of the unfilled matrix, ACS Rubber Div., Cleveland, OH, 10/95. Paper 109.
141. Hay, J. N., Physical ageing in polymer blends, Chapter 13, *The Glassy State in Foods*, Nottingham Univ. Press, 1993.
142. Brown, R. P., Physical Testing of Rubbers, 1995, Chapter 9.
143. Payne, A. R., and Scott, J. R., *Engineering Design with Rubber*, Maclaren and Sons, 1960.
144. Nicholson, J., and Wilson, A. D., *Glass Ionomer Cement*, Brit. Technology Group, Issue 3, Sept. 1988.
145. Reading, M., Calorimetry and surface measurement, AFCAT meeting 5/97, Dunkerque, France.
146. Gabo GmbH (Ahlden, Germany D29693), application note number 13.

22
Fracture Mechanics Properties

Mustafa Akay
University of Ulster at Jordanstown, Jordanstown, Northern Ireland

1 Introduction

Fracture mechanics, unlike customary impact tests, provides an ability to separate geometry from material response. However the concept still meets with scepticism and is hardly employed in the plastics industry. The standard notch impact tests are qualitative and produce at best a guide to the relative ranking of various materials for impact strength. Design against crack propagation in materials requires determination of fracture toughness, a material property that indicates the likelihood of a particular crack to cause catastrophic failure.

Fracture mechanics, which is concerned with the loads or stresses that cause a crack to propagate, was first considered by Griffith [1] in 1920 with respect to the propagation of cracks in brittle materials. He recognized that premature failure occurs in glasses due to the initiation of fracture from stress concentrating defects. It was stated that when an existing crack propagates (or extends), the reduction in the total elastic-strain energy of the stressed body balances the energy needed to create the fresh fracture surfaces. Later Irwin [2] indicated that for ductile materials, this energy balance must be between the stored strain energy and the work done in plastic deformation as well as to form new crack surfaces. He, therefore, identified an energy term, G (after Griffith), the energy release rate per unit crack area, which includes plastic work as well as surface energy. Irwin also showed that the energy approach is equivalent to a stress intensity approach. When the stress-intensity factor K exceeds a critical value K_c, then fracture occurs.

The equivalence of K and G, which strictly holds for elastic materials with linear load–deflection characteristics, is referred to as linear-elastic fracture mechanics (LEFM). Subsequently this basic concept has been modified to describe also the behavior of ductile materials. For instance, Wells [3] considered the plastic strain at the crack tip as the crack tip opening displacement (CTOD) in one such pioneering work. It was proposed that once a critical CTOD is exceeded, fracture will occur.

The significance of fracture mechanics stems from the reality that materials fail in engineering applications. The failures (typical case studies can be found in Ref. 4) range from the fracturing of a polyethylene bucket handle to the collapse of a bridge, often with serious consequences: human loss, the total or partial loss of equipment/service and their replacement/repair, and environmental damage. The solutions towards minimizing fracture-related accidents require appropriate training of engineers, research into materials and structures, the exchange of knowledge between the researchers and industry, and updating and/or introducing standards for testing and evaluation.

Fracture mechanics provides methods to evaluate the load-carrying capacities of components with the aim of minimizing catastrophic service failures originating from the flaws in the components. Whether a crack of given dimensions constitutes a hazard can be assessed from the material property of the critical stress-intensity factor K and the magnitude of applied stresses. The relationship between stress intensity K and fracture toughness K_c is similar to the relationship between stress and tensile strength. The stress intensity represents the level of stress at the tip of a crack in a specimen containing a crack, and the fracture toughness K_c is the highest value of stress intensity that the specimen can withstand at the given crack sharpness without fracturing, which is obviously dependent on the type of material. Three important parameters are, therefore, required for the prediction (assessment) of sudden failures:

1. Fracture toughness, which should be determined in accordance with internationally recognized standards
2. Crack length, which can be determined by nondestructive inspection techniques
3. Levels of stress encountered in service

Any of these parameters can be quantified from the values of the others using fracture mechanics relationships. There is, however, still a reluctance on the side of industry, particularly polymers-related industry, to use this concept. This soon becomes apparent if raw materials suppliers' literature or any of the current materials selection data bases are examined: there are no fracture mechanics data quoted, but there are values for static and impact strengths measured in a variety of manners. We are reminded [5] that traditional tests such as the tensile tests give the performance under the most favorable conditions, and as such cannot be used to indicate or assess the likelihood of unexpected brittle fractures in service. The Charpy and Izod impact tests were introduced to solve these problems in that they subject materials to a set of worst-case conditions in contrast to the (best-case) tensile tests. The specimens are notched and then struck at a high rate in bending with the notch in tension. These conditions are expected to induce brittle failure, and if a material performs well, as indicated by a high energy absorption, then it is deemed to be tough. However the recommended 0.5 mm diameter notch tip in test pieces, although it ensures reproducibility in a standard, is not sufficiently sharp and often leads to false estimates and fails to indicate the possibility of a brittle fracture in practice.

2 Fracture Mechanics Approach

It is assumed that all bodies contain cracks or flaws, and fracture is the condition when these flaws start to grow. Sufficient force needs to be applied to overcome the cohesive strength of the material and cause fracture. Although in reality, crack extension begins at much lower applied stress levels than the theoretical cohesive strength of the materials.

Flaws act as stress concentrations, magnifying the applied stresses to the theoretical cohesive strength levels at the crack tip. The molecular/atomic bonds adjacent to the crack tip can then be broken, and the crack opens.

Fracture toughness of the material, K_c, indicates the minimum level of stress intensity to cause a specimen of the material containing a crack of a certain sharpness to suffer fracture. From an energy perception, work from the release of stored elastic energy in the body of the material is required in order to facilitate crack growth. The energy release rate G must exceed a critical value G_c for fracture to occur. Alternatively, the rate at which energy is consumed in propagating a crack is called the crack resistance R. Both G and R are expressed in terms of energy per unit specimen width per unit crack extension (i.e., energy per unit fractured ligament area). These parameters are interrelated in the fracture mechanics equation for isotropic materials:

$$K_c = (EG_c)^{1/2} = Y\sigma_f a^{1/2} \tag{1}$$

where K_c is the fracture toughness (MPa m$^{1/2}$), E is the Young's modulus (GPa), G_c is the critical energy release rate per unit crack area (kJ/m^2), σ_f is the nominal applied stress at fracture (MPa), a is the crack length (or one-half of an internal crack length), and Y is the dimensionless calibration (correction) factor that accounts for the geometry of the specimen containing the flaw. Y is a function of a/W, where W is the width or depth of the specimen. The values for the Y calibration factor are available for different specimen geometries and loading modes [6,7]. For an ideal case, where an infinitely large plate containing a crack of length $2a$ is under tensile load, $Y^2 = \pi$.

Equation 1 is for plane-stress condition which exists in thin plates and fracture results in shear lips or slanted fracture surfaces. For plane-strain condition, E is replaced by $E/(1-\nu^2)$ in Equation 1, where ν is Poisson's ratio. Plane strain describes the triaxial state of stress that exists in thicker plates such that plastic deformation is constrained and fracture produces flat surfaces; see Fig. 1.

The fracture mechanics approach is well established for metals and should receive greater recognition for polymer applications with the emergence of specific standard test

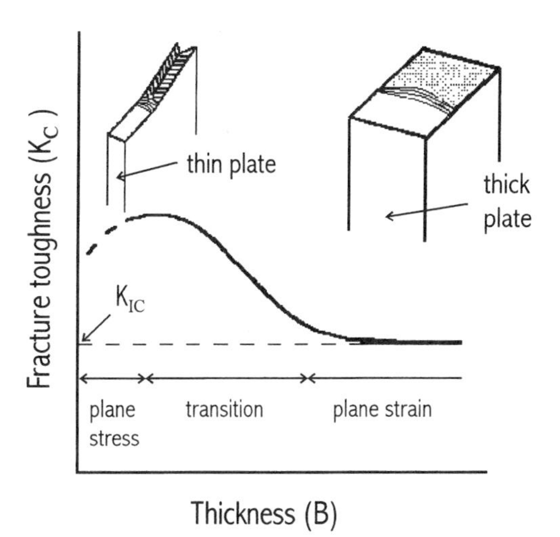

Figure 1 Effect of thickness on fracture toughness (the fracture surface sketches are more typical of metals, adopted here to indicate brittle and ductile fractures).

methods such as ASTM D5045 [8], and others in the form of test protocols from the American Society for Testing and Materials (ASTM), the European Structured Integrity Society (ESIS), and the Japanese Industrial Standardization (JIS). These proposals, however, have not yet produced an international agreement in the form of an ISO (the International Standardisation Organisation) standard.

In the following sections some of these test methods and their applications to material evaluation will be covered for different classes of polymer-based materials.

3 Plastics

3.1 Test Methods

Plain-Strain Fracture Toughness and Strain Energy Release Rate of Plastics Materials (ASTM D5045-93)

This test method is based on the ASTM E399 [9] for metals but includes certain specific considerations such as a detailed procedure for the determination of G_c for plastics, including specimen preparation details and the correction for the penetration of the loading-pin/supports into the specimen.

The properties K_c and G_c are determined under certain restrictions in order to ensure that a minimum value for these toughness parameters is obtained in a neutral environment, at room temperature, and in the presence of a sharp crack. The concept assumes linear elastic behavior, therefore certain restrictions on linearity of the load-displacement plot and specimen dimensions are placed in order to ensure validity. Specimens of sufficient thickness and notch tip sharpness are required to ensure, respectively, a plane-strain (uniform strain) condition at the crack tip, and to minimize the crack tip plastic deformation so that the plastically deformed region is small compared with crack length and the specimen thickness; although for polymers, particularly for semicrystalline thermoplastics, the microstructure is thickness sensitive, and large variations in thickness can lead to misinterpretations of the results. The microstructure of thermoplastics is also dictated by the processing direction, and accordingly the results should be identified with respect to the processing direction.

The details of the standard test pieces of three-point bend (or single-edge notched bend, SENB), and compact tension (CT) are shown in Fig. 2. Sharp notches are generated in the specimens by tapping or sliding a razor blade into the root of a sharp machined notch to produce a tip radius of approximately 10–15 μm. In both geometries the crack length should be selected so that $0.45 < a/W < 0.55$, and $W = 2B$. In fact the LEFM requirement is just that B and $(W - a)$ should be sufficient to ensure plane strain, and to avoid excessive plasticity in the ligament, respectively. Therefore, for a given B, values of $W \leq 4B$ are permitted to facilitate these requirements. It is recommended that the tests be conducted at a room temperature of 23°C and a loading rate (cross-head speed of testing machine) of 10 mm/min. Speeds of greater than 1 m/s or loading times of less than 1 ms should be avoided to reduce errors due to dynamic effects.

A typical test load–deflection trace is shown in Fig. 3. In an ideal (elastic) case the load–deflection plot produces a straight line with an abrupt drop of load to zero at the instant of crack growth initiation (crack extension), enabling a straightforward identification of the force required to initiate crack growth. Identification of where the crack growth initiates becomes unclear where the load–deflection plots produce curves. The standard recommendation (for consistency) is to define the load P_Q that gives a 5% increase in compliance C (i.e., a 5% decrease in stiffness) as a representative of the load value at the

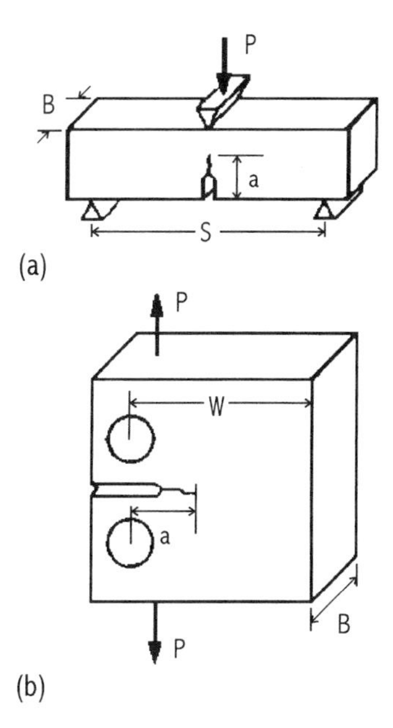

Figure 2 Specimen configuration and loading arrangements. (a) SENB; (b) CT.

onset of crack growth. The curvature in the plot is assumed to be due to crack growth and, thus, a limited amount of nonlinearity (curvature) is allowed in order to minimize the contribution of plastic deformation at the crack tip to the nonlinearity. Within the standard this limit is taken as 10%, i.e., for a valid test $P_{max}/P_Q \leq 1.1$.

A conditional fracture toughness value, K_Q, validity of which needs to be qualified, can be calculated from the following expression.

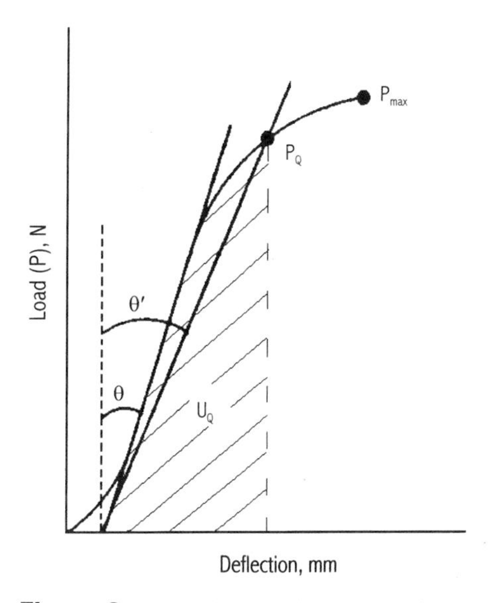

Figure 3 A load–deflection trace, where $C = \tan\theta$ and $(1.05)C = \tan\theta'$.

$$K_Q = f(x) \frac{P_Q}{BW^{1/2}} \text{ (in units of MPa m}^{1/2}\text{)} \quad (2)$$

where $0.2 < (x = a/W) < 0.8$, and $f(x)$ is the geometric calibration factor. For a CT specimen,

$$f(x) = \frac{(2 + x)(0.886 + 4.64x - 13.32x^2 + 14.72x^3 - 5.6x^4}{(1 - x)^{3/2}}$$

For an SENB specimen of $S/W = 4$,

$$K_Q = f(x) \frac{P_Q S}{BW^{3/2}} \text{ (in MPa m}^{1/2}\text{)} \quad (3)$$

where, $0 < x < 1$,

$$f(x) = \frac{6x^{1/2}[1.99 - x(1 - x)(2.15 - 3.93x + 2.7x^2)]}{(1 + 2x)(1 - x)^{3/2}}$$

Tables of the calibration factors for various a/W values are given in the designated ASTM standard. The validity of K_Q is checked by the following specimen size criteria: If B, a, and $(W - a) > 2.5(K_Q/\sigma_y)^2$, then $K_Q = K_{Ic}$.

Where σ_y is the uniaxial tensile yield stress and for polymer-based materials this is usually taken as the maximum load if a distinct yield point is not exhibited. However shear yielding in tensile tests with most polymers can be achieved by carefully polishing the specimen edges in order to remove surface blemishes and thus avoid premature failure. If yielding does not occur and brittle failure is obtained, the stress at failure should be used in the criteria which gives a conservative size value. Alternatively 0.7 times the compressive yield stress may be used. The loading time to yield (or equivalent) should be within ±20% of the loading duration in the fracture test.

Critical Strain Energy Release Rate G_{Ic}

In principle it can be obtained from K_{Ic} using the relationship

$$G_{Ic} = \frac{(1 - \nu^2)K_{Ic}^2}{E} \text{ (in units of kJm}^{-2}\text{)} \quad (4)$$

However E must be obtained at the same temperature and time conditions as the fracture test because of the viscoelastic nature of polymers. In order to avoid the associated uncertainties, it is considered preferable to determine G_Q or G_{Ic} (subject to the validity criteria of K_{Ic}) directly from the load versus deflection curve up to the same load point as used for K_Q or K_{Ic} determination (see Fig. 3). The energy should be corrected for specimen indentation (suffered under loading and support pins) by deducing the energy of indentation, U_i from U_Q. Therefore:

$$G_{Ic} = \frac{U_Q - U_i}{BW\phi} \quad (5)$$

where ϕ is the appropriate correction factor for a given a/W ratio. Tables of ϕ values, for various a/W ratios, are given in the designated ASTM standard and in other sources (e.g., Ref. 10).

The above described LEFM method is valid only if nonlinearity in deformation is confined to a small region surrounding the crack tip. Many of the engineering polymers such as high density polyethylene (HDPE), polyvinylidene fluoride (PVDF), and polyvinylchloride (PVC), however, do not exhibit linear elastic behavior at the thicknesses normally employed in service. The deviation may result from stable crack growth, viscoelasticity, or large plastic deformation at the crack tip. In such cases, material toughness to fracture may be characterized by elastic-plastic fracture mechanics, using parameters such as J (or J-integral or J-contour), which is the energy required to grow a stable crack, and the crack tip opening displacement (CTOD). These methods, strictly speaking unless modified, only apply to materials that exhibit time-independent nonlinear behaviour (i.e., plastic deformation). The critical values of CTOD and J are almost size-independent measures of fracture toughness for relatively large amounts of crack tip plasticity.

Fracture Toughness by Crack Tip Opening Displacement (CTOD)

Wells [11] first proposed the crack opening displacement as a measure of fracture toughness for tough metals such as the structural steels where LEFM is not valid. The concept became a standard in 1979 in the form of British Standard (BS) 5762 [12], which is now superseded by BS 7448: Part 1, 1991, Method of determination of K_{Ic}, critical CTOD and critical J values of metallic materials [13]. More recently, ASTM E1290–93, Crack-tip opening displacement (CTOD) fracture toughness measurement [14] was issued. It is stated that this method may be used to characterize the toughness of materials that are (a) too ductile or fail the size conditions to be tested for K_{Ic}, or (b) likely to produce unstable crack extension that would invalidate J_{Ic} tests in accordance with the requirements of ASTM E813 [15].

CTOD measurements are conducted using mostly SENB specimens fitted with a displacement gauge at the crack mouth. Three-point bend loading produces a load vs. crack mouth opening displacement plot, from which the plastic component of the crack opening displacement can be extracted. In turn, CTOD is computed by assuming that the specimen halves remain rigid and rotate about a hinge point during loading, so that CTOD can be determined from a simple geometric comparison of similar triangles. An illustration of this can be found in Ref. 16.

Fracture Toughness by *J*-integral

The contour integral [17] was applied to fracture analysis in 1968 by Rice [18] to characterize elastic-plastic material behavior ahead of a crack. This nonlinear energy release rate was expressed in the form of a line integral, which was described as the J-integral, evaluated along an arbitrary contour around the crack. The analyses [19,20] showed that J can yield a nonlinear stress intensity parameter as well as energy release rate.

Presently there are ASTM and the British Standards (BS) Institution standards that address J testing of metallic materials [13,15,21], but a specific standard for plastics does not exist. The issuing of an ASTM standard that combines E813 [15] and E1152 [21] is apparently imminent [16].

The E1152 standard applies to the entire J-crack resistance (J–R) curve, whilst E813 is concerned with the critical J value J_{Ic}, a single point on the J–R curve. This is analogous to a tensile test, from which both an entire stress–strain curve can be constructed or the yield strength determined.

Tests for the J–R curve and the determination of J_{Ic} can be conducted by employing multiple specimens or a single specimen. With the first method, identical multiple specimens are tested to different load levels and then unloaded. The extent of stable crack growth is marked by staining or fatigue cracking after the test to facilitate the crack extension measurements. Multiple measurements enable the construction of a load vs. crack-extension curve. The single-specimen test technique relies on the determination of the change in compliance with the growing crack. The crack length is computed at regular intervals during the test by partially unloading the specimen and measuring the compliance, as illustrated in Fig. 4. The specimen, as would be expected, becomes more compliant as the crack grows. The associated standards provide polynomial expressions that relate a/W to compliance. Relatively deep cracks ($0.5 \le a/W < 0.7$) are required, since the compliance variations are not sufficiently pronounced for $a/W < 0.5$.

An alternative single specimen technique to measure the crack length changes is based on monitoring the change in electrical resistance (a drop in electrical potential) that accompanies the loss of ligament area as crack grows.

The *J–R* curve is constructed by calculating the corresponding *J* values for each crack extension, using the equation

$$J = J_e + J_P \tag{6}$$

where J_e (the elastic component of J) $= [K^2(1-\nu^2)]/E$, J_p (the plastic component of J) $= (\eta U_p)/[B(W-a)]$, where U_p is the area under the load–displacement curve due to the plastic deformation, $W-a$ is the initial ligament area, and η is a dimensionless geometric constant; $\eta = 2$ for a SENB specimen and $\eta = 2 + 0.522[(W-a)/W]$ for a CT specimen.

The details of the determination of J_Q, a provisional J_{Ic}, from the *J–R* curve is given in ASTM E813 [15]. If the standard procedure is followed, $J_Q = J_{Ic}$ subject to the following size requirement.

$$B;\ (W-a) \ge 25J_Q/\sigma_y \tag{7}$$

J-integral test methods have been applied to polymers [10, 22–24] and the drafting of standard test methods by ASTM [16] and ESIS [25] suitable for J_{Ic} measurements of plastics have been under consideration for some time. The current common practice is to employ ordinary SENB or CT specimens and estimate *J* from the load *P* vs. displacement Δ plots so that

$$J = \left[\eta \int_0^{\Delta} \frac{P(d\Delta)}{B(W-a)}\right] \tag{8}$$

J–R curves for polymers are constructed mostly employing the multiple specimen technique. As previously described, the specimens are loaded to various displacements, unloaded, and fractured at low temperatures. Crack extensions *da* are measured optically

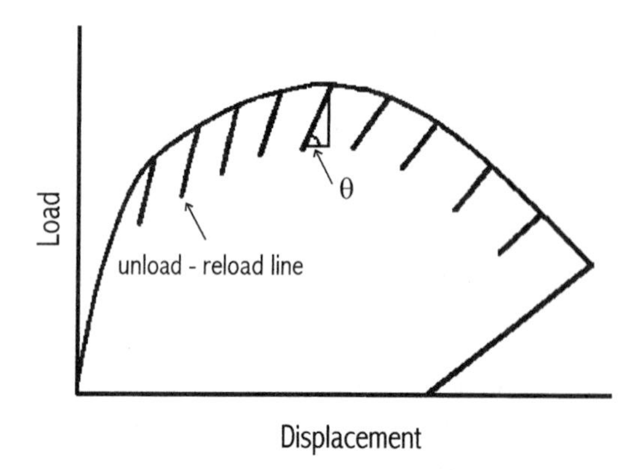

Figure 4 An illustration of partial unloading for the compliance measurement, where $C^{-1} = \tan\theta$.

for each specimen and thus a crack growth resistance curve (*J–da* plot) is constructed. The J_Q or J_{Ic} values are extracted from the *J–da* plot, by fitting a power law or straight line equation to the data as described in ASTM E813 [15].

The single-specimen technique of unloading and reloading to measure changes in material compliance can also be applied to polymers for *J*-integral testing. However, if the viscoelastic effects are significant during the period of unloading and reloading then this can produce a hyteresis loop rather than an unload–reload line shown in Fig. 4. The work by Schapery [26,27] has introduced a viscoelastic *J*-integral but has received limited interest in practice. The viscoelastic influences are obviously most pronounced at the glass transition region. Accordingly fracture tests should be conducted at temperatures well outside the glass transition temperature T_g range, preferably well below T_g in order to minimize rate effects.

Fatigue Crack Propagation

The coverage so far has been for monotonic crack growth and fracture by the application of increasing loads. Cracks that do not propagate under a static (monotonic) load may grow under an alternating load of the same magnitude. When the crack growth reaches a critical length (i.e., K_{Ic} condition) after a number of cycles N of loading then fatigue fracture occurs.

In 1960, Paris and Erdogan [28] showed that the growth of a crack on each fatigue loading cycle is a function of the range of stress-intensity factor ΔK during a loading cycle in a power-law form:

$$\frac{da}{dN} = C\Delta K^m \tag{9}$$

where $\Delta K = K_{max} - K_{min}$, and C and m are empirical material constants that depend on material properties, test frequency, applied mean load, etc. K_{max} and K_{min} are the maximum and the minimum stress-intensity factors, respectively.

Subsequent work has shown that most materials produce a fatigue crack growth behaviour of the form shown in Fig. 5, where the Eq. 9, popularly referred to as the Paris law, only holds in the region II.

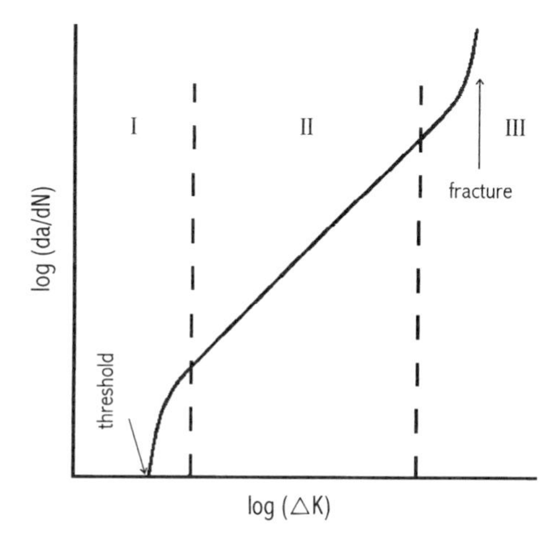

Figure 5 Fatigue crack growth rate versus stress intensity range.

ASTM E647–93, Measurement of fatigue crack growth rates [29] describes how to determine da/dN as a function of ΔK experimentally. The crack is grown by cyclic loading, and K_{min}, K_{max}, and crack length are monitored throughout the test. The standard qualifies that the behavior of the specimen, during the test, should be predominantly elastic and accordingly requires that the initial pretest ligament length for a compact-tension specimen should be

$$(W - a) \geq \frac{4}{\pi}\left(\frac{K_{max}}{\sigma_y}\right)^2 \qquad (10)$$

Although there are no specific requirements on the specimen thickness, bearing in mind the sensitivity of fracture toughness to thickness it should be representative of the section thickness of the component.

The crack length is measured at regular intervals in the course of the test by one or a combination of optical, unloading compliance, and electric-potential drop techniques. The optical method can only enable the detection of crack growth on the surface, and thus any crack propagation by tunnelling ought to be checked and corrected for. On the other hand, the interruption of cyclic loading for the unloading–loading compliance measurements can lead to creep deformation or to environmental stress cracking in an aggressive environment if the interruption times exceed 10 minutes, which is unlikely. It should be possible to perform each unloading compliance measurement within a minute.

There are two different fatigue test procedures in ASTM E647 [29] depending on the crack growth rates. The K-increasing test procedure, a constant load-amplitude test, is suitable for crack growth rates da/dN greater than 10^{-5} mm/cycle. In this procedure, the amplitude of the load is kept constant, and ΔK increases with crack extension. The K-decreasing test is preferable for crack growth rates below 10^{-5} mm/cycle when near threshold data are required. The rate of load shedding should be gradual in the K-decreasing method. This can be met by limiting the normalized K-gradient,

$$C = \frac{1}{K}\frac{dK}{da}$$

to a value of $-0.08\,\text{mm}^{-1}$.

The ASTM standard recommends a procedure for assessing the occurrence of any history effects during a K-decreasing test, since the previous cycles produce larger plastic zones that can retard crack growth during the succeeding cycles. Therefore the data obtained with the K-decreasing schedule is repeated by reversing to K-increasing and continuing with the crack propagation until the growth rate is well out of the threshold region. The segments of da/dN vs. ΔK curve obtained in these two manners should match.

The stress intensity amplitude is calculated by

$$\Delta K = \Delta K_0 \exp[C(a - a_0)] \qquad (11)$$

where ΔK is the targeted stress-intensity factor range, ΔK_0 is the initial ΔK value at the start of the test, a is the current crack length, a_0 is the initial crack length, and C is the K-gradient parameter.

da/dN values can be extracted from the a versus N curves. The ASTM standard suggests two methods of computing the derivatives. The secant method involves simply calculating the slope of the straight line connecting two adjacent data points on the a versus N curve, i.e.,

$$\left(\frac{da}{dN}\right)_{\bar{a}} = \frac{a_{i+1} - a_i}{N_{i+1} - N_i} \tag{12}$$

Since the computed da/dN is an average rate over the $(a_{i+1} - a_i)$ increment, the average crack length $\bar{a} = (a_{i+1} + a_i)/2$ may be used to calculate ΔK.

The second method, the incremental polynomial method, involves fitting a second-order polynomial to a series of successive data points and solving for the derivative mathematically. Further details and a listing of a Fortran program for the curve-fitting and da/dN computation are given in ASTM E647 [29] document.

Component-Specific Fracture Tests

There are a number of standards specific to assessing fracture behavior of various thermoplastic pipes: see BS 3505, 1986, Unplasticised polyvinylchloride (uPVC) pressure pipes for cold potable water: Method for determination of fracture toughness. Appendices C and D of this standard specify a method for the determination of fracture toughness of uPVC pipes (diameter ranging from 40 to 400 mm) after exposure to an aggressive environment at approximately 20°C. The test setup is well illustrated/described in the textbook by Anderson [16]. A 30 mm long ring section is removed from the pipe and is submerged in dichloromethane for 15 minutes, the specimen is removed from the liquid chemical, and the surface is inspected for bleaching or whitening. A segment is cut out of the ring opposite where the notch inside the pipe would be, producing a C-shaped (or arc-shaped) section. A sharp notch ($\simeq 50\,\mu$ radius) is machined on the inner surface and along the full length of the pipe section to a depth of 25% of the pipe wall thickness. The notched pipe test piece in the form of a C-profile cantilever is then loaded by supporting one and hanging a dead weight from the other of the two free edges of the C-profile (inducing bending moment at the notch region). The loading is to be maintained for 15 minutes or until visible cracking or complete fracture occurs, recording whichever period is shorter. The fracture toughness of the material, with failed specimens, can be computed from the applied load and the notch depth by means of standard K_I formulae. If no cracking is observed during the 15 minute test, the toughness may be quantified by testing additional specimens under higher loads. The standard includes a semiempirical size correction for small pipes and high-toughness materials that do not behave in an elastic-like manner.

Another is ISO 3127, 1994, Thermoplastic pipes—Determination of resistance to external blows—round-the-clock method. Test piece pipe lengths 200 mm long, conditioned at 0°C, are subjected to blows from a falling weight impact striker. The pipes with outside diameters 40 mm are hit by a single blow. Pipes with greater diameters are hit by multiple blows (the number of blows depending on the diameter) around the pipe circumference (hence the round-the-clock method). A parameter known as the true impact rate (TIR) of a batch of specimens is determined. The TIR is the total number of failures divided by the total number of blows, expressed as a percentage, and should not exceed 10%.

Another is ASTM F1474–93, Slow crack growth resistance of notched polyethylene plastic pipe. This is a qualitative screening criterion. Its remit is to determine the resistance to slow crack growth of polyethylene pipes in terms of time to failure of the pipe specimen containing axial notches under hydrostatic pressure. The test is appropriate to pipes of wall thickness > 5 mm. Test pieces are pipe sections with lengths of three times the pipe's outside diameter, containing four external V notches equally spaced around the pipe circumference. The notches are axial slots of length equal to the pipe outside diameter and notch depth of 20% of the pipe wall thickness. The specimen is filled with water for the test, immersed in a water tank at 80°C, and allowed to condition for a period of 24 h for pipes of wall thickness up to 25 mm, and 48 h for greater wall thicknesses, subsequently subjected to hydrostatic pressure by being internally pressurized with water up to the desired test pressure within 30 to 40 s at a uniform rate. The time to failure is recorded to the nearest 1 h.

It is important to remember that all these tests should be preceded by an appropriate risk assessment in order to implement all the reasonable *health and safety measures*. For instance the use of such toxic chemicals as dichloromethane is the subject of COSHH (Control of Substances Hazardous to Health) regulation. Other activities where latent energy is involved, as in the application of hydrostatic pressure, and scattering of specimen debris from a fracture test, also pose a degree of hazard. Obviously the full circumstances are best known to the actual people engaged in any specific work, and therefore they are best placed to conduct a risk assessment and establish appropriate health and safety measures/practices.

Notch-Impact Test

The notch-impact test is very popular throughout the plastics industry because it is a straightforward test and well covered by all the standards organizations. See ASTM D256–93a Pendulum impact resistance of notched specimens of plastics; ISO 179, 1993, and BS 2782, Part 3, Method 359, 1993, Determination of Charpy impact strength; ISO 180, 1993 and BS 2782, Part 3, Method 350, 1993, Determination of Izod impact strength; DIN 53 453, Testing of plastics; impact flexural test. These tests describe impact testing of notched polymeric specimens in Charpy mode, when the specimen is broken in three-point bending, and in Izod mode, if the specimen is mounted as a notched cantilever beam supported at one end and struck by the pendulum at the other free end. The normalized fracture energy is known as the impact strength. The data obtained are not true material parameters but only apply for the specific specimen size and geometry and testing conditions employed. It is, thus, inappropriate to use data from these idealized tests for the prediction of impact behavior of finished components. The suitability of a test for an application needs to be ascertained. Kakarala and Roche [30,31] have evaluated various impact test methods using a range of unfilled thermoplastics such as polypropylene (PP), acrylonitrile-butadiene-styrene (ABS), nylon (RIM), polyurethane (PU) (RIM), and reinforced thermosets such as PU (RRIM, i.e., reinforced reaction injection molding) with 20% glass flake, polyester with 28% glass fiber, and vinyl ester with 65% glass fiber, which are materials considered for automotive applications. The part of the work [30] concerned with the quality of test output showed that the results from different tests mostly do not correlate. Instances of correlation existed only in cases where there was a similarity in test stress states and measured characteristics. The study [31] on the applicability of the various test methods for predicting in service performance showed that the material impact performance ratings vary between different test methods and that it would require a number of test methods in combination for a meaningful material characterization. It was recommended that the material screening and selection for impact performance should be made using test methods with stress state and measured parameters that match the stress state, controlling variable, and failure limit of the intended application.

A comparison [32] of standard (ASTM and DIN) Charpy notch-impact test results with those of critical strain energy release rates G_c, obtained in accordance with LEFM criteria under the condition of small-scale yielding, were made for various unfilled polypropylenes, blends of PP with ethylene-propylene random copolymer (EPR) and also with the maleated version of EPR, and rigid particle filled grades of both PP and its blends. Only in the case of polypropylene with various rigid filler content was there reasonable correlation between the two methods. However with the rest of the material systems, there was no agreement between the Charpy and the fracture toughness data. The conclusion was that Charpy tests are not reliable indicators of the relative toughness of these materials and thus are of little value for material selection and design.

Part of the problem is that the standard notch-impact test methods specify blunt notches but as would be expected the fracture energy decreases as the notch becomes sharper, as can be seen in Fig. 6. Note that materials vary in their notch sensitivity, making any ranking of the materials for impact strength very arbitrary. Notwithstanding, more reliable and fundamental information can be obtained from these popular notch-impact tests, particularly with the advent of instrumented impact testing. The impact tests also enable the simulation of the high loading rates that materials encounter in many applications. Accurate force and energy information can be measured from instrumented impact tests for the determination of LEFM fracture mechanics parameters. Obviously sharp notches/cracks need to be introduced into test pieces by, for instance, lightly tapping a razor blade into the machined notch tip. Under impact loading rates, the extent of plastic deformation at the tip of a sharp notch would be smaller than that experienced under static loading rates, rendering the LEFM approach more appropriate.

A test may typically involve the extraction of force P and energy associated with the initiation of crack extension from an instrumented impact load–deflection trace (see Fig. 7) for a test conducted in three-point bending mode (Charpy arrangement). The associated fracture stress can be calculated from

$$\sigma = \frac{3PS}{2BW^2} \tag{13}$$

where S is the loading span.

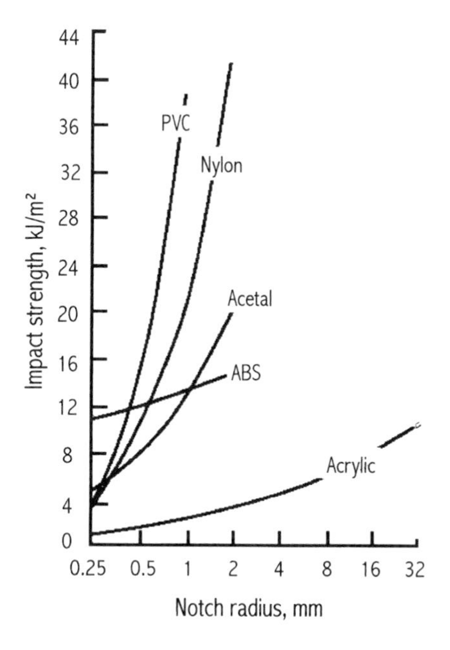

Figure 6 Effect of notch radius on the Izod strength of various engineering plastics. (From Ref. 33.)

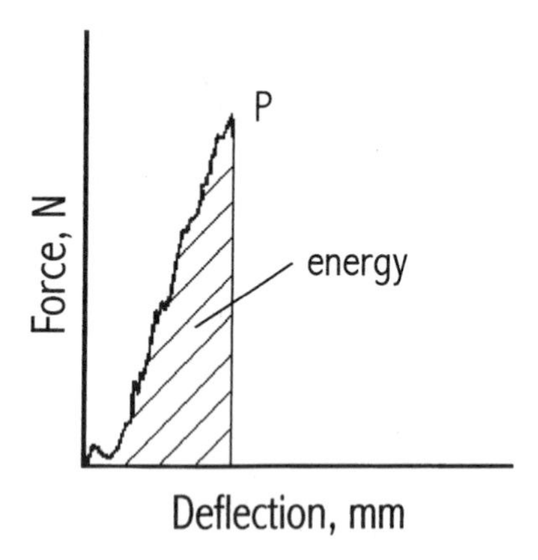

Figure 7 A typical instrumented impact load–deflection trace.

Normally a series of identical specimens with different notch lengths a are tested. Based on $K_c = Y\sigma a^{1/2}$ relationship, σY is plotted against $a^{-1/2}$, and the slope of the best straight line fit to the points is an estimate of the critical stress-intensity factor K_c.

Similarly G_c, the critical energy release rate, can be calculated from the slope of a best fit straight line to a plot of the impact fracture energy data against $BW\phi$ for the test specimens with a range of notch depths. The data for the geometric calibration factors Y and ϕ are available for various a/W and S/W ratios [6,10] and is given here in a graphical form, Fig. 8.

Further Considerations

Some of the difficulties as well as the exceptions to the above covered, mostly standard, approaches will be indicated in this section.

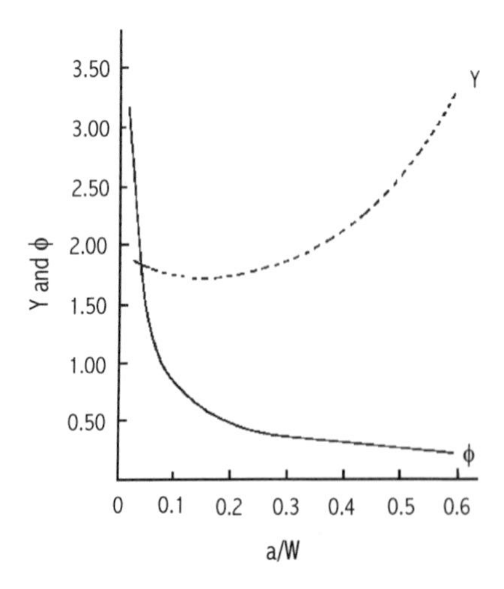

Figure 8 The variation of the calibration factors Y and ϕ with (notch length)/(specimen width) ratio.

Specimen types and crack length measurements. Compact tension (CT) and single-edge notched bend (SENB) specimens are most widely used in fracture toughness testing of plastics, although there are other standardized specimen configurations, such as the middle or center tension (MT) and disk-shaped compact-tension and arc-shaped specimens (see Ref. 16 for detailed illustrations). There are also additional, not standard, specimens that are adopted by researchers. These include double-edge-notched tensile bars [34], tapered double cantilever beams [35], double torsion specimens [36], and side-grooved SENB specimens [37] (to reduce the possibility of plastic deformation at fracture). Furthermore, there is a double-notch four-point bend specimen [38] in order to study the crack tip process zone such as tearing of the ligaments spanning the crack faces in ductile plastics. Since the two notches cannot be exactly identical, one will propagate as an unstable crack, leaving behind the other notch with a process zone at its tip.

A comparison of the most popularly used SENB and CT specimen types indicates that for plastics the SENB specimen is preferable: SENB specimens require less machining than CT specimens. The geometric condition imposed by the standards, for example $2B \leq W \leq 4B$, results in a more practical SENB specimen size. Specimens are usually taken from either injection molded plaques or extruded sheets of 2–5 mm thickness. Even a molding thickness of $B = 5\,\text{mm}$ produces a CT specimen of very small width. Handling and further machining of such small specimens can become tedious. Furthermore, loading pinholes with small diameters ($\simeq W/4$) may cause high pin penetrations during the loading. There is, therefore, a considerable amount of work when the stipulation of $W/B \leq 4$ is ignored [23,39–43]. The work [23] on PVC and polycarbonate (PC) has shown that for some materials this is not so critical: varying the specimen thickness B from approximately 3 mm to 23 mm at a fixed CT specimen width value of $W = 50\,\text{mm}$ has not indicated any significant variation in fracture toughness measurements at 25°C.

SENB specimens offer further flexibility in that they are suitable for static as well as impact three-point bending tests, where the loading span S can be adjusted.

Crack length measurements. Accuracy of crack length measurements is a major factor in obtaining reliable fracture toughness data, particularly in procedures such as J-integral, CTOD, and fatigue crack growth, where the crack growth needs to be measured at regular intervals. The work of the ESIS Task Group on Polymers and Composites, to develop a standardized testing procedure for evaluating J–R curves for plastics, has identified crack length measurements in tested specimens as an area that requires resolution [44]. A number of approaches have been examined to isolate/identify the experimental crack: cryogenic or room temperature fracturing of the crack-containing tested specimens; fatigue fracturing after the test; injection of ink into the crack to mark the crack front; measurement of the crack length from polished sections under load, etc.

Identification of crack initiation, and measurement of crack length extension during a test are also critical. Methods employ rapid photography [45,46], electric resistance (electric potential difference) [47–49], laser-based interferometric strain/displacement gauges [50], ordinary strain gauges [46], and COD clip-on gauges [51]. More sophisticated techniques have also been employed, for instance, a combination of micro focus radiography and acoustic emission (two real-time methods) [52].

It should be emphasized that together with the implementation of a sharp initial crack into the specimen, the identification of the onset of the crack growth and monitoring of the extent of crack propagation are very important. However, in spite of a considerable concentration of work, there is still concern over the reliability of these measurements.

Crack closures. The crack closure concept [53] describes the possibility and consequences of contact between crack surfaces at low load levels during fatigue testing. It is thought that when a specimen is cyclically loaded within the stress-intensity range of K_{min} and K_{max}, the crack faces make contact below a stress-intensity level, which is greater than K_{min}. It is assumed, therefore, that a lower portion of the fatigue cycle does not contribute to fatigue crack growth, and this may be accounted for by defining an effective stress-intensity range ΔK_{eff} and using it in the Paris–Erdogan equation.

The mechanisms of fatigue crack closure have been identified for metals. These are illustrated by Anderson [16], who has also expanded upon the effects of crack closure on the fatigue properties of metals. The phenomenon can also manifest itself in polymer-based materials because of the plastic deformation, viscoelasticity, residual stresses, and environmental conditions.

Another fatigue testing related consideration is that in an ideal test $dK/da = 0$, in other words a constant amplitude loading is maintained. A real structure, however, experiences a spectrum of stresses in service, which can result in overloads. A detailed analysis of the subject of variable amplitude loading and the associated concepts are covered by Anderson [16].

Slow crack growth. The fracture mechanics approach in an appropriate form is the only meaningful way to quantify the fracture behavior of materials, although the best results can only be achieved if there is a full awareness of the limitations of the test methods, data analyses, material considerations, and environmental conditions. Polymers, as indicated earlier, are viscoelastic in nature and therefore depending on the service temperature will exhibit time-dependent stress–strain behavior to varying degrees. Accordingly in polymers, even under static loading, slow crack growth can occur for stress levels below those associated with K_{Ic} if the conditions favor viscoelastic response, e.g., at temperatures near the T_g of the material, or due to environmental stress cracking. Normally K_{Ic} indicates that fracture will not occur in a material with a given crack size so long as the stress levels sustained are such that the stress intensity factor does not exceed K_{Ic}.

An estimation of slow crack growth rate and, in turn, a prediction of failure times can be made from a power law equation, which relates the crack velocity to the stress intensity factor K, similar to the equation governing the crack growth under fatigue cycling:

$$\frac{da}{dt} = AK^n \tag{14}$$

where A and n are constants.

Detailed derivations of the relationship similar to Eq. 14 can be found in the literature [16,54,55]. Double-torsion specimens [56,57] are convenient, because of their geometry and the mode of loading, for the study of crack growth rates under a fixed load, where G and K remain constant, i.e., independent of the crack length.

3.2 Materials Evaluation

Thermoplastics

Limited fracture data (room temperature measurements) for some of the thermoplastics are given in Table 1. There is good agreement between the values from different sources. Fracture performances of certain materials, however, are quite sensitive to variations in molecular weight and residual stresses. The variations between the data for uPVC (unplasticized polyvinyl chloride) is assumed to be due to differences in K values or ISO viscosity numbers (a technological term indicating molecular weight). The sensitivity of PC (polycarbonate) to frozen-in residual stresses is also clear when the data for as-extruded and quenched specimens are compared. Figure 9 clearly shows the influence of frozen-in residual stresses on fracture behavior: in the presence of compressive residual stresses at the surface layers of notched specimens, a much greater degree of plastic deformation (a plane stress condition) occurs at a given specimen thickness. Side grooving of SENB specimens results in lower G_{Ic} and J_{Ic} values due to inducement of more brittle fracture.

One of the ongoing endeavors is to improve the fracture performance of plastics. Such improvements have been achieved by the processes of copolymerization (e.g., ABS (acry-

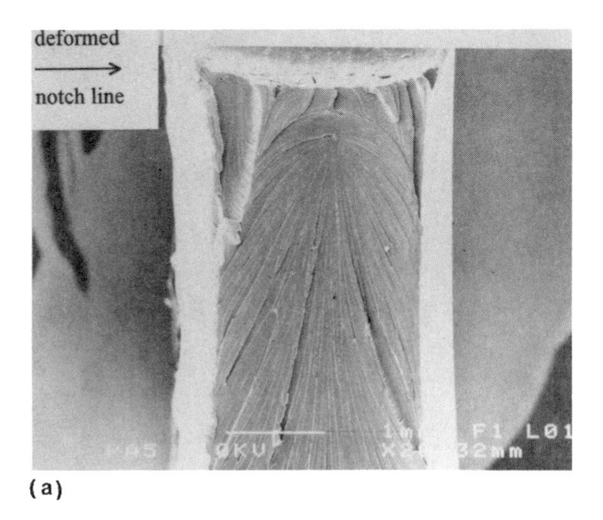

(a)

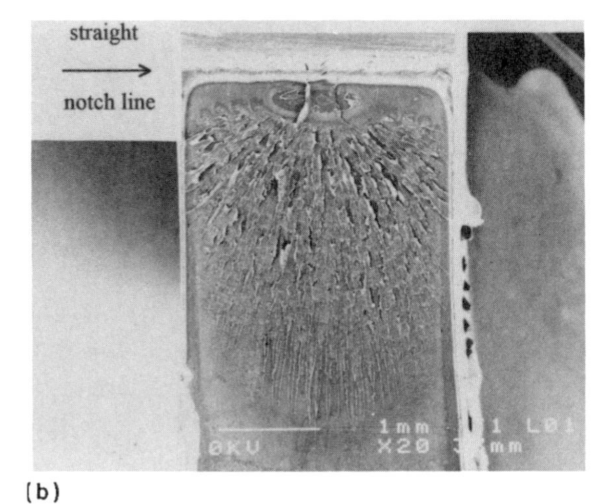

(b)

Figure 9 SENB impact fracture surfaces for (a) residual stress containing; (b) residual stress free polycarbonate specimens. (From Ref. 71.)

Table 1 Fracture Toughness Data for Thermoplastics

Material	Test type/method	Specimen type/dimension (mm)	Test speed	Fracture parameter	Ref.
1. HDPE	static/ASTM E813	SENB; $B = 27$, $W = 51$, $S = 204$	1 mm min^{-1}	$J_{Ic} \simeq 1.8\,kJm^{-2}$	58
2. HDPE	static	SENB; $a/W = 0.6$, $S/W = 4$		$J_{Ic} \simeq 2.8\,kJm^{-2}$	59
3. HDPE	static	SENB (side-grooved) $a/W = 0.6$, $S/W = 4$		$J_{Ic} \simeq 2.2\,kJm^{-2}$	59
4. HDPE (Rigidex002-55)	static/ASTM E399	SENB; $B = 10$, $W = 26$, $S/W = 8$, $a/W = 0.5$	5 mm min^{-1}	$K_{Ic} \simeq 1.6\,MPam^{1/2}$	60
5. HDPE (Rigidex006-60)	static/ASTM E399	SENB; $B = 5$, $W = 12$, $S/W = 4$, $a/W = 0.5$	1 mm min^{-1}	$K_{Ic} \simeq 1.2\,MPam^{1/2}$	60
6. HDPE (BP Rigidex002-55)	impact	SENB; $B = 9$, $W = 6$, $S = 41$		$G_{Ic} \simeq 6.4\,kJm^{-2}$	37
7. HDPE (Rigidex002-55)	impact	SENB (side-grooved) $B = 9$, $W = 6$, $S = 41$		$G_{Ic} \simeq 1.9\,kJm^{-2}$	37
8. HDPE (Rigidex006-60)	impact	SENB; $B = 9$, $W = 6$, $S = 41$		$G_{Ic} \simeq 7.4\,kJm^{-2}$	37
9. HDPE (Rigidex006-60)	impact	SENB (side-grooved) $B = 9$, $W = 6$, $S = 41$		$G_{Ic} \simeq 2.1\,kJm^{-2}$	37
10. uPVC	static/ASTM E399			$K_Q \simeq 6.3\,MPam^{1/2}$	23
11. uPVC		Four-point bending	0.013 min^{-1}	$K_{Ic} \simeq 3.5\,MPam^{1/2}$	61
12. Polycarbonate	static/ASTM E399	CT		$K_Q \simeq 3.1\,MPam^{1/2}$	23
13. Polycarbonate	impact	SENB	1 ms^{-1}	$K_c \simeq 2.8\,MPam^{1/2}$	5
14. Polycarbonate	impact	SENB		$G_c \simeq 4.8\,kJm^{-2}$	62
15. Polycarbonate (as extruded)	static	SEN (tension)	12 mm min^{-1}	$K_Q \simeq 3.6\,MPam^{1/2}$	63
16. Polycarbonate (quenched)	static	SEN (tension)	12 mm min^{-1}	$K_Q \simeq 6.5\,MPam^{1/2}$	63
17. Polystryene	impact	SENB		$G_c \simeq 8.8\,kJm^{-2}$	62

18. HIPS (7% rubber)	static	SEN (tension)	5×10^{-3} sec^{-1}	$K_{Ic} \simeq 1.5$ MPam$^{1/2}$	64
19. ABS	impact/ASTM E813	SENB; $B = 4$, $W = 10$, $L = 80$	1 ms^{-1}	$J_{0.2} \simeq 2.5$ kJm^{-2}	65
20. PMMA	impact	SENB		$G_c \simeq 1.3$ kJm^{-2}	62
21. PMMA	impact	SENB; $B = 9$, $W = 6$, $S = 41$		$G_{Ic} \simeq 0.7$ kJm^{-2}	37
22. PMMA (bone cement)	static	CT; $B = 64$, $W = 63.5$		$K_{Ic} \simeq 1.6$ MPam$^{1/2}$	66
23. PMMA	impact	SENB	1 ms^{-1}	$K_C \simeq 1.6$ MPam$^{1/2}$	5
24. Nylon 6,6	impact	SENB		$G_c \simeq 5.3$ kJm^{-2}	62
25. Nylon 6.6 (Zytel 101)	impact	SENB; $B = 6.4$, $W = 12.7$, $S = 89$		$G_{Ic} \simeq 3.1$ kJm^{-2} $K_{Ic} \simeq 3.0$ MPam$^{1/2}$	67
26. Nylon 6,6	impact	SENB		$K_c \simeq 3.2$ MPam$^{1/2}$	5
27. Nylon 6,6 (Maranyl A100)	static/ASTM E399	CT; $B = 4$, $W = 35$	10 mm min^{-1}	$K_Q \simeq 4.2$ MPam$^{1/2}$	31
28. Polyacetal (Delrin 500)	impact/ASTM E399	SENB; $B = 6.3$, $W = 25.4$, $S = 102$	1 ms^{-1}	$K_{Ic} \simeq 3.1$ MPam$^{1/2}$	68
29. Polyacetal	impact	SENB	1 ms^{-1}	$K_c \simeq 2.7$ MPam$^{1/2}$	5
30. PP (ICI GWM22)	impact	SENB; $B = 3.6$, $W = 10$, $S = 40$	1 ms^{-1}	$K_c \simeq 0.9$ MPam$^{1/2}$ $G_c \simeq 1.2$ kJm^{-2}	69
31. Unfilled PES (Victrex 4800G)	static	SENB; $B = 12$, $W = 3.4$, $S = 70$	0.5 mm min^{-1}	$K_{Ic} \simeq 2.7$ MPam$^{1/2}$ $G_{Ic} \simeq 3.1$ kJm^{-2}	70
32. Unfilled PES (Victrex 4800G)	impact	SENB; $B = 12$, $W = 3.4$, $S = 70$	1 ms^{-1}	$K_{Ic} \simeq 2.2$ MPam$^{1/2}$ $G_{Ic} \simeq 3.3$ kJm^{-2}	70

lonitrile-butadiene-styrene), EPDM (ethylene-propylene terpolymer), rubber dispersion into plastics as in HIPS (high-impact polystyrene), and a variety of polymer blendings. Fracture behavior of these modified material systems has received considerable attention [57,65,72–79]. Some of the specific material systems examined include blends of PC/ABS [65,72,73], ABS/TPU (thermoplastic polyurethane) [65] and PP (polypropylene)/EPR (ethylene propylene rubber) [75], RIM (reaction injection molding) block copolymers of polyamide-6/polyether [40], rubber toughened polymethyl methacrylate (PMMA) [76], styrene-acrylonitrile copolymer (SAN) [77], polycarbonate/polyester blends [78], and cross-linkable epoxy thermoplastics (CET) [79].

One of the attractions of the fracture mechanics approach is that the results are geometry independent. This, of course, is subject to conducting tests with valid specimens that meet the dimensions criteria set out in the standards. Considerable work has been conducted with various materials to verify how appropriate these specimen size criteria are for specific polymers and to ascertain how sensitive the test results are to changes in specimen dimensions and to deviations from the criteria. The specimen thickness is the geometric variable that is considered most, since it governs whether the fracture is plane-strain or not. The effect of specimen thickness B on fracture parameters, e.g., K, G, J, and CTOD, has been examined for uPVC [23,61], polycarbonate [23,63], HIPS [64], HDPE [37,59,60,], ABS [65], and ABS/PC blends [74]. The effect of a/W on fracture toughness properties is also considered for some of these materials, e.g., uPVC [25], HDPE [59,60], polyacetal [68], and polyether sulphone (PES) [70].

The viscoelastic nature of the polymeric materials has necessitated studies that examine the influence of the rate of loading and the test temperature on the fracture properties. The influence of impact speed on K_{Ic} and G_{Ic} has been evaluated for PC [5], PMMA [5,73], HDPE [46], PS [80], MDPE [81], uPVC [82], nylon 6,6 [67,5], and polyacetal [5,68,74]. Some of these studies [82, 60] also include the influence of temperature on the properties.

Viscoelasticity can also manifest itself in the form of slow crack growth [5, 83], which is particularly relevant in applications where the material is under stress for long periods of time, as in utility transportation pipes.

Slow crack growth is accentuated if the material is subjected to a hostile environment that can promote environmental stress cracking, although improvements also result in the presence of some liquids, for instance, the toughness of PMMA increases considerably when fracture occurs in water [84].

Time dependency of fracture is more obvious under fatigue cycling. Studies on fatigue crack propagation include the effects of cross-linking for PMMA and PS [85], molecular weight and material composition for PMMA [85,86], ABS and polystyrene acrylonitrile [87].

Fracture toughness studies on notched, representative pipe sections have been conducted in order to simulate in-service conditions and also to check the extent of correlation with small test pieces. Such works are conducted on small diameter (8–12 mm) tubes of polyamides 11 and 12 for air brake and fuel line systems [88], and on uPVC [89] pressure pipes.

Thermosetting Plastics

Static and impact fracture data (room temperature measurements) mainly for epoxy resin systems is presented in Table 2. These relatively brittle materials require modification by, for instance, blending with a suitable rubber or thermoplastic to improve fracture toughness. Such improvements, however, depend not only on the types of materials but also on the composition and therefore the type of blend structure (e.g., continuous-discrete phases, co-continuous phases, phase-inverted). Table 2 shows that improvement has been achieved in epoxy resin (listed 8) as a result of blending with 10% styrene-acrylonitrile copolymer (SAN), but increasing SAN content beyond 10% reverses the trend, and at a 20% addition, a clear deterioration in the property is indicated due to a change in the state of the phases [93]. Rigid fillers are added to thermosetting plastics to control cost and modify processing and/or end-use behavior. Contrary to expectation the addition of rigid particulate fillers to brittle thermosetting plastics can also result in increases in K_{Ic} and G_{Ic}. A review of this subject, including the influence of such parameters as the filler volume fraction, particle size, and particle-matrix adhesion on fracture toughness can be found in Ref. 95.

The data in Table 2 show that the fracture properties of the toughened epoxy systems, the same as ductile thermoplastics, can be rate dependent. A comparison of the static and impact results shows that lower values were obtained under impact conditions, which is more realistic, since at lower rates of loading, with a given specimen size, the possibility of violating plane-strain (brittle failure) condition increases. The impact tests are also more representative for cases where materials encounter high loading rates in applications, either by design or accident. Tests over a temperature range [91,92] show that the fracture toughness remains almost temperature independent for basic epoxy resin thermosets at temperatures below T_g.

Properties of the thermosetting polymers are controlled by the molecular composition and structure such as the stoichiometric ratio between the reagents, and the cross-link density of the network. Accordingly, fatigue crack propagation of an Epon 828/MDA epoxy has been examined [96] with respect to amine/epoxy ratio and cross-link density.

4 Polymer Matrix Composites

4.1 Discontinuous Fiber Reinforced Composites

There are no standard test methods specific for discontinuous fiber (or short fiber) reinforced thermoplastics. It is also not clear whether a geometry-independent fracture parameter can be measured for these nonuniformly inhomogeneous materials. However in spite of these reservations there has been considerable work conducted towards characterizing short fiber composites for fracture toughness using the standard and other procedures outlined in the previous sections. The investigators have recognized that fracture mechanics data provide much more reliable information than the customary alternative tests for material selection and also a service performance indicator for components.

The properties of short fiber composites, including fracture toughness, are a function of fiber content and fiber length and orientation distributions. These microstructural parameters usually show a variation through the thickness of a specimen, particularly the fiber orientation as dictated by the melt flow behavior. For instance, it is well known that in injection moldings there is at least a three-layered arrangement with respect to fiber orientation in the form of two surfaces, where the fibers are predominantly longitudinal with the mold fill direction (MFD) and a core, where the fibers are transverse to the MFD. The position of a crack and/or the direction of the applied load with respect to the microstructure can influence the outcome of the mechanical property measurements considerably [69,97]. The interaction of a propagating crack with fibers may result in such energy dissipation mechanisms (toughening mechanisms) as the fiber-matrix interface debonding, fiber pull-out, and matrix fracture along fiber avoidance paths, in addition to straightforward deformation and fracture of the fibers and the matrix. Energy dissipa-

Table 2 Fracture Toughness Data for Thermosetting Plastics

Material	Test type/method	Specimen type/dimensions (mm)	Test speed	Fracture parameter	Ref.
1. Epoxy resin (DGEBA) cured with piperidine	static	CT; $B = 10$, $W = 30$	12.5 mm min^{-1}	$G_{Ic} \simeq 0.23\,kJm^{-2}$	90
2. Epoxy resin (DGEBA) cured with piperidine	impact	Single-cantilever (sharp-notch Izod); $B = 10$, $W = 12.5$, $L = 64$	1 ms^{-1}	$G_{Ic} \simeq 0.29\,kJm^{-2}$	90
3. Modified epoxy resin (1) with 8% CTBN	static	CT; $B = 10$, $W = 30$	12.5 mm min^{-1}	$G_{Ic} \simeq 2.5\,kJm^{-2}$	90
4. Modified epoxy resin (1) with 8% CTBN	impact	Sharp-notch Izod; $10 \times 12.5 \times 64$	1 ms^{-1}	$G_{Ic} \simeq 1.1\,kJm^{-2}$	90
5. Epoxy resin (DGEBA) cured (120°C) with piperidine	static	CT	1 mm min^{-1}	$K_{Ic} \simeq 0.8\,MPam^{1/2}$	91
6. Modified epoxy resin (5) with 15% CTBN rubber	static	CT	1 mm min^{-1}	$K_{Ic} \simeq 2.3\,MPam^{1/2}$	91
7. Filled epoxy resin (hydantoin based) and cured (130°C) with an anhydride (HY925) (40% by vol. silica particles)	static	SEN (tension)		$K_{Ic} \simeq 2.3\,MPam^{1/2}$	92
8. Epoxy resin (DGEBA) cured (200°C) with 4,4′ DDM	static/ASTM E399	SENB; $B = 10$, $a = 3.3$	1.3 mms^{-1}	$K_{Ic} \simeq 1.1\,MPam^{1/2}$	93
9. Epoxy resin (8) modified with 10% by wt. SAN copolymer	static/ASTM E399	SENB; $B = 10$, $a = 3.3$	1.3 mms^{-1}	$K_{Ic} \simeq 1.6\,MPam^{1/2}$	93
10. Phenolic resin (BPJ2027L) cured (50°C) with 5% Plencat 10	static	SENB; $B = 6$, $W = 12$, $a/w = 0.5$, $S = 4W$	10 mm min^{-1}	$K_{Ic} \simeq 0.8\,MPam^{1/2}$ $G_{Ic} \simeq 0.2\,kJm^{-2}$	94

tion associated with these fracture processes have been estimated in isolation for ideal cases [98,99]. The energy associated with the fiber pull-out is much greater than the energy of debonding at a ratio equivalent to approximately $3E_f/\sigma_f$, where f describes fiber. To maximize, therefore, energy absorption during a fracture process, fibers with lengths approximately the same as the critical fiber length, $l_c = \sigma_f d/\tau$ (where σ_f and d are fiber tensile strength and diameter, and τ is the interface shear strength or the matrix yield strength) should be oriented at right angles to the crack surface in order to produce maximum fiber pull-out. In practice, the microstructure can be rather complex, and therefore it is difficult to quantify the extent of debonding, pull-out, etc. and make an accurate estimate. More reliably, tests should be conducted on specimens with similar microstructural arrangements as the finished part [100].

Short fiber reinforced polymer composites can undergo a considerable amount of slow crack growth, due to various microscopic failure processes, before crack grows in an unstable manner. Therefore the main fracture toughness parameter K_c or K_Q may not be appropriate by itself to characterize the total fracture behavior. Accordingly the concept of crack growth resistance curves (*R*-curves) has been adopted by various workers in conjunction with both the stress-related parameter K and the energy-related parameters such as J, G, and the specific work of fracture (denoted by w_f). However the work [101–103] on short fiber reinforced epoxy composites has shown that the R-curve based on stress-intensity factor did not lead to a size-independent material parameter because the results depended on the initial crack length. Yet it has also been shown for various fiber reinforced thermoplastics [104] that if the specimens were to be cut out of molded relatively large plaques rather than using as-molded dumbbells, the measured toughness K_Q would be size-independent. This is because of the greater uniformity of the fiber orientation distribution along the crack path in the cut test pieces.

The two fracture mechanic parameters, K_c and G_c, were also used to characterize the fracture properties of extrudates with relatively good unidirectional fiber alignment [43]. The critical strain energy release rate was determined by using the following form of the concept [105], which makes it more appropriate where the crack propagation is stable:

$$G_c = \frac{P_c^2}{2B}\left(\frac{dC}{da}\right) \tag{15}$$

where P_c is the load at the onset of crack propagation, B is the specimen thickness, C is the specimen compliance, and a is the crack length. G_c is calculated from experimental measurements of dC/da (with the aid of a suitable displacement gauge such as a linear variable differential transformer, LVDT).

While the calculation of G_c is independent of material properties, in determining the critical stress intensity factor K_c, the anisotropic elastic properties of the extrudates has to be taken into account. For the compact tension specimen geometry, as indicated before, K_c is given by $K_c = Y[Pc/(BW^{1/2})]$. Y is a geometry-dependent parameter, and for isotropic materials [6] it is a function of a/W. However for highly anisotropic materials as in some extrudates, Y should strictly be recalculated [106,107] to allow anisotropy in Young's moduli. Often such an adjustment to Y has been ignored where the longitudinal to transverse Young's moduli ratio is as high as approximately two [43].

Carling and Williams [24], working with a variety of glass fiber reinforced thermoplastics using CT specimens of different sizes (50×50 mm, 100×100 mm) and a SENB specimen of 20×100 mm, showed that J-integral tests work well with these materials. The value of J_c was reasonably independent of specimen type and size. The specimens, thicknesses varying from 3 to 10 mm, were cut from injection-molded plaques using the portion of the plaque away from the gate where the fiber orientation distribution attains a uniform pattern. The tests were conducted on an Instron testing machine with a cross-head speed of 5 mm min^{-1}. The J values were calculated from

$$J_c = \frac{\eta U}{B(W - a)} \tag{16}$$

where η is the calibration factor ($\eta = 2$ for SENB, and $\eta = 2.38$ for CT with $a_0/W = 0.5$), U is the deformation energy, and $W - a$ is the ligament length.

The J tests were conducted using multiple specimens. Each specimen was loaded to a predetermined load and then unloaded. The tested specimens were treated with a dye penetrant and broken open to measure the crack length. J_c values were taken to be the 'J-axis' intersect values from J–da plots. There was apparently no evidence of crack blunting at the notch tip. A drawing of a blunting line based on Eq. 17 below was not attempted, since using a bulk property of the material (i.e., the yield strength σ_y) to determine the extent of blunting, in this case due to microprocesses at the crack tip, was deemed to result in rather unrealistic estimates for such reinforced materials.

$$J = 2\sigma_y(da) \tag{17}$$

J_c can also be evaluated [108,109] using the locus of the crack initiation points; see Fig. 10, obtained from identical specimens that differ only in initial notch length. The accuracy of the technique depends on the accuracy of the identification of crack initiation. Points A, B, C, D in Fig. 10 denote crack initiations of the specimens with initial notch depths of a_1, a_2, a_3 and a_4. The line ABCD is the locus of the crack initiation points. J_c is determined from the crack-initiation energy differences between the specimens such that J_c is equal to the area OAD in Fig. 10 divided by the ligament area $B(a_4 - a_1)$. This method aims to eliminate the dependency of the critical displacement at which the fracture initiation occurs on a/W.

Another approach to fracture behavior characterization is by measuring the work of fracture [110] (i.e., energy absorbed during the creation of fracture surfaces following the

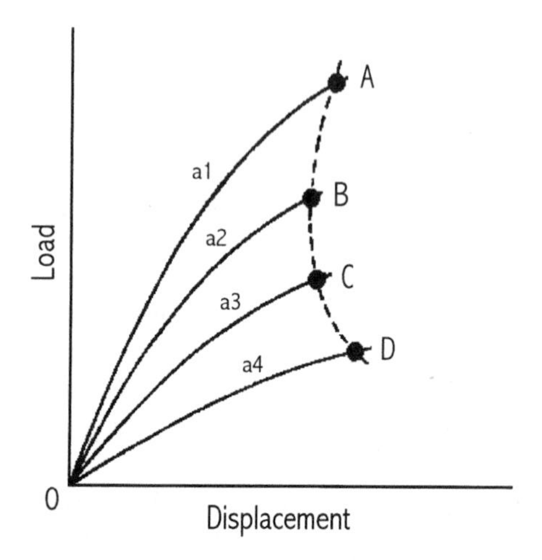

Figure 10 Schematic (load) versus (load-point displacement) traces for specimens with different initial crack lengths. (From Ref. 108.)

complete failure of the specimen). Such data can be conveniently obtained from Izod and Charpy type impact tests. Obviously greater accuracy can be achieved using instrumented equipment, otherwise the energy values indicated also represent some extraneous events. A specific work of fracture term is defined as

$$w_{\mathrm{f}} = \frac{U_{\mathrm{T}}}{B(W - a_0)} \tag{18}$$

where U_{T} is the total energy required to drive the crack through the whole specimen, and a_0 is the initial crack length.

Composites show extended damage zones (e.g., fiber-matrix debonding) in front of crack tips comparable to plastic zones of homogeneous materials. This necessitates the above indicated switch from LEFM to J-integral and R-curve concepts. There has been, however, some satisfactory work conducted where the damage zone has been treated in the same way as the plastic zone in metallic materials as far as the fracture mechanics approach is concerned [111].

One of the pioneers of fracture mechanics studies, Irwin [112], estimated the extent of plastic zone development at crack tip. The radius r_{y} of this plastic zone is given by

$$r_{\mathrm{y}} = \frac{1}{2\pi}\left(\frac{K_{\mathrm{I}}}{\sigma_{\mathrm{y}}}\right)^2 \tag{19}$$

More accurate K_{I} values, thus, can be determined by using an effective crack length in the calculations, $a_{\mathrm{ef}} = a_0 + r_{\mathrm{y}}$, which is slightly longer than the actual initial crack length a_0.

Similarly for the composites, a damage zone (or the process zone) dimension r_{p} can be incorporated into the fracture mechanics equation, Eq. 1. An extensive coverage of the damage or the energy dissipation zone for short fiber reinforced thermoplastics can be found in Ref. 113. Equation 19, however, has also been used [114] to improve the linearity of the plots of $(\sigma_{\mathrm{f}} Y)$ vs. $1/a^{1/2}$, particularly for tough thermoplastic matrix composites.

Composites specimens are also prone to being damaged by, for instance, fiber dislodgement during cutting from a larger piece and/or machining for a notch. This is a mixed blessing: the machining induced damage in fact renders the crack tip sharp and obviates any further sharpening with a razor blade, consequently similar fracture toughness results have been obtained [104,114,115] with specimens containing ordinary machined notches and those with razor or fatigue-crack sharpened notch tips. However the size of damage needs to be determined and added on to the initial crack length for more accurate calculations [116,117]. It is not easy to measure the extent of damage at the notch tip without resorting to x-ray measurements or an alternative nondestructive testing method. An estimate of the machining-induced damage size may be readily made by including data from unnotched specimens in a plot of impact energy vs. $BW\phi$ for the determination of G_{c} by using identical SENB specimens each with a different notch length. This should enable the extraction of a ϕ value that corresponds to the unnotched specimen. A value of a (which indicates the extent of damage) can then be read from the appropriate calibration curve of ϕ vs. a/W; see Fig. 8. Further details of this method are presented elsewhere [69,116].

The presence of unintentional damage in short fiber reinforced composites contributes to the relationship that appears to exist between the test results obtained with unnotched tensile tests and notched fracture toughness tests. A correlation is possible, since both these specimens fail in a similar manner where the failure is predominantly associated with the fiber–matrix interphase with little damage or yielding of the matrix beyond the immediate crack surface. The specimens fail in such a manner that the cracks grow along a fiber-avoidance path, i.e., a lateral diversion (damage) equivalent to a fiber length accompanies the forward progress of the crack. An assumption that the local strength in the failure zone is approximately equal to the ultimate tensile strength (UTS) produces, via Eq. 19, a process (damage) zone size of a fiber length long, for various short fiber reinforced thermoplastics [104]. This is realistic, since the crack has to progress by going around the fibers, which are roughly perpendicular to the crack surface. Accordingly the following relationship [104] holds for such materials:

$$l_f^* = r_p = \frac{1}{2\pi}\left(\frac{K_Q}{\text{UTS}}\right)^2$$

Rearranging we obtain

$$K_Q = (\text{UTS})(2\pi l_f^*)^{1/2}$$

Thus K_Q may be predicted from UTS and an effective fiber length representative of the longer population of the fibers, l_f^*.

A correlation between K_Q and UTS has been shown by others, also for short fiber reinforced thermoplastics [69], and more generally for a variety of composite systems, embracing unidirectional, 0°/90° woven and nonwoven, and random chopped fiber arrangements [118].

A comprehensive coverage of the theoretical considerations of the mechanical toughness of short fiber reinforced thermoplastics can be found in Ref. 103. Theoretical work usually concentrates on ideal material arrangements where equal fiber lengths are unidirectionally oriented with respect to the applied load. The outcome is often difficult to translate to real materials, where often the microstructure is much more complicated with respect to fiber length and orientation distributions. There are, however, semiempirical formulae which include microstructural features in the estimation of fracture parameters. In one such model [119] the composite fracture toughness $K_{Q,c}$ is assumed to be proportional to the toughness of the matrix $K_{Q,m}$ through the relationship of $K_{Q,c} = MK_{Q,m}$, where M is a microstructural efficiency factor, M incorporates the deformation characteristic of the matrix material a, and the relative effectiveness of all the energy-absorbing mechanisms n, so that $M = a + nR$, where R represents fiber volume fraction and fiber orientation. For accurate estimations, R needs to be determined for each case. This can be rather tedious: for instance in injection moldings, it requires a clear identification of the surface and core layers, and fiber volume fraction and fiber orientation determinations in these layers, necessitating the use of rather sophisticated image analysis equipment—in other words the medicine is worse than the pain!

Materials Evaluation

The evaluations of short fiber reinforced composites for fracture behavior includes studies with respect to material parameters such as the fiber content [42,103,120], fiber length and orientation [69,97,103,115], and fiber bundling [115,120]; testing conditions such as the temperature [24,42,121,122] and the loading rate [121,123,124]; and fractography [42,43,103,123]. The influence of some of these parameters, e.g., fiber content, rate of loading, and the test temperature on the fracture toughness of composites can be presented in the form of property maps [125,126].

It is worth stating that fractography can provide valuable information to help identify the events that have resulted in the fracture of a material. The only standard on the subject is ASTM C1256–93, Standard practice for interpreting glass fracture surface features [127].

Fracture toughness values (room temperature measurements) are given in Table 3 for some thermoplastics. Inclusion of fibers mostly improves fracture toughness, particularly as the fiber alignment becomes perpendicular to the crack surface. With some polymers, where for instance there is strong intermolecular H bonding, the test data are quite sensitive to the moisture content of specimens. The moisture absorption, as a result of matrix plasticization and the matrix–fiber interface deterioration, causes a reduction in mechanical strength and stiffness. Fracture toughness of wet specimens for glass fiber reinforced nylon 6, 6, however, has been shown to depend on the rate of loading [115]. The property improves with wet specimens under impact loading but deteriorates under long loading times (slow crack growth condition). The extent of deterioration is much less with the long fiber containing feed stock (usually in the form of 10 mm long pellets cut from thin pultruded rods) since a greater percentage of the fibers still remain longer than the critical fiber length compared with the short fiber (up to 1 mm long) containing granules, under wet conditions.

Fatigue crack propagation (FCP) studies cover materials from general purpose variety, e.g., glass fiber reinforced polypropylene [129], to high-performance variety such as carbon fiber reinforced PEEK [128,130]. The resistance of these materials to FCP improves with increasing fiber content but is, as would be expected, strongly affected by fiber orientation [129]. Testing conditions such as the fatigue frequency also influences the results.

4.2 Continuous Fiber Composites

Crack propagation through complex materials such as composites can be in many forms, and this presents the need to identify a particular failure mechanism and, in turn, to determine fracture toughness parameters that describe the material performance under that particular form of failure. Delamination is one such clearly identifiable type of failure, and as a consequence the application of fracture mechanics to delamination has received formal acceptability. The associated standard is ASTM D5528–94a, Mode I interlaminar fracture toughness of unidirectional fibre-reinforced polymer matrix composites. Submissions are being evaluated from such standardization groups as the ESIS and the JIS as well as the ASTM towards the establishment of an international agreement in the form of an ISO standard for mode I (crack opening mode) interlaminar fracture testing [131]. The type of initial crack (i.e., starter crack) to be employed in the test pieces is one of the main differences between these groups of expertise. Various forms of starter cracks, e.g., using only film inserts or insert and a precrack, in double cantilever beam (DCB) specimens, and their suitability have been covered in the literature [131,132].

ASTM D5528 [133] standard describes a test procedure for the determination of the strain energy release rate under mode I loading, G_I, for unidirectional composite laminates. Fig. 11 shows the different modes of loading that can be applied to a crack. Linear elastic behavior is assumed in the calculation of G_I when the damage zone and/or the nonlinear deformation at the delamination front are small compared with the smallest specimen dimension, which is the specimen thickness for the DCB test pieces. The details of the DCB specimens (h [note that in the ASTM DCB specimen the thickness is denoted by h but otherwise by $2h$] = 3–5 mm, $B = 20$–25 mm, and L [minimum] = 125 mm) and the types of loading fixtures (piano hinges and end blocks) attached to these specimens are shown in Fig. 12. The loading fixtures and the specimen surface should be lightly scrubbed with sandpaper and then wiped clean with a volatile solvent such as acetone prior to being bonded. If the bond failure occurs, it may be necessary to refer to ASTM D2651 [134] for a

Table 3 Fracture Toughness Data for Short Fiber Reinforced Thermoplastics

Material	Production/ crack direction	Test type/method	Specimen type/dimensions (mm)	Test speed	Fracture parameter	Ref.
1. 33% by wt GF[a] reinforced nylon 6,6 (Zytel 70G33)	injection; T*	static/ASTM D5045	SENB; $B = 3$, $W = 12$, $S = 48$	2.5 mm min^{-1}	$K_c \simeq 8.5\,\mathrm{MPam}^{1/2}$	100
2. 50% by wt GF reinforced nylon 6,6 (Maranyl A690)	injection; T	static/ASTM E399	CT; $B = 4$, $W = 24$	10 mm min^{-1}	$K_Q \simeq 9.4\,\mathrm{MPam}^{1/2}$	97
3. 50% by wt GF reinforced nylon 6,6 (Maranyl A690)	injection; L*	static/ASTM E399	CT; $B = 4$, $W = 24$	10 mm min^{-1}	$K_Q \simeq 8.6\,\mathrm{MPam}^{1/2}$	97
4. 50% by wt GF reinforced nylon 6,6 (Maranyl A690)	injection; T	impact	SENB; $B = 4$, $W = 10$, $S = 40$	1 ms^{-1}	$K_c \simeq 5.9\,\mathrm{MPam}^{1/2}$ $G_c \simeq 5.0\,\mathrm{kJm}^{-2}$	69
5. 50% by wt GF reinforced nylon 6,6 (Maranyl A690)	injection; L	impact	SENB; $B = 4$, $W = 10$, $S = 40$	1 ms^{-1}	$K_c \simeq 3.7\,\mathrm{MPam}^{1/2}$ $G_c \simeq 2.2\,\mathrm{kJm}^{-2}$	69
6. 50% by wt GF reinforced nylon 6,6 (Verton RF70010)	injection; T	impact	SENB; $B = 4$, $W = 10$, $S = 40$	1 ms^{-1}	$K_c \simeq 5.2\,\mathrm{MPam}^{1/2}$ $G_c \simeq 4.8\,\mathrm{kJm}^{-2}$	69
7. 30% by wt GF reinforced PET (Rynite 530)	injection; T	static/ASTM D5045	SENB; $B = 3$, $W = 12$, $S = 48$	2.5 mm min^{-1}	$K_c \simeq 7.3\,\mathrm{MPam}^{1/2}$	100
8. 33% by wt GF reinforced nylon 6,6	injection; L	static	SENB	5 mm min^{-1}	$J_c \simeq 5\,\mathrm{kJm}^{-2}$	24
9. 30% by wt GF reinforced PEEK	injection; L	static	CT; $B = 3$, $W = 29$	1 mm min^{-1}	$K_c \simeq 6.3\,\mathrm{MPam}^{1/2}$	121
10. 30% by wt CF[b] reinforced PEEK (Victrex)	injection; T	static/ASTM E399	SENB; $B = 64$, $W = 13$, $S = 52$	10 mm min^{-1}	$K_{Ic} \simeq 8.6\,\mathrm{MPam}^{1/2}$	128
11. 30% by wt. GF reinforced PEEK	injection; L	static	CT; $B = 3$, $W = 29$	10^3 mm min^{-1}	$K_c \simeq 1.9\,\mathrm{MPam}^{1/2}$	121
12. 25% by wt. GF reinforced POM[+] (Celcon GC-25)	extrusion; L	static	CT; $B = 3$, $W = 50$	1 mm min^{-1}	$K_c = 5.3\,\mathrm{MPam}^{1/2}$ $G_c \simeq 4\,\mathrm{kJm}^{-2}$ (stable crack growth)	43
13. 25% by wt. GF reinforced POM (Celcon GC-25)	extrusion; T	static	CT; $B = 3$, $W = 50$	1 mm min^{-1}	$K_c \simeq 6.2\,\mathrm{MPam}^{1/2}$ (unstable crack growth)	43

*L and T denote longitudinal and transverse notch directions to the extrusion direction or injection mould-fill direction.
[+]POM is polyoxy methylene.
[a]GF denotes glass fiber; [b] CF denotes carbon fiber.

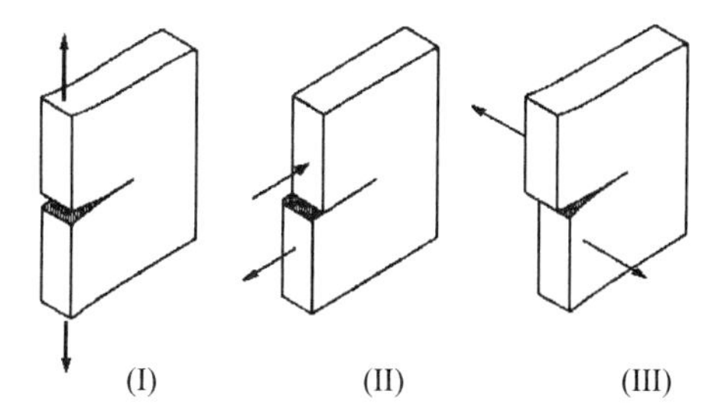

Figure 11 An illustration of the loading modes I, II, and III.

more sophisticated surface preparation procedure. A nonadhesive insert (film with thickness $\leq 13\,\mu m$) is placed at the midplane of the laminate during layup to act as an initiation site for the delamination. For epoxy-resin matrix composites with cure temperature $\leq 177°C$, polytetrafluoroethylene (PTFE) film, and higher cure temperature matrices such as polyimides, bismaleimides, PEEK, or other high melting point thermoplastic matrices, a thin polyimide film is recommended as the insert film. Nonpolymeric inserts such as aluminum foil are not recommended to avoid problems arising from possible folding or crimping at the cut edges of these alternatives. For further details refer to the ASTM D5528 document. The specimen is loaded in a displacement control mode with a constant displacement rate in the range from 0.5 to 5.0 mm min^{-1}. Slower speeds would allow crack propagation to be followed and recorded more easily. The crack length a or the delamination front as it extends along the edge of the specimen is measured using a suitable travelling microscope. A thin layer of white coating (e.g., Tippex) can facilitate the measurement. Load P versus crack opening displacement d at the point of load application is plotted; see Fig. 13. The onset of crack extension from the tip of the insert a_0, and subsequent increments of crack propagation a_i, are to be recorded regularly on this plot. When the delamination growth reaches at least 25 mm ahead of the crack-starter insert, the specimen should be unloaded.

Initiation G_{Ic} values corresponding to the crack length a_0 and subsequent propagation G_{Ic} values can be determined from the load–displacement plots, enabling the construction of an R-curve, i.e., G_{Ic} as a function of growing crack length. There are different ways of identifying the crack initiation from the load–displacement plots: the point of deviation from linearity, and the point on the curve where the initial compliance C_0 has increased by 5% (see Fig. 13). However, if the 5% offset compliance line intersects the curve at a larger

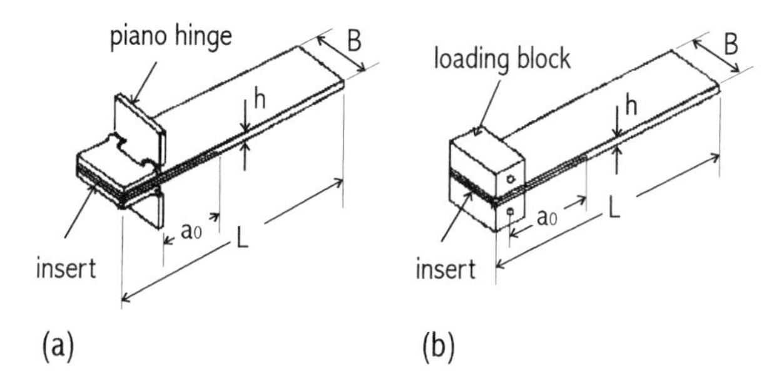

Figure 12 Details of specimen geometry and loading arrangements for DCB tests. (Note that the specimen thickness is designated by h in the ASTM standard, but otherwise identified as $2h$).

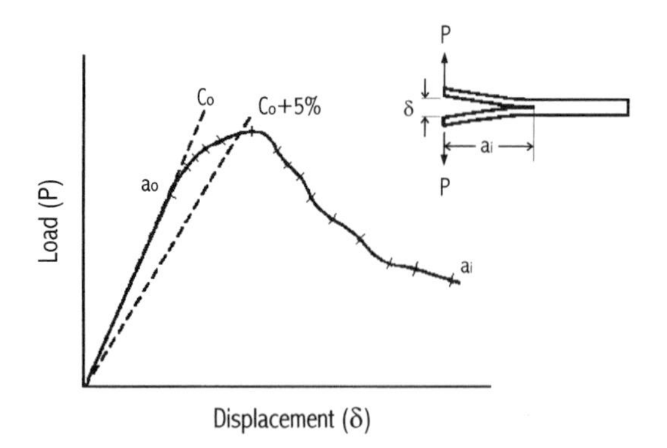

Figure 13 A typical load–displacement trace from DCB test. Measurements of delamination growth during the test are indicated (as ticks) on the trace.

displacement value than the maximum load value, the maximum load and corresponding displacement are to be selected. A visual initiation value for G_{Ic} should also be estimated based on the load and displacement data that correspond to the first sighting at which a crack is observed to move from the insert on either of the specimen edges using a travelling microscope. The point of nonlinearity yields a lower bound value for G_{Ic}. This is supported by a recent round-robin study of G_{Ic} using 60% by volume carbon fiber reinforced PEEK (polyether ether ketone). However the study also indicates that the nonlinearity criterion leads to a greater scatter in data than the 5% offset criterion [135].

In the ASTM DCB test, as the delamination grows from the insert, a resistance type fracture behavior develops such that the calculated G_{Ic} initially increases monotonically and then stabilizes with further delamination growth, producing the type of *R*-curve shown in Fig. 14. The main reason for the observed progressive resistance to delamination is the development of fiber bridging [136], which results from growing the delamination between two 0° unidirectional (UD) plies (not an occurrence between plies of dissimilar orientation). Consequently in 0° UD laminates, G_{Ic} values calculated for delamination propagation beyond the end of the embedded insert are questionable, and an initiation value of G_{Ic} measured immediately ahead of the inserted crack starter is preferred by the ASTM. However, as indicated before, there are alternative views [131], and these promote testing from a precrack generated ahead of an insert (which may introduce fiber bridging

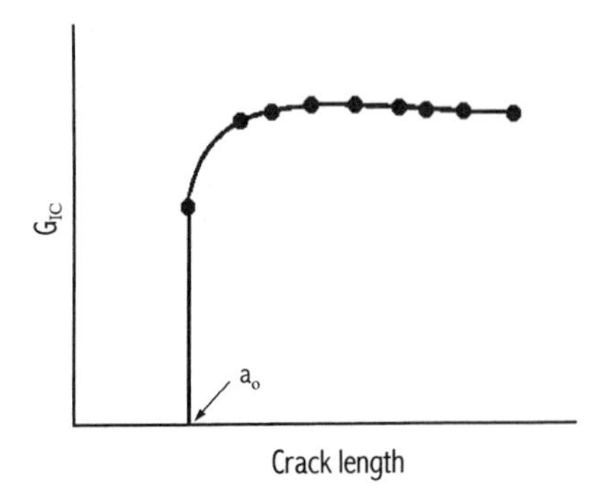

Figure 14 A typical *R*-curve from a DCB test.

to the specimen prior to the measurements). One of the arguments against the use of inserts is that it is not always practical to introduce an insert, e.g., with specimens taken from component moldings. Other drawbacks of inserts reported [132] are that the induced defect should be representative of natural cracks and should be as sharp as possible. It is also indicated that for certain cases in fact more conservative values of fracture toughness are obtained from precracks. Obviously there is scope for further work.

The standard identifies three methods of computing G_{Ic}, as follows.

Modified Beam Theory (MBT)

The simple beam theory expression for the strain energy release rate of a perfectly built-in (i.e., clamped at the delamination front) double cantilever beam is

$$G_I = \frac{3P\delta}{2Ba} \tag{20}$$

Equation 20 can be derived from the simple beam theory for a cantilever and linear elastic considerations. For a cantilever beam of thickness t and length a (i.e., one of the two halves of the DCB specimen with a crack length a) under an end load of P, the beam-end deflection $\delta/2$ (equivalent to one-half of the load point displacement of the DCB specimen) is given by the beam theory

$$\frac{\delta}{2} = \frac{Pa^3}{3EI} \qquad \text{where} \qquad I = \frac{Bt^3}{12}$$

For an increment of crack growth at a given P and δ in a DCB specimen, the work done for a displacement $d\delta$ arising from a fracture (delamination) area change dA is $dU_1/dA = P(d\delta/dA)$. The corresponding change in the stored energy U_2 is

$$\frac{dU_2}{dA} = \frac{d}{dA}\left(\tfrac{1}{2}P\delta\right)$$

or

$$\frac{dU_2}{dA} = \frac{1}{2}\left(\frac{Pd\delta}{dA} + \frac{\delta dP}{dA}\right)$$

By definition,

$$G_I = \frac{dU_1}{dA} - \frac{dU_2}{dA} \tag{21}$$

Substituting we obtain

$$G_I = \frac{Pd\delta}{dA} - \frac{1}{2}\left(\frac{Pd\delta}{dA} + \frac{\delta dP}{dA}\right)$$

$$G_I = \frac{1}{2}\left(\frac{Pd\delta}{dA} - \delta\frac{dP}{dA}\right)$$

In terms of compliance C,

$$\delta = CP$$

Differentiating with respect to A,

$$\frac{d\delta}{dA} = P\frac{dC}{dA} + C\frac{dP}{dA}$$

Substituting in G_I,

$$G_{\rm I} = \frac{1}{2}\left(P^2 \frac{dC}{dA} + PC\frac{dP}{dA} - \delta\frac{dP}{dA}\right)$$

Since $\delta = CP$,

$$G_{\rm I} = \frac{1}{2} P^2 \frac{dC}{dA}$$

or in terms of the beam width B and crack length a,

$$G_{\rm I} = \frac{P^2}{2B}\frac{dC}{da} \tag{22}$$

Substituting $\frac{\delta}{2} = \frac{Pa^3}{3EI}$ in $C = \frac{\delta}{P}$ gives

$$C = \frac{2}{3}\frac{a^3}{EI} \tag{23}$$

and $dC = \frac{2}{3}\frac{(3a^2 da)}{EI}$ or $\frac{dC}{da} = \frac{2a^2}{EI}$

substituting in $G_{\rm I}$ (i.e., Eq. 22) gives

$$G_{\rm I} = \frac{P^2}{2B}\frac{2a^2}{EI} = \frac{P^2a^2}{BEI} \tag{24}$$

Substituting for EI from the beam theory equation of $\delta/2 = Pa^3/3EI$ gives the equation for $G_{\rm I}$:

$$G_{\rm I} = \frac{3}{2}\frac{P\delta}{Ba}$$

In practice the Eq. 20 will overestimate $G_{\rm I}$ or underestimate the compliance, since the beam is not perfectly held and therefore may rotate or tilt. A means of correcting for this effect is to treat the DCB as containing a slightly longer delamination, $a + |\Delta|$. Δ can be determined experimentally by plotting the cube root of compliance, $C^{\frac{1}{3}}$, as a function of delamination a. According to Eq. 23, for an ideal setup, the plot should produce a straight line that passes through the origin, but the experimental data from the DCB tests in fact produce a negative intercept, Δ, for the delamination along the x-axis.

Therefore the Mode I interlaminar fracture toughness is computed using the following *modified beam expression.*

$$G_{\rm I} = \frac{3P\delta}{2B(a + |\Delta|)} \tag{25}$$

Compliance Calibration Method

Alternatively a general form of Eq. 23 can be adopted such that $C = Ka^n$. The exponent n can be determined from the slope of a straight line plot of $\log(C_i)$ vs. $\log(a_i)$. The value of n is substituted in Eq. 20 so that

$$G_{\rm I} = \frac{nP\delta}{2Ba} \tag{26}$$

Modified Compliance Calibration Method

G_I is expressed in terms of the compliance by using the term A_1, which defines the slope of the line generated by plotting the thickness-normalized delamination length a/h as a function of $C^{1/3}$ so that

$$A_1 = \frac{a/h}{C^{1/3}}$$

The combination of Eqs. 23 and 24 gives G_I in terms of C:

$$G_I = \frac{P^2}{2B}\frac{3C}{a} = \frac{3P^2C^{2/3}C^{1/3}}{2Ba} \tag{27}$$

By substituting for $a = A_1hC^{1/3}$ in Eq. 27, the following expression can be obtained for G_I which is used for calculations in this third method.

$$G_I = \frac{3P^2C^{2/3}}{2A_1Bh} \tag{28}$$

ASTM round-robin testing [137] has shown that the modified beam theory (MBT) method (Eq. 25) yielded the most conservative values of G_{Ic}. Accordingly this method is recommended.

The extent of errors in G_I estimations due to the linear beam theory assumption increases as the displacements get larger. The ASTM standard recommends the inclusion of correction parameters to the above-described methods to account for large displacements and also for the end block effects. Details are given in ASTM D5528 [133]. These corrections are small for short delamination lengths in 3 mm thick specimens of 60% by volume carbon fiber containing UD composites, but they may be larger for thinner or more flexible specimens or for long delamination lengths.

One of the alternatives to the ASTM procedure is the area method [138,139], where the specimen is periodically unloaded and reloaded in the course of delamination. Normally the load–displacement curve is nonlinear on the loading portion and linear and passes through the origin on the unloading portion. The strain energy release rate can be estimated from the incremental areas dU under each load–displacement curve such that

$$G_I = \frac{dU}{Bda} \tag{29}$$

A comparison of the above-covered analytical methods has been made using data for a unidirecitonal glass fiber reinforced phenolic resin composite [94].

Note that in all these tests it is recommended that the fiber volume content v_f should be recorded, preferably for each specimen. Fiber content of composite can be determined by the removal of matrix via burn off (ASTM D2584) [140] or by chemical digestion (ASTM D3171) [141] using e.g. hot nitric acid or sulphuric acid or potassium hydroxide containing chemicals depending on the matrix type. Both these methods are tedious and indeed can be hazardous. A more simple determination is based on measuring the specimen thickness and using the formula

$$v_f = \frac{\text{bulk density of the fibers in the laminate}}{\text{density of the fiber}}$$

The bulk density $= n(\text{FAW})/h$, where FAW (in g/cm^2) is the fiber areal weight, i.e., the weight (mass) of fibers in a unit area of a single ply n is the number of plies, and h (in cm) is the laminate thickness. Therefore

$$v_f(\%) = \frac{n(\text{FAW})(100)}{h(\text{FD})} \tag{30}$$

where FD (in g/cm^3) is the fiber density.

The values for FAW and FD can be found in the material suppliers' data sheets. The thickness h measurement requires greater care with specimens having a bag side (usually produces a rougher surface than the tool side).

Mode II Shear Fracture Test

Susceptibility to interlaminar failure is a major weakness of advanced laminated composite materials. It can occur by in-plane shearing (i.e., sliding) (mode II), and out-of-plane shearing (i.e., tearing) (mode III) as well as by tensile (mode I) deformation. Mode II loading is of particular interest, as G_{IIc} values have been shown to correlate with compression after impact data [142,143], which is required for such purposes as civil aircraft certification.

Considerable information exists for mode II testing, but there is no standard method of testing available. Any standardization has to resolve what type of specimen to adopt and also decide on the following, irrespective of the type of specimen selected [132, 144]: the type of starter defect (initial crack), specimen dimensions, the influence of the friction between the crack faces, which definition of G_{IIc} (i.e., identification of appropriate data points on the load–displacement curves), and which data analysis. A review paper [145] describes various test specimens and data analysis methods. Three of these test pieces, namely end notched flexure (ENF), stabilized end notched flexure (SENF), and end loaded split (ELS) (where shear stresses are generated when the arms of a DCB specimen are bent down in the same direction), are the main candidates for standardization [132]. These specimen types are contrasted in Ref. 132. The ENF and SENF specimen tests are conducted on a standard three point flexure test setup and enable the determination of the critical strain energy release rate G_{IIc} for the initiation of a delamination under mode II loading. Alternatively the ELS specimen enables both initiation and propagation values of G_{IIc} to be determined.

A brief description of the ESIS testing protocol [146] for the measurement of G_{IIc} values using the ELS specimen will be given here. Typical ELS specimen geometry and Mode II test setup are shown in Fig. 15. The specimen dimensions are approximately a free length (L) of 100 mm, width (B) 20 mm, and thickness ($2h$) 3 mm for 60% by volume CF or 5 mm for 60% by volume GF reinforced composites. PTFE film of less than 15 μm thickness should be inserted to act as starter crack (defect). A mold release agent should be used with the alternative films to PTFE. The advantage of this geometry is that crack propagation is stable for $(a/L) > 0.55$, so initiation and several propagation values of G_{IIc} can be obtained from each specimen.

During the test, the load application must remain vertical, therefore the clamped end of the specimen should be free to slide on bearings as shown in Fig. 15. The specimen is loaded at 1 to 5 mm/min. Large displacements should be avoided such that $(\delta/L) < 0.2$.

A continuous plot of load P against displacement δ should be recorded, together with measurements of crack length. A layer of white coating along the specimen edge may help crack length measurements during the test. The occurrence of any permanent deformation should be checked by unloading the specimen before the crack reaches approximately 10

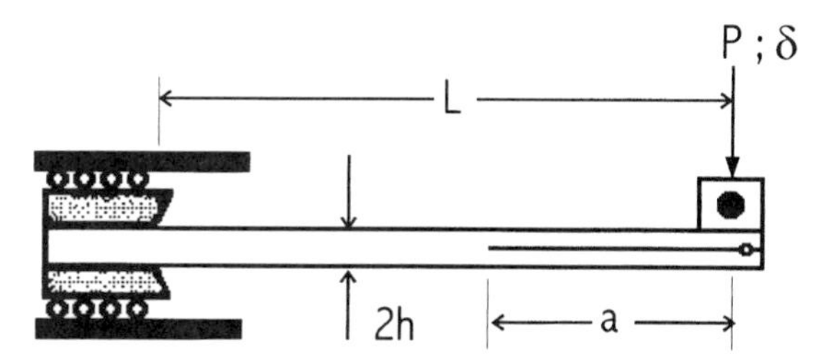

Figure 15 Specimen geometry and test setup for a mode II (ELS) test.

mm off the clamped end, and the trace should return to the origin in the absence of permanent deformation. G_{IIc} initiation values may be determined from one of the three load values corresponding to (a) nonlinear point, i.e., where the load–displacement plot deviates from linearity, (b) point of visual exhibition of delamination onset, or (c) point of the intersection of the 5% compliance offset line with the P–δ trace or the maximum load value if the intersection is at a larger displacement than the maximum point on the trace.

G_{IIc} initiation values are based on the initial (starter) delamination length, whereas for the propagation values of G_{IIc} the corresponding growth in delamination length is added to the initial crack length and the extended length is used in calculations. The energy release rate can be computed using the expressions in the following.

Beam Theory

The DCB and the ELS specimens consist of two halves, each a cantilever beam. In the DCB test, both halves are subjected to a load P, whereas in the ELS test the load P is shared, i.e., in this case a load of $P/2$ applies to each cantilever arm.

$$G_{II} = \frac{3P^2a^2}{16BEI} \tag{31}$$

or by substituting $I = Bh^3/12$,

$$G_{II} = \frac{9P^2a^2}{4B^2Eh^3} \tag{32}$$

Both these equations require a value of the elastic modulus E. This may be obtained from a standard flexure test, or alternatively the ELS specimens may be tested as above (loaded and unloaded within the elastic region) with the specimen changed in reverse such that the starter film is not within the free length. The slope P/δ of the linear P–δ line would allow a value of EI to be determined, since $EI = (Pa^3)/(3\delta)$. This value of EI can be directly used in Eq.31.

Corrected (Modified) Beam Theory

A correction Δ may be introduced into the beam theory expressions to account for any rotation of the crack tip. It has been shown [147] that a good estimate of Δ_{II} for mode II ELS loading can be made from the correction value Δ_I corresponding to mode I tests such that $\Delta_{II} = 0.49(\Delta_I)$.

Thus the modified beam expressions become

$$G_{II} = \frac{3P^2(a + |\Delta_{II}|)^2}{16BEI} \tag{33}$$

or in terms of h,

$$G_{\mathrm{II}} = \frac{9P^2(a + |\Delta_{\mathrm{II}}|)^2}{4B^2Eh^3} \tag{34}$$

Experimental Compliance Method (for Linear Elastic Behavior)

Equation 22 gives an expression of the energy release in terms of the compliance as

$$G_{\mathrm{II}} = \frac{P^2}{2B}\left(\frac{dC}{da}\right)$$

dC/da can be estimated from an experimental plot of C vs. a^3. The compliance C values for various extensions of delamination can be determined from loading and unloading curves, provided these remain linear. Using the following generalized form of Eq. 23.

$$C = C_0 + ma^3$$

where m is the slope of the plot of C vs. a^3.

Substituting for $dC/da = 3ma^2$ in the above expression of G_{II} gives

$$G_{\mathrm{II}} = \frac{3P^2ma^2}{2B} \tag{35}$$

G_{IIc} can be calculated by substituting the appropriate load values (discussed previously) into the G_{II} expressions for those three methods. Slender specimens have to be subjected to large loads and, in turn, large displacements to effect crack propagation, particularly in mode II tests. G_{II} computations need to be corrected for large displacements and for end blocks (loading fixtures). Necessary expressions for these corrections are given in the ESIS protocol [146].

The reporting of these tests should include starter defect type and thickness, loading rate and temperature, the load displacement curve, the method of calculation of the fracture toughness parameter G, and such material variables as fiber volume fraction.

Clearly both mode I and mode II tests are important in the evaluation of fiber reinforced composites. In structures, however, delaminations grow under a mixture of mode I and mode II loading, and also there are obvious advantages in test methods where both mode I and mode II data can be obtained simultaneously with a single test. One such mixed mode test is also covered in the ESIS protocol [146] and is briefly outlined below.

Mixed Mode I/II (ADCB)

The critical strain energy release rate, for initiation and propagation of a delamination failure, can be measured under mixed mode (I/II) loading using the asymmetrical double cantilever beam (ADCB) arrangement. The test setup is the same as that shown in Fig. 15, except the loading and the displacement are in the opposite direction, i.e., the arms of the cantilever beams are lifted up at the free end giving rise to both crack opening (peel) and sliding shear within the induced starter crack. With this set up the ratio of the modes is fixed, nominally $G_{\mathrm{I}}/G_{\mathrm{II}} = 4/3$. Accordingly it is also known as the fixed ratio mixed mode (FRMM) test [148]. The details of the test procedure are similar to those of mode II. The data analysis is also based on the previously described concepts.

Beam Theory

In Eq. 24, replacing P by $P/2$, since the load P is shared between the two arms of the DCB specimen, results in the following expression for G_{I}:

$$G_{\mathrm{I}} = \frac{P^2 a^2}{4BEI} \tag{36}$$

or in terms of h,

$$G_{\mathrm{I}} = \frac{3P^2 a^2}{B^2 E h^3} \tag{37}$$

G_{II} values are determined using Eqs. 31 and 32.

Corrected Beam Theory

The value of Δ obtained in the mode I tests is used to correct the expression for G_{I}, for example,

$$G_{\mathrm{I}} = \frac{3P^2 (a + |\Delta_{\mathrm{I}}|)^2}{B^2 E h^3} \tag{38}$$

The mode II correction, $\Delta_{\mathrm{II}} = 0.49(\Delta_{\mathrm{I}})$, therefore, is

$$G_{\mathrm{II}} = \frac{9P^2 (a + |\Delta_{\mathrm{II}}|)^2}{4B^2 E h^3]} \tag{39}$$

Experimental Compliance

Equation 35 applies in this case as well:

$$G_{\mathrm{II}} = \frac{3P^2 m a^2}{2B} \tag{40}$$

The strain energy release rate obtained under the mixed mode conditions can be split into mode I and II components using the relationships of $G_{\mathrm{I}}/G_{\mathrm{II}} = 4/3$ and $G_{\mathrm{I}} + G_{\mathrm{II}} = G_{\mathrm{I/II}}$. Therefore $G_{\mathrm{I}} = 0.57(G_{\mathrm{I/II}})$ and $G_{\mathrm{II}} = 0.43(G_{\mathrm{I/II}})$.

In contrast to the above described ADCB test, *the mixed mode bending (MMB) test* enables the rates of $G_{\mathrm{I}}/G_{\mathrm{II}}$ to be adjusted and therefore a combination of G_{I} and G_{II} values to be determined. A plot of G_{I} vs. G_{II} can be drawn from the various pairs of data to produce a failure locus [148]. The MMB test [149,150], which simply combines the DCB and ENF loadings, allows any combination of mixed mode loading from pure mode I to pure mode II to be tested with the same test specimen configuration. The MMB test fixture is illustrated in Fig. 16. The ratio of the modes, $G_{\mathrm{I}}/G_{\mathrm{II}}$, is controlled by varying the lever load point position c. Pure mode II loading is achieved at $c = 0$, i.e., by moving the load point to the specimen mid-span (this is equivalent to the ENF test setup). For pure mode I loading, the loading lever is removed and the specimen loaded by pulling up the hinge directly in DCB fashion. In a mixed mode situation, the contribution of mode I increases as the value of c increases and vice versa. The equation relating $G_{\mathrm{I}}/G_{\mathrm{II}}$ ratio to specimen and loading lever dimension can be found in Ref. 148. Static tests are conducted at a rate of 0.5 mm min^{-1} and the load–displacement response recorded. The critical load P_{c} for the G_{c} calculations may be taken as the load where the load–displacement curve deviates from linearity [151]. Similar to DCB specimens, film inserts are employed to act as a delamination starter. For instance, Teflon and Kapton films are used for epoxy matrix and PEEK matrix laminates, respectively.

Mode III (Tearing Shear) Fracture Toughness

The mode III toughness has been shown to be as important as the other two modes for delamination between angle plies [152]. Test method development for this mode of loading

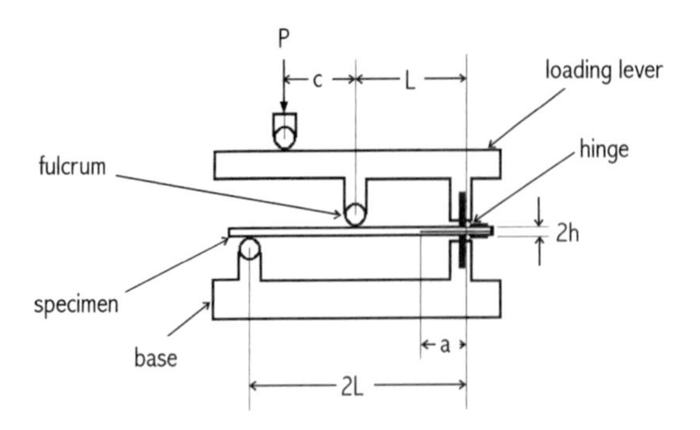

Figure 16 Mixed mode bending (MMB) test setup.

has received comparatively little attention. The specimen types used for mode III tests include a split cantilever beam (SCB) [153] and the crack rail shear [154].

In split cantilever beam tests, a strip of laminate of approximately 13 mm width, containing an insert crack starter, is bonded between two parallel aluminum bars. A mode I precrack can be introduced by clamping the specimen and then driving a razor blade wedge down the insert crack. In order to monitor crack length extensions during the test, the laminate edge is coated with white typewriter correction fluid, and a strip of graph paper is bonded near the edge of the specimen. The specimens are loaded at a constant displacement rate of 0.5 mm min^{-1} in a direction parallel to the plane of the crack and normal to the length of the specimen to produce the out-of-plane tearing. G_{IIIc} values for crack growth initiation and propagation can be computed, from the load displacement crack length data, using either the beam theory (Eq. 20) or the area method (Eq. 29).

Specimen configuration for the crack rail shear test is identical to the ASTM D4255 standard [155] specimen for the two-rail shear test. Analytical expressions for strain energy release rate determination are derived by a strength of materials approach [154].

Notch-impact tests [150–154,156–158], and tests using CT specimens [159] (see Section 3) have also been employed in the evaluation of continuous fiber composites, particularly those with woven [159] and nonwoven [157,158] fabric reinforcements, and 3D composites [156].

An alternative [160] approach independent of LEFM parameters has been developed such that the damage stages are quantified by the use of composite mechanics, while degradation of the structural behavior is quantified by finite element analysis. A global strain energy release rate parameter has been, however, employed to identify a "critical damage amount."

Materials Evaluation

Fracture toughness data (room temperature measurements) for various continuous fiber composites under different loading modes are presented in Table 4. The data from different sources are reasonably consistent for a given material system with the same fiber content. Small differences may result from the variations in the degree of cross-linking for thermosets and the degree of crystallinity for the thermoplastic matrices.

Specimen geometry independence of the experimental measurements has been examined by, for instance, conducting DCB (mode I) and ENF (mode II) tests on specimens with different thicknesses [162]. The values are virtually independent of thickness for CF/epoxy resin laminates. A magnitude of increase comparable with the standard deviation was indicated in both G_{Ic} and G_{IIc} propagation with increasing thickness in CF/PEEK laminates. This increase was suggested to be due to the introduction of multiple cracks in the thicker specimens during the precracking procedure, which may result in an additional resistance to fracture propagation.

The influence of test conditions such as test temperature and speed of testing, which influence specimen stiffness and the viscoelastic nature of the matrix, have been extensively studied: Mode I delamination resistance of both tough (e.g., PEEK) [161,163,169] and relatively brittle (e.g., epoxy [161,163,169] and polybismaleimide [161]) matrix composites increases with temperature, often to the extent that mode I loading ceases to be the lowest energy failure mode for the tougher matrix composites. Under mode II conditions the material is less sensitive to the temperature variation [161,163]. Studies [169,170] over a wide range of specimen testing speeds have indicated that the fracture toughness of carbon fiber reinforced epoxy matrix composites do not exhibit any rate sensitivity. In the case of more ductile matrices such as PEEK, both the rate sensitivity [169] and independence [170] have been reported. At room temperature the fracture toughness of CF/PEEK (APC-2) has indicated [169] a monotonic decrease beyond an approximately 5 mm min^{-1} shear displacement rate, and a nearly linear elastic response at rates ≥ 20 mm min^{-1}.

Other variables investigated with respect to fracture toughness include processing parameters, e.g., production schedules and postcure cycles [166,171] for thermosets, and cooling rates for thermoplastics [171] matrix composites, and moisture content. A study [171] on composites based on epoxy and vinyl ester matrices has shown that whilst matrix plasticization improves mode I fracture toughness, mode II fracture toughness deteriorates due to interface degradation. The sensitivity of G_{IIc} to water absorption has been demonstrated even for matrices that absorb very small amounts of water such as polypropylene [168].

Materials evaluation further includes fractography [151,158,165,167,172,173] and fatigue fracture toughness. Fatigue crack propagation tests were performed and the threshold values G_{th} were identified for unidirectional CF/epoxy and CF/PEEK laminates under mode II loading using ENF specimens [173], UD graphite fiber/bismaleimide and graphite fiber/PEEK under mixed mode bending test [174], and woven GF/polyester [175] under mode I (DCB) and mode II (ENF) conditions.

Injection-molded plaques of liquid crystalline polymers exhibit a multilayered structure through the thickness of a molding, where in each layer there is a high degree of orientation. Accordingly the fracture test methods of continuous fiber composites can be adopted for the testing of liquid crystalline polymers [176].

5 Other Polymeric Materials

5.1 Adhesives

Adhesives are increasingly being employed in the construction of aerospace, land and sea transportation vehicles. Where such applications are structural (load-bearing), then fracture resistance and also durability of performance (e.g., environmental stress cracking under hygrothermal conditions) are included as design criteria. Detailed specific information on the subject can be found in the literature: these include a coverage of surface and interfacial phenomena [177], fracture mechanics and mechanics of failure of adhesive bonds [178,179], the effects of various operating (service) environments on the performance of joints [179,180]. Adhesive fracture testing is also documented [179,180]. Some of the standard methods include the following. ASTM D3433–93, Fracture strength in cleavage of

Table 4 Fracture Toughness Data (Initiation Values) for Some Continuous Fiber Composites

Material	Specimen type/dimension (mm)	Test speed	Fracture parameter	Ref.
1. Unidirectional 0° (UD) carbon fiber (CF)/ epoxy resin (Fibredux (6376C) ($v_f \simeq 0.65$)	DCB (mode I); ELS (mode II); 2h (thickness) = 4, B(width) = 20	2 mm min^{-1}	$G_{Ic} \simeq 0.31\,kJm^{-2}$, $G_{IIc} \simeq 0.76\,kJm^{-2}$	161
2. UD CF/epoxy (Ciba-Geigy) ($v_f \simeq 0.60$)	DCB; ENF; $2h = 3.2$, $B = 20$	1 mm min^{-1}	$G_{Ic} \simeq 0.12\,kJm^{-2}$, G_{Ic} (propagation $\simeq 0.2\,kJm^{-2}$, $G_{IIc} \simeq 0.44\,kJm^{-2}$	162
3. UD CF/epoxy (T300/914) ($v_f \simeq 0.6$)	DCB; $2h = 5$, $B = 20$	2 mm min^{-1}	$G_{Ic} \simeq 0.18\,kJm^{-2}$, G_{Ic} (propagation) $\simeq 0.18\,kJm^{-2}$	163
4. UD CF/epoxy (Fibredux 376C)	DCB; $2h = 4$, $B = 25$	2 mm min^{-1}	$G_{Ic} \simeq 0.27\,kJm^{-2}$	164
5. UD CF/epoxy (1M6/6376)	DCB; $2h = 4$, $B = 20$	2 mm min^{-1}	$G_{Ic} \simeq 0.18\,kJm^{-2}$, G_{Ic} (propagation) $\simeq 0.63\,kJm^{-2}$	163
6. UD graphite epoxy	SCB; $2h = 12.7$, $B = 12.7$	0.5 mm min^{-1}	$G_{IIIc} \simeq 1.2\,kJm^{-2}$	153
7. Plain weave (PW) Kevlar 49/epoxy (Epon) ($v_f \simeq 0.45$)	DCB; $2h = 5$, $B = 25$	30 mm min^{-1}	$G_{Ic} \simeq 1.85\,kJm^{-2}$	165
8. PW glass fiber (GF)/epoxy (Epon) ($v_f \simeq 0.45$)	DCB; $2h = 5$, $B = 25$	30 mm min^{-1}	$G_{Ic} \simeq 1.67\,kJm^{-2}$	165
9. UD CF/rubber-toughened polybismaleimide (Technochemie)	DCB; $2h = 2$, $B = 20$		$G_{Ic} \simeq 0.18\,kJm^{-2}$	161

10. UD GF/polyester	DCB		$G_{Ic} \simeq 0.1\,kJm^{-2}$	166
11. UD GF/modar (i.e., modified acrylic)	DCB		$G_{Ic} \simeq 0.3\,kJm^{-2}$	166
12. 3D carbon–carbon composite	SENB; $B = 10$, $W = 10$, $S = 55$	$3\ ms^{-1}$	$K_{Id} \simeq 8.3\,MPam^{1/2}$ $J_{Id} \simeq 3.7\,kJm^{-2}$	156
13. UK CF/PEEK ($v_f = 0.6$)	DCB; $2h = 3.2$, $B = 20$	$1\ mm\ min^{-1}$	$G_{Ic} \simeq 1.4\,kJm^{-2}$ G_{Ic} (propagation) $\simeq 1.7\,kJm^{-2}$ $G_{IIc} \simeq 2\,kJm^{-2}$	162
14. UD CF/PEEK (ICI APC2) ($v_f \simeq 0.65$)	DCB; ELS	$2\ mm\ min^{-1}$	$G_{Ic} \simeq 2.3\,kJm^{-2}$ $G_{IIc} \simeq 2.3\,kJm^{-2}$	161
15. UD CF/PEEK (APC2)	DCB; $2h = 4.4$, $B = 20$	$2\ mm\ min^{-1}$	$G_{Ic} \simeq 1.8\,kJm^{-2}$	164
16. UD CF/PEEK (AS4/APC2) ($v_f \simeq 0.6$)	DCB; $2h = 5$, $B = 20$	$2\ mm\ min^{-1}$	$G_{Ic} \simeq 1.46\,kJm^{-2}$ G_{Ic} (propagation) $\simeq 2.4\,kJm^{-2}$	163
17. UD CF/PEEK (AS4/APC2) ($v_f \simeq 0.7$)	DCB; ELS; $2h = 3.2$–5.3, $B = 25$, $L = 140$	$2\ mm\ min^{-1}$	$G_{Ic} \simeq 1.2\,kJm^{-2}$, G_{Ic} (propagation $\simeq 1.9\,kJm^{-2}$ $G_{IIc} \simeq 2.1\,kJm^{-2}$ G_{IIc} (propagation) $\simeq 2.1\,kJm^{-2}$	167
18. UD GF/polypropylene (E-glass/PP Plyton ZM4350 PA), ($v_f \simeq 0.35$)	ENF; $2h = 4.7$, $B = 25$, $L = 140$	$2.5\ mm\ min^{-1}$	$G_{IIc} \simeq 2.1\,kJm^{-2}$	168
19. Swirl GF mat/NBC (i.e., nylon block copolymer) ($v_f \simeq 0.09$)	SENB; $B = 4$, $W = 12$, $S = 70$	$3.7\ ms^{-1}$	$G_d \simeq 13\,kJm^{-2}$ $K_d \simeq 4.7\,MPam^{1/2}$	158

adhesives in bonded metal joints. This test method involves cleavage testing of bonded specimens such that a crack is made to extend by a tensile force acting in a direction normal to the crack surface. The property G_{Ic} characterizes the resistance of a material to slow-stable or run-arrest fracturing in a neutral environment such that the crack tip plastic zone is acceptably small not to violate plane strain state near the crack front.

Approved test specimens are shown in Fig. 17. The tapered (or contoured) double-cantilever beam (TDCB) (developed by Mostovoy and Ripling [35] offers the advantage that G_{Ic} can be estimated (see below equation) independently of the crack length variation, which can be difficult to identify and measure accurately. The tests are conducted in the standard laboratory atmosphere (i.e., $23 \pm 2°C$ and $50 \pm 5\%$ RH) and at a testing speed of approximately 2 mm/min. A saw-toothed load–displacement trace is produced by the TDCB specimen after an initial linear rise in load to the onset of first crack initiation (first peak). This indicates that crack propagation occurs intermittently in a slip-stick manner. The peaks in the saw-toothed trace indicate crack initiation (start) and the nadirs crack arrest. Accordingly two fracture toughness parameters, G_{Ic} and G_{Ia}, can be determined. G_{Ic} can be calculated from the load value to start crack and G_{Ia} from the arrest load, using the following Eqs. 41 and 42.

If the arms of the tapered double-cantilever beam (TDCB) specimen behave in a linear elastic manner, then from Eq. 22,

$$G_I = \frac{P^2}{2B}\left(\frac{dC}{da}\right)$$

For a TDCB, it has been shown [181] that

$$\frac{dC}{da} = \frac{8m}{EB}$$

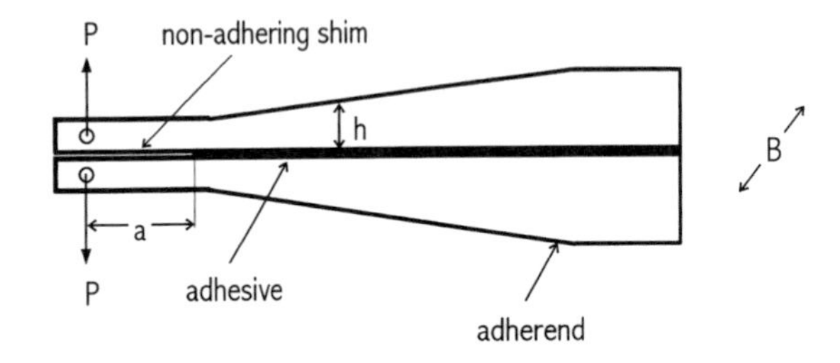

(a)

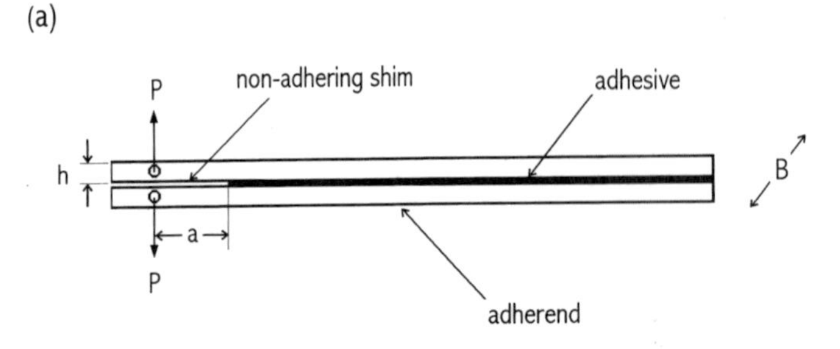

(b)

Figure 17 Tapered (a) and flat (b) double-cantilever beam fracture specimens for adhesives.

Therefore

$$G_{Ic} = \frac{4P^2_{(max)}m}{EB^2} \tag{41}$$

and

$$G_{Ia} = \frac{4P^2_{(min)}m}{EB^2} \tag{42}$$

where m is a geometry factor:

$$m = \frac{3a^2}{h^3} + \frac{1}{h}$$

where h is the height of the beam at the respective crack length a. The height is varied in a TDCB specimen so that m remains constant and, in turn, the value of dC/da is constant. Thus the values of G_{Ic} and G_{Ia} will be independent of crack length.

ASTM D5041–93b, Fracture strength in cleavage of adhesives in bonded joints. This test is for the determination of fracture strength in the cleavage of adhesive bonds when tested on reinforced plastic specimens (may be used for ordinary plastics of certain rigidity). It provides an effective means of screening structural adhesives where this is not possible with simple lap shear testing.

The test setup consists of an assembly of two panels of substrate (50–150 mm width and 150 mm height) that sandwich a core consisting of a layer of adhesive (across the full width of the substrate and 50 mm height) and a sheet of removable shim (a release agent coated steel) to control the thickness of the adhesive and to provide a gap (crack) in the assembly upon being removed following the curing of the adhesive. The bonded assembly is vertically positioned in the testing machine and a wedge (with an included angle of 45° and a tip radius of 0.02 mm) is placed above it in the gap between the substrate panels. The wedge is driven into the gap at a speed of 127 mm min^{-1} until the assembly fails and the load–deformation trace is plotted. There are two regions of interest on these traces, which are failure initiation and failure propagation. The first significant drop in load is normally used for the failure initiation point. The failure propagation covers the events beyond this point. During the test energy is absorbed by the bonded assembly as the wedge separates the two adherents. It is possible to determine two energy values: failure initiation energy E_i (the area under the load–displacement curve from the start of the test to the point where first significant load drop occurs), and failure propagation energy E_p, which is equivalent to the remaining area under the complete failure load–displacement curve. It should be noted that the preparation of the surfaces prior to bonding should either be in accordance with the adhesive manufacturer's instructions or with ASTM D2093 [182] for plastic surfaces and ISO 4588 [183] for metals. A similar test to the above is recommended for adhesives by ISO under impact loading conditions.

ISO 11343, 1993, Adhesives—Determination of dynamic resistance to cleavage of high strength adhesive bonds under impact conditions—Wedge impact method: The method is mainly aimed for the characterisation of metal substrates suitable for automotive applications. An instrumented impact testing machine (pendulum type) of 50–300 J and 3–5.5 ms^{-1} maximum capacity is required for this test. A blunt wedge of approximately 1 mm tip radius and included angle of 8° is impact driven into a bonded joint at 3 ms^{-1} for aluminum alloy adherends and 2 ms^{-1} for steel adherends. The impact event should be recorded in the form of a force–time or force–displacement plot. The material is characterized in terms of the average cleavage (i.e., separation of adherends) resistance, expressed as average force and/or energy per unit specimen width (i.e., kN/m or J/m).

As indicated before, crack growth can occur at G_I values less than that of G_{Ic} under cyclic or sustained load, which can be accelerated under certain deleterious environments. Therefore in order to determine and predict the environmental durability of adhesively bonded materials, perhaps the test methods described by ASTM D3433 [184] should be adapted to test appropriately conditioned specimens. Alternatively, a qualitative test method should be used to assess service performance: ASTM D3762–79 (reapproved 1993), Adhesive-bonded surface durability of aluminium (wedge test): The test is qualitative and can also be used for other metals and plastics depending on the thickness and rigidity of the adherends. Specimens consist of two substrate strips of 25.4 × 203 mm with a layer of adhesive of 25.4 × 152 mm between them. A wedge is placed in between the substrate strips at the unbonded end of the assembly and positioned so that the end and sides of the wedge are approximately flush with those of the specimen. Measure the initial crack lengths a_i, and then place the specimen, with the wedge in place, in an environmental chamber, set at a suitable combination of temperature and humidity, for hot/wet conditioning. The specimens are then brought out of the chamber at regular time intervals and the crack extension Δa is measured. At the conclusion of testing, the specimen is prised open to note the failure mode, i.e., adhesive failure at the interface or cohesive failure within the body of the adhesive.

A further relevant ASTM standard is ASTM D3807–93, Strength properties of adhesives in cleavage peel by tension loading (engineering plastics-to-engineering plastics). The test allows a comparison to be made of bonded engineering thermoplastics strips (25.4 × 177.8 mm) for cleavage/peel strength. The specimen is loaded under tension (crack opening of the unbonded end) at 12.7 mm min^{-1}. A trace of load values should be recorded against the distance peeled for approximately 50 mm of cleavage/peel after the initial load peak, and the average load (in kN per meter width of specimen) required to separate the adherents determined.

Typical values for the fracture toughness G_c of various adhesives under different modes of loading are given in the literature [179,180] including fatigue fracture data [180]. The references also provide data on work of adhesion for various interfaces and environments.

5.2 Rubber

Rivlin and Thomas [185,186] recognized the limitation of the LEFM concept for rubber-like materials. The reduction in stored elastic energy during a crack propagation may not just be balanced by an increase in surface energy, and there may be other energy absorbing processes. Consequently they introduced an alternative based on the "tearing energy," which is the energy required per unit thickness of a specimen to effect a unit increase in crack length. The tearing energy T involves all the deformation in the crack tip region as the crack propagates, and it has been shown [185,186] to remain proportional to the increase in crack length. Therefore T is a true material property (similar to G in plastics and metals), independent of specimen geometry and the manner of loading, but, as in fracture toughness of polymers, it can show rate and temperature dependency [187].

Assume a tear crack growth of da under a fixed deformation condition; then the associated decrease in elastic strain energy is

$$-\left(\frac{dU}{da}\right)_l = TB \tag{43}$$

where B is the thickness of the strip of specimen. The suffix l indicates that the specimen is held under constant deformation, and hence no external work is involved.

In a recent review paper [188], commonly used test specimens and the relevant expressions for determining the tearing energy of rubbers are given in terms of the strain (stored) energy density (i.e., strain energy divided by specimen volume). The specimens include single edge notched (SEN) tensile strips, trouser tear piece, center-notched tension, and pure shear specimens. For the simple case of trouser tear, where the elongation ε in the specimen is minimal (i.e., the deformation (stretch) ratio $\lambda = 1 + \varepsilon \simeq 1$) and the tear propagates down the central axis of the test piece, the equation for T can be derived as follows. The work done (or energy expended) by a force F in generating a tear of da length in the specimen is given by $dW = F(2da)$, since for every da increase in tear length, the jaws (clamps) of the tensile machine move apart by $2da$ in the case of trouser tear specimens. Since T = (energy)/(Bda), therefore $T = dW/(Bda)$, or substituting $dW = 2Fda$ gives

$$T = \frac{2F}{B} \tag{44}$$

Equation 44 becomes $T = F/B$ for specimen geometries where the jaw separation is roughly equal to the tear propagation length.

There are various standard methods for determining the tear energy (or the tear strength) and other crack growth parameters for rubbers under both static and cyclic loading conditions. Some of these standards are discussed here.

ASTM D624–91, Tear strength of conventional vulcanized rubber and thermoplastic elastomers: The test method describes procedures for measuring the force required to rupture completely or tear the specified test pieces. The recommended test pieces are two types of crescent-shaped strips containing a razor nick to induce tearing, a 90° angle specimen (chevron shape) without a nick, and a trouser tear piece (a rectangular strip of 2 × 15 × 150 mm, with a 40 mm long slit centrally cut from one end of the strip to produce two legs). The specimens are punched out of sheets or plaques of material using suitably sharp steel cutting dies. The tests are conducted on tensile testing machines at jaw separation speeds of 50 mm min^{-1} for the trouser test piece and 500 mm min^{-1} for the other types of specimens. The tests are normally conducted at the standard conditions of temperature and humidity, but additional conditioning may be required for moisture sensitive materials. The tear strength T_s in kNm^{-1} is determined using the expression

$$T_s = \frac{F}{B} \tag{45}$$

where B is the specimen thickness and F is the maximum force indicated during the sudden tearing that crescent and angle specimens experience. The trouser tear tests do not yield a simple single peak in the force vs. jaws separation curve; instead a saw-toothed curve of varying degrees of roughness is produced, reflecting that the tear failure is not sudden but occurs progressively in a start–stop fashion, causing the force to fluctuate. Accordingly the ASTM recommendation is to use an average value of force in Eq. 45. The employment of Eq. 44 is not recommended by the ASTM, unless the legs of the trouser specimen are reinforced with fabric to ensure that $\lambda \simeq 1$ and a groove is introduced in the specimen to provide a path of least resistance for the tear propagation down the central axis of the specimen.

Other standards similar to the above are as follows. ISO 34, 1994 (≡ BS 903–A3, 1995), Rubber, vulcanized or thermoplastic—Determination of tear strength, Part 1, Trouser, angle and crescent test pieces. ISO 816, 1983, Rubber, vulcanized—Determination of tear strength of small test pieces (Delft test pieces). Here, the force required to tear across the width of a small test piece ($2 \times 9 \times 60$ mm) containing a 5 mm wide slit in the center of the specimen is measured.

Tests that aim to determine the extent of crack growth, particularly under cyclic loading, include ASTM D813–95, Rubber deterioration—Crack growth; ISO 133, 1983 (≡ BS 903—All, 1985 (1992)), Rubber, vulcanized—Determination of crack growth (De Mattia). Both these standards recommend the use of De Mattia tester for flex-cracking tests of molded specimens with pierced grooves. The flexing (bending) axis is to be parallel to the groove. The crack (cut) length should be measured at frequent intervals to determine the crack growth rate.

Another cyclic flexing test is described in: ASTM D3629–94, Rubber property-cut growth resistance. T-shaped grooved specimens are pierced, mounted into slots on the periphery of the rotating disk, and then bent or flexed as they strike against deflector bars at a controlled frequency. Tests are conducted within entirely closed chambers; therefore the test temperature and atmosphere can be conveniently set. The cut growth, in a given atmospheric condition, may be recorded as the number of cycles required to reach a certain cut length.

Fatigue Crack Propagation (FCP)

Rubber is employed in numerous dynamic engineering applications, e.g., tires, springs, v-belts, and flexible tubings, where the appropriate fracture data should be of considerable value. Although there is no standard method for determining FCP for rubbers, there is significant evidence that the relationship between rate of crack growth, da/dN, and tearing energy is independent of specimen geometries [189,190].

For fatigue tests a SEN tension specimen is most suitable [189], and the associated tearing energy equation is shown [190] to be

$$T = 2k(w_0)a \tag{46}$$

or more specifically, $T_{max} = 2k(w_{max})a$, where w_0 is the strain energy density and w_{max} is its value at $\lambda = \lambda_{max}$, a is crack length, and k a constant related to λ by $k = \pi\lambda^{-1/2}$.

The actual fatigue testing should, therefore, be conducted in two stages [188]: w_0 should be determined first, and then FCP should be measured as a function of the number of fatigue cycles N.

SEN tensile strips (approximately $2 \times 25 \times 105$ mm) are tested under strain control. Initially the unnotched rubber specimens should be conditioned, particularly filled ones, by being repeatedly loaded and unloaded to the maximum strain to reduce the Mullins effect (i.e., to obtain an equilibrium level of strain softening). Strain energy density is obtained by measuring the area under the unloading curve (i.e., strain energy) of the conditioned specimen and dividing the strain energy by the specimen volume.

A sharp notch of approximately 1 mm is cut with a razor blade into the specimen. The notch is extended under fatigue by approximately 0.1 mm, prior to the onset of the fatigue test proper. During the fatigue tests the crack length extension is measured periodically after a suitable number of cycles, and at each measurement da/dN and corresponding T values are determined. The test is continued until the crack has grown to approximately 20% of the specimen width. The results are plotted in the form of $\log(da/dN)$ vs. $\log(T)$. It is possible to identify [191] four regions in a typical fatigue plot for materials such as natural rubber. These indicate two extreme values of T, i.e., the minimum value of T, $T_{(threshold)}$, below which no mechanical crack growth should occur, and a maximum value of T, $T_{(catastrophic)}$, at which unstable crack growth is initiated. In the intermediate region the plot (da/dN vs. T) shows two distinct linear relationships. The most pronounced one corresponding to higher tear energy values shows the customary power law (Paris–Erdogan equation) relationship

$$\frac{da}{dN} = BT^{\beta} \tag{47}$$

where B and β are constants depending on the material type. The values of β vary between 2 and 6 [192].

Fatigue curves, threshold tear energy values, and ozone crack growth (which may be significant below the threshold level) data for various rubbers/elastomers are presented in the review article by Seldén [188]. In the same paper an estimation of fatigue life from crack growth data is also outlined. Briefly, using Eq. 47 and substituting for T from Eq. 46 gives

$$\frac{da}{dN} = B(2kw_0 a)^{\beta} \tag{48}$$

Equation 48 can be rearranged and integrated between an initial crack length a_0 and a final crack length a_f, where unstable crack growth initiates.

$$\int_{a_0}^{a_f} \frac{da}{a^{\beta}} = \int_0^{N_f} B(2kw_0)^{\beta}\, dN$$

The integration yields the fatigue lifetime (i.e., the number of cycles to failure) N_f to be

$$N_f = \frac{1/a_0^{\beta-1} - 1/a_f^{\beta-1}}{(\beta - 1)B(2kw_0)^{\beta}} \tag{49}$$

Since $a_0 \ll a_f$, then Eq. 49 can be simplified to

$$N_f = \frac{1}{(\beta - 1)B(2kw_0)^{\beta} a_0^{\beta-1}} \tag{50}$$

Accordingly the fatigue lifetime of rubber with a given defect size a_0 may be estimated.

5.3 Cellular Materials

Most polymers and rubbers can be produced in a cellular form (i.e., foams). The "flexible" foams are characterized in a similar manner to rubbers by tear energy measurements [193]. There is also a specific standard for conducting tear strength tests for flexible foams: BS EN ISO 8067, 1995 (≡ ISO 8067, 1989), Flexible cellular polymeric materials—Determination of tear strength. This standard method is specified for flexible foams of thickness greater than 24 mm. The test pieces are rectangular parallelepipeds of 25 × 25 × 125 mm with a 50 mm longitudinal slit cut centrally from one end. The tests are to be conducted at least 72 h after manufacture in a standard laboratory atmosphere. The specimen legs are clamped and pulled apart at 50–500 mm min^{-1} so that the direction of the applied force is perpendicular to the tear width. The cut may be maintained in the center of the block while tearing, with light touches of a razor blade edge. The tear is allowed to propagate for approximately 25 mm and the maximum force F associated with the tear is recorded. The tear strength R is expressed as $R = F/d$ (in N/m), where d is the original specimen thickness (i.e., tear width).

By contrast, the "rigid" polymer foams exhibit linear-elastic behavior up to fracture in tension and therefore can be evaluated in terms of the parameters G_{Ic} and K_{Ic}. Although there are no standard procedures for this, Fowlkes [194] has given the experimental details of the compliance method of determining G_{Ic} for a rigid polyurethane (PU) foam. Briefly, the method uses Eq. 22:

$$G_{Ic} = \frac{P_c^2}{2B}\left(\frac{dC}{da}\right)_{a=ac} \tag{51}$$

The following specimens are used to conduct the experimental measurements of compliance C vs. crack length a, and the load to crack growth initiation P_c.

The double cantilever beam (DCB) specimen with side grooves to control the path of the crack propagation is used. The side grooves and the starter crack are generated using a suitable saw. The starter crack is extended with a razor blade. For the compliance calibration curve (i.e., C vs. a) six specimens with different starter crack lengths are tested for the compliance measurements. The data produces the following relationship for rigid PU:

$$C = 0.00165a^3 \qquad \text{or} \qquad \frac{dC}{da} \simeq 0.005a^2$$

Substituting in Eq. 51 gives

$$G_{Ic} = \frac{P_c^2}{2B}\, 0.005a_c^2 \tag{52}$$

P_c and a_c values are determined by loading DCB specimens of a given starter crack length and plotting load against displacement. The plot indicates that the crack advances in multiple cycles, each cycle comprised of the initiation of unstable crack growth, crack arrest, and slow crack growth. The crack lengths corresponding to these phenomena are measured via a scale attached to the side of the specimen. By substituting the appropriate values of load and crack length into Eq. 52, three fracture toughness parameters G_a (arrest), G_{sg} (slow growth), and G_{Ic} (unstable growth) can be obtained at each extension of the crack length.

The alternative specimens are center notched (CN), double-edge notched (DEN) and single-edge notched (SEN) tensile specimens. Unlike DCB tests, these tests produce a single load–displacement peak, indicating that these specimens fail spontaneously from the starting crack with no slow crack growth. The formulae used to compute G_{Ic} are [195]:

$$G_{Ic} = \left(\frac{P}{B}\right)^2 \frac{1}{EW}\left(\tan\frac{\pi a}{W} + 0.1\sin\frac{2\pi a}{W}\right) \tag{53}$$

for the CN and DEN specimens, and

$$G_{Ic} = \left(\frac{P}{B}\right)^2 \frac{1}{EW}\left(7.59\,\frac{a}{W} - 32\left(\frac{a}{W}\right)^2 + 117\left(\frac{a}{W}\right)^3\right) \tag{54}$$

for the SENB specimen. E is Young's modulus in these equations.

The values for the initial crack length and the maximum load at failure were substituted for a and P respectively in Eqs. 53 and 54 for the calculations of G_{Ic}.

The results [194] from the various specimens are in good agreement: the maximum value being obtained from DCB, and the minimum from DEN tests.

K_{Ic}, J_{Ic} and δ (i.e., COD) have also been employed in fracture toughness determination of foams using expressions appropriate for plastics [196,197]. The studies have included the effects of specimen geometry (e.g., a/W), loading rate, and foam cell size [196], and foam density [197] on the properties.

A comprehensive account of the fracture toughness of cellular materials including foams, honeycombs, woods, and cancellous bone is presented by Gibson and Ashby [198]. They have put emphasis on establishing and verifying relationships between the fracture parameters and the density of the material rather than the microstructural parameters such as the cell wall thickness. This is a practical approach since the density data for these materials are readily available or can be easily measured.

For example the dependence of fracture toughness of a cellular material, $K_{Ic}{}^{*}$, on its density ρ^{*} is shown in a simple manner as follows: Let the toughness (i.e., the fracture energy per unit area) of the solid material be G_{Ics} and that of the foam $G_{Ic}{}^{*}$. Assume an open-cell foam of cell edge length l and edge thickness t. A crack, advancing over an area l^2 in an open-cell foam, breaks (on average) one cell wall of cross section t^2. From an energy balance,

$$G_{Ic}{}^{*}l^2 = G_{Ics}t^2 \tag{55}$$

Using the fracture mechanics equation for the foam, $(K_{Ic}^{*})^2 = EG_{Ic}^{*}$, and given that $E^{*} \propto E_s(t/l)^4$, and combining with Eq. 55, we obtain

$$(K_{Ic}{}^{*})^2 = E_s\left(\frac{t}{l}\right)^4 G_{Ics}\left(\frac{t}{l}\right)^2$$

Substituting $K_{Ics}{}^2 = E_sG_{Ics}$ and $\rho^{*}/\rho_s \propto (t/l)^2$(appropriate for open cells) gives

$$K_{Ic}^{*} = CK_{Ics}\left(\frac{\rho^{*}}{\rho_s}\right)^{3/2} \tag{56}$$

Equation 56, where C is a constant, relates the fracture toughness of the foam to that of the solid polymer and their respective densities. Similar relationships have been also shown for the closed cell foams and honeycombs [198]. The estimates according to Eq. 56 compare very well with experimental data obtained for various types of rigid foams [198].

5.4 Textiles, Films, and Coated Fabrics

There is no standard methodology of determining the fracture toughness of these materials. The thickness requirement and the test geometry (CT and SENB) recommended in the standards appropriate for plastics would obviously lead to problems when applied to thin materials. Attempted measurements have selected special specimen geometries and certain fracture parameters in order to minimize/counter the effect of plane stress/plastic deformation. Accordingly the J-integral [199,200], and the essential work of fracture [201] parameters have been adopted. In one of these studies [199], center-notched rectangular strips of polyethylene film are tensile tested with the specimen being clamped along its long edges in order to generate a state of biaxial stress in the specimen. Typical specimen dimensions are 250 mm width, 150 mm length (with a gauge length of 50 mm), 20 μm thickness, and notch (crack) length > 50 mm. The specimens are clamped in grips with silicone rubber facing, and testing is conducted at approximately 2.5 mm min^{-1}. The J-integral is obtained using the expression of $J = -(1/B)(dw_T/da)$, where B is the initial film thickness and w_T is the total work done during the test, i.e., $w_T = \int[(\text{force}) \cdot$

(displacement)]. J_{Ic} is taken as the value of J at the onset of crack growth as indicated on the J_R curve (i.e., J versus *da* plot).

SEN tension rectangular strips of approximately 10 mm width and 130 mm length have also been used [200] in J-integral measurements of various polymer films intended as magnetic media (including syndiotactic polystyrene, polyethylene terephthalate, Trycite, and Kapton). These specimens are tabbed with cardboard and tested at approximately 1 mm min^{-1}. For accurate measurements of crack (tear) growth initiation and the subsequent tear propagation, video imaging (preferably with polarized light) is employed in these studies [199,200].

The existing test standards on the resistance of these materials to crack propagation concentrate on determining the tear strength values. Some of these standards are specific to plastics film and sheeting [202–208], fabrics coated with rubber or plastics [209–212], non-woven textile fabrics [213–216], and woven textile fabrics [217–218].

The force required to propagate a starter tear in the specimen is recorded in the course of the standard test procedures. The tear strength is expressed either as the maximum value (alternatively the average value) of the tearing force recorded or as the maximum force (alternatively the average force) per specimen thickness (in kN/m). The information from these standard tests, while useful for quality control and acceptance testing, does not produce a fundamental material property suitable for design applications.

Acknowledgments

I wish to express my very sincere gratitude to Mrs. Myrtle Young for the typing and Dr. David Barkley for his help with the illustrations.

References

1. Griffith, A. A. (1920), *Phil. Trans. Roy. Soc., A221*, 163.
2. Irwin, G. R. (1948), *Trans. Am. Soc. Metals, 40*, 147.
3. Wells, A. A. (1963), *Brit. Welding J., 10*, 563.
4. Jones, D. R. H. (1993), *Engineering Materials 3*, Pergamon Press.
5. Williams, J. G. (1990), *Prog. Rubber Plast. Tech., 6*, 2, 174.
6. Brown, W. F., and Srawley, J.E. (1966), ASTM STP No. 410, p1.
7. Rook, D. P., and Cartwright, D. J. (1976), *Compendium of Stress Intensity Factors*, HMSO.
8. ASTM D5045. 1993. Plane-strain fracture toughness and strain energy release rate of plastics materials.
9. ASTM E399. 1983. Plane-strain fracture toughness of metallic materials.
10. Williams, J. G. (1984), *Fracture Mechanics of Polymers*, Ellis Horwood.
11. Wells, A. A. (1961), *Proceedings of the Crack Propagation Symposium*, Vol. 1, Paper 84, Cranfield, U.K.
12. BS 5762. 1979. Crack opening displacement (COD) testing.
13. BS 7448. 1991. Determination of K_{Ic}, critical CTOD and critical J values of metallic materials.
14. ASTM E1290. 1993. Crack-tip opening displacement (CTOD) fracture toughness measurement.
15. ASTM E813. 1989. J_{Ic}, a measure of fracture toughness.
16. Anderson, T. L. (1995), *Fracture Mechanics (Fundamentals and Applications)*, 2d ed., CRC Press.
17. Eshelby, J. D. (1956), *Progr. Solid State Phys., 3*, 79.
18. Rice, J. R. (1968), *J. Appl. Mech., 35*, 379.
19. Hutchinson, J. W. (1968), *J. Mech. Phys. Solids, 16*, 13.

20. Rice, J. R., and Rasengren, G. F. (1968), *J. Mech. Phys. Solids*, *16*, 1.
21. ASTM E1152. 1987. Determining *J–R* curves.
22. Jones, R. E., and Bradley, W. L. (1989), ASTM STP 995, Vol. 1, 447.
23. Cayard, M. (1990), "Fracture Toughness Testing of Polymeric Materials," Ph.D. thesis, Texas A & M University.
24. Carling, M. J., and Williams, J. G. (1987), *Proceedings of the 6th ICCM and 2nd ECCM*, I.C., London, Vol. 3, p. 317.
25. Hale, G. E. (1990), *European Conference on Fracture—ECF8, Torino, Italy. Proceedings*, Vol. 1, p. 135.
26. Schapery, R. A. (1984), *Internat. J. Fracture*, *25*, 195.
27. Schapery, R. A. (1990), *Internat. J. Fracture*, *42*, 189.
28. Paris, P. C., and Erdogan, F. (1960), *J. Basic Engineering*, *85*, 528.
29. ASTM E647. 1993. Measurement of fatigue crack growth rates.
30. Kakarala, S. N., and Roche, J. L. (1987), Instrumented impact testing of plastics and composite materials, ASTM STP 936, p. 144.
31. Roche, J. L., and Kakarala, S. N. (1987), Instrumented impact testing of plastics and composite materials, ASTM STP 936, p. 24.
32. Jancar, J., and Dibenedetto, A. T. (1994), *Polymer Engineering and Science*, *34*, No. 24, p. 1799.
33. *Engineered Materials Handbook*, Vol. 2: *Engineering Plastics*, ASM International, 1988.
34. Faucher, B. (1994), *J. Testing Evaluation*, *22*, No. 1, 30.
35. Mostovoy, S., and Ripling, E. J. (1966), *J. Appl. Polym. Sci.*, *10*, 1351.
36. Kies, J. A., and Clark, A. B. J. (1969), *Proc. 2nd International Conference on Fracture*, Paper 42, Brighton, U.K.
37. Hodgkinson, J. M., Chow, K. H. L., and Williams, J. G. (1991), *Proc. 8th International Conference on Deformation, Yield and Fracture of Polymers*, Churchill College, Cambridge, U.K., p. 43/1.
38. Sue H. J., and Yee, A. F. (1993), *J. Material Sci.*, *28*, 2975.
39. Akay, M., and O'Regan, D. F. (1995), *Polymer Testing*, *14*, 149.
40. Karger-Kocsis, J., Daugoard, J., and Amby, L. (1992), *Int. J. Fatigue*, *14*, No. 3, 189.
41. Hargarter, N., Friedrich, K., and Catsman, P. (1993), *Composites Sci. Techn.*, *46*, 229.
42. Spahr, D. E., Friedrich, K., Schultz, J. M., and Bailey, R. S. (1990), *J. Material Sci.*, *25*, 4427.
43. Hine, P. J., Duckett, R. A., and Ward, I. M. (1993), *Composites*, *24*, No. 8, 643.
44. Hale, G.E. (1991), *Proc. 8th International Conference on Deformation, Yield and Fracture of Polymers*, Churchill College, Cambridge, U.K., p. 23/1.
45. Akay, M., and O'Regan, D. (1995), *Polymer Testing*, *14*, 149.
46. Dear, J. P., and MacGillivray, J. H. (1990), *European Conference on Fracture-ECF8, Torino, Italy, Proceedings*, Vol. 1, p. 169.
47. Baker, A. (1985), ASTM STP 856, American Society for Testing and Materials, Philadelphia, p. 394.
48. Stalder, B., Beguelin, P., Roulin-Moloney, A. C., and Kausch, H. H. (1989), *J. Materials Sci.*, *24*, 2262.
49. Cudre-Mauroux, N., Kausch, H. H., Cantwell, W. J., and Roulin-Moloney, A. C. (1991), *Internat. J. Fracture*, *50*, 67.
50. Sharpe, W. N., Jr., and Bohme, W. (1994), *J. Testing Evaluation*, *22*, No. 1, 14.
51. ASTM E399-83, American Society for Testing and Materials, Philadelphia.
52. De Kalbermatten, T., Jaggi, R., Flueler, P., Kausch, H. H., and Davies, P. (1992), *J. Materials Sci. Letter*, *11*, 543.
53. Elber, W. (1970), *Engineering Fracture Mechanics*, *2*, 37.
54. Williams, J. G., and Marshall, G. P. (1975), *Proc. Royal Soc., London, A342*, 55.
55. Schapery, R. A. (1975), *Internat. J. Fracture*, *11*, 141.
56. Matthews, F. L., and Rawlings, R. D. (1994), *Composite Materials: Engineering and Science*, Chapman and Hall.

57. Kinloch, A. J., and Young, R. J. (1983), *Fracture Behaviour of Polymers*, Applied Science Publishers.
58. Strebel, J. J., and Moet, A. (1992), *J. Materials Sci.*, *27*, 2981.
59. Chung, W. N., and Williams, J. G. (1990), *European Conference on Fracture-ECF8, Torino, Italy, Proceedings*, Vol. 1, p. 129.
60. Chan, M. K. V., and Williams, J. G. (1981), *Polymer Engineering Sci.*, *21*, No. 15, 1019.
61. Birch, M. W., Taylor, M. D., and Marshall, G. P. (1983), *Plastics and Rubber Processing and Applications*, *3*, No. 3, 281.
62. Plati, E., and Williams, J. G. (1975), *Polymer Engineering Sci.*, *15*, No. 6, 470.
63. Mallick, P., and Jennings, J. (1988), *SPE ANTEC '88*, Society of Plastics Engineering, New Orleans, p. 583.
64. Yap, O. F., Mai, J. W., and Cotterell, B. (1983), *J. Materials Sci.*, *18*, 657.
65. Seidler, S., and Grellmann, W. (1995), *Polymer Testing*, *14*, 453.
66. Sih, G. C., and Berman, A. T. (1980), *J. Biomed. Materials Res.*, *14*, 311.
67. Adams, G. C. (1988), *SPE ANTEC '88*, Society of Plastics Engineering, New Orleans, p. 1517.
68. Goolsby, R. D., and Chunliang, L. (1987), ASTM STP936, American Society for Testing and Materials, Philadelphia, p. 351.
69. Akay, M., and Barkley, D. (1991), *J. Materials Sci.*, *26*, 2731.
70. Whitehead, R. D. et al. (1987), *Plastics and Rubber Processing and Applications*, *8*, 115.
71. Akay, M., and Ozden, S. (1996), *Plastics, Rubber and Composites Processing and Applications*, *25*, 138.
72. Seibel, S. R., Moet, A., Bank, D. H., and Sehanobish, K. (1995), *SPE ANTEC '95*, Society of Plastics Engineering, New Orleans, p. 3966
73. Beguelin, Ph., and Kausch, H. H. (1991), *Proc. 8th International Conference on Deformation, Yield and Fracture of Polymers*, Churchill College, Cambridge, U.K., p. 22/1.
74. Lu, M.-L., Chiou, K.-C., and Chang, F.-C., (1996), *Polymer*, *37*, No. 19, 4289.
75. Martinatti, F., and Ricco, T. (1994), *Polymer Testing*, *13*, 405.
76. Lovell, P. A., et al. (1991), *Proc. 8th International Conference on Deformation, Yield and Fracture of Polymers*, Churchill College, Cambridge, U.K., p. 34/2.
77. Rink, M., et al. (1990), *European Conference on Fracture-ECF8, Torino Italy, Proceedings*, Vol. 1, p. 201.
78. Moskala, E. J. (1991), *Proc. 8th International Conference on Deformation, Yield and Fracture of Polymers*, Churchill College, Cambridge, U.K., p. 51/1.
79. Sue, H.J., et al. (1994), *Colloid Polym. Sci.*, *272*, 456.
80. Mills, N. J., and Zhang, P. S. (1989), *J. Materials Sci.*, 2099.
81. Greig, J. M., and Skiplorne, I. C. (1985), *Proceedings of the PRI International Conference on Impact Testing and Performance of Polymeric Materials*, University of Surrey, U.K., p. 11/1.
82. Moore, D. R., Prediger, R., and Stephenson, R. C. (1985), *Plastics and Rubber Processing and Applications*, *5*, 335.
83. Chung, W. N., and Williams, J. G. (1991), *Proc. 8th International Conference on Deformation, Yield and Fracture of Polymers*, Churchill College, Cambridge, U.K., p. 28/1.
84. Josserand, L., Schirrer, R., and Davies, P. (1995), J. *Materials Sci.*, *30*, 1772.
85. Michel, J. C., et al. (1991), *Proc. 8th International Conference on Deformation, Yield and Fracture of Polymers*, Churchill College, Cambridge, U.K., p. 120/1.
86. Clark, T. R., Hertzberg, R. W., and Mohammadi, N. (1991), *Proc. 8th International Conference on Deformation, Yield and Fracture of Polymers*, Churchill College, Cambridge, U.K., p. 31/1.
87. Bucknall, C. B., and Faitrouni, T. (1991), *Proc. 8th International Conference on Deformation, Yield and Fracture of Polymers*, Churchill College, Cambridge, U.K., p. 30/1.
88. Echalier, B. (1990), *Proceedings of the PRI seminar on Impact Testing of Plastics*, Regent's College, London.
89. Marshall, G. P., and Birch, M. W. (1982), *Plastics and Rubber Processing and Applications*, 369.

90. Ting, R. Y., and Cottington, R. L. (1980), *J. Appl. Polymer Sci.*, *25*, 1815.
91. Kinloch, A. J., Gilbert, D. G., and Shaw, S. J. (1986), *J. Materials Sci.*, *21*, 1051.
92. Cantwell, W. J., and Roulin-Moloney, A. C. (1987), *Proceedings of the 6th ICCM and 2nd ECCM, I.C., London*. Vol. 3, p. 460.
93. Zheng, S., et al. (1996), *Polymer*, *37*, No. 21, 4667.
94. Charalambides, M. N., and Williams, J. G. (1995), *Polymer Composites*, *16*, No. 1, 17.
95. Jackson, G. V. and Orton, M. L. (1995), Chapter 9 in *Particulate-Filled Polymer Composites* (R. Rothon, ed.), Longman Scientific and Technical.
96. Kim, S. L., et al. (1978), *Polymer Eng. Sci.*, *18*, No. 14, 1093.
97. Akay, M., O'Regan, D. F., and Bailey, R. S. (1995), *Composites Sci. Tech.*, *55*, 109.
98. Kelly, A. (1970), *Proc. Roy. Soc. London*, *A.319*, 95.
99. Wells, J. K., and Beaumont, P. W. R. (1988), *J. Materials Sci.*, *23*, 1274.
100. Huang, D. D. (1995), *Polymer Composites*, *16*, No. 1, 10.
101. Gaggar, S., and Broutman, L. J. (1975), *J. Comp. Mat.*, *9*, 216.
102. Agarwal, B. D., and Giare, G. S. (1981), *Fibre Sci. Technol.*, *15*, 283.
103. Lauke, B., Schultrich, B., and Pompe, W. (1990), *Polym.-Plast. Technol. Eng.*, *29*, Nos. 7 & 8, 607.
104. Mandell, J. F., Darwish, A. Y., and McGarry, F. J. (1981), ASTM STP734, American Society for Testing and Materials, Philadelphia, p. 73.
105. Irwin, G. R., and Kies, J. A. (1954), *J. Welding*, *33*, 193.
106. Sweeney, J. (1986), *J. Strain Anal.*, *21*, 99.
107. Sweeney, J. (1988), *Int. J. Fract.*, *37*, 233.
108. Kim, B. H., and Kim, H. S. (1989), *Polymer Testing*, *8*, 313.
109. Kim, B. H., and Joe, C. R. (1987), *Polymer Testing*, *7*, 355.
110. Harris, B. (1972), *Composites*, *3*, 152.
111. Owen, M. J., and Bishop, P. T. (1972), *J. Phys. D: Appl. Physics*, *5*, 1621.
112. Irwin, G. R. (1964), *Appl. Mats. Res.*, *3*, 65.
113. Lauke, B., and Pompe, W. (1986), *Composites Sci. Tech.*, *26*, 37.
114. Adams, G. C., and Williams, J. G. (1985), SPE ANTEC '85, Society of Plastics Engineering, New Orleans, p. 361.
115. Carling, M. J., and Williams, J. G. (1990), *Polymer Composites*, *11*, No. 6, 307.
116. Akay, M., and Barkley, D. (1987), *Polymer Testing*, *7*, 391.
117. Karger-Kocsis, J. (1990), *Polymer Bulletin*, *24*, 341.
118. Harris, B. (1986), *Engineering Composite Materials*, Institute of Metals.
119. Friedrich, K. (1985), *Composites Sci. Technol.*, *22*, 43.
120. Kim, J.-K., and Mai, Y.-W. (1993), *Composites Sci. Technol.*, *49*, 51.
121. Karger-Kocsis, J., and Friedrich, K. (1986), *Polymer*, *27*, 1753.
122. Miwa, M., Ohsawa, T., and Tsuji, N. (1979), *J. Appl. Polymer Sci.*, *23*, 1679.
123. Friedrich, K. (1989), *Application of Fracture Mechanics to Composite Materials* (K. Friedrich, ed.), p. 425, Elsevier Science Publishers.
124. Vu-Khanh, T., and Fisa, B. (1986), *Polymer Composites*, *7*, No. 5, 375.
125. Friedrich, K., and Karger-Kocsis, J. (1987), *6th ICCM and 2nd ECCM, I.C., London, Proceedings*, Vol. 3, 327.
126. Wells, J. K., and Beaumont, P. W. R. (1982), *J. Material Sci.*, *17*, 397.
127. ASTM C1256. 1993. Interpreting glass fracture surface features.
128. Akay, M., and Aslan, N. (1995), *Proc. Instn. Mech. Engrs.*, Vol. 209, Part H: J. Engineering in Medicine, p. 93.
129. Fedrizzi, F., Martinatti, F., and Ricco, T. (1995), *Proceedings of the 3rd IOM International Conference on Deformation and Fracture of Composites*, University of Surrey, U.K., p. 362.
130. Friedrich, K., et al. (1986), *Composites*, *17*, 205.
131. Brunner, A. J., et al. (1996), *Seventh European Conference on Composite Materials (ECCM-7), London*. Vol. 2, p. 3.

132. Davies, P., et al. (1996), *Proceedings of the Seventh European Conference on Composite Materials (ECCM-7), London*. Vol. 2, p. 9.
133. ASTM D5528. 1994. Mode I interlaminar fracture toughness of unidirectional fiber-reinforced polymer matrix composites.
134. ASTM D2651. 1995. Preparation of metal surfaces for adhesive bonding.
135. Davies, P. (1996), *Applied Composite Materials*, *3*, 135.
136. Johnson, W. S., and Mangalgari, P. D. (1987), *ASTM J. Composite Technology and Research*, *9*, 10.
137. O'Brien, T. K., and Martin, R. H. (1993), *ASTM J. Composite Technology and Research*, *15*, No. 4.
138. Smiley, A. J., and Pipes, R. B. (1987), *J. Composite Materials*, *21*, 670.
139. Hibbs, M. F., Tse, M. K., and Bradley, W. L. (1987), ASTM STP937, American Society for Testing and Materials, Philadelphia, p. 115.
140. ASTM D2584. 1994. Ignition loss of cured reinforced resins.
141. ASTM D3171. 1990. Fiber content of resin-matrix composites by matrix digestion.
142. Masters, J. E. (1987), *Proceedings of 6th ICCM and 2nd ECCM I.C.*, London, Elsevier, Vol. 3, p. 96.
143. Recker, H. G., Altstadt, V., and Strangle, M. (1992), *Proc. 37th Int. SAMPE Symposium*, p. 493.
144. Davies, P., et al. (1990), *Composites Sci. Tech.*, *39*, 193.
145. Carlsson, L. A., and Gillespie, J. W., Jr. (1989), Chapter 4 in *Application of Fracture Mechanics to Composite Materials* (K. Friedrich, ed.), Elsevier.
146. European Structural Integrity Society, Polymers and Composites Task Group, "Protocols for Interlaminar Fracture Testing of Composites," draft by P. Davies, May 1992.
147. Wang, Y. and Williams, J. G. (1992), *Composites Sci. Tech.*, *43*, 251.
148. Kinloch, A. J., Wang. Y., Williams, J. G., and Yayla, P. (1993), *Composites Sci. Tech.*, *47*, 225.
149. Reeder, J. R., and Crews, J. H., Jr. (1990), *AIAA Journal*, *28*, No. 7, 1270.
150. Reeder, J. R., and Crews, J. H., Jr. (1991), *Proceedings of ICCM/8*, Hawaii, p. 36-B-1.
151. Reeder, J. R. (1993), ASTM STP1206, American Society for Materials and Testing, Philadelphia, p. 303.
152. Wang, S. S. (1983), *J. Comp. Mat.*, *17*, 210.
153. Donaldson, S. L. (1987), *Proceedings of the 6th ICCM and the 2nd ECCM, I.C.*, London, Vol. 3, p. 274.
154. Becht, G., and Gillespie, J. W., Jr. (1988), *Composites Sci. Tech.*, *31*, 143.
155. ASTM D4255. 1994. In-plane shear properties of composite laminates.
156. Martin, E., et al. (1986), *Composites Sci. Tech.*, *26*, 185.
157. Fejas-Kozma, Zs., and Karger-Kocsis, J. (1994), *J. Reinforced Plastics Composites*, *13*, 8.
158. Karger-Kocsis, J. (1992), *J. Appl. Polymer Sci.*, *45*, 1595.
159. Czigany, T. (1995), *Periodica Polytechnica Ser. Mech. Eng.*, *39*, Nos. 3–4, 189.
160. Chamis, C. C., Murthy, P. L. N., and Minnetyan, L. (1996), *Theoret. Appl. Fracture Mechanics*, *25*, 1.
161. Hashemi, S., Kinloch, A. H., and Williams, J. G. (1987), *Proceedings of the 6th ICCM and the 2nd ECCM, I.C.* London, Vol. 3, p.254.
162. Davies, P., et al. (1991), *Composites Sci. Technol.*, *43*, 129.
163. Davies, P., and de Charentenay, F. X. (1987), *Proceedings of the 6th ICCM and the 2nd ECCM, I.C.* London, Vol. 3, p. 284.
164. Hashemi, S., Kinloch, A. J., and Williams, J. G. (1989), *J. Material Sci. Letters*, *8*, 125.
165. Wang, Y., Li, J., and Zhao, D. (1995), *Composites Engineering*, *5*, No. 9, 1159.
166. Davies, P. (1992), *Proceedings of the European Conference on Composites Testing and Standardisation*, p. 405.
167. Dyson, N., Kinloch, A. J., and Okada, A. (1994), *Composites*, *25*, No. 3, 189.
168. Davies, P. (1996), *J. Composite Materials*, *30*, No. 9, 1004.

169. Chapman, T. J., Smiley, A. J., and Pipes, R. B. (1987), *Proceedings of the 6th ICCM and the 2nd ECCM, I.C.* London, Vol. 3, p. 295.
170. Blackman, B. R. K., et al. (1995), *Proceedings of the 3rd IOM International Conference on Deformation and Fracture of Composites*, University of Surrey, U.K., p. 277.
171. Davies, P., et al. (1993), *Proceedings of ICCM/9*, Madrid, Spain, Vol. VI, p. 308.
172. Gilchrist, M. D., and Svensson, N. (1995), *Composites Sci. Tech.*, *55*, 195.
173. Hojo, M., Matsuda, S., and Ochiai, S. (1996), *Proceedings of the ECCM-7*, London, Vol. 1, p. 81.
174. Sriram, P., Khourchid, Y., and Hooper, S. J. (1993), ASTM STP1206, American Society for Testing and Materials, Philadelphia, p. 291.
175. Martin, R. H. (1995), *Proceedings of the 3rd International Conference on Deformation and Fracture of Composites*, University of Surrey, U.K., p. 297.
176. Brew, B., et al. (1991), *Proc. 8th International Conference on Deformation, Yield and Fracture of Polymers*, Churchill College, Cambridge, U.K., p. 18/1.
177. Kinloch, A. J. (1980), *J. Mater. Sci.*, *15*, 2141.
178. Gent, A. N. (1974), *Rubber Chem. Technol.*, *47*, 202.
179. Kinloch, A. J. (1982), *J. Mater. Sci.*, *17*, 617.
180. Corish, P. J. (ed), (1992), *Concise Encyclopedia of Polymer Processing and Applications* (chapter on bonding), Pergamon Press.
181. Mostovoy, S., Crosley, P.B., and Ripling, E.J. (1965), *J. Mater. Sci.*, *2*, 661.
182. ASTM D2093. 1993. Preparation of surfaces of plastics prior to adhesive bonding.
183. ISO 4588. 1989. Adhesives-preparation of metal surfaces for adhesives bonding.
184. ASTM D3433. 1993. Fracture strength in cleavage of adhesives in bonded metal joints.
185. Rivlin, R. S., and Thomas, A. G. (1953), *J. Polymer Sci.*, *10*, 291.
186. Thomas, A. G. (1955), *J. Polymer Sci.*, *18*, 177.
187. Hamed, G. R. (1991), *Rubber Chem. Technol.*, *64*, 493.
188. Seldén, R. (1995), *Progress in Rubber and Plastics Technology*, *11*, 56.
189. Gent, A. N., Lindley, P. B., and Thomas, A. G. (1964), *J. Appl. Polym. Sci.*, *8*, 455.
190. Lake, G. J. (1983), *Prog. Rubber Technol.*, *45*, 89.
191. Ellul, M. D. (1991), in *Engineering with Rubber —How to Design Rubber Components* (A. N. Gent, ed.), Hanser Verlag, Germany.
192. Royo, J. (1992), *Polymer Testing*, *11*, 325.
193. Gent, A. N., and Thomas, A. G. (1959), *J. Appl. Polymer Sci.*, Vol. II, No. 6, 354.
194. Fowlkes, C. W. (1974), *Int. J. Fracture*, *10*, No. 1, 99.
195. Srawley, J. E., and Brown, W. F., Jr. (1964), ASTM STP No. 381, p. 189.
196. Zenkert, D., and Backlund, J. (1989), *Composites Sci. Tech.*, *34*, 225.
197. McIntyre, A., and Anderton, G. E. (1979), *Polymer*, *20*, 247.
198. Gibson, L. J., and Ashby, M. F. (1988), *Cellular Solids: Structure & Properties*, Pergamon Press.
199. Tielking, J. T. (1993), *Polymer Testing*, *12*, 207.
200. Wu, S., Bubeck, R. A., and Carriere, C. J. (1995), ANTEC '95, p. 3898.
201. Hashemi, S. (1993), *J. Materials Sci.*, *28*, 6178.
202. ISO 6383–1. 1983 (≡ BS2782, Part 3, Method 360B, 1991), Determination of tear resistance of plastics film and sheeting by the trouser tear method.
203. ISO 6383–2. 1983 (≡ BS2782, 3, 360A, 1991), Determination of tear resistance of plastics film and sheeting by the Elemendorf method.
204. BS2782, Part 3, Method 360C. 1991. Determination of tear resistance of plastics and sheeting by the initiation method.
205 ASTM D1938. 1994. Tear propagation resistance of plastic film and thin sheeting by a single-tear method.
206. ASTM D1922 1994. Tear propagation resistance of plastic film and thin sheeting by pendulum method.
207. ASTM D1004. 1994. Initial tear resistance of plastic film and sheeting.

208. ASTM D2582. 1993. Puncture-propagation tear resistance of plastic film and thin sheeting.
209. ISO 4674. 1977. Fabrics coated with rubber plastics—Determination of tear resistance.
210. ISO 4646. 1989. Fabrics coated with rubber or plastics—Low temperature impact test.
211. BS 3424, Part 5. 1982. Coated fabrics—Determination of tear strength.
212. ASTM D2724–87. 1995. Test methods for bonded, fused and laminated apparel fabrics.
213. ISO 9073-4. 1989. (≡ BS EN29073, 4) Textiles—Non woven fabrics—Determination of tear resistance.
214. ASTM D5733. 1995. Tearing strength of non woven fabrics by the trapezoid procedure.
215. ASTM D5734. 1995. Tearing strength of non woven fabrics by falling-pendulum (Elmendorf) apparatus.
216. ASTM D5735. 1995. Tearing strength of non woven fabrics by the tongue (single rip) procedure (constant-rate-of-extension tensile testing machine),
217. ISO 9290. 1990. Textiles—woven fabrics—Determination of tear resistance by the falling pendulum method.
218. ASTM D4533. 1991. Trapezoid tearing strength of geotextiles.

23
Friction

Ivan James
Forncet, Wem, Shropshire, England

1 Introduction

Friction can be defined as resistance to motion, whether it be the motion of a solid through a fluid or one solid sliding over another. It is this latter aspect that is covered in this chapter.

In order to have friction between two surfaces it is necessary that there be a compressive normal force between the surfaces, and in most circumstances frictional force is proportional to this normal force. It is convenient, therefore, to take the ratio of frictional force to normal force and call it the "coefficient of friction." This term is universally used, even in circumstances where the frictional force is not quite proportional to the normal force, and it is then customary to speak of an increasing or a decreasing coefficient of friction with increasing load.

However, the situation is complicated by the fact that the applied mechanical load is not always the only normal force between the surfaces, and in some circumstances the normal force is completely unknown. For example, electrostatic attraction may enhance the normal force, or plasticizer blooming to the surface may give rise to some measure of adhesion. In some model trains, magnets are used to enhance the normal force, so increasing traction while avoiding increasing the mass of the traction unit. Similarly, when O-ring seals are used, the normal force is unknown and is critically dependent on the geometry. The force to pull out a stopper from a bottle is similarly dependent on geometry and cannot easily be related to normal force.

These examples illustrate a fundamental difficulty when measuring friction, namely that it is not possible to control completely all the variables that affect the measurement. Indeed the parameters may even change during measurement. For example, friction generates heat, thus bringing about a temperature change; wear alters the texture of the surfaces and may also give rise to deposits, which again influence friction.

In fact, the situation is more complex. If wear takes place, then energy is used in breaking up the surfaces, but this appears in the measurement as work done against friction and the figure obtained for the coefficient of friction is enhanced. This type of ambiguity makes it difficult not only to measure friction but also to define what it is we are measuring. A good example of this is so-called static friction.

Static friction is sometimes defined as the force needed to start the motion between two surfaces. It is clear, however, that if a shear force is applied to two contacting surfaces and no motion takes place it may be because the surfaces are adhering and thus it is necessary to distinguish between friction and adhesion. A car licence holder on a windscreen illustrates the dilemma perfectly. There is a very strong resistance to shear, but is this entirely due to friction?

If such a licence holder is placed on the underside of a horizontal glass plate the normal force is nominally zero, but there is still a strong resistance to shear, which is then clearly due to adhesion. The dilemma as to what is the true normal force remains, however, since air pressure plays some part and there may also be some element of normal force that is due to capillary attraction. Nevertheless, in spite of these ambiguities, it is reasonable to define shear adhesion as the shear force needed to move two surfaces in contact when the directly applied normal force is zero. Because the very term adhesion carries with it the implication that an adhesive has been used, the natural adhesion between surfaces, so closely related to friction, is sometimes termed "stiction." Then stiction may be defined as the frictional force encountered between two surfaces in contact when the applied normal force is zero. Even this is not an adequate definition, however, since it is often found in practice that stiction builds up with time, and thus it is necessary also to state the time that the surfaces have been in contact.

If static friction is defined as the force needed to establish motion between two surfaces, then experimentally it is not possible to distinguish between stiction, or adhesion, and friction. In some circumstances this may be an appropriate measurement, but it is important to realize that a wide range of values may be obtained, since the result depends on the time available to establish adhesion.

Another approach to static friction is to measure friction at a number of very low velocities and then to extrapolate to zero velocity. This gives a "static friction" value for zero contact time and is likely to be much lower than the former method, although it may be argued that extrapolation in these circumstances is wholly inappropirate and that no real measurement of static friction has been made at all. However, there is no doubt that this approach to measurement has given a much greater insight into the nature of the friction of rubber-like polymers.

The magnitude of the frictional resistance in any particular circumstance depends on two factors, firstly the adhesion between microprojections or asperities on the surfaces and secondly the energy lost through hysteresis when the polymer surface is disturbed, say by the ploughing action of hard asperities on the track. Under dry conditions the shear mechanism predominates, and under lubricated conditions the hysteresis mechanism assumes an increasing importance. Prediction of friction values is difficult, but in general low-modulus materials in dry conditions give higher levels of friction than high-modulus materials. In wet conditions it is the hysteresis level which is important, and it frequently happens that polymers that show high levels of friction when dry show very low friction values when wet.

Grosch [1] established the relationship between velocity and temperature when carrying out friction tests under lubricated conditions and showed also the influence of the glass transition temperature of the polymer. His work is illustrated in Figs. 1a and 1b, which show the effect of glass transition temperature, velocity, and ambient temperature on the friction of elastomers when the adhesion element is deliberately kept to a minimum. Note that in these plots velocity is plotted in terms of its logarithm and extrapolatuion to zero is therefore not possible. Also, the terms high and low are relative, so that a temperature of 20°C, say, would be classed as a high temperature when compared to a temperature of, say, −10°C. Similarly for glass transition temperature: a temperature of −5°C would be a high glass transition temperature in comparison with one of −40°C.

In most practical applications the range of velocities encountered is relatively small, and what makes elastomers so useful is their rising friction/velocity characteristic, which inhibits slipping. It may be worth pointing out, therefore, that under some lubricated test conditions, rubber compounds with a relatively high glass transition temperature may exhibit a falling friction/velocity curve, particularly at lower temperatures. For example, this type of compound may perform satisfactorily in the summer but not during a very cold winter. The effect is illustrated in the region u-v in Figs. 1a and 1b. The answer to the problem is to use a polymer with a lower T_g. Even though a single point test may show a lower coefficient of friction, the rising friction/velocity curve gives a more stable performance in service.

These figures also illustrate the importance of matching test conditions to the conditions encountered in service. This reasoning applies also to factors not referred to in fig. 1, such as the surface condition of the materials being tested, a point emphasised by James and Mohsen [2]. Similarly, a method of bringing an Official vinyl Composition Tile to a standard condition is described in ASTM D4103–90 [3].

2 Measurement of Friction

There are three main ways in which friction is measured. In the first of these a fixed velocity of separation is imposed on the surfaces under test and the frictional force opposing motion is measured directly. In the second method the aim is to measure static friction or friction close to zero velocity. A steadily increasing force is applied to the stationary

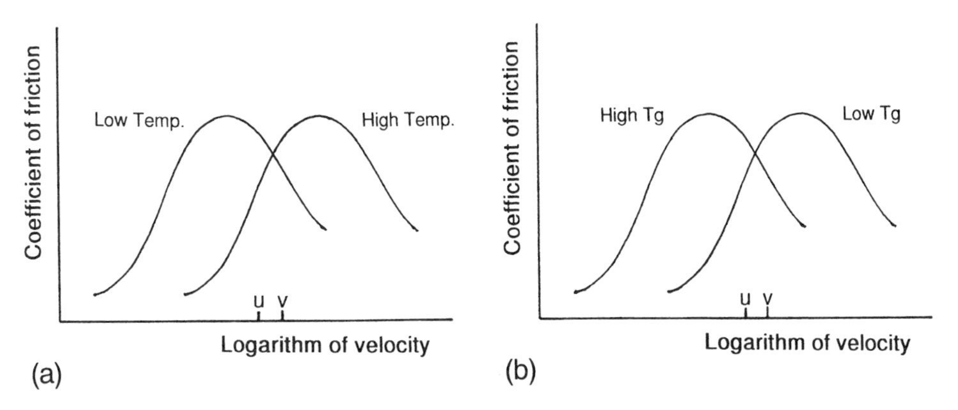

Figure 1 (a) Effect of temperature on the friction of elastomers under lubricated conditions. (b) Effect of glass transition temperature on the friction of elastomers under lubricated conditions.

surfaces until breakaway is detected, when the applied force is recorded. The third method is based on energy changes. A measured degree of potential energy is directly converted into kinetic energy, at which point the surfaces are brought into contact. The loss of energy brought about by friction is then monitored in some way.

There are literally hundreds of test methods utilizing these principles, and it is not possible to refer to them all in detail. However, once the principle involved in each class of measurement is understood, the suitability of any particular test method may be easily assessed.

Difficulties sometimes arise because in the laboratory it is convenient to standardize as many parameters as possible, whereas in service normal load, velocity, surface condition, and temperature may not be constant at all. Tires are a good example. It is possible to test tire compounds in the laboratory under a wide range of conditions, but ultimately the product itself will need to be tested using an instrumented trailer or similar method. Another good example would be bottle closures, where the force needed to pull out the closure depends as much on the geometry as on the coefficient of friction. the normal force is not known and cannot be specified, and so the only sensible test is a direct test on the product. Tests of this kind will not be discussed since the range of products is too large.

Figure 2a represents, in a basic form, the elements required in an apparatus that measures friction directly (i.e., excluding energy absorption methods).

A/B represents the interface between the two surfaces being considered, weight W applies the normal load, drive motor D supplies the motion, and a load cell, represented by spring S, measures the frictional force. Consideration of this basic apparatus leads to important design considerations.

At first spring S is being stretched, and there is little relative motion between surfaces A and B. If there is a falling friction/velocity characteristic (or high static and low dynamic friction) then the frictional force is, at first, relatively high. Spring S continues to stretch until limiting conditions are reached when motion is induced between surfaces A and B. As velocity increases, friction falls and the spring force then drags surface A forward at an increasing velocity. The spring force is thus reduced below equilibrium level and a slow buildup begins again. This jerky motion is termed stick-slip and gives characteristic saw-

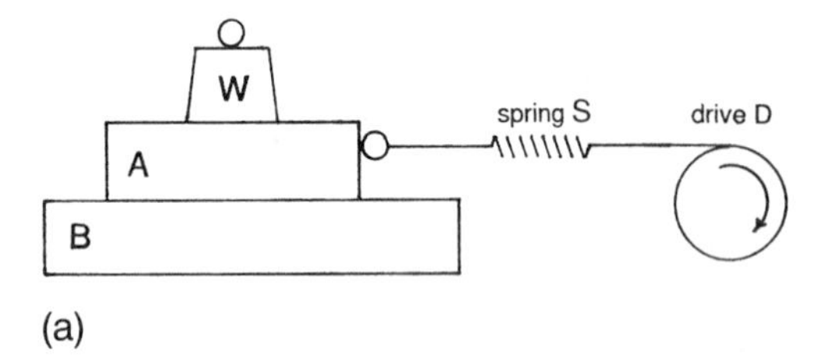

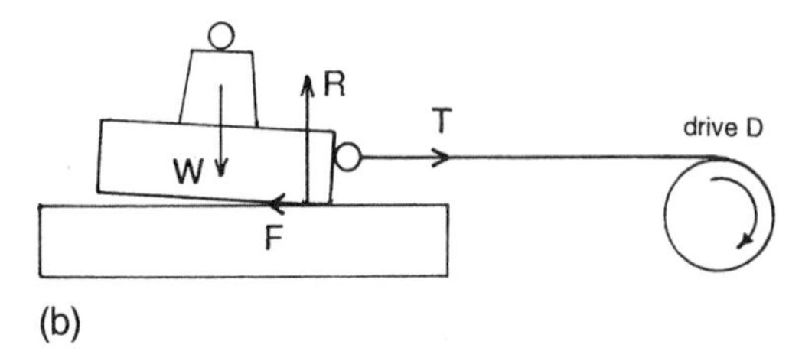

Figure 2 (a) Elements of simple friction apparatus. (b) Forces involved.

toothed records of frictional force. To minimize stick-slip the whole apparatus has to be stiffened. The load cell needs to be of high modulus, the drive motion has to be direct, rather than through a wire, and the framework should not allow vibration to take place. In practice this means a machine capable of handling a load some 20 times greater than the maximum frictional force being measured. Even if these precautions do not entirely eliminate stick-slip, it is usually possible to reduce the width of the saw-toothed band to an acceptable level.

Another important consideration is the alignment of the drive and the line of action of the load cell with the plane of friction. Figure 2b shows the forces acting in a simple arrangement where this precaution has been omitted.

Since the driving force T and the frictional force F are not in line, they form a clockwise couple, which has to be balanced by an anticlockwise couple. The other two forces acting are the weight W, providing the normal load, and the reaction R. They can provide an anticlockwise couple only if surface A tilts so that the reaction force acts forward of the weight W. When surface A tilts, the reaction takes place close to the forward edge of surface A. (If the surface A is a soft polymer, then the shift in the position of the reaction force is likely to be taken up by distortion of the polymer, with a consequent redistribution of pressure, rather than by an observable tilt.)

These points are discussed in BS 903, Part A61, 1994 [4], where it is pointed out that it is advisable to design the apparatus so that the drive and the line of action of the load cell both lie in the plane of the contacting surfaces. An alternative arrangement is to make one of the samples, say A in our diagram, in the form of a hemisphere, so that the contact area remains constant, even if R and W are not in line.

An apparatus incorporating these principles has been described by James and Newell [5]. The drive, load cell, and plotter are all provided by a tensile test apparatus, and the friction surfaces are placed vertically in the machine so that the contact plane contains both the line of action of the load cell and the drive. The advantages are that a good range of velocity is available and the high stiffness of the tensile test apparatus avoids stick-slip. The contacting surfaces can be placed in an environmental chamber, so giving a wide temperature range, and test piece geometry can be readily changed so that tests can be adapted for products or parts of products.

3 Plastics/Rubber Differences

In general, rubbers, in dry conditions, exhibit a higher coefficient of friction than plastics. This means that at high velocities rubbers tend to develop more heat than plastics. Consequently, the range of test velocities for rubbers is usually limited, in order to avoid a temperature buildup. On the other hand, many plastics exhibit low friction and are capable of being used at much higher velocities, for example in bearings, and test machines for these materials differ considerably from those for rubbers, usually involving a rotary rather than a linear device. As a general rule, test equipment designed for the measurement of friction of rubbers can also be used for plastics materials, but test equipment designed for plastics can be used for rubbers only if there is a method of reducing or limiting the velocity. Foams and sponges differ in that, in general, the shear forces they can withstand are smaller than is the case with solid materials, so that test loads are smaller and the range of velocities is also small. Equipment designed for testing solid rubbers can often be adapted for sponges and foams.

4 Plastics Test Methods

ISO 6601, 1987 [6] lists the test parameters involved and outlines the various geometries that are available. Some of these are shown in Fig. 3. The linear track apparatus shown at (1) is suitable for both rubbers and plastics, a typical example being the apparatus described by James and Newell [5], but perhaps the most well known apparatus of this type is that described in ISO 8295 [7], a test method for plastics film and sheeting, said to be suitable for measuring either static or dynamic friction.

Griffin [8] has described a test for small cylindrical plastic test pieces, and Mustafa and Udrea [9] used a rotating steel disc and a stationary plastic test piece. Bailey and Cameron [10] used a ball and peg machine, and West and Senior [11] a hemispherically ended pin contacting a flat plate. A similar device using a steel pin running on a plastic disc was used by Jost [12]. A ring on disc machine was described by Bielinski et al. [13], and Baudet and Di Bernado [14] used a solid wheel with controlled slip. The fundamentals of friction measurement do not change, and some of the older methods have now become standards, for example Westover and Vroom's [15] apparatus is described under ASTM D3028–95 [16]. It is likely that newer methods will incorporate some element of computer control, as in Benabdallah's apparatus [17].

5 Rubber Test Methods

As has been previously explained, rubber test methods differ in that, to avoid heat buildup, the velocity range is restricted. Many methods were described by James [18] in a review covering pedestrian friction and more recently by Mitsuhashi [19] in a review covering both friction and abrasion. The two most versatile testers are those described by James and Newell [5] and more recently by Roberts [20]. Both avoid the misalignment problem spoken of in BS 903, part A61 [4] and both incorporate automatic control of the load arm.

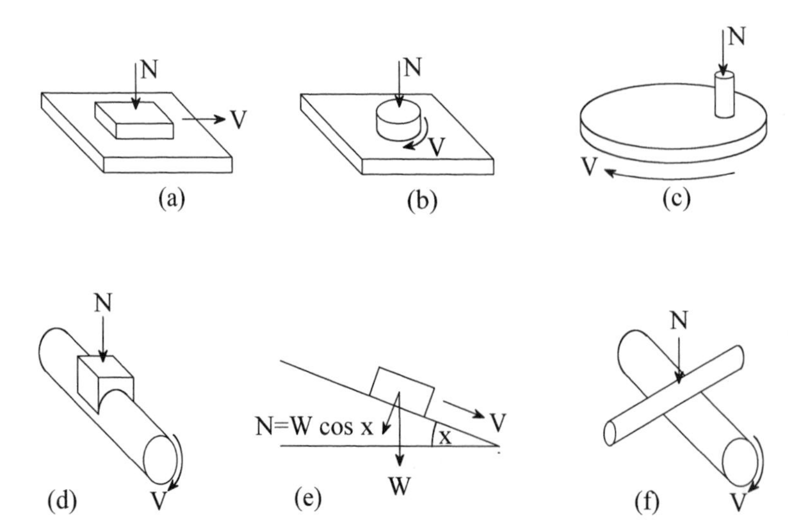

Figure 3 Different geometries of friction test apparatus.

Energy absorbent methods have been widely used by the footwear and flooring industries, for example the apparatus described by Wilson and Mahoney [21]. Energy acquired by a trolley running down an incline is absorbed by the braking action of a heel on the horizontal test surface at the base of the incline. A similar test appears in BS 7044 [22] for artificial sports surfaces.

Flooring is an important use of rubber and plastics materials, and many friction tests for flooring have been devised. Perhaps the most widespread of these are swinging pendulum tests, all loosely related to the original apparatus introduced by Sigler et al. [23]. The best of these is the Skid Tester designed by the Transport and Roads Research Laboratory of the U.K., originally described by Giles et al. [24]. The Leroux [25] pendulum device is similar in principle. Apparatus of this type is particularly valuable when testing in wet conditions, as it is capable of imitating the aquaplaning effect, which is so dangerous to pedestrians.

Other equipment, working at lower velocities and with a flat contact, does not imitate aquaplaning and is more suitable under dry conditions. The best known example is the Tortus apparatus designed by British Ceram Research Ltd. and described by Brough et al. [26]. This is a four-wheeled trolley electrically propelled across the floor. A cylindrical rubber piece contacts the floor, and the frictional resistance is monitored via a spring assembly coupled to a transducer. The FSC 2000 [27] is very similar. Portable equipment of this type is useful in that it enables tests to be made on installed floors, but in the laboratory different tests are sometimes used.

The oldest of these is the machine devised by S. V. James [28], widely favored in the U.S.A. and incorporated into ASTM F489 [29] and ASTM 2047–82 [30]. The Shoe and Allied Trades Research Association in the U.K. devised and instrumented a rig capable of testing whole shoes [31], and many laboratories utilize some form of ramp test in which test personnel walk on a test surface placed on the ramp, making a judgment at which angle of inclination they feel insecure. The best known tests of this type are DIN 51097 [32] and DIN 51130 [33]. More useful is the Rapra modification of these tests, in which shod personnel walk on water wet surfaces. Incidentally, DIN 51097 [32] is the only suitable test for barefoot applications of flooring materials.

6 Testing Methods for Foams and Sponges

As mentioned earlier, the main problems encountered when testing foams and sponges are their compressibility and their low inherent strength.

Thus a pin on flat geometry is unsuitable, since the compression under the pin would lead to high "ploughing" forces, certainly influencing test results and perhaps causing tearing of the foam. To avoid these difficulties it is advisable to use a flat on flat geometry of comparatively large area, restricting the range of normal loads to that which does not cause damage to the test piece. Similarly, test velocities should be kept low.

With these constraints, any of the linear flat on flat machines can be used for testing sponges or foams. Previous warnings about keeping the line of action of the force measuring equipment within the plane of friction are particularly relevant, since considerable distortion of the test piece may otherwise occur.

References

1. Grosch, K. A. (1968), *Proc. Roy. Soc.*, *A274, 21.*
2. James, D. I., and Mohsen, R. (1982), *Polymer Testing*, *3*(2).

3. ASTM D4103–90, Standard practice for preparation of substrate surfaces for coefficient of friction testing.
4. BS 903, Part A61, 1994, Determination of the frictional properties of rubber.
5. James, D. I., and Newell, W. G. (1980), *Polymer Testing*, *1*(1).
6. ISO 6601, 1987, Plastics—Friction and wear by sliding—Indication of test parameters.
7. ISO 8295, 1995, Plastics—Film and sheeting—Determination of the coefficients of friction.
8. Griffin, G. J. L. (1971), *Wear*, *17*, 399.
9. Mustafa, M. R., and Udrea, C. Dec. 1969), *Plast. Mod. Elast.*, *21*(10), 114.
10. Bailey, M. W., and Cameron, A. (Aug. 1971), *Wear*, *21*(1), 43.
11. West, G. H., and Senior, J. M. (Jan. 1972), *Wear*, *19*(1), 37.
12. Jost, H. (March 1969), *Plast u. Kaut.*, *16*(3), 175.
13. Bielinski, D., Janczak, K., Janczak, T., and Slusarski, L. (1991), *Polim. Tworz. Wielk.*, *36*(10), 380.
14. Baudet, P., and Di Bernado, C. (1991), ACS Rubber Div., Toronto, Paper 70.
15. Westover, R. F., and Vroom, W. I. (Oct. 1963), *SPE Journal*, *19*(10), 1093.
16. ASTM D3028–95, Test method for kinetic coefficient of friction of plastic solids.
17. Benabdallah, S. M. H. (1990), *Polymer Testing*, *9*(3), 175.
18. James, D. I. (1980), *Rubber Chem. Tech.*, *153*(3).
19. Mitsuhashi, K. (1989), *Int. Polym. Sci. Technol.*, *16*(2), T31–48.
20. Roberts, A. D. (1993), ACS Rubber Div. Spring Conference Proceedings Paper 61.
21. Wilson, A., and Mahoney, P. (March 1972), *Rubber World*, *165*(6), 36.
22. BS 7044, Part 2, Section 2.2 (1990), Methods for determination of person/surface interaction.
23. Sigler, P. A., Geib, M. N., and Boone, T. H. (May 1948), Nat. Bureau of Standards Research Paper RP189, *J. of Research*, *40*(5), 339.
24. Giles, C. G., Sabey, B. A., and Cardew, K. H. F. (1964), Road Research Technical Paper No. 66, HMSO.
25. Leroux, M. (1957), *Rev. Gen. des Routes et des Aérodromes*, *310*(43).
26. Brough, R., Malkin, F., and Harrison, R. (April 1979), *J. Phys. D.*, *12*, 517.
27. Slip Stop FSC 2000. Manufactured by Sellmaier Electronic Vertrieb GmbH, München, Germany.
28. James, S. V. (1944), *Soap Sanit. Chem.*, *20*, 111.
29. ASTM F489–96, Standard test method for static coefficient of friction of shoe sole and heel materials as measured by the James machine.
30. ASTM 2047–82, Test method for static coefficient of friction of polish-coated floor surfaces as measured by the James machine.
31. SATRA.PM 144, 1992, Machine available from RDM Test Equipment, Bishops Stortford, England.
32. DIN 51097–92, Testing of floor coverings: Determination of anti-slip properties: Wet-loaded barefoot areas: walking method—Ramp test.
33. DIN 51130–92, Testing of floor coverings: Determination of anti-slip properties: Workrooms and work areas with increased risk of slip: Walking method—Ramp test.

24
Thermal Properties

David Hands
Consultant, Sutton Farm, Shrewsbury, England

1 Introduction

This chapter deals with the measurement of thermal conductivity, thermal diffusivity, and specific heat. Other properties that are sometimes included under the umbrella term "thermal properties" are dealt with in other parts of this volume. In most cases it does not matter whether the sample is a rubber or a plastic, the experimental techniques are the same.

2 General Theory

2.1 Equation of Conduction of Heat

For the flow of heat in one direction, the heat flux J_u is related to the temperature gradient $\partial\theta/\partial x$ by Fourier's law

$$J_u = -K \frac{\partial \theta}{\partial x} \tag{1}$$

where K is the thermal conductivity. The minus sign indicates that the heat flows in the opposite direction to the temperature gradient. The form of Eq. 1 implies that heat conduction is a random process. If energy were propagated without scattering, then the heat flow would depend on the temperature difference between the end faces of the specimen instead of the temperature gradient [1].

The general equation from which the time-dependent temperature distribution may be calculated is obtained from Eq. 1 and the equation of continuity

$$\frac{\partial J_u}{\partial x} = -\rho c \frac{\partial \theta}{\partial t} \tag{2}$$

where ρ is the density, c is the specific heat, i.e., the heat capacity per unit mass, and t is the time. The equation of continuity is an expression of the conservation of energy. The heat flux can be eliminated between Eqs. 1 and 2 to give

$$\frac{\partial^2\theta}{\partial x^2}+\frac{1}{K}\frac{\partial K}{\partial\theta}\left(\frac{\partial\theta}{\partial x}\right)^2=\frac{1}{\alpha}\frac{\partial\theta}{\partial t} \tag{3}$$

where $\alpha = K/\rho c$ is the thermal diffusivity. Eq. 3 is the equation of conduction of heat, in the absence of heat generation and convection, for heat flow in one direction. If the conductivity is independent of temperature it reduces to

$$\frac{\partial^2\theta}{\partial x^2}=\frac{1}{\alpha}\frac{\partial\theta}{\partial t} \tag{4}$$

which is the equation usually referred to. It has been shown that for rubbers, and for plastics except at melting transitions, the conductivity term in Eq. 3 is very small, and Eq. 4 is adequate for most heat flow calculations [2] .

When the temperature distribution does not change with time Eq. 4 reduces to Laplace's equation

$$\frac{\partial^2\theta}{\partial x^2}=0 \tag{5}$$

The parameter α was called the thermal diffusivity by Kelvin and the thermometric conductivity by Maxwell, but Kelvin's expression has been generally adopted. It measures the change in temperature that would be produced in unit volume of the substance by the quantity of heat that flows in unit time across unit area of a layer of the substance of unit thickness with unit temperature difference between its faces [3]. It is the parameter that determines the non-steady state temperature distribution in the absence of heat generation and convection and is therefore essential for transient heat flow calculations.

In the above discussion the specific heat referred to is that at constant volume per unit mass c_v whereas in practice the sample is at constant pressure. However, a body containing temperature gradients usually also contains internal stresses, and c_p is not quite correct either [4], but it is more appropriate than c_v. However, the difference between the two principal specific heats is small for polymers [5], hence this is not a significant point. In the following sections, c_v is used in theoretical arguments, and c_p is used elsewhere, unless otherwise stated.

2.2 Units

A variety of units are used in the literature for thermal properties, and this can be a nuisance when different sets of results have to be compared, or when values from an older publication are being used for calculations. Prior to the adoption of the SI system, the two most common units for thermal conductivity were the cal/cm s °C and the BTU in/ft^2 h °F. There are two units of length in the imperial unit, because area is measured in square feet and thickness in inches, and this inconsistency is a potential pitfall for the unwary. A self-consistent conductivity unit, the BTU/ft h °F, is obtained if the temperature gradient is measured in °F/ft instead of °F/in, but this is not as common. For diffusivity the c.g.s. unit is the cm^2/s and the imperial unit is the ft^2/h. The SI unit for conductivity is the W/mK, and the unit for diffusivity is the m^2/s. For polymers it is more convenient to use a submultiple of the diffusivity unit, the mm^2/s, because this eliminates a factor of 10^{-6}. Conversion factors are given in Table 1.

Table 1 Conversion Factors for Some Thermal Conductivity Units

	W/mk	cal/cm s°C	W/cm°C	kcal/m h°C	BTU in/ft^2 h°F
W/mk	1	0.00239	0.01	0.86	6.93
cal/cm s°C	419	1	4.19	360	2900
W/cm°C	100	0.239	1	86	693
kcal/m h°C	1.16	0.00278	0.0116	1	8.06
BTU in/ft^2 h°F	0.14	0.000345	0.00144	0.12	1

2.3 Fundamental Significance of Diffusivity

Since $\alpha = K/\rho c$, diffusivity is often regarded as just a mathematical parameter rather than a fundamental material property. However, the fundamental significance of diffusivity can be seen if we think wholly in terms of energy.

The heat capacity at constant volume per unit volume is $(\partial u/\partial \theta)_V$, where u is the internal energy per unit volume. This is also equal to the product of the density and the heat capacity per unit mass, hence $\rho c = (\partial u/\partial \theta)_V$. Obtaining from this an expression for the temperature gradient gives $\partial \theta/\partial x = (1/\rho c)(\partial u/\partial x)$, and substituting this into Eq. 1 gives

$$J_{\mathrm{u}} = -\alpha \frac{\partial u}{\partial x} \tag{6}$$

Thus diffusivity is the parameter relating energy flux to energy gradient, whereas conductivity relates the energy flux to the temperature gradient. The origin of the units, mm^2/s, which are perhaps meaningless in themselves, now becomes apparent.

2.4 Heat Transfer at Interface

Many heat transfer problems involve a resistance to heat transfer across the interface between the sample surfaces and the heating or cooling medium. Such a resistance may be caused by a stationary surface film (of gas or liquid) of small but indeterminate thickness through which the temperature gradually changes from that of the solid to that of the bulk fluid. Heat transfer across a solid/fluid interface depends on wetting, film thickness, and temperature gradients in the film. If the fluid is moving, it will also depend on whether the flow is laminar or turbulent, which in turn is determined by velocity and viscosity. Similarly heat transfer across the interface between two solids depends on adsorbed gases, pressure, surface finishes, hardness, and trapped fluids [6, 7].

A surface heat transfer coefficient h can be defined as the quantity of heat flowing per unit time normal to the surface across unit area of the interface with unit temperature difference across the interface. When there is no resistance to heat flow across the interface, h is infinite. The heat transfer coefficient can be compared with the conductivity: the conductivity relates the heat flux to the temperature gradient; the surface heat transfer coefficient relates the heat flux to a temperature difference across an unknown distance. Some theoretical work has been done on this subject [8], but since it is rarely possible to achieve in practice the boundary conditions assumed in the mathematical formulation, it is better to regard it as an empirical factor to be determined experimentally. Some typical values are given in Table 2. Cuthbert [9] has suggested that values greater than about 6000 W/m^2 K can be regarded as infinite. The spread of values in the Table is caused by mold pressure and by different fluid velocities. Heat loss by natural convection also depends on whether the sample is vertical or horizontal. Hall et al. [10] have discussed the effect of a finite heat transfer coefficient on thermal conductivity measurement.

Table 2 Typical Values of Surface Heat Transfer Coefficient h

System	h, W/m^2K
Polymer/mold	500–inf.
Heating or cooling with water	200–inf.
Heating or cooling with air	4.5–90
Fluid bed	560

3 Thermal Conductivity Measurement

In this section the methods of measuring conductivity are described. The methods discussed have been chosen to illustrate the basic principles of conductivity measurement and the various ways of minimizing the major sources of experimental error. The experimental methods can be divided into two groups: steady state and transient.

3.1 Steady State Methods

Under steady state conditions, i.e., when the temperature at any point does not change with time, the temperature distribution in the sample is governed by Laplace's equation (Eq. 6). The sample geometries are chosen so that the temperature is a function of only one coordinate, and simple analytical solutions to Laplace's equation can be used. There are two geometries that satisfy this condition: the parallel-faced slab with heat flow normal to the surfaces; and the hollow cylinder with radial heat flow. In the latter case, although the heat flow is now in two dimensions the temperature is a function of only one coordinate, namely the radius, because of symmetry. The condition is also satisfied by the sphere, but that is not relevant to the present discussion. The first case is illustrated in Fig. 1 which shows a parallel-faced section of thickness x and cross-sectional area A.

One face is maintained at a uniform temperature θ and the other at a uniform temperature of $\theta - \Delta\theta$. The rate of heat flow q is in the x direction only and is normal to the two faces. The conductivity is obtained from Eq. 1:

$$K = \frac{qx}{A\Delta\theta} \tag{7}$$

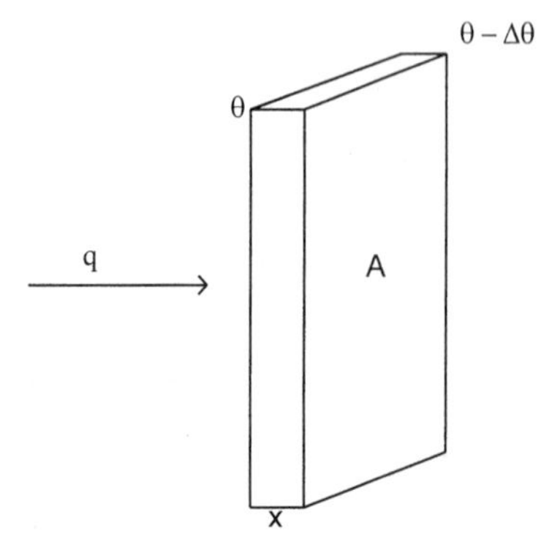

Figure 1 Flat slab.

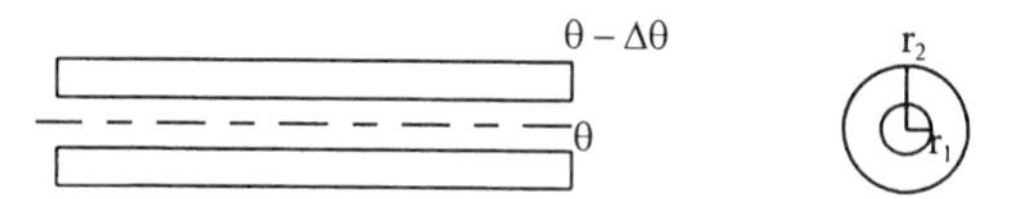

Figure 2 Coaxial cylinder.

The coaxial cylinder geometry is illustrated in Fig. 2. The cylinder is of length L, outside radius r_2 and inside radius r_1. The temperature difference between the inner and outer surfaces is $\Delta\theta$. Eq. 1 can be integrated to give

$$K = \frac{q}{2\pi L \Delta\theta} \log_e\left(\frac{r_2}{r_1}\right) \tag{8}$$

It is necessary to construct the apparatus so that the experimental conditions agree with the boundary conditions for simple analytical solutions. Failure to achieve this perfectly gives rise to three major sources of error.

The first major source of error is a result of a failure to obtain a normal or radial heat flow, depending on the geometry. Satisfying this boundary condition is very difficult in practice. The problem has been tackled in three ways: the flux is constrained by guard rings; or the heat loss from the edges of the specimen is minimized and ignored; or the heat loss is estimated and allowed for in the calculations.

The second source of error is that the surfaces of the sample are not necessarily at the same temperature as the walls of the apparatus, i.e., there is a resistance to heat flow across the interface between the sample and the apparatus. In the majority of cases the temperatures of the cell walls are measured, and this gives a conductivity value that is too low.

The third source of error results from failing to satisfy the condition that the major surfaces must be isothermals. The size of this error depends on the degree of non-uniformity.

The more important methods for measuring thermal conductivity are discussed below.

The Unguarded Hot Plate

The modern unguarded hot plate is based on the well known Lees' disc method [11]. The general arrangement is shown schematically in Fig. 3. Recommendations are given in BS 874 [12] and ASTM C518 [13].

In this method two identical samples are placed on either side of an electrically powered heat source and sandwiched between two heat sinks. The heat sinks are controlled at a given temperature, usually by circulating a liquid from a constant temperature bath, and a known power is supplied to the heat source. In order to ensure that the surface temperatures are uniform, the surfaces of the hot and cold plates are made from a metal

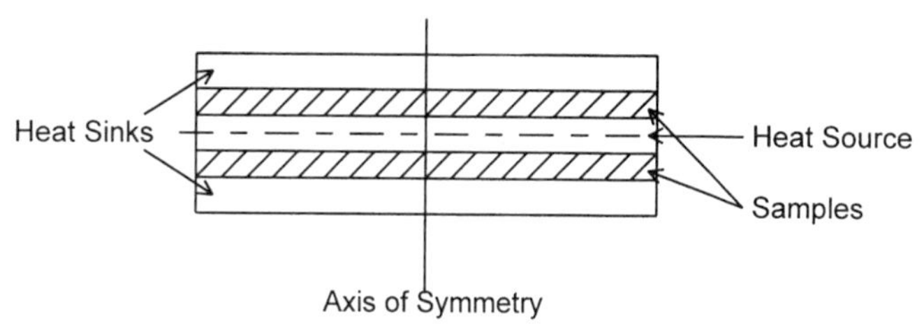

Figure 3 Unguarded hot plate.

with a high thermal conductivity such as copper, brass, or aluminum. The apparatus is allowed to reach steady state, and the temperature drop across the sample is measured, usually with thermocouples. The conductivity is calculated using Eq. 7.

In order to reduce the heat lost from the exposed edges, the samples and the heat source are made as thin as possible, and the apparatus is surrounded with a material of low thermal conductivity such as vermiculite. However, the side losses may still be a significant fraction of the total heat input, and this must be allowed for in the calculations. The heat loss may be evaluated by using a pair of samples of known conductivity; the conductivity of the calibration material would have been measured in an absolute instrument such as a guarded hot plate (see below). Ideally the conductivity of the calibration samples should be similar to that of the test samples.

Heat loss is a function of sample thickness, hot and cold face temperatures, and ambient temperature. These factors must be taken into account by making calibration measurements over a range of thicknesses and temperatures. If the losses are known to say 10%, and only represent 10% of the heat input, the error in the calculated conductivity is only 1%. A further refinement is to use heat flow meters on one or both sides of the sample [13–15]. Calibration of heat flow meters is discussed in ASTM C1132 [16] and by Scott [17].

By using thin specimens and thin heater plates with a large surface area, the side losses may be reduced to negligible proportions. However, the preparation of such specimens becomes increasingly difficult the larger the area and the smaller the thickness.

The samples can be square or round, and the most convenient size lies usually in the range from 50 to 200 mm across (edge or diameter) by 3 to 6 mm thick. The area and thickness of such samples can easily be measured with the required precision. The most important requirements for the samples are that the faces should be parallel and as flat as possible, since errors due to imperfect contact between the samples and the hot and cold plates can be very large. As an example, an air film of thickness 0.03 mm on one side of a 3 mm thick sample of a typical solid plastic would cause an error approaching 10% if the temperature drop across the film were ignored. It is customary to reduce such errors by coating the samples with a thin film of liquid of relatively high conductivity (compared with air) such as glycerol, or with a suitable grease. A correction for the resulting small temperature drop in such films may be obtained by making a measurement with just the contacting liquid between the hot and cold plates. Alternatively, measurements can be made on samples of different thicknesses. Assuming that the resistance at the interfaces is constant it can then be eliminated from the calculations for the conductivity of the test sample.

Temperatures (invariably of the plates rather than the samples) are normally measured with fine wire thermocouples in conjunction with a voltmeter with a resolution of 1 μV. The heater power can be obtained from measurements of the current and voltage. Both temperature and power measurements are capable of high precision. A stable power source is a necessity; batteries can be used, but the many admirable transistorized power supplies now available are more convenient and do not suffer from the long-term drift associated with batteries. This is particularly important when measurements are being made on low conductivity materials such as foams, because steady state conditions may not be reached for many hours after energizing the heater.

The Guarded Hot Plate

An improvement on the unguarded hot plate is the guarded hot plate. This is the most accurate method available for solid materials (including foams) and is recommended by

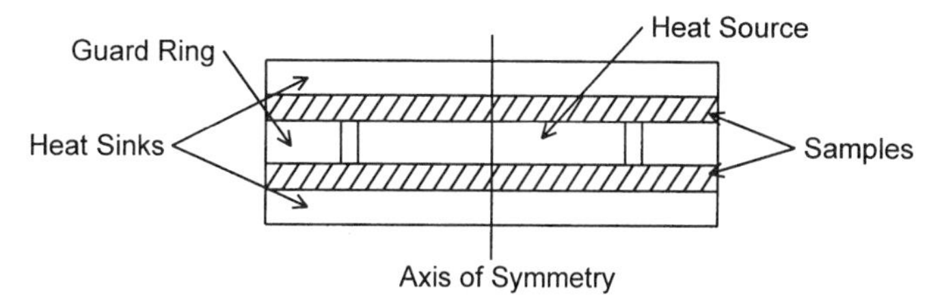

Figure 4 Guarded hot plate.

the standards organizations [18–22] for measurements on low conductivity materials. The general arrangement is shown schematically in Fig. 4.

In this method the heat source is surrounded by a guard heater that has an independent power supply. The power to the guard is adjusted so that there is no temperature difference between the heat source and the guard. The heat from the heat source thus flows normally through the samples, and the heat lost from the exposed edges comes from the guard. The whole apparatus is surrounded by a low-conductivity material such as vermiculite. With thick samples it may be necessary to provide an additional guard heater around the exposed edges in order to reduce the heat losses to a minimum.

The standard organizations recommend various dimension ratios for the guard width g, heater side length (or diameter) $2s$, and sample thickness l. If an accuracy of 1% is to be achieved, the limiting values of the ASTM specification should be used [23] and these are given in Table 3

The heat sinks are controlled at a given temperature, usually by circulating a liquid from a constant temperature bath, and a known power is supplied to the heat source. The power supplied to the guard heater is adjusted to equalize the temperatures of the guard and heat source. The apparatus is allowed to reach steady state, and the temperature drop across the samples is measured, usually with thermocouples. The conductivity is calculated from Eq. 7.

It can take a few hours to reach steady state, and in the case of low conductivity materials such as foams, it can take a few tens of hours. For this reason, power supplies that are stable over long time periods are essential. The rate of temperature change is slow and therefore difficult to detect. The selection and construction of the thermocouples is very important, and it is necessary to measure the thermal e.m.f.s with a resolution of 1 μV. As temperature differences across the samples of as little as 10°C or less may be used, variations in the cold plate temperatures must be reduced to a minimum, and a thermostat bath temperature of at least ±0.1°C is required.

The apparatus dimensions are usually 300 mm square, and the sample thickness range is from 6 mm to 50 mm. There is a gap between the heat source and the guard heater of

Table 3 Dimension Ratios for Guarded Hot Plate

Ratio	BS 874	ASTM C177
$\frac{2(g+s)}{l} \geq$	6	9
$\frac{g}{l} \geq$	1	1.5
$\frac{s}{l} \geq$	2	3

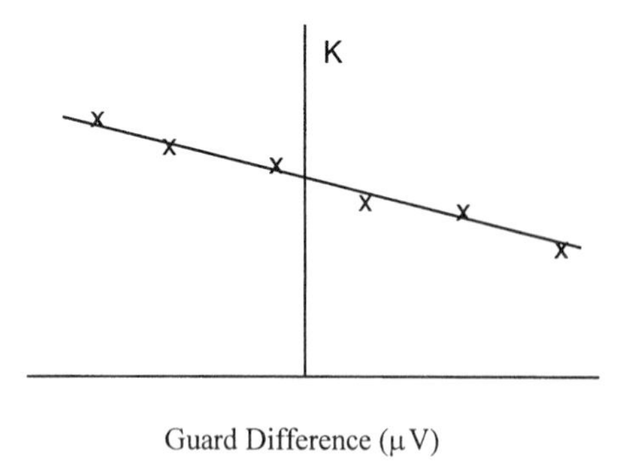

Figure 5 Thermal conductivity against guard difference.

about 2 mm. The area used in Eq. 7 for calculating the conductivity is determined from the center of this gap and is fractionally larger than the actual area of the heater plate.

A temperature difference of the order of 0.05°C between the guard and the heater can have a significant effect on the calculated conductivity value. In operation, unless the guards are controlled automatically, it is tedious to balance exactly the guard and center temperatures. It is easier to make a series of measurements with slightly different guard temperatures and plot a graph of apparent conductivity against the temperature difference between guard and heater plates. The power supplied to the central heater must be maintained constant throughout the series of measurements, and after each guard adjustment the apparatus must be left for steady state to be reestablished. This is not a measurement which can be hurried. Over small temperature differences the plot is linear, and the correct conductivity is given by the intercept at zero temperature difference. A typical plot is shown in Fig. 5.

The virtue of this refinement in technique is that, with experience, the slope of the plot indicates whether the apparatus is responding properly or not. It is also a measure of the efficiency of the plate, a high slope being undesirable because it indicates a high lateral conductivity between the central heat source and the guard heater.

Errors caused by resistances to heat flow across the interfaces between the conductivity cell surfaces and the sample surfaces are more serious when measurements are being made on solid materials rather than on foams, because the resistance to heat flow is a larger fraction of the resistance of the sample. The errors can be minimized by wetting the surfaces with a heat transfer grease or oil. Measurements can also be made on samples with different thicknesses as described in the previous section.

Quasi–Steady State Methods

ASTM D2214 [24] describes the Cenco-Fitch method. In this method a thin sample is placed between a heat source and a copper block heat sink. The heat source is held constant at the temperature of boiling water. The temperature of the copper block is measured as a function of time and plotted on log linear paper. The conductivity of the sample is obtained from the slope

$$K = -2.303\,\frac{hMc}{mA} \tag{9}$$

where h is the thickness of the sample, M is the mass of the copper block, c is the specific heat of copper, m is the slope, and A is the area.

A method based on a similar principle was developed by Eiermann and coworkers [25] to cover an extended temperature range from −190°C to +90°C. The apparatus consisted of two heat sources and two thin samples on either side of a copper heat sink, the temperature of which was monitored and which was surrounded by an adiabatic shield to minimize the heat loss. Power was supplied to the heaters in order to cover the temperature range at a slow rate. The instantaneous heat flow rate was determined from the rate of temperature rise of the heat sink

$$q_1 + q_2 = Mc\,\frac{d\theta}{dt} \tag{10}$$

where q_1 and q_2 are the heat flow rates across the two samples, M is the mass of the heat sink, c is its specific heat, and $d\theta/dt$ the rate of temperature rise. Since $q = -KA(\Delta\theta/x)$,

$$K = -\frac{Mc}{A(\Delta\theta_1 + \Delta\theta_2)}\,\frac{d\theta}{dt} \tag{11}$$

The heat capacity of the samples was involved in the analysis of the results because the experiment was dynamic. This was eliminated by repeating each experiment with a heat sink of different thermal capacity. Thermal resistance was eliminated by making measurements in a helium atmosphere and then in a nitrogen atmosphere. Thus, to obtain conductivity values for one material four separate measurements were required. Apparatus of this type has also been used by Ott [26].

Coaxial Cylinder Method

Methods based on the coaxial cylinder geometry have the advantage that the exposed areas, from which heat can be lost, are much smaller, compared with the surface area of the heat source, than is the case with the hot plate methods. Thus the effects of lateral heat flow can be minimized more easily without introducing guard rings, although guard rings can be used if necessary. The essential features of a cylindrical conductivity cell are shown schematically in Fig. 6.

The apparatus comprises a rod-shaped heat source surrounded by a tubular sample inside a cylindrical heat sink. The heat sink and the heat source have the same axis. The temperature of the heat sink is controlled, usually by circulating a liquid from a constant temperature bath, and a known power is supplied to the heat source. The temperature difference between the heat source and the heat sink is measured after steady state conditions have been established. The conductivity is obtained from Eq. 8.

The apparatus is usually constructed with a large length-to-diameter ratio so that the heat lost from the ends is small compared with the heat transferred radially through the sample. The major source of error is not heat loss from the ends but resistance to heat flow across the interfaces between the sample and the heater and heat sink, because it is very difficult to ensure that the sample is a perfect fit in the apparatus [27]. This method is more

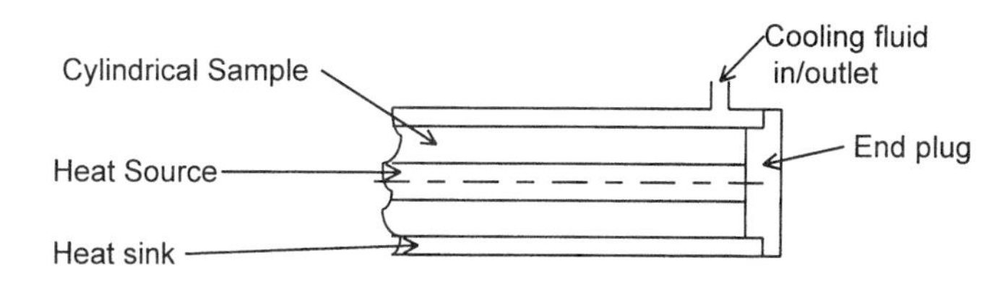

Figure 6 Coaxial cylinder apparatus.

suitable for measurements on viscous liquids, which can be poured or pumped into the apparatus; the measurement will be accurate provided that the viscosity is high enough to prevent convection. This has been used for investigating the effect of pressure on the conductivity of molten polymers [28, 29]. The main problem with this type of apparatus is loading and unloading the sample [30].

3.2 Transient Methods

The time-dependent temperature distribution in a transient experiment is governed by Eq. 4, and usually the related parameter, thermal diffusivity, is obtained. However, under certain circumstances the solution to the heat equation contains the thermal conductivity as well as the thermal diffusivity, and by choosing a suitable method the diffusivity can be eliminated from the answer. The more important methods are the line and plane source heater methods and are described below. These are not Standard methods, but they can be used where speed is more important than absolute accuracy, to give a conductivity value more quickly than the Standard methods. They can also be used to compare a range of materials.

Line Source Method

The most common transient technique that gives thermal conductivity directly is the continuous line source method. For a line source heater of infinite length in an infinite mass of sample material, the temperature at a distance r from the line source is given by [31]

$$\theta = \frac{q}{4\pi K} \int_a^\infty \frac{e^{-u}}{u} \, du \tag{12}$$

where $a = r^2/4\alpha t$, and q is the rate of heat generation per unit length. For small values of a the integral reduces to $[\log_e(4\alpha t/r^2) - \text{const}]$. Thus if the diffusivity and conductivity are constant over the temperature range $\Delta\theta$, then

$$\theta_2 = \theta_1 = \Delta\theta = \frac{q}{4\pi K} \log_e \frac{t_2}{t_1} \tag{13}$$

where θ_1 is the temperature at r at time t_1 and θ_2 is the temperature at time t_2.

Rearranging gives

$$K = \frac{q}{4\pi\Delta\theta} \log_e \frac{t_2}{t_1} \tag{14}$$

Hence the conductivity can be obtained directly by measuring q and the change of temperature with time at a known point.

It is necessary to make small corrections for finite heater wire diameter, finite wire length, and finite sample mass. These have been discussed in detail for liquids by Horrocks and McLaughlin [32], who found that the corrections were only significant for short heating times. This technique has been used with polymers [33–37], and has also been developed in the form of a conductivity probe [38–40] for making single point measurements on, for example, foams.

Plane Source Method

An alternative to a line source is to use a plane source of constant flux J. If the sample is assumed to be a semiinfinite solid ($0 \leq x < \infty$), then the temperature at a point x and time t is given by [41]

$$\theta = \frac{2J}{K}\sqrt{(\alpha t)}\, ierfc\, \frac{x}{2\sqrt{\alpha t}} \tag{15}$$

At the point $x = 0$,

$$\theta = \frac{2J}{K}\sqrt{\frac{\alpha t}{\pi}} \tag{16}$$

The temperature at the surface, $x = 0$, and at a point in the sample, $x = l$ say, are measured as functions of time, and from this both K and α can be determined. The method was developed by Harmathy [42] for measurements on building materials, and in his paper he also discussed the limitations on the technique imposed by the finite size of the samples.

4 Thermal Diffusivity Measurement

To measure diffusivity it is only necessary to know the change in temperature with time at three collinear points in the direction of heat flow. For conductivity on the other hand, although the temperature gradient can easily be obtained, the heat flux cannot be measured at a point; the total power to the heat source is known, but it is difficult to control or calculate the movement of the energy. Thus diffusivity is easier to measure; but it has received less attention, although it is the essential parameter for transient heat flow calculations. The scatter in the diffusivity data in the literature is not as great as for conductivity (possibly because not as many people have tried to measure it!).

4.1 Basic Principles

Boundary Conditions

For certain boundary and initial conditions analytical solutions to Eq. 4 can be obtained. The majority of diffusivity measurement methods are all based on such solutions. The experimental conditions are matched to these mathematical conditions as closely as possible, and the appropriate solution is used to give a value for the diffusivity. The experiment can be repeated at different temperatures in order to obtain the temperature dependence of the diffusivity. This type of experimental procedure has been criticized [43] because if the diffusivity changes with temperature then almost invariably the conductivity is also temperature dependent, and Eq. 3, which would not have given an analytical solution, should have been used instead of Eq. 4. However, Hands and Horsfall [44] have shown that, except near melting transitions, thermocouples sensitive to 0.002°C would be needed to detect the effect of the conductivity term in Eq. 3. Hence, generally speaking, the simpler equation is adequate for diffusivity measurement and for the majority of heat flow calculations.

Diffusivity measurement methods based on analytical solutions to Eq. 4 have all had the same initial condition that the whole sample is at a constant uniform temperature. But three different types of boundary conditions have been employed: first, the sample surface is subjected to a step change in temperature; second the surface is subjected to a linear rate of temperature rise; and third, the surface is subjected to a periodic temperature fluctuation.

Infinite and Non-infinite Solids

The analytical solutions to the heat equation are for one dimensional problems; this implies that the heat flow is in one direction only, or that the equation reduces to one dimension because of symmetry, i.e., the temperature is a function of one length variable only. This condition is satisfied for three sample geometries: the infinite flat slab, the infinite circular cylinder, and the sphere. An infinite slab is a parallel faced slab of finite thickness and infinite length and breadth. An infinite cylinder is a cylinder of finite diameter and infinite length.

Real samples are obviously not infinite, and the practical definition of an infinite solid depends on how much error can be tolerated. For example, if one lateral dimension of the slab is reduced from infinity to five times the thickness, or the length-to-diameter ratio of a cylinder is reduced to 5:1, then an error of about 1% is introduced if lateral heat flow is ignored.

For samples in the form of cubes, rectangular parallelepipeds (bricks) and short right circular cylinders, where lateral heat flow cannot be ignored, the infinite solid solutions can still be used [45]. For a brick, by considering heat transfer through each pair of faces in turn (imagining the lateral dimensions to be infinite) three solutions are obtained. The final solution is the product of these three. Similarly the solution for a short cylinder is given by the product of the solutions for an infinite slab and an infinite cylinder.

This approach is based on separation of variables, and it can be shown by substitution that a product of solutions is itself a solution. Separation of variables is only valid for certain boundary conditions and imposes restrictions on geometry that are, however, satisfied by the examples under consideration.

Sources of Error

In the experimental techniques, which will be described below, there are a number of recognized sources of error. These have not always been avoided, and although the measurement is basically straightforward the results in the literature are not always reliable. The significance of the errors will be discussed in connection with the methods of measurement and only a summary is given here.

1. Assuming an infinite surface heat transfer coefficient between the sample and the heating or cooling medium.
2. Ignoring lateral heat flow when it is significant.
3. Conduction along the thermocouple leads.
4. Assuming diffusivity is temperature independent.
5. Ignoring the thermal expansion of the samples.

4.2 Quenching Methods

Thermal diffusivity has usually been measured using a quenching method, i.e., the solid sample at a uniform temperature is immersed in a temperature-controlled bath at a different temperature. The rate of change of temperature at the center is then monitored with an embedded thermocouple. The sample dimensions are usually chosen so that lateral heat flow can be ignored and regular sample geometries, i.e., "infinite" flat slabs, "infinite" cylinders, or spheres, are used.

Quenching methods are based on the assumption that the thermal diffusivity is constant over the experimental temperature range. To obtain the temperature dependence of the diffusivity, the total temperature range is covered in a series of steps which are small enough for this assumption to be valid. The size of the step temperature change is a compromise between this requirement and the need to make accurate temperature difference measurements. The value usually chosen is about 5°C. In some experiments this point has been ignored and the measurement made with a much larger step. For example, the sample may be conditioned at room temperature and then immersed in boiling water. An average value for the diffusivity is obtained, and the error depends on the temperature dependence of the diffusivity and hence changes from material to material.

Two methods have been used to obtain a value for the diffusivity from the experimental results. For the geometries in question the analytical solutions to the heat equation, which are in the form of infinite series, can be conveniently represented graphically. In order to give the greatest amount of information with the minimum number of curves, the graphs are presented in terms of dimensionless groups. The first such collection of graphs was published by Williamson and Adams [46]. By reference to such graphs a value for α can be obtained.

The second method is based on the rapid convergence of the series solutions. For example, after about 10% of the time for the center temperature to reach 99% of the surface temperature has elapsed, the second term in the series is about 1/2% of the first term. Thus after a certain time, only the first term is relevant. Now the dimensionless temperature is always an exponential function of the time; for example, consider an infinite slab of thickness $2a$. The first term of the series for the dimensionless temperature at the center is given by

$$Y = \frac{\theta_1 - \theta}{\theta_1 - \theta_0} = \frac{4}{\pi} \exp\left(-\frac{\pi^2 \alpha t}{4a^2} \right) \tag{17}$$

where θ_1 is the temperature of the heating or cooling medium and θ_0 is the initial uniform temperature. For long times this is the only significant term, and the diffusivity can be obtained from the slope of a graph of $\log_e Y$ against t.

4.3 Linear Heating Method

The second boundary condition to be considered is that in which the sample surface is subjected to a linear rate of temperature rise. A method based on this has been developed by Shoulberg [47] for diffusivity measurements on polymer melts. He used two discs of his material with a thermocouple sandwiched between them; the diameter-to-thickness ratio was such that the sample sandwich could be regarded as an infinite flat slab. The sample completely filled the cavity in an aluminum block and was melted in the apparatus. The aluminum block was heated electrically, and the power was adjusted to give an approximate linear rate of temperature rise. Under his experimental conditions this lasted for about 30°C.

For this boundary condition the solution to the heat equation for the temperature at the center is [48]

$$\theta = kt - \frac{ka^2}{2\alpha} + \frac{16ka^2}{\alpha\pi^3} \sum_{n=0}^{\infty} \frac{(-1)^n}{(2n+1)^3} \exp\left(\frac{-(2n+1)^2 \pi^2 \phi}{4} \right) \tag{18}$$

where k is the linear rate of temperature rise. After a long time (in Shoulberg's case about 12 min) the summation term is negligible and the temperature difference between the surface and the center becomes constant. The diffusivity is then given by

$$\alpha = \frac{ka^2}{2\Delta\theta} \tag{19}$$

where $\Delta\theta$ is the temperature difference between the surface and center of the sample, and a is the semithickness of the sample sandwich, i.e., the thickness of one of the discs. The temperature range was covered in a series of 30°C, and using Eq. 19 he obtained an average value for the diffusivity for each step.

4.4 Periodic Heating Method

A method for measuring diffusivity of solid polymers based on this type of boundary condition has been developed by Berlot [49, 50], and by Gehrig et al. [51]. A disc sample of thickness $2a$ is held at a uniform temperature and then a sinusoidal temperature fluctuation of angular frequency ω is imposed on the outer surfaces. The amplitude ratio and phase of the temperature at the center are monitored with a thermocouple. Under these conditions the amplitude ratio A and phase ϕ are given by [52]

$$A = \left(\frac{32}{\cosh 2ka + \cos 2ka} \right)^{1/2} \tag{20}$$

$$\phi = \arg\left(\frac{1}{\cosh ka(1+i)} \right) \tag{21}$$

$$k = \left(\frac{\omega}{2\alpha} \right)^{1/2} \tag{22}$$

k can be obtained from Eq. 20 or Eq. 21 and hence α from Eq. 22.

4.5 Continuous Heating Method

If the temperature dependence of diffusivity is taken into account, then the temperature range from ambient up to, say, 250°C can be covered in one experiment. With a quenching method a true step change in temperature is difficult to achieve, because liquids capable of withstanding high temperatures tend to have high viscosities and this results in large temperature gradients close to the sample surface. Another objection to a quenching method is that the predominance of the quenching temperature throughout the experiment coupled with the extremely large initial temperature difference could give rise to computational difficulties. These problems, inherent in the quenching method, do not occur in a continuous heating method.

A method based on this principle was developed by Hands and Horsfall [2] and has been further developed by Smith [53]. The apparatus is shown schematically in Fig 7. Two disc-shaped samples are placed together and a thermocouple is sandwiched between them for monitoring the center temperature. The outside temperatures are measured with two further thermocouples that are in intimate contact with the surfaces of the sample. The three thermocouple junctions lie on the axis of symmetry. The sample sandwich is contained in a brass ring and two brass end plates that contains electrical heaters. The samples are restrained by high temperature rubber O-rings so that measurements on plastic samples can be taken into the melt region, and the end plates are allowed to move against springs. The change in thickness is measured and used in the calculations.

Enclosing the samples allowed measurements to be made on molten samples and eliminated errors caused by ignoring thermal expansion. In quenching methods, for example, the appropriate sample dimension is measured at room temperature, and this is the value used in the calculations. Since diffusivity depends on the square of the thickness, any error caused by ignoring expansion is automatically doubled. This error can be a few percent depending on the temperature range covered.

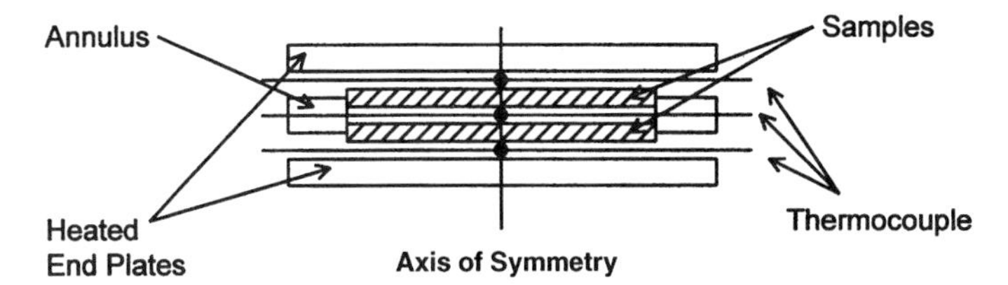

Figure 7 Schematic diagram of apparatus for continuous heating method.

In use the heaters are energized and the thermocouple outputs monitored at 20 s time intervals. The maximum temperature range is covered in about 20 min. The heat equation is solved numerically [54] to give thermal diffusivity as a function of temperature.

4.6 Pulse Heating Method

There is an ASTM Standard for this method [55]. If a disc-shaped sample is irradiated on one surface for a short time using a flash tube or a pulsed laser, then the curve of temperature against time for the back face depends on the thermal diffusivity of the sample and the heat losses. If the pulse time τ is very much shorter than the time for the pulse to pass through the sample, then the heat losses can be ignored. Cape and Lehmann [56] have given as a criterion for this condition

$$\frac{a^2}{\alpha\pi^2} \geq 10\tau \tag{23}$$

where a is the thickness of the disc. Under these conditions the ratio of the temperature to the maximum temperature for the back face is given by [57]

$$\frac{\theta}{\theta_{\text{max}}} = 1 + 2\sum_{n=1}^{\infty}(-1)^n \exp\left(-\frac{n^2\pi\alpha t}{a^2}\right) \tag{24}$$

The temperature as a function of time on the back face is usually measured with a radiation pyrometer. Analysis of the results has been discussed by Taylor [58] and by Parrott and Stukes [59]. This is possibly the most accurate method currently available for measuring diffusivity of solid materials.

5 Specific Heat

Specific heat c is the heat capacity C per unit mass or per unit volume; usually the term refers to the mass specific heat. Heat capacity is defined by

$$C = \frac{dQ}{dT} \tag{25}$$

where Q is the heat energy absorbed and T is the temperature. The heat capacities of greatest interest are those at constant pressure C_P and at constant volume C_V. Substituting into Eq. 25 for the heat energy absorbed, from the first law of thermodynamics, assuming that the work done is of a mechanical nature only, the heat capacities are given by

$$C_\text{P} = \left(\frac{\partial U}{\partial T}\right)_\text{P} + P\left(\frac{\partial V}{\partial T}\right)_\text{P} \tag{26}$$

and

$$C_V = \left(\frac{\partial U}{\partial T}\right)_V \tag{27}$$

where U is the internal energy of the system, P is the pressure, and V is the volume.

The heat capacity at constant volume is of more interest from a theoretical point of view because it is directly related to the internal energy of the system. However, it is almost impossible to measure C_V directly. To obtain a value for C_V it is necessary to measure C_P and to calculate C_V from the formula [60]

$$C_P - C_V = TVB\beta^2 \tag{28}$$

where B is the bulk modulus and β is the volume expansion coefficient. In terms of the principal specific heats this is

$$c_P - c_V = \frac{TB\beta^2}{\rho} \tag{29}$$

Putting in some typical values, the difference between the principal specific heats for polymers at room temperature is of the order of 0.004 J/gK, which is usually small enough to be ignored [5]. Typical specific heat values for solids, including polymers, range from 0.4 J/gk to 4 J/gK. An extensive collection of data is given in the review by Wunderlich and Baur [61].

5.1 Adiabatic Calorimeters

One of the most precise ways of measuring specific heat is by means of an adiabatic calorimeter. It is also the most direct method, since a small measured amount of heat is applied to the specimen and its resultant temperature rise recorded. It is adiabatic since the temperature of the system is allowed to rise, but no heat exchange with the surroundings is permitted. This is normally achieved by surrounding the test cell with a jacket that is maintained at the same temperature as the sample throughout the measurement to within very close limits (of the order of 0.01°C or less). Having reduced the heat losses to a negligible level, all the thermal energy goes into raising the temperature of the sample and the sample holder, and the specific heat is simply calculated. The thermal capacity of the sample holder is determined in a separate experiment.

A number of workers have described adiabatic calorimeters and the constructional and experimental difficulties [62–75].

5.2 Drop Calorimeters

Drop calorimeters are widely used because of their simplicity. A specimen, often contained in a metal capsule, is heated to some appropriate constant temperature in an oven or furnace and allowed to drop into liquid in a stirred calorimeter. The temperature rise of the calorimeter is monitored, and from this the specific heat can be calculated. The thermal capacity of the calorimeter must be determined in a separate experiment, and heat losses or gains to or from the environment must be allowed for.

ASTM C351 (1982) [76] describes a drop calorimeter. This Standard is for insulating materials and can be used with polymers.

Sometimes the drop calorimeter and adiabatic methods are combined, and the heated sample is dropped into and adiabatic container. Griskey and Hubbell [77] describe the use of this method for measurements on methacrylic polymers in the temperature range 120 to 300°C.

5.3 Scanning Calorimeters

The usual scanning techniques are differential thermal analysis (DTA) [78] and differential scanning calorimetry (DSC) [79–81]. In these methods it is assumed that the heat loss from the calorimeter is a function of temperature only. By comparing the rate of heat input and temperature rise for a polymer sample with that of a standard, usually synthetic sapphire, the specific heat of the polymer can be obtained.

DTA measures the difference in temperature between the sample and a standard for the same rate of heat input. DSC compares the rate of heat inputs for the same rate of temperature rise. The latter is easier to analyze as it gives a direct measure of the rate of heat input. The method is based on the assumption that the samples are so small that thermal equilibrium is obtained almost immediately. For polymers this is not correct, and errors from this source are discussed by Strella and Erhardt [82]. Richadson [83–86] and Laye [87] have discussed methods of calibrating the DSC to improve the accuracy of the results. The method is also outlined in ASTM E1269 [88].

References

1. Kittel, C. (1966), *Introduction to Solid State Physics*, 3d ed., John Wiley, p. 168.
2. Hands, D., and Horsfall, F. (1977), *Rubber Chemistry and Technology*, *50*, 253.
3. Carslaw, H. S., and Jaeger, J. C. (1959), *Conduction of Heat in Solids*, 2d ed., Oxford Univ. Press, p. 9.
4. Parrott, J. E., and Stukes, A. D. (1975), *Thermal Conductivity of Solids*, Pion, p. 5.
5. Choy, C. L. (1975), *Journal of Polymer Physics*, *13*, 1263.
6. Howard, J. R. (1975), *Engineering*, *215*, 220.
7. Rhee, B. O., Hieber, C. A., and Wang, K. K. (1994), *Antec*, *1*, p. 496.
8. McAdams, W. H. (1975), *Heat Transmission*, 3d ed., McGraw-Hill.
9. Cuthbert, C. (1954), *Transactions of the Institution of the Rubber Industry*, *30*, 16.
10. Hall, J. A., Ceckler, W. H., and Thompson, E. V. (1987), *Journal of Applied Polymer Science*, *33*(6), 2029.
11. Lees, C. H. (1898), *Philosophical Transactions of the Royal Society*, *A191*, 399.
12. BS 874 (1988), Methods for determining thermal insulation properties: Part 2, Section 2.2, Unguarded hot plate.
13. ASTM C 518 (1991), Test method for steady-state heat flux measurements and thermal transmission properties by means of the heat flow meter apparatus.
14. ISO 8301 (1991), Thermal insulation—Determination of steady-state thermal resistance and related properties—Heat flow meter apparatus.
15. ISO 2581 (1975), Rigid cellular plastics—Determination of apparent thermal conductivity by means of a heat-flow meter.
16. ASTM C1132 (1995), Practice for calibration of the heat flow meter apparatus.
17. Scott, J. A., and Bell, R. W. (1994), *Journal of Thermal Insulation*, *18* (Oct), 146.
18. ASTM C177 (1985), Test method for steady-state heat flux measurements and thermal transmission properties by means of the guarded-hot-plate apparatus.
19. BS 874 (1986), Methods for determining thermal insulation properties: Part 2, Section 2.1, Guarded hot plate.
20. ASTM C1044 (1985), Practice for using the guarded-hot-plate apparatus in the one-sided mode to measure steady-state heat flux and thermal transmission properties.
21. ISO 8302 (1991), Thermal insulation—Determination of steady-state thermal resistance and related properties—Guarded hot plate.

22. ISO 8990 (1994), Thermal insulation—Determination of steady-state thermal transmission properties—Calibrated and guarded hot box.
23. Parrott, J. E., and Stukes, A. D. (1975), *Thermal Conductivity of Solids*, Pion, p. 19.
24. ASTM D2214 (1970), Test method for estimating the thermal conductivity of leather with the Cenco-Fitch apparatus.
25. Eiermann, K., Hellwege, K. H., and Knappe, W. (1961), *Kolloid Z.*, *174*, 134.
26. Ott, H. J. (1981), *Plastics and Rubber Processing Applications*, *1*, 9.
27. Kline, D. E. (1961), *Journal of Polymer Science*, *50*, 441.
28. Lohe, P. (1965), *Kolloid Z.*, *203*, 115.
29. Gobbe, C., Bezin, M., Gounot, J., and Dehay, G. (1988), *Journal of Polymer Science: Polymer Physics*, *26*(4), 857.
30. Lane, K. Private communication.
31. Carslaw, H. S., and Jaeger, J. C., *Conduction of Heat in Solids*, p. 261.
32. Horrocks, J. K., and McLaughlin, E. (1963), *Proceedings of the Royal Society*, *A273*, 259.
33. D'Eustachio, D., and Schreiner, R. E. (1952), *Journal of the American Society of Heating and Ventilating Engineers*, *58*, 331.
34. Vos, B. H. (1956), *Applied Scientific Research*, *A5*, 425.
35. Oehmke, F., and Wiegmann, T. (1994), *Antec II*, 2240.
36. Yen, C. L., Tseng, H. C., Wang, Y. Z., and Hsieh, K. H. (1991), *Journal of Applied Polymer Science*, *42*(5), 1179.
37. Lobo, H., and Cohen, C. (1990), *Polymer Engineering and Science*, *30*(2), 65.
38. Mann, G., and Forsyth, F. G. E. (1956), *Modern Refrigeration*, *188*, June.
39. Hooper, F. C., and Lepper, F. R. (1950), *Heating Piping and Air Conditioning*, *20* (August), 129.
40. Hooper, F. C., and Chang, S. C. (1952), *Heating Piping and Air Conditioning*, *22* (October), 125.
41. Carslaw, H. S., and Jaeger, J. C. (1959), *Conduction of Heat in Solids*, p. 75.
42. Harmathy, T. Z. (1964), *Journal of Applied Physics*, *35*, 1190.
43. Martin, B. (1970), *Polymer*, *11*, 287.
44. Hands, D., and Horsfall, F., *Polymer*, *12*, 145.
45. Carslaw, H. S., and Jaeger, J. C., *Conduction of Heat in Solids*, p. 33.
46. Williamson, E. D., and Adams, L. H. (1919), *Physical Review*, *14*, 99.
47. Shoulberg, R. H. (1963), *Journal of Applied Polymer Science*, *7*, 1597 .
48. Carslaw, H. S., and Jaeger, J. C. (1959), *Conduction of Heat in Solids*, p.104.
49. Berlot, R. (1966), *Plastiques Modernes et Elastomeres*, *18*, 231.
50. Berlot, R. (1969), *Plastiques Modernes et Elastomeres*, *19*, 117.
51. Gehrig, M., Gramespacher, H., Innerebner, F., and Meisaner, J. (1991), *Kunststoffe German Plastics*, *81*(8), 30.
52. Carslaw, H. S., and Jaeger, J. C., *Conduction of Heat in Solids*, p. 105.
53. Smith, D. I. (1987), Ph.D. Thesis, University of Bradford.
54. Horsfall, F. (1976), M.Sc. thesis, University of Bradford.
55. ASTM E1461 (1992), Test method for thermal diffusivity of solids by the flash method.
56. Cape, J. A., and Lehman, G. W. (1963), *Journal of Applied Physics*, *34*, 1909.
57. Parker, W. J., Jenkins, R. J., Butler, C. P., and Abbott, G. L. (1961), *Journal of Applied Physics*, *32*, 1679.
58. Taylor, R. (1965), *British Journal of Applied Physics*, *16*, 509.
59. Parrott, J. E., and Stukes, A. D., *Thermal Conductivity of Solids*, p.37.
60. Zemansky, M. W. (1957), *Heat and Thermodynamics*, 4th ed., McGraw-Hill, p. 251.
61. Wunderlich, B., and Baur, H. (1970), *Advances in Polymer Science*, *7*, 151.
62. Southward, J. C., and Brickwedde, F. G. (1933), *Journal of the American Chemical Society*, *55*, 4378.
63. Bekkedahl, N., and Matheson, H. (1935), *Journal of Research of the National Bureau of Standards*, *15*, 503.

64. Scott, R. B., Myers, C. H., Rands, R. D., Jr., Brickwedde, F. G., and Bekkedahl, N. (1945), *Journal of Research of the National Bureau of Standards*, *35*, 39.
65. Williams, J. W., and Daniels, F. (1924), *Journal of the American Chemical Society*, *46*, 903.
66. Dole, M., Hettinger, W. P., Jr., Larson, N. R. Wethington, J. A., and Worthington, A. E. (1951), *Review of Scientific Instruments*, *22*, 812.
67. Worthington, A. E., Marx, P. C., and Dole, M. (1955), *Review of Scientific Instruments*, *26*, 698.
68. West, E. D., and Ginnings, D. C. (1958), *Journal of Research of the National Bureau of Standards*, *60*, 309.
69. Richardson, M. J. (1965), *Transactions of the Faraday Society*, *61*, 1876.
70. Aukward, J. A., Warfield, R. W., Petree, M. C., and Donovan, P. (1959), *Review of Scientific Instruments*, *30*, 597.
71. Tautz, H., Gluck, M., Hartmann, G., and Leuteritz, R. (1963), *Plastik und Kautschuk*, *10*, 648.
72. Hellwege, K. H., Knappe, W., and Wetzel, W. (1962), *Kolloidzeitschrift*, *180*, 126.
73. Wunderlich, B., and Dole, M. (1957), *Journal of Polymer Science*, *24*, 201.
74. Bowring, R. W., Garton, D. A., and Norris, H. F. (1960), UKAEA Report AEEW-R. 38, HMSO.
75. Dainton, F. S., Evans, D. M., Hoare, F. E., and Melia, T. P. (1962), *Polymer*, *3*, 263
76. ASTM C351 (1992), Test method for the mean specific heat of thermal insulation.
77. Griskey, R. G., and Hubbell, D. O. (1968), *Journal of Applied Polymer Science*, *12*, 853.
78. Slade, P. E., Jr., and Jenkins, Ll. T. eds. (1966), *Techniques and Methods of Polymer Evaluation*, Vol. 1, *Thermal Analysis*, Arnold.
79. Watson, E. S., O'Neill, M. J., Justin, J., and Brenner, N. (1964), *Analytical Chemistry*, *36*, 1233.
80. O'Neill, M. J. (1964), *Analytical Chemistry*, *36*, 1238.
81. O'Neill, M. J. (1966), *Analytical Chemistry*, *38*, 1331.
82. Strella, S., and Erhardt, P. F. (1969), *Journal of Applied Polymer Science*, *13*, 1373
83. Richardson, M. J., and Burrington, P. (1974), *Journal of Thermal Analysis*, *6*, 345.
84. Richardson, M. J. (1976), *Plastics and Rubber: Materials and Applications*, *1*(3–4), 162.
85. Richardson, M. J. (1984), *Polymer Testing*, *4*(2–4).
86. Richardson, M. J. (1989), *Comprehensive Polymer Science* (G. Allen and J. C. Berington, eds.), Vol. 1, Chapter 36, Thermal Analysis.
87. Laye, P. G. (1980), *Analytical Proceedings*, *17*(6), 226.
88. ASTM E1269 (1995), Test method for determining specific heat capacity by differential scanning calorimetry.

25
Electrical Properties

Cyril Barry
Consultant, Hay-on-Wye, Hereford, England

1 Synopsis

The electrical properties of polymers that are of general interest are those related to the behavior of the polymer as an electrical insulating material. It is, however, true that such properties may be important in assessing the behavior and suitability of a material in applications that may not be electrical in nature (for example the dissipation factor may be useful to detect the presence of an unwanted contaminant in a polyolefine).

These properties may be conveniently divided into two categories:

1. Those dependent on the mobility of electrically charged molecular entities (e.g., ions or dipoles) within the material. These movements do not generally lead to irreversible changes in the material. The important properties are volume resistivity, surface resistivity, insulation resistance, permittivity, and dissipation factor.
2. Those relevant to the ability of the material to withstand electrical stress without degradation leading to the loss, permanent or temporary, complete or partial, of its electrical insulating character. This group comprises mainly breakdown voltage, tracking resistance, arc resistance, and resistance to electrical discharges.

An important subdivision of these categories identifies those properties that are significant in determining the behavior of the material in generating or retaining unwanted electrostatic charges. These will be considered separately, for two reasons:

1. The methods may be very specialized (e.g., chargeability and charge decay).
2. The determined value of the property may be such that some experimental errors are not significant, so that a simplified procedure or lower accuracy may be acceptable (e.g., resistivity).

In preparing this chapter, the aim of the author has been to provide a guide to the standardized practical methods. Since electrical measurement is the province of the International Electrotechnical Commission (IEC), IEC standards have been preferred to ISO when both are in existence. Reference has also been made to a number of review papers with extensive bibliographies. If more fundamentally oriented discussion is required, the first volume of the handbook edited by Barnikas and Eichhorn is recommended [1].

In most cases, a brief outline of the experimental procedure is given. This is not intended as a set of instructions for making the measurement: when a measurement is planned, the standard specification for the procedure should be consulted for details of the procedure.

2. Resistivity

2.1 General

The theory of electrical conduction in polymeric materials is extremely complex, and the phenomena are still incompletely understood. The apparent current between electrodes separated by a polymeric material is neither constant in time nor proportional to the applied potential. Most aspects of the subject are discussed in a review paper [2]. This review also presents an extensive bibliography. From the practical point of view, there are several published standards dealing with both volume and surface resistivity [3–5]; in addition, several standards deal with specialized materials from the viewpoint of electrostatics (see the section on electrostatics). Some of the latter may well be withdrawn as the IEC/ISO situation on electrostatic matters is rationalized.

2.2 Practical Considerations

After application of a field, the current into the electrodes usually decreases steadily over a range of several decades, with a time constant that increases with the duration of the measurement, reaching a steady state after a period that may be minutes or hundreds of hours. Conventionally, the apparent resistance is measured either 60 or 120 seconds after the initial application of electric stress.

The current at any time is usually linearly related to the applied field ("ohmic conduction") for values of the field that are low in comparison to that which might be employed in use of the material as electrical insulation.

The measurement of resistivity can always be resolved into the measurement of a resistance in conjuction with the use of an electrode system defining the ratio of the measured resistance to the required resistivity. This ratio is frequently referred to as the "cell constant" of the electrode system.

Other than the time of application of the electic field, there are four main aspects to the accurate measurement of resistivity:

1. Accurate measurement of resistance
2. Accurate knowledge of electrode geometry
3. Elimination of series resistances (contact or lead)
4. Elimination of spurious parallel resistances

These four features are not all relevant in the same degree to all values of resistivity.

2.3 Resistance Measurement

Although accurate measurement of resistance is conventionally achieved by a bridge procedure, in commercial practice this is rarely used. Most resistance measuring instruments ("resistance meters") rely either on measuring the current caused by a defined potential or the potential necessary to cause a defined current. In either case, the instrument will indicate directly the potential/current ratio as a resistance value, or the current/potential ratio as a conductance.

The great majority of measurement in the polymer testing field are on high-restivity materials (measured resistance $\geq 10\,\mathrm{k}\Omega$), and most instruments will be fixed potential devices with current measurement. Multiple ranges for the instrument will usually be achieved through changes in the sensitivity of the current-measuring section. Such instruments usually have an accuracy of about $\pm 2\%$ at "full scale" (i.e., minimum resistance: the scale will often be reversed, and usually nonlinear).

Where low-resistivity materials are tested, fixed current/potential measuring devices are more appropriate.

Electrode Geometry

This rarely presents problems, since engineering tolerances can be much closer than necessary for the required accuracy in resistivity. The electrodes are often of a circular, concentric design but may also be linear and parallel. The ratio of resistivity/resistance is defined by the electrode dimensions and is often referred to as the "cell constant." For volume resistivity it has the dimension of a length (electrode area/electrode spacing). For surface resistivity it is a dimensionless number (electrode length/electrode spacing).

Errors Due to "Stray" Impedances: Three- and Four-Electrode Systems

The discussion of the elimination of these errors is also relevant to the determination of permittivity and dissipation factor.

Series Resistance Errors

When low resistance values ($\leq 10\,\mathrm{k}\Omega$) are being measured, series resistance errors may arise from the leads connecting the electrodes to the "resistance meter" or from the contact resistances between the electrodes and the test material. Both may be eliminated by using "four-electrode" measuring systems where the electrodes and leads used for current and potential connection are completely separate. Obviously, the meter must be one designed for four-terminal measurement and will employ the "fixed current/measured potential" system.

The "cell constant" of the electrode system is calculated from the dimensions of the "inner" electrode pair (see Fig. 1).

Parallel Conductance Errors

For high resistivity measurements ($R \geq 10\,\mathrm{k}\Omega$) the greatest sources of error arise from parallel conductances, either surface leakage currents or "fringing field" currents caused by field distortion at the edges of the electrodes. These may be minimized by the use of guard electrodes (often known as three-electrode systems). Field distortion errors cannot be completely eliminated in this way, but they can be reduced to negligible proportions by good guard and electrode design.

The guard electrode consists of a third electrode mounted as close as possible to the electrode from which the current is to be measured. It is maintained at a potential as close to the potential of the latter as possible, but not in electrical contact with it, so that the

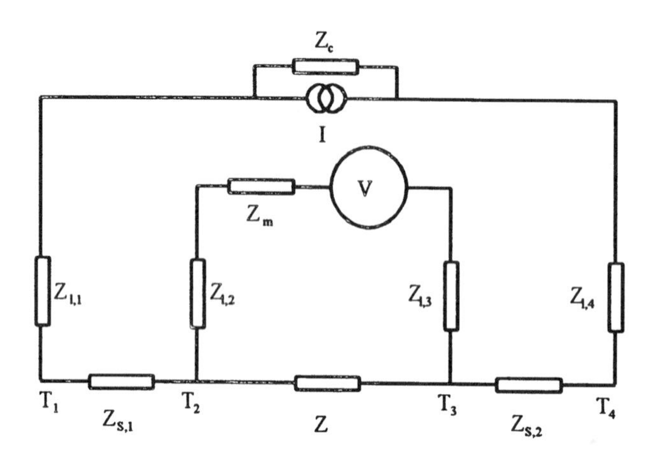

Figure 1 Four-terminal equivalent circuit where I is the current source, internal impedance Z_c; V, voltmeter, internal impedance Z_m; $Z_{l,n}$, lead and contact impedances; $Z_{s,n}$, simple impedances outside test region; and Z, desired impedance.

guard draws no current from the electrode, and any current drawn by other parts of the circuit is drawn from the guard (see Fig. 2). Similarly, the proximity of the guard, and its potential, reduce the tendency of the electrode edges to distort the interelectrode field.

2.4 Volume Resistivity

Volume resistivity values for the complete range of known materials cover a wider range than any other physical property: for conducting materials, values may be as small as $10^{-6}\,\Omega$m, and for the most highly insulating materials, greater than 10^{16}, a range of over 10^{22} to 1. The polymeric materials in common use occupy the range from about 1 Ωm to the highest values, although the greatest interest is in values greater than about $10^4\,\Omega$m. Many materials are anisotropic, and in such cases the apparent resistivity will be direction dependent. For these materials it will often be necessary to use nonstandard specimens and electrode systems. General guidance on the principles to be followed will be found in IEC 93 [3].

Measuring Instrument

For most materials the volume resistivity will be greater than $10^9\,\Omega$m, and the measured resistance using the electrodes illustrated in Fig. 3 will be above about $10^9\,\Omega$. Measurement in this range will require the use of an "electrometer" instrument drawing "bias" current less than about 10^{-12} A. Such instruments are almost invariably provided with a third terminal for connection to a guard electrode (see Fig. 2). The initial charging current for

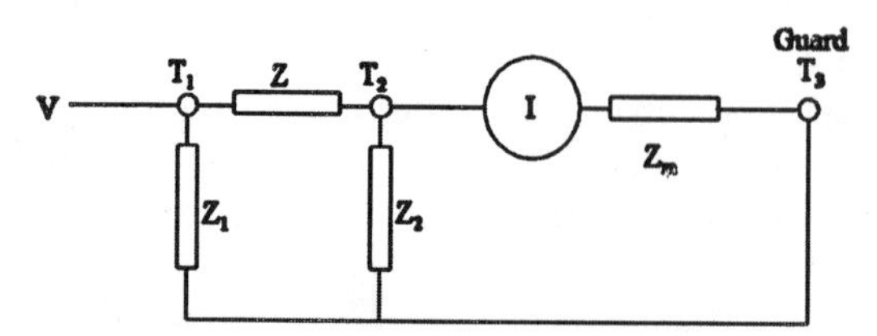

Figure 2 Three-terminal equivalent circuit where Z is required impedance; $Z_{1,2}$, stray impedances; and Z_m, current meter impedance.

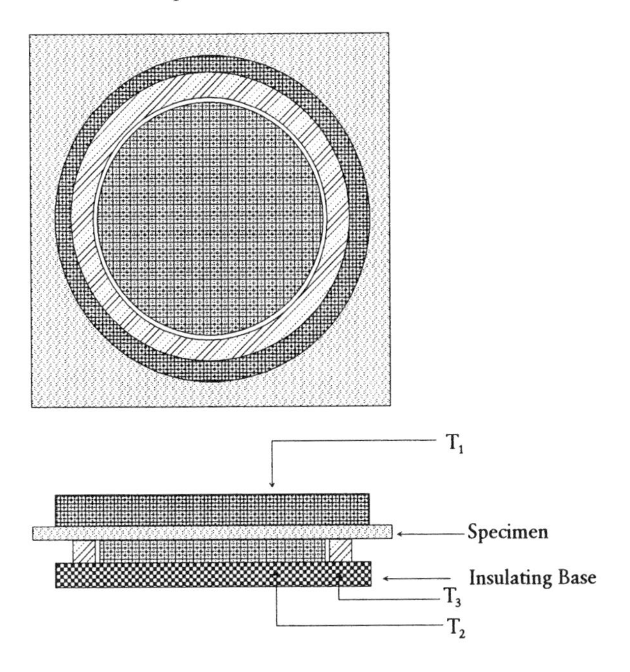

Figure 3 Electrodes for volume resistivity.

the capacitance associated with the specimen may be several orders of magnitude greater than the final measured current. For this reason, most measuring instruments are provided with an an initial charging position on the connecting switch, in which the input circuit of the electrometer section is bypassed.

A possible source of error, only significant for the measurement of extremely high values of volume resistivity ($\geq 10^{13}\,\Omega$m) using older instruments, arises from the self capacitance of the (electrode + specimen) system. If the charging time constant of this through the input resistance of the current meter is comparable with the measurement time for the resistivity, very large errors may be caused.

Electrode System

Generally, the most convenient electrode system will be a coaxial assembly similar to that shown in Fig. 3. In this assembly, the electrodes T_2, T_3 are usually mounted rigidly on the fixed insulating base, T_1 being demountable for insertion of the specimen. The assembly may be employed for sheet material in thicknesses greater than about 25 μm, with an upper limit of about 3 mm. The relation of resistivity to resistance is given by the equation

$$\rho = \frac{R \cdot \pi a^2}{t} \tag{1}$$

where a = central electrode radius and t = specimen thickness. For more accurate work, if required, a correction for fringing field may be added to a, approximately equal to $\frac{1}{2}$ the interelectrode gap.

Convenient (guarded) electrode diameters are specified in BS 2782 [4] as 50 and 150 mm, using a guard gap of about 1 mm, the unguarded electrode being about 10 mm larger than the guard.

In many cases, it is necessary to apply "intimate" electrodes to the surfaces of the specimen. These are usually of thin (6–25 μm) aluminum foil in conjunction with a conductive lubricant. Other intimate electrodes may be used, e.g., vacuum deposited aluminum or conductive paint. Detailed advice will be found in BS 2782 [4].

Specimen

In most cases, the specimen will be a sheet of material, 1 to 3 mm thick, at least 10 mm larger than the diameter of the unguarded electrode. The sheet may be the material as supplied, or one fabricated for the purpose.

Since the resistivity of most materials is strongly affected by water content, conditioning of the test specimen after preparation and before application of the electrodes is important. The standard or controlled atmosphere for conditioning is specified in the material standard. Testing should be completed either in the conditioning atmosphere or as quickly as possible after removal of the specimen from it.

For sheet material and plastic films thinner than 25 μm, wound capacitors may be used, with thin aluminum foil (usually 6 μm) as electrode [6]. Particular attention must be paid to the tightness of wind of the capacitor: in many cases it is better to wind on a large, removable mandrel, flattening the winding after removal. In this case the winding should be clamped and compressed between rigid endplates after any conditioning. The capacitance of the winding should be measured and the resistivity calculated from the equation.

$$\varepsilon \times \rho = C \times R \tag{2}$$

where ε is the permittivity of the material and C and R are the capacitance and resistance of the winding.

Outline Procedure

1. The specimen is prepared in accordance with the particular test procedure being employed and is then conditioned as necessary.
2. Any necessary intimate electrodes and contacting medium are applied. If the intimate electrodes are applied by vacuum metallizing, the specimen is conditioned after application.
3. The specimen is mounted in the specimen holder/electrode assembly and connections made to the measuring instrument.
4. Test potential is applied via the "charging" position of the instrument switch for at least about 10% of the specified electrification time (usually about 5 seconds).
5. The instrument connection switch is moved to the "read" position and the range switch set to an appropriate level at least 10 seconds before reading the resistance at the specified electrification time. The indicated resistance will not usually be constant but will steadily increase, at a rate dependent on the test material.
6. If necessary, the capacity of the specimen/electrode assembly is measured.
7. The resistivity is calculated using Eq. 1 or 2.

2.5 Surface Resistivity

General

While the volume resistivity (or rather volume conductivity) of a material is a true measure of the charge transport through the bulk of the material, surface resistivity is a composite property, made up of transport through the bulk, as well as along the surface of the material, the relative proportions being indeterminate in the general case. However, the apparent ability of a material to conduct charge along its surface is extremely important in many applications.

The surface resistivity of a material is defined as the resistance between electrodes connected to the surface, when the surface between the electrodes is a square. For this reason, surface resistivity is often expressed in units of "ohms per square," the resistance of the square being independent of the size of the square. The correct unit for expression of surface resistivity is, however, "ohms", since it is calculated from the resistance of a rectangle by dividing the product of the resistance and the electrode width by the length between electrodes.

Electrodes

The electrodes and specimen for surface resistivity ρ_s measurement may be in one of three forms.

A strip of sheet material, the linear electrodes being longer than the width of the material. In this case,

$$\rho_s = R\frac{w}{1} \quad (\Omega) \tag{3}$$

where l is the distance between electrodes, w is the width of the strip, and R is the resistance in ohms.

Although in principle, a four-electrode system may be used to eliminate series resistance (connection) errors, this will rarely be found beneficial in measuring the surface resistivity of polymers.

A guard electrode may be placed on the opposite side of the specimen from the electrodes: alternatively, the specimen may be supported on the flat surface of a material of very high resistivity, such as P.T.F.E.

A circular, concentric electrode configuration is often convenient, and suitable arrangements are specified in BS 2782, Method 231A (See Fig. 4). The resistivity is calculated using the equation

$$\rho_s = \frac{2\pi R}{\log_e[D_2/D_1]} \tag{4}$$

This electrode assembly has the advantage that a sheet of material of indeterminate size (larger than the electrodes) can be used, the measurement being unaffected by the size of the sheet.

Parallel, linear electrodes of defined length and spacing may be used on a sheet of test material larger than the electrode assembly. The method is not recommended, since a simple expression relating resistivity and resistance is not possible, the proportionality factor being dependent on the distances between electrodes and sheet boundaries. If the procedure is used, the factor for specified electrode and sheet dimensions should be determined empirically from measurements on material whose resistivity is known from measurements using the two methods above.

Measuring Instrument

In general, the instruments used for volume resistivity measurement will be satisfactory for surface resistivity. Charging current will rarely be a significant factor.

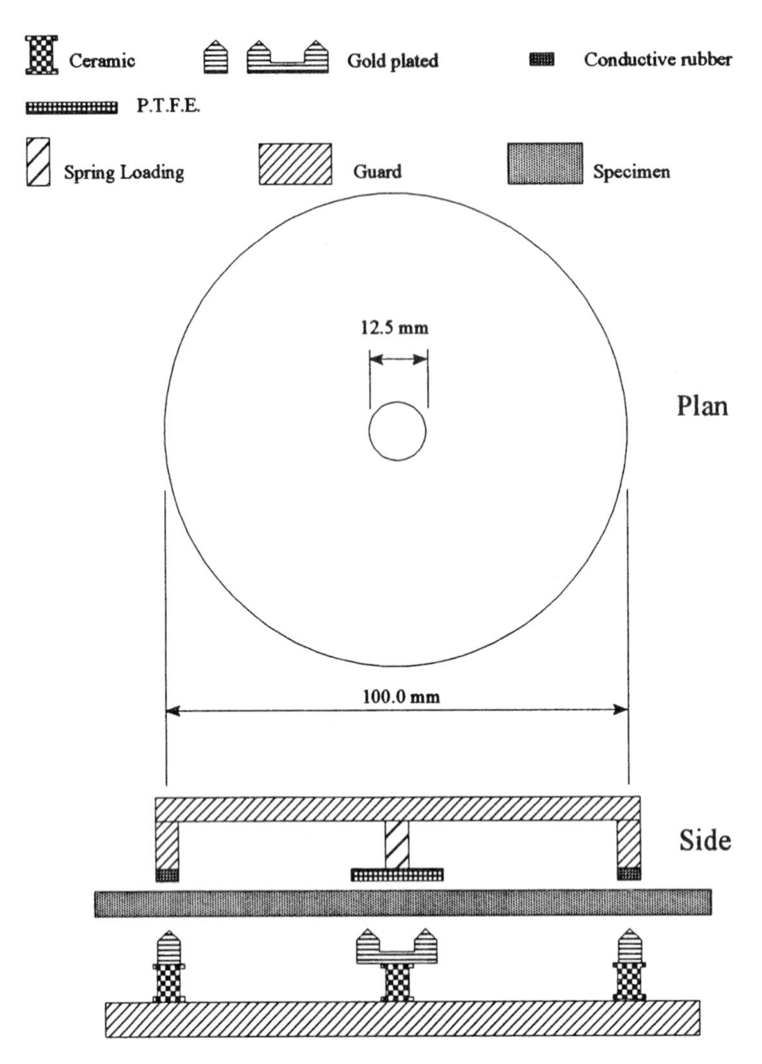

Figure 4 Surface resistivity electrodes.

Outline Procedure

The procedure outlined for volume resistivity will usually be satisfactory.

2.6 Insulation Resistance

The insulation resistance is not a "true" property of the material. It is determined by measuring the resistance between two electrodes applied to the specimen, with no attempt to separate conduction through the material from that along its surface, nor to calculate to a standard configuration from the electrode dimensions. Nonetheless, it is a useful, rapidly and easily measured property in the context of quality control. Standard electrode configurations for various purposes are recommended in IEC 167 [7].

The resistance meter suggested for volume resistivity will be appropriate for this measurement, as will the outline procedure. It should be noted, however, that many of the electrode systems used require different forms of test specimen, and some kind of machining of the material is often specified.

3 Permittivity and Dissipation Factor

The permittivity and dissipation factor of polymeric materials are important properties in themselves for many applications. In addition, the dissipation factor is often used as a criterion in measuring the concentration of undesirable contaminants or impurities. For these purposes, measurements must be made at frequencies over a very wide range, from below power frequencies (~50 Hz or lower) to microwave frequencies (~300 MHz up). Standard methods have been developed throughout this range, and the more widely invoked will be briefly described.

3.1 General

The IEC standard, Publication 250 [8], consists of an outline of the general principles of measurement methods for use up to 300 MHz. Unfortunately, it gives very little detailed instruction on procedures and is in serious need of revision. Work on such revision has been in hand for several years. A similar ASTM standard, D150 [9], is also a very useful source of background information. The following table summarizes the relevant standards of IEC, BSI, and ASTM.

Standard	Frequency	Temperature	General coverage
IEC 250 [8]	1 Hz–300 MHz		General theory
IEC 377-1 [10]	> 300 MHz		General theory
IEC 377-2 [10]	> 300 MHz		Resonance methods
ASTM D150 [9]	1 Hz–300 MHz		General theory
ASTM D1531 [11]	1 kHz–1 MHz		Thin sheet material (liquid immersion)
ASTM D2149 [12]	< 10 MHz	< 500°C	Ceramics
ASTM D2520 [13]	Microwave	< 1650°C	Ceramics
BS 2067 [14]	10 kHz–100 MHz		"Hartshorn & Ward"
BS 7663 [15]	50 Hz–1 MHz		A. "Lynch method"
BS 4542 [16]			B. Liquid immersion
			C. As ASTM D1531

Very valuable discussion and a useful comprehensive bibliography will be found in a review paper by Bartnikas [17].

3.2 Definitions

For all real electrical insulating materials, the alternating current through a capacitor using the material as dielectric is different in phase from the voltage across the capacitor by an angle less than 90°, and therefore mathematically the impedance of the capacitor is treated as a complex quantity. The angle for a capacitor with vacuum dielectric is 90°, and therefore the permittivity (see below) is itself a complex number.

Permittivity — The definitions of IEC 50 [18], the International Electrotechnical Vocabulary, are abstruse and of little use in every day work. The notes to the definition of "Relative permittivity" are more valuable: "In the case of ... alternating fields of sufficiently low frequency, the relative permittivity ... is equal to the ratio of the capacitance of

a capacitor in which the space between and around the electrodes is entirely and exclusively filled with the dielectric to the capacitance of the same configuration of electrodes in vacuum."
"In practical engineering, it is usual to employ the term *permittivity* when referring to *relative permittivity*."

Dissipation Factor — The numerical value of the ratio of the imaginary to the real part of the complex permittivity. The heat generated in a capacitor is equal to the dissipation factor multiplied by the product of current and voltage in the capacitor.

3.3 Measuring Instruments for Capacitance and Dissipation Factor

Although the discussion below refers throughout to calculations based on capacitance ratios and the measurement of capacitance, the instrument used (especially in the case of direct-indication equipment) need not measure or indicate capacitance directly. Measuring equipment may be used that displays, for example,

Capacitance and dissipation factor
Capacitance and equivalent series resistance
Capacitance and equivalent parallel resistance
Impedance and dissipation factor
Admittance and dissipation factor
Reactance and series resistance (real and imaginary parts of complex impedance)
Susceptance and parallel conductance (real and imaginary parts of complex admittance)

These representations are all equivalent and interconvertible: however, the values of capacitance in series and parallel representations are not generally exactly equal, especially if the dissipation factor is greater than about 0.1. In such cases the convention is to use the value appropriate to the parallel representation. A full discussion and explantion will be found in IEC 250 [8] and/or ASTM D150 [9].

Much of the modern direct indication equipment measures admittance or impedance by calculating the complex ratio of complex current and voltage, using sophisticated computing techniques. Apparatus of this kind is often equipped to accept inputs through 3, 4, or 5 terminals, so that errors due to series and/or parallel stray impedances can be eliminated.

Equipment employing such techniques is often restricted in its use of connection cables, in that the computing algorithms associated with the internal calculations require specific properties (capacitance, inductance, etc.) of the connection system. It may be necessary to adjust preset controls for "input short circuit" and "input open circuit" conditions.

Equipment of the above type is very convenient for measurement at audio and low radio frequencies, but for use at power frequencies a bridge instrument of the "Schering" or "transformer" design is more approprate. The former should be equipped to control a guard electrode (e.g., "Wagner earth" circuit): instruments are today available in which the guard control is automatic, not requiring a separately balanced section of the Schering bridge itself.

3.4 Methods for Frequencies Between 50 Hz and 1 MHz

BS 7663 [15]

BS 7663 was developed by BSI Committee PRI/25 (Electrical properties of rubbers and plastics) to provide detailed guidance for measurements on polymeric materials in this frequency range. It comprises 3 procedures: Methods A, B, and C.

BS 7663, Method A

This is a development of the Lynch method [16] in which it is not necessary to measure accurately the value of any capacitance. The specimen is inserted between electrodes that do not make any direct contact with it. The spacing between the electrodes can be adjusted and measured by a micrometer. It is necessary that the thickness of the test specimen be known with appropriate accuracy.

The capacitance meter is balanced and/or read. The specimen is removed and the capacitance reset to the original value by adjusting the micrometer. The permittivity can be calculated from the specimen thickness and micrometer adjustment. The dissipation factor can be calculated from these readings and the initial dissipation factor of the (specimen + electrode) assembly.

The electrode assembly is shown schematically in Fig. 5. Its dimensions may be changed to suit particular circumstances.

Test Specimen

For the electrode assembly illustrated, a suitable test specimen is a flat sheet of uniform thickness ($\pm 1\%$ is usually specified) between 1 and 3 mm, and linear dimension about 6×6 cm or 6 cm diameter.

Outline Procedure

Before making any measurements it is necessary to establish the calibration of the micrometer used for thickness measurement in terms of that of the micrometer electrode adjustment. If test specimens are sufficiently uniform in thickness, it may be possible to use the latter for thickness measurement.

1. Prepare and condition the specimen.
2. Measure the thickness of the test specimen at at least five positions uniformly distributed over its surface. Calculate the mean thickness.
3. Adjust the setting of the micrometer electrode so that the electrodes are just in contact. Record the micrometer reading m_0.
4. Adjust the setting of the micrometer electrode so that the specimen can be easily inserted.
5. Connect the instrument to the electrode assembly as appropriate for making "open circuit" and "closed circuit" adjustments. Adjust the preset controls.
6. Insert the specimen. Measure and record the capacitance C and dissipation factor D_1.
7. Record micrometer reading m_1.
8. Remove specimen: adjust micrometer so that the instrument indication of capacitance is returned to its original value C. Record the dissipation factor value D_2. Record the micrometer reading m_2.
9. Calculate permittivity and dissipation factor from the equations

$$\varepsilon = \frac{t}{t - (m_1 - m_2)} \qquad \delta = (D_1 - D_2)\frac{m_2 - m_0}{t - (m_1 - m_2)}$$

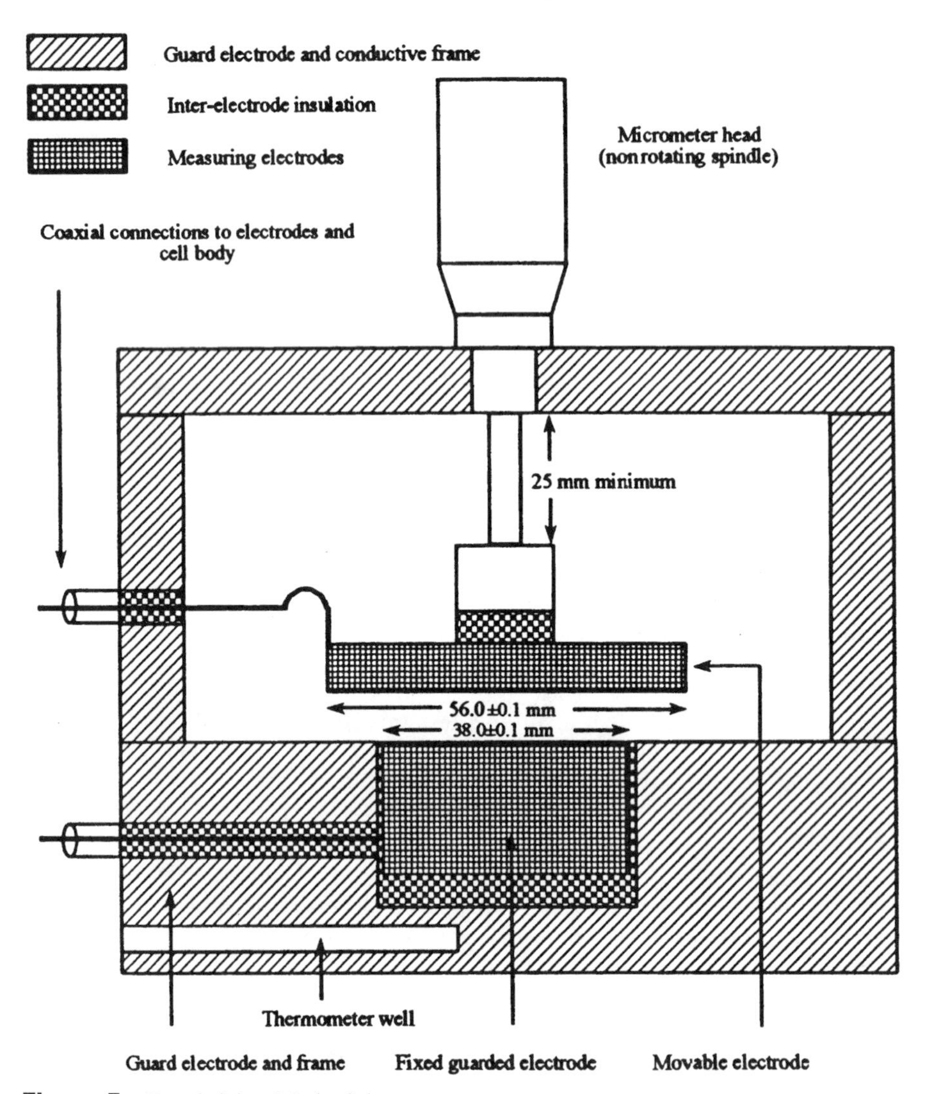

Figure 5 Permittivity, Method A.

Method B

Method B is a liquid immersion method, the capacitance of the cell (air-filled) being measured with and without the specimen between the electrodes. These measurements are repeated with the cell filled with a liquid of known permittivity. The permittivity and dissipation factor of the specimen can be calculated from the four values of capacitance and dissipation factor without knowledge of the thickness of the specimen. It is also possible to calculate the actual thickness of the material between the electrodes. Comparison of this with the geometrical thickness enables an estimate to be made of the depth of surface profile of the material. An example of a suitable cell is shown in Fig. 6.

Test Specimen

Generally, the test specimen of Method A will be suitable for Method B.

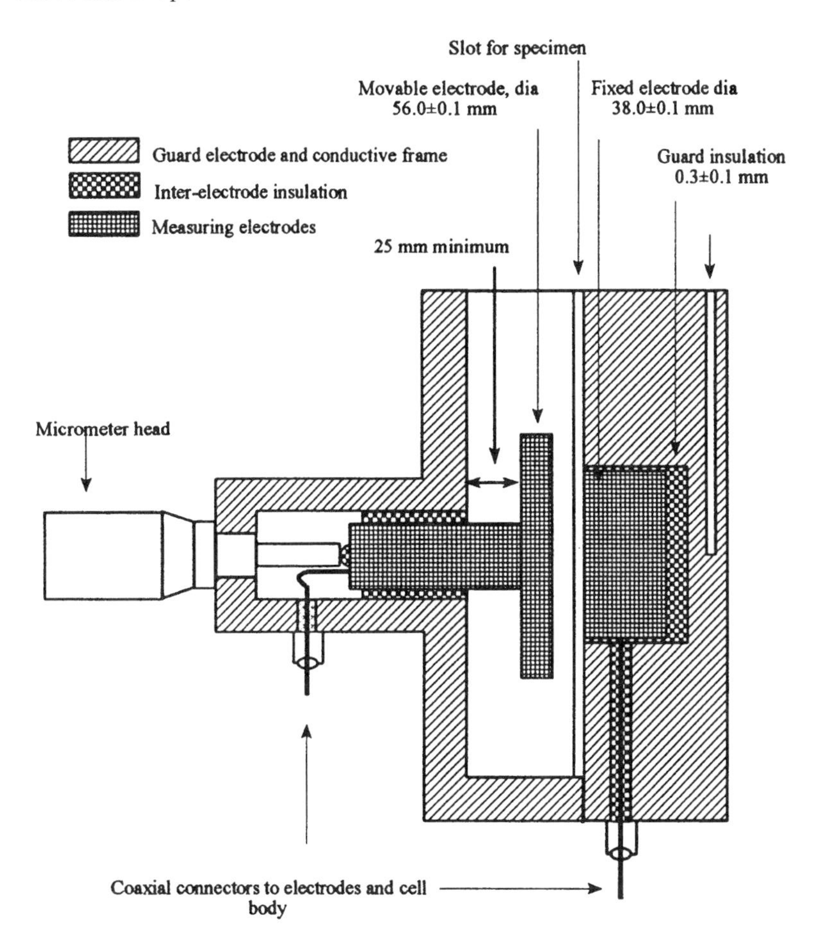

Figure 6 Permittivity, Method B.

Outline Procedure

Before use, the electrode assembly should be cleaned with deionized water, followed by acetone, and dried at 50°C.

The following procedure is extracted from BS 7663 (steps 2–8).

1. Adjust the micrometer electrode so that the test specimen can be easily inserted into the interelectrode gap.
2. Connect the leads to the cell and follow the manufacturer's instructions as to the initial bridge calibration. Record the values of capacitance and dissipation factor as C_1 and D_1 with air between the electrodes.
3. Insert the test piece between the electrodes. Record the values of capacitance and dissipation factor as C_2 and D_2. Record the cell temperature to ±0.1°C.
4. Remove the test piece from the cell and fill the cell with the fluid chosen. Measure the capacitance and dissipation factor and record the values as C_3 and D_3.
5. Insert the test piece into the cell. Measure the capacitance and dissipation factor and record the values as C_4 and D_4. Remove the specimen from the cell immediately after this measurement.

6. Repeat this procedure for a further two test pieces.
7. Calculate the values of permittivity and dissipation factor as shown below.
8. Calculate the mean values of material permittivity and dissipation factor.

$$\varepsilon = 1 + \frac{C_4(C_2 - C_1)(C_3 - C - 1)}{C_1(C_4(C_3 - C_1) - C_2(C_4 - C_3))} \tag{6}$$

$$\delta = \left[\frac{C_2 C_4(C_3 - C_1)}{(C_4(C_2 - C_1) - C_2(C_4 - C_3))}\right]\frac{(D_2 - D_1)}{(C_2 - C_1)} \tag{7}$$

Cylindrical Materials

Comparison of the expressions for the reciprocal of the capacitance of multilayer capacitors with flat electrodes and cylindrical electrodes shows close formal similarity:

$$\frac{1}{C} = \frac{\sum \frac{d_i - d_{i-1}}{\varepsilon_i}}{\varepsilon_0 A} \quad \text{(planar case)} \qquad \frac{1}{C} = \frac{\sum \frac{\log_e(r_i) - \log_e(r_{i-1})}{\varepsilon_i}}{\varepsilon_o L} \quad \text{(cylindrical case)}$$

where d_{i-l} and d_i are the distances from one electrode to the sides of layer i, ε_i is its permittivity, and A is the electrode area in the planar case. For the cylindrical case r_i is the outer radius of layer i, and L is the electrode length.

Based on this formal identity of calculation equations, it has been possible to specify a functionally similar cell in BS 7663 for use with cylindrical specimens of polymeric materials. This is shown schematically in Fig. 7. In use it is not as versatile as the cell of Fig. 6, and a separate cell is best used for each wall thickness/diameter combination. In all other respects, its use parallels that of the flat sheet system.

Method C

Method C of BS 7663 is identical with ASTM D1531 [11]. Its use is restricted to thin sheet materials where a liquid of very similar permittivity is available. The capacitance of the cell is measured when filled with the liquid, with and without the specimen in place between the electrodes. From the small change in capacitance, the difference in permittivity can be calculated accurately without very accurate knowledge of the thickness of the specimen. The specimen holder/electrode system is shown schematically in Fig. 8.

Outline Procedure

Method C is of rather restricted applicability, since its use depends on the availability of a low loss liquid of comparable permittivity to that of the material under test. In use it is similar to Method B and reference should be made to BS 7663 for details of the method.

BS 2067 [14]

This procedure (a method for frequencies between 10 kHz and 100 MHz) is mentioned largely for historical reasons. It is usually known as the Hartshorn and Ward method. The method must be considered obsolete, since the necessary equipment is no longer obtainable commercially: it is, however, still invoked in a number of standards for polymeric materials.

Like most of the older standard methods, BS 2067 employs dual electrode systems: intimate electrodes in contact with the specimen, and secondary rigid electrodes carrying the connections to the measuring device. Such methods suffer from the dangers of poor contact and/or contamination through the use of adhesive materials.

The methods of BS 7663 do not rely on contacting electrodes and therefore are not subject to these potential errors.

For most purposes, Method A of BS 7663 is equally suitable to BS 2067, and it is expected to replace BS 2067 for use in standard specification in due course.

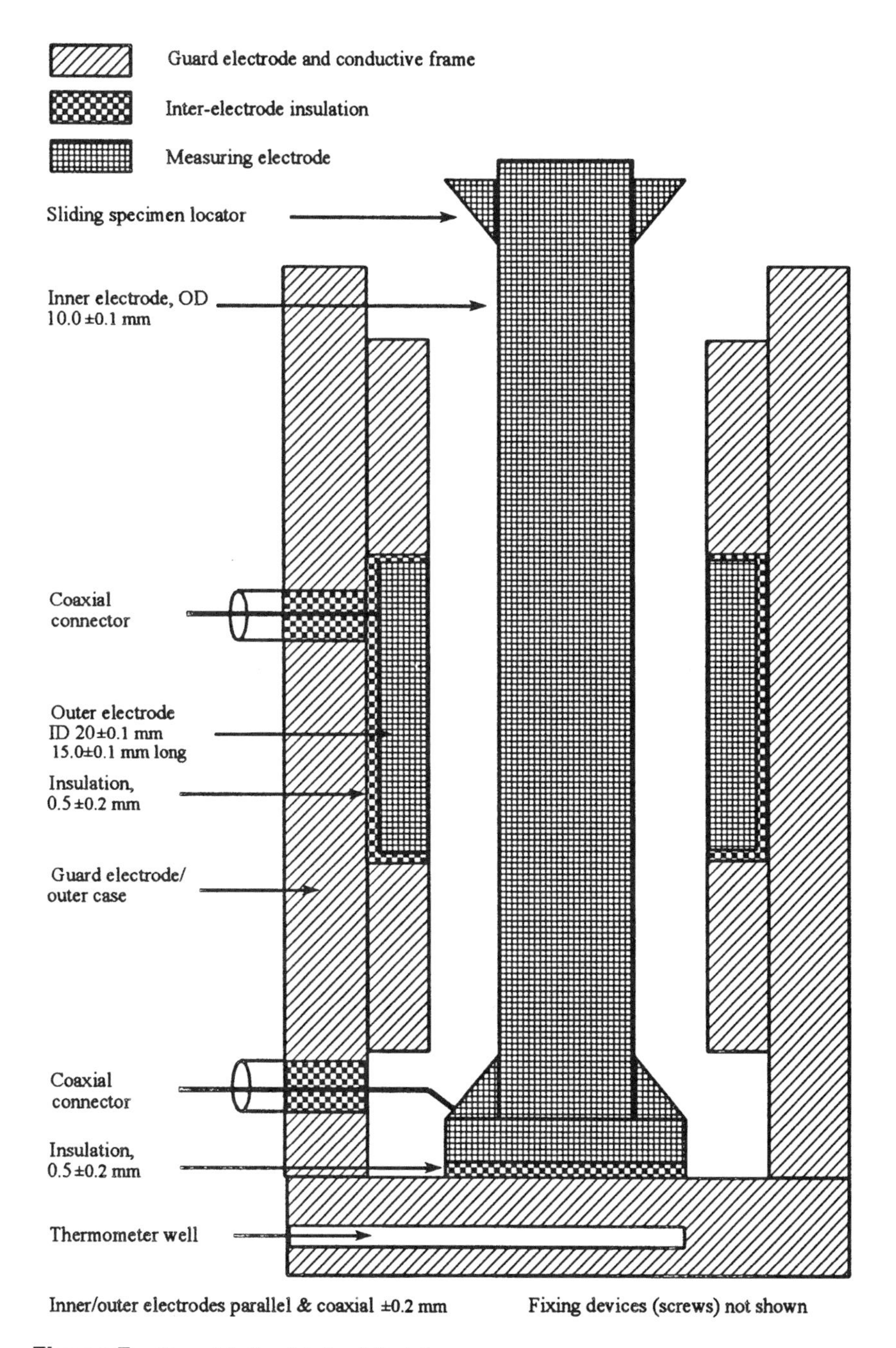

Figure 7 Permittivity, Method B, tubes.

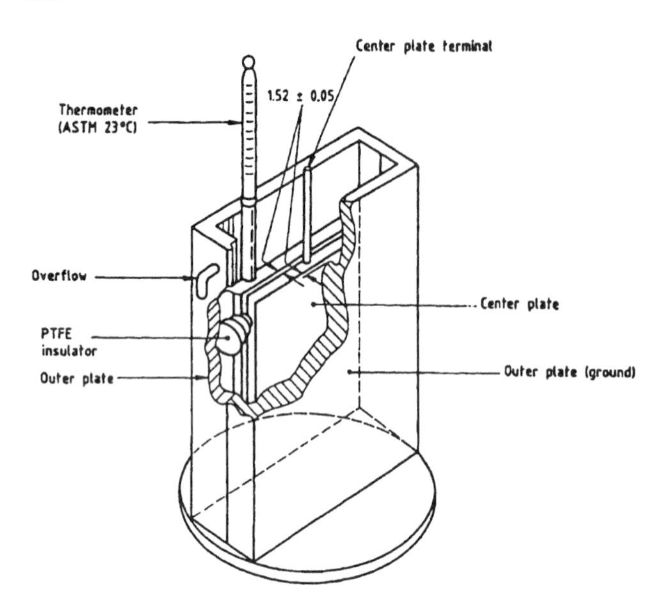

Figure 8 Permittivity, Method C.

3.5 Other Procedures

Methods using wound capacitors with aluminum foil electrodes are often used for thin polymeric film materials. This procedure requires careful control of drying and winding processes: in view of the limited applicability, the procedure will not be considered further. Reference should be made to BS 7290 [6].

3.6 Methods for Frequencies Above 300 MHz

Electrical insulation for use at frequencies above about 300 MHz (microwave frequencies) requires dielectric losses to be controlled at as low a value as possible, and the values of permittivity and dissipation factor are of vital importance. However, the determination of these properties in this frequency range involves the use of specialized microwave techniques of no interest at lower frequencies or for other applications.

The general principles involved are set out in IEC 377, Part 1 [10]. Resonant cavity techniques are specified in Part 2 of this standard, Further parts specifying other procedures may be developed.

Resonant Cavity Methods

Part 2 of IEC 377 describes resonant cavity procedures and incorporates details of the design and construction of the resonators employed and of the specimens for them. Specimens may take the form of disk, rod, tube, or plate, dependent on the type of resonator.

The procedure consists of scanning the frequency response of a loaded resonator, whose unloaded characteristics are known. The permittivity and dissipation factor are calculated from the shift in peak frequency (or wavelength) and the Q-factor.

Instructions are given for several resonators, covering the frequency range from 100 MHz to >30 GHz.

Type of cavity	Frequency	Shape	Remarks
Reentrant cavity	100 MHz to 1 GHz	Disk	$\varepsilon_r \leq 10$
Coaxial resonator	1 GHz to 3 GHz	Tube	
Cavity (closed)	1 GHz to 30 GHz	Disk, rod	
"Open cavity"	> 3 GHz	Disk	$\varepsilon_r > 5$
Optical resonator	> 30 GHz	Plate, sheet	

4 Electric Strength (Breakdown Voltage)

The international standard (which is almost universally accepted) for measurement of breakdown voltage of solid materials is IEC 243 [19]. In this standard, the breakdown voltage is defined as the voltage at which the insulating property is lost under prescribed test conditions, and the electric strength as the quotient of this and the distance between the applied electrodes. Electric srength measurements are usually made as part of the quality control of polymeric materials intended for electrical insulation, although breakdown voltage is also an important factor in the generation of unwanted electrostatic charge.

The electric strength is generally highly dependent on the interelectrode distance, the surrounding medium and, to a lesser extent, the electrode area. For this reason it is preferable to report breakdown voltage, specimen dimensions, and test conditions rather than simply electric strength.

4.1 "Intrinsic" Electric Strength

The apparent electric strength of a material is greatly influenced by the presence or otherwise of electric discharges in the medium surrounding the electrodes. In some research work, great emphasis has been placed on the apparent electric strength in the absence of such discharges, this being considered to be a true property of the material, generally referred to as the "intrinsic" electric strength. Such measurements have been made using specially designed electrodes embedded or recessed into the material specimen, shaped for the purpose [20]. Such measurements are rarely employed in routine work.

4.2 Electrode Area Effects

The area of material under electrical stress has a great effect on the breakdown voltage obvserved. In some cases it is possible to calculate the expected dependence, in other cases it is not: in all cases, great caution should be observed in making these calculations, which should always be based on as large a body of experimental data as is possible.

In simple terms, the area under stress is usually not equal to the area actually between the electrodes, since at electrode edges the value of electric field is usually increased, and consequently the effective electrode area will be greater than the simple geometric value. This effect is enhanced if the surrounding medium is ionized by the higher field, since then the electrode may be surrounded by a region of high effective electrical conductivity, thereby extending the electrode into the surrounding medium. In many cases, the effective electrode area may be many times the physical value. Clearly, this area enhancement will be voltage dependent and therefore different for each individual breakdown voltage measurement, so that no simple treatment can be possible.

However, particularly for materials of relatively low electric strength and thin film materials, it is often found that the great majority of the breakdown locations are actually between the flat areas of the electrodes. In these cases a relatively simple treatment for the calculation of the area dependence of breakdown voltage may be used.

In addition, a wide field of application of polymeric materials is in sheet or tubular material for electrical insulation, where the area exposed to electric stress is very large. A small failure frequency on "proof test" may be commercially significant. Here, the individual values of voltage at which breakdown occurs are by definition much lower than the average breakdown voltage of a small test area. At the same time, the failures causing concern are obviously located within the actual area of the electrodes.

Two procedures are useful. The first makes no assumption about the actual statistical distribution of the observed breakdown voltage values, other than that the known data are similar to those that might be obtained in further tests.

At each voltage level of interest, the probability of failure p_1 (i.e., the fraction of specimens with a value of breakdown voltage below this level, corrected if necessary for the number of specimens) is calculated, the electrode area being a_1. To calculate the probability of failure p_2 at the same voltage, with a different area a_2, the following equation is used:

$$\log\left[\frac{(1-p_1)}{(1-p_2)}\right] = \frac{a_2}{a_1} \tag{8}$$

The second procedure makes use of a statistical distribution function, the Weibull distribution. Breakdown voltages are often statistically distributed according to the function.

$$\log\left[\frac{1}{(1-p)}\right] = \left[\frac{V}{\alpha}\right]^{\beta} \tag{9}$$

where p is the cumulative probability of a breakdown below a voltage V, β is the "shape parameter," and α is the "location" parameter, equal to the voltage at which the probability of breakdown is $1 - 1/e$ (i.e., $\log(1/(1-p) = 1)$.

Computer software is available to facilitate this analysis of experimental data. A typical graphical display of the results is shown in Fig. 9.

Using the Weibull equation, it is easily shown that

$$V_{n,0.5} = n^{-1/\beta} \cdot V_{1,0.5} \tag{10}$$

where $V_{n,0.5}$ is the median breakdown voltage for area n and $V_{1,0.5}$ is the median value for unit test area. It must be emphasised that this analysis can only be carried out if the experimental data is found to conform to the Weibull equation.

A discussion of the derivation of the failure probability from the numbers of failures (particularly for relatively small numbers) will be found in Refs. 21 and 22. This probability is not generally equal to the ratio of the number failed n_F to the total number n_T but to the "median rank" probability. A widely accepted close approximation to this is

$$p = \frac{n_F - 0.3}{n_T + 0.4} \tag{11}$$

4.3 Test Parameters

The breakdown voltage is influenced by four principal factors:

1. The electric strength of the material on a microscopic scale (a scale small enough not to be influenced by "defects" or regions of lower electric strength)

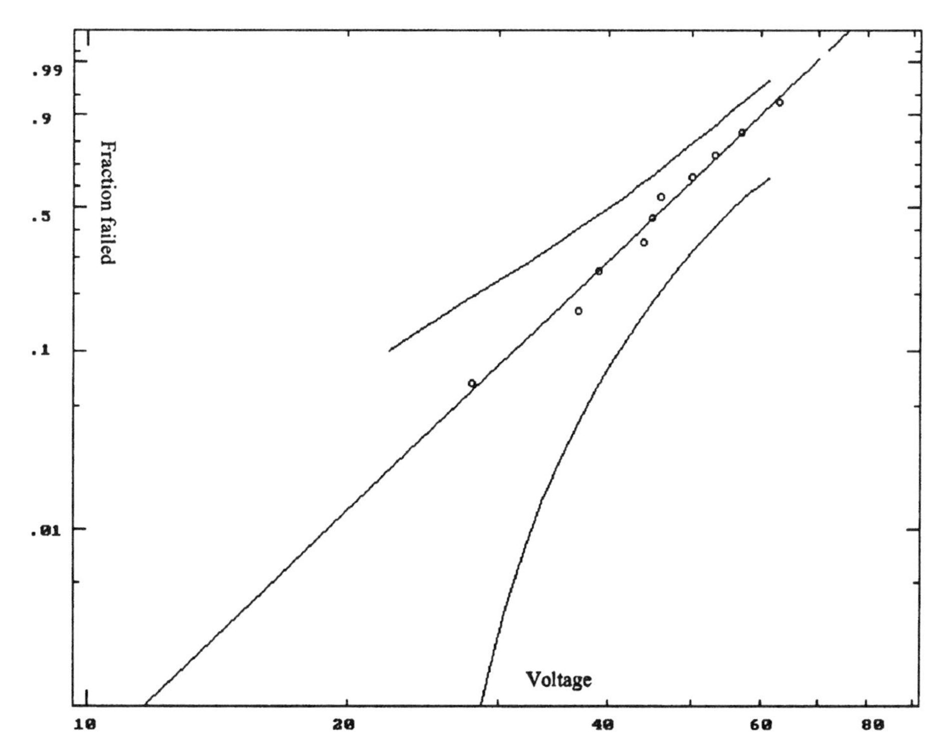

Figure 9 Weibull plot of breakdown voltage (showing upper and lower 95% confidence intervals).

2. The population density and properties of such defects
3. The electrode and specimen configuration
4. The time scale of the test

This leads inevitably to the necessity of defining carefully the test parameters, if satisfactory precision of the test result is to be obtained; these are

1. Electrode dimensions and configuration
2. Application of voltage
3. Specimen dimensions

4.4 Electrode Dimensions and Configuration

IEC 243 lists a number of electrode configurations:

Normal to surface

Symmetrical (25 mm diameter)—see Fig. 10a (IEC 243, para. 4.1.1.2, also ASTM D149[23])

Asymmetrical (25 mm on 75 mm diameter)—see Fig. 10b (IEC 243, para 4.1.1.1), probably more widely used than any others.

Symmetrical (6 mm diameter)—see Fig. 11 and 12 (IEC 243, para 4.1.2).

Specialized electrodes for tubular materials and moulded/cast materials (IEC 243, para 4.1.6 and IEC 455, part 2.1[24])

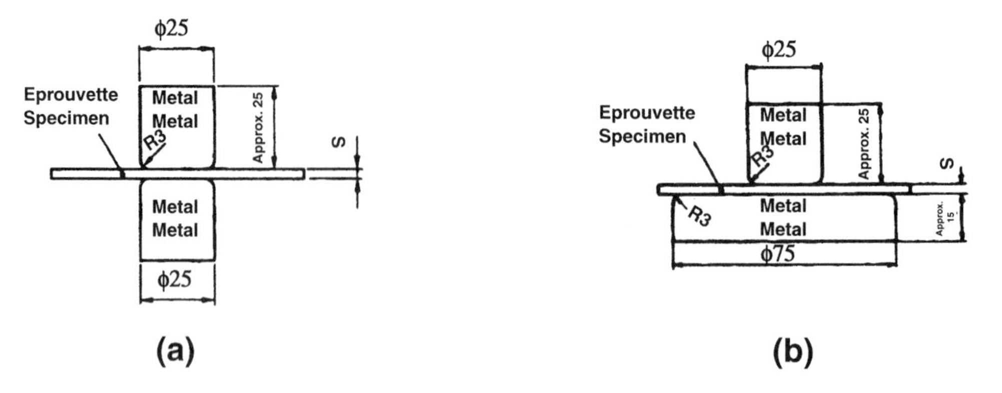

Figure 10 BDV electrodes for sheet (IEC 243).

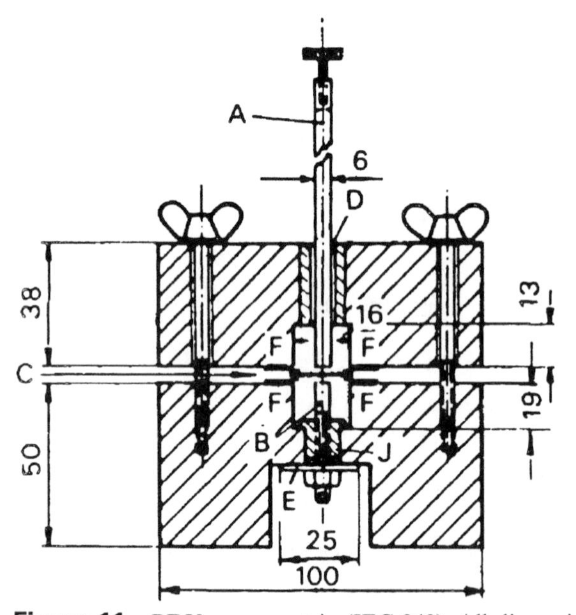

Figure 11 BDV, narrow strip (IEC 243). All dimensions in millimeters.

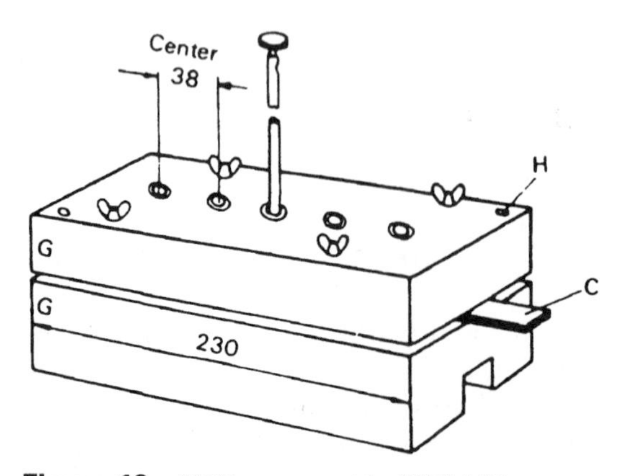

Figure 12 BDV, narrow strip (IEC 243).

Parallel to surface
  Surface and interior of material (IEC 243, para 4.2.1.1, 4.2.1.2, 4.2.2)
  Along laminae, within material (IEC 243, para 4.2.3)

For most materials, the test is conducted either in air or with the electrodes and specimen immersed in "transformer oil" conforming to IEC 296 [25].

It should be noted that the results obtained from tests using different electrode systems are not directly comparable.

4.5 Application of Voltage

For most materials, the applied voltage is of power frequency (48–62 Hz). For very thin materials, e.g., plastic films, direct voltage is often applied, since damage to electrodes caused by the breakdown is much more easily limited.

The voltage is applied in an increasing manner, either at a uniform rate or in steps of about 10% until breakdown occurs. For the uniform rate of rise, a rate is selected such that the majority of the breakdowns occur in 10 to 20 s. A stepwise mode of voltage increase is widely used in testing polymeric materials, particularly when it is necessary to test thick specimens. In this case, successive voltages are applied for periods of 20 s, increasing between steps by about 5–10% in accordance with a specified table. The starting voltage is 40% of the expected breakdown value. Other rates of increase are specified, but are not widely used.

For alternating supplies, the conventional method is to use a motor-driven autotransformer. In this case it is necessary to ensure that the steps caused by the turns of the autotransformer are less than 2% of the result. Since most variable autotransformers are wound with 100 turns, it is necessary that the result be greater than 50% of the maximum output, necessitating a further stae of adjustable transformer ratio if a wide range of materials is to be tested.

Current-limiting resistors are normally employed to minimize electrode damage. It is also recommended that the supply voltage be disconnected in not more than two cycles after breakdown.

Much of the automatic equipment in use today employs electronic generation of the output voltage, a negative feedback variable gain amplifier being used to make the amplitude follow an applied ramp waveform. This has the advantage that more rigorous limitation of energy dissipation at breakdown is possible.

For direct voltage supplies, BS 7506 [39] recommends the use of resistance/capacitance charging network to generate a suitable voltage ramp (see Fig. 13). IEC 243–2 [19] covers other requirements for direct voltage testing.

4.6 Specimen dimensions

Materials of thickness less than 3 mm are usually tested as supplied. Thicker materials are usually reduced to 3 mm by machining one side: for asymmetrical electrodes the machined face is in contact with the larger electrode.

The results from tests using different thicknesses are not directly comparable.

4.7 Indication/Recording of Breakdown Voltage

Direct visual observation of a moving pointer instrument is not recommended, since this leads to variable errors through the variable speed of response of the human eye. For the

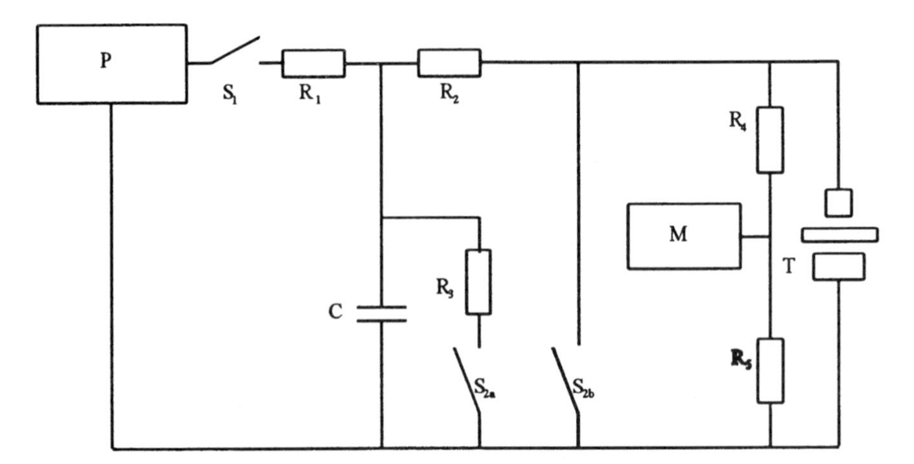

Figure 13 BDV measurement circuit (BS 7506). P, power supply (20 kV, 5 mA); R_1, charging resistor, 7 MΩ; R_2, protective resistor, 50 MΩ; R_3, discharging resistor, 0.1 MΩ; R_4, potential divider 500 MΩ; R_5, potential divider, 500 Ω; M, recorder, 10 mV; C, timing capacitor, 7 μF; T, test electrodes and specimen; and S, control switch, S_1, S_{2a}, S_{2b} ganged for safety. S_1 to open when S_{2a}, S_{2b} close, and vice versa. The switches must control a few mA at 20 kV.

recommended time to breakdown of 10–20 s, the response time constant of the indicator should be not more than 0.25 s. This is faster than most moving coil indicators: even so, it entails indicator lag errors of up to 2.5%. It is also recommended that the indicator should retain the peak value reached by the test voltage. It is essential that the indicated value be not affected by any voltage transient generated at the instant of breakdown. The writer has found a fast potentiometric pen recorder (driven by an electronic measuring device) to be most satisfactory, but wholly electronic indicators are available in commercial equipment and are also acceptable.

4.8 Outline Procedure

On account of the wide diversity of experimental features, such as electrode geometry, surrounding medium, and rate of voltage rise, it is rather difficult to give an outline procedure that will be generally applicable. The following points may be helpful:

1. For a material whose properties are not well known, preliminary measurements should be made to establish the expected mean or median breakdown voltage. This should be used to calculate the rate of voltage increase or start point of voltage steps to be used.
2. Ensure that an adequate number of specimens is prepared: if unexpectedly high dispersion of test data or "outlying" individual values should be found, extra measurements may be called for.
3. Test the electrical properties (resistivity and/or permittivity) of the immersion liquid to be used. Contamination of the liquid can often lead to erroneously high breakdown voltage values.
4. Condition the samples as specified, and carry out the test as quickly as possible after removal from the conditioning atmosphere. Allow bubbles to disperse if the test is to be in a liquid immersion medium.

5 Tracking

Tracking is the progressive formation of conducting paths produced on the surface of an electrically insulating material as the result of the combined effects of electric stress and electrolytic contamination. The subject is discussed at considerable length by K. N. Mathes [26]. A comprehensive bibliography will be found in this paper.

In the assessment of tracking resistance, the procedure involves establishing conditions likely to lead to the development of permanent surface damage to the material by the surface discharges. These conditions are generally referred to as "scintillation": there is still not universal agreement as to what constitutes satisfactory scintillation, and no instrumental method of its assessment exists at present. However, once the condition has been observed, there is little difficulty in recognizing acceptable scintillation conditions.

Two standards are widely used in the specification and assessment of resistance to tracking of polymeric materials employed in the construction of equipment such as switchgeat, connectors, and printed circuit board assemblies.

IEC 587 [27] Relatively resistant materials for use under severe conditions. Also known as the "inclined plane tracking test"
IEC 112 [28] Relatively less resistant materials.

5.1 IEC 587

The electrode/specimen assembly used by IEC 587 is shown schematically in Fig. 14. Since the test involves the use of high voltages at high current levels and noxious decomposition products may be formed, it is essential to conduct the test in an appropriate enclosure.

The test specimen is a sheet of material at least 50×120 mm and more than 6 mm thick. One side of the specimen is prepared by lightly abrading with fine silicon carbide. This test specimen is mounted in a position inclined at 45° with the prepared side facing downwards.

A contaminant solution (0.1% NH_4Cl + 0.02% nonionic wetting agent) is made to flow at the prescribed rate onto a filter pad clamped between the upper electrode and the prepared surface.

Voltage is applied through a current limiting resistor to the stainless steel electrodes either at a constant value (2.5 to 4.5 kV) until the current reaches a specified value (usually 60 mA) or until the track developed reaches a specified length. A set of specimens (usually 5) can be tested simultaneously, the voltage supply being automatically disconnected from a specimen reaching the required end point.

Test voltage can also be applied in a stepwise increasing fashion (1.0 to 6.0 kV) using the current criterion for end point.

The material is classified according to the time taken to reach the criterion.

Outline Procedure

1. Prepare an adequate number of test specimens. Mount the first batch.
2. Establish the contaminant flow for each specimen.
3. Select the correct series resistors and adjust the test voltage. Check that satisfactory scintillation is being obtained. Maintain the test condition for 6 hours.
4. Measure depth of erosion on each specimen.
5. If further testing at other voltages levels is required, it will be necessary to use fresh specimens and fresh filter paper pads.

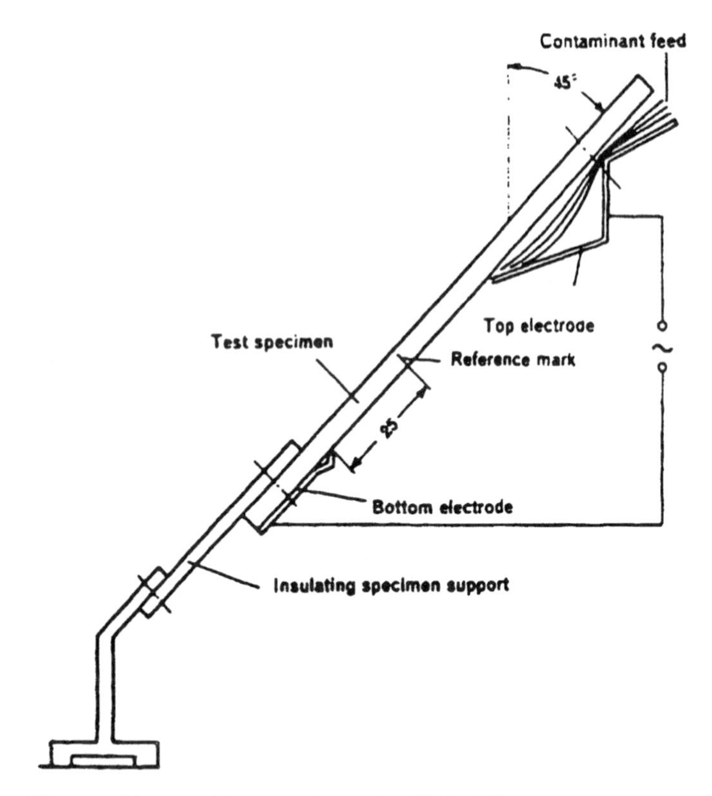

Figure 14 Tracking test electrodes (IEC 587).

Classification is based on the test voltage and the number of specimens reaching the end point in less than 6 hours.

5.2 IEC 112

The basis of this test is that a fixed voltage (100 to 600 V) is applied along the surface and a contaminant solution (0.1% NH_4Cl+ ionic wetting agent) dripped on the surface at a prescribed rate. The test assembly, which uses platinum electrodes, is shown schematically in Fig. 15. The short circuit current which may be drawn from the electrodes is limited to 1.0 ± 0.1 A by a series resistor.

Since the voltage and current levels are potentially dangerous, and noxious products may be formed, it is desirable that the test be conducted in a suitable enclosure. The atmosphere in the enclosure must be free from draughts.

The test procedures of IEC 112 produce one of two results: PTI or CTI. The Comparative Tracking Index (CTI) is the voltage at which the tracking current reaches a defined level before 50 drops of contaminant have been dripped onto the surface.

The Proof Tracking Index specifies a proof test where a number of specimens of the material are tested at a prescribed (proof) voltage, and each must withstand 50 drops without failure.

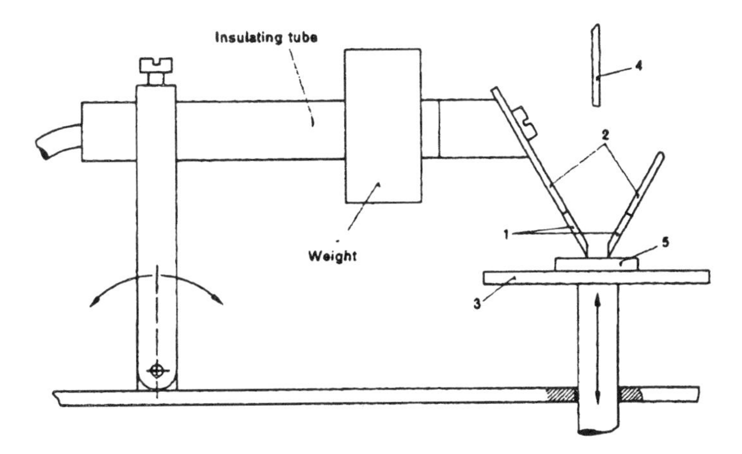

Figure 15 Tracking test electrodes (IEC 112).

Outline Procedure (CTI)

1. The test may be carried out on any flat piece of material, larger than 15 × 15 mm and of thickness > 3 mm. If the surface is specially prepared for the test, this should be reported. If more than one test is to be carried out on one test piece, the test sites must be separated by an adequate distance to avoid cross contamination between tests. Prepare an adequate number of test pieces.
2. The electrodes must be cleaned before every test.
3. Apply the electrodes to the surface with the specified force (1.0 ± 0.1 N).
4. Set the voltage at the selected level (100–600 V in 25 V steps) and drip contaminant solution on the surface at the prescribed rate until the specimen fails (current > 0.5 A for 25 s) or 50 drops have been applied. Test five specimens.
5. If no failure takes place, test a further five specimens at a higher voltage. The CTI is the highest voltage at which no failure takes place within 50 drops.

Proof Tracking Test

This test is carried out as for the CTI, steps 1–4, the voltage being set at the required proof level. No failures are permitted from five specimens.

5.3 Other Test Methods

Several other related properties appertaining to tracking, arc resistance, and surface discharge resistance have standardized test methods [29–32]. These procedures are either very specialized or else controversial and are not likely to be of general interest. For this reason, no detail is given: a discussion of arc resistance test methods will be found in the paper by Mathes [33].

6 Electrostatic Properties

On account of the widespread use of polymeric materials in modern society, and with increasing awareness of the significance of electrostatic effects in causing both safety and equipment damage hazards, the electrostatic behavior of polymers is today considered to be of great importance. In consequence, a large number of standards, international, national, and industry-specific, has been published. Unfortunately, many are of limited applicability, and it has become all too easy to obtain misleading results.

In order to resolve this situation, the International Electrotechnical Commission has set up a specialized Technical Committee (TC101) having a "pilot safety" function, this signifying that other international standards committees (including those of the ISO) should seek guidance from TC101 in matters concerned with electrostatics. A number of standards are in the course of preparation.

In BSI, a two-part standard, BS 7506 [34], is now available. The first part is a general guide to the basic principles of electrostatics; the second is a compendium of test methods for the more important electrostatic measurements. These are of two types:

Properties	Charge decay
	Resistivity
	Chargeability
	Shielding performance
	DC breakdown voltage
Effects	Electric field
	Potential
	Charge
	Charge density

The methods for measurement of the magnitude of effects are employed in the property measurement methods, and the latter will be outlined below.

6.1 Charge Decay

Two modes of charge decay exist:

1. The surface of the material may become charged (for example, by contact with a different material). The charge is evident from the associated external electric field and may give rise to hazardous discharges. The charge decays by conduction through the surface or bulk of the material.
2. A conducting object in contact with the surface of the material, but otherwise insulated, may become charged. Charge decays by conduction through the contact with the material and thence through its bulk or surface.

Measurement methods for charge decay by both modes are included in BS 7506–2 [35]. Since quite different results may be obtained, it is important to establish which of the two methods is relevant. However, the methods are formally similar, only differing in the mode of application of the charge. In both cases, the decay is monitored through measurement of the associated electric field.

6.2 Resistivity

The measurement of resistivity in work related to electrostatics is simplified considerably by the restricted range of values that is associated with significant effects. In most cases, values of resistance below 1 MΩ are associated with charge decay times of about 10^{-3} s or less (for example, the capacitance of a human body is usually about 300 pF: time constant $= CR \approx 300\ \mu s$) so that electrostatic charge remaining on an article for a significant time is unlikely. Conversely, resistance values of $10^{10}\ \Omega$ or more will be associated with time constant of 1 s or longer, and electrostatic charges are likely to remain for a time of practical significance.

This resistance range of 10^6 to $10^{10}\,\Omega$ will probably be associated with resistivity values of 10^5 to $10^9\,\Omega$m (to within a factor of 10 or so). Below this range, the material is effectively conductive, and the exact value is not significant: similarly, above the range, the material is effectively an insulator, and again the exact value is not significant. Fortunately, in the range 10^5 to $10^9\,\Omega$m no particular precautions against stray series or parallel impedances are necessary, and the required accuracy can be achieved by simple means, often a simple multimeter.

Similar considerations apply to the determination of surface resistivity.

Several standardized methods exist that may ultimately be withdrawn as the pilot/safety role of IEC TC101 (Electrostatic measurements) is implemented in IEC and ISO [36].

6.3 Chargeability

Electrostatic charges are produced on material surfaces whenever dissimilar materials are brought into contact and subsequently separated: this contact is most often achieved by rubbing together.

Charge may dissipate over and/or through a material during and after rubbing: it is important, therefore, that measurements be made promptly at the end of rubbing actions. Determination of the rate of charge decay after rubbing is in many cases necessary to establish whether measurements are affected by charge dissipation and provide information for compensation.

Chargeability may be expressed either in terms of the charge generated on individual articles or in terms of charge per unit mass, area, or volume. Four standard methods of measurement are described:

Rubbing tests
Charging during walking on flooring
Charging of sheet or web materials moving over rollers
Charging during sliding and flowing of materials or articles

Of these, the first three are more relevant to testing of sheet materials or finished products, the last to the testing of materials in powder form or finished products.

The charge produced is measured either by dropping one of the materials directly into a device known as a "Faraday pail" [37] or by observing the electric field in the vicinity of one immediately the surfaces are separated. For further details, reference should be made to BS 7506 [38].

6.4 Breakdown Voltage

The breakdown voltage of a material is often significant in electrostatics, since the material may break down itself under the influence of the potential associated with the electrostatic charge, so limiting the potential that can be achieved. Values of breakdown voltage up to about 10 kV (DC) are likely to be of interest: in this range, no particular precautions or complex equipment (such as might be needed for values up to 100 kV) will be necessary. BS 7506 suggests [39] a simple, low-current circuit in which the voltage rise is controlled by a capacitance charging system. The current is limited to a safe value and switch interlocking (for safety) is shown (see Fig. 13).

References

The following abbreviations are used for publications of

American Society for Testing Materials (Philadelphia, USA)	ASTM
British Standards Institution (London, UK)	BS
International Electrotechnical Commission (Geneva, Switzerland)	IEC
International Organisation for Standardisation (Geneva, Switzerland)	ISO

Please note that all relevant parts of BS 2782 are also published as parts of BS 903 (Physical testing of rubber). The individual identifiers of the parts are different, but each document carries both sets of titles and numbers.

1. Bartnikas and Eichorn, eds., Engineering Dielectrics, Vol. II, Part A, ASTM, 1983.
2. H. J. Wintle, Engineering Dielectrics, Vol. II, Part B, Bartnikas and Eichorn, eds., ASTM, 1984, Ch. 3.
3. IEC Publication 93, Methods of test for volume resistivity and surface resistivity of solid electrical insulating materials (1993).
4. BS 2782, Methods of testing plastics: Part 2, Electrical properties. Method 230A. Determination of volume resistivity.
5. BS 2782, Methods of testing plastics, Part 2, Electrical properties. Method 231A. Determination of surface resistivity.
6. IEC Publication 674, Plastic films for electrical purposes. part 2, Methods of test. Also published (dual numbered) as BS 7290 (1990).
7. IEC Publication 167, Methods of test for the determination of the insulation resistance of solid insulating materials (1964). ISO 2951, BS 2782, Method 232, and BS 903, Part C5 are identical with IEC Publication 167.
8. IEC Publication 250, Recommended methods for the determination of the permittivity and dielectric dissipation factor of electrical insulating materials at power, audio and radio frequencies, including metre wavelengths (1969).
9. ASTM Method D150, Methods of test for A-C loss characteristics and dielectric constant (permittivity) of solid electrical insulating materials.
10. IEC Publication 377, Recommended methods for the determination of the dielectric properties of insulating materials at frequencies above 300 MHz. Part 1. General (1973). Part 2. Resonance methods (1977). Also published, dual-numbered as BS 7737 (1995).
11. ASTM Method D1531, Method of test for dielectric constant and dissipation factor of polyethylene by liquid displacement procedure.
12. ASTM Method D2149, Dielectric constant (permittivity) and dissipation factor of solid ceramic dielectrics at frequencies to 10 MC and temperatures to 500°C.
13. ASTM Method D2520, Complex permittivity (dielectric constant) of solid electrical insulating materials at microwave frequencies and temperatures to 1650.
14. BS 2067, Determination of power factor and permittivity of insulating materials (Hartshorn and Ward method) (1953).
15. BS 7663, Methods of test for determination of permittivity and dissipation factor of electrical insulating material in sheet or tubular form (1993).
16. BS 4542, Method for determination of loss tangent and permittivity of electrical insulation in sheet form (Lynch method). This method has been incorporated into BS 7663.
17. R. Bartnikas, Engineering Dielectrics, Vol. II, Part B, Bartnikas and Eichorn, eds., ASTM 1984, Ch. 2, pp. 52–157.
18. IEC Publication 50, Chapter 212, International Electrotechnical Vocabulary: Insulating solids, liquids and gases (1990).
19. IEC Publication 243, Methods of test for electric strength of solid insulating materials. Part 1. Tests at power frequencies. Part 2. Additional requirements for tests using direct voltage. Part

3. Additional requirements for impulse tests. Also published as BS 2918, Part 1, Part 2, and Part 3 (1994).
20. R. Bartnikas, Engineering Dielectrics, Vol. II, Part B, Bartnikas and Eichorn, eds., ASTM 1984, Ch. 3, pp. 162–165.
21. J. Jacquelin, IEEE Transactions on Electrical Insulation, 28, 168 (1993).
22. J. C. Fothergill, IEEE Transactions on Electrical Insulation, 25, 489 (1990).
23. ASTM Method D149. Dielectric breakdown voltage and dielectric strength of electrical insulating materials at commercial power frequencies.
24. IEC Publication 455–2.1, Solventless polymerisable resinous compounds used for electrical insulation. Part 2, Methods of test Section 2.1, Materials other than coating powders. Also published (dual numbered) as BS 5664–2.1.
25. IEC Publication 296, Specification for unused mineral insulating oil for transformers and switchgear.
26. K. N. Mathes, Engineering Dielectrics, Vol. II, Part B, Barnikas and Eichorn, eds., ASTM 1984, Ch. 4, pp. 283.
27. IEC Publication 587, Evaluating resistance to tracking and erosion of electrical insulating materials used under severe ambient conditions (1984). Also published as BS 5604 (1986).
28. IEC Publication 112, Determining the comparative and proof tracking indices of solid insulating materials under moist conditions (1979). Also published as BS 5901 (1980).
29. IEC Publication 1302, Electrical insulating materials—Method to evaluate the resistance to tracking and erosion—Rotating wheel dip test.
30. IEC Publication 1072, Methods of test for evaluating the resistance of insulting materials against the initiation of electrical trees.
31. IEC Publication 343, Recommended test methods for determining the relative resistance of test methods to breakdown by surface discharges.
32. ASTM Method D495, Standard method of test for high voltage, low current arc resistance of solid electrical insulating materials.
33. K. N. Mathes, Engineering Dielectrics, Vol. II, Part B, Bartnikas and Eichorn, eds., ASTM 1984, Ch. 4, pp. 266–268.
34. BS 7506 Methods for measurements in electrostatics. Part 1. Guide to basic electrostatics (1995). Part 2. Test methods (1996).
35. BS 7506, Methods for measurements in electrostatics. Part 2. Test methods (1996), Clause 6.
36. For example: ISO 1853, Conducting and antistatic rubbers, vulcanised or thermoplastic—Measurement of resistivity (in revision–DIS issued 1996). BS 2050, Electrical resistance of conducting and antistatic products made from flexible polymeric material. BS 2044, Determination of resistivity of conductive and antistatic plastics and rubbers (laboratory methods). An IEC standard is in draft giving comprehensive guidance. This is at present at the Committee draft stage—15D(Secretariat)37, Methods of test for the resistance of solid materials used to control static electricity.
37. BS 7506, Methods for measurements in electrostatics. Part 2. Test methods (1996), Clause 5.3.
38. BS 7506, Methods for measurements in electrostatics. Part 2. Test methods (1996), Clause 9.
39. BS 7506, Methods for measurements in electrostatics. Part 2. Test methods (1996), Clause 11.3.2.

26
Optical Properties

Roger Brown
Rapra Technology Ltd., Shawbury, Shrewsbury, England

1 Introduction

Of all the physical properties, optical testing of polymers is probably the least different from testing other materials. Consequently, in-depth treatment of the procedures used can be found in specialized books and published literature on optical properties and their measurement. Here, the properties of most interest to polymers will be outlined, and the relatively few standards that have been formulated in the polymer industries will be considered.

Optical properties are of most relevance for plastics, although color and gloss can be important for any of the material classes for products where aesthetic considerations are of interest. In fact for most plastics products it is the aesthetic properties, such as surface texture, color, gloss, and reflectance, that are important. Light transmission characteristics and refractive index measurements are needed for only a relatively few products such as lenses and windows where the transparency or translucency of the material is a paramount consideration. This is unfortunate, as light transmission and refractive index are a good deal easier to define and and to correlate with visual experience than the other properties mentioned. Meeten [1] has edited a detailed text on the optical properties of polymers.

ISO 31/V1 (1980), Quantities and units of light and related electromagnetic radiation [2], is a table giving the international symbols and units relating to light and other radiations.

Fibers represent a special case, as such optical properties as refractive index and birefringence are important not so much for their influence on the appearance or performance of the product but as an aid to fiber identification. Fiber optical properties are considered in Chapter 19. Although similar identification techniques are applicable to transparent plastics in general, such tests are not widely used outside the forensic science laboratory. Infrared and ultraviolet absorption are more often used for polymer identification, but these techniques, and others generally considered such as chemical analysis, are outside the scope of this book. Similarly, methods of studying changes in the refractive index of solutions as an aid to following chemical reactions in solution, or methods of molecular weight determination, will not be considered. Changes of color on exposure to light or heat are dealt with in Chapter 28.

The primary optical properties considered here are

Refractive index
Birefringence and photoelastic properties
Light transmission and haze
Gloss
Color

It should be noted that most optical properties are greatly affected by the condition of the surface, so that the method of preparation of test pieces is most important and needs to be carefully controlled if comparative results are required. To obtain a very high standard of surface finish, molding against platens of borosilicate glass without a release agent has been recommended.

2 Refractive Index

2.1 Terms and Definitions

The ISO definition [2] of refractive index is given in terms of the velocity of light: "Refractive index is the ratio of the velocity of electromagnetic radiation *in vacuo* to the phase velocity of electromagnetic radiation of a specified frequency in the medium." Although this is accurate, most people find the following ray definition below more helpful. When a ray of light passes from one isotropic medium to another, the sine of the angle of incidence bears a constant ratio to the sine of the angle of refraction (both measured with respect to the normal) for all angles of incidence. This ratio depends not only on the two media concerned but also on the wavelength of the light used and on temperature. Refractive index is the term used for this ratio when light passes from a vacuum (or from air for less accurate work) into a more dense medium; it is always greater than unity.

Refractive index is dimensionless. For anisotropic materials the state of polarization of the light (and its direction, where appropriate) must be defined relative to a reference axis in the sample. It is then customary to quote two refractive indices; additionally, the maximum difference between the two indices measured in two mutually perpendicular directions is termed the birefringence of the material.

2.2 Methods of Measurement

General methods of measuring refractive index, not confined to plastics, can be found in older textbooks on optics [3]. Well-known methods are those depending on matching the unknown solid with known liquids, using the Becke line technique [4], dispersion staining [5], techniques based on the Abbe refractometer [3], and the measurement of the ratio of real depth to apparent depth [3].

The two methods recommended in ISO 489 [6] are the Abbe refractometer technique and the Becke line technique.

The test piece for the refractometer method should be about 12 × 6 × 3 mm with one flat face and one truly perpendicular surface, these two surfaces intersecting along a sharp line without a bevelled or rounded edge. For anisotropic materials specimens should be made with their polished surfaces parallel and perpendicular to the direction of orientation. The test piece is attached to the prism of the refractometer with a drop of liquid of refractive index higher than the test piece by at least 0.01 and this liquid should not soften, attack, or dissolve the plastics material. A table of suggested liquids is given in the specification, and measurements are carried out using white light. The instrument gives the refractive index for the sodium [D] line to a precision of about 0.001.

The alternative measurement in ISO 489, the Becke line method, is much more useful in that it can be used with powdered or granulated transparent material, or indeed with any small chip of material taken from a larger specimen. A microscope having a magnifying power of at least 200 diameters is required, together with a range of liquids of known refractive index. If an Abbe refractometer is available but we have only small chips of material rather than parallel-sided test pieces, then the refractometer can be used to calibrate test liquids for use with the Becke line method. The test pieces should have a thickness significantly less than the working distance of the 8 mm microscope objective, and linear dimensions sufficiently small, and so distributed, that simultaneous observation of approximately equal areas of sample and surrounding field is possible.

The material under test is mounted in a liquid of known refractive index and examined in monochromatic light with the condenser adjusted to give a narrow axial beam. When the test pieces and the liquid have different refractive indices, each particle is surrounded by a narrow luminous halo (the Becke line), which moves as focus is adjusted. If the focus is lowered, then the Becke line moves towards the medium having the lower refractive index. The test is repeated with particles mounted in other immersion liquids until a match is found or until the index of the test sample lies between two known indices in the series of liquid standards. If the Becke line phenomenon does not appear, then the refractive index of the material being examined is equal to that of the immersion liquid.

ISO 489 recommends a test temperature of 20 ± 0.5°C and gives no conditioning. A revision being processed gives a temperature of 23 ± 2°C and 50% humidity for conditioning and testing.

The draft revision has additional details of procedure for thin films, which are more difficult to measure, anisotropic materials, and translucent materials. It also has diagrams of the experimental arrangements.

The British standard, BS 2782, Method 531A [7], is identical to ISO 489.

ASTM D542 [8] includes only the refractometer method, and this is similar to the ISO procedure. The specified conditions are 23 ± 2°C and 50 ± 5% r.h. In cases where the material has a high thermal coefficient of refractive index, the temperature has to be accurately controlled at 23 ± 0.2°C.

In general, these methods have changed very little since the nineteenth century, so that the older references still have value. Wiley and Hobson's review [9] covers refractometer methods and immersion methods, whereas Brown, McCrone et al. [10, 11] deal with dispersion staining. Billmeyer [12], Ellis [13], and Schael [14] deal specifically with polymer films, where it must be borne in mind that, as in the case of fibers, birefringence may result from the processing conditions.

3 Birefringence and Photoelastic Properties

3.1 Definition

Birefringence is a measure of optical anisotropy. It is defined as the maximum algebraic difference between two refractive indices measured in two perpendicular directions. Alternatively, the velocity of the light in the two directions is different.

Optical anisotropy can be induced by mechanical stressing. Many plastics are isotropic until stressed elastically, others have permanent birefringence induced by processing, for example from drawing of fibers or stretching of film during extrusion. The phenomenon of stress-induced birefringence has been much utilized as a means of analyzing stresses and determining residual stress and degree of orientation in molded components. This requires a model of the part in a transparent material.

3.2 Methods of Measurement

Birefringence can be measured by determining the refractive index in two directions, but a more convenient way is by means of a polarizing microscope. Detailed descriptions can be found in standard textbooks [15], but the following indicates the principle. Light waves are transverse waves, vibrating at right angles to the direction in which the light is travelling, and the vibrations are symmetrical about the direction of travel. A polarizer is a device that transmits only light vibrating in one plane. Thus light emerging from the polarizer is plane polarized. A similar polarizer mounted with its plane of polarization at right angles to the first will absorb all of the light. Only if the two directions of polarization are parallel will light pass through both filters. This second piece of polarizing material is known as the analyzer. The two arrangements described above are referred to as crossed polars and open polars respectively.

Consider now a birefringent object mounted between crossed polars. The simplest case is that where a rectangular object, having its planes of polarization parallel to its sides, is mounted with one of these planes parallel to the plane of the incident light (Fig. 1). Light from the polarizer passes through the birefringent test piece and is absorbed by the analyzer, so that the field is dark. Similarly, if the test piece is rotated through 90° the field would remain dark. If the test piece is mounted at 45° to the plane of polarization, however, the object stands out brightly on a dark field. The original plane-polarized beam may be considered to be resolved into two components A and B corresponding to the principal planes of the test piece. These two components traverse the test piece at different velocities so that the two emergent beams A1 and B1 are out of phase. On reaching the analyzer both components A1 and B1 may be resolved into components parallel to and at right angles to the analyzer direction. Components parallel to the analyzer direction pass through and, since these are out of phase, interference results. Because the phase difference varies with wavelength, brilliant interference colors can be obtained. The effect on these colors of retardation wedges or compensators enables an estimate of birefringence to be made.

If a polarizing microscope is available, the experimental technique involved is relatively straightforward. Many of the difficulties arise from a lack of understanding of the physical principles behind the method, and initial study of the theoretical principles of polarizing microscopy is well worth while. For studying stress distributions a component can be placed in a light box with polarizing plates at each end.

Fibers sometimes present difficulty in that there is a different orientation at the skin and in the core of the fiber. In these cases it must be borne in mind that the Becke line

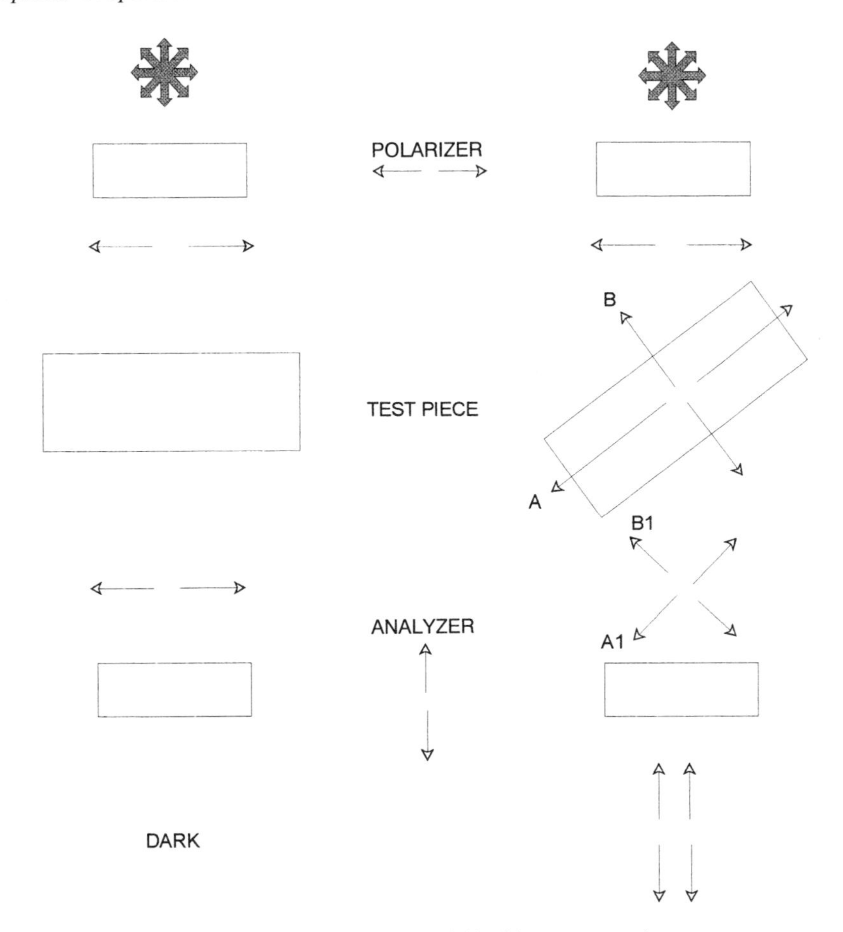

Figure 1 Interaction of polarized light and birefringent test piece.

method gives the refractive index of the skin, whereas birefringence measured by a compensator method is that of the fiber as a whole. Methods of analyzing the optical properties of fibers are discussed in detail by Barakat and Hamza [16].

4 Light Transmission and Haze

4.1 Terms and definitions

Luminous transmittance, haze, and see-through clarity are closely linked properties that together constitute the degree of transparency of a material. Total luminous transmittance is the ratio of the transmitted luminous flux to the incident luminous flux when a parallel beam of light passes through a test piece. Regular transmittance is the ratio of the undiffused transmitted flux to incident flux and is generally taken to be the flux that deviates from the incident beam by no more than 2.5%. Haze is then the transmitted flux that deviates from the incident beam by more than 2.5%. Haze is the cloudy appearance caused by forward scattering at imperfections within or at the surface of the material, and these can be filler particles, impurities, minute bubbles or voids, surface roughness, or regions of different refractive indices brought about, say, by crystallization. Abrasion, weathering, absorption of water, and even temperature change can all bring about a change in the haziness of a material. See-through clarity is less easy to define precisely but is the ability of the test piece to transmit image forming light. Clarity does not necessarily correlate with low haze, and it may be a function of light scattered over a small angle as opposed to the larger angles of scattering associated with haze [17, 18]. BS 4618, Section 5.3 [19], gives useful background material and the principles of experimental methods.

As an illustration of these properties, glazing material for windows and diffusing material for lighting fixtures must both have a high light transmittance, but the former must be free from haze and very transparent, while the latter must have maximum diffusion and minimum transparency.

The most common light sources used in illumination work and photometry are CIE (Commission International de l'Eclairage) Sources A, D_{65} and C [20, 21]. A corresponds to a tungsten lamp and the others to sunlight. C is less close to sunlight than D_{65}.

4.2 Total Luminous Transmittance

An international method for luminous transmittance was under consideration for a very long time before ISO 13468, Part 1, was published in 1996 [22]. Part 2 of this standard was still in draft form at the time of writing. Both parts use an integrating sphere to collect the light passing through the test piece, Part 1 uses a single beam of light and Part 2 a double beam.

The schematic arrangement of a single beam instrument is shown in Fig. 2. A parallel beam of light enters the sphere and reflects from the wall of the sphere to be measured by the photodetector. A measurement is made with the test piece over the compensation port to determine the incident light and then with the test piece over the entrance port to determine the light transmitted by the test piece. Detailed requirements for the instrument are given, and it is specified that the the output of the combined light source and photodetector correspond to the CIE standard colorimentric observer specified in ISO/CIE 10527 and the CIE standard illuminant D_{65} specified in ISO/CIE 10526.

In the double-beam test, Fig. 3, the light source produces two parallel beams of monochromatic light of the same wavelength and approximately the same radiant flux.

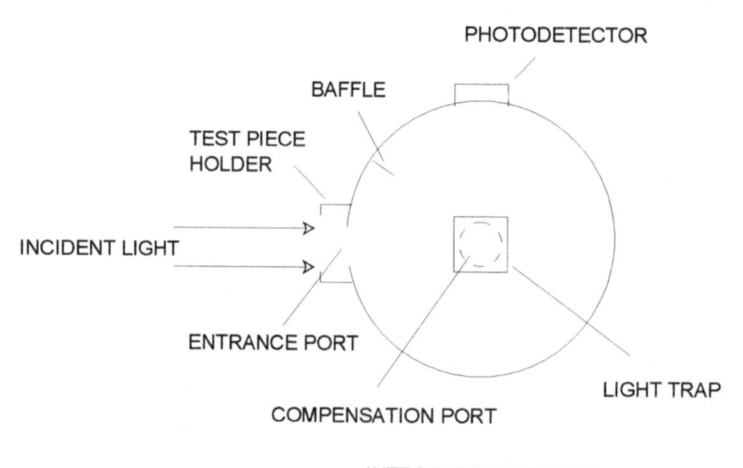

Figure 2 Principle of single-beam instrument.

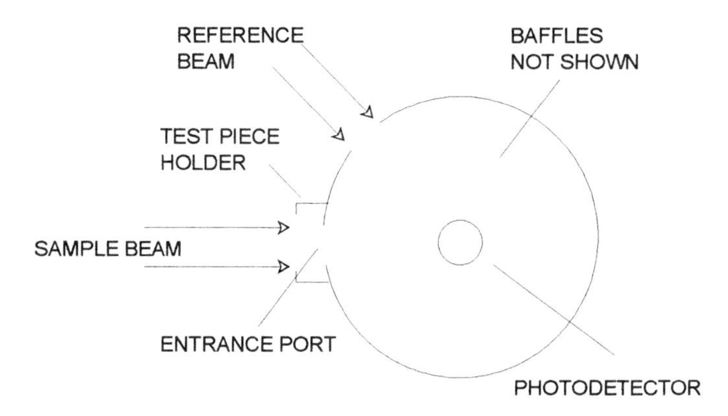

Figure 3 Principle of double-beam instrument.

Two measurements are taken, with the test piece over the entrance port, and with no test piece, to determine the transmitted and incident light. Measurements are taken over a range of wavelengths at 5 nm intervals to yield the spectral transmittance at each wavelength. The total luminous transmittance is calculated from an equation involving the sums of the incident and transmitted light for the wavelengths between 380 and 780 nm.

The British standard (also known as BS 2782, Method 532A) is identical with ISO 13468, Part 1. The ASTM method, D1003 [23], covers both luminous transmittance and haze measurements. It uses a single-beam integrating sphere, but of different geometry, and some variation of geometry is allowed, which reflects the different designs of commercially available integrating spheres. The ASTM method also differs in that it specifies the light source and photodetector corresponding to the 1931 CIE standard colorimetric observer with either CIE standard illuminant C or A.

Transmission tests are specified for some products, for example corrugated sheeting as in BS 4154 [24] and BS 4203 [25]. A light box is used capable of holding a 600 mm square test piece. The light source is an array of fluorescent tubes, and a photocell measures the transmitted light. The transmittance is given as the ratio of photocell outputs with the specimen in and the specimen out. In both standards an integrating light box procedure is given, and in BS 4203 there is also a slightly simpler direct transmission procedure.

4.3 Haze

An international method is currently at draft stage [26] that specifies an integrating sphere and light source essentially the same as in the single beam method for luminous transmittance. Four measurements are made, of the incident light, the total transmitted light, the light scattered by the instrument, and the light scattered by the test piece, from which the haze is calculated. The total luminous transmittance is of course also obtained.

ASTM D1003 also obtains the haze from four measurements but, as mentioned above, has a different sphere geometry. The British standard BS 2782, Method 521A [27], uses the pivotable integrating sphere geometry (as included in ASTM D1003). Many hazemeters use geometry complying with ASTM and BS, and not surprisingly there is some objection to the change proposed by ISO. Unpublished comparisons indicate that the two geometries differ in haze results by 3 to 5% of the value.

Diffusion of light is sometimes required to be measured on products. In BS 4154, in addition to the transmission test, a test for light diffusion is given based on a simple slit diffusion photometer.

4.4 See-Through Clarity

There is no method universally accepted for measuring see-through clarity. The method outlined in BS 4618, Section 5.3 [19], is based on the work of Webber [17]. Sets of charts (Snellen charts) are viewed through the film under test. Each chart consists of sets of parallel lines that are vertically perpendicular, each group of lines having a different separation. The angle subtended at the eye by the minimum resolvable separation between the lines is noted and compared with the angular resolution without a film. The difference between the two values gives a measure of clarity. The angular resolution with a film present depends on the distance between chart and sample and this must be specified or recorded.

ASTM 1746 [28] claims that clarity correlates with regular transmittance but then concedes that the correlation is poor for highly diffusing materials. The method given uses a light source, a system of apertures and lenses, and a photo receptor arranged so that regular transmittance is measured. It would not seem unreasonable to obtain regular transmittance from measurements with an integrating sphere.

5 Gloss

5.1 Definitions

Gloss can be defined as the degree to which a surface simulates a perfect mirror in its capacity to reflect incident light, i.e., how much of the light is reflected with reflectance angle the same as the incident angle as opposed to being scattered. It can also be defined as the amount of light scattered within a small solid angle around the mirror reflection direction. Specular gloss is the term used for the proportion of light reflected in the mirror direction. In most practical glossmeters, specular gloss is the property determined and is expressed as the ratio of the light reflected from the test piece and the light reflected from a gloss standard.

Small differences in optical geometry can cause large variations in results from one type of instrument to another, and in particular the result will depend on the angle of incidence used. A great difficulty with gloss measurements lies in correlating measurements with visual impressions; this is discussed by many people, for example Hammond [29]. Knittel [30] points out that an observer, in assessing visual gloss, turns the sample over using many angles of incidence and reflection, and his judgement is therefore based on a whole series of observations.

Because there is no absolute standard for gloss, existing standards (usually optically flat black glass plates) are assigned values that depend on the refractive index of the glass employed and that may vary from one test to another.

5.2 Methods of measurement

An international gloss method for plastics has not yet been produced. The method in BS 2782, Method 520A [31], specifies a 60° geometry for general use, a 20° geometry for high gloss materials, and an 85° geometry for low gloss materials. Although the geometry of source and receiver is defined with close tolerances on the angles, the actual instrument to be used is not described in great detail.

It is noted that the spectral characteristics of the light source are not critical unless highly chromatic low-gloss materials are tested or specular reflectance produces an obvious shift in the color of the incident light. It is also noted that if transparent materials are tested it must be possible to measure up to 200 gloss units.

The primary standard of gloss is a highly polished, plane, black glass surface with a refractive index of 1.567 for the sodium D line to which is assigned the arbitrary value of 100 gloss units for each of the three geometries. Working standards are calibrated against the primary standard and may be of of ceramic tile, vitreous enamel, opaque glass, or a similar material.

There is also a British standard for coated fabrics, BS 3424, Method 31 [32], which uses the same geometries as the plastics method.

ASTM D2457 [33] now covers plastics generally in addition to films and is essentially equivalent to the British standard except that it uses 60°, 20°, and 45° geometries. The first two geometries are as in ASTM D523, Test method for specular gloss, which also has the 85° geometry. Indeed, the British method was revised in order to align it with ASTM D523.

Despite the existence of the standard methods, differences can occur between different gloss meters. Experts suggest that the lack of specification of the characteristics of illumination and of the photodetector is a source of more discrepancies than is indicated in the standards; it is suggested that the illumination should be CIE illuminant C and the detector filtered to respond as the human eye. Other points are that for some materials the maximum specular reflection does not occur at the incidence angle, the test temperature and meter orientation should be the same as that for calibration, and there have been differences between working standards calibrated in different countries.

6 Color

Color is not an absolute property but depends very much on the surface condition of the test piece and the viewing conditions. Even to specify daylight or artificial light is not enough, for the color of the sun changes as it gets lower in the sky and the colors of all objects in our world change with it. Two colors that match under a particular set of lighting conditions may not match in another. Similarly, two moldings made of the same plastics material may appear to be of different colors if their surface finish is different. These problems are not peculiar to the polymer industry, as anyone who has taken paint samples to the daylight to compare will confirm. Discussion of color testing must of necessity cover a much wider field than polymer testing alone. Furthermore, it is a very complicated subject and there is no possibility of it being given proper consideration in this book.

The appearance of a color, being dependent on even subtle changes in how it is illuminated and in the response of the observer, means that to make objective measurements of color these factors have to be tightly controlled. The human eye is an extremely sensitive (but perhaps not consistent) observer, and any instrument to measure color satisfactorily has to match this sensitivity.

The CIE system for expressing color in objective terms was first presented in 1931 and has survived with relatively few modifications. The starting point is a standard observer, or rather a suitably filtered photocell, and a standard white light source. White light or a color can be obtained by additively mixing three colors, and the proportions of each needed relative to the proportions for white light to produce a color are referred to as tristimulus values. These tristimulus values define the color response as seen by the standard observer.

More generally, a color is uniquely defined for a defined observer and a defined light source by three coordinates that are mutually perpendicular. There are various color coordinate systems recommended by the CIE that are represented by notations such as X, Y, Z, or L, a, b, depending on the method of calculation and are detailed in CIE publication 15 [34]. The most usual standard illuminant is D_{65}, as mentioned under Light Transmission earlier.

Tristimulus colorimeters work on the principle of using three filters to read tristimulus values directly and their accuracy is largely dependent on the accuracy of adapting the photocell/filters combination to the characteristics of the defined standard observer and to maintaining this with time. The spectrophotmetric method involves measuring the relative intensity of each spectral color with a spectrophotometer and computing the tristimulus values. This is the most accurate method and also generally the most expensive. Of course, the alternative to instruments to measure color coordinates is for an observer to match colors by eye.

An international standard method for measuring color of plastics got as far as having a number but was never published. The standards for paint could be considered models, but these are currently being revised, and it is expected that a plastics standard will copy the revised paint documents. The current methods for paint are ISO 7724, Parts 1–3 [35–37]. Part 1 covers the principles of colorimetry, specifying the color coordinates and illuminant to be used. Part 2 specifies procedures for color measurement and allows spectrophotometer and tristimulus colorimeters to be used. Part 3 gives procedures for calculating small color differences using the CIELAB formula between samples from measurements of their color coordinates. Methods have also been published for textiles covering similar ground as ISO 105 J01–3 [38–40].

A procedure for visual comparison of the color of paints is given in ISO 3668 [41]. The simple basis is that an observer compares the test panel with standard color samples, but to achieve reproducible results the illuminant and viewing conditions have to be carefully controlled and the observer carefully selected, as detailed in the standard.

ASTM has a whole range of colorimetric test procedures for paint that can be found in volume 6.01 of the ASTM standards publications. There were British standards for plastics covering yellowness and color of near white materials but these were made obsolete.

Changes in color after exposure to weathering (see Chapter 28) are most accurately measured by instrumental methods as discussed above, although the weathering test methods for polymers do not at present have polymer-specific standards to reference. Estimates can also be made by using simple grey scales and visual observation to give a rating for magnitude of change

Instrumentation and the associated software for measuring and matching color has advanced rapidly in recent years as the instruments have become more accurate, smaller, portable, more user friendly, and more affordable. Comment is given by Klefinghaus [42] on recent developments in color measurements, and descriptions of instruments available can be found in *Modern Plastics International* [43, 44]. The goal is to correlate with visual assessment, but as Mulholland reports [45] the color equations are not perfect. Abrams [46] points out the objectivity of instrumental methods and some of the problems of correlation with visual assessments and gives advice on the selection and use of samples for matching.

References

1. Meeten, G. H. (1986), *Optical Properties of Polymers*, Elsevier
2. ISO 31/6 (1980), Quantities and units of light and related electromagnetic radiations.
3. Nelkon, M. (1955), *Light and Sound*, 2d ed., Heinemann.
4. James, D. I. (1974), *RAPRA Members Journal*, May, 134.
5. McCrone, W. C. (1963), ASTM STP, No. 348, p. 125.
6. ISO 489 (1983), Plastics, Determination of the refractive index of transparent plastics.
7. BS 2782, Method 531A (1992), Determination of the refractive index of transparent plastics.
8. ASTM D542 (1995), Standard test method for index of refraction of transparent organic plastics.
9. Wiley, R. H., and Hobson, P. H. (1948), *Analy. Chem.*, *20*(6), 520.
10. Brown, K. M., and McCrone, W. C. (1963), *Microscope*, *13*(11), 311.
11. Brown , K. M., McCrone, W. C., Kuhn, R., and Forlini, F. W. (1963), *Microscope*, *14*(2), 39.
12. Billmeyer, F. W. (1974), *J. Appl. Phys.*,*18*(5), 431.
13. Ellis, R.H. (1957) Review of Scientific Instruments, 28(7), 557
14. Schael, G. W. (1964), *J. Appl. Polymer Sci.*, *8*(6), 2717.
15. Hartshorne, N. H., and Stuart, A. (1970), *Crystals and the Polarizing Microscope*, 4th ed., Elsevier.
16. Barakat, N., Hamza, A. A., (1990), *Interferometry of Fibrous Materials*, Adam Hilger.
17. Webber, A. C. (1957) *J. Opt.l Soc. Am.*, *47*(9), 785.
18. Miles, J. A. C., and Thornton, A. E. (1962), *Brit. Plast.*, *35*, 1, 26.
19. BS 4618, Section 5.3, Optical properties.
20. ISO/CIE 10526 (1991), CIE standard colorimetric illuminants.
21. ISO/CIE 10527 (1991), CIE standard colorimetric observers.
22. ISO 13468-1 (1996), Determination of the total luminous transmittance of transparent materials.
23. ASTM D1003 (1995), Haze and luminous transmittance of transparent plastics.
24. BS 4154 (1985), Corrigated plastics translucent sheeting made from thermosetting resin.
25. BS 4203 (1980), Extruded rigid PVC corrugated sheeting.
26. ISO CD 14782, Determination of haze of transparent materials.
27. BS 2782, Method 521A, Determination of haze of film and sheet.
28. ASTM D1746 (1992).
29. Hammond, H. K. (1987), ASTM Standardization News, February.
30. Knittel, R. R. (1962), *Materials, Research and Standards*, *2*(3), 180
31. BS 2782, Method 520A (1992), Determination of specular gloss.
32. BS 3424, Part 28 Method 31 (1995), Determination of specular gloss.
33. ASTM D2457 (1990), Specular gloss of plastic films and solid plastics.
34. CIE Publication No. 15, Colorimetry, official CIE recommendations.
35. ISO 7724, Part 1 (1984), Paints and varnishes—Colorimetry—Principles.
36. ISO 7724, Part 2 (1984), Paints and varnishes—Colorimetry—Colour measurement.
37. ISO 7724, Part 3 (1984), Paints and varnishes—Colorimetry—Calculation of color difference.
38. ISO 105 J01 (1989), Textiles—Method for color and color difference.
39. ISO 105 J02 (1987), Textiles—Method for instrumental assessment of whiteness.
40. ISO 105 J03 (1985), Textiles—Calculation of color difference.
41. ISO3668 (1976), Paints and varnishes—Visual comparison of the color of paints.
42. Klefinghaus, K. (1995), *Masterbatch 95, Basel, Proceedings*, Paper 19.
43. Leaversuch, R. D. (1996), *Mod. Plast. Int.*, *26*, 10, p107.
44. Grande, J. A. (1995), *Mod. Plast. In*t., *25*, 12, p85.
45. Mulholland, B. M. (1995), *ANTEC 95, Boston, Proceeedings*, Vol 3.
46. Abrams, R. L. (1995), *ANTEC 95, Boston, Proceedings*, Vol 3.

27
Testing for Fire

Keith Paul
Rapra Technology Ltd., Shawbury, Shrewsbury, England

1 Fire

Designers have to ensure that a large number of parameters are correct if their product is to be both satisfactory in use and marketable. These parameters may include mechanical strength, chemical resistance, aesthetic appearance, etc. In contrast, a product may never be involved in a real fire, but if the product is exposed to a fire and its fire performance is inadequate, then the result may be loss of life and/or property. For this reason, it is essential that fire behavior should be included in all design criteria lists. It may subsequently be eliminated if it is considered to be irrelevant, but if it is ignored by oversight then the path may be open for disaster.

Fire is a highly exothermic process and may be defined as the oxidative degradation of materials, which generates heat and light. Products of partial combustion include smoke and gases. Smouldering combustion may be indicated by glowing, but most fires show flaming combustion.

Real life fires start when a material is heated and its gaseous decomposition products mix with air (oxygen) to ignite and burn. Fires may pass through several different phases with different temperatures and oxygen concentrations (fire models); see Fig. 1. Real-life fires have been subdivided into various types (ISO TR9122–1 [1]). These are shown in Table 1.

2 Use of Fire Tests

A number of possible answers exist to the question "why carry out a fire test"? including direct legal requirements for duty of care or product liability reasons, by a purchase specification, etc., but ultimately the single answer must be to improve safety by reducing fire hazards. To achieve this, it is essential that the fire test carried out and the performance

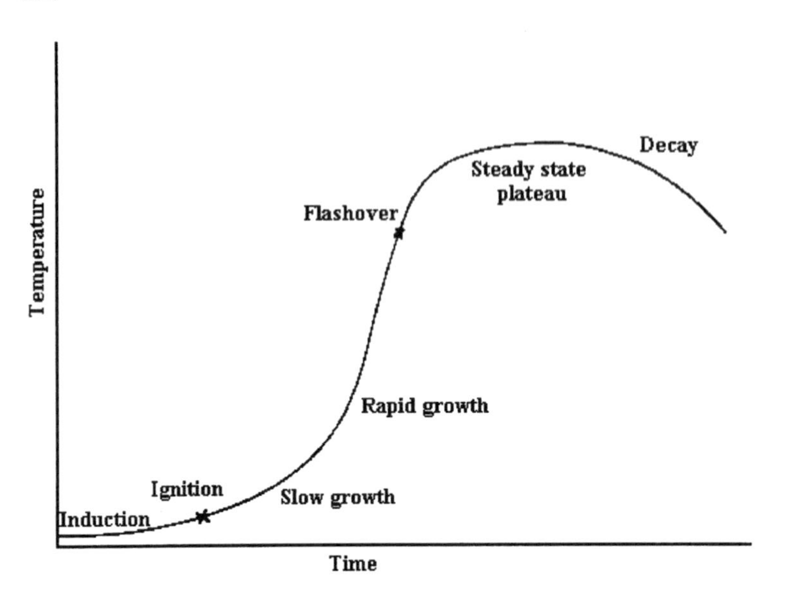

Figure 1 Stages of fire growth and development. Many fire tests apply to ignition and preflash-over period only.

requirements achieved be relevant to both the product and its end use. As a test for specification purposes, it should be reproducible, but it should be acknowledged that a highly reproducible test may be misleading if it is not relevant to the product and the end use situation.

On a macro scale, fire safety was originally improved by the use of fire breaks, i.e., gaps that contained fire within defined groups of buildings, etc. The next step was the use of fire-resistant floors, walls, and ceilings, which contained fire within a single building and then within a defined zone within a building (compartmentalization). (*Note*: The term fire resistance has a very specific meaning and is the ability of a panel to prevent the passage of flame, heat, and smoke.)

Table 1 Summary of Fire Classifications

Classification	Oxygen (%)	CO_2/CO ratio	Temperature (°C)	Irradiance (kW/m^2)
(i) Smoldering (nonflaming)	<21	N/A	400–1000	N/A
(ii) Oxidative decomposition (nonflaming)	5–21	N/A	<500	<25
(iii) Pyrolytic decomposition	<5	N/A	<1000	N/A
(iv) Developing fire (flaming)	10–15	100–200	400–600	20–40
(v) Fully developed (flaming, low ventilation)	1–5	<10	600–900	>40–70
(vi) Fully developed (flaming, well ventilated)	5–10	<50	600–1200	50–150

The next stage was to control the flammability of the surfaces of rooms using reaction-to-fire tests such as flame spread, fire propagation, heat release, etc. More recently, steps have been taken to control the flammability of the contents of rooms, etc. The latter concept has sometimes been contested, although it is logical because the fire boundaries of compartments have to be designed to withstand the fire generated by the contents. These steps obviously apply to buildings, etc., but parallel situations arise in many other applications. Transport is an obvious situation, but other and perhaps less obvious situations arise within business machines, domestic appliances, furniture, etc., all of which may comprise a chamber or series of chambers containing potentially flammable items and possibly also ignition sources.

It must be acknowledged that even if fire starts with a particular material or product, it is unlikely to restrict itself to that item and will accept any organic material as a potential fuel. Thus fire will spread via the more easily ignited items, whether they are ceilings, walls, contents of rooms, stored goods, electric or service systems, decoration, or complete items and products. Fire is a highly exothermal process and once started, it can grow very rapidly until it is restricted by lack of fuel and/or oxygen (air). For these reasons it is essential to carry out a fire hazard assessment and to consider the product itself and its end use environment as part of the design process. When selecting materials, it is important to realize that the use of flame-retarded materials, even those of greater inherent smoke and toxic gas yields, can reduce the overall fire hazard compared to non-flame-retarded products because of their lower burn rate. A mixture of flame-retarded materials and non-flame-retarded products may result in a greater hazard than either generate separately, because the fire generated by the more flammable items may be sufficient to overcome the flame retardancy of the less flammable items. A further consideration is that while materials in isolation may have excellent fire behavior, they may behave in a hazardous manner when used in combination. The role of surface coverings and fillings in upholstered furniture composites and in sandwich panels are good illustrations of this.

3 Types of Fire Test

There are a very large number of fire tests, which may be divided into various categories, see Table 2.

The entries in the table are often regarded as separate properties, and published standards may imply this belief. However most of these parameters are interrelated, and

Table 2 Sub-division of Fire Tests

Ignition Flame spread Heat release		Reaction to Fire
Smoke Toxic gasses corrosion	Fire effluent	Reaction to Fire
Fire resistance		

failure to acknowledge this can result in misleading conclusions and hence potentially unsafe situations (Fig. 2).

3.1 Ignition and Flame Spread

Ignitability, when related to a specific ignition source, determines the probability of a fire starting. This is very important, since without ignition there can be no fire. However, protection against identified ignition sources cannot guarantee fire will not occur because of the possibility of secondary ignition and it is necessary to consider the consequences of ignition and fire growth. Flame spread may be considered as a series of stepwise ignitions caused by radiant heat and pilot flames, both of which may be generated by the flame front itself.

Many current flammability tests are based on perceived hazards and often involve flames applied to products or materials. Such tests use flames of different type, size, and duration, and test specimens of different size, shape, and orientation, although considerable rationalization has taken place recently. For historical reasons, they are often industry and/or product based. The majority of these tests essentially determine whether a product will sustain combustion away from the ignition source. When assessed in terms of fire hazard analysis, the effective differences between these flames may be minor compared to the actual fire scenario.

3.2 Reaction-to-Fire Tests

Note: Ignition and simple flame spread are also reaction-to-fire characteristics but are considered separately for convenience. Reaction-to-fire tests tend to be applied to products and are considered in a separate section.

Rate of heat release is arguably the most important property of all. If the heat generated by a burning material is greater than the amount of heat needed to cause ignition, then the fire will sustain itself (heat out is greater than heat in). The rate of

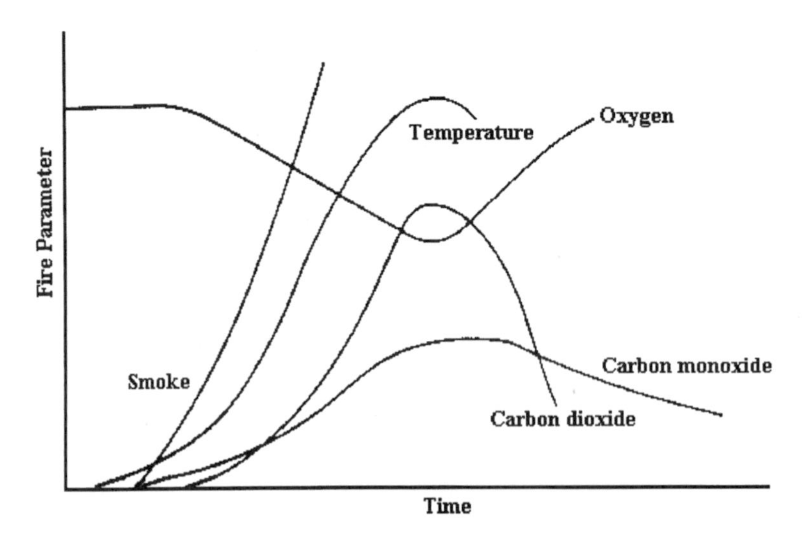

Figure 2 Fire growth showing relationship between heat, smoke, and fire gas generation rates for a single product burn.

heat release is the driving force of a fire. Tests used for wall surfaces of buildings and ships are often complex and involve exposure to radiant heat and flames to simulate the performance of walls and ceilings when exposed to real fire.

Reaction-to-fire tests, including rate of heat release, smoke, and toxic gas tests, are often empirical in nature and may measure different parameters in different ways. Direct correlations between national tests are unsatisfactory, but when they are used with national building codes they produce similar material product rankings and improved fire safety. Other major industrial groups, e.g., aircraft industry, power cable manufacturers, automobile manufacturers, warship constructors, etc., have developed their own tests, although in some cases the international nature of their use has led to a degree of test and specification harmonization. Many building type tests were developed at a time when building products were based on wood, plaster, etc., and thus they may be less suitable for testing thermoplastics, and deficiencies with standard tests inevitably lead to developments. Smoke and irritant, toxic, and corrosive gases are collectively termed "fire effluent" and depend on the materials burning, the rate of burning fire model (or temperature and oxygen availability) and the fire environment.

3.3 Fire Resistance Tests

Fire resistance is the ability to prevent the passage of flame heat, smoke, and fire gases in a defined (often developed) fire environment. Polymer products are rarely required to meet fire resistance requirements on their own, although they may be used in composite systems. Fire resistance tests of a similar nature are specified in many countries and by the ISO.

3.4 Large-Scale Tests

The use of large-scale tests designed to replicate the end use scenario may overcome many deficiencies of small-scale tests and are quite feasible for items such as upholstered furniture, but they are impractical for very large scenarios. They are also relatively expensive and limited in their scope. Examples of large-scale fire tests used for control purposes include the California TB133 test, which determines rate of heat release, smoke, and carbon monoxide tests for upholstered seating, the U.K. vandalized prison mattress test (FTS 15) and the large vertical cable tray test (IEC 332–3). While the older type of room-corridor facility has been largely replaced by the ISO room calorimeter and open calorimeter tests, concern has been expressed that these may well be too well ventilated to enable small room burning conditions to be replicated and could be misleading for smoke and toxic gas determinations.

4 Factors Affecting Future Fire Tests

It is desirable for many reasons to rationalize the current fire testing system and to reduce its diversity as well as to improve its relationship to real fires and hence its reliability. A number of current activities have this intention, although they approach it by different routes. These include voluntary industry-based developments, legislative controls, scientific developments, etc.

4.1 International Activities

The various international standards organizations (e.g., the ISO, IEC, IMO, CEN, CENELEC) develop harmonized standards, but in addition to this formal standards activity, other activities also promote the harmonization of standards. The increasing ease of transport and communication has led to a degree of harmonization within some fire testing and specification areas. In some cases, a major group has essentially taken the lead and has been followed by others. The U.S. Federal Aviation Authority (FAA) controls U.S. civil air activities, and participants in that major market are required to meet FAA requirements. Thus the aircraft regulations of many countries and airlines closely resemble the FAA system, which includes ignition and rate of heat release tests for aircraft components. Fire smoke and toxic gas tests may additionally be applied. A similar example exists in the automobile industry, where Motor Vehicle Safety Standard 302 has been adapted, albeit sometimes in a modified form, by the automobile industry, by national standards bodies, and by the ISO. Following serious fires and a major research programme, the power cable industries have also harmonized a number of fire tests within IEC.

Recent developments in the maritime industry include proposed specifications for rate of heat release and smoke requirements for fire resisting bulkheads for large high-speed surface craft. These requirements are in addition to existing fire resistance requirements (flame, heat, and smoke penetration). They are an alternative to previous noncombustible requirements that conflicted directly with weight considerations and indicate a fire engineering approach to fire safety.

4.2 Scientific Tests

The majority of current fire tests are empirical in nature and have often arisen from a desire to reproduce and simplify a perceived end use situation. Inevitably as scientists begin to develop theories, their models are a combination of theory, practice, empiricism, etc., and as such are less than perfect. However, theoretical models require better tests that measure basic properties.

The recent development of rate of heat release tests in a form that can be routinely used by test houses represents a major step forward. Probably the best known instrument is the cone calorimeter (ISO 5660–1), which determines the rate of heat release using the oxygen consumption technique. The OSU calorimeter (ASTM E906) uses a direct heat measuring system and is used in regulations by the FAA.

Applications for rate of heat release data include the specification of internal panels for passenger aircraft cabins and draft specifications for materials for inside USN submarines and for high speed surface craft (IMO). Cone calorimeter data have also been used with mathematical models to predict flame spread in small rooms, and this approach forms part of proposed applications for high-speed surface craft. A number of other applications have been suggested for bench scale rate of heat release data including the ranking of materials for applications such as bulk storage, cable coverings, upholstered furniture, railway carriage materials, etc., and doubtless these applications will grow as fire models improve and become better validated.

A test protocol has recently been published in which the toxic potency of fire gases is determined for different fire models. The test is based on a tube furnace, and the air flow, temperature, and rate of sample introduction are controlled to give steady state burning at predetermined fire models, e.g., developing fire, developed, low ventilation developed high temperature fire, etc. This is important since the toxic potency of materials can vary with different fire scenarios (Purser et al.) [2].

4.3 Fire Safety Engineering

Fire predicting models, combined with tests to determine fundamental fire properties, have led to the concept of fire safety engineering, in which the fire safety of a complete situation can be assessed and defined in engineering terms rather than in an empirical, prescriptive manner. The success of fire modelling and fire safety engineering will directly depend on proper validation of models and acceptance by fire experts and authorities. An important aspect of the fire safety engineering process is that it can include a wide range of safety measures including the use of active and passive detectors, sprinklers, etc.

5 Ignition and Flammability Tests for Plastics

It is sometimes convenient to refer to fire tests for plastics, rubbers, and textiles, and while tests exist that are specifically intended for individual materials, a large number of fire tests are intended for products. For example, the Motor Vehicle Safety Standard is intended for materials used inside motor vehicles irrespective of their type. Standards directories list a considerable number of fire tests for plastics and textiles but relatively few for rubber, although in practice a number of the plastics tests are used with rubbers.

Standard fire tests are published by international, European, and national organizations. The ISO issues general standards and has been divided into a number of subgroups, e.g., TC20, Space vehicles, TC22, Road vehicles, TC38, Textiles, TC45, Rubbers, TC61, Plastics, TC92, Building materials, products and structures, and TC188, Small craft boats.

The IEC deals with electrotechnical products while the IMO deals with tests and specifications for maritime applications.

The CEN deals with European Standards, which are then issued by national standards organizations. The equivalent of the CEN for electrical products is the CENELEC.

It is important to realize that when a European standard is issued, it replaces the equivalent national standard. This has led to confusing situations if the European standard only replaces part of a larger national standard, and a single document may then be replaced by several standards and amendments.

5.1 Guidance Documents

A number of guidance documents have been published, including ISO 3261 [3], a French/English list of defined fire terms. BS 4422 [4] and ASTM E176 [5] give a glossary of basic definitions of fire terms.

BS 6336 [6] is a most important standard and discusses the need carefully to access the limitations of small-scale tests in order to avoid possible incorrect or misleading interpretations. Large-scale tests are frequently useful in establishing the performance of a product or material in a specific environment; suitable small-scale tests can then be selected to provide quality assurance and control.

BS 476, Part 10 [7] is a guide to principles and applications of fire testing as applied to materials, composites, and products used in building construction. The BS 476 series of tests are not dealt with in detail here, as they are not primarily intended for building products, which will include plastics and textile products.

ISO TR10840 [8] is entitled Plastics—Burning behaviour—Guidance for development and use of fire tests, and is self explanatory. The report itself contains a brief introduction to fire, an outline of the various types of fire test, and a list of the various ISO and IEC tests. The design of fire tests, aspects of fire hazard assessment, and specific problems associated with the fire testing of plastics are briefly discussed.

ISO TR10353 [9], Plastics—Survey of ignition sources, used for national and international fire tests, defines the contents of this report. Approximately 140 ignition sources are tabulated in terms of burner type, dimensions, and operating conditions, and application time. The test method, purpose of use, and equivalent tests are also tabulated in groups relating to their country of origin. The report concludes (correctly in the writer's opinion) that while a large number of ignition sources are used in standard tests, many of the sources bear little relevance to ignition sources found in real-life scenarios; also that many tests provide insufficient information to ensure reproducibility between laboratories. The report recommends that attempts should be made to reduce the number of ignition sources used in standard tests, that sources should be relevant to end use hazards, and that they should be properly defined.

ISO 10093 [10], Plastics—Fire Tests—Standard ignition sources, specifies a range of laboratory ignition sources for use in fire tests on plastics. The sources vary in intensity and area of impingement and may be used to simulate the initial thermal abuse to which plastics may be exposed in certain actual fire risk scenarios.

5.2 Ignition and Flammability Tests

ISO 871 [11] details a determination of ignition temperature using a hot-air furnace. This test determines the temperature at which plastics begin to decompose to flammable gaseous products. The original (1968) version of this test has recently been replaced by a version of the Setchkin test described immediately below.

ASTM 1929 [12] details a method to determine the ignition properties of plastics. It is a vertical furnace and is used with a forced air flow introduced into the outer chamber (Fig. 3). Thermocouples positioned above and below the specimen holder are used to determine:

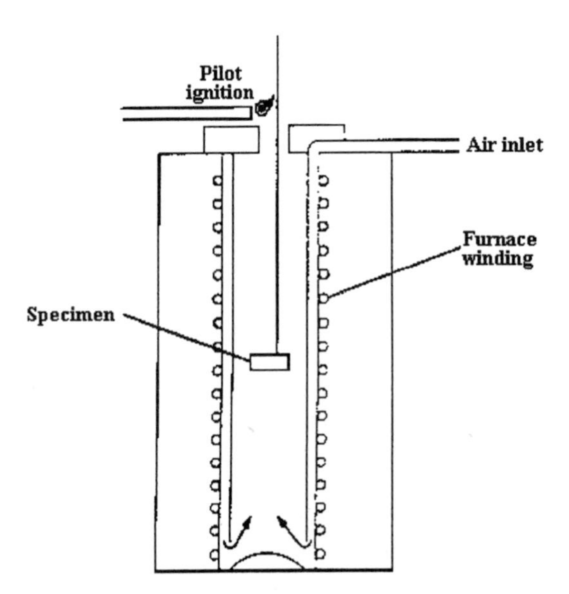

Figure 3 Schematic diagram of test apparatus for determining flash (or pilot) and self-ignition temperatures (ASTM D1929).

1. *Flash ignition temperature.* The lowest temperature of air passing around the specimen at which a sufficient amount of combustible gas is evolved to be ignited by a small external pilot flame.
2. *Self-ignition temperature.* The lowest temperature of air passing around the specimen at which, in the absence of an ignition source, the self-heating properties of the specimen lead to ignition or ignition occurs of itself, as indicated by an explosion, flame, or sustained glow.
3. *Self-ignition by temporary glow.* A special case of self-ignition temperature where, in some cases, slow decomposition and carbonization of the specimen result only in glow of short duration at various points without general ignition.

ISO 1182 [13] is essentially a simplified version of ASTM D1929 in which the furnace is held at 750°C. Specimens are classed as combustible if ignited. This is indicated by a 50°C temperature rise or by flaming for more than 10 s. The test is applied to building materials and is unlikely to be passed by plastics. Similar tests are defined by ISO and ASTM standards.

ISO 4589 [14, 15], the oxygen index test, gives good repeatability and is useful for quality control purposes. It does not purport to predict performance of a material in real fire conditions and indeed must not be used for this purpose, nor does it necessarily indicate relative merits of, say, flame-retardant additives in a polymer under conditions other than those specified in the test method; see Fig. 4.

Essentially, the test determines the minimum oxygen concentration necessary to just support flaming combustion of the material under certain conditions. The test piece is clamped vertically at its base and supported in a glass chimney of specified dimensions. Different sized specimens are defined for self supporting, cellular and flexible film or sheet materials. The upper edge of the test specimen is ignited and it then burns in a candle-like manner. A mixture of oxygen and nitrogen is metered through the base of the chimney.

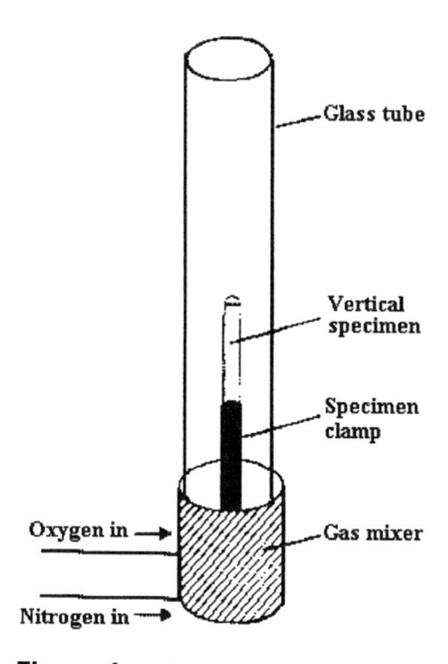

Figure 4 Schematic diagram of oxygen index test (ISO 4589).

The oxygen concentration in the gas mixture that enables the specimen to burn for a specified distance and time is determined, and this critical oxygen concentration is reported. ASTM D2863 [16] is essentially similar. ISO 4589, Part 1 [14] is a guidance document, and Part 2 [15] is the ambient temperature test method.

ISO 4589 Part 3 [17] determines the temperature index, which is similar to the oxygen index but is the temperature at which the test material will burn within defined limits in air.

ISO 12992, 1995 [18], determines the vertical flame spread of plastics sheets and films of less than 3 mm thick when exposed to a propane flame 40 mm long. The flame spread rate is defined by time required for the flame seat to move between defined markers 200 mm apart. The test provides a measure of the burning characteristics of a material but results from specimens of different thicknesses are not comparable, and results are not relevant for specimens that distort out of the flame. The standard includes results for six materials indicating average flame spread rates, repeatability, and reproducibility.

ISO 1210, 1992 [19], Fig. 5, determines the behavior of vertical and horizontal specimens when exposed to a small flame. The vertical burning test classifies materials as FV0, FV1, or FV2. A small flame is applied to the base of the vertical specimen for a specified

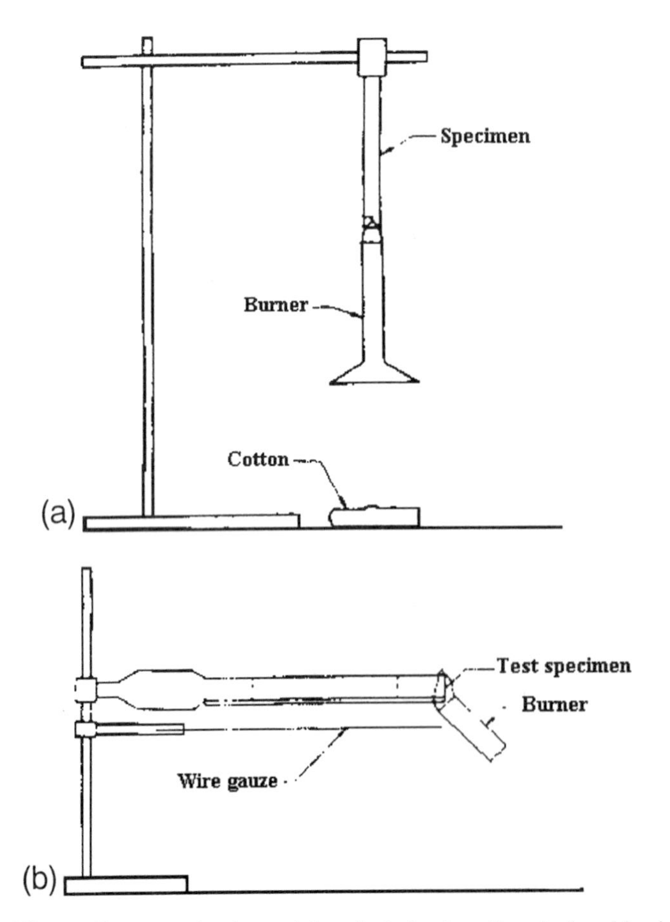

Figure 5 Tests for determining the behavior of vertical and horizontal specimens when exposed to a small flame (ISO 1210).

time (10 seconds) and the burn time noted. The flame is then reapplied for a 10 second period. Apart from the specimen not burning to the upper clamp, the classification is based on maximum limits for the individual burn time of each specimen and the total burn times for all specimens. Specification limits for FV0 classifications are less than for FV1 limits. Classification FV2 is similar to that of FV1 except that occasional burning drops are permitted to ignite cotton lint placed beneath the test specimen.

For the horizontal test, the specimen is exposed to the 25 mm flame for 30 seconds and the specimen classified FH1, FH2, FH3, or FH4, depending on the rate and extent of flame spread. All tests use test specimens of 125 × 13 × thickness mm.

These two tests are similar to those defined in BS 2782, Part 1, Method 140A [20]. This test is similar to the UL94 [21] Vertical burning (UL94, V0, V1, V2) and horizontal burn tests (UL94HB). The horizontal test is similar to ASTM D635 [22], while the vertical test is defined in ASTM D3801 [23].

ISO 9773, 1990 [24] determines the burning behavior of flexible vertical specimens in contact with a small flame ignition source, Fig. 6. The test specimen comprises a sheet 200 × 50× > 0.1 mm shaped around a mandrel 12.7 mm diameter. The upper end is clamped (sealed) and the lower end may be a closed cylinder or a flared skirt. The lower edge of the specimen is exposed to two successive three second flame applications. Appendix A of ISO 9773, 1990(E), gives criteria for the classification of materials VF0 to VF2. This is similar to BS 2782, Part 1, Method 140B [25] and to ASTM D4804 [26].

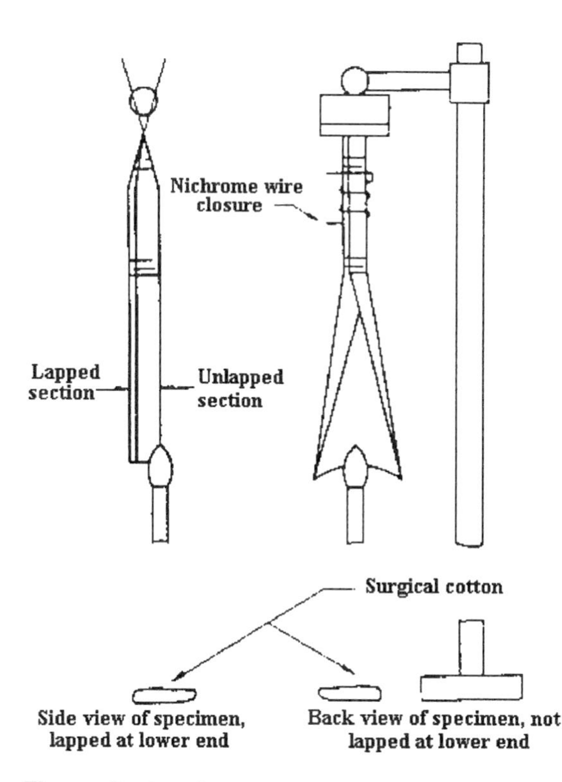

Figure 6 Test for determining the burning behavior of flexible vertical specimens in contact with a small flame source (ISO 9773).

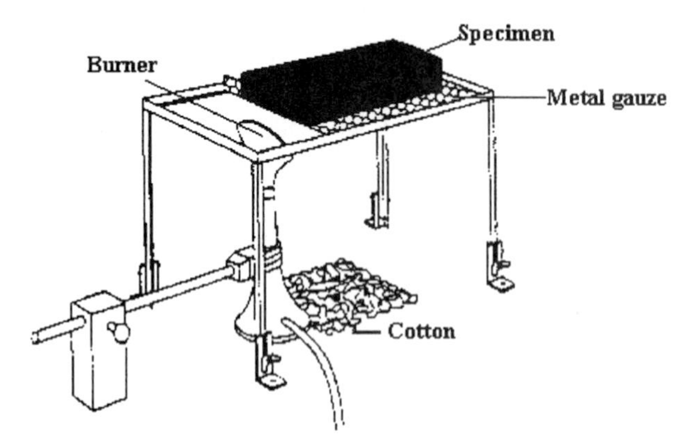

Figure 7 Test for determining the horizontal burning behavior of small cellular specimens subjected to a small flame (ISO 9772).

ISO 9772 1994 [27] determines horizontal burning characteristics of small cellular specimens subjected to a small flame. The ends of pieces of cellular materials, 150 × 50 × 13 mm, are supported horizontally on a wire grid and are exposed to a flame 38 mm high, from a wing tap burner for 60 seconds. The burn distance, time, and burn rate are determined. The Appendix gives a classification for cellular materials as FH1, FH2, or FH3; see Fig. 7.

This test is similar to BS 4735 [28] except that the latter only requires materials to be conditioned at 23°C and 50% rh, whereas ISO 9722 additionally tests materials tested for 168 hours at 70°C. ISO 9722 is essentially similar to the UL94 test for cellular materials.

ISO 10351 1992 [29] tests the combustibility of specimens using a 125 mm flame source; see Fig. 8. This method specifies a small-scale screening procedure for comparing the relative burning characteristics of small specimens of plastics and their resistance to burn through when exposed to a medium energy (500 W) flame. It is applicable to both solid and cellular materials. It is intended for quality assurance and as an aid to the preselection of materials and uses an ignition source that is approximately 10 times more severe than that of ISO 1210, 1992. Methane gas is specified, although propane or butane may also be used. Test specimens comprise vertical strips 125 × 13 × 3 mm and the 125 mm flame is applied in five, 5 second stages at 5 second intervals for the burn test. For the penetration test, the flame is applied beneath a horizontal plaque 150 × 150 × 43 mm in the same way as for the burn tests. This test is similar to the UL94 5V test, and to BS 2782, Part 1, Method 140C [30].

It is a feature of these tests with vertical specimens that a piece of surgical cotton is placed below the specimen to detect the formation of burning drips and that specimens are conditioned for 48 hours at 23°C, 50% rh, and after for 168 burns at 70°C.

ISO 181 (1981) [31] A test piece, not less than 80 mm long, 10 mm wide, and 4 mm thick, is weighed and clamped horizontally so that its edge can be brought into contact with an incandescent silicon carbide rod, 8 mm in diameter, maintained at a temperature of 950°C. The incandescent bar is held in contact with the test piece with a force of 295 mN for a period of 3 minutes. The bar is removed, any flame is dry-extinguished, and the test piece is reweighed. The difference between the original and final lengths is termed the flame spread.

IEC 707, 1981 [32]. This standard comprises two sections. The first part exposes a horizontal strip specimen 125 × 10 × 4 mm to an incandescent bar heated to 950°C, and

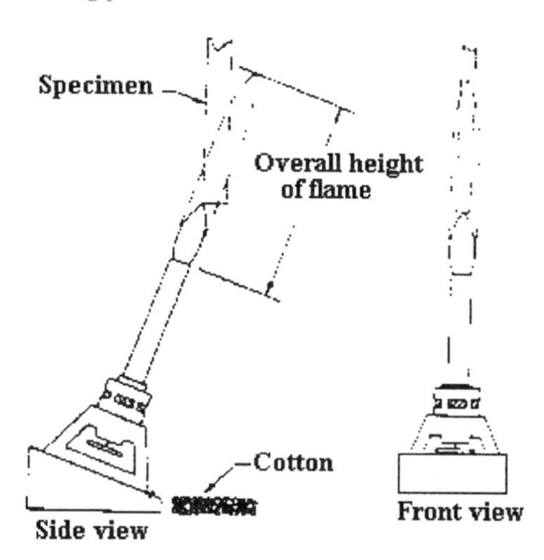

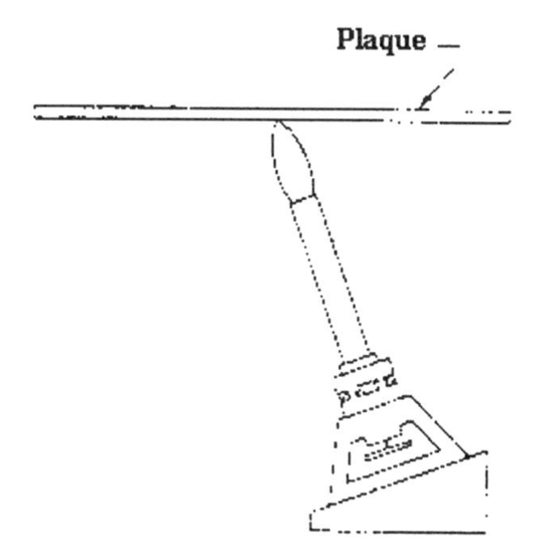

Figure 8 Schematic diagram of test for comparing the relative burning characteristics of plastics and their burn-through resistance when subjected to a medium energy flame (ISO 10351).

classifies materials as BH1, BH2, or BH3. The second part of this test exposes vertical and/or horizontal specimens to a small flame and is essentially similar to the ISO 1210 test. Specimens are classified FV0, FV1, and FV2. This test is essentially the same as BS 6334 [33].

It should be noted that the small flame vertical and horizontal burn tests (usually referred to in its UL 94 format), together with the oxygen index test, are probably the most widely used small flame tests for plastics materials. They are also used with rubber materials.

ASTM D5207 [34] gives a method of calibrating the 20 and 100 mm flames used in the above standards, while ASTM D5025 [35] gives detailed specifications for the burners used.

A series of flammability tests and guidance for their use when applied to electrical components is given in the IEC 695–2 series of tests (individually referenced below). However they include a number of different ignition sources which could be applied to many plastics materials, since these are frequently used by the electrical industry. These standards are similar to those described above but are important because each standard

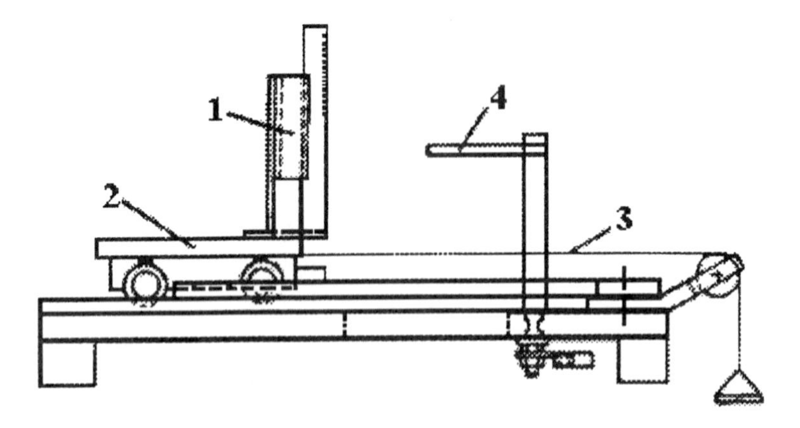

Figure 9 Schematic diagram of glow wire test (IEC 69 5–2–1). 1. Test specimen, 2. Test specimen mounting, 3. Device for applying 1 newton load. 4. Hot needle probe.

gives details of the ignition source including drawings of its construction, details of its operation and calibration, and recommended guidance for its use.

IEC 695–2–1, Part 1, Section 1 [36], is a glow wire test (Fig. 9) in which a mandrel preheated to a specified temperature between 550°C and 960°C is pushed against the test specimen with a force of 1 newton for 30 seconds. The test is intended to simulate the effect of glowing elements.

IEC 695–2 Section 20 [37], is a hot-wire coil ignitability test and is applied to solid insulating materials. It is intended to provide a comparison of the behavior of various materials according to the time taken to ignite the specimen from heat applied from an electrically heated coil wound around the specimen. The test specimen is 125 × 13 × 3 mm, and the spacing between the wire coils is 6.35 mm. The power dissipation is 0.26 W/mm, and the specimen ignition time (up to a maximum of 120 seconds) is recorded.

IEC 695–2–2 [38] describes the needle flame test in which a flame 12 mm long generated by a tube 0.5 mm diameter is applied for a predetermined time of 5, 10, 20, 30, 60, or 120 seconds to the surface of a specimen or product. The test is intended to simulate the type of hazard that may result from fault conditions within electrotechnical equipment. The specification defines the method of test and gives guidance on its use.

IEC 695–2–4/2 [39] defines a premixed propane flame of 500 W. The flame with an overall height of 115–135 mm is positioned so that the distance between the burner and the specimen is 55 mm. The flame is calibrated using a copper cylinder fitted with a thermocouple and inserted at a specified position in the flame. This specification defines the burner and flame but is not a method of test. It is intended for testing electrotechnical products.

IEC 695–2–4/1 [40] gives detailed requirements for the production of a 1 kW (nominal) premixed propane flame burner. The flame is approximately 175 mm long and is applied to specimens 100 mm from the burner top. The flame is calibrated using a copper cylinder fitted with a thermocouple and inserted at a specified position in the flame. This specification defines the burner and flame but is not a method of test. It is intended for testing electrotechnical products.

IEC 332–1 [41] tests electrical cables under fire conditions. A premixed gas flame, approximately 175 mm long is positioned so that the distance between the top of the blue cone of the flame and the vertical cable specimen is 10 mm. The flame is held in position for a period of time that is dependent on the mass of the cable. The test specification requires that the charred or affected portion of the cable shall not have reached within 50 mm of the upper cable clamp which was 475 mm from the flame impingement region. This test is essentially similar to BS 4066, Part 1 [42].

IEC 332–2 [43]. This is similar in principle to IEC 332–1, but is applied to small conductors that melt when the flame of Part 1 is applied. Propane gas is used to generate a 125 mm diffusion flame, which is applied for 20 seconds. The flame flows around the specimen. Test requirements are that the flame shall not have spread to within 50 mm of the upper cable support clamp. BS 4066, Part 2 [44], is essentially similar.

IEC 332–3 [45] and BS 4066, Part 3 [46] are not described in detail since they are large specialized tests carried out on bunched wires or cables simulating a real-life installation. The test involves a vertical duct about 3 m high and simulates actual cable ladder installations. The base of the cable is exposed to a gas burner and the vertical flame spread determined.

ISO 3795 [47] is a relatively simple method of test for materials used in road vehicles, etc. A horizontal piece of material $100 \times 350\times < 13$ mm is supported between two open-ended metal clamps and on a wire grid. A flame is applied to the end of the material and the burning time, distance, and rate determined. ISO 3795 is a method of test, but other performance specifications may require results of less than 100 mm/min, 80 mm/minute, or no sustained burning; see Fig. 10.

This test was based on the MVSS 302 specification. Many national standards organizations and motor vehicle manufacturers have a similar test, but careful study is required because many differ in important details.

BS 2782, Method 140D [48]. A strip of thin PVC of the prescribed dimensions is clipped to a semicircular rack and one end of the strip is subjected to the flame from a specified volume of ethanol. The test result is reported as the distance over which the strip is burned or charred under these conditions; see Fig. 11.

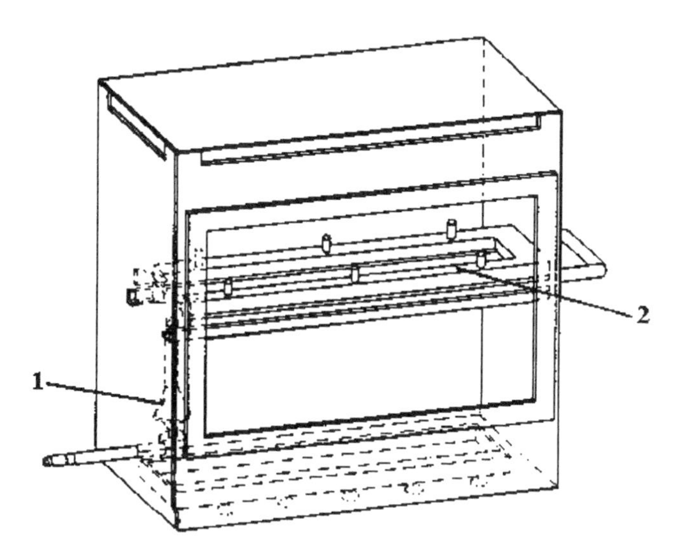

Figure 10 Schematic diagram of apparatus used to determine the flammability of materials used in the interior of vehicles showing burner (1) and specimen clamp (2) (ISO 3795).

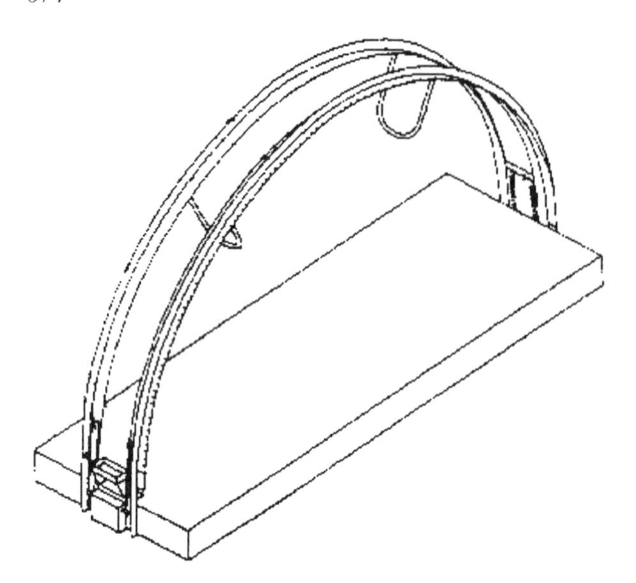

Figure 11 Schematic diagram of test apparatus to determine the flammability of thin PVC sheeting (BS 2782, Method 140D).

BS 2782, Method 140E [49]. In this test, the test piece of prescribed dimensions is supported on a simple brass and wire frame so that its major plane is at 45° to the horizontal. The flame from burning a specified amount of alcohol is allowed to impinge on the center of the sheet, and the result is expressed in terms of the amount of charring of the surface of the test piece and the duration of flaming or glowing; see Fig. 12.

ASTM D3014 [50]. In this test, a vertical specimen 254 mm long and 13 mm square is mounted in a chimney. A small flame is applied to the base of the specimen for 10 seconds, and the time for flames to extinguish, the weight loss, and the flame height recorded.

6 Ignition and Flammability Tests for Rubbers

In contrast to the considerable number of fire tests for plastics, there are relatively few fire tests specifically for rubbers as such. There are a number of tests for rubber products including cable insulation, hoses of various types, and cellular products. In some cases, e.g., cellular products, the test relates to both cellular plastics and rubbers, e.g., BS 4735 and ISO 3582, Horizontal burning characteristics when subjected to a small flame; or BS 5111, Determination of smoke generation. For convenience, these have been described in the section dealing with plastics tests. Other tests for rubber products include ISO 8030 [51], Flammability of rubber hoses for underground mining, ISO 3401 [52], Conveyor belts—Flame retardation specification and test method, and BS 5173, Part 103 [53], Fire resistance of plastics and rubber hoses and hose assemblies, and linings of hoses, etc.

It is not proposed to describe all these tests because they primarily relate to products rather than materials. Typically, the test specimen is exposed to a specified flame, and the flammability of the product is defined in terms of burning distance and/or time. In some tests, the hose may be filled with water, fuel, oil, or air at a specified pressure, and the ability of the hose to resist the flame exposure without leaking is determined. Details of these product tests may be found in the relevant lists of the ISO, EN, BS, etc., standards.

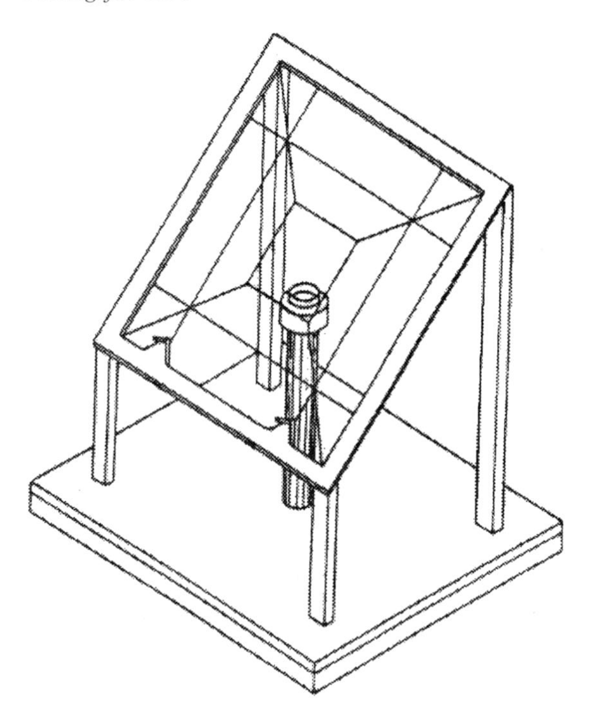

Figure 12 Schematic diagram of test to determine flammability of small inclined test piece exposed to an alcohol flame (BS 2782, Method 140E).

7 Ignition and Flammability Tests for Textiles

As for plastics, there are a number of different fire tests for textiles, but a significant proportion of these concern textile products as opposed to textiles alone.

Textiles themselves are typically tested to BS 5438 [54] and this method is used in a number of product test specifications. Examples of products which comprise textiles on their own or in conjunction with other materials include protective clothing, upholstered furniture, bedding, tents, and marquees. These may sometimes use BS 5438(43) to specify a part of the product, but they are frequently specified according to a special test. In certain cases, these tests may be directly required by legislation, e.g., children's nightwear.

With the exception of glass and mineral fiber products, and, like rubbers and plastics, textiles are largely hydrocarbon polymers and as such have a strong tendency to ignite and burn from a small flame. Textiles are essentially sheets of woven, knitted, or sometimes randomly orientated fibers and may be directly used on their own or in combination with other materials, e.g., coated fabrics, or as reinforcement, e.g., in rubber hoses. Other examples of textile products are upholstered furniture and protective clothing.

The testing of many of these products follows the patterns set for rubbers and plastics, where product tests have frequently been developed within the relevant industry and may be simplified, standardized versions of real-life hazards. It is proposed to deal only with the relatively simple textile fire tests and not to detail the many and varied product tests.

BS 5438, with its counterparts BS EN ISO 6940 [55] and 6941 [56], is probably the most widely used small flame test for textiles, and its various parts are used for both ignition and flame spread tests. It is more convenient first to consider the British Standard in its original form before the introduction of the ISO and EN tests and specifications. The latter are discussed later. The test specimen, a vertical strip of fabric mounted on a series of pins or spikes, is held in front of a metal frame. The specimen size varies with the purpose of the test and also, to a lesser extent, between the different test specifications. For the determination of minimum ignition time (Test 1), a specimen size of 200 mm high × 160 mm wide is used, for limited flame spread tests (Test 2), the specimen size is 200 mm high × 160 mm wide; and for flame spread tests it is 160 mm wide × 560 mm long (Test 3); see Fig. 14. Specimens are cut in both the warp and weft directions of the fabric and can be tested with face and surface ignition; see Figs 13 and 14.

The ignition source consists of a small premixed butane/air flame adjusted to give a 40 mm height in the vertical orientation. Butane flows through a jet and draws air in via radical inlets, and the premixed fuel is burned at a 6 mm diameter nozzle. The flame may be positioned at the lower edge or on the face of the fabric in different test methods.

To determine the ignition time, the specimen is exposed to flames of progressively increasing duration from 1 to 20 seconds. The minimum time to cause ignition is thereby determined for the edge and face of the test fabric.

To determine minimum flame spread, the flame is applied for 10 seconds or another specified time, the duration of flames and the glow are noted and also whether or not burning reaches the upper or vertical edges of the specimen. The occurrence of flaming debris is also noted.

A similar procedure is used to determine flame spread, but in this case results are defined as the times at which three horizontal threads, positioned 1 mm from the fabric face at 220, 370, and 520 mm above the burner flame, burn through.

BS 5438 was amended in 1990, and sections dealing with the determination of the minimum ignition time (Tests 1A and 1B) and the flame spread (Tests 3A and 3B) were incorporated into similar specifications such as BS EN ISO 6940 [55] and BS EN ISO 6941 [56]. The test to determine minimum ignition time is given in BS EN ISO 6940 and flame spread in BS EN ISO 6441. The face ignition procedure for limited flame spread (Test 2A) was used as BS EN 532 [57]. It is anticipated that these standards will eventually be

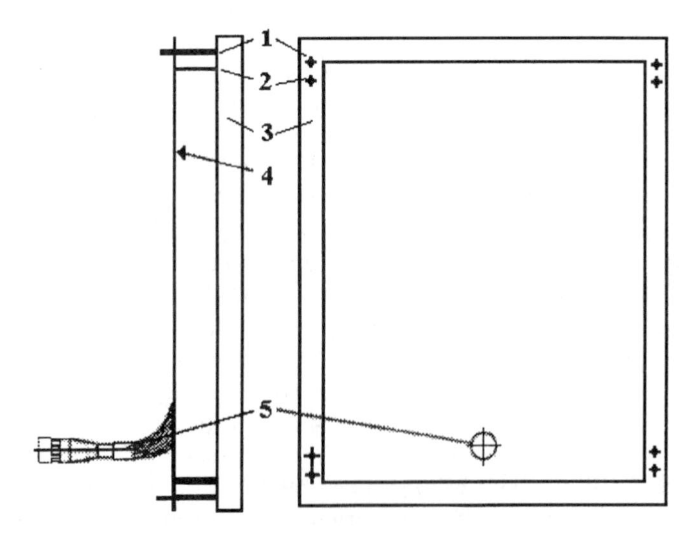

Figure 13 Schematic test to determine the ignition of vertical textile specimens. Face ignition shown (BS EN ISO 6940). 1, 2, 3, Mounting pins and frame. 4. Test specimen. 5. Burner position.

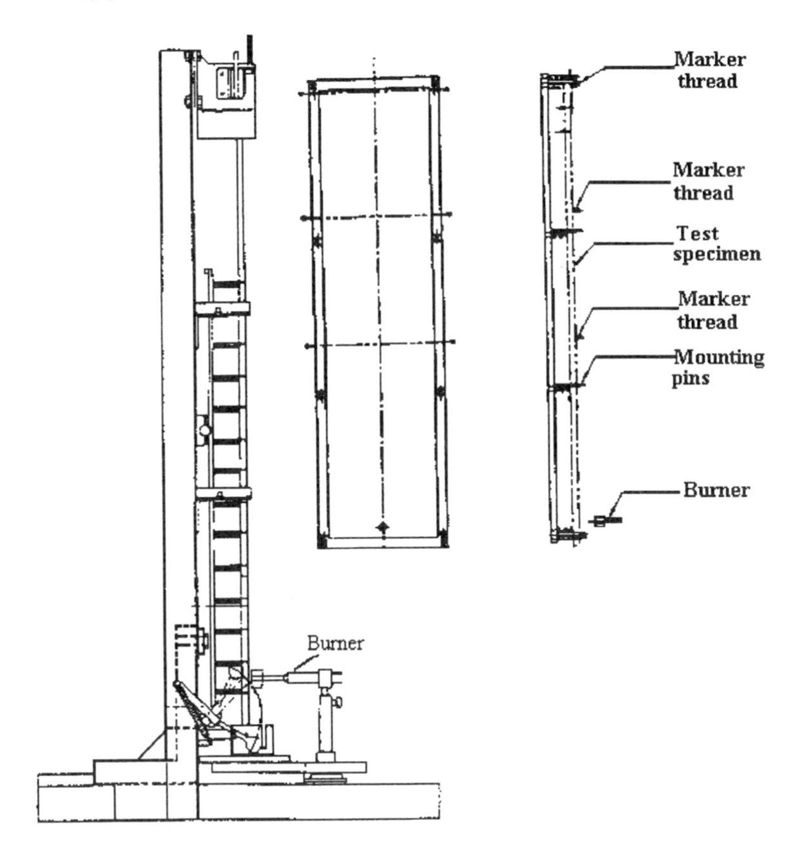

Figure 14 Schematic test to determine the flame spread of vertical textile specimens (BS EN ISO 6941).

replaced by a new set of ISO/EN standards that will have a similar scope to the original version of BS 5438.

BS PD 2777 [58] contains information about the development of BS 5438, while Appendix C of BS 5438 gives advice on the selection of test methods and acceptance criteria.

BS 5438 was first issued (albeit with a diffusion flame) in 1976 and has formed the basis of a number of specifications including BS 5722 [59] for children's sleepware and dressing gowns, BS 5867 [60], Type A for curtains, and BS 6341 Type F [61] for camping tents. Large tents, e.g., circus marquees, are tested to BS 7157 [62], in which a three-dimensional test specimen representing the internal roof corner of a tent is exposed to the No. 6 or No. 7 crib of BS 5852 [63]. The No. 6 crib (64 g of wood sticks) is used for general purpose applications while the No. 7 crib (125 g of wood sticks) is used for marquees with higher risks, e.g., cooking, heaters, etc. BS 5438 is also used to specify curtains and drapes for use in some public buildings, but the large-scale curtain test, used as an ad hoc test and considered as a potential ISO test, may be abandoned.

BS 3119 [64] applied a small flame to a vertical strip of fabric. Specification requirements were given in BS 3120 [65], but these tests have now been superseded by BS 5438 and are only mentioned here for reference purposes. A vertical strip test in which a small flame is applied to the base is used by the aircraft industry to define fabrics, curtains, etc. This test, defined in various ways, is referred to as EN 2310 [66] but has also other designations in different countries.

BS 5438 forms the basic test method referred to in BS 6249 [67] and EN 532 [57], which are used to specify protective clothing. Other tests for protective clothing include BS EN 531 [68], which relates to workers' protective clothing other than that used by fire fighters and welders.

EN 367 [69] specifies a method for comparing heat transmission through materials used in protective clothing; see Fig. 15. The horizontal specimen is exposed to an incident heat flux of 80 kW/m^2 generated by a calibrated Meker propane burner beneath the specimen. A copper calorimeter is positioned on the upper surface of the specimen, and the time in seconds is determined for the temperature of the calorimeter to rise by 24°C.

EN 366 [70] contains two test methods. The first (Method A) allows visual assessment of any changes that take place when the specimen is exposed to radiant heat, while Method B determines the protective effect of the test material. The source of radiant heat is a series of six silica carbide rods, and the test heat flux is chosen according to the intended end use of the material from the range of low intensity (5–10 kW/m^2), medium intensity (20 and 40 kW/m^2), and high intensity (80 kW/m^2). A heat transmission factor is calculated from the temperature rise of a calorimeter positioned against the reverse side of the specimen.

Upholstered furniture is a specialized application that combines a number of materials, including textiles as coverings, possible interliner fabrics, flexible, resilient foam fillings, and various waddings and nonwoven fiber paddings. Fire tests for upholstered furniture are considered here because of their increasing importance and its polymeric base. The essence of fire tests for upholstered furniture is that the complete upholstered composite, i.e., covering fabric, interliner, filling, etc., is tested as a single test specimen. This is essential because of the interaction between the different components. It is quite possible to combine fabrics with fillings and for the composite to burn to completion, whereas each individual component would resist the same ignition source. Seating composites are typically tested in an L-shaped form where the crevice simulates the more

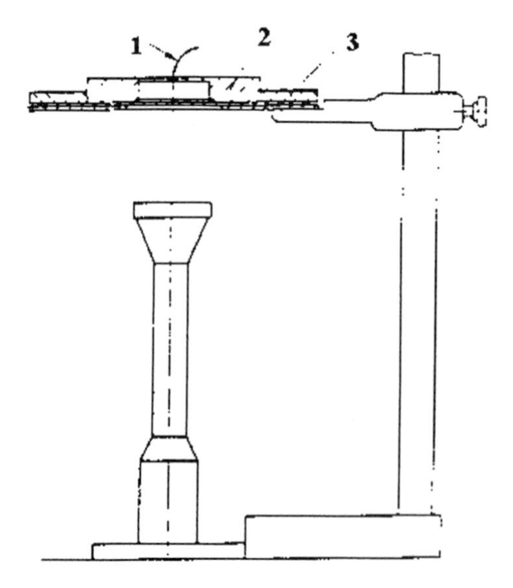

Figure 15 Schematic test to compare heat transmission through materials used for protective clothing (EN 367). 1. Thermocouple, 2. Calorimeter block, 3. Test specimen.

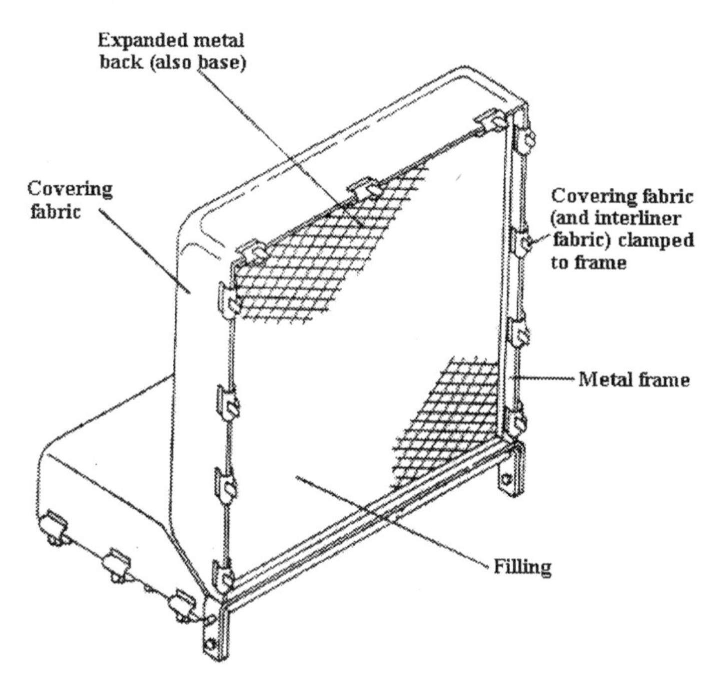

Figure 16 Test apparatus for determining the ignitability of upholstered seating composites using cigarette (EN 1021-1) and match flame equivalent (EN 1021-2) ignition sources applied to the junction between the vertical and horizontal parts of the test specimen.

critical geometry of actual furniture; see Fig. 16. Mattresses are typically tested as flat, horizontal specimens. Test specimens are relatively large, e.g., 450 × 250 mm at least 75 mm thick. The latter is important because it enables progressive smouldering reactions to occur and simulates the end use application. Typical ignition sources include a burning cigarette (ISO 8191–1 [71] and EN 1021–1 [72] for seating and EN 597–1 [73] for mattresses) and a small gas flame approximately 35 mm high (ISO 8191–2 [74] and EN 1021–2 [75] for seating and EN 597–2 [76] for mattresses). In the U.K., larger ignition sources comprise gas flames and wooden crib for upholstered seating (BS 5852) [63] and BS 6807 [77] for mattresses. Both of these standards include tests for actual products, while BS 6807 also includes a test for the mattress with known bedclothes.

BS 7157 [78] is used in the U.K. Bedding is tested using the same series of ignition sources as defined in EN 1021–1, EN 1021–2, and BS 5852, but the bedding assembly is tested over a simulated noncombustible mattress.

Other ignition sources have been developed for upholstered furniture tests and include a 100 g paper bag used to test continental railway seats [79] and a large kerosene oil burner for passenger aircraft seats (FAR Pt 25, Appendix F, Part 2) [80].

BS 4970 [81] uses a hot (900°C) metal nut to simulate the effect of a hot coal falling onto textile floor coverings, while ISO 6925 [82] uses a small (150 mg, 6 mm dia) methenanine as the ignition source.

ISO DIS 9239 [83] determines the surface spread of flame of textile floor coverings using a radiant heat test and is similar to the test defined in ASTM E648 [84].

Woven textiles of the type used to reinforce conveyor belting for use in mines and for other critical applications are tested using a spirit burner. The test BS 3289 [82] can be applied to the woven textile reinforcement as well as to the conveyor belt itself. This standard includes the spirit burner test described in ISO 340 [52], a large drum friction test, and a large gas burner test in which a multijet burner (450 mm square) is placed below the belt in a simulated tunnel.

8 Reaction-to-Fire Tests for Ignition and Spread of Flame

Reaction-to-fire tests determine the specified parameter and are applied to materials or products. As such they are not specific to plastics, rubbers, or textiles and have therefore been considered separately.

ISO 5657 (1982) [86], although primarily intended for building materials and products, is useful in that horizontal specimens (165 mm square with a 140 mm exposed central disc) are exposed to a radiant heat flux of from 10 to 50 kW/m^2 with or without a pilot ignition flame. The time to ignition is recorded, and results for specimens that do not ignite within 15 min are reported as "no ignition." The importance of this test is that specimens are exposed to high heat flux levels and thermoplastic materials do not melt away from the ignition source. The test apparatus, without the pilot ignition flame, is used as the fire model in the ISO dual-chamber smoke test, discussed later.

Many of the tests used for testing plastics were originally designed for other materials. For example, BS 476, Part 7 (1987) [87] determines surface spread of flame for building materials. A test piece 95 mm wide and 300 mm long and not more than 25 mm thick is mounted so that the width is vertical and the long axis is virtually at right angles to a gas-fired radiant panel, so that the radiant intensity reaching the test piece is within the limits specified by the test method. As soon as the test piece is in position, a gas flame is applied to the hotter end for 1 minute. Measurements of the time of the flame spread along the test piece are made over a period of 10 minutes or until the flame has reached the far end of the test piece, whichever is the shorter. The flame spread so measured is used to classify the test material into one of four categories, depending on flame spread after 1½ minutes and final flame spread.

A similar but smaller radiant panel test for building materials is being developed within the ISO as ISO 5658 [88] and within the IMO as Resolution A516 [89] for merchant shipping applications These have not been described here because they are primarily used for building products.

9 Rate of Heat Release

There are few accepted rate of heat release performance specifications, although this parameter is arguably the most important for determining fire behavior of products and materials. Although many rate of heat release tests have been suggested, few have been developed to full scientific or regulatory standards. However, a number of heat release apparatus exist and are reviewed by M. Jannssens and R. Minne [90] in ISO TC92/SC1/104 and also by Brabauskas [91].

ISO 5660–1 [92] defines the cone calorimeter, which is probably the most widely used of the rate of heat release test methods; see Fig. 17. A horizontal 100 mm square specimen is exposed to a radiant heat flux of 10 to 100 kW/m^2 with a spark ignition system. The effluent is drawn through a duct fitted with sensors for determining temperature, gas flow rate, and oxygen concentration. These data enable the rate of heat release to be determined using the oxygen consumption method. For most plastics, rubbers, and natural materials, the amount of heat produced per unit mass of oxygen consumed is approximately the

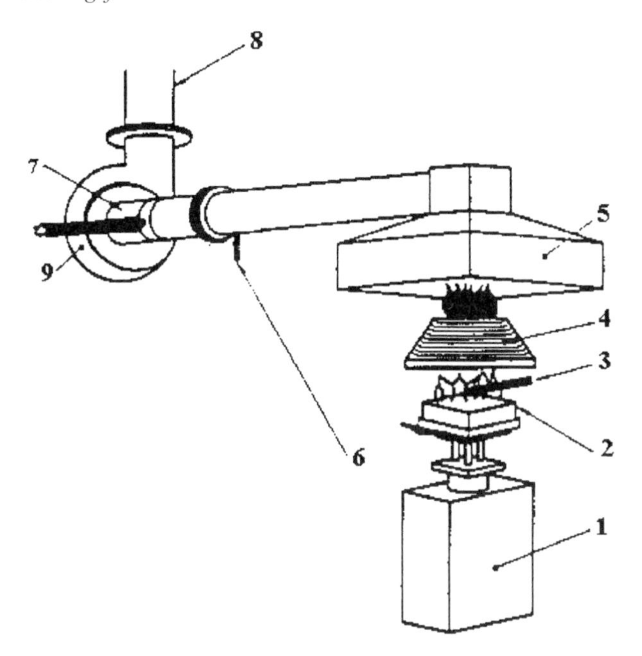

Figure 17 Schematic diagram of cone calorimeter for determining rate of heat release (ISO 5660–1) and smoke (draft ISO 5660–2). 1. Load cell. 2. Test specimen. 3. Spark igniter. 4. Conical radiant heater. 5. Exhaust hood. 6. Gas sampling probe. 7. Laser/photocell smoke determining system. 8. Pressure (velocity) measurement. 9. Exhaust fan.

same. The cone calorimeter determines rate of heat release and the effective heat of combustion. The basic test is being drafted as a smoke test (ISO 5660–2). Many cone calorimeters are also instrumented to determine carbon monoxide and carbon dioxide yields, although this is not a standard test procedure.

Simplified cone calorimeters have been suggested for quality assurance testing in which mass/loss and/or temperature measurements replace the more complex oxygen consumption calorimetry. Published specifications for the cone calorimeter include

ASTM E1354 [93], which determines heat release and smoke for building products
ASTM E1474 [94] for upholstered composites
The CBUF [95] program developed a procedure for upholstery composites

Specifications for cone calorimeter heat release data include the MIL–STD–2031 [96] for products used in U.S. Navy submarines and the IMO draft [97] for materials used in high-speed surface craft. More recently, cone calorimetry heat release data have been used to assess the fire loading of railway vehicles [98].

ASTM E906 [99] defines the OSU calorimeter. The apparatus consists of an insulated box containing a vertical specimen, a parallel electric radiant heater, and a pilot ignition device. Air at a controlled rate flows through the box, and the inlet and outlet temperatures are recorded. ASTM E906 also records the temperature of the box wall to compensate for the nonadiabatic characteristics of the apparatus. The box is calibrated using a preset gas flame. The vertical specimen size is 150 mm square, and the incident heat flux has a maximum value of 100 kW/m^2. Tests with a horizontal specimen, 110×150 mm, which involve the use of aluminum foil to reflect heat onto the specimen, are apparently not completely satisfactory. Modified equipment has been used with the oxygen consumption technique. The basic OSU calorimeter is used by the FAA [100] to specify sidewalls, ceilings, partitions, storage bins, etc., in the interiors of passenger aircraft. Tests are carried out at 35 kW/m^2, and the maximum rate and total heat release are limited for a specified period. It should be noted that the FAA-specified calorimeter and procedure [100] differ somewhat from that of ASTM E906.

Although the Factory Mutual Apparatus [101] has never been proposed as a standard test method, it is briefly mentioned here since it has been used for basic research work by several organizations and contains a number of novel features. A 100 mm square specimen is mounted in a vertical chimney and heated (up to 65 kW/m^2) by banks of quartz lamps mounted outside the chimney. Preheated air or oxygen/nitrogen mixtures are passed up the column at a constant rate. The apparatus has been mainly used to determine the fundamental fire properties of materials, although products such as cables have also been tested.

ISO 1716 [102] specifies a test method for determining the calorific value of non-metal containing building materials and uses a high-pressure bomb in a water jacket immersed in a calorimeter vessel containing water. The specimen is contained in a crucible in the bomb, which is pressurized with oxygen. The specimen is ignited electrically and the peak temperature of the water jacket determined. The gross or net calorific potential is then calculated using specified formulae. Composite specimens are tested after first being separated into separate compartments. In contrast to the previous heat release test methods, this method determines the heat of total combustion.

10 Smoke Tests

Smoke tests may use photometric or gravimetric systems to determine optical density or mass density, respectively, and may be applied to static or dynamic smoke tests. The majority of smoke tests apply an optical measuring system to static smoke tests in which the smoke is generated, stored, and measured within an enclosed cabinet. A number of more recent smoke tests meter smoke dynamically as it passes through a duct above the fire model.

BS 6401 [103] and ASTM E662 [104] are based on the NBS smoke test and are essentially the same test although some differences occur; see Fig. 18. A 75 mm square specimen of up to 25 mm thick is combusted in a vertical orientation at 25 kW/m^2 incident heat flux. Tests are carried out with and without a series of pilot flames along the lower edge of the specimen. The smoke is contained in a cabinet of 0.51 m^3 and measured using a vertical photomultiplier/lamp system. The results are typically expressed as specific optical density, which relates the optical density of the smoke to the volume of the cabinet, the length of the smoke-measuring path, and the area of the specimen exposed in the test. Other methods have been used in which specific optical density is related to the mass of the specimen combusted and/or to time.

NES 711 [105] uses the NBS test cabinet but involves a number of fundamental modifications, i.e., the cabinet includes a smoke stirring fan, a different pilot burner is used, the first 4 minutes of the test are under nonflaming conditions and the remainder under flaming conditions, and the results are expressed as an index calculated from the time for the smoke transmittance to reduce to 70, 40, 10%, and the minimum value. In practice, this means that the smoke index is biased against materials that generate large amounts of smoke in the early part of the test.

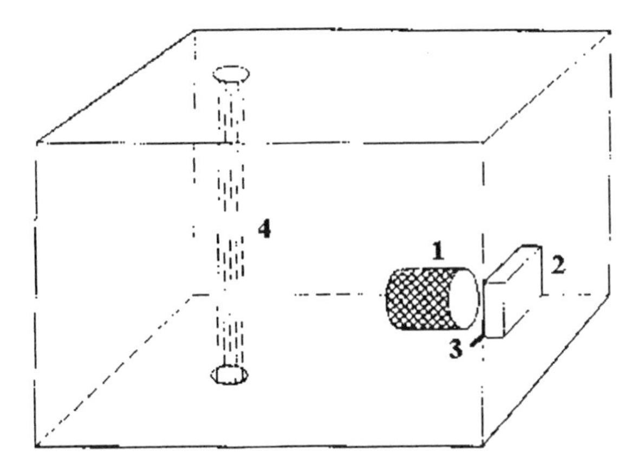

Figure 18 Schematic diagram of NBS smoke test ASTM E662). 1. Radiant heater. 2. Vertical test specimen. 3. Pilot ignition flames. 4. Photocell/light path.

prEN 2824 [106] and prEN 2825 [107] are based on ASTM E662. prEN 2824 defines the test apparatus, prEN 2825 the smoke measuring test, and prEN 2826 the measurement of toxic gases. The latter is described in the next section.

ISO 5659–2 [108] determines the optical density of smoke generated and measured in a single test chamber. The test cabinet and smoke measuring equipment is that of the earlier NBS test defined as ASTM E662 and BS 6401. This ISO test uses a horizontal fire model in which the standard test specimen (75 mm square) is supported on a load cell. The specimen is exposed to radiant heat from a conical radiator positioned above the specimen holder; see Fig. 19. Although a range of heat flux values can be used, the standard specifies that tests should be carried out at 25 kW/m^2 with and without a pilot ignition flame, at 50 kW/m^2 without a pilot flame. Calculation of results is by the method defined in the earlier tests.

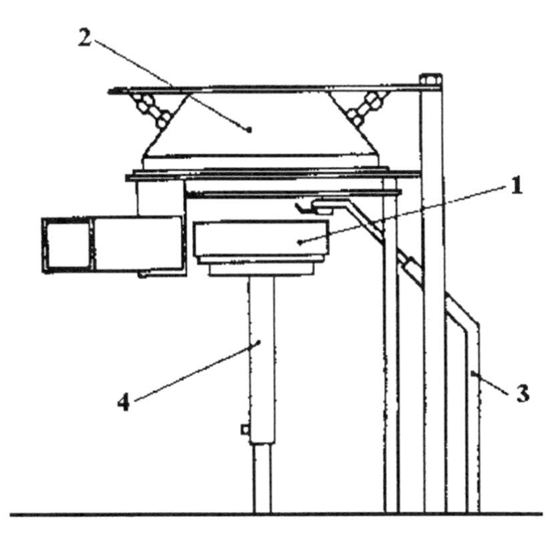

Figure 19 Schematic diagram for cone radiator specimen holder system used inside NBS smoke test cabinet for smoke test (ISO 5659–2). 1. Test specimen holder. 2. Cone radiant heater. 3. Igniter. 4. To load cell.

Part 3 of this standard refers to a draft method in which smoke is measured dynamically. The equipment is similar to that of Part 2, except that a small hood is mounted over the radiator and test specimen. The hood is connected to a duct that passes through the roof of the test cabinet. A smoke measuring system and extraction fan are attached to this duct and the test operated with the chamber cabinet open.

Part 1 of this standard contains guidance for the measurement of smoke in tests.

ISO TR 5924 [109] The dual chamber test was developed for use with building products. It uses the ISO ignitability apparatus but without the pilot flame ignitor as the fire model. Smoke may be generated at 10, 20, 30, 40, and 50 kW/m^2 and passes from the combustion chamber and flows into a larger chamber where it is stirred with a fan and determined using a horizontal photocell/light system with a path length of 360 mm. The horizontal specimen is 165 mm square but is masked to expose a 140 mm diameter zone to the radiator.

ASTM D4100 (1982) [110] describes the Arapahoe Smoke Box, which exposes a small strip of material to a gas flame for a specified material. The smoke is drawn through a fine glass filter and is expressed as the percentage mass of soot to the mass of the material combusted.

Draft DIN 53436/DIN 53437 [111] and NFT 51–073 [112] are dynamic furnace tests in which smoke is generated in an airstream through a tubular furnace and measured using a photocell/light system. The DIN furnace is more frequently encountered as a toxicity test and is considered in the next section.

BS 5111 [113]is a relatively simple test in which a foamed specimen cube 25 mm on a side is burned in a gas flame in a cabinet measuring 308 × 308 × 924 mm high. The maximum smoke density (as percent obscuration) is determined using a horizontal photocell/lamp system in the upper part of the cabinet. The maximum obscuration and the time at which it occurs are reported.

ASTM D2843 [114] uses the same cabinet and test method as in BS 5111 but is applied to a wider range of materials. A 25 mm square specimen is burned, and the smoke obscuration/time curve is reported together with the maximum smoke obscuration amid the area under the smoke obscuration/time curve.

IEC 1034–1 and 1034–2 [115] define a smoke test in which cables arranged on a horizontal tray are burned in a 3 meter cube; see Fig. 20. The fire source is a tray containing 1 liter of alcohol spirit, and the smoke is stirred within the test room by a fan. The density of the smoke is determined by a horizontal photocell/lamp light path across the upper part of the test chamber. Although this test is defined as a cable test, it is also used to determine the smoke produced by a number of materials and products. In its original form it was developed to test materials used in underground railway vehicles and stations and was then standardized for these applications as BS 6853, Appendix B [116]. Part of this specification refers to indicative smoke tests using flat plaques of plastics and rubber materials and is used for a comparison and initial selection programme.

Modified Flammability Tests. A number of smoke tests exist in which the smoke produced by a flammability or heat release test is determined using photocell/light systems positioned in the smoke streams.

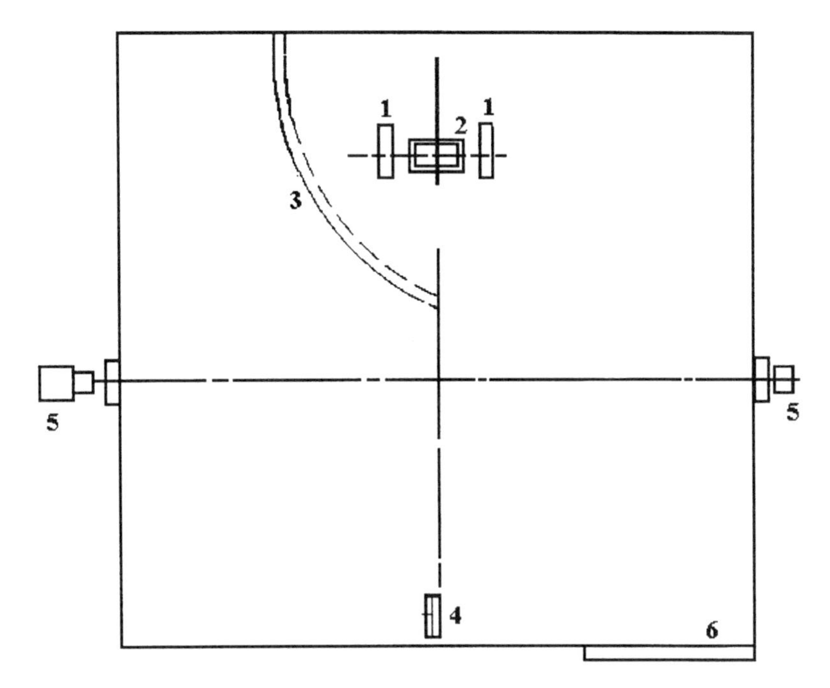

Figure 20 Plan view of three meter cube smoke test (IEC 1034–1 and 1034–2). 1. Specimen supports. 2. Alcohol tray. 3. Draught screen. 4. Mixing fan. 5. Photo cell/lamp system. 6. Entry door.

11 Corrosivity of Fire Gases

A number of tests have been developed to determine the corrosivity of fire gases. These have resulted from damage caused to building materials and components, to machinery, and metallic products, and to electrical and electronic installations.

Corrosivity tests frequently involve heating materials in a horizontal tubular furnace and determining the corrosivity by dissolving the gases in water and determining pH, or acids equivalent (e.g., HCl), but more recent tests involve assessing the effect of corrosive gases on electrical circuits.

IEC 754–1 [117) heats 0.15 to 1 gram of material at a rate of 20°C/minute to 800°C and then for 20 minutes at 800°C in a horizontal-like furnace. Air at a preset rate is passed through the tube and the effluent passed through bubblers containing demineralized water. The contents of the bubblers are titrated with silver nitrate solution and the halogen acid per gram of sample determined. The accuracy of this method is such that it should not be used to determine halogen acid concentrations of less than 5 mg/g. For this purpose and for assessing materials described as "zero halogen", IEC 754–2 [118] is recommended.

IEC 754–2 [118] is generally similar to Part 1 of this standard, but the material is inserted into a furnace that has already been heated to more than 935°C. The specimen mass is 1 gram, and the test is continued for 30 minutes. The fire effluent is absorbed in bubblers, and the pH value and conductivity of the absorber liquid are determined after the test. The results are expressed as a weighted pH and conductivity values; see Fig. 21.

BS 6425, Parts 1 and 2 [119], are similar to IEC 754, Parts 1 and 2, respectively. Other corrosivity tests have been published that are essentially similar to the above test methods,

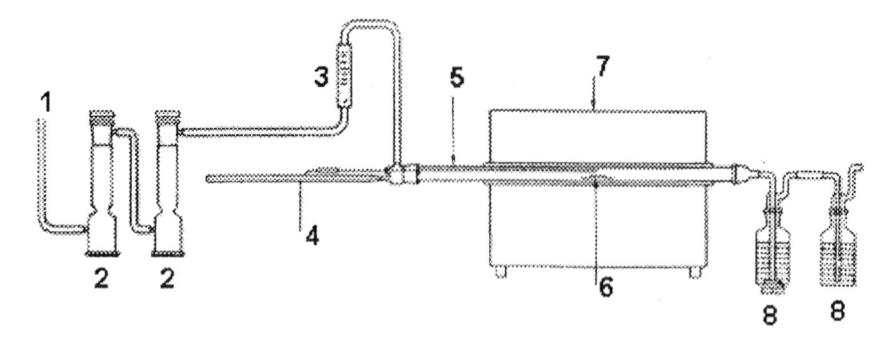

Figure 21 Schematic diagram of apparatus to assess the corrosivity of fire gases by measuring pH and conductivity (IEC 754–2). 1. Air inlet. 2. Air cleaning filters. 3. Flow meter. 4. Thermocouple. 5. Quartz glass tube. 6. Test specimen in boat. 7. Furnace. 8. Gas absorbing solutions.

but they differ in experimental details, for example test temperature, duration, airflow rate, and determination of acidity, e.g., methyl orange/red indicator.

A different approach is used in UTE C20–453 [120], where specimens are burned in an enclosed cabinet and the change in resistance of a copper wire or other electrical circuits contained in the cabinet measured. Tests of this type appear to be more realistic, since they directly assess the corrosive effects of fire gases on actual components, but the acid gas type of test described above is simpler.

ISO 11907–2, 1995 [121] is a static method of test used to determine the corrosivity of fire gases. The specimens (600 mg) in the form of granules or chips are heated with an electric resistance wire (800°C) in a crucible. The effluent is contained within a closed cabinet of 20 liters volume maintained at 50°C and 65% rh. The corrosion detector consists of a resistance etched copper plated laminate, and the corrosivity is assessed as the variation in electrical resistance due to attack on the copper circuit. Two operating procedures are given, one for the condensing mode, in which the corrosion sensor is water cooled to 40°C, and one for the non-condensing mode, without water cooling to the sensor; see Fig. 22.

ISO 11907–3 [122] consists of a tube furnace and quartz tube similar to that of DIN 53436 [111]. The test specimen is placed in a boat and the hot furnace (600°C) traversed along it. Air flows through the furnace tube and over the corrosivity detector, which consists of an etched copper printed circuit board. The printed circuit board is cooled to 10°C to cause the corrosive vapors to condense. The corrosivity is expressed as the percentage change in resistance relative to the original value.

ASTM D5485 [123] is based on the cone calorimeter. Fire effluent generated by the cone fire model is drawn through an 11.2 liter chamber containing a corrosion detecting target. The target consists of two electric elements, one of which is protected while the other is exposed to the fire gases. The two circuits form part of a bridge circuit. The change in resistance is used as a measure of the corrosivity of the fire gases.

12 Toxicity Tests

ISO TC92/SC3 have set up a series of working groups to consider fire models (WG1), chemical analysis methods (WG2), biological assay (WG3), guidance documents (WG4) and bioanalytical methods (WG5). These working groups have not developed specific fire

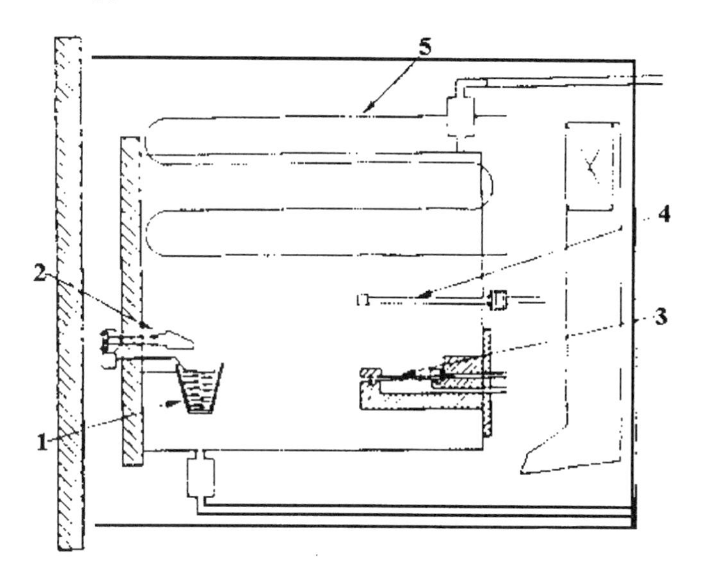

Figure 22 Schematic diagram of apparatus to determine the corrosivity of fire gases in a static exposure chamber (ISO 1197–2). 1. Ignition source. 2. Test specimen dispenser. 3. Corrosion detector. 4. Thermocouple. 5. Heater.

toxicity tests but have produced state-of-the-art reports and recommendations for each of the specialized topics. These include six parts of ISO TR 9122.

Part 1 [1]	General Information and guidance.
Part 2 [124]	Guidance for bio assays to determine the acute inhalation toxicity of the effluents (basic principles, criteria, and methodology)
Part 3 [125]	Methods for the analysis of gases and vapors in fire effluents
Part 4 [126]	Fire model (furnaces, and combustion apparatus used in small scale testing
Part 5 [127]	Prediction of toxic effects of fire effluents
Part 6 [128]	Guidance for regulators and specifiers on the assessment of toxic hazards in fires in buildings and transport

ISO DIS 13344 [129] concerns the determination of the lethal toxic potency of fire effluents. This standard gives methods of calculating toxic potencies from analytically determined fire gas concentrations.

BS PD 65031 (1982) [130] reports on the toxic hazard in fire and makes recommendations concerning problems related to the development of tests for the measurement of toxicity and combustion of products used in buildings, contents of buildings, and transport. BS PD 65031 contains background information useful to those concerned with fire-fighting, the handling and use of flammable materials, and carrying out the fire tests and interpreting their results.

ASTM E800 [131] gives details of the methods of test and analytical procedures to be used in determining the main gas produced in fires.

ISO 9122–4 [124] describes the assessment of hazards caused by toxic fire gases and the development of suitable test methods. The main tests considered include the NBS Protocol [132] and DIN 53436 [111] tube furnace.

In the NBS Protocol [132] the test material (about 8 grams) is decomposed in a small furnace heated to 25°C above and then to 25°C below the material self-ignition tempera-

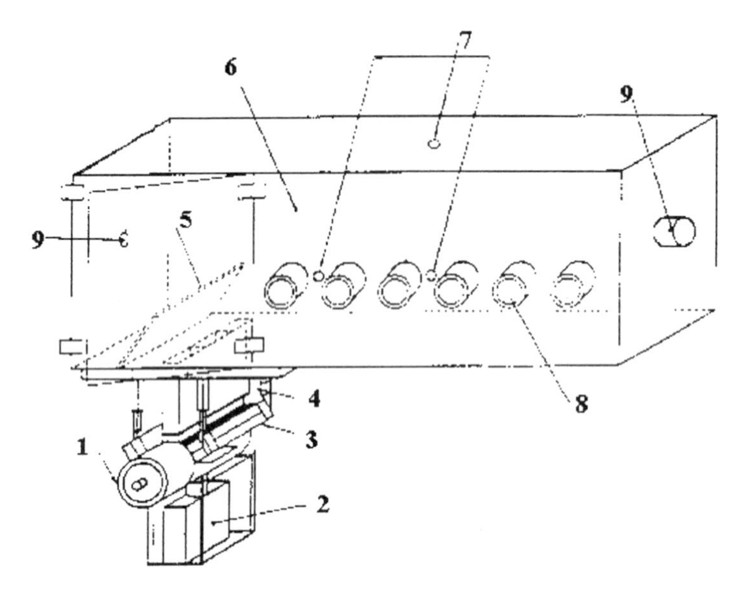

Figure 23 Schematic diagram of U.S. radiant smoke toxicity apparatus. 1. Specimen combustion cell. 2. Load cell. 3. Radiant heater. 4. Chimney. 5. Shutter. 6. Gas mixing chamber. 7. Gas sampling ports. 8. Animal exposure ports (if used). 9. Gas circulating port.

ture. The furnace is mounted in the base of a cabinet of approximately 0.2 m volume. Fire gases circulate around the cabinet and are determined analytically or by animal protocols.

The U.S. Radiant Smoke Toxicity Apparatus [133] consists of the test cabinet of the NBS Protocol [128], but the test specimen, 76 × 125 × 51 mm thick, is decomposed by a radiant heater fitted beneath the smoke box; see Fig. 23. The maximum heat flux is 50 kW/m^2. This equipment was developed following criticism of the fire model of the NBS Protocol.

DIN 53436 [111] describes a 40 mm diameter tubular furnace in which the specimen is heated under predetermined conditions of temperature, gas flow rate, and oxygen concentration. The furnace moves over the specimen in the opposite direction to the airflow. Materials may be typically tested at 300, 400, 500, and 600°C in nitrogen and air. The fire gases are typically diluted with air prior to entering an animal exposure chamber and/or are determined analytically. Tests at temperatures between 200°C and 1000°C have also been reported; see Fig. 24.

The French test NFX 70-100 [134] is similar in principle to the IEC 754–1 corrosivity test and is increasingly applied to materials used for railway transport. Air is passed through the preheated furnace at 2 liters/minute, and the fire effluent generated by heating 1 g of material at 200, 400, 600, and 800°C is passed through bubblers; individual gases are determined by various analytical methods.

NES 713 [135] is a relatively simple test in which, typically, 2 g of material are burned by a gas flame in a polypropylene-lined cabinet of about 1–1½ m. Simple fire gases (carbon monoxide, carbon dioxide, hydrogen cyanide, hydrogen fluoride, hydrogen chloride, hydrogen bromide, nitrogen oxides, sulphur dioxide, formaldehyde, acrylonitrile, phenol, and phosgene) are determined using chemical reagent tubes. Test results are expressed as an index that is the sum of the ratios of each gas concentration (expressed as volume ppm per 1 m^{3}per 100 g sample) to the concentration required to cause death in

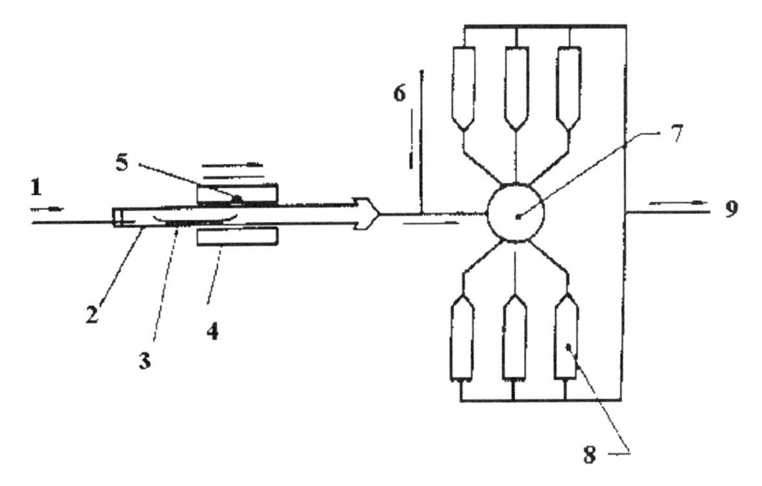

Figure 24 Schematic diagram of smoke toxicity apparatus (DIN 53436). 1. Combustion air. 2. Quartz glass tube. 3. Test specimen in boat. 4. Travelling furnace. 5. Thermocouple. 6. Dilution air (for animal exposures). 7. Gas mixing chamber (if used). 8. Animal exposure tubes (if used). 9. To gas analyzers. *Note*: Items 6, 7, and 8 not always used.

30 minutes. In the U.K. this test is applied to materials for use in warships and fighting vehicles and occasionally for other applications.

prEN 2826 [136] is based on the NBS Smoke test (ASTM E662) and determines simple fire gases directly using chemical reagent tubes or indirectly by specific ion electrodes. Limits are specified for smoke and gas concentrations for the combustion products, which are generated by exposing a 75 mm square sample to a radiant heat flux of 25 kW/m with and without pilot flame ignition. Although intended for use with aerospace materials, this test has been applied to other products. prEN 2824 and prEN 2825 give details of the test apparatus and smoke measuring methods respectively.

ISO 5659–2 is a development of this test and has been adopted as a draft IMO smoke and toxicity test [137]. The fire model consists of a conical radiator mounted above the specimen. Tests are carried out at 25 kW/m^2 with and without a pilot ignition device and at 50 kW/m^2 without pilot ignition. The concentrations of specified gases in the test cabinet are determined at specified times and compared to limiting values.

13 Large-Scale Fire Tests

Detailed discussion of large-scale fire tests is outside the scope of this book, and those requiring any such fire tests to be carried out are advised to get in touch with specialist organizations. A useful appraisal of large-scale tests and especially on the need carefully to design the tests to meet the end use environment, and also to establish the limitations of such tests, is given in BS 6336 [6].

Large standard tests are used for building products, and examples include the Fire Resistance Test (ISO 834) [138] for walls, ceilings, etc. This requires furnaces that can heat panels to temperatures of up to 1200°C and specimens of 2½ × 2½ m for wall panels and 2½ × 4 m for floors. Other examples of large-scale tests include ISO 9705 [139], which evaluates wall and ceiling systems in a small room ventilated by an open door (Fig. 25) and the furniture calorimeter NT Fire 032 [140]. The latter two tests use the oxygen consump-

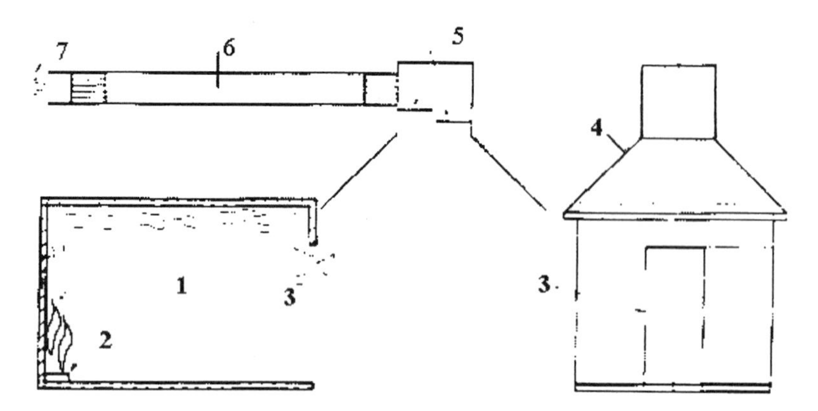

Figure 25 Schematic diagram of room calorimeter test (ISO 7905). 1. Room. 2. Gas burner ignition source. 3. Room exit door. 4. Hood. 5. Fire gas mixing baffles. 6. Gas sampling, temperatures, and velocity probes, smoke measuring sensors. 7. Exhaust fan. *Note*: Furniture calorimeter is similar but without room. Test specimen is burned directly under hood (NT Fire 032).

tion technique, and the ducting and instrumentation resemble that of a large-scale cone calorimeter. Additional sensors are used to determine smoke and toxic gases.

ASTM E603 [141] contains a Guide for Room Fire Experiments and gives advice on the methodology, instrumentation, test protocols, etc. for use in the full-scale testing of room linings and contents.

ASTM E1537 [142] and ASTM E1590 [143] test upholstered chairs and beds in an instrumented test room. Rate of heat release, smoke and carbon monoxide, and mass loss limits are specified for these full-scale product tests carried out for furniture for high-risk public area usage. These tests actually assess the fire performance of a composite construction, including design effects in a simulated end use situation, and are based on the California State tests.

ASTM 3894 [139] is the standard that tests the flammability of the specimen arranged as two walls only, each 1220 mm × 610 mm high, or two walls plus a ceiling 1220 mm × 610 mm high, or two walls plus a ceiling 1220 mm square, which are exposed to a premixed propane flame positioned in the lower corner. A specified temperature/time calibration is specified, and results are assessed in terms of flame spread. Thermocouples are positioned at specified locations on the specimen surface, and the times to maximum temperatures and the maximum temperatures are recorded.

13 References

1. ISO TR 9122–1(1989), Toxicity testing of fire effluents, Part 1, General.
2. Purser D. A., et al., *Flame Retardants 94, BPF Conference*, Interscience London, 1994.
3. ISO 3261 (1975), Fire test vocabulary.
4. BS 4422, 1987, Glossary of terms associated with fire.
5. ASTM E176 (1981), Terminology relating to fire standards.
6. BS 6336, 1982, Guide to the development and presentation of fire tests and their use in hazard assessments.
7. BS 476, Part 10 (1989), Guide to the principles and applications of fire testing (for building materials and structures).

8. ISO TR10840 (1995), Plastics—Burning Behaviour—Guidance for development and use of fire tests.
9. ISO TR10 353 (1992), Plastics—Survey of ignition sources used for national and international fire tests.
10. ISO 10093 (1994) Plastics—Fire tests—Standard ignition sources.
11. ISO 871 (1996), Plastics—Determination of ignition temperature using a hot air furnace.
12. ASTM D1929 (1977), Test method for the ignition of plastics.
13. ISO 1182 (1990), Fire tests, building materials, non-combustibility tests.
14. ISO 4589 (1996), Plastics—Determination of burning behavior by oxygen index, Part 1, Guidance.
15. ISO 4589 (1996), Plastics—Determination of burning behavior by oxygen index, Part 2, Ambient temperature test.
16. ASTM D2863, Measuring the minimum oxygen concentration to support candle-like combustion of plastics (oxygen index).
17. ISO 4589 (1996), Plastics—Determination of burning behavior by oxygen index, Part 2, Elevated temperature test.
18. ISO 12992 (1995), Plastics—Vertical flame spread determination for film and sheet.
19. ISO 1210 (1992), Plastics—Determination of flammability characteristics in the form of a small flame.
20. BS 2782, Part 1, Method 140A (1992), Determination of the burning behavior of horizontal and vertical specimens in contact with a small flame ignition source.
21. UL94, Tests for the flammability of plastics materials for parts in devices and appliances, Underwriters Laboratories USA.
22. ASTM D635 (1990), Standard test method for rate of burning and/or extent and time of burning of self supporting plastics in a horizontal position.
23. ASTM D3801, Method for measuring the comparative extinguishing characteristics of solid plastics in a vertical position.
24. ISO 9773 (1990), Plastics—Determination of burning of flexible vertical specimens in contact with a small flame.
25. BS 2782, Part 1, Method 140B (1993), Determination of the burning behavior of flexible specimens in contact with a small flame ignition source.
26. ASTM D4804, Test method for determining the flammability of non-rigid solid plastics.
27. ISO 9772 (1994), Cellular plastics, Determination of the horizontal burning characteristics of small specimens subjected to a small flame.
28. BS 4735 (1984), Laboratory method of test for assessment of the horizontal burning characteristics of specimens no longer than 150 mm × 50 mm × 13 mm (nominal) of cellular plastics and rubbers subjected to a small flame.
29. ISO 10351 (1992), Method of testing plastics, Part 1, Thermal properties, determination of the combustibility of specimens using a 125 mm flame source.
30. BS 2782, Method 140C (1993), Determination of combustibility of specimens using a 125 mm flame source.
31. ISO 181 (1981). Plastics—Determination of flammability characteristics of rigid plastics in the form of small specimens in contact.
32. IEC 707, 1981, Determination of the flammability of solid electrical insulating materials when exposed to an igniting source.
33. BS 6334 (1993), Method of test for the determination of the flammability of solid electrical insulating materials when exposed to an ignition source.
34. ASTM D5207 (1991), Standard practice for calibration of 20 and 125 mm test flames for small-scale burning tests on plastics materials.
35. ASTM D5025, Specification for a laboratory burner used for small scale burning tests on plastics materials.
36. IEC 695–2–1, Fire hazard testing, Glow wire test and guidance.
37. IEC 695–2–20, Fire hazard testing. Hot wire coil ignitability test.

38. IEC 695–2–2 (1991), Fire hazard testing, test methods section, needle flame test.
39. IEC 695–2–4/2, Fire hazard testing, Test methods Section 4, sheet 2,500 W nominal gas flame and guidance.
40. IEC 695–2–4/1, Fire hazard testing. Test methods Section 4/sheet 1, 1 kW nominal premixed test flame and guidance.
41. IEC 332–1 (1979), Tests on electrical cables under fire conditions, Part 1, Tests on a single vertical insulated wire or cable.
42. BS 4066, Pt. 1, 1995, Amd 8539, 1992, Tests on electrical cables under fire conditions. Part 1, Method of test on single vertical insulated wire or cable.
43. IEC 332-2 (1994), Tests on electric cables under fire conditions, Part 2, Method of test on a single vertical insulated wire or cable.
44. BS 4066, Pt. 2 (1989), Method of test on a single vertical insulated cable.
45. IEC 332–3 (1994), Tests on electric cables under fire conditions, Part 3, tests on bunched wires or cables.
46. BS 4066, Pt. 3, 1994, Method of test on bunch wire or cables.
47. ISO 3795 (1989), Road vehicles and tractors and machinery for agricultural and forestry—Determination of burning behavior of interior materials 1989.
48. BS 2782, Part 1, Method 140D (1987), Flammability of a test piece 550 mm × 35 mm of thin PVC sheeting (laboratory method).
49. BS 2782, Part 1, Method 140E, 1988, Flammability of a small inclined test piece exposed to an alcohol flame (laboratory method).
50. ASTM D3014 (1989), Standard test method for flame height, time of burning and loss of weight of rigid, thermoset cellular plastics in a vertical position.
51. ISO 8030 (1995), Rubber and plastics hoses, method of test for flammability.
52. ISO 346, Conveyor belts, Flame retardation—Specifications and test method.
53. BS 5173, Part 103, Section 103.6., Determination of ignitability of lining (of gas hoses).
54. BS 5438 (1995), Flammability of textile fabrics when subjected to a small igniting flame applied to the face or bottom of vertically oriented specimens.
55. BS EN ISO 6940 (1995), Textile fabrics—Burning behavior, Determination of ease of ignition of vertically orientated specimens.
56. BS EN ISO 6941 (1992), Textile fabrics—Burning behavior, Measurement of flame spread properties of vertically orientated specimens.
57. BS EN 532 (1992), Protective clothing, Protection against heat and flame, Test method for limited flame spread.
58. BS PD2777 (1994), Fabric flammability burning accidents and the relevance of BS 5438.
59. BS 5722 (1991), Specification for flammability performance fabrics and fabric assemblies used in sleepwear and dressing gowns.
60. BS 5867, Part 2 (1993), Flammability requirements (for textiles).
61. BS 6341, Type F, Specification for fabrics for camping tests.
62. BS 7157 (1989), Method of test for ignitability of fabrics used in large tented structures.
63. BS 5852 (1990), Methods of test for assessment of the ignitability of upholstered seating by smouldering and flaming ignition sources.
64. BS 3119, 1959, Specification for method of test for flameproof materials (withdrawn).
65. BS 3120 (1959), Specification for performance requirements of flameproof materials for clothing and other purposes (withdrawn).
66. EN 2310 (1990), Test methods for flame resistance rating of non-metallic materials.
67. BS 6249, Materials and material assemblies used in clothing for protection against heat and flame.
68. BS EN 531 (1995), Protective clothing for industrial workers exposed to heat excluding fire fighters and welders clothing.
69. BS EN 367 (1992), Protective clothing—Protective against heat and fire—Method of determining heat transmission on exposure to flames.

70. BS EN 366 (1993), Protective clothing—Protection against heat and fire—Method of test: evaluation of materials and materials assemblies when exposed to a radiant heat source.
71. ISO 8191–1 (1987), Furniture—Assessment of the ignitability of upholstered furniture—Part 1, Ignition source: smouldering cigarette.
72. BS EN 1021–1 (1994), Furniture—Assessment of ignitability of upholstered furniture. Ignition source smouldering cigarette.
73. BS EN 597–1 (1995), Furniture—Assessment of ignitability of mattresses and upholstered bed bases, Ignition source: smouldering cigarette.
74. ISO 8191–2 (1988), Furniture—Assessment of the ignitability of upholstered furniture—Part 2—Ignition source: match flame equivalent.
75. BS EN 1021–2 (1994), Furniture—Assessment of ignitability of upholstered furniture, Ignition source: match flame equivalent.
76. BS EN 597–2 (1995), Furniture—Assessment of ignitability of mattresses and upholstered bedbases, Ignition source: match flame equivalent.
77. BS 6807 (1996), Method of test for the assessment of the ignitability of mattresses, divans and bedbases with primary and secondary sources of ignition.
78. BS 7175 (1989), The ignitability of bedcovers and pillows by smouldering and flaming ignition sources.
79. UIC Instruction Sheet 564.2VE, Appendix 3, Testing of the fire performance of seats. Union Internationale des Chemins de Fer, Paris.
80. FAR, Part 25. Appendix F. Part 2. Code of federal regulations. Aeronautics and space 1990, U.S.A.
81. BS 4790 (1987), Method for determination of the effects of a small source of ignition on textile floor coverings (hot nut method),
82. ISO 6925 (1982), Textile floor coverings, Burning behavior, Tablet test at ambient temperature.
83. ISO DIS 9239 (1995), Floor coverings—Determination of critical heat flux using a radiant heat energy source.
84. ASTM E648 (1986), Test for critical radiant, Flux of floor covering system using a radiant heat energy source.
85. BS 3289 (1995), Specification for textile carcase conveyor belting for use in underground mines (including fire performance),
86. ISO 5657 (1986), Reaction to fire ignitability of building materials.
87. BS 476, Part 7 (1993), Method for the classification of the surface spread of flame of products.
88. ISO 5658, Fire tests—Reaction to fire lateral surface spread of flame on building products with specimen in vertical configuration.
89. IMO Resolution A516, Recommendation of fire test procedures for bulk head surface and deck finish materials.
90. Jannssens, M., and Minne, R., ISO TC92/SC1, 104.
91. Babrauskas, V., NBS1R 82–2611, U.S. Bureau of Standards, NIST Washington, 1982.
92. ISO 5660–1 (1993), Fire test—Reaction to fire—Part 1, Rate of heat release from building products (cone calorimeter method),
93. ASTM E1354 (1990), Standard test method for heat and visible smoke release for materials and products using an oxygen depletion calorimeter.
94. ASTM E1474 (1992), Standard test method for determining the heat release rate of upholstered furniture and mattress components or composites using a bench scale oxygen consumption calorimeter.
95. Fire safety of upholstered furniture, Final report of CBUF programme. Interscience, London, 1995.
96. MIL–STD–2031 (1991), Fire and toxicity test methods and qualification procedures for composite material systems used in hull machinery and structural applications inside navy submarines (U.S.A.).

97. Draft specification for bulkheads and fixed furniture for high speed surface craft 1995. (Draft IMO/Finnish and National Maritime Administration specification),
98. GM/RT 2125 (1996), Fire Performance requirements for railway vehicles.
99. ASTM E906 (1983), Standard method for heat and visible smoke release rates for materials and products.
100. DOT/FAA/CT–96/15, *FAA Fire Test Handbook*, FAA, Washington, 1990.
101. Tewarson, A., and Pion, R. F., *Combustion and Flame*, *26*, 85 (1976).
102. ISO 1716 (1973), Building materials determination of calorific potential.
103. BS 6401 (1983), Method of measurement in the laboratory of specific optical density of smoke generated by materials.
104. ASTM E662 (1979), Test for specific optical density of smoke generated by solid materials.
105. NES 711 (1981), Determination of smoke index of the products of combustion from small specimens of materials.
106. prEN 2824, Burning behavior, Determination of smoke density and gas components in the smoke of materials under the influence of radiating heat and flames, Test equipment apparatus and media.
107. prEN 2825, Burning behavior determination of smoke density and gas components in the smoke of materials under the influence of radiating heat and flames—Determination of smoke density.
108. ISO 5659–2 (1994), Plastics—Smoke generation—Part 2, Determination of specific optical density.
109. ISO TR5924 (1989), Fire Tests—Reaction to fire—Smoke generated by building products (dual chamber test),
110. ASTM D1400 (1982), The Arapahoe smoke test apparatus.
111. DIN 53436/DIN 53437, Tubular furnace for the thermal decomposition of plastics.
112. NFT 51–073, Method of measuring the optical density of smoke.
113. BS 5111 (1983), Laboratory method for testing a 25 mm cube of specimens of low density material to continuous flaming conditions.
114. ASTM D2843 (1977), Standard test method for density of smoke from burning or decomposition of plastics.
115. IEC 1034 (1990), Measurement of smoke density of electric cables burning under defined conditions. Part 1, Test apparatus (1990), Part 2, Test procedures and requirements (1991),
116. BS 6853, Appendix B (1987), Code of practice for fire precautions in the design and construction of railway passenger rolling stock.
117. IEC 754–1, Tests on gases evolved during combustion of materials from cables, Part 1, Determination of amount of halogen and gas.
118. IEC 754–2, Tests on gases evolved during combustion of materials from cables, Part 2, Determination of degree of acidity of gases evolved during the combustion of materials taken from electric cables by measuring pH and conductivity.
119. BS 6425, Parts 1 (1990) and 2 (1993), Tests on gases evolved during the combustion of materials from cables, Part 1, Determination of amount of halogen acid gas, Part 2, Determination of acidity of gases by measuring pH and conductivity.
120. UTE C20 0 453 (1976), Determination of corrosivity of fumes.
121. ISO 11907–2 (1995), Plastics—Smoke generation—Determination of corrosivity of fire effluents, Part 2, Static method.
122. ISO DIS 11907–3 (draft), Plastics burning behaviour, Determination of the corrosivity of fire effluents, Part 3, Dynamic test.
123. ASTM D5485, Standard test method for determining the corrosivity of combustion products using the cone corrosimeter.
124. ISO TR 9122–2 (1990), Toxicity testing of fire effluents, Part 2, Guidelines for biological assays to determine the acute inhalation toxicity of fire effluents (basic principles, criteria and methodology),

125. ISO TR 9122–3, 1933, Toxicity testing of fire effluents, Part 3, Methods for the analysis of gases and vapours in fire effluents.
126. ISO TR 9122–4, Toxicity testing of fire effluents, Part 4, The fire model/ furnace and combustion apparatus used in small scale testing.
127. ISO TR 9122–5, 1993, Toxicity testing of fire effluents, Part 5, Prediction of toxic effects of fire effluents.
128. ISO TR 9122–6 (1994), Toxicity testing of fire effluents, Guidance for regulators and specifiers on the assessment of toxic hazards in fires in buildings and transport.
129. ISO DIS 3344 (1994), Determination of the lethal toxic potency of fire effluents.
130. BS PD 6503, Part 2 (1988), Report and recommendations on the development of tests for measuring the toxicity of combustion products.
131. ASTM E800 (1981), Guide for the measurement of gases present or generated during fires.
132. Levin, B. C., NBSIR82/2532, U.S. Department of Commerce, NIST Washington D.C., 1982.
133. Alexeeff, G. V., and Packham, S. C., *J. Fire Sci.*, *2*(5) (1984),
134. NFX 70–100 (1986), Fire tests analysis of pyrolysis and combustion gases—Tube furnace method.
135. NES 713 (1985), Determination of the toxicity index of the products of combustion from small samples of materials.
136. prEN 2826, Burning behavior determination of smoke density and gas components in the smoke of materials under the influence of radiating heat and flames—Determination of gas concentrations in the smoke.
137. IMO FP 39/19, Annex 3, Draft MSC resolution, Interim standard for measuring smoke and toxic products of combustion.
138. ISO 834 (1975), Fire resistance tests on elements of building construction.
139. ISO 9705 (1988), Fire tests—Full scale room test for surface products.
140. Nord test NT Fire 032, Upholstered furniture burning behaviour. Full scale test.
141. ASTM E603 (1977), Guideto room fire experiments.
142. ASTM E1537 (1995), Standard test method for fire testing of real scale upholstered furniture items.
143. ASTM E1590 (1995), Standard test method for fire testing of real scale mattresses.
144. ASTM D3894, Evaluation of fire response of rigid cellular plastics using a small corner configuration.

General References

Mackower, A. D., *Flammability of Solid Materials, A Guide to the relevant British Standards*, BSI.
Troitzsch, J. (1982), *Plastics Flammability Handbook, Principles, Regulations, Testing and Approval*, Hanser, MacMillan.
Hilardo, C. J., *Flammability Handbook for Plastics*, Technomic Publishing.
Drysdale, D. (1985), *Fire Dynamics*, John Wiley.
Babrauskas, V., and Grayson, S., *Heat Release in Fires*, Elsevier Applied Science, 1992.
Fire, F. L., *Combustibility of Plastics*, Van Nostrand Reinhold, New York, 1991.

28
Weathering

Dieter Kockott
Consultant, Hanau, Germany

A great number of organic materials are used outdoors. Resistance to weathering is therefore a very important factor influencing the usefulness of these materials. The demands placed on resistance to weathering thus differ depending on the prevailing climatic conditions where the materials are put to use. We can check the resistance to weathering in a manner very similar to general practice by exposing a sample to that climate to which the material will be exposed during subsequent use. Natural weathering in various climates is certain to be too expensive. In the course of time, Florida has come to be considered as a reference climate owing to its moist, warm weather, followed by Arizona, with its hot, dry climate. One result of the world's increasing economic integration is that it is often impossible for producers to know where their materials will be used in the end. In this case, it is important to know how the material will react to extreme climates. In a limited, local market, such as that in the moderate climate zone of Central Europe, outdoor weathering will often be preferred to obtain a realistic idea of the material's behavior and to keep the demands placed on resistance to weathering from driving up costs unnecessarily.

For economic reasons, however, it is not possible to wait years for the results of outdoor weathering. Thus there is a need for accelerated weathering in the laboratory. This is intended to identify and measure the same changes in the properties of polymeric materials that would occur under real conditions during the normal service life of these materials. This is a very high goal to aim at, since accelerated time testing generally poses the danger of initiating degradation processes other than those occurring in general practice, apart from the desired acceleration of the aging processes that do occur naturally; or, by the same token, changes occurring in the materials outdoors may not take place during accelerated weathering, or not to the same extent. This chapter will present various methods of outdoor weathering and explain the fundamental possibilities for acceptably accelerating weather-related aging processes, as well as which methods and equipment are currently available for this purpose. In addition, support will be offered for making decisions regarding which accelerated methods of testing are suitable for a given problem.

1 General Effects of Solar Radiation of Polymeric Materials

In connection with the aging of polymeric materials, only that radiation absorbed by the material that substantially causes the material to heat up is of interest. Only a small portion of the absorbed radiation gives rise to primary photochemical processes, such as the formation of excited states, radicals, and chain scission. For the chain scission of plastics, quantum yields in the range of 10^{-2} to 10^{-5} have been measured. This means that, when 100 to 100,000 polymer molecules have absorbed radiation, only one will respond with scission.

Figure 1 shows, in tabular form, the major reactions and their causes, as well as their interactions. The primary degradation processes in a polymer matrix are largely the result of ultraviolet radiation, owing to its greater quantum energy. Dyestuffs and pigments also absorb light in the visible spectrum, otherwise they would have no color. They are thus in principle also sensitive to radiation in the visible range. The portion of ultraviolet light in the solar radiation is irrelevant to the heating of an irradiated sample. This is caused solely by the absorption of visible and infrared radiation. The primary aging processes are essentially unrelated to temperature. Many secondary degradation processes, however, do depend on temperature. Therefore, the heating of a sample as a result of irradiation should not be left out of consideration in an aging test. The heating of irrradiated polymers is also the cause of processes that are not directly related to photochemistry but do have an effect on the course of aging. The diffusion and evaporation of additives that are not chemically combined in the polymer but are instead in solution or suspension, is

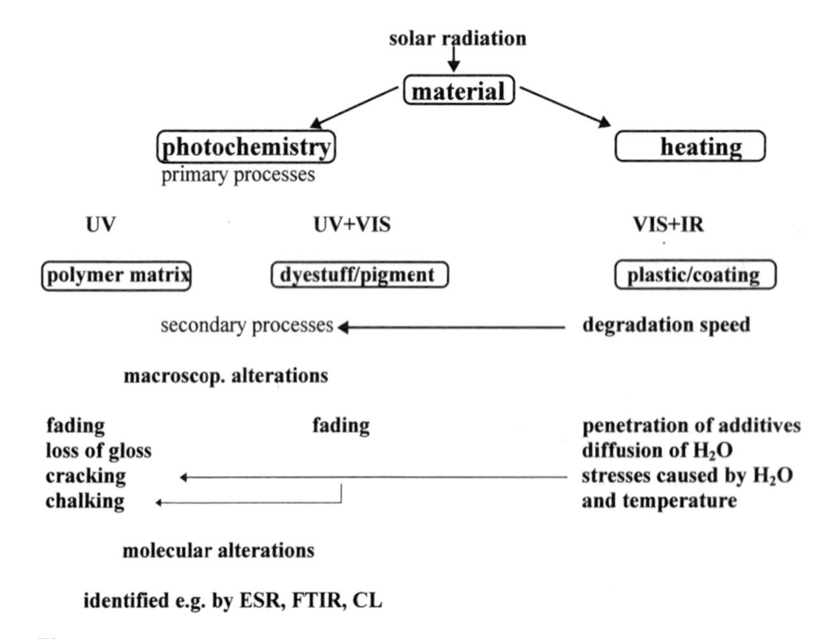

Figure 1 Principal reactions and interactions.

greater at higher temperatures. The loss of softeners and the concomitant reduction of flexibility is another typical example.

The macroscopic changes in the polymer are based on molecular changes that can be identified using various analytic methods. Figure 1 mentions but a few examples. Sensitive analytic methods make it possible to detect changes in the material at the molecular level long before they become observable to the naked eye, for example, loss of gloss, discoloration, or cracking. In principle, this is a way to enable early detection of weathering-induced damages. The decisive question, however, is whether and under what experimental conditions the results of the measurements taken by analytic methods correlate with the macroscopic changes in the material. Work is currently being done to this effect in various places. This is not the topic of the present chapter.

1.1 Radiation

Important Parameters for Characterizing Radiation

The radiant flux $\phi = Q/t$, measured in W, indicates how much radiant energy is released by the source of radiation per unit time.

The irradiance $E = \phi/F$, measured in W/m^2, indicates the amount of radiant flux F striking a surface per unit of area.

For both of the parameters mentioned, it is necessary to indicate the spectral range in which the measurements were taken or for which the values were calculated. If we turn our attention to narrow wavelength intervals, then we obtain the spectral irradiance, $E_\lambda = \phi_\lambda/F$, measured in W/m^2nm. This indicates the strength of the radiant flux striking a surface, F, in the spectral range $\Delta\lambda$ (usually 1 nm). Plotting spectral irradiance E_λ in a number of equally spaced wavelength intervals yields the spectral distribution of the radiation. The irradiance in the wavelength range between λ_1 and λ_2 then becomes

$$E = \int_{\lambda_1}^{\lambda_2} E_\lambda d\lambda \tag{1}$$

If radiation with an irradiance E acts upon the surface of a material for a period of time t, then a specific amount of radiant energy will impinge upon the surface in this time. This parameter is called the radiant exposure H

$$H = \int_{t_1}^{t_2} E\, dt \tag{2}$$

and is stated in J/m^2.

In common parlance, H is also occasionally referred to as "dose." When observing very narrow wavelength intervals, we obtain the spectral radiant exposure, H_λ

$$H_\lambda = \int_{t_1}^{t_2} E_\lambda\, dt \tag{3}$$

which is stated in J/m^2nm.

It is only reasonable to indicate the radiant exposure when the relevant wavelength range is also mentioned. Figure 2 is a summary of the terms mentioned above.

relative spectral distribution

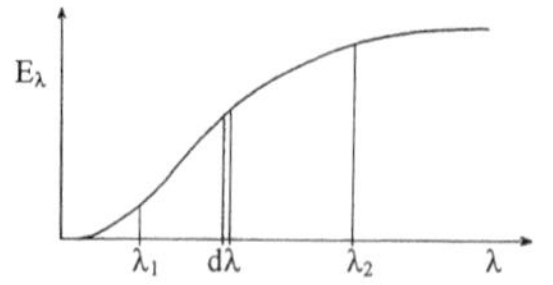

spectral irradiance

E_λ in W/m² nm

spectral radiant exposure

$H_\lambda = \int_{t_1}^{t_2} E_\lambda \, dt$ in Ws/m² nm = J/m² nm

irradiance between λ_1 and λ_2

$E = \int_{\lambda_1}^{\lambda_2} E_\lambda \, d\lambda$ in W/m²

radiant exposure between λ_1 and λ_2

$H = \int_{t_1}^{t_2} E \, dt$ in Ws/m² = J/m²

Figure 2 Important terms characterizing radiation.

The radiant exposure H only gives evidence of the *impinging* radiant energy. It tells us nothing about the radiant energy *absorbed* by the material. An ideal white body with an absorption ratio of $\varepsilon = 0$ absorbs no radiant energy at all, since it will reflect all of the impinging radiation!

The Absorption Ratio, Spectral Sensitivity, and Activation Spectrum of Materials

A portion of the radiation impinging on a material is absorbed. The absorption ratio $\varepsilon(\lambda)$, which is generally dependent on the wavelength, indicates the ratio of absorbed to impinging radiation. The greater portion of absorbed radiation causes the irradiated material to heat up. This will be discussed in Section 2. Although radiation absorption is a necessary precondition for aging—no aging without absorption—it is not alone sufficient. Therefore the absorption ratio is not capable of describing the behavior of a material with respect to the change in it useful properties as a result of optical radiation. To this end, it is better to use the spectral sensitivity s_λ of the material. This indicates which portions of the radiation spectrum cause a specified property of the material to change, for example, discoloration. Figure 3 shows the spectral sensitivity s_λ of polypropylene and polyethylene as an example of the change in optical density at 5.3 μm [1].

If we know s_λ, we can also calculate the effective irradiance E_{eff} for the specified material and the observed property change,

$$E_{\text{eff}} = \int_{\lambda_1}^{\lambda_2} E_\lambda s_\lambda \, d\lambda \tag{4}$$

and from this, the effective radiant exposure H_{eff}:

$$H_{\text{eff}} = \int_{t_1}^{t_2} E_{\text{eff}} \, dt \tag{5}$$

Figure 4 shows this in the form of a graph. The crosshatched area is a measure of the effective irradiance E_{eff}. Radiation outside the crosshatched spectral range contributes

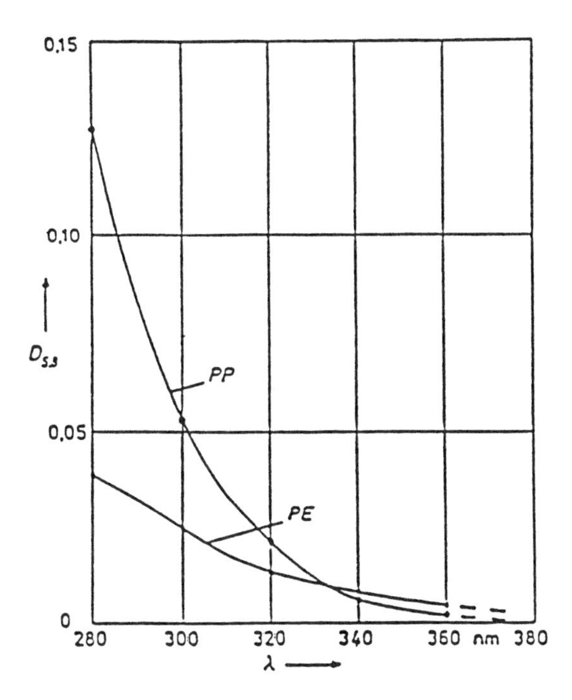

Figure 3 Spectral sensitivity of polypropylene and polyethylene.

nothing to the aging process under observation here. Hence the polymer does not have to be protected against this radiation.

This mathematical approach enables the aging behavior of a material with a spectral sensitivity s_λ, when acted upon by radiation with a spectral distribution $E_{\lambda,1}$, to be predicted as an initial approximation when the aging behavior of the material when exposed to radiation with the spectral distribution $E_{\lambda,2}$ is known. In doing so, we assume that the aging will be proportional to the impinging effective radiant exposure H_{eff}. This calculation only takes account of the effects of the radiation and disregards all other influencing factors, such as temperature.

The spectral sensitivity s_λ is a material-specific quantity and refers to a corresponding observed property change in a polymer. It does not depend on the source of radiation used for measurement. In general practice, however, we often need to know only how a polymer will act in the presence of a specific type of radiation, such as solar radiation. Then it will be sufficient to break down the solar radiation into its spectral portions and study which property changes occur in which narrow spectral range as a result of the irradiation of the polymer. The result is called the "activation spectrum" [2, 3]. This represents the product $E_\lambda s_\lambda$ (cf. Eq. 4 and Fig. 4). From this we can directly derive the spectral ranges of the radiation, which are responsible for the degradation. Efforts at achieving improved stability must therefore concentrate primarily on this wavelength region.

One way of ascertaining the activation spectrum is by using a monochromator to break down the incident radiation into its spectral portions, which then strike various places on a sample. The spectral irradiance on the sample is relatively minor when this method is used, however, resulting in extremely long irradiation periods. Therefore, a variate difference method employing a number of cut-off filters with a steep absorption edge is frequently used (Fig. 5). In the visible and in the infrared range, all filters exhibit

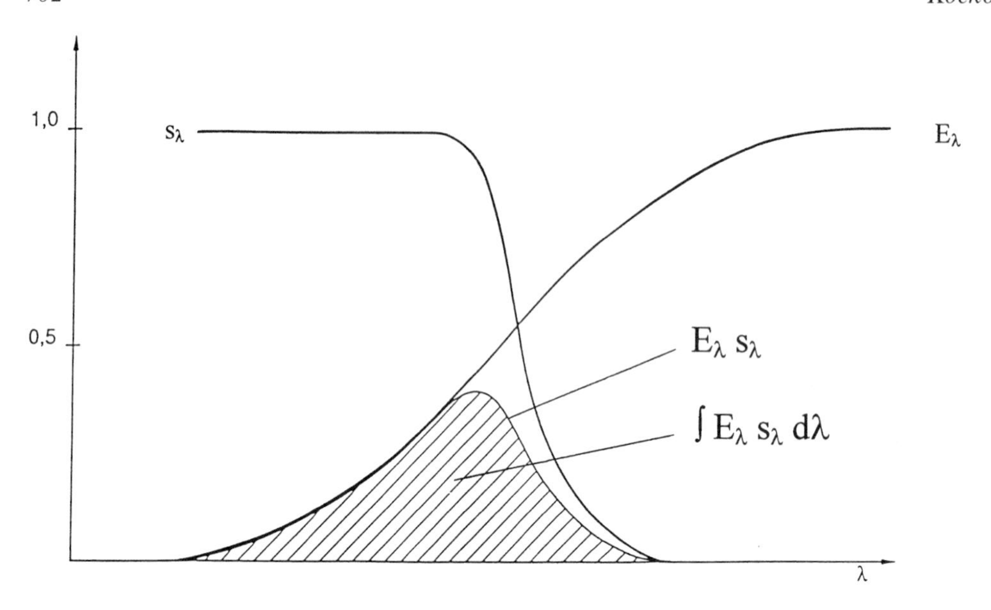

Figure 4 Drawing of spectral irradiance E_λ weighted with spectral sensitivity s_λ.

equally high transmittance; in the ultraviolet range, they differ with respect to the position of the absorption edges. These are roughly 10 nm apart. The filters are laid next to one another directly on the material being tested. The test specimen so prepared is exposed to solar radiation or a source of artificial radiation. The difference in the property change of the material behind filters with neighboring absorption edges then characterizes the property change in the spectral range between the neighboring filters. Figure 6 shows, by way of example, the activation spectrum for the yellowing of polycarbonate as a result of irradiation with artificial global radiation. In this example, radiation between 300 and 320 nm proves to be especially damaging. using a polyarylate as an example, Fig. 7 shows that activation spectra can also be found—for the same source of radiation—whose maximum is of a considerably longer wavelength, lying in the range of 340–360 nm.

All examples thus far refer to unpigmented materials. However, a pigment can change the activation spectrum, either by absorbing damaging radiation or as a result of its photoactivity [4].

Figure 8 shows the activation spectrum of a polyester film. The change in the film's yellowness index was measured step by step after it had been irradiated with monochromatic radiation. Here we see that it yellows at wavelengths of less than 350 nm, while wavelengths of more than 350 nm cause bleaching. Hence we are confronted with two competing processes. Irradiation with polychromatic radiation, such as solar radiation, triggers both processes. The yellowing observed in general practice is produced by the sum of the combined effects of yellowing and bleaching. This example very clearly shows the kind of errors that can occur when only certain portions of the solar radiation spectrum are used for artificial weathering.

The activation spectrum very closely approximates the characterization of the wavelength-related degradation behavior of a polymeric material under the influence of the radiation of a given radiation source. At the same time, it tells us the spectral range in which the material must have additional protection to lengthen its service life.

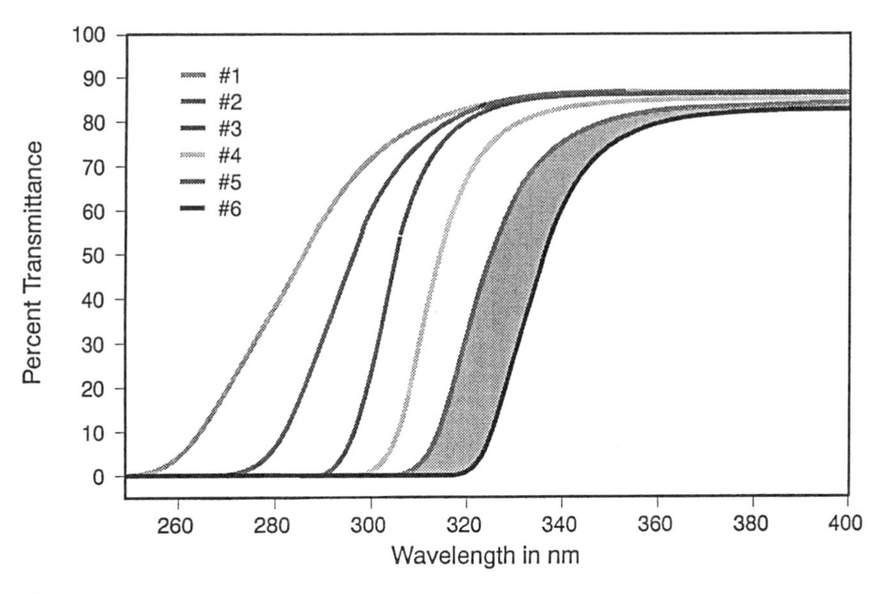

Figure 5 Transmittance of filters being used for measuring activation spectra.

1.2 Heat

Heating an Irradiated Object

An irradiated object absorbs a portion of the radiation and converts it to heat. It thus becomes warmer than its surroundings. The temperature increase depends on properties specific to the material (for example, the absorption ratio, ε_λ, or the thermal conductivity) and on the spectral irradiance of the impinging radiation. As an initial approximation, we can assume that all the absorbed radiation contributes to heating. The absorbed irradiance that can lead to heating is then

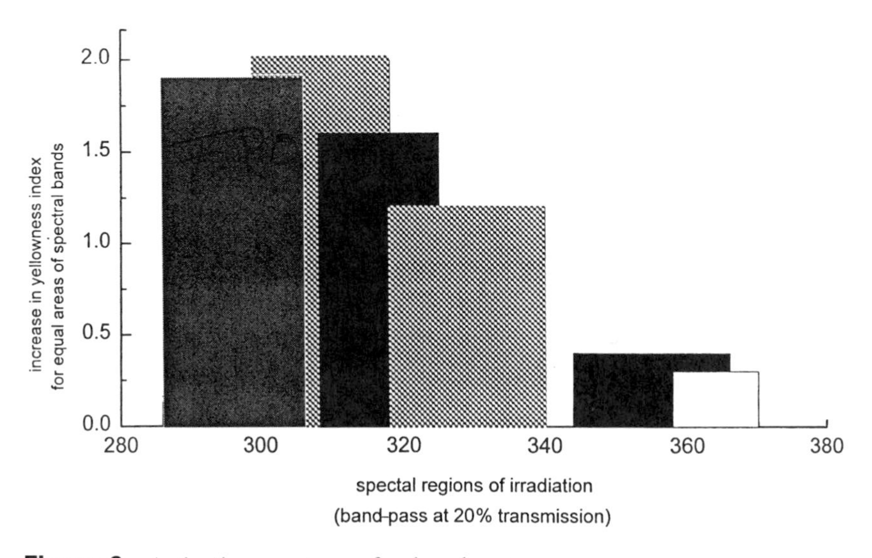

Figure 6 Activation spectrum of polycarbonate.

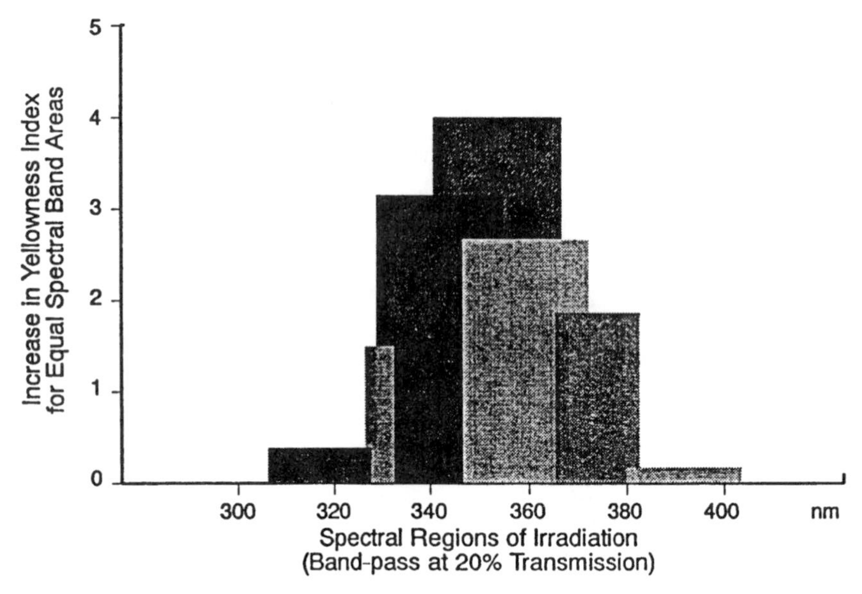

Figure 7 Activation spectrum of polyarylate.

$$E_{\mathrm{abs}} = \int E_\lambda \varepsilon_\lambda \, d\lambda \tag{6}$$

The absorption ratio ε_λ of various materials differs greatly. However, knowing the upper and lower limit is sufficient for making an estimate. Under otherwise identical conditions, an object whose absorption ratio is 1 will exhibit the greatest rise in temperature compared to its surroundings, while one with an absorption ratio of 0 will heat up least. In materials testing, black and white panel thermometers have proved to be useful in

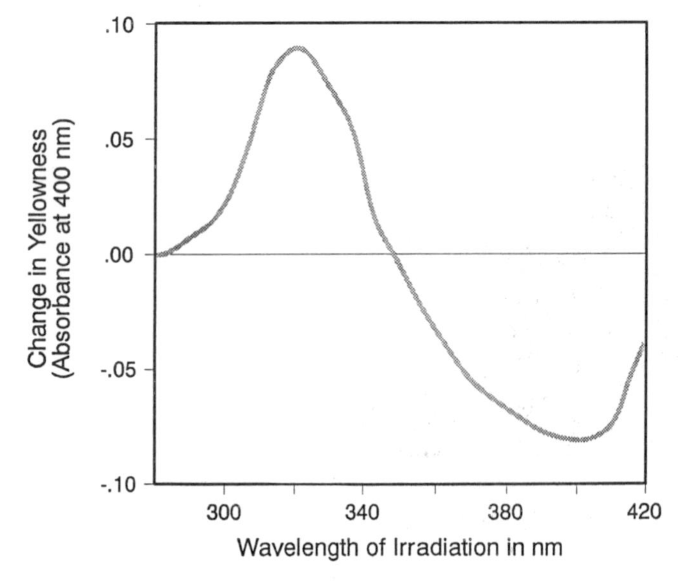

Figure 8 Activation spectrum of polyester.

measuring surface temperature. These thermometers consists of thin metal plates that are coated with a special black or white lacquer and whose temperature can be measured with the aid of a temperature sensor on the back. Alongside the conventional black panel thermometers, there is also a recent type whose back is provided with thermal insulation, which is called a black standard thermometer [5]. The absorption ratio of both lacquers is shown in Fig. 9 as a function of wavelength. The black lacquer is a technically adequate approximation of an ideal black body, for which $\varepsilon = 1$. The white lacquer is far removed from an ideal white body, whose $\varepsilon = 0$. However, it is typical of white industrial surfaces. The temperature of irradiated colored surfaces is between that of the black and the white panel thermometers [6].

Figure 10 shows the temperature of variously colored PVC foils during outdoor weathering at an angle of 45° to the south with the back covered and uncovered—and in a weathering device under the weathering conditions in the illustration [6]. It compares the difference in temperature between the colored samples and a black sample. Figure 11 demonstrates how the difference in temperature of the variously colored samples depends on the irradiance. The higher the irradiance, the greater the difference in temperature. Hence selecting the appropriate irradiance in a weathering device, including a simulation of global radiation, can heat the colored samples much the same as outdoor weathering would.

Emission and Absorption of Longwave Radiation

Every body whose temperature is above absolute zero emits electromagnetic radiation. The radiant exitance $M = \Phi/F$ (Φ = radiant flux, F = surface from which the radiation emanates) and the spectral distribution of this radiation depend on the temperature. For the so-called black-body, which *per definitionem* absorbs all impinging radiation and converts it into heat, the Boltzmann law makes the radiant exitance $M = \Phi/F$,

$$M = \sigma T^4 \qquad \text{where } \sigma = 5.67 \cdot 10^{-8} \quad \frac{\text{W}}{\text{m}^2\ \text{degrees}^4} \tag{7}$$

Every body exchanges heat with its surroundings. This takes place through heat conduction and radiation. In this chapter, we would like to consider only the radiation. Every body constantly emits radiation and receives radiation from its environment. At thermal

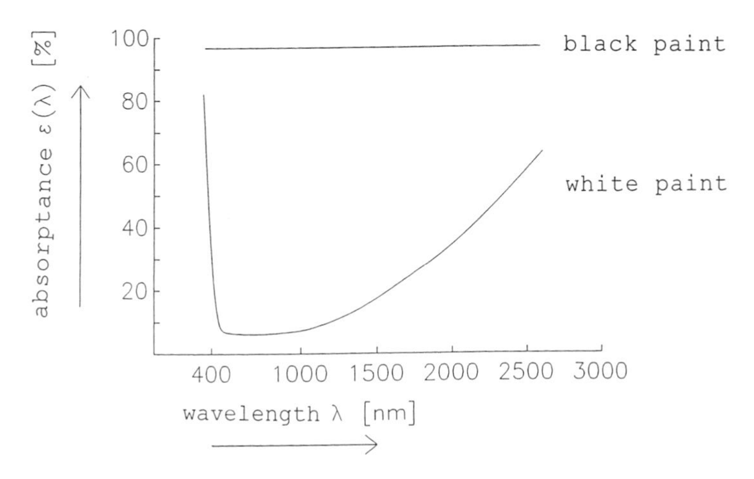

Figure 9 Spectral absorptance of paints.

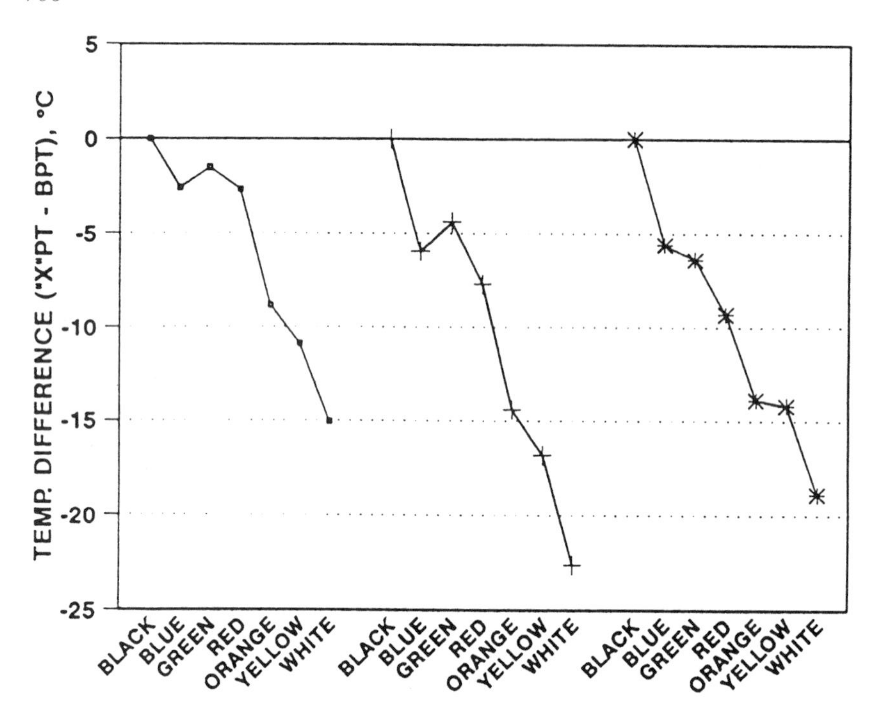

Figure 10 Temperature differences of colored samples outdoors and in a Xe weatherometer. BPT = 70°C, spectral irradiance at 340 nm: 0.55 W/m^2nm.

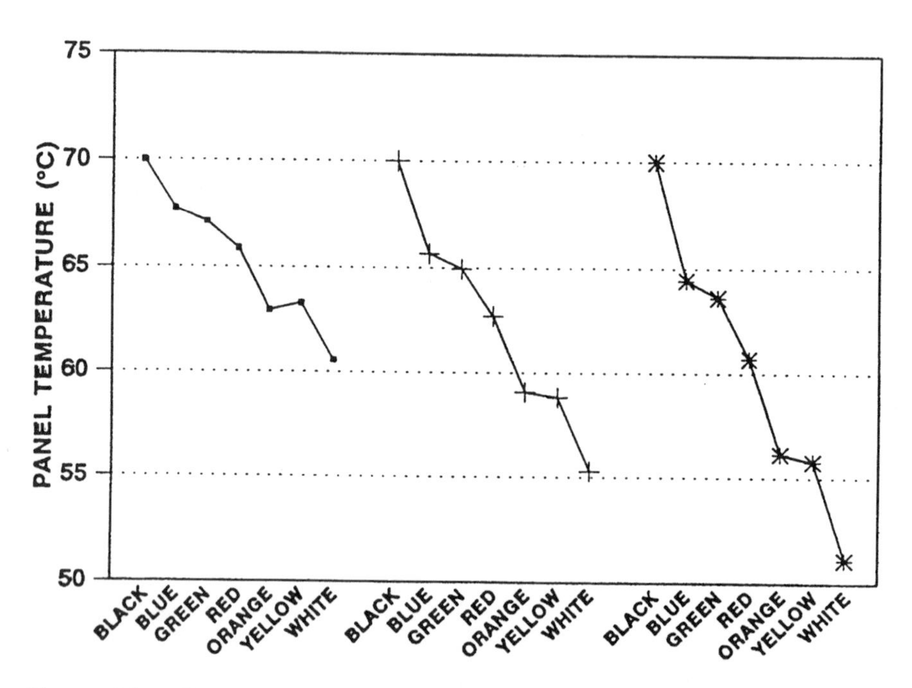

Figure 11 Effect of increasing irradiance on temperature of colored samples in a Xe weatherometer. BPT = 70°C.

equilibrium, the energy radiated per unit of surface and time is just equal to the energy absorbed from the surroundings. When there is a difference in temperature, $T_1 - T_2$, the energy flowing to the colder body per unit of surface and time amounts to

$$\Delta M = \sigma(T_1^4 - T_2^4) \tag{8}$$

(cf. Fig. 12).

Example: $T_1 = 340\text{K}\,(67^\circ\text{C});\ T_2 = 300\text{K}\,(27^\circ\text{C})$
$$\Delta M = 5.67 \cdot 10^{-8}\,(340^4\text{–}300^4)\ \text{W/m}^2$$
$$\Delta M = 298\ \text{W/m}^2$$

This example describes, as a technically adequate approximation, the radiation conditions between the black floorboard of a car, say, at 27°C, and that of the black road surface at 67°C on which the car is moving or parked. The resulting irradiance of around 300 W/m² on the car's floorboard is *not* negligible in comparison to the irradiance of the solar radiation acting with a maximum irradiance of 1000 W/m² on the car from above. Both types of radiation are in quite different spectral ranges, and heat radiation has, of course, nothing to do with primary photochemical processes in organic materials. It does, however, affect the temperature in the car's interior.

So far, we have only spoken of the ideal black body, with an absorption ratio of $\varepsilon = 1$. Real bodies have an absorption ratio of $\varepsilon < 1$, whereby ε is also dependent on the wavelength. Following Kirchhoff's Law, the radiant emittance of a real body is $M = \varepsilon M_S$, in which M_S is the radiant exitance of a black body at the same temperature.

Radiation Balance and Energy Balance

In the following discussions, it will be very useful to distinguish between solar radiation and terrestrial radiation. Radiation from the sun corresponds roughly to the radiation of a black body at a temperature of approx. 6000K, while that of the thermal radiation of terrestrial bodies and the lower atmosphere is approximately 300K. Figure 13 shows the spectral distribution of both types of radiation on a log-log scale. The spectral areas are almost entirely separated from one another. It is common in meteorology to designate these as shortwave (solar) and longwave (thermal) radiation.

Here we will discuss only the integral values of the appropriate spectral areas. The radiation balance Q of a horizontal surface is

$$Q = (G - R) + (A - E) \tag{9}$$

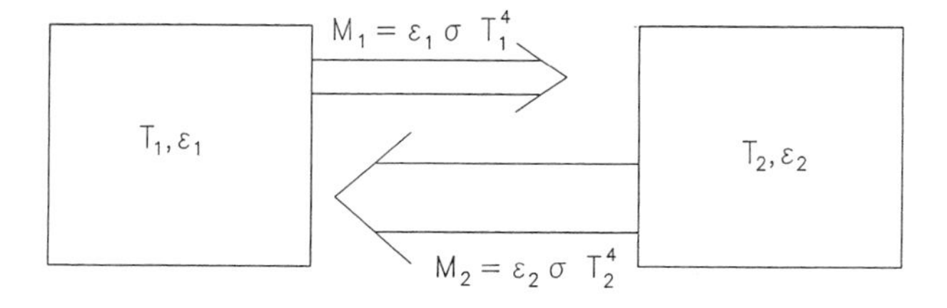

Figure 12 Radiant exitance M_1 and M_2 of two bodies with temperatures T_1 and T_2 and absorptances ε_1 and ε_2.

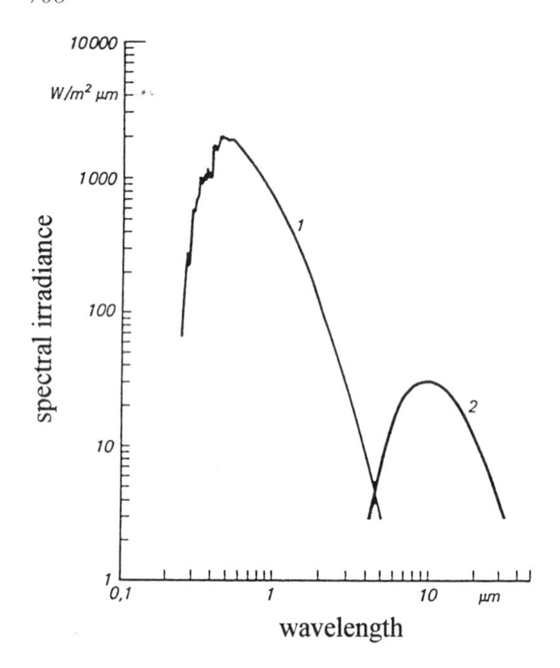

Figure 13 Spectral irradiance. Curve 1: Extraterrestrial solar radiation. Curve 2: Terrestrial radiation of a black body of temperature 300K $\approx$ 27°C.

in which

G = global radiation, i.e., the total of direct and diffused solar radiation
R = global radiation reflected from the surface
A = thermal radiation from the atmosphere directed downwards
E = thermal radiation of the body.

Figure 14 shows the daily progression of the individual members of the radiation balance of the earth's surface on a cloudless summer's day. The global radiation G and the reflected global radiation R follow the position of the sun. The thermal radiation A of the atmosphere and of the earth E change but little in the course of the day. The radiation balance is positive during the daytime, owing to the sun's rays. At night, it is negative because the thermal radiation directed away from the earth is greater than the thermal radiation sent from the atmosphere down to earth.

Aside from the radiation, there are yet other factors that contribute to the energy balance of a plane surface:

$$Q + K + H + V + P = 0 \tag{10}$$

in which

Q = radiation balance
K = thermal flux density from the interior of the body to its surface
H = flux density of the sensible heat from the atmosphere to the surface as a result of molecular and convective heat conduction
V = flux density of latent heat as a result of condensation and evaporation
P = flux density of the heat carried to the surface by precipitation

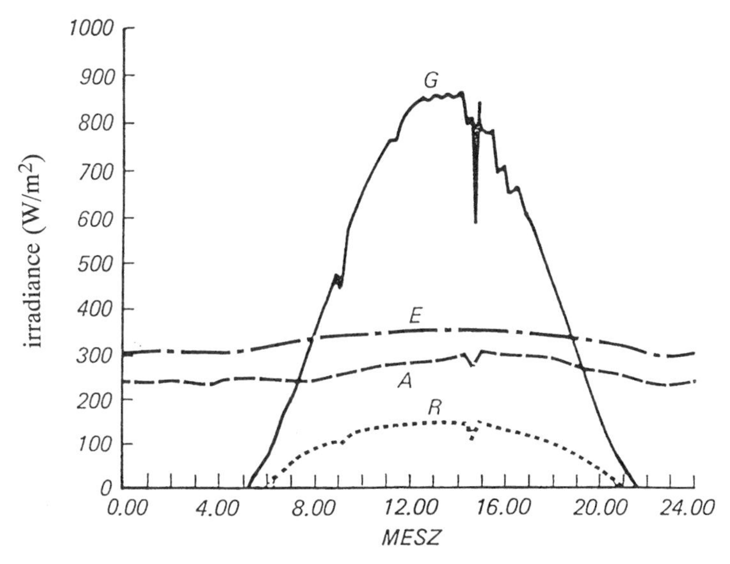

Figure 14 Diurnal variation in individual terms of radiation balance on earth's surface (example).

Figure 15 shows the daily progression of the individual members of the energy balance for the same cloudless summer's day as in Fig. 14. The greater part of this day's incoming radiant energy is used to evaporate water.

Details on how to calculate the individual members of the radiation balance and the energy balance can be found in the new VDI guideline 3789, Part 2 [7].

2 The Dependence of Aging on Temperature

Many secondary degradation processes of organic materials following the primary photochemical step depend on the temperature. The speed of a degradation reaction usually

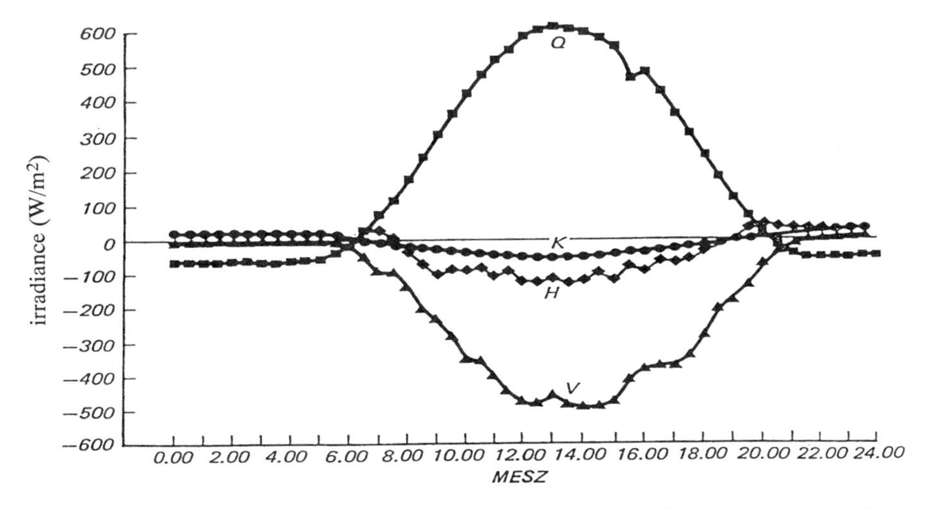

Figure 15 Diurnal variation in individual terms of energy balance on earth's surface (example).

increases with increasing temperature. A rule of approximation from the field of reaction kinetics states that an increase in temperature of 10°C roughly doubles the reaction speed. In many cases, we can observe no linear dependence of photochemical degradation on temperature; instead, change in a property depends but little on temperature initially. However, from a certain (material-specific) temperature on, it increases rapidly [8].

An example taken from general practice is shown in Fig. 16. Here, it is a matter of a reduction in the tensile strength of industrial yarn as a result of artificial weathering [9]. The weathering conditions in both parts of the figure differ only in the temperature of the surrounding air, and thus also in the temperature of the samples. At an air temperature of 30°C (diagram on the right), the sequence, with respect to aging stability, is polyamide, polypropylene, polyester. At an air temperature of 60°C (diagram on the left), not only is the degree of aging greater, as could be expected, but the sequence is also in part reversed. It now becomes polyamide, polyester, polypropylene. This means that, when the test conditions for accelerated weathering are set up, not only the radiation but also the temperature must correspond to that occurring where the material will be used in general practice. Otherwise we may well obtain very unrealistic information on the aging behavior.

The dependence of the speed of a chemical reaction on temperature is described by the Arrhenius equation:

$$K(T) = Ae^{-E/RT} \tag{11}$$

where

$K(T)$ = speed constant
A = material-specific factor
E = activation energy
R = gas constant
T = absolute temperature

From Eq. 11, it follows that

$$\ln K(T) = \frac{-E}{RT} + \ln A \tag{12}$$

If the degradation reaction follows the Arrhenius equation, plotting $\ln K$ over $1/T$ must yield a linear relationship. Figure 17 shows an example in which not $\ln K$ but rather the $\ln t$ of the testing period t is plotted up to the attainment of a specific degree of discoloration [10]. This example also shows that a linear extrapolation of the values measured at specific temperatures to other temperatures is only possible to a limited extent. In particular, a linear relationship will be unreliable if, for instance, a second degradation mechanism is

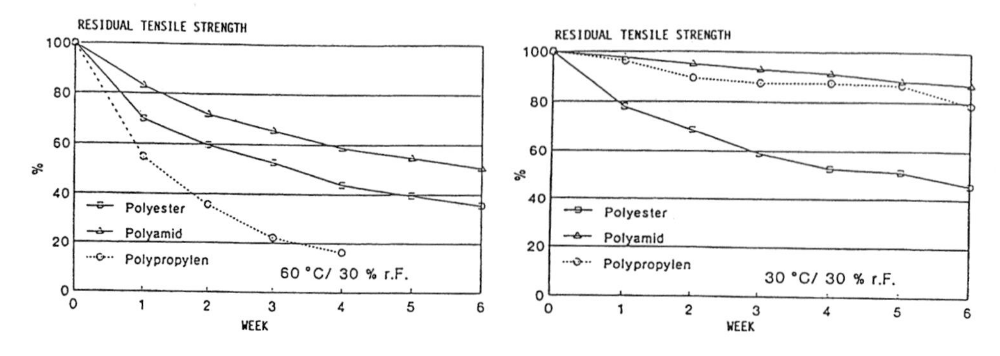

Figure 16 Decay of residual tensile strength of technical yarns at 30°C and 60°C.

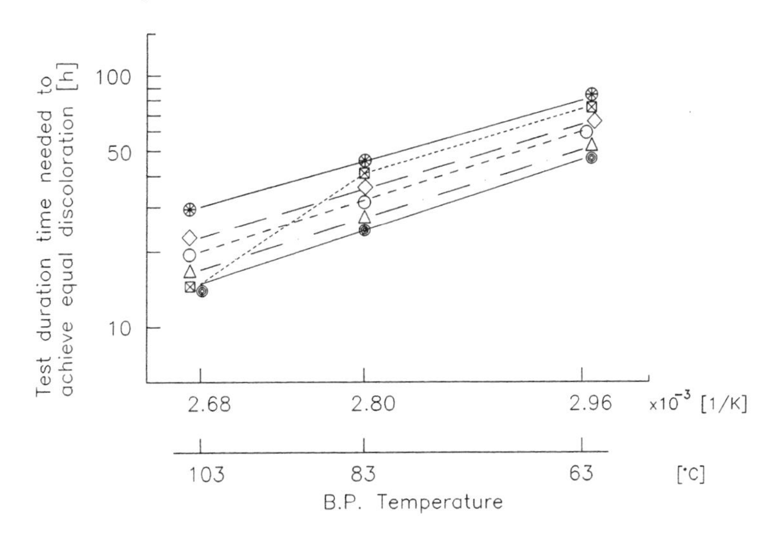

Figure 17 Temperature effect on light degradation speed of seat fabrics.

triggered at a specific temperature. In degradation processes based on photochemical oxidation, the supply of oxygen in the material is also significant. The diffusion speed, $D(0_2)$ of the oxygen also depends on the temperature.

As far as amorphous and semicrystalline polymers are concerned, the glass transition temperature T_G also plays a special part in this context, since the mobility of the chains above T_G increases precipitously while at the same time the oxygen diffusion coefficient distinctly changes. The effect this has on the speed of degradation depends on the material and the specific degradation process. The example in Fig. 18 shows the change in the glass transition temperature of a clear PE lacquer as a function of time when sorted at a temperature of 80°C [11]. This state of affairs applies in principle to natural and artificial weathering as well, since the irradiation causes the samples heat up in comparison with their surroundings. When the testing temperature is the same as the glass transition temperature, small differences in temperature during weathering cause great differences in the results of weathering. The differing glass transition temperatures of two identical samples that only differ with respect to their thermal pretreatment is also one of the causes of the difference in weathering results despite identical testing parameters (cf. Section 9 as well).

3 Rain/Humidity

Water is ubiquitous in our environment, whether in the form of humidity or of rain and dew. All materials used outdoors are exposed to these influences. It is expedient to distinguish two mechanisms by which water affects organic materials. On the one hand is the purely mechanical stress of the material due to swelling and shrinking resulting from the varying availability of water in the surroundings; on the other is the chemical reaction of the polymeric organic materials with the water, for example, during hydrolysis or the more roundabout route involving the formation during irradiation of OH and HO_2 radicals, which in turn react with the organic material. First, as to swelling and shrinking:

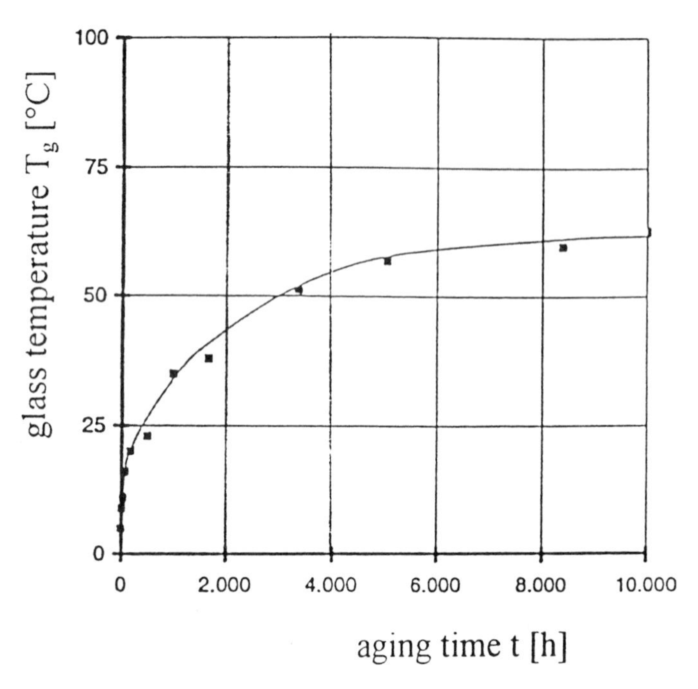

Figure 18 Glass temperature T_g as function of aging time t. Temperature 80°C, PE-C clear coat.

When parts made of synthetic materials or their coatings absorb water from humid air or are directly moistened by rain or dew, they tend to swell. A subsequent dry period implies that water will be released. When the surface layers dry out, this should cause a contraction in volume; however, this is partially prevented by the fact that the underlying layers are still swollen. This results in tensile stresses on the surface which can lead to cracking [12]. Figure 19 is a schematic diagram of the compressive and tensile stresses occurring when a plate with a thickness d absorbs and releases water. When the water is absorbed, compressive stresses prevail on the outside and tensile stresses on the inside; when water is released, compressive stresses prevail on the inside and tensile stresses on the outside, accordingly. These stresses disappear when moisture equilibrium is reached. In this context, it is important to point out that the swelling and shrinking in themselves already subject the material to considerable stress. Solar radiation increases this only inasmuch as a large number of photochemical aging processes cause the surface of the material to become brittle and thus increase its tendency to crack under the tensile stresses that appear during the dry period.

As we mentioned at the beginning, however, the water can also be directly involved in the degradation reaction in a chemical sense. An example of this is the chalking of TiO_2-pigmented coatings or synthetic materials. Experience has long shown that TiO_2 systems are subject to chalking, i.e., the binder or synthetic material is degraded by the action of the weather and the TiO_2 particles are exposed on the surface, where they form a dull white layer that can be wiped off. We know from experience that the chalking becomes more intense the more water is available on the surface. In a dry atmosphere, no chalking can be observed. The details of the individual photochemical processes were cleared up several years ago [13]. They result from the combined effects of three meteorological factors: shortwave radiation, water, and oxygen in the air.

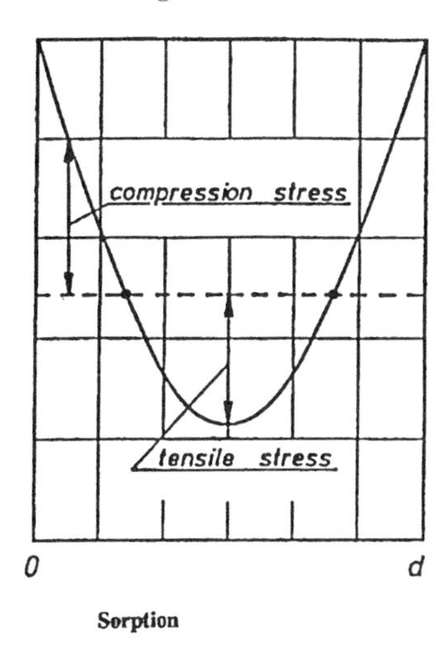

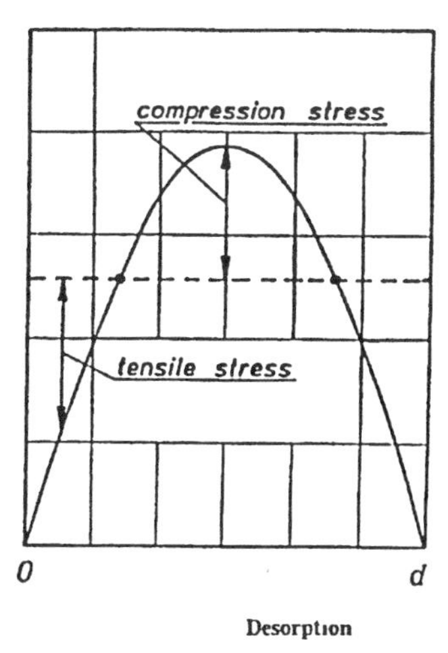

Figure 19 Tensile stress and compression stress during water absorption and desorption.

4 Industrial Pollutants

Industrial pollutants are playing an ever greater part in our environment—unfortunately, one might well add. To start with, they may react directly with organic materials, leaving aside other meteorological factors. However, the true problems regarding materials do not occur as a rule until several meteorological factors act together. An example is the formation of atmospheric SO_2 and its effects on organic coatings in combination with water and the ultraviolet radiation from the sun. The effects of this combination of SO_2, water, and ultraviolet radiation is responsible for the rapid discoloration. Not until it is irradiated does the watery SO_2 solution on the surface of the samples cause dilute sulfuric acid to form, which in turn causes discoloration as a result of its reaction with the pigments. The effects of nitrogen oxides on polymers in the dark and when subjected to solar radiation are discussed in [14].

Testing with increased concentrations is a good method for determining the influence of industrial pollutants on the weather resistance of polymers. Schulz and Trubiroha [15] give some results of the action of pollutants and radiation on automotive coatings.

5 Outdoor Weathering

The information provided by outdoor weathering is correct only in regard to the behavior of the specimen at the respective site during or after the exposure period. All conclusions regarding its behavior when put to practical use rest on the assumption that the climatic conditions it will then encounter will be the same as those during the outdoor weathering. Experience has shown, however, that extreme weather conditions can indeed occur in certain years despite uniform long-term averages. Periods of outdoor weathering lasting one year or less are also affected by seasonal factors during exposure. Measuring and recording all essential meteorological parameters during outdoor weathering is thus decisive. Only then is it possible to draw well-founded conclusions or make comparisons to short-term weathering in devices in which the same exposure parameters can be set and measured.

5.1 Measuring the Exposure Parameters

Commercial weathering stations measure both the usual meteorological data, such as temperature, relative humidity, and direction and speed of the wind, and the following data, which are important for the aging of polymeric materials.

Duration of Rain

As far as the aging of polymeric materials is concerned, the duration of rain is more important than the amount. It makes a great difference if a certain amount of rain falls, say, in a matter of a few minutes in a sudden shower, or in the form of a drizzle lasting several hours. The depth of penetration into the material, and thus the influence on the weathering behavior is much greater in the second case than in the first!

The duration of rain is measured as follows: Both ends of a strip of cotton approximately 5 cm long are fastened to a glass plate. The glass plate is slanted 45° to the horizontal. The electrical conductivity of the cotton fabric is constantly measured. When it rains, the conductivity increases, and these periods of increased conductivity are recorded. In order to be sure that only the periods of rain are recorded, and not any periods during which the cloth is covered with dew, the bottom of the glass plate is enclosed by a cover beneath it in which a 25-watt light bulb burns continuously. The heating of the glass plate prevents dew from forming. When it rains, an additional 40-watt bulb switches on that ensures that the cotton fabric dries quickly when the rainfall stops, so that only the effective duration of the rain is recorded.

Duration of Humidity

The duration of humidity is the most important of all measured data regarding humidity. It indicates how long the surface of the sample was moist. Apart from the duration of rain, it also includes periods of dew and the period of dampness after the rain stops. The measuring apparatus is in principle the same as that used for the duration of rain, but without the sources of heat.

Black Panel or Black Standard Temperature, or if Applicable, White Panel or White Standard Temperature

(Cf. Section 2.1).

Total Irradiance and Total Radiant Exposure of Global Radiation

The total irradiance, i.e., the amount of energy per unit time and surface element which reaches the earth's surface as a result of direct sunshine and diffuse celestial radiation in the entire wavelength range of the radiation, is usually measured with a solarimeter. Its sensor consists of one black and one white surface, whose temperature difference it measures. The difference in temperature is a measure of the total irradiance E. From E, we can then calculate the total radiant exposure, $H = \int E\,dt$, which strikes the samples during weathering. It is usually stated in MJ/m^2.

Irradiance and Radiant Exposure in the Ultraviolet Range

Ultraviolet radiation is one of the primary causes of aging in organic materials. Measuring the ultraviolet radiation thus tells us much more about the stress on the material than measuring the total radiation. The irradiance in the ultraviolet range, E_{UV}, is usually measured with a device equipped with a photocell receiver and a broadband ultraviolet filter. This combination makes the measuring device sensitive to the spectrum between 300 and 400 nm. A diffusing cap ensures that the evaluation of the radiation takes proper account of the cosine response. We can calculate the radiant exposure in the ultraviolet range from E_{UV} by using the formula $H_{UV} = \int E_{UV}\,dt$. It is usually stated in kJ/m^2.

Some weathering stations offer an additional service consisting of supplementary measurements, for example, the pH value of the rain and/or dew, and the concentration of pollutants, such as O_3, SO_2, and NO_X.

5.2 Typical Arrangement of Samples

The samples are usually exposed at an angle of 45° to the equator, occasionally at 5°, seldom at the angle of the respective latitude. ASTM G7 and ASTM D1435 describe the most commonly used types of sample layout. It is advisable to expose a reference material of known aging behavior along with the sample material. In this way, extreme weather fluctuations can be identified and taken into consideration. In the following, the most important sample arrangements are listed:

Weathering without back coating
Weathering with back coating (Fig. 20)

Under the same exposure conditions, samples with a back coating will become warmer than those without. They also remain moist for a longer time. Samples sensitive to temperature and humidity will thus yield different weathering results. The choice of an appropriate arrangement for the samples depends on the intended use of the material being tested. At all events, the type of the sample arrangement must be precisely indicated in the test report.

Standard black box

The samples are fastened to the top of an open black box in such a way that they form the upper edge of the black box. The samples are *not* covered. This arrangement can be used to test lacquers, coatings, and decorative elements for use on the outside of car bodies. The

Figure 20 Oudoor weathering.

black box is intended to simulate the car. The temperature of the samples is higher than in the standard arrangement.

Weathering under glass
Black box with glass cover

Here the samples are arranged *inside* a black box. This type of weathering arrangement was developed to simulate conditions inside a car. It therefore primarily serves to test materials and components for the interior decoration of cars.

BBUGVACT (black box under glass variable angle controlled temperature)

This is an advanced development of the black box with glass cover (see ASTM G24). The glass used in BBUGVACT is a type of safety glass used in car making, and the temperature of the air in the black box is kept constant at 95°C by a thermostatically controlled electric heater from 6:00 a.m. to 6:00 p.m. The results, for example, those from the car seat covers, exhibit a good correlation with results obtained in general practice.

5.3 Accelerated Outdoor Weathering

To accelerate the aging processes in outdoor weathering, an array of mirrors that focuses the solar radiation on the samples and thus increases the irradiance on the samples by a factor of roughly 8 has been in used for some time. Figure 21 shows such an apparatus,

Figure 21 Accelerated outdoor weathering.

which has been set to follow the position of the sun. The samples are cooled by a stream of air and, if necessary, sprayed with water at regular intervals. It is only reasonable to use this type of mirror layout in areas where the direct solar radiation is not weakened by clouds or haze. This main area of application is therefore in desert climates, such as Arizona. This testing method is described in detail in ASTM D4364, ASTM E838, and ASTM G90.

6 Accelerated Weathering in Devices

In an ideal world, accelerated weathering would cause all aging processes in polymers to take place exactly as in general practice, but much, much faster, in order to enable us to identify the changes in the material after a short period of time. In principle, this ideal can be achieved, but only in part. It is indeed technically possible, if quite expensive and complicated, to simulate all meteorological parameters for an observed climate zone as they occur in the course of time. However, this will not accelerate the degradation processes, since these will still take place at the same rate as they do naturally. Instead, we restrict ourselves to a simulation of the significant parameters—radiation, heat, rain/humidity—in the weathering devices, whereby we change their development over time in comparison to natural weathering.

First, let us consider how the meteorological parameters are technically simulated.

6.1 Sources of Radiation

The most decisive technical problem with respect to weathering devices is to achieve an adequate simulation of solar radiation. When artificial weathering first began in the 1920s, the only source of intense radiation available was the *carbon arc*. The spectral energy distribution of this radiation bears but a faint similarity to that of solar radiation.

In the 1950s, this was replaced by *xenon lamps*, making a great technical advance. Xenon lamps are gas discharge lamps. Their radiation spectrum is quite similar to that of solar radiation, except that they give off radiation of less than 300 nm, which does not occur in the solar radiation reaching the earth's surface, and that they also emit considera-tion infrared radiation. Both must be eliminated by suitable filters. The target spectral distribution that must be simulated for the purposes of materials testing is specified in CIE Publication No. 85, Table 4.

In the course of time, two systems of devices have come to the fore: those using air-cooled and those using water-cooled xenon lamps. In principle, air or water cooling is irrelevant to the radiation emitted by a xenon lamp. This does not change the spectral energy distribution of the radiation. The method of cooling does, however, affect the respective filter system and the overall design of the devices. In the water-cooled devices, the source of radiation consists of the xenon lamp itself and two concentrically arranged absorption filters, between which the cooling water flows (Fig. 22). Apart from its cooling function, the water also functions as a filter in the infrared range, which is quite desirable. The two cylindrical filters enable different qualities of glass to be combined in order to generate different spectral energy distributions of the radiation, for example, sunshine outdoors or through window glass. Figures 23 and 24 compare the spectral energy distribution of the radiation with two commonly used combinations of filters to global radiation outdoors (CIE Publication No. 85, Table 4) and through window glass.

The latest generation of air-cooled xenon lamps use selective reflection filters that, along with absorption filters, eliminate the infrared radiation (Fig. 25). The infrared

Figure 22 Xe burner and filter system.

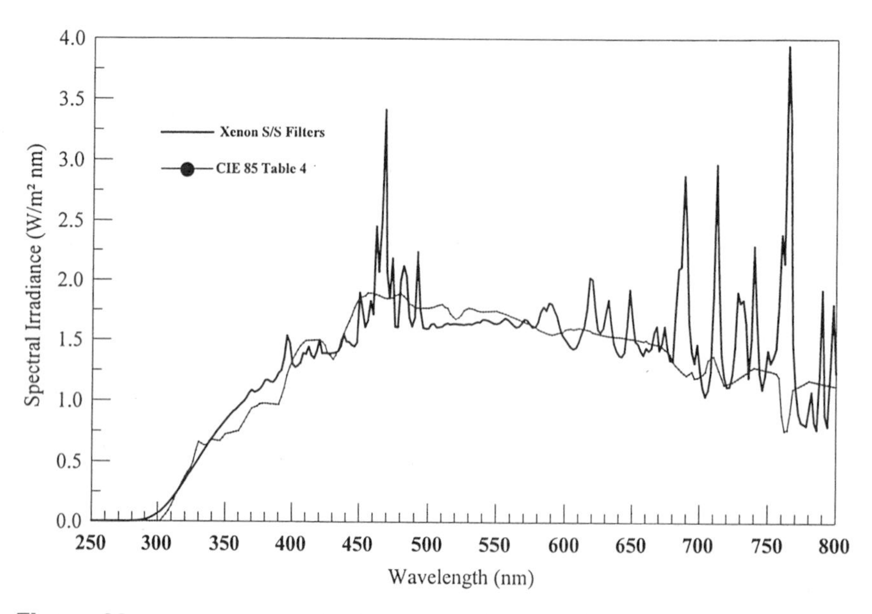

Figure 23 Xenon arc with S/S filters vs. global radiation CIE 85, Table 4.

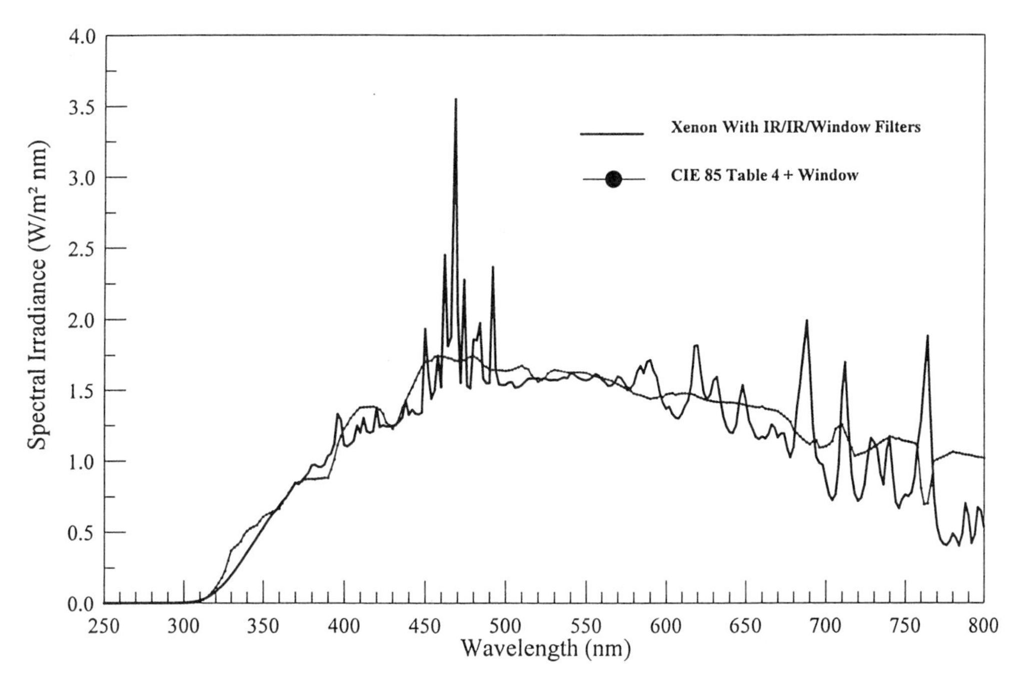

Figure 24 Xenon arc with IR/IR/window filters vs. CIE 85, Table 4 + window.

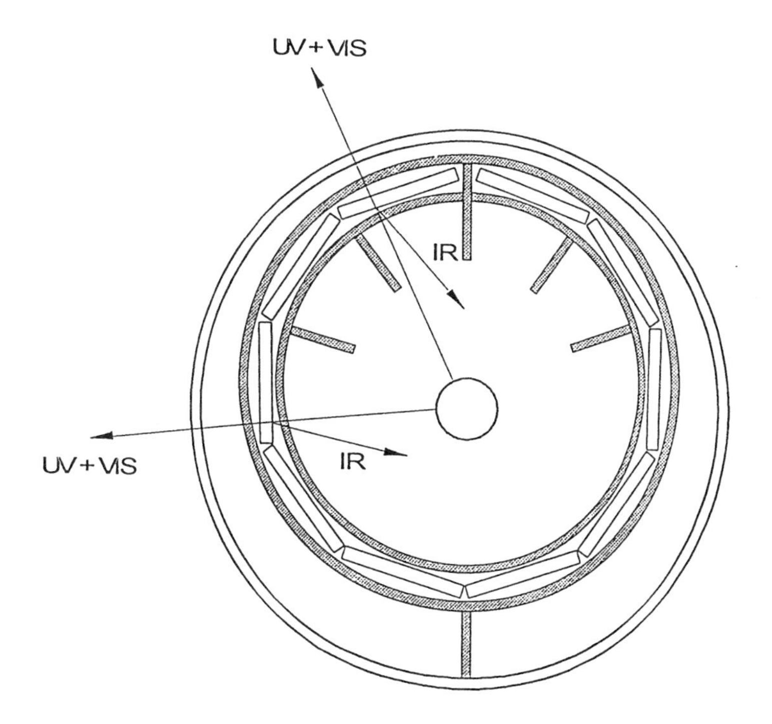

Figure 25 Xenochrome filter system.

radiation is reflected by the reflection filters and absorbed by black absorbers, which in turn are cooled by the cooling air. The reflection filters allow the ultraviolet and visible light to pass through them and enter the test cabinet through the outer cylinder. There are two models of reflection filter, which differ in their transmittance to ultraviolet radiation and with which the spectral distribution of solar radiation through window glass or outdoors can be simulated. In addition, there are special models for special applications.

Although the filter systems for the various types of devices are built differently, they serve the same purpose, viz., the simulation of global radiation in conformance with CIE Recommendation No. 85. The better this goal is achieved, the smaller are the differences between the various filter systems. Hence today's standards or test specifications indicate only the physical conditions of testing, supplemented by examples of suitable filter systems only if necessary. The manufacturer of the devices is responsible for providing the appropriate filter systems. It is important that users of the testing devices employ the right filter system for their problems and indicate this in all test reports. It is not enough to state merely "Tested in an XY weathering device"; an indication of the type of filter system used is also imperative. Otherwise, professional communication with other testing agencies becomes impossible.

Together with an appropriate filter system, the xenon lamps in use today offer the best simulation of global radiation in the ultraviolet and visible range. Apart from their many technical advantages, such as the fact that their radiant power can be controlled within broad limits by changing the electric power without altering the spectral distribution of the radiation, they also have one disadvantage: their yield of usable radiation is relatively low. A large part of the input electric power must be given off in the form of heat. In laboratory devices, this is easy to control and thus acceptable. With large systems, however, such as those used to irradiate entire vehicles, this becomes a technical and economic problem. Therefore, *metal halide lamps* are used in large solar simulation systems. These consist of lamps that send out a multilinear spectrum that can be considered to be a continuum for the purposes of materials testing. The yield of usable radiation from these lamps is approximately three times that of xenon lamps. Figure 26 shows a comparison of the spectral distribution of an optimized metal halide lamp with that of global radiation in conformance with CIE No. 85.

Metal halide lamps have technical peculiarities, however, that must be taken into account when laying out and operating weathering systems or laboratory devices for weathering. Three of these should suffice to give an idea of the problems: first, the dependence of the spectral distribution of the radiation on the temperature of the lamp. Therefore measures must be taken to ensure that the design provides as constant a temperature as possible in the vicinity of the lamp. The second peculiarity comes from the same physical cause: the dependence of the spectral distribution of the lamp on the electric power supply. Thus the ability to set a desired irradiance on an object by changing the electric power supply is narrowly limited. Lowering the irradiance becomes possible when filters are used that do not depend on wavelength, such as close-meshed wire grating, or when the distance of the lamp from the object is increased. The third peculiarity is that, compared to xenon lamps, the sample strew with respect to the spectral distribution varies from one lamp to another of the same type. At the present time, this can be partially ameliorated by measuring the lamps and selecting one accordingly. The lamp units (Fig. 27) are arranged in such a way that they can be shifted and equipped with various neutral filters to reduce the radiant flux, so that an identical irradiance can be set to strike the surface of a car, for example.

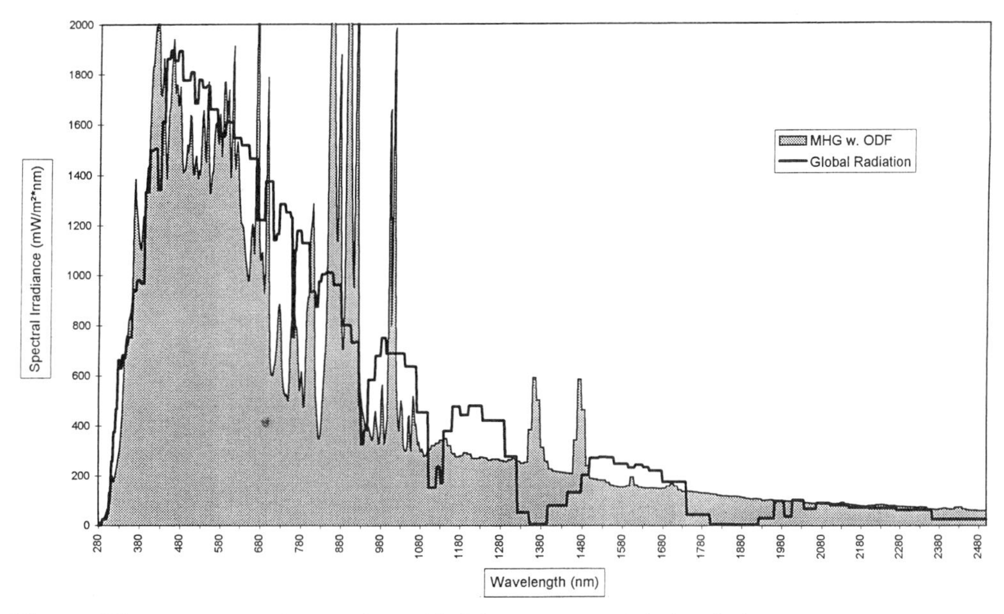

Figure 26 Spectral irradiance of metal halide system vs. global radiation.

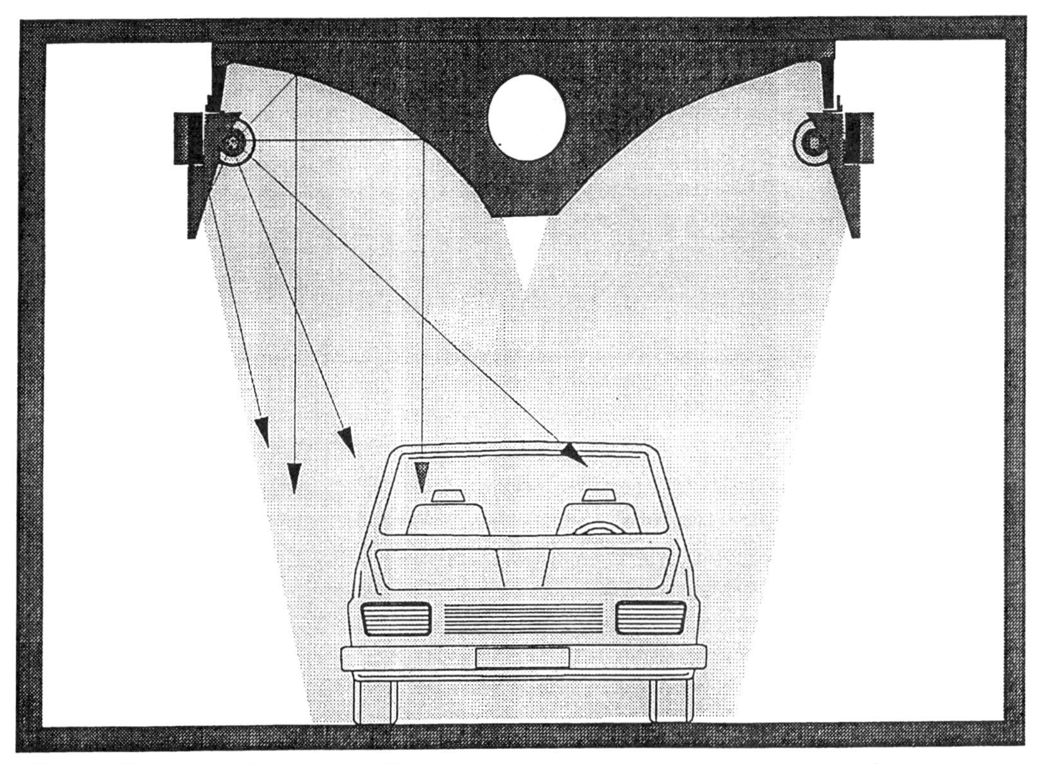

Figure 27 System SolarModul 4000.

Not only car makers are interested in testing large components, but also suppliers. For economic reasons, however, they will hardly be likely to operate large solar simulation systems for entire vehicles. Thus there is a demand for laboratory testing devices in which large components can be tested under the same conditions as those in large simulation systems. By way of example, Fig. 28 shows a schematic diagram of such a laboratory testing device with a test surface of $1.6 \times 0.6\,m$.

Another development also attempts to get around one of the drawbacks of the xenon lamp, its relatively low yield of radiation. *Fluorescent lamps* have a distinctly higher radiation yield. The decisive drawback to all fluorescent lamps, however, is their geometrically large expansion at low radiance. Image-forming systems for generating the required high irradiance on the samples are thus difficult to contrive. Even if fluorescent lamps with interior reflectors are packed as densely as possible, it is impossible to generate a sufficiently high irradiance on the samples in the ultraviolet and visible spectral range. By selecting suitable fluoresent materials, however, the fluorescent radiation can be limited to the ultraviolet range, where it will emit an irradiance on the samples that is comparable to that measured outdoors. Figure 29 shows the spectral irradiance of two commonly used lamps in comparison with global radiation. The UVB-313 lamps formerly in common use show a considerable amount of radiation in spectral ranges that do not occur in solar radiation. They cannot be recommended for aging tests on polymers. The new UVA-340 lamps provide a good simulation of solar radiation in the spectral range of $300 \leq \lambda \leq 350\,nm$. *Nota bene*: This technology only simulates global radiation at the shortwave end of the ultraviolet spectrum. The radiation of ultraviolet fluorescent lamps contains only small portions in the visible and infrared ranges. This heats up the samples until they are only slightly warmer than the surrounding air, in contrast to conditions outdoors. This has the advantage of keeping the samples moist for a longer period of time after they have been covered with dew, for instance, which is especially important with respect to the corrosion of the undersurface of a coating. On the other hand, the mechanical stresses resulting outdoors from the heating and cooling of the upper absor-

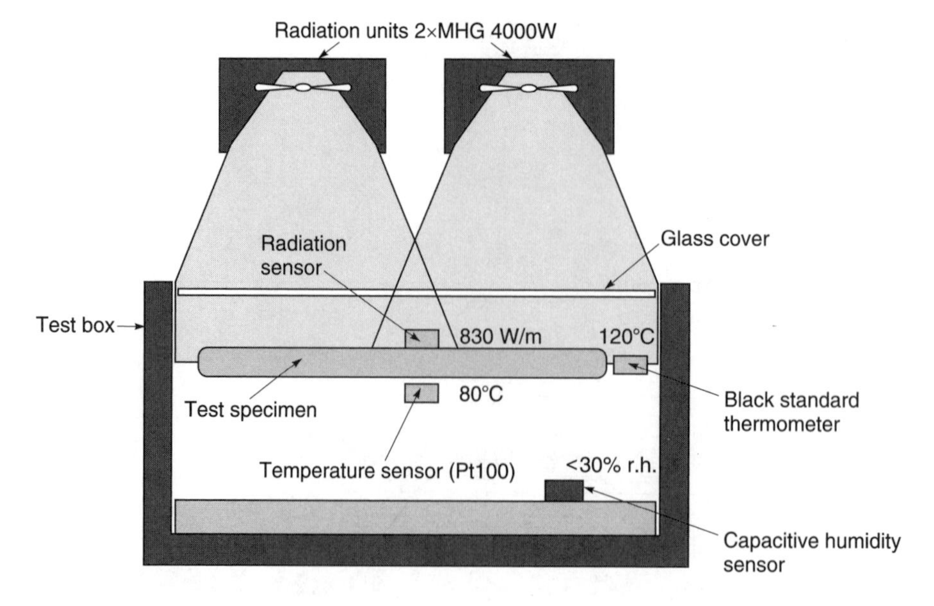

Figure 28 SolarClimatic 1600.

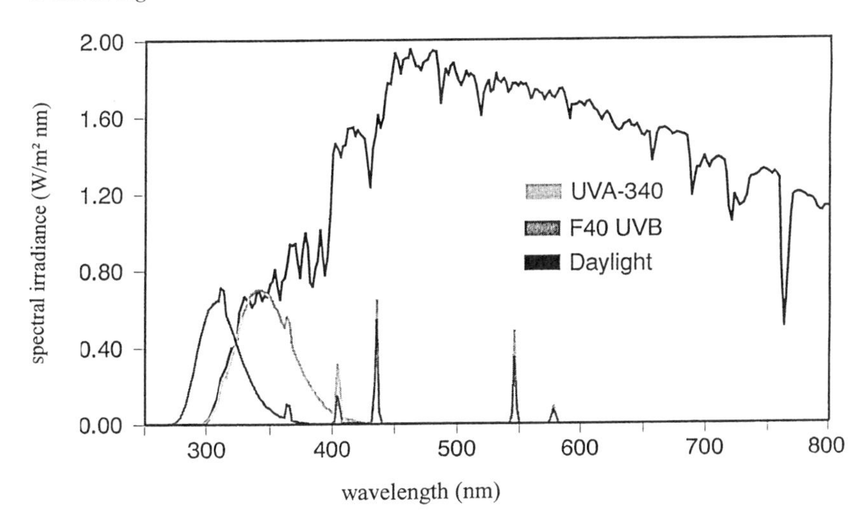

Figure 29 Spectral irradiance of two fluorescent lamps vs. daylight (Miami).

bent layer do not appear, which is a definite disadvantage. This is important in relation to cracking, among other things.

Devices using fluorescent lamps are frequently used for purposes of quality control in comparison to a standard. Figure 30 shows a schematic diagram of a fluorescent lamp device taken from ASTM G 53–93. The samples are at a distance of approximately 5 cm from the fluorescent lamps. The air in the test cabinet can be heated up. The water on the

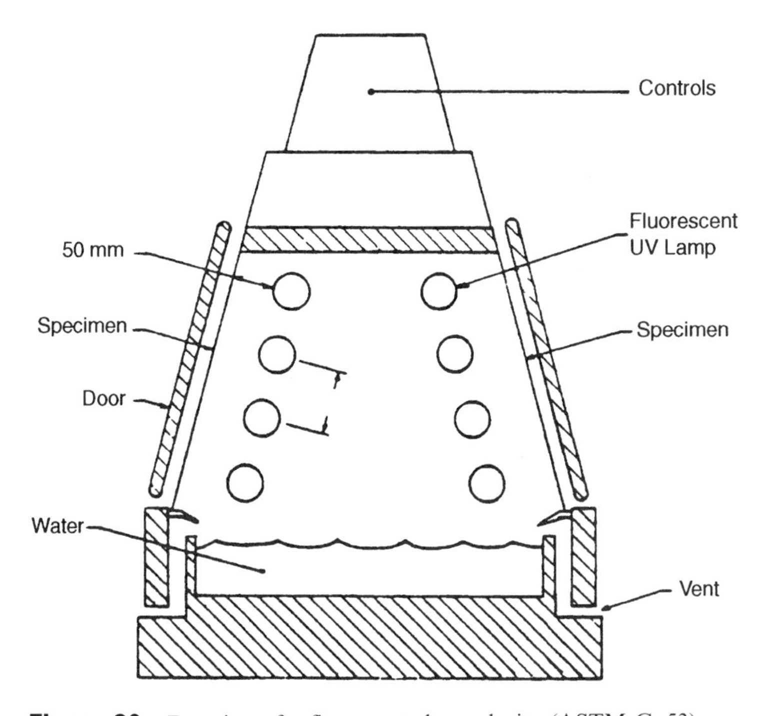

Figure 30 Drawing of a fluorescent lamp device (ASTM G–53).

floor, which can also be heated up, humidifies the test cabinet. It can even produce dew on samples with good thermal conductivity, such as lacquered metal sheets. Modern fluorescent lamp devices are also equipped with a contrivance for spraying the samples. However, whoever uses these devices in general practice must be aware of the technical limitations mentioned above. To what extent these are of importance, however, depends on the problem and the material being tested, especially its spectral sensitivity.

Important prerequisites for reliable test results are uniform and constant conditions in all weathering devices. The improvement of this aspect runs like a thread through all the technical developments made in the devices for the past 40 years. In modern weathering devices, the irradiance is continually measured on the surface of the samples and the electric power supply of the lamp readjusted if, for instance, the irradiance should sink because the lamp and/or the filter should be old or dirty. Two approaches have been taken as far as metrology is concerned. In some devices, the spectral irradiance is measured and controlled at a specific point in the spectrum (narrowband measurement), usually at 340 or 420 nm (Fig. 31). In others, the irradiance is measured and controlled in the range between 300 and 400 nm (broadband measurement) (Fig. 32). There has been much discussion regarding the advantages and disadvantages of each method. The better the aging behavior of the lamps and filters, and there has been considerable progress made in this respect, the smaller are the differences between them. Each type of data on irradiance can be converted to the other if the spectral distribution of the radiation is known. The measuring and control unit used in recent fluorescent lamp devices is calibrated to a spectral irradiance of 340 nm and thus conforms to the narrow-band measurement mentioned above.

The radiation measuring unit built into the weathering device must be calibrated at regular intervals. Two methods are commonly used for this purpose: either a calibrated xenon lamp is used, which generates an irradiance indicated in the calibration certificate on the surface of the samples at a specific level of electric power, or the radiation measuring device must be recalibrated by the manufacturer at regular intervals.

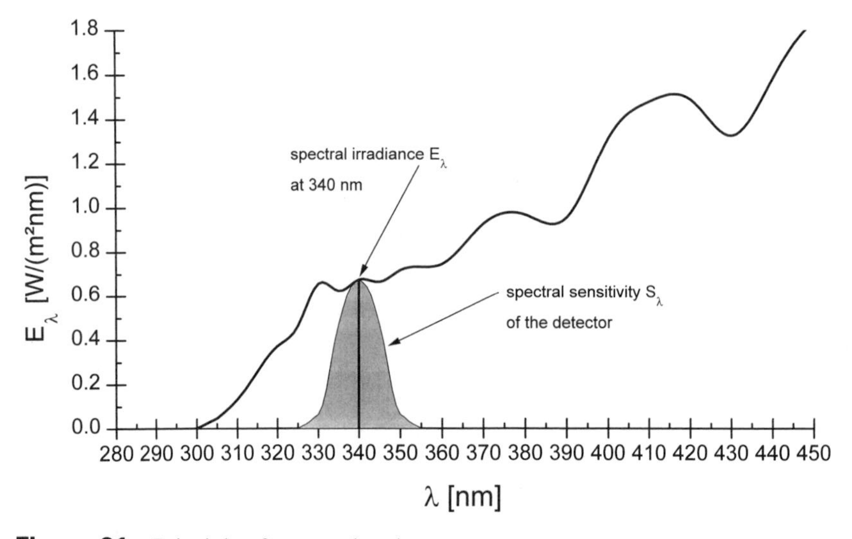

Figure 31 Principle of narrow band measurement.

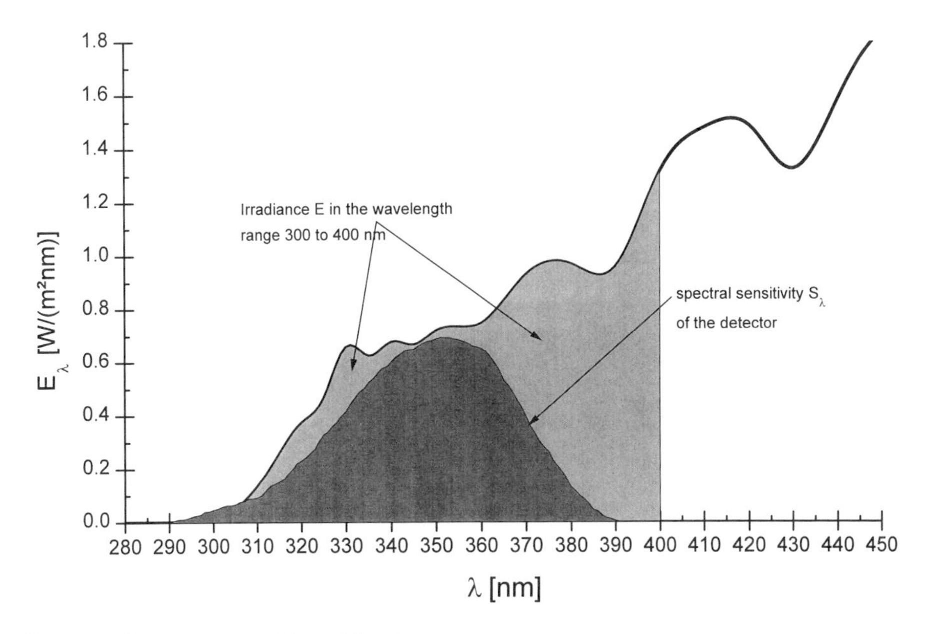

Figure 32 Principle of broad band measurement.

6.2 Heat

Weathering devices that simulate global radiation generate so much radiant energy that the test cabinet and the samples must be cooled by means of a stream of air. The air in the laboratory is used to this end. Only in special cases are additional heating or cooling units necessary. In weathering devices relying on fluorescent lamps, the temperature of the samples is roughly equal to the temperature of the air in the test cabinet, since the radiant energy, which is transmitted only by the ultraviolet radiation, is so low as to be negligible. If higher sample temperatures are desired, the air in the test cabinet is heated.

There are two important adjustment parameters in weathering devices:

Black and White Panel Temperature

This topic has already been discussed in Section 1.2.

Temperature in the Test Cabinet

The effects of temperature on the aging behavior of polymeric materials became especially clear when materials for the interior equipment of motor vehicles were tested. Since these are often exposed to solar radiation at high temperatures, the test conditions for artificial weathering must be selected accordingly. A specific black panel temperature is set in the weathering devices.

Experience has shown that not only the black panel temperature but also the temperature in the test cabinet can affect the test results. This may at first be rather surprising, but it becomes understandable when we observe more closely the effects of the addition and removal of heat on a sample in a weathering device. The surface temperature of an irradiated sample depends on the one hand on the properties of the sample itself (material, type of surface) and on the other, on the conditions of exposure. Here we will take the sample and how it is held in the sample holder as given and direct our attention exclusively to the conditions of exposure. By way of example, we will observe the black panel thermometer. The temperature of the black panel thermometer T_{BPT} is a function of the irradiance E, the temperature T_{air} of the air in the test cabinet, and the speed v_{air} of the air in the test cabinet, including the turbulence, the angle of flow, etc.

$$T_{BPT} = f(E, T_{air}, v_{air}) \tag{13}$$

At a constant irradiance, T_{BPT} still depends on T_{air} and v_{air}. Hence it is possible to set one and the same T_{BPT} by selecting differing T_{air} and v_{air}. Or, to put it differently, T_{air} and v_{air} can differ in the devices despite the same T_{BPT} in weathering devices No. 1 and No. 2. As long as we direct our attention exclusively to the black panel thermometer, this is of no consequence. However, the heat balance, and thus the temperature of the samples, does differ as a result of the various materials and surface structures in weathering devices 1 and 2. This leads inevitably to different weathering results. In terms of device technology, this implies that T_{air} and v_{air} must be stipulated in addition to T_{BPT}. T_{air} is easier to measure and control than is v_{air}. Therefore modern weathering devices allow t_{air} to be set and controlled independently of T_{BPT}, within certain limits. Consequently, newer testing standards stipulate the temperature of the air in the test cabinet along with the black panel temperature.

6.3 Rain and Humidity

Most weathering devices enable the samples to be sprayed with rainwater at specified intervals for various periods of time. To this end, they generally allow the temperature of the rainwater to be adjusted. The quality of the rainwater is decisive. Impurities collect on the surface of the samples and mimic a change of chalking in the course of weathering, for instance, at highly polished surfaces. Thus the use of distilled water or at least deionized water is recommended, and is even expressly prescribed in many standards.

The air in the test cabinet is moistened either by mechanically or ultrasonically atomizing the water, or by evaporation. The relative humidity is measured and controlled at a representative site inside the test cabinet. In doing so, it must be noted that relative humidity is only one adjustment parameter for the weathering device. It tells us nothing about the relative humidity on the irradiated surface of the sample or in its immediate vicinity, since the samples have a higher temperature than the surrounding air in the test cabinet owing to the absorbed radiation. This also applies, of course, to samples in outdoor weathering or in normal use.

7 Possibilities and Limits of Time Acceleration During Accelerated Weathering

The demand for increasingly shorter testing periods can be understood from a commercial point of view. This is limited by factors specific to the different materials, however. Leaving aside the analytic methods for characterizing the aging of polymeric materials mentioned in Fig. 1, which are not the topic of this chapter, the only remaining possibility for reducing the time required for testing is to intensify one or more of the significant meteorological parameters. However, the consequence is that the degradation processes during accelerated weathering do not necessarily take the same course as in practical use. Fig. 33 lists the common methods for accelerating degradation processes, as well as possible future developments. These will require the spectral distribution of solar radiation across the entire spectrum, or only in the ultraviolet range, to be as good as possible—not forgetting the restrictions mentioned in Section 6.1. The decisive reason for the reduction of the time spent on tests in accelerated weathering devices is the fact that simulated solar radiation acts on the samples at the maximum irradiance E for 24 hours a day. The specified calculated acceleration factor is a purely mathematical quantity yielded by the quotient of the radiant exposure (dose) received by the samples, $H = \int E\,dt$, and the radiant exposure in Florida over a period of one year.

The calculated acceleration factor does not take account of the effects of other parameters, such as temperature. The tendency here is to follow a similar procedure and select as a rule the maximum temperature occurring in general practice as the test temperature, which is then allowed to prevail for 24 hours a day, however. This yields an additional acceleration factor that is difficult to estimate, since it is specific to the material and also depends on the property under observation in a given material. Multiplying both acceleration factors ushers us into the realm of experimentally observed acceleration factors, which average out at 10–15 but have a wide margin of error both upward and downward. Let it be stated once again for purpose of clarification: The mathematical estimate mentioned here cannot substitute for determining the acceleration factor of a given material experimentally, observing the change in the property and stipulating the exposure parameters.

In modern xenon weathering devices, the irradiance on the samples can be increased to nearly twice the maximum irradiance of solar radiation. A short time ago, devices appeared on the market that are capable of increasing the irradiance on the samples to three times the irradiance that would otherwise commonly prevail. To this end, the infrared portion of the radiation is kept low and/or the samples are cooled to ensure that the black panel temperature does not rise despite the increased irradiance. This new way of reducing the time of accelerated weathering has yielded very encouraging results, in particular in the textile industry and in the Japanese car industry. There is certainly a need for further systematic studies undertaken in general practice to determine the experimental conditions and material-specific prerequisites that will allow widespread application of this method.

	calculated accel. factor compared to Florida	**remarks**
radiation (provided $E_{\lambda,sample} = E_{\lambda,sun}$)		
irradiance $E_{sample} = E_{sun,max}$ 24 h/day	≈5	**applied for many years**
increase E_{sample} about 3-fold without increase of T_{sample}	≈3	**in test at present**
shift of E_λ towards shorter wavelength		**attention:applicable for specific use only**
heat		
$T_{sample} \approx T_{practice,max}$ 24h/day	≈2 - ? depending on material	**applied for many years**
increase T_{sample}		**attention: other degradation processes may occur**

Figure 33 Chances and limits of accelerated weathering.

8 Guide to Selecting Appropriate Methods of Testing

Figure 34 lists the three most important tasks of accelerated weathering. The easiest of these is to compare the polymers being tested with a standard agreed between supplier and purchaser, or within a company, whose aging behavior is well known. This method is often used to ensure the quality of running production. Or the purchaser stipulates a minimum requirement, for example, a specific degree of light-fastness, color difference (ΔE) or chalking after being exposed to a specified level of radiant exposure. The most difficult task is to predict the service life of a new or modified material. In the simplest case, relevant standards or in-company testing regulations are already in place. Often, however, existing standards must be modified or new test conditions developed for a particular test. Figures 35 and 36 illustrate this problem. Figure 37 is a schematic diagram of the material-specific dependencies of the degradation processes, which should, if possible, be known before test methods for special materials or applications are developed. These are

The spectral sensitivity s_λ of the material or alternatively its activation spectrum $s_\lambda E_\lambda$, that is, the product of the spectral sensitivity and the spectral distribution of the radiation, which is easier to determine experimentally. This allows us to identify those portions of the radiation spectrum responsible for the property change Δ under consideration.

goals of accelerated weathering

quality control
(comparison to a general or internal standard)

control of minimum requirements

lifetime prediction

Figure 34 Guide to suitable test procedures.

materials without knowing about application, e.g.:	**general test procedures,** **ISO 4892, ASTM G26,G53** **ASTM D2565** **ISO 11341, ASTM D4587**
materials for specific application **e.g.:automotive interior material**	**specific test procedures** **ISO 105-B06, SAE J-1885**
e.g.:automotive exterior material	**SAE J-1960**
semifinished products or component parts **e.g. PVC sheet roofing**	**ASTM D4434**
complete cars or component parts **e.g. dash boards**	**DIN 75220**

Figure 35 Guide to suitable test procedures.

measurement of environmental conditions at the place of use as function of time or max./min.

radiation: E_λ; E_{total}, E_{UV}, E_{340}; $H = \int E dt$

temperature: T_{sample}, T_{air}, $T(d)$
black panel/black standard temperature
white panel/white standard temperature

rain: t_{rain}

humidity: r.h., $t_{humidity}$

Figure 36 Guide to suitable test procedures.

property change Δ

$\Delta = f(\lambda)$ **spectral sensitivity** s_λ

activation spectrum $s_\lambda \cdot E_\lambda$

$\Delta = f(E)$ **t = const**
T = const

$\Delta = f(T)$ **E = const**
t = const

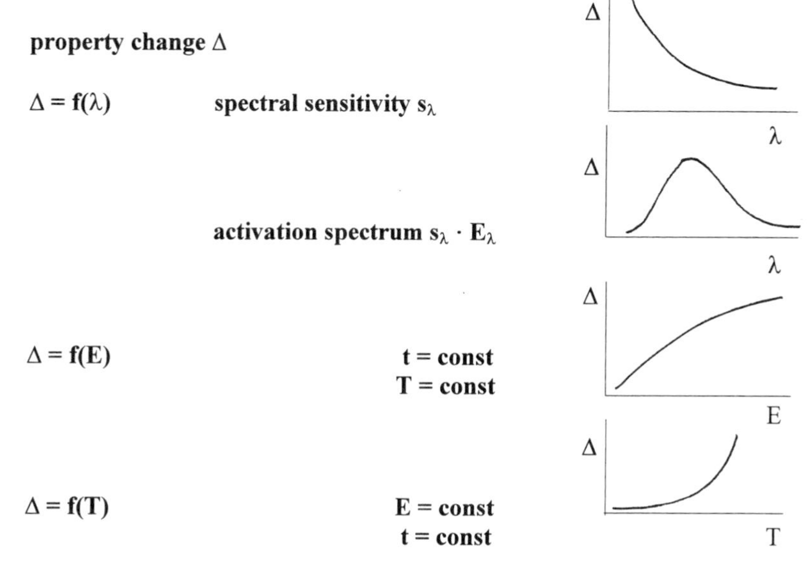

Figure 37 Guide to suitable test procedures.

From this, we can estimate how the change in the spectral distribution of natural or artificial radiation will affect the aging behavior.

As an initial approximation, we assume that the property change Δ is proportional to the impinging radiant exposure $H = \int E\,dt$. This is not always accurate, especially for a high irradiance *E*. Hence this point must be examined before considering a test in which the irradiance would be tripled, for instance.

Dependence of the property change Δ on the temperature. This point has already been discussed in Section 2.2.

9 Final Remarks

To conclude, let us consider one more point taken from daily practice: When problems occur related to reproducing the results of accelerated weathering, or to its agreement with the long-term behavior of a coating, the blame is often placed solely on the accelerated weathering test employed or the weathering device used in it. This may well be the case in some instances. However, it is often forgotten that the preparation of the samples and their prehistory, which is often unknown, can have a substantial effect on the test results. Even if the samples are guaranteed to come from the same batch, it is still worth investigating how the samples were stored or treated between the time of sampling and the start of testing. As a rule, property changes due to weathering depend upon processes that take place on the surface of the sample and in a thin layer immediately beneath the surface. These can easily have been damaged in any number of ways during preparation or storage, as a result of exposure to heat or irradiation, for instance. Nonuniform conditions during the production of the specimens can also make the results of weathering tests difficult to reproduce.

References

1. Trubiroha, P. (1978), *Chemie-Kunststoffe-Aktuell*, Sondernummer, 10, Donauländergespräch Wien, pp. 25–29.
2. Searle, N. D. (1985), *Seventh International Conference on Advances in the Stabilization and Controlled Degradation of Polymers*, Luzern, May 22–24.
3. Searle, N. D. (1987), *Internationales Symposium "Bewitterung"*, 28/29 September, Essen.
4. Andrady, A. L., Torikai, A., and Fueki, K. (1989), *J. Appl. Polym. Sci.*, *37*, 4, 935–946. Andrady, A. L., Fueki, K., and Torikai, A. (1990), *J. Appl. Polym. Sci.*, *39*, 3, 763–766.
5. Boxhammer, J., Kockott, D., Trubiroha, P. (1993), Black standard thermometer—Temperature measurement of polymer surfaces during weathering tests, *Material-Prüfung*, *35*, 143–147.
6. Fischer, R. M., and Ketola, W. D. (1994), Surface temperatures of materials in exterior exposures and artificial accelerated tests, *Accelerated and Outdoor Durability Testing of Organic Materials*, ASTM, STP 1202.
7. VDI 3789, Part 2 (1994), Environmental meteorology. Interactions between atmosphere and surfaces. Calculation of short-wave and long-wave radiation.
8. Trubiroha, P. (1981), Hanau Symposium, May 5.
9. Tabor, B. J., and Wagenmakers, J. C. (1992), *Melliand Textilberichte*, *1*, 9–14.
10. Watanabe, Y. (1989), *International Symposium: Automotive Test Procedures for Interior Trim Materials*, August 17–18, Williamsberg, U.S.A.
11. Peters, H. G. (1996), *Korrelation zwischen Kurzzeitprüfung und Langzeitbewährung von Beschichtungen*, I-Lack 4, 64, Jahrgang.
12. Lehmann, J. (1991), Lecture Deutsche Physikalische Ges., Meeting, March 5.
13. Volz, H. G., Kaempf, G., and Klaeren, A. (1976), *Farbe + Lack*, *82*, 805. Kaempf, G., Volz, H. G., and Klaeren, A. (1978), *Defazet*, *32*, 288.
14. Jörg, F., Schmitt, D., and Ziegahn, K.-F. (1984), *Staub-Reinhalt. Luft*, *44*, 276.
15. Schulz, U., and Trubiroha, P. (1996), Predicting water spot and acid etch resistance of clearcoats by artificial weathering: Acid dew and fog test, *Durability Testing of Nonmetallic Materials*, ASTM STP 1294 (Robert J. Herling, ed.), American Society for Testing and Materials.

List of Specifications on Weathering

AATCC–016–93–94	Colorfastness to light.
AATCC–111–90–94	Weather resistance.
AATCC–169–90–94	Weather resistance of textiles: Xenon lamp exposure.
ASTM C 217–90	Test method for weather resistance of natural slate.
ASTM C 732–87	Test method for aging effects of accelerated weathering on latex sealing compounds.

ASTM C 734–93	Test method for low-temperature flexibility of latex sealing compounds after artificial weathering.
ASTM C 793–91	Test method for effects of accelerated weathering on elastomeric joint sealants.
ASTM C 1257–94	Standard test method for accelerated weathering of solvent-release-type sealants.
ASTM D 358–88	Specification for wood to be used as panels in weathering tests of coatings.
ASTM D 518–91	Test method for rubber deterioration—Surface cracking.
ASTM D 750–95	Test method for accelerated weathering test conditions and procedures for bituminous materials (carbon-arc method).
ASTM D 750–95	Test method for rubber deterioration in carbon-arc weathering apparatus.
ASTM D 822–96	Practice for conducting tests on paint and related coatings and materials using filtered open-flame carbon-arc light and water exposure apparatus.
ASTM D 904–94	Exposure of adhesive specimens to artificial (carbon-arc type) and natural light.
ASTM D 1006–92	Practice for conducting exterior exposure tests of paints on wood.
ASTM D 1014–88	Method for conducting exterior exposure tests of paints on steel.
ASTM D 1150–87	Method for single and multi-panel forms for recording results of exposure tests of paints.
ASTM D 1435–85	Practice of outdoor weathering of plastics.
ASTM D 1654–92	Method for evaluation of painted or coated specimens subjected to corrosive environments.
ASTM D 1669–89	Method for preparation of test panels for accelerated and outdoor weathering of bituminous coatings.
ASTM D 1670–90	Test method for failure end point in accelerated and outdoor weathering of bituminous materials.
ASTM D 2249–94	Test method for predicting the effect of weathering on face glazing and bedding compounds on metal sash.
ASTM D 2565–92	Practice for operating xenon-arc type light exposure apparatus with and without water for exposure of plastics.
ASTM D 2898–86	Test method for accelerated weathering of fire-retardant-treated wood for fire testing.
ASTM D 3361–87	Practice for operating light- and water-exposure apparatus (unfiltered carbon-arc type) for testing paint, varnish, lacquer, and related products using the dew cycle.
ASTM D 3424–92	Test methods for evaluating the lightfastness and weatherability of printed matter.
ASTM D 4141–87	Practice for conducting accelerated outdoor exposure tests of coatings.
ASTM D 4303–93	Test methods for lightfastness of pigments used in artist's paints.
ASTM D 4329–92	Practice for operating light- and water-exposure apparatus (Fluorescent UV—Condensation type) for exposure of plastics.
ASTM D 4364–84	Practice for performing accelerated outdoor weathering of plastics using concentrated natural sunlight.

ASTM D 4459–93	Practice for operating an accelerated lightfastness xenon-arc type (water-cooled) light-exposure apparatus for the exposure of plastics for indoor applications.
ASTM D 4587–91	Practice for conducting tests on paint and related coatings and materials using a fluorescent UV-condensation light- and water-exposure apparatus.
ASTM D 4798–88	Test method for accelerated weathering test conditions and procedures for bituminous materials (xenon-arc method).
ASTM D 4799–88	Test method for accelerated weathering test conditions and procedures for bituminous materials (fluorescent UV condensation method).
ASTM D 4909–89	Test method for color stability of vinyl-coated glass textiles to accelerated weathering.
ASTM D 5031–89	Practice for testing paints, varnishes, lacquers, and related products using enclosed carbon arc light- and water-exposure apparatus.
ASTM D 5071–91	Practice for operating xenon arc-type exposure apparatus with water for exposure of photodegradable plastics.
ASTM D 5105–90	Practice for performing accelerated outdoor weathering of pressure-sensitive tapes using concentrated natural sunlight.
ASTM D 5208–91	Practice for operating fluorescent ultraviolet (UV) condensation apparatus for exposure of photodegradable plastics.
ASTM D 5722–95	Standard practice for performing accelerated outdoor weathering of factory-coated embossed hardboard using concentrated natural sunlight and a soak-freeze-thaw procedure.
ASTM E 782–95	Exposure of cover materials for solar collectors to natural weathering under conditions simulating operational mode.
ASTM E 881–92	Exposure of solar collector cover materials to natural weathering under conditions simulating stagnation mode.
ASTM E 1596–94	Standard test methods for solar radiation weathering of photovoltaic modules.
ASTM F 1164–88	Test method for evaluation of transparent plastics exposed to accelerated weathering combined with biaxial stress.
ASTM G 11–88	Test method for effect of outdoor weathering on pipeline coatings.
ASTM G 23–92	Practice for operating light- and water-exposure apparatus (carbon-arc type) for exposure of nonmetallic materials.
ASTM G 24–87	Practice for conducting natural light exposures under glass.
ASTM G 26–92	Practice for operating light-exposure apparatus (xenon-arc type) with and without water for exposure of nonmetallic materials.
ASTM G 53–91	Practice for operating light- and water-exposure apparatus (fluorescent UV-condensation type) for exposure of nonmetallic materials.
ASTM G 9–94	Practice for performing accelerated outdoor weathering of nonmetallic materials using concentrated natural sunlight.
prEN 513–93	Unplasticised polyvinylchloride (PVC-U) profiles for the construction of windows; determination of the resistance to artificial weathering.

EN 1056–96	Plastics piping and ducting systems—Plastics pipes and fittings—Method for exposure to direct (natural) weathering.
ENV 1170–8–96	Test method for glass-fibre-reinforced cement—Part 8, Cyclic weathering type test.
prEN 2591–320–97	Aerospace series—Elements of electrical and optical connection—Test methods—Part 320, Simulated solar radiation at ground level.
prEN ISO 105–B03-96	Textiles—Tests for colour fastness—Part B03, colour fastness to weathering: Outdoor exposure (ISO 105–B03, 1994).
prEN ISO 105–B04-96	Textiles—Tests for colour fastness—Part B04, Colour fastness to artificial weathering; Xenon arc fading lamp test (ISO 105–B04, 1994).
EN ISO 877–96	Plastics—Methods of exposure to direct weathering, to weathering using glass-filtered daylight, and to intensified weathering by daylight using fresnel mirrors (ISO 877, 1994).
prEN ISO 11341–97	Paints and varnishes—Artificial weathering and exposure to artificial radiation—Exposure to filtered xenon-arc radiation (ISO 11341, 1994).
EN ISO 12017–96	Plastics—Poly(methyl methacrylate) double- and triple-skin sheets—Test methods (ISO 12017, 1995).
ISO 2810–74	Paints and varnishes, Notes for guidance on the conduct of natural weathering tests.
ISO/DIS 3917–97	Road vehicles—Safety glazing materials—Test methods for resistance to radiation, high temperature, humidity, fire and simulated weathering (Revision of ISO 3917, 1992).
ISO 4607–78	Plastics, methods of exposure to natural weathering.
ISO/DIS 4665–96	Rubber, vulcanized and thermoplastic—Resistance to weathering (Revision of ISO 4665–1, 1985, ISO 4665–2, 1985, and ISO 4665–3, 1987).
ISO 4892–94	Parts 1–4, Plastics—Methods of exposure to laboratory light sources.
ISO/DIS 9370–96	Plastics—Instrumental determination of radiant exposure in weathering tests—General guidance and basic test method.
ISO 11341–94	Paints and varnishes—Artificial weathering and exposure to artificial radiation—Exposure to filtered xenon-arc radiation.
ISO/FDIS 11507–96	Paints and varnishes—Exposure of coatings to artificial weathering—Exposure to fluorescent UV and water.
ISO 12017–95	Plastics—Poly(methyl methacrylate) double- and triple-skin sheets—Test methods.

29 Lifetime Prediction

Roger Brown
Rapra Technology Ltd., Shawbury, Shrewsbury, England

1 Introduction

The object of any degradation test, and also tests of such time-dependent mechanical properties as creep and stress relaxation, is directly or indirectly to predict service life. If time were not important, tests could be made under the expected service conditions and the test continued for the desired lifetime. With expected lifetimes sometimes measured in tens of years, this is clearly not a viable option in most cases, and some form of extrapolation, and usually acceleration, will be needed.

An indirect indication of service life is obtained simply by comparison of the performance of materials under given test conditions, the one which shows the smaller change being deemed to perform better. If one material is a "standard" with known service performance, an estimate can be made of the other material's expected performance. Particularly with accelerated tests, this is in fact a dangerous assumption, because the differences seen under the test conditions may not be similar to the differences realized in practice. To make a direct estimate of service life it is necessary to apply some form of extrapolation technique to measured data.

For tests made under unaccelerated conditions it is a matter of extrapolating to longer times, which means obtaining a function for the change of the parameter(s) of interest with time.

By definition an accelerated test requires that the degrading agent or agents be present at a level higher than that to be seen in service. The general procedure is to measure the degree of degradation by change in selected properties of interest as a function of time of exposure to the degrading agent and, unless there is previous knowledge, it is necessary to carry out tests at a number of levels of the agent. There are then two stages to modelling the degradation process:

Obtaining a function for the change of the parameter(s) of interest with time
Obtaining a function for the rate of change of the parameter(s) with the level of the degrading agent

Using these relationships the change in the property for longer times and lower levels of the degrading agent can be predicted. Clearly, the success of the process is critically dependent on the validity of the models used. Whilst a number of models applicable to polymers have been known for a long time, they are in practice relatively infrequently applied, and the majority of accelerated durability tests carried out are used on a comparative basis only. There are a number of reasons for this, not least that there is a lack of evidence for the universal validity of the models, and that the behavior found for many materials is very complicated. It is also a fact that the generation of data over sufficient times and levels of agent is an extremely time-consuming and expensive process.

2 Standardized Procedures

International and national standards for predicting lifetimes of polymeric materials are mostly conspicuous by their absence. The only aspect to have been standardized is application of the Arrhenius approach (see below) to evaluating accelerated test results involving the effect of temperature. IEC 216 [1] is a guide to evaluating the thermal endurance of electrical insulating materials, and ISO 2578 [2] applies the same principle to determining time/temperature limits to plastics. In both cases the accent is more on finding maximum service temperatures than on extrapolating to normal ambient temperature. Use of the same Arrhenius relationship is also in the course of being standardized for rubbers [3].

There are of course many standard methods for carrying out natural or accelerated tests and time dependent mechanical property tests and for measuring the properties used to monitor degradation. These can be found in this volume and in other standard texts [4,5] and their use for weathering tests has been reviewed [6].

3 Models for Change of Parameter with Time

The change in parameter with time may take several forms, and the form may vary with the level of the degrading agent as well as with the parameter chosen. Indeed, using too great an acceleration, resulting in changing the degradation mechanism and hence the rate of change, is a common pitfall. The difference in degree of change with different monitoring parameters should also be emphasized, and the best practice is to use properties of direct relevance to service.

Some possible forms of the change of parameter with time are shown in Fig. 1. The easiest form to handle, a linear relationship, is unfortunately frequently not found because of the complicating effect of several factors. There may be an induction time, due for example to protective additives, or an initial nonlinear portion while equilibrium conditions are reached. A chemical reaction may produce a linear change, whilst a physical effect may be logarithmic and the two may occur together. An autocatalytic reaction will show an increasing rate after a period of time. The cyclic trace shown in Fig. 1 is a real example of the modulus of a rubber after aging at an elevated temperature with firstly the effect of curing more, then softening, and finally becoming brittle.

In some cases it may be possible to transform a curve to linear form, for example by taking logarithms, or a relatively simple relation can be found to fit. With composite curves it may be justifiable for the end purpose intended to deal only with one portion,

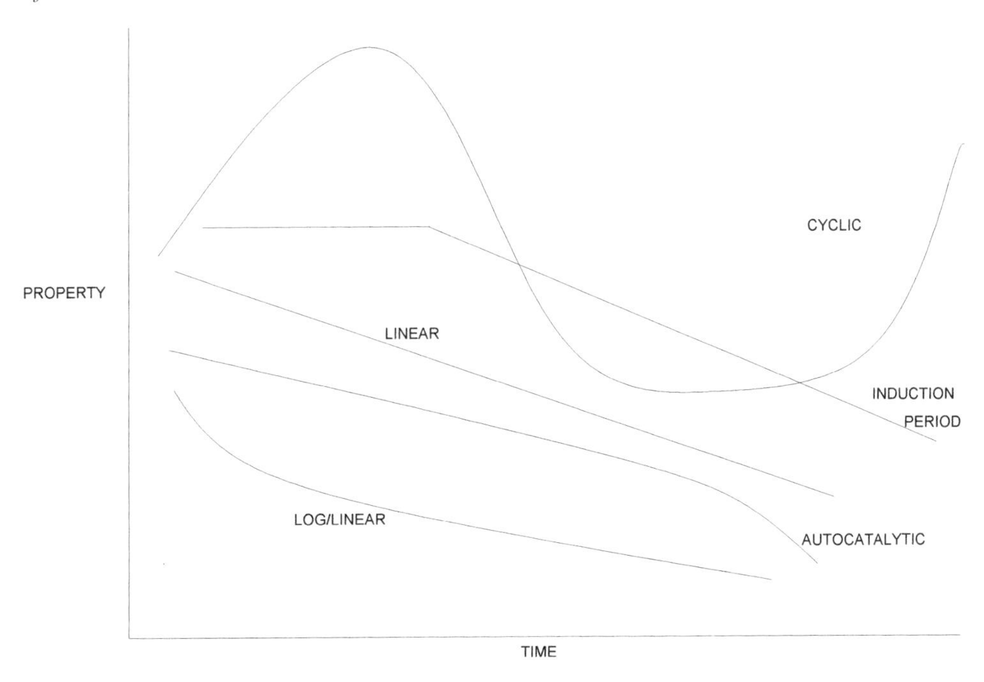

Figure 1 Change of property with time.

for example by ignoring what happens before an equilibrium condition is reached. The cyclic rubber modulus example was dealt with by considering the fairly linear softening section on the basis that curing would not happen at normal ambient temperature and useful life had been exceeded when brittleness set in.

It is common practice when similar materials are being compared to ignore the shape of the curve and to take the time for the property to reach some percentage, say 50%, of its initial value. This may be expedient but is clearly less satisfactory than modelling the curve and could be extremely misleading if materials with substantially different curves are compared. If a time/temperature shift method is used to model the effect of temperature (see below), no function to describe the change of property with time need be assumed.

4 Time-Dependent Mechanical Properties

4.1 Creep and Stress Relaxation

When a polymer is subjected to a stress or strain, the observed stiffness changes, an increase in strain is seen in a creep test, and a reduction of stress in a relaxation test. The changes will be composed of physical and chemical effects, with physical processes dominating at lower temperatures and shorter times and chemical or degradation effects dominating at higher temperatures and longer times.

If we consider the case of tests at ambient temperatures and moderate times, then the effects measured are principally physical. Creep and stress relaxation tests under these conditions may need extrapolating to longer times. Ignoring any possible degradation effects, it is commonly found with rubbers that a plot of modulus against log of time will yield a linear relationship, which makes extrapolation very easy. With plastics, the log linear plot generally does not yield a straight line, but a log log plot may do, or extrapolation can be achieved by fitting a relation to the curve. It should be emphasized that it is generally dangerous to extrapolate in this way greater than one decade in time. At longer times, other processes may come into play, including the effects of chemical aging.

Creep or stress relaxation tests may be made at a series of different stresses or strains to create a family of curves. A relation where strain is a function of stress and time will exist between the curves, and this relationship can be represented as a surface in three dimensions. Computer techniques allow this surface function to be generated, including smoothing of the experimental data.

When creep or stress relaxation tests are made at elevated temperatures, aging effects are generally present, and extrapolation to lower temperatures and longer times may be made using the techniques discussed later. It may be noted that it is fairly common to make creep tests on plastics at different temperatures, but for short times, and extrapolate using a time-temperature shift procedure.

4.2 Fatigue

Fatigue can be defined as the decrease in load-bearing capacity with time under load. Under constant load conditions this is termed static fatigue or creep rupture, and under cyclic or intermittent load dynamic fatigue.

The most simplistic approach to making predictions of fatigue life is to carry out tests at a series of stress or strain levels and construct a curve of cycles or time to failure against stress or strain. A limiting stress or strain below which the fatigue life is very long may be found as illustrated in Chapter 10, or extrapolation can be made to lower stresses by finding a relation to fit the curve. This can be successful with rubbers for modest extrapolations, but with plastics there is generally a strong probability of results being misleading.

Considering static fatigue of plastics, service failures are almost inevitably brittle, whereas failures from short-term tests at higher stresses are often ductile. Extrapolations from tests giving ductile failures will result in an overestimate of long-term performance. At some point there will be a ductile–brittle transition that may be seen as a "knee" in the lifetime curve, beyond which the decline in strength accelerates. The transition is shifted to lower times by

Temperature
Stress concentrations
Effect of liquids (environmental stress cracking)
Dynamic stressing

Rupture starts at stress-concentrating defects, but the growth of cracks in plastics is complicated. Using a test piece with a sharp notch produces a large stress concentration and helps to obtain brittle fractures. However, the fatigue life is then based on the crack propagation rate rather than including the time to crack initiation. Time to failure in a product will depend on the severity of stress-raising features.

The same difficulties apply to dynamic fatigue tests, but additionally for all materials there is a danger of running at frequency and stress conditions that cause temperature rise in the test pieces. Whilst the increase in temperature will tend to decrease fatigue life, in plastics at least the shape of the fatigue life–stress curve can change so as to give increased estimates of life at low stresses. The fatigue life will also be affected by the shape of the waveform used, a square wave being the most severe, and by the mean stress during the cycle, a positive mean stress reducing fatigue life.

It will also be appreciated that at longer times and with higher temperatures there will be aging effects that generally reduce fatigue resistance.

A further difficulty with making predictions from fatigue data is that the spread of failure times is generally large and not usually normally distributed. In consequence, the measured data has relatively large uncertainties attached to it that are magnified considerably by any extrapolation process.

Considering these difficulties in making predictions from short-term tests on plastics, it is not surprising that for design purposes reliance is often put on the measured long-term data that exist for some materials and accepted maximum allowable strains are used.

4.3 Abrasion

Laboratory abrasion tests are notorious for not correlating with service. The abrasion process is complicated, and the rate of abrasion is very much dependent on the particular conditions. Laboratory tests cannot often properly represent the conditions in service, and hence correlation is difficult if not impossible to find. This does not mean that correlations never exist, but these will be application and test specific. The usual approach is to seek a correlation for the particular circumstances and to test materials on a comparative basis, rather than to predict service wear rate in absolute terms.

5 Environmental Degradation Tests

5.1 Models for Effect of Level of Degradation Agent

When the form with time has been established and a suitable measure to represent that form selected, the relation with the level of the degradation agent is needed to allow extrapolation to the service level. Generally, measures need to be made at several levels to establish a model with reasonable confidence. Typically five levels are considered satisfactory, but it should be noted that when extrapolation is to be made over several decades of time the uncertainty of the prediction will be large, even if the measured data looked very consistent. Estimates of uncertainty should always be made.

It is feasible to make an empirical fit to a graph of change against level, although it can be dangerous to do this with no theoretical justification. In cases with multiple degrading agents, and hence a complicated relation, it could be the only option, but normally an established form with theoretical justification is fitted if possible. A number of models that have been used are considered below.

Before applying any model it is essential to have confidence that the input data are valid. There are repeated warnings in the literature of data being invalid because obtained at such accelerated levels of the degradation agent that reactions occur that do not do so at lower levels.

5.2 Arrhenius Relationship

The best known and most widely used model is the Arrhenius relationship, which in particular is applied to the permanent effects of temperature as the degrading agent.

The Arrhenius relationship is

$$K(T) = A \exp \frac{-E}{RT}$$

Thus

$$\ln K(T) = \frac{-E}{RT} + C$$

Where

$K(T)$ is the reaction rate for the process
E is the reaction energy
R is the gas constant
T is absolute temperature
C is a constant

A plot of $\ln K(T)$ against $1/T$ should yield a straight line with slope E/R, which can with caution be extrapolated as shown in Fig. 2.

The Arrhenius relation is generally the first choice to apply to the effects of temperature, but no general rule can be given for the measure of reaction rate (change of parameter with time) to be used with it. Very frequently the time taken to a given percentage of the initial value is chosen.

When a form of the change of parameter with time other than linear is proposed, a power law is usually tried first

$$f(X) = X^n$$

Combining these gives the Avrami equation,

$$X = X_0 \exp\left[-At^n \exp\left(\frac{En}{RT}\right)\right]$$

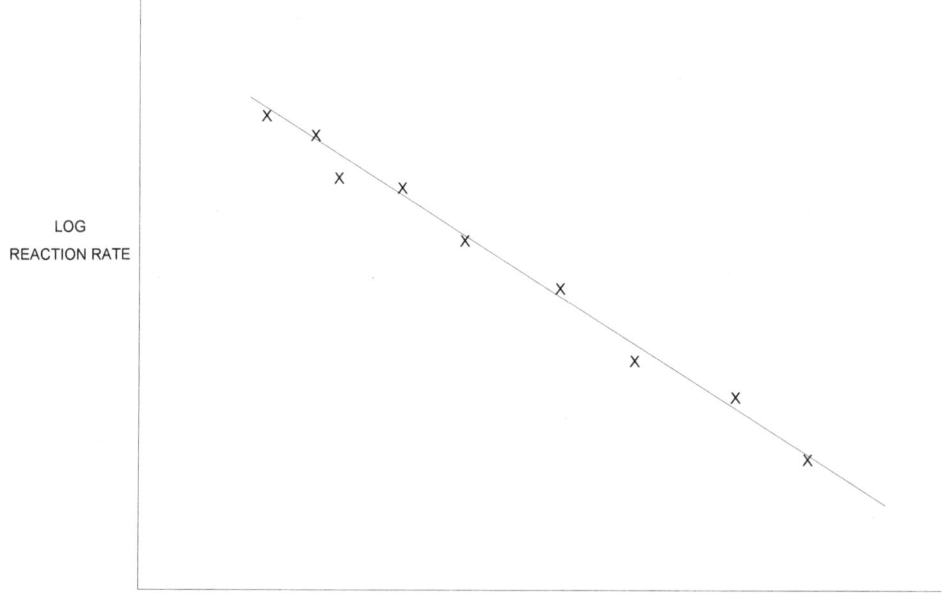

Figure 2 Arrhenius relationship.

An example of the application of this model has been given by Kantor and Rakova for a polyimide film [7].
A power law can be applied to describe compression set of rubbers [8], but if modelling is required to higher levels of set it is necessary to use an exponential or logarithmic function that is self-limiting between 0 and 100%.

Clearly other forms of reaction rate may be more appropriate in particular cases as outlined earlier.

There are occasions when the Arrhenius equation does not give a straight line and hence there is clear indication that predictions from it will not be valid. An alternative expression that has improved the line in certain cases is

$$\ln K = \ln K_0 + \frac{B(T_0 - T)}{10}$$

Where

K_0 is the reaction rate at a reference temperature T_0

This was found by Benson [9] to describe more successfully the thermal degradation of polyethylene film.

Predictions using these models take no account of differences in the oxygen diffusion situation in the test pieces compared to the product, and this can have an enormous effect on service performance [10]. Prediction of the oxidized layer is discussed by Verdu [11].

5.3 Time/Temperature Shift

An alternative to constructing the Arrhenius plot $\log(K)$ against $1/T$ is to shift the plots of parameter against time along the time axis to construct a master curve. Use can be made of the WLF equation

$$\log(a_T) = \frac{-c_1(T - T_0}{c_2 + T - T_0}$$

where

a_T = shift factor
c_1 and c_2 = constants
T_0 = reference temperature

The shift factors to align the plots at different temperatures with the plot for the chosen reference temperature are determined, and then these are fitted to the WLF equation to find the constants. The equation can then be used to predict the response at service temperatures.

This technique has the advantage that no particular measure of the reaction rate has to be chosen nor any form assumed for the change of parameter with time, but it can only be used if the curves at different temperatures are of the same form. In principle, other relationships between the shift factors and temperature could be fitted on an empirical basis but, with no theoretical justification, particular caution would be advised with extrapolation.

It would be of considerable interest to have comparisons of predictions made by Arrhenius and by time/temperature shift, but no examples have been found.

5.4 Artificial Weathering

Weathering is clearly a more complicated case than heat aging alone because there are temperature effects added to the light, and probably other agents such as moisture and ozone as well. Not surprisingly, there is no very widely accepted relationship equivalent to the Arrhenius. The result is that many workers have developed empirical relations that are usually only shown to be applicable to a narrow range of materials and conditions.

An attempt can be made to combine the various effects. In a number of cases at least, the rate of degradation can be considered as proportional to the intensity of light and hence to radiation dose:

$$X = X_0 + bD$$

where

D = radiation dose

This can be combined with Arrhenius for temperature effects to give a relation of the form

$$X = X_0 D \exp \frac{-E}{RT}$$

Rather than linear, other proposals for the relation between property and radiation dosage include a power law and an exponential relationship

$$X = X_0 + bD^n$$

or

$$X = X_0 + a \exp(D)$$

Note that dosage has been used, although this is substituted by time of exposure by many workers.

The relationship of degradation with light intensity is certainly nearer to linear than is the case with temperature. This implied time scale adds considerably to the effort needed to collect data at a series of acceleration levels, and in fact it is extremely uncommon for this to be done.

Another form used is [12]

$$K = 10^{b(D-a)}$$

where

K is the ratio of properties X/X_0

The constant a represents an induction period before degradation starts, which is commonly found.

A further approach is to start with a relation of the form [13]

$$X = f(y_1 + y_2 + y^3 \cdots + y_n)$$

where y_1 etc. represent the various factors or agents that may cause degradation, e.g., dosage, intensity, temperature, humidity.

Regression analysis techniques are then used to find the significant agents and produce a model for the particular data in question. This is essentially an empirical approach, and alternatively a mathematical form for the data could be found by curve fitting without consideration of the effect of individual agents. Extrapolation is then particularly dangerous.

This form of approach is probably the only way to deal with multiple agents, for example where contaminants are added to the normal weathering factors of light, heat, and water, and it can be extended to include synergistic effects. However, this quickly becomes very complicated and attempts at such analysis are not common.

In many weathering trials, results are only available for one set of conditions, and strictly no extrapolation can then be made for temperature and the degradation is assumed to be independent of light dosage rate. The temperature conditions in weathering cabinets can be quite high, and if the temperature of service were more modest the predictions could be very misleading.

A further factor, which accounts for a great deal of the difficulties of correlating accelerated and natural weathering, is the differences in spectral distribution of the light sources. The degree of degradation is critically dependent on wavelength, but natural sunlight varies in intensity, and spectral distribution and several different lamps are used in weathering cabinets. This is not taken account of in the usual models but in principle could be incorporated into a multifactor regression analysis.

5.5 Ionizing Radiation

The first assumption can be that degradation is independent of dose rate. However, acceleration levels can be very high, and this is a prime reason why in practice it is often found that the effect of a given dose decreases with increased dose rate. The limiting factor is the rate of oxygen diffusion. Recommended practice is to test at two or more dose rate levels to determine the magnitude of the effect.

When heat and radiation are considered together, the two effects will be additive. However, there can also be a synergistic factor [14].

5.6 Effect of Liquids

The case of liquids is simpler than weathering in that it is essentially a two agent situation, liquid and temperature. However it is generally necessary to take account of the rate of diffusion of the liquid into the material, which may be slow in relation to the time scale of an accelerated test. Also, it is necessary to consider that there may be physical change (swelling) of the polymer as well as chemical degradation.

When equilibrium absorption is attained well within the time of the experiment, the situation is similar to heat aging; the form of change with time has to be modelled to a degree of degradation specified, and then it is sensible to use an Arrhenius relation to account for temperature change. Clearly it is advantageous to work with thin test pieces so that equilibrium is obtained quickly, but this is not always possible, and extrapolation to thicker products may be needed.

Diffusion in the unsteady state before equilibrium is described by Fick's second law:

$$\frac{dC}{dt} = D\frac{d^2C}{dx^2}$$

where

C = volume
t = time
D = diffusion coefficient
x = thickness

As a general rule the time to equilibrium is proportional to the square of thickness. When the diffusion coefficient is known, a suggested estimation of the time to reach equilibrium to a depth b is

$$t = \frac{b^2}{2D}$$

Both the diffusion coefficient and the solubility coefficient vary with temperature in accordance with an Arrhenius relationship. The diffusion coefficient increases with temperature, but the solubility coefficient increases for gases and decreases for vapors. For a full treatment of absorption, a text on mass transport should be consulted.

The change in property measured up to, and past, equilibrium absorption is likely to show a marked change in the shape of the degradation curve. This is very noticeable in compression stress relaxation results on some rubbers in water at ambient temperature when a rise in modulus can be seen due to swelling at times greater than one year. Serebryakov et al. [15] demonstrate a similar type of effect for the strength of polystyrene.

For the particular case of water at less than 100% humidity, the amount of water absorbed at equilibrium is dependent on humidity. In some, but by no means all, cases it may be simply linear, and the relation needs to be known if performance at different humidities is to be estimated. Zinbo and Golovoy demonstrate the relationships for polycarbonate [16]. Again, absorption will change with temperature.

It becomes apparent that to transpose data from different humidities, temperatures, thicknesses, and varying levels of approach to equilibrium can be very involved. Further, whilst the transport relationships apply to the uptake of fluid, the effect on properties at times below equilibrium can never be simple because of the varying concentration with thickness.

Where the liquid causes environmental stress cracking, the situation is one of fatigue, which is accelerated by the presence of the stress cracking agent. Quite often a limiting strain will be found, and the limit is very similar for agents of different severity, but of course the time scales may be very different. The cracking of rubbers by ozone is a similar situation, and plots can be made of time to cracking against both strain and ozone concentration. Most commonly, materials are subjected to a high level of acceleration at one set of conditions with the aim of selecting a material that is resistant or protected such that no cracking occurs.

5.7 Dynamic Conditions

All the treatments discussed above have been concerned with static conditions, i.e., where in the accelerated tests the level of the degrading agents has been held constant throughout one exposure, and any extrapolation to service implicitly assumes that conditions there will also be constant. In real life however it is much more likely that service conditions will be cyclic. Generally, therefore, further approximations have to be made.

Most commonly the worst situation is assumed, for example with temperature the reaction rate will be something like doubled for a 10°C rise and lower temperatures will be relatively insignificant. With natural weathering the conditions change daily and geographically, and this is the basic reason why light dosages should be used rather than time. Even with temperature it is theoretically possible to estimate an equivalent "dose," i.e., the temperature that represents the mean of temperatures encountered, duly weighted for their degradation effects. With fluids, contact can be intermittent with drying out possible. The total chemical effect is likely to be less for less contact, but there may be effects of expansion and contraction. Such conditions would best be modelled in the accelerated experiment.

In some simple cases an additive approach can be successful. The service life is divided into stages, for example moderate exposure for one year, severe exposure for one month, and low exposure for five years. The predicted effects for the three periods can be summed and the condition of the product at the end, and hence residual lifetime, estimated.

6 Limitations

From the foregoing it is clear that prediction from accelerated tests is at best a hazardous procedure. To minimize the limitations in any particular case it is essential to design the accelerated trials to simplify and ease as far as possible the prediction process. Of the limitations the most important can be summarized as

Statistical uncertainty due to quality and number of test results.

Quality of accelerated data in terms of test conditions being sufficiently valid to relate to service conditions

Validity of extrapolation procedure

The first is a matter of minimizing variability, maximizing the quantity of data, and minimizing the degree of extrapolation needed. The quality of accelerated data in relation to service is dependent on a considerable number of factors, many of which have been mentioned above and must be systematically addressed when designing the trials. The validity of the extrapolation model is likely to be better the more proven the procedures used and the smaller the degree of extrapolation. The validity of the data and the validity of the model become interwoven. Generally, it is not unreasonable to argue that the model is valid if the data shows a good fit, but this provides absolutely no evidence that the data is valid or even relevant when applied to the intended service conditions.

7 Simulated Design Life Exposure of Products

The usual approach to accelerated durability trials is either to make a simple comparison of materials or, by extrapolation, to predict performance at longer times under less severe conditions.

An alternative to predicting performance is to simulate a given design life by accelerated exposure and then to assess performance under service conditions. This approach can particularly be used with whole products where the end performance assessment can be made by operating the product. It is further particularly suited to cases where, after the accelerated exposure, it is possible for the product to be subjected to abnormally harsh conditions and these are used for the end assessment.

The approach makes the assumptions that a valid extrapolation procedure is known and that the necessary input data has been determined to define the accelerated simulation of design life. Whilst this may seem to be a typical chicken-and-egg situation, it is possible for the data needed to be known to a sufficient approximation (perhaps with a safety factor) for the important material, but there need be no data on how the actual product manufactured from it may perform after aging.

The process can be illustrated by an extremely simplistic example. It can be claimed that via the Arrhenius relation reaction rates for mechanical properties of a rubber are known at least approximately. Hence, for example, a tire (which is a complicated structure) could be given a simulated lifetime of, say, 10 years by heat aging and then actually run under heavy loading conditions.

It might be found that although the mechanical properties after 10 years were reasonable, they deteriorated very rapidly in heavy service because the protection additives against high running temperatures had been depleted. Hence one would be advised not to use tires stored for long periods even if they seemed intact.

In practical applications of the approach, mechanical fatigue may be carried out simultaneously with the aging, and the end assessment may involve multiple agents. A problem can be that with several materials making up the product, the simulated design life aging, in being correct for one material, over-ages another, which then is the element to fail.

References

1. IEC 216, 1990, Parts 1, 2 and 3, Guide for the determination of thermal endurance properties of electrical insulating materials.
2. ISO 2578, 1993, Determination of time-temperature limits after prolonged exposure to heat.
3. ISO CD 11346. Lifetime predictions using Arrhenius plots.
4. Brown, R. P. ed. (1988), *Handbook of Plastics Test Methods*, Longman Scientific and Technical.
5. Brown, R. P. (1986), *Physical Testing of Rubber*, Elsevier Applied Science.
6. Brown, R. P. (1991), *Polym. Test.*, *10*, 1, 3.
7. Kantor, L. A., and Rakova, V. G. (1991), *Int. Polym. Sci. Tech.*, *18*, 10, T/71.
8. Budrugeac, P. (1992), *Polym. Deg. & Stab.*, *38*, 165.
9. Benson, G. W. (1986), *J. Plast. Film Sheet.*, *2*, 7, 180.
10. Malek, K. A. B., and Stevenson, A. (1992), *J. Nat. Rubb. Res.*, *7*, 2, 126.
11. Verdu, J. (1993), *Int. Pulym. Sci. Tech.*, *20*, 1, T/64.
12. Peeva, L. B., and Tsveteva, V, A. (1986), *Die Ange. Makro. Chem.*, *141*, No. 2293, 95.
13. Hamid, S. H., and Prichard, W. H. (1991), *J. App. Polym. Sci.*, *43*, 651.
14. Ito, M. (1987), *Int. Polym. Sci. Tech.*, *14*, 7, T38.
15. Serebryakov, G. A., Sitimov, S., Aristov, V. M., and Khukmatov, A. I. (1991), *Int. Polym. Sci. Tech.*, *18*, 10, T/75.
16. Zinbo, M., and Galovoy, A. (1992), *Polym. Eng. Sci.*, *32*, 12, 786.

30
Permeability

David Hands
Consultant, Sutton Farm, Shrewsbury, England

1 Introduction

1.1 Basic Theory

No polymeric material forms an impervious barrier to gas or vapor molecules. The transmission of such molecules through a membrane is a result of intermolecular spaces, pinhole defects, or porosity, or some combination of these three structural features of the material. In this chapter we are restricting the term permeation to the movement of gas or vapour molecules through molecular scale voids. The quantity of permeant is usually measured as the volume at STP for a gas, or the mass in the case of a vapor.

Permeation of gases and vapors takes place in three stages. In the first stage the permeant dissolves into the surface of the material. The second stage is the diffusion of the dissolved molecules through the material under the action of a concentration gradient. The third stage is the evaporation of the dissolved molecules from the low pressure surface of the sample.

The solubility of the permeant in a polymer is defined as that volume (or mass) which dissolves in unit volume of the material under applied unit pressure. It is described by Henry's law,

$$c = SP \tag{1}$$

where c is the concentration, i.e., the volume at STP of the gas per unit volume of polymer, S is the solubility coefficient, and P is the applied pressure. Departures from Henry's law are discussed below.

The diffusion process is described by Fick's first and second laws. Fick's first law relates the flux to the concentration gradient:

$$J_{\mathrm{n}} = -D \frac{\partial c}{\partial x} \tag{2}$$

Where J_{n} is the permeant flux, i.e., the rate of flow per unit area, D is the diffusion coefficient or diffusivity, and $\partial c/\partial x$ is the concentration gradient.

Equations 1 and 2 can be combined to give

$$J_{\mathrm{n}} = -Q \frac{\partial P}{\partial x} \tag{3}$$

where $Q = DS$ is the permeability and $\partial P/\partial x$ is the pressure gradient.

Fick's second law, from which the time dependent concentration distribution can be calculated, is obtained from Eq. 2 and the equation of continuity

$$\frac{\partial J_{\mathrm{n}}}{\partial x} = -\frac{\partial c}{\partial t} \tag{4}$$

The equation of continuity is an expression of the conservation of mass. The flux can be eliminated between Eqs. 2 and 4 to give

$$\frac{\partial c}{\partial t} = D \frac{\partial^2 c}{\partial x^2} + \frac{\partial D}{\partial c}\left(\frac{\partial c}{\partial x}\right)^2 \tag{5}$$

If the diffusivity is independent of concentration, this reduces to

$$\frac{\partial c}{\partial t} = D \frac{\partial^2 c}{\partial x^2} \tag{6}$$

Equations 4 to 6 can also be written in terms of pressure instead of concentration.

Assuming that Q is independent of pressure and that the sample is homogeneous, when steady-state conditions have been achieved Eq. 3 can be integrated to give

$$J_{\mathrm{n}} = Q \frac{\Delta P}{h} \tag{7}$$

where ΔP is the pressure drop across the sample and h is the sample thickness. If a gas mixture is being used, this equation must be used with the partial pressure drop for each component to give the flux for each component.

If Q changes with pressure, Eq. 7 would give an average value for the permeability. If there is any doubt as to the homogeneity of the material it is usual to calculate the gas transmission rate, or permeance, for the sample under test rather than to calculate a permeability. The gas transmission rate (GTR) is the flux per unit pressure drop

$$\mathrm{GTR} = \frac{J_{\mathrm{n}}}{\Delta P} \tag{8}$$

The use of laminates in packaging is common, and it is useful if a permeability for the laminate can be calculated from the permeabilities of the constituent layers. For a laminate of n separate layers the permeability is given by

$$\frac{1}{Q_{\mathrm{L}}} = \frac{1}{h} \sum_{i=1}^{n} \frac{x_i}{Q_i} \tag{9}$$

where Q_{L} is the permeability of the laminate, h is the thickness of the laminate, Q_i is the permeability of the ith layer, and x_i is the thickness of the ith layer. This equation can be used if, and only if, the permeabilities are independent of pressure. If any of the permeabilities are pressure dependent the final value would depend on the order of the layers, and a more complicated calculation would be required.

The mechanism of diffusion in polymers, and the effects of varying the environmental factors on the values of the material properties, are discussed in a number of reviews [1–8].

1.2 Nonlinear Behavior

The steady-state equation for gas transmission only applies when the gases are sparingly soluble in the polymer ($<0.2\%$), and when there is no chemical association. This is true for the air gases at pressures below 10^5 Pa, but often not true for water and organic vapors, which can have pressure-dependent permeabilities. Vapors may also act as plasticisers and so increase diffusion.

Five types of sorption behavior have been described in the literature [8,9] under isothermal conditions. Type I is the ideal behavior described by Henry's law, in which the quantity of gas that dissolves in the polymer is directly proportional to the applied pressure.

Type II behavior occurs when there is a preference for polymer penetrant pairs to form, or when there are sites in the material that preferentially absorb the penetrant. As an example, consider a black filled rubber and a gas that has an affinity for carbon black. Under these conditions the rate of sorption decreases with increasing pressure. At higher pressures, when all the sites are occupied, small quantities of the gas will then dissolve randomly in the continuous phase and the system behaves ideally. It follows from this that diffusion does not depend on the total concentration but on the much smaller concentration of gas in the continuous phase.

Type III behavior occurs when there is a preference for penetrant–penetrant pairs to form. In this case, solubility increases with increasing pressure, and diffusion decreases with increasing concentration, because penetrant pairs or clusters are less mobile than isolated molecules. This type of behavior is also associated with swelling of the sample.

Type IV behavior is a combination of Type II behavior at low pressures and Type III behavior at high pressures. Some authors [10,11] have also described a dual mode sorption, which is a combination of Type I and Type II.

For systems where Henry's law is not valid, sorption must be measured directly [12–17]. This is usually done gravimetrically. The diffusion coefficient can be obtained from the time taken to reach equilibrium. To obtain the effect of concentration on diffusion, the measurement has to be repeated at different pressures; in the case of a liquid the measurement is repeated at different vapor pressures. It is important to take into account the increase in the sample dimensions caused by swelling and buoyancy; previous swelling history may also be important.

1.3 Effect of Temperature

For all polymer–permeant systems the effects of temperature on the permeability, the diffusivity, and the solubility are usually described by Arrhenius equations [18]:

$$Q = Q_0 \exp\left(-\frac{E_P}{RT}\right) \tag{10}$$

$$D = D_0 \exp\left(-\frac{E_D}{RT}\right) \tag{11}$$

$$S = S_0 \exp\left(-\frac{E_S}{RT}\right) \tag{12}$$

where Q_0, D_0, and S_0 are constants for a particular system, E_P is the activation energy for the permeation process, E_D is the activation energy for diffusion, E_S is the heat of solution, R is the gas constant, and T is the absolute temperature.

Since $Q = DS$ it follows that

$$E_P = E_D + E_S \tag{13}$$

For both gases and vapors the diffusivity increases with increasing temperature. The solubility increases with temperature for gases but decreases for vapors. The change in the heat of solution is usually smaller than the change in the activation energy for diffusion. It follows from this that permeability increases with increasing temperature other than in exceptional circumstances.

1.4 Units

A bewildering variety of units have been used in the literature, and Huglin and Zakaria [19] reported 29 different units for permeability. Pauly [18] gives useful conversion factors as well as tables of selected values.

The SI unit for gas transmission is $m^3/(m^2\, s(Pa/m))$. This can be rearranged to give $m^2\, s^{-1}\, Pa^{-1}$ or, if the pressure is expressed in terms of newtons, $m^4\, s^{-1}\, N^{-1}$. The SI unit for vapor permeability is $g\, m^{-1}\, s^{-1}\, Pa^{-1}$ or $g\, m\, s^{-1}\, N^{-1}$.

The SI unit for diffusivity is $m^2\, s^{-1}$.

2 Gas Permeability

In this section well-established techniques used for measuring permeability and gas transmission rate are described.

The various methods that have been described in the literature, and in the standard specifications, have some features in common in the design and operation of the gas transmission cell. The main differences between them are in the methods used for measuring the quantity of gas that has permeated through the sample. The methods described in the ASTM specification have been discussed by Demorest [20]. The sample holder is divided into two parts by the sample. Usually the low-pressure side is evacuated, and the high-pressure side is held at a pressure of one atmosphere. The sample is supported against the high pressure on a porous substrate; this can be a piece of filter paper, a disc of wire mesh, or a disc of sintered metal. Leakage from the sides is prevented by O-ring seals. The general arrangement is illustrated in Fig. 1. Other methods allow measurements to be made on products such as hoses [21].

The quantity of gas permeating through the sample is measured and plotted against time (Fig. 2). When steady-state conditions have been established, the volume of gas permeating through the sample per unit area is given by [22]

$$\frac{V}{A} = \frac{Q\Delta P}{h}\left(t - \frac{h^2}{6D}\right) \tag{14}$$

where V is the volume of gas permeating through the sample expressed at STP, A is the area of the sample, Q is the permeability, ΔP is the pressure drop across the sample, h is the sample thickness, and t is the time.

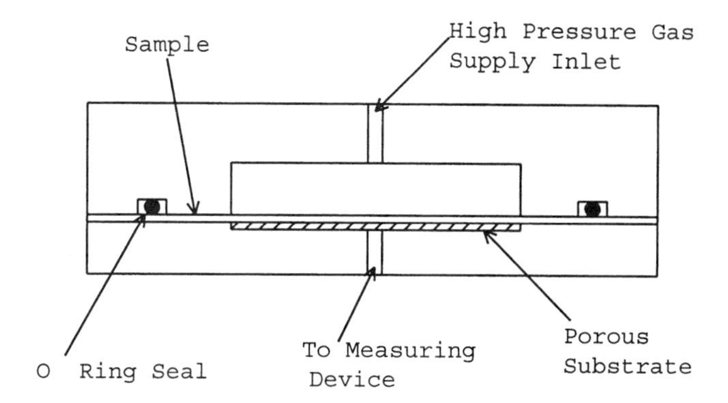

Figure 1 Schematic diagram of permeability apparatus.

Permeability can be obtained from the slope of a graph of V/A against t. This graph has an intercept on the time axis of $h^2/6D$, and from this the diffusivity can be obtained [23]. As a rule of thumb, steady state conditions are reached after about two or three times the intercept time. This method is usually more accurate for permeability than for diffusivity because it is more accurate to measure a slope than an intercept. The accuracy in measuring diffusivity, is improved if a thicker sample is used, but the experiment would then take very much longer, because the time to reach steady-state conditions depends on the square of the thickness.

It is essential in measurements of this nature to condition the samples in a vacuum before starting the experiment in order to remove the air gases from the sample. Failure to do this will not affect the final slope of the curve, but the intercept will be meaningless. As an alternative, the sample can be conditioned in the test gas at, say, one atmosphere. After equilibrium has been attained, the pressure in the high pressure side can then be increased to, say, two atmospheres.

It is absolutely essential when measuring permeability that steady state be established, i.e., when equal quantities of permeant enter and leave the sample. Under these conditions

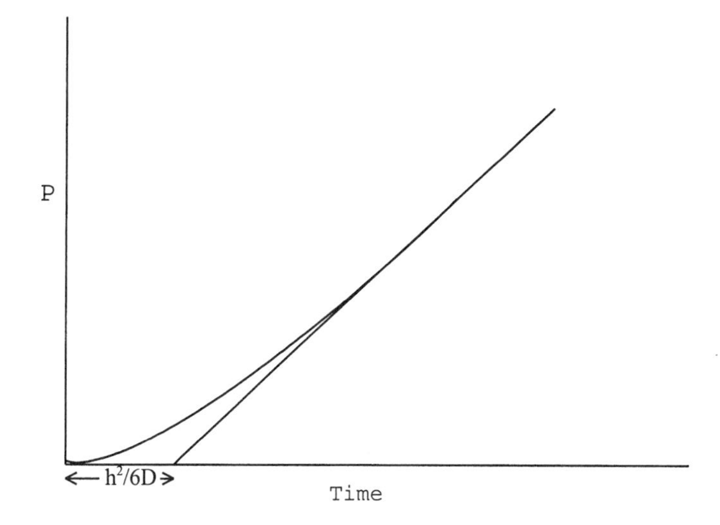

Figure 2 Typical graph of permeating species against time.

the concentration gradient does not change with time, and Eqs. 7 and 8 can be used. Before steady state conditions are achieved, more permeant dissolves into one face of the polymer sample than evaporates from the other. The first stage of the permeation process (solubility) is dominant initially, but as the concentration gradient increases, the second stage (diffusion) becomes increasingly important. The attainment of steady state may take anything from a few hours to many days. If the measurement is made before steady-state conditions have been achieved, then the calculated permeability will be smaller than the correct value.

The above requirement illustrates a problem that can occur when permeability data obtained from measurements on thin films is used for calculations on thick films. The calculations are valid if, and only if, the problem is a steady-state problem. In order to estimate whether or not steady state conditions have been achieved, it is necessary to know the diffusivity.

2.1 Manometric Method

The manometric method is described in the BS [24] and ASTM [25] and ISO [26] standard specifications. The quantity of gas that permeates through the test piece in a given time is measured as a change in pressure and volume. As described in the previous section, the test piece forms a barrier between two chambers in a gas transmission cell. A constant high pressure (usually 1 atmosphere) is maintained in one chamber and a low pressure (usually a vacuum) is initially established in the other chamber. The test piece is supported against the high pressure by a porous substrate. A mercury manometer is coupled to the low-pressure chamber to measure the variation in pressure and volume over a specified test time. A plot of mercury displacement against time gives the characteristic curve illustrated in Fig. 2. The slope increases until steady state conditions are established. As the pressure increases in the low-pressure side of the apparatus, and approaches that in the high-pressure side, the slope decreases. Steady state occurs at the point of inflexion in the curve. However, in practice steady state is marked by a lengthy section of the curve, which is sensibly a straight line. It is thus not necessary to prolong the experiment in order to obtain the whole curve; it is sufficient to wait until the curve becomes a straight line. As noted above, it is essential that steady state conditions have been achieved, and in experiments where the rate of change is slow the appearance of the curve can be deceptive. It is always better to make a few more measurements than to terminate the experiment prematurely.

From Eq. 8 it can be shown that the instantaneous gas transmission rate is given by

$$\mathrm{GTR} = \frac{273}{AT\Delta P}\left(P\frac{\partial V}{\partial t} + V\frac{\partial P}{\partial t}\right) \tag{11}$$

where A is the permeated area, T is the absolute temperature, ΔP is the pressure difference across the test piece, P is the instantaneous pressure in the low pressure side of the apparatus, and V is the instantaneous volume. Equation 11 may be written in terms of mercury displacement:

$$\begin{aligned}\mathrm{GTR} &= \frac{273}{AT(P_{\mathrm{H}} - P_0 - h)}\left[(P_0 - h)a\frac{dh}{dt} + (V_0 + ah)\frac{dh}{dt}\right] \\ &= \frac{273}{AT(P_{\mathrm{H}} - P_0 - h)}[V_0 + aP_0 + 2ah]\frac{dh}{dt}\end{aligned} \tag{12}$$

where V_0 is the initial volume between the test piece and the datum line in the manometer, P_0 is the initial pressure in the manometer, P_H is the high pressure, h is the mercury displacement from the datum line, dh/dt is the rate of change of mercury displacement, and a is the cross-sectional area of the measuring capillary.

BS 2782 uses Eq. 12 to measure the gas transmission rate at a single point chosen on the steady-state portion of the plot of mercury displacement against time. ASTM D1434 uses an approximation to the integrated form of Eq. 12 for an extended time interval.

In practice it is essential that the manometer and mercury be thoroughly clean to eliminate stick-slip effects in the movement of the mercury meniscus. A tipping manometer is frequently used to transfer mercury from the reservoir into the arms of the manometer after evacuation. During the tipping operation it is essential that the mercury not reach the stopcock, because this is always greased and the resulting contamination will spread rapidly through the manometer [27].

A different type of manometer, which eliminates the need for tilting the apparatus, has been designed [28]. In this manometer there is a direct connection between the measuring capillary and the mercury reservoir, and this permits the mercury to remain in the manometer while it is being evacuated or aerated.

Leaks in the high-pressure chamber are not important if a large reservoir of gas is used; the important thing is that the pressure remain constant. There must be no leaks in the low-pressure chamber. Leaks can occur at the interface between the test piece and the metal case of the transmission cell, or at the various joints in the manometer, and some simple procedures for detecting leaks are given in the relevant standards.

The volume of the low-pressure chamber can usually be varied to alter the sensitivity of the apparatus. This can also be achieved by using a larger permeated area, or by using an inclined measuring capillary.

Readings of mercury displacement may be made from a scale mounted behind the capillary or, more accurately, with a travelling microscope. ASTM D1434 also makes provision for the use of a calibrated resistance wire inserted into the capillary, thus allowing a continuous measurement of displacement as a function of time. It is essential to chose an alloy for the resistance wire that is not affected by mercury.

A close tolerance on test temperature is required, and therefore the apparatus is usually placed in a temperature-controlled chamber.

For full details of the test procedures, the relevant standards should be consulted.

2.2 Constant Volume Method

This method has been described in a number of research papers and is used in a test standard for rubbers [29, 30].

If the volume on the low-pressure side of the test piece is kept constant, Eq. 11 becomes

$$\mathrm{GTR} = \frac{273}{AT\Delta P}\left(V\frac{dP}{dt}\right) \tag{13}$$

where V is the volume between the test piece and the measuring device. The permeability may then be obtained from $\mathrm{GTR} = Q/h$, where h is the thickness. An estimate of the time required to reach steady-state is obtained from the relationship $t = h^2/2D$ if the value of the diffusivity, D, is approximately known.

The onset of steady-state is indicated by an inflexion in the curve of pressure against time, similar to the results from the manometric method, and is illustrated in Fig. 2.

Strictly speaking, beyond the inflexion point conditions are again transient, because the increasing pressure continuously reduces the flux from the sample. However, since the pressure change is very much smaller than the pressure drop across the sample, this perturbation is usually negligible, and after the inflexion point the curve of pressure against time is sensibly a straight line.

The use of pressure transducers instead of a manometer to measure the pressure change makes the apparatus very much easier to use [31]. It is also safer, because mercury is highly toxic. The use of transducers has the additional advantage that measurements can be made at much higher pressures, and hence the effect of pressure on the permeability can be measured [32]. Two transducers are required for the high- and low-pressure sides of the apparatus, or one transducer and a differential transducer.

The high-pressure chamber is flushed out with the test gas prior to pressurization. If the pressure in the low-pressure side is anything other than a vacuum, then the low pressure side including the ancillary pipework must also by purged with the test gas to prevent bidirectional gas permeation. Time must also be allowed for the air gasses to come out of solution in the sample. Leaks in the low-pressure chamber manifest themselves as a falling off in the gradient of the pressure–time curve. The test piece is supported on a porous substrate. This can be a disc of filter paper, a fine wire mesh, or a disc of sintered metal. To prevent the apparatus from acting like a gas thermometer, it is essential that the test cell, especially the low-pressure side, be kept at a constant temperature.

2.3 Constant Pressure Method

ASTM D1434 [25] provides for a second method of measuring the gas transmission rate of plastics, namely that at constant pressure. This technique also forms the basis of a rubber test standard [29]. In this case the pressure is kept constant in the low-pressure chamber and the volumetric change in the permeated gas is measured. Equation 11 reduces to

$$\mathrm{GTR} = \frac{273}{AT\Delta P}\left(P\frac{\partial V}{\partial t}\right) \tag{14}$$

where P is the atmospheric pressure.

The same basic design of gas transmission cell is used as in the manometric and constant volume methods. The test piece forms a barrier between the high- and low-pressure chambers. The test gas is maintained at atmospheric pressure on the low-pressure side of the test piece and at a higher pressure on the high-pressure side. The volumetric change is registered by the displacement of a slug of manometer fluid, which does not dissolve in the test gas, in a measuring capillary, which can be mounted horizontally or vertically. This displacement may be measured on a scale or with a travelling microscope. The bore of the capillary must be uniform to a close tolerance. Three diameters are suggested, 1 mm, 0.5 mm, and 0.25 mm, for use with materials ranging from high to low permeability.

The volumetric displacement is plotted against time, and the slope of the linear steady-state portion is used in the calculation of gas transmission rate and permeability coefficient.

As in the previous method, it is essential that both test chambers and associated pipework be flushed out with test gas unless air is being used. This method is highly susceptible to temperature change, the capillary acting as a gas thermometer, and the standards recommend a maximum variation in the test temperature of ±0.1°C in order to reduce this effect to a minimum. It also helps if the volume of the low pressure chamber is kept to a minimum.

An adjustable measuring manometer can also be used. The volumetric change is registered as a change in the height of the balanced liquid columns.

2.4 Carrier Gas Methods

Carrier gas techniques are commonly referred to as "dynamic" because gas is permitted to flow across each side of the test specimen, at equal pressure. The gas transmission cell is similar to those of the previous methods, in that the test piece forms a barrier between two chambers. In the most basic of carrier gas methods the test gas flows at a constant rate through one chamber and a second gas, the carrier, flows through the other chamber at a constant rate. Test gas permeates through the sample and is swept away to a detector by the carrier gas. The detector may be of the absorptiometric [33] or of the thermal conductivity type [34,35], although other detectors have also been used.

An important feature is the lack of need for a mechanical support for the test piece, because both gases are held at the same pressure, usually 1 atmosphere. The difference in test gas partial pressure (the driving pressure) across the sample is also the same throughout the measurement, because the concentration of permeant in the carrier gas is kept small by suitable choice of flow rate.

Another feature of the technique is that small differences in partial pressure can be established across the polymeric membrane while still maintaining a constant hydrostatic pressure of 1 atmosphere. Thus the service conditions of some polymers used in the packaging industry, for example food packaging, can be approximated in the laboratory. These small partial pressure differences are achieved by passing the carrier gas through both chambers with the permeant introduced as an impurity into one of the gas streams. As an example the permeation of sulphur dioxide at partial pressure differences of less than 100 Pa through a polymeric membrane has been demonstrated using an ultraviolet detector [36].

Gas leakage is less of a problem with the use of a carrier gas, although the problem of diffusion of air gases through seals is still present when small permeation rates of such gases are being measured. Suitable choice of sealing materials should minimize this.

A logical development of the carrier gas technique has been the introduction of a gas chromatographic column between the transmission cell and the detector [37–39]. This permits gas mixtures to be used for the permeant and their individual transmission rates to be measured. It should be noted that no single gas chromatographic column and detector combination is suitable for all gases. The system must be chosen to suit the permeant and the carrier gases being used. In addition, a pulsed input to the column is required if separation of components is achieved by elution development. This is the most common and versatile method of separation.

2.5 Other Methods

Several techniques have been developed for detecting the quantity of gas, and in some cases the type of gas, that has permeated through the sample. Perhaps the most important of these from the research point of view is the use of a mass spectrometer as the detector [40,41]. In this case a pressure differential may be maintained across the test piece by coupling the low-pressure chamber of the transmission cell directly to the inlet manifold of the instrument. The mass spectrometer offers high sensitivity and the ability to measure the permeation rates of several gases simultaneously.

Some methods that are specifically for the detection of oxygen have been used: the colorimetric method [42]; the coulometric method [43]; and another based on the phenomenon of radiothermoluminescence of polymers, as a function of oxygen content [44]. The latter was used in a transient experiment and hence measured the diffusivity, but if the solubility is known, or can be measured, the permeability can be calculated.

If a radioactive isotope of a test gas exists, a small, but known, quantity can be introduced into the test gas. Making the reasonable assumption that the two isotopes permeate at the same rate, the total quantity can be inferred by measuring the radioactive component [45,46].

3 Vapor Permeability

Methods used to measure the permeability of polymers to vapors fall broadly into two categories, namely those for water vapor and those for volatile liquids. In both cases the most common measuring technique is the weight change or gravimetric method. Several test standards are based on this. Other methods include techniques similar to those used for gases.

3.1 Water Vapour

The transmission cell used in the gravimetric method is relatively simple, consisting of a lightweight metal dish with a lid (Fig. 3). Depending on the procedure, a desiccant, such as calcium chloride, or water is introduced into the dish and the test piece is sealed onto the lip of the dish using wax, which has a very low permeation rate to water vapor. The assembly is then placed in an atmosphere of controlled temperature and humidity. The weight gain or loss is measured at regular intervals; if this is plotted against time the curve shape resembles that for gas transmission as in Fig. 2. When the weight change becomes constant with time, the rate of weight change is used to calculate the quantity of water vapor that permeates through unit area of the test piece in unit time. The intercept method should not be used because Henry's law is not valid [47].

In all cases the desiccant in the dish effectively reduces the water vapor pressure to zero on one side of the test piece, while the water methods raise the vapor pressure to a

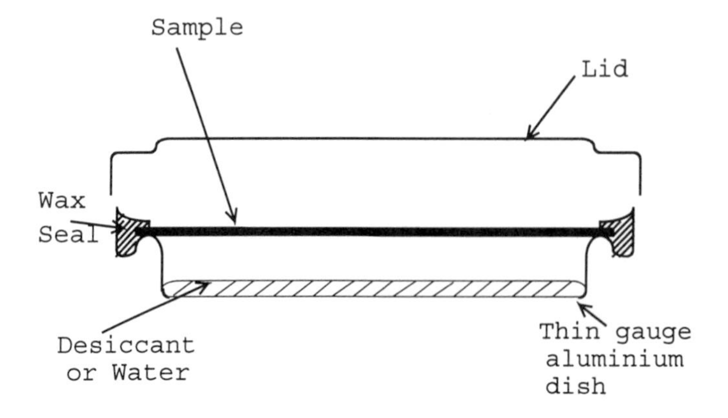

Figure 3 Water vapor transmission apparatus.

maximum for a given test temperature. The water contact method, specified in ASTM E96, also tests for capillary transfer as well as diffusion.

The experimental techniques are described in BS 2782 [48], BS 3177 [49], ASTM E96 [50], and ISO 2528 [51]. Separate test standards [52,53] accommodate thicker sheet materials and rigid cellular materials.

The use of wax to seal the test piece into the dish requires care and dexterity in order to avoid minute leaks. Such leaks are difficult to detect, and it is useful to test the sealing technique using a metal template instead of the polymeric sample. It is also sensible to make measurements on a number of samples of the same material at the same time. If one of the seals is leaking, the transmission rate will be significantly higher than the others and the results can be discarded.

An alternative to the dish method is that which uses the test material in the form of a sachet [54]. This is suitable for thermoplastic materials in thin sheet form, which can be heat sealed. The transmission of vapor through the seals must be considerably less than through the sheet.

Two different definitions of water vapor permeability are used in the BS and ASTM standards, and this can cause confusion. BS 3177 and BS 2782 define the water vapor transmission rate as the mass transfer rate of water vapor per unit area (g/m^2 24h) (see Section 1), but they call it "permeability." When the material is homogeneous and the transfer rate is inversely proportional to thickness, BS 3177 defines an "equivalent permeability" as the "permeability" multiplied by the thickness of the test piece in thousandths of an inch (g mil/m^2 24h).

The ASTM standards adopt definitions that are consistent with the equivalent definitions for gas transmission. Water vapor transmission rate is the mass transfer rate of water vapor per unit area (g/m^2 24h). Permeance is the ratio of the water vapor transmission rate to the difference in vapor pressure between the surfaces of the test piece measured in mm of mercury; this unit is known as the metric perm (g/m^2 24h mmHg). This is equivalent to the gas transmission rate. Permeability is the product of the permeance and the thickness of the test piece, assuming that the permeance is inversely proportional to thickness for homogeneous materials; this unit is known as the perm-centimetre (g cm/m^2 24h mmHg). Since the adoption of SI units, the water vapor permeability may also be expressed in the units of microgram meter per newton hour ($\mu g\, m\, N^{-1}\, h^{-1}$ or $\mu g\, m\, m^{-2}\, Pa^{-1}\, h^{-1}$).

Refinements of the weighing technique have been made by enclosing the dish and a sensitive balance in a temperature- and humidity-controlled cabinet. This eliminates the need to remove the cell from its controlled atmosphere for weighing. However, this is still a time-consuming test.

Electrical and chemical detection systems have also been employed. In addition, radioactive labelling of water and its subsequent detection with a liquid scintillation counter have been demonstrated. ASTM F1249 uses an infrared sensor [55].

3.2 Volatile Liquids

The vapor transmission rates of other volatile liquids can also be measured from weight change data [56–58]. A suitable vessel is illustrated in Fig. 4. The test piece is sealed into the vessel by a screw-on lid having an aperture that defines the permeated area. To prevent distortion of the material, a metal ring separates the lid from the specimen. The ring and lid are free to rotate with respect to one another on ball bearings. The vessel is fitted with inlet and outlet ports to allow replenishment of the test liquid. This is particularly important if the liquid is made up of more than one volatile component having different per-

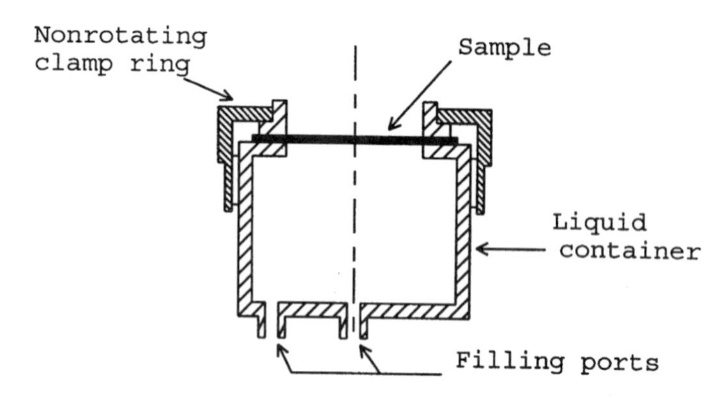

Figure 4 Volatile liquid vapor transmission cell.

meation rates. Thus the composition may be maintained constant throughout the test. When testing, the vessel is usually inverted so that the polymer and liquid are in contact.

The result is quoted as a transmission rate with the test conditions stated. It should be remembered that some polymers swell when in contact with organic liquids, hence the thickness and the permeability are not constant throughout the test. Steady-state conditions must also have been achieved before the transfer rate is calculated.

There are also standard specifications for testing the permeation of liquid chemicals through protective clothing [59], and for the permeation of vapor through the walls of a hose [60,61].

As an alternative to the gravimetric method of monitoring the permeation, other methods make use of the manometer and gas chromatography. A mass spectrometer could also be used.

References

1. Comyn, J., ed. (1985), *Polymer Permeability*, Elsevier.
2. McGregor, R. (1974), *Diffusion and Sorption in Fibres and Films*, Vol. I, Academic Press.
3. Vieth, W. R. (1991), *Diffusion in and Through Polymers*, Hanser.
4. Stastna, J., and De Kee, D. (1995), *Transport Properties in Polymers*, Technomic.
5. DiBenedetto, A. T. (1963), *Journal of Polymer Science: Part A*, *1*, 3459, 3477.
6. DiBenedetto, A. T., and Paul, D. R. (1964), *Journal of Polymer Science: Part A*, *2*, 1001.
7. Paul, D. R., and DiBenedetto, A. T. (1965), *Journal of Polymer Science: Part C Polymer Symposia*, *10*, 17.
8. Naylor, T. de V. (1989), in *Comprehensive Polymer Science* (G. Allen, and J. C. Berington, eds.), Vol. 2, Chapter 20, 643.
9. Rogers, C. E. (1985),in *Polymer Permeability*, Elsevier, Chapter 2.
10. Petropoulos, J. H. (1970), *Journal of Polymer Science: Part A2*, *8*, 1797.
11. Paul, D. R., and Kemp, D. R. (1973), *Journal of Polymer Science: Polymer Symposia*, *41*, 79.
12. Volabuyev, P. V., Kupryazhkin, A. Ya., and Suetin, P. Ye. (1972), *Vysokomolekulyarnye Soyedineniya*, *A14*(2), 489.
13. Davis, H. A. (1983), *Review of Scientific Instruments*, *54*, 1412.
14. Wissinger R. G., and Paulaitis, M. E. (1987), *Journal of Polymer Science: Polymer Physics*, *25*(12), 2497.
15. Toi, K., Ito, T., Shirakawa, T., and Ikamoto, I. (1992), *Journal of Polymer Science: Polymer Physics*, *30*(6), 549.

16. Booth, J. R., and Holstein, T. J. (1993), *Journal of Thermal Insulation, 16*, 246.
17. Stewart, M. E., Hopfenberg, H. B., and Koros, W. J. (1987), *Journal of Applied Polymer Science, 34*, 721.
18. Pauly, S. (1989), *Polymer Handbook*, 3d. ed. (J. Brandrup, and E. H. Immergut eds.).
19. Huglin, M. B., and Zakaria, M. B. (1983), *Die Angewandte Makromolekulare Chemie, 117*, 1.
20. Demorest, R. L. (1992), *Journal of Plastic Film and Sheet, 8*(2), 109.
21. ISO 4080 (1991), Rubber and plastics hoses and hose assemblies—Determination of permeability to gas.
22. Crank, J., and Park, G. S., eds. (1968), *Diffusion in Polymers*, Academic Press.
23. Daynes, H. A. (1920), *Proceedings of the Royal Society, A97*, 286.
24. BS 2782, Part 8, Method 821A (1993), Determination of the gas transmission rate of films and thin sheets under atmospheric pressure—Manometric method.
25. ASTM D1434 (1982), Test methods for determining gas permeability characteristics of plastic film and sheeting.
26. ISO 2556 (1974), Plastics—Determination of the gas transmission rate of films and thin sheets under atmospheric pressure—Manometric method.
27. Hems, G. (1987), Private communication.
28. Linovitski, V. (1977), *Kunststoffe, 67*(8), 433.
29. BS 903, Part A30 (1996), Determination of permeability to gases.
30. ISO 2782 (1995), Rubber, vulcanized or thermoplastic—Determination of permeability to gases.
31. Pye, D. G., Hoehn, H. H., and Panar, M. (1976), *Journal of Applied Polymer Science, 20*, 1921.
32. Izydorczyk, J., Podkowska, J., and Salwinski, J. (1990), *International Polymer Science and Technology, 17*(8), T/69.
33. Landrock, A. H., and Proctor, B. E. (1952), *Modern Packaging, 25*, 131.
34. Pasternak, R. A., Schimscheimer, J. F., and Heller, J. (1970), *Journal of Polymer Science: Part A2, 8*, 467.
35. Ziegel, K. D., Frendsdorff, H. K., and Blair, D. E. (1969), *Journal of Polymer Science: Part A2, 7*, 809.
36. Davis, E. G., and Rooney, M. L. (1975), *Journal of Applied Polymer Science, 19*, 1829.
37. Caskey, T. L. (1967), *Modern Plastics, 45*(4), 148.
38. Kapanin. V. V., Lemanik, O. B., and Reithninger, S. A. (1974), *Vysokomolekulyarnye Soyedineniya, A16*(4), 911.
39. Senich, G. A. (1981), *Polymer Preprints, 22*(2), 343.
40. Invashchenko, D. A., Krotov, V. A., Talakin, O. G., and Fuks, Ye, V. (1972), *Vysokomolekulyarnye Soyedineniya, A14*(9), 2109.
41. Eustache, H., and Jacquot, P. (1968), *Modern Plastics, 45*, June, 163.
42. Speas, C. A. (1972), *Packaging Engineering*, October, 78.
43. ASTM D 3985 (1995), Test method for oxygen gas transmission rate through plastic film and sheeting using a coulometric sensor.
44. Kiryushkin, S. G., and Gromov, B. A. (1972), *Vysokomolekulyarnye Soyedineniya, A14*(8), 1715.
45. Kirshenbaum, A. D., Strong, A. G., and Dunlap, W. B., Jr. (1954), *Rubber Age*, March, 903.
46. Algar, M. M., and Ward, M. J. (1987), *Journal of Plastic Film and Sheet, 3*(1), 33
47. Barrie, J. A., Newman, A., and Sheer, A. (1974), in *Permeability of Plastic Films and Coatings* (H. B. Hopfenburg, ed.), Plenum Press, New York, p. 167.
48. BS 2782, Part 8, Method 820A (1992), *Determination of water vapour transmission rate—Dish method.*
49. BS 3177 (1959), Method for determining the permeability to water vapor of flexible *sheet material used for packaging.*
50. ASTM E96 (1994), Test methods for water vapor transmission of materials.
51. ISO 2528 (1995), Sheet materials—Determination of water vapor transmission rate—Gravimetric (dish), method.

52. BS 4370, Part 2, Method 8 (1973), Methods of test for rigid cellular materials—Measurement of water vapour transmission.
53. ISO 1663 (1981), Cellular plastics—Determination of water vapor transmission rate of rigid materials.
54. BS 2782, Part 8, Method 822A (1992), Determination of water vapor transmission rate—Sachet method.
55. ASTM F1249 (1990), Test method for water vapor transmission rate through plastic film and sheeting using a modulated infrared sensor.
56. ASTM D814 (1995), Test method for rubber property—Vapor transmission of volatile liquids.
57. BS 903, Part A46 (1991), Method for determination of the transmission rate of volatile liquids.
58. ISO 6179 (1989), Vulcanized rubber sheet, and fabrics coated with vulcanized rubber—Determination of transmission rate of volatile liquids (Gravimetric techniques),.
59. BS EN 369 (1993), Protective Clothing. Protection against liquid chemicals. Test method. Resistance of materials to permeation by liquids.
60. BS 5173, Part 103, Section 103.12 (1988), Determination of vapor transmission of liquids through walls.
61. ISO 8308 (1993), Rubber and plastic hoses and tubing—Determination of transmission of liquids through hose and tubing walls.

31
Adhesion

Roger Brown
Rapra Technology Ltd., Shawbury, Shrewsbury, England

1 Introduction

Adhesives and adhesion are very large subjects that are by no means restricted to polymers, although a great many adhesives are polymeric materials. There are numerous test methods for characterizing and measuring the performance of adhesives, but adhesives as a material type (even if restricted to polymeric adhesives) are outside of the scope of this book. Hence this chapter makes no attempt to cover the field in general but is restricted to the few adhesion tests that have been developed and standardized within the solid polymer industries.

Polymers, and rubbers in particular, are frequently used as composites with other solids, for example in tires, belting, and coated fabrics; indeed a class of polymer considered in this book is coated fabrics. It is because of the manufacture of this type of product that adhesion test methods were developed within the industry. These tests can be classified as adhesion to metals, adhesion to fabrics, and adhesion to cord, which can be traced to rubber-to-metal bonding, belting, coated fabrics, etc., and tires respectively. When considering the use of these methods, the limitations of their purposes should be borne in mind, and for a fuller understanding of the relevance of adhesion tests, as well as for alternative test methods, it is suggested that reference should be made to the great volume of literature from the adhesives and coatings industries.

2 Adhesion to Metals

2.1 General

Rubber is bonded to metal during processing to form a variety of products, and in most cases a very strong bond is necessary for the product to perform satisfactorily. It is usually desirable to measure bond strength by testing the actual product, but this is not always possible or convenient and, particularly for evaluating bonding systems, there is a need for tests using standard laboratory prepared test pieces. Whether the product or a test piece is used, the bond should be strained in essentially the same manner as would occur in service, although this may be complex rather than, for example, in simple tension or shear.

Four geometries have been standardized as illustrated in Fig. 1 they can be described as peel, tension, shear, and a variation on tension using conical end pieces.

With peel and direct tension tests, failure tends to occur in the rubber if the bond strength is high. It can be argued that if the bond is stronger than the rubber it is strong enough, but this attitude assumes that failure would be similar with another mode of straining and may not allow discrimination between a good and a very good bond. The tension test with cone-shaped metal end pieces was developed to encourage failure at the interface between rubber and metal because of a stress concentration at the tips of the cones.

The place the failure occurs can be as significant as the numerical value of the bond strength, and hence it is usual to report the type of failure. The bond can be made up of one or more adhesive layers, and symbols commonly used are as follows:

R = failure in the rubber.
RC = failure at the interface between the rubber and the cover cement.
CP = failure at the interface between the cover cement and the primer cement.
M = failure at the interface between the primer cement and the metal.

In practice it is not always possible to distinguish between RC and CP, and in many cases a single coat bonding system may have been used.

2.2 Peel Tests

The international method for rubbers is given in ISO 813 [1] and uses a 90° peel geometry with a test piece, 6 ± 0.1 mm thick and 25 ± 0.1 mm wide, bonded to a 1.5 mm thick metal strip along 25 mm of its length. The rubber is peeled at 50 ± 5 mm/min using (preferably) a low-inertia tensile machine, having first started to strip the rubber from the metal using a sharp knife. This rather dubious procedure of cutting at the bond line is intended to lessen the probability of failure in the rubber, and the standard states that if the rubber starts to

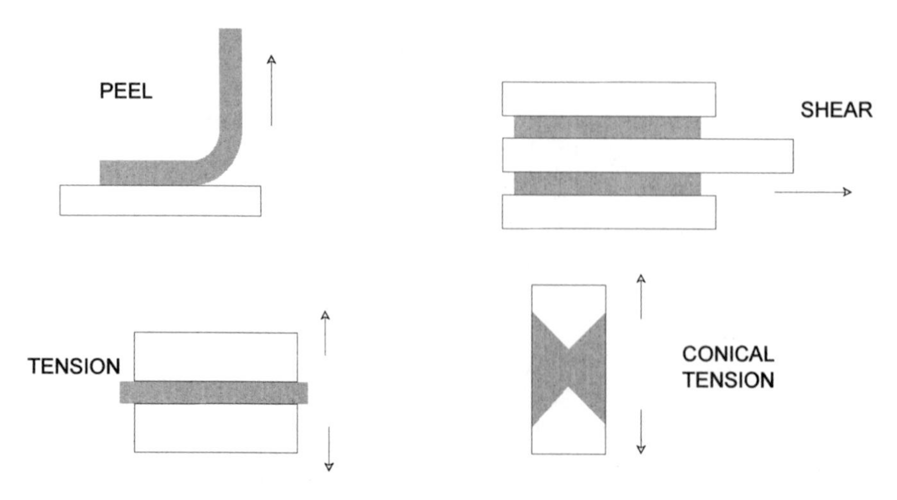

Figure 1 Rubber-to-metal adhesion geometries.

tear during the test it shall be cut back to the metal. The maximum force during stripping over the 25 mm length expressed per mm of width is taken as the bond strength. It is suggested that an autographic recording of the strength over the 25 mm length be taken, but the peaks and troughs in force that are likely to occur are not taken account of.

The measured adhesion strengths from peel tests are lower than those from tension tests, and the theoretical aspects of this have been discussed by Kendall [2]. Choosing 90° for the peal test is to some extent arbitrary, and Gent and Kaang [3] have investigated the effect of peel angle for adhesive tapes. Orthman [4] studied the ISO method and found that a great deal of the scatter in results could be attributed to the direction of peeling relative to any pattern on the metal plate and proposed a modified test piece with which peeling could take place in two opposite directions.

The equivalent British Standard is BS 903, Part A21 [5], Method B, which is essentially identical to ISO 813. The ASTM equivalent, D429 [6], Method B is very similar but uses slightly different test piece dimensions, which are attributable to being direct metric equivalents of imperial sizes, and test speeds.

These standard methods are simple procedures intended for quality control and the comparison of bonding systems, but they could be readily used as the basis for wider investigations such as the effect of peeling angle, test speed, etc., or adhesion to other rigid substrates.

2.3 Tension Tests

The international standard method is given in ISO 814 [7] and uses a disc test piece 3 ± 0.1 mm thick and between 35 and 40 mm in diameter, bonded to metal plates that are a little less in diameter than the rubber. The slightly smaller size of the metals is intended to prevent the rubber tearing from the edges of the metals during test.

The bonded test piece is separated at 50 mm/min in a tensile machine and the maximum force recorded. The result is then expressed as the maximum force divided by the cross-sectional area of the disc. With such tension tests, great care must be taken to ensure that the test piece is accurately aligned so that the tension is uniformly distributed over its cross section during test, as any misalignment will introduce a peeling action. As the rubber deforms under the tensile force, shear forces are introduced, and the measured bond strength depends on the thickness of the rubber disc, the strength increasing with decreasing thickness.

The equivalent British Standard, BS 903, Part A21, Method A [5] is essentially identical to ISO 814. The ASTM equivalent, D429, Method A [6] is very similar but, as for the peel tests, uses slightly different test piece dimensions.

A method using conical metal end pieces is standardized in ISO 4600 [9]. The test piece diameter is 25 mm, the cone angle 45°, and the distance between the tips of the cones 121 mm. The bonded test piece is separated at 50 mm/min, and the result is simply expressed as the maximum force recorded. BS 903, Part A40 [10] is identical, and ASTM D429, Method C [6] is very similar, but the distance between the cones is 11.5 ± 1.2 mm.

The stress distribution is not even, with this test piece, and the action is not pure tension but involves peel and shear forces. The test was investigated by Painter [11] who showed that the stress is concentrated at the tips of the cones. He found that failure occurred at the interface rather than in the rubber, and the measured strengths were lower than with a plain disc test piece of similar diameter, more in line with the results of peel tests. It is probable that the result is sensitive to geometry, particularly the cone angle. An earlier draft of ISO 5600 put a tolerance of ±0.1 mm on the tip separation, which could imply that this dimension is also critical.

2.4 Shear Tests

The international standard method for shear adhesion is now incorporated into the standard for shear modulus, ISO 1827 [12], which has been discussed in Chapter 15. The quadruple element test piece provides a stiff construction that will will remain in alignment under the high stresses that can be expected when the test is continued to failure point to measure adhesion. The test piece is strained at 50 mm/min, and the maximum force is recorded in line with the other adhesion-to-metal tests. The result is expressed as the maximum force divided by the total bonded area of one of the double sandwiches.

2.5 Dynamic and Impact Tests

With very good bonding systems it is often difficult to discriminate between the systems because of failure in the rubber, and yet in service differences in performance are quite clear. High-speed deformations (impact) or repeated application of stress (fatigue or dynamic stressing) can occur [8,14], but neither of these factors is considered in the standard methods discussed above. For the effect of impact, Buist and Naunton [8] preferred falling weight apparatus, with each test piece receiving a single blow. With the particular case of automobile bumpers in mind, Given and Downey [15] developed a high speed test using a double element shear test piece and a sophisticated servohydraulic universal test machine.

Impact methods can be used to test fatigue resistance of bonds by making repeated blows, but this is not very convenient. The use of the Goodrich Flexometer (see Chapter 15) with a modified test piece holder to fatigue bonds in tension was able to discriminate between bonding systems that appeared equal in the standard tension test [8]. Other procedures tried are the Goodrich Flexometer in compression prior to making the standard tension, a slow-speed cycling test in shear [14], and fatiguing in bend with a modified Rotoflex machine [16].

The modern approach is to use a universal tensile machine that can make tests under various straining modes at a range of strain rates. The disadvantage is the cost of tying up such an expensive apparatus for long periods. Many ad hoc rigs have been devised to test bonded components, often using pneumatic or hydraulic cylinders, and such a device could be arranged to test standard test pieces.

A BRMA publication [17], which gives recommendations for testing rubber to metal bonded components in general, suggests conditions for carrying out dynamic tests. Aubrey et al. [18] made a systematic study of the most commonly used methods and developed a procedure for industrial laboratories to predict strength and lifetime of bonded components.

In service, rubber-to-metal bonds are often required to withstand harsh environments, so it is logical that tests for strength will be required after exposure to simulated aging conditions. Any of the environmental exposure tests (see Chapter 13) can in principle be applied, and it is generally preferable to subject the bond to stress during the exposure.

2.6 Adhesion to and Corrosion of Metals

In the tests considered above, the object was to measure the degree of adhesion of test pieces that were intended to be well adhered. The opposite situation can be encountered with some rubber compounds that may cause corrosion of, and tend to stick to, metal surfaces with which they are in contact, and corrosion can even be caused to a metal in close proximity but not touching the rubber, i.e., a case of adhesion occurring when it is not wanted. Although this is not a very widespread problem, there has been sufficient concern, particularly for some military applications, for tests to be devised to assess the relative degree of corrosion and adhesion caused by different compounds.

Most tests are based simply on placing the rubber in contact with metal under load, aging for a period under specified conditions, and assessing corrosion and adhesion by visual inspection. The international standard method is ISO 6505 [19] in which rubber strips are sandwiched between the metals of interest (usually copper, brass, aluminum, and mild steel) under a load of 10 kg and clamped. The sandwich is normally aged under relatively dry conditions, for example 7 days at 70°C, and then visually examined for signs of adhesion or corrosion. It has proved rather difficult to obtain good reproducibility, and it is essential that great care be given to the preparation of samples, in particular as regards cleanliness.

The same method is given in BS 903, Part A37 [20], which also contains a second method for assessing the degree of corrosion when the rubber is not in contact with the metal. Zinc is used as the standard metal, as this is fairly readily corroded. A strip of zinc and the rubber test piece are both suspended over distilled water in a stoppered container maintained at 50°C. After a period of three weeks the corrosion products are removed from the zinc by immersion in chromium trioxide solution, and the loss in weight is used as the measure of degree of corrosion. Even more care has to be taken than in the contact method to avoid contamination and to obtain reproducible results.

3 Adhesion to Fabrics

3.1 General

Rubbers and plastics are used with textile fabrics to produce composites in such products as belting and hose, and also as a coating on the fabric to form materials that find application in many areas such as waterproof covers and as a replacement for leather.

Tests for adhesion are carried out in peel or direct tension, peel being the most common, although tension tests are particularly useful for thin coatings where the rubber is too thin or too weak to carry out a peel test successfully. Tests have been standardized in the rubber industry, where the aim was probably at such products as belting, and also in the coated fabric industry.

3.2 Peel Tests

The international method, ISO 36 [21], for adhesion strength of rubber to fabrics uses a 25 mm wide strip test piece, long enough to permit separation over at least 100 mm. The fabric and rubber are separated by hand over a length of about 50 mm, and the two ends are placed in the grips of a tensile testing machine. The grips are separated at a rate of 50 ± 5 mm/min so as to give a rate of ply separation of 25 mm/min.

The geometry of the test is shown in Fig. 2. The angle between the two gripped legs of the test piece is 180°. The unstripped portion of the test piece is left to find its own level during the test, but variation in the angle a will affect the measured result. This angle depends on the relative stiffness of the plies B and C; the greater the stiffness ratio B/C the nearer the angle approaches 180°. The plies should separate at a sharp angle, but this will depend on the thickness and stiffness of the plies. The standard suggests that the thickness

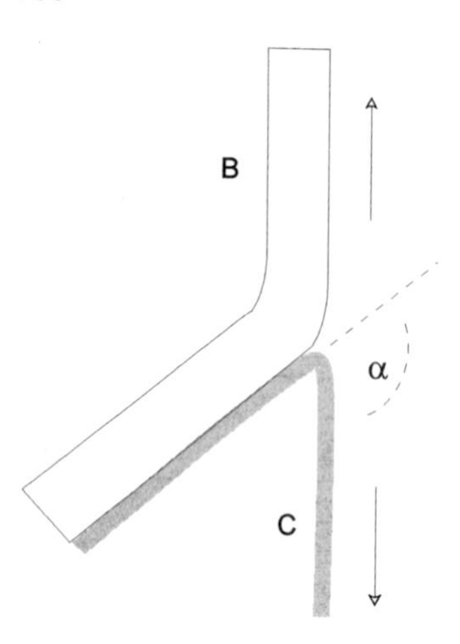

Figure 2 Rubber-to-fabric peel test.

should be reduced if necessary, so that the line of separation of the plies lies as closely as possible to the plane of the axis of the legs of the test piece held in the grips. It would seem better to restrain the unpeeled portion A so that α is either 90° or 180°. Variations of the peel test can be applied to rings and directly to peel the cover from a helical wound hose.

The stripping force is recorded continuously, which results in a trace containing a series of peaks and valleys. It should be noted that the peaks and especially the troughs are affected by the dynamic response of the test machine. For this reason, only measuring systems having very low inertia should be used for this test. There are various ways in which an adhesion value can be obtained from this trace but the standard procedure now adopted is given in ISO 6133 [22]. This gives three procedures, one for traces having less than five peaks, another for 5–20 peaks, and another for more than 20 peaks. For less than five peaks the median of them all is taken, for 5 to 10 peaks the median of the peaks in the central 80% of the trace is taken, and for more than 20 peaks the trace is divided into tenths by nine lines, the peak nearest to each line noted and the median of these taken. A draft revision of ISO 6133 has further cases: those in which there are many peaks that are not easily distinguished or counted, and those in which the trace is undulating. In these cases the mean is reported, taken as the midpoint between maximum and minimum deflections. The simplest procedure, and it is a reasonable one, is to take the mean of the force trace, and in practice many workers do this by drawing a line through the trace using a transparent rule, whatever the number of peaks. Others have preferred to take some measure from the lowest troughs on the basis of recording the worst case.

The equivalent British Standard, BS 903, Part A12 [23], is identical with the ISO. ASTM D413 [24] contains a method similar to ISO 36 but with provision for holding the unpeeled portion so that there is 180° separation. The mean force level is taken as the adhesion strength. ASTM D413 also details methods for peeling the strip with the unpeeled portion held so as to give 90° separation, for rings of up to 100 mm diameter, and it gives a simple dead load method whereby a mass, large enough to cause peeling, is hung from one leg of the test piece and the rate of separation noted. Tensile machines are common enough that there would seem to be little use for the dead load type of procedure.

In such products as belts, essentially the same method is used to separate different plies of rubber when it can be especially difficult to obtain interfacial failure. Loha et al.

[25] successfully used test pieces including a perforated metal sheet at the interface to measure rubber-to-rubber adhesion strength.

Separate peel methods have been developed for coated fabrics. Apart from the need to cater for thin weak coatings, the only reason for there being separate methods is the fact that coated fabrics are considered in different committees. The international method is ISO 2411 [26], the British standard is BS 3424, Part 7 [27], and the ASTM standard is ASTM D 751 [28]. The test piece is a strip, but the width used varies, and a portion is peeled back by hand in order to facilitate gripping as for the rubber/fabric methods. If the coating has lower strength than the bond, the problem is overcome by sticking two strips back to back or using a piece of fabric as a backing.

In the ISO standard, one method is given using a tensile machine in what might be called the conventional manner. The separation speed is 100 mm/min and the median of the peaks taken. Clearly, results will not be comparable with the rubber standards, and the differences in procedure serve to illustrate the arbitrary nature of these standards. Three alternative procedures are given in the British standard, using dead weights, a tensile machine, and a spring balance. The dead weight and spring balance methods simply reflect a wish to have methods requiring minimal equipment, and it is a little surprising that they still exist.

3.3 Direct Tension Tests

A direct tension method [29,30] is said to measure more nearly the true adhesion between fabric and rubber and to be particularly useful for discriminating between adhesive systems. The main objection to the method in practice is that the preparation of test pieces is rather difficult.

The standard method is given in ISO 4637 [31] and in BS 903, Part A 27 [32]. The test piece consists of two metal cylinders, between which the composite to be tested is cemented in a similar manner to the method discussed earlier for rubber-to-metal bonds. The piece of fabric/rubber under test is a square of side approximately 32 mm and hence larger than the metal cylinders. The metals are gripped in a tensile machine and separated at a rate of 50 mm/min, and the maximum force is recorded. The most important part of the test is the preparation of the metal/rubber/fabric test piece, and unless very careful preparation of the metals is carried out, failure occurs at the metal surface. Considerable detail on surface preparation is given in the standard. After machining the ends are lapped and degreased with trichloroethylene, whilst the test piece is wiped with a solution of ammonium hydroxide in acetone. The assembly is cemented together with a cyanoacrylate adhesive using a special jig.

3.4 Dynamic Tests

A dynamic ply separation test is really a fatigue test on the rubber/fabric composite to weaken the bond or to determine the number of cycles until the bond fails. In principle any flexing test (see Chapter 15) could be used, but there has been little standardization for this particular product type. A notable exception is the Scott flexer detailed in ASTM D430 [33], which is specifically intended for testing ply separation of belts and tires, etc. The relative bond strengths of different rubber/textile systems may be different in dynamic and static tests, and hence it would seem important to assess any composite that will be subjected to fatigue in service by a dynamic method.

4 Adhesion to Cord

The adhesion of textile or metal cord to polymers is essentially restricted to rubbers used in tires and to some extent in belting. The basic form of static test consists of measuring the force to pull a cord out of a block of rubber into which it has been vulcanized, and it is apparent that the result is critically dependent on the efficiency with which the test piece was molded. The measured force is also dependent on the amount that the rubber deforms during the test.

The original method, the H-pull or H-block test, was standardized by the ASTM together with a variant, the U test. In the former, two blocks of rubber are gripped in the tensile machine, and in the latter, a loop of cord is hooked onto one of the grips. A number of variations on the procedure for producing the test piece, and on test piece and supporting jig geometry, have been tried, and it took some time before agreement was reached on international standards. The H-pull method for textile cord was eventually published as ISO 4647 [34] and the method for steel cord as ISO 5603 [35]. In the steel cord method the block of rubber is held in one grip and the cord held in the other grip. BS 903, Part A48 [36] is identical to ISO 4647 and Part A56 [37] to ISO 5603.

Factors affecting the measured adhesive strength and improvements to the standard methods have been discussed by, for example, Hicks et al. [38], Skobrik [39], and Campion and Redmond [40]. ISO 5603 contains two methods of molding, the second being of ASTM origin and the first resulting from the work of Campion in particular.

One of the main points of debate with the above methods is the stress distribution due to gripping the rubber block. Nicholson et al. [41,42] used a test with two cords embedded in the block of rubber and avoided holding the block in one grip of the testing machine. Further analysis was made by Brodsky [43], who used three cords. Ellul and Emerson [44,45] used cords embedded in steel enclosed rubber cylinders with alternatively hot and cold bonding. Ridha et al. [46] have calculated the stress fields in tire cord adhesion test pieces, and Mollet [47] has compared the various methods.

Tires operate in a dynamic situation and are continuously fatigued in use, so that bonds should be assessed dynamically. Although methods have not been standardized, a variety of procedures have been reported [48–55]. Some workers have used the same or a similar test piece as in static tests and applied a cyclic tensile stress on strain, whilst others have used some form of fatigue tester operating in compression/shear to stress or cord rubber composite repeatedly or even to flex samples in the form of a belt.

5 Blocking

Blocking can be described as the unintentional adherence between materials and thus, like the adhesion and corrosion of rubbers to metals discussed above, is a case of measuring adhesion with the hope of finding a low answer. The problem is most usually found with thin sheets of material, and standards have been produced for two cases, plastic film and sheeting and coated fabrics.

Two methods for plastics film are given in ISO 11502 [56], one qualitative and one quantitative. Both consist of keeping pieces of the material in contact under specified conditions of time, pressure, and temperature. In the qualitative method, the test pieces are separated manually in a peeling mode, and it is noted whether a metal bar of defined mass is lifted before separation occurs or whether there is damage to the test piece surfaces. In the qualitative method the test pieces are separated in direct tension with a tensile machine.

The specified conditions for holding the test pieces in contact are not the same for the two methods, and the different geometries of separating the test piece ensure that the two are not comparable. Clearly, the degree of adhesion is likely to depend on the conditions of pressing the two pieces together, and it is easy to envisage other arrangements for carrying out the separation. These methods simply represent arbitrary procedures that have become established.

The method for coated fabrics given in ISO 5978 [57] is based on the same principle, but only a manual separation procedure is given and the result is a subjective judgement of the difficulty of getting separation.

6 Print Adhesion

A method for print adhesion given in BS 2782, Method 826A [58] is not an adhesion test in the normal sense but a type of abrasion test in which the print on plastics sheet is rubbed in a specified manner with an abrading fabric and number of cycles to cause damage to the print noted.

7 Interlaminar Adhesion

In the broad sense, interlaminar adhesion means the adhesion between any laminations of similar or dissimilar materials and hence could be taken to include just about any adhesion test. One particular type for fiber-reinforced plastics, which is known as interlaminar shear strength, is normally considered as a short beam flexural test (see Chapters 9 and 18). A method peculiar to laminated plastics tube is given in BS 2782, Method 346A [59] and called cohesion between layers of laminated tube. A sample of tube is subjected to compression to induce shear forces between layers and delamination observed by eye.

8 Nondestructive Tests

To be able to estimate bond strength by a nondestructive method is extremely attractive, especially for quality control purposes. It is one aspect of nondestructive testing that is considered in detail in Chapter 32. The possibility of using ultrasonics for this application has been recognized for a long time [60], and the basic technique has been applied to such components as rubber/metal engine mountings [61]. However, the difficulty and the expense of ultrasonics and other promising techniques such as radiography, holography, and thermography have in practice largely restricted nondestructive methods to detection of flaws and disbonds in fiber-reinforced composite structures and tires. In addition to the descriptions of techniques and literature references in Chapter 32, reviews of applications in the polymer industry have been published [62–65] that contain many hundreds of references.

References

1. ISO 813, 1986, Determination of adhesion to metal—One plate method.
2. Kendall, K. (1973), *J. Adhesion*, *5*, 1, 77.
3. Gent, A., and Kaang. S. Y., (1987), *J. Adhesion*, *24*, 2–4, 173.
4. Othman, A. B. (1988/9), *Polym. Test.*, *8*, 4, 289.
5. BS 903, Part A21, 1989, Determination of rubber to metal bond strength.
6. ASTM D429-82, Adhesion to rigid substrates.

7. ISO 814, 1986, Determination of adhesion to metal—Two plate method.
8. Buist, J. M., and Naunton, W. J. S. (1950), *Trans. IRI*, *25*, 378.
9. ISO 5600, 1986, Determination of adhesion to rigid materials using conical shaped parts.
10. BS 903, Part A40, 1988, Determination of adhesion to rigid materials using conical shaped parts.
11. Painter, G. W. (1959), *Rubber Age N.Y.*, *86*, 262.
12. ISO 1827, 1991, Determination of modulus in shear or adhesion to rigid plates—Quadruple shear method.
13. BS 903, Part A14, 1992, Determination of modulus in shear or adhesion to rigid plates—Quadruple shear method.
14. Buist, J. M., Meyrick, T. J., and Stafford, R. L. (1956), *Trans. IRI*, *32*, 149.
15. Given, D. A., and Downey, R. E. (1975), *Polym. Sci. Tech.*, 9A, ACS Symposium, Philadelphia, PA., p. 315.
16. Beatty, J. R. (1973), ACS Div. Rubb. Chem. Fall Meeting, Preprint 42.
17. BRMA Publication (1982), Testing of rubber to metal bonded components.
18. Aubrey, D. W., Southern, E., Harman, A. D. (1985), PED Review Meeting, Loughborough, Paper 52.
19. ISO 6505, 1984, Determination of adhesion to, and corrosion of, metals.
20. BS 903, Part A37, 1987, Assessment of adhesion to and corrosion of metals.
21. ISO 36, 1993, Determination of the adhesion strength of vulcanised rubbers to textile fabrics.
22. ISO 6133, 1981, Analysis of multi-peak traces obtained on determinations of tear strength and adhesion strength.
23. BS 903, Part A12, 1975, Determination of the adhesion strength of vulcanised rubbers to fabrics (ply adhesion),
24. ASTM D413–82, Adhesion to flexible substrate.
25. Loha, P., Bhowmick, A. K., and Chakravarty, S. N. (1987), *Polym. Test.*, *7*, 3, 153.
26 ISO 2411, 1991, Fabrics coated with rubber or plastics—Determination of the coating adhesion.
27. BS 3424, Part 7, 1982, Coating adhesion strength.
28. ASTM D751–89, Methods of testing coated fabrics.
29. Borroff, E. M., and Wake, W. C. (1949), *Trans. IRI*, *25*, 199.
30. Meardon, J. I. (1962), *Rubb. and Plast. Weekly*, 142, 107.
31. ISO 4637, 1979, Determination of rubber to fabric adhesion—Direct tension method.
32. BS 903, Part A27, 1986, Determination of rubber to fabric adhesion (direct tension),
33. ASTM D430 1973, Rubber deterioration—Dynamic fatigue.
34. ISO 4647, 1982, Determination of static adhesion to textile cord—H-pull test.
35. ISO 5603, 1986, Determination of adhesion to wire cord.
36. BS 903, Part A48, 1984, Determination of static adhesion to textile cord in H-pull test.
37. BS 903, Part A56, 1989, Determination of adhesion to wire cord.
38. Hicks, A. E., Chirico, V. E., and Ulmer, J. D. (1972), *Rubb. Chem. and Tech.*, *45*, 1, 26.
39. Skolnik, L. (1973), ACS Div. of Rubb. Chem. Fall Meeting, Denver, Paper 13 (Oct.).
40. Campion, R. P., and Redmond, G. B. (1975), *Rubb. Ind.*, *9*, 1, 19.
41. Nickolson, D. W., Livingston, D. I., and Fielding-Russell, G. S. (1978), *Tire Sci. Technol.*, *6*, No. 1.
42. Fielding-Russell, G. S., Livingston, D. I., and Nickolson, D. W. (1980), *Rubb. Chem. Technol.*, 53, No. 4.
43. Brodsky, G. I. (1984), *Rubb. World*, *190*, No. 5.
44. Ellul, M. D., and Emerson, R. J. (1987), ACS Rubber Division Meeting, Cleveland, Paper 36.
45. Ellul, M. D., and Emerson, R. J. (1987), ACS Rubber Division Meeting, Cleveland, Paper 37.
46. Ridha, R. A., Roach, J. F., Erickson, D. E., and Reed, T. F. (1981), *Rubb. Chem. Technol.*, *54*, No. 4.
47. Mollet, J. R. (1982), International Rubb. Conference, Paris, June 2–4, Paper 012.
48. Lunn, A. C., Evans, R. E., and Ong, C. J. (1981), ACS Meeting, Cleveland, Oct. 13–16, Paper 20.

49. Bourgois, L., Davidts, J., and Schittescatte, M. (1973), *Text. Inst. Ind.*, *11*, No. 1.
50. Campion, R. P., and Corish, P. J. (1977), *Plast. Rubb. Mat. Appln.*, *2*, No. 2.
51. Kachur, V., and Weaver, E. J. (1976), ACS Meeting, Minneapolis, April, Paper 2.
52. Hewitt, N. L. (1985), *Rubb. Plast. News*, *15*, No. 5.
53. Vorachek, J. J., Causa, A. G., and Fleming, R. A. (1976), ACS meeting New Orleans, Oct. 7–10, Paper 60.
54. Bourgois, L. (1978), ASTM Special Publication 694.
55. Panchuk, F. O., Semak, B. D., Grabov, I. F., and Oplochenko, N. A. (1990), *Int. Polym. Sci.*
56. ISO 11502, 1995, Plastics film and sheeting—Blocking resistance.
57. ISO 5978, 1990, Rubber and plastics coated fabrics—Determination of blocking resistance.
58. BS 2782, Method 826A, 1992, Determination of adhesion of print on plastics.
59. BS 2782, Method 346A, 1984, Cohesion between layers of laminated tube.
60. Heughan, D. M., and Sproule, D. O. (1953), *Trans. IRI*, *29*, 5, 255.
61. Preston, T. E. (1970), *British J. of N.D.T.*, March, 17.
62. Trivisonno, N. M. (1985), *Rubb. Chem. Technol.*, *58*, No. 3.
63. Bergen, H. (1981), ACS Rubb. Div. Meeting, Minneapolis, Paper 44.
64. Scott, I. G., and Scala, C. M. (1982), *UDT Int.*, *15*, No. 2.
65. Reynolds, W. N. (1990), Rapra Review Report, No. 30.

32
Nondestructive Testing

Xavier E. Gros
Independent NDT Centre, Bruges, France

1 Introduction

The high strength-to-weight ratio of composites make them well suited to the manufacture of items that require a combination of lightness and strength. Included are products required by the aerospace industry, such as helicopter rotor blades and aircraft parts and sport articles, such as golf clubs and tennis rackets. Where composites are used in the transport industry, multiple advantages, such as fuel economy, weight reduction, and a longer life-cycle, are gained. Recent advances in composite technology have resulted in tougher thermoplastic resins for improved impact resistance and the employment of ceramic components to improve performance in high-temperature applications. Ceramics are less complex to fabricate than metals and have high corrosion and wear resistance, high temperature resistance, and low electrical and thermal conductivity.

Because of improvements in performance, composite technology is fast gaining around, but flaws or defects do still occur that need to be detected and accurately quantified in order to predict their effect on the structural integrity of a component. Although the ability to detect, locate, and quantify defects in composite materials has greatly improved over the years, due to their inherent anisotropy and heterogeneity, composites remain difficult to inspect with conventional nondestructive testing (NDT) methods. For example, the high density of composites causes absorption and scattering of heat, x-rays, and ultrasounds. Therefore the performance of certain inspection techniques can be greatly compromised. With an augmentation of the use of composites throughout industry, and because there is limited knowledge of the behavior of these new materials during in-service operations, an increase in terms of inspection frequency, defect monitoring, and maintenance is required to meet stringent safety and economic regulations. Defect characterization is also necessary to add understanding to the types of damage incurred with use. This chapter deals with nondestructive testing (NDT) techniques used in the assessment of the structural integrity of composite materials. An overview of existing NDT techniques and methods is presented, their principles described, and their main and potential applications discussed. In addition to the conventional techniques of ultrasound or radiography, other less traditional techniques such as magnetic particle tagging or shearography are considered.

Industrial concern for quality and process control, as well as in-service examination, has led to an increased demand for composite reliability inspection techniques. New and more sophisticated NDT techniques have been developed to meet this demand. However, there is still the view that the tried and tested conventional methods are generally more acceptable, their limitations understood and appreciated. "Despite the development of new NDT techniques for composite examination, conventional NDT techniques will remain favoured to NDT engineers as they are expected to be accepted more readily [1]." The introduction of innovations, in addition to highly proven and accepted reliabilities in flaw detection, is significantly desirable to composite manufacturers and users.

It is expected that composite nondestructive evaluation (NDE) will continue to undergo significant improvements and will become an increasingly important discipline. The principal objective of NDT is not to find defects but to ensure that there are none present. Moreover, NDT can be used to establish process control and facilitate the correct choice of manufacturing process. For these reasons NDT techniques must be efficient and produce results rapidly and unambiguously. Ultrasound is the most widely used technique in industry, but other techniques described in this chapter present comparable or better results in certain applications than ultrasonic testing. Each NDT technique has its own strengths and limitations depending on the type of material tested and the type of flaw being sought out; thus more than one technique is usually required for a full inspection of a material and to achieve optimum material evaluation. The sensitivity of each NDT technique greatly depends upon the equipment used, the procedure selected, and the material tested. For example, in the case of eddy current inspection, the sensitivity is determined by probe geometry and frequency. Ultrasound is best suited to check for material integrity, while eddy current can provide information about fiber alignment. For all techniques, instrument calibration is usually required to establish the limits of defect detectability, resolution, and sensitivity. Realistic defects built into calibration pieces should also be part of the calibration procedure, and reference standards should be of the same material as that examined. Critical flaw size, crack growth, and other flaw detection criteria are multiple parameters to be defined prior to inspection in order to select an appropriate testing method, and for the evaluation of the remaining life time of a component. For example, it is necessary for flaws smaller than 50 μm to be detected in ceramic components for good quality assurance [2]. Acceptance/rejection criteria should be established prior to an inspection, taking into account expertise from NDT engineers, structural mechanics engineers, designers, and quality assurance specialists. These criteria should define a maximum acceptable defect size for a specific structure. As yet there are no well-based acceptance/rejection criteria for NDT of composites, so the techniques described in this chapter are of interest for composite examination in laboratory or field tests. Only after such tests would the NDT engineer be able to select an appropirate inspection method and acceptance/rejection criteria. Other parameters that will enter into consideration are budget allocation for NDT, inspection speed, and equipment cost. Significant savings can be realized by the use of appropriate NDT methods and properly designed and managed inspection procedures. Savings through waste reduction, increased product reliability, and improvement in productivity and quality of the finished product can be achieved.

The inspector of composites must be familiar with the many types of construction materials and fabrication methods and must be aware of the type of defects expected as well as the limitations of each NDT technique [3]. For composite materials themselves, their physical degradation is dependent upon their composition, the type of defect, and the operating environment. Accurate defect detection and quantitative information enable an assessment to be made of the remaining life expectation of the material. To achieve this, quality control should occur during all phases of manufacture and during in-service operations.

As new composite materials are being produced, NDT techniques are becoming more sophisticated in the accurate detection and quantification of defects. Recent developments in high-resolution imaging systems coupled to NDT apparatus provide defect visualization and facilitate its location and signal interpretation. In addition, the development of new processes in artificial intelligence, and the adaptation of data fusion to NDT, greatly improves the NDE of composites and materials in general [4].

Evaluating composites nondestructively not only includes the use of an adapted NDT technique but also encompasses the disciplines of physics, material science, statistics, and computing. The capabilities of NDE in assessing the structural integrity of composites continue to grow, and proper employment of each technique remains a real challenge.

2 Defects in Composite Materials

Fracture and failure mechanisms in composites are complex, can be difficult to detect, and can lead to structural degradation and shorten structure life. Despite high-quality raw materials, a properly mixed resin, a clean manufacturing environment, and correctly designed artefacts and molds, flaws may occur due to modified mechanical, thermal, chemical, or electrical parameters. Although effective reinforcement is achieved with a good bonding between the fibers and the matrix, excessive fiber–matrix bonding can cause a brittle material such as ceramic to become more brittle as the strong fiber–matrix bonding causes cracks to propagate in the direction perpendicular to the fiber–matrix interface [5]. Reifsnider reported that a reduction in stiffness of up to 30% can occur in quasi-isotropic graphite epoxy laminate subject to tensile loading [6]. Harris *et al.* [7] showed that tension fatigue cycling on carbon fiber reinforced plastics (CFRP) produces a reduction in Young's modulus and strength, and that transverse compression causes minor permanent changes in shape, matrix plastic deformation, local bending of fibers, and fiber breakage in both 0° and 45° plies. For glass fiber reinforced plastics (GFRP), there appeared to be no visible signs of fiber breakage of matrix cracking but a reduction in strength and fracture toughness [7]. Moreover, stress corrosion of GFRP causes a small overall decrease in strength and toughness [7]. An exhaustive list of the types of defects as well as their effects on the structural integrity of composites can be found in the literature [8].

Despite an increase in life cycle, strength, and stiffness, composite materials may be subject to manufacturing or in-service damage that may cause a reduction in static strength and weaken structural integrity. The more likely events of occurrence of some of the most widely spread flaws, and their effect on the structural integrity of composites, are described next.

2.1 Manufacturing Flaws

Manufacturing flaws could be avoided by following a quality control procedure and setting up more stringent regulations or monitoring of manufacturing processes. Djordjevic and Green summarized the parameters and properties that can be nondestructively monitored for process control of organic matrix composites [9]. These include but are not restricted to cure conditions, temperatures used, viscosity, porosity, fiber orientation, lamination, and material density. According to Rebello and Charlier, cracks, porosity, and inclusions are the three main types of flaw indentified in ceramic composite manufacture [10]. Irregularities in composite makeup can appear at almost any stage of the process. In addition to those briefly described, the following details more general flaws that occur in composites.

Fiber/Matrix Distribution

The distribution of fibers and matrix is critical to the performance of the composite structure [11]. Steiner showed that matrix-starved areas could be found around the edges of composites, whereas matrix-rich regions usually occur in the center. It was demonstrated by Hancox that with pre-impregnated fibers, if pressing is undertaken before the resin can flow or after it has gelled, errors in fiber and resin volume fraction occur [12]. This results in a reduction in mechanical properties. Improper temperature can produce excessive tightening of the epoxy material out of the composite, resulting in epoxy-starved regions within the composite.

Fiber Misalignment

Misalignment of fibers can be brought about at several stages in manufacture. Incorrect weaving of the fibers may cause changes in volume fraction, thus preventing adequate packing of fibers, while layup errors may cause ply misalignment. Movement of fibers may occur during fabrication when the resin melts, or its viscosity is reduced, due to inappropriate internal or external heating temperatures. Hand layup fabrication, using dry fiber and a liquid resin, is the most common situation in which fiber misalignment can appear. Such flaws are also common in matched dye moldings using preimpregnated fiber. Faults in this situation arise in the early stages of high-performance CFRP production such as during the manufacture of turbine blades. Fiber misalignment results in a reduction of directional strength and stiffness, eventually leading to failure. In these failure situations, it was noted that the longitudinal tensile modulus decreases very rapidly with increasing angle between the fiber and loading axis [12]. If fibers in unidirectional composites or laminate sheets are out of alignment by even a few degrees, reductions in the strength and modulus will occur.

Fiber–Resin Ratio

Inappropriate temperatures and pressures or inaccurate assembly can cause evacuation of a significant amount of resin from the composite during manufacture, inducing an incorrect fiber–resin ratio. Poor fiber–resin ratio inevitably ends in a reduction in the mechanical properties of the composites structure.

Material Thickness

Because composites are often assembled in layers, variations in material thickness can occur and need to be accurately quantified. Thickness variations in composite structure are often associated with differences in the resin content of the laminate. This type of manufacturing error can lead to, for example, a loss of balance of a structure or a lack of flexibility.

Bonding Defects

Poor bonding is a common defect found in all types of fiber reinforced materials and is due to incorrect adhesive curing time or to the presence of contaminants on the adherend surface, affecting adhesion and leading to a weak adhesive–adherend interface [13]. Other causes include the failure of the resin to wet the fibers because of surface contamination, insufficient resin, incorrect processing temperature, or an inability to bond to the fibers. Poor bonding may sometimes appear as abnormally large internal voids. As a result of their formation, a reduction in interlaminar shear strength and stiffness occurs because stress can no longer be transferred effectively into and between fibers [13]. The mechanical strength of an adhesive bond depends on cohesion and adhesion. Bond failure is due to stress forces exceeding the material strength at its weakest point. The difference between cohesion and adhesion, as defined by Munns *et al.*, is that cohesion refers to binding properties within the adhesive itself, while adhesion refers to the bonding phenomenon (i.e., how well the adhesive attaches itself to another surface) [14]. These authors demonstrated that defects within the adhesive layer causes a reduction in cohesive strength, while defects within the adherend–adhesive interface causes a reduction in adhesive strength [14].

Foreign Inclusion

Foreign particles, or material inclusions, can be present during composite manufacture. A source of such contamination can be from plastic carrier film and delivery release paper of impregnated composite plies that remain on the material during processing, thus causing inclusions. Another cause of contamination can happen during the curing process. Peel plies are used to prevent bonding of the laminate to the mold, and sometimes peel ply fragments can be present within the laminate bulk and can lead to a reduction in bonding between plies. Foreign inclusions may affect the structural integrity of a composite, depending upon their location.

Porosity

Porosity and voids within the material of a composite are caused through air bubbles or volatiles becoming trapped between fibers or combined into the resin due to pressure loss during the cure cycle. Void and pore formation is considered one of the most common faults occurring in fiber reinforced plastic manufacture. If pores form along the interlaminar interfaces, a reduction in static shear strength occurs that renders the composite more susceptible to impact damage and reduces its fatigue life [15]. Voids, of greater volume than pores, and caused by gases trapped in the adhesive layer, tend to concentrate between two boundaries or laminae, causing a line or planar fault. Hagemaier and Fassbender estimated that interlaminar shear strength in CFRP decreases by 7% for each 1% of void content up to at least 4% and that a void content in excess of 2% is considered unacceptable in secondary structures [16]. Voids within composite structures may act as stress raisers, or contribute to premature failure because of the presence of unsupported regions of fiber [12]. The presence of voids in the bulk matrix leads to an overall reduction in cured resin properties, which may be reflected in lower transverse, flexural, shear, and compression performance. Voids also increase composite susceptibility to chemical attack and reduce the electrical properties of graphite reinforced plastics. It has been shown that increasing void content decreases compressive failure strain in laminates exhibiting negligible weaviness and also reduces interlaminar shear strength significantly [17]. Skin-to-core voids may occur at the edges of doublers or during chem-milled steps, when the adhesive bridges across a discontinuity [3].

Incompletely Cured Matrix

Incomplete matrix curing is due to an incorrect cure cycle and may occur due to inaccurate mixing of compounds or temperature or pressure variations during the manufacturing process. Cure failure can also occur in the adhesive layer and develop into a more significant fault, thus reducing further the structural integrity of a material.

Matrix Cracking

Although matrix cracking occurs during in-service operation due to impact or excessive strain, it can also happen during manufacture, mainly due to differential thermal contraction. In [0.90 $\pm$ 45] graphite epoxy composites, matrix cracking occurs throughout the thickness of plies inclined in orientation to the principal load axis. The cracks tend to form along the direction of the fiber [6]. A reduction in stiffness caused through matrix cracking can appear before a structure has completed less than 20% of its expected life span. Reifsnider reported that micro-cracks found clustered in zones, centered around transverse cracks in adjacent plies, are formed under tensile loading, in addition to the major matrix cracks created through the thickness of off-axis plies [6].

Delamination and Disbond

Delaminations are planar defects occurring within the composite and are either due to contamination during layup or can appear following impact damage during in-service operation. They can be described as the splitting between laminae and laminate and are considered one of the most important flaws affecting composites. Delamination can be defined as a lack of adhesion between two layers of wound fibers, while a disbond can be characterized as a lack of adhesion between two dissimilar materials. The most common and effective initiators of delamination are tensile interlaminar stresses. Interlaminar shear stresses greater than a threshold stress value initiate deliminations, which reduce compressive strength and fatigue life [6].

Very common in composite laminates that have free edges is edge delamination, occurring due to an increase in tensile loading [6]. On compression loading of glass fiber epoxy composites in excess of that tolerated by the structure, delamination first appears near the surface, towards the inside [18]. Chester and Clark demonstrated that no delamination takes place between plies of the same orientation in graphite epoxy laminates, and the delamination spreads out along the fiber direction of the ply furthest from the impact location [19]. Delaminations and disbonds do not transfer interlaminar shear stresses correctly and can be the cause of rapid fracture. This produces a reduction in local cross-sectional bending stiffness and locally eccentric loading, which introduces transverse shear forces of sufficient magnitude to provoke composite failure [20]. Heslehurst and Scott reported that the fracture process of delamination involves interlaminar cracking between two anisotropic fiber reinforced plies, and that the propagation of delaminations in graphite–epoxy composites is sensitive to several parameters such as toughness, stacking sequence, delamination size, location, and environmental effects [21]. Disbonds can also occur due to lack of adhesive, or the presence of an uneven layer of adhesive, or from the appearance of a contaminant, such as grease, on an adherend [13]. Zero volume disbonds occur between the adherend and the adhesive zone where there is contact but not mechanical or chemical bonding [14]. Disbonds reduce laminate strength and can favor moisture ingression and subsequently ensuing corrosion.

2.2 In-Service Defects

From research in fracture mechanics, it is clear that damage initiation in composites is influenced by fiber matrix assembly, stacking sequence, curing cycle, and the environment in which the structure operates. Several mechanisms of in-service degradation have also been identified and include static overload, impact, fatigue, lighting strike, overheating, creep, and hygrothermal effects [15]. These may lead to fiber breakage, fiber–matrix interface breakage, moisture ingress, bond failures, delamination, and through-thickness cracks, with most defects occurring at the leading edge and at the side closures of honeycomb structures [22]. Some of these effects are dealt with individually in the following subsections.

Water Ingress

Honeycomb structures can be susceptible to water intrusion, which may affect, for example, the weight and balance of an aircraft, and in the long term induce corrosion. During winter, or at high altitude in the case of aircraft, trapped water in such structures may freeze, and the subsequent expansion of ice causes cracks, breakage of the honeycomb cells, and disbonds. Moisture absorption in unidirectional carbon fiber epoxy–matrix composites causes swelling and degradation of epoxy film adhesive joints.

Impact damage

Impact damage is the primary cause of in-service delamination in composites and can be brought about by dropping a tool during maintenance, hailstones, bird strikes, a stone, runway debris, thrown tire treads, or baggage loading. In addition to delamination, impact damage can also lead to matrix cracking and fiber fracture, which cause reductions in static residual strength. Assessing accurately the extent of impact damage is necessary in order to evaluate the residual strength of a structure.

Fibers absorb part of the energy during an impact, but they also distribute some of the load internally. This excess energy can induce cracking and delamination. Impact damage detection and quantification can be a difficult task, especially with low-energy impacts, where most of the damage is not readily visible on the external surface. Barely visible damage (also referred as blind side impact) produces visible cracking of the matrix solely on the back side of the component and cannot be detected with conventional rapid visual examination. Theoretical and experimental studies about the failure modes and the damage susceptibility of graphite fiber reinforced composites subjected to low-energy impact have been discussed by several authors [23,24,25]. It was shown that due to low strain to failure, graphite sandwich panels were much more susceptible to impact damage than glass–epoxy panels, which did not suffer from fiber fracture [24]. For graphite honeycombe core, low-compressive buckling strength is a factor affecting impact resistance, core crushing being initiated immediately after impact. As graphite composites absorb energy, progressive fracturing along planes defined by the laminate fiber orientation is brought about. Additional damages, such as transverse cracking, penetration, fiber breaking, and structural failure, are all results of impacts of varying energy [25]. It was demonstrated that for a given mass, bird impacts are more severe—in terms of damage produced—than impacts from ice, and that the projectile–target contact stress is independent of the projectile radius and is a function of velocity and material properties only [25]. In similar studies, it was observed that the extent of local damage increases with the magnitude of applied axial compressive strain, and that the damage in the high axial stiffness region precipitated failure at an applied strain well below the design strain level [20]. Cantwell and Morton observed a reduction in residual and tensile and compression strength of up to 50% in CFRP laminates [26]. Contact mechanics have shown that tensile stresses are generated at the periphery of the area of contact, which at sufficient impact velocities are large enough to cause failure at the fiber–matrix interface. In thin CFRP composites, damage is initiated in the lowest ply due to the flexural action of the beam, whereas in thicker, stiffer materials damage occurs in the uppermost plies. For example, impacts on metal fiber composites produce buckling of the fibers, fracture of the matrix, or splitting of the matrix–fiber interface. CFRPs exhibit low susceptibility to short-time impact loads compared to isotropic materials.

Impact damage in laminated composites has been widely studied; a review of an analytical and experimental approach to damage modelling has been made by Chester and Clark [19], while Girshovich *et al.* discussed ultrasonic and radiographic results obtained from inspections of impact damage in 48-ply thick graphite/epoxy laminates [27]. Moreover, Chester and Clark demonstrated that the combination of NDT techniques and theoretical and experimental modelling can provide information to facilitate prediction of the effects of impact damage on the static compression strength of graphite epoxy laminates. Damage development in plates of glass fiber reinforced polypropylene has been studied by Reed and Bevan, who show that impact damage apears to be progressive. Its initiation could not be detected by freeze-frame photography or visual examination [28]. However, microscopic observation found that the initial damage mechanism occurring was matrix cracking at the fiber–matrix interface, the crack propagating along the fiber, and initial damage occurring when the transverse tensile stress reached a critical value. Ultrasonic C-scans were used to quantify impact damage in graphite/epoxy thin (0.254 cm) and thick 2.54 cm) laminates [33]. This study showed that the thin/thick laminate impact behavior transition was dependent on the thickness of the specimen as well as its flexural stiffness. Thin specimens are more vulnerable to delamination and matrix cracking at the back surface, while the thicker the specimen, the greater the threshold energy (due in part to a greater local or hertzian stiffness).

Fiber Breakage

Broken fibers, a result of impact damage, cause a reduction in tensile strength and thus shorten the life of composite laminates. Under cyclic loading, fiber fractures begin within the first third of the expected life of cross-ply laminates [6]. If carbon or boron fibers are used, these tend to be more brittle, and cracking may occur more easily during onset of impacts, which reduces the structural integrity of composites made with such materials.

Lightning Strike

In the case of aerospace vehicles, damage is possible due to lightning strike. Although lightning damage is usually confined to the surface layers of stricken composites, corrosion, disruption of the ground path, and a breakdown of electrical conductivity can result. Lightning can vaporize resin matrices at the point of discharge and cause charring of nonmetallic honeycomb cores, which leads to extensive and serious structural damage [29].

Corrosion

Corrosion can be brought about as a result of water ingress or the coupling of anodic metals and carbon fiber composites in the absence of a barrier coating [29]. If the adhesive layer is improperly coated or sealed, it absorbs moisture and may corrode. The core material in honeycomb sandwich materials is principally subject to bondline corrosion [29]. Coupling a cathodic material such as carbon with an anodic metal can cause galvanic corrosion if a protective barrier is not present.

Coating Deterioration

Natural weathering such as wind and rain erosion can eventually lead to failure of protective coatings of composites including paints, and the underlying fiber–resin material, if left exposed, is then vulnerable to continued weathering [29]. Water penetration and UV radiation are other factors damaging composites.

3 Optical Testing

Optical testing is probably the most common method for materials evaluation and metrology. Optical methods include visual inspection and microscopy and the use of more recent technology such as optical fiber sensors, photoelastic coatings, moiré fringes, holographic methods, shearography, and laser optical triangulation. These techniques are described in detail in the next paragraphs.

3.1 Visual Inspection

Visual inspection is best suited in rapid, noncontact survey of specific areas and is ideally followed by more sophisticated methods to determine the extent of visual damage. Visual inspection is an important and simple technique by which gross defects, discoloration and external corrosion can be identified. The efficiency of such visual tests can be improved by the use of facilitating tools such as a light source, a magnifying glass, and a boroscope. Other apparatus are also available such as the profile projector of Walker and McKelvie [30]. Used to examine complex parts, this instrument has a resolution of 2.5 μm.

In the aerospace industry, the immobilization of planes is costly, so cheap, portable methods for rapid in-service inspection are preferred. Automatic or semiautomatic systems that reduce inspection time and allow rapid scanning of the whole surface of a vehicle are favored. The condition and continued airworthiness of a composite structure is visually verifiable for the greater part, although where there is cause for concern inspections should be followed by a tap test or an intensified visual inspection if necessary, to assess the nature and the extent of damage [3]. Personnel involved in the visual analysis of structure integrity should have good eyesight, and inspection should be performed under good lighting and on relatively clean surfaces. One of three tests can be carried out:

A quick visual inspection with time limitation
A long visual inspection without time limitation
A detailed visual inspection using facilitating apparatus and without time limitation.

Such tests can identify incorrect component dimensions, lack of adhesive, lack of filleting, edge voids and porosity at the manufacturing level, throughout the cross section of a laminate after the part has been cured [29]. Other in-service damages detectable include paint failure, lighting strikes, and peeled surfaces.

Sophisticated visual inspections use video technology and image processing recognition systems for improved signal interpretation. A video detection system consisting of video acquisition and processing hardware was developed by Cauffman *et al.* [31]. This system is capable of 360° flaw detection and measurement of axial and circumferential fiber alignment. A desktop computer based system with three cameras provides 360° coverage. Illumination symmetry obviates calculation of two 2-D Fourier transforms, normally required during homomorphic image enhancement. This preprocessing step improves the performance of the pattern matching algorithm, which determines braids per inch and fiber alignment. Braid patterns extracted from the captured image are subsequently adjusted to compensate for individual lens focus and actual camera angle. The resultant robust system is transferable to 2-D braiding, weaving, and other fiber processes.

Among the multiple methods for testing rubber tires, visual examination remains the most popular. Interest in infrared thermography is increasing for the detection of delamination between the rubber and the metallic reinforcement fibers. Another promising technique for on-line inspection at the manufacturing level uses eddy currents, but additional research is still required in order to estimate its full potential. Specially designed apparatus have been developed to test both bias and radial tires (e.g., the NDT-II tire tester from Hawkinson) in the treading process, prior to and after buffing. Such a system is estimated to test a tire in less than a minute and carry out electronic and visual inspection in one operation. It can be used for in-service testing of tires to detect nails, holes, cuts, tears, and bad repairs. Bad casing can also be detected before curing loses appear as well as inner liner porosity. Stress related cracks of plastic materials due to poor design or chemical damage are usually identified by visual inspection. Special tests (such as the melt flow index and the intrinsic viscosity tests) have been developed which, by comparing the molded sample to the original design, provide a measure of the amount of degradation of polymers that may have occurred from the molding process, thus causing a change in the molecular weight of the material.

3.2 Microscopy

Optical microscopy, scanning electron microscopy (SEM), and transmission electron microscopy (TEM) can be used to study the mechanism of adhesion between fibers and matrix, for surface and failure analyses. The limitations of all microscopy methods are the small size of the sample required and the coating of samples for EM with a conductive surface such as gold or carbon, which is necessary to ensure that an image is obtained of the sample surface. Optical microscopy requires little sample preparation, and the resolution appears to be limited to 1.0 μm; curved surfaces cannot be studied [32]. Detection of impacts of 0.4 J in 8 plies ±45° CFRP laminates by optical microscopy was achieved by Cantwell [26]. Figure 1 is a micrograph (optical microscopy) showing breakage of carbon fibers, metal fibers, and the matrix on the surface of a CFRP composite subjected to a 5.0 J impact.

Acoustic microscopy is based on the principle of sending high-frequency ultrasonic waves (10–100 MHz) into a material and detecting on the back side the transmitted waves using a laser beam ripple. These waves are attenuated by variations in structural properties or the presence of flaws [34]. The scanning acoustic microscope (SAM) is appropriate for flaw detection in ceramics and ceramic–matrix composites, but it is limited to the detection of surface defects and subsurface defects with a spatial resolution of 1.0 μm [34]. A low-frequency SAM (3–10 MHz) is available, and it is better suited for internal damage assessment. Yen and Tittman used SAM to visualize damage to composite materials during the carbonization process [35]. Due to their high sensitivity to surface breaking cracks, Rayleigh waves were generated on the surface of the composite, and subsequent study of the resultant micrographs showed that transverse crack constituted the predominant form of damage occurring during pyrolysis.

The SAM is a reflection mode technique that uses a transducer with an acoustic lens to focus the wave below the surface; the transducer is scanned across the sample to create an image [36]. Images from SAM are formed through the interaction of sound waves by direct reflection of the longitudinal waves with the sample, or by mode conversion to surface acoustic waves that propagate and scatter. High-resolution images of the material surface

Figure 1 Micrograph (optical microscopy) showing breaking of carbon fibers, metal fibers, and matrix on a CFRP composite subject to a 5 J impact.

are produced. Several different versions exist, for example, analysis at specific depths can be achieved by C-SAM, designed for moderate component penetration. In addition, the scanning laser acoustic microscopy (SLAM) is a transmission mode instrument that creates true real-time images of a sample through the entire thickness by detecting ultrasounds by a laser that scan the material surface [37]. Experimental results of SLAM in the projection of impact damage at a microscopic level, and C-mode SAM in the provision of a 3-D distribution of the damaged areas, have been described in the literature by Wey *et al.* (1992) [37].

A new tool for cost-effective microscopy and surface texture analysis is parallel confocal microscopy. Confocal microscopy delivers high depth-of-field optical images by constructing a 3-D image from a series of sequential scanned slices. This technique uses pinholes localted at the focal plane of a lens to ensure that all the light used to form an image comes from the plane on the specimen surface, which is at the focal plane of the lens. Scanning along both the length and the height of the specimen allows building of complete images. Real-time height maps of the area under examination can also be generated, allowing surface texture analysis of a composite. Textiles require multiple parameters to be tested, such as tensile strength and elongation, length, and resistance to abrasion of textile fabrics and plastic-coated materials. The elasticity of textile fibers can be measured using a ductilometer; yarns and textiles are stretched at a uniform velocity to their breaking point, measuring their load density and elasticity. Devices have also been designed to measure the properties of contractile force of synthetic fibers when heated. Visual inspection remains one of the principal techniques for quality checking of textiles and for identifying causes of defects in woven and knit cloth. Scanning electron microscopy allows detection and enlargements of disfigurements of textile fibers and clinging materials on the surface of yarn and woven cloth. Other techniques, such as infrared spectrometry, x-ray analysis, and thermal analysis can also be applied to textile testing.

Polymeric products such as paper, leather, rubber, plastic, resin, and textiles are regularly tested to ensure high quality and performance. Defect analysis and charcterization of textile products on the basis of the final product use, and identification of fiber species and accurate measurements of the fiber diameter, are important aspects of quality control. Traditionally, fiber diameter was determined by manual visual inspection. Computerized fiber identification and new measurement systems, taking into account the specific gravity of the fiber and its weight fraction, have been developed for improved accuracy and reduction of testing time. Measurement of fibers diameter using an optical fiber diameter analyzer helps to determine its best use in textile manufacturing. Microprojection is also used for measuring fiber diameter for core testing. Testing and analysis of the structure of fibers and their physical characteristics are among the regularly tested textile parameters by projection microscopy. Also checked is the moisture content of the textile fiber and the yarn construction such as its linear density, ply of yarn, thickness, and twist of yarn in fabric (BS2864, 1984, 1990). Additional tests include measurement of the electrostatic properties of textile fabrics as recommended by the British standards BS2050, 1978, BS6524, 1984, and BS EN 100015, 1992. Among the methods employed to measure these parameters are dispersive infrared spectroscopy, Fourier transform infrared spectroscopy, microscopy, and ultraviolet and visible spectrophotometry. Visual quality control for production of textile composite materials can also be achieved using automated image processing. Images of the surface of a composite from a microscope coupled to a computer allow the extraction of important features such as yarn orientations. Further image analysis can also be used for 3D mapping of yarn, which helps the determination of matrix pocket locations.

3.3 Optical Fiber Sensors

In situ embedded optical fiber sensors, also known as smart sensors, have been developed for monitoring and control of composites during both fabrication and in-service operation. They have the advantages of combining small size and light weight with a wide operating temperature range, compatibility with composites, and immunity to electromagnetic interference and corrosion. In the aerospace industry, optical fiber time domain sensors measure the time-of-flight or time delay of optical signals propagating into embedded optical fibers. Deformation of the structure into which the fibers have been incorporated also produces a distortion of the optical fiber, which in turn causes a change in the time of flight of the optical signal. Quantitative information regarding the structural integrity of a composite can be obtained by relating the deformation of the embedded optical fibers. Because the optical refractive index is correlated to the state of cure of a composite, continuous monitoring of changes in the optical refractive index can allow information to be gathered during the curing process; the determination of strain and temperature changes using optical fiber time domain strain and temperature network have been described by Lou *et al.* [38]. Embedded optical fibers can also be used to assess and map damage growth in composites; however it is necessary to embed different kinds of sensors, as there are multiple types each designed to detect only specific parameters.

For optimum performance as strain and temperature sensors, embedded optical fibers should be mounted between two collinear plies and be aligned with the reinforcing fibers. Maximum sensitivity is achieved when fibers are embedded as close to the surface of maximum tensile strain as possible, and sandwiched orthogonally between a pair of collinear plies [39]. Ideally, each fiber should be able to provide a well-behaved, linear response and a sufficiently sensitive signal. Measures proved that embedded optical fibers do not compromise the strength of composites, nor increase their damage vulnerability, but improve the damage resistance of the material [39]. Of the available types, carbon coated optical fibers (12 μm diameter) are more compatible, in term of adherence, with epoxy–carbon composites, and they improve the damage tolerance of the material [40]. Embedded Michelson fiber optic strain gauges can detect acoustic emission signals directly correlated with the formation of matrix cracks and interlaminar debonding [39]. The relative strength and weakness of embedded fiber optic strain rosette sensors and Fabry-Perot–based fiber optic interferometric strain sensors have been compared by Measures, who showed that the latter were superior to both Michelson type optical fibers and simple fiber optic strain rosettes [39].

Due to their high sensitivity to strain, temperature variation, vibration, and acoustic waves, embedded extrinsic Fabry-Perot interferometric optical fiber sensors have been developed to detect delamination, based on changes in the acoustic properties of the materials before and after delamination [41]. Impact events and corrosion cracking generate ultrasonic waves, which can be characterized using elliptical core fiber sensors [40]. The in-line Fabry-Perot interferometer seems well suited for the local detection of shear waves and the characterization of impact-induced damage.

An increased confidence in terms of maintenance could be achieved by the use of such complete monitoring systems, thus allowing longer periods between inspection.

3.4 Photoelastic Coating

This method consists of bonding to the surface of a composite structure a thin sheet (0.25 mm) of photoelastic material, so that the bonded interface is reflective [34]. When the material become stressed, surface strains are transmitted to the coating and induce a fringe pattern to be produced that is recorded and analyzed using a reflection polariscope. This method is useful in the detection of localized strain concentration areas and damage propagation in composites.

A comparable technique to that of the photoelastic coating is brittle lacquer testing. In this technique, a coating of a thin brittle lacquer layer is applied to the composite surface [30]. Applying a load onto the coated surface will cause cracking of the lacquer, thus revealing areas of excessive strain.

3.5 Moiré Fringe Methods

Moiré fringe methods make use of the moiré effect, which is a pattern formed by interference between two sets of regularly spaced gratings [30]. With the shadow moiré technique, a grating located close to the surface to be tested is illuminated obliquely, and lines of the grating merge with the lines projected onto the surface to create a series of fringe patterns. The spacing of the contour fringe pattern is constant on a defect-free area but becomes modified in the presence of a flaw. The separation S of the moiré fringes is given by

$$S = \frac{d}{\theta} \tag{1}$$

where d is the line spacing of the grating and θ the angle of intersection of the gratings. As one can see, the spacing of the grating is a limiting factor in the accuracy of this technique. This low-cost technique is used for the measurement of complex shapes, out-of-plane deformations, and in-plane strain. Moiré interferometry is useful in detecting surface variations on plastics and polymers, but its sensitivity to vibration and dust thus are limiting factors to its use in industrial applications.

3.6 Laser Holography

Laser holography, also known as hologram interferometry, generates fringe patterns similar to contour maps that show variations on the surface of a material. Each fringe represents a differential surface movement of half a wavelength, and any changes in the fringe patterns relates to the presence of subsurface flaws. External applications of heat, loads, or vibrations are applied to the specimen under test, and deformations in fringe pattern due to the presence of flaws are recorded. The principle of this technique is based on the superposition of a coherent light source, reflected from an object, and a reference laser beam (Fig. 2). The light scattered from the surface of the specimen is received by a recording medium. The interference pattern recorded on film is called a hologram and

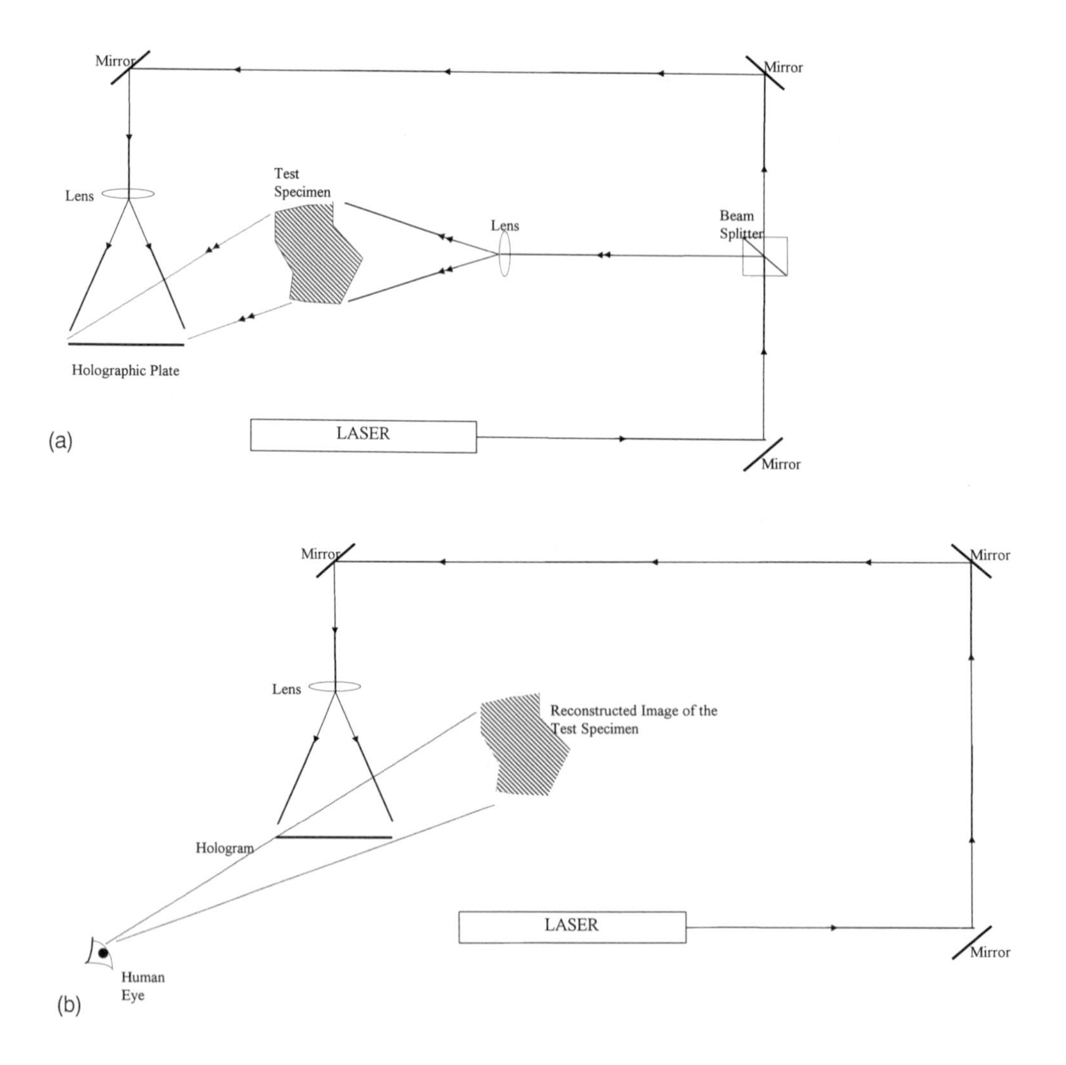

Figure 2 Schematic of a laser holographic system; (a) image formation, (b) image reconstruction.

contains the wave pattern characteristic of the structure. When materials are under stress, strain displacements occur at the proximity of flaws, producing a deformation in the fringe pattern. These interference fringes are a measure of the amount of surface deformation between the initial object and the object measured at a later time [42]. Evidence of the deformation is given by comparing the original hologram of the structure not under stress with one obtained when an external load is applied. This technique is also known as double-exposure holography, in which two holograms are superimposed on the same holographic plate; each one captures the material structure in a different state separated by a fixed time interval, with the intensity and the extent of the pattern anomaly being proportional to the defect extent. However, this technique can be difficult to use in the case of thermal application; holography, which detects the differences in heat conduction occurring on the surface of the material, reamins a cumbersome technique largely limited to laboratory examination despite several improvements. Bruce and Clark found that defects deeper than 2.0 mm were not detectable in CFRP when using temperature as a variation parameter [15]. In addition, vibration and mechanical loading can be difficult to implement on complex structures, they are insensitive to subsurface damage, and interpretation of fringe patterns can be subjective.

Applications

In certain circumstances, holography can be an effective NDT technique; the cellular structure of honeycomb materials can be accurately examined, and areas of poor adhesion between cover ply and honeycomb structures can be detected. Holographic inspection of GFRP honeycomb structures appeared to be a sensitive and repeatable method for disbond detection, and both defect size and depth can be estimated by experienced operators [43]. However, sensitivity is reduced as defect depth increases, and tightly pressed shallow flaws are difficult to detect by means of thermal stressing. Tay *et al.* used double-exposure holographic interferometry to detect subsurface flaws in GFRP with a 10.0 mW He–Ne laser of 632.8 nm wavelength and successfully detected localized thinning in woven GFRP plates [44]. Delamination in composites can also be detected as bonding strength affects the fringe pattern, while acoustic imaging indicates a perfect bonding. Matrix cracks, fiber breaks, and delamination can be detected by laser holography, but subsurface matrix cracking and internal flaws will remain undetected. Holography failed to detect inclusions located by acoustics, largely because the determination of the border of the fringe pattern, and therefore the extent or depth of damage, can be difficult to ascertain. The appearance of the fringe is influenced by the bonding structure and material strength, leaving defects undetected [45]. Moreoever, black and specular surfaces are difficult to inspect because illumination of the specimen cannot be uniform (a solution is to spray the panel with a diffuse white coating) [46]. Cracking, voids between plies, and debonding after thermal fatigue were found by Delgrosso and Carlson during the inspection of jet engine composite fan blades made of boron filaments coated with silicon carbide [47].

Vibration is the major problem affecting the accuracy of laser holography. The development of a portable holographic interferometry testing system that overcomes the vibration isolation problem associated with laboratory holography is described by Heslehurst *et al.* [48]. Using their system, adhesively bonded joint weakening was determined both during fabrication and in an in-service period of a composite structure. The system is built so that both the laser and the optical equipment are mounted as one unit, and the reference and object beam are one and the same. Although it is actually a qualitative method, research is currently being carried out to determine quantitatively the correspondence between the in-plane load and the material properties corresponding to out-of-plane deformation measurements [48].

3.7 Acoustic Holography

Ultrasonic waves can be used to create real-time acoustic holograms of the interior of a structure. The hologram formed of an object inside an opaque material is then transformed into an optical hologram, thus creating a visible image of the object [49]. Scott and Scala used a 5.0 MHz transducer to detect disbonds caused after impacts between a graphite–epoxy skin and a honeycomb core, and to determine the extent of the impact damage [50]. This highly sensitive technique can scan large areas at low cost and size defects with an accuracy of 3.0 mm, but its usefulness remains under consideration [49]. A limitation is that sound fields must be kept absolutely constant during scanning, a task that can be time consuming.

3.8 Television Holography

Conventional laser holography is a slow and cumbersome process, which limits its use for on-line inspection of composites. An alternative holographic technique, known as television (TV) holography, or electronic speckle pattern interferometry, offers the advantage of recording holograms on the photosensitive surface of a video camera which, after processing, can be displayed on a TV monitor. Holograms are formed by coherent light illuminating a diffusely reflecting surface; the result is a speckle pattern. The speckle field is recorded by the video camera together with a reference beam, before and then after deformation. Despite the poorer quality of the holograms, this technique is faster than conventional laser holography and produces real-time images. However, the coarse, speckled pattern displayed on the TV monitor may be difficult to interpret, and defect location can be a problem. Image processing software has been developed to overcome this problem, and Vikhagen and Løkberg developed an image processing algorithm based on phase-shifting procedure to compensate for the poor quality of the fringe patterns [45]. The best results of TV holography were obtained in the detection of debonded areas, but components with irregular shapes and surfaces remain difficult to inspect, and considerable phase gradients, not related to flaws, can occur, thus introducing false indications [45].

3.9 Shearography

Bruce and Clarke defined shearography as a real-time wavefront division technique that produces images of the difference in surface displacement rather than absolute measurements of the displacement profile [15]. The component inspected is illuminated with a laser beam and is visualized using a real-time image-shearing camera that produces a couple of sheared images (speckle interference patterns) in the image plane (Fig. 3). When placed under stress, the component is deformed and the presence of a flaw produces a displacement in speckle patterns. The principle can be described as follows; a laser illuminates the surface of the test material, and the reflected light is detected by a shearographic camera that produces two overlapping slightly offset images. The images combine to form light intensity variations that result from the complex addition of the two coherent wave fronts. A reference video image is stored by computer, and every new video image, at a different strain state, is subtracted from the reference image. The difference reveals the strain patterns caused by the presence of a flaw [51]. Shearography is a noncontact technique,

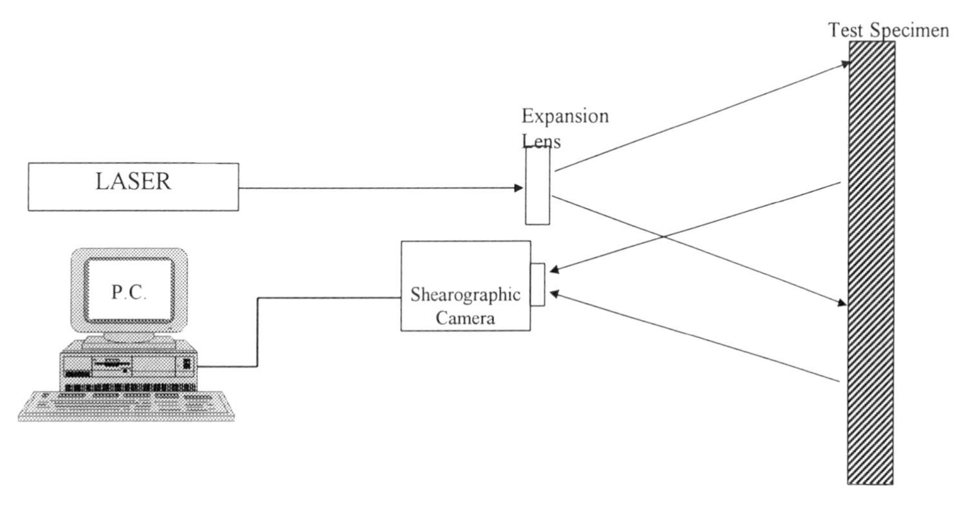

Figure 3 Schematic of a shearographic system.

and the signal output is in the form of an image on a video monitor, facilitating interpretation and sizing; although the specimen is required to be under stress during testing. Because it is not a point-by-point scanning procedure, shearography is well suited for rapid inspection of large areas. It is less sensitive to environmental disturbances than conventional holography, its sensitivity depending upon the angle of the shearing prism employed, as well as on object-to-camera distance.

Shearography has been used in the inspection of large turbine engine composite thrust reversal panels, and in the detection of collateral damage, including delamination, in graphite-epoxy composite panels with honeycomb core [52]. Shearography can detect disbonds, delaminations, and impact damage in composites and honeycomb structures. Disbonds as small as 0.64 cm in diameter have been detected in composites, and the system is regularly used to inspect components such as helicopter rotor blades, flaps, honeycomb panels, and missile fuselages. The sensitivity of inspection results depends upon the stressing technique applied, which itself is chosen as a function of the material composition and geometry. The size of the damaged area can be estimated from the size of the anomalous shearographic fringe pattern.

A large selection of shearographic equipment is now commercially available, and shearographic cameras can weigh as little as 141.8 g [53]. Generally, large structures such as aircraft are placed in a test chamber for pressure reduction stressing or thermal stressing. Shearography is sensitive to submicroscopic deformations in the surface of a test object, and video images of surface strains with a sensitivity of 0.1 microstrain can be obtained [53].

3.10 Laser Optical Triangulation

The triangulation approach is useful in the testing of three-dimensioanl shapes to determine variations in depth by measuring scattered light from a surface illuminated by two overlapping spots [30]. Such systems have been developed to measure tire tread depth but suffer from poor resultion and dynamic range. For these reasons, laser based systems have been developed for improved precision and accuracy.

In laser optical triangulation, illumination of the measured surface is achieved by a collimated laser beam. Irregularities in surface height cause the imaged spot to shift due to parallax. A lens focuses the illuminated spot onto a detector that records the spot position in its image plane [54]. Output data is in a two-dimensional height format, which provides quantitative and qualitative relief information. Robust and accurate systems remain difficult to design and suffer from electronic noise (e.g., slot noise from detected photons or Johnson noise from resistive elements); and surface reflectivity can introduce fundamental limitations to accuracy. The technique is used in surface profiling (e.g., detailing surface height/roughness), gauging (e.g., thickness, diameter measurements), and defect detection such as cracks, pitting, and corrosion in any type of composite, rubber, or ceramic. The main characteristics of well designed systems are high speed, high accuracy and noncontct testing. Certain systems have been built with the possibility of acquiring 20,000 height readings per second, with an accuracy of 5.0 μm [54].

3.11 High-Speed Optical Scanning

High-speed optical scanning, known as D-Sight, uses a white light source mounted below a CCD camera and a retroreflective screen to reveal surface variations of a structure [55]. The surface tested is illuminated with a light source. Variations on the surface of the material tested disperse the light into a retroreflective screen. The pattern formed on the screen is unique and specific to a particular angle of the original light source. The dispersed light is back-reflected onto the tested surface of the composite, which in turn reflects it back. One observer placed off-axis from the primary light source can see the pattern of light from the screen through the surface. Deformations on the surface of the composite appear as black and white variations. These variations are picked up by the CCD camera and displayed on a TV monitor. This very simple technique can be used to detect impacts, disbonds, and delaminations on composites with reflective surfaces.

4 Mechanical Impedance Testing

The principle of the following techniques is based on the analysis of frequency release following a mechanical force.

4.1 Tap Test

The tap test, or coin tapping, is made by tapping the surface of a structure with a metallic pin and listening for a change in sound abnormally caused by the presence of a flaw. Sounds produced by tapping over defect-free regions are usually shorter in duration and more intense than over flawed areas. The kinetic energy released during a coin tap test has been measured to approximately 2 mJ, which is very much less than the energy required to cause damage [56]. The damaged areas have more energy at low frequencies than those produced over intact areas; the generation of a dull sound when striking an area contaning a defect is due to the absence of high frequency sounds. For example, voids and disbonds sound hollow in comparison with a defect-free area, though flaws that bring about an increase in material density, such as a dent, a crushed core, or an adhesive-filled area, can be mistaken for flawless areas [32]. Coin tapping does not necessitate the purchase of any expensive equipment, but it does require expertise, because interpretation of sound information can be subjective, especially with structures of complex geometry. The sensitivity of the tapping technique decreases as flaw depth increases. it is inadequate for inspection of thick laminates, does not provide quantitative information, and is limited to the detection of laminar flaws such as disbonds. Moreover, the detection of any flaw should be followed by a more sophisticated method for accurate sizing.

An automated version of the coin tap technique allowing the production of C-scan images has been developed by Cawley and Adams [56]. Their system compares the time history, or frequency spectrum, of the measured pulse with a reference standard obtained from a defect-free structure. Detection of 10 mm diameter defects in 1 mm thick CFRP skin has been achieved with this technique, while manual coin tapping detected only 20 mm diameter defects. Still, defect detection with the automated system is greatly dependent upon the threshold value selected.

4.2 Acoustic Impact Technique

The acoustic impact technique, a method analogous to the coin tap test, involves striking the surface of a structure with a constant force and recording the time history of the sound impulse; time domain anlaysis then allows identification of defective areas [57]. Impact damages of 2 and 5 J were detected by this technique in 2 mm thick unidirectional prepreg graphite–epoxy panels by Haque and Raju [57], who also reported that the value of the force amplitude increases when the size of delamination, and the impact energy, decreases, while the value of the pulse width decreases when the size of delamination and impact energy diminishes. However, no significant difference was noted between force and time response to identify moisture damage in laminated composites. The acoustic impact technique is used to detect disbonds, impact damage, interlaminar porosity, crushed cores, and moisture in honeycomb laminated structures.

Adams and Cawley described the development of a tapometer for analysis of the frequency content of the force amplitude [58]. They determined that the sound resulting from tapping a structure is mainly emitted at the frequencies of the major structural modes of vibration, and that the characteristics of the tapping depend on the local impedance of the structure and on the impacter.

Pattern recognition techniques can be used to analyze the frequency spectrum of a pulse force. The surface of the structure is tapped at a velocity of 0.1–0.5 m/s, during which a force is generated between the head of the probe delivering the taps and the surface of the structure at the time of impact. The force causes the probe head to decelerate as it springs away from the surface. A force–time pulse output is displayed on an oscilloscope and compared with the standard pulse from a defect-free component. A tapometer allows both time-domain and frequency-domain analysis methods, provides quantitative measurements, and is less subjective than manual coin tapping. However, no flaw depth information is obtained, small voids are not detectable, and certain coatings must be removed from the surface of a component before a tap test can be performed (e.g., coatings containing silicon or ceramic fillers) [3].

The point impedance of a structure Z is function of the harmonic force F and the sound velocity v of the structure at the same point; it is given by [59]

$$Z = \frac{F}{v} \tag{2}$$

This relationship has been exploited for NDT, where impedance transducers have been developed that measure the elastic strain at two different sections of a vibrating rod excited at one end and in contact with a composite structure at the other end. Certain impedance transducer systems measure the potential difference between a top driving piezoelectric element and a receiving piezoelectric element; the relationship between these voltages is dependent on the impedance. It was reported by Cawley that a 3 dB difference in impedance between defect-free areas and defective zones is measurable in stiff CFRP structures [59]. Transducer output is a complicated nonlinear function of structural impedance, and a spectrum of the pulse response is obtained by Fourier transformation of the signal originating from the piezo-transducer attached to the component under inspection [60]. The spectrum is normalized to the corresponding quantity derived from the pulse used for excitation of the rod. Analysis of the transfer functions after a number of load cycles allows estimation of the deterioration of the composite.

5 Vibration Measurements

Vibration measurements are based on the resonant frequencies emitted by structures excited by a vibrational impulse. Monitoring changes in resonant frequency provides damage assessment of a selected structure area. Cawley and Adams showed that both defect detection and quantification can be performed by measuring changes in the vibration frequencies of a material [61], and Scott and Scala used vibration frequencies from 20 Hz to 20 kHz to evaluate remaining material life [50]. Access to any single point on a structure and a short inspection time are the principal advantages of this technique. However, it is recommended to carry out tests at constant temperature, as variations in temperature can influence the signal output.

6 Dielectric Monitoring

Dielectric tests are used to monitor curing of composites and organic resins. Dipoles contained in organic resins will change position in an alternating electric field, and a measure of the energy lost due to dipole movement can inform on the physical nature of the curing resin and on pressure and temperature changes [32]. For example, in a hard resin matrix, cured or melted, the dipoles will remain immobile. The test is performed by making a capacitor using the composite under test as a dielectric. The dielectric constant and the energy dissipation, which are established by changes in the resin during the curing process, are measured. The capacitor is formed using the resin matrix as a dielectric medium, by inserting a section of thin metal foil into the composite to make a small condenser. A similar method to the dielectric test, known as ion graphing, involves the continuous measurement of DC resistance of composite resins during cure, which changes with the fluidity of the resin.

Computer controlled systems have been developed that enable investigation of the spatial distribution of impedance within resins and in turn the identification of flaws [62]. Water trapped within the composite will cause dielectric permittivity to increase, thus revealing its presence. Detection of voids, variation in the cross section of the adhesive layer, ingress of moisture into a joint structure, and characterization of cure in epoxy resin can be achieved with high frequency dielectric spectroscopy (10^{-2}–10^{+9} Hz) [63].

7 Infrared Thermography Techniques

The technique of infrared thermography relies upon the detection of infrared radiation emitted from the surface of a structure. An infrared scanning unit converts electromagnetic thermal energy radiated from an object into electronic video signals and produces color-coded maps of isotherms [64]. Differences in thermal waves on the surface of a material can be detected that will make certain flaws visible and allow detection of flaws with low-temperature differentials with respect to the surrounding area within

±0.1°C. Due to the extreme sensitivity of the equipment, vibrations, background noise, and changes in temperature should be avoided during testing as they may affect signal output. Enhancing thermographs through post signal processing is often required to facilitate defect detection. Detailed information on NDE thermal methods can be found in the literature [65,66].

Infrared radiation (IR), part of the electromagnetic spectrum, is emitted by all objects. Its wavelength is longer than that of visible light but shorter than that of microwave radiation and therefore a special device is required for its detection. Detection occurs in wavelengths with a high transmission band such as 3–5 μm or 8–12 μm. Orlove recommends the longer band of wavelengths for outdoor work due to its superior atmospheric transmission qualities and its freedom from solar reflection [67]. Cooling of the solid-state detector that converts infrared radiation to an electrical signal is often required, and it is achieved through maintaining cryogenic temperatures using liquid nitrogen. Alternative electric cooling systems using the Peltier effect or the Stirling effect have been developed, which eliminate the inconvenience of the cryogenic gases.

Thermal waves have relatively shallow penetration into a material; surface flaws tend to cause heat to diffuse at a different rate from that of the surrounding material [68]. In thermographic testing, two techniques exist, the single-sided technique and the through-transmission or double-sided test. In the former, which is more sensitive to flaw depth indication, only the test area scanned is heated (Fig. 4), while the double-sided technique uses the transmission characteristics of a heat pulse passing through a thin section of composite material [64]. Temperature patterns at the surface of a composite are produced

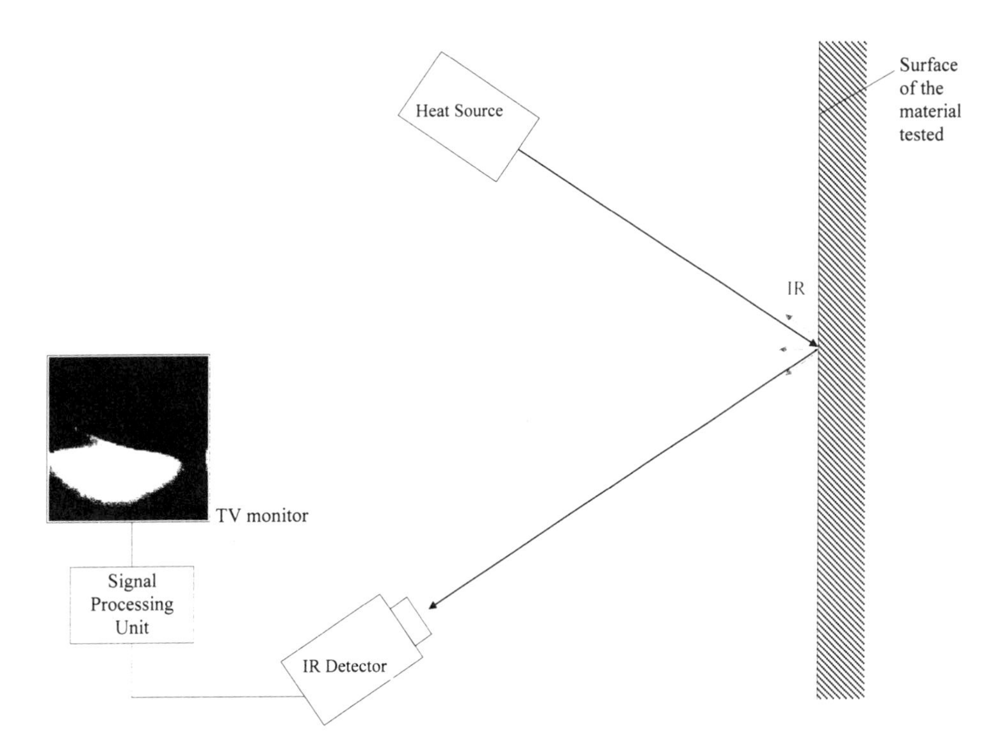

Figure 4 Schematic of the single-sided thermographic technique.

by inducing a temperature differential using heating or cooling systems. Sophisticated technique use flash lamps to send a brief heat pulse to the component inspected. With certain techniques, a laser is scanned over the surface tested to produce heating that is in phase with the pixel being scanned by the detector.

During heat dissipation, an increase in temperature occurs across cracked surfaces. Delaminations and voids therefore show up as hot spots, preventing conduction of the heat across the surface and through the solid material to the other side. In CFRP material the thermal conductivity in the lamellar plane is nine times that in the through thickness direction (dominated by the polymer matrix), and defects deep in the material tend to remain undetected owing to this transverse diffusion [64]. Recent progress has been made to try to overcome the limitations of infrared thermography for composites testing [69,70]. As CFRP composites are highly anisotropic, fiber layup can be detected mainly because the coefficient of thermal expansion is anisotropic. The thermal conductivity of CFRP and GFRP composites with perpendicular fibers has been estimated at 0.8 W/mk and 0.4 W/mk respectively, while steel is 46.0 W/mk [71]. Smith and Clayton showed that a cyclic tensile stress of 70 MPa exhibits a thermoelastic effect of 0.2°C in CFRP materials [71]. The greater the stress range and the thicker the composite, the higher the temperature rise.

Despite its high investment cost, infrared thermography is better suited to monitor low thermal conductivity composites. Evaluation of filler distribution close to the surface in polymeric matrix materials in graphite fiber cured parts can be determined, as absolute thermal diffusivity is sensitive to fiber concentration.

Space-domain processing techniques have been developed to compensate for fluctuations that occur in the heat source, while time-domain image processing was developed to reduce thermal noise caused by variations in surface emissivity or by unwanted reflections [72]. The surface temperature history of a material is given by

$$\Delta T = \frac{Q}{e\sqrt{\pi t}} \tag{3}$$

where ΔT is the temperature increase of the front side, Q the density of the pulsed energy, e the layer effusivity, and t the time. The layer effusivity is given by

$$e = \sqrt{k\rho c} \tag{4}$$

with

$k =$ the thermal conductivity
$\rho =$ the material density
$c =$ the specific heat capacity

The thermal diffusivity of the material is also in terms of k, ρ, and c:

$$\alpha = \frac{k}{\rho c} \tag{5}$$

Balageas *et al.* showed that measuring both the apparent effusivity of a composite and the thermal resistance between two layers can be a useful exercise in the determination of the depth at which delamination has occurred in carbon–epoxy laminates [73]. The heat flow in a single layer of graphite–epoxy composite with a constant flux applied to one surface is given by [74]

$$\frac{\partial T(x, t)}{\partial t} = \alpha \frac{\partial^2 T(x, t)}{\partial x^2} \tag{6}$$

where $\partial T(x, t)$ is the temperature in function of the material thickness x and the time t.

Transient thermography, also known as pulse video thermography, uses a heat source such as a xenon flash or a pulse laser to induce a temperature differential within a material. The technique can reveal flaws at depth because the heat pulse spreads out as it penetrates into the specimen, so that late frames are affected by breaks in diffusion paths at great depth [75]. With conventional thermography, defects at only a few microns below the surface of the specimen are detectable, and different heat emissivity, due to the presence of dissimilar materials such as composite and metal, causes difficulties in defect detection and masking of defects due to the high emissivity of metal. A solution to overcome this problem is to cover the material tested with a black coating.

With lock-in thermography, the component is heated gently using a harmonically modulated frequency source ranging from 0.03 to 3.75 Hz, and amplitude and phase images are produced [76]. An amplitude equation image calculation allows the cancellation of different reflectivity components and eliminates reflected ambient radiation and thus selects out bright spots. Phase image processing further cancels out multiplicative signal components, including emissivity, and reveals the internal structure of the test component in detail [76]. Lock-in thermography eliminates the drawbacks of conventional infrared thermography by recording frequency controlled and capturing four correlated thermal images for every one wavelength of energy from the illumination source [77]. The results are averaged over time, with defect-free composites giving a uniform image while defective materials produce waves of reflected energy that interfere with incoming waves at the surface, resulting in a kind of standing wave pattern that leaves spots on the surface.

7.1 Vibrothermography

Vibrothermography is a variant of infrared thermography and involves measurements of temperature changes occurring in materials subjected to vibrations. High-frequency, low-amplitude vibrations (10–20 kHz) are used to induce localized heating in a composite in such a way that flawed areas are set into local resonance [66]. Real-time thermography is then applied to detect and measure temperature variations occurring on the material surface. Vibrating materials at their resonance frequency allows detection of zero-volume matrix shear cracks in CFRP and GFRP composites [50].

7.2 Applications

Infrared thermography is a technique that finds many applications and is well accepted in the aerospace industry, partly because materials with low thermal conductivity, such as CFRPs or ceramics, are well suited to infrared examination. It was reported that thermographic inspection of an entire aircraft can be accomplished within 20 minutes [67], and as early as 1973 the Lockheed C-5 was completely inspected using thermography [78]. In the case of field inspection of aircraft, access is limited to external surfaces, thus requiring the use of portable single-sided methods and allowing areas up to 1 m^2 to be inspected simultaneously [79]. Infrared thermography is progressively replacing radiography in the location of water ingress in honeycomb structures, and aerospace companies are using it extensively for the diagnosis of this problem [67]. After aircraft landing, ice in areas of water ingress remains frozen while the rest of the aircraft is still warm, thus favoring its detection by IR thermography. Laser heating thermographic techniques are used in the inspection of carbon fiber brakes, aerodynamic control surfaces, and fuselage panels [67]. Cockpit windows made of Perspex with small filament heating elements to prevent icing can be thermographically inspected for discontinuity or satisfactory current flow over the entire window [67].

Thermal methods can be used for the evaluation of fiber content in graphite epoxy composites, and in the evaluation of fiber orientation in extruded or molded parts. Maldague used a 0.5 W argon laser beam to point heat the surface of a composite inspected, while an infrared camera recorded the thermal pattern; in the case of unidirectional graphite epoxy, an elliptical pattern can be observed [78]. The ratio between the x and y axis of the ellipse is related to the square of the diffusivity ratio along longitudinal and transverse directions and thus to the fiber orientation. It was also reported that the detection of delamination may be affected by the surface temperature distribution only during a specific period, and that a smoothing effect may reduce defect detectability in isotropic materials; defects of a radius-to-depth ratio of less than or close to unity are difficult to detect [78]. Other applications include measurement of the thickness of coatings on composites and determination of the individual thickness of two layers of paint on a polymer substrate [60]. Connolly and Copley have used thermography to identify oxidation damage occurring between the silicon carbide coating and the carbon fibers of a composite material [80]. Shiratori *et al.* developed a computational hybrid system that consists of an infrared thermal video camera, by which the temperature distribution of the body surface can be measured, and image processing software, which transforms the thermographs generated into an image displaying flaws more clearly, as raw thermographs are complicated to interpret since they reflect the influence of heating or cooling systems [81]. During the analysis the lower surface of CFRP or GFRP honeycomb structures with aluminum or aramide cores were heated up to 70°C. The heat is conducted through the material to the upper surface, from which it is detected. A large quantity of heat is necessary for detection through thick-sheet constructions.

One major problem is that heated structures cool down during inspection, so that defects disappear rapidly. Thus storing of thermographs is essential. It was found that the relationship between the size of a hot spot on a thermograph and the size of a defect is a complex function of the defect depth and the thermal properties of the composite material inspected [82]. Impact energies of 6.00 J on 1.14 mm thick graphite–epoxy plates, heated to 15°C above room temperature with a 1 second pulse from one 1000 W projector, have been detected by thermography [73].

The capabilities of externally applied thermal field (EATF) thermography have been described by McLaughlin [83]. The technique involves the application of external heat to a surface and the study of the resulting transient thermal pattern. External heat can be applied to the structure using light bulbs, infrared lamps, radiant heaters, or xenon flashes. Pulse lasers are also useful with thin laminates and coatings, as they generate high intensity heat for a short duration. An evaluation of different heat sources and infrared equipment has been made by Jones and Berger [82]. This study showed that the most uniform heating is achieved with a quartz halogen lamp. Flaws located 10% below the surface of graphite–epoxy lamiantes are easily detectable, but small defects produce low thermal contasts, so that deep, small flaws may not generate sufficient contrast to be seen. Using EAFT, surface cracks can be detected by heat conduction parallel to the observed surface only if heat conduction occurs laterally in a direction perpendicular to the crack [83]. For example, in 1–8 mm thick glass–epoxy laminates with an in-plane thermal conductivity of 0.5 W/mk, a heat flux of 1 solar constant is adequate to detect delaminations that are at least 30 mm in diameter and situated at up to $\frac{3}{4}$ of the depth of the laminate from the observed surface [83]. However, the detection of large delaminations close to the surface is easier than that of deeper defects of the same size. It was also shown that thermal characteristics of flaws affect their detectability, for example a damaged region with a fractional conductivity of 1 will not be detectable by EAFT. Paints and other coatings may generate thermal patterns during testing that are due to thickness variations in the paint layer, occurring if the coating has a thermal conductivity different from the substrate. Stress-generated thermal fields (SGFT) are due to cyclic loading, and flaws produce heat dissipation that can be detected by thermography, but thermographic images may be distorted by anisotropy in thermal conductivity [50].

A typical pulse-echo thermal wave imaging system uses a pulsed heat source of 6.4 kJ pulses of 2 ms duration, a thermographic camera, and an image processing software unit (Fig. 4) [68]. When the energy from the lamps is absorbed at the surface of the material, a plane thermal wave pulse is induced into it. When this pulse is reflected from a subsurface boundary between layers having different thermal properties (e.g., corroded parts, disbonds), the reflected portion propagates back to the surface, where it modifies the surface temperature distribution [68]. Thermal waves are monitored by a camera that follows the time-dependent surface distribution of the IR emission from the surface. When uniform heating of the surface cannot be achieved due to coating variations, processing of thermographic signals is required.

It was reported that thermography can detect corrosion in aircraft lap slices, in the vicinity of wing fasteners, disbonds between the fiber glass skin and a foam core in an aircraft structure, impact induced delaminations in ceramic composites, and subsurface impact damage in multiply graphite fiber reinforced polymer [68]. Figure 5 shows a thermograph of a section of a dashboard airbag made of PUR foam layers sandwiched between a 1.0 mm thick thermoplastic foil and a thermoplastic carrier, revealing 12 mm diameter indents. Infrared thermography can also detect delamination, voids, missing layers, and cloth joints in axisymmetric composite structures made by filament winding and hand layup techniques [64]. Where ceramics are concerned, due to the low thermal conductivity coefficients of these materials, only near-surface defects are detectable by thermography [10]. Plastic identification can be performed by Fourier transform infrared analysis. With this technique, infrared radiation is beamed to the sample, and the absorption or transmission wavelength is measured and computer processed. A spectrum can be produced that can be referred to as the fingerprint of the plastic tested. Analysis of this spectrum allows identification of materials such as polycarbonate, acetal, polypropylene, and other plastics. Further spectrum analysis and data processing allow the identification of contaminants that may be present within the plastic.

Vibrothermography shows high potential in the detection of impact damage near the surface plies of a composite. Henneke and Tang have related the temperature variations of a material under fatigue loading to the thermoelastic constant, the local absolute temperature, and the change in principal strains [65]. The temperature change in principal strains can be related to thermoelastic data, and thermoelastic strain analysis can be used to detect manufacturing flaws such as nonuniform fiber and matrix distribution. A modulated optical reflectance thermal wave method has been developed to examine silicon wafer semiconductors [65]. This method uses a laser beam to generate a thermal wave on the surface of a component that will modulate the refractive index of the material tested. The refractive index variations are detected by a probe beam reflected from the specimen surface. Although the thermal conductivity of GFRP is about 20 times lower than that of CFRP, vibrothermographic detection of 15 J impacts on GFRP composites was achieved by Potet *et al.*, but impact location was only roughly delineated, and heat source mapping could not be performed [66].

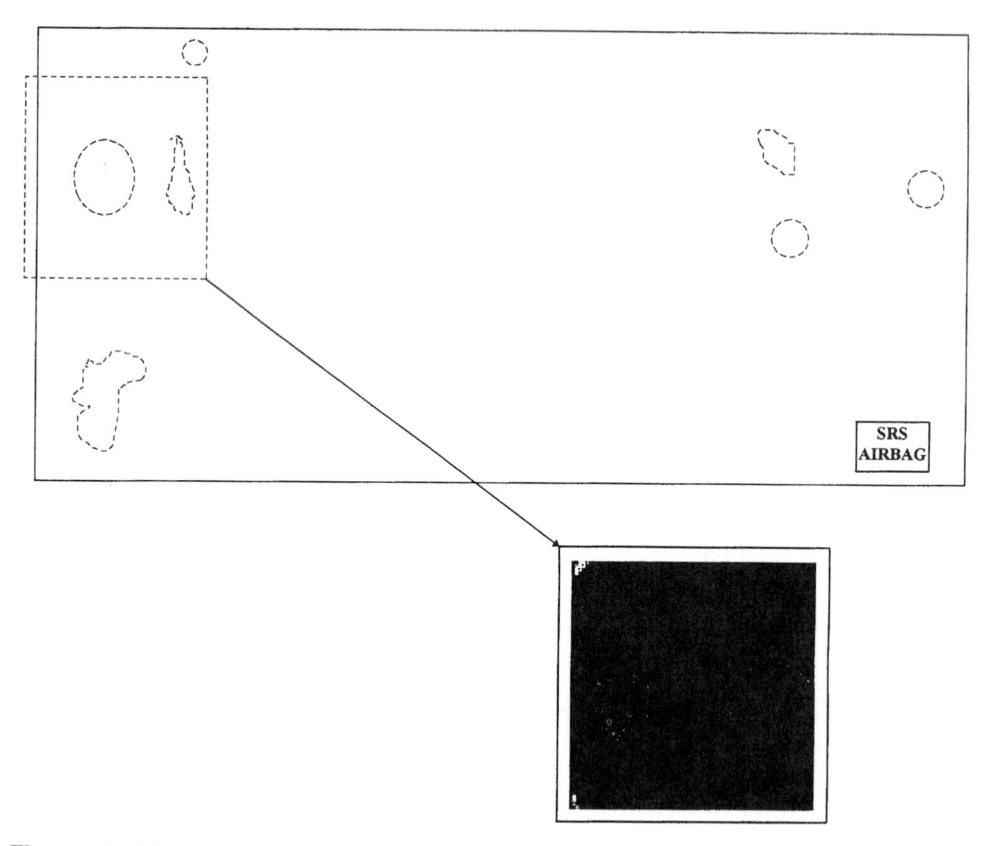

Figure 5 Thermograph of a section of a dashboard airbag revealing indentation damage.

7.3 Summary

Thermography is well suited to detect defects that run parallel to the surface inspected, and automatic infrared thermography examinations using image reconstruction and segmentation can be an economic solution in the case of repeated inspections of large components. Despite this, manual thermography lacks reproducibility, and quantitative signal interpretation can be difficult [84]. The high price of infrared thermographic equipment and their relative limitations in detecting flaws in composite materials are factors that compromise widespread adoption by industry. The main attraction of infrared thermogrpahy is that an easy-to-interpret color coded image is produced as a signal output. However, image processing is often required in order to improve the quality of the thermograph and to help in its interpretation. Orlove states that in some cases, thermography is proving to be faster and more cost effective than traditional nondestructive testing methods and presents new developments in infrared thermographic equipment [67]. Although disbond detection is good in graphite–epoxy laminates, cured adhesive is detectable only with difficulty, and defect detection in glass–epoxy structures is poor due to the low emissivity and thermal conductivity of this material. Hot spot details can be smeared due to too high a surface emissivity, and heat flow parallel to high conductivity fibers results in great perturbations of the thermal dissipation patterns. Spraying the surface inspected with a flat enamel results in a uniform surface emissivity and avoids false

Table 1 Advantages and Limitations of Infared Thermography

Advantages	Limitations
Safe technique	High price equipment
Portable equipment	Lack of reproducibility
Non-contract	Difficult quantitive signal interpretation
Non-invasive	Image processing required for image enhancement
Real-time imaging	Efficiency is function of material type, defect size and location
Speed of inspection	High emissivity materials difficult to inspect
Digital format archiving	Edge effect causes false indication
Outdoor IR detected free form solar radiation interference	No defect depth information

indications due to edge effects and variations in surface emissivity [50]. Defect detection is a function of defect type, size, depth, and initial temperature rise. Thermographic performance is also affected by the thermal properties of the composite inspected, and by cell size in honeycomb structures. The resolution of thermographic technique remains inferior to that of ultrasound, and it provides only an estimate of defect depth. Thermography is unlikely to reveal information not detectable by ultrasound, but it remains a noncontact method with multiple advantages, but also limitations which are summarized in Table 1.

8 Liquid Penetrant Testing

Liquid penetrant testing, principally used to detect surface breaking damage, porosity, and edge delamination, is rarely employed for composite testing because of the cleaning process involved following an inspection, and also because it may cause damage to certain composites by infiltration of the liquid dye. If complete removal of the dye is not possible, adhesion problems may result [85].

Penetrant testing is a time-consuming process composed of six phases (Fig. 6) and includes

Cleaning of the surface inspected
Spraying of a liquid penetrant onto the material surface
Removal of excess liquid
Application of a developer over the surface
Inspection of the component
Cleaning to remove the dye after inspection

A penetrant dye is sprayed onto the surface of the composite inspected. The liquid infiltrates cavities and is made visible by applying a developer to the surface. Defect size is indicated by the color intensity and the rapidity of infiltration of the liquid penetrant. Two types of penetrant, visible and fluorescent, can be used. Fluorescent dyes are visible under UV light, while the visual dyes are visible under normal light. The choice of the penetrant is function of the material tested and the sensitivity required. Automatic dye penetrant inspection systems exist that transfer and apply the dye, apply the revealing agent (or developer), and finally clean the part after testing. However, only specific shapes can be inspected automatically, and manual inspection is still required for most structures.

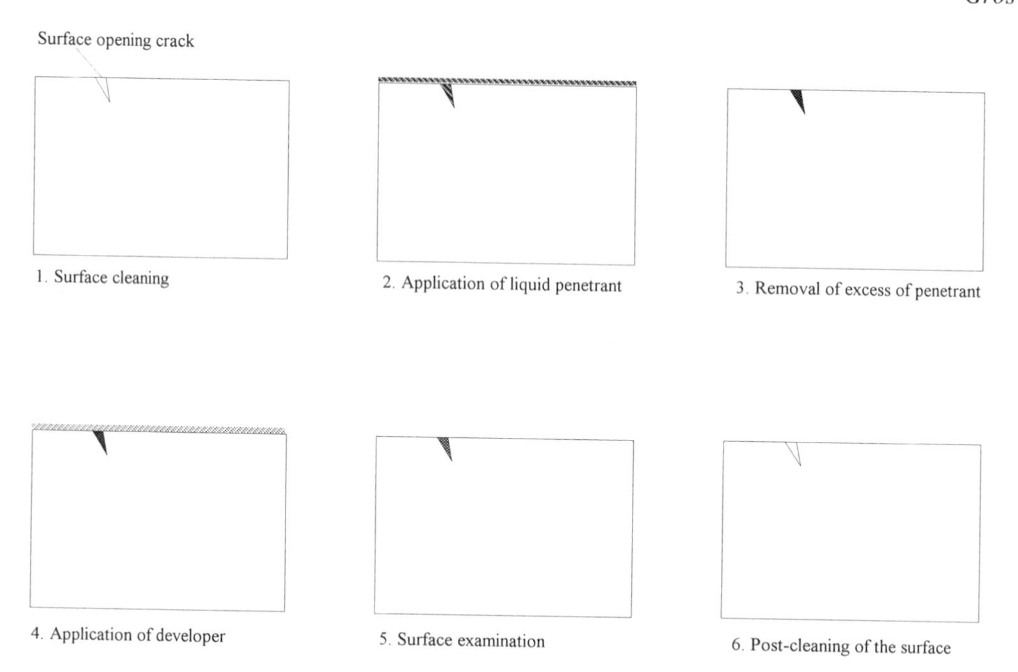

Figure 6 The six phases of liquid penetrant testing.

9 Microwave Inspection

The basic principle of microwave testing is based on the transmission of a microwave into a dielectric material; the recorded amplitude/phase of the transmitted or reflected wave is analyzed and processed to create a 2D or 3D image of the specimen inspected. Microwaves penetrate deep inside dielectric materials, are sensitive to the presence of dissimilar layers in composites, and can detect and locate irregularities such as voids, delamination, porosity, and moisture content. At frequencies ranging between 10 and 20 GHz, microwaves are useful for obtaining accurate thickness variation measurements in the range of a few micrometers [86]. The technique is sensitive to geometrical and dimensional variations of a material. With no attenuation due to internal scattering, this noncontact technique requires access to one side only and is suitable for the inspection of thick composites. The technique is used for in-process control during composite manufacture (e.g., state of cure) and for material characterization and classification. Moreoever, Ganchev *et al.* showed that the polarizability of microwave signals enables the study of fiber bundle orientation or misalignment during manufacturing, and provides information about cut or broken fiber bundles [86].

Microwave detection can be performed from the penetrating side (pulse-echo mode) or from the opposite side (through-transmission mode). The through-transmission technique can determine the state of resin cure in boron fiber laminates. However, it is not suitable for laminates with 0°/60° filament orientation [87]. The propagation of microwaves (wavelength: 10–10^{-3} m) is governed by Maxwell's equations of the magnetic field [88]. Maxwell's equations are

$$
\begin{aligned}
\nabla E(r,t) &= -\frac{\partial B(r,t)}{\partial t} \\
\nabla H(r,t) &= J(r,t) + \frac{\partial D(r,t)}{\partial t} \\
\nabla D(r,t) &= \rho(r) \\
\nabla B(r,t) &= 0
\end{aligned} \tag{7}
$$

where H is the magnetic field strength, B is the magnetic flux density, E the electric field intensity, J the current density, D the displacement current, and ρ the electric charge density.

Microwaves sources include electron beams (e.g., magnetron, klystron), semiconductors (e.g, Gunn diode, transistor) and masers (microwave amplification by stimulated emission of radiation) [88]. The depth of penetration d of microwaves into a dielectric is a function of the loss factor δ in the dielectric material and the frequency f of operation (or vacuum wavelength λ_0) [89] and is given by

$$
d = \frac{\lambda_0}{\pi\sqrt{\varepsilon}\tan\delta} \tag{8}
$$

where ε is the real part of the dielectric function at the microwave frequency.

Although laboratory test equipment can be expensive, it is claimed that simply designed and portable hardware for real-time, on-line examinations can be developed at relatively low cost [90]. Moreover, there are no environmentally hazardous or undesirable by-products associated with microwave testing [90]. The disadvantages of this technique are that microwaves cannot penetrate inside conductors or graphite composites, and results from 2D raster scans can sometimes be ambiguous to interpret.

An increase in resolution can be obtained using very high frequency (100 GHz) radiation, but this tends to increase inspection cost greatly. Open-ended waveguides can be used to transmit and receive microwaves, emitted in frequencies ranging from 225 MHz to 100 GHz, and these present the advantages of combining improved near-file resolution and depth range potential (single-sided technique) [89]. During emission and reflection of microwaves, a standing wave field is created that depends upon the reflectivity of the material tested and the stand-off distance, varying from 0 to 3 mm. Variations in signal amplitude are a function of the interaction of the emitted radiation with the material. This technique is sensitive to small variations in the multilayer system of a laminate. The microwave absorption can be mapped using a microwave detector or, as described by Diener, by using a lock-in thermographic camera and the through transmission technique [89]. However, the open-ended technique provides better resolution and higher contrast than lock-in thermography in the detection of contamination in GFRP [89]. Material anisotropy can be investigated using a rotating waveguide to detect fiber-induced anisotropic fields in glass fiber reinforced injection molded polymers [89].

Applications of microwave testing have included thickness measurements of plastic composites, ceramics, and dielectric materials; delamination, void detection, fiber bundle orientation, and fiber breakage as well as moisture content, impact damage detection, and evaluation of curing. Correlation of measured microwave data to physical and mechanical attributes of a structure are also possible [91]. Thickness variation detection, an accurate measurement of coating thickness, and measurement of disbond thickness as little as a few microns using frequencies in the order of 10 GHz have been achieved [91]. Because microwaves are absorbed and scattered by water molecules, detection of moisture content in composites is possible. Microwave testing has also been used in 38 mm thick GFRP for the detection of flaws embedded 19 mm within the sample [91]. The technique can be easily adapted for on-line measurement, allowing process control during the manufacturing of composites.

10 Millimeter Wave Inspection

Microwaves in the higher frequency ranges, 30–300 GHz, are called millimeter waves. The use of wavelengths shorter than those of conventional microwaves gives better defect contrast as well as higher spatial and depth resolution. The millimeter wave technique is based on amplitude measurements of back-scattered and forward-scattered radiation. This technique uses an open-ended waveguide to transmit a signal that penetrates dielectric materials, and a voltage related to the magnitude of the reflection coefficient at the waveguide aperture is recorded during scanning of the surface of the composite. The reflected signal is related to the dielectric properties of the material under test, and its analysis provides information on its microstructure. Radford *et al.* described the examination of the microstructural aspects of damage initiation and growth in glass fiber–epoxy composites subjected to low-energy impacts [92]. The technique has also been used to find image defects in Kevlar–epoxy composites [93]. Millimeter-wave examination is used for monitoring damage accumulation in composites for lifetime prediction, and to detect subsurface damage such as voids and disbonds.

11 Magnetic Resonance Imaging

Magnetic resonance imaging (MRI), also known as nuclear magnetic resonance (NMR), uses the nuclear magnetic resonance of protons to produce proton density maps or images. Its principle is based on the magnetization of nuclear magnetic moments contained in certain atomic nuclei. When placed in a strong magnetic field, the nuclear moments can only take up certain discrete orientations, each orientation corresponding to a different energy state, and the nuclear moment deflects with a specific period. Transitions between these energy levels can be induced by the application of characteristic radio frequencies.

MRI is useful for the detection of surface damage, but internal voids are more difficult to detect and require direct imaging of the polymer matrix using multiple pulse dipolar decoupling and novel coil geometries [94]. This method is applied to the detection of solvent distribution and diffusion parameters in swollen polymers, the detection of inhomogeneities in mobile polymers, the observation of fillers in ceramic materials, and the localization of the extent of polymerization in acrylic systems [94]. It is also used to monitor the curing process, measure moisture content, determine fiber volume fraction, and detect jet fuel ingress in PEEK composites [36].

12 Acoustic Emission

Acoustic emission is a passive technique that records sound generated by a physical event such as crack growth, fiber breakage, or delamination. Each type of flaw produces a specific acoustic emission signal, which is referred to as a signature and is used to determine the type of flaw that has been detected. Acoustic emission events are usually displayed on the cathode ray tube (CRT) of an oscilloscope and can be related to the position of a transducer, placed at a specific location on the structure tested. Acoustic emission signals are characterized by their amplitude, duration, rising time, stress delay, energy, and counts (i.e., number of signals crossing a threshold value). The relationship between acoustic emission parameters and the structural integrity of a component must be known for correct signal interpretation. Among the factors to be considered the following have been identified by Scott and Scala [50]:

The condition of the fiber–resin interface
The extent of the damage
The environmental conditions during testing

Acoustic emission occurs in events (or bursts) of continuous counts, and the cumulative number of counts indicates the extent of damage and the percentage of ultimate load carried by a structure. The time for the signal to reach the different transducers can be used to estimate defect position within the material [95]. Low-frequency transducers monitor emission from a large area, while high-frequency transducers monitor emission from a local area. Thus the source of emission can be precisely localized using multiple high-frequency sensors, while a more general indication of the integrity of the whole structure can be obtained with low-frequency sensors. A limitation of the method is that only growing flaws are detected. Its advantages are its suitability for detection of crack initiation and for continuous in-situ monitoring of damage. The various microfracture modes can be identified from the analysis of amplitude distribution and frequency. Analysis of the Kaiser effect, an irreversible acoustic emission wave, is useful in the detection of subcritical flaw growth, but analysis of the Felicity effect is preferred for composites inspection [34].

Acoustic emission has been successfully applied to the detection and location of impact-damaged areas in CFRP composites under tension cyclic loading and residual strength in composites prediction using long-duration events [96]. It appears that both acoustic emission RMS voltage and event duration data are useful parameters for the quantitative evaluation of the degree of damage in CFRP and that the value of the Felicity ratio is very sensitive to the stacking sequence [96]. The Felicity ratio is defined as the ratio between the load at which acoustic emission occurs over the previous maximum stress. Bailey *et al.* described the use of acoustic emission to locate and monitor impact damage progression in 16-ply graphite fiber reinforced epoxy composites and showed that acoustic emission can detect damage at very low stress levels [97]. Yen and Tittman monitored acoustic emission generated by carbon–carbon composites during pyrolysis to investigate the extent of porosity within the material, the amount of acoustic energy released being proportional to the size of the defect [35]. Three different measurement thresholds were used to distinguish among sources of different amplitudes (e.g., delamination cracks from fiber–matrix cracks). They found that the temperature onset for fiber–matrix interfacial cracking was about 400°C. Acoustic emission testing can give an indication of the quality of adhesive bond by taking into account the number of oscillations above a specific threshold [14].

Acoustic emission is best suited to detect disbonds between thin composite skins and a foam or honeycomb structure. Williams and Lee observed that the acoustic emission stress delay in GFR polyesters can be correlated with the ultimate and tensile strength of a structure [98]. This method can also be used to monitor the resin-cure process during manufacture of composites [99]. Shrinkage and hardening of the resin cause acoustic emission due to microcracking in the laminate as a result of thermoelastic stress. Melve and Moursund described the use of acoustic emission for the testing of GFRP pressured pipes transporting water based media for cooling or fire fighting systems on offshore platforms [100]. Lack of adhesion, interface cracks between adhesive and adherents, undercured adhesive, porosity, wet surface, and misalignment of adherents have all been successfully detected. Although commercial acoustic emission field equipment can be used without on-site modification, noise generated by the operating environment (e.g., electromagnetic noise, vibration noise) may be detected by the sensors as they pick up sounds in the range 50 kHz–1 MHz (100–300 kHz for composites) [100].

A state-of-the-art acoustic technique to detect faults in rubber tires has been developed by Dunlop SP and uses a microphone that picks up the road noise of a tire. Preprocessing of the acoustic signal helps in producing a spectrum, which is used as the input data for a neural network. The neural network monitors the sound and raises an alarm as soon as a change in road noise that indicates a fault is detected. Research on the use of acoustic impedance spectroscopy to measure the progress of moisture uptake, electromigration, and corrosion within plastics as well as to monitor the cure process of polymers will provide useful real-time qualitative and quantitative information. Fiber reinforced polymers are materials well suited for acoustic emission monitoring of damage occurring due to tensile or thermal loading. The technique was successful in determining the initiation and crack growth in SiC fiber reinforced Pyrex glass and polymers.

13 Electrical and Electromagnetic Testing

Although ultrasound is the most widely used technique in industry, other techniques, described in this section, present results often comparable to, if not better in certain applications, than ultrasonic testing. Carbon and graphite materials have good thermal conductivity and fair resistance to thermal shock and are electrically conductive. Impregnation of carbon and graphite with metals (e.g., copper, silver) and resins reduce the porosity and increase the materials' strength. These properties favor the use of electromagnetic inspection techniques. This section reviews some electrical and electromagnetic nondestructive testing techniques used for composite inspection.

For metal or resin impregnated carbon/graphite materials, the extent of polymerization, cure cycles, and degradation can be assessed by measuring electrical resistance of the resin [101]. For example, the internal viscosity of a polymer can be determined by monitoring the electrical resistivity as a function of temperature. Electrical and electromagnetic properties of composites as well as NDT techniques are described next.

13.1 Electromagnetic Emission

Electromagnetic emission testing is based on the principle that materials under loads generate electromagnetic field pulses in the radio frequency at the moment of defect and crack propagation. Gordeev *et al.* observed that dielectric transformation processes occur due to mechanical deformation of a structure [102]. These processes produce a vibrational motion of electrostatic charge in the material surface, which in turn creates an electromagnetic field within the material. This electromagnetic field, stable over time, varies when in close proximity to flawed regions, thus allowing defects such as cracks, disbonds, and foreign inclusions to be mapped. The characteristics of the electromagnetic emissions can also be used to predict the ultimate compression and bending strength of the test objects. This electromagnetic emission phenomenon is similar to acoustic emission and presents several advantages for composite examination: the electromagnetic field enables high penetration into the material, the Kaiser effect is absent, and early stages of crack initation can be observed. A system called ÉKhO, based on registration of the electromagnetic emission when rising from the structure under test, was developed by Tomsk

Polytechnical University in Russia for on-site composite inspections. This equipment has the advantage of being portable and does not require external operating power. Despite the existence of this system, methods using electromagnetic emission are not very widespread, mainly due to the lack of appropriate measuring apparatus. Another limiting factor is that the material tested must be under stress, which may be constraining, especially with large composite structures.

13.2 Electrical Impedance Monitoring

Measure of the electrical properties of a CFRP material employed as an electrode can provide information about the structural integrity of a composite, allowing detection of impact damage, microcracks, moisture content, and delamination. Delamination shows a capacitive behavior easily detectable with impedance measurement. The material tested is placed outside a cell filled with an electrolyte fluid (K_2SO_4), over a round opening. The counter electrode, made of CFRP, is fixed inside the cell. The system, composed of two electrodes (The CFRP specimen as the test electrode and a one-way electrode) is polarized with a small DC potential. An AC potential overlies this DC potential to measure the electrical impedance and phase values at different frequencies (10^{-1}–10^{5} Hz) using an impedance spectrum analyzer. The result is an impedance spectrum with a phase spectrum. The impedance spectrum of the speciment tested is compared with one from a reference standard. Changes in impedance are related to damage in the material.

13.3 Magnetic Particle Tagging

Magnetic particle tagging is a new concept for process control in composite fabrication and involves the addition of small amounts of magnetic particles (5 nm in size) to the composite matrix in order to facilitate composite assessment using conventional electromagnetic devices. In his excellent paper, Clark claimed that this technique can reveal important fabrication parameter such as reinforcement ratio, fiber layout, material composition, state of cure, and void content [103]. Moreover, activation of the magnetic particles by an external electromagnetic source allows selective curing, self-repair, and even the measurement of reinforcement to matrix bond strength. Process control magnetic particle tagging finds applications in adhesive joint monitoring by measuring the layer of the applied adhesive. This is done by mixing ferromagnetic particles with the adhesive and comparing the signal output from the measuring apparatus with a reference standard. The detection of any signal change is representative of a varition in joint width, or the presence of voids. Other applications include control and qualification of polymeric adhesive joining systems, monitoring and control of liquid-to-solid ratios in material processing, and characterization of powder processing techniques such as blending, compaction, and extrusion. The advantage of magnetic particle tagging is that slight electromagnetic properties are conferred to otherwise nonconducting materials, without affecting their insulating characteristics. Ferromagnetic particles are available at low cost, do not affect the performance of composite materials, or matrix properties, and are easily detectable with conventional NDT apparatus. In addition, several benefits can be gained, which include monitoring of the fabrication process and in-service inspection by eddy current.

13.4 Eddy Current Inspection

High-frequency magnetic fields cause eddy currents to flow in conductive materials, or composites containing conductive fibers or a certain amount of graphite. Mainly limited to the inspection of metallic components, eddy currents has been successfully employed to locate and characterize defects in composite materials, particularly CFRP composites [104–108]. However, eddy current NDE of composite materials remains complicated due to the anisotropic nature of their electrical conductivities.

Principle

Limited at first to the inspection of metallic components, it was later demonstrated that high-frequency magnetic fields causing eddy current production could be applied to the testing of composites containing conductive fibers, or a certain amount of graphite [108,109]. Eddy current testing has been successfully applied in the detection and characterization of defects in carbon fiber reinforced plastics (CFRP) panels, helicopter rotor blades, truck tires, and more [110–112].

The principle of eddy current testing is based on the interaction between an electrical current induced in a material and the structure of the material tested. A primary magnetic field is generated in the vicinity of a coil excited by an alternating current (Fig. 7). The magnetic flux density Φ associated with the primary magnetic field is proportional to the magnitude I of the electric current within the coil and to the number N of turns in the coil:

$$\Phi = NI \tag{9}$$

When the coil is brought into the proximity of a conductive material, eddy currents are induced within this material, which according to Lenz's law generate a secondary magnetic field ϕ_s opposed in direction and magnitude to the primary magnetic field ϕ_p. Both magnetic fields are linked by the equation

$$\phi_s = \mu \times \phi_p \tag{10}$$

A phenomenon known as the skin effect is used to describe the depth of penetration of the eddy currents, which is proportional to the current frequency (f in Hz), the electrical conductivity (σ in mhos/m) and the magnetic permeability (μ in H/m) of the material tested. The standard depth of penetration (d in m) can be calculated using the formula

$$d = \frac{1}{\sqrt{\pi f \mu \sigma}} \tag{11}$$

The distribution of eddy current in a material varies exponentially, and the current density (J_x in A/m^2) at a depth x can be calculated by

$$J_x = J_0 \exp\left(\frac{-x}{\sqrt{\pi f \mu \sigma}}\right) \tag{12}$$

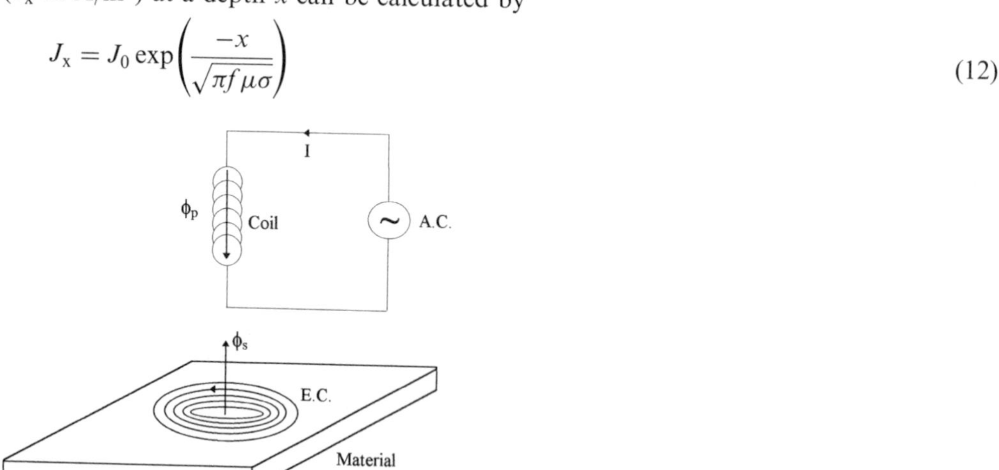

Figure 7 Schematic of an eddy current circuit.

where J_0 is the current density at the surface of the material. And the flow of induced currents is given by

$$\nabla^2 J = \mu\sigma \tag{13}$$

Additional information about eddy current theory can be found in the literature [113].

The eddy current path will be altered by the presence of flaws in the material tested, and signal variation can be recorded and analyzed on the cathode ray tube of an oscilloscope. During inspection, the operator scans the sensor over the surface of the component inspected and analyzes any changes in probe impedance. Eddy current testing is subject to edge effects and probe liftoff. Indeed, near the edges of a component, the eddy current signal can be distorted, thus providing misleading information. Similarly, signal variation can occur if the distance between the scanning probe and the inspected surface is not kept constant, also leading to interpretation error. Because expertise may be required to interpret such analogue eddy current signals, color-coded visualization techniques have been developed to facilitate signal interpretation [114].

The electrical conductivity of carbon fiber composites is affected by the fiber type, density, and waving pattern. Several authors have demonstrated electrical conductivity in carbon fiber composites and measured resistance for a range of composite types. For example, carbon fiber–epoxy has a resistivity ranging from $5000\,\mu\Omega \cdot \text{cm}$ to $20{,}000\,\mu\Omega \cdot \text{cm}$ [104,105,110]. It has also been shown that carbon fibers in matrix composites produce a transverse electrical conduction path [112]. Eddy currents flow along fibers and pass from one fiber to another at the points of fiber contact, as shown in Fig. 8. The longitudinal and transversal resistivity of carbon fiber reinforced epoxy resin for a volume fraction of 50% is $0.009\,\Omega \cdot \text{cm}$ and $0.5\,\Omega \cdot \text{cm}$ respectively [112]. Further detailed infomration on the electrical resistance of resins and reinforcement fibers can be found in the article by Summerscales [101]. Knowing the resistivity ρ in $\mu\Omega \cdot \text{cm}$ one can calculate the electrical conductivity in %IACS using the equation [115]

$$\sigma = \frac{172.41}{\rho} \tag{14}$$

Due to their low electrical conductivity, high frequencies are required for eddy current examination of composites. For instance, a 1.0 mm thick CFRP composite requires a skin depth frequency varying from 12.6 to 50.6 MHz [104]. Prakash estimates that eddy current

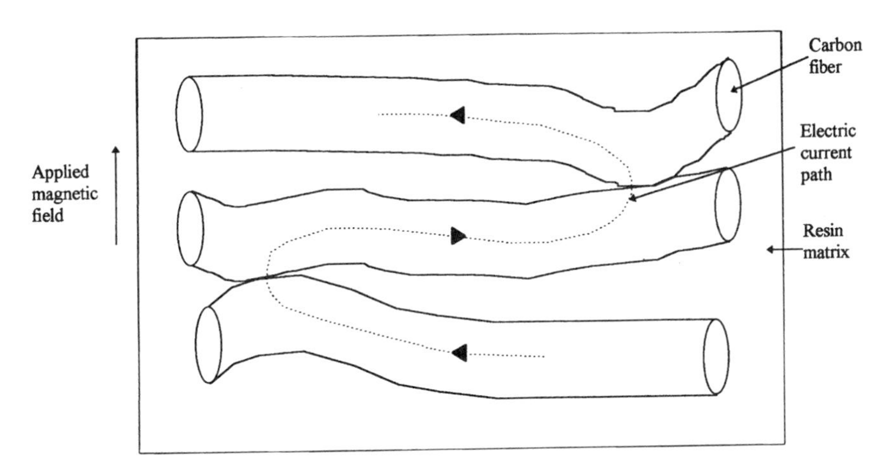

Figure 8 Schematic of an electrical current induced in a CFRP material. (Modified from Ref. 52.)

testing of composites is limited to material having at least a 40% carbon fiber volume fraction [116]. Eddy current sensitivity increases when probe size diminishes and when the skin depth is less than the crack depth. To achieve maximum sensitivity, Valleau suggests that for the detection of shallow back-surface breaking faults in a panel of thickness T, the skin depth of the frequency used for the inspection must be of the order T [104]. As the resistive component of probe impedance is sensitive to the presence of cracks, it is recommended to use high frequencies in order to achieve good spatial resolution. Ferrite probe cores are best suited to direct the magnetic field at high frequencies and for maximum flux penetration. They also offer the advantages of being fairly inexpensive and widely available off the shelf; they can be custom made and they have a relatively high initial permeability with low change.

C-scan eddy current instruments displaying color-coded images require a full scan of the surface to produce a complete visual assessment of the specimen, to facilitate signal interpretation, and to reduce false interpretation. Menu-driven software in a point-and-click environment facilitates manipulation, data storage, and redaction of inspection reports. Inspection can be either manual or automated using a computer-controlled frame scanning apparatus.

Application of Eddy Current Testing of Composites

Comparison of the impedance signal obtained from a reference material with that of the material under test can provide information on the type of defect detected, establishing fault signatures that enable eddy currents to be used for a wide variety of inspection tasks. Indeed, fiber damage with or without matrix cracking can be detected by eddy currents [104] and variations in fiber volume fraction revealed [112]. Determination of fiber orientation can be achieved using an appropriate coil transducer with a horizontal axis and polar plots of the impedance change, indicating the principal fiber direction [115]. Good correlation between volume fraction and eddy current parameters was identified by Dingwall and Mead at a frequency of 10–12 MHz [117]. However, their system did not work for cross-ply laminates, and signal analysis appeared to be difficult. Using phase-sensitive equipment operating at 500 kHz, Hagemaier and Fassbender established that excellent correlation between eddy current conductivity readings and resin content was evident in graphite fiber composites [16].

Hashimoto *et al.* used eddy currents with an optimal frequency ranging between 0.5 and 4.0 MHz, for quality control of the graphite core internal structures of high-temperature, gas-cooled reactors [109]. The technique was remotely applied in a radioactive environment and detected cracks of 0.3 mm wide, 1.0 mm long, and 1.0 mm deep. Lange and Mook demonstrated not only that the inspection of composites using eddy currents is based on the conductivity of the fibers but also that the properties of the matrix could be studied by analyzing the capacitive effects when using high frequencies (10 MHz) [111]. This allowed a measure of the anisotropic electrical properties of composites as well as an estimation of their degradation stage. Other research has been performed to evaluate the eddy current method in monitoring the electrical conductivity of carbon–carbon composite materials during high-temperature pyrolytic processes [108]. It was demonstrated that the frequency of the peak eddy current dissipation was proportional to the product of the material thickness and its electrical conductivity.

A lack of reproducibility, of ultrasounds for the detection of disbonds on helicopter rotor blades made of glass fiber reinforced titanium or carbon fibers, made eddy current inspection an alternative technique with high reproducibility [119]. A longer life-cycle is achieved with composite rotor blades, and certain blades have demonstrated a service life in excess of 5000 hours [87]. Helicopters operate in temperatures from −40°C to +70°C, which enhances the need for thorough testing. Figure 9 shows a plan view of the eddy current inspection result of a helicopter rotor blade. Disbonds are color coded for faster defect location (orange on the figure) and a superimposed contour map gives information on disbond height. The detection of low-energy impact damage (0.5 J) in CFRP materials has been achieved, and the sensitivity of the technique was far better than that of any other widely used NDT technique, such as ultrasound, radiography, or infrared thermography [120]. Impacts on a CFRP panel can be identified through peaks of eddy current signal amplitude on Fig. 10. Electromagnetic testing can also be used to detect the presence of water trapped in a structure, as its presence affects the resistivity of the composite. The effect of moisture content on the electrical resistivity of carbon reinforced epoxy resins was studied by Belani and Broutman [121]. Their results showed that the electrical resistance transverse to the fiber direction increases proportionally to the volume of water ingress. Further research showed that certain insulating materials, such as rubber, may become electrically conductive when a sufficient concentration of carbon black is used in their manufacture, thus allowing inspection of tires [122].

Valleau developed a high frequency (> 15 MHz) eddy current C-core probe suitable for nondestructive testing of graphite composites [104]. This multifrequency eddy current NDT system allows real-time scanning, data collection, and C-scan display. Probe lift off is computer controlled, and an expert system determines the type of faults detected based on the analysis of the signal output. Fault type characterization was achieved with an accuracy of over 95%, and misidentification occurred principally due to improper calibration. C-core probes are anisotropic and quite sensitive to faults along the axis, but they are blind to faults off the axis and tend to deform circular faults into elongated shapes (due to the eddy current patterns). However, Valleau demonstrated that a more accurate image of the defect detected can be obtained by mathematically combining two measurements from C-core probes [104]. Vernon and Gammell reported that maximum defect resolution is achieved when the radius of a coil is three times the skin depth, and the skin depth about eight tenths the material thickness, thus restricting the use of conventional probes [123]. Indeed, they designed high-frequency probes (1.5–27 MHz) to encompass the poor electrical conductivity of composites (e.g., 0.1% IACS for graphite epoxy) and in order to assess the integrity of conducting fibers in a nonconducting matrix.

Eddy currents have also been used to measure the volume fraction of titanium alloy reinforced with silicon carbide fibers [124]. Beissner studied the effect of abnormal microstructure on eddy current probe response, a change in response occurring due to the movement of fibers within rows leading to changes in volume fraction.

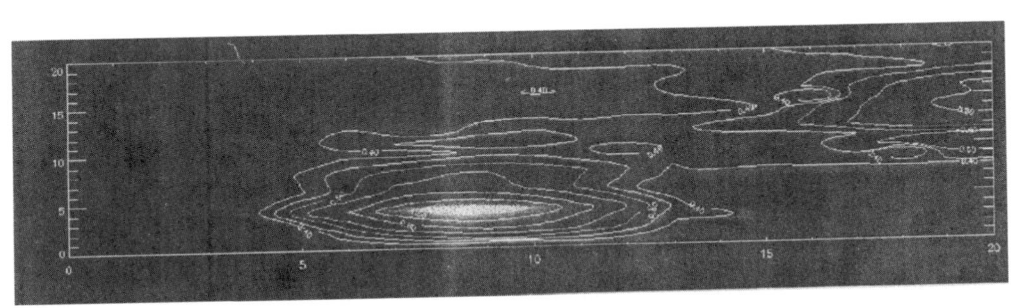

Figure 9 Disbonds on a helicopter rotor blade detected by eddy currents (view from above).

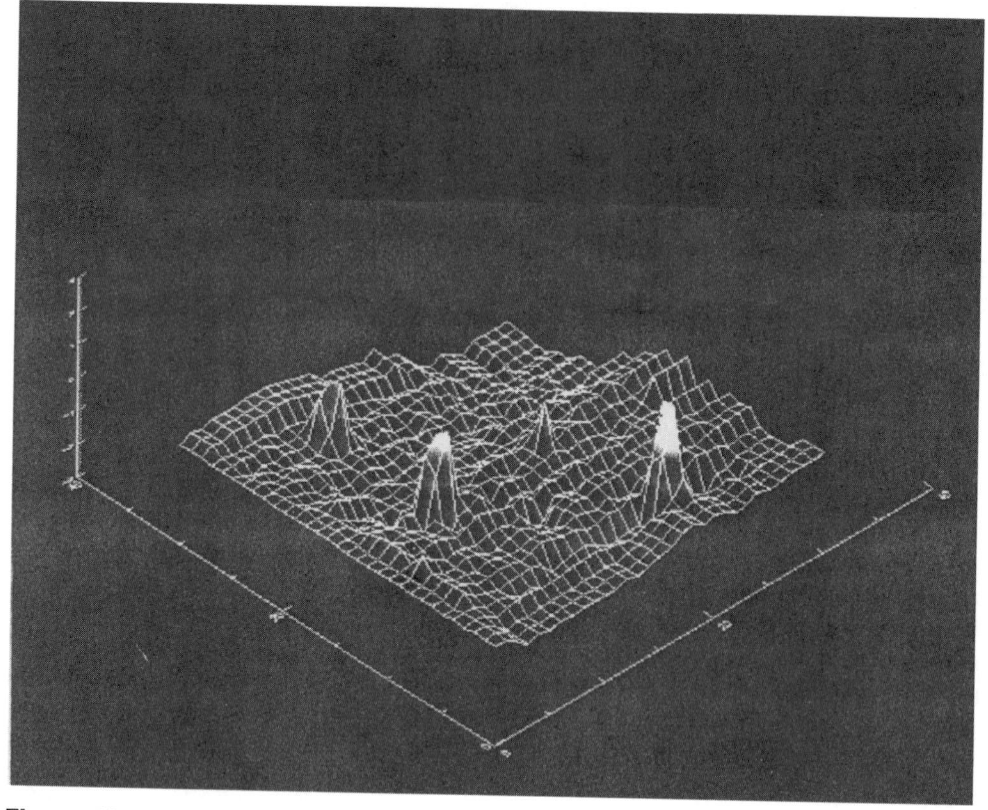

Figure 10 Visualization of low-energy impacts detected by eddy currents on a CFRP panel.

Eddy current testing is best suited to the detection of surface and subsurface defects, and further research is required fully to exploit its potential in composite evaluation.

13.5 Magnetooptic Eddy Current Imaging

An eddy current technique known as magnetooptic imaging (MOI) has been developed to test aging aircraft [77]. Real-time eddy current images of both surface braking fatigue cracks and hidden multilayer cracking or corrosion can be produced. The system uses frequencies ranging from 6 to 100 kHz, and inspection is five times faster than conventional eddy currents. Magnetooptic imaging can be used to inspect composites containing fine magnetite (Fe_3O_4) particles or with embedded wire screens such as copper or aluminum. A magnetooptic sensor (MOS) made of thin bismuth-doped iron garnet film exhibits a uniaxial magnetic anisotropy and can retain its established magnetization after removal of an applied magnetic field. The principle of this technique consists of sending normally incident, linearly polarized light transmitted through a MOS with fixed magnetization, and then viewing the reflected back light of this same region through an analyzer. This reveals the focal state of magnetization in a region of the sensor. This state of magnetization can be seen as a high contrast dark or bright area. This property allows the creation of images of the normal component magnetic field associated with eddy currents [77].

13.6 Summary

Electromagnetic testing of composites presents several advantages over other conventional means of inspection: reproducibility, portability, and competitive price. They are also truly quantitative [125]. Moreover, such techniques can be applied for process control at manufacturing level and in-service composite evaluation.

It was demonstrated that eddy current inspection of composite materials is a possible low-cost alternative to other more expensive NDT techniques, such as laser generated ultrasound or shearography [110]. Operator dependence can be minimized by the use of automated or semiautomated eddy current systems that display color-coded images of the region inspected, thus facilitating signal interpretation and defect characterization. However the resolution of the eddy current technique is influenced by fiber distribution and is limited to the inspection of composites containing conductive fibers such as graphite or boron. Still, the electrical conductivity of carbon fiber composites is affected by the fiber type, density, and weaving pattern. For example, the conductivity of unidirectional CFRP perpendicular to the fiber axis increases with frequency and varies with the fiber volume fraction. Eddy current is useful for the inspection of CFRP materials and, combined with the magnetic particle tagging process, will enable both manufacturing process control and in-service structural assessment of nonconducting composites. One can see that this small alteration in composite fabrication process will, in the long term, greatly improve composite quality, increase product reliability, facilitate NDE, favor defect monitoring, and thus reduce cost and minimize waste. Nevertheless, other NDT techiques are required to assess fully the structural integrity of composite materials, and electromagnetic techniques have the potential to complement other means of inspection.

14 Ultrasonic Examination

Ultrasonic examination is currently the most commonly used NDT method for inspection of composites. It presents desirable features such as providing information about defects situated deeply inside a material, but equally this method has several limitations. Flaws modify ultrasonic parameters such as wave velocity, refraction, reflection, scattering, and intensity, thus affecting the efficiency of ultrasound in defect location. In order to fully understand the concept of ultrasonic testing, it is necessary to use some mathematics, which will be kept to a minimum. The principle, advantages, and limitations of ultrasonic NDT techniques for composite inspection are described next.

14.1 Theory of Ultrasonic Testing

A transducer containing a piezoelectric crystal converts electric energy into mechanical vibrations in a frequency range inaudible by the human ear. The ultrasonic waves are transmitted through the material inspected and are reflected onto the back surface of the specimen. The reflected waves are then received by the same or a different transducer and analyzed on the CRT of an oscilloscope. A smooth clean surface and a couplant, either a gel or water (or even water immersion in a tank), are usually required between the transducer and the material to facilitate penetration of the ultrasonic signal within the structure. Signal output can be a time-based graph known as A-scan or B-scan; an attenuation of the signal is related to the presence of internal defects (Fig. 11).

Attenuation is due to scattering of ultrasonic waves by the fibers and by absorption into isothermal resin. However, attenuation is reduced with the through-transmission method and by the use of low frequencies. Defects such as voids. resin rich or resin starved

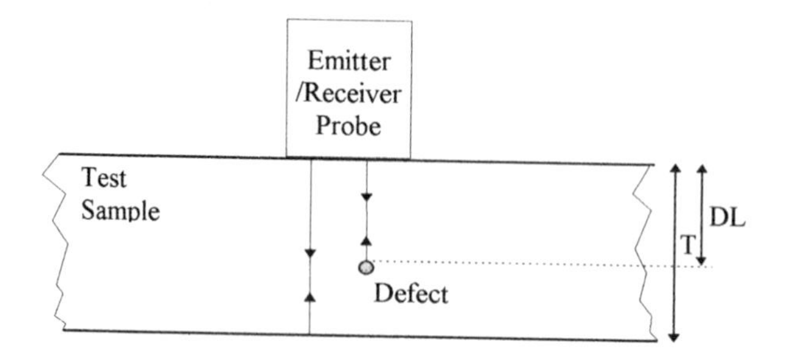

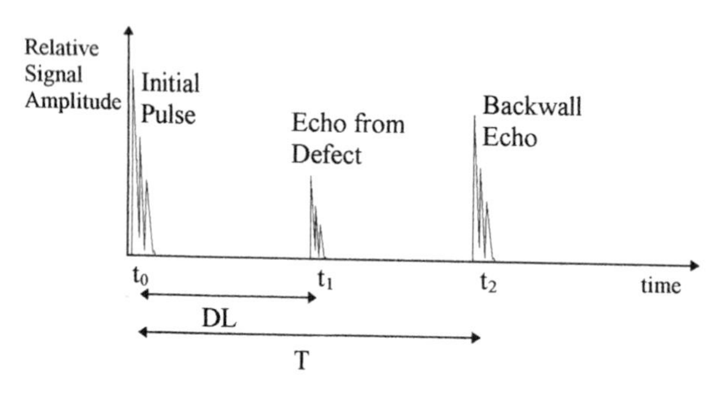

Figure 11 Schematic ultrasonic A-scan of a pulse-echo inspection.

areas, and delaminations are easily detected as they produce signal attenuation, and such techniques are suitable for the measurement of dispersed defects [34]. Experiments performed by Williams *et al.* showed that the attenuation of longitudinal waves propagating perpendicular to the plies was sensitive to the interlaminar quality of a component and could therefore be used to identify damage [126].

From an A-scan, material thickness T or defect location DL can be determined using the equations

$$T = v\,\frac{t_2}{2} \tag{15}$$

$$DL = T\,\frac{t_1}{t_2} \tag{16}$$

where v is the sound velocity in the material and t_i the time at the instant i (see Fig. 11). A B-scan gives a cross-sectional view along a scan line by displaying signal amplitude as a function of wave transmit time and position. In addition, certain ultrasonic systems produce C-scans, which use multiple scans to generate a plan view of a material. Recent equipment can present results as color-coded maps of the inspected surface.

The two major techniques in ultrasonic testing are the pulse-echo mode and the through-transmission mode. In pulse-echo mode, the same transducer is used to emit and receive ultrasonic waves and requires access to only one side of the structure inspected (Fig. 12). The pulse-echo technique is effective on the near side skin laminate only, and the sensitivity decreases as a function of depth [29]. With the through-transmission mode, access to both sides of a structure is required as an emitter probe is placed on one side and a receiver on the other (Fig. 13). This technique measures the signal amplitude of ultrasonic waves transmitted through the material tested and is more sensitive to small defects than the pulse-echo mode. The detection of flaws throughout the whole depth of a

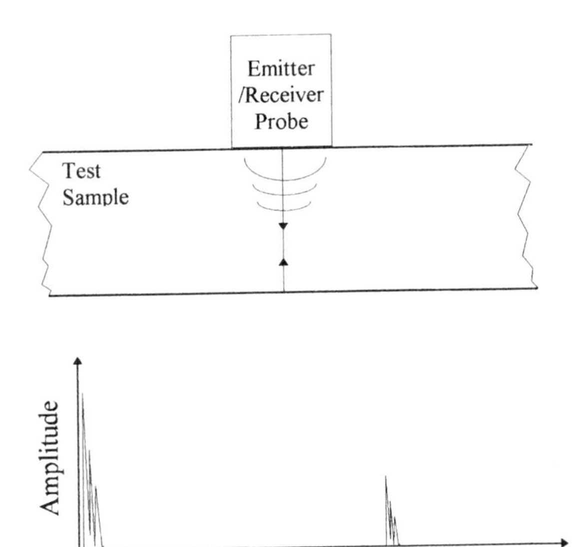

Figure 12 Schematic of the pulse-echo single-probe ultrasonic technique and associated A-scan.

laminate is possible. For example, with the through-transmission approach, the wave travels only once through the material, therefore allowing the inspection of thick composites. This provides a better signal-to-noise ratio as the ultrasonic wave is less attenuated.

The pitch-catch technique is an alternative method that consists of placing at an angle both transducer and receiver on the same side of the specimen inspected. Although dry coupled ultrasonic probes are being developed, the low frequencies used with such equipment require a certain amount of signal processing to facilitate interpretation. The main advantage is that no couplant is required so that scanning time is reduced.

Commonly used frequencies for ultrasound scanning range from 20 kHz to 25 MHz, and the ultrasonic wavelength λ is given by

$$\lambda = \frac{v}{f} \tag{17}$$

where v is the wave velocity in a given material and f the transmitting frequency. Resolution cannot be increased by the use of high-frequency focused transducers, as composite materials tend to filter high frequencies. Reference standards with embedded defects made of a material similar to that tested are required to calibrate the ultrasonic instrument prior to inspection.

Ultrasonic inspections are usually carried out with the sound beam propagating in a direction normal to the axis of the reinforcement fibers, resulting in a relatively simple directional propagation of the sound beam [127]. The nominally acceptable level of randomly distributed porosity present in graphite–epoxy material results in a large variation in ultrasound transmission, which can mask defect indications at frequencies above approximately 0.2 MHz. For this reason these composites are most effectively inspected at relatively low frequencies, where the lateral variation is minimal. The ultrasonic beam pattern can be divided into two regions outside the transducer; these are called the near field (or Fresnel zone) and the far field (or Fraunhofer zone). The near field length, noted N, can be calculated using the formula

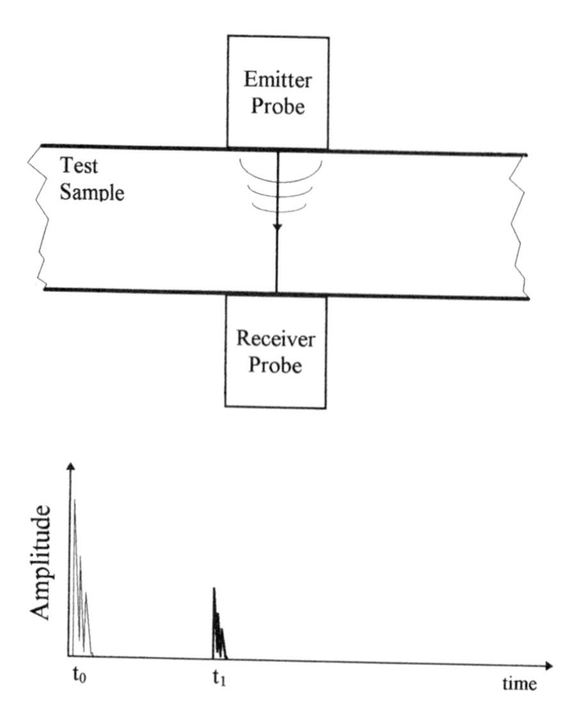

Figure 13 Schematic of the through-transmission ultrasonic mode and associated A-scan.

$$N = \frac{FD^2}{4v} \tag{18}$$

where D is the probe diameter. In the near field, the amplitude of the beam varies similar to an interference pattern, while in the far field it varies according to the inverse square law. Although Wong *et al.* demonstrated that sizing of delaminations in thin composites is possible as long as the reflected echoes from the delaminations can be separated from each other and from the transmission pulse, sizing in the near field zone is not recommended [128].

Compressional wave probes send ultrasonic waves that propagate vertically through the specimen. With angle probes, both reflected and refracted compressional and shear waves are generated. The incident α or refracted angle β can be calculated using Snell's law:

$$\frac{\sin\alpha}{\sin\beta} = \frac{v_1}{v_2} \tag{19}$$

where v_i is the wave velocity in the medium i (Fig. 14). Material thickness T can be determined by calculating the half-surface distance (HSD) using (Fig. 15a)

$$HSD = T \times \tan\alpha \tag{20}$$

and defect location l using (Fig. 15b)

$$DL = l \times \cos\alpha \tag{21}$$

Thickness measurements of for example 0.5 mm composite materials within accuracies of ±0.03 mm or less can be troublesome using conventional ultrasonic techniques. The use of a zero interface probe can provide accurate results, as such transducers are specifically

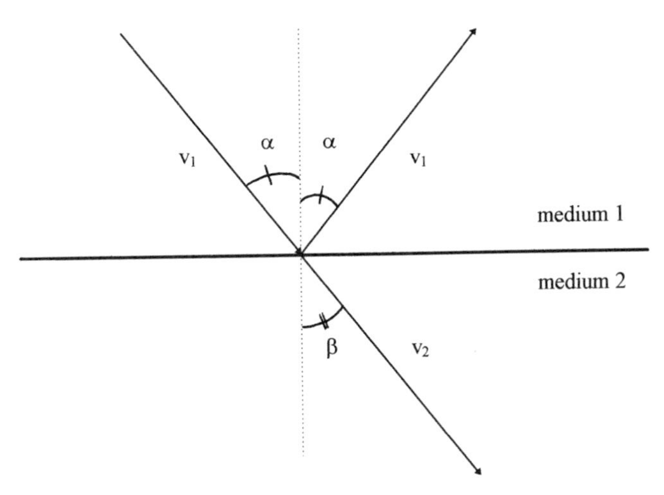

Figure 14 Ultrasonic shear waves at the interface between two media ($v_1 > v_2$).

designed for testing composite material. The delay line is acoustically matched to the composite material and yields maximum coupling.

Ultrasonic spectroscopy is based on analysis of the received signal in the frequency domain. The frequency domain signal is obtained by taking the fast Fourier transform of the time domain signal. The method is based on the comparison of the spectral response of a bonded object with a disbonded reference standard. For inspection of composites thicker than 13 mm (e.g., honeycomb core structures), frequencies ranging from 0.5 to 5.0 MHz are used [129]. For thicknesses between 2.5 and 13.0 mm, 5–20 MHz are used, and for thin composites of less than 2.5 mm, 30–75 MHz [129]. Hagemaier recommendeds to inspect thin laminates with a 10 MHz ultrasonic wave, while lower frequency (2.25 MHz) waves are preferable for thick laminates [16]. Honeycomb parts are difficult to immerse; instead, a water spray system is best suited. Despite high attenuation, a sharper defect definition can be obtained with higher frequencies than with lower frequencies [50]. Still, confusion may arise from attenuation, depending upon the degradation process and the material tested.

Other ultrasonic techniques have been developed to overcome the limitations of conventional ultrasonic testing and are discussed next.

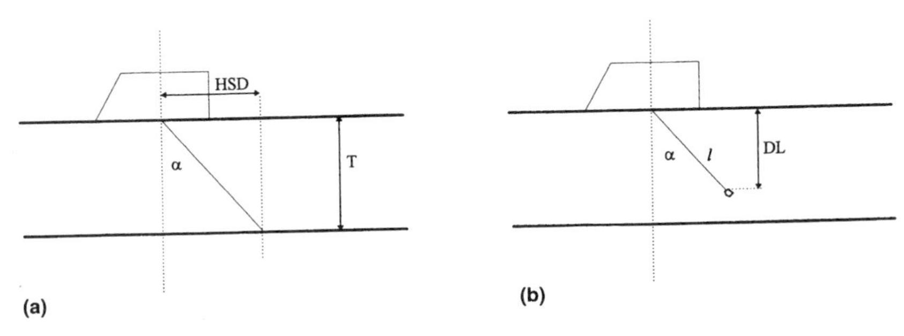

Figure 15 Diagrams illustrating the calculation of (a) material thickness T, and (b) defect location l.

14.2 Noncontact Ultrasonic Inspection

Because of the use of a couplant is not recommended for porous materials testing, ultrasonic techniques requiring no couplant have been developed.

Air-Coupled Ultrasound

Noncontact, or air-coupled, ultrasound uses piezoceramic/epoxy transducers in a through-transmission mode. The principle of an air-coupled ultrasonic transducer can be found in the literature and will not be detailed here [130]. Because the signal attenuation is too high (e.g., attenuation coefficient for an epoxy composite with organic fiber reinforcement is 3.5 dB/cm), air-coupled ultrasonic testing is performed at low frequencies (40–50 kHz) [13]. However, this makes the detection of small flaws difficult. Large-diameter transducers allow rapid inspection, but the sensitivity and spatial resolution are low. Only a small fraction of the energy enters into the solid, and the thicker the material, the lower the fundamental resonant frequency.

Although the pulse-echo mode is possible, this requires a degree of signal processing, and the Lamb wave method is preferred. Indeed, incline insonification permits better transitional parameters to be achieved than for normal incidence.

Laser-Generated Ultrasound

Laser-generated ultrasound typically uses a CO_2 to Nd:YAG laser, emitting a pulsed beam of 150 mJ, to generate longitudinal ultrasonic waves within a material. Another laser coupled to an optical interferometer is used for the detection of ultrasonic waves from the same surface. The detector laser must have a stable frequency, such as an Nd-YAG, and emit pulses of 50 μs long at a repetition rate of 50 Mhz [132]. Laser generated ultrasound in the frequency range of 1–10 MHz can be detected using a Fabry-Perot interferometer using a 400 mW argon-ion laser source [133]. Because the direction of wave propagation is insensitive to the orientation of the laser beam, this technique can be used to inspect complex structural geometries. Due to a small signal amplitude, specially designed detectors are required, and in the case of powerful lasers special precautions may be needed (e.g., the laser energy deposition must be low enough not to damage the composite). Detection is performed by interferometry and optical beam deflection; the detector must be carefully aligned with the detection surface for optimum resolution, and defect visibility needs to be enhanced using digital filtering. An image resolution of ±1.0 mm can generally be obtained.

Despite their price, laser ultrasonic systems such as the LUIS developed by UltraOptec can generate A, B, and C scans and are effective in detecting flaws at the bondline of contoured adhesively bonded honeycomb structures at a distance of up to 2.4 m without the use of a couplant [134]. Laser ultrasound testing allows rapid inspection of large polymer matrix composites at a speed exceeding conventional ultrasonic testing ($5.5\ m^2/h$), produces real-time easy-to-interpret signals (provided adequate signal processing software is available), and is superior to conventional pulse-echo ultrasound in the case of contoured parts [132]. Because it is noncontact, laser ultrasound can be used to inspect parts moving on a conveyor, composites that cannot be immersed, and parts heated to elevated temperatures [135]. At high frequencies laser ultrasound is limited by the laser pulse duration and the depth of penetration of the laser light. There is no restriction on the type of material to be inspected; e.g., polymer–matrix composites (graphite, Kevlar and glass-epoxy), thermoplastics, painted metal, nonmetallic structures, or honeycomb structures. The deeper the light penetration, the wider the observed ultrasonic displacement pulse [135]. However, focusing could damage materials that are strongly absorbing, such as polymer matrices.

The main application of laser-generated ultrasound is rapid examination of aircraft and automotive inspection at manufacturing level. It also finds application in the visualization of laminar defects in 1–20 mm thick CFR materials, can identify damage through separate ply layers, can resolve individual cells of honeycomb structures, and is useful for high-temperature coating monitoring [136]. Djordjevic and Green developed a new laser system that generates discrete frequency ultrasound rather than the broad frequency bands usually generated [9]. Their system has a flat frequency response with no phase shift over the bandpass range of interest (10 kHz–10 MHz), providing maximum amplitude calibration at the beginning and end of a test and instantaneous calibration at the moment of measurement. The pulsed laser beams are delivered to the composite surface through a linear array of optical fibers, and the reflected beams are detected with a second linear array of optical fibers. This allows monitoring of mechanical properties as a function of depth.

A disadvantage of laser ultrasound is that the ultrasound is generated at a single point and beam diffraction occurs, making signal detection and interpretation difficult [137]. A phased array of sources and detectors helps in controlling beam size, shape, and direction by focusing the beam at fixed depth.

Electromagnetic Acoustic Transducer (EMAT)

The electromagnetic acoustic transducer (EMAT) generate ultrasounds by applying a magnetic field to a material with an electrically conductive surface. Radially polarized shear waves (frequencies 1–8 MHz) are generated without physical contact with the test specimen. At normal incidence to an adherent-adhesive interface, a shear wave is a more sensitive probe of interfacial shear stiffness than a longitudinal wave [138]. The main advantages of EMATs are that no couplant is required, and they are useful for high-temperature testing. However, their use is limited to the inspection of electrically conducting materials, and the efficiency of ultrasound generation and detection decreases rapidly with liftoff distance.

14.3 Oblique Incidence Testing

Ultrasonic Back-Scattering

Oblique back-scattering, also known as polar back-scattering, is based on the principle that specular reflection along the fiber direction occurs within a composite if insonified at an angle. Ultrasonic beams (1–25 MHz) enter the material at an angle other than normal to the surface, but normal to the fiber direction in the ply or plies tested [139]. Beam energy is returned only when a properly oriented scatter is present, and it is affected by layered heterogenous composites. Signal levels are reduced due to symmetric spacing of fibers, and cracks will act as reflectors, thus producing a large signal. Pulse and spatial averaging are usually used to improve the signal-to-noise ratio.

Ultrasonic back-scattering requires access to one side only of the inspected component and is well suited for detection of matrix cracking, porosity, fiber orientation, and fiber misalignment in individual plies [34]. An approximation of the defect shape, location, and orientation can also be calculated. The technique is useful for characterization of matrix cracking within the composite plies [84]. If small amounts of porosity are to be detected, smooth surfaces are required. This technique allows the detection of fiber weaviness, translaminary cracks, and ply end discontinuity [49]. Ultrasonic oblique insonification can also provide quantitative information about defects, adhesion, and the elastic properties of composites [140]. Ultrasonic wave attenuation is affected by material composition, density, porosity, and the presence of anomalies. Fiber orientation, misalignment, and stacking can be determined by measuring the spatial distribution intensity of ultrasonic back-scattered signals for a constant angle of incidence [99]. Bashyam described an oblique incidence immersion mode technique used for the generation and measurement of surface acoustic wave (SAW) velocity that, when combined with other measurements, allows an estimation of the stiffness coefficient of ceramic matrix composites such as CAS-Nicalon [141]. Using a pair of 50 MHz broad-band 50 mm focused transducers, his technique allowed characterization of fibers and matrix–fiber interface degradation in a quantitative manner.

Oblique ultrasonic waves sent to a composite at frequencies that excite plate wave modes induce the leaky lamb wave phenomenon. When the leaky Lamb wave is generated, the specular reflection is distorted. When the specular reflection and the leaky Lamb wave interfere, a phase cancellation occurs, and two components are generated with a phase between them. Because each type of defect has a unique response, this technique can be used to determine material eleastic constants and to estimate the volume content of resin as well as porosity content. Detection of transverse cracking and delamination in a 24-layer unidirectional graphite-epoxy laminate has also been reported [140], and oblique incidence back-scattering techniques give accurate fiber orientation of the first composite layers [15].

Lamb Waves Technique

If the thickness of a material is comparable to the wavelength used, Lamb waves are produced within this material. Ultrasonic lamb waves propagate along the plane of plate-like structures, and two transducers set at an angle to normal in a pitch-catch configuration scan the line between them. The pitch-catch mode has been designed to facilitate detection of the reflected ultrasonic wave. Generation of lamb waves can be a time-consuming process, as in-plane and out-of-plane displacements must be calculated for each material tested in order to obtain optimum inspection results. And accurate calculation of the angle between the two probes is also necessary.

Stretching and distortion of the tone burst envelope of the transmitted wave can make signal interpretation difficult. Another problem involved with Lamb waves is the angle of defect edges relative to the direction of propagation, which can influence the magnitude of reflection of the signal. Highly attenuative materials produce a loss of signal, and the distance between the transducer and the test piece should be kept constant, and as short as possible, for improved signal-to-noise ratio. Due to the high sensitivity of the technique, small variations in material thickness affect the signal output [142]. This technique is based on the principle of oblique incidence of an ultrasonic pulse (0.1 to 15 MHz). Refracted incident waves induce guided waves along the material, resulting in leakage of radiation into the immersion fluid. Phase cancellation occurs due to interference between the reflected and the leaky Lamb wave, thus creating a null zone. Detection of delamination, porosity, and inclusion is made through recording the null zone. The leaky Lamb wave receiver is positioned at the null zone and the leaky wave components. Tone burst signals are induced in the composite to establish a steady-state condition in the laminate. The reflection coefficient and the dispersion curve can be analyzed from the received signal. Variation in amplitude changes or spectral response at a tone burst frequency that induces Lamb waves in composites can be due to the presence of flaws within the composite. However, because of its high sensitivity, small variations in elastic properties may induce unwanted noise, thus making signal discrimination difficult, and making the use of signal processing techniques necessary to facilitate signal interpretation and defect detection. The advantages of this techique are its simplicity and the fact that no particular type of transducer is required. Its problems are the necessary immersion of samples and the difficulty in signal interpretation [143].

Farlow and Hayward described the development of an automated air-coupled Lamb wave scanner for rapid inspection and imaging of defects in carbon fiber composites [142]. It can detect disbonding of the fiber–matrix interface, broken or missing fibers, and changes in elastic properties of the adherend, which may indicate a change in thickness (i.e., a void) [14].

14.4 Time-of-Flight Diffraction

Time-of-flight diffraction (TOFD) systems measure the time required for an ultrasonic wave to reach a flaw and return to the transducer. Defect depth and visualization of defect regions can be obtained with this technique. Incident ultrasonic energy propagates differently within a material as follows:

It is reflected as compressional waves
It is mode converted to shear waves
It may pass around the defect without interacting
It is scattered from the defect face
It may diffract at the tips of the crack.

Diffracted waves from crack tips are detected by a received probe, and a measurement of ultrasonic path length allows calculation of defect depth. Although TOFD can achieve great accuracy, recognizing diffracted echoes from other echo types and noise can be difficult, thus necessitating some signal filtering. Pulses of 100 ns have been used to detect 1.0 mm diameter delamination in graphite–epoxy laminates with a depth accuracy of ±0.2 mm [144]. Determination of the elastic constants of a composite can be achieved through measurements of the velocity of longitudinal and shear waves. Delamination of less than 1 mm diameter can be detected by TOF in graphite epoxy laminates. Ultrasonic TOF scans from the rear surface of a composite are used to determine the depth of penetration of lightning damage by measuring the thickness of the remaining undamaged material [15]. However, TOF measurements are not sufficient to detect the presence of foreign materials in laminates, because TOF could be affected by both velocity and thickness changes [145].

14.5 Transducer Resonance Technique

Transducer resonance techniques, such as the Fokker bond tester, are used to determine adhesive-bond quality and to detect disbonds in composites. A standing wave is generated in a material when the thickness of the material is equal to an integral number of half the wavelength. An unbonded material gives rise to a higher resonant frequency than that of the bonded structure. Analysis of the wave behavior is performed to determine the unbound/bound areas [22]. The Fokker bond tester operates in the range 100 Hz–6 MHz and requires a couplant and a reference standard for calibration. It is a slow, manual scanning process that can be extremely time-consuming in the case of large structure inspection. This instrument was primarily designed for metal matrix composites and has limited effectiveness with honeycomb sandwich structures [29]. Variations of frequency and amplitude vibrations are observed and related to the cohesive strength of the bond between laminates and honeycomb structures [49]. Focused beam transducers have a better lateral resolution than ordinary transducers, and a greater signal-to-noise ratio. it can predict cohesive strength when the thickness of the bondline is accurately known and the adhesive has a low specific stiffness, but it is unsuitable to assess most commonly used adhesives of specific stiffness $\geq 4.5 \times 10^4$ GN/m^3 [14].

Another transducer resonance system is the Fokker Stack Tester, developed to examine ply orientation and stacking sequence in the different layers of CFRP composites. This computer-controlled system displays the signal output in a graphical format. It consists of passing a probe into a rivet or fastener hole and scanning the material surface by a focused laser beam. The intensity of the reflected light is correlated to the orientation of each ply and can provide information on the ply number, the thickness of each ply, and fiber orientation.

14.6 Acoustoultrasonic Testing

Acoustoultrasonic testing was developed by the NASA-Lewis Research Center to determine the interaction between defects and the microstructural environment in a composite. Repeated ultrasonic pulses are sent to a material and interact within the material microstructure along its path. Modified stress waves due to microcracking are detected by an acoustic emission transducer and analyzed [84]. The presence of flaws will affect transmission of the acoustic wave, and damaged areas produce a decrease of the high-frequency energy and a slowing of the wave propagation velocity. Examination of the amplitude, frequency, and time parameters provides qualitative information on the integrity of a structure and can be correlated to residual strength/stiffness of the specimen through additional experimental measurements [146].

By using the stress wave factor and sound velocity in the material, the interlaminar shear strength of fiber composite laminates can be determined [49]. The stress wave factor is a measure of the efficiency of transmission of ultrasonic stress waves; the higher the material strength, the more efficient it is at transferring stress waves. The stress wave factor can be used to detect microporosity and inconsistencies in the fiber–resin ratio, and defects such as delamination, stacking faults, matrix cracking, and disbonds. However, the stress wave factor is not useful in the detection of faults in fiber reinforced composite materials due to high scattering, caused by the uncertain quality of coupling between transducers and the test piece. The technique can be used to measure the interlaminar shear strength of fiber reinforced composites in honeycomb panels and laminated structures. Acoustoultrasounds have also been used to assess and monitor damage under compressive loads in SiC/CAS-II ceramic composites [146].

14.7 Pulse Compression Technique

The Barker code pulse compression technique is a special ultrasonic testing method that compresses the ultrasonic pulse through the use of a special class of binary code called the Barker code [49]. The pulse compression techique can be used to overcome the limitations of the traditional pulse-echo techique and is useful in testing materials that cause high attenuation.

14.8 Synthetic Aperture Focusing Technique

The synethic aperture focusing technique (SAFT) uses a scanning transducer to collect and process ultrasonic waves into a unit known as an aperture, by shifting adjacent waves with respect to the middle element of the aperture, summing the waves point by point across their length, and then placing the summing vector at the center of the chosen aperture [147]. One single transducer, emitting a high broad beam, moving along the surface of the composite at a constant speed, is used, and multiple A-scans are stored and coherently processed.

Although SAFT requires a great deal of signal processing, the technique can determine the size, shape, orientation, and location of a defect [148]. However, Thompson *et al.* demonstrated that SAFT tends to produce an exaggeration of defect width in thick Plexiglas composites but provides correct defect depth estimation, with results lying between 1% of the actual defect depth [149]. It was also shown that composite anisotropy tends to defocus the synthetic aperture, but this does not affect the sizing.

14.9 Applications of Ultrasonic Testing of Composites

Ultraound is well suited to detect defects such as delamination oriented perpendicular to the direction of the ultrasonic wave. Evaluation of the integrity of bonds between dissimilar materials of low and high acoustic impedance, such as plastic, glass fiber, rubber, or epoxy bonded to metal, glass, or ceramic, can be performed ultrasonically. Analysis of the reflected ultrasonic wave from the interface between two materials is used for structural bond integrity assessment. Such assessment is very good with metal-to-metal or plastic-to-plastic materials having similar acoustic impedance. With dissimilar materials, i.e., materials with different acoustic impedances such as in plastic-to-metal bonding, signal interpretation is more difficult. The technique consists of determining the phase and amplitude of the reflected ultrasonic wave at the boundary between two materials. The phase of the echo will be reversed when the order of relative acoustic impedance is reversed, and a lack of bonding will produce a composite/air boundary that results in a negative echo, while a solid bond inverts the echo to positive [150]. Ultrasonic immersion C-scanning (5–25 MHz) can detect impacts on composites, but a computerized system would be required to provide information about the distribution and size of the associated individual delaminations [27]. Delaminations closer to the impacted surface cause reflection of the ultrasonic wave, a phenomenon known as the shadow effect. This effect tends to limit the detectability of delamination by ultrasound to damage close to the impacted surface or within the boundaries of the immediate subsurface. Detection of impacts outside these areas necessitates the use of another NDT technique such as radiography [27]. Ultrasound is insensitive to fiber breakage, and quantification of delamination areas can be difficult. Relatively minor delaminations tend to produce ultrasonic information similar to that from areas of more signfiicant delamination [104]. Jones describes an automated water-spray through-transmission ultrasonic system developed for the inspection of large aircraft parts [151]. At 5 MHz, this system easily detected large planar discontinuities, such as adhesive voids and delamination, which cause signal attenuation. Foreign material inclusions are detectable by through-transmission, as they do not bond into the laminate, or are very attenuative. However, certain fabrics such as Mylar can be difficult to detect, as their acoustic impedance is very close to that or composite. Ultrasonic examination can also be used to inspect for skin-to-core disbonds in aircraft assemblies to ensure the integrity of the adhesive bondline [152]. The method for multiple reflections, based on the evaluation of multiple ultrasonic signals reflected from the interfaces of bonded materials, is useful to test bonded metal-to-plastic composites when access to only one side is possible [153]. Ultrasonic evaluation of reinforced plastics and ceramics (25 MHz) is used for quality sorting of materials and assessment of the physical properties of materials.

Kaczmarek described a polar back-scatter technique to measure ply-by-ply delamination and progression when the delamination areas propagate conically through the thickness of the composite [154]. Crack detection occurs because a wave-scattering center, causing signal attenuation, is created by transverse cracks if the damage dimension is of the same order of magnitude as the wavelength. This technique, based on the maximum amplitude measurement in a thin time gate (time-of-flight), allows the production of multiple C-scans in a single pass. However, immersion of the specimen in a water tank and the use of 20 MHz focused transducers generating longitudinal waves are still not sufficient to detect matrix cracking. No depth information was obtained about delaminations with pulse echo and/or through transmission testing. Long focal lengths of 75–150 mm can be used in water immersion to produce a narrow, collimated ultrasonic beam in the material to obtain good lateral resolution [155]. With planar laminated specimens and delamination type defects, reflections are produced that are out of phase with the input waveform. By digitizing the entire waveform, recording of the amplitude at each point corresponding to an interface in the material is possible, and imaging of delamination between each layer can be achieved [155]. Although defect features can be displayed on a ply-by-ply basis, the lower ply delaminations are not completely imaged, as they are shadowed from the ultrasonic energy by the delaminations located in previous interfaces. Closely spaced delaminations can still be resolved using data processing software and utilizing gating at various frequencies. The ply-by-ply imaging enables the investigation to follow growth of delaminations through the damaged area. The assembly of these images allows one to see the cumulative effect of damage through the material and to recognize any significant patterns that may exist [155].

Schwartz suggests the use of the pulse-echo technique for the inspection of glass fiber bonds and delaminations in fiberglass laminates up to 19 mm thick, the through-transmission technique to inspect sandwich constructions, and thick fiber glass laminates and the resonant frequency for detection of disbonds in glass–fiber–metal bonds [87].

Steiner describes the use of a multiaxis robotic system to scan graphite/epoxy laminates ultrasonically and then digitize the ultrasonic signal to produce computer images of the area inspected with a resolution of 0.1 mm [11]. Transducer frequencies ranging from 15 to 25 MHz, resulting in a wavelength of between 0.075 and 0.05 mm in graphite–epoxy composites, were used to monitor fiber/matrix distribution and detect porosity, delamination, and impact damage. Fiber/matrix distribution can be analyzed with ultrasound because fibers reflect the ultrasonic wave while the matrix tends to absorb some of the energy. From 0.17 to 3.57% porosity has been detected in 2.00 mm thick composites; high-porosity composites produce a reduction in the back echo. Ultrasonic C-scan analysis allowed the detection of a 15 J impact energy in a 32-ply graphite epoxy laminate [11]. Chester and Clark developed an immersion ultrasonic TOF C-scan system (2.5–5.0 MHz) producing real-time color-coded maps to monitor damage growth in composite laminates subject to impacts [19].

The absorption of ultrasonic waves in the adhesive layer significantly reduces the amplitude of the echoes from the back surface, thus limiting the detectability of a disbond in that region. Filtering and signal processing are required to reduce the masking effect of front adherent reverberations. New surface flaws (large flaws or delamination) may reflect nearly all of the ultrasonic energy and prevent the detection of deeper defects (also known as hidden or shielded damage) [129]. In this case, multiscan systems are required to obtain pulse-echo images from each side of a component simultaneously. With C-scans, if limits of depth range are set too near material interfaces, images will have nonrelevant interface echoes that mask defects and materials with inconstant thickness limiting the effectiveness of C-scans. In the case of adherend–adhesive–adherend material, reflection of the ultrasonic wave occurs at each interface, so if a defect lies in the path of an ultrasonic beam an additional reflection occurs that can then be visualized on an oscilloscope [14]. Short inspection wavelengths are required to detect thin-layer disbonds between adherend and adhesive, and oblique incidence waves are better suited than normal waves for the detection of interfacial imperfections [14].

Because ultrasonic attenuation rises sharply with void content, it can be used to estimate porosity level; an attenuation of 1.0–1.5 dB·mm^{-1} at 10.0 MHz was reported [15]. Attenuation of ultrasonic waves due to scattering is proportional to the size of the void; the greater the attenuation, the larger the void. However, with prepreg materials there is little variation in the fiber reinforcement volume fraction, thus causing an abnormal attenuation and indicating porosity where there is none. Voids can be detected and evaluated with an accuracy of 0.5%, provided that their size is greater than the ultrasonic wavelength used [16]. Williams and Lee use through thickness attenuation of ultrasonic waves in composites to determine residual static and fatigue strengths and observed that the ultrasonic attenuation of glass fiber reinforced polyester composites can be correlated with the ultimate strength of a structure, independent of flaw type [98]. A correlation between ultrasonic attenuation and porosity-void content was observed by Olson *et al.* [17].

Recent experiments by Dr. Bar-Cohen *et al.* have shown that ultrasonic oblique insonification can be used to characterize thermal damage to composites [156]. Using an inversion technique based on a micromechanical model, the reflected ultrasonic signals are analyzed to determine the overall laminate stiffness constant before and after loading. Another technique developed by the NASA to encompass the limitation of pulse-echo ultrasonic and photomicroscopic methods is diffuse-field acoustoultrasonic coupled vibration damping [157]. Both NASA techniques are complementary and are used to assess microstructural damage accumulation in ceramic matrix composites.

At manufacturing level, chemical reactions between the matrix and the fiber produce an interface zone of different mechanical properties from the two phases producing it [158]. The load of a composite is usually transferred through the interface between the matrix and the fiber, and the toughness of the composite is determined. Karpur *et al.* measured ultrasonically the shear stiffness coefficient of the interface in fiber reinforced metal matrix and ceramic matrix composites [158]. They claim that the significance of the quantification of the shear stiffness coefficient of the interface is that the elastic property of the interface can be used as a basis for composite life prediction.

Because ceramic matrices may be porous, a carbon or boron layer is often deposited between the fiber and the matrix to improve matrix toughness by deflecting cracks and preventing premature fiber fracture [159]. Oxidation damage caused by diffusion of oxygen though matrix pores and cracks occurs at this interface and results in a loss in strength and stiffness. Oxidation damage and the initiation of oxidation in ceramic matrix composites can be determined by measuring the changes in ultrasonic velocity and elastic moduli. Chu *et al.* also showed that the interfacial moduli determined from the measured composite moduli can be used as a quantitative measure of damage severity [159].

Thompson *et al.* described a series of ultrasonic techniques used for in-situ measurements of elastic constants on thick-walled submersible vessels [149]. The elastic constants can provide information about fabrication errors such as wavy fibers and fiber disbonds. Elastic constant measurements can be performed using Rayleigh or Lamb wave modes, or by using angle beam techniques. It was shown that the effect of the anisotropy increases with increasing flaw depth, thus producing poorer accuracy in defect sizing measurements in scans along and normal to the fiber direction [149].

Image processing techniques are often used to improve the signal-to-noise ratio in the case of graphical signal output. Contrast enhancement, edge detection, image sharpening, thresholding, and filtering are all part of today's ultrasonic inspection systems. However, image enhancement is usually performed after completion of an inspection. It was estimated that an accuracy of over 90% could be achieved for ultrasonic detection of large voids and delamination if image enhancement is performed [34]. The difference in acoustic impedance between air and solid materials makes honeycomb structures difficult to inspect due to ultrasonic propagation, mainly at cell wall locations, and pulse-echo inspection is unreliable [13]. Ultrasonic oblique incidence techniques are used to inspect the adhesive–adherend interface in the order of 2 μm by measuring variation in the material reflection coefficient [13]. Ultrasonic wave dispersion can also be used to evaluate moisture content in composites and was successfully applied by Sachse *et al.* in the inspection of boron–epoxy materials [160]. The quality of butt fusion welds in polyethylene plastic pipes used in the gas and water industries is usually inspected by visual examination to detect misalignment and gross flaws. Recent studies have shown that ultrasonic creeping waves can be used to inspect the fusion area beneath the outer weld bead in polyethylene pipes [161].

In terms of equipment, McDonnell Douglas proposes an automated ultrasonic scanning system (AUSS) with 9 axes of motion, a 7 axes ultrasound system (ADIS-II) and a portable inspection system (MAUS II) for aerospace composite structure inspection [36]. The ultrasonic resin analyser (URA 2002A from Quatro Technologies) can be used to measure the resin content in polymer based composites [10]. The UCMS-200 ultrasonic cure monitoring system from Micromet Instruments utilizes measurements of the ultrasonic sound speed to monitor changes in the viscosity/rigidity of composites.

Focused and unfocused back-scatter ultrasonic B- and C-scans have been studied through experiments by Moran *et al.* for the inspection of graphite–epoxy composites with ply cracks [139]. A high-frequency focused probe (25 MHz) is recommended in order to achieve good image sensitivity. Back-scatter B-scan produce high noise scattering and poor image sensitivity, while C-scans produce better results provided the C-scan time gate is properly positioned relative to the surface roughness signals. Calibration procedures are required to evaluate the dispersive effect of an ultrasonic wave inside a composite material in order to perform a reliable examination. Focused transducers can be used to reduce the attenuation of ultrasonic waves in heterogenous and anisotropic materials. Moreover, it is suggested to use narrow ultrasonic beams (high-frequency, focused transducer, large-diameter transducer) to achieve a good lateral resolution in composites [162].

14.10 Outlines

Despite its extensive use and success in industry, ultrasound is not fully convenient in inspecting composite materials; similar inspection results could certainly be achieved with other, less costly, NDT techniques. While some find ultrasonic signal interpretation easy, calibration can be time-consuming and complicated with composites. The severity of the damage can be difficult to assess, and water immersion of composite structures is not always possible. If a gel couplant is used, cleaning of the component is required prior to repair. Pulse-echo requires precise entry angle alignment, suffers from increased complexity in signal processing, and is poor with complex shaped composites. Manual point-by-point scanning is slow, and wave scattering and wave resonance in adhesively bonded composites mask flaw echoes, so advanced signal processing may be required to improve the signal-to-noise ratio. It was reported that when the duration of the transducer impulse response approaches the acoustic transit time across the layers, signal interpretation can be difficult [163]. For these reasons, computer controlled scanners and data handling and signal processing systems have been developed to reduce human intervention and inspection costs [151]. For example, for a scan of curved parts with conventional ultrasound, a robotic computer controlled device and a water tank or water spray system are required.

The major problem with ultrasonic NDE is that the ultrasonic wave velocity behaves differently in composites than in homogeneous materials and is therefore difficult to use precisely as it changes from point to point [162]. Couplants cannot be used with certain structures such as satellites as they may affect the adhesion of rain-erosion resistant coatings. Ultrasonic testing is useful for detection of delamination in CFRP materials, but with lack of reproducibility [32]. The through-transmission technique is difficult to implement for in-service inspection, as access to both sides is often impossible. Planar flaws such as delamination, interlaminar porosity, voids, and foreign inclusions are usually oriented normal to the ultrasonic beam, making them more readily detectable by ultrasound. Through-transmission is superior to pulse echo to detect surface flaws, but this method does not give depth information about the defect. Availability of fast, inexpensive PCs favors the use of several ultrasonic imaging systems suitable for a range of budgets. System performance is affected by stiffness and precision of the scanner, bandwidth, noise level, and resolution of the instrumentation. Those interested in knowing more about ultrasonic nondestructive evaluation of CFRP can find detailed information in the publication by Edmund and Henneke [164].

In summary, it can be said that ultrasonic testing is a well-established method but difficult to implement with composites, mainly because of their heterogeneity, anisotropy, and multiple layers. The repeatability of the technique is also often limited; signal interpretation remains uncertain or difficult, and further improvements are required. Complex geometries and nonplanar surfaces are difficult to inspect and tend to complicate signal interpretation, defect detection, and ultrasonic inspection in general.

15 Radiography

Radiography is a well-established technique in medicine that also finds applications in materials testing as it provides visual information on internal flaws. The principle of this technique and those of other x and gamma radiation methods are described next.

15.1 Principle

Radiography relies on the absorption of electromagnetic radiation (x-rays) while passing through a material. Penetrating radiation passes through a material and the emerging radiation is recorded on a film (Fig. 16). The amount of radiation absorbed by the material is proportional to its thickness and density (e.g., less radiation will be absorbed by a void, resulting in a dark spot on the radiograph). Cracks are detectable principally when viewed from a direction parallel to their plane; cracks perpendicular to the beam may remain undetected. The minimum detectable size for a foreign particle is a function of the x-ray absorption coefficient and density difference between inclusion and matrix. Radiographic films show variations of grey levels proportional to the amount of radiation absorbed; the

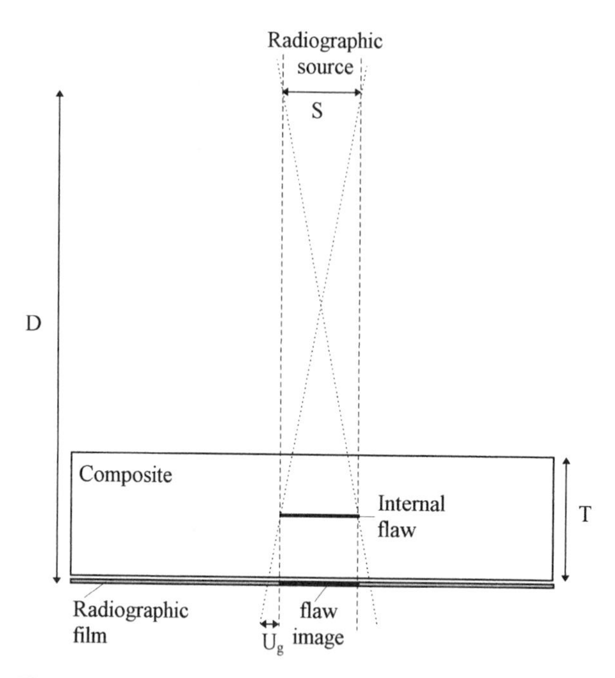

Figure 16 Schematic of film radiography.

lighter the shade of grey, the larger the amount of absorbed radiation. The quality of a radiograph can be quantified by measuring its geometric unsharpness U_g which is proportional to the source size S, the material thickness T, and the source-to-film distance D. It is given by the equation (Fig. 16)

$$U_g = \frac{ST}{D - T} \tag{22}$$

X-rays or gamma rays can be employed in radiography. X-rays are generated in an x-ray tube from the interaction of electrons and a metallic target made of tungsten, platinum, or gold. Gamma rays on the other hand are natural radiation emitted from an isotope such as cobalt 60 or iridium 192. X radiation is usually preferred in NDE, as it is more sensitive than gamma, and its emission is more controllable.

Because graphite–epoxy composites structures are relatively low absorbers of x-rays, low kilovoltage machines with beryllium windows are preferred. The greatest sensitivity is achieved with low energy x-rays, because they are more attenuated than those of higher energy (Fig. 17). A 30 seconds exposure using a 10–25 kV and 25 mA x-ray source is a good compromise. The presence of glass and boron fibers helps in radiographic examination of composites. However, radiography of composites can be enhanced by the injection of a radiopaque substance (e.g., iodine, bromine, chlorine) into the composite. This substance is made apparent on the radiograph, allowing an accurate assessment of damage extent.

Radiographic film has an excellent spatial resolution but requires film processing, which can either be costly or time consuming. The development of a solid state detector has favored the development of filmless systems.

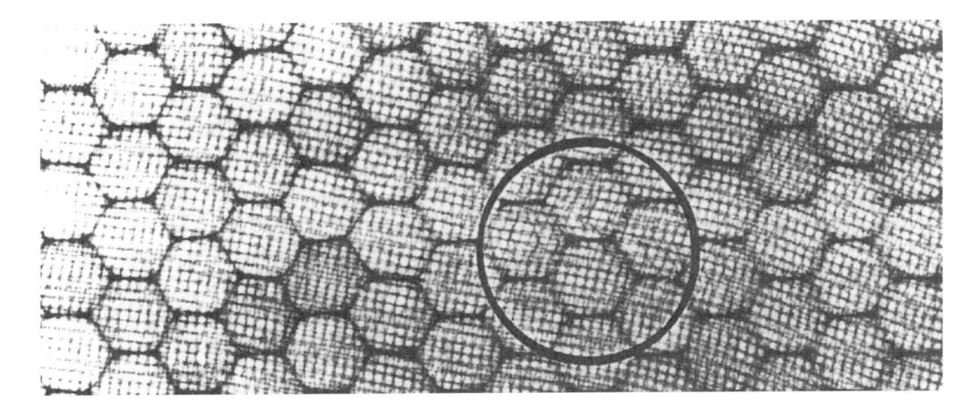

Figure 17 X-radiography showing damaged fiber in a honeycomb cell composite after an impact of 2 joules.

15.2 Real-Time Radiography

Real-time radiography (RTR) is used to produce instantaneous radiographs that are displayed on a TV monitor. Unlike conventional radiography the detector is not a film but an image intensifier screen, a fluorescent screen, or a scanning linear array sensitive to radiation. The radiation is converted into a digital signal, which is displayed on the screen and recorded onto an optical disk. The development of new x-ray detectors using charge coupled devices (CCD) allows better contrast and computer-aided defect recognition, their principal advantages being instantaneous generation of radiographs, lack of film processing cost, and automation. Real-time radiographic systems can be fully automated to inspect parts passing on a conveyor during manufacture. Despite its rapidity over conventional film radiography, RTR remains a technique not readily accessible to all NDT users. Indeed, only when large-scale inspections are performed on a regular basis, will the cost of the equipment become less than that of a conventional film radiographic system. Moreoever, image processing (IP) software is also often required. For example, at low energies several image intensifier tubes are not sensitive enough to detect the emerging radiation, and IP software can provide images of quality surpassing those obtainable on film. Computerized image processing systems enable easier and faster structural assessment of composites. The differential imaging procedure consists of substracting one radiograph of a sample taken with a radiopaque contrast liquid from that of a conventional radiograph. The difference between the two images tends to show clearer details unavailable on each separate image before processing. Differential imaging of radiographs taken at different energies, or averaging a large number of images, is also possible and improves inspection results [165]. Other advantages include increased contrast sensitivity, more convenient data storage and transfer, and improved implementation of IP algorithms.

Also known as fluoroscopy, RTR is used to detect exaggerated resin-rich/starved areas, to determine ply orientation, and to detect cracks oriented parallel to the x-ray beam with a depth greater than 3% of the total part thickness [32]. It is particularly applicable to bonded honeycomb structures (e.g., helicopter rotor blades, tail rotor blades, panels) in the detection of hidden foreign objects, mismatch, or misalignment. Real-time radiography can be incorporated as an on-line system to monitor the fabrication of composite parts during all the process cycles. This can provide instant feedback on operations such as layup, consolidation, and curing or weaving [166]. Jensen uses an energy of 35–75 kV at 4.4 mA for assessment of the adhesive line of a windmill blade of glass fiber reinforced polyester and disbonds and voids in the polyurethane foam core of a sandwich panel (19 kV and 4.0 s) [165].

15.3 Neutron Radiography

Neutron radiography, developed using thermal neutrons from a nuclear reactor, and although it remains unsafe, costly, and not portable, is successful if the detection of water trapped in honeycomb cells and useful in inspecting bond lines and composites at proximity to metals [50]. Because x-ray films are not directly sensitive to neutrons, an additional neutron-sensitive screen is often required, or neutrons are converted to electrons by passing through a gadolinium metal film.

Neutrons are more strongly absorbed by adhesives than x-rays and therefore are best suited for disbond detection and to determine resin content (e.g., a change in matrix volume of 1% can be detected) [14]. Organic materials containing hydrogen, such as epoxy, exhibit good radiographic contrast, and a change in material density due to voids or inclusions can be seen [50]. However, defect indiation can be awkward, as low absorption inclusions produce images similar to voids.

15.4 Computer Tomography

Computer tomography (CT), also called computerized axial tomography (CAT), takes one or several radiographic slices from many different viewing directions of an object and combines them into a cross-sectional array. The principle of this technique is based on a collimated beam of x-rays that passes through the test piece, which is rotated through an angle of 180°. The emergent radiation is received by a row of detectors that measure the attenuation of the beam intensity in a finite number of angular increments. Computed tomography presents significant advantages over conventional radiography. Image interpretation can be greatly simplified, as there is no image overlap, and by stacking a series of 2D CT images, one can reconstruct a 3D picture of the inspected specimen. This 3D reconstruction of the test object allows a complete examination of a structure. It was reported that tomographic images are of much better quality than those obtained with radiography, and density differences up to 0.1% can be observed [18]. Moreover, the contrast resolution of CT has been estimated to be 0.1 to 0.2% better than that of x-ray films, and structural noise is often absent [167]. An additional feature of CT imaging is its faster speed than conventional radiography, as no film processing is required. While 50–75 μm voxels are too coarse for damage quantification in composites, 25 μm voxels are obtainable from a commercial CT scanner [168]. High-resolution CT is also known as x-ray tomographic microscopy (XTM) and is capable of detecting matrix cracking and fiber fracture in composites [168]. Despite certain advantages, CT remains expensive, and in the case of large specimens, data collection can be time-consuming and costly. Figure 18 is a

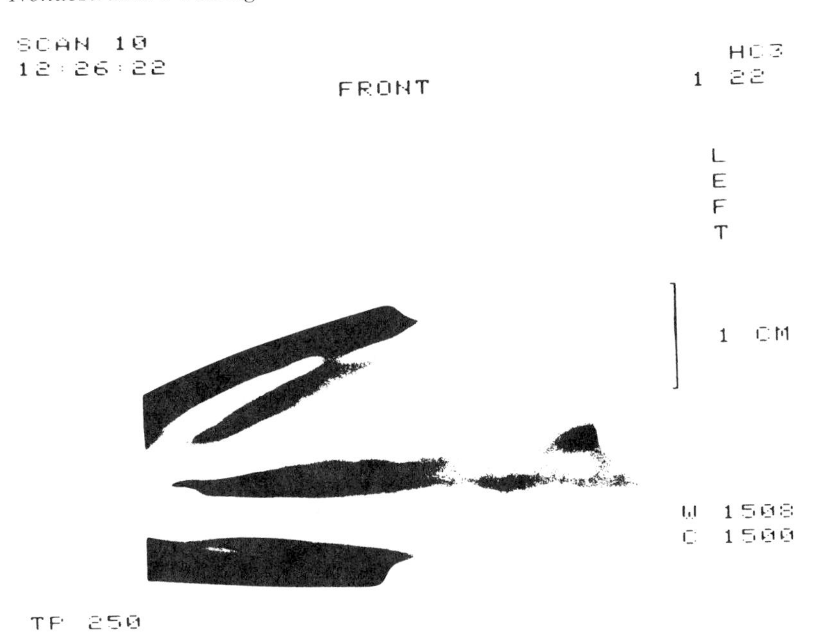

Figure 18 CT radiograph of a section of carbon fiber helicopter rotor blade with light delamination.

CT radiograph of a section of carbon fiber helicopter rotor blade showing light delamination.

Delamination, cracks, and density variations in graphite epoxy composites can be visible on the same image with a single scan using CT [167]. This technique is very useful in the detection of fiber pull-outs in ceramic matrix composites and can quantitatively measure 3D density distribution in materials [166]. Detection of objects as small as 1 μm has been reported in silicon carbide materials, and determination of material density differences within an accuracy of about 0.1% has been achieved [36]. However parts made of materials exhibiting regions of different compositions (e.g., with Si_3N_4 ceramic turbine rotors) are difficult to inspect, and the use of two different energy levels is necessary. This technique will become more cost-effective as computer power increases and computer prices diminish. It has good repeatability, can be adapted to inspect any geometry, and being digital it can be fully automated and used in combination with expert systems.

15.5 X-Ray Refractography

X-ray refractography can be performed where x-rays pass through a collimation slit and penetrate a composite sample. Absorption by the sample is monitored using an x-ray absorption detector, while refracted rays pass through a refraction slit positioned at an angle. The emergent refracted radiation is then collected by a refraction detector and visualized on a monitor. The signal received can be defined as the relative difference between refraction level and absorption level [169]. The physics of this technique is analogous to optical light refraction (e.g., Snell's law). X-ray beam deflection occurs at very small angles and, due to a smaller difference in density, radiation refraction is greater in a debonded area than in a defect-free region. Comparison of a standard reflected reference beam allows the detection of flaws in composites using an x-ray refraction detector. X-ray refractography is a real-time scattering technique that can be used to determine the amount of inner surfaces and interfaces of nanometer dimensions, due to the short x-ray wavelength of around 0.1 nm [169]. X-ray refractography finds applications in integral analytical determination of mean pore size in ceramics and average single fiber debonding.

15.6 Positron Annihilation

Positron annihilation uses the principle where positively charged electrons emitted by radioisotopes (positrons) enter a material and annihilate with conduction to produce gamma rays [170]. Emergent radiation is then detected and a radiograph produced. With resin based composites, the lifetime of positrons within the material can be observed and is governed by the formation of positronium. Polar molecules such as water affect the distribution of positron lifetime, therefore permitting moisture detection. This technique is useful in providing information on degradation caused by fatigue and moisture ingress. However, it remains a laboratory technique rarely used in industry.

15.7 X-Ray and Gamma Ray Back-Scattering

Gamma and x back-scattering techniques are based upon the Compton photon back-scatter effect. Collimated low-energy gamma rays or X-rays are emitted and beamed at the inspected material. The rays become scattered back toward the detector in direct proportion to the mass of the material in front of the probe. A scintillation crystal detector is used to convert the back-scattered protons into an electrical signal that can be related to material thickness, provided that the material density is constant. Back-scattered x-rays can be detected by an array of scintillation detectors, and the position of the detector relative to the sample relates to different depths in that sample [84].

Although the results of back-scatter techniques are affected by surface roughness, information on material density change, voids, and inclusions can be obtained in laminated structures such as pressure vessels and rocket motor cases. The detection of tight delaminations with gaps less than 50 μm in width have been made [84]. Although back-scatter imaging tomography systems suffer from slow data acquisition, they do provide significant capabilities in NDE of multilayered composites (e.g., inspection of inner layers that have been shielded by outer coverings and structures) [171].

15.8 Applications

There are contradictory opinions on the suitability of x-rays to detect defects in composite materials. Scott and Scala [50] found radiographic techniques unsuited to the detection of many defects, while Harris [172] identified voids but could not detect thermal cracks. On the other hand, Prakash [173] suggested that thermal cracks are readily detectable, while Salkind was unable to detect voids or cracks [174]. Case-to-liner and liner-to-propellant bonds in solid rocket motors are currently inspected by passing x-rays tangentially to the bonding surfaces [153]. Radiography can be used to detect crushed cores normally associated with impact damage and condensed cores resulting from excess foaming adhesive or associated with core slippage during the cure cycle [3]. Other applications include the detection of voids in foam adhesives, skin-to-core voids, poor adhesive filleting, and bulk density gradients in ceramics [10].

Low-kilovoltage radiography is useful in the inspection of honeycomb structures for core or adhesive filler defects and accurately to detect water ingress in honeycomb structures. The radiographic detection of 4.5 J impacts in 6 mm thick 6-ply laminate was reported [16]. Low-energy radiography (< 15kV) can also pinpoint localized variations in fiber density in thin composites. Microfocus systems can produce high spatial resolution radiographs by means of geometric magnification, but delamination remains difficult to detect due to the limited performance of the tube and, in the case of RTR, of the image intensifier tube. Delamination is difficult to detect with radiography as there is no density variation between delaminated and defect-free areas, and x-radiography cannot detect fiber breaks except in the case of boron fiber [34]. A radiopaque penetrant, such as diiodobutane (DIB), is often used to increase image resolution and to give a general view of a damaged area near the composite surface [27]. The detection of delamination does require the use of radiopaque liquid, which means that the defect has to be open to the surface to enable the injection of the liquid. Moreover, Girshovich *et al.* showed that this method was not sufficiently reliable nor comprehensive [37]. Polyethylene pipes can be inspected using radiographic intensities of between 16 kV and 26 kV and allow the detection of cold weld in 5–50 mm diameter pipes [161].

Three-dimensional spatial distribution of damage during deformation can be imaged by XTM, thus allowing identification of the mechanisms that control the macroscopic response of a composite to mechanical or thermal stresses [168]. SiC/Al matrix composites are best studied at energies close to 21 keV, while SiC/Ti_3Al and nickel based composites require energies of up to 31 keV [168]. Microfocus x-rays can detect defects in the order of microns using small focal spots (10–50 μm compared to 1–4 mm with conventional x-rays), thus producing a low geometric unsharpness at up to 200 times magnification. Microfocus x-radiography is suitable for the detection of critical flaw sizes of 20–50 μm in ceramics using 30–60 kV, a beam current of 0.25–0.32 mA, a 10 μm focal spot, and a molybdenum anode [10].

Real-time radiographic systems, such as the x-star from OIS, have been developed for helicopter companies to inspect complete rotor blades up to 10 meters in length [175]. Such systems use a 160 kV constant potential dual-focus tubehead, a 220 mm low-absorption aluminum input window image intensifier tube, and an automatic manipulator movement. One of the limitations is that the component inspected has to be removed from the helicopter prior to inspection.

Examples of the use of x-ray tomography to study damage in graphite/epoxy laminates subjected to impact, tension, and compression loading are given by Stock *et al.* [168]. Using computer tomography, Bathias *et al.* detected damage in 16-ply glass–epoxy plates of dimensions equal to 250 × 50 × 11 mm, and a [0/0/45/45/ − 45/ − 45/90/90] stacking sequence [18]. This example showed that transverse cracks ±45° or ±90° are associated with delamination initiation. Tomography is also useful for thick structures such as filament wound tubes and rocket cases. However, computer tomography is adequate for the identification of cracks in adjacent fibers of the lower ply of composites. Low energy (1.33 J) impact damage was not apparent in radiography of 8 ply ±45° CFRP laminates and barely detectable in 8 plies 0° ± 45° samples [26].

X-rays proved inaccurate in determining resin content, while thermal neutron radiography showed good results (an increase in absorption of neutrons occurs due to an increase in resin content). Indeed, neutron radiography can assess resin content with a resolution of ±1% of the material thickness [99]. However, this technique is impractical and costly for large-structure examination [16].

A study of radiographic parameter variations due to low-energy inspection of graphite–epoxy composites has been made by Fassbender and Hagemaier [176]. This study demonstrated that radiographic films exhibited higher subject contrast at 25 kV than at

50 kV, and that type AA film can be exposed at kilovoltages not practical for type M films. This results in high-contrast radiographs and short exposure times. Fluorometallic intensifying screens produce up to a 40% enhancement of exposure compared to lead screens.

15.9 Concluding Remarks

Radiography can provide information regarding internal and subsurface defects, provided the correct technique is used and that defect orientation is prone to detection by radiography. X-ray radiography is usually sensitive to 1–2% changes in material thickness or density, and a spatial resolution of 1 μm has been reported [34]. The major limitations of radiography are the orientation of the defect within the composite, which means difficulty in detecting surface damage, the large initial investment required, the limited portability, the use of hazardous radiation, and the need to use a radiopaque substance.

16 Conclusion

Most defects in composites can be detected and quantified using existing NDT methods. However, the accuracy in defect quantification is still greatly dependent upon the technique used and operator experience. Material composition, environmental factors, specimen location, and access are other significant parameters to take into account before an examination. Moreover, certain factors must be considered preceding the selection of an NDT method, which include but are not limited to ease of use, portability, safety, ease of signal interpretation, rapidity, reliability, reproducibility, accuracy, the material to be tested, the type of defect to be detected, and cost.

The need for improved understanding and further developments in material characterization methods for novel materials such as metal–matrix composites has been highlighted by Bruce and Clarke [15]. However, no NDT technique alone is suitable for detection and quantification of flaws in composites. Eddy current for example is suitable for the estimation of reduction in tensile strength, while ultrasound is particularly sensitive to evaluate loss of compression strength in laminates. Whichever the NDT technique used, a knowledge of structural design and the material is necessary. Optimum nondestructive examination could be achieved through a better understanding of the failure mode and crack propagation in composite materials. Radiography, ultrasonic, and infrared testing measure the amplitude of transmission of the respective energy used and provide efficient measurements about the structural integrity of a composite; but they are imprecise for detailed microstructural evaluations.

Because the structural integrity of composites relies mainly on the interfacial bond strength of organic resins, their strength is diffused throughout the structure; thus special care must be given to their inspection [29]. For efficient NDE, each instrument must be carefully calibrated and the output signal correctly interpreted. The use of color-coded images or computer-aided interpretation systems are preferable as they facilitate signal analysis. Training of the NDT inspector is also necessary prior to inspection.

In order to facilitate in-service inspection, nondestructive testing considerations should be kept in mind throughout the design and process stages. In 1995, Olson *et al.* recognized that NDE techniques, at their present stage of development, did not have the capability fully to interrogate all the features in polymer composites necessary for quantifying material quality [17]. However, efforts are being made in the NDT industry to develop techniques adapted to assess the structural integrity of current and future composite materials.

References

1. Sterke, A. (1994), Inspection of adhesively bonded joints in glass-reinforced plastic pipe systems, Proc. of the Inspection Structural Composites Conf., paper no. 14.
2. Nonaka, T., Hayakawa, Y., Takeda, S., and Nishimori, H. (1989), Ultrasonic imaging systems for composites and ceramics, *Materials Evaluation*, *47*, 5, 542–546.
3. Seidl, A. L. (1994), Inspection of composite structures, *SAMPE. J.*, July/Aug. 1994, *30*, 4, 38–44.
4. Gros, X. E., Strachan, P., Lowden, D. W., and Edwards, I. (1995), A Bayesian approach to NDT data fusion, *Insight*, May 1995, *37*, 5, 363–367.
5. Chung, D. D. L. (1994), Carbon Fiber Composites, Butterworth-Heinemann, pp. 200–212.
6. Reifsnider, K. L. (1989), Service induced damage in composite structures, Handbook of Composites, Vol. 2, Structures and Design (Herakovich, C. T., and Tarnopol'skii, Y. M., eds.), pp. 231–262.
7. Harris, B., Chen, A., Coleman, S. L., and Moore, R. J. (1990), Residual Strength and Toughness of Damaged Composites, Proc. of the 4th Europ. Conf. on Composite Materials, Developments in the Science and Technology of Composite Materials, pp. 157–162.
8. Kelly, A., ed. (1994), *Concise Encyclopedia of Composite Materials*, Pergamon.
9. Djordjevic, B. B., and Green, R. E. (1994), Non-contact Ultrasonic Techniques for Process Control of Composite Fabrication, Proc. of the Conf. on NDE Applied to Process Control of Composite Fabrication, Oct. 1994, pp. 57–76.
10. Rebello, J. M. A., and Charlier, J. (1993), Some aspects of NDT applied to ceramics, *European J. of NDT*, April, 4, 150–158.
11. Steiner, K. V. (1992), Defect classification in composites using ultrasonic nondestructive evaluation techniques, Damage Detection in Composite Materials, ASTM, STP 1128, pp. 72–84.
12. Hancox, N. L. (1983), High performance composites with resin matrices, Handbook of Composites, Vol. 4, Fabrication of Composites (Kelly, A., and Mileiko, S. T., eds.), pp. 29–38.
13. Cawley, P. (1992), NDT of adhesive bonds, Flight-Vehicle Materials, Structures, and Dynamics, Vol. 4, Tribological Materials and NDE, pp. 349–363.
14. Munns, I. J., and Georgiou, G. A. (1995), Non-destructive testing methods for adhesively bonded joint inspection—A review, *Insight*, Dec. 1995, *37*, 12, 941–952.
15. Bruce, D. A., and Clarke, B. (1994), Non-destructive evaluation of composite materials for aerospace, Proc. of the Inspection of Structural Composites Conf., paper no. 1.
16. Hagemaier, D. J., and Fassbender, R. H. (1979), Nondestructive testing of advanced composites, *Materials Evaluation*, June 1979, *37*, 6, 43–49.
17. Olson, B. D., Lamontia, M. A., Gillespie, J. W., and Bogetti, T. A. (1995), The effects and non-destructive evaluation of defects in thermoplastic compression-loaded composite cylinders, *J. Thermoplastic Composite Materials*, Jan. 1995, *8*, 109–136.
18. Bathias, C., and Cagnasso, A. (1992), Application of X-ray tomography to the nondestructive testing of high performance polymer composites, Damage Detection in Composite Materials, ASTM, STP 1128, pp. 35–54.
19. Chester, R. J., and Clark, G. (1992), Modelling of impact damage features in graphite/epoxy laminates, Damage Detection in Composite Materials, ASTM, STP.
20. Rhodes, M. D., Williams, J. G., and Starnes, J. H. (1976), Effect of low velocity impact damage on the compressive strength of graphite-epoxy hat-stiffened panels, Report NASA TM X-73988 (TN D-8411), December 1976.
21. Heslehurst, R. B., and Scott, M. (1990), Review of defects and damage pertaining to composite aircraft components, *Composite Polymers*, *3*, 2, 103–133.
22. Hagemaier, D. J., and Fassbender, R. H. (1985), Ultrasonic inspection of carbon-epoxy composites, *Materials Evaluation*, April 1985, *43*, 556–560.

23. Greszczuk, L. B., and Chao, H. (1976), Impact damage in graphite-fiber-reinforced composites, Proc. of the 4th Conf. on Composite Materials: testing and design, AST Pub. 617, May 1976, pp. 389–408.
24. Oplinger, D. W., and Slepetz, J. M. (1975), Impact damage tolerance of graphite/epoxy sandwich panels, Foreign Object Impact Damage to Composites, ASTM STP 568, pp. 30–48.
25. Preston, J. L., and Ccok, T. S. (1975), Impact response of graphite-epoxy flat laminates using projectiles that simulate aircraft engine encounters, Foreign Object Impact Damage to Composites, ASTM STP 568, pp. 49–71.
26. Cantwell, W. J., and Morton, (1985), Detection of impact damage in CFRP laminates, *Composite Structures*, *3*, 3–4, 241–257.
27. Girshovich, S., Gottesman, T., Rosenthal, H., Drukker, E., and Steinberg, Y. (1992), Impact damage assessment of composites, Damage Detection in Composite Materials, ASTM, STP 1128, pp. 183–199.
28. Reed, P. E., and Bevan, L. (1993), Impact damage in a composite material, *Polymer Composites*, Aug. 1993, *14*, 4, 286–291.
29. Seidl, A. L. (1995), Inspection of composite structures Part II, *SAMPE J.*, Jan./Feb. 1995, *31*, 1, 42–48.
30. Walker, C. A., and McKelvie, J. (1987), Optical methods, Non-destructive testing of fibre-reinforced plastics composites, Vol. 1, pp. 105–149.
31. Cauffman, S., Brown, R., and Bartos, A. (1994), Video monitoring of a large-scale commercial three-dimensional braider for advanced composite preform process control, Proc. of the Conf. on NDE Applied to Process Control of Composite Fabrication, Oct. 1994, pp. 127–142.
32. Wegman, R. F. (1989), Nondestructive test methods for structural composites, Sampe Handbook 1.
33. Marshall, A. P., and Bouadi, H. (1993), Low-velocity impact damage on thick-section graphite/epoxy laminated plates, *J. Reinforced Plastics Composites*, Dec. 1993, *12*, 12, 1281–1294.
34. Daniel, I. M. (1992), NDE of composite materials, Flight-Vehicle Materials, Structures, and Dynamics, Vol. 4, Tribological Materials and NDE, pp. 313–342.
35. Yen, C. E., and Tittman, B. R. (1995), Fiber-matrix interface study of carbon-carbon composites using ultrasonics and acoustic microscopy, *Composites Engineering*, *5*, 6, 649–661.
36. MacRae, M. (1993), Details without damage: NDT methods "find faults" with composite materials, *Advanced Composites*, *8*, 4, 28–34.
37. Wey, A. C., and Kessler, L. W. (1992), Quantitative measurement of delamination area in low-velocity impacted composites using acoustic microscopy, Review of Progress in QNDE, Plenum Press, Vol. 11, pp. 1413–1419.
38. Lou, K. A., Zimmerman, B., and Yaniv, G. (1994), Combined sensor system for process and in-service health monitoring of composite structures, Proc. of the SPIE, Vol. 2191, Smart Sensing Processing and Instrumentation, pp. 32–45.
39. Measures, R. M. (1991), The detection of damage and the measurement of strain within composites by means of embedded optical fiber sensors, Review of Progress in QNDE, Plenum Press, Vol. 10B, pp. 1247–1258.
40. Voet, M. R. H., Vertongen, M. C. M., Boschmans, L. M. H., and Mertens, M. M. J. (1993), Overview of new damage techniques in composite materials for aeronautics using fibre optic sensors, Proc. Inter. Symp. on Intelligent Instrumentation for Remote and On-site Measurements, 6th TC-4 Symp., pp. 127–132.
41. Murphy, K. A., Schmid, C. A., Tran, T. A., Carman, G., Wang, A., and Claus, R. O. (1994), Delamination detection in composites using optical fiber techniques, Proc. of the SPIE, Vol. 2191, Smart Sensing Processing and Instrumentation, pp. 227–231.
42. Rosenthal, D., and Trolinger, J. (1995), Holographic nondestructive testing, *Materials Evaluation*, Dec. 1995, *53*, 12, 1353–1355.
43. Zhu, J. (1992), Optical nondestructive examination for honeycomb structure, Proc. of the Non-Destructive Testing 92 Conf., pp. 399–403.

44. Tay, T. E., Shang, H. M., and Lwin, M. (1993), Application of holographic interferometry in the detection of flaws in composite plates, Proc. of the Inter. Conf. on Advanced Composite Materials, pp. 683–686.
45. Vikhagen, E., and Løkberg, O. J. (1990), Detection of defects in composite materials by televsion holography and image processing, *Materials Evaluation*, Feb. 1990, *48*, 2, 244–248.
46. Sheldon, W. H. (1978), Comparative evaluation of potential NDE techniques for inspection of advanced composite structures, *Materials Evaluation*, Feb. 1978, *36*, 2, 41–46.
47. Delgrosso, E. J., and Carlson, C. E. (1977), Holographic inspection of jet engine composite fan blades, *Automot. Eng.*, Dec. 1977, *85*, 62–65.
48. Heslehurst, R. B., Baird, J. P., Williamson, H. M., and Clark, R. (1993), Portable holographic interferometry testing system: Application to crack patching quality control, Proc. of the Inter. Conf. on Advanced Composite Materials, pp. 687–692.
49. Raju, P. K. (1993), Ultrasonic characterization of composites, Proc. of the 1993 National Conference on Noise Control Eng., May 1993, pp. 441–446.
50. Scott, I. G., and Scala, C. M. (1982), A review of non-destructive testing of composite materials, *NDT International*, April 1982, 75–86.
51. Newman, J. W. (1991), Production and field inspection of composite aerospace structures with advanced shearography, Review of Progress in QNDE, Plenum Press, Vol. 10B, pp. 2129–2133.
52. NASA (1994), Successful shearographic tests were made on honeycomb core composite panels, *NASA NDE Working Group Newsletter*, Jan. 1994, *2*, 1, p. 6.
53. Newman, J. W. (1991), Shearographic inspection of aircraft structure, *Materials Evaluation*, Sept. 1991, *49*, 9, 1106–1109.
54. McCullough, R. W., Bondurant, P. D., and Doyle, J. L. (1995), Laser-optical triangulation systems provide new capabilities for remote inspection of interior surfaces, *Materials Evaluation*, Dec. 1995, *53*, 12, 1338–1345.
55. Hageniers, O. L., and Karpala, F. (1995), Composite NDE—A rapid scan technology, ASNT Spring Conf., 20–24 March 1995, pp. 194–196.
56. Cawley, P., and Adams, R. D. (1989), Sensitivity of the coin-tap method of nondestructive testing, *Materials Evaluation*, May 1989, *47*, 5, 558–563.
57. Haque, A., and Raju, P. K. (1995), Sensitivity of the acoustic impact technique in characterizing defects/damage in laminated composites, *J. Reinforced Plastics Composites*, March 1995, *14*, 3, 280–296.
58. Adams, R. D., and Cawley, P. (1986), Low-velocity impact inspection of bonded structures, Proc. Int. Conf. Structural Adhesives in Engineering, pp. 139–142.
59. Cawley, P. (1985), The operation of NDT instruments based on the impedance method, *Composite Structures*, *3*, 3–4, 215–228.
60. Busse, G. (1994), Nondestructive evaluation of polymer materials, *NDT&E International*, *27*, 5, 253–262.
61. Cawley, P., and Adams, R. D. (1979), Vibration technique for non-destructive testing of fibre composite structures, *J. Composite Materials*, *13*, 161–175.
62. Joshi, S. B., Hayward, D., Wilford, P., Affrossman, S., and Pethrick, R. A. (1992), A method for the non-destructive investigation of adhesively bonded structures, *European J. of NDT*, April 1992, *1*, 4, 190–199.
63. Banks, W. M., Hayward, D., Joshi, S. B., Li, Z. C., Jeffery, K., and Pethrick, R. A. (1995), High frequency dielectric investigations of adhesive bonded structures, *Insight*, Dec. 1995, *37*, 12, 964–968.
64. Muralidhar, C., and Arya, N. K. (1993), Evaluation of defects in axisymmetric composite structures by thermography, *NDT&E International*, *26*, 4, 189–193.
65. Henneke, E. G., and Tang, B. S. (1992), Thermal methods of NDE and quality control, Flight-Vehicle Materials, Structures, and Dynamics, Vol. 4, Tribological Materials and NDE, pp. 223–245.

66. Potet, P., Jeannin, P., and Bathias, C. (1987), The use of digital image processing in vibrothermographic detection of impact damage in composite materials, *Materials Evaluation*, April 1987, *45*, 4, 466–470.
67. Orlove, G. (1992), Infrared thermal imaging in the aerospace industry, *Sensors*, Feb. 1992, 37–42.
68. Favro, L. D., Kuo, P. K., Thomas, R. L., and Shepard, S. M. (1993), Thermal wave imaging for aging aircraft inspection, *Materials Evaluation*, Dec. 1993, *51*, 12, 1386–1389.
69. Sakagami, T., Ogura, K., Burleigh, D. D., and Spicer, J. W. M. (1996), Recent progresses in NDT techniques using infrared thermography, *J. JSNDI*, Nov. 1996, *45*, 11, 788–797.
70. Okamoto, Y. (1996), Nondestructive evaluation technique by means of infrared thermography, *J. JSNDI*, Nov. 1996, *45*, 11, 780–787.
71. Smith, G. M., and Clayton, B. R. (1994), Thermography and other NDE techniques for wind turbine blades, Proc. of the Inspection Structural Composites Conf., paper no. 7.
72. Cielo, P., Maldague, X., Deom, A. A., and Lewak, R. (1987), Thermographic nondestructive evaluation of industrial materials and structures, *Materials Evaluation*, April 1987, *45*, 4, 452–460, 465.
73. Balageas, D. L., Deom, A. A., and Boscher, D. M. (1987), Characterization and nondestructive testing of carbon epoxy composites by a pulsed photothermal method, *Materials Evaluation*, April 1987, *45*, 4, 461–465.
74. Zalameda, J. N., and Winfree, W. P. (1991), Thermal imaging of graphite/epoxy composite samples with fabricated defects, Review of Progress in QNDE, Plenum Press, Vol. 10A, pp. 1065–1072.
75. Shelley, T. (1988), Flashes show defects in depth, *Eureka Transfer Technology*, June 1988, 25–28.
76. Jeffreys, J. (1994), The impossible comes into view, *Eureka Transfer Technology*, Oct. 1994, 40–43.
77. Fitzpatrick, G. L., Thome, D. K., Skaugset, R. L., Shih, E. Y. C., and Shih, W. C. L. (1993), Magneto-optic/eddy current imaging of aging aircraft: A new NDI technique, *Materials Evaluation*, Dec. 1993, *51*, 12, 1402–1407.
78. Maldague, X. P. V. (1993), Nondestructive Evaluation of Materials by Infrared Thermography, Springer-Verlag.
79. Hobbs, C. (1992), The inspection of aeronautical structures using transient thermography, Proc. of the Conf. on NDT for Corrosion in Aerospace Structures, Royal Aeronautical Society, pp. 6.1–6.10.
80. Connolly, M., and Copley, D. (1990), Thermographic inspection of composite materials, *Materials Evaluation*, Dec. 1990, *48*, 12, 1461–1463.
81. Shiratori, M., Qiang, Y., Takahashi, Y., and Ogasawara, N. (1994), Application of infrared thermography to detection of flaws in honeycomb sandwich constructions, *JSME International Journal*, Series A, *37*, 4, 396–402.
82. Jones, T., and Berger, H. (1992), Thermographic detection of impact damage in graphite-epoxy composites, *Materials Evaluation*, Dec. 1992, *50*, 12, 1446–1453.
83. McLaughlin, P. V. (1988), Defect detection and quantification in laminated composites by EATF (passive) thermography, *Review of Progress in QNDE, 7B*, 1125–1132.
84. Jones, T. S., and Berger, H. (1989), Application of nondestructive inspection methods to composites, *Materials Evaluation*, April 1989, *47*, 4, 390–400.
85. Reynolds, W. N. (1986), Nondestructive testing techniques for metal-matrix composites, Harwell Report No. AERE-R-13040, UK.
86. Ganchev, S. I., Carriveau, G., and Qaddoumi, N. (1994), Microwave detection of defects in glass reinforced polymer composites, Proc. of the SPIE, Advanced Microwave and Millimeter-wave Detectors Conf., Vol. 2275, pp. 11–20.
87. Schwartz, M. M. (1992), Composite Materials Handbook, 2d ed., McGraw-Hill.
88. Summerscales, J. (1990b), Microwave Techniques, Non-destructive testing of fibre-reinforced plastics composites, Vol. 2, Elsevier Applied Science, pp. 361–412.

89. Diener, L. (1995), Microwave near-field imaging with open-ended waveguide—Comparison with other techniques of nondestructive testing, *Research in NDE*, *7*, 2/3, 137–152.
90. Zoughi, R. (1995), Microwave and millimeter wave nondestructive testing: a succinct introduction, *Research in NDE*, *7*, 2/3, 71–74.
91. Zoughi, R., and Ganchev, S. I. (1994), Application of microwave nondestructive testing and evaluation to thick composites, Proc. of the Conf. on NDE Applied to Process Control of Composite Fabrication, Oct. 1994, pp. 77–89.
92. Radford, D., Ganchev, S., Qaddoumi, N., Beauregard, G., and Zoughi, R. (1994), Millimeter wave nondestructive evaluation of glass fiber/epoxy composites subjected to impact fatigue, Proc. of the SPIE, Advanced Microwave and Millimeter-Wave Detectors Conf., Vol. 2275, pp. 21–26.
93. Gopalsami, N., Bakhtiari, S., Dieckman, S., Raptis, A., and Lepper, M. (1994), Millimeter-wave imaging for non-destructive evaluation of materials, *Materials Evaluation*, March 1994, *52*, 3, 412–415.
94. Jackson, P., Barnes, J. A., Clayden, N. J., Carpenter, T. A., Hall, L. D., and Jezzard, P. (1990), Defect detection in carbon fibre composite structures by magnetic resonance imaging, *J. Materials Science Letters*, *9*, 10, 1165–1168.
95. Fowler, T. J. (1977), Acoustic emission testing of fiber reinforced plastics, Proc. of the ASCE Fall Convention and Exhibit, Oct. 1977, Paper 3092.
96. Kwon, O., Lee, J. H., Ahn, D. M., and Yoon, D. J. (1994), Acoustic emission monitoring of tensile testing of low velocity impact-damaged CFRP laminates, Proc. of the 12th Acoustic Emission Symp., Progress in Acoustic Emission VII, Oct. 1994, pp. 505–510..
97. Bailey, C. D., Hamilton, J. M., and Pless, W. M. (1979), Acoustic emission of impact-damaged graphite-epoxy composites, *Materials Evaluation*, May 1979, *37*, 5, 43–48.
98. Williams, J. H., and Lee, S. S. (1985), Promising quantitative nondestructive evaluation techniques for composite materials, *Materials Evaluation*, April 1985, *43*, 4, 561–565.
99. Bar-Cohen, Y. (1986), NDE of fiber-reinforced composite materials—A review, *Materials Evaluation*, March 1986, *44*, 3, 446–454.
100. Melve, B., and Moursund, B. (1994), Acoustic emission testing of glass fibre reinforced pipes on offshore platform, Proc. of the Inspection Structural Composites Conf., paper no. 13.
101. Summerscale, J. (1990), Electrical and magnetic testing, Non-Destructive Testing of Fibre-Reinforced Plastic Composites, Vol. 2, Elsevier Applied Science, pp. 253–297.
102. Gordeev, V. F., Eliseev, V. P., Malyshkov, Y. P., Chakhlov, V. L., and Krening, M. (1994), Apparatus for quality control of non-metallic materials and object using electromagnetic emission characteristics, *Russian J. Nondestructive Testing*, *30*, 4, 278–283.
103. Clark, W. G. (1994), Magnetic particle tagging for process control in composite fabrication, Proc. of the Conf. on NDE Applied to Process Control of Composite Fabrication, Oct. 1994, pp. 199–211.
104. Valleau, A. R. (1990), Eddy current nondestructive testing of graphite composite materials, *Materials Evaluation*, Feb. 1990, *48*, 2, 230–239.
105. Gros, X. E., and Lowden, D. W. (1995), Electromagnetic testing of composite materials, *Insight*, April 1995, *37*, 4, 290–293.
106. Vernon, S. N. (1987), Eddy current nondestructive inspection of graphite epoxy using ferrite cup core probes, Report NSWC-TR-87-148, Sept. 1987, Naval Surface Warfare Center, Silver Spring, MD, USA.
107. Gerasimov, V. G., Malov, V. M., Efraimov, L. Y., and Evdokimov, P. A. (1993), Eddy-current inspection of carbon components, *Russian J. NDT*, *29*, 9, 643–649.
108. Gvishi, M., Kahn, A. H., and Mester, M. L. (1992), Eddy current testing of carbon-carbon composites, Review of Progress in QNDE, Plenum Press, Vol. 11, pp. 289–297.
109. Hashimoto, M., Nakamura, H., Sugiura, T., Miya, K., and Kaneki, K. (1988), Eddy current testing of graphite material, Proc. of the 9th Inter. Conf. on NDE in the Nuclear Industry, April 1988, ASM Pub., pp. 385–389.

110. Gros, X. E. (1995), Advantages and limitations of non-destructive eddy current testing of composite materials: From rotor blades to dashboards, *Engineering Plastics*, *8*, 6, 410–425.
111. Lange, R., and Mook, G. (1994), Structural analysis of CFRP using eddy current methods, *NDT&E International*, *27*, 5, 241–248.
112. Owston, C. N. (1976), Eddy current methods for the examination of carbon fibre reinforced epoxy resins, *Materials Evaluation*, Nov. 1976, *34*, 11, 237–244, 250.
113. Rudlin, J. R. (1989), A beginner's guide to eddy current testing, June 1989, *British J. NDT*, *31*, 6, 314–320.
114. Lowden, D. W., and Gros, X. E. (1994), Visualising defect geometries in composite materials, Proc. of the Inter. Symp. "Advanced Materials for Lightweight Structures", ESTEC, March 1994, pp. 683–686.
115. Davis, J. M., and King, M. (1994), Mathematical formulas and references for nondestructive testing—Eddy currents, 1st ed., Art Room Corporation.
116. Prakash, R. (1980), Non-destructive testing of composites, *Composites*, *11*, 217–224.
117. Dingwall, P. F., and Mead, D. L. (1976), Non-destructive inspection and volume fraction determination of CFRP using an eddy current method, June 1976, Dept. of Defence, Report RAE-TR-76078.
118. Gros, X. E. (1997), Eddy current: The future of composite materials evaluation? *J. JSDNI*, October 1997, *46*, 9, 642–648.
119. Lowden, D. W. (1992), Quantifying disbond area, Proc. of the Inter. Symp. "Advanced Materials for Lightweight Structures," ESTEC, March 1992, pp. 223–228.
120. Gros, X. E. (1996), Characterization of low energy impact damage in composites, *J. Reinforced Plastics Composites*, March 1996, *15*, 3, 267–282.
121. Belani, J. G., and Broutman, L. J. (1978), Moisture induced resistivity changes in graphite reinforced plastics, *Composites*, Oct. 1978, *9*, 4, 273–277.
122. Gros, X. E. (1996), Some aspects of electromagnetic testing of composites, *Insight*, July 1996, *38*, 7, 492–495.
123. Vernon, S. N., and Gammell, P. M. (1985), Eddy current inspection of broken fiber flaws in non-metallic fiber composites, Review of Progress in QNDE Plenum Press, Vol. 4B, pp. 1229–1237.
124. Beissner, R. E. (1992), Theory of eddy currents in metal matrix composites, Review of Progress in QNDE, Plenum Press, Vol. 11, pp. 225–232.
125. Auld, B. A. (1992), Electromagnetic methods in NDE, Flight-Vehicle Materials, Structures, and Dynamics, Vol. 4, Tribological Materials and NDE, pp. 213–221.
126. Williams, J. H., Nayeb-Hashemi, H., and Lee, S. S. (1979), Ultrasonic attenuation and velocity in AS/3501-6 graphite/epoxy fiber composite, NASA Report No. CR3180.
127. Hawkins, G. F., Sheaffer, P. M., and Johnson, E. C. (1991), NDE of thick composites in the aerospace industry—An overview, Review of Progress in QNDE, Plenum Press, Vol. 10B, pp. 1591–1597.
128. Wong, B. S., Tan, K. S., and Tui, T. G. (1994), Ultrasonic testing of solid fiber-reinforced composite plate, *SAMPE J.*, Nov./Dec. 1994, *30*, 6, 36–40.
129. Michaels, T. E., and Davidson, B. D. (1993), Ultrasonic inspection detects hidden damage in composites, *Advanced Materials Processes*, 3, 34–38.
130. Hutchins, D. A., Wright, W. M. D., Scudder, L. P., Mottram, J. T., and Schindel, D. W. (1994), Air-coupled ultrasonic testing of composites, Proc. of the Inspection Structural Composites Conf., paper no. 8.
131. Solokhin, N. V., and Sumbatyan, M. A. (1993), Ultrasonic through-transmission technique for composite materials, *European J. NDT*, Jan. 1993, *2*, 3, 91–93.
132. Padioleau, C., Bouchard, P., Heon, R., Monchalin, J. P., Chang, F. H., Drake, T. E., and McRae, K. I. (1993), Laser ultrasonic inspection of graphite epoxy laminates, Review of Progress in QNDE, Vol. 12, pp. 1345–1352.
133. Dewhurst, R. J., He, R., and Shan, Q. (1993), Defect visualization in carbon fibre composite using laser ultrasound, *Materials Evaluation*, Aug. 1993, *51*, 8, 935–940.

134. Chang, F. H., Drake, T. E., Osterkamp, M. A., Prowant, R. S., Monchalin, J. P., Heon, R. Bouchard, P., Padioleau, C., Froom, D. A., Frazier, W., and Barton, J. P. (1993), Laser ultrasonic inspection of honeycomb aircraft structures, Review of Progress in QNDE, Vol. 12, pp. 611–616.
135. Monchalin, J. P. (1993), Progress towards the application of laser-ultrasonics in industry, Review of Progress in QNDE, Vol. 12, pp. 495–506.
136. Ringermacher, H. I., and McKie, A. D. W. (1995), Laser ultrasonics for the evaluation of composites and coatings, *Materials Evaluation*, Dec. 1995, *53*, 12, 1356–1361.
137. Addison, R. C., Ryden, H. A., and McKie, A. D. W. (1991), Laser-based ultrasonics for the inspection of large area Gr/Epoxy composites, Review of Progress in QNDE, Plenum Press, Vol. 10A, pp. 485–492.
138. Dixon, S., Edwards, C., and Palmer, S. B. (1995), Experiment to monitor adhesive cure using electromagnetic acoustic transducers, *Insight*, Dec. 1995, *37*, 12, 969–973.
139. Moran, T. J., Crane, R. L., and Andrews, R. J. (1985), High-resolution imaging of microcracks in composites, *Materials Evaluation*, April 1985, *43*, 4, 536–540.
140. Bar-Cohen, Y., Mal, A. K., and Lih, S. S. (1993), NDE of composite materials using ultrasonic oblique insonification, *Materials Evaluation*, Nov. 1993, *51*, 11, 1285–1296.
141. Bashyam, M. (1995), Ultrasonic technique to measure stiffness coefficients of CMC and its implications on characterizing material degradation, *Composites Engineering*, *5*, 6, 735–742.
142. Farlow, R., and Hayward, G. (1996), An automated ultrasonic NDT scanner employing advanced air-coupled 1-3 connectivity composite transducers, *Insight*, Jan. 1996, *38*, 1, 41–50.
143. Karpur, P., Benson, D. M., Matikas, T. E., Kundu, T., and Nicolaou, P. D. (1995b), An approach to determine the experimental transmitter-receiver geometry for the reception of leaky lamb waves, *Materials Evaluation*, Dec. 1995, *53*, 12, 1348–1352.
144. Bar-Cohen, Y., Arnon, U., and Meron, M. (1978), Defect detection and characterization in composite sandwich structure by ultrasonics, *SAMPE J.*, *14*, 1, 4–9.
145. Hsu, D. K., and Hughes, M. S. (1993), Ultrasonic nondestructive evaluation of foreign objects in composite laminates, Proc. of the Inter. Conf. on Advanced Composite Materials, pp. 699–705.
146. Tiwari, A., Lesko, J., and Henneke, E. G. (1995), Real-time acousto-ultrasonic (AU) nde technique to monitor damage under compressive loads, ASNT Spring Conf., 20–24 March 1995, pp. 202–203.
147. Ghorayeb, S. R., Hughes, M. S., Holger, D. K., and Zachary, L. W. (1991), A preliminary application of SAFT on composites, Review of Progress in Quantitative Nondestructive Evaluation, Vol. 10A (Thompson, D. O., and Chimenti, D. E., eds.), Plenum Press, pp. 1027–1032.
148. Hao, C., Ming, B., and Sheng, Z. C. (1992), Ultrasonic synthetic array imaging technique applied to composite NDT, Non-destructive testing '92, pp. 823–830.
149. Thompson, R. B., Thompson, D. O., Holger, D. K., Hsu, D. K., Hughes, M. S., Papadakis, E. P., Tsai, Y. M., and Zachary, L. W. (1991), Ultrasonic NDE of thick composites, Proc. of the ASME Conf. on NDE, Enhancing Analysis Techiques for Composite Materials, NDE-Vol. 10, pp. 43–57.
150. Panametrics (1992), Phase shift test bond integrity, Technical Sheet.
151. Jones, T. S. (1985), Inspection of composites using the automated ultrasonic scanning system (AUSS), *Materials Evaluation*, May 1985, *43*, 5, 746–753.
152. Lord, R. J. (1985), In-service nondestructive inspection of fighter and attack aircraft, *Materials Evaluation*, May 1985, *43*, 5, 733–739.
153. Rogousky, A. J. (1985), Ultrasonic and thermographic methods for NDE of composite tubular parts, *Materials Evaluation*, April 1985, *43*, 4, 547–555.
154. Kaczmarek, H. (1995), Ultrasonic detection of damage in carbon fibre reinforced plastics, *J. Composite Materials*, *29*, 1, 59–95.
155. Buynak, C. F., Moran, T. J., and Donaldson, S. (1988), Characterization of impact damage in composites, *SAMPE J.*, March/April 1988, *24*, 2, 35–39.

156. NASA (1994), Ultrasonic oblique insonification used to evaluate thermal damage to composites, NASA NDE Working Group Newsletter, Oct. 1994, Vol. 2, No. 4.
157. NASA (1994), NDE methods successfully applied to ceramic matrix composites, NASA NDE Working Group Newsletter, Oct. 1994, Vol. 2, No. 4.
158. Karpur, P., Matikas, T. E., and Krishnamurthy, S. (1995a), Ultrasonic characterization of the fiber-matrix interphase/interface for mechanics of continuous fiber reinforced metal matrix and ceramic matrix composites, *Composites Engineering*, *5*, 6, 697–711.
159. Chu, Y. C., Lavrentyev, A. I., Rokhlin, S. I., Baaklini, G. Y., and Bhatt, R. T. (1995), Ultrasonic evaluation of initiation and development of oxidation damage in ceramic-matrix composites, *J. American Ceramic Society*, *78*, 7, 1809–1817.
160. Sachse, W., Ting, C. S., and Hemenway, A. (1979), Dispersion of elastic waves and the nondestructive testing of composite materials, ASTM STP 674, pp. 165–183.
161. Munn, I. J., and Georgiou, G. A. (1995), Ultrasonic and radiographic NDT of butt fusion welds in polyethylene pipe, Plastic Pipes IX Conf., Scotland, UK, 18–21 Sept. 1995.
162. Rose, J. L. (1985), Ultrasonic wave propagation principles in composite material inspection, *Materials Evaluation*, April 1985, *43*, 4, 481–483.
163. Freemantle, R. J., Challis, R. E., and White, J. D. H. (1994), Limitations to ultrasonic testing of bonded structures, Proc. of the Inspection Structural Composites Conf., paper no. 17.
164. Edmund, G., and Henneke, I. I. (1990), Ultrasonic nondestructive examination of advanced composites, Non-Destructive Testing of Fibre Reinforced Plastics Composites, Vol. 2, Elsevier Applied Science.
165. Jensen, T. H. (1994), Filmless X-ray evaluation of structures made of light materials, *NDT&E International*, *27*, 2, 89–96.
166. Yancey, R. N., and Mitchell, C. R. (1994), Nondestructive evaluation of composites using digital radiography and computed tomography, Proc. of the Conf. on NDE Applied to Process Control of Composite Fabrication, Oct. 1994, pp. 111–114.
167. Armistead, R. A., and Yancey, R. N. (1989), Computed tomography for the nondestructive testing of advanced engineering materials, *Materials Evaluation*, May 1989, *47*, 5, 487–491.
168. Stock, S. R., Breunig, T. M., Guvenilir, A., Kinney, J. H., and Nichols, M. C. (1992), Nondestructive X-ray tomographic microscopy of damage in various continuous-fiber metal matrix composites, Damage Detection in Composite Materials, ASTM, STP 1128, pp. 25–34.
169. Hentschel, M. P., Ekenhorst, D., Harbich, K. W., and Lange, A. (1994), New X-ray refractography for nondestructive investigations, Proc. of the Int. Symp. on Advanced Materials for Lightweight Structures, March 1994, pp. 661–664.
170. Smith, R. L. (1994), Techniques for the determination of impact damage and uptake moisture in composite materials, Proc. of the Inspection Structural Composites Conf.
171. Cordell, T. M., and Bhagat, P. K. (1991), Air force requirements for NDE of composite materials, Proc. of the ASEM Conf. on NDE, Enhancing Analysis Techniques for Composite Materials, NDE-Vol. 10, pp. 67–75.
172. Harris, B. (1980), Accumulation of damage and non-destructive testing of composite materials and structures, *Annales de Chimie–Science de Matériaux*, Vol. 5, pp. 327–339.
173. Prakash, R. (1990), Eddy-current testing, Non-Destructive Testing of Fibre-Reinforced Plastic Composites, Vol. 2, Elsevier Applied Science, pp. 299–326.
174. Salkind, M. J. (1976), Early detection of fatigue damage in composite materials, *J. Aircraft*, *13*, 764–769.
175. Hastings, K. P. (1994), Real-time radiography gains a firm foothold in aerospace composite component inspection, *Insight*, May 1994, *36*, 5, 314–315.
176. Fassbender, R. H., and Hagemaier, D. J. (1983), Low-kilovoltage radiography of composites, *Materials Evaluation*, June 1983, *41*, 6, 831–838.

Index

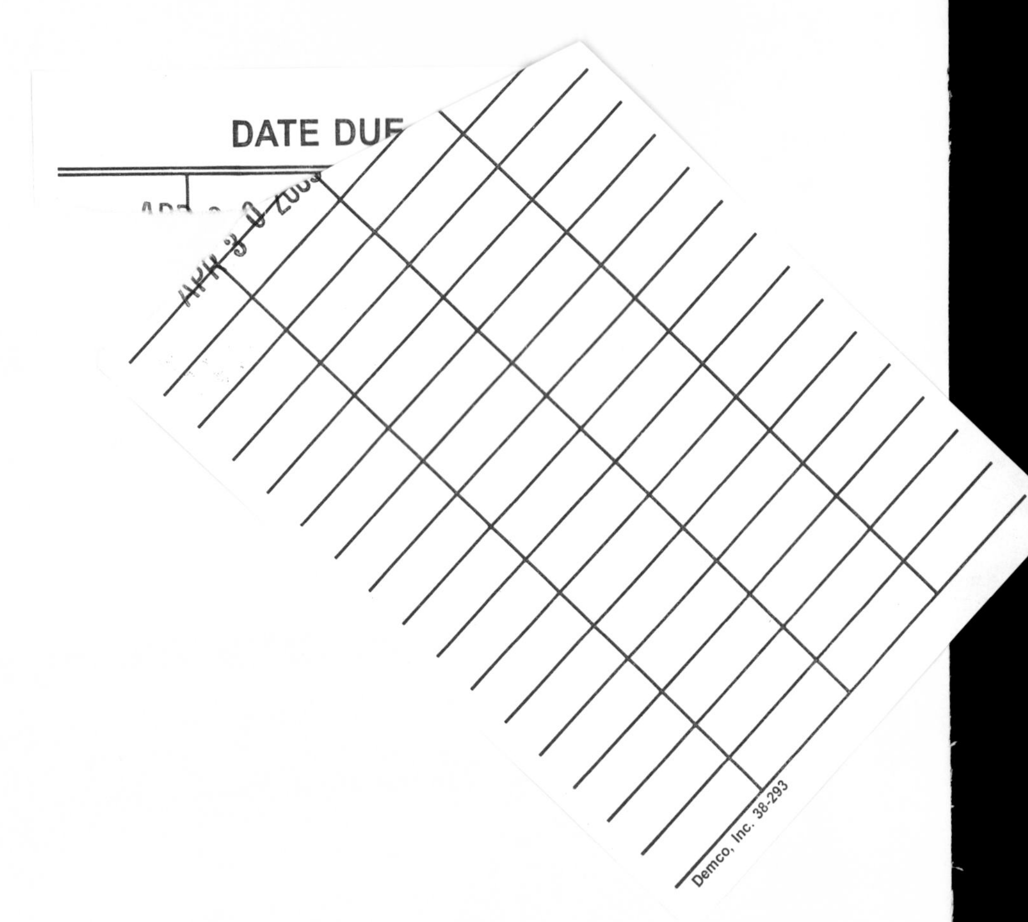
DATE DUE
APR 3 0 2003
Demco, Inc. 38-293

DEPAUL UNIVERSITY LIBRARY
3 0511 00698 4280